NEUROLOGICAL ANATOMY
In Relation to Clinical Medicine

NEUROLOGICAL ANATOMY

In Relation to Clinical Medicine

THIRD EDITION

A. Brodal, M.D.

Professor Emeritus of Anatomy, University of Oslo, Norway
Formerly Resident of the University Neurological
*   and Psychiatric Clinics in Oslo*

New York   Oxford
OXFORD UNIVERSITY PRESS
1981

Library of Congress Cataloging in Publication Data

Brodal, Alf, 1910–
  Neurological anatomy in relation to clinical
medicine.

  Bibliography: p.
  Includes index.
  1. Neuroanatomy.   2. Nervous system—Diseases—
Diagnosis.   I. Title.   [DNLM: 1. Nervous systems
—Anatomy and histology Neurological manifestations.
WL101 B864n]
QM451.B7853   1980   611'.8   80-11701
ISBN 0-19-502694-2

Preface

THE SCOPE and aim of the third edition of this book are the same as those outlined in the preface to the second edition (see below), to which the reader is referred for a general orientation. It should be emphasized that the book is not intended to serve as a systematic textbook of neuroanatomy. In general, purely topographical relations have accordingly been little dealt with. For information on this subject the reader should consult some of the available atlases, for example DeArmond, Fusco, and Dewey's (1976) or Nieuwenhuys, Voogd, and van Huizen's (1978).

In the ten years that have passed since the publication of the second edition, our knowledge in all fields of the neurological sciences has increased with an ever-accelerating speed. This is the case not least in neuroanatomy, due first and foremost to the development of several new methods, to be considered in Chapter 1. The new data have given us a picture of the structure of the central nervous system which encompasses a complexity that we should not have imagined ten years ago. This is true also in the fields of neurophysiology, neurochemistry, neuropharmacology, and behavioral and communication research.

Only a fraction of the new anatomical information can as yet be satisfactorily correlated with observations in clinical neurology; consequently only a few of the findings can be of immediate value in practical medicine. However, there are other reasons why it is important for the clinical neurologist to be aware of essential results of recent research. An insight into present-day knowledge of the structure and function of the nervous system will aid in evaluating the frequent deviations—seen in the clinic—from the textbook descriptions of symptoms, signs, and syndromes. Second, insight of this kind may initiate thinking about problems and may foster ideas about how to explain certain symptoms and how to elaborate refined testing methods of neurological disorders in man.

Today it is virtually impossible for a single author to cover the entire field of neuroanatomy in a satisfactory way. I am very grateful, therefore, for the assistance offered by my son, Assistant Professor *Per Brodal*, M.D., and by my friends and colleagues, Assistant Professor *Kirsten Kjelsberg Osen*, M.D., and

Professor *Eric Rinvik,* M.D., all of the Anatomical Institute, University of Oslo. They have assisted me in the revision of Chapters 2, 3, 4, 8, and 9. Some chapters or parts of chapters have been entirely recast. Some 75 new figures have been added, while some of the old ones have been deleted.

The amount of new data from research in recent years is enormous. What should be included for thorough consideration in the third edition of a book like this, what might be treated more superficially, and what can be neglected? To some extent, decisions about these matters must be subjective. To give a balanced presentation of the entire field is virtually impossible. Much relevant information has certainly not been considered as fully as some readers might have wanted. Other readers, on the other hand, may find that too many details have been discussed. It is inevitable that the numbers of pages and references have increased considerably. It has not been possible to include references to all papers relevant to a particular item. The choice of authors to be quoted has of necessity often been arbitrary.

Several friends and colleagues have read through parts of or complete chapters and given valuable advice concerning special subjects about which they have firsthand knowledge. Professor Emeritus of Neurosurgery, K. Kristiansen, M.D., and Associate Professor of Neurology, B. Vandvik, M.D., University of Oslo, have given special consideration to the clinical questions treated in the book. I am greatly indebted to all these friends and colleagues for advice and constructive criticism. However, the responsibility for errors and omissions that are to be found in the text rests entirely with myself and my coauthors.

The expert technical assistance given by the artists of the Anatomical Institute, the late Mrs. Nanti Stang-Lund and Mrs. Kari Øztürk, and the photographer, Mr. E. Risnes, is gratefully acknowledged. Valuable assistance has further been given by the Institute's librarian, Miss Wenche Sandberg. I am deeply indebted to Miss Oddlaug Gorset for indefatigable and conscientious help during the preparation of the third edition in typing the manuscript and secretarial assistance of various kinds. It is hardly an exaggeration to say that the preparation of the third edition would not have been possible without the assistance of the collaborators mentioned above. My sincere thanks, therefore, go to the present head of the Anatomical Institute, my close collaborator for many years, Professor Fred Walberg, M.D., for putting the technical facilities of the institute at my disposal and for assistance of many kinds during the work.

The careful scrutiny of the English of the manuscript by Miss Ellen Johannessen is greatly appreciated. I would also like to thank the editors of journals and many colleagues in various countries who have given permission to reproduce illustrations from their publications.

It is a pleasure to extend my thanks to the staff of Oxford University Press for an agreeable collaboration and for their efforts to make this a better book.

Finally, I would like to thank my wife for a lifelong enduring patience and forbearance during my preoccupation with the present and other works, and for encouragement and valuable criticisms.

OSLO   A. Brodal
September 1979

From the Preface to the Second Edition

Of all the natural phenomena to which science can turn
its attention, none exceeds in its fascination the working
of the human brain. Here, in a bare two-handsful of
living tissue, we find an ordered complexity sufficient
to embody and preserve the record of a lifetime of
the richest human experience. We find a regulator and
co-ordinator of the hundreds of separate muscle systems
of the human body that is capable of all the delicacy
and precision shown by the concert pianist and the surgeon.
Most mysterious of all, we find in this small sample of
the material universe the organ (in some sense) of our
own awareness, including our awareness of that
universe, and so of the brain itself.

D. MacKay (1967)

SOME TWENTY YEARS have passed since the first edition of this book appeared.
During this time there has been a steadily increasing activity in all fields of
neurological research, which has furnished us with a wealth of new information on
the structure and function of the nervous system. As a consequence of our increased knowledge, many points of view have had to be modified. Some traditional concepts have had to be discarded; new ones have been created. For example, there is no longer any justification for retaining the traditional concept of the
"extra-pyramidal motor system," as will be discussed in Chapter 4. On the other
hand, the neurological literature has been enriched with concepts such as the "ascending reticular activating system," "the limbic system," and others. In order to
proceed in research we obviously have to create concepts as tools in the formula-

tion of working hypotheses. However, we are often inclined to forget the limited validity of such concepts, particularly when a concept is verbalized in a simple and catching fashion. This tendency of the human mind has been aptly characterized and ridiculed by Goethe:

> Denn eben wo Begriffe fehlen,
> da stellt ein Wort zur rechten Zeit sich ein.
> Mit Worten lässt sich trefflich streiten
> mit Worten ein System bereiten.*

The neurological literature furnishes numerous examples of the truth of this dictum. Much confusion and unnecessary disputes have arisen because a sufficiently clear distinction has not been made between facts and hypotheses, between observations and interpretations. There is little doubt that many concepts which are at present accepted and in general use will in the future have to be considerably modified or even abandoned. The speed with which new information is forthcoming makes it more important than it was some decades ago to keep incessantly in mind the provisional tenability of current concepts. This, however, is not as easy as it may sound.

The author has felt this very strongly in the preparation of a second edition of this book. Largely due to the accumulation of new data our views are today on many points quite different from those accepted twenty years ago. It has therefore been necessary to rewrite entirely several chapters and sections of the first edition. For the same reason it has been difficult to decide what to include in the text, and to make clear what may be considered as established and what is still hypothetical. The main criterion in the selection of data to be presented has been to retain the original aim of the book: to bridge the gap between basic neurological sciences and clinical neurology, with particular reference to correlations between the structure of the nervous system and its functions as revealed in man under normal and pathological conditions.

Like the first edition, the present does not pretend to be a systematic textbook on neurological anatomy, nor is it a textbook of clinical neurology. Some subjects are discussed in considerable detail, others are treated only briefly, and still others are not dealt with at all. Attempts have been made to include anatomical data which may be of interest from a clinical point of view. However, certain subjects are selected for more comprehensive consideration, in order to convey an impression of contemporary problems in the field of neurological research, even if the subjects are not of immediate practical interest. Results of neurophysiological research have been incorporated insofar as they can be correlated with structural features and can be understood by a writer who is not a neurophysiologist. More specific neurophysiological problems are not considered. The extensive new data

---

*For just where fails the comprehension,
 A word steps promptly in as deputy.
 With words 't is excellent disputing;
 Systems to words 't is easy suiting.

  Goethe: *Faust,* First Part, Scene IV.
  Translated by Bayard Taylor. Strahan & Co.,
  Publishers, London, 1871.

in neurochemistry and neuropharmacology will be mentioned only to a limited extent. This is due to two circumstances. In the first place the author's knowledge of these subjects is too limited to permit a fuller treatment. Second, at the present state of research, data from these fields can only in few instances be definitely correlated with structural features. Likewise, most of the attempts made by workers in the field of neurocommunication to prepare models or theories of the brain or parts of it are not considered. Interesting as they may be, most of the models and theories are still of rather limited interest from the clinician's point of view.

With the enormous literature in the neurological sciences of today it is virtually impossible to give references to all relevant publications, nor would this be practical. In the selection of references to original publications, preference has in general been given to recent papers (in them interested readers will find references to previous publications). Furthermore, attempts have been made to give credit to authors who have been the first to report a new observation. Finally, where reviews or monographs are available, these are mentioned. However, the selection of references will of necessity be to some extent arbitrary, and it cannot be avoided that some papers which ought rightly to have been included are left out.

OSLO   A. Brodal
May 1968

Contents

NEUROLOGICAL ANATOMY

In Relation to Clinical Medicine

1

Introduction, Methods, Correlations

BEFORE DEALING with different parts of the nervous system certain generalizations with regard to its anatomy must be considered, since this will facilitate an understanding of the material dealt with in this book. It is necessary first of all to look briefly at the methods we have at our disposal for the investigation of the anatomy of the nervous system and especially its fiber connections.

Structure and function. Anatomy is one of the fundamental branches of medicine, but anatomy alone, without relation to function, is a somewhat barren study, for it is function which primarily interests us in medicine. Some knowledge of the structure of organs and organisms is a necessary prerequisite for the understanding of their function. In fact, a knowledge of structure will often yield information on functions which have not been elucidated by other data. It is now recognized almost as an axiom that where structural differences are found there are also functional differences and vice versa. To take a simple example from the nervous system, one can, as is well known, differentiate between the somatic and visceral efferent nuclei in the brainstem. On some occasions it is not clear that there are structural differences, but this may be due to the inability of our present methods to bring more subtle differences to light.

In neurology a knowledge of structural features is perhaps more important for an understanding of function under normal and pathological conditions than in any other branch of medicine. Symptoms produced by diseases of the nervous system can be multifarious and involve different functions and components of functions, according to which different structures are affected. By clinical examinations we attempt to analyze the functional conditions and to understand the extent to which normal functions have been altered. The better our knowledge of the anatomy of

the nervous system, the more easily we can infer from these functional disturbances and symptoms which structures are affected. However, the *topographical
diagnosis,* the determination of the site of the lesion, is not, as one would first
imagine, of importance only in those cases in which surgical procedures may be
applied. It may also be of great importance in the *etiological diagnosis,* the determination of the nature and perhaps the cause of the lesion. Some diseases of the
nervous system affect selectively, and in some cases exclusively, certain anatomical and functional units, while in other cases the localization of the lesion is quite
irregular.

The determination of the site of the lesion is therefore an important link in
understanding the nature of the disease. The better and more detailed our knowledge of the anatomy of the nervous system, the better equipped we are to interpret
the clinical findings and the closer we can come to an exact diagnosis. Details of
the clinical phenomena which at first may seem meaningless and unintelligible
can, with a widened knowledge of the structural foundation, become intelligible
and give valuable information.

Methods of studying the structure of the nervous sytem.   The older anatomists who studied the nervous system depended on dissection and macroscopic
findings, and by these primitive methods succeeded in demonstrating certain fundamental features, but it was not until the development of microscopic anatomy
that any considerable clarity was achieved. It is surprising to realize that microscopic study of animal and human organisms did not begin until 1840. The German anatomist Schwann demonstrated about this time that animal organisms, like
plants, are built up of cells. During the past century the cell theory has, as we
know, gone through a striking development. The discovery (about 1850) that
animal cells could be stained after first being fixed, i.e., killed by coagulation of
the cell albumen, and that the different parts of the cells absorbed stains to various
degrees, was another epoch-making event. The first stains to be used, first and
foremost carmine, gave rather unsatisfactory pictures, however, when compared
with present-day standards. Only with the introduction of hematoxylin (about
1890) was it really possible to undertake a finer structural analysis of cells. Numerous other methods then followed. *Weigert* devised a method for selective
staining of myelin sheaths, and another for glial tissue. The introduction by *Golgi*
of his method of impregnating nerve cells with silver nitrate opened new possibilities for study of the finer structure of the nervous system. The *Nissl* method, introduced about 1890, is also of decisive importance. It is based on the staining
with basic aniline dyes of specimens which are fixed in alcohol and remains the
routine method in investigations of the cytology of the nervous system. About the
same time *Marchi's method of demonstrating degenerating myelin sheaths* was developed, and numerous other methods subsequently evolved. Of particular importance is the development of *silver impregnation methods* which permit one to trace
the degenerating unmyelinated axons and terminals.

A number of *histochemical methods* have been devised by which various
chemical constituents in the nerve cells can be determined and the occurrence of
certain enzymes, for example cholinesterases, can be demonstrated. The *use of radioactive isotopes* in morphological studies has created possibilities for studying

details in the development of the various nuclei of the nervous system and in addition has given information on metabolic processes in nerve cells and glial cells. However, the most important new research tool for the morphologist has been the *electron microscope*. In the thirty years it has been in use, a wealth of new data on the finest structural details of nerve cells and fibers and glial cells has been revealed, which has immensely increased our understanding of the nervous system. In recent years methods have been developed which utilize the anterograde and retrograde axonal transport of macromolecules for the study of fiber connections. Some of these methods will be briefly reviewed below.

The study of normal preparations.   Many features of the fiber connections of the nervous system have been elucidated by the study of normal preparations in which the fibers are followed microscopically, partly by staining of the myelin sheaths and partly by *impregnation of the cell body and processes with silver and gold salts*. The latter methods have given useful information, especially about the lower, simply constructed organisms, but in many such cases it is impossible to state definitely where the fibers end. One of these methods, the *Golgi method,* has in recent years experienced a renaissance. This is largely a consequence of progress in modern electrophysiology of the nervous system, where it is often of importance to know the distribution of axon collaterals and dendrites in evaluating one's findings. Qualitative and quantitative ultrastructural investigations of the central nervous system have also proved more rewarding when correlated with parallel light microscopic Golgi studies. (Several papers illuminating present-day use of the Golgi method are found in the book *Golgi Centennial Symposium: Perspectives in Neurobiology,* 1975.) With the Golgi method only random cells are impregnated; that method has for decades been the only means of visualizing the course and type of all processes of a cell. Figure 1-1 shows an example taken from the vestibular nuclei. However, in recent years the technique of iontophoretically injecting single cells by means of micropipettes has made it possible to visualize a single cell and its axon and dendrites.

Several substances have been used as markers, such as Procion yellow, cobolt ions, and tritiated amino acids. Figure 1-2 shows an example of a nerve cell injected in this way with ³H-glycine. The amino acid has been incorporated into protein, transported intraneuronally, and is made visible by autoradiography. With double-barreled pipettes one of them can be used for recording the electric potential changes in the cell, and the method thus permits the precise identification of a particular cell whose behavior has been studied physiologically. For recent exhaustive reviews of the problems of intraneuronal transport and the marking of single cells, see Kreutzberg and Schubert (1975) and Schubert and Holländer (1975). Dendrites and axons (and their collaterals) of single cells may also be studied after injection of horseradish peroxidase (see below) in the perikaryon (see, for example, Jankowska, Rastad, and Westman, 1976; Snow, Rose, and Brown, 1976; McCrea, Bishop, and Kitai, 1976).

The so-called *myelogenetic method,* introduced by Flechsig, makes use of the fact that the different fiber systems are myelinated at different times in embryonic life and during the first period after birth. This method is hardly ever used today.

The study of pathological conditions in humans and experimental investigations on animals.   By far the greatest part of our present-day knowledge of the fiber connections has been attained by the study of pathological conditions of

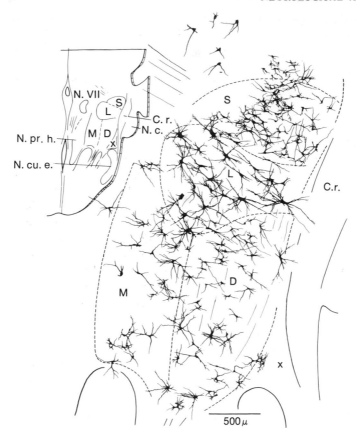

FIG. 1-1 Drawing of a Golgi preparation from the lateral vestibular nucleus in the cat illustrating the various sizes of cells and orientation of dendrites in the four main vestibular nuclei as seen in a horizontal section (cf. inset above). *S, L, M,* and *D:* superior, lateral, medial, and descending vestibular nuclei. From Hauglie-Hanssen (1968).

the ''degenerations'' which occur after lesions in the nervous system. Cases from human pathology with lesions in different sites may be used for study, but it is the study of experimentally produced lesions in animals which has proved to be the more fruitful. In experimental investigations one usually has the advantage of being able to vary the test conditions at will, which in human cases is, of course, impossible. Furthermore, in human pathology the lesions are often large and diffuse and involve several different regions so that definite conclusions cannot be drawn. On the other hand, many small lesions suitable for study are undoubtedly present in humans; but, being so small, they often do not give rise to symptoms and are therefore not investigated. There is *no essential difference between experimentally produced lesions and those which occur in human cases.* The secondary changes which accompany the lesions are also similar, and the methods used in their study are the same. One must, of course, exercise care in applying to human beings the results achieved by experiments on animals. Since postmortem changes occur very rapidly in the brain, a major problem in the application of most anatomical methods to human material is the difficulty of obtaining properly fixed material. The use of brain biopsies circumvents this difficulty but has its obvious limitations.

Gudden's method, secondary atrophy in newborn animals. About 1870 the German anatomist B. Gudden discovered that if one injured, for example, the cerebral cortex of a newborn rabbit and then examined the brain after 7 to 8 weeks or somewhat longer, the fiber systems which originated in the part of the cortex concerned had completely disappeared, having been resorbed, and the nuclei to which the affected tracts sent their fibers were diminished and atrophied. Originally Gudden was of the opinion that this atrophy, which he found after lesions in newborn animals, affected only the fiber systems that originated in the injured or destroyed parts. Later it was realized that this was not the case (von Monakow, about 1880). The fiber tracts which terminated in the injured or destroyed regions also degenerated. After a lesion of the cerebral cortex there also appeared atrophy and wasting of the fibers that terminate there and of their nuclei of origin, e.g., certain thalamic nuclei.

Secondary atrophy in grown animals and human beings. It is characteristic that such a marked wasting as is observed in newborn animals does not appear in fully grown animals. Some atrophy and decrease in volume of the affected systems do occur, but not until a much later date and to a lesser degree. Secondary atrophy in fully grown animals can, therefore, also be used for the study of fiber connections, but this method is somewhat coarse. In such cases, as a rule, myelin

FIG. 1-2 The soma of a motoneuron in the spinal cord of the cat has been injected iontophoretically with ^{3}H-leucine. Thirty minutes after the injection, labeled proteins are visualized autoradiographically in the dendrites. ×320. From Schubert (1974).

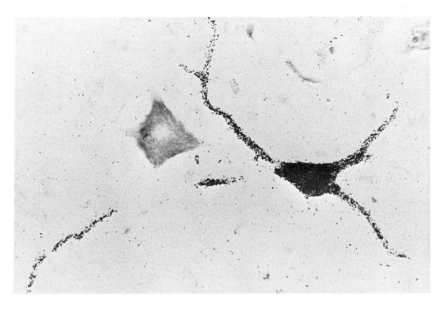

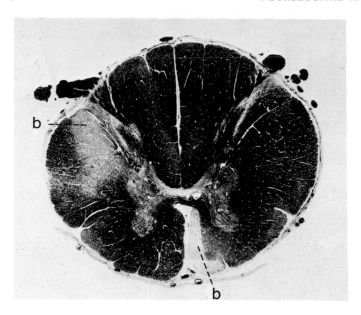

FIG. 1-3 Photograph of a myelin sheath stained preparation of the thoracic spinal cord from a patient who had a cerebral hemorrhage several years before he died. The light areas (*b*) in the lateral and ventral corticospinal tracts are due to the disappearance of the myelin sheath around the degenerated nerve fibers of the tracts.

sheath preparations are used, where the healthy, myelinated systems are stained a bluish color, while the affected systems, in which not only the axons but also the myelin sheaths have disintegrated, are pale. After old hemiplegias with hemorrhage into the internal capsule, one will thus find in myelin sheath preparations of transverse sections of the spinal cord the affected pyramidal tract as two pale areas in the remaining bluish-black stained white matter (Fig. 1-3). The coarser features of the fiber connections are thus brought to light, but this method is not suitable for detailed studies.

> The difference just mentioned, between mature and newborn individuals, appears not only in animals but also in humans. It is a well-known fact that a cerebral injury to a newborn baby or a very young child causes bigger and more visible secondary changes in the brain than the same injury to an adult. A hemorrhage into the internal capsule, for example, which in adults causes essentially only a slight reduction in volume of the pyramidal tract, as a rule in the newborn baby leads to retardation of growth of the whole side of the brain concerned and of the contralateral cerebellar hemisphere. Some experimental findings have been interpreted as showing that brain damage occurring early in life is less debilitating than comparable damage later in life. However, a critical consideration of the problem by Isaacson (1975a) does not lend much support to this notion. Isaacson's review is a thoughtful reminder of the complexities involved in the evaluation of early brain damage.

Most of the methods of investigation mentioned so far are based on the examination of large numbers of altered cells or fibers. This means that a considerably widespread injury is necessary and a great number of neurons has to be damaged. Owing to this fact detailed studies cannot be made with these methods.

The classical neuron theory. At this stage some comments on the neuron theory are appropriate. Introduced about 1890, the theory was strongly supported by His and Ramon y Cajal and had the endorsement of the most prominent anatomists and neurologists in the latter part of the last century—Forel, Koelliker, von Monakow, Waldeyer, and others. The most important assertion of the neuron theory is that nerve tissue, like other tissues, is built up of individual cells which are genetic, anatomical, functional, and trophic units. The neuron, a nerve cell with its processes, is the structural unit of nervous tissue, and the neurons are the only elements in the nervous system that conduct nervous impulses. The dendrites and soma are receptive—that is, they are acted upon by impulses from other neurons—while the axon transmits impulses arising in the neuron to its terminals ("dynamic polarization" of the neuron). The other types of cells, the various glial types, the ependyma, the epithelium in the choroid plexus, and the connective tissue cells, serve other functions. This classical neuron theory has been a central point in our interpretation of the nervous system and has proved very useful as a working hypothesis.

For several decades the proponents of the neuron theory were opposed by other researchers (often called "reticularists"), who, largely on the basis of silver-impregnated material, maintained that there is a continuity between nerve cells by means of fine fibrils, while according to the neuron theory the termination of an axon on a cell is a mere contact. The pros and cons of the neuron theory were carefully evaluated by Cajal in a monograph which appeared after his death and was translated into English about 20 years later (Cajal, 1954). The controversy was definitely settled in favor of the neuron theory by electron microscopic studies. Although there are many different morphological varieties of nerve terminals by which the final branches of an axon or its collaterals may establish contact with other nerve cells or their dendrites, there is always a clear separation between the elements belonging to the two neurons. Figure 1-4 shows an example of the most

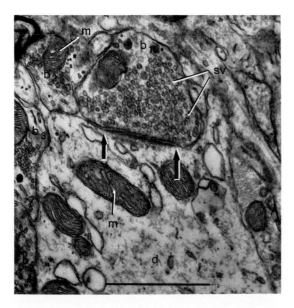

FIG. 1-4 Electron micrograph showing synaptic contact of terminal bouton (b_1) with a thin dendrite (d). Note membrane thickenings (*between arrows*) and subsynaptic condensation of material in the dendrite. To the left three other boutons $(b_2, b_3,$ and $b_4)$ in contact with the dendrite, but the synaptic complexes are not included in the section. *m:* mitochondria; *sv:* synaptic vesicles. Scale line 1μm.

common type of contact between the terminal swelling (terminal bouton, end foot, bouton terminal) of an axon and a nerve cell as seen in the electron microscope. There is a cleft of approximately 200Å (Å − Ångström unit: 1/10,000 of a micron) between the bouton and the nerve cell. (Optic wavelengths are now usually given in nanometers, nm. One nm = 10 Ångström units. In this text the old designation will be used.) At certain places along the area of contact, specializations are seen (arrows in Fig. 1-4). These indicate the areas where transmission of impulses from the bouton to the cell is believed to occur and thus represent the sites of *synapses*. Because of their importance for the understanding of nervous function the synapses will be considered separately below.

It should be emphasized that not only is the neuron a structural unit; it also, in most cases, behaves as a trophic unit, as defined by the neuron theory. This is very obvious in lesions involving neurons and, as will be seen, this fundamental fact makes possible the study of fiber connections in the nervous system in a precise manner.

The synapse.   The term "synapse" was coined by Sherrington in 1897, referring to the site of contact between two neurons where transmission of the nervous impulse occurs, i.e., as a functional term implying, however, that there is a structural basis for the phenomenon of impulse transmission. In light microscopic preparations a variety of types of endings of nerve fibers has been observed. The most common type of presynaptic structure is the *terminal bouton,* referred to above, which appears as a small spherule or bulb at the end of axonal branches and collaterals (see Fig. 1-7). Such boutons may be found on the surface of cell bodies (perikarya or soma), on dendrites, and on axons. Accordingly, one speaks of *axosomatic, axodendritic,* and *axoaxonic synapses.* The boutons vary in size, being commonly 0.5–3 μm in diameter. By special staining methods it can be shown that, at least in certain places, the entire surface of the cell body and its dendrites is densely beset with boutons (Fig. 1-5). The electron microscope has made possible a minute study of the many varieties of contacts which occur between nerve cells and has clearly shown that a contact between two nerve cells

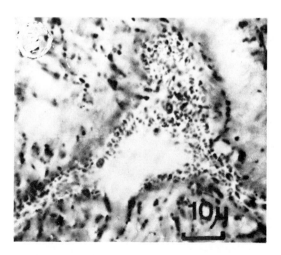

FIG. 1-5 Photomicrograph of a ventral horn cell of the cat lumbar spinal cord, stained with a special chrome silver method for demonstration of terminal boutons. Parts of the surfaces of dendrites and soma of a cell are in focus and are seen to be studded with terminal boutons of somewhat varying sizes. From Illis (1973a).

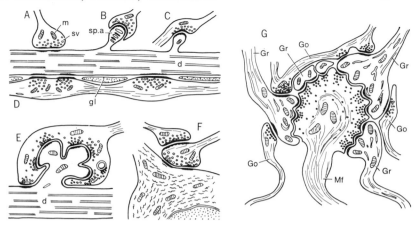

FIG. 1-6 Diagram of different types of synaptic contact as seen in electron micro-graphs (cf. text). Abbreviations: *d:* dendrite; *gl:* glial cell; *m:* mitochondrion; *sp. a:* spine apparatus; *sv:* synaptic vesicles. To the right (*G*) an example of a complex synapse, showing some of the elements in a cerebellar glomerulus. A mossy fiber ending (*Mf*) establishes synaptic contact with dendritic branches of several granule cells (*Gr*). Collaterals of Golgi cells (*Go*) end on the latter dendrites. Diagram *G* adapted from Szentágothai (1965a).

seen in sections prepared by classical methods does not necessarily represent a synapse.

A prototype of a synaptic contact is shown in Fig. 1-4. The presynaptic ter-minal, the *bouton,* is richly provided with mitochondria (indicating presumably a high rate of oxidative phosphorylation). Especially at the place of contact with the postsynaptic structure (cell body or dendrite) the bouton contains a number of small vesicles, 300–600Å in diameter, the so-called *synaptic vesicles.* In certain boutons special vesicles with a "dense core" are seen and are generally assumed to be related to biogenic amines. Sometimes "complex vesicles" and fine fila-ments are seen as well. At the site of contact the membranes of the presynaptic and postsynaptic structures show specializations in the form of electron-dense con-densations of material. In the *synaptic cleft* between these two regions, usually 150 to 250Å wide, there is often a condensation. In some synapses, for example in the cerebral cortex, this material appears as a series of fine filaments bridging the synaptic cleft. Organelles of different kinds have also been described beneath the postsynaptic membrane thickening (see Gray and Guillery, 1966; De Robertis, 1967; and the reviews of D. G. Jones, 1975, 1978).

A number of variations of the prototype of synapse described above have been encountered. Some are shown in Fig. 1-6. A bouton may be attached to a cell body or a dendrite (Fig. 1-6A). On the dendrite, boutons often establish synaptic contact with the spines (Fig. 1-6B) often found on dendrites of nerve cells. These spines were recognized by early workers in Golgi preparations, but by many they were previously thought to be artifacts. The bouton may cover the en-tire spine. In certain cells, for example, the pyramidal neurons of the hippocam-pus, a very large excrescence from the dendritic stem splits into several spinelike tips, and the whole structure is enclosed by a single large bouton (Fig. 1-6E) with a number of synaptic complexes between bouton and spine. The terminal branches of a dendrite may protrude into the bouton. An axon running along a dendrite and lying in close apposition to it may show several sites of synaptic contact, so-called *boutons en passage* (Fig. 1-6D). A passing fiber may contact a spine (Fig. 1-6C). In all these types of synapses, synaptic vesicles occur in the presynaptic element.

On the postsynaptic side various specializations have been described. One of these, a "spine apparatus" (see Fig. 1-6B), appears as a series of alternating, parallel-oriented plates of dense material and elongated vesicles in the spines (Gray, 1959). Or there may be no postsynaptic thickening but a flattened "subsynaptic sac" (Gray) or a "subsynaptic web" (De Robertis). On the basis of certain differences in the structure of the presynaptic and postsynaptic thickenings and the dimensions of the cleft, a distinction has been made between synapses of type 1 and type 2 (Gray, 1959). Type 1 has a thick postsynaptic thickening, type 2 a thin one. An often used distinction is one proposed by Colonnier between symmetrical and asymmetrical synapses. In the latter the postsynaptic membrane is considerably thicker than the presynaptic; in the symmetrical synapses the two membranes are of apparoximately equal thickness. Furthermore, in the asymmetrical synapses, the terminal contains round vesicles; in the symmetrical the vesicles are more or less flattened.[1] It should be noted that transitional types of synapses are frequent. (For recent surveys of the morphological varieties and details in synaptic structure, see, for example, *Structure and Function of Synapses,* 1972; Pfenninger, 1973; Jones, 1975, 1978.) In certain regions complex synaptic arrangements are found, for example in the so-called glomeruli. Figure 1-6G shows a diagram of a cerebellar glomerulus, to be described in Chapter 5. The morphological basis of presynaptic inhibition has been claimed to be a contact between one bouton and another as shown in Fig. 1-6F.

Often a particular synaptic variant is found only in certain regions or nuclei. It is currently generally accepted that places of contact, provided with synaptic vesicles and membrane specializations, are the actual sites where chemically mediated impulse transmission occurs. The *transmitter substance,* which is an essential link in the process, is assumed to be bound to the synaptic vesicles. On arrival of an impulse at the bouton, the transmitter presumably passes the presynaptic membrane and enters the postsynaptic cleft, to be bound to a receptor substance at the postsynaptic membrane.

The permeability of this membrane is then altered with consequent depolarization if the impulse has an excitatory action on the postsynaptic structure, or hyperpolarization of the membrane with a raising of the threshold of excitability if the action is inhibitory. The transmitter is then rapidly destroyed by enzymatic action, and the cell is ready to receive another impulse. Or the transmitter may again be taken up by the terminal. Several pharmacologically active substances are present in nervous tissue, and between 10 and 20 of them have been claimed to act as transmitters, being probably confined to specific neurons. *Acetylcholine* and *noradrenaline* appear to be definitely established as transmitter substances. Among other putative transmitters may be mentioned *GABA* (γ-aminobutyric acid), *dopamine, serotonin* (5HT, 5-hydroxytryptamine), *glycine,* and *glutamic and aspartic acid.* Acetylcholine is generally said to act as an excitatory transmitter (possibly also glutamic and aspartic acid), while GABA and glycine are considered inhibitory. The monoamines noradrenaline, dopamine, and serotonin may have an exictatory effect at some sites, an inhibitory effect at others.

In spite of intensive investigations in recent years there are still numerous unsolved questions concerning transmission at chemical synapses and factors that influence it. It is not clear, for ex-

---

[1] It has been claimed that round and flattened vesicles are typical of excitatory and inhibitory synapses, respectively. Although this may be true in certain situations, there is no doubt that the shape of the synaptic vesicles is influenced by the kind of fixation used. It is certainly not permissible to draw conclusions about synaptic functions from vesicle types in the presynaptic part of a synapse.

ample, whether a single neuron may release different transmitters at different terminals, or even at the same terminal (for a review, see Burnstock, 1976). It has been shown in invertebrates that a neuron that acts on two different cells may excite one and inhibit the other. (Both terminals are provided with the same transmitter, acetylcholine, indicating that the different responses depend on the receptor properties of the cells.) These uncertainties should induce caution in attempts to explain many findings concerning the workings of the nervous system. For some data on synaptic transmission and unsolved problems see, for example, *Neurotransmitters* (1972), *Neurotransmitters and Metabolic Regulation* (1972), and Krnjević (1974).

In spite of intensive research in this field in recent years, there are still many unsolved problems. The organization of the nervous system is extremely complex, and generalizations may be misleading. Thus, synapses operating by *electrical transmission,* known for some time to exist in invertebrates, are present also in some vertebrates (see Gray, 1966; Hinojosa and Robertson, 1967). In these *electrotonic synapses* there is a very close contact between the pre- and postsynaptic membrances and usually no synaptic vesicles. Such "gap junctions" have been carefully studied by Sotelo and Palay (1970), Korn, Sotelo, and Crepel (1973), and others. For reviews see Pappas and Waxman (1972), Pappas (1975), and Sotelo (1975).

In recent years it has been shown that in addition to the types of synapses described above (axosomatic, etc.) there are other types. Thus *dendrodendritic synapses* have been observed in various regions of the central nervous system (for example, the retina, the olfactory bulb, the superior colliculus, the thalamus, and the cerebral cortex). When two dendrites contact each other synaptically, as judged from their morphology (the distribution of synaptic vesicles, membrane specializations, and other features) a particular dendrite may have some synapses that are presynaptic, others that are postsynaptic, in relation to the other dendrite. Findings like these are difficult to reconcile with the classical neuron doctrine (as will be considered later).

Almost every nucleus appears to have its own individuality with regard to cell types, synaptic pattern, etc. Despite recent advances in all fields of neurological research, we are still far from a complete understanding of the finer organization of the nervous system.

The basis of experimental studies of fiber connections.   As indicated in a preceding section, the neuron behaves as a "trophic unit." Until quite recently all methods used to study fiber connections were based on this fact. If an axon is transected, its peripheral parts, including its terminal ramifications and boutons, and the myelin sheath undergo degeneration. As a common denominator for these changes the term *anterograde degeneration* is often used, "anterograde" referring to the direction of impulse conduction in the axon. However, the parts of the neuron proximal to the site of transection are affected as well. These changes, called *retrograde* involve the proximal part of the axon, the perikaryon, and its dendrites. Figure 1-7 illustrates the main features of this process. It further shows that the neuron receiving its afferent input from a damaged axon may be affected. These changes are called *transneuronal*. The histological methods which can be used for the demonstration of these changes differ, depending on which element one wants to study.

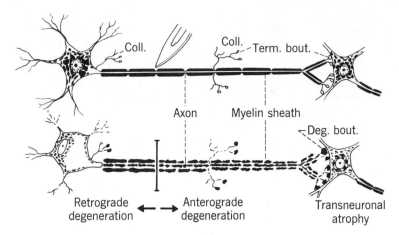

FIG. 1-7 Diagrammatic representation of the changes in the neuron which follow transection of its axon. Above, the normal situation; below, the changes as they appear some days after the lesion, central to the cut (retrograde degeneration) and peripheral to it (anterograde degeneration). Below, to the right, the transneuronal changes are indicated. (See text for description.)

Anterograde degeneration.   The changes that appear in the part of the nerve fiber which is separated from the perikaryon are often collectively referred to as *Wallerian degeneration.* The degeneration also affects the myelin sheaths of transected fibers. The degenerating myelin sheaths can be demonstrated, for example with the so-called *Marchi method,* introduced by the Italian, Marchi, about 1890. The method is based on the fact that degenerating myelin becomes impregnated by an osmic acid solution when it has previously been treated with a mordant, e.g., potassium bichromate. These changes, as a rule, reach their maximum 10 to 20 days after injury. The degenerating myelin sheaths appear as black dots or somewhat longer cylindrical particles (Fig. 1-8). The normal myelinated fibers are stained light yellow and provide a favorable background. Gradually the lipid globules are converted to fatty acids, and finally they are completely resorbed.

The *Marchi method* gives best results within a definite period of time after the injury (see above), even though in experiments with animals degeneration products can be seen at the end of one year (Glees, 1943). Marion Smith (1951) has shown that the Marchi method may also be used with human material for tracing degenerating fibers at intervals up to 13 months after division of the fiber tracts.[2] It should be noted that the optimal time for the use of the Marchi method varies with fiber systems and animal species. All methods of investigation are hampered by certain sources of error, and the Marchi method is no exception. (For a detailed study of the artifacts occurring with the Marchi method, see M. Smith, 1956b.) Used critically, the Marchi method has given much valuable information about a number of connections in the nervous system.

As stated above, most terminal axonal branches end as globule-shaped swell-

[2] According to Marion Smith (1956a; see also Smith, Strich, and Sharp, 1956) the Marchi method can be used with advantage on human material fixed in formalin.

ings called *terminal boutons*. With *silver impregnation methods* the terminal boutons may appear as ringlike figures of different sizes or as solid globules, depending on the method used. When the distal part of a transected axon disintegrates, its
terminal boutons will also degenerate. They usually swell, become irregular, and
show an increased argyrophilia (see Fig. 1-7). These processes occur in the course
of a few days. Consequently, it should be possible to determine exactly where the
fibers of a transected tract terminate by mapping the distribution of degenerating
boutons and the finest terminal axonal branches. However, in such studies it is important to consider that there are wide variations among normal boutons with
regard to their appearance in silver-impregnated sections, and that there are variations between regions or nuclei, and even within subdivisions of one nucleus, as,
for example, in the inferior olive (Blackstad, Brodal, and Walberg, 1951; Walberg, 1960). Furthermore, the speed of degeneration is not the same for all fiber
tracts or for fibers of different calibers. Negative results may be misleading because of an inappropriate survival time.

Several methods are available for the study of degenerating boutons and
fibers. With some of them (e.g., the method of Glees, 1946), normal as well as
degenerating fibers and boutons are impregnated. This makes it difficult and very
laborious to distinguish the degenerating particles, especially when they are few.
However, extremely fine degenerating fibers can be identified (see Fig. 1-9C). The
silver impregnation method worked out by Nauta and Gygax (1954) and modified
by Nauta (1957) has been much used. In successful preparations only degenerating
axons and branches are impregnated. These black degenerating structures contrast
clearly with the yellow tone of the normal structures, which provide a favorable
background. A nucleus or part of a nucleus which is the site of termination of a

FIG. 1-8   A Marchi section from the 2nd cervical segment of a cat in which 15 days
previously the lateral funiculus had been transected in the lumbar cord. Degenerating ascending fibers, interrupted by the lesion, are seen as black particles in the lateral funiculus. The degeneration in the ipsilateral dorsal funiculus is due to damage
inflicted on dorsal roots at the level of operation.

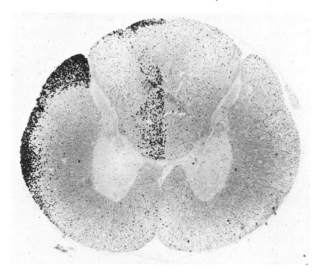

transected fiber bundle can often be distinguished very clearly even when examined at low power of the microscope (Fig. 1-9A). This method is also very well suited for tracing the course of bundles of degenerating fibers, particularly when sections are cut in the plane of the fibers. Where the degenerating fibers branch and terminate, rows of fine black particles indicate the site of terminal branches. Quite often such fibers are seen almost to encircle a perikaryon (Fig. 1-9B) or to run along the proximal dendrites, which can usually be seen.

It is important to be aware of the fact that the various silver impregnation methods differ with respect to the impregnation of degenerating fibers and of terminals (see Heimer, 1967, for a survey). For this reason negative findings may not be decisive. Disputes have occasionally arisen over divergent findings because the investigators have not been aware of the differences just mentioned.

It was originally assumed that the Nauta method did not demonstrate degenerating boutons. However, comparative studies with the methods of Glees and

FIG. 1-9 Photomicrographs from sections treated with the silver impregnation methods of Nauta (1957) and Glees (1946).

A: From a transverse Nauta-impregnated section through the lateral (*L*) and medial (*M*) vestibular nuclei in a cat, 10 days after transection of the vestibular nerve (*N VIII*). ×20. The terminal areas (*arrows*) are clearly indicated by the presence of darkly stained degenerating fibers. From Walberg, Bowsher, and Brodal (1958).

B: From a Nauta-impregnated section of the spinal cord of a cat 6 days after a lesion of the contralateral primary sensorimotor cortex. Degenerating fibers and small fragments, some of which are presumably terminal boutons, are found close to the cell bodies. ×330. From Nyberg-Hansen and Brodal (1963).

C: From a Glees-impregnated section of the spinal trigeminal nucleus in the cat 5 days after a lesion of the frontoparietal cerebral cortex. Extremely fine degenerating fibers, some of them indicated by arrows. ×530. From Brodal, Szabo, and Torvik (1956).

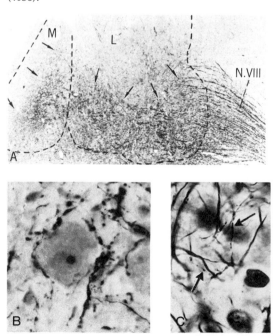

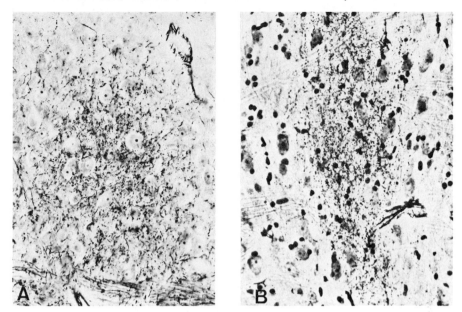

FIG. 1-10 Photomicrographs showing the circumscribed area of degenerated fibers and terminals that may be seen in silver-impregnated sections of the pontine nuclei following minute lesions of the cerebral cortex.
A: Small patch of the pontine nuclei in the cat 6 days after a lesion of the visual cortex. Silver impregnation method of Fink and Heimer (1967). ×300.
B: A small patch in the pontine nuclei of the monkey 6 days after a small lesion of the motor cortex. Method of Wiitanen (1969). Courtesy of Per Brodal.

Nauta have shown that it does (e.g., Walberg, 1964), although it is usually impossible to decide whether a particular black body represents a degenerating bouton or a fragment of a fine degenerating fiber. Electron microscopic studies of material impregnated by the Nauta method have proved unequivocally that degenerating terminal boutons can be made visible with this method (Guillery and Ralston, 1964; Lund and Westrum, 1966; and others).

Numerous modifications of the Nauta method have been developed. Among these the procedures described by Fink and Heimer (1967) appear to have been most used. With these methods the terminal boutons are usually clearly impregnated, while the terminal fibers are less conspicuous. For this reason the field of termination of a degenerating fiber bundle can be more precisely outlined than with the Nauta method. Figure 1-10 shows examples. (For an account of the use of silver impregnation methods, see *Contemporary Research Methods in Neuroanatomy,* 1970.)

Silver impregnation methods have been used by some workers on human material with satisfactory results. (The greatest difficulty is to obtain well-fixed material.) Degenerative changes in transected fibers and in boutons are usually seen as early as 4 days after the lesion, or even earlier, but it should be remembered that in human material the optimal periods may be different from those in animals.

When critically used, silver impregnation methods give information on many details of interest. Obviously, these methods may be used not only following transections of axons but also when the perikarya of axons are destroyed by lesions of a nucleus.

As stated above, the final proof that contacts observed under the light microscope represent true sites of synapses can only be obtained in electron microscopic studies, since with the silver impregnation methods it is impossible to decide whether a black particle found attached to a cell is a bouton. Furthermore, a thin glial sheet, not visible in the light microscope, may be interposed between the bouton and the cell or its dendrite. It is fortunate, therefore, that degenerating nerve fibers and terminal boutons can be identified with the electron microscope (Gray and Hamlyn, 1962; Colonnier and Gray, 1962; and others). In most places where they have been studied, degenerating boutons appear in the electron micrographs as shown in Fig. 1-11. The bouton appears to be somewhat reduced in size, its matrix has a finely granular appearance, and the mitochondria show fragmentation with loss of structural details. The synaptic vesicles are closely packed. These changes give the degenerating bouton an electron-dense appearance. At early stages the sites of synaptic contacts can still be recognized (arrows in Fig. 1-11). These early changes may be apparent in boutons of many fiber systems 2 to 5

FIG. 1-11 Electron micrograph showing a degenerating bouton (*db*) in the lateral vestibular nucleus of the cat 3 days after transection of the vestibular nerve. The bouton is dark and heavily changed (cf. Fig. 1-4), but its site of synaptic contact (*between arrows*) with a dendrite (*d*) is still visible. *b:* normal boutons; *m:* mitochondria. Scale line 1 μm.

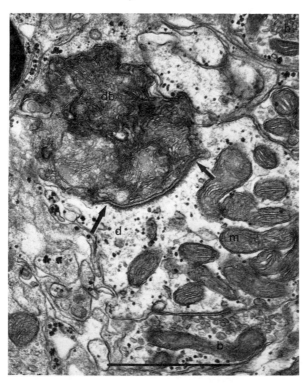

days after the axons have been cut, and the boutons may disappear rather rapidly (see, for example, Mugnaini, Walberg, and Brodal, 1967). In other instances degenerating boutons may be identified in electron micrographs for considerable periods after the lesion (e.g., 112 days in the inferior olive, Walberg, 1965a).

> In addition to this "dark type" of degenerating terminal boutons, other types have been described. The most common is the "filamentous type" where the bouton is not electron-dense but is swollen and contains an increased amount of filamentous material running in different directions. At later stages the filamentous boutons may show an increased electron density. A third, pale (electron-lucent) type as well as other types of degenerating boutons have been described (see Grant and Walberg, 1974, for a review). The causes of these variations in appearance are not known, but they induce caution in the interpretation of boutons as degenerating.

Electron microscopic studies of degenerating boutons open up new possibilities for establishing the synaptic relationships of fiber systems in great detail. Not only can it be determined whether a contact between two elements is a synapse, but information can also be obtained on other problems which cannot be solved with the light microscope. For example, it is possible to determine whether there are synaptic contacts with the finest dendritic terminals. It is essential in such studies that one knows exactly where to look for the alterations. Therefore, a precise determination with silver impregnation methods of the area in which the fiber bundle under study terminates is a necessary prerequisite (see Alksne, Blackstad, Walberg, and White, 1966).

Retrograde changes. Changes in the perikaryon and the central part of the axon after this has been cut. Originally, it was erroneously assumed that no significant changes occur in the *central part of the axon* after this has been cut. The changes in the myelin sheaths are not conspicuous, and the changes in the axon have been particularly difficult to clarify and the subject of much dispute.

> Thus it had been maintained that the retrograde changes in the axon do not extend farther toward the center than to the next proximal node of Ranvier, and it had been assumed that the proximal part of the axon undergoes only a slow atrophy. Subsequent studies, however, indicated that this was not so. Cowan, Adamson, and Powell (1961) found that in the pigeon retrograde degeneration of the axon begins close to the perikaryon and spreads centrifugally. The same conclusion was reached by Grant and Aldskogius (1967) in a study with kittens. The degenerative changes in the central part of a transected axon may be visualized with the Nauta method. Even the dendrites are affected (Cerf and Chacko, 1958; Grant and Aldskogius, 1967). The changes in the central part of the axon require a somewhat longer period of time to develop than the anterograde degeneration in the distal part (see Powell and Cowan, 1964). (This is of practical importance, as the choice of too long a survival period in studies of efferent fiber connections may give erroneous results because of simultaneous impregnation of fibers degenerating in the retrograde direction.) For references to the retrograde degeneration of axons (or indirect Wallerian degeneration), see Grant (1970, 1975), Aldskogius (1974), and Grant and Walberg (1974).

Retrograde cell changes in peripheral neurons. The changes that appear in the *perikaryon of a cell* after an injury have been of great value in the study of connections of the nervous system, particularly in determining the site of origin of fiber tracts. Even though the recent methods, based on retrograde axonal transport of macromolecules, have proved to be superior in many respects (see below), the older methods deserve attention, since they form the basis for much of our fundamental knowledge of the organization of the nervous system. Proper evaluation of findings made with them is only possible with some knowledge of the weaknesses in what they may yield.

Our first accurate knowledge of the changes that occur in the perikarya of nerve cells after transection of their axons came from Nissl. At the end of the nineteenth century he (1892) found that after division or excision of a nerve, characteristic structural changes appeared in the cells of origin of its motor fibers. In rabbits in which he evulsed the facial nerve, the cells of the facial nucleus showed a tigrolysis (dispersion of the Nissl bodies into dustlike basophilic material) that began centrally and, in the course of a few days, extended throughout the cytoplasm, leaving only a brim of fine tigroid granules in the periphery. The central parts of the cytoplasm acquired a "milky" appearance. The nucleus was displaced to the periphery, opposite the axon hillock and often was somewhat flattened. The cells appeared more round than usual, apparently because of some swelling. These changes are fully developed after 7 to 10 days, when the changed cells can clearly be distinguished from the normal ones. Nissl called this stage *primäre Reizung* (primary irritation). Corresponding changes are seen in the motoneurons of the spinal cord following transection of their axons (Fig. 1-12). The discovery of these cell changes by Nissl (in 1892) meant that one had a new means of determining the source of origin of fibers in the peripheral nerves.

Nissl did not carry his study of cell changes beyond the stage of *primäre Reizung*. When, about 1900, several authors began a renewed study of the cell changes Nissl had described, they made the interesting discovery that the facial cells, after having reached the *primäre Reizung* stage, could behave somewhat differently (Fig. 1-13). Either they underwent an increased swelling and disintegrated, or a restoration took place in the following manner; the tigroid substance reappeared, first in the central parts of the cell; the cell then became smaller and

FIG. 1-12 Photomicrographs of motor ventral horn cells in the adult monkey. *A:* normal; *B:* 6 days following transection of axons. All cells present typical acute retrograde changes. ×68. From Bodian (1947).

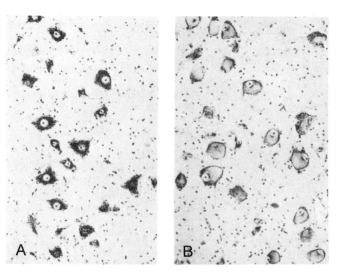

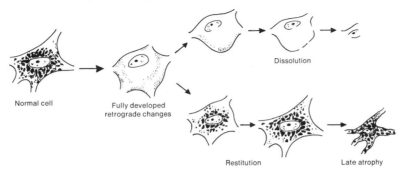

FIG. 1-13 A diagram of the usual course of the retrograde cell changes in peripheral motor neurons. To the left a normal cell. From the stage of the fully developed
retrograde changes the cell may disintegrate (*above*) or recover, but then frequently
the cell atrophies later on (cf. text).

took its normal contour again and the nucleus resumed its central position. After a
considerable time, however, these restored cells could be attacked by a slow atrophy with an increasing diminution and darker color. It appears from later studies
that this occurs when the axon of a regenerating cell does not establish functional
contact with the muscle.

These retrograde cell changes which are seen in the cell's perkaryon after
an injury to its axon appear not only in the motor cells in the ventral horns of the
spinal cord and in the motor cranial nerve nuclei, but are also seen in the cells of
spinal and cranial nerve ganglia. In these instances the cell changes appear distinctly only after division of the peripheral branch, while division of the central
branch which goes to the central nervous system does not produce clear-cut
changes.

Retrograde cell changes in central neurons. Originally it was believed
that *primäre Reizung,* as Nissl had described it in the facial nerve cells, appeared
in the same manner in all other types of neurons, but it has since been shown that
this is not the case. On the contrary, the *early changes after division of the axon
can vary considerably.* Following transection of their axon, some cells do not
swell but become smaller. The nucleus is not always peripherally displaced. And,
more important, dissolution of the Nissl bodies—chromatolysis or tigrolysis—
does not always occur, and when it does, it may not, as is usual, start centrally.[3]
These variations in the early stages of the retrograde reaction occur not only in peripheral neurons (see, for example, Cammermeyer, 1963) but also in neurons referred to as central—that is, neurons that do not send their axons out of the central
nervous system. An important difference between peripheral and central neurons
appears at a later period. *In central connections some of the affected cells disin-*

---

[3] It is unfortunate that many authors have used the terms ''chromatolysis'' and ''tigrolysis'' to
designate the retrograde cellular changes considered here. Chromatolysis, as mentioned, may be absent
in the retrograde reaction. On the other hand, chromatolysis is a feature of many manifestations of
pathological changes in nerve cells. Terms such as ''retrograde reaction'' or ''axon reaction'' are to be
preferred when reference is made to the alterations following axon transection.

tegrate acutely and none of them are restored. Those that do not disintegrate undergo a slowly progressing atrophy, which after some months may become extreme. There are variations among nuclei and among species with regard to the cellular changes and their time course, and one cannot, therefore, expect to find the typical picture of *primäre Reizung* in all neuron systems.[4] For example, in the facial nucleus of the mouse a peripheral chromatolysis may be seen 12 hours after section of the nerve (Cammermeyer, 1963), while in the human spinal cord some cells showing the classic condition of *primäre Reizung* may be seen as late as 10 months following a chordotomy (Marion C. Smith, personal communication). It is often difficult to distinguish between less characteristic reactions and normal variations. Sometimes shrunken, dark cells have been interpreted as affected by acute retrograde changes, but it has been conclusively shown that this cell condition arises as an artifact (Cammermeyer, 1960, 1962, 1972, 1975). However, where the *early* changes are clear-cut and typical, they can be used in the study of fiber connections if one is careful in interpreting the findings.

> Many studies have been made to clarify the process going on in the cell during the acute retrograde changes. In recent years numerous studies of the chemical (including enzymatic) and ultrastructural changes have been performed (see the extensive review of Lieberman, 1971). These subjects will not be considered here. The results of histochemical as well as electron microscopic studies are not all concordant as to details. This may presumably be due, at least in part, to the fact that different animal species and different cell types have been used. For a short account of the main points concerning the axon reaction, particularly chromatolysis, see Torvik (1976).

Because of the variety in pictures of the early retrograde changes and the resulting uncertainty, another procedure has been used to a great extent, namely

---

[4] Little is known about the reasons for these variations among neurons. It appears that in general an axonal lesion is less deleterious the greater its distance from the perikaryon. In some instances it appears that preservation of collaterals proximal to the injury to the axon may prevent the axon reaction. The age of the animal may play a role (see also below). Only speculations can be made about the "signal" for the axon reaction (see Cragg, 1970).

FIG. 1-14   Photomicrograph of a Nissl preparation showing retrograde loss of cells and cellular atrophy in a limited part (A) of the thalamus in the monkey, following extirpation of a cortical area. From Le Gros Clark (1937).

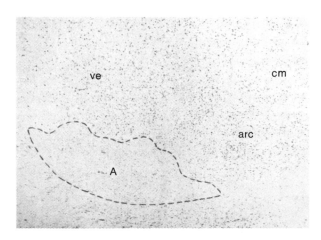

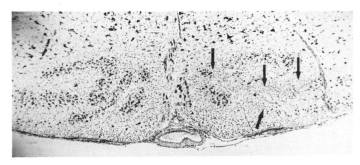

FIG. 1-15 Photomicrograph of a Nissl-stained section through the inferior olive of a rabbit in which a lesion of the cerebellar cortex was made when the animal was 11 days old. (Killed 8 days later.) The complete disappearance of the nerve cells in circumscribed parts of the contralateral olive is visible (*between the arrows*) (cf. text). From Brodal (1940b).

the examination of the *late retrograde changes, often called retrograde atrophy*. The animal is allowed to live for a longer time after the lesion is made, for weeks or months. Then, as mentioned above, as a rule, most of the cells whose axons are severed have degenerated so that the number of cells is reduced, and most of the remaining nerve cells will be atrophied. Often there is proliferation of glial cells. In many instances the parts which are affected in this manner stand out distinctly from the surrounding normal regions, and it is possible to map out the affected areas fairly accurately. (Figure 1-14 shows an example.)

By making use of the late retrograde cell atrophy, less equivocal changes may sometimes be obtained than by studying the early cell changes. In some cases, however, neither the early nor the late changes give sufficiently clear-cut pictures. This is the case with the inferior olive. A systematic investigation of the retrograde changes in the inferior olive following lesions of the cerebellum (Brodal, 1939) made it clear that the cells of the olive in very young rabbits and cats (8–15 days old at operation) behave in a manner different from those in adult animals. In the course of 4–8 days the cells show characteristic retrograde degeneration (while in the adult the changes are equivocal). Furthermore, after a few days more, all the cells disintegrate completely. The area of the olivary band which sends its fibers to the ablated part of the cerebellum then stands out as a zone free of nerve cells, with some increase in the glial cells (Fig. 1-15). Corresponding observations have been made in the pontine nuclei.

This procedure, called a *modified Gudden Method* (Brodal, 1939, 1940a), has been used with advantage in our laboratory in studying the origin of a number of fiber connections. Usually the acute changes are clearer than in adult animals. Figure 1-16 shows cells undergoing acute retrograde changes in the red nucleus of the kitten following a lesion of the spinal cord. Nerve cells are apparently more susceptible to damage to their axon in very young animals than in adult animals (see, for example, La Velle and La Velle, 1958, 1959). Also compatible with this view is the recent demonstration that in very young kittens, not only the perikaryon but also the central part of the axon and even the dendrites of the affected nerve cells degenerate rapidly (Grant and Aldskogius, 1967; Grant, 1970).

When using the modified Gudden method it is important to be aware that the time parameters vary between species and between fiber systems. At a certain stage of maturation the cells change their mode of reaction to the adult pattern. Consequently, if the animal is only a little too old, the advantages of the susceptibility of the young cells are lost. It is especially fortunate that with this method retrograde changes can usually be identified, not only in large cells but also in small ones (Fig. 1-16). This has made it clear that the efferent fibers of many nuclei come from cells of different sizes. For example, the vestibulospinal tract is made up of axons from small as well as giant cells in the nucleus of Deiters (Pompeiano and Brodal, 1957a). The drawbacks of the method are that the very young animals are relatively less resistant to operative procedures than adult ones, that dimensions are smaller, and that sterotaxic coordinates are difficult to establish.

Transneuronal changes.   All the cell and fiber changes that we have so far discussed can be explained by the neuron theory. There are, however, other changes occurring in the nerve cells which are not easily explained. In certain cases, after a fiber system is severed, changes appear in the cells to which its fibers conduct their impulses. The neurons themselves are not directly injured, but are affected nevertheless. Since the deleterious effect of the lesion is transmitted over a synapse, the cell changes are called *transneuronal* or *transsynaptic* (Fig. 1-17). These are seen most clearly in the cells of the lateral geniculate body (Fig. 1-17) after a lesion of the retina. (The ganglion cells in the retina send their axons to the lateral geniculate body where they terminate on its cells and dendrites.) In human pathology atrophy of the pontine nuclei following long-standing lesions of the cerebral cortex or cerebral peduncle (with damage to the corticopontine fibers) has been frequently observed and interpreted as being transneuronal. In other instances the changes observed have been less clear-cut. However, when more accurate methods are used, in which cell size, nuclear size, and nucleolar size are measured and counts of cells are made (Cook, Walker, and Barr, 1951), it has been possible to identify transneuronal changes in a series of nuclei following damage to their main afferent systems. As a rule the changes develop rather slowly. For example, following destruction of the cochlea in cats, changes were first noticed in the cochlear nuclei after 60 days (Powell and Eulkar, 1962). However, the time required for such changes to occur varies with the species. Changes in the lateral geniculate body following transsection of the optic nerve appear much more rapidly in the monkey (Glees and Le Gros Clark, 1941; Matthews,

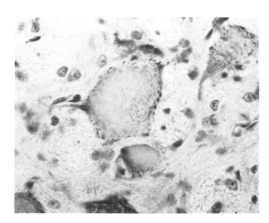

FIG. 1-16 Acute retrograde cellular changes in large and small cells in the red nucleus of a kitten following a partial transection of the cord at C_{2-3} (age at operation 8 days, killed 10 days later). From Pompeiano and Brodal (1957c).

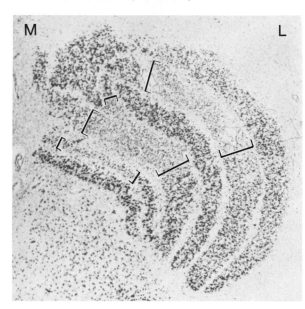

FIG. 1-17 Photomicrograph of a Nissl preparation showing transneuronal disappearance of the nerve cells in 3 of the laminae of the lateral geniculate body of a monkey after a lesion of the retina. From Le Gros Clark and Penman (1934).

Cowan, and Powell, 1960) than in the cat and rabbit (Cook, Walker, and Barr, 1951). The cellular changes consist of a moderate shrinkage of the cell, the nucleus, and the nucleolus, with some reduction in the Nissl substance. Eventually the cells may disappear completely. Just as is the case with retrograde cellular changes, the transneuronal changes occur more rapidly in young animals than in adult ones (Torvik, 1956a; Matthews and Powell, 1962). The same appears to be the case in man.

It appears that the transneuronal changes are essentially an atrophy of disuse. It is to be expected, therefore, that their intensity in any nucleus will depend on the degree of functional deafferentation. Cells which receive afferent inputs from many sources may, therefore, suffer relatively little if one afferent contingent is cut. This seems to be true in general, but, as elsewhere, one has to take into account variations in the morphology and function of the various nuclei, the age at which the damage is made, and species differences. The mechanism of transneuronal degeneration is poorly understood. For a discussion of this problem and a review of transneuronal degeneration, see Cowan (1970) and *Neuron–Target Cell Interactions* (1976, p. 355 ff.).

The transneuronal degeneration considered above may be referred to as *anterograde*. In recent years convincing evidence of the occurrence of *retrograde transneuronal degeneration* has appeared: neurons whose axons end on cells affected by primary retrograde changes (due to damage to their axon) may themselves be subject to degeneration.

The phenomenon of secondary retrograde degeneration has been particularly fruitfully studied in the medial nucleus of the mammillary body (Cowan and Powell, 1954; Bleier, 1969; and others). In young animals this second-order retrograde degeneration may proceed to complete cell loss (Bleier, 1969). Even a third-order retrograde transneuronal degeneration has been observed (in the ventral tegmental nucleus). Retrograde transneuronal changes appear to occur in neurons that are functionally particularly "dependent" on being able to act on the neurons which have suffered from primary retrograde changes. This may explain why the phenomenon is most often observed in neuronal pathways that have rather limited connections with other parts of the nervous system. For a discussion, see Cowan (1970).

The phenomena of transneuronal anterograde and retrograde degeneration are not easily compatible with the classical neuron doctrine and the notion that the neuron is a trophic unit (see below).

At this point mention should be made of a process that has recently received attention, the so-called *transganglionic degeneration*. The term refers to the degenerative changes that may be provoked in the central branches of primary sensory neurons by transection of their peripheral branches. Since the degenerative process in the neuron spreads across the ganglion (transganglionic), the sites of termination of the central branches of the primary sensory fibers may be determined with considerable precision, and the central projections of peripheral nerve branches may be studied. Promising results have been reported, particularly in studies of the central projection of branches of the trigeminal nerve (Grant and Arvidsson, 1975; Arvidsson and Grant, 1979; and others). A refinement of the method has been found by making use of the transganglionic transport of horseradish peroxidase (Grant, Arvidsson, Robertson, and Ygge, 1979), as will be considered later in this chapter.

Recent anatomical methods for the study of fiber connections by means of axonal transport of macromolecules. Until a few years ago the methods described above were the only ones that could be used for experimental studies of the connections of the nervous system. They are all based on the study of degenerative changes that occur in neurons when they are damaged; that is, they necessitate that lesions be made somewhere in the nervous system in order to interrupt fiber tracts or to destroy nuclei. Following the discovery that in neurons there occurs a transport of material from the perikaryon toward the peripheral axonal branches, and also from the latter back to the axon, methods have been developed which make it possible to visualize anatomically both anterograde and retrograde axonal transport (for a brief historical review, see Cowan and Cuénod, 1975). These methods, based on physiological properties of neurons, have opened up new perspectives for precise tracking of fiber connections of the nervous system and have been extensively used in the last few years.[5]

If tritiated amino acids are injected into a group of nerve cells, they are taken up by the neuronal somata, incorporated into proteins, and transported centrifugally in the axon to its terminal branches. (The most commonly used tritiated amino acids are leucine and proline; lysine, asparagine, adenosine, and others have also been used.) As regards the rate of transport, a distinction is commonly made between a slow flow (1–10 mm/day) and a fast flow (100–300 mm/day), although a number of intermediate rates occur. The presence of radioactive proteins in cells, fibers, and terminals is visualized by autoradiography (sections are covered with a photographic emulsion, which is bombarded by particles emitted from the isotope present within the axon and its endings). Thus an image of the tissue elements containing radioactive proteins is obtained on the photographic emulsion

---

[5] The various methods, their biological basis, the technical procedures, problems of interpretations, and sources of error are exhaustively treated in *The Use of Axonal Transport for Studies of Neuronal Connectivity* (1975).

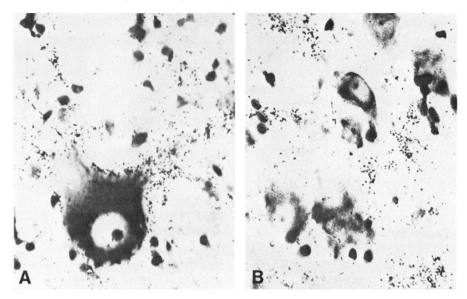

FIG. 1-18 After injection of a small amount of tritiated leucine in the cerebellar nucleus interpositus in the cat, the labeled protein has been transported anterogradely in the axons to their terminals. In autoradiographs the labeled terminals appear as black dots. The fibers from the nucleus interpositus are found to end on somata and dendrites of the cells of the nucleus ruber (A), while in the nucleus ventralis lateralis of the thalamus (B) the terminal labeling is found only in the neuropil. ×500. From Grofova and Rinvik (1974).

when it is developed; such elements appear as fine granules in axons and terminals. In this way it is possible to trace with great precision the terminal distribution and the course of the axons from a particular injection site. Figure 1-18 shows an example of the use of the autoradiographic method for determining the terminal distribution of a projection. In Fig. 1-19 the terminal distribution of climbing fibers in the cerebellar cortex is seen (see also Chap. 5).

There are several technical difficulties associated with the processing of the material, and the interpretation of autoradiographic findings is not always simple. An important source of error in interpreting labeled terminals with this method is the fact that there is usually a slight degree of radioactivity in the tissue, which gives a background of finely granular particles in the autoradiographs. Used critically, however, the technique represents an immense improvement in our armamentarium for qualitative and quantitative studies of the nervous system.

The fast rate of transport generally gives best results when the terminals of axons are to be identified. The experimental animal is then killed a few hours after the injection (depending on the length of the fiber system under study). When tracing of axons is desired, it is better to use the slow transport flow; then longer animal survival times are needed. Many factors are of importance for obtaining clear results (see Graybiel, 1975b, for a review). The concentration and the degree of spreading of the injected amino acid fluid and other factors affect the size of the "effective injection site." In some regions one amino acid appears to be preferentially taken up by a certain type of cell but not by another (Künzle and Cuénod, 1973). Further, it appears that even if tritiated amino acids may be taken up by intact axons, axons do not incorporate them into proteins (see Cowan and Cuénod, 1975). While labeled proteins may leave the axon terminals and be taken up

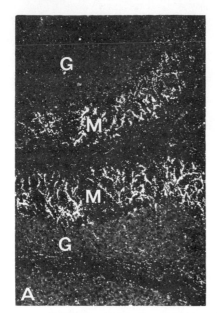 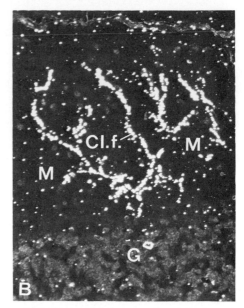

FIG. 1-19 Autoradiograph from the cerebellar cortex of the cat, after injection of tritiated amino acids in the inferior olive. In the sections, cut perpendicular to the cerebellar folia, labeled fibers can be seen entering from the granular layer (G) and continuing in the molecular layer (M) as climbing fibers (Cl.f.) along the dendrites of Purkinje cells. From Kawamura and Hashikawa (1979). *A:* ×60. *B:* ×270.

by postsynaptic neurons and thus be transported transneuronally (Grafstein, 1971; for further references, see Graybiel, 1975b), uptake of amino acids does not occur from terminals (Cowan et al., 1972; Holländer, 1974b). There is so far no autoradiographic evidence of retrograde transport of labeled proteins toward the soma, which, if it did occur, would represent a serious source of error in the interpretation of findings with this method. The radioactive proteins can be visualized in the electron microscope and have been found to be present in the terminals (Hendrickson, 1969; and others). Thus a precise determination of the synaptic relations of labeled fibers is possible. There are still many unsolved questions with regard to the anterograde transport of labeled proteins, its mechanism, and its practical application in the tracing of fiber connections.

Although our knowledge of many points is still incomplete, it is apparent that in many respects the method of tracing fiber connections in the anterograde direction by means of injection of tritiated amino acids has advantages over the use of silver impregnation methods. Most important, fibers that do not originate at the site of injection but only traverse the region will not take up the tritiated amino acid and be labeled, whereas damage to "fibers of passage" is inevitable when a lesion is made to trace degenerating fibers. Such fibers may erroneously be interpreted as originating in the area of the lesion. Furthermore, since the track of the injection cannula is very fine, one will usually avoid damage to vessels, with consequent bleeding, and distortion of the normal topography by glial reaction will be minimal. The confusion of anterograde and retrograde degenerating fibers, which may be a problem in degeneration studies in the case of two reciprocally connected regions, is no problem with this method. Finally, in some situations, the method may be more effective in revealing projections than the degeneration methods. However, like all methods, the autoradiographic method has its disadvantages and limitations (discussed recently by Cowan and Cuénod, 1975), which warn against its uncritical use. The possibility of transneuronal transport, mentioned above, is one of the sources of error.

In the study of fiber connections by autoradiography, as described above, *endogenous* proteins serve as markers. However, *exogenous* proteins may also be used for the study of efferent projections, when they are transported in an anterograde direction. One of these markers is horseradish peroxidase (HRP). This was originally assumed to be transported only retrogradely from axon terminals (see below); but later studies showed that HRP, when taken up by a perikaryon, may be transported in the anterograde direction along the axon to the terminals.[6]

The method of tracing efferent *anterograde* axonal transport of HRP is still in its beginning (for references see LaVail, 1975). It has, however, been applied to several fiber systems (see, for example, Lynch, Gall, Mensah, and Cotman, 1974). Further, it may be used for the study of the projection of Purkinje cells to the cerebellar nuclei (see Walberg, Brodal, and Hoddevik, 1976; and others). It has been used with success for tracing postganglionic sympathetic fibers from the superior cervical ganglion, for example to the ciliary processes in the eye (Brownson, Uusitalo, and Palkama, 1977). It appears from the studies performed so far that there are variations among fiber systems, for example with regard to rates of flow and the appearance of the labeled axons. There are also species differences. The presence of HRP in the terminals can be confirmed with the electron microscope. In most instances this anterograde transport presents no serious difficulty in the interpretation of findings made with retrograde tracing with HRP. The anterograde transport results in the presence of labeled fine terminals in the neuropil; the retrograde transport gives rise to coarse granulations in somata and dendrites of neurons.

Anterograde axonal flow is the basis for the tracing of efferent connections from a brain region or a particular nucleus. *The naturally occurring retrograde axonal flow may also be used for anatomical tracing of fiber connections,* particularly the *determination of the site of origin of a projection.* (For a historical review, see Kristensson, 1975.) Exogenous macromolecules injected at the site of termination of nerve fibers may be taken up by the terminals and transported retrogradely to the perikaryon where they can be identified under the microscope. Most commonly used as a tracer substance has been the enzyme *horseradish peroxidase* (HRP).[7] It is visualized in the tissue by an enzyme-histological technique, and in the light microscope the reaction product (with diaminonobenzidene) appears as brown granules in the somata and dendrites of cells whose axons have taken up the enzyme. The labeling is clearly seen with dark-field illumination. (The HRP reaction products in the nerve cells can also be identified in the electron microscope.) Figure 1-20 shows examples of retrogradely labeled cells.

The rate of the retrograde axonal transport varies somewhat between species and fiber systems, but is generally of the order of 60–100 mm/day. In the course

[6] As mentioned above, intracellular HRP injections may give information as to the course of dendrites, axons, and collaterals of single cells and permit an anatomical study of particular cells that have been recorded from in electrophysiological studies.

[7] In addition to the commonly used chromogen, 3,3'-diaminobenzidene tetrahydrochloride (DAB), others have been tried. In the oxidized state DAB turns brown. Mesulam (1976, 1978) has, for example, described procedures in which other benzidene compounds are used as chromogens. The reaction product is then blue and offers better contrast than what can be achieved with DAB. It appears that, at least in some systems, these procedures may disclose connections that are not visualized with DAB. (See also Deschênes, Laundry, and Labelle, 1979.)

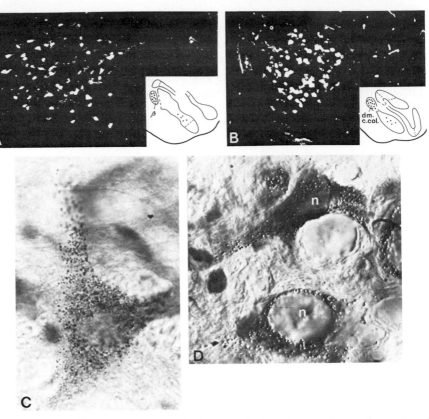

FIG. 1-20   After injection of a minute amount of a solution of HRP, the enzyme is taken up by the terminals and transported retrogradely in the axons to the somata and dendrites of the neurons. The reaction products (with diaminobenzidine) can be seen in the light microscope and with dark-field illumination. The photomicrographs show labeled cells in nuclei projecting to the cerebellum after HRP injections in the cerebellar uvula (*A* and *B*) and paramedian lobule (*C* and *D*) of the cat.

A and *B:* Most of the neurons in two small cell groups of the inferior olive (*A:* nucleus β, *B:* dorsomedial cell column) are labeled. Insets show the position of the labeled groups in transverse sections of the olive. Dark-field illumination. ×54. From A. Brodal (1976).

C: A labeled neuron in the lateral reticular nucleus, showing abundant granules when examined in the light microscope. Section weakly stained with cresyl violet. ×1400. From P. Brodal (1975).

D: Two labeled neurons from the pontine nuclei as seen with interference microscopy. The nucleus (n) is free of granules. ×1100. From Hoddevik (1975).

of a few days the enzyme disintegrates, apparently by the activity of lysosomes. Depending on the length of the axons under study (and other factors), a survival time of 24 hours to 2–3 days is usually best suited in studies of this kind. The intensity of cell labeling and the number of cells labeled in the nucleus of origin of a fiber tract whose terminals are exposed to HRP depend on many factors, among them the amount and concentration of the HRP solution injected. Even under apparently identical conditions, the degree of spreading at the injection site is variable, apparently depending on many factors, so far largely unknown.[8] With microinjections or by iontophoresis it is possible to deposit HRP in a very minute piece of tissue, but then often only a few labeled cells are found in the nucleus of origin.

The uptake of HRP by the nerve terminals appears to occur by pinocytosis. It was originally assumed that uptake occurred only at the terminals. However, it has been shown that HRP can be taken up by nerve fibers when they are damaged (de Vito, Clausing, and Smith, 1974, vagus nerve in the cat and monkey; Kristensson and Olsson, 1976, sciatic nerve in the mouse). This appears to apply to central nerve fibers as well (Halperin and LaVail, 1975; and others).

The method of determining the origin of fibers by means of the retrograde transport of HRP offers several advantages to the study of retrograde cellular changes following damage to axons. There is no serious destruction of tissue; only a few passing fibers will be destroyed; and identification of labeled cells is more reliable than the evaluation of retrograde cellular changes or retrograde cell loss. With this method many hitherto unknown areas of origin of afferent fibers to a nucleus have been revealed. For example, when the paramedian lobule of the cerebellum is ablated, retrograde cell loss in the inferior olive (see Fig. 1-15) is found only in a restricted part of the principal olive (Brodal, 1940b). After HRP injection in the same lobule (see Fig. 5-17), however, labeled cells are present in four separate regions of the olive (Brodal, Walberg, and Hoddevik, 1975).[9] Similar findings have been made in other fiber systems. On the other hand, a negative result is not decisive, since it appears that a fiber system identified with other methods may escape recognition with the HRP method (Nauta, Pritz, and Lasek, 1974; Holm and Flindt-Egebak, 1976; and others). This may indicate that not all kinds of cells are capable of HRP uptake by their axon terminals, but other explanations may be suggested as well (see footnote 8). It should be noted that when the diaminobenzidine method is used, some neurons exhibit an endogenous peroxidatic activity (Wong-Riley, 1976).

It follows from the description above that the HRP method permits precise determination of particular cells that send their axon to another cell group. However, this method, like all others, has its sources of error and limitations (for recent accounts see Cowan and Cuénod, 1975; LaVail, 1975; see also footnote 8). Even though axons of passage may take up HRP when damaged, this is usually no serious problem on account of the fine caliber of the injection needle. Nor is there

---

[8] The greatest difficulty in the evaluation of the results obtained with the retrograde HRP method is the determination of the area from which uptake has taken place, since this area, as judged by its color under the light microscope, begins to shrink some hours after the injection. Some authors have found evidence that uptake by terminals occurs only at the point where the tip of the injection needle has been. Other authors have concluded that the entire visible stained area is to be considered the area from which uptake of HRP has occurred. (Such differences may be related to differences in the texture of the neuropil.) The density of its terminal axonal field will influence the amount of HRP available to a particular cell. A spreading of the injected HRP solution via the cerebrospinal fluid to uninjected areas of brain surface may give erroneous results.

[9] The reasons for the absence of visible retrograde changes in three of the areas are not clear. The most likely explanation appears to be that, since the olivocerebellar axons branch (as has been concluded from physiological studies; see, for example, Armstrong, Harvey, and Schild, 1971, 1974), the cells in these areas of the olive will not react with discernible retrograde changes when only one of their collateral branches is destroyed.

any convincing evidence that the HRP reaction product, transported retrogradely to a cell soma or its dendrites, is transported transsynaptically to terminals of other cells that establish synapses with labeled cells. However, as mentioned above, at the injection site HRP is taken up by the somata of nerve cells and may be transported anterogradely in their axons.

The problem of retrograde axonal transport in general is of interest from a clinical point of view. Experimental studies have confirmed the old view that *tetanus toxin* reaches the central nervous system via nerve fibers that innervate the tetanus-infected wound. When experimentally injected tetanus toxin is radioactively labeled (with ^{125}I), it can be identified in sections by means of autoradiography. Tetanus toxin may, therefore, be used as a marker for retrograde axonal transport (see Schwab, Agid, Glowinski, and Thoenen, 1977).

In addition to toxins, *neurotrophic viruses* may also be transported from parts of the body to the central nervous system. This applies, for example, to the *herpes simplex virus*. When the cornea is affected, the virus may spread centrally and affect the brain, first reaching the ganglion cells in the semilunar ganglion. From here the virus spreads to the trigeminal nuclei.[10] It may, in addition, spread further within the nervous system via fiber connections. This spreading has been concluded to occur by way of retrograde axonal transport of the virus. The presence of the virus can be ascertained with electron microscopy, and to some extent also by light microscopy. From a recent study by Bak, Markham, Cook, and Stevens (1977), it appears that certain classes of neurons do not take up the virus at their terminals, whereas others do. Experimental studies of the spread of neurotrophic viruses may prove of great interest for the understanding of the pathology of virus-dependent infections of the brain (encephalitides, probably also multiple sclerosis).

Studies of the transganglionic transport of HRP (referred to in footnote 10) may be expected to give much new and important information. Such studies make it possible to determine precisely the site of termination of primary sensory fibers (somatosensory, vestibular, and others) in the dorsal horn or primary sensory nuclei. The central sites of termination of sensory afferents from cutaneous areas supplied by different peripheral nerves may be determined and compared. Comparisons between the central projections of muscular and cutaneous peripheral nerves may be made. This method (Grant, Arvidsson, Robertson, and Ygge, 1979) makes it possible to map the central distribution of peripheral nerves morphologically, while previously such information could only be obtained by electrophysiological methods.

It will be evident from the account in this chapter that today several methods are available for anatomical studies of nerve cells and their interconnections. Furthermore, in many instances it is possible to make *combined use of two methods* in the same experiment.

---

[10] This view is in agreement with the recent experimental findings of Grant, Arvidsson, Robertson, and Ygge (1979). After exposing the cut sciatic nerve of the rat to HRP, the authors observed labeled ganglion cells in the corresponding ganglia and labeling in the dorsal horn. (The procedure used may supplement information obtained with the method of transganglionic degeneration, mentioned above.)

The cells of origin of a certain projection (for example, the pontocerebellar) may be determined by means of the HRP method (labeling of pontine cells after HRP injections in the cerebellum). The sites of ending of afferents to these cell groups (for example, from the cerebral cortex) may be studied with the methods of anterograde degeneration or anterograde transport of labeled amino acids or of HRP. In this way a precise correlation of afferent and efferent connections of even a small cell group may be possible. For studies of reciprocal connections (for example, corticothalamic and thalamocortical) the injection (into either the cortex or the thalamus) of a combined solution of HRP and a tritiated amino acid may be very useful (see, for example, Jacobson and Trojanowski, 1975b). Since HRP or radioactively labeled cells and boutons can be identified in the electron microscope, it is further possible to study synaptic relations this way. Electron microscopic examination of particular cells identified in Golgi sections may be made, and studies of their contacts with normal or degenerating boutons (after experimental lesions) seem possible (see Blackstad, 1975a, b).

It is a common phenomenon that a nucleus sends fibers to more than one region. Studies of its efferent fibers by autoradiography or degeneration methods usually do not allow reliable conclusions as to which cells within the nucleus give rise to the various efferent projections. The retrograde HRP method is valuable in this situation because it may demonstrate, for example, that axons of cells of different types have different targets, or that a particular projection arises only from a subdivision of a nucleus. Examples will be mentioned in later chapters.

A special problem arises when cells may be assumed to supply more than one target area because their axons give off collaterals or branches. (This is the case, for example, within the thalamocortical projection, see Chap. 2, and the olivocerebellar projection, see Chap. 5.) Physiologically, such branchings or collateralizations may be inferred to be present from studies of the antidromic potentials recorded from the cells of origin. It is now also possible to demonstrate such a situation anatomically, since a cell can be labeled retrogradely with different markers that can be distinguished in the sections and are injected at two different sites of axonal endings from the same nucleus.

A combination of ordinary HRP with tritiated HRP (^{3}H-HRP) has been suggested (Geisert, 1976). HRP combined with tritiated bovine serum albumin is another procedure (Steward, Scoville, and Vinsant, 1977). With the combination of HRP, injected at one site, and injection at another site of tritiated HRP that has had its enzymatic activity eliminated (^{3}H apo-HRP), Hayes and Rustioni (1979) obtained promising results. Kuypers, Bentivoglio, Van der Kooy, and Catsman-Berrevoets (1979) used two substances that show yellow-green and orange fluorescence, respectively. In the future such double-labeling techniques will presumably bring forth valuable information.

It is obvious that the new methods of tracing fiber connections by utilizing the physiologically occurring axonal transport hold great promises for precise mapping of the connections of the nervous system, as discussed by Cuénod and Cowan (1975). However, it is fitting to quote the same authors (Cowan and Cuénod, 1975, p. 19) when they say, in concluding their account of the HRP method: "At the same time it should be emphasized that neither the development of this new method, nor that based on the anterograde transport of isotopically labeled materials or exogenous macromolecular markers, renders any of the more established techniques redundant. As in the past, the elucidation of the connections of any given region of the nervous system will depend on the effective application of *the most appropriate methods,* whether they be old or new."

FIG. 1-21 Photomicrograph of the locus coeruleus of a 7-day-old rat in sagittal section, treated according to the method of Falck and Hillarp to visualize noradrena-
line-containing cells and fibers. Moderately fluorescent bundles of fibers with
varicose-like enlargements are seen to leave the locus coeruleus, containing noradrenergic nerve cells. The fibers pass in the rostral direction. ×190. From Olson and
Fuxe (1971).

Finally, another new approach should be mentioned, namely the *mapping of
cells and fibers according to their monoamine content*. The monoamines dopamine, noradrenaline (norepinephrine), adrenaline (epinephrine), and 5-hydroxy-
tryptamine (serotonin) can be visualized in the fluorescence microscope with a
technique worked out by Falck and Hillarp (Falck, 1962; Falck, Hillarp, Thieme,
and Torp, 1962). It is based on the observation that the amines react with formaldehyde vapors to form fluorescent products, and cells, axons, and terminals
containing the monoamines can be seen (with the usual filters, 5-hydroxytryp-
tamine shows a yellow, the others a green fluorescence). The introduction of this
method has given rise to an avalanche of papers. Certain nuclei and fiber tracts
have been shown to be characterized by a high content of one or another of the
monoamines. About ten groups of noradrenergic neurons have, for example, been
identified in the brainstem, the best defined being that shown in Fig. 1-21, the
nucleus locus coeruleus (from which a number of noradrenergic connections have
been traced; see Chap. 6). Among dopaminergic projections, that from the substantia nigra to the striatum should be mentioned (see Chap. 4). The method has
also been applied to experimental studies involving either the destruction of fibers
or cell groups or systemic application of chemical compounds that influence the
production or transport of monoamines. It is further possible to combine staining
for monoamines with electron microscopy, to use exogenous markers, or to make
use of immunohistochemical methods (for a review, see Hökfelt and Ljungdahl,
1975). Reference will be made later to results obtained with the method of Falck
and Hillarp.

Essentially, the problems that are amenable to studies with this and related
methods relate closely to neurochemistry and transmitter distribution. It may be
mentioned that by this method and its modifications, some heretofore unrecognized fiber connections have been revealed, as for example the noradrenergic

projection from the locus coeruleus to the cerebellum and other parts of the brain. The noradrenergic innervation often appears as a widespread distribution of branching varicose terminals. It has been hypothesized that this branching may reflect not only a termination of axons from the nucleus locus coeruleus on neurons, but also a central noradrenergic influence on the cerebral microcirculation.

Physiological methods of tracing fiber connections. It is seen from the preceding account that a number of anatomical methods are at our disposal for studying fiber connections of the nervous system. However, information concerning these fiber connections can also be obtained from physiological studies. Thus the effects observed when various parts of the nervous system are stimulated may be examined and conclusions drawn concerning the pathways followed by the impulses, particularly when comparisons are made with results in similar experiments in which certain fiber tracts have been transected.

A more direct study of fiber connections has been made possible by the development of electrophysiological techniques. By means of an amplifier-cathode-ray oscillograph the action potentials arising during activity of nervous structures can be recorded, and the influence of artificial (e.g., electrical) or natural stimulation (e.g., of receptors of sense organs) on the electrical activity of the terminal area of a certain fiber system can be studied. This *method of recording evoked potentials* has yielded important information with regard to several fiber connections, and it will be referred to frequently in the following chapters. The method has an advantage in that it may be applied in many cases where it is difficult or impossible to place lesions which will give unequivocal evidence when anatomical methods are employed. It should, however, be stressed that anatomical control is necessary to make sure that the electrodes have been properly placed. It should also be realized that it is usually difficult or even impossible to decide whether the impulses recorded have traversed one synapse or more on their way from the point of stimulation to the site where they are recorded.

In recent years improvements in electrophysiological techniques have made it possible to record the action potentials of single fibers or cells (single-unit recordings) by means of microelectrodes. Important information has been obtained in this way. The recording of antidromic potentials, which occur in nerve cells on stimulation at the site of their axon terminals, has been particularly useful for correlations with anatomical data. The localization of the cells is determined histologically, and a precise correlation between the sites of origin and termination of fibers is possible. This procedure has been used, for example, by Armstrong, Harvey, and Schild (1974) to construct a detailed map of the olivocerebellar projection (recording of antidromic responses in the olive; see also Chap. 5).

The neuron doctrine in need of revision. As alluded to in the preceding text, the classical neuron theory (p. 9) can no longer be upheld in its original form. Data from recent anatomical (particularly electron microscopic), electrophysiological, and biochemical research make this clear. So far it is, however, difficult to formulate a theory that incorporates the many new data in a coherent picture.

According to the classical theory, the neuron is dynamically polarized; that is, it is a one-way information-transmitting system, the dendrites and soma surfaces being receptive, and the axon transmitting messages from the neuron to other cells. It has further been generally assumed that this message transmission requires the propagation of "spike" action potentials. Neither view is now tenable. Likewise, the notion of the neuron as a trophic unit needs modification. In the following only some pertinent data, particularly anatomical ones, will be mentioned.

Since the early days of neurobiology it has been common to distinguish two fundamentally different kinds of neurons. The *Golgi-type I* cell is provided with a long axon that branches only near its point of termination and serves impulse transmission over long distances (so-called projection neurons, for example corticospinal neurons). The *Golgi II type* neuron has an axon that arborizes abundantly close to its exit from the soma, establishes contact with cells in the immediate vicinity (so-called association neurons, some of which act as internuncial cells), and takes care of the "local" impulse traffic. The Golgi II cells occur in a variety of forms, differing with regard both to the degree and extent of their axonal ramifications and to the length, thickness, and orientation of their dendrites. This differentiation between cells of type I and II is, however, not as distinct as was formerly assumed. Thus axons of type I cells often give off an abundant number of collaterals along their course (thus being able to influence cells situated along their trajectory). Typical Golgi II cells have repeatedly been found to give off, in addition to their "local" branches, a long axon (e.g., probably all Golgi II type cells in the reticular formation; Scheibel and Scheibel, 1958; Leontovich and Zhukova, 1963). It is of further interest that in many regions dendrites may be packed in dense bundles (e.g., in the spinal cord, the reticular formation, and the neocortex; see Scheibel and Scheibel, 1975b, for a brief review). Such arrangements of dendrites are certainly of functional importance.

The heaviest blow to the classical neuron doctrine is probably the demonstration of the common occurrence of dendrodendritic synapses (dendrosomatic and somatosomatic synapses have likewise been described, see Racik, 1975, for a review). These synapses may be of the chemical type, with synaptic vesicles, or of the electrotonic type, appearing morphologically as gap junctions. Where two dendrites establish contact, synapses of the two kinds may occur in close vicinity to each other. As mentioned previously, synapses between two dendrites may often be reciprocal, one dendrite being presynaptic to the other at one synaptic site, postsynaptic at another. Physiological bioelectrical studies of the functioning of dendrodendritic synapses are beset with great difficulties, and many details remain to be clarified.

The physiological significance of electrotonic synapses and electrotonic couplings between neurons is imperfectly understood. It appears that such couplings permit a synchronization of the firing of neurons of a nucleus; also, a correlation of the occurrence of gap junctions with a tendency to synchronous neural firing has been found in several nuclei (for example, in the inferior olive).

Another point on which our concepts of the nervous system have been changed is the following: It appears to be established that the propagation of spike action potentials is not the only possible mode of message transmission between neurons, but that graded changes in potential in one neuron can synaptically influ-

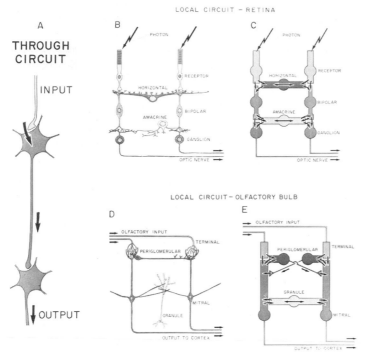

LOCAL CIRCUIT – RETINA

LOCAL CIRCUIT – OLFACTORY BULB

FIG. 1-22 Diagram to illustrate principal differences between simple "through" pathways (A) and local neuronal circuits (B to E). Dendrodendritic interactions in the retina are shown above (B and C), in B as evidenced by the arrangement of the cellular elements, while C shows the routes for interneuronal transfer of information (*arrows*). D and E show corresponding diagrams for the olfactory bulb. From Schmitt, Dev, and Smith (1976), based on Shepherd (1974).

ence the electrical activity in other neurons. This is pertinent to the understanding of, among other things, the functional role of dendrodendritic synapses.[11]

The new data give us a far more complex picture of the organization of the brain than can be accounted for by the classical neuron doctrine. The possibilities for mutual interactions between nerve cells seem to be almost limitless. This applies particularly to short-axon cells of Golgi's type II and to interactions between such cells. These interactions will occur by means of what may be called neuronal *"local circuits."* Schmitt, Dev, and Smith (1976) present an illuminating analysis of the new data.[12] A "local circuit" is described (see Schmitt et al., p. 115) as one "whose components may include many neurons joined through dendrodendritic junctions or may involve regions along individual dendrites or even patches in the neuronal membrane." The diagrams of Fig. 1-22 show two

[11] The new data have made it possible to shed some light on the function of axonless cells (amacrine cells in the retina, granule cells in the olfactory bulb, and others); see Shepherd, 1974; see also Fig. 1-22.

[12] The relative role of electrotonic and chemical synaptic transmission, the influence of the environment (ions, hormones, transmitters), and the problem of molecular transfer between neurons are likewise discussed. In recent years the functional relations between glial cells and nurons have received increased attention (see *Neuron-Target Cell Interactions,* 1976, p. 322, for references; see also Illis, 1973b).

examples. Schmitt, Dev, and Smith emphasize the role that such local circuits must be assumed to play, particularly in higher nervous functions. In this connection they point to the fact that the number and proportions of neurons that have exclusively local contacts increase in phylogeny and reach their peak in the human brain (both in absolute and relative numbers; for references see Rakic, 1975).

The data considered above are not compatible with the idea of the dynamic polarization of the neuron. The concept of the neuron as a trophic unit is likewise scarcely acceptable as it was originally formulated. Increased knowledge of anterograde and retrograde transneuronal changes (see p. 24) and particularly the finding that molecules may be transferred transsynaptically from one neuron to another indicate that a single neuron is dependent on other neurons for its viability. The significance of the interneuronal molecular transport is not clarified. One role may be to function as a biochemical signaling device between cells (for information on this point, see *Neuron-Target Cell Interactions,* 1976).

Regeneration in the central nervous system. Recovery after lesions.   Research in recent years gives us a far less rigid picture of the structure and function of the central nervous system than that prevailing only some ten years ago. Physiological, biochemical, and, to a lesser extent, anatomical studies have revealed a high degree of *plasticity in the nervous system;* that is, it is capable of a considerable degree of *remodeling,* both functionally and structurally. In general, this makes it easier for us to understand much of what we experience of human development and behavior in daily life.

The recognition of the plasticity of the nervous system has further had consequences in our understanding of the capacity of the nervous system to regain functions that have been disturbed or lost after an injury or a disease. The literature on the latter subject and on neuronal plasticity is overwhelming. Only some salient points can be touched upon here.[13]

It has long been known that after damage to the central nervous system, particularly traumatic or vascular, for example a cerebral stroke, initial and often severe functional defects will eventually be reduced, and a more or less complete restitution takes place.

New nerve cells are not formed in the developed nervous system, and transected central nerve fibers have not been convincingly shown to regenerate, as peripheral nerve fibers do.[14] The functional recovery, accordingly, must depend on other mechanisms. Improvement during the first days and weeks is generally assumed to be due largely to resorption of blood, decrease in the initial edema, and to the fact that fibers that have suffered from compression or anoxia regain their conductive capacity.[15] However, the later recovery, which is seen when the local

---

[13] For recent comprehensive reviews, see, for example, *Development and Regeneration in the Nervous System,* 1974; *Functional Recovery after Lesions of the Nervous System,* 1974; *Plasticity and Recovery of Function in the Central Nervous System,* 1974; *Outcome of Severe Damage to the Central Nervous System,* 1975; *Recovery from Brain Damage. Research and Theory,* 1978.

[14] Some recent studies suggest, however, that some central monoamine neurons have a regenerative capacity corresponding to that of peripheral neurons (for a recent account, see Svendgaard, Björklund, and Stenevi, 1975). Regeneration in peripheral neurons will not be considered here.

[15] In addition to the pathological changes mentioned above, a brain damage may produce others that contribute to functional disturbances, such as changes in cerebrospinal fluid pressure and contamination of the fluid, supersensitivity of denervated neurons, glial reactions, and retrograde changes in other regions due to destroyed axons passing through the lesion.

reparative processes of the tissue are probably completed (in the course of a few months), and which may continue for years, is more difficult to explain. Much speculation has centered on this problem, and a final solution is not yet available. The idea that nondamaged parts take over functions originally subserved by the destroyed brain regions—that is, they *function vicariously*—is scarcely satisfactory as a general explanation (see Goldberger, 1974a, for a discussion of theoretical aspects). In recent years considerable evidence has appeared which suggests that what may be called *functional reorganization* is the most important factor. *Denervation hypersensitivity* (see Chap. 11) has been assumed to play a role in such reorganization, but little is known about its role in the recovery of nervous function. Numerous observations indicate that *collateral sprouting* may be an important basis for functional reorganization.

The first demonstration of this phenomenon was made by Liu and Chambers (1958) in the cat. They transected a series of dorsal roots on one side, except one, for example L_7, with the consequence that the terminal ramifications of the transected roots in the dorsal horns degenerated. After 270 days, when the debris of the degenerated terminals had been resorbed, they cut the remaining root (L_7) as well as the corresponding root on the other side. After a survival time of a few days they mapped the distribution of degeneration in the spinal gray matter on both sides with a silver impregnation method. It turned out that on the side where the neighboring dorsal roots had been transected at the first operation the distribution of degeneration resulting from the section of a single root was more massive and extended over more spinal segments than on the other side. The terminals of the transected root had obviously spread into territories that previously had been occupied by terminals of fibers from the adjoining dorsal roots; that is, they had formed new terminal branches. Such occurrence of *collateral sprouting* has subsequently been demonstrated in several regions, for example in the hippocampus (Lynch, Stanfield, and Cotman, 1973; Zimmer, 1973, 1974b), in the superior colliculus (Lund and Lund, 1971), and in the spinal cord following transection of descending tracts (Liu and Chambers, 1958; Murray and Goldberger, 1974) or of dorsal roots (Illis, 1964b). It appears that when a cell group in the central nervous system loses some of its afferent fibers, collaterals of other preserved fibers will invade the area.

This, however, does not necessarily imply that the sprouting fibers establish synaptic contact with the denervated neurons. And if so, are they functionally active?

It has been shown in several instances that sprouting fibers do form synapses. Studies of such problems require meticulous and systematic electron microscopic investigations. It is pertinent to choose the pioneer studies of Raisman (1969a, 1969b) and Raisman and Field (1973) as an example (Fig. 1-23). They studied the medial septal nucleus of the rat, which receives two main contingents of afferents (see Chap. 10), one from the hippocampus via the fornix (fimb in Fig. 1-23) and the other from the hypothalamus via the medial forebrain bundle (MFB in Fig. 1-23). It was first shown that in the normal animal there are two main types of terminals, axosomatic and axodendritic. The fornix fibers terminate on dendrites only, the hypothalamic fibers on somata as well as dendrites (Fig. 1-23A). When one of the afferent fiber systems was transected, the fibers and boutons belonging to it degenerated and disappeared in the course of some 6 weeks. The other fiber

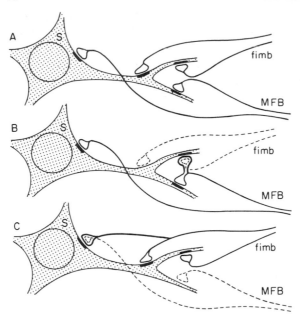

FIG. 1-23   Diagrams to illustrate the formation of new synaptic contacts due to collateral sprouting. *A* shows the normal situation, in which fibers from the medial
forebrain bundle (MFB) terminate in the medial septal nucleus with boutons on the
cell soma (S) and on dendrites, whereas the fimbrial fibers (fimb) are restricted in
termination to the dendrites. *B* shows the situation several weeks after a lesion of
the fimbria. The medial forebrain bundle fiber terminals extend across from their
own sites to occupy the vacated synaptic sites, forming double synapses. Degenerated connections are shown as broken lines, presumed plastic changes in heavy
black lines. *C* shows the situation several weeks after a lesion of the medial forebrain bundle. The fimbrial fibers now give rise to terminals occupying synaptic sites
on the soma, presumably those vacated as a result of the lesion of the medial
forebrain bundle. From Raisman (1969a).

system was then transected. The fibers and boutons belonging to this system then
degenerated and could be identified electron microscopically after a few days. It
turned out that changes had occurred in the synaptic patterns. When the fornix afferents had been transected first, the degenerating hypothalamic fibers were found
to contact an unusually large number of dendrites (Fig. 1-23B), whereas after
long-term lesions of the hypothalamic fibers, the fimbrial fibers made contact with
the somata (Fig. 1-23C). The findings strongly suggest that synaptic sites of boutons belonging to fibers that had been transected at the first operation had been occupied by boutons belonging to the other fiber system. *Collateral sprouting thus
had resulted in the establishment of new synapses at the vacated synaptic sites.*

In a continued study Raisman and Field (1973) brought forward further interesting details about collateral sprouting (in the medial septal nucleus of the rat). It
appears that it is chiefly axon terminals close to the denervated site which are
stimulated to form new synapses (see also Zimmer, 1974a). The process of reinnervation appears to follow a rigid time course, and as the degenerating terminals
are removed, the newly formed boutons occupy the vacant sites (see also Illis,
1973a, b).

It appears from the literature that all regions and fiber systems of the nervous system are not identical with regard to collateral sprouting. Some evidence indicates that the process is rather specific. Sprouting collaterals from a particular source may invade one synaptically denervated area but not another (see Murray and Goldberger, 1974; Lynch, Stanfield, Parks, and Cotman, 1974). It appears that sprouting starts early and may be complete already after 7 days in some regions. Zimmer (1974b) found the changes to be permanent (in the fascia dentata of the rat).

There are still many open questions concerning the phenomenon of collateral sprouting and its functional importance. In some instances there is evidence from electrophysiological studies that the newly formed synapses are indeed functioning (see for example, Lynch, Deadwyler, and Cotman, 1973; Steward, Cotman, and Lynch, 1974; Murakami, Fujito, and Tsukahara, 1976; see also Wall and Egger, 1971). However, inevitably there will be a *loss of specificity* when a certain group of neurons is activated by afferents other than those in the normal condition (for some considerations, see Isaacson, 1976). What will happen if the sprouting fibers operate with another transmitter than the original one?[16] Changes have been described in dendrites of denervated cells (even in second-order, transneuronal denervation; see, for example, Globus and Scheibel, 1967a; Valverde, 1968) which may influence the final result. It is an open question to what extent special factors, for example the ''nerve-growth factor'' (NGF) may influence sprouting.[17]

So far, relatively few parts of the central nervous system have been studied with regard to collateral sprouting. Most of the studies have been performed on rats. It is probable that conditions are in principle similar in man, but no conclusive evidence is available.[18] Finally, it cannot *ipso facto* be assumed that the results of collateral sprouting following lesions of the nervous system are beneficial. It is quite conceivable that the sprouting may result in disturbances. However, it is also conceivable that the plasticity of the nervous system will permit changes in the initial pattern of collateral sprouting which will make this more useful in the course of time, for example, by disappearance of initially formed synaptic contacts, or by some synapses becoming ineffective, even though they are still present. (For some comments on plasticity of synapses, see Cragg, 1974.)

There is considerable evidence that *neuronal connectivity may be modifiable.* It may change in response to influences from the surroundings and be dependent

---

[16] Little is known about the functional consequences. Collateral sprouts have been found, particularly in the autonomic system (see Chap. 11), which belong to fibers equipped with a different transmitter than the terminals of the degenerating ones. In the septum of the rat the adrenergic fibers, entering it via the medial forebrain bundle, appear to substitute for the nonadrenergic afferents entering via the fornix when this is cut (Moore, Björklund, and Stenevi, 1971). According to Raisman (1969b), large, dense-core vesicles are present chiefly in septal afferents entering via the medial forebrain bundle.

[17] No mention has been made of the factors that are at work in neuronal sprouting. Despite some positive findings, for example of the importance of the target (manufacture of ''sprouting factors'' by denervated cells?), little is known about a number of neurobiological and biochemical (including normal) factors that seem to be involved (see *Neuron-Target Cell Interactions*, 1976, for a recent account).

[18] In a recent study Goldman (1978) reports sprouting of nerve fibers in the rhesus monkey. After prenatal removal of the dorsolateral frontal cortex on one side, a projection was traced autoradiographically from the intact frontal cortex to the contralateral caudate nucleus. This crossed corticocaudate projection was not found in normal animals. The observation is taken as evidence that neuronal rearrangements can occur also in the primate telencephalon in response to lesions.

on the functional activity of neighboring cells. Support for this view may be found in studies of the changes that occur in the nervous system of newborn and very young animals as a consequence of alterations in stimulation from the environment. This has been studied especially for the visual system (for references and brief review, see Blakemore, 1974). When kittens are deprived of light from birth, overt morphological changes appear in the visual cortex and the superior colliculus (such as defective development of dendritic spines). The appearance of direction-selective units in the visual cortex has been shown to be dependent on visual experience (see Chap. 8). Such environmental influences on the development are most pronounced during a certain "sensitive period" (in kittens starting at about three weeks of age, with a maximum in the fourth and fifth week and then declining gradually), but, as stated by Blakemore (1974), the modifiability may well persist in some rudimentary fashion throughout life.

The explanation of functional recovery on the basis of collateral sprouting presupposes that morphological changes occur in neural connectivity and synaptic patterns. The structural alterations are correlated with functional changes. However, evidence is accumulating that factors that do not require overt morphological changes are at work in functional recovery of the nervous system. The term *"unmasking"* has been used for a process whose essential feature is that *previously unrecognized pathways become functioning*. In a series of elegant experiments Wall and his collaborators (see Wall, 1980, for a recent review) have brought forward interesting new information.

In physiological studies they have found in several situations that, when a particular nervous connection is completely interrupted, and one would expect total loss of the corresponding function, there is, nevertheless, some remaining function. Briefly stated, after partial destruction of the afferent input to a region (for example, the dorsal column nuclei; see Dostrovsky, Millar, and Wall, 1976) the innervated zone of the nuclei, as determined physiologically, expanded in the course of days and weeks into the denervated zone. Cells that had lost their input as a consequence of the lesion began to respond to input via intact afferents from other regions. Wall assumes from this and other findings that some connections—in this case input pathways—are normally suppressed by the activity in the main routes. When the major input fails, the normally present inhibition of the alternative (or supplementary) pathway is lost, and it starts functioning.

There appears to be little doubt that "unmasking" of nervous connections must be taken into consideration in any attempts to explain plasticity in the nervous system. "Unmasking" may function as a supplement to collateral sprouting. Whether it is universally valid remains to be seen. It is of interest, not least from the therapeutic point of view, that certain drugs may influence synaptic transmission. Experimental studies show that "unmasking" may be influenced by drugs. As mentioned by Bach-y-Rita (1980), it is conceivable that some drugs may produce "unmasking."

To demonstrate such normally inactive fibers anatomically is difficult. There is, however, some evidence of their presence in the dorsal column nuclei in the cat (Hand, 1966; Rustioni and Macchi, 1968) and in rat. In the latter, Grant, Arvidsson, Robertson, and Ygge (1979) have recently made interesting observations using transganglionic transport of HRP. They found that central processes of spinal ganglion cells belonging to the sciatic nerve, in addition to their main (and well-established) termination in the ipsilateral gracile nucleus, send some fibers to both cuneate (forelimb-related) nuclei and to the contralateral gracile nucleus.

It appears from present-day knowledge *that whereas the main cell groups and major fiber connections in the nervous system are present at birth, the finer and more detailed patterns develop largely postnatally.* It is these, making up the local neuronal circuits (see p. 37), which can be influenced environmentally by proper use and stimulation. *Learning of any kind may be assumed to require a capacity of the brain to modify its circuitry,* and to alter and adjust synaptic connections, certainly functionally, but probably also structurally. Animal experiments and investigations in children have shown the importance of the surroundings with regard to type and degree of stimulation in early years of life. The capacity to learn remains throughout life (although it decreases gradually). It appears reasonable to assume that *the improvement that occurs after lesions of the nervous system is in essence a learning process,* albeit in a defective nervous system. If so, retraining after cerebral damage must be assumed to be of importance, and most researchers who have studied this problem are of the opinion that this is indeed the case. Consequently, active therapeutic measures to improve the deficiencies following damage to the brain (be they motor, sensory, or revealed as defects in language or in other "higher" functions) are now common, even though their theoretical basis and the contribution of special factors, such as collateral sprouting, are still conjectural.

It is a common experience that the degree of recovery from an apparently identical damage may vary from patient to patient. In attempts to explain such differences the existence of *individual variations* between patients is presumably an important factor, as emphasized by Geschwind (1974a), among others, in an extensive review. Such individual differences may in part be explained by differences in anatomical structures of the brain (see Geschwind, 1974a, p. 503), but, in addition, at least concerning mental functions, the premorbid functional capacity of the brain (its "quality") seems to be of importance. Isaacson (1975a), concludes that the effects produced by brain damage depend on the genetic endowment, developmental history, and the age at which the lesion occurs and states emphatically: "The variability of behavior after brain damage in adult or infant is understandable once we give up the absurd view that the structure of each and every brain is identical." It should finally be recalled that, as with all kinds of diseases, the *patient's motivation* to recover is of overwhelming importance for the results of any retraining therapy. In recent years increasing attention has been devoted to the importance of psychological factors in the retraining and rehabilitation of brain-injured patients. For a lucid and thoughtful review, see Bach-y-Rita (1980).

The recovery that may occur after brain damage can be amazing (see Bach-y-Rita, 1980). However, closer analysis will reveal that it is scarcely ever complete, even though it may appear so to an observer. There are always minor sequelae that an observant patient will notice, but which are not obvious to others and may not be detected in a routine examination, not even with sophisticated test procedures.[19]

---

[19] An example: following a left-sided hemiparesis the right-handed patient (Brodal, 1973) noted changes in his handwriting. (These have not completely disappeared in the course of seven years.) This, in addition, illustrates that functions that are generally considered to be governed by one half of

Although our knowledge concerning what really goes on during recovery after lesions of the nervous system is only in its infancy, it is clear that the problems are extremely complex. Only a few aspects have been dealt with here. The information from recent years has given rise to a therapeutic optimism and to a general view on the processes of recovery which may be summarized by the statement of Geschwind (1974a, p. 468): "While we have been accustomed in the past to thinking about the effects of a 'fixed lesion of the nervous system,' it may in fact be the rule that a sequence of changes follows all lesions, and that perhaps one never achieves an equilibrium. There are immediate changes, and changes occurring over seconds, hours, days, weeks, months, and indeed years."

The scope and application of the methods.   Mention was made toward the beginning of the chapter that the same anatomical methods are applied in the study of fiber connections in experimental animals and in cases derived from human pathology. Attention was also directed to the frequently very extensive lesions met with in human material, which make such cases of limited value. Another circumstance which often reduces the value of human material for study of fiber connections is that the time interval which has elapsed from the development of the lesion until the death of the patient does not permit use of the method which would have been appropriate for the particular problem.

The most reliable and exact information concerning the fiber connections of the nervous system has come from experimental investigations. In animal studies the investigator himself can determine the conditions of the experiments, and he may reproduce series of more or less exactly similar cases. Especially in regard to details of the fiber connections, most of our knowledge is obtained by such studies. During the last twenty years many experimental investigations of fiber connections have been performed, and much new light has been thrown on hitherto obscure relationships between various parts of the nervous system. It should again be emphasized that great care must be taken when findings in animals are to be applied to man. This holds true particularly when the experiments have been performed on lower animals. However, monkeys have in recent years been rather extensively employed as experimental animals, and in some cases even apes, which are closely related to man in many respects. The correspondence between the experimental findings in animals and the data obtained from studies of human beings is very close in many instances. Examples of this will be mentioned later.

Even a cursory examination of neurological periodicals of recent years will show that experimental investigations make up the bulk of the papers published. This is a reflection of the fact that what is needed at present is a more accurate knowledge of the details of structure and function of the nervous system, not least a more minute mapping out of the vast number of different fiber systems, their exact origins, their detailed patterns of organization, their modes of ending, and their functional qualities. This goal can only be reached by extensive experimental investigations, although additional information from research along other lines

---

the brain only, are influenced from the other half as well, although to a lesser degree. The same is true particularly for mental functions, often said to be dependent on one cerebral hemisphere only (cp. Chap. 12).

may contribute to the ultimate verification of the results derived from experiments. In order to clarify details, extensive series of experiments usually have to be performed. Recent achievements from other fields of neurology (such as data obtained in clinical and biochemical studies) also have to be considered in the planning, performance, and interpretation of the experiments. Due attention must be paid to special features, such as cytoarchitecture (e.g., when making lesions in the cerebral cortex), chemoarchitecture, phylogenesis, and ontogenesis. The amount of exact knowledge derived from recent studies is enormous and has made modern neuroanatomy an altogether different subject from the old, mainly morphological, descriptive anatomy. There is reason to assume that in the years to come the new research tools will make it possible to considerably increase our understanding of the complex functions of the nervous system. This will undoubtedly have consequences for the diagnosis and treatment of disorders of the nervous system.

2

The Somatic Afferent Pathways*

THE DESIGNATION "somatic afferent pathways" comprises several fiber systems and nuclei, partly differing in regard to anatomical features, partly also in regard to function. The term chosen as a heading of this chapter, it is admitted, is not unequivocal, and, in order to avoid misunderstandings some preliminary remarks are necessary, which will delimit more exactly what is inherent in the concept as used here.

The different sensory modalities. Most parts of the human body are equipped with some sort of receptors, although, as is well known, normally most impulses from the viscera are not consciously perceived. However, impulses from the viscera (often called interoceptive) play an important part in visceral function, being necessary for the manifold visceral reflexes. These visceral afferent impulses, and the fiber tracts they traverse, will be treated in another chapter. Here our interest will be focused on the pathways followed by somatic afferent impulses arising in the skin, muscles, ligaments and joints, and fascia, more precisely called *general somatic afferent,* in contrast to the special somatic afferent impulses originating in some of the sense organs (eye, ear). The general somatic impulses can further be differentiated into *exteroceptive* and *proprioceptive*. The former arise from receptors developed from the integument, and mediate sensations of touch, light pressure, cold, warmth, cutaneous pain, and some more complex sensations; the adequate stimuli are changes in the environment. The latter are mediated by receptors developed in the muscles, fascia, ligaments, and joints, i.e., the deeper somatic structures, and their adequate stimuli are changes taking place

*Revised by Per Brodal and Eric Rinvik.

within the body.[1] Only the exteroceptive and some proprioceptive impulses, their receptors, the tracts they use, and their relay and terminal stations, will be considered in the present chapter. Common to the sensory modalities to be treated in this chapter is that they *reach the level of consciousness* and are perceived by the individual, giving him information of changes occurring in the external world in immediate contact with the body (exteroceptive) and of the changes in the deeper somatic structures (proprioceptive). Some impulses presumably originating from visceral receptors, such as those producing sensations of hunger, thirst, feeling of well being, and others, are, however, also consciously perceived but will not be considered in this connection. The pathways for sensory impulses passing to the cerebellum are also omitted (cf. Chap. 5 as concerns these). The muscle spindles and tendon organs, belonging to the proprioceptors, are treated in connection with the peripheral motor neurons in Chapter 3, although the role of muscle spindles in kinesthesia is treated here together with other receptors involved in kinesthesia.

Our knowledge of the pathways followed by impulses aroused by the stimulation of the different types of receptors has not been obtained solely by anatomical investigations. Only through a combination of anatomical investigations with clinical observations and animal experiments has it been possible to approach the solution of the many problems in the mechanisms of the sensory apparatus. Even though considerable advances have been made, there are still many open questions.

The receptors. Different types of sensory receptors have long been known to exist, in the skin as well as in the deeper structures, and their functional significance has been inferred partly from purely morphological data, partly also from physiological investigations. Starting with the formulation by Johannes Müller in 1840 of his "law of specific irritability," the opinion was held by many that receptors of different types give rise to sensations peculiar to each of them, regardless of the sort of stimulus applied to them. Furthermore, it was generally accepted that each type of receptor is specifically adapted to react to a definite kind of stimulus, which represents, physiologically speaking, the "normal" stimulus for that type of receptor. In other words, its threshold is low for the appropriate sort of stimuli, high for other types. Although this concept of receptor specificity has been seriously questioned (see, for example, Calne and Pallis, 1966), it is generally supported by studies of single afferent fibers, showing that at least the majority of the receptors have such a high degree of selective specificity that under natural circumstances they will convey information only of a particular type (see Iggo, 1977).

One of the problems in the study of *cutaneous sensation,* and of sensation as a whole, has been to elucidate correlations between the different structural types of receptors and the various kinds of sensations which are perceived. When one attempts to analyze this correlation, several difficulties become apparent. These are partly purely technical, but other circumstances have frequently caused confusion, such as the failure to realize that what is perceived by the mind in normal

---

[1] A special category of proprioceptive impulses, namely those originating in the vestibular part of the inner ear, is discussed in Chap. 5 on the cerebellum and Chap. 7 on the VIIIth cranial nerve.

life is not the stimulus of a single receptor or frequently not only a single type of receptor. What the individual experiences is a complex impression, resulting from a spatial and temporal summation of stimuli of different kinds. This applies especially to sensations usually described as *discriminative:* the capacity to recognize differences between the objects in contact with the skin, such as their size, form, texture, surface characteristics, etc. Another source of confusion may be the choosing of unappropriate test methods.

As concerns the morphology of cutaneous receptors, some brief data will be sufficient. The receptors are commonly subdivided into *free or unencapsulated nerve endings and encapsulated endings.*[2] Free nerve endings are present nearly everywhere in the body. In the skin all over the body nerve endings are found emerging from subepithelial nerve nets to the deeper layers of the stratum germinativum, where they terminate with arborizations between the epithelial cells (Fig. 2-1A). In the connective tissue of the subcutis and the corium many fine terminal nerve fibers are also found. Many of these free endings probably are supplied by unmyelinated axons. They are invested with Schwann cells, except near the terminal tip, where they appear to be "naked" (Fig. 2-1D,E). The nerve fibers encircling the hair follicles are fine and form a dense network on their surface, although they are derived from myelinated fibers (Fig. 2-1C).

A somewhat specialized type of free ending is the so-called *disc of Merkel.* The nerve fiber, after having branched several times, ends eventually in concave, flattened, disc-like formations, each in close contact with a single, enlarged epithelial cell of a special structure (Fig. 2-1A,F). These discs occur in man and appear to be particularly numerous in glabrous skin of distal parts of the extremities; but they are found also in hairy skin and in the skin of the lips and external genitals.

The *encapsulated nerve endings* are enclosed within a covering of connective tissue. They are found most frequently and are most elaborate structurally in regions with a highly developed sensibility. They are also found more frequently in primates than in subprimates.

Most elaborate are the *Pacinian corpuscles* (Fig. 2-1A). They may attain a length of 1 to 4 mm and may thus be visible to the naked eye as white, egg-shaped bodies. They are covered by a capsule of connective tissue rich in fibrils, arranged in concentric lamellae. This capsule encloses a protoplasmic bulb, consisting of a large number of cytoplasmic lamellae, separated by fluid spaces. In the center is a single terminal nerve fiber, running through the corpuscle. This nerve fiber belongs to those of large caliber, and loses its myelin sheath when entering the corpuscle. In the skin the Pacinian corpuscles are found in the subcutaneous layer (Fig. 2-1A), especially abundant on the tips of the fingers and toes, the palms, and soles. (Corpuscles of the Pacinian type are also found in ligaments, the periosteum, the peritoneum, mesenteries, the pancreas, and other viscera.)

Between this elaborate type of encapsulated nerve ending and the free nerve

---

[2] For accounts of morphological and functional aspects of cutaneous receptors see the Ciba Foundation Symposium: *Touch, Heat, and Pain,* 1966. An extensive account of all aspects of cutaneous sensation can be found in the monograph by Sinclair (1967). For more recent accounts the reader is referred to *Somatosensory and Visceral Receptor Mechanisms,* 1976; *Handbook of Sensory Physiology,* Vol. II/2: *Somatosensory System,* 1973; *Sensory Functions of the Skin in Primates,* 1976; *Somatic and Visceral Sensory Mechanisms,* 1977; Chouchkov: *Cutaneous Receptors,* 1978.

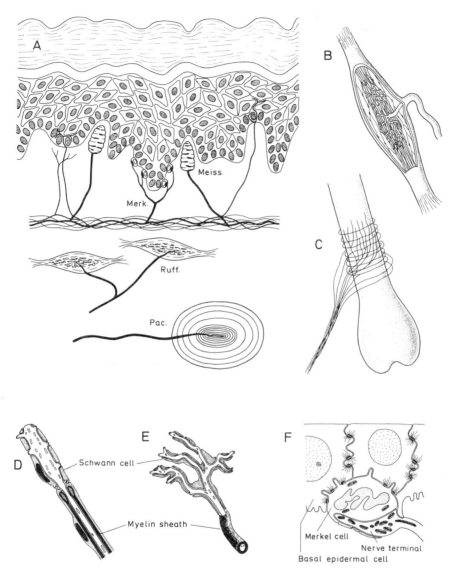

FIG. 2-1 Diagrammatic drawings of various types of cutaneous receptors. For description, see text. *A:* Typical distribution of receptors in hairless skin. Ruffini endings are found in the dermis, while the Pacinian corpuscles occur subcutaneously *B:* Drawing of a Ruffini ending (redrawn from Bannister, 1976). The capsule has been opened, and the nerve fibers can be seen to course between the collagen fibrils. *C:* The arrangement of myelinated nerve fibers around a hair root. *D,E:* Drawings of free nerve terminals, based on electron micrographs (redrawn from Andres and von Düring, 1973). The nerve fiber is enclosed by Schwann cells except at small regions near its end. At such places the axoplasm shows an increased granularity ("receptor matrix"). *F:* Drawing based on an electron micrograph of a Merkel's disc (redrawn from Iggo and Muir, 1969).

endings several transitional types exist, making any classification arbitrary. Fairly characteristic are the *corpuscles of Meissner,* found in the dermal papillae (Fig. 2-1A). They are elongated or ovoid, with their longitudinal axis perpendicular to the surface. Structurally they possess a thin connective tissue sheath, and flattened cells of an epitheloid type are found in the interior, subdividing it into transversely arranged, incompletely separated small chambers. Between these cells the entering nerve fibers wind, forming a network. There may be 2 to 6 fibers derived from axons of the thick myelinated type. Meissner corpuscles are especially numerous on the hairless volar surface of the fingers, toes, hands, and feet.

Several other types of encapsulated receptors have been described in various animal species. Among those said to occur in man are the corpuscles of Ruffini and the Krause end bulbs. The latter occur in primates usually in the lips, tongue, conjunctiva, and so forth, and have a few lamellae of connective tissue around the nerve terminal. According to Iggo (1977), the Krause end bulbs are equivalent to the Meissner corpuscles in glabrous skin (Fig. 2-1). The Ruffini ending is found in the dermis in hairy and glabrous skin (Fig. 2-1A,B), and the nerve terminals are intimately associated with collagen fibrils in the capsule merging with the dermal collagen (Chambers, Andres, von Düring, and Iggo, 1972). It is supplied by a large myelinated axon. Although there appear to be all kinds of transitions from encapsulated to free sensory endings, those mentioned above are usually recognized as distinct types by morphologists (see Andres and von Düring, 1973).

The morphological diversity of the nerve endings has naturally made it seem reasonable to allot to each of them a special function. The well-known fact that cutaneous sensibility is punctuate in nature makes it a likely assumption that the position of the various types of end organs is the factor determining the place and number of cold, warm, touch, and pain spots, the four sensory modalities regarded by most investigators as being the primary qualities of cutaneous sensibility.

Following a study by Woollard (1935) continued investigations by Woollard and Weddell appeared to support this conception. These investigations were performed on normal individuals and on patients with nerve lesions, as well as on animals. After mapping out and marking cold spots, touch spots, etc., the corresponding areas were subsequently studied with regard to the presence of nerve fibers and endings, especially with the use of intravital methylene-blue staining. It was found that *each physiologically determined spot usually includes more than one terminal corpuscle.* Thus the touch spots contain two or three Meissner corpuscles in the dermal papillae. On the finger pads they are most densely placed, approximately ten groups per mm² (Weddell, 1941a). Furthermore, the fibers covering the hair follicles are important touch receptors in the hairy parts of the skin. In the ear of the rabbit a single nerve fiber was seen to supply some 300 hairs (Weddell, 1941b). The receptors for warmth were assumed to be end bulbs resembling the Ruffini type, usually situated deep in the corium. The receptors for cold were thought to be the Krause endings.

Continued studies in the early fifties made it questionable whether a strict correlation was tenable between the morphology of the nerve endings and their function. Thus, in regions of the body like the ear, which are sensitive to touch, warmth, and pain, only two types of nerve endings occur, namely fibers ending freely and nerve fibers encircling the hairs (Sinclair et al., 1952; Hagen et al., 1953). In fact, it has been maintained by Weddell and others that there is no specificity among the receptors and that sensory perception depends only on spatial and temporal patterns of nerve impulses. This argument appeared to be particularly strong with regard to pain perception, as will be discussed later. However, more recent studies, recording the activity of single sensory nerve fibers have clearly shown that each fiber as a rule is activated only by certain kinds of stimuli applied to the skin area it innervates—that is, they show a high degree of specificity. This point will be treated further in the following account of the cutaneous receptors, the joint receptors, and the receptors for vibratory sensibility. The cutaneous receptors are generally subdivided into *mechanoreceptors, thermoreceptors,* and *pain receptors* or *nociceptors.*

Mechanoreceptors are characterized functionally as having a low threshold to mechanical stimulation of the skin. There are different kinds of mechanoreceptors. The fibers coming from the nerve plexuses surrounding the hairs respond to slight movements of a hair and adapt rapidly (see Iggo, 1966, 1974). The potentials led off from an afferent fiber stop when the movement of the hair ceases. The receptive field—the skin area from which the fiber can be excited—is smallest on the distal parts of the body, for example the dorsum of the fingers. In glabrous skin, rapidly adapting receptors with similar functional properties, probably Meissner corpuscles, are found (see Iggo, 1977). There are also, however, slowly adapting touch receptors in the skin. By single-unit recordings some receptors (type I) have been found to show a persistent discharge for 5 min or longer if the mechanical effect on the skin is continued. These receptors have been identified as Merkel's discs (Iggo and Muir, 1969) and are found in the dermis (Fig. 2-1A,F). Another type of slowly adapting receptor (type II) has a resting discharge that increases when a mechanical stimulus is applied to the skin. These receptors, which differ also in other respects from the first type, for example concerning receptive fields, are found in the dermis and appear to be Ruffini endings (Chambers et al., 1972). The slowly adapting receptors are particularly sensitive to stretching of the skin in certain directions, which is in accord with the structure and location of the Ruffini endings (see above, and Fig. 2-1B).

Elegant studies by Vallbo and Johansson (1978; see also Johansson and Vallbo, 1979) on the hairless skin of the human hand reveal the presence of receptors with the same functional properties as those described in animals. These authors recorded the activity of single nerve fibers in the median and ulnar nerves and determined their receptive fields and the kind of stimulus required to excite them. Four types of mechanosensitive units were described: two with rapidly adapting responses and two with slowly adapting responses. Within each group, one had small receptive fields, the other large. The units with rapidly adapting responses and small receptive fields (RA units) were assumed to end in Meissner corpuscles; those with small receptive fields and slowly adapting responses (SA I) most likely end in Merkel's discs. Both types of units showed a much higher density in the finger tips (140 and 70/cm², respectively) than in the palm (25 and 8/cm², respectively). Among those with large receptive fields, those with a slowly adapting response (SA II) were assumed to innervate Ruffini corpuscles, whereas nerve fibers with a rapidly adapting discharge (PC units) appear likely to end in Pacinian corpuscles. The latter two types of receptors showed about the same density in the finger tips as in the palm.

Thus four or five particular types of receptors appear to be definitely concerned in responses to what are usually called tactile stimuli.[3] There is, however,

---

[3] An interesting observation has been made in human studies (Hensel and Boman, 1960): the mechanical stimulus required for a threshold sensation is of the same order as that for setting up a single impulse in a single fiber. That a single impulse is consciously perceived has not been proved,

reason to believe that free nerve endings may also be involved. Thus the cornea contains only free nerve endings, although it responds both to tactile stimuli and to pain (Lele and Weddell, 1956). Furthermore, the response of tactile receptors may be influenced by changes in temperature, and receptor units have been observed which respond to temperature as well as mechanical stimuli (see Iggo, 1977).

Special *thermoreceptors* have so far not been identified structurally, but they are almost certainly not encapsulated, and a combined physiological and ultrastructural study indicates that cold receptors are free endings of thin myelinated fibers (Hensel, 1973). Sensations of warmth and cold are not related to the absolute temperature of the skin, as everybody knows from personal experience (e.g., lukewarm water may feel cold or warm according to the temperature of the hand when it is put into the water). The sensations of warmth and cold appear to depend on the transfer of heat to and from the skin, respectively (Lele, Weddell, and Williams, 1954). This is in agreement with the results of recent neurophysiological studies. Thus single afferent fibers have been found which respond to local application of either cold or warm stimuli. These units often show a spontaneous discharge, which in the case of a "cold" fiber increases when the skin temperature is lowered, and ceases when the temperature is raised. The "warmth" fibers behave in the opposite way. The adequate stimulus thus is a *change* in temperature, not the absolute temperature. The temperature range for maximum discharge is between 38 and 43°C for the "warmth" fibers, between 16 and 27° for the "cold" fibers (Hensel, Iggo, and Witt, 1960; and others). Interestingly, normal skin temperature is about 33°C, at which temperature the warm fibers are inactive and the cold fibers only slightly active (Iggo, 1977). A change of skin temperature of about 0.2°C is sufficient to cause a considerable change in the discharge of "cold" or "warmth" fibers. This corresponds with the thresholds of temperature sense in man, an observation of some interest since most studies on the temperature receptor units have been made in experimental animals. On recording from single fibers in the superficial branch of the radial nerve in man, "cold" fibers have been identified which behave like those studied in animals (Hensel and Boman, 1960).

The problems related to *pain receptors* and the perception of pain have attracted much interest, presumably on account of their clinical importance. Several difficulties arise in studies of this subject, particularly that of producing painful stimuli without concomitant excitation of receptors for other stimuli. Although much remains to be understood, considerable progress has been made during recent years with regard to the anatomical structures and physiological mechanisms involved in pain perception. (For a historical account of changing views, see Keele, 1957.) It appears to be generally accepted that the pain receptors are fine, freely ending fibers in the skin and other organs. Support for the idea that sensations of pain are mediated by thin nerve fibers is gained from studies in conscious human subjects using electrical stimulation of peripheral nerves and electrical or natural stimulation of the skin supplied by the nerve(s). Thus, when the nerve is stimulated directly with a low intensity, only the thick, myelinated af-

---

however. If so, it would be most likely to occur with mechanoreceptors in distal parts of the extremities, e.g., finger tips (Vallbo and Johansson, 1976). Single impulses in unmyelinated (C) fibers do not appear to be consciously perceived (Torebjörk and Hallin, 1974).

ferent fibers are activated, giving rise to a tactilelike sensation but no pain. Increasing the strength of electrical stimulation so that the thinnest myelinated fibers are also activated causes pain to be felt. A further increase in stimulus strength, which activates the unmyelinated (C) fibers, increases the sensation of pain (Heinbecker, Bishop, and O'Leary, 1933; Collins, Nulsen, and Randt, 1960; and others).[4] Compression of the nerve blocks rather selectively the activity in the thick, myelinated fibers (Torebjörk and Hallin, 1973), and this was used to show that activity in thin fibers (C group) is sufficient to cause pain (Price, 1972; Torebjörk and Hallin, 1973). Conversely, injections of a local anesthetic agent block mainly the conduction in the thinnest fibers (C group), and reduces pain sensibility much more than tactile sensibility (Torebjörk and Hallin, 1973). Furthermore, recordings from single C fibers in human nerves make it likely that they are activated rather selectively by painful stimuli (Torebjörk and Hallin, 1974).

Subjectively, there are many varieties of pain, a further indication of the complexities involved. However, it is generally agreed that as concerns pain from the skin, a distinction can be made between two types. One type is aroused by superficial penetration with a fine sharp needle, is abrupt in onset, hurts little, may be accurately localized, and disappears when the stimulus ceases. The other is felt when the needle is pressed deeper into the skin, is perceived after a short latent period, is more intense and diffuse, and has a tendency to outlast the stimulus. Both types of pain, often referred to as "pricking" and "stinging," "fast" and "slow" or "first" and "second" pain, respectively, can be elicited from most parts of the body. Among other things, experiments with graded electrical stimulation of peripheral nerves and selective blocking of different fiber groups, make it highly probable that the "fast" pain is mainly transmitted in thin, myelinated fibers (group Aδ), whereas the "slow" pain is provoked by activity in the thin, unmyelinated fibers. (Dyson and Brindley, 1906. For a study of "slow" pain, see Sinclair and Stokes, 1964.)

Judged from recording of activity in single afferent fibers in various animals, there appear to exist two principal types of nerve fibers sensitive to painful stimuli (for reviews see Burgess and Perl, 1973; Lynn, 1977). One type responds to intense mechanical stimulation of the skin but not (or very much less) to heating or irritant chemicals. These fibers appear to belong to the Aδ group. The other type, which probably belongs to the C group, is activated by a variety of sensory stimuli, such as strong heating or cooling, strong mechanical stimulation, and irritant chemicals. The first type has been called *mechanical nociceptors,* the latter *polymodal* (Bessou and Perl, 1969) or *thermal nociceptors* (Iggo, 1977). In man, afferent C fibers with properties similar to the "polymodal nociceptors" have been described (Torebjörk and Hallin, 1974). Most pain-producing stimuli appear to cause some tissue destruction. It is not known whether different types of stimuli, such as mechanical squeezing, heat, and cold, excite pain receptors through a

---

[4] According to their conduction velocities, the nerve fibers are subdivided into three groups, A, B, and C. The myelinated A fibers fall into four groups, partly overlapping, α, β, γ, and δ with decreasing conduction velocities and diameters. The B group covers myelinated efferent autonomic fibers, the C group unmyelinated fibers. In myelinated fibers the conduction velocity, as measured in meters per second, is approximately the figure obtained when the diameter of the fiber is multiplied by 6 (Hursh, 1939). However, there appear to be some species and other differences (see McLeod and Wray, 1967, for some data and references).

common factor. However, several mediators capable of activating nociceptors have been found, such as bradykinin and some of the prostaglandins (see Lynn, 1977, for a review). Although all afferent units sensitive to painful stimuli appear to be slowly adapting, the "polymodal" nociceptors with unmyelinated afferent fibers also sensitize; that is, they show a more vigorous response when repeatedly stimulated (see Perl, 1976). It should be stressed here that many C fibers are activated by stimuli other than painful ones, and not all Aδ fibers appear to be concerned with pain transmission. Thus, specific warmth receptors are most likely supplied by unmyelinated axons, and cold receptors by thin, myelinated ones (see Hensel, 1973; Darian-Smith and Johnson, 1977, for reviews). In animal experiments a large proportion of the C fibers are activated by apparently nonpainful, mechanical stimuli and may be categorized as low-threshold mechanoreceptors. However, so far, experiments in man indicate that very few C fibers are activated by low-intensity mechanical stimuli and that most afferent C fibers supply nociceptors (van Hees and Gybels, 1972; Torebjörk and Hallin, 1974).

> The *sensation of itching* is believed by several authors to be mediated by the "pain fibers" of the C group, because, among other reasons, in clinical cases it is often seen that loss of touch sensibility does not prevent itching, but in an area insensitive to pain, itching does not occur (see Arthur and Shelley, 1959; Rothman, 1960). This is supported by recent recordings from single C fibers after histamine was pricked into human skin (van Hees and Gybels, 1972; Torebjörk and Hallin, 1974). There appear to be itch spots as well as touch and temperature spots. While injection of histamine in the skin is known to produce itching, liberation of histamine does not appear to be the essential factor in the production of itch (see Arthur and Shelley, 1959).

The *Pacinian corpuscles,* referred to above, are found in the subcutaneous tissue and in many other places. In some respects they may therefore be grouped with cutaneous receptors. Since they are supplied by a single nerve fiber and, furthermore, can be readily identified on account of their size, they have proved useful for studies of certain general features of the function of receptors, especially because they may be isolated with their nerve fiber and examined in vitro. When a steady pressure is applied to a Pacinian corpuscle, it responds with a rapidly adapting discharge. Whereas these corpuscles have in the course of time been assumed to be receptors for a variety of stimuli, it appears now to be established that they essentially record *vibration.* Psychophysical studies by Talbot, Darian-Smith, Kornhuber, and Mountcastle (1968) have shown that the Pacinian corpuscles are responsible for the perception of high-frequency vibration (>100 cycles/sec), whereas receptors (not yet identified, but presumably Meissner corpuscles) more superficially in the skin appear to be responsible for perception of low-frequency vibration (<100 cycles/sec). It is a further point of interest that the Pacinian corpuscles, like most receptors, show regressive changes with advancing age (Cauna and Mannen, 1958) since vibratory sensibility is likewise reduced. The afferent fibers conduct quite rapidly (in accordance with the relative thickness of the myelinated fibers supplying the corpuscles, some 9 to 16μm). The functional importance of these vibration receptors is not yet entirely clear, but conceivably they may, together with other receptors, give information about moving stimuli on the skin. In clinical neurology examinations of "vibratory sensibility" are often performed routinely by means of tuning forks, or with special instruments (pallesthesiometers).

It remains to consider *"joint sensibility"* or *kinesthesia,* which is probably a better term in view of present knowledge. As is well known, we are able to recognize even very small movements of joints (for example, if a joint is moved passively while the subject closes his eyes), and to perceive its position even when it has not been moved for a long time (see, for example, Horch, Clarke, and Burgess, 1975). With time, various receptors have been regarded as responsible for awareness of position and movements of joints, but the problem is apparently not yet fully understood (see Matthews, 1977, for a comprehensive and critical review). The discovery before the turn of the century of elaborate sense organs in the muscles led to the assumption, unchallenged for many decades, that these organs (muscle spindles and tendon organs in particular) were of prime importance for kinesthetic sensation. Later, mainly in the 1950s and 1960s, experiments with local anesthesia of joint capsules led to the view that mainly or solely receptors in and around the joints could account for "joint sense" (see Skoglund, 1973). It was also argued by some that, since the information from the muscle spindles depends not only on the length of the muscle but also on the level of fusimotor activity, muscle spindles were not able to record the absolute position of the joints.[5] Recent experiments, however, have clearly shown that muscle receptors are able to give information of joint position and movement. Further, there is some doubt as to whether joint receptors alone can give all information needed. Consequently, receptors both in the joints and in the muscles have to be considered in the following. The muscle spindles and Golgi tendon organs will be treated fully in Chapter 3, and only their role in kinesthesia will be briefly treated here, after a description of sense organs in and around the joints.

The joints are amply innervated, but "articular neurology" has, until recently, attracted little interest. Most studies have been made in animals, particularly on the knee joint of the cat (Gardner, 1944, 1967; Skoglund, 1956; Freeman and Wyke, 1967; Clark, 1975; Clark and Burgess, 1975; Grigg, 1975; Millar, 1975; Grigg, Harrigan, and Fogarty, 1978; and others). A monograph by Poláček (1966) gives an exhaustive account of joint innervation in man as well as in several mammals and birds. The few studies performed in the monkey and man indicate that conditions are essentially the same as in the cat (Stilwell, 1956, 1957a, b; Jackson, Winkelmann, and Bickel, 1966; Keller and Moffet, 1968; Grigg and Greenspan, 1977).

The joints are partly supplied by separate nerve branches, and partly by branches from nerves supplying adjacent muscles. Recent studies of the anatomy and physiology of the endings of these nerve fibers agree on the main points. Several morphological types of receptor endings have been found in the joint capsules and in relation to them. On account of similarities with structures found in other tissues they have often been referred to as Ruffini endings, Pacinian corpuscles, etc. However, it is advisable to avoid such analogies, which may be misleading, and instead, to use neutral designations for the various types, as done by Freeman and Wyke (1967). The latter authors group the joint receptor endings in four categories.Each type appears to have its particular functional properties, but they

---

[5] However, this is not necessarily true; it may well be imagined that the central nervous system can compare the afferent signals from the muscle and the fusimotor activity and thereby determine the actual length of the muscle (see Matthews, 1977).

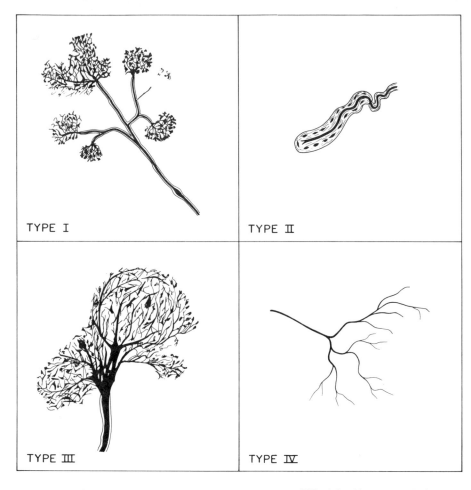

TYPE I

TYPE II

TYPE III

TYPE IV

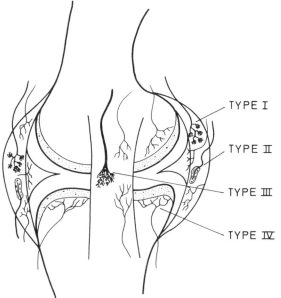

TYPE I

TYPE II

TYPE III

TYPE IV

FIG. 2-2 Above, semischematic illustration of the four joint receptors usually distinguished (based on Skoglund, 1956; Poláček, 1966; Freeman and Wyke, 1964). For description see text. To the left, a diagram of the knee joint, showing the distribution of the various receptor types in the capsule and ligaments of the joint. The menisci are free from nerve fibers except at their attachment to the fibrous capsule.

are all tension receptors, giving information of the tension in various parts of the capsule and ligaments. (Fig. 2-2).

The *type I receptors* (Fig. 2-2) are ovoid corpuscles with a thin connective tissue capsule and are supplied by a small myelinated fiber (5 to 8 μm) which arborizes within the capsule. Type I receptors occur almost exclusively in the fibrous joint capsule (a few in the extrinsic ligaments) and act as slowly adapting mechanoreceptors (stretch receptors). Both structurally and functionally these receptors resemble closely the Ruffini endings in the dermis (see Fig. 2-1B). They respond with a sustained discharge to continuous stimulation, the impulse frequency depending on the position of the joint and the speed of movement to or from their "best position." Each receptor is effective within certain angles of the range of movement in the joint. The changing frequencies of the impulses from these receptors thus may signal the direction and speed of movement and the position of the joint (Skoglund, 1956; however, see below).

Another type of receptor, called *type II* (Fig. 2-2), is about twice as large as type I and is supplied by a somewhat thicker myelinated fiber (8 to 12 μm) which usually ends as a single terminal within a rather thick laminated capsule. These receptors (which resemble the Pacinian corpuscles) occur only in the fibrous joint capsule and have been shown to be rapidly adapting mechanoreceptors. They are very sensitive to rapid movements starting from any position of the joint. They have therefore been termed "acceleration receptors" (Skoglund, 1956).

The *receptors of type III* (Fig. 2-2) are the largest ones. Each is supplied by a thick, myelinated fiber that branches profusely. These receptors (which resemble the Golgi organs) do not occur in the joint capsule but only in extrinsic and intrinsic ligaments. They adapt very slowly and have a high threshold. Their function is unclear. Their high threshold presumably makes them unsuitable for accurate recording of position, and a suggested protective function is not corroborated by studies in conscious human subjects (Petersén and Stener, 1959).

The *type IV receptor* (Fig. 2-2) of Freeman and Wyke (1967) is represented by plexuses of fine, unmyelinated fibers which occur in the fibrous capsule, the ligaments, the subsynovial capsule, and fat pads. Most authors have not been able to find them in the synovial membrane (see, however, Goldie and Wellisch, 1969). These fibers are interpreted as pain receptors.[6] Stimulation of joint nerves at a strength sufficient to activate type IV receptors gives contraction of all muscles around the joint, thereby immobilizing it. On the other hand, distending the knee joint with intra-articular fluid without provoking pain brings about a strong quadriceps inhibition, presumably by activating a large number of type I and II receptors (de Andrade, Grant, and Dixon, 1965). This phenomenon may be an important factor in the quadriceps wasting so often seen in patients with knee joint disorders.

In addition to these four types of afferent endings, the joints receive sympathetic vasomotor fibers. These degenerate after extirpation of the appropriate sympathetic ganglia (Samuel, 1952).

It appears from these studies that the joints are supplied with several receptors, each of which plays a particular role in informing the central nervous system about the position and movements of any joint at any moment.[7] As mentioned above, however, several recent studies have raised the suspicion that joint receptors are not by themselves able to mediate the information needed for all aspects of kinesthesia (see Burgess and Clark, 1969; Clark and Burgess, 1975; Horch, Clark, and Burgess, 1975; Millar, 1975; Grigg and Greenspan, 1977; and others).

[6] Some thin myelinated and unmyelinated fibers appear to be involved in cardiorespiratory reflexes evoked by gentle passive manipulation of joints in decerebrate animals (for a review, see Coote, 1975).

[7] Information from the joints is carried to the cerebral cortex chiefly (but apparently not solely) by way of the dorsal columns—medial lemniscus and by a pathway ascending in the dorsal part of the lateral funiculus (see p. 80). Such information is also of importance in the regulation of motor impulses to the muscles, where it must cooperate with afferent impulses from muscle spindles, tendon organs, and other receptors (see Chap. 3).

Thus, there appear to be very few slowly adapting joint receptors that are active when the knee and elbow joints are held in mid-range positions. Furthermore, even some earlier studies, using local anesthesia of skin and joints, have shown that the perception of movement is not completely lost, particularly not for active movements. This observation has later been confirmed and extended (Gandevia and McCloskey, 1976). The presence of virtually normal "joint sense" after total hip joint replacement has also been taken to indicate that joint receptors are not the only ones capable of giving rather precise information of joint position and movements (Grigg, Finerman, and Riley, 1973). Finally, studies in conscious human subjects have shown that vibration of muscles (regarded as a way of selectively stimulating muscle spindles) gives illusions of movement and a false impression of the joint position (Eklund, 1972; Goodwin, McCloskey, and Matthews, 1972). It seems likely that both the primary and secondary afferents of the spindle contribute to this effect. The Golgi tendon organs are presumably also excited when the vibration leads to contraction of the muscle.

It appears from the discussion above to be clearly established that muscle receptors (muscle spindles and possibly Golgi tendon organs) and not only joint receptors play a role in kinesthetic sensation. The relative contribution of each receptor type, however, has not been worked out. Apparently, various joints differ in this respect; in distal joints (finger and toes) the joint receptors are probably more important than in more proximal ones (such as knee and hip). There also appear to exist considerable individual variations (see Gandevia and McCloskey, 1976).

The preceding simplified account of receptors that respond to various kinds of stimuli to skin, joint, and deeper tissues (Pacinian corpuscles) makes it clear that we are, at least to a certain extent, justified in relating a particular morphological type of receptor to a particular type of stimulus. It is obviously no longer tenable, as previously maintained by Weddell and others, that there is no specificity among the receptors, and that sensory perception depends only on spatial and temporal patterns of nerve impulses, although such factors are of importance in the coding of information. Furthermore, it ought to be kept in mind that, although there is much evidence in favor of specificity of peripheral receptors, this situation is not necessarily matched by equally specific channels to the cerebral cortex, as will become evident from what follows later. Melzack and Wall (1962) emphasized the confusion that has arisen in premature attempts to correlate anatomical, physiological, and psychological observations. Furthermore, it is often overlooked that the selection of sensory "modalities" to be studied is to a large extent arbitrary. As Calne and Pallis (1966, p. 737) phrased it, "Though many sensations may readily be placed into broad categories labelled touch, pain, warmth and cold, the vocabulary of even the most articulate is clearly inadequate to describe the innumerable gradations of sensation which fail to fall into these convenient but quite arbitrary taxonomical pigeon-holes."

In spite of many morphological and functional differences between the receptors considered in this chapter, it nevertheless appears quite likely that they have certain basic characteristics in common, for example with regard to the elements which are involved in the transfer of energy from the non-nervous to the nervous elements. Particular attention has been directed to the membranes at these sites. (For some data on these subjects see the Ciba Symposium on *Touch, Heat, and Pain,* 1966.)

Much still remains to be done before we can understand how the many more or less specific receptors provide us with that precise information, both with regard to the qualities of stimuli and their spatial localization, which enables us to be aware of even subtle changes in the environment and in our body. The problems concerned in the central transmission of this information to the "highest levels" are no less complex.

Reference should be made here to a theory which has been much debated, namely *Head's theory of protopathic and epicritic sensibility* (Rivers and Head, 1908; Head, 1920). Experiments in nerve division which were performed by Head and Rivers on themselves formed the basis for the theory. Following the division of a cutaneous nerve of the forearm they found an area of completely abolished superficial sensibility, surrounded by a narrower zone, where pain sensibility was preserved and extreme grades of temperature were recognized, whereas touch, discrimination, and the perception of slighter differences of temperature were abolished (the intermediate zone). Furthermore, the pain elicited from this zone was abnormally intense, irradiating, and could not be accurately localized. In order to explain this a dual innervation was postulated, consisting of a *protopathic* system of primitive character, subserving pain and extreme temperature differences, yielding ungraded, diffuse impressions of an all or none type, and an *epicritic* system concerned in the mediation of smaller temperature changes, touch, and especially discrimination of all types. The epicritic system was believed to be a phylogenetically younger acquisition. The intermediate zone, as well as some phenomena observed during nerve regeneration, was assumed to be caused by a wider distribution of the protopathic system fibers in the peripheral nerves.

> The fact that an intermediate zone is regularly found between the normal parts of the skin and the totally anesthetic area is agreed upon by all observers. This has been emphasized, for example, by Wollard, Weddell, and Harpman (1940), who in the forearm of human beings, by anesthetizing the lateral and medial cutaneous nerves on successive days, found that the adjacent margins for anesthesia to touch of the two nerves coincided, whereas the area anesthetic to pain was approximately 1 cm smaller for each nerve at the border of contact, owing to overlap in the distribution of pain fibers. However, Head, and Rivers's opinion that tactile sensibility is entirely lacking in this zone has not been uniformly verified by other investigators. For example, Trotter and Davies (1909) concluded from experiments similar to those performed by Head and Rivers that the intermediate zone was not anesthetic to touch, but only hypoesthetic, the hypoesthesia increasing gradually when passing from the normal skin to the anesthetic area. Likewise they maintained that the changes in pain and temperature perception in the intermediate zone may be explained as due to a hypoesthesia to these qualities, and the same applies to two-point discrimination. They found it unnecessary to resort to Head's hypothesis to explain these phenomena as well as those observed during regeneration.

In a critical review, Walshe (1942a) analyzed the implications of Head's theory, and pointed to the many difficulties which oppose its acceptance (see also Semmes, 1969; and Henson, 1977). However, in spite of much evidence against the theory, it still seems to have its proponents.

The afferent sensory fibers. The impulses from the receptors travel centrally through the afferent sensory fibers, having their perikarya in the spinal ganglia. The fibers found in the peripheral nerves are of varying caliber, ranging from very fine unmyelinated to thick fibers.[8] The larger afferent fibers belong to

---

[8] In all peripheral nerves, fibers of sympathetic origin are also present, and in mixed and motor nerves motor fibers are also found.

the large spinal ganglion cells, the finest unmyelinated to the smallest. In general, pure cutaneous nerves and branches are richer in fine unmyelinated fibers than are the motor or mixed nerve trunks. For example, in the deep branch of the ulnar nerve in man the proportion of unmyelinated to myelinated fibers is 0.69:1, whereas the corresponding proportion in the superficial branch is 2.12:1 (Ranson, Droegemueller, Davenport, and Fisher, 1935). The fibers supplying the Pacinian corpuscles, some of the joint receptors as well as the Meissner corpuscles, the muscle spindles, and tendon organs are relatively thick and myelinated, whereas the fibers in the epidermis and subepidermal layer of the dermis are mostly very fine, in keeping with the fiber distribution in the deep and superficial nerves. As mentioned above, to some degree fibers of particular sizes are related to various kinds of receptors. This question has been tackled by recording the potentials set up in nerve fibers following stimulation of different types, since the velocity with which an impulse is conducted in a nerve fiber is related to the diameter of the fiber; the thicker the fiber the faster the conduction velocity (see Gasser, 1935). In large, myelinated fibers the impulses are propagated with a speed of some 75 to 100 m/sec; in the finest unmyelinated fibers the velocity is only 1.5–0.3 m/sec (see Footnote 4, p. 88). From several such studies it appears that there is not a very close correlation between fiber size and sensory modalities perceived (see Melzack and Wall, 1962; Iggo, 1966). Thus the responses not only to painful but also to tactile and temperature stimuli appear to be transmitted centrally by fibers of different diameters, and fibers with very similar size may have different functional properties (see, for example, Light and Perl, 1979b). Valuable information comes from recording of potentials from single fibers in response to stimuli of different kinds, and many such studies have now been performed in man (see *Sensory Functions of the Skin in Primates with Special Reference to Man*, 1976). It appears that receptors that can be fairly well defined, such as Meissner corpuscles, Merkel discs, muscle spindles, and joint receptors, have fibers with a relatively limited range of thickness, and this also seems to hold true for the afferent fibers from the functionally different types of free ending receptors, as discussed above. However, high-threshold mechanoreceptors of the skin appear to be an exception to this rule, since they have myelinated afferent fibers with conduction velocities from 5 to over 40 m/sec (see Burgess and Perl, 1973, for a comprehensive review).

The dorsal roots and afferent fibers in the ventral roots.   The somatic afferent fibers, discussed above, which convey impulses from skin, muscles, tendons, and so forth, enter the spinal cord in the dorsal or posterior roots and have their pseudounipolar perikarya in the spinal ganglia. In addition, there is now considerable evidence that many unmyelinated afferent fibers enter the spinal cord through the ventral roots. This problem will be considered below. The afferent fibers from the face pass through the trigeminal nerve and have their cell bodies in the semilunar ganglion (cf. Chap. 7). The fibers composing the dorsal roots are of a highly variable thickness, ranging from heavy fibers with a thick myelin sheath, up to 20 μm in diameter, to very fine, unmyelinated fibers of a diameter even less than 2 μm. The total number of fibers in each dorsal root is considerably larger than that in the corresponding ventral root. (For data on the fiber spectra in the various dorsal roots of man see Rexed, 1944.) It appears that fibers coming from different regions of the territory supplied by one dorsal root (a dermatome, see below) are not systematically arranged within the root. However, microelectrode recordings indicate that fibers which converge on a particular cell in the gray matter are grouped together in "microbundles" (Wall, 1960).

The fibers in the dorsal roots enter the spinal cord in the dorsolateral sulcus as a series of slender bundles. It is customary to distinguish here between two groups of fibers. Those in the smaller *lateral group* are of the thin type, mainly unmyelinated. Shortly after their entrance into the spinal cord, they dichotomize into an ascending and a descending branch, both equipped with collaterals distributed in a ventral direction and reaching the gray matter of the dorsal or posterior horn into which both branches also soon terminate. Most ascending and descending branches do not extend for more than one or two segments of the spinal cord, but some reach considerably longer, probably at least 3–6 segments rostrally and 4–5 caudally (see Imai and Kusama, 1968; Wall and Werman, 1976). Some of the thinnest are contained in a distinct, easily recognizable area, the *zona terminalis,* the *tract of Lissauer,* or dorsolateral fasciculus (Fig. 2-3B).

The fibers composing the *medial group* are larger and thicker, and most of them are myelinated. They enter the white matter closely medial to the dorsal horn and, like the fibers in the lateral group, they dichotomize in the medullary substance into an ascending and a descending branch.[9] Some of the ascending fibers are very long and can be followed in the cranial direction to the caudal part of the medulla oblongata, where they terminate in the nuclei of the posterior funiculi, *nucleus gracilis,* and *nucleus cuneatus.* These fibers give off only scanty collaterals during their course, predominantly near their origin. Other fibers in the medial group have only short ascending and descending branches, richly endowed with collaterals. These enter the gray matter of the dorsal horn, ultimately establishing synaptic contact with nerve cells in its different groups.

It appears now to be firmly established that *afferent fibers enter the cord in the ventral roots.* Although there has long been some evidence suggesting this, it was apparently the electron microscopic finding of a surprisingly high proportion of unmyelinated fibers in the ventral roots in several mammals including man that led to a systematic investigation of this problem (see Coggeshall, Applebaum, Frazer, Stubbs, and Sykes, 1975). In man, on an average 28% of the axons in the ventral roots are unmyelinated, a figure very close to that found in the animals investigated. Apparently, in the cat up to half of these fibers are preganglionic autonomic fibers (see Applebaum, Clifton, Coulter, Vance, and Willis, 1976), but this proportion seems to vary considerably between various roots. The evidence that many of the ventral-root unmyelinated fibers are afferent, comes from a variety of experiments: e.g., cutting the central roots and looking for fibers degenerating proximal to the cut, and fibers surviving distal to it. More conclusive evidence comes from experiments with injections of HRP into the spinal cord after severance of the dorsal roots proximal to the ganglion. Many labeled ganglion cells were found in such cases (Maynard, Leonard, Coulter, and Coggeshall, 1977; see,

---

[9] The fundamental discovery that dorsal root fibers dichotomize in an ascending and a descending branch was first made by the Norwegian explorer Fridtjof Nansen (1886) in myxine (hagfish).

however, Yamamoto, Takahashi, Satomi, and Ise, 1977). When the ventral root was also cut prior to injection, "almost no labeled cells" were found (Maynard et al., 1977). Physiological studies have so far been performed to a limited extent. Most afferent fibers in the sacral ventral roots of the cat were found to have receptive fields in pelvic visceral organs, but about one-third were activated from somatic structures. The thresholds of the latter group were high as a rule, suggesting that these fibers may be involved in transmission of painful stimuli (Clifton, Coggeshall, Vance, and Willis, 1976).

Some myelinated afferent fibers in the ventral roots have been demonstrated as well, but their number appears to be so small that it is doubtful that they have any functional significance (see Loeb, 1976).

It has been suggested that the afferents entering the ventral roots may be responsible for persistence of pain in some patients after dorsal rhizotomy, but other possibilities have to be taken into account as well in cases of chronic pain. Furthermore, there appears to be no evidence of pain produced by stimulation of ventral roots in man (see White and Sweet, 1969).

Some features of the organization of the spinal cord. Termination of dorsal root fibers.   The structure of the gray matter of the cord as seen in transverse sections is not identical all over. Various cell groups stand out fairly distinctly; for example, groups of multipolar cells in the ventral horn and (in segments Th_1–L_2) the column of Clarke at the medial aspect of the dorsal horn (see Fig. 3-1) while other groups are less clearly outlined. By making use of thick Nissl-stained sections Rexed (1952, 1954) has shown that on the basis of its cytoarchitecture the gray matter of the spinal cord of the cat can be subdivided into ten, largely horizontal, zones or laminae (Fig. 2-3A). Most of these are present throughout the cord, although with minor variations between levels. In fiber-stained preparations most of the zones likewise have their characteristic features (see Nyberg-Hansen and Brodal, 1963, for some data). The assumption that the zonal subdivision of the spinal gray matter reflects functional properties is borne out by a number of data, physiological as well as anatomical. For example, the fiber systems descending to the cord from the cerebral cortex and nuclei in the brainstem end preferentially in particular zones (see Chap. 4). Other data will be mentioned later in this and other chapters. Apart from bearing witness to a detailed functional and structural differentiation in the gray matter, Rexed's (1952, 1954) studies have provided us with an exact and much needed reference map which permits precise statements as to points studied, anatomically and physiologically, in the spinal gray matter.[10]

From a functional point of view the neurons of the gray matter may be subdivided into three kinds. One kind, found in lamina IX and the intermediolateral cell column (see Figs. 2-3 and 11-1), sends axons out of the cord into the ventral roots. Another type of cell gives off long ascending axons to supraspinal levels. A third group consists of cells whose axons remain in the cord. These cells may be referred to as *internuncial cells* or *interneurons*. The classical interneuron, the

---

[10] A corresponding map for man has not yet been made. It appears likely that the principal organization will turn out to be as in the cat.

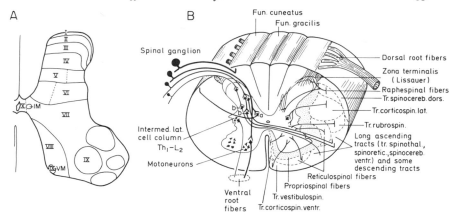

FIG. 2-3 A: Diagram of a transverse section through the 6th cervical segment of the spinal cord of the cat, illustrating the cytoarchitectonic subdivision of the gray matter into approximately horizontally arranged laminae (I to X) according to Rexed (1954). LM and VM: lateromedial and ventromedial groups of motoneurons. B: Schematic drawing of an isolated segment of the spinal cord, showing the location of some cell groups and the main ascending and descending tracts. Not all tracts are included, and the borders between adjacent tracts are not sharp. a: Neuron in Clarke's column, giving origin to the dorsal spinocerebellar tract; b: neurons giving off spinoreticular and spinothalamic fibers.

Golgi type II cell, with a short axon that does not leave the gray matter and arborizes extensively in the vicinity of the cell body, appears to be rare in the spinal cord (Scheibel and Scheibel, 1966a). Accordingly, cells identified physiologically as interneurons in certain spinal reflex pathways have recently been shown with intracellular injection of HRP to send axons into the white matter after giving off several local collaterals (see, for example, Czarkowska, Jankowska, and Sybirska, 1976). The topography of the internuncial cells has been worked out in detail only for a few types. (Some features of functional interest will be described in Chap. 3.) However, the great number of internuncial cells provide convincing evidence of a complex organization of impulse pathways in the gray matter of the cord, as does the fact that the dendrites of nerve cells in the cord may extend for considerable distances from the perikaryon (Aitken and Bridger, 1961; Sprague and Ha, 1964; Testa, 1964; Scheibel and Scheibel, 1966a; Szentágothai, 1967; Brown, Rose, and Snow, 1977a; and others). These and other data argue against correlating various laminae too specifically with particular functions.

Many axon collaterals of internuncial cells remain in their own segment of the cord, and are concerned in the transmission of incoming impulses to cells of the other two types at the same level. It is important to realize, however, that probably the majority of cells give off collaterals which extend up or down for several segments (see Chap. 3) and thus establish an extensive network of interconnections between neighboring segments and provide possibilities for interaction between these. Some of these axons of internuncial cells course within the gray matter. Others, however, enter the white matter, descend or ascend in this for various distances, and then enter the gray matter again. Most of these intersegmental fibers are found close to the gray matter, where they make up the massive *fasciculi proprii* (sometimes called propriospinal fibers) (see Fig. 2-3B). An impression of their abundance is obtained when in an experimental animal a couple of segments of the cord are isolated by transverse lesions above and below the segments, and the corresponding dorsal roots are cut. After some time the dorsal root fibers and the ascending and descending long fibers passing through the segments will degenerate, and the remaining fibers should all represent segmental interconnections. Such fibers occur in all three funiculi of the cord (see Tower, Bodian, and Howe, 1941; Anderson, 1963). It should be especially noted that the tract of Lissauer, or dorsolateral fasciculus, situated peripherally to the dorsal horn and lateral to the entering dorsal root fibers (Fig. 2-3B), contains a large number of thin fibers which are intrinsic, i.e., they originate in the gray matter of the cord and re-enter this after having coursed in the tract for one or two segments. In the cat some 75% of all fibers of Lissauer's tract are of this type (Earle, 1952; see also Szentágothai, 1964a). However, in the monkey cervical cord La Motte (1977) concluded on the basis of autoradiographic studies that 50% of the fibers in the tract of Lissauer are primary afferents. An exhaustive review of the fasciculi proprii in man has been given by Nathan and Smith (1959; see also La Motte, 1977).

In addition to connections between neighboring segments, axons of cells on one side cross the midline, making possible a cooperation between the right and left halves of the cord. Axons of such cells, which are particularly abundant in lamina VIII, pass in the ventral and dorsal commissures.

> The location of cells giving rise to long ascending and descending propriospinal fibers, interconnecting cervical and lumbar levels of the cord, has recently been mapped with the horseradish peroxidase method in cat and monkey (Burton and Loewy, 1976; Molenaar and Kuypers, 1978; Matsushita, Ikeda, and Hosoya, 1979). Such cells are particularly numerous in lamina VIII and adjoining parts of VII, but are present also in lamina I and V. Degeneration experiments indicate that such fibers terminate mainly in lamina VIII and the adjoining part of lamina VII, but to a limited extent also in lamina IX (Barilari and Kuypers, 1969; Matsushita and Ikeda, 1973; Matsushita and Ueyama, 1973). Physiological experiments indicate that the long propriospinal fibers mediate mono- and disynaptic excitatory and inhibitory effects on motoneurons (Jankowska, Lundberg, Roberts, and Stuart, 1974; and others). Such interconnections between cervical and lumbar levels of the cord are presumably of importance for coordinated activity in fore- and hindlimb muscles during walking. Whether, in addition, the long propriospinal fibers arising in lamina I and V are concerned with modulation of sensory transmission is not known.

As mentioned above, some dorsal root fibers turn cranially on entering the cord and ascend in the dorsal columns. These can be followed quite easily as a separate fiber system (see below). The distribution of the other, larger contingent of root fibers, which enters the gray matter directly, has been studied in Golgi preparations and experimentally with the Marchi and silver impregnation methods, and more recently with methods making use of the axonal transport of various substances. It has even been possible to inject horseradish peroxidase into single dorsal root afferents coming from physiologically identified receptors (see Brown, 1977; Ishizuka, Mannen, Hongo, and Sasaki, 1979; Light and Perl, 1979b), thereby permitting a detailed description of their sites of termination in the spinal cord gray matter (see Fig. 2-5). This has so far only been achieved with a limited number of myelinated afferent root fibers. However, the site of termination of fibers of different calibers can at least to some degree be determined from degen-

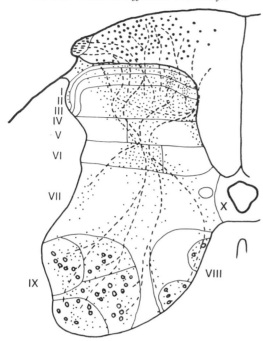

FIG. 2-4 Drawing from the 6th lumbar segment in the spinal cord of the cat to show the sites of termination of degenerating fibers 5 days following transection of the ipsilateral root of L$_6$ as seen in Nauta preparations. Degenerating fibers in the dorsal column are indicated by coarse dots, those in the tract of Lissauer by intermediate-sized dots. Small dots indicate sites of termination. Fibers of passage are shown as dashed lines. The laminae of Rexed are indicated (see text). From Sprague and Ha (1964).

eration studies (see La Motte, 1977), and the thick dorsal root afferents ending among the motoneurons can be identified as primary muscle spindle afferents (see Iles, 1976; Brown and Fyffe, 1978). Furthermore, by applying horseradish peroxidase to the proximal part of cut dorsal rootlets, the distribution in the cord of fibers of different sizes can be elegantly visualized (see Light and Perl, 1979a).

As seen in Fig. 2-3, some dorsal root fibers pass through the gray matter and end on the large ventral motor horn cells in lamina IX. In silver impregnation studies contacts have been seen on the somata and on proximal dendrites, confirming Cajal's classical Golgi observations, as will be further discussed in Chapter 3. Collaterals of dorsal root fibers have further been found to contact cells of the column of Clarke (Szentágothai and Albert, 1955; Liu, 1956; Grant and Rexed, 1958; and others). Other root fibers and collaterals end in the dorsal horn and the "intermediate zone." According to Sprague (1958) and Sprague and Ha (1964), in the lumbar cord there are terminations in all laminae (Fig. 2-4), but especially in laminae IV and VI, in the latter mainly centrally. In the brachial cord Sterling and Kuypers (1967a) have made largely corresponding findings, but stress the scarcity of endings in the lateral parts of laminae V and VI. The terminal branches show a tendency to run in the same direction as the dendrites of the cells in the particular laminae Sensory afferents appear to a large extent to end on internuncial cells, but also on cells giving rise to ascending fibers (see below). It has been known for a long time that a few dorsal root fibers cross the midline to terminate in the contralateral dorsal horn (Schimert, 1939; Sprague and Ha, 1964; and others). It is of interest that, according to recent combined anatomical and physiological studies, such fibers appear to come particularly from nociceptors in the

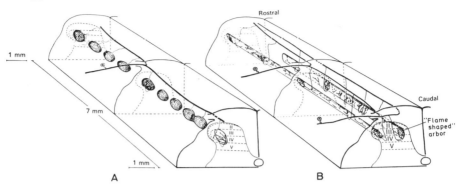

FIG. 2-5 Schematic drawings of the cat lumbosacral dorsal horn, showing the three-dimensional arrangement of axon collaterals from two types of cutaneous receptors. *A:* the arborizations of a dorsal root fiber from a type I slowly adapting receptor. *B:* the arborization of two dorsal root fibers from hair follicle receptors. See text for description. The figure illustrates that axons from functionally different receptors have different modes of termination in the cord (from Brown, Rose, and Snow, 1977b, 1978).

skin and to terminate in lamina I and V (Light and Perl, 1979a, b). As will be discussed later in this chapter, many cells in these laminae send their axons into the spinothalmic tract of the opposite side. Thus, the crossing dorsal root fibers may conceivably contribute to a "pain pathway" ascending ipsilateral to the afferent input.

A detailed Golgi study of the orientation and distribution of dorsal root afferents and of dendritic fields in the gray matter of the cord was published by Scheibel and Scheibel in 1968. Many of their observations have recently been confirmed and extended by use of intra-axonal injection of horseradish peroxidase (Brown, Rose, and Snow, 1977b, 1978; Ishizuka, Mannen, Hongo, and Sasaki, 1979; Light and Perl, 1979b; Light, Trevino, and Perl, 1979).

Thus, Brown and co-workers were able to show that hair follicle afferents form the "flame shaped" arbors in the dorsal horn described in earlier Golgi studies (Fig. 2-5B). One hair follicle afferent coursing in the dorsal column gives off collaterals to the dorsal horn which penetrate to lamina IV or V and then turn dorsally, making abundant synaptic contacts in lamina III and to a lesser extent in IV and II. The collaterals from one axon form a sheet that is narrow mediolaterally but extends rostrocaudally for at least 7 mm in the adult cat. Other types of cutaneous afferents appear to have different types of aborizations; for example, the afferents from types I (Fig. 2-5A) and II slowly adapting receptors arborize extensively in the transverse plane in the dorsal horn, but very little rostrocaudally (see Brown, 1977, for a review).

The existence of *primary afferents terminating in the substantia gelatinosa* (Rexed's lamina II) was long disputed but has now been firmly established experimentally in various mammals (including primates) by the use of silver impregnation methods, autoradiography, and electron microscopy (Ralston, 1965, 1968b; Heimer and Wall, 1968; Réthelyi and Szentágothai, 1969; Coimbra, Sodré-Borges, and Magalhães, 1974; La Motte, 1977; Ralston and Ralston, 1979). The interpretation of the findings of various authors is complicated by the use of different survival times and often by lack of quantitative data, but the number and distribution of such terminals appear to differ among various species (see Heimer and Wall, 1968; Ralston, 1971b; Ralston and Ralston, 1979).

It has generally been held on the basis of Golgi studies that dorsal root afferents to the substantia gelatinosa are of two kinds: thick, myelinated afferents entering from the ventral aspect, making the characteristic "flame-shaped" arbors and thin myelinated and unmyelinated fibers entering from the dorsal aspect (Scheibel and Scheibel, 1968; Réthelyi and Szentágothai, 1973; and others). Autoradiographic and degeneration studies in the monkey (La Motte, 1977; Ralston and Ralston, 1979), however, indicate that few thick fibers terminate in lamina II. Accordingly, as mentioned above, Brown, Rose, and Snow (1977b) concluded on the basis of horseradish peroxidase injections of single dorsal root axons, that the "flame-shaped" arbors terminate mainly in lamina III and only to a very limited extent reach lamina II. After extensive and careful studies with horseradish peroxidase applied to cut dorsal rootlets, Light and Perl (1979a, b) came to corresponding conclusions. Thin, myelinated fibers were found to terminate in lamina I, the inner part of lamina II, and lamina III. Unmyelinated fibers appeared to terminate mainly in the outer part of lamina II, while lamina III was the only one among the dorsalmost three laminae to receive terminations of thick, myelinated fibers.

In the course of time, different opinions have been held about the functional role of the substantia gelatinosa. It appears to play a special role in the processing of sensory information and is not, as was formerly believed, the site of the first synapse in the "pain pathway." It has become increasingly clear during recent years that the sensory information entering the cord through the dorsal roots is subject to considerable modification in the dorsal horn and that this modification will vary under different conditions. According to Golgi, silver impregnation, and electron microscopic studies (Pearson, 1952; Szentágothai, 1964a; Scheibel and Scheibel, 1968; Ralston, 1968a, b, 1971b; Sugiura, 1975; Mannen and Sugiura, 1976; Réthelyi, 1977; La Motte, 1977; Ralston, 1979; Ralston and Ralston, 1979; and others) *the main features of the organization of the substantia gelatinosa* may be summarized as follows: the small cells of the substantia gelatinosa are arranged with their dendrites extending rostracaudally and dorsoventrally, constituting thin sheets oriented sagittally (this holds true also for the neurons of lamina III). The dorsal root afferents to the substantia gelatinosa show a similar distribution. In addition, the neurons in the substantia gelantinosa receive afferents from the brainstem, particularly the nucleus raphe magnus and the adjacent retricular formation (Basbaum, Clanton, and Fields, 1978). The fine, largely unmyelinated axons of most of the substantia gelatinosa cells follow a tortuous course toward the white matter, giving off several collaterals in lamina II and III. Some collaterals probably also terminate in lamina I. They descend or ascend for one to four segments in the zona terminalis and the adjacent dorsolateral funiculus before they return to the substantia gelatinosa. Some axons cross the midline to terminate in the substantia galatinosa of the other side. A few cells have axons that ramify extensively in the vicinity of the cell body, apparently without sending any branches to the white matter. According to recent HRP studies, a small number of cells in lamina II and III in the rat send their axons to the lateral cervical nucleus and the lower brainstem (Giesler, Cannon, Urca, and Liebeskind, 1978) and in the monkey even to the thalamus (Willis, Leonard, and Kenshalo, 1978). Although the latter cell type may be directly involved in sensory transmission, the exact way in which the rest of the substantia gelatinosa neurons can modify the activity of cells projecting rostrally to the brainstem and thalamus remains enigmatic.

Cells in lamina III may act on the dorsally directed dendrites of the large cells in lamina IV and thereby influence sensory transmission to higher levels, since many of these cells project to the lateral cervical nucleus. However, experiments with intracellular injection of horseradish peroxidase indicate that the dendrites of lamina IV cells as a rule do not extend into lamina II (Brown, Rose, and Snow, 1977a). On the other hand, it appears from Golgi studies (Scheibel and Scheibel, 1968; Sugiura, 1975) and experiments with intracellular injection of horseradish peroxidase (Bennett, Hayashi, Abdelmoumene, and Dubner, 1979; Light, Trevino, and Perl, 1979) that cells in lamina II may influence neuronal structures in lamina I and III by way of axon collaterals. Furthermore, Mannen and Sugiura (1976) showed in Golgi reconstructions cells in lamina III with a few axon collaterals apparently ending in lamina IV. It still remains unclear how cells in lamina II and III influence cells in deeper laminae, like V and VII, known to contain a large proportion of the neurons involved in pain transmission (see later). On the basis of what is said above, however, it seems reasonable to assume that it happens by way of several interneurons coupled in series.

It has been suggested that the substantia gelatinosa neurons modify the activity of dorsal root afferents by way of presynaptic inhibition (presumably taking place in lamina II itself or in lamina III). The gate control theory of Melzack and Wall (1965), among other things, postulated a differential effect of thick and thin primary afferents on the substantia gelatinosa cells, the latter acting, as it were, as a gate for sensory signals. Activity in thick dorsal root fibers was supposed to close the gate by activating the substantia gelatinosa cells and thereby increasing presynaptic inhibition of thick and thin primary afferents, whereas thin fibers would inhibit the gelatinosa cells and thus open the gate. However, some of the basic assumptions of Melzack and Wall (1965) have turned out to be incorrect (see Nathan, 1976; Wall 1973, 1978, for reviews). Axoaxonic synapses, usually considered to be the sites of presynaptic inhibition, have been described repeatedly in the substantia gelatinosa, although the "vast majority of synapses are axodendricic" (Ralston, 1971b).

Duncan and Morales (1978) counted the number of various types of synaptic contacts in the cat's substantia gelatinosa. Ninety-seven percent of all synapses examined appeared to be axodendritic; the rest were equally divided between axoaxonic, dendrodentritic, and dendroaxonic ones. Very few of the axoaxonic contacts appeared to have relation to dorsal root fiber terminals; the presynaptic element in such contacts contained flattened vesicles, and the postsynaptic profile appeared to be the primary afferent. (See also Ralston, 1979, Ralston and Ralston, 1979, for quantitative data from the monkey.)

On the whole, the present knowledge of the anatomical organization of the substantia gelatinosa appears insufficient to allow an understanding of the mechanisms by which it may modify sensory transmission. Physiological studies indicate that such affects may be very complex. Denny-Brown, Kirk, and Yanagisawa (1973) found evidence for inhibitory and facilitatory effects after selective severance of parts of the zona terminalis in the monkey. For example, the receptive fields of dorsal root fibers could be drastically altered by such procedures (see footnote 13).

It has recently been possible to record from single cells in laminae I, II, and III (Hentall, 1977; Light, Trevino, and Perl, 1979; Wall, Merrill, Yaksh, 1979). It is of particular interest that some cells show a prolonged discharge, sometimes for minutes, after a single electrical stimulus to their peripheral receptive field.

The types of transmitter substances released in the substantia gelatinosa by dorsal root afferents have attracted much interest (see, for example, *Neurobiology of Peptides,* 1978, *Chemical Pathways in the Brain,* 1978, and Emson, 1979). Histochemically, two types of peptide-containing small dorsal root ganglion cells have been identified (Hökfelt, Elde, Johansson, Luft, Nilsson, and Arimura, 1976). One type contains substance P, the other somatostatin. Substance P is found particularly in lamina I and II, while somatostatin is concentrated in lamina II. The amount of substance P in the dorsal horn decreases after dorsal rhizotomy (Hökfelt, Kellerth, Nilsson, and Pernow, 1975), and there is also other evidence indicating that substance P may act as an excitatory transmitter for thin dorsal root afferents. There is some evidence that the amino acid glutamate could be a transmitter for other primary afferents (Johnson, 1977). The finding of opiate-binding receptors in the dorsal part of the dorsal horn may be of particular importance for the role of the substantia gelatinosa in pain transmission (see La Motte, Pert, and Snyder, 1976). The in vitro liberation of substance P from primary afferent terminals is inhibited by opiates (Jessel and Iversen, 1977). Enkephalin, a peptide occurring naturally in the brain, shows a distribution similar to that of opiate receptors in the cord and has an opiatelike effect on pain transmission.[11] Furthermore, it appears to bind to the same receptors as morphine. The amount of enkephalin is, unlike substance P, not altered by dorsal rhizotomy. It has therefore been suggested that enkephalin acts as a transmitter for small cells in the substantia gelatinosa, and that the latter inhibit substance-P-containing dorsal root afferents presynaptically. Other terminals in lamina I, II, and III, some of them axoaxonic, probably release GABA (gamma amino butyric acid; Wood, McLaughlin, and Vaughn, 1976). Using immunocytochemical methods, the GABA-synthesizing enzyme glutamic acid decarboxylase (GAD) was shown to be localized in presynaptic boutons engaged in axoaxonic contacts (Barber, Vaughn, Saito, McLaughlin, and Roberts, 1978). Such GAD-positive terminals were often presynaptic to dorsal root afferent terminals, and, at the same time, presynaptic to a nearby dendrite. Thus, the same transmitter possibly mediates both postsynaptic and presynaptic inhibition.

As referred to above, physiological studies have produced evidence for the view that there are functional differences between the laminae in the dorsal horn. Wall (1967) studied units in laminae IV to VI following application of cutaneous and joint stimuli and found that units in all three laminae respond to cutaneous stimuli, but only those in lamina VI are sensitive to movement. Furthermore, there appears to be a mediolaterally arranged somatotopic pattern within lamina IV (Wall, 1967; Brown and Fuchs, 1975), but the cells, which have small receptive fields, respond to different cutaneous stimuli, such as hair movement, touch, and cooling of the skin. It was concluded that the cells in lamina V were activated mainly via axons of cells in lamina IV. The cells in lamina V receive converging inputs of different modalities, and many are activated by painful stimuli (Pomeranz, Wall, and Weber, 1968; Besson, Conseiller, Hamann, and Maillard, 1972; and others). Of particular interest is the fact that many cells receive converging inputs from visceral and somatic structures (Selzer and Spencer, 1969a, b; Pomeranz, Wall, and Weber, 1968). These findings thus agree with the anatomical data, as do some other physiological studies (see Sprague and Ha, 1964, and Wall, 1973, for some data and references). It appears, however, that the situation is rather complex, and that no lamina can be related to a particular sensory "modality." Lamina I may come nearest an exception from this rule, since many cells there are activated exclusively by painful stimuli (Christensen and Perl, 1970; and others).

The segmental sensory innervation. The dermatomes. Before following the pathways for impulses entering in the dorsal roots in their further routes within the central nervous system, a particular aspect of the peripheral distribution of the

---

[11] However, the correspondence between distribution of opiate receptors and enkephalin is not without exceptions (see, for example, Simantov, Kuhar, Pasternak, and Snyder, 1976).

sensory fibers will be dealt with. In the peripheral distribution the originally segmental origin of the body is revealed, in the same manner as for the fibers of the
ventral roots (see Chap. 3). Each dorsal root is composed of sensory fibers from
those regions of the skin, muscles, connective tissue in ligaments, fascia, tendons,
and joints, from bones, and also from viscera which are developed from the same
body segment (somite) as the corresponding segment of the spinal cord. (It should
be noted that, strictly speaking, segments cannot be distinguished within the spinal
cord. A spinal cord segment is by definition that part of the cord which gives rise
to those root fibers which unite to form a pair of spinal nerves.) The main features
of the segmental innervation have been made clear by careful dissection, in following the fibers from the roots in their course in the plexuses and further in the
peripheral nerves (Bolk and others). This is most easily performed in the case of
the cutaneous fibers. *The segmental innervation of the skin, the dermatomes,* may,
however, be studied more exactly by other methods. For example, a continuous
number of dorsal roots may be cut, leaving intact only a single root in the midst of
the area. The peripheral distribution of fibers from this root will then be apparent
as a zone of retained sensibility within a larger anesthetic area. By this ''method
of remaining sensibility'' Sherrington proved experimentally in monkeys that
neighboring dermatomes overlap to a greater or lesser extent. This method has
been applied also in man, most extensively by Foerster in cases where several dorsal roots have had to be sectioned on account of pain, for example, from neuromas
following amputations. A third method, introduced by Dusser de Barenne, utilizes
local application of strychnine to one or more dorsal roots, resulting in a hypersensitivity in the area of distribution. The distribution of the herpetic vesicles in cases
of herpes zoster and the sensory disturbances in clinical cases of lesions of the
dorsal roots or of the spinal cord have also given valuable information for the
study of the distribution of the dermatomes. Finally, the distribution of the vasodi-
lation following irritation of the dorsal roots may be utilized (antidromic impulses,
cf. Chap. 11), and local anesthesia of dorsal root ganglia may be used as an experimental procedure in healthy subjects. This has been done by Keegan and Garrett
(1948) whose diagrams of the dermatomes, reproduced in Figs. 2-6 and 2-7, are
based on a large series of cases of dorsal root compression by herniated discs.[12]

The maps of the dermatomes in man worked out by different methods are not
concordant in all respects. However, the main principles are identical. Thus there
is a *considerable degree of overlapping between neighboring dermatomes* (not
shown in Figs. 2-6 and 2-7).[13] For example, the 4th thoracic dermatome is cov-

[12] In contrast to the dermatome charts of many other authors, the diagrams of Keegan and Garrett
show the dermatomes as continuous zones extending from the spine to the distal parts of the limbs.
They show only one axial line (the line where non-neighboring dermatomes meet), the ventral one,
while some authors have advocated the presence of a dorsal axial line as well. The paper of Keegan
and Garrett (1948) should be consulted for particulars and for a historical account on the dermatomes.

[13] There is evidence that the skin area innervated by one dorsal root may be approximately twice
the size of that revealed by the classical method of sectioning three neighboring roots on either side of
that under study. Such a wide area of innervation was revealed in monkeys, when, in addition to cutting of neighboring roots, the spinal cord was transected just above or hemisected below the test segment (Denny-Brown, Kirk, and Anagisawa, 1973). Apparently, only impulses from the central and
most densely innervated part of the peripheral field of an isolated dorsal root are consciously perceived
and give rise to reflexes, for example, on scratching of the skin. According to Denny-Brown et al., this
appears to be caused by inhibitory effects mediated by the lateral division of the zona terminalis.

ered in its upper half by the 3rd, in its lower half by the 5th thoracic dermatome. A practical consequence of this arrangement is that even a complete lesion of a single dorsal root will, as a rule, not be followed by a sensory loss, at least not definitely. This statement, however, requires certain qualifications. Considerable individual variations exist with regard to details in the areas occupied by different dermatomes. Still more important is the fact that *the borders of the dermatomes are not exactly the same for touch as for pain and temperature. The dermatomes are somewhat more extensive in regard to touch than to pain and temperature,* i.e., the "touch" fibers belonging to a dorsal root overlap to a greater extent with those from the neighboring roots than do the fibers conveying sensations of pain and temperature. This can be seen in Fig. 2-8, showing the findings in one of Foerster's cases. Thus a careful investigation of cutaneous sensibility in cases where a single dorsal root has been interrupted will frequently reveal a limited

FIG. 2-6 The dermatomes on the trunk and upper extremity (see text). From Keegan and Garrett (1948).

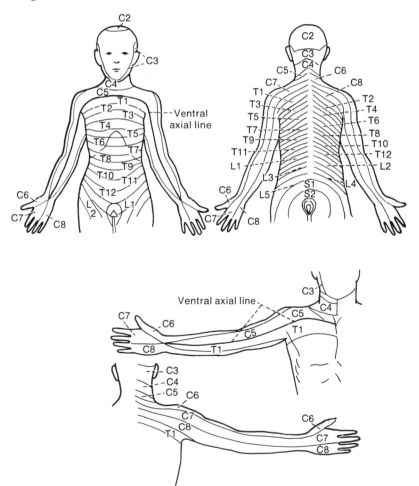

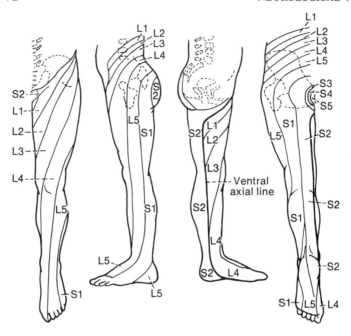

FIG. 2-7   The dermatomes on the lower extremity. From Keegan and Garrett (1948).

zone of hypalgesia, or more rarely even of analgesia, whereas there may be no loss of the sense of touch. It is on this account especially important to test pain sensibility very carefully in searching for monosegmental sensory loss, e.g., in suspected cases of herniated intervertebral disc.

Where a sensory loss of the segmental, dermatomal type is encountered, it is of course of interest to determine the level of the lesion as exactly as possible, and usually one will resort to a comparison with diagrams such as those presented in Figs. 2-6 and 2-7. However, certain principal features in the dermatomal arrangement ought to be remembered, in order to enable the examiner to decide at least approximately which dorsal roots are involved in the lesion. From Figs. 2-6 and 2-7 it can be seen that, broadly speaking, the skin of the lateral (radial) side of the arm, forearm, and hand with the thumb belongs to the 5th and 6th cervical dermatomes, the medial (ulnar) side to the 8th cervical–1st thoracic dermatomes. The nipple is located in the region of the 4th to 5th thoracic dermatomes, the umbilicus in the 9th to 10th. The inguinal sulcus falls within the 1st lumbar dermatome, the anterior aspect of the thigh and the knee is covered by the 3rd to 4th lumbar, whereas the heel is found in the area of the 5th lumbar and 2nd sacral dermatome, and the posterior aspect of the thigh and upper calf belongs to the 1st to 2nd sacral dermatomes. The anogenital region is supplied by fibers from the 3rd to 5th sacral roots. (For a clinical study on the segmental sensory innervation, see Hansen and Schliack, 1962.)

The segmental distribution discussed above refers to the exteroceptive fibers. But the visceral afferent, as well as the proprioceptive and other afferent fibers, are also segmentally distributed. There is still some uncertainty about certain details in this arrangement, but some principal features are known. Concerning the visceral afferent fibers the reader is referred to Chapter 11. With regard to *the proprioceptive fibers* some data are worth mentioning. As the afferent fibers from the muscles follow the motor fibers, the segmental distribution of muscle sensibility will not agree fully with the segmental cutaneous pattern. (The segmental motor innervation of muscles is treated in Chap. 3.) With respect to the skeleton and the joints it is generally assumed that the segmental distribution is even more different from the dermatomes than is the muscular innervation. Probably the segmental areas of the skeleton of the limbs have retained more of their original pattern, partly extending as narrow longitudinal zones through the entire length of the extremity. It is probable that the deep pains, usually dull and difficult to localize, which occur in many cases of damage to the dorsal roots, may be due to irritation of the afferent fibers from the bones and joints. These pains also have a tendency to irradiate upward or downward in the limbs.

In the dorsal roots the different kinds of sensory fibers treated above are intermingled, and an interruption of a dorsal root consequently will affect all sensory modalities. Upon entering the spinal cord, however, the fibers separate, and the impulses conveyed by them take different courses. This circumstance favors the analysis of lesions of the sensory systems, facilitating as it does in many instances an exact focal diagnosis.

As referred to above, most dorsal root fibers enter the gray matter. The cells within this which give off long ascending axons transmitting impulses to the brainstem and thalamus are commonly spoken of as "secondary sensory neurons," even if presumably many of them are not secondary. The tracts into which the long ascending fibers are grouped are named according to their destination as spinocerebellar, spinoreticular, spinovestibular, spinotectal, spino-olivary, etc. Most of them will be considered in other chapters. Here we are concerned primarily with those ascending pathways which appear to be essential for the transmission of sensory impulses which reach the level of consciousness. One such pathway is the spinothalamic tract. Another important one is the pathway established by the axons of *primary* sensory neurons which, as mentioned previously, ascend in the dorsal funiculi. In the following discussion the latter will be considered first. In addition the more recently discovered spinocervicothalamic tract, present in many animals and probably also in man, will be treated. Figures 2-3B, 2-10, and 2-12 are simplified diagrams showing some of the ascending spinal pathways.

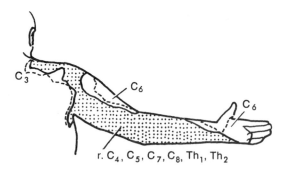

FIG. 2-8 The 6th cervical dermatome (C₆) after resection of the dorsal roots of neighboring segments of the cord (C₄, C₅, C₇, C₈, Th₁ and Th₂). Sensibility to pain and temperature is abolished in the area limited by the broken line. In the stippled area tactile sensibility is also lacking (see text). From Foerster (1936a).

At the spinal level there is no clear segregation between "secondary sensory" fibers passing to different end stations, for example the thalamus and the reticular formation. (For recent reviews of somatosensory pathways, see Boivie and Perl, 1975; Lynn, 1975; Brown and Gordon, 1977; and Webster, 1977.)

The fibers in the dorsal funiculi and the medial lemniscus. The large myelinated dorsal root fibers, which ascend to terminate in the nuclei of the dorsal funiculi in the medulla oblongata (see Figs. 2-10 and 7-4) make up a considerable part of the medical group of the dorsal root fibers. At their entrance these fibers are situated immediately medial to the dorsal horn. During their ascending course, however, they are steadily pushed in a more medial direction, because the fibers entering at succeeding rostral levels intrude between the ascending fibers and the dorsal horn. As a consequence of this the fibers occupying the most medial part of the medial funiculus gracilis in the upper cervical region will belong to the sacral dorsal roots, then follow the fibers from the lumbar dorsal roots—i.e., the fibers from the lower extremity are found most medially. The fibers belonging to the upper extremity are found most laterally in the funiculus cuneatus, close to the dorsal horn; the fibers from the upper cervical roots are found more laterally than those from the lower roots. Of the thoracic fibers approximately the lower six occupy the lateral part of the funiculus gracilis, the upper six the medial part of the funiculus cuneatus.[14]

The disposition of the fibers of the dorsal funiculi has been established by experimental investigations and studies of human pathological cases. Even if there appears to be some degree of overlapping between fibers from adjacent roots (Walker and Weaver, 1942), *the ascending fibers of the dorsal funiculus are somatotopically organized* (Fig. 2-9), reflecting in their arrangement the original segmental pattern of the organism. *The fibers terminate in the nuclei of the dorsal funiculi, following the same principle,* as shown in the monkey (Ferraro and Barrera, 1935b; Walker and Weaver, 1942) and in the cat (Glees, Livingston, and Soler, 1951; Hand, 1966) and as confirmed in single-unit recordings by Gordon and Paine (1960), Kruger, Simonoff, and Witkowsky (1961), and others. However, more recent anatomical and physiological studies have shown that, although the arrangement of ascending fibers in the funiculus gracilis is segmental at lumbar levels, considerable re-sorting of the fibers takes place as they ascend. Thus, at cervical levels the fibers from different segments are organized topographically so that those from the foot are situated together, apart from those from the leg, and so on (Whitsel, Petrucelli, Sapiro, and Ha, 1970).

The dorsal column fiber population is, according to recent studies, far more heterogeneous than appears from the above. Thus, a number of primary afferent fibers coursing in the dorsal columns terminate at spinal levels (*propriospinal fibers*). In addition to primary afferents, ascending in the dorsal columns, there are also many axons of *second-order neurons* in the dorsal horn which terminate in

[14] Some of the fibers from the cervical and 4th to 5th upper thoracic roots terminate in the external cuneate nucleus, from which their impulses are conveyed to the cerebellum, as mentioned in Chapter 5. According to recent studies with the horseradish peroxidase method some cells in the external cuneate nucleus project to the thalamus (Boivie, Grant, Albe-Fessard, and Levante, 1975, monkey; Fukushima and Kerr, 1979, rat).

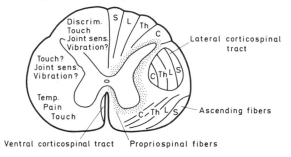

FIG. 2-9 A diagram of the spinal cord showing the segmental arrangement of the nerve fibers within some major tracts. On the left side are indicated the sensory "modalities" that appear to be mediated via the main ascending pathways situated in the dorsal, dorsolateral, and ventral fasciculus. Fibers ascending in the dorsolateral fasciculus, conveying the types of sensory information indicated, would belong to the spinocervical tract but also to nonprimary afferents to the dorsal column nuclei and some other nuclear groups in the medulla, projecting to the thalamus (nucleus z, external cuneate nucleus, see text). It should be emphasized that the diagram is highly simplified since it appears that most sensory "modalities" are served by fibers widely scattered in the spinal cord white matter and not restricted to only one of the classical sensory pathways. Note the broad zone close to the gray matter occupied by propriospinal fibers. *C:* cervical; *L:* lumbar; *S:* sacral; *Th:* thoracic.

the dorsal column nuclei. Furthermore, the existence of *descending* fibers in the dorsal columns, arising in the dorsal column nuclei, should be kept in mind when one evaluates the effects of lesions or stimulation of the dorsal columns. Finally, it should be noted that the funiculus gracilis and cuneatus differ in important respects with regard to fiber composition, particularly fibers carrying information from joint and muscles.

The *primary afferent fibers coursing in the dorsal columns* and terminating at spinal levels (see Horch, Burgess, and Whitehorn, 1976) are of various calibers. The thinnest fibers appear from physiological studies to ascend for only 2–3 segments (A δ group), whereas thicker fibers (in the A α range) may pass for 4–12 segments. Some of these fibers apparently leave the dorsal columns to take another course to the brainstem, for example fibers from slowly adapting joint receptors in the hindlimb (see below). The existence of propriospinal fibers coursing in the dorsal columns has long been known (see Nathan and Smith, 1959), but only recently has the existence of spinal cord *second-order neurons* projecting to the dorsal column nuclei been firmly established. Uddenberg (1968) found evidence of synaptically activated units located deep in the cat dorsal columns, and after injecting horseradish peroxidase into the dorsal column nuclei of the monkey, Rustioni (1976) described a large number of labeled cells in the spinal gray matter. They were most numerous in lamina IV at cervical and lumbar levels, but some labeled cells were also found more ventrally in the dorsal horn, and some even in the lateralmost part of the ventral horn ("spinal border cells").

The existence of *long, descending fibers in the dorsal columns* has been firmly established in the rat, cat, and monkey with the anatomical techniques making use of anterograde and retrograde axonal transport (see Burton and Loewy, 1977). The cell bodies are located in the dorsal column nuclei, and the axons terminate somatotopically in the cord, apparently most densely in lamina V. This lamina contains a large number of neurons sensitive to noxious stimuli. However, it is still conjectural whether descending fibers in the dorsal columns contribute to the effects of dorsal column electrical stimulation on pain transmission (see Foreman, Beall, Applebaum, Coulter, and Willis, 1976; Burton and Loewy, 1977).

The funiculus gracilis and cuneatus differ in important respects with regard to fiber composition. Thus, primary afferents from receptors in joint capsules and in muscles in the forelimb ascend in the funiculus cuneatus to reach the cuneate nucleus (Oscarsson and Rosen, 1963; Clark, Landgren, and Silfvenius, 1973), whereas most such fibers from the hindlimbs leave the funiculus gracilis, probably before reaching low thoracic levels (Burgess and Clark, 1969; Whitsel, Petrucelli, and Sapiro, 1969; Clark, 1972). Thus in the cat, Burgess and Clark (1969) found that only 10% or less of the afferents in a particular knee joint nerve could be antidromically activated by stimulation of the dorsal columns at high cervical levels. Furthermore, almost all such fibers were rapidly adapting; that is, they were sensitive to movement but not to position. Conditions appear to be quite similar in the monkey since it appears that most afferent fibers from receptors in "deep" tissue of the leg leave the dorsal columns before they reach cervical levels (Whitsel, Petrucelli, Sapiro, and Ha, 1970).

The *classical view,* based mainly on clinical experience, has been that the *impulses ascending in the fibers of the dorsal columns mediate sensations of touch, deep pressure, vibratory sense, and sense of position of joints and are particularly important for sensory discrimination* (see below).

However, during the last few years this view has been seriously challenged (see Wall, 1970), the critics arguing that in the clinical cases said to demonstrate the sensory deficits of dorsal column lesions, the damage nearly always includes other structures as well. Histological controls of the lesions are naturally seldom available. Cook and Browder (1965) reported some cases of dorsal chordotomy—that is, section of the dorsal columns—in which almost no permanent loss was found in the abovementioned sensory qualities. The question of the functional role of the dorsal column–medial lemniscus pathway will be discussed in connection with the consideration of the symptoms following dorsal column lesions. Suffice it here to say that, although the matter is far from fully clarified, several careful studies in primates indicate that *the dorsal columns mediate sensory signals necessary for rather complex discriminative tasks.* For example, they appear to be necessary to compare the magnitude of pressure upon the skin, to judge the actual distance between two stimuli (as distinct from the awareness of two stimuli), to adjust the finger grip when an object is slipping, and so on. On the basis of examination of three patients· with lesions of the dorsal columns, Wall and Noordenbos (1977) concluded that "From a clinical point of view, deficiency in detecting direction of movement on the skin and figure writing seem to be the most useful and simplest tests" (to detect lesions of the dorsal columns). *The dorsal columns appear not to be critically important for joint sensation and sense of vibration,* although refined tests may well be thought to demonstrate some role of the dorsal columns in these aspects of sensation as well, presumably first and foremost in the ability to judge the rate of movement (see Vierck, 1978a, for a critical review). The lack of obvious lasting deficits after lesions of the dorsal columns may appear puzzling in the light of the functional properties of fibers in the dorsal columns and of neurons in the dorsal column nuclei, as will be considered below. It should be recalled, however, that most routine clinical tests give information about very crude sensory functions, which may presumably be served by a small fraction of the neurons and nerve fibers engaged in transmission and analysis of sensory impulses. For example, Dobry and Casey (1972) found that cats had a reduced capacity to distinguish between surfaces with different roughness only when more than 90% of the fibers in the dorsal columns were transected. For the complex sensory tasks served by the dorsal columns, a very large number of fibers are presumably needed.

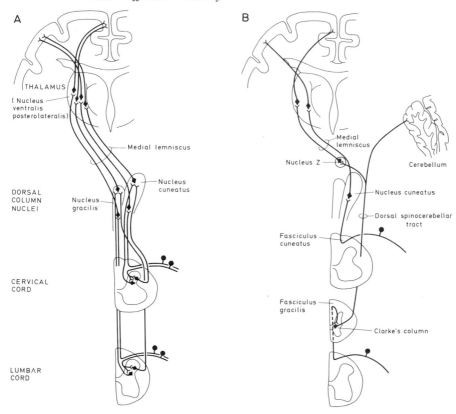

FIG. 2-10 Simplified diagrams which show the main anatomical features of the dorsal column-medial lemniscus pathway. A: The main routes taken by fibers carrying impulses arising in low-threshold cutaneous receptors and rapidly adapting deep receptors. Note that apart from the massive input of primary afferent fibers to the dorsal column nuclei, many afferents arise from second-order sensory neurons, located in the dorsal horn, which ascend in the dorsal columns and in the dorsolateral fasciculus. B: A diagram of the main routes taken by fibers carrying impulses from slowly adapting receptors in joints and muscles, presumably responsible for "joint sense." Note that such impulses from the hindlimb and forelimb follow different routes. See text.

Anatomical studies show that the *dorsal column nuclei* may be subdivided into parts that differ cytoarchitectonically and with regard to the distribution of dorsal root fibers, secondary afferents from the cord, and other afferent contingents. Physiological studies, furthermore, indicate that these anatomical differences are reflected by functional differences.

On the basis of Nissl and Golgi material, Kuypers and Tuerk (1964; see also Hand, 1966) described two main cell types in the cat dorsal column nuclei. Large, round cells with richly branching dendritic trees are concentrated in the middle dorsal portion of the nuclei and are arranged in clusters ("cluster zone"), while medium-sized cells with long, radiating dendrites are particularly common in the rostral and ventral parts of the nuclei ("reticular zone"). Despite some differences, conditions appear to be principally similar in primates. For example, in the nucleus cuneatus of the monkey a pars rotunda appears to correspond to the "cluster region'' of the cat (Ferraro and Barrera, 1936; Rustioni, Hayes, and O'Neill, 1979). The pars rotunda in the monkey receives dorsal root afferents from segments supplying distal parts of the extremities (Shriver, Stein, and Carpenter, 1968), and in the cat, dorsal root afferents terminate with the highest density among the cell clusters (Kuypers and Tuerk, 1964; Keller and Hand, 1970). The overlap between the terminal area of fibers from neighboring roots is smaller in the ''cluster region'' and the pars rotunda than in other parts of the nuclei. On the other hand, axons from second-order neurons in the cord which ascend in the dorsal columns and the dorsolateral funiculus (Rustioni and Molenaar, 1975, cat; Rustioni, Hayes, and O'Neill, 1979, monkey) terminate chiefly in the rostral part of the dorsal column nuclei—that is, in the ''reticular zone'' (Rustioni, 1973, cat; Nijensohn and Kerr, 1975, monkey). The ''reticular zone'' appears also to be the main area of termination of fibers from the cerebral cortex (Kuypers and Tuerk, 1964) and from the gigantocellular nucleus of the reticular formation (Scheibel and Scheibel, 1958; Odutola, 1977; Sotgiu and Marini, 1977). Thus, it appears from anatomical studies that the cluster zones are the recipients of dense and precisely organized primary afferent projections, particularly from distal parts of the extremities, whereas the ''reticular zones'' are sites of convergence of inputs from various sources. A large part of their inputs from the spinal cord appears to consist of primary and ''nonprimary'' afferents with receptive fields in more proximal parts of the body and with more diffuse patterns of termination. The two subdivisions of the dorsal column nuclei, furthermore, differ with regard to their efferent connections. We shall return to this below, after consideration of the functional properties of neurons in the dorsal column nuclei.

The highly simplified account of the anatomy of the dorsal column nuclei given above makes it clear that their organization is far more complex than was assumed some years ago, when it was believed that they were simple relay nuclei: afferents from the dorsal columns establish synaptic contact with cells in the nuclei, and these cells give off their axons to the medial lemniscus. In general, physiological studies are in good agreement with the anatomical data.

Recordings from single units in the dorsal column nuclei after stimulation of various receptors show that a large proportion of the neurons are activated only from one particular kind of receptor; that is, they are ''modality-specific'' (Perl, Whitlock, and Gentry, 1962; Gordon and Jukes, 1964a; Winter, 1965; and others). Furthermore, the receptive field for touch—that is, the cutaneous area which influences a single cell—is usually small (''place specificity''), although it varies in size in different parts of the body. The receptive field is smallest in the digits and peripheral parts of the extremities (Kruger, Siminoff, and Witkowsky, 1961; Perl et al., 1962; Winter, 1965; and others). The spatial precision is further increased by lateral inhibition (inhibition of the ''surround type,'' see Fig. 2-16) of cells supplied by cutaneous receptors; that is, a stimulus applied near but outside the receptive field for the unit may inhibit the cell (Gordon and Paine, 1960; Perl et al., 1962; Gordon and Jukes, 1964a; Bystrzycka, Nail, and Rowe, 1977). However, afferent spatial facilitation may also occur (Gordon and Paine, 1960; Gordon and Jukes, 1964a). Neurons with the above-mentioned properties appear to be particularly abundant in the ''cluster zones.'' Most units appear to be activated from low-threshold mechanoreceptors in the skin, such as hair follicle receptors, Meissner's corpuscles, Merkel's discs, Ruffini endings, and Pacinian corpuscles. (The latter type is, strictly speaking, not located in the skin, see Fig. 2-1A.) Some units are activated by movements of joints. In most cases the responses are rapidly adapting; that is, the units signal *changes* of stimulation more than the steady state. Units responding with a slowly adapting discharge to movements of joints, presumably of prime importance for kinesthetic sensation, are found in the cuneate nucleus but are much more infrequent in the gracile nucleus. (The question of pathways for signals informing about joint position will be returned to separately below.) The reported preponderance of units activated from hair follicle receptors may not be representative of conditions in man, since nearly all studies of functional properties of dorsal column units have been done in the cat and rat.

A recent study of responses of cat cuneate neurons to natural stimulation of the foot pad described in detail three classes of neurons (Douglas, Ferrington, and Rowe, 1978). One cell type adapts slowly and has other properties suggesting that it receives input from Merkel's discs. The impulse frequency of such cuneate neurons follows approximately linearly the degree of indentation of the skin in its receptive field; that is, these neurons "could contribute discriminative information about indentation intensity" (Douglas et al., 1978). The two other types of cuneate neurons adapt rapidly; both are particularly sensitive to vibration in their receptive fields. One is excited particularly by vibratory stimuli below 80 Hz, the other by frequencies above 80 Hz. The properties of the two types of cells suggest that they receive input from Meissner's and Pacinian corpuscles, respectively (see pp. 51–54). Douglas and co-workers (1978) concluded that "Although the activity of cuneate neurons provides a poorer signal of vibratory frequency than that of primary fibers their properties are nevertheless consistent with their being able to signal the information which is utilized for subjective frequency discrimination."

Cells in the dorsal column nuclei with input from only one kind of receptor in a very small peripheral field seem capable of transmitting very precise information about the type of stimulus and its location. However, it should be emphasized that such knowledge about the properties of the units in the dorsal column nuclei does not necessarily give insight into how this information is used by the central nervous system, or the symptoms that occur after lesions of the dorsal columns and their nuclei of termination.

The functional properties described above as typical of the neurons in the "cluster zones" of the dorsal column nuclei—that is, small receptive fields and specificity with regard to receptor types that excite them—are not found in all dorsal column nuclei neurons. Other cells, predominantly located in the "reticular zones," have wide receptive fields and are activated by various types of stimuli, apparently also painful ones (Gordon and Jukes, 1964a; Angaut-Petit, 1975). That some dorsal column neurons have these properties is not unexpected in light of the properties of afferents from second-order neurons in the spinal cord (Anguat-Petit, 1975).

Many cells in the dorsal column nuclei appear from physiological studies to be interneurons (Andersen, Eccles, Schmidt, and Yokota, 1964b). They are found chiefly in the "reticular zones" and receive converging excitatory inputs from the cerebral cortex and second-order sensory neurons in the spinal cord and may, by way of local axons, inhibit the cells projecting to the thalamus or other regions. In accordance with the physiological data, Blomqvist and Westman (1976) have described in Golgi material from the feline gracile nucleus neurons whose axons ramify after a short distance and end with fine branches within the nucleus. They also found initial collaterals from axons of cells projecting into the medial lemniscus.

Because of the changed points of view on the functional role of the dorsal column nuclei outlined above, it may be appropriate to consider some aspects more specifically. One item of interest concerns the *pathways in the cord and relay stations in the medulla oblongata for impulses arising in muscles and joints,* particularly impulses from slowly adapting receptors, which are crucial for awareness of joint position. (Rapidly adapting receptors can obviously only signal changes in position and the speed with which the changes take place.) The relation of the dorsal column–medial lemniscus pathway to "joint sense," or kinesthesia, is more complex than formerly held, when it was thought that the dorsal columns were solely responsible for carrying information about joint position. As mentioned above, fibers carrying impulses from slowly adapting receptors in joint capsules and muscles are virtually absent from the funiculus gracilis at cervical levels, but they are present in the funiculus cuneatus. Instead of coursing in the dorsal columns, many of the fibers activated from hindlimb muscle and joint receptors ascend in the dorsolateral funiculus above lumbar levels. At least some appear to be collaterals of the dorsal spinocerebellar tract and to terminate in nucelus z, situated just rostral to the gracile nucleus (see Fig. 7-14). On the other hand, it is well established that neurons excited by fast-conducting afferents from mucles and joints in the forelimbs (presumably involved in kinesthetic sensations) are located in the cuneate nucleus. On the basis of the anatomical features described above for pathways presumably involved in kinesthetic sensation, Brown and Gordon (1977) state: ". . . one would expect that the arms and hands would be more severely disabled than the legs by damage to the dorsal columns. . . . Charcot (1881) seems to have appreciated that, though there was always sclerosis of the cuneate fascicle in arm ataxia, sclerosis of the cervical gracile fascicle did not necessarily accompany leg ataxia in human patients."

Several studies have revealed neurons activated from receptors in joints and muscles in the "reticular" part of the cuneate nucleus (Rosén, 1969; Clark, Landgren, and Silfvenius, 1973; Rosén and Sjölund, 1973), whereas Millar (1979) in a careful study found most of the cuneate cells excited from the elbow joint nerve in the cat to be located in the middle, "cluster" zone of the nucleus. Many of these cells had additional cutaneous receptive fields around the cubital fossa. The convergence of joint and cutaneous afferents is interesting in light of evidence that both types of information may contribute to kinesthetic sensation in man (see Marsden, Merton, and Morton, 1977).

Neurons receiving corresponding information from the hindlimbs are found in nucleus z, first described by Brodal and Pompeiano (1957). From studies of anterograde degeneration after lesions of various parts of the cord it appears that nucleus z receives its main spinal input from the dorsolateral fasciculus (Rustioni, 1973, cat; Rustioni, Hayes, and O'Neill, 1979, monkey). Many, but not all, such afferents appear to be collaterals from the dorsal spinocerebellar tract arising in Clarke's column (Johansson and Silfvenius, 1977a). It is conceivably of importance for their role in kinesthetic sensation that cells in Clarke's column receive monosynaptic excitation from slowly adapting joint afferents in the hindlimbs (Lindström and Takata, 1972; Kuno, Muñoz-Martinez, and Randić, 1973) in addition to the well-known input from muscle spindles (see Chap. 5). Further studies by Johansson and Silfvenius (1977b), with extracellular recordings from cells in nucleus z of the cat, confirmed that many cells were excited on stimulation of ipsilateral group I muscle afferents, as first shown by Landgren and Silfvenius (1971) and further showed that some were excited from low-threshold afferents from the skin, and a few from low-threshold joint afferents. The assumption that nucleus z is an important relay site for transfer of proprioceptive information to the cerebral cortex is supported by the fact that a large proprotion of its cells can be antidromically activated from the thalamus (Johansson and Silfvenius, 1977a).

Another aspect of the function of the dorsal column nuclei concerns the role played by their *afferents from the cerebral cortex*. As mentioned above, these fibers follow the pyramidal tract. They have been shown in experimental studies to end in the nuclei gracilis and cuneatus (Chambers and Liu, 1957; Walberg, 1957a, in the cat; Kuypers, 1958b, in the monkey and chimpanzee; Kuypers and Lawrence, 1967, in the monkey). This projection is somatotopically organized and mainly crossed; fibers from the arm area of the sensory cerebral cortex end in the cuneate nucleus, those from the leg area in the nucleus gracilis. The second somatosensory region also projects in the same manner to these nuclei (Levitt, Carreras, Liu, and Chambers, 1964).

Electron microscopic studies show that the cortical fibers end with uniformly small boutons, whereas the dorsal column fibers have boutons of various sizes (Walberg, 1966). The presence of axoaxonic contacts (Walberg, 1965b, 1966) is of interest in the light of physiological evidence of presynaptic inhibition from the cortex of dorsal column fibers (Andersen, Eccles, Schmidt, and Yokota, 1964a).

After injections of horseradish peroxidase in the dorsal column nuclei, the origin of afferents from the cortex was mapped in detail (Weisberg and Rustioni, 1976, 1977). In the monkey and cat, labeled cells were found in layer V in the trunk and forelimb and hindlimb regions of area 4 and S I. A smaller number occurred in S II, in agreement with the degeneration studies (Levitt et al., 1964). In the monkey (Weisberg and Rustioni, 1977) but not in the cat (Weisberg and Rustioni, 1976), labeled cells were also found in the supplementary motor cortex (the medial part of area 6) and in area 5.

Several physiological studies have demonstrated that the *sensorimotor cortex influences the transmission of impulses from the dorsal column nuclei*, as was assumed on the basis of the anatomical data described above. This is only one of many observations (perhaps so far the most extensively studied one) that have drawn our attention to the central control exerted on receptors and sensory impulse transmission. (For reviews see Towe, 1973; Brown and Gordon, 1977.) Other examples will be mentioned later; suffice it here to mention that impulse transmission in the dorsal column–medial lemniscus pathway can be influenced also at the thalamic level by way of corticothalamic fibers and, furthermore, that the secondary afferents reaching the dorsal column nuclei from neurons in the cord are subject to descending influences from the cerebral cortex and brainstem at the segmental level. The overall effect of cortical stimulation on sensory transmission through the dorsal column nuclei appears to be inhibitory (Magni, Melzack, Moruzzi, and Smith, 1959; Towe and Jabbur, 1961; Gordon and Jukes, 1964b; Winter, 1965). However, some cells are facilitated, and some receive both excitatory and inhibitory influences from the cortex (Towe and Jabbur, 1961; Gordon and Jukes, 1964b; Winter, 1965). Most authors agree that the inhibitory influences are mediated via the pyramidal tract fibers to these nuclei. According to Jabbur and Towe (1961) the facilitatory influences are likewise mediated via the pyramidal tract (however, according to these authors, some inhibitory influences may be relayed through the reticular formation).

The functional role of the cortical connections to the dorsal column nuclei is not fully understood, but Gordon and Jukes (1964b) suggested that the inhibitory influences might serve to increase the lateral inhibition taking place in the nuclei (see above), thereby sharpening the spatial pattern of the sensory signals. One might also imagine that the overall transmission through the nuclei may be depressed or enhanced according to the need of the organism; for example, "an interesting sight or smell would be accompanied by suppressed transmission in other systems" (see Brown and Gordon, 1977).

The transmission of sensory impulses through the dorsal columns appears to be suppressed before and during movements, as shown by mass recording from the medial lemniscus in the cat (Ghez and Pisa, 1972; Coulter, 1974; and others) and monkey (Dyhre-Poulsen, 1978). Although not directly shown, it seems very likely that this effect is mediated by descending connections from the cerebral cortex to the dorsal column nuclei. The suppression is limited to sensory signals from a region close to where the movement is taking place; for example, movement of the left arm has no effect on the transmission from the right arm. Psychophysical experiments in man show that, from about 200 msec before and during a movement, the threshold is raised for various types of stimuli applied to the skin of the moving part (Coquery, 1978; Dyhre-Poulsen, 1978). Dyhre-Poulsen also showed that the threshold is raised in relation to both rapid (ballistic) and slow movements. The functional significance of this phenomenon is not understood. It might conceivably be a result of increased lateral inhibition, which would be of importance during exploratory movements, or it might represent a suppression of cutaneous sensation in order to favor transmission of proprioceptive signals. (For a wealth of information on the relation between movement and sensation, see the proceedings of the symposium *Active Touch. The Mechanism of Recognition of Objects by Manipulation,* 1978.)

As discussed above, the dorsal column nuclei are not homogeneous cytoarchitectonically and with regard to the distribution of afferent connections. This appears to be the case also with their *efferent connections,* although there appear to exist species differences that make it difficult to draw general conclusions. (The ending in the thalamus of fibers from the dorsal column nuclei will be considered below; here only some points on their origin will be discussed. Furthermore, some connections to cell groups apart from the thalamus are briefly mentioned here.) Studies using various anatomical techniques in the cat have shown that most of the neurons projecting to the thalamus are relatively large and located mainly in the "cluster zones" (see Kuypers and Tuerk, 1964; Berkley, 1975; Hand and van Winkle, 1977; Blomqvist, Flink, Bowsher, Griph, and Westman, 1978; and others). In the dorsal column nuclei of the monkey, however, there are apparently no such clear-cut regional variations in the proportin of cells projecting to the thalamus and in the size of projecting and nonprojecting cells (Rustioni, Hayes, and O'Neill, 1979). According to Rustioni and co-workers, the ratio of labeled to unlabeled cells in the rostral "reticular" part of the dorsal column nuclei is 1:1 after large injections of HRP in the thalamus of the cat, whereas after corresponding injections in the monkey the ratio is 5:1. Since it appears that the "reticular zones" in cat and monkey receive similar afferent inputs, different from those reaching the "cluster" regions, it seems possible that the information conveyed to the thalamus by the medial lemniscus is more varied in the monkey than in the cat. For example, information reaching the dorsal column nuclei from second-order sensory neurons (nonprimary afferents), some with wide receptive fields and responding to various stimuli, seems likely to be forwarded to the thalamus via these nuclei.

In the cat, neurons, in the "reticular zones" which do not project to the thalamus appear either to be interneurons or to project to certain parts of the brainstem or the spinal cord. Thus, there are connections, arising mainly in the "reticular zones," to the cerebellar anterior lobe and the dorsal and medial accessory olives (see Chap. 5). Some fibers have, furthermore, been traced to the contralateral inferior and superior colliculus, the red nucleus, the pretectum, the magnocellular part of the medial geniculate body, the suprageniculate nucleus, Forel's field, and the zona incerta (see Hand and van Winkle, 1977; Berkley and Hand, 1978b; Blomqvist et al., 1978). Although less extensively studied, conditions in primates appear to be similar (Bowsher, 1961; Burton and Loewy, 1977; Boivie, 1979). However, neither Bowsher (1958, 1961) nor Boivie (1979) mention terminal degeneration in the red nucleus after electrolytic lesions of the dorsal column nuclei in rhesus monkeys.

As mentioned above, the axons of a great number of the cells of the nuclei gracilis and cuneatus form the *medial lemniscus* or mesial fillet ascending in the brainstem and ending in the thalamus. There is general agreement among anatomists that all fibers of the medial lemniscus cross the midline in the medulla, in man as in the monkey (Rasmussen and Peyton, 1948; Matzke, 1951; Glees, 1952; Bowsher, 1958; Boivie, 1979; and others). Furthermore, the fibers do not appear to give off collaterals to the reticular formation (in contrast to the spinothalamic fibers). However, the *segmental somatotopic organization present in the dorsal columns and their nuclei is maintained in the second link of this sensory pathway, the medial lemniscus and its terminal nucleus.* In the medulla the fibers of the medial lemniscus, after crossing, are found to occupy a triangular area dorsal to the pyramidal tract (Figs. 2-11, 7-4 and 7-7). Within this area the fibers from the gracile nucleus are situated ventrolaterally, those from the cuneate nucleus dorsomedially. Within these main subdivisions there is apparently also a finer arrangement according to the segmental pattern, as shown by the investigations of Ferraro and Barrera (1936) in the monkey and of Walker (1937) in the chimpan-

FIG. 2-11 Diagram of a section through the lower part of the medulla oblongata, indicating the somatotopic arrangement of the fibers in the medial lemniscus, the spinothalamic tract, and the corticospinal tract. *A:* arm; *F:* face; *L:* leg; *T:* trunk.

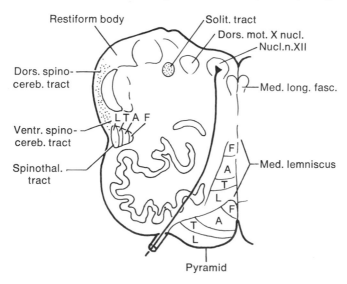

zee. This arrangement is also retained in the further course of the lemniscus, probably also here with a certain degree of overlapping, and the fibers from the main sensory trigeminal nucleus in the pons, which join the lemniscus, are found in the extreme dorsomedial part of its field. Conclusive evidence that the same arrangement is valid in man has apparently not been brought forward, but it appears probable that conditions are similar to those in monkeys. The segmental arrangement of the fibers of the medial lemniscus is represented diagrammatically in Fig. 2-11.

Further along the tract a certain rotation takes place, so the fibers which originally were placed ventrolaterally finally occupy the lateral position, whereas the originally dorsomedial fibers from the cuneate nucleus will be found medially. In this order the fibers enter the ventral nucleus of the thalamus, more precisely the nucleus ventralis posterior lateralis (VPL). The trigeminal fibers enter most medially, in a distinct portion of the nucleus, usually called the nucleus ventralis posterior medialis (VPM). From these thalamic regions the impulses are conveyed mainly to the somatosensory areas in the cerebral cortex. The endings of the fibers of the lemniscus and the thalamocortical projection will be described below.

The spinothalamic tract.   In addition to the dorsal column–medial lemniscus route, the spinothalamic tract represents a second pathway for sensory impulses from the body which are consciously perceived. It differs from the first system in certain respects, anatomically as well as functionally, and on the whole our knowledge of it is less (see Fig. 2-12 for a very simplified diagram).

As referred to above, a number of cells in the gray matter give off axons which ascend on both sides of the cord. Those ending in the thalamus make up the *spinothalamic tract*. (Sometimes a distinction is made between a ventral and a lateral spinothalamic tract, but this distinction is open to criticism; see below.) It should be noted that the spinothalamic tract, defined as the assembly of fibers which arise in the cord and end in the thalamus, does not occupy a well-delimited region in the spinal white matter. On the contrary, *its fibers are mixed with other ascending projections from the cord* referred to above. Furthermore, some of the spinothalamic fibers give off collaterals to certain nuclear regions during their course, for example the reticular formation. These features have until recently made it difficult to study certain aspects of the anatomy of the spinothalamic tract, for example to determine its cells of origin. However, the use of retrograde axonal transport of horseradish peroxidase has dramatically changed this situation in the last few years. With regard to the distribution in the cord of cells projecting to the thalamus, conditions appear to be similar in rat, cat, and monkey (certain differences exist, however; these will be considered below).

The vast majority of labeled cells in the spinal cord are found contralateral to the thalamic injection of horseradish peroxidase (Albe-Fessard, Boivie, Grant, and Levante, 1975; Trevino, 1976; Willis, Leonard, and Kenshalo, 1978, monkey; Carstens and Trevino, 1978a, cat; Giesler, Menétrey, and Basbaum, 1979, rat), as one would have expected on the basis of degeneration studies and clinical data. However, somewhat unexpectedly, a conspicuous group of labeled cells is found ipsilaterally in the C_2 segment, concentrated in the lateral part of laminae VII–VIII (Trevino, 1976; Carstens and Trevino, 1978b; Giesler et al., 1979). The contralateral labeled cells are found mainly in laminae I, V, VII, and VIII, but the

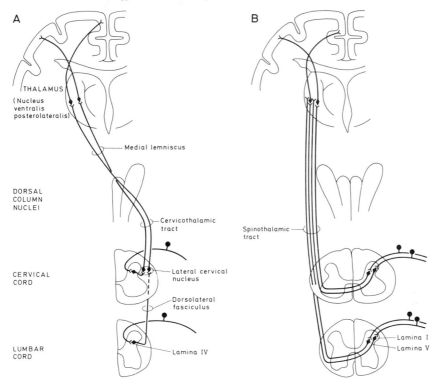

FIG. 2-12 Simplified diagrams which show the main anatomical features of the spinocervicothalamic (A) and the spinothalamic (B) tracts. The termination of spinothalamic fibers in other parts of the thalamus than the VPL is not shown.

mutual distribution within these laminae varies between cervical and lumbar levels of the cord and between species.[15]

Thus, in the monkey, at cervical levels almost all the labeled cells are found in laminae I and V (some are located in parts of laminae IV and VI adjoining lamina V), whereas at lumbar levels a fair number of labeled cells occur more ventrally in laminae VII and VIII as well (Trevino, 1976). In the cat and rat, as compared with the monkey, relatively more cells are found in laminae VII and VIII and fewer in lamina V; in the cat, labeled cells in lamina V appear to occur only at cervical levels (Carstens and Trevino, 1978a).

The anatomical data correspond well with the distribution of spinothalamic cells as determined electrophysiologically with antidromic activation from the thalamus (Willis, Trevino, Coulter, and Maunz, 1974; Price and Mayer, 1975, monkey; Trevino, Maunz, Bryan, and Willis, 1972, cat). That a relatively small number of spinothalamic units were revealed in lamina I by physiological methods

[15] By the use of more sensitive incubation methods for horseradish peroxidase a few cells in the substantia gelatinosa (lamina II) and in lamina III have been found to project to the thalamus in the monkey (Willis, Leonard, and Kenshalo, 1978). Parts of the rat trigeminal complex assumed to correspond to laminae II and III likewise send some fibers to the contralateral thalamus (Fukushima and Kerr, 1979).

is probably due to the fact that small cells are more difficult to identify than large ones.

It has been known for a long time (Petrén, 1910; and others) that in man lesions of the anterolateral funiculus of the cord give rise to loss of sensations of pain and temperature on the contralateral side of the body. Since some sensibility to touch and deep pressure remains after total interruption of the dorsal columns, and anterolateral cordotomy generally produces some tactile deficit (Foerster and Gagel, 1932), it has further been concluded that the spinothalamic tract is also involved in conveying impulses perceived as touch and deep pressure.

Physiological studies are the whole in accord with the clinical data. Thus, after transection of both medial lemnisci and one anterolateral funiculus in monkeys, leaving only one spinothalamic tract for conduction, single-unit responses have been studied in the ventral posterior thalamic nucleus, where many of the spinothalamic fibers end (see below). Units were found which were excited by displacement of hairs (touch), light mechanical distortion of the skin, distortion of muscle–fascial relations, joint rotation, pressure, heat, or stimuli of a tissue-damaging nature (Perl and Whitlock, 1961). More decisive and detailed information about the functional properties of spinothalamic cells has recently been obtained by studying, in the primate, spinal cord single units identified by antidromic activation from the thalamus (Willis, Trevino, Coulter, and Maunz, 1974; Applebaum, Beall, Foreman, and Willis, 1975; Willis, Manuz, Foreman, and Coulter, 1975). Many such cells are characterized as being excited by tactile stimuli as well as intense mechanical stimuli ("wide dynamic range neurons"), although some appear to be activated only by one type of stimuli. These functional classes of spinothalamic neurons appear to be differentially distributed among Rexed's laminae. Thus, cells in lamina I as a rule are activated solely by intense mechanical stimulation or painful heating, whereas cells in lamina V as a rule are influenced by light tactile stimulation and by more intense noxious stimuli. Spinothalamic cells receiving input from deep structures (mainly studied with regard to muscular afferents) appear to be located particularly in laminae V, VII, and VIII—that is, on the whole more ventrally than the other groups. Such "deep" cells often have additional cutaneous receptive fields (Willis, Trevino, Coulter, and Maunz, 1974; Foreman, Schmidt, and Willis, 1979), which may be of relevance to the cutaneous hyperesthesia often present in patients with disease processes in deep structures. Some spinothalamic cells activated by rotation of joints, also show a slowly adapting discharge when the joint is held in a fixed position (Willis, Trevino, Coulter, and Maunz, 1974). Such cells may therefore forward information about joint position. It is of some interest in this connection that, in order to produce total loss of hindlimb position sense in the monkey, it seems necessary to cut the contralateral anterolateral funiculus in addition to the ipsilateral dorsal column and dorsolateral funiculus (Vierck, 1966).

In accordance with their functional properties the "wide dynamic range" cells receive converging inputs from thick, myelinated (A α,β) and unmyelinated (C) fibers (Foreman, Applebaum, Beall, Trevino, and Willis, 1975). Since the discharge frequency of such cells increases with the stimulus intensity, they seem capable of transmitting information about both light, tactile stimuli and painful stimuli of various intensities in their receptive fields.

As with the lemniscal system, receptive fields of spinothalamic neurons are smaller on peripheral than on proximal parts of the extremities (Applebaum, Beall, Foreman, and Willis, 1975). The units with a high stimulus threshold (presumably activated only by noxious stimuli) have smaller receptive fields than units with a low threshold. Thus, units in lamina I may be of particular importance for localization of painful stimuli.

Although there is little reason, from clinical experience, to doubt that the spinothalamic tract is involved in mediating sense of temperature, units responding to changes of skin temperature below the noxious level have been hard to find in animal experiments (Willis et al., 1974). The reason for this is not known, but it is possible that axons of such units terminate at levels below the thalamus. Norsell (1979) found defects in the cat's ability to detect small changes in skin temperature after lesions of the middle part of the lateral funiculus, apparently in agreement with the suggestions of Foerster and Gagel (1932) and Gildenberg (1972) that thermosensitive fibers are located in the dorsal part of the ventral quadrant in man. However, the results of Norsell (1979; see also Norsell, 1980) indicate that fibers carrying information about temperature must be widely spread in the spinal white matter (including the dorsal columns) since no partial lesion produced a complete thermosensory defect.

Inhibition of the surround type, which is so characteristic of the lemniscal system, appears to be less common in the spinothalamic pathway. However, some units are subject to a peculiar form of inhibition on stimulation of the corresponding receptive field on the other side of the body (Willis et al., 1974).

It appears from electrophysiological studies that spinothalamic axons of neurons in lamina I are thinner than those arising from neurons in lamina V. This difference was used by Mayer, Price, and Becker (1975) to investigate whether activity in units of the wide dynamic range type is sufficient to cause pain to be felt in man. They stimulated the anterolateral quadrant in human subjects prior to chordotomy at an intensity sufficient to evoke activity in fibers from lamina V cells but not in fibers from lamina I cells. Such relatively low-intensity stimulation was reported by the patients as giving rise to "burning" or "hot" pain, but some reported "visceral" pain or pain "like needles." The pain could be located to, for example, a single finger or the entire leg. Furthermore, the intensity of pain felt increased with increasing stimulus intensity.

On the basis of clinical data, Foerster and Gagel (1932) suggested that within the spinothalamic tract there is an arrangement according to sensory qualities: the fibers mediating sensations of pain and temperature appear to occupy the dorsolateral part of the anterolateral funiculus, and those conveying sensations of touch and deep pressure are found in the ventromedial part. However, no such localization was found in the monkey by Applebaum, Beall, Foreman, and Willis (1975) when leading off in the white matter from spinothalamic fibers with known peripheral input. In particular, fibers mediating noxious stimuli were found both in the lateral and in the ventral part of the spinothalamic tract. These data appear to argue against the existence of two functionally different parts of the spinothalamic tract (often referred to as the paleo- and neospinothalamic tracts; see Semmes, 1969; Kerr, 1975; Boivie, 1979). However, even though functionally different spinothalamic fibers, originating to some extent in different laminae of the cord, are not segregated in the spinal white matter, they may well terminate in different parts of the thalamus (see below). This question appears not to be settled; the use of small, localized horseradish peroxidase injections in various parts of the thalamus has led to different conclusions (Carstens and Trevino, 1978a, cat; Giesler, Menetrey, and Basbaum, 1979, rat).

It appears from clinical experiences with operative section of the anterolateral funiculus that the spinothalamic fibers cross the midline shortly after their origin. Foerster (1936a) concluded that the crossing is completed in the segment above the entrance of the dorsal root fibers, since the analgesia and thermal anesthesia after the operation extend rostrally to the caudal limit of this dermatome. This view gained full support when operative section of the spinothalamic tracts for cure of otherwise intractable pain became more generally used. This operation, usually called *chordotomy*, is applied especially in cases of cancer in the pelvic organs or lower parts of the abdominal viscera. As a result of the operation the patient may be able to spend his last months of life in a tolerable condition.[16] Postmortem findings have been reported in several cases in which the sectioned fibers have been followed by the Marchi method. From these (Foerster and Gagel, 1932; Hyndman and von Epps, 1939; Weaver and Walker, 1941; Gardener and Cuneo, 1945; and others) and from experimental studies (Morin, Schwartz, and O'Leary, 1951), it is learned that the spinothalamic tract presents a segmental lamination. The fibers crossing at a particular level are added to those coming from more caudal parts of the cord at the inner side of the tract. Thus at higher levels the longest fibers, from the sacral segments, will be found most superficially, followed inward by fibers originating at successively more rostral levels (Fig. 2-9).[17] The arrangement is not strictly a question of depth, as in addition the lowest fibers are displaced more dorsally during their upward course. Thus *the fibers in the spinothalamic tract are somatotopically arranged*. This arrangement is also of diagnostic importance, and it is of practical relevance when a chordotomy is to be performed. If the section of the anterolateral funiculus is not made sufficiently deep, the deepest, i.e., shortest, fibers will escape destruction, and the upper border of the ensuing analgesia will be found several segments below the expected level. In order to avoid failure due to a too superficial section of the tract, therefore, many neurosurgeons prefer to transect the spinothalamic tract some six segments higher than the upper border of the painful region.[18]

It appears that the somatotopic arrangement is retained during the further course of the tract in the medulla and pons, although at rostral levels the lamination has not been observed as clearly as in the spinal cord (Walker, 1940a, 1942a; Weaver, Jr. and Walker, 1941; Gardner and Cuneo, 1945). In Fig. 2-11 the segmental arrangement in the medulla is indicated. At this level the tract occupies a

[16] For surveys of the operative procedure, its indications, contraindications, and results, see White and Sweet (1955, 1969) and Nathan (1963).

[17] It may be mentioned that fibers of the other long ascending and descending tracts in the spinal cord are organized in a similar manner, the longest fibers being situated most superficially. Besides being applicable to the fibers in the dorsal funiculi and the spinothalamic tracts, this principle is valid for the corticospinal tract, the spinocerebellar tracts, and others.

[18] The arrangement of the fibers has made it possible in some cases to produce an analgesia restricted, for example, to the chest or the upper extremity by performing a chordotomy limited to the ventral part of the anterolateral funiculus (Hyndman and von Epps, 1939; Jenkner, 1961).

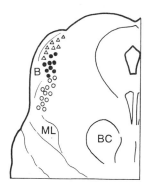

FIG. 2-13 Drawing showing the position of the spinothalamic tract and its somatotopic pattern in the mesencephalon according to Walker (1943). *Triangles:* lower extremity; *dots:* upper extremity; *circles:* face; *B:* brachium quadrigeminum inferius; *BC:* brachium conjunctivum; *ML:* medial lemniscus.

rather narrow zone dorsal to the inferior olive (see also Figs. 7-4 and 7-7). Most investigators who have followed the fibers in Marchi preparations comment upon the reduction of fibers in the tract during its rostral course. Presumably this is due to the filtering off of fibers which belong to other ascending systems, and which have been cut simultaneously with the spinothalamic tract. Furthermore, the tract appears to be considerably better developed in man and chimpanzee than in the monkey and is even less massive in the cat (see Mehler, 1966a). In the cat, Marchi studies have given meager results, while silver impregnation methods have made it possible to study the tract in some detail in this animal (see below). However, in man it has been identified in Marchi preparations in the mesencephalon underneath the brachium of the interior colliculus. Also at this level the fibers in the tract appear to be somatotopically arranged (Walker, 1942a, 1943; Morin, Schwartz, and O'Leary, 1951), at least in man and monkey, as seen in Fig. 2-13. Ventral to the spinothalamic fibers is a zone of fibers from the trigeminal nucleus, and ventral to this the medial lemniscus ascends.

From the mesencephalon the spinothalamic fibers proceed to the thalamus, where the majority end in three regions: the ventral posterior nucleus (VPL), the posterior complex, and parts of the intralaminar nuclei (see below). Like the fibers of the medial lemniscus, those of the spinothalamic tract end in a somatotopic pattern in the VPL (see Fig. 2-12).

Just as the nuclei of the dorsal columns receive fibers from the cerebral cortex which may influence their activity, so *the neurons giving rise to the spinothalamic tract are subject to influences from "higher levels."* Hagbarth and Kerr (1954) demonstrated that the central transmission in the spinothalamic tract could be inhibited by stimulation of the sensorimotor areas of the cerebral cortex, the anterior lobe of the cerebellum, and the reticular formation (see also Lindblom and Ottosson, 1957). The effect was interpreted as being caused by a postsynaptic action on the second-order sensory neurons. Later it was found that single units in the dorsal horn and sensory pathways responding to natural stimulation could be both inhibited and facilitated (Hagbarth and Fex, 1959). Furthermore, at least some of the supraspinal inhibition is presynaptic, as shown by Andersen, Eccles, and Sears (1964a) and Carpenter, Lundberg, and Norsell (1963). Coulter, Maunz, and Willis (1974) were able to show that stimulation of the monkey's sensorimotor cortex can selectively inhibit the responses of spinothalamic tract cells to tactile stimuli without affecting the responses to noxious stimulation. Other studies also indicate that the descending influences on sensory transmission in the spinothalamic tract are very complex and specific.

There are a number of *descending pathways to the spinal cord which may take part in the modulation of ascending activity in the spinothalamic tract.* Fibers of the pyramidal tract, particularly those arising in the *somatosensory cortex,* terminate in cord laminae containing spinothalamic neurons (see Chap. 4) and may be responsible for the effects of cortical stimulation on sensory transmission. Other pathways, synaptically interrupted in the brainstem, may also be involved. The descending fibers from the dorsal column nuclei, mentioned earlier in this chapter, may be of particular relevance since they originate in parts of these nuclei that receive afferents from the cerebral cortex and terminate in laminae IV, V, and possibly lamina I (Burton and Loewy, 1977). Whether these fibers also contribute to the inhibition of transmission produced by stimulation of the dorsal columns in animals and man does not appear to have been studied yet.

Spinal projections from the *periaqueductal gray,* the *raphe nuclei,* and parts of the *reticular formation* have recently received much attention because of their relation to natural pain-inhibiting mechanisms and the suppression of pain in animals and man after electrical stimulation of these or nearby brainstem sites. The physiological roles of these descending pathways, some of them probably involving endogenous morphine-like substance, appear to be very complex and far from fully understood. Only some main points will be mentioned here. For more complete information the reader is referred to review articles by Mayer and Price (1976), Hughes and Kosterlitz (1977), Fields and Basbaum (1978), and the Neurosciences Research Program bulletins *Pain* by Kerr and Casey (1978) and *Neurobiology of Peptides* by Iversen, Nicoll, and Vale (1978). Chapter 6 should be consulted for a more complete account of the connections of the reticular formation, raphe nuclei, and the periaqueductal gray substance.

Descending fibers from the gigantocellular reticular nucleus of the medulla oblongata have been traced with a variety of anatomical techniques. They descend in the ventral quadrant of the cord and distribute only to the ventral cord laminae VII and VIII, where they may conceivably influence some spinothalamic tract cells. However, their effect on motor activity appears to be more important. On the other hand, fibers arising in the nucleus raphe magnus and an adjacent part of the reticular formation, descending in the dorsolateral fasciculus, may have more direct access to pain-transmitting neurons (see Dahlström and Fuxe, 1965; Basbaum, Clanton, and Fields, 1978; Martin, Jordan, and Willis, 1978). Thus, according to studies using small medullary injections of radioactively labeled amino acids, such fibers terminate chiefly in the ipsilateral dorsal horn laminae I, II, and V and medial parts of laminae VI and VII (Basbaum, Clanton, and Fields, 1978). The projection patterns of the nucleus raphe magnus and the adjacent reticular formation appear not to be quite identical. A spinal projection from the lateral part of the periaqueductal gray substance and the adjoining part of the reticular formation (nucleus cuneiformis) to the spinal cord has recently been described in the monkey by Castiglioni, Gallaway, and Coulter (1978) in a study with the horseradish peroxidase technique. In the cat, however, such connections appear not to exist, as judged from studies using horseradish peroxidase (Kuypers and Maisky, 1975) and autoradiography (Edwards, 1975). Nevertheless, the periaqueductal gray can apparently influence neurons in the dorsal horn indirectly via projections to the nucleus raphe magnus (see also Chap. 6).

Electrical stimulation of the periaqueductal gray substance or the nucleus raphe magnus and nearby medial brainstem structures has repeatedly been shown to produce profound analgesia in various animals (Reynolds, 1969; Mayer, Wolfe, Akil, Carder, and Liebeskind, 1971; Oliveras, Besson, Guilbaud, and Liebeskind, 1974; Oliveras, Redjemi, Guilbaud, and Besson, 1975) and in man (Hosobuchi, Adams, and Linchitz, 1977; Richardson and Akil, 1977a, 1977b).

The analgesic effect appears to be mediated, at least in part, by descending fibers to the spinal cord, as indicated by the facts that electrical stimulation of the periaqueductal gray inhibits lamina V cells in the cat (Liebeskind, Guilbaud, Besson, and Oliveras, 1973) and that spinothalamic tract cells in the monkey dorsal horn can be inhibited by stimulation of the nucleus raphe magnus (Willis, Haber, and Martin, 1977). Furthermore, Willis and co-workers (1977) showed that the analgesic effect was abolished by cutting the dorsolateral fasciculus, where, as mentioned above, the raphe fibers to the cord run. These fibers contain serotonin (5-hydroxytryptamine), which may function as a transmitter at their terminations in the dorsal horn. When the synthesis of serotonin is inhibited by systemic administration of certain drugs, the analgesic effect of brain stimulation is markedly reduced (Akil and Mayer, 1972).[19]

To what degree descending fibers from medial brainstem regions act on spinothalamic neurons directly, via interneurons, or presynaptically on primary afferent fibers, is not known, but it seems likely that all modes may be of importance (see Martin, Haber, and Willis, 1979). The complexity of the neuronal mechanisms involved is illustrated by experiments showing that stimulation of the nucleus raphe magnus rather selectively inhibits dorsal horn neurons with a noxious peripheral input, whereas cells with a tactile input are mainly unaffected (Willis, Haber, and Martin, 1977). Iontophoretic application of serotonin in the cord appears also to give rather selective effects (but with a very long latency) on spinothalamic tract cells. Thus, cells of the ''wide dynamic range'' and ''high threshold'' variety (both types responsive to noxious stimuli) were inhibited, while cells with a proprioceptive input were excited (Jordan, Kensalo, Martin, Haber, and Willis, 1979).

The discovery of endogeneous substances (endorphins, enkephalins) with opiatelike effects (see Hughes and Kosterlitz, 1977) has led to an enormous research activity to reveal their sites of action and physiological significance. Somewhat earlier, specific binding sites (receptors) for morphine had been demonstrated in many parts of the brain; particularly high levels of such receptors are found, among other places, in the periaqueductal gray substance and in the spinal cord dorsal horn, as mentioned earlier in this chapter. Interestingly, the level of enkephalin is also high in these sites, and it appears that systemically administered morphine exerts its action both in the periaqueductal gray substance and in the dorsal horn, imitating the effect of the endogenous enkephalin. The effect of brain stimulation on pain transmission seems likely to involve liberation of enkephalin, since the opoid antagonist naloxone is reported to reverse the analgesic effect in man (Hosobuchi, Adams, and Linchitz, 1977; Richardson and Akil, 1977b).

As discussed above, there is now much evidence that the brain possesses a complicated system that can effectively control the information ascending in the spinothalamic tract and, particularly important from a clinical point of view, effectively suppress pain transmission. The mechanisms by which this system is activated under normal circumstances is not known, but it may be of relevance that the periaqueductal gray receives fibers ascending with the spinothalamic tract (Kerr, 1975; Trevino, 1976; and others). It has been suggested that, for example,

---

[19] There is evidence that some descending fibers mediating supression of pain transmission are not serotonergic (see Fields and Basbaum, 1978; Carstens, Klumpp, and Zimmermann, 1980).

acupuncture exerts its action via such ascending connections that activate the pain-modulating descending systems.

The very simplified account above leaves out a number of still controversial or unresolved questions concerning modulation of pain transmission. It should also be recalled that the periaqueductal gray substance and raphe nuclei have important connections with parts of the brain other than the spinal cord (see Chap. 6), to be considered when their functional roles are evaluated. We shall return later to the question of the pathways for transmission of the sensation of pain.

The spinocervicothalamic tract. Although the nucleus had apparently been noted by some anatomists (for example, Ranson, Davenport, and Doles, 1932), it was only after the description of Rexed and Brodal (1951) that the *lateral cervical nucleus* was recognized as a specific entity. It has since been rather extensively studied. The nucleus consists of a slender column of multipolar nerve cells situated in the lateral funiculus in segments C_1 and C_2, just ventrolateral to the dorsal horn. It is present in the cat and some other animals but may be present in another form or completely absent in still other species. Ha and Morin (1964) claim that it is present but very small in man, whereas Truex, Taylor, Smythe, and Gildenberg (1970) found the lateral cervical nucleus to be present in 9 out of 16 human specimens. It appears to be consistently present in the monkey, but is probably smaller than in the cat (see Ha and Morin, 1964; Mizuno, Nakano, Imaizumi, and Okamoto, 1967). The possibility cannot be excluded, however, that in the monkey and man some of the cells of the lateral cervical nucleus are incorporated in the adjacent doral horn (Brown and Gordon, 1977).

The assumption that the nucleus sends its fibers to the cerebellum (Rexed and Brodal, 1951) has been disproved (Grant, Boivie, and Brodal, 1968). Lesions in the upper brainstem result in marked retrograde cell loss in the contralateral nucleus (Morin and Catalano, 1955). Its efferent fibers cross in the upper cervical region and ascend with those of the medial lemniscus (Busch, 1961). Studies using degeneration techniques (Boivie, 1970; Ha, 1971) and retrograde transport of horseradish peroxidase (Trevino, 1976; Blomqvist, Flink, Bowsher, Griph, and Westman, 1978; Craig and Burton, 1979) have shown that most of the neurons in the lateral cervical nucleus project to the contralateral thalamus and terminate in a somatotopic manner in the nucleus ventralis posterolateralis, and in certain other nuclear groups (see below). Figure 2-12A gives a schematic representation of the spinocervicothalamic tract.

The main source of afferents to the lateral cervical nucleus is the spinal cord, as first shown with silver impregnation methods by Brodal and Rexed (1953) and later confirmed by others (Ha and Liu, 1966; Craig, 1978). Brodal and Rexed (1953) also showed that fibers from all levels of the cord terminate in the nucleus, and that the fibers are not primary afferents but come from cells located in the gray matter of the cord. The fibers ascend in the dorsolateral part of the lateral funiculus and terminate with a profuse plexus of fine collaterals throughout the nucleus. The investigators furthermore concluded that the cells of origin are not, or at least not exclusively, located in Clarke's column. These conclusions, reached on the basis of degeneration techniques, have been confirmed and extended in recent studies with retrograde transport of horseradish peroxidase. Thus, after injections of the nucleus cervicalis lateralis in the cat and dog, most labeled cells are found in the ipsilateral lamine IV of the dorsal horn, with a few scattered ones in more ventral laminae (Craig, 1978). This is also in general agreement with physiological observations (Eccles, Eccles, and Lundberg, 1960; Bryan, Trevino, Coulter, and Willis, 1973; Bryan, Coulter, and Willis, 1974), although the physiological studies have indicated a more widespread origin. This might possibly due due to inclusion of cells not projecting to the lateral cervical nucleus in the physiological samples. By means of intracellular injection of horseradish peroxidase in spinocervical cells, their dendritic arborizations and axonal trajectory have been studied in great detail (Brown, Rose, and Snow, 1977a). Spinocervical cells give off several collaterals before the axons enter the ipsilateral dorsolateral funiculus; and further collaterals, which course back to the dorsal horn, are given off as the fibers ascend. Thus, such cells must obviously serve segmental functions as well as the function of transmitting information to the lateral cervical nucleus (see Brown, Rose, and Snow, 1977a).

"Descending" connections to the lateral cervical nucleus have been traced anatomically from the "reticular" parts of the dorsal column nuclei (Craig, 1978; Burton and Loewy, 1977). The functional significance of these connections is not clear, but they presumably serve to modulate the flow of afferent impulses in the cervicothalamic tract. Stimulation of the cortical sensorimotor region appears as a rule to inhibit transmission in the spinocervicothalamic pathway, but the effect is rather selective. Thus, the imput from nociceptors to spinocervical cells is strongly inhibited, whereas the input from hair follicle afferents is unaffected (Brown, Coulter, Rose, Short, and Snow, 1977).

The original demonstration by Rexed and Ström (1952) that the spinocervical tract mediates impulses of cutaneous origin has been confirmed and extended in subsequent physiological studies, using extracellular and intracellular recordings. Cells of origin of the spinocervical tract receive monosynaptic excitatory input from thick, myelinated afferents of cutaneous origin and polysynaptic excitatory input from thin afferents (including unmyelinated fibers) from muscle and skin (see Brown, 1973; Boivie and Perl, 1975, for reviews). Many spinocervical cells receive convergent inputs from different types of receptors, such as hair receptors and receptors requiring direct and more intense stimulation of the skin. Cells in the lateral cervical nucleus have corresponding physiological properties (see Craig and Tapper, 1978; and others). Thus, in carnivores three-quarters of the lateral cervical nucleus cells can be excited by movement of hairs, but other cells require strong pressure or squeezing of the skin to be activated. Although earlier studies indicated an input from joint and muscles, this has not been verified by direct recordings from cells of the lateral cervical nucleus.

As with the other ascending sensory pathways, receptive fields of the spinocervicothalamic fibers are smaller on peripheral than on proximal parts of the body. Although it has not been apparent in anatomical studies, recent physiological results indicate that there is a somatotopic pattern within the lateral cervical nucleus: the hindlimb is represented dorsolaterally and the forelimb ventromedially; the face is represented in the medialmost part of the nucleus (Craig and Tapper, 1978).

The spinocervicothalamic tract thus seems capable of transmitting somato-

topically organized information of both light cutaneous and noxious stimuli. Its exact functional role in the overall transmission of sensory signals is not clear, however. Owing to the mixing of spinocervical fibers with fibers belonging to other ascending and descending tracts (dorsal spinocerebellar tract; nonprimary afferents to the dorsal column nuclei; corticospinal, rubrospinal, and raphespinal tracts), it would be almost impossible to analyze its functional role on the basis of lesion experiments.

The thalamus and the thalamocortical projections.   The third neuronal link in the long ascending somatic sensory fiber systems is made up of the neurons of those nuclei of the thalamus which receive the spinothalamic and medial lemniscus fibers, and whose axons transmit the impulses to the cerebral cortex. However, these fiber systems constitute only a restricted part of the extensive thalamocortical projection, as several of the other thalamic nuclei also project onto the cerebral cortex. It will be appropriate to consider the thalamus and its connections in general before treating this third link in the sensory systems in more detail.

Some of the pioneers of neuroanatomy (particularly Gudden, von Monakow, and Nissl) ascertained that extensive parts of the thalamus undergo atrophy after removal of the cerebral cortex, and they were also able to indicate definite nuclei which were connected with different parts of the cortex. Around the turn of the century the major ascending afferent connections to the thalamus had been described, and it was realized that this nuclear complex represents the final link in most afferent fiber systems transmitting impulses to the cerebral cortex. In the light of present-day knowledge the thalamus must be looked upon as being composed of a number of subdivisions which are functionally dissimilar, even if the functional role and the connections of many of the subdivisions are so far not known. The thalamus or parts of it will have to be considered in many chapters of this book. However, it is practical at this juncture to review briefly the thalamus as a whole, leaving special parts for discussion in subsequent chapters. Only aspects related to the somatic sensory pathways under consideration in this chapter will be dealt with in any detail here.

On dissection of the brain, the two egg-shaped thalami, separated by the third ventricle, can each be subdivided into three gray masses or nuclei: the anterior, the medial, and the lateral thalamic. (The latter extends posteriorly to include the pulvinar.) They are separated by laminae of white matter which in transverse sections appear as a Y, the *internal medullary lamina*. Under the microscope it is possible to distinguish within each of the three main nuclei various subgroups which differ in their cytoarchitecture and in their myeloarchitecture as well as with regard to the fiber pattern, as seen in silver-impregnated sections, and often with regard to their glial architecture. Furthermore, several small cell groups are found within the internal medullary lamina, often collectively referred to as the *intralaminar nuclei*. Close to the midline are the so-called *nuclei of the midline*. Along almost the whole external (lateral) surface of the thalamus is a thin layer of cells, the *reticular nucleus of the thalamus*, which is separated from the main body of thalamic nuclei by another thin lamina of white matter, the *external medullary lamina*. The main nuclear groups as well as some of the subdivisions and their afferent fiber contingents are shown in the diagram of Fig. 2-14. On the whole, the

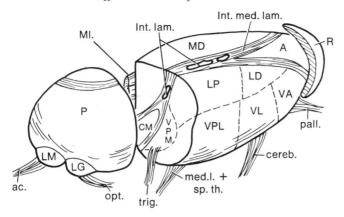

FIG. 2-14 Diagram of a three-dimensional reconstruction of the right human thalamus seen from the dorsolateral aspect. The posterior part is separated from the rest by a cut to display some features of the internal structure. Only the rostral tip of the reticular nucleus is included. The main afferent contingents to some of the major nuclei are indicated. Abbreviations for thalamic nuclei: *A:* anterior; *CM:* centromedian; *Int. lam:* intralaminar; *LD* and *LP:* lateralis dorsalis and posterior; *LG:* lateral geniculate body; *LM:* medial geniculate body; *MD:* dorsomedial; *MI:* midline; *P:* pulvinar; *R:* reticular; *VA:* ventralis anterior; *VL:* ventralis lateralis; *VPL* and *VPM:* ventralis posterior lateralis and medialis.

principal pattern of the thalamus appears to be the same in most animals. However, there are conspicuous differences as concerns the relative development of some of the cell groups. For example, the dorsomedial nucleus (MD), the dorsal nuclei (the nuclei lateralis dorsalis and posterior, LD and LP), the pulvinar, and the centromedian nucleus (CM) are particularly well developed in primates, including man.

Several authors have proposed parcellations of the thalamic nuclei and assigned particular names to each of the subdivisions. The criteria used for definition of the subdivisions have varied. On a purely descriptive anatomical largely cytoarchitectonic and myeloarchitectonic basis a number of minor units can be distinguished. The most commonly accepted scheme is still that suggested by Walker (1938c) as shown in Fig. 2-15 and adopted in Fig. 2-14. This is based chiefly on studies of the thalamus of the monkey and the chimpanzee. Hassler (1959) has extended this scheme for the thalamus of man and undertaken a far-reaching subdivision, distinguishing a large number of nuclei and areas. Somewhat different parcellations of the human thalamus have been made by Sheps (1945), Dekaban (1953), and Kuhlenbeck (1954), although Van Buren and Borke (1972) in their extensive study of the normal and pathological human thalamus adopted mainly the terminology suggested by Hassler (1959). Some authors have taken into consideration embryological criteria, while still others base their subdivisions on what is known of the fiber connections or on results of physiological studies. (For a brief review of various classifications of thalamic nuclei, see Ajmone Marsan, 1965). It appears likely that when we eventually possess sufficient knowledge of the fiber connections and the functional aspects of the barious groups of the thalamus, a rational subdivision may be arrived at. However, much remains to be clarified before this goal is achieved. An essential prerequisite will be a detailed mapping of all fiber connections of the thalamic nuclei. This is, however, an extremely difficult task, since many subdivisions are very small and can scarcely ever be damaged in isolation. Furthermore, any lesion of one of these small nuclear groups is apt to interrupt fibers passing through or near them. Physiological recordings may contribute to identifying certain connections, but again sources of error have to be considered, such as concomitant stimulation of passing fibers. The introduction during recent years, however, of autoradiographic techniques and the method of retrograde transport of horseradish peroxidase (see Chap. 1) has overcome several of these problems and greatly increased our knowledge of the fiber connections of the various thalamic subnuclei. Still, numerous questions remain, and it is therefore not deemed necessary to discuss the minutiae of the thalamic connections. The main connections of the various thalamic nuclei will be presented in their proper context in different chapters (in particular, Chapters 4, 5, 6, 8, 9, 10, and 12). Here only the thalamic nuclei concerned in the mediation of somatic sensibility will be dealt with. However, it seems appropriate to first mention some main features in the general organization of the thalamic connections.

In the presentation of the thalamic nuclei it will be practical to take as the starting point the fact that a large number of them give off *fibers to the cerebral cortex* (thalamocortical fibers). These nuclei are often said to be "cortical dependent" because they show retrograde cellular changes or cell loss following ablations of various parts of the cerebral cortex (see Fig. 1-14). As first studied in detail by Le Gros Clark and Walker most of these thalamocortical projections show a precise topical relationship between the particular thalamic nucleus and the corresponding cortical area. Among the "cortical-dependent" thalamic nuclei is the *anterior thalamic nucleus,* in the gross anatomical sense (A, Fig. 2-14). It may be subdivided architectonically into three parts, all of which send their fibers to the medial surface of the cerebral hemisphere (see Chap. 10). Of the *medial group* of macroscopical anatomy only its largest subdivision, the *nucleus medialis dorsalis* (MD), sends fibers to the cortex, more precisely to various regions of the frontal lobe, and chiefly the orbital cortex (Fig. 2-14; see also Chap. 12). Within the *lateral cell mass* certain parts give off fibers to the cortex. Proceeding in an anteroposterior direction the *inferior or ventral part* of the nucleus may be subdivided into three units (Fig. 2-14): the ventralis anterior (VA), the ventralis lateralis (VL; see Chap. 5), and the ventralis posterior. This again consists of three parts: the ventralis posterior lateralis (VPL), the ventralis posterior medialis (VPM) and the ventralis posterior inferior (VPI). These three nuclei (except apparently parts of VPI and parts of VA) all project to the cerebral cortex, the VL to the precentral, the VPL and VPM to the postcentral cortex (Fig. 2-15). As mentioned previously, VPL is the main final relay station in the medial lemniscus system, while the VPM receives fibers from the trigeminal sensory nuclei. The *dorsal part of the lateral nuclear mass* is subdivided into the nuclei lateralis (LD) and lateralis posterior (LP). The former appears to have cortical projections, at least chiefly, to the posterior part of the cingulate gyrus and probably part of the parietal cortex. The latter (LP) projects chiefly to the parietal cortex (Fig. 2-15). The pulvinar sends fibers to the parietotemporal cortex and will be dealt with more

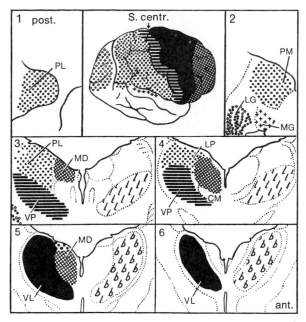

FIG. 2-15 Walker's diagram of the thalamocortical projection in the chimpanzee. 1–6: Representative sections through the thalamus, in a posteroanterior sequence. Corresponding areas of the thalamus and cerebral cortex marked with identical symbols. *LG:* lateral geniculate body; *LP:* nucleus lateralis posterior; *MD:* nucleus medialis dorsalis; *MG:* medial geniculate body; *PL* and *PM:* nuclear pulvinaris lateralis and medialis; *VL:* nucleus ventralis lateralis; *VP:* nucleus ventralis posterior; *l:* the area of termination of the fibers of the medial lemniscus and the spinothalamic tract. *b:* the area of termination of the fibers of the brachium conjunctivum. Slightly altered from Walker (1938c). Subsequent studies have necessitated certain alterations as to details in the diagram (see text).

fully in Chapter 12. Finally, the *lateral and medial geniculate bodies* (LG and MG), being stations in the pathways for vision and hearing, respectively, have well-defined cortical projection areas (Fig. 2-15) (see Chaps. 8 and 9 for details). Of particular relevance for the question of thalamic relay of somatic sensory information is the so-called *posterior complex* of the thalamus. The term was introduced by M. Rose (1935) and J. E. Rose (1942) and referred to an ill-defined cellular grouping in the caudal thalamus at the meso-diencephalic junction. Renewed interest in this thalamic area followed the work of Rose and Woolsey (1958), which suggested that this area might project diffusely upon the auditory and adjoining cortex, and the physiological study of Poggio and Mountcastle (1960), which suggested that the area might be of importance in central pain mechanisms. This notion received considerable support in the many experimental anatomical studies showing that the posterior thalamic region represents an important area of termination of spinothalamic fibers (see below; for references see Boivie 1971b, 1979, and Jones and Burton, 1974). Careful examination of well-prepared thionine-stained sections of normal brain has made it possible to define the borders of the posterior complex more precisely (Moore and Goldberg, 1963;

Rinvik, 1968a; Jones and Powell, 1971; Heath and Jones, 1971a; Robertson and Rinvik, 1973; Jones and Burton, 1974; Burton and Jones, 1976; Boivie, 1978, 1979). It has become clear from these and other studies that the posterior complex is not a homogeneous structure, and that it rather consists of several more or less distinct cellular groups, including the suprageniculate and limitans nuclei, the magnocellular division of the medial geniculate body, and portions of the pulvinar as this is delimited by Olszewski (1952). The posterior complex also includes an area of mixed cell types intercalated between the ventroposterior nucleus and the nucleus lateralis posterior, and it extends medially to the centrum medianum. Studies on the fiber connections of this thalamic area have further underlined its heterogeneity. Thus, in a detailed autoradiographic study in the monkey, Burton and Jones (1976) have shown that the medial part of their posterior nucleus projects upon the retroinsular field lying posterior to the second somatic sensory area (see below), whereas the lateral part of the posterior nucleus projects upon post-auditory cortical areas. The suprageniculate and limitans nuclei, on their part, send fibers to the granular insular area of the cortex.

It is characteristic that thalamic nuclei that give off fibers to a particular part of the cerebral cortex generally receive abundant afferents (*corticothalamic fibers*) from the same cortical areas, so that intimate reciprocal connections are established (Rinvik, 1968b, c, d, 1972; Jones and Powell, 1968c; Petras, 1972; Robertson and Rinvik, 1973; Diamond, Jones, and Powell, 1969; Niimi, Kawamura, and Ishimaru, 1971; Kawamura, Sprague, and Niimi, 1974; Holländer, 1974b). The detailed arrangements within these fiber connections indicate that functional correlations between interconnected regions of cortex and thalamus must be very close. The reciprocity between cerebral cortex and a particular cell group in the thalamus has been particularly well demonstrated in studies combining the use of retrograde transport of HRP and of autoradiographic visualization of anterogradely transported tritiated proteins (Jacobson and Trojanowski, 1975b; Trojanowski and Jacobson, 1975). These studies have, furthermore, emphasized some interesting differences in the general organization of the reciprocal connections between the cerebral cortex and various thalamic nuclei. It thus appears that all corticothalamic fibers arise from cells located in the infragranular layers of the cerebral cortex, whereas the termination of the thalamocortical fibers within the different cortical layers varies according to the thalamic nucleus of origin (Jacobson and Trojanowski, 1975b).

As can be seen from the enumeration above, the quantitatively largest nuclei of the thalamus, together making up the bulk of its entire mass, are closely related to the cerebral cortex. (For a review of the literature up to the time of the introduction of the techniques of axonal transport, see Macchi and Rinvik, 1976.) For several of the minor cell groups, which have commonly been considered not to be "cortical-dependent," data had suggested that they send at least some efferents to the cortex (see Murray, 1966; Scheibel and Scheibel, 1967). These suggestions have later been confirmed in studies in which the technique of retrograde transport of horseradish peroxidase has been used. Results have shown that the intralaminar nuclei of the thalamus project diffusely, but sparsely, upon the cerebral cortex, in addition to their dense and well-known projection upon the striatum (Jones and Leavitt, 1974; Macchi et al., 1977, Robertson, 1977; Kievit and Kuypers, 1977).

In this context it is interesting that the reciprocal connections—that is, corticofugal projections upon the intralaminar thalamic nuclei—appear to be more massive (Petras, 1972; Rinvik, 1972).

It should be noted that most of the thalamic nuclei that send massive projections to specific parts of the cortex and receive fibers from corresponding cortical regions are acted upon by afferent impulses from fairly well-characterized extrathalamic sources (somatic sensory, cerebellar, pallidal, hypothalamic), as will be considered in this and following chapters (see also Fig. 2-14).

It has for a long time been customary to speak of some thalamic nuclei as *specific*, in contrast to others which are termed *nonspecific*. This distinction is made largely on the basis of functional, electrophysiological observations. Common to the "nonspecific" thalamic nuclei is that on electrical stimulation evoked potentials occur over wide cortical areas of the two hemispheres, appearing after a long latency and showing the phenomenon of "recruitment." These cortical responses markedly contrast with the sharply localized and rapidly occurring potentials which follow stimulation of the "specific" thalamic nuclei. Most of the "nonspecific" nuclei are situated in or near the midline or in the internal medullary lamina (see Fig. 2-14 and Fig. 6-12). The reticular thalamic nucleus (R, Fig. 2-14) is also generally considered to be "nonspecific." The "nonspecific" thalamic nuclei have been considered to form part of the morphological substratum of the "ascending activating system." Although the lumping together of certain thalamic nuclei as "nonspecific" may serve a didactic purpose, it has become more and more obvious as new anatomical and physiological data accumulate that a separation between "specific" and "nonspecific" thalamic nuclei is an oversimplification. These problems and the "nonspecific" thalamic nuclei will be considered in Chapter 6, on the reticular formation.

As will appear from the above, the thalamus is a complex of nuclei that differ with regard to their connections, and accordingly they are functionally dissimilar. In fact, one or more thalamic cell groups are related to such diverse functions as vegetative reactions, motor functions, maintenance of the conscious state and its variations, the senses of hearing and vision, and the mediation of exteroceptive and proprioceptive sensations that are perceived consciously. Only the role of the thalamus as the last subcortical station for the pathways from the spinal cord to the cortex will be considered in this chapter.

The thalamic terminations of the medial lemniscus and the spinothalamic tracts. As referred to previously, it is unanimously agreed that the nucleus ventralis lateralis posterior (VPL) of the thalamus receives fibers from the medial lemniscus (Fig. 2-10). The pathway is somatotopically organized, with fibers from the gracile nucleus ending most laterally and those from the cuneate nucleus ending in the larger medial part of VPL. Medial to the latter nucleus, in the nucleus ventralis medialis posterior (VMP), are found the fibers arising from the main sensory nucleus of the trigeminal complex (see Chap. 7). Until the beginning of the 1970s it was further generally accepted that VPL also represented the major area of termination of the spinothalamic fibers. More careful examination of this fiber projection during the last few years has seriously challenged this concept. It is now clear that the medial lemniscus and the spinothalamic tract display several differences in their distribution in the thalamus and in other ways that are of interest with regard to the types of peripheral stimuli which give rise to impulses in each of them.

All the *fibers of the medial lemniscus* appear to be crossed. According to several earlier investigators, their thalamic terminal area is restricted to VPL (Gerebtzoff, 1939, in the rabbit; Matzke, 1951, in the cat; Le Gros Clark, 1936a; Bowsher, 1958; Mehler, Feferman, and Nauta, 1960, in the monkey; Walker, 1938a, in the chimpanzee; for previous studies see these articles). In retrospect, however, it is noteworthy that Bowsher (1961) in the monkey, and Lund and Webster (1967a) in the rat found some lemniscal fibers to terminate in the magnocellular part of the medial geniculate body—that is, in a part of the posterior complex of the thalamus (see above). Renewed investigations on the projections of the dorsal column nuclei in the cat (Boivie, 1971a; Jones and Burton, 1974) and in the monkey (Boivie, 1978) have established that the medial part of the posterior thalamic nucleus (POm) represents an area of termination of the lemniscal fibers in addition to VPL. It should be emphasized, however, that the lemniscal projection to POm is sparse and that, furthermore, it has so far not been shown to be somatotopically organized, as is the major lemniscal input to VPL (Boivie, 1971a, 1978). It is not yet clarified whether the dorsal column projection to POm consists of collaterals of the lemniscal input to VPL, or whether the two projection systems arise from separate cell groups in the dorsal column nuclei. In any case, judging from the caliber of the degenerating fibers, it appears that the lemniscal fibers to POm are thinner than those to VPL (Boivie, 1978). In Golgi studies Scheibel and Scheibel (1966e) have demonstrated that the axonal ramifications of the fibers of the medial lemniscus end as regularly arranged bushy arbors (see Fig. 2-17). This has been confirmed in experimental studies in which small fields free from degeneration alternate with patches of very dense degeneration in VPL following incomplete lesions of the dorsal column nuclei (Boivie, 1978). Following total destruction of the nuclei, however, the ensuing degeneration was evenly distributed throughout most of the monkey's VPL (Boivie, 1978). In experimental electron microscopic studies in the cat (Ralston, 1969) it has been shown that the degenerating lemniscal boutons belong to the largest bouton category seen in VPL (Ralston and Herman, 1969). Indirect evidence suggests that the lemniscal fibers contact both thalamocortical projecting neurons and local interneurons.

The exact area of termination of the *spinothalamic fibers* has been a subject of much controversy in the literature. Some spinothalamic fibers ascend on the ipsilateral side of the cord, but most cross the midline. All earlier investigators emphasized that a substantial number of the spinothalamic fibers end in VPL, as shown in experimental anatomical studies in the rabbit (Gerebtzoff, 1939), in the cat (Getz, 1952; Anderson and Berry, 1959), in the monkey (Le Gros Clark, 1936a; Chang and Ruch, 1947; Mehler, Feferman, and Nauta, 1960; Bowsher, 1961), in the chimpanzee (Walker, 1938a), and in man (Bowsher, 1957). In addition, it appears that spinothalamic fibers supply other thalamic cell groups, among them some situated more posteriorly. There has been some lack of clarity concerning the latter terminations, in part owing to differences in nomenclature and difficulties in distinguishing particular cell groups in this region. More recently it has become quite obvious, however, that this area of termination of spinothalamic fibers lies within the posterior complex of the thalamus (see above). Using silver impregnation methods in the monkey, Mehler, Feferman, and Nauta (1960) and Bowsher (1961) traced spinothalamic fibers, degenerating as a consequence of spinal cord lesions, to the magnocellular part of the medial geniculate body and to the suprageniculate nucleus (Bowsher, 1961). These fibers are also described in the rat (Lund and Webster, 1967b). In addition to the above-mentioned thalamic nuclei, most investigators have described an important input from the spinal cord to some of the intralaminar thalamic nuclei, in particular to the nucleus centralis lateralis (CL) (for particulars and discussion see Mehler, Feferman, and Nauta, 1960; Mehler, 1966a, b, 1969; Jones and Burton, 1974; Boivie, 1979). However, the detailed area of termination of spinothalamic fibers within CL seems to vary in different species (Jones and Burton, 1974; Boivie, 1979).

During the last few years, however, our concept of the spinothalamic projection to the ventral thalamus has undergone a radical change. In a careful experimental study in the cat, Boivie (1971b) concluded that, contrary to all reports in the literature, the spinothalamic fibers to not terminate in VPL at all, *as long as the ascending fibers from the lateral cervical nucleus are not included in the spinal cord lesion.* When the lateral cervical nucleus is selectively damaged, extensive terminal degeneration is seen in VPL (Boivie, 1970). When the lateral funiculus is lesioned at levels caudal to the lateral cervical nucleus in the cat, the degenerating spinothalamic fibers can be followed to POm, CL, and to an area lying at the transition between VL and the oral part of VPL. This area obviously corresponds to the nucleus ventralis intermedius of Hassler (1959) and Mehler (1971) and is included in Olszewski's (1952) oral part of VPL (VPLo). The findings of Boivie (1971b) in the cat were later confirmed by Jones and Burton (1974), who emphasized that there is virtually no overlap in the termination of the medial lemniscus and the spinothalamic tract in the ventral thalamic nuclear complex (VPL and VL). It should be noted, however, that the spinothalamic distribution to the transitional area between VL and rostral VPL (Boivie, 1971b, 1979; Jones and Burton, 1974) appears to coincide with the area of termination of fibers arising from the nucleus z of Brodal and Pompeiano (1957), as shown in an experimental degeneration study in the cat (Grant, Boivie, and Silfvenius, 1973). This pathway is known to relay impulses from group I muscle afferents from the cat's hindlimb (Landgren and Silfvenius, 1971).

Much of the controversy in the literature concerning the distribution of spinothalamic fibers can be explained on the basis of species differences. A recent, careful investigation of the termination of spinothalamic fibers in the monkey appears to have shed considerable light on this problem (Boivie, 1979). This author could show that the spinothalamic fibers terminate more profusely in POm in the monkey than in the cat and that the pattern of spinothalamic degeneration in the intralaminar nuclei differs somewhat in the two species. Furthermore, contrary to the findings in the cat (Boivie, 1971b; Jones and Burton, 1974), the spinothalamic fibers in the monkey appear to terminate in the whole VPL (Boivie, 1979). However, the pattern of termination in VPL of spinothalamic fibers differs markedly from that of the medial lemniscus. The spinothalamic fibers are very unevenly distributed in the monkey's VPL, with very sparse degeneration in the central parts of the nucleus and small clusters of dense degeneration in the outskirts of the forelimb and hindlimb representation areas—that is, along the borders of VPL (Boivie, 1979). A somatotopic pattern is seen in the projection upon the ventral thalamus but not in the projections to POm. An area of dense degeneration is also seen in the transitional zone rostral to VPL (V.im., see above) as found in the cat (Boivie, 1971b; Jones and Burton, 1974).

Although it may at first glance seem difficult to reconcile the contradictory observations made on the termination of spinothalamic fibers in the cat and the monkey, a closer examination of the available evidence may shed some light on this problem. As discussed by Boivie (1979), there are striking similarities between the pattern of spinothalamic tract degeneration in the monkey's VPL and the pattern of cervicothalamic tract degeneration in the cat's VPL. Thus, the combined cervicothalamic and spinothalamic tract projection areas in the cat's ventral thalamus (Boivie, 1970, 1971b) appear to correspond to the total spinothalamic tract projection in the monkey (Boivie, 1979). The studies of Boivie suggest that in carnivores and primates the spinothalamic and cervicothalamic tracts are linked together in the evolutionary process. It is conceivable that the clustered spinothalamic projection to VPL in the monkey represents a component that is homologous to much of the cervicothalamic projection in the cat (Boivie, 1979).

Textbooks of neurology generally describe two spinothalamic tracts, one ventral and one lateral, attributing different functions to the two. Although Mehler (1966, 1969) reported that the two divisions of the spinothalamic tract showed differences in their distribution in the thalamus, recent anatomical investigations in the monkey (Kerr, 1975; Boivie, 1979) indicate that fibers ascending in the ventral funiculus of the spinal cord generally end in the same regions of the thalamus, as do the more laterally ascending fibers. Anatomically speaking, there is only one spinothalamic tract, located mainly in the lateral funiculus but extending well into the ventral one (Boivie, 1979). In support of this view, Applebaum, Beall, Foreman, and Willis (1975) in a physiological study in the monkey found no anatomical segregation of functional categories among spinothalamic fibers in the upper lumbar segments except for a superficial location in the spinal cord of all spinothalamic fibers activated by deep receptors. On the other hand, it does not seem farfetched to assume that the distribution of spinothalamic fibers to three different thalamic areas must have a functional significance (see Jones and Burton, 1974).

It should be clear from the brief outline presented above that renewed anatomical investigations during the last few years have led to a thorough revision of several of our traditional concepts concerning the organization of the two major ascending pathways to the thalamus, the medial lemniscus and the spinothalamic tract. As mentioned previously, also our idea that the dorsal column nuclei receive only primary afferent fibers ascending in the dorsal funiculi can no longer be upheld (see p. 74).

The anatomical studies carried out during the last few years also emphasize that the fibers of the medial lemniscus and the spinothalamic tract by and large end in different thalamic areas, although some overlap is obvious in POm and probable in peripheral parts of VPL (see above). It is important to realize, however, that, although impulses from the dorsal column nuclei are mainly relayed in VPL, it is an oversimplification to consider VPL a pure relay nucleus in the medial lemniscus–sensory cortex pathway. This erroneous view is, unfortunately, all too often expressed in current textbooks of neurology. It should be recalled that, in addition to afferents from the dorsal column nuclei and the spinal cord, VPL receives fibers from several other sources, such as the mesencephalic reticular formation, the midline thalamic nuclei, and the medial forebrain bundle (Scheibel and Scheibel, 1966e), as well as from the reticular nucleus of the thalamus (Scheibel and Scheibel, 1966d, 1970, 1972a; Minderhoud, 1971; see Chap. 6). In addition to these structures, VPL receives a very substantial input from the somatosensory cortical areas (Rinvik, 1968b, c, 1972; Jones and Powell, 1968c). In this context it is worthwhile to recall that Ralston (1969), from his experimental electron microscopic studies in the cat, concluded that only about 8% of the boutons in VPL belong to lemniscal afferents.

All the above-mentioned anatomical observations are highly relevant to the interpretation of the often contradictory reports in the literature concerning the physiological properties of cells in the thalamic areas which receive dorsal column or spinal cord afferents. Much of this controversy can be explained by the fact that most earlier investigators thought that fibers ascending in the dorsal columns and the anterolateral columns subserved different, quite specific, sensory modalities from the periphery.

As has been mentioned earlier (pp. 78–93), however, many physiological observations have made it clear that the dorsal columns and the spinocervical tract handle similar information from the periphery, notably from all receptor types in the cat's skin and subcutaneous tissue with the exception of the sensitive thermoreceptors. It should be particularly noted that nociceptive stimuli are relayed in both pathways (for a review see Brown, 1973).

In the 1950s and early 1960s Mountcastle and collaborators made a series of investigations of the properties of single cells in VPL. It appeared from these studies that stimulus—as well as place—specificity is largely maintained in VPL (and indeed even in the sensory cortex, see below).

> Thus single units of the VPL were found to respond to only one particular type of stimulus. Of about one thousand units studied in the nucleus in the unanesthetized monkey, 42% responded only to stimulation of skin receptors (gentle mechanical stimuli), 32% responded to mechanical distortion of fascia or periosteum, and 26% to joint movements (Poggio and Mountcastle, 1963). All units responded only to stimulation within a restricted region on the contralateral side of the body. The receptive fields were small on the fingers and toes (about 0.2 cm²), considerably larger on proximal parts of the limbs. Most of the joint units were characterized by being excited by rotation of the particular joint in one direction only. Many of them, however, signaled not only movements but also certain steady joint positions. (For various reasons the units recorded from were taken to be activated by fibers of the medial lemniscus.) An analysis of the positions within the VPL of the different types of units confirmed the somatotopic pattern described anatomically and as mapped physiologically in the monkey and other animals (Mountcastle and Henneman, 1949, 1952; Rose and Mountcastle, 1952; Gaze and Gordon, 1954; Kruger and Albe-Fessard, 1960; Poggio and Mountcastle, 1960; and others). The precise mapping of the anatomical position of the units recorded from, furthermore, enabled Poggio and Mountcastle (1963) to show some other interesting features. The distal portions of the body dispose over a much greater volume of thalamic tissue than do the proximal portions. Neurons related to deep receptors are commonly located in the dorsal part of the nucleus, while neurons activated from skin are located ventrally.

In recent years, however, views on the properties ascribed to cells in VPL in earlier physiological studies have been modified. Poggio and Mountcastle (1963) commented on the effect of barbiturate anesthesia on the properties of VPL cells, and it has become increasingly clear that the percentage of VPL neurons with small peripheral receptive fields, and the size of such fields, is significantly altered by the general state of anesthesia and is also dependent on the anesthetic agent used (Jabbur, Baker, and Towe, 1972; Lanoir and Schlag, 1976, for a review). In the chronic, awake, and freely moving cat, Baker (1971) found a large number of cells in the rostral VPL with much larger peripheral receptive fields than those described by Poggio and Mountcastle in the anesthetized cat (1960) or in the paralyzed, nonanesthetized monkey (1963). These units displayed peripheral receptive field properties very similar to those of single cells in the cat's posterior thalamic complex (PO, see above) as found by Poggio and Mountcastle (1960). In a detailed physiological investigation of the properties of VPL cells in the paralyzed, nonanesthetized monkey, Loe, Whitsel, Dreyer, and Metz (1977) found that cells that are responsive to deep pressure and movements of limbs are preferentially located to the rostral and caudal ends of VPL. In the central parts of the nucleus the neurons are responsive to cutaneous tactile stimuli (Fig. 2–22). The basic body representation in VPL described by Loe et al. (1977) is in general agreement with that found by Poggio and Mountcastle (1963) although some significant differences are noted. According to Loe et al. (1977), the body representation in VB can be regarded as a series of nearly identical horizontal two-dimensional maps stacked on top of each other. Each of these horizontal maps receives a highly organized input which, proceeding from lateral to medial, describes a characteristic sequence of receptive fields. These data are in conflict with earlier physiological findings suggesting that VPL neurons situated in the dorsal or ventral part of a lamella received their input from different regions of the same dermatome (Mountcastle and Henneman, 1952; Poggio and Mountcastle, 1963). The observations of Loe et al. (1977), however, are in remarkable agreement with anatomical studies showing that a very restricted part of the monkey's somatosensory cortex has reciprocal connections with a lamella of VPL cells that extends from the dorsal to the ventral borders of the nucleus (Jones and Powell, 1970c; Jones and Leavitt, 1974; Strick, 1975; DeVito and Simmons, 1976; see below).

The medial part of the posterior complex of the thalamus (POm) represents a terminal area of both lemniscal fibers and, more importantly it seems, spinothalamic fibers. The physiological properties of the cells in this region have not been investigated as extensively as those in VPL, but it appears that they respond to a large variety of peripheral stimuli, including light, mechanical touch (Poggio and Mountcastle, 1960; Perl and Whitlock, 1961; Curry, 1972; Pugh and Wagman, 1977; Guilbaud, Caille, Besson, and Benelli, 1977). Since the spinothalamic tract is traditionally considered to relay pain impulses, considerable attention has been devoted to the reports by Poggio and Mountcastle (1960) that 60% of all PO units in the unanesthesized cat reacted only to noxious stimuli. These observations have been confirmed by Guilbaud et al. (1977), whereas Pugh and Wagman (1977) found only 15% of PO cells to be activated by noxious stimuli. The percentage of PO units responsive to painful stimuli apparently decreases under barbiturate anesthesia (Perl and Whitlock, 1960; Curry, 1972). Obviously, the available physiological evidence suggests that POm plays a role in central pain mechanisms. In VPL, on the other hand, only a few nociceptive cells are found, although some recent observations in the rat indicate that there may be more such neurons than was previously thought (Mitchell and Hellon, 1977).[20]

---

[20] It is at present difficult to assess the functional significance of the termination of spinothalamic fibers in restricted areas of CL (see Boivie, 1979), since no physiological responses registered have

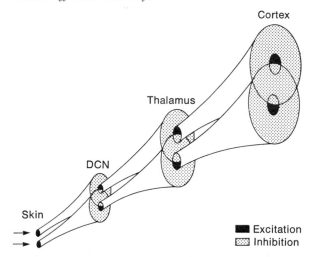

FIG. 2-16   A schematic drawing illustrating the role inhibition is thought to play in discrimination in the somatic afferent system. Each stimulus delivered to the skin activates a group of cells of the dorsal column nuclei, and inhibits those which surround that excited group. This pattern is repeated at each successive level of the system. When two stimuli are delivered and brought closer together the surround inhibition will delay fusion of the excited zones set up by the two stimuli, thus preserving the two peaks of activity at the cortical level, an event thought to be essential to the perception that two stimuli have been delivered. More complicated patterns of central excitation and inhibition will accompany more complicated peripheral stimuli, but the same organization should contribute to pattern and contour recognition. *DCN:* dorsal column nuclei. From Mountcastle and Powell (1959b).

In recent years *single-unit recordings have been made in the thalamus of human beings,* primarily as an aid in locating the points of stereotaxically placed electrodes used in the treatment of parkinsonism. Single units in the VPL have been found to respond to light touch, pressure, and joint movement (Albe-Fessard, Arfel, and Guiot, 1963; Gaze et al., 1964; Albe-Fessard et al., 1966; Jasper and Bertrand, 1966; Bates, 1973). A somatotopic pattern corresponding to that in the monkey has been found. Responses to joint movements are obtained more anteriorly than responses to tactile stimuli (Jasper and Bertrand, 1966). On stimulation of points in the VPL by implanted electrodes, conscious patients experience highly localized sensations on the contralateral side of the body, described as tingling, pins and needles, numbness, or electricity (Ervin and Marks, 1960; Hassler, 1961) while modality-specific sensations do not occur. With certain stimulus frequencies, pain sensations have, however, been elicited from the basal part of the nucleus (Hassler, 1961).

It appears from the data reviewed above that the specificity of stimulus type

been confined to such cells. Recordings from the intralaminar-medial thalamus, including CM and Pf as well as CL, have consistently shown that the neurons have large receptive fields, often including both sides of the body. They respond to a variety of stimuli, including noxious ones. However, on the basis of available experimental and clinical evidence, Mehler (1974) argued against a role played by the intralaminar thalamic nuclei in pain perception. With regard to the CL nucleus, it is in any case interesting that, in addition to a heavy projection on the striatum (see Chap. 4), it has reciprocal connections with the cerebral motor cortex (Petras, 1972; Rinvik, 1972; Jones and Leavitt, 1974; Strick, 1975).

and of site of stimulation, characteristic of the two somatic sensory pathways to the thalamus, is largely preserved within the thalamus. Furthermore, within the thalamic relay, the VPL, lateral inhibition occurs as it does at the first stations in the pathways and serves to increase sharpness in spatial localization (see Fig. 2-16 and explanation in the legend). The neurons in the VPL are subject to influences from the corresponding part of the cortex by way of corticothalamic fibers; and fibers from other thalamic nuclei and other sources are likely to play an important role.[21] These processes are obviously very complex and not yet fully understood. Golgi and electron microscopic studies of VPL have revealed the intricacy of its cellular and synaptic organization (Pappas, Cohen, and Purpura, 1966; Scheibel and Scheibel, 1966e; Jones and Powell, 1969d; Ralston and Herman, 1969; Scheibel, Davies, and Scheibel, 1972a; Scheibel, Scheibel, and Davies, 1972); Spacek and Lieberman, 1974 (Fig. 2-17). For example, the functional role played by the large number of dendrodendritic synapses in the VPL is at present purely conjectural. Obviously, the VPL is concerned with what is generally spoken of as discriminating modalities of sensation, but the complexity of its cellular and

---

[21] For further details on this subject, the reader is referred to *Corticothalamic Projections and Sensorimotor Activities* (1972).

FIG. 2-17 Arrangement of some of the components making up the presynaptic and postsynaptic neuropil of the *VB* nucleus. *1:* the diffuse net or matrix, generated by spinothalamic and reticular projections; *2:* the bushy arbors produced by terminal fibers of the medial lemniscus, arranged in an onionskin distribution around the hilus of entry; *3:* transverse discoid and more diffuse terminal patterns of corticothalamic and striatothalamic reflux systems; *4:* some aspects of the postsynaptic thalamocortical units, showing the grouping of axons from adjacent neurons, the marked overlap of dendrite domains, and the relation of somata to presynaptic arbors (schematic sagittal sections). From Scheibel and Scheibel (1966e). The term *VB* (ventrobasal nucleus) used above is often employed as a common designation for *VPM* and *VPL*.

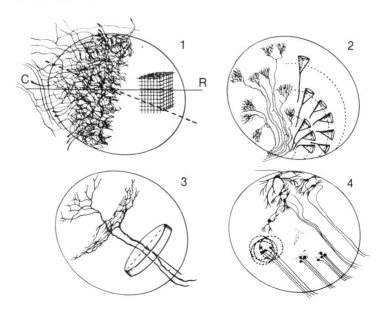

synaptic organization as revealed in Golgi and electron microscopic studies clearly indicates that the nucleus is more than a relay in the transmission of sensory information to the somatosensory cortex. Physiological studies during the past few years have also implicated a role for the VPL in the genesis of cortical rhythmical activity (for a review, see Lanoir and Schlag, 1976).

The somatosensory cortical areas. It is a matter of taste how one defines a somatosensory cortical area. If this term is applied to all cortical areas from which potentials can be recorded following stimulation of various somatic receptors, a very extensive part of the cortex should be considered sensory. On the other hand, one could use the term in a restricted sense, to cover only areas of the cerebral cortex which are end stations of fiber tracts which transmit responses to sensory stimuli via thalamic relays directly to the cortex. Until some years ago this definition appeared rather clear. Among such sensory cortical areas one would list, for example, the striate area, since this appeared to be the only cortical area receiving impulses from the eye via the lateral geniculate body. The cortical areas which are the sites of termination of fibers coming from those subdivisions of the thalamus where the somatosensory pathways treated in this chapter (and those from the trigeminal nuclei) terminate would likewise by covered by the definition. However, it appears that the criterion mentioned above is becoming more equivocal than was formerly assumed. For example, fibers from the lateral geniculate body do not supply the striate area only, but the two surrounding visual areas (18 and 19) as well (see Chap. 8). The cortical projections of the thalamic nuclei which receive the somatic afferent pathways are likewise less simple than formerly believed, and it seems certain that there is not only one but at least three cortical areas which are closely related to the reception of somatosensory messages.

From clinical and physiological observations it has been known for a long time that the postcentral gyrus in man (or the corresponding part of the brain in other species) is a main somatosensory cortical area. In 1941 Adrian demonstrated that, following peripheral somatic stimuli, responses could be picked up in the cat's brain not only in this cortical region but, in addition, in another smaller area situated beneath the lower end of the classical sensory area. The area described by Adrian is generally referred to as the *second somatosensory area,* in contrast to the classical or *first somatosensory area.* Both areas show a somatotopic pattern. Further studies have disclosed an *additional area on the medial aspect of the cerebral hemisphere* from which somatotopically localized responses can be recorded following stimulation of peripheral nerves or dorsal roots. However, it has become evident that *none of these areas is purely sensory.* From all of them motor effects can be obtained on electrical stimulation. Among the pioneer studies in this field are those of Dusser de Barenne (1933) and Dusser de Barenne, Garol, and McGulloch (1941). It has therefore become common to speak of *sensorimotor cortical areas.* Since certain parts of each area appear to be predominantly either motor or sensory, Woolsey (1964) has suggested a nomenclature which takes this into account. In the designation of an area, capital letters are used to indicate functions dominant to functions indicated by lower case letters (see Fig. 2-19). The first somatosensory area (in the postcentral gyrus) is labeled Sm I; the second somatic (sensorimotor) area is called Sm II. The third sensory area is labeled Ms II, since

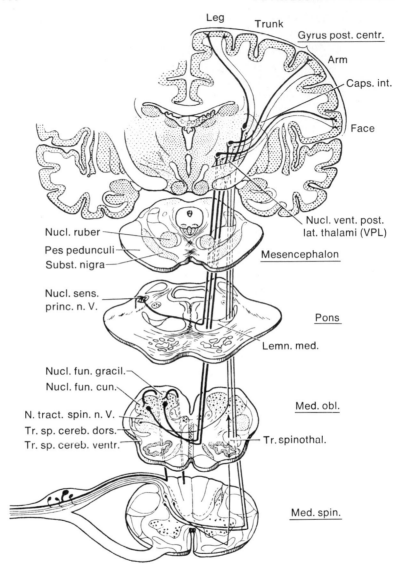

FIG. 2-18 A simplified diagram of the main features in the somatosensory pathways. Redrawn and slightly modified from Rasmussen (1932).

it appears chiefly to be related to motor functions.[22] The precentral area, often spoken of as the "motor" area (see Chap. 4), is designated Ms I, since it is also concerned in sensory functions. In the following the easier terms, "first" and "second" somatosensory areas (S I and S II), will be employed, but it should be recalled that these areas are not purely sensory. Even though some of the areas

[22] Ms II is often spoken of as the supplementary motor area (see Chap. 4). According to Blomquist and Lorenzini's (1965) studies in the squirrel monkey, a *supplementary sensory* area is found caudal to the extension of Sm I on the medial aspect of the hemisphere. There would thus be altogether five sensory areas (see also Penfield and Jasper, 1954).

listed above are as yet not known sufficiently to be of use in the evaluation of sensory disturbances in clinical neurology, the available data indicate that conditions in man are essentially the same as in experimental animals. It should, however, be emphasized that extensive parts of the cortex outside these areas are involved in the elaborate processes which take place in the brain when somatosensory stimuli are perceived.

The first somatosensory area, S I or Sm I, is fairly well known. Cytoarchitectonically it consists of three longitudinal zones in the postcentral gyrus (Brodmann's areas 3, 1, 2, area retrocentralis, see Fig. 12-2). In man this area anteriorly reaches the depth of the central sulcus where it merges with a transitional field, area 3a, which separates area 3 proper from the first motosensory cortex, Ms I (Brodmann's area 4). It should be noted that area 3 proper is often named area 3b and it is what most authors refer to in the monkey simply as area 3. The first somatosensory area, S I, covers the entire post-central gyrus to the postcentral sulcus,[23] where area 2 borders on Brodmann's areas 5 and 7. On the medial surface

[23] The same subdivision into areas, 3, 1, and 2 is found in the monkey and the cat. Particularly in the latter animal there has been some dispute as to the anterior border of the first somatosensory area. Some physiologists have maintained that the border between the "motor" and the "sensory" cortex corresponds to a "dimple" which is found rather constantly in the posterior sigmoid gyrus in the cat. This shallow sulcus is then taken to correspond to the central sulcus in monkey and man. The cytoarchitectonic studies of Hassler and Muhs-Clement (1964) support this notion. However, with regard to the amount of corticofugal fibers descending to certain nuclei in the brainstem, the border between an anterior, amply projecting and a posterior, relatively sparsely projecting, region appears to correspond to the cruciate sulcus (for some data see P. Brodal, 1968a; Rinvik, 1968b). This problem is of some practical importance for the interpretation of anatomical and physiological observations on this part of the cortex in the cat.

FIG. 2-19 Diagram of the brain of the monkey showing the general arrangement of localization patterns within the first and second sensorimotor (*Sm I* and *Sm II*), the first motosensory (*Ms I*) and the supplementary motor (*Ms II*) regions. Question marks suggest possible incomplete information for caudal borders of *Sm I*. From Woolsey (1964).

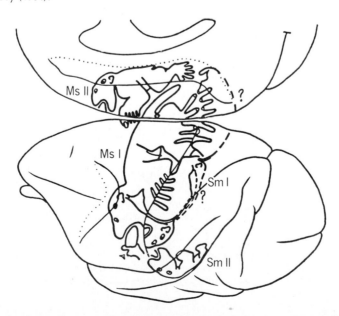

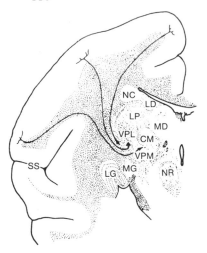

FIG. 2-20 Diagram showing the somatotopic pattern in the projection from the ventral posterior nucleus of the thalamus (*VPL* and *VPM*) onto the postcentral cortex in the macaque monkey. *NC:* caudate nucleus; *NR:* red nucleus; *SS:* Sylvian fissure. Other abbreviations as in Fig. 2-14. From Walker (1938a).

of the hemisphere it is found in the paracentral lobule. The entire region is distinguished by a distinct lamination with prominent well-developed granular layers (2nd and 4th layers) whereas the 5th layer is faintly developed (Fig. 8-7b). This is a feature typical of cortical areas predominantly receptive in function (cf. Chap. 12). Anatomically, cortical areas 3, 1, and 2 have been shown in several animal species to receive their thalamic afferents from the ventral posterior thalamic nucleus, VPL and VPM; see Fig. 2-14, (Polyak, 1932; Le Gros Clark and Boggon, 1935; Le Gros Clark, 1937; Walker, 1934, 1936, 1938b; Le Gros Clark and Powell, 1953; Chow and Pribram, 1956; Roberts and Akert, 1963; Pubols, 1968), where the medial lemniscus and the cervicothalamic tract terminate together with some of the fibers of the spinothalamic tract (see above). After a restricted lesion in this part of the cortex a sharply circumscribed patch of retrograde cell loss occurs in the nucleus ventralis posterior. The conclusion that this suggests the existence of a very sharply localized pattern within this part of the thalamocortical projection has been fully substantiated in further studies on anterograde fiber degeneration after small electrolytic lesions in various parts of the ventral posterior nucleus of the thalamus (Jones and Powell, 1969c, 1970b; Hand and Morrison, 1970, 1972) and in studies using the method of retrograde axonal transport of HRP (see Chap. 1) (Ralston and Sharp, 1973; Jones and Leavitt, 1973, 1974; Whitsel, Rustioni, Dreyer, Loe, Allen, and Metz, 1978; Lin, Merzenich, Sur, and Kaas, 1979).[24] In addition to providing new information on the organization of the thalamic projections to somatosensory cortical areas (see below), the use of modern techniques for studying axonal transport has confirmed the existence of a clear somatotopic arrangement within the projection. An informative study on this subject is that by Welker and Johnson (1965) in the raccoon. Their combined ana-

[24] It is of interest that several of these authors have reported that cells in one of the intralaminar thalamic nuclei, the nucleus centralis lateralis (CL)—which has widespread cortical projections—also projects upon Sm I (Ralston and Sharp, 1973; Jones and Leavitt, 1974; Lin et al., 1979) and the first motosensory cortex, Ms I (Jones and Leavitt, 1974). These observations are of interest since CL represents one major site of termination of spinothalamic fibers (see p. 101).

tomical and electrophysiological study demonstrates a remarkable degree of precision in this part of the thalamocortical projection. The main principles of the projection are shown in Fig. 2-20. It will be seen that regions of the VPL which receive "hindlimb" fibers project onto the upper part of the cortex of the central region, near the medial surface, while those regions which receive "forelimb" fibers project more laterally and lower. Fibers from the VPM, receiving secondary trigeminal fibers (Chap. 7), reach the lowest part of the central cortical region. This indicates that *there is within the first sensory area a somatotopic pattern, with the hindlimb represented uppermost, the fact lowermost.* More precise information has been obtained in clinical and experimental physiological studies.

The somatotopic pattern in the *postcentral gyrus in man* has been studied particularly by recording the sensations which a patient experiences when this part of the cortex is electrically stimulated during exposure for brain surgery. Extensive studies have been published by Foerster (1936a) and Penfield and his collaborators (see Penfield and Boldrey, 1937; Penfield and Rasmussen, 1950; Penfield and Jasper, 1954). Most frequently the patients describe sensations such as itching, tickling, tingling, numbness, or a feeling of pressure, although more accurate testing conditions have revealed a much greater incidence and variety of more natural sensations (Libet, Alberts, Wright, Lewis, and Feinstein, 1975). It should be noted, however, that sensations of pain have not been elicited by stimulation of the postcentral gyrus in man (Foerster, 1936a; Penfield and collaborators; Libet et al., 1975). When weak currents are applied to the postcentral gyrus, the various sensations that are elicited are sharply localized, for instance to one of the fingers.[25] In this manner it has been established that in the postcentral gyrus in man, the calf and foot are "represented" on the medial surface of the hemisphere, then follow the foci for thigh, abdomen, thorax, shoulder, arm, forearm, hand, digits, and then the face region (Fig. 2-21). Lowest on the medial surface the foci for the bladder, rectum, and genital organs are found. Electrical stimulation produces sensations in the corresponding organs.[26] Occasionally the sensations provoked by electrical stimulation of the cortex also appear in the corresponding region on the homolateral side of the body. This most readily and frequently happens as concerns the face, especially the oral region. In the larynx, pharynx, rectum, and genitalia the sensations appear to occur bilaterally. Bilateral sensations are more difficult to produce in the extremities, especially their distal parts. As referred to above, localized sensations follow irritation of the cerebral cortex are not strictly limited to the areas 3, 1, and 2 in the postcentral gyrus. Identical sensations may arise from irritation of the posterior part of the precentral cortex (Penfield and Boldrey, 1937, Libet, 1973, and others).

Figure 2-21 illustrates *another important feature in the somatosensory area I: the different parts of the body are "represented" in territories of very different*

[25] It is appropriate to recall, however, that the precise somatic site to which a sensation is referred after stimulation of a given point of the postcentral gyrus can at times shift or change in size, probably as a function of previous stimuli to the same point or to other points in the vicinity (Libet, 1973).

[26] Erickson (1945) has reported a case of nymphomania, dut to a hemangioma in the sagittal sulcus, and concluded that the nymphomania was due to irritation of the sensory focus for the genital organs. After removal of the tumor the nymphomania, as well as the accompanying motor seizures, disappeared.

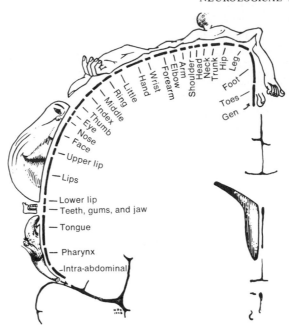

FIG. 2-21   Diagram showing the relative size of the parts of the central cortex from which sensations localized to different parts of the body can be elicited on electrical stimulation in man. From Penfield and Rasmussen (1950).

sizes. Thus the "face area" is large, the "hand area" likewise, while the proximal parts of the limbs occupy relatively small territories. As will be recalled, the same principle is valid for the "representation" of various bodily regions in the VPL. Thus the amount of tissue within the sensory cortex which is allotted to a particular region of the body is related to the importance of that region in somatic sensation and not to its absolute size.

A number of experimental studies in animals have given results concordant with those in man. The classical studies of Woolsey, Marshall, and Bard established the general principles in the discrete localization of tactile sensibility in the postcentral gyrus in the monkey (1942) and chimpanzee (1943). With the animals in deep anesthesia, these authors mapped out, point by point, the evoked cortical potentials when various parts of the skin were stimulated. The somatic receiving area determined in this way was found to correspond well with the histologically delimited areas 3, 1, and 2. The pattern of "representation" corresponds to that found in studies of human cases. The cortical projection areas of different dermatomes were found to vary widely in size. Furthermore, the mapping of evoked potentials indicated that the dermatomes are "represented" by overlapping bands of various widths which parallel each other and are arranged roughly at right angles to the central fissure. Corresponding results were obtained in Adrian's (1941) studies in the rabbit, cat, dog, and monkey. The results of these pioneer studies were subsequently confirmed by others, for example Woolsey and Fairman (1946), Mountcastle and Henneman (1952), Benjamin and Welker (1957), Celesia (1963), and Blomquist and Lorenzini (1965). A particularly discrete localization has been described in the raccoon (Welker and Seidenstein, 1959), which has an unusually well-developed cutaneous digital sensibility. Several authors have found responses to peripheral nerve stimulation or to natural somatic stimulations in front of the areas 3, 1, and 2, in the agranular "motor" cortex (Malis, Pribram, and Kruger, 1953; Benjamin and Welker, 1957; Woolsey and collaborators, see Woolsey, 1964; Albe-Fessard and Liebeskind, 1966; Asanuma, Stoney, and Abzug, 1968; Zimmerman, 1968; and others). These responses are preserved even when the sensory cortex is ablated. The latter findings, and corresponding findings in man, raise questions of details in the thalamocortical projections to the first somatosensory area. It is possible that the somatosensory— evoked cortical responses recorded in the "motor" cortex have been relayed in area 3a, lying at the transition between S I proper and M I (see below and Chap. 4). It is worth mentioning, however, that sensory input to the motor cortex appears to vary significantly with the level of anesthesia and to show considerable species differences as demonstrated in a comparative study by Goldring, Aras, and Weber (1970). The conclusion reached by these authors is that the human motor cortex plays a less important role in the integration of sensory input from the periphery than does the motor cortex in cat and even in monkey.

The use of microelectrodes for recording the activity of single cortical cells brought forward further data of great interest for our understanding of the function of the first somatosensory cortex (areas 3, 1, and 2). In a series of pioneering experiments on this topic in the monkey it was shown that over 90% of the neurons in area 2 are related to receptors of the deep tissues of the body, particularly of the joints, whereas most of the cells in area 3 are activated only by cutaneous stimuli (Mountcastle and Powell, 1959; Powell and Mountcastle, 1959b). Area 1 occupies an intermediate position with regard to the modality specificity of its neurons. This functional gradient corresponds to cytoarchitectonic transitions across the somatosensory region (Powell and Mountcastle, 1959a). The modality specificity appears to be present throughout vertically oriented columns of cells extending through all layers of the cortex. These columns are supposed to be capable of integrated activity of a rather complex order, which is independent of lateral spread of activity within the gray matter (Powell and Mountcastle, 1959b, and earlier papers). (A similar arrangement has been found and investigated in greater detail in the visual cortex, see Chap. 8; and also Chaps. 7 and 12).

The specificity of the cortical neurons in S I is especially well illustrated by Mountcastle and Powell's (1959a) observations of joint receptors in the cortex. Just as in the thalamus (see above) there are cortical units which respond to movements of a particular joint in one direction only. Furthermore, when the joint is kept in a steady position, the unit goes on discharging for a long time (several seconds). When the movement is continued in the excitatory direction the rate of discharge increases. Obviously the sensory cortex thus gets very precise information on the position and movements of the joints, and it appears that this information (together with that resulting from deformation of fascia) is the basis of what is usually referred to as "position sense." This agrees with conclusions reached in clinical and clinico-experimental studies (Stopford, 1921; Provins, 1958; and others) about the importance of an intact innervation of the joints for the position sense. Whether muscle receptors contribute substantially to the appreciation of position appears not to be quite settled (Mountcastle and Powell, 1959a; see also below). The information arriving from the cutaneous receptors to the first somatic sensory cortex appears to be quite specific as well, since there are in the cat (Mountcastle, 1957), as well as in the monkey (Powell and Mountcastle, 1959b), rapidly adapting units which respond to the displacement of hairs and slowly adapting ones which are excited by pressure of the skin.

Subsequent microelectrode studies undertaken by other investigators confirmed the observations made by Powell and Mountcastle and correlated the cytoarchitectonic differences within S I with a segregation of submodalities (Werner and Whitsel, 1968; Whitsel, Dreyer, and Roppolo, 1971; Dreyer, Loe, Metz, and Whitsel, 1975). However, these and other observations raised several questions concerning the "representation" of the body in S I. With the method of surface recording of evoked cortical potentials in the monkey it was seen that the distal parts of the extremities are "represented" more anteriorly, closer to the central sulcus, that is, in area 3, whereas the dorsum of the back is "represented" posteriorly, in area 2 (Fig. 2-19, Woolsey, Marshall, and Bard, 1942). In microelectrode studies, however, it has been shown that each part of the body is "represented" in a rostrocaudal cortical strip extending across all the cytoarchitectonic areas (areas 3, 1, and 2) comprising the entire S I (Powell and Mountcastle, 1959b; Whitsel et al., 1971; Werner and Whitsel, 1968, 1973; Pubols and Pubols, 1971, 1972; Dreyer et al., 1975). Consequently, it has been suggested that the "representation" of the body in S I is simply a representation of dermatomes extending across the rostrocaudal dimension of this cortical area; these representations are numbered along its mediolateral axis (Powell and Mountcastle, 1959b; Werner and Whitsel, 1968, 1973). Thus, from the brief outline presented above it is apparent that two different views have prevailed with regard to how the body is "represented" in S I: on one hand the "homunculus" or "simiunculus" concept (Fig. 2-19) and on the other hand the rostrocaudally oriented "isorepresentation" strips. It is difficult, however, to reconcile completely either concept with several physiological observations that have been made. Thus, many authors have observed some degree of somatotopic organization within the rostrocaudally oriented strips, and discontinuities and multiple "representations" within a single somatotopic map (for a detailed review of the literature on this topic, see Werner and Whitsel, 1973; and Merzenich, Kaas, Sur, and Lin, 1978).

Some recent physiological observations may, however, explain many of the controversial reports in the literature concerning the organization of the first somatosensory cortex. Contrary to the generally accepted notion that there is one total body "representation" that covers all cytoarchitectonic areas in S I, Paul, Merzenich, and Goodman (1972) reported that in the macaque the entire hand is "represented" within area 3 and there is a second complete cutaneous "representation" of the hand within area 1. There is possibly a third "representation" of the hand's deep modalities within area 2 and a fourth noncutaneous hand "representation" within area 3a. Later, Krishnamurti, Sanides, and Welker, (1976) reported a complete body surface "representation" within area 3b of the slow loris.

A very detailed mapping study of S I with microelectrode multiunit recording technique has recently been undertaken by Merzenich, Kaas, Sur, and Lin, (1978) in the owl monkey.[27] These

____

[27] In the owl monkey the central fissure is very shallow, and all four cytoarchitectonic areas (areas 3a, 3b, 1, and 2) are exposed on the cerebral surface, thus greatly facilitating the electrophysiological exploration of cellular activity in individual cortical areas.

authors have shown that there may be as many as four separate "representations" of the body within the classical first somatosensory cortical region. There is one large body surface "representation" that is coextensive with area 3b, and a second smaller body surface "representation" within area 1. Area 3a lies outside the cutaneous sensory strip. Finally, there is a systematic "representation" of deep body structures caudal to area 1, probably coextensive with area 2. Merzenich et al. (1978) have emphasized, however, that in neither area 3b nor area 1 is there a systematic representation of dermatomes and that the body surface "representations" in these areas are better considered as composites of somatotopically organized regions rather than as simple continuous simiunculi. Another interesting observation reported by Merzenich et al. (1978) is that the relative proportion of cortex devoted to the "representation" of various body parts differs in areas 3b and 1.

On the basis of their observations Merzenich et al. (1978) suggested that the first somatosensory region of primates should be named the parietal somatosensory strip and that areas 3b and 1 should be named *S I proper* and *posterior cutaneous field,* respectively. The reader should consult their article for further details and for an informative discussion of the various concepts of the organization of the first somatosensory area that prevail in the literature.

The physiological observations made in microelectrode studies of S I have prompted a reinvestigation of the organization of the thalamocortical projection to this cortical area. Silver impregnation studies of degenerating fibers after small electrolytic lesions of VPL and VPM had established that these thalamic nuclei project heavily upon areas 3b, 1, and 2 in the cat (Jones and Powell, 1969c; Hand and Morrison, 1970, 1972) and in the monkey (Jones and Powell, 1970a).

In the cat the thalamocortical fibers appear to be evenly distributed to the three cortical areas, whereas in the monkey area 3b receives a much greater number of thalamic afferents than do areas 2 and 1 (Jones and Powell, 1969c, 1970a; Jones and Burton, 1976).[28] Furthermore, the thalamocortical fibers to the latter areas are thin and possibly collaterals of the much thicker fibers to area 3b (Jones and Powell, 1970a, b), supporting the suggestions advanced by Le Gros Clark and Powell (1953) and Roberts and Akert (1963) in their retrograde cellular degeneration studies. In their recent study in the owl monkey, Lin, Merzenich, Sur, and Kaas (1979) have shown that injections of horseradish peroxidase into the same body "representation" part in either of the two cutaneous cortical fields (areas 3b and 1) labeled sheets of cells in the same region of the ventral posterior thalamic nucleus (VP). Furthermore, Lin et al. (1979) suggested on the basis of their observations that as many as one-third of the neurons in VP project to both area 3b and area 1, but that a significant number of VP cells project exclusively to one or the other of these two cortical areas. An important still unanswered question is how ascending lemniscal and spinothalamic fibers interact with these two types of thalamocortical projection neurons.

The observations made by Lin et al. (1979) imply that there is a single body "representation" in VP (Loe et al., 1977, see above) which provides input to two body surface "representations" in S I which differ in several important details (Merzenich et al., 1978, see above).

Furthermore, in their anatomical investigation Lin et al. (1979) found that cytoarchitectonic area 2, to which information from deep body tissues is relayed (see

---

[28] The detailed autoradiographic investigation by Jones and Burton (1976) in the monkey has also revealed distinct differences in the *laminar* distribution of thalamic afferents to the cytoarchitectonic subareas of S I. Thus, for instance, in areas 1 and 2, cortical lamina IV is practically devoid of thalamic afferents, whereas this lamina represents a major field of termination of thalamocortical fibers in area 3b.

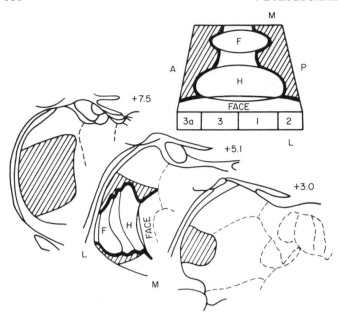

FIG. 2-22 Schematic diagram illustrating the principles in the organization of the thalamocortical projection from VP to SI in the monkey. Three stereotaxic frontal levels (anterior +3.0, +5.1, and +7.5) through the thalamus are shown. The inset shows the unfolded postcentral gyrus with its four cytoarchitectonic areas 3a, 3, 1, and 2. *H* and *F* indicate the "representation" of the hand and foot, respectively in VP and SI. *M:* medial; *L:* lateral; *A:* anterior; *P:* posterior. Oblique lines indicate zones in the thalamus and in the postcentral gyrus where cells respond predominantly to stimulation of deep body tissues (so-called "border zones"). The clear areas in VP and in SI are zones where cells respond predominantly to peripheral cutaneous stimulation (so-called "core zones"). Modified from Whitsel et al. (1978).

above), receives afferents from thalamic cells that lie mainly outside the borders of VP itself. These thalamic neurons have particularly been localized to the thalamic area named nucleus ventralis intermedius (Walker, 1938c; Hassler, 1959; Mehler, 1971) or pars oralis of VPL (VPLo, Olszewski, 1952). Similar observations appear to have been made by Whitsel, Rustioni, Dreyer, Loe, Allen, and Metz (1978) in a horseradish peroxidase study in the macaque monkey. On the basis of their observations these authors concluded that the central part of VP projects upon the "cutaneous" cortical zones (areas 3b and 1), whereas the extreme caudal and rostral parts of VP and its border zones project upon area 2 and 3a, respectively (see Fig. 2-22).[29] As emphasized by Whitsel et al. (1978), this organization of the thalamocortical projections would preserve the highly ordered spatial relations between neurons with different submodality and receptive field properties, which was shown to exist in VP (Loe et al., 1977).

In recent years particular attention has been devoted to area 3a, which in several physiological studies has been shown to receive information from group I muscle afferents and other deeply

[29] These observations support previous fiber degeneration studies showing that degenerating fibers to area 3a are particularly abundant when the rostralmost part of VP and the thalamic area immediately rostral to VP (VPLo) are damaged (Jones and Powell, 1969c, 1970a, b).

situated receptors (Mountcastle and Powell, 1959; Powell and Mountcastle, 1959b; Oscarsson and Rosén, 1966; Werner and Whitsel, 1968; Landgren and Silfvenius, 1969; Philips, Powell, and Wiesendanger, 1971; Paul, Merzenich, and Goodman, 1972; Yumiya, Kubota, and Asanuma, 1974; Tanji, 1975; Heath, Hore, and Philips, 1976). As mentioned above (p. 80) and also considered in Chapter 4, it has for some years been known that group I muscle afferents in the cat are relayed via nucleus z of Brodal and Pompeiano (1957) in the brainstem to the ventral thalamus, particularly to an area lying at the transition between VP and VL (Landgren and Silfvenius, 1971; Grant, Boivie, and Silfvenius, 1973). This thalamic area appears to correspond to the nucleus ventralis intermedius (Walker, 1938c; Hassler, 1959; Mehler, 1971) in the monkey and is identical to the rostral half of VPLo (Olszewski, 1952). Although detailed studies on the origin of thalamic afferents to area 3a have not yet been undertaken, there is some evidence that V.im. (VPLo) may project to this cytoarchitectonic division of S I (Jones and Powell, 1970a, b, 1973; Strick, 1975; Kalil, 1976; Whitsel et al., 1978). This would account for the reports that group I muscle afferents are relayed through this thalamic area in the cat (Landgren and Silfvenius, 1971; Grant, Boivie, and Silfvenius, 1973) and possibly in the monkey (Jones and Burton, 1974; Boivie, 1978). It should be noted, however, that this thalamic region represents the area of termination not only of spinothalamic fibers (Boivie, 1971b, 1979; Jones and Burton, 1974) but also of other fiber systems, most notably cerebellar afferents (Mehler, 1971).

The question of whether thermal and painful stimuli are relayed to S I is still debated. In man, stimulation of the postcentral gyrus has never elicited definite sensations of pain (Penfield and Boldrey, 1937; Penfield and Rasmussen, 1950; Foerster, 1936; Libet, 1973). Large ablations of the first and second (see below) somatosensory cortical areas in the cat (Berkley and Parmer, 1974) and the monkey (Porter and Semmes, 1974) have shown that the animals' sensitivity to temperature and noxious stimuli is largely unaffected. On the other hand, Powell and Mountcastle (1959b) found a few cells that responded to painful stimuli in the monkey's S I; and tooth pulp stimulation in the cat (Vyklidey, Keller, Brozek, and Butkhugr, 1972) and the monkey (Hassel, Biedenbach, and Brown, 1972) results in evoked cortical responses in S I. In this context it is relevant to ask whether thalamic nuclei other than VP project upon S I, since spinothalamic fibers, conveying painful stimuli, seem to terminate mainly outside VP proper, most notably in the posterior group of thalamic nuclei (PO) and in the intralaminar nucleus centralis lateralis (CL) (see p. 707). As mentioned previously, studies using the technique of retrograde axonal transport of horseradish peroxidase have shown that cells in CL project upon wide cortical areas, including S I (Ralston and Sharp, 1973; Jones and Leavitt, 1973, 1974; Macchi, Bentivoglio, Atena, Rossini, and Tempesta, 1977; Pearson, Brodal, and Powell, 1978). On the basis of experimental and clinical evidence, however, Mehler (1974) found it difficult to imply an important role for CL in pain perception. The cells in PO, on the other hand, have for some time been known to respond to noxious stimuli (see p. 704), and the cortical projection of PO has been extensively studied during the last few years. In a few horseradish peroxidase studies some labeled cells have been localized to PO after injections of the enzyme in S I (Ralston and Sharp, 1973; Pearson, Brodal, and Powell, 1978). In anterograde degeneration studies, however (Heath and Jones, 1971a; Graybiel, 1973) no evidence was found for a projection from PO to S I in the cat. A similar conclusion was arrived at in a detailed autoradiographic investigation in the monkey (Burton and Jones, 1976). Thus, the anatomical data so far corroborate most physiological and clinical reports that appear to minimize the role of S I in pain perception. On the other hand, however, results of animal experiments support the conclusion, drawn on the basis of the studies mentioned on the preceding pages, that the first somatosensory area is important for sensory discrimination. According to Orbach and Chow (1959), ablation of the postcentral gyrus in monkeys results in marked disturbances of tactile form and roughness discrimination. Other data may be found in the report of Semmes and Mishkin (1965). In a more recent investigation Randolph and Semmes (1974) studied the differential contribution of the cytoarchitectonic areas of S I to the monkey's ability to discriminate objects by touch. The authors found that lesions restricted to area 3 resulted in a severe impairment of all tactile discrimination tasks tested. Animals with lesions limited to area 1 or area 2 were significantly impaired in tasks involving discrimination of "texture" and of "angles," respectively. Randolph and Semmes (1974) could also show that the input to the hand region of area 3 was normal after an area 1 lesion, and vice versa. These findings are remarkably concordant with the findings made in microelectrode studies of S I and the recent reports of a multiple body "representation" in S I (see above). It should be noted, however, that the tactile discrimination ability appears to be only transiently lost after lesions of the first and second somatosensory areas (Norsell, 1967a, b, 1980). Several explanations can be advanced for understanding this phenomenon (for a comprehensive discussion, see Semmes, 1973), among these the existence of association and commissural fiber connections of the somatosensory cortical areas (see below).

Reference has on several occasions been made to a *second somatosensory cortical area, S II (or Sm II)*. This cortical area was discovered by Adrian (1941), in the cat, where it lies in the anterior ectosylvian gyrus. A second somatosensory area was later discovered in several other species, including the monkey, by Woolsey and his colleagues (see Woolsey, 1952, 1958) and man (Penfield and Rasmussen, 1950). In the monkey S II lies largely on the upper bank of the Sylvian fissure adjacent to the insula (Fig. 2-19). Its position in man is rather similar to that in the monkey. In all animal species studied a somatotopic pattern has been determined by studying the evoked potentials set up after peripheral stimulation (Woolsey and Fairman, 1946; Knighton, 1950; Benjamin and Welker, 1957; Welker and Seidenstein, 1959; Celesia, 1963, and others), but the localization appears to be less discrete than in S I. The face is "represented" rostrally, the hindleg most caudally (Fig. 2-19). In a detailed microelectrode mapping study in unanesthetized monkeys, Whitsel, Petrucelli, and Werner (1969) confirmed the somatotopic organization in S II but they also emphasized the extensive dermatomal overlap. Whitsel et al. (1969) also confirmed previous observations of a bilateral body "representation" in S II, in contrast to the contralateral "representation" in S I. The ipsilateral responses in S II to peripheral stimulation are not as marked as the contralateral ones, but this can be explained by the level of anesthesia, which apparently affects the ipsilateral responses more readily than the contralateral ones (Whitsel et al., 1969).

Single units have been identified that react specifically to touch and vibration. Stimulation of Pacinian corpucles gives rise to potentials more marked in S II than in S I (McIntyre, 1962). Even stimulation of a single corpuscle may be recorded in S II (McIntyre et al., 1967). It appears that impulses from muscles and deep tissues (periosteum, fascia, etc.) reach both area S II and S I (Mountcastle, Covian, and Harrison, 1952; Landgren and Wolsk, 1966; Whitsel, Petrucelli, and Werner, 1969). Movements of joints do not evoke responses in S II of the cat (Andersson, 1962) or the monkey (Whitsel et al., 1969). In a single unit study in the cat, Carreras and Andersson (1963) identified individual neurons in S II that responded to various types of peripheral stimuli, both cutaneous and deep tissue, including nociceptive stimuli. Such polysensory neurons appeared to be diffusely distributed in S II, but with an increasing gradient toward the posterior border zone of S II (Carreras and Andersson, 1963). In the unanesthetized monkey, on the other hand, Whitsel et al. (1969) found that the neurons with polysensory modality convergence and the cells responding to nociceptive stimuli were restricted to a well-localized region at the posterior margin of S II. Some cells in this region also displayed wide cutaneous (often discontinuous and asymmetrical) receptive fields. Conversely, in the anterior part of S II most neurons responded to gentle tactile stimuli and a minority to stimulation of deep tissues, such as periosteum and fascia (Whitsel et al., 1969).

According to Andersson (1962), in the cat the afferent impulses to S II are mediated via the dorsal columns, via the spinocervical tract, and via fibers that ascend in the ventral funiculi. The responses relayed to S II by the ventral funiculi appear only on strong mechanical stimulation of large receptive fields (Andersson, 1962) and resemble those localized to the posterior margin of S II in the monkey (Whitsel et al., 1969).

In man electrical stimulation of the second somatosensory area gives rise to sensations that resemble those occurring on stimulation of the first somatosensory area (Penfield and Rasmussen, 1950; Penfield and Jasper, 1954; Penfield and Faulk, Jr., 1955; Libet, 1973). Feelings of tingling, numbness, and warmth are usually referred to a particular region on the contralateral side of the body but may be felt bilaterally. (Penfield and Faulk quite often observed abdominal and epigastric sensations, and there was a concomitant change in gastric motility.) Since the area is partly hidden in the fissure of Sylvius it is not easily accessible.

Conflicting results have been reported over the years concerning *the thalamic relay for the sensory impulses reaching S II,* and only recently has the nature of this thalamocortical projection been established. Since a considerable number of cells in S II have modality and peripheral receptive field properties unlike those of S I, but similar to those of PO (see above), it appeared reasonable to suggest that there was a dual scheme in the thalamocortical somatosensory pathway in which VP and PO projected upon S I and S II, respectively. In a retrograde cellular degeneration study in the cat, however, Macchi, Angeleri, and Guazzi, (1959) concluded that VP projects upon both S I and S II. This opinion was supported with some reservation by Guillery, Adrian, Woolsey, and Rose (1966) who recorded localized responses in both somatosensory areas after focal stimulation of the cat's VP. The authors (Guillery et al., 1966), however, also claimed that PO contributed to the projection onto S II, contrary to the conclusions arrived at in a retrograde cellular degeneration study in the oppossum (Diamond and Utley, 1963). In detailed fiber degeneration studies in the cat (Jones and Powell, 1969c; Hand and Morrison, 1970, 1972) and the monkey (Jones and Powell, 1970a) it was established that VP projects upon both S I and S II. This conclusion has later been confirmed in studies with the technique of retrograde axonal transport of horseradish peroxidase (Jones and Leavitt, 1973; Ralston and Sharp, 1973). In an elegant investigation using two different labels, Hayes and Rustioni (1979) could furthermore show that some cells in the cat's VP have dichotomizing axons, sending one branch to S I and another to S II. Other cells in VP appear to project exclusively to one or the other somatosensory area (Hayes and Rustioni, 1979).

The site of cortical projection of cells in PO appears also to have been established in recent years. Destruction of the cat's PO results in fiber degeneration in a band of cortex that includes the insular and suprasylvian fringe auditory fields and overlaps a part of the posterior fringe of S II where the polysensory neurons are localized (see above) (Heath and Jones, 1971a, b; Graybiel, 1973). Thus, in the cat at least, PO does not appear to project upon S I and S II proper (Jones and Powell, 1973). This has been confirmed in a detailed autoradiographic study in the monkey (Burton and Jones, 1976) in which it was shown that PO projects upon the retroinsular area. According to Burton and Jones (1976), this area corresponds to the posterior region of S II as defined in the microelectrode investigation of Whitsel et al. (1969) (see above) and should not be considered to belong to S II proper.[30] The anterior part of S II as defined by Whitsel et al. (1969) should be considered S II proper according to Burton and Jones (1976) since it receives afferents from VP, as shown in the latter authors' autoradiographic investigation.

From the available evidence it thus *appears that VP is the main source of thalamic afferents to both S I and S II*. All the above-mentioned investigations have emphasized, however, that the projection from VP to S II is less extensive than to S I, and the topographical organization less precise.

It should particularly be noted, however, that, although PO does not appear to project—to a significant degree at least—upon either S I or S II, both these cortical areas have been shown to project massively back upon PO as well as VP in cats and monkeys (for particulars on this subject see Rinvik, 1968b, c, 1972; Jones and Powell, 1968c, 1970c, 1973; Jones and Burton, 1974).

Functional deficits after ablations of S II have been found by some authors and denied by others. Although the acute effects of damage to S II alone, or of this in combination with damage to S I, can be notable after some time, remarkable recovery of sensory discrimination is quite often seen (see Semmes, 1973, for a critical discussion). Among the many possible explanations for this phenomenon, we shall only draw attention to the *extensive associational and commissural connections* of the two somatosensory areas. These have been studied in great detail in the cat and the monkey by Jones and Powell (1968a, b; 1969a, b; and other authors) and the findings appear in principle to be similar in the two species. Only some main points will be briefly mentioned (Fig. 2-23). For details the reader should consult the original reports. Each of the cytoarchitectonic subdivisions of S I projects upon the ipsilateral S II, which in turn sends fibers back to areas 3a, 3b, 2, and 1 of S I. In addition to sending associational fibers to S II, S I projects heavily upon area 5 of the parietal lobe and the motor cortex (area 4) and less profusely upon the supplementary motor area of the same hemisphere. S II, on the

---

[30] It is interesting that cats (Berkley and Parmer, 1974) and monkeys (Porter and Semmes, 1974) are able to respond to noxious and thermal stimuli as long as the cortex lying posterior to S II is undamaged. This cortex would include the retroinsular area in which Burton and Jones (1976) reported that fibers from PO terminate.

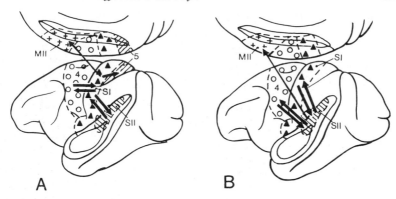

A   B

FIG. 2-23 A schematic diagram summarizing the ipsilateral cortical connections of S I (*A*) and S II (*B*). Reciprocal connections join S I and S II to one another and to the motor cortex (area 4). Each area has an additional projection to the supplementary motor area (M II) but S I alone sends fibers to area 5 of the parietal cortex. See also Fig. 4-14. Modified from Jones and Powell (1973).

other hand, does not send fibers to area 5 but only to motor and supplementary motor cortical areas. On the afferent side areas S I and S II seem to receive associational fibers only from the motor cortex (area 4). As far as the *commissural connections* are concerned, it has been shown that S I projects to both S I and S II of the opposite hemisphere, whereas S II has connections with its counterpart only. Neither somatosensory cortical area appears to project to motor or parietal cortex of the opposite hemisphere. One finding deserves to be particularly mentioned. From earlier studies it appeared that the regions of S I and S II related to the distal parts of the extremities did not receive fibers from the somatosensory areas of the opposite hemisphere (Ebner and Myers, 1962, 1965; Pandya and Vignolo, 1968; Jones and Powell, 1968b, 1969b). In a more recent investigation in the monkey, Shanks, Rockel, and Powell (1975) could show that all cytoarchitectonic subdivisions of S I receive commissural fibers from the opposite postcentral gyrus but that the intensity of the distribution of such fibers varies, being particularly heavy at the boundaries between cytoarchitectonic subdivisions. In view of the recent physiological findings of multiple body "representations" in the postcentral gyrus (see above) it may be necessary to reinvestigate the occurrence of commissural connections between cortical areas "representing" the distal parts of the extremities.

The sensory function of the supplementary motor area (Ms II of Woolsey) is relatively little known. Some observations have been mentioned above. According to Penfield's experience, stimulation of this region in conscious man gives rise to sensations in the head and abdomen and a sense of palpitation. Most often the patient describes it as a general sensation (Penfield and Jasper, 1954). There is some evidence that also in man the "sensory" part of the supplementary area is situated posteriorly to the "motor."

What is known so far of the somatosensory cortical areas makes it clear that they are not functionally equivalent. Presumably each of them has its specific tasks. Even though they all appear to be cortical receiving stations for somatosensory responses, they differ with regard to the types of stimuli which activate them, as well as in certain features of their organization. However, *they all appear to be relatively elementary links in the neural mechanisms which are related to somatosensory perception.* Disturbances in such functions are best studied in man, and clinical studies show that sensory perception may suffer even if the somatosensory areas are intact (see, for example, Weinstein et al., 1958). This is the case also in the monkey and chimpanzee (Ruch, Fulton, and German, 1938; Bates and Ettlinger, 1960; and others). The part of the cortex which appears to be specially important for sensory perception and discrimination is the *superior parietal lobule,* which in Brodmann's map (Fig. 12-2) corresponds to his areas 5 and 7 (Foerster calls Brodmann's area 7 area 5b). The cortex of these areas has a well developed layer IV, and there are many large pyramidal cells in the "efferent" layers III and V (see Chap. 12). According to Foerster (1936a), on *stimulation of the cortex in his area 5* localized sensations do not occur. Sensations are of the same type as on stimulation of the postcentral gyrus, but they are more or less diffusely projected on the entire contralateral half of the body. Furthermore, stronger currents are required, and the sensations show a greater tendency to occur bilaterally. According to these findings there appears to be no topical representation of somatic sensibility in this region.

It must be assumed that many other cortical regions in addition to area 5 are essential for the proper perception and interpretation of the large variety of sensory stimuli which reach the cerebral cortex. Furthermore, an integration of somatosensory information with optic, acoustic, and other messages will be of importance. In spite of some physiological studies on functional interrelations between various cortical areas, and their importance for processes such as perception, the question of how the brain handles all this information is still largely in the realm of speculation, and the available data give no clues of practical value in clinical neurology. Other problems concern the collaboration between the two hemispheres in sensation and the relation between "sensory" and "motor" performances. Some clues may be gotten from a study of commissural and associational connections in general (to be discussed in Chap. 12) and of the somatosensory areas in particular (see above).

The pathway for pain. On account of its practical importance, it is deemed appropriate to consider the problem of central pain transmission separately. It may at once be stated, however, that in spite of considerable progress in the last decade, knowledge on this subject is still fragmentary. In the following discussion some of this information will be considered, while only a few of the many hypothetical conceptions of the matter will be mentioned.

It may be appropriate to emphasize a few facts at the outset. In the first place, pain is a subjective experience which requires the presence of consciousness. Subjectively, pain may be of many kinds. The appreciation of pain and the evaluation of its intensity is subject to considerable variations from time to time in the same individual, depending on psychological and other factors; pain tends to increase if attention is directed to it, to decrease when attention is diverted from it.

Various stimuli—mechanical, thermal, chemical, and others—may produce pain. As discussed at the beginning of this chapter, the receptors for cutaneous pain are generally assumed to be free nerve endings, and there is no direct evidence that the sensation of pain can be produced by stimulation of encapsulated nerve endings. However, free nerve endings are certainly also engaged in the reception of stimuli other than painful ones. There is now much evidence, including experiments in human volunteers, that the peripheral nerve fibers transmitting information about painful events are of the thinly myelinated (Aδ) and unmyelinated (C) type. Furthermore, the first category of fibers appears to convey the sensation of "fast" pain while the second conveys "slow" pain. Activity restricted to thick, myelinated fibers has not been reported to produce painful sensations in man. There is rather good evidence that in the spinal cord the substantia gelatinosa is particularly closely related to the transmission of painful impulses, but its role in this process is still not clear. It may be considered fairly certain that most of the fibers which transmit the painful impulses rostrally in the cord are found in the ventrolateral funiculus (cf. the discussion below on chordotomies), even though there may be other subsidiary pathways.

When we try to follow the paths taken beyond the spinal cord by impulses that give rise to the conscious appreciation of pain, we meet with perhaps greater problems than those concerning the preceding links in the "pain pathways." All too often we have been inclined to think only of the spinothalamic tract in this connection (perhaps on account of the results of chordotomies) and to forget that in the ventrolateral funiculus the spinothalamic fibers run intermingled with many others, among them spinoreticular and spinotectal fibers, which will be interrupted in chordotomies. If "pain impulses" ascend in the ventrolateral funiculus, they might, therefore, as well continue in these tracts as in the spinothalamic. Whether they do has to be decided by physiological experiments and by clinicopathological studies. Both approaches meet with difficulties. For example, it is scarcely possible to be sure in experimental animals that one produces pure pain-eliciting stimuli. In human material reliable information can only be expected from patients with small and discrete lesions (which are rare) and in whom a precise clinical examination has been made during life. With regard to neurophysiological observations it should be recalled that what concerns us here is the *conscious appreciation of pain*. Animal experiments in which abolition of reflexes elicitable on pain-producing stimuli has followed destruction of certain regions of the brain are accordingly of limited interest.

One of the most debated questions has been *whether the appreciation of pain involves the cerebral cortex,* more particularly the postcentral gyrus. There is an extensive literature on this subject which goes back to the end of the nineteenth century. In spite of this, the problem is not solved. (For brief reviews, see Marshall, 1951, and the Neurosciences Research Program Bulletin *Pain,* 1978.) On electrical stimulation of the postcentral gyrus in conscious human beings, the patients very rarely report feeling pain. For example, in Penfield and Boldrey's (1937) large material only 11 of more than 800 responses to electrical stimulation of the postcentral gyrus were described as pain. Correspondingly, in the monkey, Mountcastle and Powell (1959b) found a few units in the first somatosensory area which responded only to nociceptive stimuli. Following cortical ablations in man most authors have observed that the patient has retained his sensibility to painful stimuli. Others have found a reduction in pain sensibility or an elimination of preexisting pain, for example phantom limb pain. As emphasized by Lewin and Phillips (1952), who described three cases of the latter type, such observations do not prove that the sensory cortex is involved in the conscious appreciation of pain, and they carefully state that "all we may conclude is that in these patients spontaneous pain is associated with activation of the sensory cortex." In a critical study of 18 patients with cortical wounds, Marshall (1951) reached a similar conclusion. He found impairment of pain and temperature sensibility in 11 of these patients. In two cases there was impairment of temperature sensibility but preservation of pain sensibility. Even though the cerebral cortex, presumably especially the postcentral gyrus, may thus be engaged in the appreciation of pain under certain conditions, it does not seem to be essential. However, there is considerable evidence that the *thalamus* is an important structure. This is evidenced both by clinical and experimental observations.

Single units responding to stimuli that are presumably pain-producing (tissue-destructive) have been found repeatedly in the posterior complex (PO, see p. 150; Poggio and Mountcastle, 1960; Whitlock and Perl, 1961; Guilbaud, Caille, Besson, and Benelli, 1977; see, however, Curry, 1972). Poggio and Mountcastle reported that nearly 60% of the PO neurons could be activated only by noxious stimuli and from large parts of the body.[31] Albe-Fessard and Bowsher (1965) found similar units in some of the nonspecific thalamic nuclei, including the centromedian nucleus. Although anatomical studies indicate that there is no termination of spinothalamic fibers in the CM (Mehler, Feferman, and Nauta, 1960; Boivie, 1979), it appears to receive afferents from parts of the reticular formation (see Chap. 6). The other thalamic regions where nociceptive units have been found are, however, supplied with spinothalamic fibers. In addition, pain impulses appear to reach other regions of the brain as well, such as the periaqueductal gray substance, the parafascicular nucleus, and the optic tectum (the colliculi). Furthermore, nociceptive impulses have been shown to be particularly potent in influencing the ascending activating system, presumably by way of spinoreticular fibers (see Chap. 6).

Fiber connections from the reticular formation ascend particularly to the "nonspecific" (but also to some extent the specific) thalamic nuclei (see Chap. 6) as well as to the superior colliculus and the periaqueductal gray. The colliculi, particularly the superior (see Tarlov and Moore, 1966), project to the central gray and to certain thalamic nuclei, among them the PO complex (see also Altman and Carpenter, 1961; Benevento and Fallon, 1975). When taken together, these and other anatomical and physiological observations strongly suggest that "pain impulses" may ascend along routes other than the spinothalamic tract. Clinical data support this view. In order to relieve unbearable pain, transection of the spinothalamic tract in the mesencephalon (mesencephalic tractotomy) has been performed. Selective sections according to the somatotopic pattern (see Fig. 2-13) have re-

---

[31] Since such units are not likely to provide information concerning the location of the stimulus, Poggio and Mountcastle suggested as a possibility that simultaneous stimulation of mechanoreceptors may be responsible for the localization of the stimulus, an assumption which receives some support from observations in humans whose lemniscal system has been interrupted. In these patients the capacity to localize a painful stimulus is poor. However, some spinothalamic units, particularly those located in lamina I of the dorsal horn, have small receptive fields (see p. 87).

sulted in analgesia in restricted parts of the body (Walker, 1942b, 1943). Some neurosurgeons have used a stereotactic approach (mesencephalotomy). There has often been immediate relief of pain and a more or less complete hemianalgesia or hypalgesia on the opposite side of the body and face. However, after some time spontaneous pain and pain sensibility appear often to return to some extent. This has been observed even if the lesion, in addition to the spinothalamic tract, has included part of the medial lemniscus (Torvik, 1959). Since the spinotectal fibers ascend close to the spinothalamic in the mesencephalon, the two tracts are likely to be interrupted together. Searching for pathways which may be responsible for transmitting pain impulses following mesencephalic tractotomies or stereotactic mesencephalotomies, several authors have suggested a spinoreticulothalamic route as the most likely. Bowsher (1957, 1976) is even inclined to associate this pathway with diffuse, slow pain, the spinothalamic tract with fast, localized pain (see, however, Boivie and Perl, 1975). The fact that the spinothalamic and reticulothalamic fibers in part end in the same thalamic nuclei (see Chap. 6) introduces further problems.

As referred to above, there is physiological evidence that noxious stimuli give rise to potentials in the PO group of the thalamus. One would expect, therefore, that this part of the thalamus might be especially closely related to pain perception in man. However, the evidence on this point is meager. Some information has been obtained from stimulations of thalamic points during stereotactic operations for the relief of thalamic pain and from the results of stereotactic destruction of parts of the thalamus.

Cooper (1965a) reports that relatively large lesions within the specific nuclei which project to somesthetic cortex "do not produce appreciable lasting objective sensory deficit on the contralateral side of the body." On the other hand, other authors (Hassler and Riechert, 1959; Bettag and Yoshida, 1960; Mark, Ervin, and Hackett, 1960; and others) have obtained relief of spontaneous pain, with or without concomitant analgesia or hypalgesia, following stereotactic destruction of the posterior part of the nucleus ventralis posterior. Only in relatively few cases has postmortem control of the lesions been made and, as a rule, the lesions have extended somewhat more rostrally than to the region presumably corresponding to the PO group (as evidenced by the ensuing sensory loss of other qualities than pain, sometimes somatotopically localized). Mark, Ervin, and Yakovlev (1963) have undertaken a correlation of anatomical sites of the lesions with clinical and experimental findings in stereotactic thalamotomy. They concluded that lesions restricted to the "sensory relay nuclei" of the thalamus result in profound sensory loss with little pain relief. In cases where good pain relief with little sensory loss had been obtained, the lesions included the parafascicular and intralaminar nuclei. (Lesions of the dorsomedial and anterior thalamic nuclei were followed by a pronounced change in affect.) Hassler and Riechert (1959) found loss predominantly of pain and temperature sensibility following destruction of the basal parts of what appears to correspond to VPL (Hassler's small-celled V.c.pc.). On stimulation of this region the patients reported pain sensations. However, Hassler (1960, 1966a) concluded that this is not the only thalamic nucleus related to pain. Other attempts to locate the thalamic structures involved in pain perception have aimed at determining the site of the lesion in cases of spontaneous "thalamic pain." In some cases of this kind a circumscribed lesion of vascular origin has been found in the caudalmost part of the VPL (Hoffmann, 1933; Garcin and Lapresle, 1954; and others). It is an apparent paradox that a destruction of this part, presumably involved in transmission (and integration?) of pain sensations gives rise to pain. Several theoretical explanations have been set forth (see, for example, Hassler, 1966a), but a satisfactory theory is still lacking.

The only conclusion which can apparently be drawn, so far, concerning the thalamus and pain is that in addition to the PO group, and presumably to some extent the VPL, other thalamic regions are probably involved, perhaps particularly some of the "nonspecific" nuclei. Central "pain transmission" does not appear to be a function of the spinothalamic tract only, but is apparently taken care of by several other fiber connections as well, not least, perhaps, the spinoreticulothalamic pathway. The central processes related to pain perception appear to be extremely complex. It may indeed be misleading to consider "pain sensibility" a sensory quality of a kind similar to, for example, tactile and joint sensibility.[32] This is suggested especially by the fact that pain sensations are far more closely linked with emotion than any other sensation, a well-known experience from daily life. Objective evidence for this close correlation comes from both experimental and clinical experience. Of particular interest are observations of patients subjected to so-called prefrontal leucotomy (see Chap. 12) since, following this operation, a preexisting pain may still be present, but it does not hurt any more; it is not of concern to the patient. Similar effects have been described following cingulumotomy (Foltz and White, 1962). (In this connection it may be recalled that the cingulate gyrus receives thalamic afferents from the anterior thalamic nucleus, the frontal cortex from the dorsomedial nucleus.) Some aspects of visceral pain will be considered in Chapter 11.

Examination of somatic sensibility.   Before describing the clinical symptoms resulting from lesions of the somatosensory system, some remarks on the examination of sensory changes are appropriate. A complete examination of sensibility, aiming at a thorough mapping out of losses or changes of the different sensory modalities, is usually a tiresome and lengthy procedure for the patient (as well as for the examiner), and it ought, if possible, to be performed in several steps, the first examination aiming only at a rough orientation. As the examination requires the co-operation of the patient, the results may be severely invalidated if the patient is tired. Without the friendly co-operation of the patient nothing can be achieved. Furthermore, it is of supreme importance for the reliability of the results obtained that the patient is of a normal level of intelligence, as clearly his mental equipment to a large extent determines his capacity not only to analyze his sensations but also to describe them properly. In mentally defective persons even the most thorough examination of the sensory functions will be futile, not least on account of their great suggestibility.

It is important to realize that, in the examination of sensory functions, *what is registered are the subjective sensations which the patient experiences.* They are the result, not only of the stimulation of one or more types of receptors, but also of the analysis and integration of the impressions perceived, a process which must be assumed to take place at higher levels of the central nervous system, presumably first and foremost in the cerebral cortex. Saying, for example, as is often done, that the dorsal columns convey impulses of vibratory sense, joint sensibil-

---

[32] The problems of pain have been discussed at length in a number of publications, and theories abound. Among some relevant references the following may be mentioned: Lewis, 1942; *Pain* (Vol. 23 of *Ass. Res. Nerv. Ment. Dis.*), 1943; White and Sweet, 1955, 1969; Keele, 1957; Noordenbos, 1959; *Pain* (Vol. 4 of *Advances in Neurology*), 1974; *Advances in Pain Research and Therapy* (Vols. 1 and 2), 1976, 1979. A theory based on anatomical and physiological observations (Albe-Fessard and Delacour, 1968) appears to give a rational explanation of some features concerning central pain mechanisms. (See also Albe-Fessard, 1968.)

ity, and two-point discrimination does not imply that each of these modalities is mediated by its separate receptors and fiber systems.[33] On the contrary, it is extremely probable, and for certain modalities practically certain, that the *sensations experienced result from a simultaneous stimulation of several types of receptors,* and that just this simultaneous stimulation is responsible for the complex character of the sensation. A loss of a certain type of sensibility on this account cannot always be interpreted as indicating a lesion of a definite part of the anatomical substratum.

As will be further elucidated below, more refined methods of examination may reveal changes of sensibility which are not detected by the usually applied routine methods. By grading the intensity of the stimuli, information can be obtained of altered threshold values of the receptors, and reduction of the territory to which the stimulus is applied may disguise slighter reductions of sensibility, otherwise not detected and not observed by the patient himself. The more refined the methods applied, the more complex the matter of sensibility appears to be, as has been seen from the data presented in preceding sections of this chapter.

These preliminary considerations ought to be borne in mind when the symptoms produced by lesions of the somatic sensory "systems" are considered. The following description will be limited to the central parts of these "systems." The symptoms following lesions of the visceral sensory "system" are treated in Chapter 11.

Symptoms following lesions of the dorsal roots. The pathological processes affecting the *dorsal roots* may, like damage to many other parts of the nervous system, give rise to irritative symptoms and to deficiency phenomena.

Irritative symptoms may appear when the dorsal roots are the subject of traction, compression, or other mechanical influences, when they are the seat of inflammatory changes, or when their blood supply is interfered with. Most frequently the irritation betrays itself as pain, "radicular pain," or "root pain," a pain which is more or less exactly limited to the dermatome belonging to the affected root, eventually also to the deeper structures supplied by the root. Affections of the thoracic roots therefore manifest themselves typically as girdle pains. They may vary in intensity, sometimes reaching intolerable height. If the irritation is slight, the pain may occasionally be limited only to certain minor areas within the dermatome, the so-called *maximal points of Head,* e.g., to the inguinal sulcus and the trochanter major in lesions of the first lumbar dorsal root.

A special type of root pain is the *lancinating pain,* found in some cases of tabes dorsalis. It is characterized by occurring in fits, varying in intensity and irradiation in the zone of distribution of the affected roots. The *tabetic crises,* which appear to be related to the lancinating pain, appear as periodically recurring, often excessive pains, usually localized by the patient to one of the inner organs. They are frequently accompanied by motor or secretory irritative symptoms. Most usual, perhaps, are the gastric crises, where the pain is felt in the epigastrium and is accompanied by a pronounced hyperesthesia of the skin in this region. It is generally believed that these tabetic

[33] Walshe (1942a) has emphasized this very clearly as follows: "There is no such *thing* as localization or discrimination, there are only things localized or discriminated, and we cannot conceive of an impulse that allows us to localize or to separate without making us aware of the thing localized or separated."

crises are also due to an irritation of the dorsal roots, but in this case predominantly the visceral afferent fibers are affected, in the case of the gastric crises the 6th–9th thoracic dorsal roots which supply the stomach. Consequently the hyperesthetic dermatomes are those belonging to these roots. (The visceral afferent impulses are discussed more fully in Chap. 11).

Radicular pain, however, does not necessarily accompany lesions of the dorsal roots. Occasionally the pain appears to be entirely absent in such cases, but frequently when it is absent other irritative phenomena occur, which may eventually also be present together with pain. Not infrequently, *paresthesias* are met with: sensations of numbness, pricking, tingling, or other peculiar sensations. Naturally these paresthesias will occur only as long as some of these root fibers remain intact. These, as well as the *hyperesthetic areas* which also may occur, are localized to the dermatomes supplied by the affected roots. On this account they may be a valuable guide in the diagnosis of the level of intraspinal morbid processes.

If the pathological process of the dorsal roots progresses, the fibers in the root will be more or less damaged, and finally their conductive capacity will be entirely lost.[34]

The deficiency symptoms in lesions of the dorsal roots will consequently be a diminished sensibility, a hypoesthesia, or a complete loss of sensibility, an *anesthesia,* and the distribution of these changes will correspond to the dermatomes of the affected roots, i.e., they will be *characterized by a segmental upper and lower limit.* Since the fibers conveying impulses of the various types are intermingled in the dorsal roots, *as a rule this sensory loss will comprise all sensory modalities,* superficial (cutaneous) as well as deep. However, aberrations are not infrequently encountered, insofar as the different modalities are affected to different degrees. This is usually explained by assuming that the different types of fibers display different resistance toward deleterious influences. The fibers conveying touch are said to be more resistant than those transmitting impulses of pain and temperature. *Intrathecal injections of phenol* have been employed for many years as a therapeutical measure, primarily in order to interrupt the central conduction of pain impulses. The method is based on the assumption that when the phenol solution infiltrates the roots, it destroys chiefly thin fibers, among these the "pain" fibers. Experimental electrophysiological studies appeared to lend some support to this assumption (Nathan and Sears, 1960), but pathological anatomical findings have not corroborated it (see Nathan, Sears, and Smith, 1965; Hansebout and Cosgrove, 1966). It is possible to apply the phenol solution fairly precisely in the vicinity of a particular root, and also to deposit it near ventral roots. The latter procedure has been used to counteract spasticity in paraplegic patients, on the assumption that the phenol will chiefly affect the γ-fibers. However, α-fibers are also affected (see Koppang, 1962; Pedersen and Juul-Jensen, 1965, for some data on results).

---

[34] It is noteworthy, however, that pain may be present even when the fibers in a dorsal root are completely interrupted. This "anesthesia dolorosa," with a segmental sensory loss and concomitant radicular pain, is assumed to arise when the fibers are interrupted, but the pathological process, e.g., a tumor, still exerts some irritation on the proximal stump of the root.

More important than "dissociations" of this type, however, is the fact, previously mentioned, that the dermatomes vary with regard to different qualities. *The dermatomes of touch extend over a somewhat larger territory than those of pain and temperature* (see Fig. 2-8).

As already mentioned, on this account, *interruption of one dorsal root will usually give no definite sensory loss,* although careful examination may reveal a limited zone of analgesia or hypalgesia, because of the somewhat more restricted overlap of the "pain" fibers. But it is important to remember that the *absence of a segmental sensory loss does not exclude the possibility of a lesion to a single dorsal root.* Using refined methods, however, in such instances some minor sensory changes can usually be detected.

When a segmental sensory symptom is present, it is important to remember that within the spinal canal the nerve roots (except the upper cervical roots) follow a descending course on their way to the intervertebral foramen. The more caudal the root, the longer is the distance between its point of exit from the cord to its intervertebral foramen. This is shown in Fig. 3-7, and, as explained more fully in Chapter 3, if overlooked, may give rise to diagnostic errors.

A segmental sensory loss, produced by a lesion of one or more dorsal roots, may be encountered in different pathological processes. The most typical instances are perhaps encountered in *intraspinal tumors,* for example neurinomas originating from one of the dorsal roots, and when a *protruded intervertebral disc* affects the root. In the latter case the region of impaired sensibility is most frequently localized to the 5th lumbar or 1st sacral dermatome. As a rule, an exact analysis of the distribution of the sensory loss or impairment will make it possible to decide whether it is of the radicular type or is due to a peripheral nerve lesion. The most important clue is given by the distribution of the sensory changes. Signs of motor impairment may, however, be present in both cases, but usually these are more prominent when peripheral nerves are affected. If a single mixed nerve is damaged, the motor and sensory disturbances will be confined to the structures supplied by it. The motor impairment in lesions of the dorsal roots may be due to a simultaneous affection of the ventral roots as well, caused, for example, by pressure of the tumor. In this case the paresis or paralysis will also be segmentally distributed (cf. Chap. 3). Or the disturbed motor functions may be due to affection of the long descending motor pathways, producing a monoplegia or hemiplegia according to the intensity and level of the lesion (cf. below). Finally, autonomic disturbances may be present or may be ascertained by special tests, such as the sweat test (cf. Chap. 11).

On account of their practical importance the symptoms occurring in cases of *intraspinal protrusion or herniation of an intervertebral disc* will be briefly considered. This condition is due to degenerative processes of one or more intervertebral discs with a dorsal displacement of the nucleus pulposus through defects in the annulus fibrosus. In some cases a narrow spinal canal ("spinal stenosis") caused by a combination of congenital and acquired conditions, may give rise to similar symptoms (Verbiest, 1954, 1977). In some cases a *thickening of the ligamentum flavum* (between the vertebral laminae) gives rise to similar symptoms. A thickening of the ligamentum flavum is occasionally found together with a herniated disc. The degenerative process in itself may also produce similar symptoms (osteochondrosis columnae).[35]

In the majority of cases the lumbar discs are affected, and among these again the discs between the 4th and 5th lumbar and between the latter and the 1st sacral vertebrae are most commonly those producing clinical symptoms. The symptoms can be explained on the basis of the conditions arising in the spinal canal when a protrusion is present and will differ according to whether it takes place laterally or near the midline. The localization of the symptoms will depend on which root or roots are affected. It is not unusual that more than one root is involved. The oblique course of the lumbar roots in the spinal canal (see Fig. 3-7) explains that when more than one root is affected, the roots leaving the canal at levels lower than the one primarily affected will be involved. This is seen in Fig. 2-24, where a lateral protrusion originating from the disc between the 4th and 5th vertebrae presses on the 5th root but, in addition, also affects the 1st sacral root. A median prolapse is apt to affect the roots of both sides and, if large, will give gross symptoms due to its pressure on the cauda equina. Extensive damage to the lumbar and sacral roots may be observed, making the differential diagnosis between a prolapse and an intraspinal tumor difficult.

The first sign of pressure of a protruded intervertebral disc usually comes from the dorsal root affected. This is manifested as pain and is frequently preceded by transient attacks of low-back pain. Not infrequently the symptoms begin acutely, when the patient lifts a heavy weight while he is bending forward. In ventroflexion of the back the nucleus pulposus is pushed dorsally, and the degenerated annulus may give way or rupture. The *pain* is more or less clearly localized in the area of distribution of the affected root, and in affections of the 5th lumbar and 1st sacral clinically appears as sciatica. (It should be stressed, however, that "scia-

[35] In autopsy studies of the human cord it is not uncommon to find meningeal changes in the lumbar and sacral nerve roots, in the form of arachnoidal proliferations, often combined with cyst formations (see Rexed and Wennström, 1959). These changes are most often found in elderly persons, and are commonly multiple. It is not clear to what extent they may give symptoms of "backache" or sciatica. Beatty, Sugar, and Fox (1968) report findings in a series of patients in whom clinical symptoms of lumbar and sacral root compression appeared to be due to a folding of the posterior longitudinal ligament. The cause of this may be a degeneration of a disc.

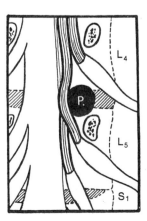

FIG. 2-24 A diagram illustrating the pressure exerted by a protruded 4th intervertebral disc. Originating laterally, the protrusion (*P*) affects the 5th lumbar root but in addition affects the 1st sacral root. Redrawn from Spurling and Grantham (1940).

tica'' may be due to several pathological conditions other than protrusion of an intervertebral disc, since the fibers proceeding from the roots into the nerve may be affected also in the pelvis and the thigh.) The pain is usually assumed to be a consequence of pressure on the nerve fibers of the root (see, however, below).

This concept has been challenged on the basis that acute peripheral compression neuropathies are usually painless. Furthermore, in experimental studies, acute compression or mechanical injury of spinal nerve roots or nerves produces only a few seconds of repetitive firing of nerve impulses (Wall, Waxman, and Basbaum, 1974; Howe, Loeser, and Calvin, 1977). According to Howe et al. (1977), however, acute compression of a *dorsal root ganglion* in cats and rabbits produces prolonged (5–25 min) periods of firing, and if the dorsal root or ganglion is the site of chronic inflammation, minimal compression will regularly produce prolonged discharges in Aβ, Aδ, and probably C fibers. In the light of the close spatial relationship between the dorsal root ganglion and the lateral part of the intervertebral disc (Lindblom and Rexed, 1948), it seems reasonable to assume that irritation of the dorsal root ganglion, be it caused by compression or by stretching (see below), is of particular importance for the pain produced in "sciatica." Further support for this notion comes from the observation of Lindblom and Rexed (1948) that in cases of herniated lumbar discs the affected dorsal root ganglia were distorted and showed various degrees of degeneration.

Characteristic of "sciatica" are the exacerbations of the pain following increase of intracranial pressure, such as occurs in sneezing, coughing, and straining. Compression of the jugular veins also usually augments the pain for the same reason. Raising of the straight leg (Lasègue's sign) increases the pain or is painful when spontaneous pain is absent, in severe cases even when the leg is raised only 20° or even less. This phenomenon has been explained as being due to stretching of the nerve fibers (when the ankle is dorsiflexed the pain is still more exaggerated). It is commonly assumed that reflex spasm of the muscles at the back of the thigh is also responsible for some of the pain. Most patients are in less pain when lying than when sitting. This has been explained by the assumption that pressure on the vertebral discs is less in the lying position than when the spine is kept erect, and thus allows the protrusion to recede somewhat. When the patient is standing, a *scoliosis* is regularly seen, most often with the tilt to the side opposite the protrusion. This, and also the common flatening of the *lumbar lordosis,* may be explained on the same basis: a slight movement of the spine to the opposite side and a slight ventroflexion may relieve the protruding force and the pressure on the root. However, in other instances the tilt is to the side of the protrusion, and other explanations must be sought. The slackening of the root passing over the protrusion (see below) during the movement to this side may be of relevance. The mechanical situation in the spinal canal in cases of protruded discs may obviously be different among patients (see Sand, 1970) depending on the site and size of the protrusion, its consistency, the degree of degeneration and flattening of the disc, concomitant alterations in the vertebral joints, and other factors.

If one root only is affected, a hypoesthesia will frequently not be discovered, except by testing very carefully for sensibility to pain (see above). The *hypalgesia,* if present, is limited to the affected dermatome. Thus in lesions of the 5th lumbar root it will be mainly found on the lateral aspect of the calf, sometimes extending toward the dorsum of the 1st toe; in lesions of the 1st sacral root the area is found on the heel, extending beneath the lateral malleolus on the lateral aspect of the foot (see Fig. 2-7 on the dermatomes). The *anklejerk* is weakened or abolished if the 1st sacral root is affected. (The kneejerk is usually influenced by lesions of the 3rd or 4th lumbar roots.) In some cases *paresthesias* occur, of the same segmental distribution as the pain and hypalgesia.

If the ventral roots are affected, a *motor impairment* will ensue. When one root only is damaged a pareses may be inconspicuous, but, like the sensory changes, its distribution will be segmental and may aid in the diagnosis. When the protrusion is larger, there will occasionally be changes in the cerebrospinal fluid at levels beneath the lesion, but on the whole the purely clinical signs are the most important for diagnosis. Myelography (radiculography) may verify the exact site of the protrusion.

It should be remembered that, although protrusion of intervertebral discs is by far the most common in the lower lumbar spine, it may occur in any intervertebral disc. Particularly those originating from the lower cervical discs should be borne in mind since they give rise to symptoms that may be interpreted as due to other affections. In the rather rare protrusions of thoracic discs, symptoms are often uncharacteristic. However, these protrusions are particularly apt to result in compression of the cord and, therefore, may require immediate operative treatment (see Kite, Whitfield, and Campbell, 1957).

Particularly with regard to the cervical spinal canal, Breig's studies (1960, 1978) are of interest. The common view has been that the spinal cord and the dura move up and down in the vertebral canal as the spine is bent forward or backward. Breig found no evidence of this. In an extensive study on autopsy material he demonstrated that in ventroflexion the cord is straight, and so are the dura and the roots (see Fig. 2-25, to the left). On dorsiflexion when the vertebral canal is shortened the spinal cord also shortens, with a consequent folding of the dura, most marked dorsally (Fig. 2-25, to the right). The nerve roots are slackened as well. Microscopic studies showed that, while the longitudinal nerve fibers of the cord are straight in ventroflexion, they assume a wavy course when the cord is shortened in dorsiflexion. As might be expected, this shortening is compensated for by an increase in the transverse direction; the cross area of all levels of the cord is larger in dorsiflexion than in ventroflexion. On lateral movements of the body the cord is shortened on the side to which the bending takes place, and straightened on the other. Obviously mutual rearrangements take place between the elements of the soft tissues of the cord during movements.

These findings have important bearing on diagnostic and therapeutic problems, as dealt with in some detail by Breig (1960, 1978). Thus in exposing the cord during operative interventions, the risk of traumatization is reduced if the cord is slack, i.e., the spine is dorsiflexed. With regard to the symptoms from the roots occurring in protrusion of an intervertebral disc, it appears that the protrusion does not necessarily have to compress the root, but when the nerve "rides" on the protrusion, a movement that causes stretching of the root may produce pain because the root is not free to stretch normally. Since the roots are solidly fixed in the intervertebral foramina the appearance of pain on Lasègue's test may in part be due to the ventroflexion of the spine which ensues when the leg is raised, with concomitant stretching of the roots. The pain on coughing and sneezing may likewise be explained in part by the straightening of the back which takes place. The observations of Breig have a bearing also on other types of diseases involving

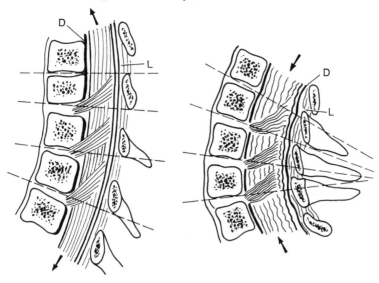

FIG. 2-25 Diagrammatic representation of the change in form of the lower cervical spine and of the concomitant changes in the cord and nerve roots occurring on bending of the spine. In ventroflexion (to the left) the cord, dura (*D*), nerve roots, and ligamenta flava (*L*) are straight. In dorsiflexion of the spine (to the right) all structures are shortened, especially the most dorsally situated ones (see text). Redrawn from Breig (1960).

the cord, as discussed by the author. More recently, however, other authors have come to conclusions differing somewhat from those of Breig with regard to movement to the spinal cord and dural sac. Thus, Adams and Logue (1971) found that gliding as well as folding and unfolding of the dural sac occurs, but they considered the gliding movement to be quantitatively most important.

It is of some interest that the spinal dura is only innervated (from the meningeal rami) on its ventral aspect (Edgar and Nundy, 1966). This explains why piercing the dura on lumbar puncture is not painful. Fine, unmyelinated nerve fibers have been said to be present in the annulus fibrosus (Roofe, 1940), but most authors agree that nerve fibers are present only in the dense connective tissue in the outermost zone of the disc (see, for example, Stilwell, 1956; Jackson, Winkelmann, and Bickel, 1966; Kumar and Davis, 1973). A recurrent branch from the 2nd lumbar nerve descends on the dorsal aspect of the vertebrae, in the posterior longitudinal ligament, and can be followed to the level of the 5th lumbar vertebra. Irritation of this branch may explain the common occurrence of low-back pain in protrusions of lumbar discs.

Symptoms following lesions of the dorsal horn and of the central gray matter of the cord. A *segmental sensory disturbance can be caused* not only by affections of the dorsal roots but *also by lesions of the dorsal horn.* This is a consequence of the anatomical arrangement whereby many of the afferent fibers from the spinal ganglia are distributed to the cells in the dorsal horn in the segment in which they enter the cord. Irritative symptoms may appear, most frequently as pain of radicular distribution, or a radicular sensory loss may be found. "Dorsal horn pains," however, occur less frequently than "dorsal root pains." Presumably symptoms of this kind, indicating a damage to the dorsal horn, arise most frequently in syringomyelia or hematomyelia. However, there is a *characteristic difference between the symptoms following lesions of the dorsal roots and dorsal horns. In the latter case the classical dissociated sensory loss may be found in which the sensibility to temperature and pain is abolished or weakened, whereas the sense of touch appears to be intact when tested with routine methods.* In lesions of the dorsal roots, as previously mentioned, all sensory modalities are affected. The explanation of the different symptoms in the two types of lesions has been thought to be found in the anatomical arrangement of the sensory fibers (Figs. 2-3 and 2-4). The fibers conveying impulses of temperature and pain are among those which enter the dorsal horn of the gray matter to terminate on its cells in the same or neighboring segments from which the spinothalamic tract takes its origin. Consequently, a lesion of the dorsal horn will affect these fibers and eventually interrupt them. On the other hand, the medial group of fibers entering the cord and ascending to the nuclei of the dorsal funiculus will escape destruction as long as the lesion is confined to the gray matter. As these fibers mediate sensations of touch, this sensory quality will not be affected. *Likewise the deep sensibility (joint sense, vibratory sensibility) is usually found to be preserved.* Furthermore, the sense of discrimination, the ability to distinguish qualities like smoothness, roughness, and others, is not affected. In this respect also, a dorsal horn lesion differs from a dorsal root lesion.

A lesion of the central parts of the gray matter of the spinal cord will produce the same symptoms as a dorsal horn lesion. In this instance the crossing fibers from the cells of the dorsal horn forming the spinothalamic tracts will be interrupted, whereas the long ascending fibers of the dorsal funiculi will remain untouched. However, the symptoms will obviously as a rule occur bilaterally.

In summary, *a sensory loss with a segmental or approximately segmental upper and lower border points to a lesion of the dorsal roots, of the dorsal horn, or in the central part of the gray matter. If the sensory loss is of the dissociated type, it indicates that the morbid process is located in the dorsal horn or in the central gray matter.* Most frequently this is seen in syringomyelia.

Evidently, a lesion affecting a dorsal horn or the central gray matter of the cord will often be of large dimensions, be it a syringomyelia, a hematomyelia, or an intramedullary tumor. The pathological process will frequently tend to invade the surrounding parts of the cord, the ventral horns ventrally, or the white matter to both sides. As all experience shows, the pathological processes arising in the gray matter of the cord are to a certain extent inclined to confine themselves first and foremost to it. This is, for example, often the case in syringomyelia. The process most frequently starts from the central parts of the gray matter, and later on extends in a dorsal direction to affect the dorsal horns, but also ventrally to the ventral horns. *Pari passu* with the progressing destruction of the anterior horn cells, a paresis and finally a paralysis will develop, affecting the muscles supplied by the diseased segments of the cord. (As described in Chap. 3 this paresis is of the peripheral type, characterized by atrophy, loss of tone, and weakened or abolished myotatic reflexes.) Syringomyelia, as mentioned, shows a predilection for starting in the upper thoracic and lower cervical segments; later on it extends rostrally and also caudally. The first clinical signs therefore are, as a rule, segmental sensory disturbances, usually followed soon after by motor disturbances, from the ulnar fingers and the ulnar side of the hand and forearm. With the progression of the anatomical changes the clinical symptoms spread to the radial side of the hand, frequently also to the forearm and arm and eventually the thorax. A similar development of the symptoms is often displayed also by intramedullary tumors.

Sensory symptoms following lesions of the anterolateral funiculi of the spinal cord. Other types of sensory disturbances will ensue if the long ascending sensory paths in the cord are damaged. The symptoms are different in lesions of the dorsal funiculi and of the anterolateral funiculi. We may begin by considering the latter, and confine ourselves to the sensory symptoms. The motor symptoms, which are frequent companions of the sensory, will be briefly treated below in connection with a discussion of transverse lesions of the cord.

From what has already been said it is evident that *a destruction of the anterolateral funiculus of the spinal cord will be followed by a loss of pain and temperature sense on the contralateral side of the body at all levels caudal to the site of the lesion,* because the tracts conveying pain and temperature sensations are made up of axons crossing the median line, arising from cells in the dorsal horn (see Figs. 2-3 and 2-12). Furthermore, since the crossing is completed over a distance of not more than two segments, the upper border of the sensory loss will correspond approximately to the lower border of the dermatome belonging to the lowest preserved segment of the cord. However, this will be true only under the presupposition that the lesion is deep enough to reach nearly to the gray matter of the cord, thus interrupting *all* fibers of the spinothalamic tract at the level in question. *If the lesion is limited to the more superficial parts of the tract, a more restricted sensory loss will ensue,* and its upper limit will be found several segments farther caudally, the more so the more superficial the lesion. This is a natural consequence of the anatomical arrangement of the fibers already described, the longest fibers, derived from the caudal levels, being situated most superficially (see Fig. 2-9). This is a point of considerable clinical importance, since ignorance of this condition may lead to a false segmental diagnosis.

To give an example: A patient complains of radicular pain at the level of the right nipple (corresponding to the 4th–5th thoracic dermatomes). He presents a hypoesthetic belt at this level with affection of all sensory modalities. In addition, however, the examination reveals a hypoesthesia to pain and temperature from the inguinal sulcus (1st lumbar dermatome) and downwards on the other, left, side. These symptoms may be due to a developing intraspinal tumor, e.g., a neurofibroma, arising from the right 4th or 5th thoracic dorsal root. This tumor causes the radicular pain, and by destruction of the fibers of the roots is responsible for the hypoesthetic belt. But the tumor has also exerted pressure on the lateral side of the cord and has damaged the most superficially situated fibers at this place, which convey impulses of pain and temperature from the lumbar and sacral segments of the other side. If the tumor is allowed to continue its growth, the hypoesthesia to pain and temperature will gradually ascend and include the succeeding lower thoracic dermatomes. Of course, in such a case, motor disturbances are also quite frequently found, e.g., paresis of the leg on the side of the lesion, and symptoms due to affection of the dorsal funiculus.

It is customary to say that lesions of the anterolateral funiculi produce loss of pain and temperature sensibility, whereas touch sensibility is preserved. However, if in these cases the superficial tactile sensibility is examined by refined methods, it usually also presents some reduction. (The same applies to lesions of the dorsal horns and the central gray matter.) This and other findings demonstrate that the anterolateral funiculi also contain fibers subserving touch sensibility. The amount appears to present some individual variation. On the basis of clinical experience it was formerly held that the fibers within the anterolateral column are not only arranged according to their segmental origin but also, to a certain extent, according to sensory qualities. The fibers conveying impulses of deep-pressure sense and superficial touch were said to be gathered predominantly in the ventral parts, whereas those mediating impulses of pain were found to be located more dorsolaterally, and the fibers for temperature sense in the dorsalmost part of the tract, closely ventral to the corticospinal tract. This has, however, not been corroborated by experimental studies in the monkey, as mentioned previously. The most pronounced and enduring defects in reactivity to pain-provoking stimuli were observed by Vierck and Luck (1979) in monkeys after severance of both the ventrolateral and the ventral funiculus. Additional lesions of the dorsal columns, Lissauer's tract, and the dorsolateral funiculus apparently did not produce long-term effects on the monkey's response to painful stimuli. Furthermore, the experience gathered from a large number of chordotomies, performed during the last 20 to 30 years, agrees with the concept of a rather widespread distribution of pain-transmitting axons in the anterior quadrant. Thus, White and Sweet (1969) stated on the basis of observations in 422 patients subjected to "spinothalamic tractotomy," that "It is now recognized that the more the incision is carried beyond the line of motor root outflow and centrally close to the ventral sulcus, the higher and more complete will be the extent of analgesia," and that "With the present-day recognition of the widespread distribution of pain-conduction fibers and the extent of transection required to ensure their complete severance, it would be more accurate to speak of anterior quadrant cordotomy than to continue referring to section of the anterolateral quadrant."

A few individuals have what is called "congenital insensitivity to pain" Postmortem examinations of the central nervous system and nerves, performed in a few such cases, have been negative. However, Swanson, Buchan, and Alvord (1965) found an absence of small cells in the dorsal root ganglia, lack of small fibers in the dorsal roots, and absence of the tract of Lissauer in a boy who had insensitivity to pain and defective temperature sensibility.

The operative section of the anterolateral funiculi, *chordotomy*, already referred to, has given valuable information concerning other aspects of pain transmission as well. Sections that include the entire anterolateral funiculus on one side have been found also to be followed by a slight reduction of the sensibility to touch. Furthermore, a slight reduction of temperature and pain sensibility has also been ascertained on the operated side. This agrees well with the anatomical finding that some of the axons from cells of the dorsal horns do not cross but ascend in the homolateral anterolateral funiculus. Some of the pain and temperature impulses thus evidently traverse the cord without crossing. Another observation is in accord with this. Following a successful unilateral chordotomy, the patient is usually free from pain for some time if the pain has been unilateral. After a varying interval of time, however, some pain usually recurs, and the pain sensibility reappears to a certain degree. The common interpretation of this phenomenon is that the normally subsidiary homolateral pain fibers gradually take over some of the functions previously performed by the crossing ones, and a practical consequence is that even with unilateral pain a bilateral chordotomy will give better results than a unilateral one. However, even after a bilateral chordotomy a certain degree of pain and temperature sensibility may reappear.

Foerster (1936a) assumed this to be due to scanty fibers in the extreme dorsal part of the anterolateral funiculus, which may escape destruction, or to some pain-conducting fibers in the dorsal funiculi (see below). The possibility also exists that some pain impulses may ascend in the sympathetic trunk, for even after a complete transverse lesion of the cord some pain sensibility may be present. These "aberrant" pathways for pain and temperature sensations appear to be better developed from the lower sacral segments. At least this may be an explanation of the fact that in lesions of the spinal cord these dermatomes—that is of the anogenital region—are frequently either entirely spared or less severely affected than the more rostral segmental areas. This applies to bilateral, but especially to unilateral, lesions of the cord. A possibility also to be taken into consideration when the lesion is not complete is that fibers from the lower sacral segments may be uncrossed to a larger extent than those from more rostral segments (compare the nearly always bilateral sensations on stimulation of the anogenital area in the first somatosensory cortical area).

It may happen that following a unilateral chordotomy the patient experiences pain in a part of the body where it had not been felt before, most often the same place on the opposite side of the body. Nathan (1963) assumed that this is due to a reference of the original pain: "Impulses arising from the neoplasm are blocked in the opposite spinothalamic tract, but they reach the neurological substratum of consciousness by other pathways." In a review of the results of chordotomy in 104 patients, Nathan (1963) found that the results were not as good as may appear from many reports. In a fair number of cases there are sensory disturbances, such as reappearance of pain and occurrence of dysesthesisas. In large materials presented by White and Sweet (1969), and others, the pain recurs in more than 50% of the patients after chordotomy. Furthermore, other complications are common, some of them being indeed unavoidable, since it is impossible to transect only the spinothalamic fibers. Thus some disturbances in micturition and defecation regularly occur, owing to the course with the spinothalamic tract of visceral afferent fibers from sacral segments, and because the descending fibers to the sacral cord related to the bladder and rectum (Nathan and Smith, 1958) may be interrupted by the operation. Motor disturbances may also appear.

During recent years the operation has been performed percutaneously by introducing a needle under radiographic control and making the lesion by radiofrequency coagulation. The patient is usually awake during the procedure, and the correct placement of the electrode can be ascertained by stimulation before making the lesion. The risk of serious complications, such as motor impairment (hemiplegia) and bladder disturbances, appears to be much less with the percutaneous method than with sectioning during open operation. Quite unexpectedly, in a few patients subjected to percutaneous chordotomy with subsequent pain relief and no hemiparesis the lesions turned out to be confined to the dorsolateral funiculus (see the Neurosciences Research Program Bulletin *Pain,* 1978).

In any consideration of the results of chordotomy the complexities in the anatomical and functional organization of the nervous system should be borne in mind. In the first place, it is obvious that section of the anterolateral funiculus interrupts ascending and descending fiber tracts other than those mediating pain. Secondly, the spinothalamic tract is certainly not the only pathway along which impulses arising on stimulation of pain receptors may be conveyed centrally. As mentioned previously, it may well be that the interruption of spinoreticular fibers is as important as the section of the spinothalamic tract, since pain stimuli are potent activators of the reticular formation (see Chap. 6).

Sensory symptoms following lesions of the dorsal funiculi. Whereas the anterolateral funiculi are engaged in the transmission of pain, temperature, and, furthermore, superficial touch and deep pressure, the other large ascending sensory system, that of the dorsal funiculi, transmits primarily different types of superficial tactile sensibility and deep sensibility. Since these tracts ascend in the spinal cord without crossing, *the symptoms ensuing on lesions of the dorsal funiculi will always appear on the side of the lesion* (in contradistinction to those following damage to the anterolateral funiculi). The results of recent anatomical and physiological studies, described in a previous section of this chapter, are on the whole in good accord with the conclusions drawn from findings in clinical studies of patients, correlated with postmortem anatomical examinations. However, as mentioned previously, it appears that at least the *lasting* symptoms reported in many patients are more pronounced than those observed in experimental animals undergoing lesions of the dorsal columns. This difference might be more apparent than real, since the morbid process in humans is very seldom limited to the dorsal columns; encroachment on the dorsolateral funiculus would be expected to increase the symptoms (presumably particularly with regard to kinesthesia). Conversely, in experimental animals the lesions would often be less than complete in order to avoid inclusion of other pathways than those located in the dorsal columns. As considered previously in this chapter, it appears that if only a very small proportion of the fibers in the dorsal columns escapes destruction, a large degree of discriminative sensibility remains. This may possibly help to explain that very slight symptoms were observed in some patients subjected to sections of the dorsal columns (Cook and Browder, 1965).

It has long been known that a morbid process limited to the dorsal funiculi will give no clear-cut loss of the simple sensibility of cutaneous touch, and it is sometimes stated that *the sense of touch as well as of pain and temperature is preserved in lesions of the dorsal funiculi*. However, concerning the touch sensibility, certain qualifications have to be made. Even if a patient with his dorsal funiculi severely damaged is able to perceive very slight cutaneous touch stimuli, a more detailed examination, including a determination of the amount of "touch spots" and threshold values, usually reveals some reduction of touch sensibility. Conditions are similar to those prevailing in the system of the anterolateral funiculi. It appears that if one of the two systems subserving touch sensibility suffers, the other cannot sufficiently take over the function of both. However, this reduction is so moderate that it is of no practical relevance. Clinically important, however, is that *lesions of the dorsal funiculi produce defects in the sense of discrimination*. The capacity of differentiating two simultaneously applied touch stimuli is reduced. Examining, for example, with Weber's test, it will be found that the distance between two points that can be appreciated as separate is enlarged. The capacity to recognize figures and letters drawn on the skin is lessened, and the patient is only able to indicate approximately which point of the body is touched. In lesions of the gracile nucleus, for example, the patient cannot tell which toe is being touched. It is, furthermore, difficult or impossible for him to decide whether the object with which he is being touched is rough or smooth, hard or soft.

The clinical observations given above are in accord with physiological studies which have demonstrated a high degree of stimulus and place specificity in the dorsal column nuclei, as described in some detail above. As we have seen, the same is true for the further links of this pathway to the cerebral cortex. The different aspects of the discrimination itself are probably purely cortical processes, based on a synthesis and integration of impressions gained from several types of receptors. It seems farfetched and illogical to postulate the existence of special paths for discrimination.

In monkeys subjected to section of the dorsal columns, the persisting symptoms are of a similar kind, but less pronounced than in human patients. For example, two-point discrimination is not significantly reduced, nor is the capacity to distinguish between surfaces differing in roughness. However, more complex sensory functions are clearly affected. Thus, monkeys show a reduced capacity to distinguish between pressure on the skin with different intensities, to judge the *distance* between two points, and to distinguish between coins of different sizes pressed against the skin (see Vierck, 1978a, for a critical review of dorsal column functions as studied by animal experiments). The deficits described above are, in fact, remarkably similar to those observed by Wall and Nordenboos (1977) in a patient with a stab wound transversely severing the spinal cord except the anterolateral part. They concluded, as cited previously in this chapter, that figure writing on the skin and examining the patient's ability to judge the *direction* of movement on the skin would be the most sensitive tests to detect lesions of the dorsal columns.

In clinical studies of dorsal column affections *vibratory sensibility* is diminished or even abolished, in agreement with the physiological observations or even abolished, in agreement with the physiological observations of units in the dorsal column nuclei responding to vibratory stimuli. However, neither in animal experiments nor in the patients of Wall and Nordenboos (1977) and of Cook and Browder (1965) were clear-cut deficits in vibratory sensibility observed. Nevertheless, considering the properties of single cells in the dorsal column nuclei, as discussed previously, it seems hard to accept the possibility that the dorsal columns are irrelevant to vibratory sensibility. Again, one would expect that they are of importance particularly for more complex, discriminatory aspects, such as distinguishing between vibration with different frequencies.

It appears from some clinical studies that pain and temperature sense also have a certain connection with the dorsal funiculi. In lesions of these a reduction of the threshold values and the number of pain spots has been noted (Foerster, 1936a), and the type of painful stimulus applied is not recognized, probably because of reduced discriminatory abilities. These data, compared with the results of studies of the sensibility in lesions of the anterolateral funiculi, tend to show that, although the two large ascending exteroceptive sensory fiber systems are functionally mainly different, both of them appear to a certain degree to be concerned not only with tactile sensibility but also with pain and temperature sensations. However, these observations in clinical cases may be opposed on the grounds that a concomitant affection of dorsal roots may explain the disturbances in pain and temperature sensation.

Furthermore, direct irritation of the dorsal columns is not painful (Nashold, Somjen, and Friedman, 1972). In fact, electrical stimulation of the dorsal columns can effectively alleviate chronic pain in some patients. The mechanism of this effect is not fully understood.

It appears from clinical experience that *lesions of the dorsal funiculi are accompanied by disturbances of coordination,* which are more conspicuous than are the more subtle disturbances of exteroceptive sensory function described above. Thus the typical *ataxia* occurring in tabetic patients is explained as being due to the degeneration of the dorsal funiculi, which is a fairly constant feature of the anatomical picture of this disease.[36] The basis for the ataxia has been assumed to be the loss of proprioceptive impulses, this loss being the consequence of the lesion of the dorsal funiculi. Clinically this is manifested as follows: *If his eyes are closed the tabetic patient is unable to tell in which position his joints and limbs are moved,* and if the limbs or parts of them are moved passively, he does not recognize whether the movement consists of flexion or extension (joint sense, position sense). If the cuneate funiculus is degenerated, the upper limbs will also be affected, and it will be difficult or impossible for the patient to estimate the weight of objects. Furthermore, he has not the normal capacity for recognizing the size and form of objects with his eyes closed (astereognosia), partly at least because he cannot properly perceive the numerous small movements, especially of the fingers, which are necessary for this purpose. In addition, the reduction of the cutaneous discriminative sense is of importance in this connection.

The so-called dorsal funiculi ataxia is revealed in a reduced capacity to perform movements smoothly and precisely. The movements become unsteady, of uneven range; the patient points now too far, now too near. It appears as if the patient tries to compensate for his loss in coordination with a surplus of muscular power. The gait is unsteady and jerky. Frequently these disturbances of coordination are more prominent if the movements are performed with the eyes closed, vision thus partly making up for the reduction in proprioceptive impulses. This is brought out also by the fact, readily observed, that the patient is able to stand in a stable position, but if he is asked to stand with his feet side by side and with eyes closed, he will usually lose his balance (Romberg's sign).

Recent animal experiments and observations in some patients with section of the dorsal columns (Cook and Browder, 1965; Wall and Nordenboos, 1977) raise some doubt that the ataxia described above is due solely to lesions of the dorsal columns. Although destruction of the dorsal columns in monkeys is followed by a severe ataxia that is most pronounced in the forelimbs (Gilman and Denny-Brown, 1966), the ataxia appears to be transitory (Vierck, 1978b). During some months after surgery a gradual return of motor function takes place, so that eventually no impairment is exhibited during running, climbing, and other motor activities (Vierck, 1978b).[37] Nor can any impairment be found in the ability to execute fast and accurate projection movements of the arm toward a target. Particularly with regard to the distal parts of the extremities, however, some deficits, lasting at least for many months, can be observed.

---

[36] However, the dorsal funiculi are not exclusively involved in the degeneration in tabes dorsalis. The pathological process is essentially an inflammation of the dorsal roots (a "posterior radiculitis"), and this causes a degeneration not only of the long ascending tracts in the dorsal funiculi but also of the short fibers, which enter the dorsal horns as well. Anatomically the latter is, however, not very conspicuous, at least not in myelin sheath preparations which clearly show the changes in the dorsal funiculi. As described previously (Chap. 1), the methods for study of myelin sheaths were known long before more modern neurohistological technique was developed, and the old thesis of the degeneration of the dorsal funiculi as the anatomical substratum of tabes has shown a peculiar tendency to dominate the minds of neurologists. It is, however, important to realize that the symptoms arising from the diseased spinal cord in tabes are not all due to the degeneration of the dorsal funiculi, even though these usually form a conspicuous feature of the clinical picture of tabes dorsalis.

[37] The mechanism underlying this remarkable degree of recovery is not known. It appears to be of critical importance that the animals are rewarded for using the affected limbs; otherwise they tend to neglect them. Factors of possible importance for restitution after damage to the central nervous system are considered in Chapter 1.

For example, the opposition of the thumb and forefinger is deficient so that the monkey uses scraping movements instead of the usual precision grip to pick up bits of food from a depression, and its ability to adjust the grip when an object is grasped insecurely is reduced. Furthermore, refined analysis reveals some impairment also of jumping movements. It is also of interest that in monkeys with complete section of the dorsal columns the threshold for detection of passive movements of the metacarpophalangeal joints increased from 5° before operation to 10° after (Schwartzman and Bogdonoff, 1969).

Joint receptors were previously regarded as the only receptors of crucial importance for kinesthesia, and it was believed that impulses arising in joints and ligaments ascend mainly or only in the dorsal funiculi. These notions together appeared to explain fully the reduced joint sense and the ataxia seen after destruction of the dorsal columns in man. As considered previously in this chapter, however, the receptors and pathways for impulses signaling joint movement and position are more diverse than was formerly held; all aspects of kinesthesia do not depend upon joint receptors, and all afferent impulses from joints destined for higher levels do not ascend in the dorsal columns.[38] It thus appears that the behavioral data from primates subjected to destruction of the dorsal columns are in accord with the recent anatomical and physiological data.

Lesions of the somatosensory tracts in the medulla oblongata, pons, and mesencephalon. Even a superficial knowledge of the structure of the brainstem will suffice to make clear that the clinical pictures resulting from lesions involving the somatosensory systems in the brainstem will be far more complex than in the cord. In the brainstem the long, ascending tracts are to a great extent intermingled between cellular masses and fiber bundles belonging to other neuronal systems, partly related to the cranial nerves. The symptoms following lesions of these structures are considered in Chapter 7. Many lesions of the brainstem are immediately fatal, because they involve regions which are essential in the regulation of cardiovascular functions and respiration, and there may be disturbances of consciousness (discussed in Chap. 6). The following remarks will be restricted chiefly to symptoms following lesions of the somatosensory pathways.

A lesion affecting *the lateral part of the medulla oblongata* may interrupt the spinothalamic tract (see Figs. 2-11 and 7-7), and thus give origin to a loss of pain and temperature sensibility on the opposite side of the body, whereas the sense of touch, discrimination, vibration, and deep sensibility will be intact if the medial lemniscus has escaped. (See Soffin, Feldman, and Bender, 1968.) Since secondary trigeminal fibers join this tract, the analgesia following a lesion at a somewhat higher level in the medulla oblongata or in the pons will also include the face on the same side as the body, and there results a complete *hemianalgesia and hemithermanesthesia.*[39] It is noteworthy, however, that a superficial lesion may

[38] As a further example of the complexity of this problem, it may be mentioned that section of the dorsal columns in monkeys appears to produce more severe proprioceptive deficits in the distal parts of the extremities, whereas lesions of the dorsolateral funiculus affect proximal joints to a higher degree (Schneider, Kulics, and Ducker, 1977). In experimental studies of the spinal course of joint afferents, as considered earlier in this chapter, attention has been focused almost entirely on afferents from the knee joint, which may possibly be regarded as a proximal joint in this context.

[39] The exact course of secondary trigeminal fibers appears to be more complicated than was previously held (see Chap. 7), and this may help to explain variations in the clinical picture of lesions similar to those treated here.

give an incomplete picture, owing to involvement of only the longer fibers which are placed superficially. Thus, for example, an analgesia from the mammary papilla downward may well be due to a lesion in the medulla oblongata.

A lesion of the brainstem will, however, only in rare cases be limited to the spinothalamic tract. Frequently the spinal trigeminal tract and its nucleus will be included (see Fig. 2-11), producing an additional thermanesthesia and analgesia in the face on the side of the lesion (*hemianesthesia alternans*). The extension of the process in a rostral direction will decide whether the crossed trigeminal area is intact or not. Symptoms indicating damage of the vagus nuclei or fibers may also appear, as homolateral paresis of the soft palate or of the laryngeal muscles, accompanying the hemianalgesia or hemihypalgesia. If the lesions injure the spinocerebellar tracts and the inferior olive there will be disturbances of coordination, as will also be the case when the dorsally placed restiform body is affected. In extensive lesions of the lateral part of the medulla oblongata the pyramidal tract may also be encroached upon, leading to some motor impairment. The most usual cause of lesions in this part of the brainstem is a vascular disorder in the territory of the posterior inferior cerebellar artery or the vertebral artery (thrombosis, embolism, or hemorrhage). In such cases all the symptoms mentioned above may be present, and in addition a Horner's syndrome on the side of the lesion may be found, evidencing damage to the descending sympathetic fibers (see Chap. 11).

A lesion affecting the *medial parts of the medulla oblongata* will give symptoms of a somewhat different type. A unilateral lesion of the medial lemniscus is usually said to be followed by an impairment or loss of deep sensibility (deep pressure, joint, and vibratory sensibility), and reduced or abolished sense of discrimination on the entire contralateral half of the body if the lesion is situated above the crossing of the fibers.[40] If the lesion is situated in the pons, the trigeminal area in the face on the contralateral side will usually be included (see Fig. 2-18). However, the lemnisci of both sides will frequently be affected simultaneously on account of their close proximity. On account of this the *symptoms mentioned are nearly always bilateral*, but often of a different distribution on the two sides. A frequent accompaniment to the sensory disturbances will be a paralysis or *paresis of the tongue of the peripheral type* on the side of the lesion, with atrophy of the tongue muscles, since the hypoglossal nerve passes immediately lateral to the medial lemniscus (Fig. 7-7). If the lesion extends farther in a ventral direction the pyramidal tract and other descending tracts will be encroached upon, and in addition there appears a *hemiplegia on the side opposite the hypoglossal lesion*. If the pathological process invades the pyramidal tract from the dorsal side, a consideration of the arrangement of the fibers (cf. Fig. 2-11) will explain that a paresis of one or both upper extremities may appear, while the lower extremities are affected slightly or not at all. Symptoms indicating a lesion in the median region of the medulla oblongata are usually due to syringobulbia or an interference with the blood supply of the anterior spinal artery.

These few remarks will suffice to give an impression of the variation and

---

[40] Since the medial lemniscus contains fibers from the lateral cervical nucleus, and the dorsal column nuclei receive afferents ascending outside the dorsal columns, symptoms would be expected to be somewhat different and more severe after lesions of the medial lemniscus than after lesions of the dorsal columns.

complexity of the symptoms which may follow lesions of the medulla oblongata. Still, the description is far from complete. With a knowledge of the anatomical structure of the medulla it will as a rule be possible to determine the site of the lesion fairly exactly. However, it is important to remember that, owing to the segmental arrangement of the fibers in the long ascending and descending tracts traversing the brainstem, *there frequently will not be a complete unilateral sensory or motor impairment*. Monoplegias and circumscribed regions of sensory loss are not rare, and a great variety of combinations of motor dysfunction (central and peripheral) and sensory defects may be encountered. Another point is also of practical importance: *the regions presenting sensory and motor impairment are often only indistinctly outlined*. As with lesions in other regions, this is no wonder, as frequently the regions totally injured are surrounded by areas which are only more or less impaired in their function on account of pressure or edema.

Above, only the deficiency sumptoms occurring in lesions of the medulla oblongata have been referred to. Without doubt these are the most important, but *symptoms of irritation of the sensory tracts* may also occur, manifesting themselves usually as paresthesias of various types. Not infrequently a peculiar form of irritative sensory symptoms occurs in lesions of the medulla oblongata. These so-called *hyperpathias* are pains of an unusual character. The intensity of the pain bears no relation to the strength of the stimulus. The slightest superficial touch may be felt as severe pain. The pain shows a tendency to irradiate in regions of the body not being stimulated, and usually lasts long after the stimulus has ceased. Similar hyperpathias may occur in lesions of the sensory system in other places, especially the thalamus (see below) and they may occasionally be observed in lesions of the spinal cord and, in rare instances, even of the peripheral nerves. Their genesis is not clearly understood.

Lesions affecting the sensory tracts in the pons and mesencephalon will give symptoms similar to those in the medulla oblongata. However, the accompanying symptoms may be of a somewhat different type on account of their spatial relationship to other tracts and nuclei (compare Figs. 7-18, 7-22, and 4-5). No detailed description of the different possibilities will be given here. It should, however, be pointed out that the *most important landmark in the diagnosis of the level of lesions of the brainstem is an eventual affection of one or more of the cranial nerves*. A simultaneously developed facial paralysis or paresis of the peripheral type (affecting upper and lower territories) points to a lesion in the middle or lower part of the pons. If the pyramidal tract is involved, a hemiplegia or hemiparesis appears on the opposite side (facial paralysis with crossed hemiparesis: Millard-Gubler's syndrome). Other possible combinations are with an abducens paralysis or a trigeminal paralysis. In pontine lesions usually the pontocerebellar system will be affected, with consequent disturbances of coordination. However, an exact focal diagnosis in pontine lesions is somewhat more difficult than in the medulla oblongata, since the function of several structures is less well known.

The same applies to lesions in the *mesencephalon*. An accompanying unilateral *oculomotor paresis* or paralysis is indicative of a lesion at the level of the superior colliculus, a paralysis of the trochlear nerve of a lesion near the inferior colliculus. Lesions in the mesencephalon are frequently accompanied by *disturbances of coordination and hyperkinesias*, tremor, or athetotic and choreatic movements. The basis of these has usually been assumed to be lesions of the brachium conjunctivum, the red nucleus, or the substantia nigra. Actually, however, our knowledge is very limited concerning the symptoms which can be referred to the individual structures in this part of the brainstem. The number of anatomically controlled cases is small. These questions will be touched on in Chapter 4.

Symptoms following lesions of the thalamus. As described previously, the medial lemniscus and most of the fibers of the spinothalamic tract terminate in the nucleus ventralis posterior (VPL). However, since the thalamic nuclei are relatively small, it is to be expected that lesions affecting only one of them will be very rare. Because several large fiber systems pass in the immediate neighborhood of the thalamus, thalamic lesions will frequently be accompanied by symptoms from other structures. Most important of these are the large tracts passing in the internal capsule close to the thalamus, the corticospinal fibers and corticofugal fibers to nuclei in the brainstem (see Chap 4). More posteriorly are the fibers to the cortex from the lateral and medial geniculate bodies (optic and acoustic radiations). Tumors especially are apt to involve neighboring structures. Among 24 patients with thalamic tumors McKissock and Paine (1958) found only 5 who had sensory impairment. The sensory modalities involved varied.

A destruction of the entire thalamic area receiving the sensory fiber systems would be expected to result in an impairment or loss of somatic sensibility in the opposite half of the body. In many of the cases of this kind described in the literature, deep sensibility and discriminating sense are found to be severely impaired, the sense of touch and temperature less so, while the perception of pain is only slightly affected (Head and Holmes, 1911–12; and others). Whether this may be explained by circumscribed affections of particular regions of the thalamus allotted to specific sensory modalities is unknown, and autopsy findings in such cases are scarce. In part the differences may be explained by different numbers of uncrossed fibers in the two large somatosensory pathways, since the qualities most severely affected are those mediated by the lemniscus, which is completely crossed.

Concerning the presence of a *somatotopic localization within the thalamus,* the few clinical findings reported are in good accord with experimentally established data. Some cases have been described in which smaller lesions within the nuclei of the thalamus have been followed by a sensory impairment limited only to parts of the body (Garcin and Lapresle, 1954; Hassler and Riechert, 1959; and others). Most frequently these localized lesions are of vascular origin (affecting the thalamogeniculate artery), and in most instances the sensibility has been retained in the face, corresponding to a sparing of the medially situated VPM. In cases of this type the limits of hypesthesia or anesthesia of the different sensory modalities are usually not quite identical. Fisher (1965) reports the clinical findings in 25 patients presenting what he calls ''pure sensory stroke.'' These are no pareses, visual defects, or other signs apart from a persistent or transitory numbness or mild sensory loss over one entire side of the body, or in part of the body. The symptoms, which may be transitory, but often recur, are probably due to transient ischemia caused by thrombosis. In one case, where autopsy was made, a lacuna, 7 mm in diameter, was found in the VPL.

A peculiar aspect of thalamic lesions is that they are frequently accompanied by spontaneous pains. These ''thalamic pains'' are often very intense, occur in paroxysms, frequently irradiate to the entire half of the body, and are usually intractable to analgesics. The pains are usually present together with a sensory loss, but they may also occur without, usually then as an initial symptom. Frequently the pains, as with lesions in the medulla oblongata, have the character of hyperpathia (in this connection often termed dysesthesia). Even slight stimuli, normally not painful, may provoke paroxysms lasting long after the application of the stimulus has ceased. The period of latency between the stimulus and the perception of pain is prolonged. No universally accepted explanation of these pains has been given. Some are inclined to ascribe them to vasomotor disturbances in the thalamus. Others assume that a disappearance of cortical inhibition on account of the affection of thalamocortical fibers or the lack of intrathalamic association may be the cause. As mentioned when discussing the central pathways for pain, "thalamic pains" most often occur with small vascular lesions. Surgical attempts to abolish them by stereotactic thalamotomy have been in part successful.

Symptoms following lesions of the somatosensory cortical areas. From what has been discussed in previous sections it will be apparent that *circumscribed lesions of the postcentral gyrus* (Brodmann's areas 3, 1, 2) *will be followed by a localized sensory loss in part of the opposite half of the body.* Disturbances of this type have been observed in smaller lesions, traumatic, neoplastic, or other kinds. Cortical ablations to remove a neoplasm or a depressed bony fragment have yielded evidence confirming the somatotopic organization of the retrocentral area. For example, a damage to or a removal of the upper part of the gyrus with the adjacent lobulus paracentralis results in a sensory loss limited to the contralateral leg, extending farther proximally the lower the lesion reaches on the lateral surface of the cortex. In some cases the observed sensory loss has presented an approximately segmental distribution resembling lesions of the dorsal roots, but not infrequently the borders of the affected area have been more circular, producing an anesthesia of the "glove" or "stocking" type.[41]

In an acute lesion of the postcentral gyrus the anesthesia at first comprises all sensory modalities, superficial as well as deep. After some time sensibility reappears to a certain extent, as a rule, beginning with a partial return of pain sensibility. In time this may be restored to such a degree that only highly refined methods will ascertain the slight persistent reduction. Somewhat later the appreciation of touch stimuli is partly regained and also the temperature sense is largely restored. However, *the sense of discrimination is usually permanently and severely impaired,* and is never, or only exceptionally, completely regained. Likewise *the vibratory and joint sensibilities are permanently reduced or practically abolished.* This is even more true of the stereognostic sense, where the lesion includes the foci for the hand and fingers.[42] Thus in cortical lesions the sensory modalities transmitted through the dorsal funiculi and the medial lemniscus are those usually most severely and enduringly affected.

However, in the restitution following a lesion of the postcentral gyrus, there

[41] Anesthesia of these distributions is most frequently seen in hysteria. The differential diagnosis will as a rule not be very difficult.

[42] It is difficult to decide whether this is a "primary" astereognosia or a secondary astereognosia, due to the abolition or reduction of the more simple sensory qualities. The affection of the latter presumably plays an important role.

is also a *regional difference*. The different parts of the body do not regain their sensibility at the same rate. The most rapid and extensive restitution takes place for the face, especially the oral region, larynx and pharynx, and the anogenital region. The neck and trunk are in a more favorable position in this respect than are the extremities, and especially as concerns the distal parts of the latter where restitution is usually far from complete. The parallelism with the possibilities of restitution after a hemiplegia (see Chap. 4) is very close. It is reasonable to suppose that the potencies of the different parts of the body as regards restitution in lesions of the sensory cortex are dependent first and foremost on the degree of bilateral "representation" in the cortex of the parts in question. The degree of the restitution corresponds also to the ease with which bilateral sensory symptoms may be elicited on cortical stimulation.

Little appears to be known of the sensory defects following *lesions of the second somatosensory cortical area (S II) in man*. Since the total area is small it will most often be affected together with other parts of the cortex, particularly S I, making analyses of the symptoms observed difficult. From animal experiments one might expect some affection of discriminating sensory qualities.

The symptoms following *lesions of the superior parietal lobule* with area 5 (Brodmann's areas 5 and 7) are likewise incompletely known. It appears, however, that no somatotopic localization is present in this part of the cortex, the disturbances affecting more or less the entire contralateral half of the body. The different sensory modalities are affected according to a principle similar to that in lesions of the postcentral gyrus. The permanent impairment is most conspicuous for the combined sensory qualities, i.e., discrimination of various types. This has been found by Foerster (1936a), while Penfield and Rasmussen (1950) did not report sensory defects following excision of most of the superior parietal lobule. Foerster's observations are compatible with Ruch, Fulton, and German's (1938) experiments on monkeys and the chimpanzee, since following lesions of the posterior parietal lobe the animal's capacity for discriminating weight, as well as roughness and geometrical forms by palpation (in darkness), is seriously, possibly permanently, impaired. However, the postcentral gyrus is also of importance in this connection. In some human cases these authors found similar disturbances. It is appropriate to emphasize, however, that there is a wide diversity of opinion in the literature concerning the nature of the deficit in somesthetic discrimination after posterior parietal lesions. In particular, nonsensory deficits such as motor retardation and spatial discrimination contribute significantly to the patient's lack of performance of somesthetic tasks (for a comprehensive discussion, see Semmes, 1973).

Apart from deficiency symptoms dealt with above, *irritative symptoms* may appear in lesions of the somatosensory areas. Real pain very rarely occurs. More often the irritation is revealed clinically as paresthesias. If the irritation (caused by a tumor, a foreign body, a bone fragment, or other causes) is limited to the postcentral gyrus, the paresthesias may be localized to a limited part of the opposite half of the body. Important diagnostic hints may thus be obtained.

However, the sensory epileptical paroyxsms or *sensory Jacksonian fits* are more conspicuous and more indicative of a process in the cortex. Just as an irritative process in the precentral gyrus may elicit a motor discharge, spreading regularly, an irritation of the cortex of the postcentral gyrus may provoke a wave of sensory irritative symptoms, traveling over the body in accordance with the somatotopic organization of the retrocentral area. For example, paresthesias of a different type may start in the fingers and proceed to the forearm, arm, and so on. In some instances only sensory symptoms of this type will occur; in others, however, the irritation is propagated also to the precentral gyrus, first and foremost the readily excitable area 4. In this case a motor Jacksonian fit (cf. Chap. 4) will follow the sensory phenomena, with clonic twitchings of small groups of muscles, spreading regularly. It is customary to speak of the sensory symptoms in these instances as a *sensory aura,* but really they are an epileptic fit originating from the postcentral gyrus. Just as a motor Jacksonian fit is often followed by a postparaxysmal paresis, an examination of the patient immediately after a sensory Jacksonian fit may in some instances reveal a postparoxysmal sensory loss in the parts of the body which have been affected. The sensory Jacksonian fits, like the motor, are of great clinical importance, since they indicate the presence of a focal lesion of some sort or another in the first sensorimotor area. However, occasionally such fits may occur without a cortical focus.

In epileptic fits originating from the superior parietal lobs, a sensory aura may occur. This usually comprises the entire contralateral half of the body and as a rule soon becomes bilateral. According to Penfield and Kristiansen (1951) an epileptic discharge starting in the parietal lobe will frequently not manifest itself until it spreads forward to the postcentral gyrus where it produces a somatic sensory aura.

Apart from certain types of seizures arising from foci in particular areas of the cerebral cortex, it is difficult to relate focal or generalized epileptic seizures to well-defined regions of the brain. The many varieties of epileptic seizures, their classification, and other problems of epilepsy will not be considered in the present text. For exhaustive treatments of this subject the reader is referred to the monographs of Penfield and Kristiansen (1951) and Penfield and Jasper (1954).

3

The Peripheral Motor Neuron[*]

IN CLINICAL neurology the examination of what is commonly called "the motor system" is a fundamental part of the analysis of symptoms in disease of the nervous system. Different types of functional disorders are observed, which, when carefully examined, may give valuable information about the localization and type of pathological process responsible for the symptoms. A distinction can be made between the peripheral motor apparatus—the peripheral motor neurons and the striated muscles of the body—and the central motor tracts and systems. The latter will be considered in the next chapter.

The motor cells of the ventral horns of the spinal cord. The peripheral motor neurons, often called *motoneurons,* have their perikarya or cell bodies in the ventral horns of the spinal cord and in the motor nuclei of some of the cranial nerves in the brainstem. The latter will be treated in more detail in Chapter 7, but principally they are of the same type as the motor cells of the ventral horns. Following the discovery of an efferent innervation of the muscle spindles, it has become customary to distinguish between large and small motoneurons, commonly referred to as α (alpha) and γ (gamma) motoneurons, respectively. We shall return to the latter below. Some of the alpha motoneurons are among the largest nerve cells present in the nervous system, and, because of this, have been more frequently used than any other cell type for the study of fine cytological changes. Like the cells of the corresponding cranial nerve nuclei, they are polygonal, multipolar, and endowed with a rich amount of tigroid granules in the cytoplasm (see Fig. 1-12A). The numerous dendrites arborize in the gray matter, mainly of the ventral horn. The axon takes a ventral course to leave the spinal cord through one of the ventral roots. The peripheral nerves then convey the axons

* Revised by Per Brodal.

to the muscles. (Not all peripheral nerves contain such somatic efferent motor fibers; most of them also have somatic afferent and visceral fibers.)

The motoneurons are found within the cytoarchitectonically defined lamina IX of Rexed (see Fig. 2-3). However, intermingled with them are a number of small cells. In fact, the number of these exceeds that of the large motoneurons by far. In certain groups the relation is 1:16.5 (Balthasar, 1952; see also Aitken and Bridger, 1961). It is generally held that some 30% of the fibers in the ventral roots supply muscle spindles. Accordingly, the majority of the small cells in lamina IX cannot be the perikarya of fibers to the muscle spindles. Some certainly are just small motoneurons; whether others are internuncials has not yet been settled, although some seem likely to be the Renshaw cells. These subjects will be considered in a later section.

The motoneurons in the ventral horns of the spinal cord and the motor cranial nerve nuclei are the "final common path" (Sherrington) for all those impulses that are transmitted to the skeletal musculature of the body. They are influenced by impulses from many sources. As stated previously, primary sensory fibers have been found to end on them. In addition they are acted upon, chiefly through internuncial cells, by fibers which descend in the spinal cord from the cerebral cortex and several nuclei in the brainstem (to be discussed in Chap. 4).

The movements that occur as the ultimate effect of the action of the motoneurons are determined by the impulses which impinge upon the motor ventral horn cells, the latter being, as it were, played upon by the different systems. The multitude of different movements that may occur normally and in conditions of disease of the nervous system are accordingly not primarily determined by the motoneurons themselves (except when these are directly involved in the pathological process). In recent years a wealth of information has accumulated, on the function of the motoneurons, as well as on other aspects of motor function, including its central control. These subjects are treated fully in textbooks of neurophysiology and will therefore be considered only briefly here, with particular reference to anatomical aspects and to the light they may shed on symptoms in neurological diseases.

The arrangement of the motoneurons. *The motoneurons* in lamina IX, like the cells in some of the motor nuclei of the cranial nerves, *are arranged in a characteristic manner*. In a cell-stained transverse section through the spinal cord several groups of cells can be distinguished in the ventral horns (see Fig. 3-1A). Actually these groups belong to longitudinally arranged columns of nerve cells. The various columns are present in certain segments of the spinal cord only, as indicated in Fig. 3-1A. When a peripheral nerve is cut, retrograde cell changes occur in the motoneurons that send their axons into the cut nerve, as mentioned in Chapter 1 (see Fig. 1-12B). In this way it has been determined experimentally, in several animals, which nuclear groups are concerned in the innervation of the various muscles. In the same manner, postmortem studies of the human spinal cord in cases of amputations of limbs or nerve injuries have given information on the conditions in man. Sharrard (1955) has published a careful study of the subject based on a correlation of the distribution of paralysis in cases of poliomyelitis with the sites of cell loss in the ventral horn. One fundamental point that has been made clear by investigations of this type is that the medial cell groups supply the muscles of the trunk and neck. Accordingly, these groups (more precisely, only the ventromedial column) are developed throughout the length of the cord, as seen in Fig. 3-1A. The lateral groups, on the other hand, are present mainly in the cervical and lumbar enlargements of the cord and have been shown to supply the muscles of the limbs. Within the various lateral cell groups of columns a further pattern is found: the cells concerned with the innervation of the distal muscles of the extremities are, as a rule, dorsal to those that send their axons to the proximal muscles. It may also be noted that the cell groups related to the distal parts of the extremities are only developed in the caudal part of the enlargements. The motoneurons are, therefore, somatotopically arranged. This pattern of *somatotopic localization* may be illustrated diagrammatically, as in Fig. 3-1B. This arrangement explains why a localized lesion of the ventral horn or of the motoneurons at a particular level (for example, in poliomyelitis) will result in a segmentally distributed paresis or paralysis (for example of the deltoid, supraspinatus, biceps, coracobrachialis, brachioradialis, and brachialis muscles when C_{5-6} is involved; compare Fig. 3-1B and the dermatome map of Fig. 2-4). Some investigators maintain that a still more detailed differentiation can be made, and that the cells of the various columns are arranged according to the function of the muscles that they supply (see Sterling and Kuypers, 1967b). Romanes (1953) concluded a critical evaluation of the subject by saying that apparently "the motor cell groups represent in part the morphological divisions of the muscles, but also in the majority of mammals have a topographical significance which is related to the joints moved, with occasional subdivisions representing single muscles."

Very detailed studies of the distribution of motoneurons innervating particular muscles may be performed by the new HRP method (see Chap. 1). Injecting the soleus and medial gastroc-

FIG. 3-1 *A:* Diagram of the human spinal cord presenting in a transverse section the position of the various cellular columns. The letters and figures indicate the segmental distribution of the cellular columns. Redrawn from Kappers, Huber, and Crosby (1936).
B: Three-dimensional diagram of the gray matter in the cervical cord, showing the somatotopic arrangement of the groups of motor ventral horn cells which supply different parts of the upper extremity. The hand should cover a larger area than shown in the diagram, since the distal parts of the upper extremity have a far ampler motor supply (cf. small motor units) than the proximal parts. See text.

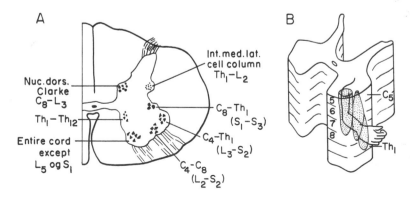

nemius muscles in cats, Burke, Strick, Kanda, Kim, and Walmsley (1977) found labeled cells in the cord to be limited to a column in L_7 and S_1. Both muscles were innervated from the same column, although there was a slight rostrocaudal difference in the distribution, the soleus motoneurons extending most rostrally. Exact measurements of the size distribution of motoneurons were also made.

The motor units. The term *motor unit,* introduced by Sherrington, has proved to be a very useful one. *A motor unit is an α motoneuron with its axon and the muscle fibers which it supplies.* When a motor ventral horn cell discharges, all muscle fibers belonging to the unit will contract. The size of the motor units varies widely, as a single ventral horn cell may supply from very few up to a thousand or more muscle fibers. The number of muscle fibers per nerve cell has been determined in animals in the following way: After removal of the spinal ganglia of the dorsal roots that supply, for instance, one of the extremities, the experimental animal is allowed to survive until sensory fibers in the appropriate spinal nerves have degenerated. Then the number of remaining nerve fibers, which presumably are motor (some also sympathetic efferent, however), is counted in the branches supplying certain muscles and the number of muscle fibers in these muscles is also counted (Clark, 1931), and thus the number of muscle fibers per nerve fiber and, accordingly, per ventral horn cell is calculated. In some muscles, particularly such small ones as the extrinsic eye muscles and the small muscles of the hand, the motor units are small. It also appears likely that in primates the motor units are smallest in muscles that are used for finely differentiated movements, such as the small muscles of the hand, the laryngeal muscles, and the extrinsic eye muscles. This has been confirmed in some instances (Feinstein, Lindegård, Nyman, and Wohlfart, 1954). The authors counted the number of muscle fibers in several muscles and the number of fibers in the nerves to the same muscles. Assuming that some 40% of the thick nerve fibers are afferent, they found the following number of muscle fibers per nerve fiber: platysma 25, first lumbrical 108, first dorsal interossus 340, anterior tibial about 600, gastrocnemius 1600 to 1900. (For some other values, see Buchthal, 1961; Mc Comas, Sica, Upton, Longmire, and Caccia, 1972). The nerve fibers of the small units are of a finer caliber than those of the larger ones. A single nerve fiber is able to supply a large number of muscle fibers because of extensive branching of the axon. This branching occurs when the nerve fiber is close to the muscle or within it (Cooper, 1929; Eccles and Sherrington, 1930). According to Sunderland and Lavarack (1953), some axons may also branch peripherally along their course.

Just how the muscle fibers that belong to the various motor units are mutually arranged within the muscle is a question of considerable interest. From physiological experiments in cat sartorius and tenuissimus muscles (Cooper, 1929), it has been concluded that the muscle fibers of one unit must be assumed to be distributed throughout the length of the muscle, arranged, as it were, in series.

The arrangement in cross section of fibers belonging to one unit has been considerably clarified in recent years. By using the glycogen depletion technique of Edström and Kugelberg (1968)[1] in various mammals, fibers belonging to one

---

[1] Single motor axons are stimulated repetitively, thereby depleting all muscle fibers innervated by this axon of their glycogen. When cross sections of the muscles are subsequently stained by the PAS method, the glycogen-depleted fibers remain unstained in contrast to the unstimulated fibers.

motor unit were found to be distributed evenly throughout about 15–20% of the muscle cross section (Edström and Kugelberg, 1968; Burke and Tsairis, 1973; and others). It has been argued that this may not be the case in man, since in muscles of patients with diseases in which the ventral horn cells are affected (amyotrophic lateral sclerosis, syringomyelia), the atrophic muscle fibers are typically found close together in groups, as seen in Fig. 3-2. Some earlier physiological studies were taken to support this subgroup concept (see Buchthal and Rosenfalck, 1973). However, studies with refined physiological techniques have made clear that the arrangement in normal human muscle does not differ markedly from that in other mammals (Stålberg, Schwartz, Thiele, and Schiller, 1976). The grouping of atrophic muscle fibers seen in patients with diseases involving the motoneuron (Fig. 3-2) is presumably due to the fact that, whereas some muscle fibers lose their innervation, the remaining normal motor nerve fibers will sprout and reinnervate the denervated fibers (Morris, 1969; Schwartz, Stålberg, Schiller, and Thiele, 1976; and others). In this way one motor axon will come to innervate muscle fibers lying close together, and at the same time the size of the remaining motor units is increased. When this motoneuron also dies, the atrophic muscle fibers will occur in groups.

The occurrence of a regular arrangement of the atrophic fibers in cases of muscular atrophies due to affection of the peripheral motoneurons is of diagnostic value, although it is not, as previously held, an expression of the normal arrangement of the motor units. Following the pioneer studies by Slauck (1921 and later) and Wohlfahrt and Wohlfart (1935), extensive research has confirmed that the microscopic changes in the muscles in disease of the type mentioned above (in which motoneuron damage is the primary affection) differ from other types of muscular atrophy, especially the progressive muscular dystrophies, where there is no regular arrangement of atrophic fibers and there is commonly a considerable increase of fat and connective tissue in the muscles (cf. the "pseudohypertrophy" of the affected muscles in cases of infantile muscular dystrophies). There are also other differential features.

However, there are a number of varieties of these diseases, which to some extent have their own particular histopathological characteristics (see Adams, 1975). This can be seen from examination of a small piece of affected muscle, which may be removed by biopsy under local anesthesia. By using histochemical techniques in addition to conventional histological staining methods, further information may be obtained. In normal human skeletal muscle, three main fiber types can be distinguished by using a variety of histochemical methods (see Brooke and Kaiser, 1974). Type 1 corresponds roughly to the "red" fibers, type 2B to "white" fibers, while type 2A is similar to the type 1 fibers in some respects (for example, in the content of mitochondrial enzymes) but similar to type 2B in others.

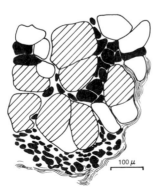

FIG. 3-2 Drawing of a primary muscle bundle in a biopsy taken from a diseased muscle in a patient suffering from progressive neuropathic peroneal muscular atrophy (Charcot-Marie-Tooth's disease). Some muscle fibers are of normal diameter (white), some are "hypertrophic" (hatched), while most fibers are atrophic (black). Note tendency of fibers at the same degree of atrophy to aggregate. From Brodal, Bøyesen, and Frøvig (1953).

In animals a close correlation between histochemical and physiological properties of muscle fibers has been worked out (see Burke and Tsairis, 1974) and seems to hold true for human muscle fibers as well (Buchthal and Schmalbruch, 1970; Warmholtz and Engel, 1972; Grimby and Hannerz, 1977). Type 1 fibers are slowly contracting, while types 2A and B presumably are fast contracting. Certain fiber types are selectively involved in some disorders (see Engel, 1965, for a review). It is also possible to obtain biopsies from motor end plates and thus to study alterations in these (see Coërs and Woolf, 1959; Engel and Santa, 1973).

The terminal arborizations of the motor nerve fibers are found in the muscles as part of the so-called *motor end plates*. Histochemical and electron microscopic studies have clarified a number of interesting structural details of the motor end plates. The main features may be summarized as follows: The end plate consists of two parts, one nervous and one muscular, separated by a cleft (20–30nm). This *myoneuronal junction* in many respects resembles a synapse in the central nervous system. The muscular part of the end plate contains several nuclei and numerous mitochondria, but lacks contractile myofibrils. The nervous part is provided with a number of grapelike swellings, much like terminal boutons. Each of these ''dips into'' the muscle fiber and contains synaptic vesicles and mitochondria. It is established that acetylcholine acts as a transmitter substance at the end plate. It enters the cleft, is bound to the muscle cell membrane, and alters the ionic permeability of this. One nerve impulse produces an end-plate potential in the muscle fiber, and when this reaches a sufficient height it gives rise to an action potential that is propagated along the fiber and elicits shortening of the myofibrils. The acetylcholine, which is liberated on the arrival of a potential in the nerve fiber, is rapidly destroyed by cholinesterase. With histochemical methods the end plates may be visualized by virtue of the presence of this enzyme. The impulse transmission at the myoneural junction may be influenced in various ways. Curare, for example, reduces the end-plate potential, and eventually prevents the appearance of an action potential. Thus a paralysis of the muscle results (compare the use of curarelike substances in anesthesia to obtain muscular relaxation). The essential disturbance in myasthenia gravis is a defective transmission at the end plate (for some data, see Desmedt, 1973). The potentials recorded in electromyography are the muscle fiber action potentials described above.

When the nerve fiber is cut, the motor end plate, as well as the fiber, undergoes degeneration. As a rule one muscle fiber is supplied by only one axon and has a single motor end plate (see Feindel, Hinshaw, and Weddell, 1952).

As is well known, motor axons can regenerate after being cut and may reestablish functional end plates with the muscle. The factors governing this synapse formation, as well as that occurring during normal development, have been studied intensively using a variety of experimental manipulations (see *The Synapse,* Cold Spring Harbor Symposium, 1976; Gordon, Jones, and Vrbová, 1976). In some way or other, the muscle cell must be able to signal whether new synapses (end plates) may be allowed to be formed. It has been assumed that the presence of acetylcholine receptors in the muscle cell membrane outside the end-plate region is closely related to this signal, since extrajunctional acetylcholine receptors are present during both normal development and regeneration of the end plates, while they disappear as soon as functional end plates are established. In a denervated muscle, electrical stimulation prevents both formation of synapses outside the end-plate region and the occurrence of extrajunctional acetylcholine receptors (see, for example, Jansen, Lømo, Nicolaysen, and Westgaard, 1973; Frank, Jansen, Lømo, and Westgaard, 1975). However, it has been recently shown that if the acetylcholine receptors are blocked by α bungarotoxin (a snake poison), the regenerating muscle nerve still makes functional end plates (Jansen and van Essen, 1975).

During postnatal development the muscle fibers undergo characteristic histochemical and physiological changes. These changes in properties of single muscle fibers (contraction speed, fatigability, maximal tension, etc.) are presumably due to changes in the activity of the motoneurons innervating them. It is, for instance, well known that the contraction speed of a fast muscle may be reduced if its nerve is cut and the muscle then innervated by the nerve to a slow muscle (see Gutman, 1976, for a review).

The motor unit undergoes a reduction in size postnatally, presumably because each muscle fiber is initially innervated by more than one motoneuron (Brown, Jansen, and Van Essen, 1976). After acquiring the adult pattern with only one axon innervating each muscle fiber (about 2 weeks postnatally in the rat), the size of the motor units appears to be constant (Kugelberg, 1976).

The function of the motor units. Electromyography. *The motor units are the smallest functional units in the locomotor apparatus.* Variations in movements—in their range, force, and type—are ultimately determined by differences in the interaction and collaboration of motor units. Information on the function of these elementary units has been obtained particularly by the use of *electromyography,* the recording of electrical changes in the muscles during contraction. This can be done by using surface electrodes applied to the skin where a muscle is situated superficially, as, for example, the dorsal interossei. When records are to be made of more deeply situated muscles, needle electrodes have to be used. These offer an advantage since one can record from a smaller total area and, accordingly, get more precise information. If the potential changes occurring in a muscle in a forceful contraction are led off, a multitude of negative and positive waves occur in the record, making it extremely complicated and difficult to interpret. If, on the other hand, the muscle is made to contract only feebly, single action potentials occur with certain intervals. Each potential is taken to reflect the activity of one motor unit. Since the potentials from different units may vary in amplitude, several unit potentials can be kept apart when they are relatively few, as in a feeble contraction.

From electromyographic studies it is learned that when a contraction gradually increases in strength, more and more units appear in the record until, finally, an interference curve appears. It is of interest to know how different motor units collaborate during a more forceful contraction. In paretic muscles in poliomyelitis, where only some of the motor units are preserved, some features can be more easily studied than in healthy muscles (Seyffarth, 1940). However, essentially the same findings are made in both cases.

In the very beginning of the contraction there is only one motor unit visible in the electromyogram, i.e., there is only one motoneuron active in the part of the muscle examined. When the strength of the contraction increases, another unit appears in addition, then a third, a fourth, and so on. This recruitment of the motor units is usually found to follow a fixed order according to their size (as will be considered later, this also relates to the size of the motoneurons). The smallest units, which are called into action first, are found histochemically to consist of type 1 (''red'') fibers; these fire with a relatively low frequency, evolve little tension, buy may go on working for a very long time. They are also the last to disappear from the electromyogram when the force of the contraction recedes. The largest motor units, consisting of type 2B (''white'') fibers, are normally recruited only when a fairly high force is required. They fire with a high frequency and evolve a high tension, but are easily fatigued. Thus, the increase in strength of a voluntary muscular contraction is, at least to a certain extent, due to the fact that more and more motoneurons discharge impulses to the muscle fibers that they supply. However, it appears from experiments in both animals and man that a great majority of the motor units are recruited at a low force. Consequently, a further increase in force must be achieved mainly by an increase in firing frequency (see Stein, 1974, for a review). In very brief and forceful contractions (ballistic movements), the recruitment order of the motor units seems to be preserved, but the larger motor units are recruited at an earlier stage of force development (Desmedt and Godaux, 1977; see, however, Grimby and Hannerz, 1977), thus making possible a more rapid increase in force.

The strength of a muscular contraction depends partly on the number of motor units being activated (recruitment) and partly on the frequency with which motoneurons send impulses to the muscle fibers they supply. It appears that recruitment is the most desirable mechanism for grading force in movements of low force, and in very brief movements in which there is time for only a few nerve impulses. However, considering the entire physiological range of forces, increase of frequency appears to be the most important factor (Stein, 1974).

The discovery of motor units, and of a method of recording their function, opened up new perspectives and made possible an understanding of many important problems. In clinical neurology, electromyography is now routinely used, since it gives valuable information, particularly in cases of muscular atrophies and dystrophies where it may aid in differential diagnosis (see Buchthal, 1962, for a review). In peripheral nerve lesions, electromyography may be used to decide, at an earlier stage than any other method, whether regeneration occurs satisfactorily. The appearance of motor units in a record from a paralyzed muscle must be taken as proof that some nerve fibers are acting and intact (having escaped destruction or having regenerated), even if no contraction can be seen. The electromyographic findings in completely denervated muscles will be referred to later. The number of motor units in some superficially located muscles has been determined by quantitative electromyography (see McComas, Fawcett, Campbell, and Sica, 1971; Sica, McComas, Upton, and Longmire, 1974; Hansen and Ballantyne, 1978; Defaria and Toyonaga, 1978). In the abductor pollicis longus muscle the normal number was estimated to be about 400 motor units, in the extensor digitorum brevis about 200, and in the thenar muscles innervated by the median nerve about 380. In particular, quantitative electromyography appears to be the most sensitive tool for detecting denervation and for assessing the degree of reinnervation taking place in the muscles. Thus "up to 70% denervation could be detected in patients in whom the electromyographic interference pattern appeared normal" (McComas, Fawcett, Campbell, and Sica, 1971).

Electromyography has further proved to be a valuable tool in research on muscle function, for example in studies of the muscles of mastication, laryngeal muscles, extrinsic ocular muscles, muscles of the inner ear, etc., as will be discussed later. It has contributed substantially in analyses of the cooperation between various muscles or parts of muscles (see Basmajian, 1974). The method has the great advantage of being applicable to human beings, and there is an extensive literature on its application in research and in clinical diagnosis of diseases of the neuromuscular apparatus. For a wealth of information on these topics, the reader is referred to *New Developments in Electromyography and Clinical Neurophysiology,* Vol 1–3, 1973.

Although the electromyogram reflects the activity of the α motoneurons, it gives relatively crude and incomplete information on their function. Recent studies in various fields have extended our knowledge of the motoneurons and the factors that influence their activity and cooperation. Not least new light has been shed on their behavior and control in spinal reflexes. Some of these data are of importance for clinical neurology.

The receptors in the muscles.   A wealth of information on this subject is available today,[2] but in spite of this there are many unsolved problems. There are still differences of opinion among specialists in the field on several points. Here only a brief outline can be given, beginning with a presentation of the principal, well-established features. Following this, more recent data will be presented.

The skeletal musculature is provided with several types of receptors (Fig. 3-3A), which may be grouped as follows: muscle spindles, tendon organs (or Golgi organs), Vater-Pacinian corpuscles, various types of endings in the capsules of the synovial joints, and free nerve endings. Of these, only the first two will be considered here. However, other types of receptors, situated in the skin, have been shown to influence the motoneurons as well. The Pacinian corpuscles and joint receptors have been considered in Chapter 2.

Of these receptors, the *muscle spindles* are morphologically and functionally most elaborate and have the most complex organization. They were rather completely described by Ruffini in the 1890s. Muscle spindles appear to be present in all muscles of the locomotor apparatus, and in many muscles supplied by motor cranial nerves as well, such as the laryngeal muscles, the muscles of mastication, the tongue, and the extrinsic ocular muscles (for reviews, see Cooper, 1960; Barker, 1974). The relative density of muscle spindles varies widely in different muscles. Thus, in the small human abductor pollicis brevis there are 80 spindles (Schulze, 1955–56), in the large latissimus dorsi 368 (Voss, 1956). An estimate of the number of spindles per gram of muscle weight gives the values of 29.3 and 1.4, respectively, for these two muscles. Thus, *muscles used in delicate movements are far more amply supplied with spindles than muscles used in coarse movements.*

The muscle spindles have derived their name from their shape. The principal points in their organization may be summarized as follows (see Fig. 3-3B): A spindle-shaped connective tissue capsule, some millimeters long, surrounds a few, 3–8 (in man an average of 10: Cooper, 1960), slender muscle fibers. These so-called *intrafusal fibers are* attached at their extremities to the connective tissue of the capsule and separated from this by a space filled with fluid. The *intrafusal fibers are thus arranged in parallel with the ordinary, extrafusal muscle fibers* and both are ultimately attached directly, or indirectly by way of inelastic collagen fibers, to the tendon (see Fig. 3-3A). When the extrafusal fibers contract, those of the spindles will be subjected to a reduced tension, while a stretching of the muscle as a whole will increase the length and the tension of the intrafusal fibers. The central equatorial parts of the intrafusal muscle fibers are not contractile and harbor many nuclei. They are surrounded by spirally coursing nerve fibers, the *primary sensory* or *annulospiral ending,* made up of fine ramifications of a single

---

[2] The reader interested in details is referred to the following comprehensive treatments of the subject: *Mammalian Muscle Receptors and their Central Actions* by P. B. C. Matthews (1972); *Handbook of Sensory Physiology* Vol. III/2; *Muscle Receptors* (1974), and the reports from a Symposium in Tokyo in 1975, *Understanding the Stretch Reflex.*

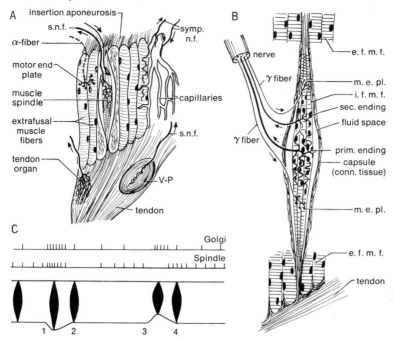

FIG. 3-3 *A:* Diagrammatic drawings of the three main receptors in skeletal muscles: a muscle spindle, a tendon organ, and a Vater-Pacinian corpuscle (*V-P*). Note that the spindle is arranged in parallel with the extrafusal muscle fibers, while the tendon organ is arranged in series with them. This has important consequences for their function.
B: Very simplified diagram of a muscle spindle to show its principal components (see text). Abbreviations: *e.f.m.f.,* extrafusal muscle fibers; *i.f.m.f.,* intrafusal muscle fibers; *m.e.pl.,* motor endings.
C: A diagram illustrating the response of a Golgi tendon organ and a muscle spindle ending during stretch and contraction of the muscle. The muscle is drawn below in black, between lines indicating its insertion. Action potentials are shown in two top lines. Further explanation in text. Fig. C from Granit (1955a).

relatively thick afferent fiber (12–14 μm in the cat). A *secondary sensory ending* or *flower-spray ending*, coming from a thinner afferent fiber, may be present at one or both sides of the annulospiral ending (Barker, 1948). In addition, the muscle spindles receive a *motor innervation*. This consists of relatively fine nerve fibers, the previously mentioned γ fibers (or fusimotor fibers), which come from the ventral horn. They terminate in the distal cross-striated parts of the intrafusal fibers, which are contractile. It follows from this that a contraction of the distal parts of the intrafusal fibers in response to impulses in the γ fibers will result in a stretching of the central "sensory" part of the intrafusal fibers, with a consequent stimulation of sensory endings, just as will a stretching of the entire muscle.

The *tendon organs* are much simpler than the muscle spindles. In essence, a tendon organ consists of a group of branches of a large myelinated nerve fiber (12–18 μm, Cooper, 1960; Schoultz and Swett, 1972) which terminates with a spray of fine endings between bundles of collagenous fibers of tendons, usually near the musculotendinous junction (Fig. 3-3A). A single fiber may supply more than one tendon organ. These are usually covered by a delicate connective tissue capsule. To understanding the function of the tendon organs, it is essential to be aware of the fact that they are *arranged in series with the extrafusal muscle fibers*. Whether the muscle contracts or is stretched, the tendon organ will be stimulated, since in both cases the tension of the tendon organ will increase.

Physiological studies have confirmed the views on the function of muscle spindles and tendon organs which had been inferred from an analysis of their structure. By leading off from single fibers in dorsal roots which transmit sensory impulses from a muscle, action potentials can be recorded which are identified by their behavior as belonging to either muscle spindles or tendon organs. Figure 3-3C shows that when the muscle is stretched (1), the frequency of afferent discharge rises in both sense organs and, after a short pause, goes back to the resting values when the stretch is released (2). When the muscle is made to contract (3), the Golgi organ again responds by increasing its firing rate on account of the pull on the tendon. The muscle spindle, however, will be unloaded during the contraction and will pause. When the contraction recedes (4), the slack spindles will again be pulled upon and consequently discharge, while the tendon organs will not respond since the tension is reduced. The *tendon organs,* according to these findings, are *tension recorders,* while the *muscle spindles give information about the length of the muscle*. These essential findings were made by Matthews (1933) and later confirmed by others (see Granit, 1955a, and 1975, for reviews).

With regard to the *efferent innervation* of the muscle spindles, Leksell (1945) first succeeded in leading off compound action potentials from the thin fibers in the ventral roots (having a conduction velocity of 20–38 m/sec) and could show that stimulation of these γ fibers did not result in contraction of the muscle. However, upon stimulation of the γ fibers, afferent action potentials could be recorded from the muscle nerve. These potentials were interpreted as arising from the muscle spindles, due to the stretching of the central sensory region of the intrafusal fibers produced by the shortening of the distal, contractile parts when the γ fibers were active. Leksell's (1945) findings, which have since been repeatedly confirmed with single fiber techniques, made it clear that the *muscle spindles are subject to control from the central nervous system*. The degree of stimulation of a muscle spindle's γ fibers will regulate its sensitivity to stretch since contraction of the distal contractile parts will lengthen the central sensory part and reduce its threshold for stimulation. The γ neurons are played upon by impulses from various sources, which thus may influence the activity of the spindles (see below).

From intracellular recordings, Eccles, Eccles, Iggo, and Lundberg (1960) concluded that *the γ neurons lie intermingled with the motoneurons that supply the same muscle*. Anatomical demonstration of the γ neurons was lacking until Nyberg-Hansen (1965b), by means of the modified Gudden method (Brodal, 1940a), succeeded in demonstrating them. Following transection of a peripheral nerve there are, among the large motoneurons that present retrograde cellular changes, a number of small ones that display the same alterations and, accordingly, must be assumed to have had their axons interrupted. Burke and co-workers (1977) found that horseradish peroxidase was transported retrogradely from the muscles to both small and large cells in lamina IX. The labeled cells showed a bimodal distribution with regard to soma diameter. The smallest group, having diameters between 18 and 38 μm, was assumed to consist of γ neurons (see also Pellegrini, Pompeiano, and Corvaja, 1977). After injections of the soleus and medial gastrocnemius muscles in the cat, these authors noted that up to 96% of the cells were labeled in the cell column supplying these muscles, indicating that few, if any, interneurons are present in lamina IX.

The very schematic account given above of the muscle spindles and their function summarizes approximately what was known around 1950. Later research has shown that the spindles are in fact far more complex with regard to both structure and function.

Although it has long been known that the muscle spindles are not randomly distributed in the muscle, it has more recently become clear that their distribution is related to the histochemical properties of the muscle fibers. Thus, in the cat the spindles are found almost exclusively among fibers of the "red," or slow twitch, fatigue-resistant type (see Richmond and Abrahams, 1975). As mentioned above, such muscle fibers belong to small motor units, responsible for the execution of, for example, delicate movements requiring relatively little force, in which one would think that the muscle spindles are particularly important. "White" parts of the muscles are thought to be practically devoid of muscle spindles. The functional properties of the muscle spindles appear to depend upon whether they are located in the superficial or deep part of the muscle, the superficial ones being less sensitive to stretch of the muscle than the deep ones (see Meyer-Lohmann, Riebold, and Robrecht, 1974). Muscle spindles in various parts of the muscle will, therefore, presumably not behave in the same way, and may be believed to give relatively specific information only about a small number of motor units (Binder, Kroin, Moore, Stauffer, and Stuart, 1976).

It was shown 20 years ago that most spindles contain two types of intrafusal muscle fibers (Fig. 3-4) which usually differ in diameter, among other things. With reference to the arrangement of their nuclei, the thick fibers were called *nuclear bag fibers*, the thin one *nuclear chain fibers*. It has recently become clear that the bag fibers are of two types, referred to as *bag*$_1$ and *bag*$_2$, that differ with regard to histochemistry and ultrastructure (see Barker, Banks, Harker, Milburn,

FIG. 3-4 Simplified diagram of the central region of a muscle spindle, showing the two kinds of intrafusal muscle fibers and their sensory and motor innervation. The presence of two types of bag fibers is not shown. The nuclear bag fiber in the diagram may be taken to represent a bag$_1$ fiber while the nuclear chain fiber represents both chain and bag$_2$ fibers. (Slightly changed from Matthews, 1964.)

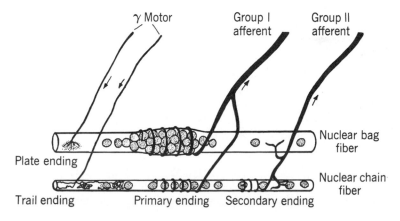

and Stacey, 1976). There is also good evidence that they are functionally dissimilar (see below). Likewise, anatomical studies on the innervation of the muscle spindles have revealed a greater complexity than was previously known. The primary sensory (annulospiral) endings are related to both types of muscle fibers, while the secondary (flower-spray) endings, mostly, but not always, appear to be present only on the nuclear chain fibers. Muscle spindles in man are in all essentials similar to those in the cat (Cooper and Daniel, 1963; Swash and Fox, 1972).

Early workers observed some features in the response of the muscle spindles to stretch which indicate that they might not be only length recorders. Jansen and Matthews (1962a) emphasized that the *potentials recorded from primary afferent fibers give information not only on the length of the muscle but also on the velocity of extension*. The former response is called the *static response* and is seen on maintained extension of the muscle. The *dynamic response* is the response of the receptor to actually being stretched. For various reasons the authors suggested that the dynamic response originates mainly in the nuclear bag fibers while the static response is due to extension of the nuclear chain fibers; both of these fiber types are supplied by the primary endings. Recent detailed physiological studies of isolated muscle spindles, particularly by Boyd and co-workers (see Boyd, 1976) are compatible with the idea that the bag_1 fibers are chiefly responsible for the dynamic response, while the bag_2 and chain fibers both give a static response. The secondary sensory endings appear, in many respects, to be functionally similar to the primary ones, but they are less sensitive to changes in length, particularly to rapid changes (see Bessou and Laporte, 1962; Matthews, 1964). They have not been studied as completely as the primary endings. Some secondary endings have been shown to have mono- and disynaptic excitatory effects on homonymous motoneurons (Kirkwood and Sears, 1974; Stauffer, Watt, Taylor, Reinking, and Stuart, 1976) and there is some indirect evidence that the secondary endings contribute substantially to the tonic stretch reflex (Matthews, 1969, 1972).

Physiological studies have further shown that *there is more than one type of γ fiber*. Jansen and Matthews (1962a, b) concluded from recordings of the activity in the muscle spindles following stimulation of the efferent fusimotor fibers in decerebrated cats that the dynamic and static sensitivity of the primary endings are independently controlled. Matthews (1962) subsequently isolated two functionally distinct types of fibers by subdividing ventral roots and named them the *static* and *dynamic* fusimotor fibers by virtue of their action on the responsiveness of the primary ending to stretching. Using the glycogen depletion technique of Edström and Kugelberg mentioned previously in this chapter, Barker, Emonet-Dénand, Harker, Jami, and Laporte (1976) were able to show that of the two types of bag fibers, only the bag_1 type is influenced by dynamic γ fibers, bag_2 and chain fibers being solely activated by static γ axons. By another approach, Boyd reached principally the same conclusions (see Boyd, 1976). However, whether the effect of static γ fibers is limited to bag_2 and chain fibers (see Boyd, Gladden, McWilliam, and Ward, 1977) or may in addition influence bag_1 fibers seems not to be settled (Barker, Banks, Harker, Milburn, and Stacey, 1976; Barker, Bessou, Jankowska, Pagès, and Stacey, 1978). The γ fibers to the spindles are provided with two kinds of terminals. One is represented by motor end plates of the kind found in extrafusal muscles (as has been long recognized); the other terminal forms an extensive and diffuse network of fine axonal fibers and small elongated endings (see Fig. 3-4). Following the notation of Barker (1962, 1966) these are commonly referred to as *plate endings* and *trail ending,* respectively. By combining various anatomical and physiological techniques, it has been established that, at least as a rule, the plate endings mediate the dynamic fusimotor effect (on the bag$_1$ fibers), while the trail endings are responsible for the static effect (on bag$_2$ and chain fibers; see Barker, Banks, Harker, Milburn and Stacey, 1976). Efferent fibers giving off branches to both extrafusal and intrafusal fibers (referred to as β fibers) have been demonstrated histologically (Adal and Barker, 1965). In some cat hindlimb muscles, about one-third of the spindles appear to receive a β axon, which almost exclusively terminates on the bag$_1$ fiber, mediating a dynamic effect (Barker, Emonet-Dénand, Harker, Jami, and Laporte, 1977; Boyd, Gladden, McWilliam, and Ward, 1977). Whether some β fibers may be static still seems to be an open question (see Laporte and Emonet-Dénand, 1976).

The efferent control of muscle spindles has been extensively studied in man after Hagbarth and Vallbo in 1968 succeeded in recording from single muscle spindle afferents in conscious human subjects. The findings are essentially in accord with those obtained in animal experiments (see Granit, 1975, for a review). However, the study of humans has yielded some information not obtained in animals. For example, humans can easily be instructed to perform various voluntary movements, or to react in particular ways to a disturbance of a movement, which, if at all possible, requires a long period of training in animals. During voluntary contraction of a particular muscle, the α and γ motoneurons are both found to be activated (Vallbo, 1970a, b; 1974a, b; and others)—the so-called α and γ coactivation of Granit (1955a). The activation of γ motoneurons during voluntary contractions will prevent shortening of the sensory region of the muscle spindle. In this way the flow of impulses from the muscle spindles exciting the motoneurons will be maintained, presumably preventing a decrease of force, with shortening of the muscle. In accordance with this, if the thinner fibers in the muscle nerve are selectively blocked (among them the γ fibers), the force and speed of movements are reduced. Both may, however, be restored by vibrating the muscle belly (Hagbarth, Hongell, and Wallin, 1970; Hagbarth, Wallin, and Löfstedt, 1975), a procedure that is known to be a powerful way of activating muscle spindle afferents (Granit and Henatsch, 1956; Matthews, 1966; Burke, Hagbarth, Löfstedt, and Wallin, 1976a, b; and others).

If vibration is applied, a peculiar reflex contraction of the muscle occurs, the *tonic vibration reflex,* characterized by its long latency (often more than 30 sec) and slow rise and fall (Hagbarth and Eklund, 1966; De Gail, Lance, and Neilson, 1966; Marsden, Meadows, and Hodgson, 1969; Burke, Hagbarth, Löfstedt and Wallin, 1976 a, b; and others). Concomitant with the contraction, the antagonists are inhibited, as is the tendon reflex in the vibrated muscle. The tonic vibration reflex may be completely suppressed voluntarily but recurs as soon as the attention of the subject is diverted to something else. Although it is clear that the spindle afferents constitute the main afferent link in the tonic vibration reflex, the central mechanisms underlying it are not understood. Due, among other things, to the long latency, it must involve a polysynaptic pathway, possibly transcortical.

The tonic vibration reflex is of some clinical relevance (see Hagbarth, 1973). In certain conditions it may be altered or abolished, as in cerebellar lesions, where it is often found to be reduced. In spastic paresis, vibration applied to antagonists often reduces the spasticity, in some cases making voluntary movements possible.

There are still unsolved problems concerning the functional organization of the muscle spindles.[3] Even the simplified account given above makes it clear, however, that the muscle spindle is an extremely complex type of receptor. The spindles inform the central nervous system of the length of the muscle as well as the speed of stretching and thus provide information that is essential for the proper activity of the spinal reflex apparatus. In addition, they contribute to muscular control by sending messages to supraspinal levels, e.g., to the cerebral cortex. We shall return to the muscle spindles when discussing the stretch reflexes and muscular tone.

The *tendon organs* have been less intensively studied than the muscle spindles. As mentioned above, these are arranged so as to record tension in the muscle, whether it results from contraction or extension of the muscle. It has, for some time, been generally held that their main task is to protect the muscle against overstretching. It appears, however, that they have a more differential function, and in fact are far more sensitive to the tension arising in the tendon on contraction of the muscle than to tension arising on stretching of the muscle. This is probably explained by the anatomy of the tendon organs. They are found at the junction between muscle and tendon, and each tendon organ (in the cat) is attached to about 10 muscle fibers (Schoultz and Swett, 1972), and these probably belong to nearly as many motor units (see Reinking, Stephens, and Stuart, 1975). It is conceivable that the force produced by, say, the activity of one motor unit will have greater influence on the tendon organs to which its muscle fibers are attached than if the same force were passively applied by stretching the muscle, which would distribute it equally to all parts of the tendon. Matthews (1972) pointed out that "Since relatively little passive tension is normally produced by stretching a muscle over its physiologically permitted range of lengths . . . for most practical purposes tendon organs may be looked upon as 'contraction receptors.' " Under certain circumstances they seem to be able to respond to the contraction of even a single motor unit (Jansen and Rudjord, 1964; Houk and Henneman, 1967; Reinking, Stephens, and Stuart, 1975; Jami and Petit, 1976). This as does also their relatively great numbers, suggests that the tendon organs may be more important in the control of the motor apparatus than hitherto considered. The number of tendon organs has been determined in some muscles and has been found to be only a little less than the number of muscle spindles (Swett and Eldred, 1960; Wohlfart and Henriksson, 1960). Information from the tendon organs is delivered not only to the cord, but also to central levels, especially the cerebellum. The spinal interneurons mediating the effect of tendon organs on the motoneurons are influenced from higher centers, for example via the corticospinal and rubrospinal tracts. (See Lundberg, 1975. This subject is treated more fully in Chap. 4).

The *Pacinian corpuscles* (discussed in Chap. 2) appear to be vibration receptors. Like the receptors of the joints (see Chap. 2), they presumably send additional information to the central nervous system, concerning what goes on during muscular contraction, and contribute to the reflex control of movements. The role of these receptors has, however, been relatively little studied so far.

---

[3] The role of muscle spindles in kinesthesia has been treated in Chapter 2.

The stretch reflex and the control of muscles. Even when it is isolated from the rest of the brain, the spinal cord is capable of mediating a number of reflexes, somatic and autonomic. The morphological basis of a nervous reflex is generally referred to as a reflex *arc,* which, in its simplest form, consists of five links: (1) *A receptor,* which reacts to the stimulus; (2) *An afferent conductor,* which transmits the impulses to the "reflex center." (The afferent conductor is an afferent sensory fiber, in most cases having its cell body in a spinal or cranial nerve ganglion.); (3) A *"reflex center,"* where the afferent message from the receptor may converge with afferent impulses from other receptors, or with afferents from other sources, which may modify the effect of the afferent impulses from the receptor; (4) *An efferent conductor,* a nerve fiber going to the effector organ; (5) *An effector,* which produces the reaction, and may be a muscle, a gland or a vessel, or may include several such components.

Reflexes vary enormously, from very complex ones, such as the reflex of swallowing in which a number of different effectors are involved, to relatively simple ones. Here we are concerned with *spinal reflexes,* and only with those in which the effectors are skeletal muscles, i.e., somatic reflexes. The fundamental principles governing the reflex activity of the cord were clarified by Sherrington. Many types of reflexes may be distinguished. *Flexor reflexes* are those in which the response is a flexion of the limbs. The most potent stimuli are nociceptive ones, and the result is a withdrawal of the limb (withdrawal reflex). In other reflexes there is an extension of a limb; for example, in the crossed extensor reflex which may accompany a flexor reflex. There are still other more complex reflexes; for example, the scratch reflex. All of these reflexes usually involve several muscles, and the reflex response may vary according to the situation (type and site of application of stimulus, intensity of stimulus, concomitant application of other stimuli, etc.). The arcs for these reflexes are of great complexity. Quite another type of reflex is the *stretch reflex, the contraction of a single muscle in response to its being stretched.* This is an elementary reflex which probably occurs in all muscles. Stretch reflexes form the basis of many so-called *postural reflexes* (see Chap. 4), which, broadly speaking, aim at maintaining the correct posture of the body and adjusting it to various demands, be these due to external forces or resulting from movements performed by the organism.

When discussing the stretch reflexes and their anatomical basis, it is appropriate to start with a relatively simple model, that of the patellar reflex, routinely tested in every clinical neurological examination. This is a *monosynaptic* reflex, the single synapse being in the spinal cord. It should be pointed out that, however useful clinically, this reflex probably has little or no functional significance, the necessary stimulus being highly unphysiological. Furthermore, the reflex does not occur in all muscles, and is entirely absent in some healthy subjects. Reflex responses to stretch of a muscle, occurring with a longer latency than the monosynaptic stretch reflex, are apparently of much greater functional significance and have received increasing attention since being first described by Hammond (1956). Less is known about the pathways involved in these polysynaptic stretch reflexes (see below). In clinical terminology the abovementioned reflexes are often referred to as *tendon reflexes* or *deep reflexes.* Neither name is really appropriate, as will

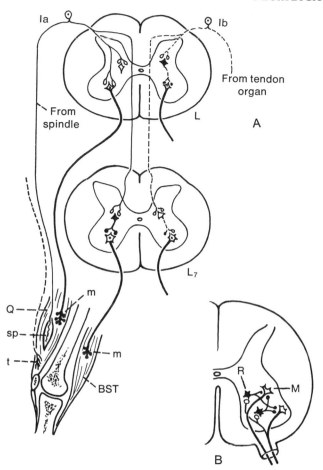

FIG. 3-5   *A:* Diagram showing (to the left) the pathways involved in a stretch reflex. Impulses arising in the muscle spindles are transmitted in group *Ia* fibers. To the right are shown (dotted) pathways arising in the tendon organs and passing in group *Ib* fibers. Excitatory cells and synapses are shown as open structures, inhibitory as black. *BST:* flexors of the knee; *m:* motor end plate; *Q:* quadriceps muscle; *sp:* muscle spindle; *t:* tendon organ (cf. text). Slightly modified from Eccles (1957).
B: A diagram of the proposed anatomical basis of the inhibition of motoneurons *(M)*, occurring through their recurrent collaterals to Renshaw cells *(R)*. Modified from Eccles (1957).

be apparent from the following. The designation *myotatic reflexes* is often used and is appropriate, but the term *stretch reflex* is apparently now most used.

Figure 3-5 shows the main features of the monosynaptic stretch reflex as deduced from experimental neurophysiological findings in the cat. As an example applicable to man we shall consider the patellar reflex with reference to the diagram. A tap on the patellar tendon stretches both the extrafusal fibers and the muscle spindles of the quadriceps (Q). The muscle spindles respond by sending a train of impulses to the cord, as described in the previous section. The afferent fibers in the dorsal roots conveying these impulses (because of their large caliber belonging in physiological nomenclature to group Ia fibers) establish synaptic contact with motoneurons which are excited and produce a contraction of the quadriceps. For this to occur there must be a concomitant reduction of tension in the antagonists to the quadriceps, the flexors of the knee (Sherrington's principle of reciprocal innervation). On the basis of physiological studies by several workers (among whom may be mentioned Lloyd, Granit, and Eccles) it has been concluded that this occurs as shown in Fig. 3-5A (to the left). The afferent spindle fiber gives off collaterals to a group of neurons in the "intermediate part" of the gray matter. These neurons are inhibitory and send their axons to motoneurons which innervate the antagonists. Consequently, the firing rate of the motoneurons contacted is reduced, and the antagonists (BST) relax.

The pathways indicated (to the left) in Fig. 3-5 are in agreement with anatomical data. As previously noted, dorsal root fibers can be traced to the large ventral horn cells by experimental methods as well as in Golgi material (see Fig. 2-3B). Whether all of these fibers are derived from muscle spindles cannot be decided, but the fibers of a single dorsal root reach ventral horn cells situated two or more segments above or below the segment of entry (Schimert, 1939; Sterling and Kuypers, 1967a; and others). This is in agreement with the physiological observation that afferents from one muscle may activate several (usually synergistic) muscles (Eccles, Eccles, and Lundberg, 1957). Of particular interest is the demonstration of ample terminations of collaterals of such fibers (Fig. 2-4) in the medial and central parts of Rexed's laminae V–VII (see Sprague and Ha, 1964; Scheibel and Scheibel, 1966a; Sterling and Kuypers, 1967a; Szentágothai, 1967), since interneurons responding to afferent impulses from muscle spindles (as well as from tendon organs, see below) were found by Eccles, Eccles, and Lundberg (1960) in this region and have since been identified anatomically (see later). Axons of cells in this region can be traced into the ventral horn, and it appears that their axons and collaterals extend for several segments up and down the cord (Szentágothai, 1951; Scheibel and Scheibel, 1966a; Mannen, 1975).

The example given above illustrates *a phasic stretch reflex,* in which the muscle spindle, by its afferent impulses, monosynaptically activates the α motoneurons to its own muscle and thus elicits a contraction of the muscle and causes relaxation of the antagonists. During contraction, the muscle spindle will stop discharging when its length is reduced (see Fig. 3-3C), the α neurons will no longer be excited, and the muscle will again relax. However, as shown by Sherrington, the stretch reflex may also be tonic. This tonic reflex seems to be mediated by a polysynaptic pathway, mentioned above. Before turning to the problem of muscle tone, this more prolonged stretch reflex will be briefly treated.

Hammond described in 1956 a dual reflex response to stretching of the biceps muscle in conscious human subjects—one with a latency of about 25 msec (the well-known monosynaptic reflex), the other with a latency of about 50 msec. The latter response has subsequently been much studied in humans, in particular by Marsden, Merton, and Morton. It is present also in muscles where the monosynaptic reflex is very weak or absent, for example the long flexor of the thumb (Marsden, Merton, and Morton, 1976a, b). In contrast to the monosynaptic reflex, it is to some degree influenced by whether the subject is instructed to resist stretch of the muscle. It occurs with both slow and fast stretching but depends upon some ongoing activity in the muscle. There is now much, although indirect, evidence in favor of the view, first suggested by Phillips (1969), that the pathway for this reflex involves the cerebral cortex (see Marsden, Merton, and Morton, 1973, 1976b; Evarts and Granit, 1976). The receptors have been shown to be the muscle spindles, although skin and joint receptors seem to be involved in a less specific way (Marsden, Merton, and Morton, 1972, 1977; Torebjörk, Hagbarth, and Eklund, 1978). It is well known that information from the muscle spindles reaches the cerebral cortex (see Chap. 2), concerning both primary and secondary endings. The disappearance of the long-latency stretch reflex in humans with lesions of the internal capsule, the dorsal columns, or the sensorimotor cortex lends further support to the conclusion that a transcortical reflex loop is involved (Adam, Marsden, Merton, and Morton, 1976; Marsden, Merton, Morton, and Adam, 1977).

The exact role played by the tonic stretch reflex in posture and postural adjustments is not yet clearly understood (Melvill-Jones and Watt, 1971; Nashner, 1973; Gurfinkel, Lipshits, Mori, and Popov, 1976; Freedman, Minassian, and Herman, 1976).

Muscle tone.   If a normal muscle is palpated when it is resting, it will be felt that it is not completely flaccid but possesses a certain degree of tension. This is also the impression gained when passive movements are made. *This normally existing condition of muscular tension is generally called muscle tone,* more properly *resting tone.* In various pathological conditions it is observed that this normal tone is changed, sometimes increased (hypertonus), in other cases reduced (hypotonus). Clinically, muscle tone is commonly examined by palpation and passive movements. However, probably two different components of muscle tone are evidenced by the two methods of examination (see below). Thus it is not uncommon to observe, for instance in capsular hemiplegias, that muscle tone is reduced when judged by palpation (the consistency of the muscles is reduced), while in passive movements it appears to be augmented, since the resistance offered to the movements is greater than normal.

There has been much debate on the problem of muscle tone, and some confusion has arisen because various definitions have been employed. The resistance offered by a muscle to being stretched may in principle be due to two factors: the inherent viscoelastic properties of the muscle and the tension set up by contraction. Both factors are certainly of importance, but their respective contributions obviously differ in various situations. The properties of the muscle alone may under certain circumstances explain much of the tension arising when it is stretched as shown by Grillner and his co-workers (see Grillner, 1974). From animal experiments, largely on decerebrated cats, it has been concluded that muscle tone is largely of a reflex nature, caused by a continuous stream of impulses from the muscle spindles, activating the α motoneurons. However, these results can hardly be applied directly to conscious human beings. They may be most relevant for the muscles maintaining the upright position, which show a continuous or intermittent (in phase with the body sway) activity in a relaxed standing subject (e.g., some of the extensors of the spine, the psoas major muscle, and the soleus muscle; see Basmajian, 1974); most other muscles usually show no EMG activity in this situation. However, even in these "postural" muscles, the activity may not be due to a segmental stretch reflex (see Gurfinkel, Lipshits, Mori, and Popov, 1976; Burke and Eklund, 1977). There is, furthermore, no evidence of tonic impulse activity in spindle afferents from relaxed forearm muscles (Vallbo, 1974a) or from relaxed leg muscles in standing subjects (Burke and Eklund, 1977). When muscles showing no activity at rest, as determined with EMG, differ with regard to resistance against passive stretch, the cause may be differences in the sensitivity of the muscle spindles (due to a different level of activity of fusimotor neurons) and/or in the responsiveness of the motoneuron pool in the spinal cord (which in turn may have various reasons). The muscle tone, as judged by palpation, may vary in different normal subjects even if they are able to relax the muscles investigated completely (see Weddell, Feinstein, and Pattle, 1944). Thus, the consistency of a muscle will probably depend not only on whether there is ongoing muscle activity but also on poorly understood properties of the muscle itself.

Since the γ neurons may alter the sensitivity of the spindles, they are of importance for muscle tone. An increased γ activity will "set" the spindles at a more active level, and so may increase the tone of the extrafusal muscles.[4] Our understanding of the role of the muscle spindles and their central connections in diseases with altered muscular tone is far from complete, despite a rapidly growing literature dealing with this problem. This seems hardly surprising, in view of the complexity of the structures involved and the limited knowledge of their role in normal control of movement and posture. Furthermore, the methods used must be indirect.

In some spastic patients, cooling of the muscles reduces the spasticity, presumably because the afferent input from muscle receptors is reduced. In other patients this treatment has no or adverse effects (see, for example, Knutsson, Lindblom, and Mårtensson, 1973). This, among other things (see Herman, 1970), has been taken to indicate that in some patients spasticity may be due to hyperactivity of the fusimotor system, while in others the flow of impulses from the muscle spindles plays no important role in causing the increased excitability of the motoneurons.

Authors studying the spindle sensitivity in spastic patients have come to somewhat different conclusions (see *New Developments in Electromyography and Clinical Neurophysiology* Vol. 3, pp. 475–588). Hagbarth and co-workers, leading off from single fibers, compared the activity of presumed primary spindle afferents during various types of passive movements and clonus in normal and spastic persons (Hagbarth, Wallin, and Löfstedt, 1973; Hagbarth, Wallin, Löfstedt, and Aquilonius, 1975). They could find no clear evidence of increased dynamic or static sensitivity of the muscle spindles in calf muscles in patients with spasticity following, for example, cerebral vascular accidents or spinal cord lesions (see, however, Jacobi, Krott, and Poremba, 1970; Dietrichson, 1971; Szumski, Burg, Struppler, and Velho, 1974). Animal experiments also indicate that there need not be a causal relationship between spasticity and fusimotor activity. Thus Gilman, Lieberman, and Marco (1974) found spasticity some time after removal of cortical areas 4 and 6 in monkeys, despite evidence of normal or even reduced spindle sensitivity. Hagbarth, Wallin, Löfstedt, and Aquilonius (1975) summed up the present state of these matters as follows: "Until further studies . . . have been made in different forms of spasticity, it remains uncertain whether an increased dynamic fusimotor drive plays an important role in the development of clonus. Internuncial neurones excitability changes in the spinal cord, loss of presynaptic inhibition on the Ia fibre terminals, and central sprouting are other plausible causes of the exaggerated dynamic stretch reflexes which, in turn, are responsible for the clonic oscillations." (For some data concerning the role of presynaptic inhibition, the Renshaw cells, and other intraspinal factors in

---

[4] The spindles have been shown to activate γ neurons, but such action does not occur monosynaptically but by way of internuncial cells, unlike the action on the α neurons, and the effect is weaker and less widely distributed (see Ellaway and Trott, 1976, 1978).

spasticity, the reader is referred to Delwaide, 1973; Herman, Freedman, and Meeks, 1973; and Veale, Rees, and Mark, 1973.)

The rigidity occurring in Parkinson's disease seems to be due to an increased central input to α and static γ motoneurons (see Dietrichson, 1971; Wallin, Hongell, and Hagbarth, 1973). During parkinsonian tremor, there appears to be a normal coactivation of α and γ motoneurons (Hagbarth, Wallin, Löfstedt, and Aquilonius, 1975), unlike the clonus beats in spastic patients, where the afferent spindle discharge decreases during shortening of the muscle, indicating lack of fusimotor coactivation.

In the execution of movements, the γ innervation is of importance in determining the state of the muscles to be used and is active in their control, while the α neurons are responsible for the contraction itself. The collaboration of α and γ neurons in the execution of movements has been the subject of several studies, and, although much still remains to be clarified, investigations in conscious human subjects have recently brought new insight (Vallbo, 1974b; Marsden, Merton, and Morton, 1972, 1976a, b; Hagbarth, Wallin, and Löfstedt, 1975).

Marsden and co-workers have, for example, provided elegant examples of how the spindle servomechanism operates during voluntary movements in man. If a continuous, slow flexion movement of the top joint of the thumb suddenly meets increased resistance, the force of contraction of the long flexor is increased with a latency corresponding to that mentioned above for the long-latency stretch reflex. (Later on there will be a voluntary response.) This may be explained as follows: During the voluntary contraction, α and γ motoneurons are activated together, as described above. The activity of both is accurately adjusted to the required force and speed of movement. If the resistance is suddenly increased, the movement will be slowed or halted, as will the shortening of the extrafusal muscle fibers, while the intrafusal muscle fibers will go on shortening owing to their continuously increasing γ input in anticipation of a steady shortening. This will lead to an increased afferent input from the muscle spindles, in turn giving higher motoneuron activity and stronger contractions to overcome the resistance. If the resistance against the movement is suddenly reduced, the opposite events take place. Furthermore, Marsden, Merton, and Morton (1972) have been able to calculate that a "misalignment" of less then 50 μm (that is, the difference between actual and anticipated muscle length) is enough to alter afferent spindle activity, and thereby the force of contraction.

As referred to previously, it appears that the central nervous system is able to control static and dynamic γ and α neurons individually. To what degree this central control is mediated by separate pathways is not known, but certain structures are found to influence mainly one type of fusimotor neuron (see *Mammalian Muscle Receptors and their Central Actions,* pp. 510–545, for a critical review and references). Thus, stimulation in the medial part of the reticular formation gives increased static spindle sensitivity, while stimulation more laterally produces increased dynamic sensitivity (Vedel and Mouillac-Baudevin, 1969).

In several instances the same structure has been found to act on both γ and α neurons, and, at least in the cat, both actions occur via internuncial cells. However, certain fiber systems appear to excite, exclusively or at least chiefly, either flexor or extensor muscles (see Chap. 4). The cooperation between γ and α neurons has been referred to as the α–γ linkage (see Granit, 1955a) or, better, α–γ coactivation (see Matthews, 1972), and evidence has been produced that the anterior lobe of the cerebellus is particularly relevant for this function (Granit, Holmgren, and Merton, 1955); but more recent experiments indicate that, although γ neuron excitability is considerably reduced after cerebellectomy, there is no disruption of the α–γ coactivation (see Gilman, 1976). Via the nuclei fastigii and interpositus the spinal regions of the cerebellum may influence several structures that are known to act on both kinds of neurons, such as the reticular formation, the red nucleus, and the vestibular nuclei. In addition, the cerebellar nuclei, presumably in particular the dentate nucleus, exert a powerful effect on fusimotor activity through the precentral motor cortex.

Thus, lesions or cooling of the brachium conjunctivum (containing the ascending fibers from the cerebellar nuclei), the ventrolateral nucleus of the thalamus, or the cortical areas 4 and 6 in monkeys produce a marked reduction in fusimotor activity, although less than after cerebellectomy (see Gilman, 1973).

The problems of muscle tone and stretch reflexes have been discussed above with reference to the muscle spindles alone. It should be mentioned, however, that other factors are also involved, but the role they play is less clear. Reference was made previously to the *tendon organs*. On stretching and, more importantly, on contraction (see above) of a muscle these will be stimulated. The impulses from the tendon organs pass via dorsal root fibers of somewhat smaller calibers than the muscle spindle afferents (in physiological nomenclature belonging to group Ib). These impulses (see Fig. 3-5A, right side) do not impinge directly upon motoneurons, but on interneurons, which send their axons to α motoneurons. These interneurons have recently been anatomically identified, as will be discussed in the following section. The main function of the tendon organs has been thought to be to prevent overcontraction of the muscles and to act as a "brake" and at the same time facilitate the antagonists. However, recent physiological studies have shown that their effects at the spinal level are not limited to the well-known inhibition of synergists and facilitation of antagonists. Watt, Stauffer, Taylor, Reinking, and Stuart (1976) found, for example, that a considerable proportion of the synergistic motoneurons were excited, while many antagonistic motoneurons were inhibited. These authors concluded, in agreement with several others, (see Reinking, Stephens, and Stuart, 1975; Lundberg, 1975; Jami and Petit, 1976) that the functions of the tendon organs are more complex and subtle than was previously believed.

In addition to impulses from the tendon organs, there is another mechanism that tends to limit the activity of the excited motoneurons, the so-called *recurrent or Renshaw inhibition*. This mechanism, illustrated in Fig. 3-5B, was constructed on the basis of physiological observations. When a motoneuron fires impulses, these will pass via its recurrent collaterals to cells with short axons situated in the ventral horn. These cells, called Renshaw cells, have an inhibitory effect on the motoneurons. The Renshaw cells, as with many other interneurons in the cord, are subjected to supraspinal control, for example from the cerebellum and the mesencephalic reticular formation (see Granit, 1975). In the monkey, a large proportion of the descending fibers acting on the Renshaw cells are said to be inhibitory (McCouch, Liu, Chambers, and Yu, 1970). It seems indeed functionally meaningful that higher levels of the nervous system are able to switch off the recurrent inhibition of motoneurons, for example during certain movements requiring more than a brief muscle contraction.

The above account of some features of what is known of the function of the motoneurons, their role in movements, and in the maintenance of muscular tonus and myotatic reflexes, shows that there are very complex problems still far from being solved. Further indications of the existing complexity come from some recent data which will be briefly outlined below.

Some features of the functional organization of motoneurons.   It appears from the preceding survey that the performance of even a small movement must be an extremely complicated act. It requires a harmonious collaboration of a number of motor units, distributed over several segments of the cord and supplying different muscles. The activity of all these units must be coordinated in time, as well as with regard to intensity; facilitation and inhibition of the units must be precisely timed. The complexity is well illustrated by the fact that part of a muscle may be employed in different, even opposed movements. Thus certain parts of the deltoid muscle may function as abductors in one position of the shoulder joint, as adductors in another position. No wonder we are still far from understanding these functions! Recent research has contributed in a general way to our comprehension of how the intimate collaboration between motoneurons and related cells can be achieved, but on the other hand, these studies have revealed a complexity in the minute organization which so far escapes analysis. Some of these data are of interest from the point of view of a general understanding of the nervous system and will therefore be reviewed.

The columnar and somatotopical arrangement of motoneurons is a fundamental feature of the organization of the spinal cord. However, this refers only to the cell bodies. Further data are needed if we are to understand how they are acted upon and cooperate.

As previously mentioned, dendrites of motoneurons may extend for some distance beyond the confines of lamina IX, into laminae VII and VIII. Most of the richly branching dendrites of the motoneurons are, however, arranged longitudinally within the cell column to which the cell belongs (Scheibel and Scheibel, 1966a), particularly in the columns innervating the trunk and proximal parts of the extremities (Sterling and Kuypers, 1967b). The dendrites of a single cell may overlap those of many hundreds of adjacent motoneurons.

The majority of the motoneuron dendrites in cat and monkey appears to be arranged into longitudinal bundles, containing from 4 to 25 closely packed dendrites (Scheibel and Scheibel, 1970). In light of some evidence that motoneurons may influence each other electrically in the cat (Nelson, 1966), it is of interest that electron microscopic studies (Matthews, Willis, and Williams, 1971) demonstrate sites with close apposition of motoneuron dendrites (gap about 18 nm). However, typical gap junctions, usually considered to be the site of electronic coupling, were not found between the motoneurons.

As to the dorsal root fibers supplying the motoneurons monosynaptically (muscle spindle afferents, perhaps others as well), it has been shown experimentally that a single dorsal root afferent may spread its collaterals over several segments (for a comprehensive review, see Réthelyi and Szentágothai, 1973). From Golgi studies of *longitudinal* sections of the cord it has been learned (Scheibel and Scheibel, 1969) that these primary afferent fibers run longitudinally for a considerable distance near or in the dorsal columns, giving off fibers at regular intervals in a ventral direction to reach the motor nuclei. These branches are found close together in so-called microbundles, the fibers of which arborize mainly in the transverse plane within the motor nuclei (see, however, Sterling and Kuypers, 1967a)—that is, at right angles to the direction of the motoneuron dendrites. The above arrangement makes it likely that afferents from one spindle may influence motoneurons in several segments of the cord. In fact, it has been shown physiologically that one primary spindle afferent fiber activates nearly all motoneurons innervating the muscle in which the spindle lies (see Henneman, 1974).

In addition to the dorsal root fibers sending collaterals that end monosynaptically on the motoneurons, a much larger number of primary afferent fibers give off a dense plexus of collaterals in the medial and central parts of laminae IV–VI, and there is another plexus in the transition between laminae VII and VIII.[5] Also within these regions, one primary afferent will usually give off

---

[5] With the use of intra-axonal injections of horseradish peroxidase in dorsal root afferents, Brown and Fyffe (1978) were able to describe in great detail the distribution in the spinal cord of fibers identified with some certainty as primary spindle afferents. The transverse collaterals from each longitudinally running stem fiber appear to make synaptic contacts in three well-defined places: in the medial part of lamina VI, in restricted parts of lamina VII (apparently in the region of the interneurons mediating the reciprocal inhibition from primary spindle afferents), and among the motoneurons in lamina IX.

collaterals at various levels, and thus may act on interneurons in several segments. Furthermore, studies of these interneurons (see also below) with the Golgi method (Scheibel and Scheibel, 1966a, b; Mannen, 1975) show that their axonal branches and collaterals extend longitudinally for 1 to 4 segments, as concluded by Szentágothai (1951) from experimental studies.

Obviously, fibers in one dorsal root, establishing contact with interneurons, may via these influence motoneurons in several segments of the cord. A similar arrangement appears to be valid for other afferents to the motoneurons, for example the corticospinal, as described by Scheibel and Scheibel (1966b). The motoneurons are thus contacted by a great number of axons and collateral branches.

Early silver impregnation studies demonstrated that the motoneurons are amply provided with terminal boutons, but for technical reasons these could be identified only on the soma and the proximal dendrites. Bouton stains which essentially impregnate mitochondria have shown that boutons are present even on fine dendritic branches. In fact, they appear to be at least as densely placed on these as on the cell body, covering 50–70% of the cell's entire surface area (Illis, 1964a in the cat; Gelfan and Rapisarda, 1964, in the dog). Electron microscopic studies in monkey and cat have likewise shown the presence of boutons on all parts of the soma and dendritic tree of motoneurons, and the density increases as one proceeds distally so that the distal branches, 0.3–1.0 μm thick, are tightly covered with boutons (Bodian, 1964, 1975; Conradi, 1969; McLaughlin, 1972a).[6] The total number of boutons on a single cell amounts to several thousand in the cat (see Illis, 1964a, for references). Using a silver impregnation method on human autopsy material, Minckler (1940) calculated that in man there are some 1475 boutons per cell, presumably much too low a figure. Electron microscopic studies show that the motoneurons are contacted by seven morphologically distinct bouton types (Bodian, 1966b; Conradi, 1969; McLaughlin, 1972a). Their distribution differs somewhat on various parts of the cell surface and between cells of different sizes. For example, the synaptic density is higher on the cell somata of the large than on those of the small cells, which is of some interest because the small cells are more excitable than the large. The size of the boutons varies from about 0.5 to 5 μm, in agreement with previous light microscopic estimates. After dorsal rhizotomies, no degenerating synaptic knobs are found in the motoneuron pools (see, however, Bodian, 1975), but a particular type of large bouton is said to disappear (Conradi, 1969; McLaughlin, 1972b). These boutons constitute only a tiny fraction of the total population contacting the motoneurons, but their disappearance is easily detected because of their conspicuous appearance. However, whether a similar loss of the much more numerous small boutons takes place would be much more difficult to establish, as pointed out by Conradi (1969).

From the Golgi studies referred to above, as well as from physiological experiments (see Henneman, 1974), it has been suggested that the primary afferent fibers terminate on widely different parts of the motoneuron dendritic tree. By iontophoretic application of cobalt chloride to cut dorsal roots, Iles (1976) was able to trace several primary afferents to their terminations in lamina IX. The fibers often ended with two or three large boutons close together on one dendrite. Both proximal and more distal parts of the dendritic tree were apparently contacted, but most boutons were found proximally. Intracellular HRP injections of fibers physiologically identified as muscle spindle primary afferents as well as of motoneurons in the same animal (Burke, Walmsley, and Hodgson, 1979) strongly indicate that these fibers terminate on proximal as well as distal parts of the motoneuron dendrites; the latter site of termination appears to be the most frequent. As mentioned above, only a small fraction of the boutons contacting motoneurons appears to be of dorsal root origin. In the cat, long descending fibers ending on motoneurons are apparently few and end only on dendrites (McLaughlin, 1972c). After lesions of the motor cortex in the monkey, Bodian (1975), by electron microscopy, found small degenerating boutons contacting motoneurons, thereby confirming previous light microscopic observations and numerous physiological reports of monosynaptic corticomotoneuronal connections in primates (see Chap. 4). Propriospinal fibers appear to terminate on cell somata as well as dendrites. Judging from the large number of normal

---

[6] These findings are of some relevance for physiological interpretations, since there has been some dispute about the functional potency of synapses on peripheral dendrites. The ample presence of such boutons is a morphological argument in favor of the view that dendritic synapses must be active.

boutons persisting in motor nuclei immediately below a cord transection, McLaughlin (1972c) concluded that the majority of cells making synapses on motoneurons must be situated in their immediate vicinity.

The terms *interneurons* and *internuncial cells* have been mentioned repeatedly. Traditionally these cells have been assumed to belong to the so-called Golgi II type, being characterized by an axon that breaks up into a rich tree of branches and collaterals close to the perikaryon. This concept has been used in the explanation of many physiological observations. However, recent Golgi studies have shown that such cells are rather rare. In several regions, for example the reticular formation (see Chap. 6), they have not been found.

In an exhaustive Golgi study of the spinal cord in a series of mammals, including monkey and man, Scheibel and Scheibel (1966a) found a small number of such cells only in the dorsal horn. In all other regions of the cord the cells with short branching axons always have, in addition, a long branch (see also Scheibel and Scheibel, 1966b). This may take a different course depending on the situation of the perikaryon. Thus many of the cells situated medially in the ventral horn send their long axonal branch across the midline, while in most other regions the long branch takes a longitudinal course and may extend through several segments. Furthermore, these branches give off collaterals along their course and thus may establish synaptic contact with a number of nerve cells. Many of these ascending or descending longer branches leave the gray matter and run in the deepest region of the white matter and thus belong to the cell category usually referred to as *propriospinal neurons* (see p. 64).

Recently, elegant experiments, combining physiological and anatomical methods, have supported this general conclusion about spinal interneurons, which was based on Golgi material (for a review, see Jankowska, 1975). After physiological identification, the neuron is injected intracellularly with the fluorescent dye Procion yellow or horseradish peroxidase. The latter has the advantages of being suitable for electron microscopical examination and of giving a better visualization of the axons (Jankowska, Rastad, and Westman, 1976; Snow, Rose, and Brown, 1976; Cullheim and Kellerth, 1976), even to their terminals. The axonal pattern of spinal interneurons in laminae V and VI monosynaptically activated from group Ia muscle spindle afferent and group Ib tendon organ afferents was recently mapped in detail by Czarkowska, Jankowska, and Sybriska (1976). Most of these cells were found to send their stem axon into the white matter, usually ipsilaterally, while giving off collaterals on their way through the grey matter, for example to the motor nuclei in lamina IX. The same appears to be the case for the interneurons in lamina VII mediating the reciprocal inhibition of motoneurons (Jankowska and Lindström, 1972) and for the Renshaw cells (see Jankowska, 1975) referred to above (Fig. 3-5B). With the Procion yellow technique the Renshaw cells were located in the ventral part of lamina VII, just medial to the motor nuclei, thus confirming what was predicted on the basis of Golgi material by Scheibel and Scheibel (1966a) and Szentágothai (1967).

The Renshaw cell hypothesis further requires the termination on these cells of recurrent collaterals of the motoneuron axons. This problem can be studied anatomically, using the Golgi method. According to Scheibel and Scheibel (1966a), there may be as many as three to four such recurrent collaterals of one axon, but many motoneurons appear to lack them (see also Prestige, 1966). As to their distribution, it appears from Szentágothai's study (1967), in which superimposed drawings of Golgi preparations were used, that the main mass of the recurrent collaterals is distributed in the region where the Renshaw cells have been identified physiologically and anatomically (see also Szentágothai, 1958). However, about 20% of recurrent collateral terminations are found within the motor nuclei (Scheibel and Scheibel, 1966a; Cullheim, Kellerth, and Conradi, 1977), and at least some terminate upon homonymous α motoneurons, as shown by intercellular injection of horseradish peroxidase and electron microscopy by Cullheim, Kellerth, and Conradi (1977).

Physiologically, two types of α neurons have been distinguished, *tonic* and *phasic* (Granit, Henatsch, and Steg, 1956; and others), the tonic ones having, among other things, a lower rate of firing than the phasic ones. The axons of the former conduct more slowly than those of the latter and presumably, therefore, are thinner. The tonic cells are assumed to be smaller than the phasic ones, and to innervate smaller motor units. Mention has also been made of the γ motoneurons which are presumably among the smallest neurons in lamina IX. A histological study of the cell sizes in this lamina reveals transitional types, and it is not possible from the size of its perikaryon to decide whether a ventral horn cell may be a phasic or tonic α motoneuron, or whether another cell may be a relatively large γ motoneuron. Working on the assumption that there is a proportionality between the size of a perikaryon and the diameter of its axon, Henneman and his collaborators, in a series of studies on the functional properties of motoneurons (see Henneman, Somjen, and Carpenter, 1965a,b) in which motoneurons were excited reflexly, concluded that the excitability of spinal motoneurons varies inversely with their size. With all types of excitatory stimuli the recruitment of the cells occurred in the order of their size (as judged from amplitudes of potentials in recordings from axons), the smaller cells being excited before the larger ones. On the other hand, ''inhibitability'' of the motoneurons is directly related to size, the largest ones being most easily inhibited. Corresponding conclusions were drawn by Kernell (1966), who studied spinal motoneurons in the cat using intracellular recordings. However, factors other than motoneuron size also appear to be important. Thus, Harris and Henneman (1977) found that small motoneurons of the same size fall within two distinct groups with regard to firing rate. As pointed out by the authors, this may partly explain the fact that metoneurons of similar size may innervate muscle fibers with widely different contractile properties, since it is well established that motoneuron firing rate influences the contractile properties of fibers.

When a muscle is made to work, this is not the result of activation of one or a few motor units; a *"pool" of motoneurons* (Sherrington) is involved. Assuming that the synaptic input to this pool in any situation is relatively equally distributed to all of its cells, which are of various sizes, the sequence of activation usually observed (γ neurons—tonic α neurons—phasic α neurons) can be explained on a morphological basis. Furthermore, the ''various cells in a pool are fired rarely, moderately, or frequently according to their size'' (Henneman et al., 1965a, p. 578).

Thus, there is, as a rule, a fixed recruitment order among the motoneurons in a motoneuron pool, as is now well established. Nevertheless, it appears that humans, with visual or auditory feedback, are able to select voluntarily, at least to some degree, among low-threshold motor units in hand muscles (see Kato and Tanji, 1972). However, this does not invalidate the general principle of a fixed order of recruitment during ordinary movements. More appropriately, such voluntary control serves as an example of the precision of the human corticomotoneuronal connections when maximally focused, presumably by way of, among other things, surround inhibition (see Phillips and Porter, 1977, for a critical review).

Symptoms following lesions of the peripheral motor neurons.   It follows from the preceding account that when all the motor ventral horn cells which supply a certain group of striated muscles of the body are destroyed, the corresponding muscle or part of a muscle will not be able to contract. The most clear-cut examples of such lesions are probably those met with in poliomyelitis. The virus of this disease affects first and foremost the motoneurons, those of the spinal cord and those of the brainstem. In cases of the latter type there will be pareses or paralyses of cranial nerves such as facial palsy, pareses of ocular muscles, etc.

In a *paralysis due to a destruction of all or practically all motoneurons supplying the muscle, not only are all voluntary movements abolished, but also no reflex contractions can be elicited.* Furthermore, since the reflex arc of the myotatic reflexes is broken, *the muscles are flaccid;* they lack their normal tonus. Their consistency is reduced, and they offer no resistance to passive movements. This flaccidity is to a certain extent characteristic of such cases, and enables one to distinguish them from paralyses due to lesions of the central motor systems (cf. Chap. 4). Another characteristic is the occurrence of a *muscular atrophy,* which is more conspicuous the more complete the paralysis and the longer its duration. The reduced volume of the affected muscles can frequently be seen with the unaided eye, for instance when it affects the small musculi interossei of the hand. As the muscle tissue atrophies there is usually some proliferation of the intermuscular and intramuscular connective tissue.

If the peripheral motor neurons supplying a muscle are not all destroyed, only a partial paralysis, a paresis, will ensue, roughly proportional to the number of cells affected. By postmortem examination of the cord in such cases, the regions supplying the paretic muscles are found to contain a certain number of nerve cells which appear to be normal and have probably been functioning. The accompanying symptoms will be the same as in cases of paralysis, but they will be less severe. This applies to the *atrophy;* to the *muscular tonus,* which will not be abolished but only reduced; and to the myotatic *reflexes,* which may be present but weakened.[7] In such cases only the muscle fibers that belong to the affected nerve cells will atrophy (see also the anatomical findings mentioned above). The movements that can be performed will be more-or-less weak, and usually also *slower* than normal. This *retardation of movements* is explained by the reduced amount of innervation. In clinical cases there is a difference between the retardation of movements seen in cases of lesions of the peripheral motor neurons and in those pareses which are due to lesions of the central motor pathways. In the latter case the retardation appears to be much greater than can be accounted for by the frequently very moderate paresis. *In peripheral motoneuron lesions the retardation approximately parallels the reduction of muscular strength.*

The symptoms described above, paresis, atrophy, hypotonus, abolished or weakened reflexes, and retardation, will of course occur not only in lesions of the perikarya of the motoneurons; an interruption of their axons in the ventral roots, the plexuses, or the peripheral nerves will have similar consequences. The interruption may in some cases be only functional, without morphological destruction of the fibers. Thus intraspinal tumors may press on one or more roots, or a periph-

---

[7] When the tendon reflexes in the leg are weak or difficult to elicit, the Jendrassik maneuver may be of value. When the individual forcefully attempts to draw his interlocked hands apart, a tap on the petellar or achilles tendon may give a marked reflex response. Several studies have been devoted to the mechanism of this facilitation of the stretch reflex occurring in the Jendrassik maneuver. Somewhat different opinions have been expressed (see for example, Gassel and Diamantopoulos, 1964; Clarke, 1967; Hagbarth, Wallin, Burke and Löfstedt, 1975).

eral nerve may suffer by anoxia due to compression or circulatory disturbances.

Occasionally, in patients suffering from damage to the peripheral motor neurons, one finds so-called *fibrillary twitchings*. These are fine, rhythmical, or more often irregular twithcings of small groups of muscle fibers visible through the skin or to be felt by palpation. They become particularly evident when the appropriate part of the body is cooled. A more adequate name, now commonly used, is *fasciculations*. They can be recorded electromyographically, but their origin is not understood. They may occur in healthy persons, but the pattern is then as a rule somewhat different from that in disease (Trojaborg and Buchthal, 1965). It was previously commonly held that the occurrence of fasciculations indicated a primary affection of the motoneurons (for example, in amyotrophic lateral sclerosis or progressive spinal muscular atrophy) and that the symptom, therefore, may help in the differential diagnosis between a "nuclear" lesion and a "radicular" lesion, where the disease process attacks the ventral roots. However, the diagnostic value of the symptom appears to be rather limited (see Buchthal, 1962).

In *denervated muscles,* i.e., muscles that have lost their nerve supply, another phenomenon may be observed in the electromyogram: small action potentials occurring with a high frequency. These *fibrillation potentials* appear to arise from single muscle fibers which have lost their motor innervation. The contraction of the fibers can neither be seen nor felt. The origin of the fibrillation is not quite settled, but it has been maintained that it is due to a "sensitization" of the muscular part of the end plate following denervation (compare Chap. 11, on the autonomic nervous system). The absence of the normal transmitter, acteylcholine, may make the muscle fiber respond to the minute amount of acetylcholine present in the circulating blood. Prostigmin, which delays the breakdown of acetylcholine, tends to increase the fibrillation in denervated muscle. The presence of fibrillation potentials in the electromyogram of a paretic or paralytic muscle may be taken as evidence that some of its nerve fibers have been interrupted, whereas the presence of true motor units indicates that some nerve fibers are intact. However, fibrillation does not necessarily indicate denervation (see Buchthal, 1962). Nevertheless, fibrillation potentials have turned out to be useful diagnostic and prognostic criteria, particularly in cases of lesions of peripheral nerves.

The clinical symptoms of lesions of the motoneurons are of the same type whether the perikarya or the axons are damaged. The impulses starting the contraction are prevented from reaching the muscle in either case. However, even if the *types* of symptoms are the same, their *distribution* may be different, and in many cases it will be *possible to determine the exact site of the lesion when attention is paid to the localization of pareses, atrophy, and changes of reflexes.* Since this point is of practical importance it will be dealt with in some detail.

The segmental motor innervation. Lesions of the brachial plexus. The fibers of the different spinal nerves (their ventral branches) from certain parts of the cord are intermingled in the formation of the plexuses, from which they continue in the various peripheral nerves. Consequently, fibers from several segments of the cord are present in most of the peripheral nerves. In the radial nerve, for example, motor fibers from the ventral horn cells of the 5th cervical—1st thoracic segments are present. Figure 3-6 shows the peripheral distribution of motor fibers

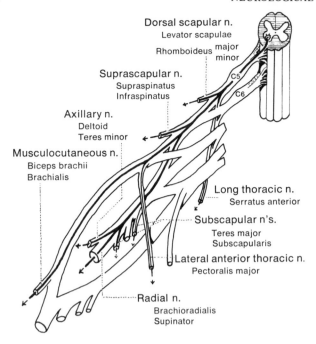

Dorsal scapular n.
Levator scapulae
Rhomboideus major
minor

Suprascapular n.
Supraspinatus
Infraspinatus

C5

C6

Axillary n.
Deltoid
Teres minor

Musculocutaneous n.
Biceps brachii
Brachialis

Long thoracic n.
Serratus anterior

Subscapular n's.
Teres major
Subscapularis

Lateral anterior thoracic n.
Pectoralis major

Radial n.
Brachioradialis
Supinator

FIG. 3-6   A diagram of the brachial plexus, showing how fibers from the 5th cervical segment are distributed to various peripheral nerves. The most important muscles supplied by these fibers are indicated. From Haymaker and Woodhall (1945).

from the 5th cervical segment. Thus the segmental arrangement of the fibers disappears during their course. However, *in the distribution of the peripheral nerves to the muscles, the segmental origin of the nerve fibers again becomes evident.* This is explained by the development of the neuromuscular system in ontogenesis. As will be recalled, the organism is, in the early stages of development, segmentally constructed. In the primitive metameres different parts are allotted to the future development of connective tissue and bones, of muscles, and of skin, and each of the metameres is related to one particular segment of the primitive spinal cord. Apparently the segmental pattern is completely lost during later development in most parts of the body. However, this is more apparent than real, and, particularly with regard to the peripheral innervation, it is actually retained in both the motor and the sensory innervation.

The segmental pattern of the striated muscles is clearly visible in the thorax with the intercostal muscles, and traces are seen also in the subdivision of the rectus abdominis muscle by its tendinous intersections. However, a relation to particular segments of the cord can also be demonstrated in other muscles. Most muscles receive their motor fibers from two segments of the cord, some smaller muscles only from one segment, and some of the larger from three or even, in some cases, from more than three neighboring segments. The various muscles get their motor (and sensory) innervation from the segments of the cord which correspond to the metameres from which they were originally developed.[8] In a lesion that comprises only a few segments of the cord or a limited number of ventral roots, the affected muscles will be those that developed from the corresponding metameres. In a lesion of a peripheral nerve, all those muscles that it supplies will be affected. The segmental innervation of the various muscles need not be memorized (complete tables are found in several textbooks, e.g., Monrad-Krohn's *Clinical Examination*), but certain principal points ought to be kept in mind.

Of most practical importance is the segmental innervation of the *upper extremity*. This may be diagrammatically regarded as developing as a lateral bud from the trunk at the level of the lower cervical segments, and consequently it is these nerves (more precisely from the 5th cervical to the 1st thoracic inclusive) which supply the muscles of the arm with motor fibers. Among these nerves the fibers arising from the 5th and 6th cervical segments are concerned mainly in the innervation of muscles situated proximally and laterally, e.g., the deltoid, supraspinatus and infraspinatus, teres major and minor, and the flexors on the anterior aspect of the upper arm, the coracobrachialis, biceps, brachialis, and brachioradialis (see also Fig. 3-6). In a lesion of the upper parts of the cervical enlargement of the cord or the 5th and 6th ventral roots these muscles will be paralytic (in addition to some smaller muscles of the neck). As will be seen, this distribution of paralysis (or paresis) does not correspond to that met with in leisons of any peripheral nerve. Some of these muscles are supplied by the axillary or circumflex nerve (the deltoid and teres minor), others by the musculocutaneous nerve (coracobrachialis, brachialis, and biceps), others by still other nerves. A paralysis of the distribution referred to above with muscular atrophy, hypotonus, and affection of the reflexes may be observed, for example, in cases of so-called *upper brachial plexus palsy* (Erb, Duchenne) which sometimes occurs in newborn babies on account of violent stretching of the plexus during delivery. (If a sensory loss is present, this will also be segmental; see description of the dermatomes, Chap. 2.) A similar distribution of the paresis or paralysis will be seen if the ventral roots of the 5th and 6th cervical nerves are damaged, for instance by an intraspinal tumor, or if the ventral horn cells of these segments are affected, such as may occur in poliomyelitis.

The extensors of the upper arm, triceps, and anconeus are supplied by the (6th) 7th and 8th cervical segments. The flexors and extensors on the forearm are mainly taken care of by fibers from the 7th and 8th cervical segments, the small muscles of the hand by the 8th cervical and 1st thoracic segments. Therefore, a lesion of the lower parts of the brachial plexus or of the ventral horn cells in the lower part of the cervical enlargement of the cord will result in paralysis or paresis predominantly of the muscles of the hand and forearm. This is seen in the so-called *lower brachial plexus palsy* (Déjerine-Klumpke), which may occur in new-born babies (although less frequently than an upper brachial plexus palsy). A localization of paresis and atrophy of this distribution is most frequently observed in cases of progressive spinal muscular atrophy, in which a slowly progressing degeneration of motor ventral horn cells occurs. The disease shows a predilection for starting in the upper thoracic and lower cervical segments of the cord, and, consequently, quite frequently the first symptoms noted are paresis and atrophy of the small muscles of the hand (the interossei and the muscles of the hypothenar and, somewhat less pronounced, of the thenar eminences). As the pathological process proceeds upwards in the cord, the muscles of the forearm and eventually those of the upper arm and shoulder girdle also become affected. In syringomyelia the same localization is also frequently observed, indicating that the pathological process is located in the lower cervical segments. A localization of the paresis of the type mentioned above

---

[8] The establishment of the segmental motor innervation has been achieved by different methods. Apart from gross dissections, the study of the retrograde changes in the ventral horn cells in cases of amputations, etc., has been utilized as memtioned in Chapter 1. Studies of cases of transverse lesions of the spinal cord have also given valuable information. Following electrical stimulation of the ventral roots during operations, the ensuing contractions of muscles may be observed and used for determining the segmental motor innervations. In recent years electromyography has been used to record the muscles activated. There appear to be considerable individual variations with regard to the segmental innervation of particular muscles, as shown for the lower limb, for example, by Thage (1965, 1974).

does not correspond to a lesion of any particular nerve. In an affection of the median nerve some of the flexors of the forearm will usually be involved, making flexion of the 2nd and 3rd fingers impossible, and the superficial muscles of the thenar eminence will contribute to making opposition as well as flexion of the thumb impossible or impaired. The interossei and the muscles of the hypothenar eminence, however, will not be affected. Lesions of the ulnar nerve are those most apt to cause diagnostic mistakes, since in those cases paresis and atrophy of the interossei and hypothenar muscles will be present. However, the thenar muscles will only be partly affected, since some of them are supplied by the median nerve (the ulnar innervating only the deep part of the flexor pollicis brevis and the adductor pollicis) and the accompanying sensory disturbance will as a rule settle the question.

In *the lower extremity* conditions are essentially similar with regard to the segmental motor innervation. The hip flexors (iliopsoas, rectus femoris, and sartorius muscles) and the adductors are innervated from the upper lumbar segments of the cord, mainly L_{2-3}, the gluteal muscles and those on the posterior aspect of the thigh from the (4th) 5th lumbar and 1st sacral segments, the muscles of the calf from the 5th lumbar–2nd sacral segments. The small muscles of the foot and the flexors of the toes are mainly supplied by the 1st and 2nd sacral segments (S_3 and S_4 are concerned in the innervation of the muscles of the pelvic diaphragm). Since localized affections of the roots of the lumbar and sacral nerves are not characteristic of certain diseases to the same extent as in the case of the cervical roots, the segmental innervation of the muscles of the lower extremity is of less importance than is that of the upper. However, the motor impairment may be an aid in the topographical diagnosis, e.g., in cases of tumors or protrusion of intervertebral discs (cf. Chap. 2). The distribution of the segmental pareses will as a rule not coincide with those due to lesions of the peripheral nerves, although the difference is on the whole not as marked as in the upper extremity.[9]

It will appear from the preceding account that an analysis of the impairment of motor function, its type as well as its distribution, may give important information with regard to the localization of a lesion. When a paralysis is found to be of the peripheral type, there is usually little difficulty in deciding whether it is due to a lesion of a peripheral nerve. In the latter case the determination of the exact place of the lesion, the segments of the cord affected, or the ventral roots involved is in some cases of practical importance, since this *segmental diagnosis* will determine at which place the operative measures have to be undertaken, for instance in cases of extramedullary tumors which lend themselves favorably to surgical intervention.[10]

In segmental diagnosis, it is necessary to consider that the spinal nerves pursue an increasingly oblique course within the vertebral canal as they approach the caudal nerves. Whereas the upper cervical nerves pass practically in a horizontal direction from the cord to their intervertebral foramina, the lower lumbar and the sacral nerves pass obliquely downward for a long distance within the dural sac before leaving the vertebral canal. This will be seen in Fig. 3-7. The explanation is found in the so-called "ascent of the medulla spinalis" in ontogenesis, since the growth of the spinal cord does not keep pace with that of the spine. If, therefore, a segmental motor impairment of the segments L_{2-4} has been ascertained, this may mean that the lesion is situated at the level of the corresponding vertebrae if the pathological process is near the intervertebral foramina, but if the cause is an in-

[9] For an exhaustive account of the distribution of pareses following lesions of particular ventral roots, see the monograph of Hansen and Schliack (1962).

[10] The condition of the reflexes may also aid in the segmental diagnosis. A list of the level of the reflex center of various reflexes is found in several textbooks, e.g., Monrad-Krohn's *Clinical Examination.*

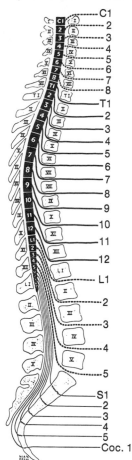

FIG. 3-7 A diagram showing the relation of the levels of the spinal cord and the spinal nerves to the corresponding vertebrae. From Haymaker and Woodhall (1945).

tramedullary lesion, e.g., a tumor or syringomyelia, the pathological focus will be found at the level of approximately the 11th to 12th thoracic vertebra. An intraspinal but extramedullary process, such as a tumor, a protruded intervertebral disc, or an arachnoiditis, may cause a distribution of this type when it is situated anywhere between the 11th thoracic and 4th lumbar vertebrae. It is particularly in the latter cases that an exact diagnosis of the level is of interest since surgical therapy must be considered. The final determination will be made by the examination of other features, such as signs from other ventral roots, accompanying sensory symptoms, autonomic disturbances, the conditions of the cerebrospinal fluid, myelography, and the history of the development of the symptoms.

Another point should also be stressed: since most muscles are innervated from more than one segment of the spinal cord, *a destruction of cells or fibers of only one segment will as a rule not betray itself clinically, or only when carefully sought.* In this case electromyographic examinations may give valuable information, as has been mentioned above, and sensory symptoms of irritation may also afford a clue (see Chap. 2).

4

Pathways Mediating Supraspinal Influences on the Spinal Cord The Basal Ganglia*

THE READER will have noticed that this book does not contain chapters with the headings "Pyramidal tract" and "Extrapyramidal motor system," found in most textbooks and used also in the first edition of the present text. This deviation from current usage is rather heretical and necessitates some comments.

"Pyramidal" and "extrapyramidal" motor systems. The cerebral cortex and a number of nuclei in the brainstem are able to influence the activity of the spinal cord since they give off fibers which either descend directly to the cord or end in other nuclei which in their turn send efferent fibers to the cord. These nuclei receive afferent fibers from other sources as well (for example, from the cerebellum), and the impulse transmission from "higher levels" to the cord via these nuclei will, therefore, be subject to modifications at those places where the pathways are synaptically interrupted. Furthermore, the cerebral cortex and all nuclei projecting, directly or via intermediate stations, to the cord will receive information from this, ultimately to a large extent derived from receptors of various kinds. Some of the fiber bundles that pass from supraspinal structures directly to the cord and may influence its activity are shown in Fig. 4-1. It seems likely, a priori, that these pathways are not functionally identical, an assumption which has been amply supported by recent neurophysiological research. It must be assumed,

* Revised by Alf Brodal and Eric Rinvik.

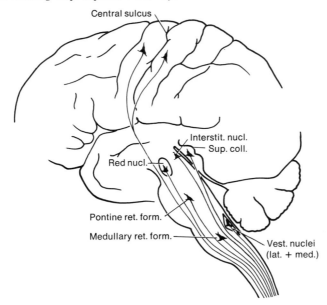

FIG. 4-1   Diagram of fiber systems descending to the cord from supraspinal levels:
the corticospinal, rubrospinal, tectospinal, interstitiospinal, pontine and medullary
reticulospinal and vestibulospinal tracts. Some minor tracts, such as the solitariospinal tract, are omitted. All of these connections may influence the activity of
the peripheral motor neurons.

further, that under normal conditions they cooperate in an integrated manner. An
analysis of this cooperation presents considerable difficulties, and much still remains to be done in this field.

It was formerly generally thought that the task of the descending fiber systems to the cord was to act on its efferent mechanisms, determining the impulse
discharges of its somatic and visceral efferent neurons. Since the 1950s it has been
conclusively demonstrated that the descending fiber systems in addition have
another important function, namely to regulate the transmission of sensory impulses from the periphery to the brain, as described in Chapter 2. This had indeed
been hypothesized by such clinical neurologists as Head, Brouwer, and others.
Furthermore, most of these fiber systems influence the reflex activity of the cord.
It is, therefore, no longer adequate—and not even permissible—to talk of the descending pathways collectively as "motor" fiber systems. The untenability of
such a concept is further borne out by the fact, referred to above, that the nuclei
giving off fibers which influence the efferent mechanisms of the cord receive ascending fibers providing them with sensory information, necessary for their proper
functioning. To deal with a "motor" function without considering its concomitant
sensory phenomena is a theoretical abstraction. As Jung and Hassler (1960, p.
864) phrase it: ". . . a 'motor system' without sensory control is a fiction, not
even a useful fiction."

The many new observations which have been provided by neuroanatomical
and neurophysiological research in the last two to three decades necessitate thor-

ough revisions of many classical concepts concerning the so-called motor systems. Unfortunately, many of these concepts have got a strong hold in our minds, in part, it seems, because they are simple and for this reason easy to use in routine work (although it is admitted that they are not adequate). Furthermore, the concepts are linked with certain names, and names and designations, as is well known, have a tendency to serve as convenient labels and help us to hide our ignorance. With reference to the subject of concern here, the terms "pyramidal" and "extrapyramidal" and "upper motor neuron" are particularly relevant, and deserve brief comment.

The *pyramidal tract* by definition is all those fibers which course longitudinally in the pyramis of the medulla oblongata, regardless of their site of origin or termination. The vast majority of these fibers come from the cerebral cortex (see below), and most of them continue into the spinal cord, making up the *corticospinal tract*. However, some fibers take off in the pyramid or at higher levels to pass towards motor cranial nerve nuclei in the brain stem. Strictly speaking these are thus not pyramidal tract fibers according to the above definition. Collectively they are often referred to as the *corticobulbar tract*. Since the latter appears functionally to correspond to the corticospinal tract, the term pyramidal tract may be—and often is—used as a common term for both. Although objections may be raised against stretching the definition in this way it may be advantageous to have a common name for the two components. The term "pyramidal tract," therefore, will be used in this sense in the following, and the names corticospinal and corticobulbar tract will be used when reference is made to the particular components. The term "corticobulbar tract" is sometimes used as including also those corticofugal fibers which pass with the corticospinal fibers and end in the bulbar reticular formation and other bulbar nuclei.

Anatomically the pyramidal tract is fairly well defined. This, however, does not justify the current use of the term *"pyramidal tract syndrome"* for a certain constellation of symptoms (see below). This term implies that the symptoms are due to damage to the pyramidal tract and only to this, while what is described as the pyramidal tract syndrome is actually those motor and reflex changes which are found after a lesion of the internal capsule or certain parts of the cerebral cortex. These, however, are never true, pure lesions of the pyramidal tract, since numerous other descending fiber projections coming from the same cortical regions as the pyramidal tract fibers and passing together with them in the internal capsule will necessarily be concomitantly damaged. The only lesions which would produce a pyramidal tract syndrome are interruptions of the fibers where they are isolated, i.e., in the pyramis of the medulla, as will be further discussed in a later section. The "pyramidal tract syndrome" is, therefore, a misnomer.

The term *"extrapyramidal"* is properly an anatomical concept and strictly speaking includes everything which is not "pyramidal." However, it has been currently used with the tacit qualification that it refers to "motor systems" (the extrapyramidal motor system), implying that the name covers all descending pathways and their intercalated nuclei, apart from the pyramidal tract, which may influence motor activities. As referred to above there are a number of such pathways, and in view of the fact that several other parts of the nervous system (such as the cerebellum) are well known to influence the function of the motor apparatus, difficulties arise as to the delimitation of the concept ''extrapyramidal'' structures. Some authors have included in it the pontine nuclei and the cerebellar nuclei, and particularly the basal ganglia are considered to form an integrative part of it. It is, however, not possible to delimit and define this ''system'' anatomically.

Clinical observations show that lesions of various regions of the brain give rise to disturbances of motor functions which differ from those of the ''pyramidal tract syndrome,'' and consist chiefly in changes in muscular tone and involuntary movements of various kinds. These clinical pictures are often lumped together as ''extrapyramidal diseases.'' They comprise those syndromes which are seen following lesions of the basal ganglia. Oscar and Cecile Vogt (1920) spoke of a ''striate system'' (including the striatum, pallidum, subthalamic nucleus, and other nuclei and tracts). As somewhat similar syndromes were found later to follow lesions of other parts of the brain as well, for example of the dentate nucleus, the concept of the ''extrapyramidal system'' expanded. In spite of the fact that these disorders may have some clinical features in common, it is difficult to see the advantages of considering them under a common heading, the more so since the ''system'' that they refer to escapes definition. The only reason for retaining the designation ''extrapyramidal diseases'' would be to avoid the problems of analyzing each particular syndrome on a rational basis.

The *''extrapyramidal system''* was in the past, by some kind of antithetic reasoning, usually thought of as something quite separate and different from the ''pyramidal system.'' It has for a long time been clear, however, that this distinction is untenable. It is impossible—and if it were possible, it would be artificial—to separate the function of the pyramidal tract from those of other descending fiber systems. In the intact being they must cooperate. Anatomically a separation is equally impossible. For example, as will be described below, those cortical regions which give rise to fibers in the pyramidal tract also give off fibers to a number of nuclei which project further caudally, and which are included in the ''extrapyramidal system.'' Furthermore, there are reciprocal connections between many of these nuclei, which bear further witness of possibilities for interaction among the nuclei and with the pyramidal tract. These multitudes of connections make it clear that in almost every lesion of the brain a number of fiber connections will be involved simultaneously, making it precarious to attribute a particular clinical feature to the lesion of a particular structure. While this may be disheartening insofar as it makes it difficult to give a certain constellation of symptoms a name which refers to a specific part of the brain, the recognition of these complexities is certainly necessary if progress is to be made in the analysis of clinical disturbances of motor functions.

The expression *''upper motor neuron''* is sometimes employed—particularly, it appears, in connection with symptoms attributed to damage to the pyramidal tract (upper motor neuron disease). Since there are many ''upper motor neurons,'' the use of the term in this sense is misleading, and it might with advantage be avoided.

It is the author's firm conviction that neither theoretically nor practically is a useful purpose served by retaining the term ''extrapyramidal.'' In the light of present-day knowledge the term, as referring to structures and functions ''other than pyramidal'' is completely meaningless, as emphasized also by others, for example Russel Meyers (1953) and Bucy (1957). Accordingly, the presentation to be given here will not follow the orthodox lines, perhaps to the disappointment of some students. In the following an attempt will be made to describe the essential points in the anatomy and to some extent the physiology of the pathways descending to the spinal cord which may influence the activity of the motor neurons. As will be seen, in spite of the accumulation of many new data which have increased our understanding of the subject, it is often not easy to make correlations with clinical findings.

Changing views on the pyramidal tract. The pyramidal tract got its name from the medullary pyramid, where its fibers run in isolation. Its definition as generally accepted at present was mentioned above. The vast majority—possibly all—of its fibers come from the cerebral cortex.

The pyramidal tract appears to be the first large fiber bundle recognized as a particular tract in the brain. It was described by Türck in 1851 as extending from the cerebral cortex to the spinal cord, and Türck also observed that in patients suffering from spastic hemiplegia a bleeding in the internal capsule is often found postmortem. In 1870 Fritsch and Hitzig demonstrated that on electrical stimulation of the frontal lobe in dogs movements of the opposite limbs can be elicited. About the same time Hughlings Jackson (1875), on the basis of clinical observations, concluded that there must be a somatotopical "localization of movements in the brain." Around the turn of the century a number of studies were undertaken which further elaborated the pioneer observations. Thus Grünbaum and Sherrington (1902; 1903) by cortical stimulations and ablations showed the existence of a somatotopic pattern in the precentral gyrus in apes. Corresponding conclusions were drawn from clinical cases with cortical lesions. In 1909 Holmes and May reported that following transection of the pyramidal tract in the upper spinal cord in monkeys retrograde cellular changes occur only in the giant cells of Betz in Brodmann's area 4. The origin of the tract appeared then to be settled. As a result of these and other studies a notion of the pyramidal tract emerged which was generally held until about some forty years ago: The pyramidal tract arises from the precentral gyrus and descends to the cord. Its fibers are thick and myelinated. In the spinal cord they establish contact with the peripheral motor neurons; the tract is concerned in the mediation of voluntary, particularly finely isolated discrete movements. According to this conception the pyramidal tract is rather simply organized and has a clearly defined function.

Doubts on the correctness of this simple view on the pyramidal tract had been voiced by students in various fields of neurological research when Lassek and Rasmussen in 1939 counted the fibers in the pyramid of man, where most of the corticobulbar fibers have left it. In silver-impregnated sections they found about a million fibers in each pyramid. More recently DeMyer (1959) reported that the number varies between 749,000 and 1,391,000, with a mean value of 1,087,200 (21 cases). Since the number of Betz cells in Brodmann's area 4 is only some 25,000 to 30,000 (Campbell, 1905; Lassek, 1940) it appears that only between 3 and 4% of the fibers in the pyramid can be axons of Betz cells, even if, as pointed out by Lassek (1941) and others, there exists no definite criterion which makes it possible to characterize a Betz cell. There are transitions to large pyramidal cells which occur in the Vth layer of area 4 as well as in neighboring areas. However, there can be no doubt that *the vast majority of pyramidal tract fibers come from cells other than the Betz cells* in the precentral gyrus. This conclusion is in agreement with observations made by some earlier authors (Wohlfahrt, 1932; Levin and Bradford, 1938). Extirpation of area 4 in the monkey results in the disappearance of only about a quarter of the fibers in the pyramid (Häggqvist, 1937; Lassek, 1942a). Further studies have shown that pyramidal tract fibers take origin also from regions of the cortex other than area 4 of Brodmann, as will be discussed later.

In the 1940s Lassek published a series of studies of the pyramidal tract, especially in man. It appears that these studies have been substantial in directing renewed interest to a fiber tract which had for a long time been considered as one of the best known and most simply organized ones in the brain. The results of anatomical, physiological, clinical, and clinicopathological studies from the last thirty-five years have necessitated a complete reevaluation of the older concepts of the pyramidal tract. Before discussing present-day views on the pyramidal tract, its gross anatomy in man will be briefly described.

The course of the pyramidal tract.   The pyramidal tract fibers descend from the cortex to the internal capsule (see Fig. 4-9), where they are found collected in a relatively small area in the posterior limb. It is worth stressing their close relationship to the thalamus medial to them, and the lentiform nucleus laterally to them, as well as the presence in the internal capsule of other fibers, such as corticopontine, corticorubral, pallidothalamic, etc.

In Foerster's diagram of the internal capsule, reproduced in many textbooks, the pyramidal tract fibers are distributed over the entire posterior limb and the genu of the internal capsule. It appears, however, from some pathological anatomical studies (Marion Smith, 1960; Hirayama et al., 1962 Brion and Guiot, 1964; Marion Smith, 1967; Beck and Bignami, 1968; Hanaway and Young, 1977; and others) that in man the fibers occupy only a relatively small area of the internal capsule, and that this area is found posteriorly in its posterior limb (see Fig. 4-9). Studies of motor responses occurring during electrical explorations of deep brain structures in stereotactic surgery have led to the same conclusion (Guiot et al., 1959; and others; for references, see Hanaway and Young, 1977). Some data indicate that there is a somatotopic localization within the tract in the internal capsule (fibers to the cranial nerve nuclei most anteriorly; then follow fibers to cervical, thoracic, lumbar, and sacral segments successively more dorsally), but the somatotopic localization is not sharp.

Concerning the gross features of the pyramidal tract in its further downward course the following may be noted. It is often stated that the pyramidal fibers occupy the middle two-thirds of the *cerebral peduncle.* However, this is not reasonable, since in man corticospinal fibers account for only about 1 million of the approximately 20 million fibers in the peduncle on each side (Tomasch, 1969). The patterns of the distribution of the various corticopontine fiber contingents in the peduncle and of the somatotopic arrangement of corticospinal fibers shown in Fig. 4-2 (based on findings in man by Foerster, 1936b; in the monkey and rat by Bar-

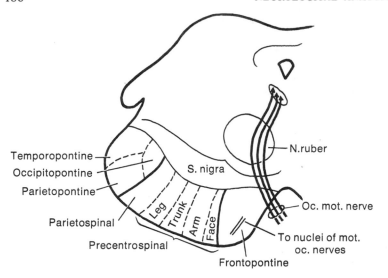

Temporopontine
Occipitopontine
Parietopontine
Parietospinal
Precentrospinal

S. nigra

Leg
Trunk
Arm
Face

N.ruber

Oc. mot. nerve

To nuclei of mot.
oc. nerves

Frontopontine

FIG. 4-2 Diagrammatic drawing of a section through the mesencephalon, indicating the position in the cerebral peduncle of the pyramidal tract fibers to the different parts of the body. Somewhat modified from Foerster (1936b) on the basis of findings made by other authors (cf. text).

nard and Woolsey, 1956, and others) are probably far less distinct than indicated. (For further data on the peduncle in man see Marin et al., 1962; Martinez et al., 1967.)

In the pons the fibers from the peduncle split into several bundles, which intrude between the large cellular masses of the pontine nuclei. The corticopontine fibers (to be considered in the chapter on the cerebellum) take off from it and end in the pontine nuclei. At the lower border of the pons the remaining fibers of the pyramidal tract unite and leave the pons as a distinct bundle, forming the *pyramid* of the medulla oblongata. Whether the fibers retain a somatotopic arrangement during their course in the pyramid has been questioned by some but advocated by other authors. Woodburne (1939) in anatomical investigations in animals (Marchi method) described a somatotopic arrangement in agreement with the conception of Foerster (1936b) as illustrated in Fig. 2-9. However, Tower (1944) and more recently Gilman and Marco (1971) are inclined to deny such an arrangement, at least in the pyramids of monkeys, on the basis of the functional defects seen following partial lesions of the medullary pyramid, and Barnard and Woolsey (1956) found no localization in the cord in monkeys. (It should be emphasized that the localization is certainly not as distinct as appears from a simplified diagram such as Fig. 2-11.)

During the course of the pyramidal tract in the brainstem some fibers are given off in order to supply the motor (somatic efferent) nuclei of the cranial nerves. It is of practical relevance that some of these fibers cross the median line, others not. This means that *some of the motor cranial nerve nuclei are influened by fibers from both cerebral hemispheres.* These conclusions are chiefly based on clinical observations. Making allowance for individual variations, it may be said that, as a rule, the hypoglossal nucleus, the nucleus of the accessory nerve, and the parts of the facial nucleus concerned with the lower part of the face appear to have purely crossed corticofugal connections. In some individuals the motor trigeminal nucleus belongs to the same group. Consequently only the corresponding muscles will be paralyzed in unilateral lesions of the corticobulbar fibers.

The *corticospinal tract* represents what is left of the pyramidal tract when it has passed the brainstem. The major part of its fibers crosses the pyramidal decussation. After having crossed, these fibers are found in the spinal cord in an area lateral to the dorsal horn as the *lateral corticospinal tract* (cf. Fig. 2-9; see also Fig. 1-2), and are presumably arranged in an orderly sequence, the shortest fibers being situated deeply, the longest most superficially. Some authors, in contrast to Foerster, maintain that in the spinal cord the somatotopic arrangement within the lateral corticospinal tract is entirely lost. The final word on this question has, however, not been said. A priori it seems reasonable that there should be some sort of somatotopic arrangement of the pyramidal fibers along their entire course, although the localization is scarcely as sharp as depicted in Foerster's diagram.

A small proportion of the pyramidal fibers descends uncrossed in the spinal cord; the percentage, however, varies individually. Some of the uncrossed fibers form a long recognized bundle, the *ventral corticospinal tract* in the ventral funiculus of the cord along the ventral median fissure (Fig. 2-9). This tract probably rarely descends below the thoracic part of the cord. In addition, however, *some uncrossed fibers join the crossed fibers of the lateral corticospinal tract.* The literature on the many individual variations of the human pyramidal tract has been reviewed by Nyberg-Hansen and Rinvik (1963). Some of these variations, particularly the ratio of uncrossed to crossed fibers, are of interest from a clinical point of view.

It is important to be aware, as shown already by Cajal, that many, if not all, corticospinal fibers give off abundant *collaterals* during their course. They pass to many nuclear masses such as the striatum, the thalamus, the red nucleus, the pontine nuclei, and the reticular formation. This has been concluded from physiological studies as well. The presence of such collaterals shows that the cortical cells that send information to the spinal cord may also influence the activity in several nuclei that via their efferent connections may act on the cerebral cortex, the spinal cord, or the cerebellum, and thus cooperate, by feedback arrangements or in other ways, with the direct corticospinal "commands" to the motoneurons. (It should be noted, however, that the various relay nuclei, for example the red nucleus, the pontine nuclei, and others, are activated from the cortex not only via collaterals of the pyramidal tract. In addition they receive direct fibers from the cerebral cortex. Functionally these direct connections may be at least as important as the collaterals from the corticospinal fibers.) In the spinal cord the corticospinal fibers likewise give off numerous collaterals. From an electrophysiological study (antidromic recordings) in the cat (Shinoda, Arnold, and Asanuma, 1976) it appears that 30% of the corticospinal fibers passing to the cervical cord give off branches to lower levels of the cord. Such axonal branchings have been demonstrated also in the monkey (Shinoda, Zarzecki, and Asanuma, 1979).

Most of our present-day insight into the anatomical and functional organization of the pyramidal tract has been derived from experimental studies of animals.

However, the comparable data available for man indicate principal similarities, even though there may be some functional differences between species. Assuming that the pyramidal tract is of particular importance for the execution of delicate voluntary movements, a question to be discussed below, one might expect the number of pyramidal tract fibers to be especially large in man, as appears indeed to be the case (Lassek and Wheatley, 1945). A discussion of the function of the pyramidal tract and of the symptoms following its destruction in man will be postponed until other descending pathways from the cerebral cortex to the spinal cord have been considered.

Origin and termination of the pyramidal tract fibers. For anatomical studies of the cortical origin of the fibers of the pyramidal tract two approaches have been used. As referred to above, one may look for the occurrence of retrograde cellular changes in the cerebral cortex following transection of the fibers. The difficulty here is that not all affected cells will necessarily present changes which are definite enough to permit recognition. Another procedure is to make lesions in the cerebral cortex and to trace the efferent degenerating fibers, either with the Marchi method or with silver impregnation methods which enable one to identify not only myelinated but also unmyelinated fibers.[1] Studies of this kind are beset with other difficulties. It is almost impossible to produce lesions restricted to particular cytoarchitectonic areas of the cortex. Neighboring areas may be involved on account of vascular changes, and microscopic identification of the part destroyed or removed is difficult on account of the ensuing distortion of the tissue. Conclusions as to the origin of fibers from a particular area must, therefore, be evaluated with some caution. In spite of technical difficulties a large number of studies devoted to the origin of the fibers of the pyramidal tract established the main points. It is obvious from these that *the area 4, the classical "motor area," is not the only source of corticospinal fibers.* In various animal species such fibers have been traced from area 6, in front of area 4 (Levin and Bradford, 1938, Russell and DeMyer, 1961, in the monkey; Minckler, Klemme, and Minckler, 1944, in man), while there appears to be no conclusive evidence that the areas in front of area 6 contribute. In the cat, however, Nyberg-Hansen (1969b), using the Nauta method, traced pyramidal tract fibers from the gyrus proreus (which appears to correspond to areas 8–12, often referred to as the prefrontal cortex, in primates). In all species studied the parietal cortex, particularly areas 3, 1, and 2 of Brodmann, the "primary sensory cortex," is an important source of pyramidal tract fibers (Levin and Bradford, 1938; Peele, 1942; Barnard and Woolsey, 1956; Kuypers, 1958 a, b, c, 1960; Russell and DeMyer, 1961; Nyberg-Hansen and Brodal, 1963; Liu and Chambers, 1964; and others). Furthermore, the parietal cortex (areas 5 and 7) sends fibers to the spinal cord. The second somatosensory area (see Chap. 2 and below) projects to the spinal cord, at least in the cat (see Nyberg-Hansen, 1969b, for references). Likewise there is in the cat a corticospinal projection from the supplementary motor region on the medial surface of the hemisphere (see below). The small contributions from the temporal and oc-

---

[1] In recent years anatomical methods based on the anterograde and retrograde axonal flow (see Chap. 1) have become available and have been used for studies of the pyramidal tract (see below).

cipital cortices in the cat, found with the method of Glees (1946) by Walberg and Brodal (1953a), could not be identified using the Nauta method (Nyberg-Hansen, 1969b). Recent studies of the distribution of labeled cells within the cortex after HRP injections in the spinal cord of the monkey (see, for example, Coulter, Ewing, and Carter, 1976) have generally confirmed the results obtained with other methods. Attempts have been made to determine the relative proportion of fibers from the various cortical regions of origin.

Quantitative assessments of degenerating fibers are difficult. Another approach to the problem is to count the number of normal fibers left after sufficient time has passed for complete disappearance of the fibers coming from an ablated cortical area. Studying monkeys surviving for one year after ablation of various cortical areas, Russell and DeMyer (1961) concluded that in this animal 31% of the descending fibers in the pyramid come from area 4, 29% from area 6, and 40% from the parietal lobe. Coulter, Ewing, and Carter (1976) found that of the cortical pyramidal cells labeled within the primary sensorimotor area (Sm I) and the adjoining cortex in the monkey following injections of horseradish peroxidase in the lumbar cord, 22% were in area 4, 48% in area 3, 17% in area 1, 7% in area 2, and 6% in area 5.

Even though the primary sensory and motor cortices are the most important sites of origin of corticospinal fibers in all mammals, there appear to be species differences. Thus areas 5 and 7 appear to give rise to more corticospinal fibers in monkeys than in the cat, whereas the reverse seems to be the case for the premotor cortex.

The view that most corticospinal fibers are axons of pyramidal cells in layer V of the cerebral cortex has recently been confirmed in studies in which cortical neurons were labeled after injections of horseradish peroxidase in the lumbar cord of cats and monkeys (Coulter, Ewing, and Carter, 1976). Large (including Betz cells in area 4) as well as smaller pyramidal cells were labeled in all parts of the primary sensorimotor cortex (see also Berrevoets and Kuypers, 1975). Jones and Wise (1977) in a careful study in the monkey further found that the cells of origin of corticospinal fibers show a tendency to be arranged in clusters of a few cells, separated by gaps, in only the deeper parts of layer V (see their paper for particulars). The studies above also confirm that the cell groups projecting to different levels of the cord show a somatotopic pattern (see below).

The *somatotopic arrangement* of the cells of origin of the corticospinal tract, corresponding in the main to the pattern of the endings of somatosensory pathways in the cerebrtal cortex (see Chap. 2), will be discussed later in this chapter (see Figs. 2-21 and 4-15).

The fibers of the pyramidal tract are of greatly varying diameter and consequently have different conduction velocities. A large number of the fibers are thin. Lassek (1942b) found that in the human pyramid, just above the decussation, 90% have a diameter of only 1–4μm, whereas not more than 1.73% are between 11 and 22μm. According to Lassek (1942b) about 60% of all pyramidal fibers are myelinated in man, while DeMyer (1959), using another method, found about 94%. The number of large myelinated fibers thus corresponds approximately to the estimated number of Betz cells in area 4, and it is also worthy of notice that, following extirpation of area 4, almost all large fibers disappear (Häggqvist, 1937; Russell and DeMyer, 1961). It appears probable indeed that the largest fibers in the tract are the axons of the largest pyramidal cells of the cortex, the Betz cells.

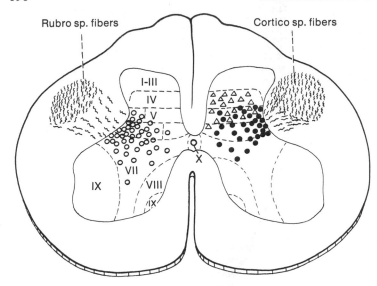

○ Sites of termination of rubro sp. fibers
● Sites of termination of cortico sp. fibers from "motor" cortex
△ Sites of termination of cortico sp. fibers from "sensory" cortex

FIG. 4-3   Diagram of a transverse section of the spinal cord of the cat at C_8 showing the location in the lateral funiculus and the sites of termination within the spinal gray matter of rubrospinal and corticospinal fibers from "motor" and "sensory" parts of the first sensorimotor region of the cerebral cortex. Note the similarities in the areas of termination of rubrospinal and corticospinal fibers from the "motor" cortex. From Nyberg-Hansen (1966a) based on experimental studies of Nyberg-Hansen and Brodal (1963, 1964).

The largest Betz cells are found in the lower part of area 4 (Lassek, 1941) which projects onto the lumbosacral cord, a fact supporting the general assumption that there is a relationship between the size of a nerve cell and the length of its axon.

The conduction velocities of corticospinal fibers (for references, see Phillips and Porter, 1977; Humphrey and Corrie, 1979) have been found to vary widely (from 7 to 70 m/sec in the cat). There are two velocity peaks within this range, at about 14 m/sec and 42 m/sec. The situation in the monkey appears to be rather similar. On the basis of such findings a distinction is sometimes made in physiological studies between "fast" and "slow" corticospinal fibers. The "fast" fibers are presumably axons of large pyramidal cells, including the Betz cells. It appears that the "fast" fibers are among those that supply the more ventral regions of the "intermediate zone" of the cord (see below and Fig. 4-3). The "slow" ones terminate more dorsally, their main function may be to influence the central transmission of sensory impulses and spinal reflexes.

By means of silver impregnation methods the *sites of termination of the corticospinal fibers* have been determined fairly precisely. Even the first studies of this kind made it clear that in the cat few, if any, of the corticospinal tract fibers establish contact with the large motoneurons. The degenerating boutons and terminal arborizations of degenerating fibers were uniformly found more dorsally, in what was usually referred to as the "intermediate zone" of the spinal gray matter and the base of the dorsal horn (Hoff and Hoff, 1934; Szentágothai-Schimert, 1941a; and others). These observations have been confirmed by Chambers and Liu (1957) and Kuypers (1958a) and agree with the results of physiological studies by Lloyd (1941) in the cat, that pyramidal tract impulses activate the motoneurons by way of internuncials. More precise information has been obtained by making use of the architectonic map of Rexed (1952, 1954). Nyberg-Hansen and Brodal (1963), using the Nauta method, found the corticospinal fibers from the "central region" in the cat to end in laminae IV to VII, chiefly laminae V to VI (Fig. 4-3), where the fibers of the lateral corticospinal tract enter the gray matter.[2] No fibers were traced to lamina IX. Studies of Golgi-impregnated sections (Scheibel and Scheibel, 1966b) show this distribution clearly (Fig. 4-4). Largely corresponding results were obtained in autoradiographic studies by Flindt-Egebak (1977). Some fibers end in corresponding regions on the other side. Further, the fibers coming

[2] In an electron microscopic study in the cat, Dyachkova, Kostyuk, and Pogorelaya (1971) found degenerating boutons on cells in the lateral part of Rexed's lamina V, mainly on their dendrites, following ablation of the contralateral sensorimotor cortex.

FIG. 4-4 Synaptic presentation of findings made in Golgi sections of the cat's spinal cord. Rexed's laminae are indicated. On the right side are seen corticospinal fibers and collaterals (A) entering the gray matter in a fan-shaped pattern, mainly within laminae V-VII. On the left side various types of cells in the gray matter are seen. These may be contacted by terminals of corticospinal fibers. Some cells (f) send their axon into lamina IX, harboring motoneurons. Note that some (e) send their axon across the midline. Interneurons, c, d, and e are situated within the terminal area of corticospinal fibers. From Scheibel and Scheibel (1966b).

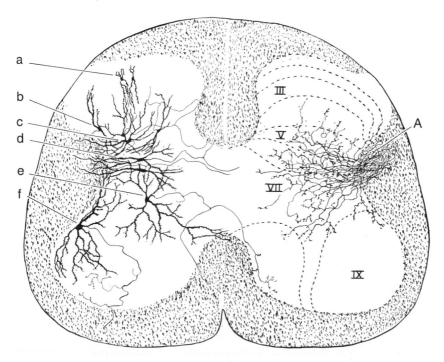

from the cortical region supposed by most authors to correspond to the precentral
"motor" cortex in monkeys and man, differ with regard to their distribution from
those coming from the "postcentral, sensory" cortex. The former fibers, which
are on the whole thicker than those from the sensory cortex, and appear to belong
to the "fast" corticospinal fibers, end chiefly in lamina VI, especially laterally,
and in the dorsal part of lamina VII (Fig. 4-3). The latter fibers terminate more
dorsally, chiefly in laminae IV and V, with some overlapping onto lamina VI.
These findings were confirmed in physiological studies (Fetz, 1968). A corresponding differential distribution appears to be present in the monkey, according
to Kuypers (1958b), Liu and Chambers (1964), and Kuypers and Brinkman
(1970). By means of autoradiographic tracing of tritiated leucine and proline,
Coulter and Jones (1977) could establish further details in the pattern. Some fibers
from cortical area 3a and in part from area 4 were found to end in the region of
motoneuron cell groups in the monkey. Further, there appear to be differences between the sites of termination of fibers from areas 3, 1, and 2 (Coulter and Jones,
1977). *Even though there is some overlapping between the terminal areas of fibers
from the "motor" and "sensory" cortex, they obviously have their preferential
sites of termination,* an observation which is of interest for functional correlations.
The corticospinal fibers from the supplementary motor area in the cat end in the
same regions as do those from the primary sensorimotor cortex (Nyberg-Hansen,
1969a).

Thus in the cat it appears that corticospinal fibers from the "central region"
can act on motoneurons only by way of internuncials,[3] as is concluded also from
electrophysiological studies (Lloyd, 1941; Hern, Phillips, and Porter, 1962; and
others). In the monkey, however, some corticospinal fibers have been found to
end on motoneurons (Hoff and Hoff, 1934; Kuypers, 1960; Liu and Chambers,
1964), and there is physiological evidence of a monosynaptic (in addition to
polysynaptic) impulse transmission from corticospinal fibers to alpha motoneurons
(see Bernhard, Bohm, and Petersen, 1953; Preston and Whitlock, 1961; Landgren, Phillips, and Porter, 1962, and below).[4] Little is known about the situation
in man, but some observations by Kuypers (1958c) and Shoen (1964) indicate that
in man some corticospinal fibers end on motoneurons.

As mentioned in Chapter 3, the motoneurons of the cord are somatotopically
arranged, those supplying the axial musculature being found most medially
throughout the cord, those for the extremities in more laterally and dorsally situated cell columns, present only in the cervical and lumbar intumescences. Furthermore, the motoneurons for the distal muscles (hand and foot) are found more dorsally and caudally than those for the proximal muscles of the extremities (cf. Fig.
3-1). In the sensorimotor cortex, the areas "representing" the face and the distal
parts of the extremities are particularly large (see below). When these regions are

---

[3] Since some dendrites of motoneurons extend rather far dorsally into the gray matter, at least
into laminae VI and VII (Aitken and Bridger, 1961; Sprague and Ha, 1964; Scheibel and Scheibel,
1966a) these findings do not definitely exclude the possibility that contacts may be established with
dendrites of motoneurons. As mentioned in Chapter 3, most of the dendrites of motoneurons take a
longitudinal course.

[4] After removal of cortical areas 4 and 6 in the monkey, Bodian (1975) identified electron microscopically some degenerating boutons on motoneurons in lamina IX. For some data on electron microscopic studies of synaptic connections of motoneurons, see Chapter 3.

damaged, the number of corticospinal fibers that degenerate is large. It is well known that a great majority of corticospinal fibers terminate in the intumescences of the cord. Kuypers and Brinkman (1970), in a careful study of the corticospinal fibers from the precentral cortex in the monkey, have made some noteworthy observations. They found that the rostral (anterior) part of the precentral cortex gives rise mainly to fibers to the ventromedial part of the "intermediate zone" of the gray matter of the cord, whereas the posterior part (nearer the precentral sulcus and within this) supplies mainly the dorsolateral parts of the "intermediate zone," that is, regions in proximity to the motoneuron pools for proximal and distal extremity muscles, respectively. Fibers terminating in the area containing motoneurons (Rexed's lamina IX), and in part presumably establishing monosynaptic contacts with the latter, are predominantly localized to areas harboring motoneurons for distal extremity muscles. These anatomical observations are in accord with conclusions drawn from physiological experiments.

As to *the pyramidal tract fibers influencing the somatic and special visceral efferent cranial nerve nuclei,* information is relatively scanty. However, in the cat, fibers from the central region have not been found to end in the motor cranial nerve nuclei but only in their vicinity (Walberg, 1957b; Kuypers, 1958a; Szentágothai and Rajkovits, 1958), while in the monkey and chimpanzee some fibers appear to end in the nuclei (Kuypers, 1960).

We will return to the function of the pyramidal tract at a later junction, but it is appropriate to mention here some electrophysiological studies which have given interesting new data. Thus it has been shown that the corticospinal fibers in general have a facilitatory action on motoneurons which supply flexor muscles, while the action on extensor motoneurons is usually inhibitory (Lundberg and Voorhoeve, 1962; Agnew, Preston, and Whitlock, 1963; Corazza, Fadiga, and Parmeggiani, 1963; and others). Furthermore, their action on the γ-neurons appears to be similar (Corazza et al., 1963; Grigg and Preston, 1971). Only studies in which precautions are taken to exclude simultaneous transmission of impulses from the cerebral cortex to the cord along other pathways are conclusive in this respect.

All authors appear to agree that in the cat the effect on both α and γ neurons is mediated via internuncial cells, in agreement with the anatomical data. The internuncials involved appear to be located in laminae V–VII (see Fig. 4-4). Internuncial cells involved in the reflex arcs for certain somatic spinal reflexes appear to be facilitated by impulses in the corticospinal tract (Lundberg, Norsell, and Vorrhoeve, 1962). In the cat the excitatory effect of corticospinal fibers on the motoneurons occurs disynaptically (Illert, Lundberg, and Tanaka, 1974), and the interneurons involved appear to be influenced both from primary spindle afferents and from other descending fiber systems (Illert, Lundberg, and Tanaka, 1975, 1976). It appears that these interneurons have important integrative functions.

On account of the monosynaptic contact of some pyramidal tract fibers with motoneurons in monkeys, studies in these species are of particular interest. A wealth of critically evaluated information can be found in the monograph of Phillips and Porter (1977).

As mentioned above, in physiological studies in the monkey, corticospinal fibers have been found to excite motoneurons monosynaptically. Inhibition, however, appears to occur disynaptically, via intercalated neurons (Preston and Whit-lock, 1961; Landgren, Phillips, and Porter, 1962; Phillips and Porter, 1964; and others). This is compatible with the demonstration that many corticospinal fibers terminate outside Rexed's lamina IX. Jankowska, Padel, and Tanaka (1976) found that the inhibition is effected via the same interneurons that mediate the inhibition of motoneurons in response to afferent volleys in Ia primary sensory fibers (see Chap. 3). Interneurons, apparently in part propriospinal neurons, appear to play important roles in the integration of impulses from various sources which finally influence the motoneurons. For a recent study see Illert, Lundberg, and Tanaka (1977).

The excitatory and inhibitory actions of corticospinal neurons on mo-toneurons and their influence on both α and γ motoneurons are obviously related to the nervous mechanisms of movement and their control—problems that have turned out to be extremely complex. For a discussion of these, the reader is re-ferred to special texts or reviews (see, for example, Granit, 1970; Brooks and Stoney, 1971; Phillips and Porter, 1977). It should again be emphasized that in these mechanisms the afferent information to regions concerned in motor functions plays an important role. Such information reaches even the cells of origin of the pyramidal tract. Some data on the influence of such sensory control will be briefly considered in the account of the motor cortex.

The rubrospinal tract.   Recent experimental studies have shown that this pathway, coming from the red nucleus, has certain features in common with the corticospinal tract. The *nucleus ruber* or *red nucleus* (Figs. 4-2 and 4-5) receives its name from being readily visible in the fresh brain by its slightly reddish color, like the red zone of the substantia nigra. It is situated in the mesencephalon at the level of the colliculi and is almost spherical. The nucleus has a high content of iron, and is covered by a sort of capsule consisting of afferent and efferent fibers. Many fibers of the oculomotor nerve traverse it (Fig. 4-5). Both with regard to fiber connections and with regard to its cellular structure the nucleus is not an en-tity. It is customary to distinguish between a caudal *magnocellular part,* contain-ing large multipolar nerve cells, and a rostral *parvicellular part,* harboring small cells of various types.

This distinction appears to be derived from the situation in man, where large cells occur only in the caudalmost part of the nucleus, even though there are small cells interspersed between the large ones. Within both parts several minor groups can be distinguished. Grofová and Maršala (1959) describe four groups in the magnocellular part of the human red nucleus which altogether contains only 150 to 200 large cells. In the phylogenetic scale the relative proportion of large cells decreases. It is unfortunate that the names "parvicellular" and "magnocellular" part have been applied also to the red nucleus of animals because, for example in the cat, there is an entirely gradual transition from caudal to rostral with regard to the occurrence of large and small cells (Brodal and Gogstad, 1954). In the monkey large cells appear to be present, although inter-mingled with small ones, only in the caudal third (Kuypers and Lawrence, 1967; Miller and Strominger, 1973).

The red nucleus has several efferent and afferent connections. Only a few will be considered here, first and foremost the projection to the spinal cord, the *rubrospinal tract* (often also called von Monakow's bundle). Since our knowledge of the connections of the red nucleus in man is relatively scanty, it will be appro-priate to consider first the data on the tract in animals.

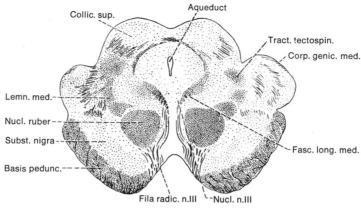

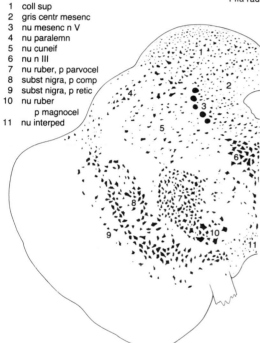

1 coll sup
2 gris centr mesenc
3 nu mesenc n V
4 nu paralemn
5 nu cuneif
6 nu n III
7 nu ruber, p parvocel
8 subst nigra, p comp
9 subst nigra, p retic
10 nu ruber
    p magnocel
11 nu interped

FIG. 4-5 Drawing above of a transverse section through the mesencephalon in man. The red nucleus is traversed by root fibers of the oculomotor nerve. Left, a diagram (from Nieuwenhuys, Voogd, and van Huijzen, 1978) of the main cellular groups at a level of the brainstem corresponding approximately to the level shown above.

After stereotactic placement of lesions in the red nucleus, degenerating fibers have been traced with silver impregnation methods to lumbosacral levels in the cat (Hinman and Carpenter, 1959; Nyberg-Hansen and Brodal, 1964), in the monkey (Orioli and Mettler, 1956; Kuypers, Fleming, and Farinholt, 1962; Poirier and Bouvier, 1966; Miller and Strominger, 1973; Murray and Haines, 1975), and in other animal species such as the rat (Brown, 1974b) and the opossum (Martin, Dom, Katz, and King, 1974). In the cat Edwards (1972) has recently traced the projection following injections of tritiated leucine in the red nucleus. The rubrospinal tract appears to be almost entirely crossed, the crossing occurring immediately after the fibers have left the nucleus. In the cord the rubrospinal fibers are found just ventral and somewhat lateral to those of the lateral corticospinal tract (Fig. 4-6). It is of interest that following interruption of the rubrospinal tract in the cat, small as well as large cells in the red nucleus show retrograde changes (Pompeiano and Brodal, 1957c) as seen in Fig. 1-16. In the monkey, likewise, small as well as large cells appear to give rise to rubrospinal fibers (Kuypers and Lawrence, 1967). This is in accordance with the presence of fibers of various diameters in the rubrospinal tract and with the findings of Padel, Armand, and Smith (1972) that their conduction velocity varies between 31 and 120 m/sec. The HRP study of Kneisley, Biber, and LaVail (1978) showed that in the monkey, as in the cat (Pompeiano and Brodal, 1957c), rubrospinal fibers originate throughout the length of the red nucleus.

Of particular interest is the demonstration of a *somatotopic pattern within the projection from the red nucleus of the cat onto the cord*. This pattern emerged from a study of the distribution of cells showing retrograde changes in the red nucleus of the cat following transections of the tract at different levels (Pompeiano and Brodal, 1957c), and was further confirmed anatomically by a study of the distribution of degenerated rubrospinal fibers following lesions of different parts of the red nucleus (Nyberg-Hansen and Brodal, 1964). As seen in Fig. 4-6, fibers ending in the cerivcal cord take origin from the dorsomedial part of the nucleus, and fibers ending in the lumbosacral cord from the ventrolateral part, from what may be referred to as a "neck and forelimb region" and a "hindlimb region," respectively, with an intermediate area representing a "trunk region." Corresponding findings have been made in the monkey by Murray and Hines (1975) and confirmed with the HRP method (Kneisley, Biber, and LaVail, 1978).

The functional validity of this pattern was demonstrated by Pompeiano (1957) on liminal stimulation of the nucleus in precollicularly decerebrated cats (see also Maffei and Pompeiano, 1962b). The somatotopic pattern has, furthermore, been confirmed by antidromic activation of rubrospinal fibers (Tsukahara, Toyama, and Kosaka, 1964; Padel, Armand, and Smith, 1972; Shinoda, Ghez, and Arnold, 1977; and others). However, in the cat many rubrospinal fibers branch and supply different levels of the cord (Shinoda, Ghez, and Arnold, 1977).

Several authors have investigated *the sites of termination of the rubrospinal fibers* in the cord (for references see Nyberg-Hansen and Brodal, 1964). In the cat the fibers do not enter lamina IX of Rexed, harboring the motoneurons, but are distributed to lamina V laterally, to lamina VI, and to the dorsal and central parts of lamina VII (Fig. 4-6). These sites of ending correspond largely to those of the corticospinal fibers from the "motor" cortex, a point of some interest (see Fig. 4-3). This pattern has been confirmed by autoradiographic tracing of the fibers in the cat (Edwards, 1972). It appears to be present in the monkey as well (Murray and Haines, 1975). Electron microscopic studies show that most terminals establish axodendritic contacts (Kostyuk and Skibo, 1975; Brown, 1974b; Goode and Sreesai, 1978).

Results of electrophysiological studies are in general agreement with anatomical findings. Thus, in agreement with Pompeiano's (1957) observations, referred to above, stimulation of the red nucleus activates contralateral flexor motoneurons, in which excitatory postsynaptic potentials can be recorded intracellularly, while contralateral extensor motoneurons give inhibitory postsynaptic potentials (Sasaki, Namikawa, and Hashiromoto, 1960). Corresponding findings have been made by Hongo, Jankowska, and Lundberg (1965), who, furthermore, concluded that both actions occur by way of internuncial cells, as would be expected from the anatomically deter-

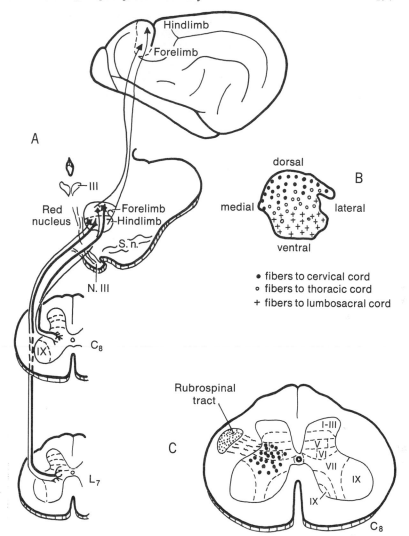

* fibers to cervical cord
∘ fibers to thoracic cord
+ fibers to lumbosacral cord

FIG. 4-6 Diagram showing the principal features in the cortico-rubro-spinal pathway as determined experimentally in the cat. *A:* The corticorubral fibers take origin in the anterior sigmoid gyrus, the "motor cortex" (above), and end in a somatotopical pattern in the red nucleus. The somatotopic arrangement is upheld in the rubrospinal projection, coming from small as well as large cells in the red nucleus. *B:* The somatotopic pattern in the red nucleus as seen in a transverse section at middle levels of the nucleus. *C:* The sites of termination of rubrospinal fibers are restricted to Rexed's laminae V to VII. Based on findings of Pompeiano and Brodal (1957c), Rinvik and Walberg (1963), and Nyberg-Hansen and Brodal (1964).

mined site of termination of the fibers. These and subsequent studies with recordings from the spinal cord (Hongo, Jankowska, and Lundberg, 1969, 1972) indicate that the fibers end on internuncial cells in Rexed's laminae VI and VII, and that these cells are (monosynaptically) activated by rubrospinal fibers. All these interneurons are also influenced from group I primary afferents from muscle or from flexor reflex afferents. The rubrospinal monosynaptic action on the interneurons is always excitatory. Various spinal reflexes are facilitated by the rubrospinal tract.

As first suggested by Granit and Holmgren (1955), the rubrospinal tract influences not only α but also γ neurons. According to Appelberg and Kosary (1963), the static γ motoneurons (see Chap. 3) of the flexor muscles are facilitated, those of the extensor muscles inhibited. In a later study, Appelberg, Jeneskog, and Johansson (1975) found that dynamic as well as static γ neurons are influenced and that this effect probably occurs via at least one interneuron.

Corresponding electrophysiological studies do not appear to have been made in the monkey, but on account of the anatomical similarities of the rubrospinal pathway in the cat and monkey, it appears likely that functional conditions are the same in the two animal species. The rubrospinal tract appears, thus, to be first and foremost a somatotopically organized fiber system involved in the excitation of α motoneurons and static γ neurons of flexor muscles. (For a review of the physiology of the red nucleus see Massion, 1967.)

In addition to the spinal cord the red nucleus supplies other regions. Most of these fibers pass to the *lateral reticular nucleus,* the *inferior olive,* and the *external cuneate nucleus,* which send their efferent fibers to the cerebellum. Some rubral efferents pass directly to the *cerebellum.* As described in Chapter 5, this projection has been found to supply only the nucleus interpositus anterior and to be somatotopically organized. This rubrocerebellar pathway thus appears to be organized like the reciprocal cerebellorubral pathway that arises from the nucleus interpositus anterior and is somatotopically organized. Only the caudal two-thirds of the red nucleus appears to be involved in these two pathways.

The afferent fibers from the nucleus interpositus anterior end preferentially on somata and proximal dendrites in the cat and rabbit (Nakamura and Mizuno, 1971) and in the opossum (King, Dom, Conner, and Martin, 1973). These data are in agreement with physiological studies. Like stimulation of the sensorimotor cortex (Tsukuhara and Kosaka, 1968), stimulation of the brachium conjunctivum or the nucleus interpositus anterior has consistently been found to produce monosynaptic excitatory potentials in cells of the red nucleus (Tsukuhara, Toyama, and Kosaka, 1964, 1967; Toyama, Tsukahara, Kosaka, and Matsunami, 1970; Anderson, 1971; and others). Many interpositus axons (up to 50, according to Toyama et al., 1970) appear to converge on a single red nucleus cell.

Still other efferents from the red nucleus pass to the *facial nucleus* (see Chap. 7, section e), the *sensory trigeminal nuclei* (see Chap. 7, section f), the *gracile* and *cuneate nuclei,* and certain parts of the *vestibular complex* (for a recent study, see Edwards, 1972). It appears from several studies that the efferent fibers from the red nucleus passing to destinations other than the spinal cord and cerebellum, originate almost exclusively in the rostral third, approximately, of the nucleus.

It has been difficult to decide whether the red nucleus gives off ascending fibers, particularly to the *thalamus,* since cerebellothalamic fibers traverse the nucleus (cf. Chap. 5 on the cerebellum). Hopkins and Lawrence (1975) made stereotactic lesions in the red nucleus in monkeys who had lived for one year after bilateral interruption of the superior cerebellar peduncle, allowing time for degeneration of cerebellothalamic fibers. They were unable to trace degeneration in the thalamus. Likewise, Edwards (1972) in an autoradiographic study found no evidence of a rubrothalamic projection in the cat, nor could it be verified physiologically (Anderson, 1971).

It is of interest here to consider some *afferent connections to the red nucleus* which may influence its action on the spinal cord. The *afferents from the cerebel-*

lum were briefly mentioned above and will be considered in the chapter on the cerebellum. Since these fibers come from the nucleus interpositus anterior and this is played upon by axons of Purkinje cells in the intermediate region of the anterior lobe of the cerebellum, it follows that the red nucleus is influenced chiefly from this cerebellar part. Of particular interest in this context is the presence of *fibers to the red nucleus from the cerebral cortex*. The discovery of the *somatotopic pattern in the red nucleus and the rubrospinal pathway* (Pompeiano and Brodal, 1957c) raised the question whether a corresponding organization is present within the corticorubral projection. In experimental studies in the cat, Rinvik and Walberg (1963) were able to demonstrate that this is indeed the case (see also Mabuchi and Kusama, 1966). The fibers come chiefly from the cat's "motor" cortex (corresponding approximately to the anterior sigmoid gyrus) and pass to the ipsilateral red nucleus. Fibers from the forelimb region of this cortical area end in the dorsomedial part of the red nucleus, i.e., its forelimb region, while those from the cerebral hindlimb region end ventrolaterally (see Fig. 4-6). In addition there are some fibers from the second somatosensory area and a region which appears to represent the supplementary motor area in the cat, as well as from the gyrus proreus of the frontal lobe (Rinvik, 1965). The fibers from the supplementary motor area are distributed bilaterally.[5] The cells of the red nucleus projecting to the cord appear to be excited monosynaptically from the cortex (Tsukahara, Toyama, and Kosaka, 1967, see also Tsukahara and Kosaka, 1968). The presence of a somatotopic localization within the corticorubral projection was later confirmed in the monkey (Kuypers and Lawrence, 1967) as well as in the chimpanzee. In the monkey most of the fibers come from the precentral gyrus, but smaller contingents were traced from the adjoining areas and from the supplementary motor cortex. There is suggestive evidence that the fibers from the latter show a somatotopic pattern similar to those from the precentral gyrus (Kuypers and Lawrence, 1967). For a recent autoradiographic study in the monkey, see Hartmann–von Monakow, Akert, and Künzle (1979).

It has often been assumed in physiological studies that the cortical activation of cells in the red nucleus occurs by way of collaterals from pyramidal tract fibers. From an anatomical point of view this is not obvious. A physiological study by Humphrey and Rietz (1976) in the monkey brought forward evidence that at least most of the fibers from the cortical arm area to the red nucleus are axons of cells other than those giving off corticospinal fibers. Corresponding conclusions were reached by Jones and Wise (1977) in an HRP study.

Other studies have brought forward further details. Whereas cerebellorubral fibers end chiefly on somata and proximal dendrites of cells in the red nucleus, the corticorubral fibers have been found electron microscopically to establish mainly axodendritic synapses in the opossum (King, Martin, and Conner, 1972), the rat (Brown, 1974a), and the rabbit (Mizuno et al., 1973). The anatomical organization of the red nucleus has been studied using the Golgi method and electron microscopically in the monkey (King, Schwyn, and Fox, 1971) and in the opossum (King, Bowman, and Martin, 1971). Different types of synaptic endings are described, and Golgi type II cells appear to occur. According to Sadun (1975), the distribution of cortical fibers within the red

---

[5] It is of functional interest that the corticofugal projections from the second somatosensory area differ from those from the primary sensorimotor area also with regard to their termination in other structures (pontine nuclei, thalamus, dorsal column nuclei, and others).

nucleus is more restricted (chiefly to lateral regions) when studied electron microscopically than when silver impregnation methods are used. When studied electrophysiologically, the somatotopic pattern in the corticorubral projection in the cat appears to be less schematic than can be deduced from anatomical experiments (Padel, Smith, and Armand, 1973). A definite somatotopic arrangement has not been demonstrated in all species studied, for example the opossum (King, Martin, and Conner, 1972).

For analyses of the functional role of the red nucleus, it is of interest that many of its cells have dichotomizing axons (Cajal) that may supply two different targets. From anatomical studies of retrograde changes in the red nucleus it was concluded (Brodal and Gogstad, 1954) that fibers to the spinal cord and to the cerebellum to a large extent are derived from the same red nucleus cells. In a physiological study in which rubral efferent cells were identified by antidromic stimulation Anderson (1971) confirmed this and, furthermore, indicated that some cells supply the spinal cord as well as nuclei in the medulla (possibly the inferior olive). The latter projection is uncrossed, in contrast to the other efferent projections, in agreement with what has been concluded from anatomical studies (see Chap. 5).

From a functional point of view it is of interest that the number of red nucleus afferents from the cerebral cortex is far smaller than the number of cerebellar afferents. In agreement with this, physiological studies have shown that the cerebellar input to the red nucleus is more powerful than that from the cerebral cortex.

It can be seen from the discussion above that the red nucleus may be influenced from many sources and that it is able to act particularly on the spinal cord and the cerebellum. It is clear that it is involved in motor mechanisms. While its collaboration with the intermediate part of the cerebellum in these functions is presumably most important and involves feedback systems, it should be noted that *in the cat and apparently also in the monkey (perhaps in the chimpanzee), the red nucleus is also a station in an indirect corticospinal pathway. Its organization shows remarkable similarities to the "precentral" component of the direct corticospinal tract.* Both fiber systems originate from largely the same cortical regions, contain fibers of various calibers, and extend throughout the cord. Both show a somatotopic pattern, and their terminal regions appear—at least in the cat where they have been properly studied in this respect—to be very similar (Fig. 4-3). Both fiber systems appear to produce predominantly facilitation of flexor α and γ motoneurons and to excite monosynaptically the same interneurons.

Little is known of the *corticorubral and rubrospinal tracts in man*. As to the latter, Collier and Buzzard (1901), from Marchi studies of human cases, suggested that the rubrospinal tract in man extends to sacral levels of the cord, but their evidence is not conclusive (for further references see Nathan and Smith, 1955a). The general opinion has for a long time been that the rubrospinal tract is rudimentary in man. However, there appears to be no good evidence for this. The only positive information of the tract is derived from a human case described by Stern (1936), in which retrograde cellular changes were found in the magnocellular part in both red nuclei following a total transverse lesion of the upper thoracic cord. It follows from this that the tract in man extends at least below this level. As described above, there are only a few large cells in the human red nucleus, and since it has—incorrectly as we have seen—been held that the tract is composed of myelinated axons from large cells only, it has been tacitly assumed that the rubrospinal tract is rudimentary in man. However, the fact that rubrospinal fibers come from small as well as large cells in the cat and also in the monkey (see above), and that it contains a considerable number of thin fibers in these species, makes it likely that the same will turn out to be the case in man when the problem is studied with proper methods and in suitable human cases. If this suggestion proves to be correct, it will have consequences for the interpretation of some clinical findings.

Corticorubral fibers in man have been described by several authors. Among several reports may be mentioned the study of Margareth Meyer (1949) on brains from patients subjected to frontal leucotomy and a corresponding study by Kanki and Ban (1952). At least part of the fibers appear to come from the precentral region. No details appear to be available concerning these connections in man, but it is clear that they descend in the internal capsule and in the cerebral peduncle to reach the red nucleus and appear to be massive (Beck and Bignami, 1968).

The vestibulospinal tract is another fairly well characterized pathway from the brain stem to the cord. It was well known to the classical neuroanatomists. After leaving the vestibular nuclei the fibers pass in a ventral direction before turning caudally and descending in the ipsilateral ventrolateral funiculus of the cord. As for the rubrospinal tract, most of our knowledge comes from animal experiments. The limited data available on the human tract are, however, in agreement with findings made in animals. In addition to this classical vestibulospinal tract there is another pathway from the vestibular nuclei to the cord, consisting of a smaller number of fibers which descend in the medial longitudinal fasciculus. In the cord these fibers are found bilaterally in the ventral funiculus, close to the midline, in the so-called sulcomarginal fasciculus. Following a suggestion of Nyberg-Hansen (1966a), the latter will be referred to as the *medial vestibulospinal tract,* while the former will be called the *lateral vestibulospinal tract.* Practically all studies of these pathways have been made in the cat.

When discussing the vestibulospinal connections it is essential to be aware of the fact that the vestibular nuclei are not an entity, but represent a complex of minor cell groups, each of which has its particular connections. These features will be dealt with in the chapters on the cerebellum and the cranial nerves. Here our concern is primarily with the vestibulospinal tracts.

The *lateral vestibulospinal tract* has been found by most authors (for references see Pompeiano and Brodal, 1957a) to take origin from the lateral vestibular nucleus of Deiters (Fig. 4-7), characterized by harboring a fair number of giant cells or Deiters' cells. Almost all authors are agreed that the tract extends to lumbosacral levels of the cord (in man as well as in the cat and monkey). In an experimental study (Pompeiano and Brodal, 1957a) in which advantage was taken of the modified Gudden method (Brodal, 1940a) it was confirmed that the tract takes origin from the lateral vestibular nucleus only. Furthermore, it could be shown that its fibers are derived not only from the giant cells but from its small cells as well, since after lesions of the tract in the cord, retrograde changes are present in cells of all types in the nucleus. This has recently been confirmed by studies of labeled cells in the lateral vestibular nucleus after HRP injections in the spinal cord in the cat (Peterson and Coulter, 1977); it fits in with the presence of thin as well as thick fibers in the tract.[6] Of particular interest was the demonstration of *a somato-topic pattern within the lateral vestibular nucleus* (Fig. 4-7). Fibers to the lumbosacral cord take origin in its dorsocaudal part, fibers to the cervical cord from its rostroventral part, with a zone of overlapping in between. Accordingly, one may speak of a "neck and forelimb region," a "trunk region," and a "hindlimb region," although the borders are not sharp. It is of interest that, according to the recent HRP study of Kneisley, Biber, and LaVail (1978), the pattern in the vestibulospinal projection in the monkey appears to be almost identical to that in the cat.

> The correctness of this pattern has further been established by the study of the distribution within the cord of fibers degenerating as a consequence of lesions restricted to different regions of the nucleus (Nyberg-Hansen and Mascitti, 1964), in electrophysiological studies by localized stimulations (Pompeiano, 1960) and by recordings in the nucleus following antidromic stimulation of the fibers in the tract (Ito, Hongo, Yoshida, Okada, and Obata, 1964; Wilson, Kato, Thomas, and Peterson, 1966). The localization is, however, not as distinct as might appear from the anatomical diagram (Wilson, Kato, Peterson, and Wylie, 1967). In agreement with the fiber spectrum of the lateral vestibulospinal tract, the conduction velocities vary considerably, from 24 to 140 m/sec (Ito et al., 1964; Wilson et al., 1966). For recent reviews of the physiology of the vestibulospinal tracts, see *Basic Aspects of Central Vestibular Mechanisms* (1972).

The terminal site within the spinal cord of the vestibulospinal fibers differs from those of the corticospinal and rubrospinal tracts. While some previous authors had indicated a termination in the ventral horn and ventral part of the "intermediate" part of the gray matter, Nyberg-Hansen and Mascitti (1964) in a more detailed study found the lateral vestibulospinal fibers to end in lamina VIII and the ventral and central parts of lamina VII (Fig. 4-7), especially on large dendrites. Only a few fibers could be traced to lamina IX containing the motoneurons. Many of the cells in lamina VIII send their axons across the midline. This may explain the occurrence of bilateral effects on stimulation of the lateral vestibular nucleus.

It has been known for some time that the nucleus of Deiters and the lateral vestibulospinal tract increase the extensor tonus of the ipsilateral limbs (for a review see Pompeiano, 1972a). The increased tonus following decerebration at levels above the nucleus of Deiters (decerebrate rigidity, to be discussed later) is to a large extent due to the unopposed action of the lateral vestibular nucleus on the cord, and disappears when the nucleus is destroyed. In keeping with this, microelectrode studies have shown that stimulation of the lateral vestibular nucleus of the cat produces excitatory postsynaptic potentials in extensor motoneurons only, while flexor motoneurons are inhibited (Lund and Pompeiano, 1965). The excitatory effect is concluded to be mediated monosynaptically. This is at variance with the lack of fibers terminating in lamina IX, but the discrepancy may perhaps be explained if a substantial number of contacts are established with the peripheral parts of dendrites from motoneurons extending into laminae VIII and VII.[7]

[6] It is of some practical importance that during its descent the tract shifts its position. While (in the cat) at cervical levels it is situated peripherally in the ventrolateral funiculus, it moves dorsomedially along the ventral median fissure as it descends (see Nyberg-Hansen and Mascitti, 1964).

[7] In an electron microscopic study in the cat, Rogers (1972), after hemisections of the thoracic cord, found degenerating boutons in the lumbar cord, mainly in Rexed's layer VIII. Some occurred on soma and large dendrites of motoneurons. However, not all the transected descending fibers were vestibulospinal.

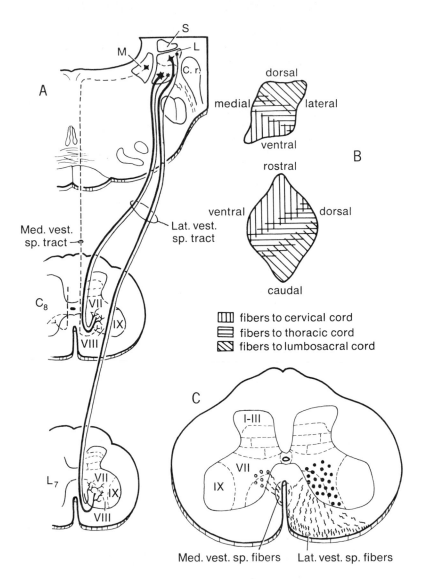

FIG. 4-7 Diagram showing the principal features in the vestibulospinal pathways as determined experimentally in the cat. *A:* The lateral vestibulospinal tract comes only from the lateral vestibular nucleus of Deiters (L) but from small as well as large cells. It is somatotopically arranged. The medial vestibulospinal tract is derived from the medial vestibular nucleus (M) and does not reach the lumbar cord. *B:* The somatotopic pattern in the lateral vestibular nucleus as seen in a transverse section (above) and in a sagittal reconstruction of the nucleus (below). *C:* The vestibulospinal tracts terminate in Rexed's laminae VII and VIII. Based on findings of Pompeiano and Brodal (1957a), Nyberg-Hansen and Mascitti (1964) and Nyberg-Hansen (1964a).

Like the corticospinal and rubrospinal tracts, the lateral vestibulospinal tract acts not only on α motoneurons but on γ neurons as well (Andersson and Gernandt, 1956; Gernandt, Iranyi, and Livingston, 1959). The action on the latter is described as being excitatory (Carli, Diete-Spiff, and Pompeiano, 1967a).[8]

Further studies have shown that conditions are less simple than appears from the discussion above. Thus, in the *hindlimb* of the cat, the motoneurons for knee and ankle extensors usually receive monosynaptic excitation, whereas those for hip and toe extensors do so only occasionally. Disynaptic inhibitory potentials are common in flexor motoneurons, but occur in some hip extensor motoneurons as well (Grillner, Hongo, and Lund, 1970). The disynaptic effects may be due to activation of interneurons in laminae VII and VIII (Lund and Pompeiano, 1968). However, motoneurons to *forelimb* muscles in the cat do not appear to be monosynaptically excited from the lateral vestibular nucleus, but polysynaptic activation is common in extensor and sometimes flexor motoneurons (Wilson and Yoshida, 1969a). These differences are only some among several that indicate differences between the motor regulation of forelimbs and hindlimbs even in a quadruped. It may be expected that the differences are more marked in primates.

The lateral vestibulospinal tract influence on γ neurons (see Pompeiano, 1972a, for a review) appears to be exerted on the static neurons only, to occur monosynaptically, and to be excitatory. Extensor and flexor γ motoneurons appear to be facilitated and inhibited, respectively (Kato and Tanji, 1971). α and γ motoneurons in a particular motoneuron pool appear to be activated in parallel from the nucleus of Deiters, and their activity appears to be closely linked.

According to a physiological study by Abzug, Maeda, Peterson, and Wilson (1974), some vestibulospinal fibers branch and supply lumbar as well as cervical segments, as has been found also for corticospinal and rubrospinal fibers (see above). This is of interest with regard to the problem of cooperation of forelimb and hindlimb, and for views on the somatotopic organization of the lateral vestibular nucleus.

While the lateral vestibular nucleus does not appear to receive fibers from the cerebral cortex it receives abundant projections from the cortex of the spinal part of the cerebellum, from the "vestibulo-cerebellum," and from the fastigial nucleus. Each of these afferent fiber contingents has its particular area of distribution within the lateral vestibular nucleus (see Fig. 7-10). These features will be considered in connection with the cerebellum. Suffice it to emphasize here the important role played by the cerebellar afferents for the function of the nucleus of Deiters.

The *other vestibulospinal pathway,* here referred to as the *medial,* is relatively modest and less well known than the lateral tract (Fig. 4-7). It appears now to be finally established that this tract takes its origin only from the medial vestibular nucleus, according to the experimental anatomical studies of Nyberg-Hansen (1964a), who also studied the termination of the fibers. Their terminal area is approximately the same as that of the lateral vestibulospinal tract (Fig. 4-7). The tract can only be traced to midthoracic levels. It appears, thus, that the medial vestibular nucleus and its spinal tract are concerned chiefly in the mediation of vestibular impulses to the neck and forelimb muscles, while the lateral nucleus and tract may influence the entire body. It is of some interest that the medial and lateral nucleus appear to be supplied by primary vestibular fibers from different parts of the vestibular apparatus (see Chap. 7). Like the lateral vestibular nucleus, the

[8] Whether there are different fibers which mediate the effect on α and γ neurons is not known. After vestibular stimulation the latter have been found to discharge at lower strength of stimulation than the α neurons. The suggestion may perhaps be ventured that this may be correlated with the fact that both large and small cells of the nucleus of Deiters send axons to the cord, and these axons have different diameters and conduction velocities.

medial nucleus receives fibers from the cerebellum, but it does not appear to receive fibers from the cerebral cortex (although it does receive fibers from the mesencephalon).[9] Physiological studies (Wilson and Yoshida, 1969b) indicate that in the cat axons descending from the medial vestibular nucleus establish monosynaptic contact with motoneurons in the upper cervical cord (but not with limb motoneurons), and that they are inhibitory. (See also Grillner, Hongo, and Lund, 1971.)

The *vestibulospinal tracts in man* are relatively little known (for a review see Nathan and Smith, 1955a). However, an analysis of the normal anatomy of the human vestibular nuclei (Sadjadpour and Brodal, 1968) shows that on practically all points they resemble those in the cat, and the same subdivisions can be identified. This makes it extremely likely that the fiber connections are also, in principle, identical. There is suggestive evidence from some observations by Foerster and Gagel (1932) that also in man there is a somatotopic organization within the nucleus of Deiters. Further support for this view has been provided by Løken and Brodal (1970).

Reticulospinal tracts. The presence of fibers passing from the reticular formation of the brainstem to the spinal cord was established by the classical neuroanatomists, such as Probst, Kohnstamm, and Lewandowsky (for references see Brodal, 1957). Later physiological studies of the effects which can be elicited by stimulation of the reticular formation have fostered renewed interest in these fiber connections and their functional role.

As will be described in the chapter on the reticular formation (Chap. 6), this is not a diffuse aggregation of nerve cells and fibers. It may be subdivided into a number of more or less well circumscribed nuclei, which in part differ with regard to their afferent and efferent fiber connections and presumably, therefore, are functionally dissimilar. As to the reticulospinal connections, these may be divided into two groups which have their particular sites of origin and differ in other respects as well. The sites of origin of reticulospinal fibers may be determined by studying the spatial distribution of cells showing retrograde changes after lesions of the cord which interrupt the reticulospinal fibers, as done in the cat by Pitts (1940) and in the monkey by Bodian (1946). These authors were, however, not concerned with certain details in the projections, which became of interest following more recent research. Taking advantage of the modified Gudden method (Brodal, 1940a), Torvik and Brodal (1957) were able to demonstrate that reticulospinal fibers are derived from small as well as large cells scattered at all levels of the medullary and pontine reticular formation. However, *there are two clearly maximal areas of origin, one in the pons, another in the medulla.* Both are restricted to approximately the medial two-thirds of the reticular formation where large cells occur. These areas are indicated in Fig. 6-4, from which it is also seen that the pontine fibers descend ipsilaterally while the fibers from the medulla are crossed as well as uncrossed. Most of the medullary fibers come from the region called the nucleus reticularis gigantocellularis, while the pontine fibers are derived

[9] Recently a third vestibulospinal pathway has been described. See footnote 11 in Chapter 7.

from the entire nucleus reticularis pontis caudalis and the caudal part of the nucleus reticularis pontis oralis. No fibers to the cord appear to come from the mesencephalic reticular formation (Nyberg-Hansen, 1965a; Edwards, 1975).

The presence of changes in small as well as large cells indicates that both types send fibers to the cord, and tallies with the presence of fibers of varying diameters within the reticulospinal tracts.[10] It is estimated that more than half of the large cells of the caudal pontine reticular nucleus project onto the cord (Torvik and Brodal, 1957). After lesions at various levels of the cord no difference in the distribution of changed cells could be ascertained. This indicates that *there is no somatotopic pattern within the reticulospinal projection.*[11] The recent physiological demonstration that a large proportion of reticulospinal neurons projecting to the lumbar or thoracic cord give off branches to the cervical cord (Peterson, Maunz, Pitts, and Mackel, 1975) is another indication that the reticular formation may be important in the coordination of activities at different spinal levels (compare similar data as concerns corticospinal and vestibulospinal fibers).

There has been some difference of opinion concerning the *course of the reticulospinal fibers in the cord,* in part, it appears, because authors who studied this subject by making lesions in the reticular formation and tracing degenerating fibers with the Marchi or silver impregnation methods have damaged fibers belonging to both components or fibers of other descending fiber tracts. These difficulties are significantly reduced if lesions are placed in those regions which are known to give off the majority of reticulospinal fibers. This was done by Nyberg-Hansen (1965a) in a study in the cat (Nauta method). It could then be settled that reticulospinal fibers descend to the lowermost levels of the cord, in contrast to conclusions made on the basis of retrograde cellular changes, where convincing alterations were not seen following lesions below the thoracic cord in the monkey (Bodian, 1946) and in the cat (Torvik and Brodal, 1957). Furthermore, the pontine reticulospinal fibers descend almost exclusively ipsilaterally in the ventral funiculus of the cord, in agreement with the results of several previous workers (for references see Nyberg-Hansen, 1965a). Some of the fibers cross in the anterior commissure of the cord before they terminate. The medullary fibers, however, descend bilaterally in the lateral funiculus (Fig. 4-8). In a careful electrophysiological study Peterson, Maunz, Pitts, and Mackel (1975) confirmed that reticulospinal fibers descending in the lateral funiculus are derived from the nucleus reticularis gigantocellularis, in agreement with the anatomical data (see Fig. 6-4). However, some fibers from this area descend in the ventral funiculus as well (where the pontine reticulospinal fibers pass). Most of the latter project as far as the lumbar cord. According to an autoradiographic study in the cat (Basbaum, Clanton, and Fields, 1978), some reticulospinal fibers from the medulla descend in the dorsolateral funiculus.

The two groups of reticulospinal fibers differ with regard to their sites of termination. Most authors have described the fibers as ending in the "intermediate zone" and the "ventral horn," for example, Kuypers, Fleming, and Farinholt (1962). Studying the terminations in greater detail, Nyberg-Hansen (1965a) could establish that in the cat the pontine reticulospinal fibers terminate more ventrally than do the medullary ones (Fig. 4-8). The former end on cells of various types in

---

[10] The conduction velocity of the reticulospinal fibers has been estimated to be from 20 to 138 m/sec (Wolstencroft, 1964). Pilyavsky (1975) separates them into two groups. According to Peterson, Maunz, Pitts, and Mackel (1975), the reticulospinal fibers descending in the ventromedial and the lateral funiculus have median conduction velocities of 101 and 70 m/sec, respectively.

[11] In electrophysiological studies (Peterson, Maunz, Pitts, and Mackel, 1975; Peterson, 1977) evidence was found for some degree of somatotopic pattern in the spinal projection from the nuclei reticularis gigantocellularis and pontis caudalis. Neurons projecting to the cervical cord were found most dorsally and rostrally. (See also Peterson, Pitts, and Fukushima, 1979.)

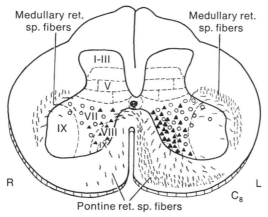

Medullary ret.
sp. fibers

Medullary ret.
sp. fibers

I-III

V

VII

IX

VIII

IX

R

L

C₈

Pontine ret. sp. fibers

▲ Sites of termination of pontine ret. sp. fibers

• Sites of termination of medullary ret. sp. fibers

FIG. 4-8 Diagram of a transverse section of the spinal cord of the cat at C₈, showing the sites of termination and the position in the cord of reticulospinal fibers from medullary and pontine reticular formation. From Nyberg-Hansen (1965a).

Rexed's (1952, 1954) lamina VIII and adjacent parts of lamina VII, while the medullary fibers end chiefly in lamina VII with a few terminations also in laminae VIII and IX. The results of Petras (1967, in the cat) appear largely to agree with this, although the sites of termination are not indicated with reference to Rexed's laminae. In the opossum it appears that the sites of origin of reticulospinal fibers (Beran and Martin, 1971) as well as their sites of termination (Martin and Dom, 1971) are in all essentials as in the cat. A short report on the labeling of reticular formation cells after injections of horseradish peroxidase in the spinal cord of the cat has been published by Kuypers and Maisky (1975).

As to *reticulospinal tracts in man,* there are indications in the literature that they are present, but no reliable information is available (see Nathan and Smith, 1955a, for a review). The failure of Foerster and Gagel (1932) to find cell changes in the reticular formation following chordotomies is not decisive. It seems a likely assumption that the pontine and medullary reticulospinal tracts differ functionally in some respects.

Although there are several electrophysiological studies bearing on the *function of the reticulospinal connections,* there is relatively little information concerning possible differences between the two components. This may be due in part to the fact that the correlation with anatomical data has not been of great concern to most physiologists studying the reticular formation. In addition, the anatomical arrangement of the reticulospinal projections often makes analyses of the effects of stimulations or ablations equivocal.

In a recent electrophysiological study Peterson, Pitts, and Fukushima (1979) found that in the medial pontomedullary reticular formation different zones are concerned in the monosynaptic activation of motoneurons supplying neck and back or limb muscles. In general, their observations are compatible with anatomical data. Polysynaptic responses were also elicited. Further data on

the reticulospinal neurons will be discussed in Chapter 6. Here it may be mentioned that Jankowska, Lund, Lundberg, and Pompeiano (1968) and others found stimulation of the medullary
reticulospinal projection area to give rise to inhibitory postsynaptic potentials in α motoneurons.
This effect is sufficiently strong to suppress decerebrate rigidity. It is also well established that the
reticular formation influences activity of the γ fibers (Granit and Kaada, 1952; and others).
Spindle activation has been observed after stimulation of the reticular formation in the mesencephalon (Granit and Holmgren, 1955) or laterally in the medulla (Diete-Spiff, Carli, and Pompeiano, 1967). Whether the activity of the muscle spindles is inhibited from the medullary reticular formation appears so far not to have been settled (see, however, Shimazu, Hongo, and
Kubota, 1962). Grillner, Hongo and Lund (1966b) found monosynaptic excitatory potentials in γ
motoneurons on stimulation of the lower brainstem. See also Bergmans and Grillner (1968).

Obviously the reticulospinal pathways are capable of influencing α as well as
γ neurons, and there is physiological evidence which supports the notion that the
medullary and pontine reticulospinal pathways are functionally dissimilar (see also
Chap. 6). Their different sites of termination in the cord are of interest from this
point of view, as will be discussed below.

In addition to acting on motoneurons, apparently via internuncials, the reticulospinal projections, like several other descending connections, have been
shown to influence the central transmission of sensory impulses (see Chap. 2).
Furthermore, they may influence various reflex pathways. In recent years several
studies have been devoted to the two subjects, and numerous bits of information
have been gathered. It is evident from Pompeiano's (1973) review of available
data, to which the reader is referred, that the reticular central control of sensory
transmission and of spinal reflexes is highly differentiated. The interneurons and
their connections are important for both functions.

In relation to the effects exerted by the reticulospinal fiber tracts on the peripheral motor apparatus it is of relevance to recall that the *reticular formation is
acted upon by fibers from other regions*. As described in Chapters 2 and 6, there is
a considerable influx of *ascending fibers from the cord*. Furthermore, there are
cerebellar afferents from the fastigial nucleus (see Chap. 5). Of most immediate
concern in the present chapter are the *corticoreticular connections*. As described
in Chapter 6, the majority of these fibers are derived from the central "sensorimotor" region. They descend in the internal capsule and cerebral peduncle and are
distributed bilaterally in the reticular formation. It is of interest that the largest
proportion of corticoreticular fibers end in the two regions of the reticular formation which give off reticulospinal fibers (see Fig. 6-4), since this makes clear that
*there are impulse pathways from the cerebral cortex, particularly its primary sen-
sorimotor region, to the spinal cord via the reticular formation*. There are also
corticoreticular fibers to the mesencephalic reticular formation (which, as mentioned above, does not give off fibers to the spinal cord).

Reticulospinal neurons, identified by antidromic stimulation, have been found
to be monosynaptically excited from the cerebral cortex (Peterson, Anderson, and
Filion, 1974; Pilyavsky, 1975) and from the superior colliculus (Udo and Mano,
1970; Peterson, Anderson, and Filion, 1974). They also respond to vestibular and
cutaneous stimulation (Peterson and Felpel, 1971). For some other data, see
Chapter 6.

Other descending tracts to the spinal cord. In addition to fibers in the
corticospinal, rubrospinal, vestibulospinal, and reticulospinal tracts, the spinal

cord receives fibers descending directly from other masses of gray matter. Most of these pathways are quantitatively less important than those mentioned, and their functional importance is less well known. They will, therefore, be treated rather briefly.

The tectospinal tract comes from the superior colliculus (Figs. 4-1 and 7-31). Its fibers cross ventral to the periaqueductal gray and descend in the brainstem, just ventral to the medial longitudinal fasciculus. In the cord the fibers travel medially in the ventral funiculus. Divergent opinions have been held concerning the origin and termination of this tract (for a review see Nyberg-Hansen, 1964b). However, it appears now to be established that the fibers come only from the superior colliculus, and that the majority end in the upper cervical segments. Only a few reach the lower cervical segments (Nyberg-Hansen, 1964b). They terminate chiefly in Rexed's laminae VII and VI, while some end in lamina VIII, through which they enter. The tectospinal fibers probably exert their action on motoneurons via interneurons. Electrical stimulation of the superior colliculus gives rise to turning of eyes and head to the contralateral side. This effect may, at least in part, be mediated via the tectospinal tract. Anderson, Yoshida, and Wilson (1971), after electrical stimulation of the superior colliculus in cats, observed short-latency excitatory postsynaptic potentials in contralateral neck motoneurons and a combination of excitatory and inhibitory potentials in ipsilateral neck motoneurons. The superior colliculus is considered in Chapter 7. Suffice it here to mention that it is influenced by the cerebral cortex, especially the visual areas in the occipital lobe, and that the cells of origin of the tectospinal tract are found in the deeper layers of the colliculus.

The interstitiospinal tract (Fig. 4-1) is another pathway to the spinal cord which has received relatively little attention in clinical neurology, although it has been described and discussed by many neuroanatomists. A review of the relevant literature has been given by Nyberg-Hansen (1966b), who in an experimental study in the cat confirmed the conclusion of most authors that the tract originates from the interstitial nucleus of Cajal in the rostral mesencephalon. The nucleus is situated just ventral to the periaqueductal gray (see Figs. 7-28 and 7-31) and separated from this by the medial longitudinal fasciculus. The fibers of the interstitiospinal tract descend to the cord in the latter bundle, chiefly ipsilaterally. They can be traced to sacral levels of the cord and terminate chiefly in lamina VIII and the adjoining part of lamina VII, but not in lamina IX. Their terminal area thus coincides approximately with that of the vestibulospinal fibers (see Fig. 4-7). The origin of spinal fibers from the interstitial nucleus has recently been confirmed with the HRP method (Kuypers and Maisky, 1975).

The interstitial nucleus of Cajal receives afferents, from various sources, among them the vestibular nuclei (see Chap. 7). Several authors have claimed that it receives fibers from the cerebral cortex (see Pompeiano and Walberg, 1957, for references).

The interstitiospinal fibers are presumably involved in the rotation around the longitudinal axis of the head and body which can be observed following electrical stimulation in the region of the nucleus, and the nucleus probably is an important link in mediating effects on the neck and body musculature in response to optic and vestibular impulses (see Hyde and Eason, 1959). The nucleus may also be of importance for certain central effects on the autonomic system (see Nyberg-Hansen, 1966b).

The *solitariospinal tract* is a minor projection to the spinal cord. It originates from the nucleus of the solitary tract, the visceral afferent nucleus of the vagus and glossopharyngeal nerves, as described in Chapter 7, section c. In a combined HRP and autoradiographic study in the cat, Loewy and Burton (1978) traced the solitariospinal fibers to the region of the motoneurons of the phrenic nerve in cervical segments 4–6, to the ventral horn in the thoracic segments, and to the intermediolateral column. For a recent HRP study of the tract in the monkey, see Kneisley, Biber, and LaVail (1978). The solitariospinal tract appears to descend predominantly ipsilaterally, and has been traced to lumbar levels. Although the presence of ascending connections from the nucleus of the solitary tract shows that it has its main function in the propagation of visceral sensory input to higher levels of the brain (see Chap. 7), it is obviously also capable of influencing spinal mechanisms, somatic and visceral. The solitariospinal fibers are presumably of importance in activating the spinal cord in response to visceral impulses entering in the vagus and glossopharyngeal nerves. The nucleus receives fibers from the cerebral cortex (see Chap. 7). These fibers may be relevant when attempts are made to explain the well-known fact that psychic stimuli may influence—inhibit or facilitate—visceral reflexes. This statement should, however, not be taken to imply that the influence of "higher levels" on visceral functions is only a matter of cortical efferents to visceral nuclei.

> Some observations bearing on this complex problem will be discussed in Chapter 11. Among the recent data of general interest is the demonstration with modern tracing methods that there are direct fibers from the hypothalamus to the nucleus of the solitary tract and also to the (parasympathetic) dorsal motor vagus nucleus and the (sympathetic) intermediolateral cell column.

It has for some time been doubted that there are fibers from the *cerebellum* directly to the spinal cord. As described in Chapter 5, however, with the use of the HRP method and in autoradiographic studies, the presence of a *fastigiospinal tract* has been demonstrated. The number of fibers is small, and they appear to supply chiefly the cervical segments of the cord.

Fibers from the raphe nuclei to the spinal cord should be mentioned. The raphe nuclei will be considered fully in Chapter 6. Suffice it to mention here that their projection to the cord appears to be modest, that most of these fibers arise from a particular part of the complex (the nucleus raphe magnus), that they descend in the dorsolateral funiculus, and that they are *serotonergic*. There is some evidence that such fibers end in the intermediolateral cell column, but other fibers appear to end in the ventral and dorsal horns. Some studies suggest that the spinal projection from the nucleus raphe magnus is involved in pain mechanisms, possibly by inhibiting the central transmission of nociceptive stimuli. The raphe nuclei receive some fibers from the cerebral cortex.

Whereas the spinal pathway from the raphe is serotonergic, there is also a *noradrenergic* spinal projection. This comes from the *nucleus locus coeruleus* (see Chap. 6) and descends in the ipsilateral ventral funiculus as far as the lumbar cord. Little is so far known of the functional role played by the noradrenergic innervation of the cord.

The olivospinal tract, mentioned in many textbooks, is probably nonexistent. At least no evidence is available from experimental studies. In man a triangular zone found in the ventral funiculus at upper cervical levels and having a light hue in myelin-stained preparations (Helweg's Dreikantenbahn, triangular bundle of Helweg) has been claimed to be an olivospinal tract, but this appears to be entirely conjectural.

The basal ganglia and some related nuclei. It is apparent from the above that there are a number of fiber bundles which descend to the cord from higher levels and thus must be able to influence its activity. The cerebral cortex may influence the cord directly via the corticospinal tract, but in addition indirectly by fibers passing from the cortex to some nuclei which sends fibers to the cord (red nucleus, parts of the reticular formation, superior colliculus, nucleus of the solitary tract, some of the raphe nuclei). Furthermore, there are other, more circuitous routes by which the cortex may play a role in acting on the spinal cord, for example by connections from the cerebral cortex to the cerebellum and from this to the vestibular nuclei and reticular formation (to be considered in the chapter on the cerebellum). Other pathways are, however, also available, namely some passing via the basal ganglia. Since the basal ganglia and their connections during the last 15 years have attracted great interest and have been studied by a large number of investigators working in different fields of the neurobiological sciences, and since the basal ganglia make up a large and characteristic part of the brain, they will be considered in some detail. While they were previously thought to represent important links in pathways leading from the cerebral cortex to the cord, and thus to be primarily "motor" in function, recent research has necessitated revisions of this concept.

In the course of time the term *"basal ganglia"* has carried different connotations. The old anatomists used it as a common denominator for all the large nuclei in the interior of the brain, including the thalamus. When the development of the brain became better known, the thalamus was excluded, while, for instance, the amygdaloid nucleus was included. There is still no generally accepted definition of what one should include in the concept "basal ganglia" although all authors consider the caudate nucleus and the lentiform nucleus with its two divisions, the putamen, and the globus pallidus as representing the main mass. The claustrum is usually included, while the amygdaloid nucleus, on account of its largely different connections and functions, is often excluded. It is common to consider the subthalamic nucleus and the substantia nigra in conjunction with the basal ganglia. This will be done also in the following account. The term *striate body* or *corpus striatum* is often used as almost synonymous with the basal ganglia and covers the claustrum, caudate, putamen, and globus pallidus. The name refers to the appearance in myelin-sheath-stained sections, where a number of myelinated fiber bundles traverse the cellular masses and give them a striated appearance. In the following a brief account of some of the main features of these nuclei will be given, before their connections are described.

The large gray nuclear masses of the *corpus striatum* are situated in the medullary layer of the hemisphere and are subdivided by fiber strands into different portions (Figs. 4-9 and 4-10). Most laterally, beneath the insula, the *claustrum* forms a thin sheet of gray substance. It is separated laterally from the cortex by a thin medullary layer, the capsula extrema, and medially from the putamen by

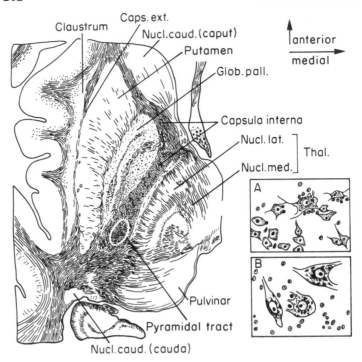

FIG. 4-9 Drawing of a horizontal section through the human brain, showing the corpus striatum (myelin-sheath staining). Redrawn from Ranson (1943). Circle indicates the approximate position of the corticospinal fibers in the internal capsule. Below to the right, representative cells from the striatum (A) and the pallidum (B). Redrawn from Foix and Nicolesco (1925).

the capsula externa (Fig. 4-9). The claustrum appears to be derived from the cerebral cortex of the insular region, and developmentally it therefore does not belong to the striate body in a restricted sense. The striate body proper, the caudate-lentiform nuclei, presents some features worth mentioning. *Both in regard to phylogenetic development and finer structure the caudate nucleus and the putamen are similar, whereas they both differ from the globus pallidus.* The latter, or *pallidum,* as it is often called, forms the medial part of the lentiform nucleus and takes its name from its pale color compared with the putamen and caudate nucleus. It can be subdivided into an outer or anterior-lateral segment and an inner or posterior-medial one (Fig. 4-9).[12] The pallidum is phylogenetically older than the two latter nuclei, and is developed from the basal plate of the primitive neural tube (in the diencephalon).

The *caudate nucleus* and the *putamen,* on the other hand, which appear later in phylogenesis, are derived from the telencephalon and increase in size corresponding to the development of the cerebral cortex to a greater extent than does the pallidum. In lower animals the putamen and the caudate nucleus are not

[12] There seems to be a general agreement that the inner segment of the pallidum in monkey and man is represented in carnivores by the entopeduncular nucleus (see Fox et al., 1966). The globus pallidus of carnivores thus corresponds to only the outer pallidal segment in man.

clearly separated by any internal capsule, since this fiber mass becomes conspicuous only with the more progressive development of the cerebral cortex. Particularly in man the internal capsule is rich in corticofugal (e.g., pyramidal) and corticopetal fibers. It is, however, bridged by strands of cells connecting the nucleus caudatus and the putamen. The pallidum represents the so-called *paleostriatum;* the putamen and the caudate nucleus together constitute the *neostriatum,* or simply the striatum. (The amygdaloid nucleus is the archistriatum.)

Subdivision in this manner is justified by the cytological structure of the palaeostriatum and neostriatum, respectively (Figs. 4-9A and B). The pallidum is composed of large, mainly spindle-shaped cells situated rather far apart, while the striatum is distinguished by densely lying small polymorph cells with interspersed larger multipolar cells. However further studies have shown that the cytology is far more complex.

In a detailed morphometric-statistical analysis of thirteen normal human brains the number of small and large striatal cells was estimated to be 110 million and 670,000, respectively, that is a ratio averaging 170:1 (Schroeder et al., 1975). The lateral and medial segments of the globus pallidus contain 540,000 and 170,000 neurons, respectively (Thörner, Lange, and Hopf, 1975). It is, however, not permissible to characterize the organization of the striatum in terms of two neuronal populations only. In detailed Golgi studies of the neostriatum several cell types may be distinguished on the basis of differences in dendritic spines and length and arborizations of axons, (Fox, Andrade, Hillman, and Schwyn, 1971; Fox, Andrade, Schwyn, and Rafols, 1971/1972; Kemp and Powell, 1971a; Fox, LuQui, and Rafols, 1974; Pasik, Pasik, and DiFiglia, 1976; DiFiglia, Pasik, and Pasik, 1977). These studies and experimental investigations on the striatonigral projections in the cat (Grofová, 1975) strongly refute the classical theory of Vogt and Vogt (1920) that only the large cells of the striatum have long projecting axons.

Golgi studies of the *globus pallidus* reveal that large oval or polygonal neurons by far outnumber the smaller neurons with few dendrites (Fox, Andrade, LuQui, and Rafols, 1974). Surprisingly, Golgi-impregnated material does not reveal significant differences between the two segments of the globus pallidus.

Electron microscopic studies, however, show clear ultrastructural differences in the neuronal and synaptic organization of the globus pallidus and the striatum (Fox, Hillman, Siegesmund, and Sether, 1966; Kemp and Powell, 1971a; Fox, Andrade, LuQui, and Rafols, 1974; Fox and Rafols, 1975; Bak, Choi, Hassler, Usunoff, and Wagner, 1975; Pasik, Pasik, and DiFiglia, 1976).

Caudally the globus pallidus is continuous with the rostral part of the pars reticulata of the substantia nigra (see below). The two nuclei show strikingly similar ultrastructural organizations of the neuropil (see below for references), possibly related to the fact that both receive massive afferent fiber projections from the striatum (Kemp 1970; Fox and Rafols 1976).

The pallidum, just like the substantia nigra and the nucleus ruber, contains a large amount of iron which can be identified histochemically. Also, the capillary bed is different, being very dense in the striatum, but not in the pallidum (Alexander, 1942a). The structural differences described above between the pallidum and the striatum correspond to functional differences, evidenced, inter alia, by the symptoms following diseases in these nuclei (see below).

The *subthalamic nucleus* or body of Luys is situated in the basal part of the diencephalon at the transition to the mesencephalon. This is an approximately ovoid nucleus, which caudally and ventrally is continuous with the substantia nigra (see Fig. 4-10). This nucleus is relatively larger in man than in other mam-

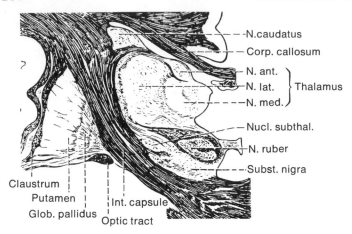

FIG. 4-10 Drawing of a myelin-sheath-stained frontal section through the human brain, showing the basal ganglia.

mals. Its cells are medium-sized, round, or fusiform, and there are regional differences in their size and concentration (Whittier and Mettler, 1949). Golgi studies of the subthalamic nucleus have disclosed at least two types of neurons (Rafols and Fox, 1977).

The *substantia nigra,* particularly well developed in man (Figs. 4-5 and 4-10), is seen on macroscopic inspection of the transected mesencephalon as a darkly colored arched stripe on both sides in the cerebral peduncle. The dark color is due to the presence in its nerve cells of densely packed abundant pigment granules (melanin) dispersed in the cytoplasm. Ventral to this dark zone (the so-called pars compacta) the substantia nigra consists of a broader zone, composed of more scattered nerve cells, which partly penetrate the bundles in the pes pedunculi ventrally. This part of the substantia nigra (the so-called pars reticulata) can also be discerned ventral to the dark zone with the naked eye. Like the pallidum and the red nucleus, it contains a certain amount of iron, particularly in the glial cells.

Many Golgi studies in the rat (Gulley and Wood, 1971; Bak et al., 1975; Juraska, Wilson, and Groves, 1977), in the cat (Rinvik and Grofová, 1970), and in the monkey (Schwyn and Fox, 1974) have shown that several types of neurons can be identified in the substantia nigra on the basis of soma size, dendritic caliber, and arborization. It is of particular interest that a large number of cells in the pars compacta send their dendrites ventrally into the pars reticulata of the substantia nigra (see Fig. 4-12). Electron microscopic studies (Rinvik and Grofová, 1970; Gulley and Wood, 1971; Schwyn and Fox, 1974) have disclosed that these dendrites are particularly densely covered by boutons, giving the neuropil of the pars reticulata a striking resemblance to the globus pallidus ultrastructurally (Fox, Andrade, LuQui, and Rafols, 1974). In addition to the similarities mentioned between the two nuclei, they often display corresponding changes in pathological conditions.

Fiber connections of the basal ganglia and related nuclei. Concerning this subject our knowledge is still rather incomplete. One reason for this is that these nuclei are difficult to examine by experimental methods, as they lie buried in the interior of the brain. It is very difficult to place a lesion in one of these structures without inflicting some damage on others and on adjacent or passing fiber tracts. Some of the various fiber connections which have been described are probably only by-passing fibers.

Human material seldom gives satisfactory results. Even if the lesions may be limited to the deeper parts of the brain without injuring superficial structures, they are, as a rule, so large that they involve different, frequently heterogeneous, nuclei. Studies of normal material are not decisive as concerns the direction of fiber tracts, and can only give crude information. Only with the introduction of silver impregnation methods for experimental studies (see Chap. 1) and the use of stereotactical placement of restricted lesions was it possible to obtain a rough idea of the basic features of the organization of the fiber connections of the basal ganglia. Later, biochemical and fluorescent histochemical techniques provided further information. Thus, convincing evidence has been provided for the existence of a prominent nigrostriatal dopaminergic projection (Andén, Carlsson, Dahlstrøm, Fuxe, Hillarp, and Larsson, 1964; Hornykiewicz, 1966; Bédard, Larochelle, Parent, and Poirier, 1969) and a serotoninergic connection from the dorsal nucleus of the raphe (Poirier, Singh, Boucher, Bouvier, Olivier, and Larochelle, 1967; Kostowski, Giacalone, Garattini, and Valzelli, 1968; Poirier, McGeer, Larochelle, McGeer, Bédard, and Boucher, 1969; Kuhar, Roth, and Aghajanian, 1971; Kuhar, Aghajanian, and Roth, 1972). More recently autoradiographic and horseradish peroxidase tracing techniques have to an unprecedented degree provided new data on the precise origin and detailed organization of the fiber connections of the basal ganglia. As usual, it has turned out that there are more connections than was previously assumed and, furthermore, that the relations between nuclei are more specific than was formerly believed. The account given below is restricted to some major points.[13]

In general it may be said that *the various subnuclei of the basal ganglia are richly interconnected*. The major afferent connections to the basal ganglia appear to be fibers from the cerebral cortex to the caudate nucleus and putamen. Only the inner segment of the globus pallidus and the substantia nigra seem to project to any significant degree to structures outside the basal ganglia proper, most notably to the thalamus. Some of the main connections are shown in Fig. 4-11.

Cortical projections to the striatum have been a matter of some dispute. It was only with the introduction of silver impregnation methods that the existence of corticostriatal fibers was unequivocally demonstrated. Using these methods in the rat (Webster, 1961; Knook, 1965), rabbit (Carman, Cowan, and Powell, 1963), cat (Webster, 1965), and monkey (Kemp and Powell, 1970), investigators have shown that *virtually all major cortical areas project upon both the caudate nucleus and the putamen*. The corticostriatal projection is topographically organized, although a considerable degree of overlap is present. The frontal lobe, including the sensorimotor areas, is related to the anterior part of the head of the caudate nucleus and the precommissural part of the putamen, the parietal cortex to the body of the caudate nucleus and the postcommissural part of the putamen, and

---

[13] For details the reader is referred to the recent reviews by Carpenter (1976), Grofova (1979), and Fonnum and Walaas (1979).

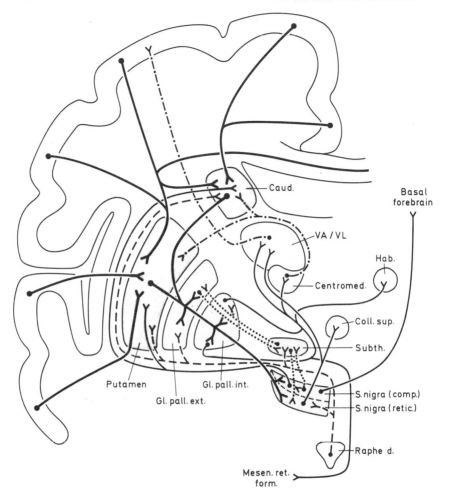

FIG. 4-11 Diagram of some of the main connections of the basal ganglia. For details see text.

the occipital visual cortex to the most posterior parts of both the caudate nucleus and the putamen. There is also a mediolateral pattern in the organization of the corticostriatal projection, which appears to differ somewhat between monkeys (Kemp and Powell, 1970) and lower mammals (Webster, 1961, 1965; Knook, 1965; Carman, Cowan, and Powell, 1963). The difference is apparently due to the development of the temporal lobe in primates.

However, at least in the monkey, recent autoradiographic investigations on the corticostriatal projections have revealed a somewhat different distribution pattern within the striatum than did the silver impregnation studies (Künzle, 1975b, 1977, 1978; Künzle and Akert, 1977).

Thus, somatic motor area 4 has been found to project bilaterally to a large anteroposterior division of the putamen. This projection is somatotopically organized; fibers from the cortical leg and tail areas terminate most rostrodorsally, those from the face area more caudoventrally

(Künzle, 1975b). On the other hand, fibers from somatic sensory areas 3a, 3b, 1, and 2 are strictly ipsilateral but have a distribution within the putamen similar to that seen in the projection from area 4 (Künzle, 1977; Jones, Coulter, Burton, and Porter, 1977). The motor and the sensory cortical areas project to nearly the entire anteroposterior extent of the putamen, but apparently only negligibly upon the caudate nucleus. (Prefrontal area 9, however, projects upon the entire rostrocaudal extent of the caudate nucleus; Goldman and Nauta, 1977.)

It appears from these autoradiographic studies on the corticostriatal projections that the caudate nucleus and the putamen cannot be considered a functional entity. It seems safe to predict that further autoradiographic—as well as other studies—on striatal afferent connections will further emphasize the necessity to differentiate between the caudate nucleus and the putamen.

The autoradiographic studies of Yeterian and Van Hoesen (1978) in the monkey have added interesting details. These authors could show that association areas of the frontal, parietal, occipital, and temporal lobes typically project upon more than one of the major subdivisions of the caudate nucleus. Furthermore, in addition to having unique overall patterns of projections to the caudate nucleus, cortical areas that have reciprocal corticocortical connections project in part to the same region of the nucleus. This observation implies that a given region of the caudate nucleus receives input not only from a particular area of the cerebral cortex, but also from all other cortical areas reciprocally interconnected with that area.[14] The abovementioned observations strongly endorse the cautious view expressed by Teuber (1976) against considering the caudate nucleus as having only either "motor" or "sensory" functions.

It has for a long time been debated whether the corticostriate fibers are axons of separate cortical cells or whether they are collaterals of other corticofugal, for example corticospinal, fibers. Although this question is far from being definitely settled, recent studies using the retrograde axonal transport of horseradish peroxidase (Wise and Jones 1977b; Jones, Coulter, Burton, and Porter, 1977) have shown that the cells of origin of corticostriatal fibers are invariably located in cortical layer V and that they appear to form a group that is distinct from the cells of origin of other subcortical projections, particularly the corticospinal (see also Chap. 4).

Next to the cerebral cortex, the *substantia nigra* is the major source of afferents to the striatum. Although several older neuroanatomists had on the basis of retrograde cell degeneration studies suggested that there is a *nigrostriatal projection,* (see Grofová, 1979, for references), this pathway could be demonstrated neither with the Marchi technique nor with the silver impregnation methods of Glees (1945) and Nauta (1954). The first convincing evidence of the existence of a nigrostriatal pathway was provided by biochemical and fluorescent histochemical techniques (Andén, Carlsson, Dahlstrøm, Fuxe, Hillarp, and Larsson, 1964; Poirier and Sourkes, 1965; Hornykiewiez, 1966; Ungerstedt, 1971). Shortly thereafter the nigrostriatal projection was demonstrated by means of modifications of the silver impregnation techniques (Moore, Bhatnagar, and Heller, 1971; Szabo, 1971; Carpenter and Philip, 1972; Ibata, Nojyo, Matsuura, and Sano, 1973; Maler, Fibiger, and McGeer, 1973; Usunoff, Hassler, Romansky, Usunova, and Wagner, 1976), by retrograde axonal transport of horseradish peroxidase (Nauta, Pritz, and Lasek, 1974; Sotelo and Riche, 1974; Kuypers, Kievit, and Groen-Klevant, 1974;

---

[14] The corticostriatal terminals originating from a particular cortical area show a patchy, striplike distribution. Strips or clusters of terminals, originating in several cortical areas, lie in intimate relation to one another within the striatum.

Miller, Richardson, Fibiger, and McLennan, 1975; Szabo, 1977), and by anterograde transport of tritiated proteins (Carpenter, Nakano, and Kim, 1976; Domesick, Beckstead, and Nauta, 1976). The nigrostriatal fibers originate mainly from cells in the pars compacta of the substantia nigra. Some neurons in the pars reticulata of the nucleus and some paranigral cell groups in the ventral and ventrolateral tegmental areas also contribute to this projection (Sotelo and Riche, 1974; Butcher and Giesler, 1977; Kocsis and Vandermaelen, 1977; Szabo, 1977).

> The fibers are topographically organized and ascend along the dorsomedial border of the substantia nigra to the lateral hypothalamus, enter the medial part of the internal capsule, and course in a dorsorostral direction to reach the head of the caudate nucleus and the rostral part of the putamen. Fibers destined for more posterior parts of the striatum leave the main bundle in Forel's field, course laterally, dorsal to the subthalamic nucleus, and penetrate the posterior part of the internal capsule and the globus pallidus.

The nigrostriatal projection is generally referred to as being dopaminergic and is thought to represent the principal source of striatal dopamine (Andén, Fuxe, Hamberger, and Høkfelt, 1966; Høkfelt and Ungerstedt, 1969; Ungerstedt, 1971). Degeneration of the nigrostriatal pathway is considered an important feature in the etiology of Parkinson's disease (Ehringer and Hornykiewicz, 1960; Hornykiewicz, 1971). This forms the rationale for the treatment of these patients with a precursor in the synthesis of dopamine—l-dihydroxyphenylalanine (l-DOPA) (Birkmayer and Hornykiewicz, 1961; Cotzias, Van Woert, and Schiffer, 1967). In recent years, however, dopamine in the central nervous system has been implicated in a variety of disorders other than Parkinson's disease.

> Of particular interest are the findings that the drugs used to treat schizophrenia act as dopamine antagonists in the brain (for reviews see Matthysse, 1973; Snyder, Banerjee, Yamamura, and Greenberg, 1974; Iversen, 1975). In this connection, however, it should be recalled that there are several dopaminergic pathways in the brain other than the nigrostriatal projection, although this seems to be the largest and so far the best-studied dopaminergic system. It is still far from clear which of the various dopaminergic pathways is crucial for the antipsychotic effects observed following a blockade of the dopamine receptors.

A third important component of striatal afferents originates in the thalamus. A *thalamostriatal projection* has been demonstrated in human neuropathological material (Vogt and Vogt, 1941; McLardy, 1948; Simma, 1951), and in experimental studies using retrogade cellular degeneration (Droogleever-Fortuyn, 1950; Droogleever-Fortuyn and Stevens, 1951; Powell and Cowan, 1954, 1956), orthograde fiber degeneration techniques (Nauta and Whitlock, 1954; Mehler, 1966), and retrograde transport of horseradish peroxidase (Jones and Leavitt, 1974; Kuypers, Kievit, and Groen-Klevant, 1974; Nauta, Pritz, and Lasek, 1974). These connections *originate in the intralaminar thalamic nuclei*—that is, the nuclei centralis lateralis, centralis medialis, paracentralis, centrum medianum, and parafascicularis—and they are topographically organized. Nuclei centrum medianum and parafascicularis are connected with more posterior parts of the striatum, whereas the other intralaminar nuclei project upon more anterior parts.

> Electron microscopic evidence (Kemp and Powell, 1971b) and recent autoradiographic investigations (Kalil, 1978; Royce, 1978a) indicate that the thalamostriatal fibers terminate in discontinuous clusters in the caudate nucleus in the same manner as the afferent fibers from the cerebral cortex. There are some discrepancies, however, in the findings reported in the two recent au-

toradiographic studies. Royce (1978a) describes a prominent projection from the nucleus centrum medianum (CM) to both the caudate nucleus and the putamen in the cat, whereas in the monkey, Kalil (1978) reports a fiber connection from CM to the putamen only. It remains an open question whether species differences may explain the discrepant results.

A fourth quantitatively important source of afferents to the striatum appears to be the *raphe nuclei,* as recently shown by several investigators using the horseradish peroxidase tracing technique (Nauta, Pritz, and Lasek, 1974; Sotelo and Riche, 1974; Miller, Richardson, Fibiger, and McLennan, 1975) or autoradiography (Conrad, Leonard, and Pfaff, 1974; Bobillier, Seguin, Petitjean, Salvert, Touret, and Jouvet, 1976).

Destruction of the dorsal nucleus of the raphe or interruption of fibers presumably arising in this nucleus results in an extensive reduction of striatal serotonin and its synthetizing enzyme, tryptophan hydroxylase (Poirier, Singh, Boucher, Bouvier, Olivier, and Larochelle, 1967; Poirier, McGeer, Larochelle, McGeer, Bédard, and Boucher, 1969; Kuhar, Roth, and Aghajanian, 1971; Kuhar, Aghajanian, and Roth, 1972). Although the highest concentrations of serotonin in the brain are found in the striatum, it appears to be located mainly in the ventrocaudal regions of the striatum (Conrad, Leonard, and Pfaff, 1974; Bobillier, Petitjean, Salvert, Ligier, and Seguin, 1975; Ternaux, Héry, Bourgoin, Adrien, Glowinski, and Hamon, 1977).

The observations mentioned above and other data underline the importance of not considering the striatum a homogeneous structure. Physiological studies have shown that stimulation of the dorsal nucleus of the raphe results in a strong inhibition in the striatum (Olpe and Koella, 1977), but so far practically nothing is known about the functional significance of the serotoninergic input to the striatum.[15]

While the striatum thus receives at least four major afferent fiber contingents, the only well-documented efferent projections of the caudate nucleus and the putamen are those to the globus pallidus and to the substantia nigra. The striatofugal fibers are very thin (average diameter, 0.6 μm) and poorly myelinated (Verhaart, 1950; Adinolfi and Pappas, 1958; Fox, Rafols, and Cowan, 1975). They collect in small bundles, the so-called pencils of Wilson (Wilson, 1914), that converge radially upon the globus pallidus and pass through and along this nucleus. Further caudally the striatofugal fibers traverse the cerebral peduncle as the ventral component of Edinger's "comb system" and then enter the substantia nigra (Nauta and Mehler, 1966; Fox and Rafols, 1975, 1976; Fox, Rafols, and Cowan, 1975). Golgi and electron microscopic studies indicate that the striatopallidal fibers are collaterals of the axons passing to the substantia nigra (Fox and Rafols, 1975; Fox, Rafols, and Cowan, 1975). There is a precise topographical organization of the striatofugal peojection in the monkey (Szabo, 1962, 1967, 1970, 1972) and in the rat (Knook, 1965), and they are similar to those in the cat (Grofová, personal communication). Fibers from the caudate nucleus project upon the dorsal parts of both segments of the globus pallidus, whereas fibers from the putamen course and terminate ventrally. There is also a distinct mediolateral orga-

[15] There is evidence suggesting that the very high concentration of acetylcholine in the striatum is contained in interneurons, since destruction of the main known afferent connections to the caudate nucleus and the putamen does not result in a reduction of striatal acetylcholine levels (McGeer, McGeer, Fibiger, and Wickson, 1971; Butcher and Butcher, 1974; McGeer, McGeer, Grewaal, and Singh, 1975; McGeer, Hattori, Singh, and McGeer, 1976).

nization of the striatopallidal projection, whereas the rostrocaudal organization is not so obvious but can be seen in horizontal sections.

The *substantia nigra* represents the other major area of termination of striatofugal fibers. Since a particularly large number of investigations has been devoted to the study of this nucleus during the last decade, it is deemed justified to consider its fiber connections in some detail (Figs. 4-11 and 4-12). During their course through the posterior limb of the internal capsule and the cerebral peduncle the fibers from the ventromedial and rostralmost part of the striatum are gradually displaced dorsally by fibers originating in more caudal and dorsal regions of the striatum. The striatonigral fibers are distributed to the pars compacta and to the pars reticulata of the substantia nigra in a distinct mediolateral topographical pattern. Fibers from the ventromedial part of the head of the caudate nucleus end most medially and those from the posterior putamen most laterally. The anteroposterior organization seen in silver-impregnated sections appears somewhat doubtful in the light of recent autoradiographic studies in the cat (Nauta, 1974) and the rat (Domesick, 1977).

The question of whether the globus pallidus projects upon the substantia nigra has been difficult to settle definitely with experimental silver degeneration techniques since the great majority of the striatonigral fibers course through the globus pallidus. Autoradiographic studies, avoiding the problem of injury to passing fibers, have, however, demonstrated the existence of a *pallidonigral projection* in the rat (Hattori, Fibiger, and McGeer, 1975) and in the monkey (Kim, Nakano, Jayaraman, and Carpenter, 1976). Studies based on the retrograde transport of horseradish peroxidase (Grofová, 1975, 1979; Bunney and Aghajanian, 1976a; Kanazawa, Marshall, and Kelly, 1976; Tulloch, Arbuthnott, and Wright, 1978) have further confirmed the existence of a pallidonigral projection and have provided new details on the precise origin and termination of these fibers. Thus,

FIG. 4-12   Diagram illustrating the principal features in the organization of the substantia nigra. Note that the dendrites of many cells in the pars compacta (SN comp.) extend into the pars reticulata (SN retic.). For details see text and Fig. 4-11.

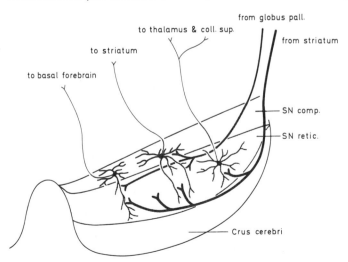

Grofová (1975, 1979) found evidence of a small projection from the entopeduncular nucleus in the cat (the feline homologue of the medial globus pallidus in man) to the pars compacta of the substantia nigra but not to its pars reticulata.

Numerous recent studies, using the horseradish peroxidase technique, have also revealed that several brain structures other than the neostriatum and the globus pallidus project upon the substantia nigra. Among these nuclei are the *dorsal nucleus of the raphe* (Bunney and Aghajanian, 1976a; Kanazawa, Marshall, and Kelly, 1976; Tulloch, Arbuthnott, and Wright, 1978), the *central nucleus of the amygdala* (Bunney and Aghajanian, 1976a), and the *bed nucleus of the stria terminalis* (Grofová, 1979; Tulloch, Arbuthnott, and Wright, 1978).

Bunney and Aghajanian (1976a) have also reported the existence of a projection from the *nucleus accumbens septi* upon the substantia nigra, confirming the autoradiographic studies of Swanson and Cowan (1975). Other investigators, however, did not observe labeled cells in the nucleus accumbens after injections of horseradish peroxidase in the substantia nigra (Grofová 1975, 1979; Kanazawa, Marshall, and Kelly, 1976; Tulloch, Arbuthnott, and Wright 1978). One possible explanation for these discrepant observations may be the difficulties in distinguishing between the dopaminergic cells in the ventral tegmental area of Tsai and the medial part of the pars compacta of the substantia nigra. On the other hand, however, it appears well established that the ventral tegmental area projects upon the nucleus accumbens (see Bjørklund and Lindvall, 1978, for references).

Another controversial question concerns the existence of a *corticonigral projection*. Although earlier investigators using the Marchi technique had described a massive corticonigral projection, the existence of this projection was questioned in an experimental silver degeneration study (Rinvik, 1966) and in an electron microscopic study (Rinvik and Walberg, 1969) in the cat. In recent years, however, several authors have described a few scattered labeled cells in the prefrontal cerebral cortex following injections of horseradish peroxidase in the substantia nigra (Bunney and Aghajanian, 1976a; Tulloch, Arbuthnott, and Wright, 1978; Grofová, 1979). In a recent autoradiographic study, Künzle (1978) described a patchy distribution of corticofugal fibers to the pars compacta of the substantia nigra from frontal areas 6 and 9 in the monkey. A third, until recently unexpected, source of afferents to the substantia nigra appears to be the *subthalamic nucleus*. This has traditionally been considered to receive afferents from the external pallidal segment and to project back upon the internal segment of the globus pallidus (see below). Several authors have, however, reported a substantial number of labeled cells in the subthalamic nucleus following injections of horseradish peroxidase in the substantia nigra (Kanazawa, Marshall, and Kelly, 1976; Tulloch, Arbuthnott, and Wright, 1978; Grofová, 1979). In a recent autoradiographic study in the cat and the monkey, Nauta and Cole (1978) confirmed the existence of a massive subthalamonigral projection.

It should be apparent from the many studies cited above that the substantia nigra receives several afferent fiber systems in addition to those from the neostriatum. Similarly, it has become clear that the substantia nigra does not send its fibers to the striatum only (see above). Thus, a *prominent nigrothalamic projection* has been demonstrated, which enables the substantia nigra to influence the cerebral motor cortex by way of the thalamus. By way of the nigrostriatal projection the substantia nigra may, however, exert a similar influence on the cerebral cortex via the route: substantia nigra–striatum–globus pallidus–thalamus–cortex.

The nigrothalamic projection has been demonstrated with silver impregnation techniques (Cole, Nauta, and Mehler, 1964; Afifi and Kaelber, 1965; Faull and Carman, 1968; Carpenter and Philip, 1972), with retrograde axonal transport of horseradish peroxidase (Rinvik, 1975), and with the autoradiographic technique (Carpenter, Nakano, and Kim, 1976). This fiber projection appears to originate mainly—if not solely—from cells of the pars reticulata of the substantia nigra and to terminate in the ventral medial (VM) and in parts of the medial areas of the ventral lateral nucleus (VL) of the thalamus. The latter thalamic nucleus (see Chap. 2) is known to project upon the cerebral motor cortex (Le Gros Clark and Boggon, 1935; Walker, 1938, 1949; Mettler, 1943, 1947; Chow and Pribram, 1956; Macchi, 1958; Strick, 1973).

Using the retrograde axonal transport of horseradish peroxidase, several authors have recently established the existence of a heavy *nigrotectal projection* (Hopkins and Niessen, 1976; Rinvik, Grofová, and Ottersen, 1976; Grofová, Ottersen, and Rinvik, 1978). This projection is bilateral, although mainly ipsilateral, and appears to originate only in the pars reticulata of the substantia nigra and to terminate in the intermediate and deep layers of the superior colliculus, as recently confirmed in an autoradiographic investigation in the monkey (Jayaraman, Batton, and Carpenter, 1977). It thus appears that the nigrotectal fibers can influence directly the cells of origin of tectospinal fibers, which are located in the deeper collicular layers (see Chap. 7 and Fig. 7-31) (Kuypers and Maisky, 1975). The physiological study of York and Faber (1977) indicates that the nigrotectal projection may be of importance with respect to visual orienting behavior.

Since both nigrotectal and nigrothalamic fibers appear to originate only in the pars reticulata of the substantia nigra, one obvious question is whether the two projections belong to separate cell populations. Physiological studies have suggested that nigrofugal fibers to the thalamus and to the superior colliculus are branches of the same stem axons (Anderson and Yoshida, 1977; Deniau, Hammond, Riszk, and Féger, 1978). In a horseradish peroxidase study in the rat, however, Faull and Mehler (1979) emphasized a separation within the pars reticulata of the cells of origin of the nigrothalamic and nigrotectal projections, although some overlap is evident. The recent elegant study of Bentivoglio, van der Kooy, and Kuypers (1979), using a retrograde fluorescent double tracing technique (cf. Chap. 1), has to a considerable degree clarified this problem. According to these authors, a great number of nigral cells in the pars reticulata project to either the superior colliculus or to the thalamus. A considerable number of cells, however, also showed double fluorescence, indicating that they have dichotomizing axons, sending one branch to the superior colliculus and the other to the thalamus. In the same investigation (Bentivoglio, van der Kooy, and Kuypers, 1979) no evidence was found of a dichotomy of the nigrostriatal axons.

In addition to the abovementioned connections, recent studies using injections of horseradish peroxidase have revealed some quantitatively minor projections from the substantia nigra to the reticular formation of the lower brainstem (Rinvik, Grofová, and Ottersen, 1976) and to the lateral part of the periaqueductal gray substance (Hopkins and Niessen, 1976; Grofová, Ottersen, and Rinvik, 1978).

Of particular interest are recent studies indicating that cells in the pars compacta of the substantia nigra project not only to the striatum but also to basal forebrain structures such as the olfactory tubercle and the amygdala and even the frontal cerebral cortex (Thierry, Blanc, Sobel, Stinus, and Glowinski, 1973; Fuxe, Hökfelt, Johansson, Jonsson, Lidbrink, and Ljungdahl, 1974; Lindvall, Björklund, Moore, and Stenevi, 1974; Avendaño, Reinoso-Suarez, and Llamas, 1976; Björklund and Lindval, 1978). There is ample biochemical, histochemical, and anatomical evidence suggesting that these nigrofugal projections to basal forebrain structures are dopaminergic (for references, see Björklund and Lindvall, 1978). It remains to be settled whether these fibers are collaterals of the dopaminergic nigrostriatal axons. A recent study in the rat indicates that separate populations of dopamine-containing cells in the pars compacts of the substantia nigra project to the neostriatum and to allocortical areas (Fallon, Riley, and Moore, 1978).

In the last few years there has been an impressive increase in the number of publications dealing with the *so-called dopamine pathways* in the brain. The great interest devoted to these projections stem particularly from the observation that the drugs used to treat schizophrenia act as dopamine antagonists in the brain (for references see Matthysse, 1973; Snyder, Banerjee, Yamamura, and Greenberg, 1974; Iversen, 1975). Although the *nigrostriatal projection* seems to be the largest—or at least so far the most extensively studied—dopaminergic pathway in the central nervous system, several lines of research during the last few years clearly suggest that dopamine-containing neurons in the mesencephalon are not localized only to the pars compacta of the substantia nigra. There appears to be a continuum of dopamine-rich neurons extending from the medially situated ventral tegmental area of Tsai through the pars compacta of the substantia nigra and even extending dorsally with scattered cells in the ventral part of the mesencephalic reticular formation (for a recent review see Bjørklund and Lindvall, 1978). The cells within this band of neurons have distinct efferent and afferent fiber connections. This fact should be borne in mind. It serves as a warning against making generalizations concerning the role of dopamine in brain functions.

It is obvious from what has been summarized above that by way of its diverse efferent and afferent fiber connections the substantia nigra may play a versatile role in several higher brain functions. Until fairly recently the only putative neurotransmitter associated with nigral function was dopamine in the nigrostriatal pathway.[16] In recent years, however, attention has been directed toward the detection of—and the role played by—other putative transmitter substances in the other nigral pathways.

In biochemical studies it has been shown that after lesions of the striatum there is a very substantial reduction of GAD—the enzyme necessary for the synthesis of GABA—in the substantia nigra (Fonnum, Grofová, Rinvik, Storm-Mathisen, and Walberg, 1974) and in the globus pallidus (Fonnum, Gottesfeld, and Grofová, 1978). These observations, together with earlier physiological studies (Yoshida and Precht, 1971; Precht and Yoshida, 1971), strongly suggest that the striatofugal projection—or at least part of it—uses GABA as the transmitter substance. The elegant immunochemical investigations of Roberts and collaborators (Ribak, Vaughn, Saito, Barber, and Roberts, 1976) support the earlier findings by demonstrating that most of the boutons in the substantia nigra are filled with the GAD-positive reaction product. They form axodendritic and axosomatic synapses.

Several recent studies indicate that another putative transmitter—an undecapeptide called substance P, first discovered by von Euler and Gaddum (1931)—is also transported in striatonigral fibers (Kanazawa, Emson, and Cuello, 1977; Hong, Yang, Racagni, and Costa, 1977). Some indirect evidence suggests that substance P and GABA in nigral afferent fibers originate in separate cell groups in the striatum (Gale, Hong, and Guidotti, 1977), but the possibility that both substances may be transported in the same axon still remains open.

According to Hattori, McGeer, Fibiger, and McGeer (1973), there is evidence suggesting that, like the striatonigral, the pallidonigral projection is GABAergic. In an elegant study these authors have further demonstrated that the pallidonigral fibers terminate mainly on dopaminergic cells in the rat's substantia nigra, whereas the striatonigral fibers probably end on nondopaminergic as well as dopaminergic nigral cells (Hattori, Fibiger, and McGeer, 1975). In physiological studies it has been shown that the striatonigral fibers contact monosynaptically the nigrothalamic and nigrotectal neurons (Deniau, Féger, and LeGyuader, 1976), whereas a possible

---

[16] In this connection it may be appropriate to call attention to the evidence suggesting that not all nigrostriatal neurons are dopaminergic (Feltz and De Champlain, 1972; Fibiger, Pudritz, McGeer, and McGeer, 1972; Ljungdahl, Hökfelt, Goldstein, and Park, 1975).

monosynaptic effect of striatonigral fibers on the dopaminergic nigrostriatal neurons is still debated (Bunney and Aghajanian, 1976 b; Kitai, Wagner, Precht, and Ohno, 1975). Kitai et al., however, found no evidence of a direct, monosynaptic effect of nigral neurons on the caudatonigral projecting cells and are inclined to believe that the dopaminergic nigrostriatal fibers may act on striatal interneurons (see also McGeer, McGeer, Grewaal, and Singh, 1975).

In a biochemical and unltrastructural study Vincent, Hattori, and McGeer (1978) have shown that the nigrotectal fibers originating in the pars reticulata are also probably GABAergic. Since some nigrotectal cells send axon collaterals to the thalamus, it seems probable that the nigrothalamic projection may also be GABAergic.

As is seen from the brief account given above, in recent years our knowledge of the complex connections and the biochemistry of the nigral projections has increased considerably. The many new data on these subjects serve as a *strong warning against considering dysfunctions of the substantia nigra in terms of nigrostriatal dopamine alone*.

The fiber connections of the subthalamic nucleus (see Fig. 4-11) have been particularly difficult to study because so many fiber systems course close to the nucleus and even traverse parts of it. Until very recently the only well-established afferent projection of the subthalamic nucleus seemed to originate in the external segment of the globus pallidus (Ranson, Ranson, and Ranson, 1941; Nauta and Mehler, 1966; Nauta, 1974). The *pallidosubthalamic projection* is topographically organized (Carpenter, Frazer, and Shriver, 1968; Grofová, 1969) and appears to use GABA as the neurotransmitter, as shown in a biochemical study (Fonnum, Grofová, and Rinvik, 1978). In a recent investigation in the cat and the monkey it has been shown by means of the technique of retrograde transport of horseradish peroxidase that several *nuclei other than the lateral pallidal segment project upon this nucleus,* notably the *dorsal raphe nucleus,* the *nucleus locus coeruleus,* the *pars compacta of the substantia nigra,* and the *pedunculopontine nucleus* (Rinvik, Grofová, Deniau, Féger, and Hammond, 1979). The functional significance of this varied input to the subthalamic nucleus is not known. It is interesting that the nigrosubthalamic projection is reciprocated by a prominent subthalamonigral projection, as recently shown in an autoradiographic study in the cat and the monkey by Nauta and Cole (1978). These authors have confirmed that the *major efferent projection from the subthalamic nucleus is destined for the globus pallidus,* as shown in previous studies (Carpenter and Strominger, 1967).

As has been mentioned above, the globus pallidus consists of a medial and a lateral segment. In an autoradiographic study in the cat and the monkey Nauta and Cole (1978) have shown that the subthalamic nucleus projects profusely upon both segments of the globus pallidus, and not only to the medial segment, as had been generally thought. The *medial segment of the globus pallidus* represents, in addition to the pars reticulata of the substantia nigra, the only part of the basal ganglia which sends fiber projections outside the basal ganglia proper. These pallidofugal fibers are arranged in two distinct bundles, the *ansa lenticularis* and the *fasciculus lenticularis*. The ansa and fasciculus lenticularis follow different courses with respect to the internal capsule. The ansa lenticularis originates mainly in the lateral part of the medial segment of the globus pallidus. The fibers course along the ventral border of the pallidum and sweep around the posterior limb of the internal capsule before entering Forel's field H. The fasciculus lenticularis arises from the dorsomedial part of the medial pallidal segment and traverses the internal capsule in several small fascicles. The ansa and fasciculus lenticularis merge in Forel's field H and then pass laterally and rostrally in the *fasciculus thalamicus*. Most of the fibers in the fasciculus thalamicus terminate in the VLo (pars oralis of the ventral lateral nucleus of the thalamus), the lateral part of the VLm (pars medialis of the VL, according to the terminology of Olszewski, 1952), and in the VApc (pars principalis of the ventral anterior nucleus), as described by Nauta and Mehler (1966), Kuo and Carpenter (1973), and Kim, Nakano, Jayaraman, and Carpenter (1976).

The pallidofugal projection to these thalamic nuclei was established by early investigators using the Marchi technique for tracing degenerating myelinated axons (Ranson, Ranson, and Ranson, 1941; Glees, 1945). The detailed description of the pallidofugal projections, however, was made by Nauta and Mehler (1966) using silver impregnation techniques in the monkey. These authors also described a substantial number of fibers that take off from the fasciculus thalamicus and terminate in the nucleus centrum medianum (CM) of the thalamus. This projection has been confirmed in autoradiographic studies (Kim, Nakano, Jayaraman, and Carpenter, 1976).

Kim et al. (1976), furthermore, described a rostrocaudal, dorsoventral, and possibly mediolateral organization in the pallidothalamic projection, although a considerable overlap was seen in the rostral and ventral thalamic nuclei. It has also been shown that the ansa and fasciculus lenticularis terminate mainly in the VApc and VLo, respectively.

The medial segment of the globus pallidus has also been reported to give origin to several smaller projections whose functional significance remains to be determined. Thus pallidofugal fibers have been followed to the pedunculopontine nucleus of the brainstem (Olszewski and Baxter, 1954), which lies partially embedded in the fibers of the superior cerebellar peduncle (Nauta and Mehler, 1966; Carpenter and Strominger, 1967). A substantial number of pallidofugal fibers have also been described to pass to the lateral habenular nucleus (Ranson, Ranson, and Ranson, 1941; Nauta and Mehler, 1966; Carpenter, Frazer, and Shriver, 1967; Kuo and Carpenter, 1973; Nauta, 1974; Kim, Nakano, Jayaraman, and Carpenter, 1976; Herkenham and Nauta, 1977).

The *claustrum* (Fig. 4-9) is a morphologically well-characterized lamina of gray matter separated from the cortex of the insula by the capsula extrema. Nothing definite appears to be known of its function, and only recently have we obtained some information on its fiber connections. While Berke (1960), who reviews the subject, did not obtain decisive results with the Marchi method in the monkey, authors using silver impregnation methods have succeeded in demonstrating a projection from the cerebral cortex. According to Carman, Cowan, and Powell (1964) and Druga (1968) the projection is very similar to the corticostriate projection insofar as all parts of the cerebral cortex project in an orderly topical pattern onto the claustrum. This projection is reciprocated by a prominent claustrocortical connection, as recently shown in a horseradish peroxidase study in the cat (Macchi, Bentivoglio, Minciacchi, Rossini, and Tempesta, 1978). It is interesting that the claustrofugal projection to the motor cortex is bilateral, with an ipsilateral prevalence. Other efferent projections of the claustrum are not yet known in sufficient details.

The account given on the preceding pages of the fiber connections of the *basal ganglia, including the subthalamic nucleus and the substantia nigra,* is far from complete. It should be clear, however, that these parts of the brain *can only to a limited extent exert an influence on nuclei that give off fibers to the spinal cord*. Actually, the only important candidate for a function like this appears to be the nigrotectal fiber projection, since this may influence the cells of origin of the tectospinal pathway (see above and Chap. 7). The *basal ganglia appear to be first and foremost concerned in the collaboration between the cerebral cortex and the thalamus*. This follows from the fact that the basal ganglia receive well-organized projections from the cortex and give off the bulk of their efferents to the thalamus. On the other hand, as has been shown above, with the use of new techniques during the last few years it has become very clear that the fiber connections of the basal ganglia are exceedingly more complex than was ever suspected. From a clinical point of view the functional role of the multifarious fiber connections is practically unknown. The widening gap between our steadily increasing knowledge of the anatomy, physiology, and biochemistry of the basal ganglia and their clinical implications is unfortunate and only too evident. It appears so far possible to correlate few clinical observations with the known organization of the basal ganglia. Reference has been made to the treatment of patients who have Parkinson's disease with the dopamine precursor *l*-DOPA. The first enthusiasm with which the therapeutic use of this drug was received has subsided considerably owing to the many short- and long-term side effects that have been observed in patients receiving *l*-DOPA treatment. (For a recent account of these problems the reader should consult *The Extrapyramidal System and its Disorders,* 1979.) We shall return to some of these questions later in this chapter. It is deemed practical to postpone a discussion of this subject until we have considered the motor cortical areas and the consequences of lesions of supraspinal structures and pathways influencing peripheral motor neurons.

The "motor" regions of the cerebral cortex.   The term "cortical motor area" is usually employed without any definition. With strong electrical stimuli movements may be elicited from many parts of the cortex. Various factors, among these the type and depth of anesthesia, influence the excitability of the cerebral cortex and determine the results of stimulation. By varying the stimulus parameters one can obtain different motor maps on the same brain, as shown by Liddell and Phillips (1950). It cannot be stressed too often that electrical stimulation is an entirely artificial method of exciting nervous tissue. However, it is so far the only method (in addition to local application of strychnine and certain other substances) which can be used for local stimulation of the cerebral cortex. If a cortical area is said to have a motor function on the basis of stimulation experiments, it must be required that motor responses are obtained with liminal stimuli. Furthermore, the defects following its destruction should affect motor functions, although these defects need not necessarily be detectable with ordinary methods of examination. (On the other hand, disturbances of motor function following damage to a part of the nervous system does not prove that this part has a "motor" function. Such disturbances may follow damage to definitely sensory structures, the most clear-cut example being perhaps the dorsal funiculi.) To be classified as a "motor" region, a cortical area should be provided with efferent fibers which directly or via a few intercalated cell groups establish a pathway to peripheral motor neurons.[17] As will be described below, the three areas considered today as motor more or less fulfill the requirements mentioned above. However, it appears that a clear definition of what we consider a motor cortical area cannot be given. Here, as in so many other instances in biology and medicine, there are no unequivocal criteria which enable us to define our subject precisely, functionally, or structurally.

For a long time it was customary to speak of one "motor region" of the cerebral cortex, situated in the precentral gyrus in monkeys, apes, and man. This region was originally assumed to correspond to area 4 of Brodmann, probably because the Betz cells of area 4 were believed to be the sole origin of the fibers of the pyramidal tract. As discussed above, the tract originates in many cortical regions in addition to area 4, both in front of and behind the central sulcus, and the original view has had to be thoroughly modified. As discussed in Chapter 2, the precentral "motor cortex" receives sensory information, and motor effects can be obtained on stimulation of the "sensory," postcentral gyrus. Accordingly one could speak of a *sensorimotor cortex,* of which the part in front of the central sulcus is preponderantly motor, the part behind the sulcus preponderantly sensory. (In Woolsey's nomenclature they are labeled Ms I and Sm I, respectively; see Fig. 2-19). Furthermore, just as there is more than one somatosensory cortical area, as described in Chapter 2, *there are at least three somatomotor areas, the first, the second, and the supplementary.* These will be considered below.

> Before the presence of the latter two areas had been established, attempts were made, particularly by Fulton and his collaborators in the 1930s to distinguish, in the precentral cortex, between a "motor" and a "premotor" area, the latter corresponding approximately to area 6 of Brodmann (see Fig. 12-2) and being functionally different from the former. Ablation of the premotor cortex resulted in the appearance of the grasp reflex (see below) and its stimulation gave rise to complex movements. With the use of stronger electrical stimuli than those required for producing discrete movements from the "motor" cortex, movements could, furthermore, be elicited from certain other regions of the cortex. These movements were slower and more complex than those obtained from area 4. Mainly, it appears, due to the general belief that the pyramidal tract came only from area 4, these other, less clearly delimited cortical regions were lumped together as "extrapyramidal motor areas" and were supposed to be essential in the production of coarse movements. These areas were particularly studied in man by the German neurosurgeon Foerster (see his account in *Handbuch der Neurologie,* 1936b) and covered extensive parts of the cortex of all four cerebral lobes.

In the course of time a voluminous literature has grown up on the motor functions of the cerebral cortex, and on the clinical consequences of lesions of the "motor" area, and heated debates have taken place. Not least have there been differences of opinion between experimental workers and clinicians. One of the most ardent spokesmen for the latter has been Sir Francis Walshe, whose eloquent writings on this subject (see, for example, Walshe, 1942b, 1943) still are fascinating reading and certainly did much to focus attention upon discrepancies between clinical and experimental observations. It would take us too far to go through the development in the views of the motor functions of the cerebral cortex. (The situation about 1944 is well summarized in *The Precentral Motor Cortex.*) In the fol-

---

[17] It would be rather far-fetched to consider, for example, all cortical areas giving rise to corticopontine fibers (see Chap. 5) as "motor," even though the pons, via the cerebellum and the projections of the latter to brainstem nuclei, may act on motoneurons.

lowing, reference will be made to only some of the earlier studies. Emphasis will be put on more recent observations on cortical motor function and their anatomical basis in man and in experimental animals. The results of the two lines of research can today be linked in a more fruitful way than was possible a few decades ago.

The first, primary, central, or Rolandic, somatomotor area, originally considered to be represented by Brodmann's area 4, belongs to the so-called agranular cortex (see Chap. 12 and Figs. 8-7 and 12-3). Area 4 is characterized by the presence of the giant cells of Betz in layer V, whereas layers II and IV are poorly developed.[18] As we have seen, the corticospinal tract receives contributions from many cortical areas other than area 4 (such as areas 6, 3, 2, 1, 5, and 7). (Furthermore, other major corticofugal projections—to the thalamus, the striate body, the red nucleus, the reticular formation, and other subcortical stations—arise from the same cortical regions.) Judged on the criterion of efferent fibers, the primary motor cortex will extend beyond the confines of area 4 frontally and will include most of the postcentral gyrus as well. This is in agreement with the results of electrical stimulation of the cortex in animals and in man. The same kinds of movements (see below) as can be elicited from the cortex immediately in front of the central sulcus and belonging to area 4 can be obtained both from the parts of the postcentral gyrus close to the sulcus and from precentral cortex belonging to area 6, although motor responses are less frequent in the postcentral than in the precentral gyrus (Clark and Ward, 1948; Woolsey et al., 1952; Welker et al., 1957; and others). Fig. 4-13 shows Penfield and Boldrey's (1937) map of their observations in man. The central or primary motor area thus is far more extensive than previously assumed. (As mentioned above, Woolsey has suggested that the precentral part may be referred to as Ms I, the postcentral as Sm I; see Fig. 2-19).

Before proceeding, it is appropriate to mention some data on the *finer structure of the primary motor cortex*. This shares main general characteristics with other parts of the isocortex (see Chap. 12). The designation "motor cortex" is not always used in the same sense. Many studies have been concerned with the sensorimotor cortex as a whole, in agreement with the recognition that sharp distinctions cannot be made between "motor" and "sensory" cortical areas. Other studies have been devoted to particular cytoarchitectonic areas. Among these, reference has been made to the "sensory" areas 3, 1, and 2 in Chapter 2; the parietal areas 5 and 7 will be briefly dealt with in Chapter 12. The following data will refer chiefly to area 4.

With regard to the *efferent connections of the motor cortex* an interesting new finding has been repeatedly mentioned in this book: corticofugal fibers to the various subcortical stations are to a considerable extent axons of different cell populations, largely concentrated to particular cortical layers. Subcortical stations (the striatum, the thalamus, the red nucleus, the pontine nuclei, the reticular formation, the inferior olive, and others), therefore, appear to receive much of their information from the cortex via separate routes, and not to be informed from the cortex only by way of collaterals of corticospinal fibers, as has often been maintained. From a functional point of view this means that the *subcortical stations, receiving*

---

[18] Area 6, like area 4, belongs to the agranular cortex but, unlike area 4, lacks Betz cells. In the areas in the postcentral gyrus, layers III and V, giving rise to long corticofugal fibers, are rather well developed.

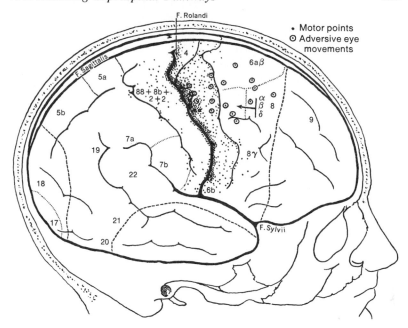

FIG. 4-13 Drawing of the right hemisphere in man showing (dots) the points from which discrete movements can be elicited on electrical stimulation. Note that points are found in the postcentral gyrus and in the entire precentral gyrus. From Penfield and Boldrey (1937).

fibers from the motor cortex, must be less closely linked with the activity of the corticospinal system than often assumed in physiological studies, although collaterals of corticospinal fibers undoubtedly pass to many of these stations.

These new data have emerged most convincingly from HRP studies, in which the types of cells and their distribution within the cortex have been mapped after injections of HRP in the subcortical target area (Berrevoets and Kuypers, 1975; Coulter, Ewing, and Carter, 1976; Jones, Coulter, Burton, and Porter, 1977; Wise and Jones, 1977b; and others).[19]

It appears from these studies that the *corticothalamic* fibers arise chiefly from cells in the deepest cortical layer, VI. Most *corticostriate fibers* (Jones et al., 1977) come from the superficial parts of layer V, whereas *corticospinal fibers* are axons of cells in deeper parts of layer V. Efferents to the *red nucleus*, the *pontine nuclei*, and the *reticular formation* take origin from levels of layer V between the deeply situated neurons projecting to the cord and the superficially situated corticostriate neurons. The latter appear to belong to the smallest pyramidal cells of layer V. On the other hand, corticospinal cells all belong to the largest pyramidal cells, including the Betz cells in area 4. The cortical efferents to the *dorsal column nuclei* come from layer V (Weisberg and Rustioni, 1976, 1977; and others).

[19] Among the efferent connections of many cortical regions are also *long ipsilateral and contralateral association fibers,* to be considered in Chapter 12. Perikarya of commissural fibers are found chiefly in the deeper half of layer III; the ipsilateral association fibers arise mainly from cells superficial to the commissural ones. For some data on this point with regard to the somatosensory areas, see Jones and Wise (1977).

The studies referred to above are of further interest with regard to the problem of the *columnar organization* of the motor cortex. As discussed in section (f) of Chapter 7 and in Chapters 8 and 12, physiological evidence indicates that in several (perhaps all?) regions of the cerebral cortex there is an organizational pattern in the form of functional columns, oriented vertically to the surface. The anatomical basis of these functional columns is still debated. However, some features in the arrangement of cortical cells that send their axons to various sites, cortical and subcortical, are compatible with the presence of a columnar organization within the motor cortex. (For a review, see Asanuma, 1975; see also Jones and Wise, 1977; and Groos, Ewing, Carter, and Coulter, 1978).

> The labeled cells occurring in the motor cortex after HRP injections in the spinal cord or other sites of terminations of corticofugal fibers are usually aggregated in clusters. In the motor cortex of the monkey, clusters of three to five or more cells are separated by gaps of various sizes (Coulter, Ewing, and Carter, 1976; Jones and Wise, 1977). The clusters tend to occur in strips, 0.5–1 mm wide. The strips are oriented mediolaterally across the cortex. This orientation may be reflected in the striplike pattern of termination of the fibers found in several of these efferent projections (see Jones and Wise, 1977, for references).

Some data on the efferents from the motor cortex to their various subcortical targets are considered when the latter are dealt with in other chapters of this book. As to the topically organized *corticothalamic projection* (see Chap. 2), it may be mentioned that in the cat, the monkey, and the chimpanzee, the fibers from the motor cortex have been found to end not only in the VL but also in some other thalamic subdivisions, including some of the intralaminar nuclei and the thalamic reticular nuclei; see reviews by Rinvik (1972, cat) and Petras (1972, chimpanzee). For an autoradiographic study in the monkey, see Künzle (1976).

A considerable proportion of *afferent fibers to the motor cortex* is derived from the thalamus, particularly the VL (nucleus ventralis lateralis). In addition, the motor cortex receives ipsilateral and commissural association fibers. It appears that the terminal fibers of at least some of the three contingents are arranged in a columnar pattern.

> The thalamocortical projection in general has been discussed in Chapter 2. As referred to there (see also Chap. 12), the pattern in this projection is less simple than previously assumed. Some data on the thalamic projection to the precentral gyrus will be referred to in Chapter 5. Suffice it to mention here that, according to HRP studies (Strick, 1975; Kievit and Kuypers, 1977), in the monkey the precentral gyrus receives afferents from other thalamic nuclei in addition to the VL, for example the VPL and some intralaminar nuclei; see Chapter 2. The projection is somatotopically arranged, but organized in a rather complex pattern. The thalamocortical fibers to the motor cortex appear to branch amply. According to the combined anatomical and electrophysiological study of Asanuma, Fernandez, Scheibel, and Scheibel (1974), the fibers supply widely separated parts of the motor cortex in the cat.

The projection from the thalamus (especially the VL) *to the motor cortex* is first and foremost a link in the route for cerebellar influences on the motor cortex. An account of the organization of the cerebellothalamic connections and some comments on their influence on the motor cortex and movements will be postponed to Chapter 5. It should be recalled that via the thalamus, regions of the brain other than the cerebellum may also influence the activity of the motor cortex. Thus the pallidum sends fibers to parts of the thalamus that project to the motor cortex.

It is important to be aware that the afferent input to the motor cortex does not come only from subcortical levels (the thalamus). It may be *influenced from various cortical regions by way of association fibers.* Some of these are intrinsic; others come from other cortical areas in the ipsilateral or contralateral hemisphere.

The *intrinsic association fibers* are restricted to a particular architectonic area and serve to interconnect, for example, regions related to different parts of the body. In area 4 of the monkey Gatter and Powell (1978) found these fibers to extend only for a few millimeters and to pass horizontally but also vertically.[20] According to the experimental electron microscopic study of Gatter, Sloper, and Powell (1978), these fibers most likely are axons of stellate cells. They establish symmetrical contacts with somata, apical dendrites, and initial segments of Betz cells and have been assumed to be inhibitory in function.

The *association connections between area 4 and other cortical regions* are not yet known in all details. It is evident, however, that this cortical area has possibilities for intimate collaboration with the first and second sensory cortex (Sm I and Sm II). In cats (Jones and Powell, 1968a) and in monkeys (Jones and Powell, 1969a; Pandya and Vignolo, 1971; Vogt and Pandya, 1978) both the first and the second somatosensory area project to area 4 (Fig. 4-14). The projections are strictly somatotopically arranged, the arm region of the sensory cortex projecting to the arm region of area 4, and so forth. Reciprocal connections (from area 4 to the somatic sensory areas) show a corresponding pattern. (The discussion above is

[20] The horizontal distribution agrees well with the dimensions of the area of monosynaptic activation of cortical cells found by Asanuma and Rosen (1973) on microstimulation (4 μA) within the cat's motor cortex.

FIG. 4-14 Schematic diagram to show the main association connections between area 4 and the first and second somatosensory areas (S I and S II, respectively) in the monkey. Note that these connections are reciprocal and somatotopically arranged. Other corticocortical connections of area 4, not included in the diagram, are mentioned in the text. *SMA:* supplementary motor area. From Jones and Powell (1969a).

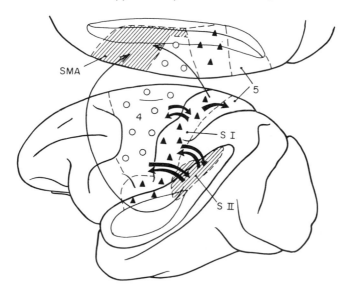

a very simplified account of a very complex pattern. There appear, for example, to be differences between subareas 3a, 3b, 2, and 1. For a recent detailed study the reader should consult Jones, Coulter, and Hendry, 1978.) The presence of association fibers to area 4 from the primary and secondary sensory areas and parietal cortical areas related to somatosensory functions (see Chap. 12) has attracted considerable interest in connection with the demonstration that impulses arising on somatosensory stimulation can be recorded in area 4 (see below). For a recent study of the association fibers from area 3 to area 4 in the cat, see Grant, Landgren, and Silfvenius (1975).[21] A projection from the posterior parietal cortex, area 5, to area 4 has recently been confirmed anatomically with the HRP method (Strick and Kim, 1978) and with recording of antidromic potentials (Zarzecki, Strick, and Asanuma, 1978).

Other ipsilateral association connections of area 4 have been described, for example from the supplementary motor area (see Jones, Coulter, and Hendry, 1978; Sloper and Powell, 1979b). Further, there are association connections with the contralateral hemisphere. It should be noted that homotopic commissural (callosal) connections between the two areas 4 are restricted to face, trunk, and proximal extremity regions, whereas (as in the sensory areas, see Chap. 2) the "representations" of distal parts of the extremities lack such interconnections (see Pandya and Vignolo, 1971).

Some information on the *synaptic organization of the motor cortex* (see also Chap. 12) has been obtained in studies with the Golgi method and electron microscopically. For a recent study of the ultrastructural features of the sensorimotor cortex of monkeys, see Sloper and Powell (1979a).

The afferents from the thalamus to area 4 (ending in layer IV and adjacent parts of layers III and V) and ipsi- and contralateral association fibers appear to establish asymmetric synapses both with spines of pyramidal apical dendrites and with somata and dendrites of large stellate cells (Sloper, 1973; Strick and Sterling, 1974; Sloper and Powell, 1979b). As mentioned above, it appears that axons of such large stellate cells (in layer IV and adjacent parts of layers III and V) establish symmetric synapses with somata, apical dendrites, and initial segments of Betz cells. It has been assumed that the large stellate cells may have an inhibitory action on neighboring cells. Axons of pyramidal cells give off recurrent collaterals that may influence the stellate cells.

In the collaboration between various cortical layers it appears that the apical and basal dendrites of the pyramidal cells may play an important role (see Gatter and Powell, 1978, for some comments). The functional significance of *dendrite bundles,* consisting of 3 to 10 closely packed dendrites, described in the human motor cortex by Scheibel and Scheibel (1978), is not clear. The recent demonstration of the occurrence of dendrodendritic synapses (see Chap. 1) in the motor cortex of monkeys (Sloper and Powell, 1978) is one of many data that give an idea of the complexity of the synaptic "machinery" of the cerebral cortex, *in casu* the motor cortex.

Many *physiological studies of the primary motor cortex* have been undertaken. A fair proportion of them have been devoted particularly to the physiology of the corticospinal neurons. This is probably in part because, of the main efferent projections from the primary motor cortex, only the corticospinal fibers have direct contact with the peripheral motor neurons. In the following we shall consider mainly the results of some of the numerous physiological studies of the motor cortex which have given information about the function of corticospinal

[21] It appears that these association fibers terminate in a pattern that is compatible with the concept of a columnar organization.

neurons. In general, these findings are in good agreement with what is known of the anatomy of the motor cortex. Although results of such studies, even in primates, cannot without qualification be applied to man, they have contributed much to a better understanding of the function of the primary motor cortex and the consequences of lesions of this in man.

Many methods have been applied. The cells of origin of corticospinal and other corticofugal fibers may be stimulated chemically or electrically. "Microstimulation" with currents as low as $5\mu A$ can be used to excite cells in deeper layers of the cortex. The corticospinal cells stimulated may be identified by stimulating them antidromically with electrodes inserted in the pyramidal tract. The intracellular insertion of microelectrodes permits the recording of responses of a single cell to the synaptic actions (excitatory or inhibitory) of the fibers ending on it. The responses to injections of transmitters and drugs can likewise be recorded. Potentials can be led off from the axons of corticospinal cells. The responses of motoneurons in the spinal cord may be recorded by intracellular electrodes inserted into these cells or by recording their activity electromyographically from the muscles they innervate. To some extent recordings may be made in awake animals, trained to perform certain movements. All these techniques are complicated and require critical evaluation of findings. For example, even with microstimulations of the cortex, several corticospinal neurons (and others) will be excited, in part due to physical spread and in part to "physiological spread," by way of collaterals of pyramidal tract cells and connections between neighboring cells, as recently discussed by Phillips and Porter (1977). It is possible here to mention only a few results of numerous recent studies.

As referred to above, in the monkey, α motoneurons can be monosynaptically excited by pyramidal tract volleys. The same appears to be the case for a fair proportion of γ motoneurons (Clough, Phillips, and Sheridan, 1971; Grigg and Preston, 1971). The cortical foci for activation of α and γ neurons to the same muscle appear to be identical (Mortimer and Akert, 1961), indicating (p. 243, loc. cit.) that " a close tie between alpha and gamma innervation exists not only at the spinal level but at the highest level as well." Whether the effects on coactivated α and γ neurons occur by identical fibers is unknown.

Problems concerning the somatotopic pattern in the motor cortex have attracted much interest in experimental studies. While the general pattern is well known (see below and Fig. 4-15), there are still questions of great functional interest that are unsolved. Although contractions of a single muscle or of a part of it may be obtained on electrical stimulation of a point in the motor cortex, this does not imply that there are rigidly fixed transmission lines from the cortical points to the motoneurons involved in the individual contractions. A large body of evidence shows that the functional organization and the somatotopic pattern of the motor cortex is fairly flexible.

It is clear from the anatomical data on the distribution of the efferents from the motor cortex that somatotopically localized motor responses, elicited on stimulation of the motor cortex, are not necessarily mediated only via the corticospinal tract.[22] Other pathways, particularly the corticorubrospinal route, may be involved. So far, however, almost all functional studies concerning the somatotopy of the motor cortex have been restricted to the direct corticospinal projection. Studies in monkeys—with monosynaptic contacts between corticospinal fibers and motoneurons—have been particularly illuminating, as fully and critically considered

[22] A somatotopic pattern is also found in the rabbit's motor cortex, even though this animal lacks a corticospinal tract (Woolsey, 1958).

in Phillips and Porter's monograph (1977). A point of interest is the question of the distribution within the cortex of pyramidal tract cells that influence a particular muscle or part of a muscle.

On the basis of experience with cortical stimulations with microelectrodes, some authors have concluded that the pyramidal tract cells influencing a particular movement are confined to a well-delineated cylindrical column of the cortex (see Asanuma and Rosén, 1972; Asanuma, 1975). Others disagree. According to Phillips and Porter (1977), "there is no experimental basis for any suggestion that the motor cortex might be built up of isolated cylindrical output columns of equal radius, one for every muscle in the body." Landgren, Phillips, and Porter (1962), recording intracellularly from motoneurons in monkeys, found that for many motoneurons there appears to be in the cortex a "best point," which consists of more than one pyramidal cell. However, the same motoneuron can also be brought to discharge from a larger surrounding area.[23] Groups of pyramidal tract cells which may activate a single motoneuron are referred to as *colonies*. In a recent study Andersen, Hagan, Phillips, and Powell (1975; see also Landgren, Phillips, and Porter, 1962) found considerable overlap between colonies projecting to motoneuron pools of individual hand muscles in the baboon. Each motoneuron receives synapses from more than one corticospinal cell (Phillips and Porter, 1964; Phillips, 1969). Corresponding findings have been made for the hindlimb in monkeys (Jankowska, Padel, and Tanaka, 1975). The colonies influencing motoneurons of distal muscles in the forelimb are more restricted than those influencing motoneurons of proximal muscles (see Phillips, 1967, 1969), as might be expected in view of the more precise somatotopic pattern in the parts of the corticospinal projection related to distal than to those related to proximal parts of the extremities. It is further in agreement with other data that the thresholds of stimulation are lowest for fingers, toes, and face, and the EPSPs are larger in the "distal" than in "proximal" motoneurons (Jankowska, Padel, and Tanaka, 1975). These monosynaptic responses appear to be mediated via the largest axons of the pyramidal tract, conducting at 60–70 m/sec.

The few data mentioned here illustrate some essential facts that are in good accord with concepts of the organization of the motor systems, formed on the basis of experience in man (to be considered below), such as the more elaborate and differentiated cortical control of distal than of proximal musculature. It has been known for a long time that sensory impulses of various kinds—visual, cerebellar, and vestibular impulses; and from skin, muscle receptors, and receptors in joints—reach the cortical areas from which corticospinal fibers arise (see Chap. 2). Afferent impulses of different kinds are related preferentially to certain minor areas of the sensorimotor cortex, and many of them are relayed via parts of the thalamus. Much of this afferent input may play a role in feedback systems and may serve to inform the motor cortex about the results of its "commands." In recent years several authors have studied these problems, not least with regard to how the single corticospinal neuron is influenced. (For a recent survey, see Phillips and Porter, 1977). In Chapter 5 some data on the cerebellar influence on the motor cortex will be mentioned. Here some questions concerning the somatosensory input from the periphery will be touched upon.

Group I impulses, derived from primary spindle afferents, influence mainly area 3a, while group II impulses, from secondary spindle afferents, influence area 4 (Hore, Preston, Durkovic, and Cheney, 1976; and others; see also Chap. 2). (In the monkey, area 3a forms a transition from area 3 to area 4 and is found in the depths of the central sulcus.) There appears to be a high degree

---

[23] In a recent physiological study in the squirrel monkey Strick and Preston (1978) found evidence suggesting that within the forearm region of area 4 there is a *double representation* of parts of the body. Thus, the concentrations of points from which hand movements could be evoked showed two spatially separated peak zones in area 4.

of specificity since some units increase their discharge in response to the shortening, others to the lengthening, and still others to both phases of the muscle (Yumiya, Kubota, and Asanuma, 1974). Units activated from joint receptors are abundant, particularly in area 2 (see Chap. 2).

In studies in which they identified the corticospinal neurons recorded from, Phillips, Powell, and Wiesendanger (1971) found pyramidal tract cells in area 3a to be activated from muscle (stimulation of muscular nerves in the baboon). Pyramidal tract cells in area 4 likewise respond to stimulation of muscular nerves (Wiesendanger, 1973). The input reaches the cortex later than that passing to area 3a (for a review, see Porter, 1976).

The pathways utilized by impulses from receptors in muscles and joints to the motor cortex are not fully known. The longer latencies for responses from stimulation of muscle receptors in area 4 than in area 3a suggest that they are mediated via different pathways, the latter presumably via the dorsal column–medial lemniscus–thalamic route (for some data, see Porter, 1976). Experiments with cooling of the dentate nucleus suggest that the responses in area 4 in monkeys are dependent on an intact cerebellum (Meyer-Lohmann et al., 1975) and may pass via this (and the VL) in the monkey (likewise in the cat; Murphy, Wong, and Kwan, 1974). According to Zarzecki, Shinoda, and Asanuma (1978), some neurons in area 3a, receiving excitatory input from Group I muscle afferents, project to area 4 (cat). The presence of such association neurons was mentioned above.

Among the many kinds of impulses which may influence the activity of the cells of origin of the pyramidal tract, some may thus activate these cells monosynaptically from the thalamus (for reviews, see Porter, 1976; Philips and Porter, 1977). Other afferents will also influence these cells either directly or only indirectly, via one or more intercalated neurons and neuronal circuits that make possible a differentiated cooperation between the incoming impulses.

The anatomical and physiological data considered above make it clear that the primary motor cortex is extremely complexly organized, but little is actually known of its intrinsic machinery. It receives information of many kinds and exerts its action not only on motoneurons in the cord but also on many subcortical nuclei. In recent studies of movements, particularly voluntary movements, conclusions about the underlying cortical mechanisms have been based mainly on data from studies of the corticospinal tract. Although this is certainly an oversimplification, it may be a useful approach to the understanding of the situation in man, with which we shall be chiefly concerned in the following. We shall begin with some comments on the stimulation of the primary motor region in man.

All students appear to agree that *on application of weak electrical stimuli to the first somatomotor region (Ms I and Sm I) discrete movements may be obtained on the opposite side of the body.* This was shown as early as 1870 by Fritsch and Hitzig in the dog and has since been amply confirmed, e.g., by Grünbaum and Sherrington (1902, 1903) in monkeys and anthropoid apes, and by C. and O. Vogt (1919) in monkeys, to mention only some of the pioneers in this field. In man similar results have been obtained (Ferrier, Horsley, Foerster, Penfield, and others).[24] It was established by early workers that *there are discrete points from which each small movement is evoked.* In this manner it is possible to construct a cerebral map of the motor "foci" for the different parts of the body (Fig. 4-15). The excitable area extends onto the medial surface of the hemisphere (the paracentral lobule). Lowermost, in part buried in the central sulcus, are found the foci for

[24] The studies in man are made in patients in whom the cortex has to be exposed as part of a therapeutic measure. Since this can usually be done under local anesthesia the neurosurgeon has the advantage that the patient can communicate with him and report his subjective sensations (see below).

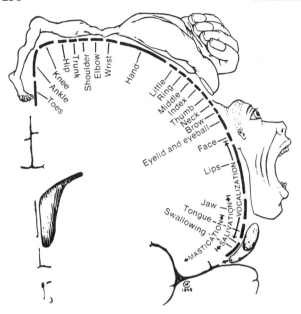

FIG. 4-15 Diagram showing the relative size of the parts of the human primary motor cortex from which movements of different parts of the body can be elicited on electrical stimulation. From Penfield and Rasmussen (1950).

the face, lips, jaw, tongue, larynx, and pharynx. The foci for the distal parts of the extremities are found closer to the central sulcus than the foci for the proximal parts (Fig. 2-19). It is important to be aware that *the cortical areas allotted to the different bodily regions are of unequal size.* Generally speaking, those parts of the body which are capable of performing the most delicate movements have the largest "representation," as shown in Fig. 4-15. (Compare, for example, the cortical "representation" of the fingers with that of the abdomen.)

The movements elicited on stimulation of the primary "motor" area in man are always contralateral as far as the extremities are concerned (see Penfield and Jasper, 1954). However, movements of the soft palate, the laryngeal muscles, and, as a rule, the masticatory muscles always occur bilaterally. Bilateral movements of the muscles of the trunk can also be provoked quite easily, while movements of the face are usually contralateral. Muscles, which under normal circumstances as a rule are active bilaterally, appear to be influenced bilaterally on cortical stimulation. These differences are probably to a large extent a functional consequence of an anatomical feature, namely a varying proportion of crossed and uncrossed corticofugal connections.

A vast number of experimental investigations of the somatotopic pattern in the motor cortex have been undertaken. Among them the extensive studies of Woolsey (1958) should be noted. The main features established by early workers in studies of animals and man have been confirmed. For an exhaustive historical account, see Phillips and Porter (1977). Several methods have been used in such studies. Lately it has been possible to make observations in unanesthetized, awake animals (see below). Many studies give information only of the somatotopic cortical pattern as reflected in the organization of the corticospinal tract.

Comparisons of results obtained in the cat with data from monkeys and man are hampered by several difficulties. One of them concerns the homology of the sensorimotor cortex in the cat and the monkey.[25] However, the cortical motor control of the musculature in the cat is finely topographically organized, as judged from the results of cortical microstimulations and recordings (cinematographic) of movements (for a recent study see Nieoullon and Rispal-Padel, 1976). For a study of the results of recordings of antidromic responses in corticospinal neurons to stimulation at various levels of the cord in the cat, see Armand, Padel, and Smith (1974).

Studies in monkeys have been particularly informative. After the presence of a somatotopic map of the motor cortex was firmly established, interest has been devoted especially to questions concerning the functional "mechanisms" at work in the production of movements on stimulation of the cortex and to analyses of factors governing movements. Only a few points from this vast field of research will be mentioned here. A critical review of the problems can be found in the monograph by Phillips and Porter (1977).

The problem of the *stability of the motor points* has attracted considerable interest. It is clear that the responsiveness of cortical cells is influenced by several factors and may show what has been referred to as "temporal fluctuations," "functional instability," etc. Nevertheless, there appears to be a rather fixed basic arrangement. The points from which a particular movement can be elicited by threshold stimulation with electrodes in the motor cortex appears to be constant, even when tested at intervals of weeks (Craggs and Rushton, 1976).[26] For further data, see Phillips and Porter (1977). These points apparently correspond to the "best points'" for a particular movement. However, the "representation" of bodily regions in the primary motor cortex should not be taken to be too rigidly fixed. As mentioned above, a particular motoneuron pool is supplied by a colony of corticospinal neurons, and there is considerable overlap between colonies projecting to motoneuron pools of individual muscles. Further, complex physiological processes, among them changing inputs to the colonies of corticospinal neurons, may alter the excitability of cortical neurons, and accordingly influence the results of stimulation. Although there appears to be a basic main line of conduction from a "best point" to the motoneurons of a particular muscle, even responses to minimal cortical stimuli are not always confined to a single joint, and even more rarely to a single muscle, as shown already in the pioneer study of

[25] It was formerly believed that the anterior and posterior sigmoid gyri (separated by the cruciate sulcus) in the cat correspond to the precentral and postcentral gyri, respectively, in the monkey. Not least on the basis of physiological studies it was later concluded that the border between "motor" and "sensory" areas in the cat is found along the so-called "postcruciate dimple" in the posterior sigmoid gyrus. In a cytoarchitectonic study Hassler and Muhs-Clement (1964) mapped the motor cortex in the cat and worked out a finer parcellation. According to them, area 6, in front of area 4, extends posteriorly and reaches the cruciate sulcus in its upper medial part. Area 4 is found both anterior and posterior to the cruciate sulcus and reaches practically to the postcruciate dimple.

[26] The situation appears to be the same in man. Bates (1960) investigated the cortical maps to electrical stimulation in patients in whom the cortex was exposed on two different occasions (at intervals of three months to more than three years). In general, identical results were obtained from stimulation of the same points on the two occasions. Bates's studies further lent support to the notion of Leyton and Sherrington (1917) that there is some degree of individuality in the cortical "representation," since the responses differ in some respects between individuals.

Chang, Ruch, and Ward (1947). As phrased by Craggs and Rushton (1976), "It seems likely that such apparent shifts of representation are caused by changes in relative thresholds of different movements subserved by overlapping groups of neurons."

It should be emphasized that *electrical stimulation of the primary motor cortex never elicits coordinated purposeful movements,* as amply documented in man, particularly by Penfield and his collaborators (Penfield and Boldrey, 1937; Penfield and Rasmussen, 1950; Penfield and Jasper, 1954). When the strength of the stimulus is increased beyond that necessary to give a small isolated contraction of a single muscle, only contractions of synergic muscles with relaxation of the antagonists occur. Another interesting observation made in studies of humans is the following: If stimulation is done while the patient performs a movement which would require the cooperation of the point stimulated, this will cause the movement to stop. The patient then often reports that he has a funny feeling, and he feels that his arm (or leg) is paralyzed.

The notion that patterns of skilled *movements* are "represented" in the motor cortex is hardly tenable. Apart from the fact that coordinated movements are never elicited on cortical stimulation, other observations argue against this view, which for a period was held by some students. Thus the results of stimulation are the same in a child of 8 years as in a man aged 60, and the same whether the man is an accomplished pianist or a manual laborer (Penfield and Jasper, 1954). On the other hand, although contractions of a single muscle may sometimes be obtained from a cortical point and although there appears to be a fairly fixed pattern of connections from a particular motor cortical point to the motoneurons supplying specific muscles or even parts of muscles, this does not permit us to conclude that *muscles* are "represented" in the motor cortex.

It has generally been believed that the primary motor cortex is an important starting point for impulses that elicit voluntary and skilled movements. In discussions of the role played by the primary motor cortex in the initiation of movements, the motor cortex has sometimes been likened to a clavichord, upon which impulses from other sources play. As we have seen, numerous other brain regions may influence the motor cortex (among them the cerebellum, the pallidum, and other parts of the cortex). Further somatosensory inputs, not least from muscles and joints, are of importance. We shall return to the primary motor cortex and the corticospinal tract below. However, first a brief account of the two other "motor cortical regions," the supplementary motor area and the second somatic sensorimotor area, will be given.

The *supplementary motor area,* described in the monkey and in man by Penfield and Welch (1951), has no clear-cut cytoarchitectonic borders. It is found on the medial surface of the cerebral hemisphere, above the cingulate gyrus, and extends forward from the "leg region" of the primary motor area where this continues onto the medial aspect of the hemisphere. In the monkey (see Fig. 2-19) its anteroposterior dimension is approximately 1 cm; in man it is larger. In the monkey, Penfield and Welch on *stimulation of this area* obtained movements of contralateral extremities, in a topographic pattern, movements of the hindlimb being elicited most posteriorly. Most often the responses involved distal musculature. Both in the monkey and in man, the threshold for the stimulation is a little higher than for the precentral and postcentral regions, and bilateral responses are frequent (there are other differences in the excitability of the two motor regions as well). The results obtained by Woolsey and his collaborators (Woolsey et al., 1952) in experiments on monkeys largely correspond to those obtained by Penfield and Welch (1951; see also Penfield and Jasper, 1954) concerning a somatotopic organization within the supplementary motor area.

However, the results in man differ in some respects from those obtained in the monkey. No somatotopic pattern has so far been demonstrated, and the motor effects usually consist in inhibition of voluntary activity or in the production of synergies resulting in the assumption of characteristic postures and in the perfor-mance of complex maneuvers. In addition, vocalization has often been observed, and sometimes pupillary dilatation or acceleration of the heart rate. Some patients report sensory experiences during stimulation.[27]

The results of *ablations of the supplementary motor area* are not entirely con-sistent. This may in part be due to the fact that the posterior part of the supple-mentary area appears to fuse with the primary motor area (the hindlimb regions of both areas are closely related). Erickson and Woolsey (1951) observed a weak and short-lived grasp reflex after unilateral involvement of the area in man. This has been confirmed by Penfield and Jasper (1954) who, furthermore, found that even after the passage of a year there is some slowness of movements on the opposite side, particularly of alternating movements. In the monkey they consider forced grasping a specific sign of removal of the supplementary area, as does also Travis (1955). While Coxe and Landau (1965) did not observe persistent changes follow-ing bilateral ablations of the supplementary motor area in the monkey, Travis (1955) in addition to the occurrence of transient grasp reflexes found disturbances in posture and tonus, but practically no paresis. There was increased resistance to passive movements of the limbs, with hypertonia in the flexor muscles. The spas-ticity demonstrated topographical localization.

The discrepancies between the observations made on the supplementary motor area in man and in the monkey are presumably due largely to method-ological difficulties. There is, however, reason to assume (Travis, 1955) that "the precentral and supplementary motor areas for the most part exert different influ-ences on the motor reflex systems." Some anatomical data support this conclu-sion. Thus the efferent projections from the two motor areas differ in certain re-spects. Further, there may be species differences. Bertrand (1956) recorded potentials in regions of the ventral horn at all levels of the cord on stimulation of the supplementary motor cortex in the monkey. He further concluded that the proportion of uncrossed fibers is greater than in the corticospinal projection. Later research indicates, however, that Bertrand's recordings were due to current spread from the supplementary to the precentral motor area. There are only a few ana-tomical studies of the efferent projection from the supplementary motor area in the monkey. It appears that efferent fibers to the spinal cord have not been convinc-ingly demonstrated (see, for example, DeVito and Smith, 1959; Wiesendanger, Seguin, and Künzle, 1974).[28] Fibers have, however, been traced to the thalamus and the striatum and to brainstem nuclei, such as the pontine nuclei and the red

[27] As mentioned in Chapter 2, there is evidence that the supplementary motor area is also sensory (Woolsey's Ms II, see Fig. 2-19).

[28] In a short report Murray and Coulter (1976) mentioned, however, that they observed labeled cells in the supplementary motor area after HRP injections in the spinal cord.

nucleus. It has, therefore, been suggested that the supplementary motor cortex exerts its influence on peripheral motor neurons via subcortical structures. This appears likely on the basis of the studies of Wiesendanger, Seguin, and Künzle (1974). With regard to the association connections the supplementary motor cortex shows some similarities with the primary motor cortex. In the monkey it receives somatotopically arranged afferents from both the first and the second sensory area (Sm I and Sm II). Unlike area 4, however, it appears not to project back to these (Jones and Powell, 1969a; Pandya and Vignolo, 1971). In addition there appear to be topically arranged afferents from area 4 and likewise a reciprocal projection back to this (Jones, Coulter, and Hendry, 1978; Sloper and Powell, 1979b). Jones et al. (1978) also described reciprocal connections with area 5. Other fibers reach the supplementary motor area from "association" areas of the neocortex (see Chap. 12).

The differences in connections between the precentral and the supplementary motor areas presumably explain the dissimilar results obtained on stimulations and ablations of the two areas. A recent study of the behavior of neurons in the supplementary motor area of monkeys during movements shows that these units differ in many respects from those in the primary motor area (Brinkman and Porter, 1979). It appears that the somatotopically organized supplementary area may be related particularly to the control of posture and some motor synergies such as the grasp reflex. Our understanding of the supplementary motor area and its function is still far from adequate.

> While the supplementary motor area has been outlined fairly precisely in the monkey with methods of stimulation, its anatomical boundaries have not yet been defined. It appears that the posterior part of it is found wihin area 4 of Brodmann, where this extends onto the medial surface of the hemisphere, while its anterior region covers part of area 6. Whether this is related to some differences in connections and functions of the forelimb and hindlimb regions of the supplementary motor area is not known, but may be considered. The fact that the hindlimb region of the supplementary area touches on the hindlimb region of the primary motor cortex may explain certain discordant results of ablation studies, particularly concerning the hindlimb.
>
> In the cat there is probably a corresponding supplementary motor area (see Woolsey, 1958). From experimental studies on the efferent projections of this cortical region it appears, however, that it projects to the cord and that the terminal area is largely the same as that of the corticospinal fibers from the sigmoid gyri (Nyberg-Hansen, 1969a). There is a topically arranged projection to the thalamus (Rinvik, 1968c), in part ipsilateral, in part contralateral (via the anterior commissure). Further, there is a bilateral topical projection to the dorsal column nuclei and the pontine nuclei, whereas the projections to these nuclei from the primary motor area are unilateral. These bilateral connections may be related to some of the functional phenomena observed.

The grasp reflex and forced grasping. As the name implies, *forced grasping* is a compulsory grasping movement and is of reflex nature. This symptom was known clinically as a sign of frontal lobe affection (Adie and Critchley, 1927) some time before it was concluded from experimental investigations that it was dependent on a lesion of area 6 (Fulton, 1934; Richter and Hines, 1934). However, it now appears well established that the responsible cortical region is the supplementary motor area. An analysis of reports attributing the symptom to lesions of area 6 indicates that in the relevant cases the lesions have in fact involved the supplementary motor area (Travis, 1955). The grasp reflex is essentially similar in man and in monkeys. It is elicited when the palmar surface of the

hand or fingers is gently touched, for example by a stick. A flexion of the fingers
then appears and the stick is held firmly.

> According to an exhaustive study of Seyffarth and Denny-Brown (1948) the adequate stimu-
> lus for the full reaction is dual. There is an initial tactile "catching phase" followed by a
> proprioceptive response, the "holding phase," due to traction on the appropriate muscles. The lat-
> ter phase is a heightened stretch reflex which is triggered by the tactile response. The grasp reflex
> is usually grouped with the postural reflexes (cf. later), since it is influenced by changes in the
> posture of the body. When a grasp reflex is present and the animal is lying on its side, the reflex is
> far more vigorous in the uppermost hand than in the other on the under side of the animal. Emo-
> tional factors also influence the reflex, anxiety, for example, increasing it. It appears that the
> reflex is entirely subcortical, and it is supposed that when it occurs it is due to loss of a normal
> inhibitory cortical influence, most likely exerted by the supplementary motor area.

It is well known that in *human infants a grasp reflex is normally present*. Its
gradual disappearance during the first months of life been explained by a development of cortical dominance over lower "centers." According to Pollack
(1960) the infantile grasp reflexes are, however, not identical with the responses
seen in adults with frontal lobe damage. It might be advisable, therefore, to reserve the term "grasp reflex" for the normal reflex in infants, and to use the term
"forced grasping" for the phenomena seen in lesions of the supplementary motor
area.

The motor functions of the *second somatic sensorimotor area* are less well
known than those of the supplementary motor area. As mentioned in Chapter 2,
this area (Fig. 2-19) was labeled Sm II by Woolsey in order to emphasize that,
although it is mainly concerned in the reception of somatic sensory impulses, it
should to a lesser degree also be considered a motor region. In pioneer studies
Adrian (1941) and Woolsey (1947) obtained motor responses upon stimulation of
the second somatic sensory cortical area. Furthermore, the motor effects were
somatotopically organized. These findings have been confirmed by Welker et al.
(1957) and others. There is evidence (Penfield and Rasmussen, 1950; Penfield and
Jasper, 1954) that in man there is a corresponding second sensorimotor area in the
upper bank of the fissure of Sylvius, but motor deficits have not been observed
following isolated ablation of this area.

Since these investigations in animals and man were carried out, subsequent
studies on Sm II have almost exclusively been concerned with its role as a sensory
cortical region (see Chap. 2). Little is known concerning the cortiofugal projections
from this area. In the cat degenerating fibers have been traced to lower levels of
the spinal cord (Nyberg-Hansen, 1969b). From an HRP study Groos, Ewing, Carter, and Coulter (1978) concluded that the fibers do not descend further than to the
upper thoracic cord. Corticospinal fibers from the second sensorimotor area have
been studied physiologically in the cat (Atkinson, Seguin, and Wiesendanger,
1974). In the monkey such fibers have been found to supply both cervical and
lumbar parts of the spinal cord (Murray and Coulter, 1976). Ample corticofugal
connections from Sm II to the dorsal column nuclei and the main sensory relay
nuclei of the thalamus have been described in the cat (Levitt, Carreras, Liu, and
Chambers, 1964; Kusama, Otani, and Kawana, 1966; De Vito, 1967; Rinvik,
1968b). Likewise there is a somatotopic projection onto the pontine nuclei
(P. Brodal, 1968b). Much remains to be done before the role of Sm II in motor

functions is understood. Atkinson, Seguin, and Wiesendanger (1974) are inclined to believe that the projection to the spinal cord is involved especially in the cortical control of somatosensory transmission and, by way of the fibers from Sm II to the pontine nuclei, in the transmission of spatial information to the cerebellum.

Movements and the pyramidal tract.   Before we go on to a discussion of clinical aspects, it is appropriate to consider some recent data on the neural circuitry involved in movements. These data have emerged mainly from experimental studies in monkeys.

The almost unlimited kinds of movement which animals and human beings are capable of, may be subdivided according to various principles into several categories. A distinction is often made (cf. Chap. 5) between "ramp movements" (where, for example, a slowly moving target is followed by one hand) and "ballistic movements" (rapid and suddenly starting and stopping movements). From the point of view of central control of movements it may be useful—as done by Phillips and Porter (1977), following Hughlings Jackson—to distinguish broadly between "more automatic" and "less automatic" performances. On one end of this scale would be the "least automatic" movements. To these belong the so-called "skilled movements," dependent on learning processes.

> Presumably such movements are based on "central programs"—that is, more or less stabilized neuronal circuits built up during life. In contrast "more automatic" movements (such as respiratory movements) are basically dependent on preformed reflex arcs of varying complexity. Under natural circumstances there is an incessant interplay and cooperation between the different categories of movements. An example of this may be found in the stepping (walking) movements. Extensive studies have clarified important points of these complex processes (see Grillner, 1975, for a review and references). It appears that locomotion is essentially generated by intrinsic spinal mechanisms responsible for the production of rhythmic movements of the legs. In addition, locomotion is normally subject to a very intricate control from "higher" regions of the central nervous system. Nevertheless, a considerable degree of well-integrated locomotor behavior is retained in *chronic* spinal animals. (For a recent study see Grillner and Zangger, 1979.)

In man, as will be discussed later, an interruption of the fibers of the internal capsule results in a clinical picture that was—and (with dubious right) still often is—referred to as the pyramidal tract syndrome (see below). From clinical experience in man the most important task of the pyramidal tract has been concluded to be to initiate and control the performance of rapid and skilled discrete movements. In addition, it is generally said to be of importance of myotatic reflexes, muscle tone, and cutaneous reflexes.

To decide whether the entire symptom complex is due to an interruption of the pyramidal tract, the consequences of such surgical interventions in monkeys and apes have been studied intensively. In general, operations on the cat have not given much information. There are small and largely transient defects in movements, and reduction of some postural reflexes occurs after section of the pyramidal tract (mainly loss of tactile placing reactions). There is some increase in the latency of the motor responses to cortical stimulation (see, for example, Laursen and Wiesendanger, 1967; Nieoullon and Gahéry, 1978). In monkeys the consequences of transection of the pyramid are also less marked than the symptoms seen in capsular hemipareses in man.

In her pioneer study in 1940, Tower found that, after transection of the medullary pyramid in monkeys, the animals were unable to perform discrete movements of the fingers. However, there was no spasticity but a hypotonia of the muscles. The abdominal and cremasteric reflexes were abolished or weakened. In chimpanzees (Tower, 1944) somewhat similar findings were made. More recently, Gilman and Marco (1971) and other authors (see their paper for references) likewise found an enduring hypotonia after transection of the pyramid in monkeys. There was no evidence of increased myotatic responses.[29] Like some other authors, they noted in addition some loss of contactual orienting responses, manifested by defects in tactile placing, grasping, and avoiding. There appears to be agreement among authors that after pyramidotomy there is defective use of fingers and toes for discrete movements. According to Lawrence and Kuypers (1968a), who studied their monkeys for several weeks, animals with a bilateral interruption of the pyramidal tract regain their capacity for independent use of their extremities and within three weeks can use their hands adequately to pick up morsels of food. However, the capacity for individual finger movements never returns. The importance of the pyramidal tract for independent finger movements was confirmed in a later study by Lawrence and Hopkins (1976). Hepp-Reymond, Trouche, and Wiesendanger (1974) have brought forward evidence (in tests of an isometric "precision grip" in awake monkeys) that the reaction time for the grip was lengthened after transection of the pyramid. (The increased reaction time was found to be due to a delay in the execution rather than the initiation of the movement.)

Increased insight into several fields of neuroscience have made us aware of sources of error in experiments such as those considered. When studied in the first days or weeks, secondary consequences of the lesion (edema, bleedings, spasms of blood vessels) may give false positive signs of defective functions. When studied in long-term experiments, the recovery processes may disguise defects originally present.

The results of transection of the pyramidal tract in monkeys are thus in accord with clinical views insofar as they indicate that the *function of the pyramidal tract is first and foremost to secure precision and rapidity of fine, skilled movements*. This is further in accord with what is now known of the anatomy and physiology of the pyramidal tract, considered in this chapter. Concerning muscular tone and reflexes there is, however, less correspondence between the changes observed in monkeys after pyramidotomy and those occurring in man after capsular hemiplegias. We shall return to this subject later.

In recent years studies of the behavior of corticospinal neurons during movements have been undertaken in awake animals. Among a wealth of data only some will be mentioned. (For an exhaustive and critical consideration, see Phillips and Porter, 1977.) When the activity of cortical neurons (many of them identified as giving rise to corticospinal fibers) is recorded, in the monkey precentral neurons

[29] In monkeys, Gilman, Marco, and Ebel (1971) found that the afferent responses of the muscle spindle primaries to passive muscle stretching (see Chap. 3) were reduced after transection of the pyramidal tract. They conclude that pyramidotomy decreases the tonic fusimotor innervation of the muscle spindles in extensor muscles. The authors suggested that the hypotonia after section of the pyramidal tract is related to that seen after ablation of the neocerebellum (more particularly the dentate nucleus, see Chap. 5). In both instances a normal activating effect of corticospinal axons on γ-neurons is reduced or abolished.

discharge before the start of the movement in which they play a role (Evarts, 1966; Porter, 1972; and others). Precentral neurons commonly become active 60–40 msec before the response (recorded mechanically and electromyographically), some as early as 140 msec. Cells in the postcentral gyrus are also active before the movement starts, but somewhat later than the precentral ones, the earliest activity appearing 60 msec before the muscular response (Evarts, 1972). Thus, the start of activity in these neurons cannot be determined by feedback influences from the periphery. Further, the pyramidal tract neurons may change their discharge according to the force needed in the prime mover muscles that are to be used (Evarts, 1968). An increase in the force required may also lead to the activation of a larger number of corticospinal neurons (Porter and Lewis, 1975). It appears that the populations of active cells change continuously with time, apparently according to modulations of contraction in the various muscles engaged in a movement. As described above, the anatomical arrangement of cells in the motor cortex and their synaptic relations are such that a single corticospinal cell may be influenced from afferent inputs from many sources, as has been shown in several physiological studies. The interaction between the various inputs to the motor cortex presumably varies incessantly during a voluntary movement.

> It should be recalled that even an apparently very simple movement involves more than one—often numerous—muscles and that each of them may have to change its speed and force of contraction in the various phases of a movement (a muscle acting as an agonist in one phase may be active as an antagonist in another phase); an afferent feedback to the motor cortex from proprioceptors in joints and muscles must be equally differentiated. Such considerations are compatible with the minute differences in the neuronal behavior of corticospinal neurons which have been recorded physiologically. See Phillips and Porter (1977) for particulars.

The patterns of activity in corticospinal neurons (and even more in the motor cortex) during voluntary movements are far from fully understood. Although refined electrophysiological studies have added much new information, the complexity seems to be overwhelming. Among the afferent inputs, much interest has been devoted to that from the cerebellum (the projection: dentate nucleus–VL of the thalamus–motor cortex). Some data on this subject will be considered in Chapter 5. A somewhat surprising finding is the demonstration that some dentate neurons are active before any activity in the motor cortex can be recorded and before the movement begins (Thach, 1970a). Units in the dentate nucleus appear to change their behavior as much as those in the motor cortex (see, for example, Grimm and Rushmer, 1974). Some comments on the role played by the cerebellum in the execution of movements will be made in Chapter 5. As is well known, after lesions of the dentate nucleus a muscular hypotonia ensues. As mentioned above, the same is the case after a pyramidotomy in the monkey (Gilman, Marco, and Ebel, 1971). In both instances there appears to be loss or reduction of the cortical facilitatory effect on motoneurons, mediated via fibers of the corticospinal tract (cf. footnote 29).

Among features that illustrate the complexity in the activity of the corticospinal tract (and the motor cortex) is the presence of activity in the motor area one or more seconds before a movement starts. This subject has been studied in several kinds of tests of movements in monkeys. For example, instructions given to a monkey concerning an arm movement he has been trained to perform lead to anticipatory activity in neurons of the motor cortex, even though overt muscular response is delayed while the monkey awaits a sensory input that triggers the movement (Tanji and Evarts, 1976). It appears from this and other experimental studies that important changes may occur in the motor cortex before it emits its commands to muscles. Psychological factors appear to influence this preparatory potential activity. Presumably the situation is essentially the same in man.

According to Deecke, Scheid, and Kornhuber (1969), EEG recordings from the scalp, preceding voluntary finger movements in man, have shown that first there appears a "readiness potential," about 850 msec before the movement starts. This potential is bilaterally symmetrical over the pre- and postcentral region. It is followed by a so-called motor potential, which shows its maximum over the contralateral precentral hand area and occurs about 56 msec before onset of the movement. For some further data on man, see Phillips and Porter (1977).

As is apparent from the data above, most studies of the neurophysiological events occurring during voluntary movements have been concerned with the role and activity of the corticospinal tract. It should again be emphasized that *studies of the corticospinal neurons can give information on only a fraction of what really goes on in the motor cortex.* Further, it should be noted that chiefly the thicker, rapidly conducting pyramidal tract fibers have been examined.[30] These, as we have seen, make up only a modest proportion of the total number. What is the role played by the many thin fibers which to a considerable extent originate in—and end in—the same areas (in the cortex and the spinal cord, respectively) as the axons of large pyramidal neurons? In reference to this problem Humphrey and Corrie (1978, p. 240) comment, "Clearly, then, additional studies of the movement-related behavior and synaptic actions of these small cells are required, if our concepts about the major functions of the pyramidal tract system are to be based—as they eventually must be—on a careful, representative study of all of its major neuronal subsets."

There is another—perhaps more important—reason that studies of corticospinal fibers can give only incomplete information about the role played by the motor cortex in movements. As repeatedly mentioned in this text, *the same cortical areas from which corticospinal fibers arise give off fibers to various subcortical stations,* such as the red nucleus, the reticular formation, and the pontine nuclei. In part these fibers are collaterals of corticospinal axons; in part the fibers are axons of other cells than those projecting to the cord. Little is so far known of the involvement of such indirect cortiscopsinal routes in various kinds of movements.

As has been mentioned in preceding sections of this chapter, descending projections from many structures in the brainstem may act on the peripheral motor neurons (α and γ motoneurons) directly. The brainstem structures may be influenced from the cerebral cortex, some indirectly, others more directly, by efferents from the cortex, for example to the red nucleus and the reticular formation. The presence of several morphological and physiological similarities between the corticospinal and the corticorubrospinal pathway, referred to earlier (p. 194 ff.), is

---

[30] Humphrey and Corrie (1978) found the modal velocities of fast and slow corticospinal fibers in the monkey to be 54 and 10–12 m/sec, respectively. Other authors have reported somewhat higher values.

especially noteworthy. As will be mentioned below when the "pyramidal tract syndrome" is discussed, there is evidence for the assumption of a very close co-operation between the corticospinal and corticorubrospinal pathways, for example in the performance of finger and hand movements.

Very little precise information appears to be available about the collaboration between the corticospinal tract and the many descending brainstem projections. Likewise, the anatomicophysiological basis of the collaboration between exquisite voluntary movements and more simple postural adjustments—between "less automatic" and "more automatic" movements—is relatively little known.

It is often thought that the central nervous activity in the performance of discrete finger movements is dependent, at least chiefly, on the first sensorimotor cortex. Bates (1957) ventured the suggestion that the fundamental patterns of discrete movements are actually organized largely at the spinal level. This view is based on the results of stimulation of the motor cortex of man and on other data (see Bates, 1963), such as studies of the types of movements the newborn is capable of. Further support comes from the observation (Bates, 1953) that there is close similarity between the movements in response to stimulation of the cortex and stimulation of the internal capsule after hemispherectomy. According to Bates's hypothesis, the role of the cortex would be chiefly to modulate the activity of the spinal patterns of movements and to impose a degree of discreteness on them. This view is in accord with experimental observations indicating that the spinal cord is capable of a considerable degree of integrated activity in the motor sphere (cf. studies of locomotion referred to above; see Grillner, 1975). Bates's hypothesis raises the interesting question of whether the learning of motor skills may be as much a task of the spinal cord as of the cortex. Recent insight into the extremely complex neuronal organization of the spinal cord is compatible with the view that the programs for learned motor activities may be "laid down" in the cord. "Higher levels" may have their importance during the learning processes, mainly by correcting and refining the first crude programs.

Symptoms following lesions of supraspinal structures and pathways influencing peripheral motor neurons. This heading covers a great variety of disorders of motor functions. Since the disturbances are not caused by damage to the peripheral motor neuron, they have in common that there is no marked atrophy of the muscles, and although there may be changes in the myotatic (stretch) reflexes these will not be abolished. Nor will muscular tonus be abolished, even if changes in this are the rule, due to alterations in the normal action of supraspinal structures on the motoneurons (α and γ). We are still far from being able to understand the mechanisms at work when motor symptoms follow lesions of supraspinal structures and pathways. It is, therefore, not possible to discuss these subjects in an entirely satisfactory fashion, and a correlation of experimental findings and observations on humans can only be made to a certain extent. For this reason it is deemed practical in the following presentation first to describe the symptoms seen in humans and then to discuss their relation to experimental observations and the attempts to explain the symptoms.

Before turning to the motor disturbances occurring in lesions at supraspinal levels it will be appropriate to devote some attention to the mechanisms in the spinal cord which ultimately mediate the supraspinal influences on the motoneurons, and to consider the consequences of lesions of the descending fiber systems within the cord.

Some features in the organization of supraspinal influences on the spinal cord. Essentially the spinal cord may be considered as a series of segments of nervous tissue charged with the nervous regulation of the corresponding segments of the body. Cooperation between segments is secured by intersegmental connections. Presumably the fact that many fibers of descending systems give off collaterals to several levels of the cord contributes to the intersegmental cooperation. Even though it appears that innervation patterns for many movements are to a large extent laid down in the cord, the integration into patterns of reaction subserving the entire organism is chiefly taken care of by descending connections from higher levels of the nervous system. In addition, the intrinsic connections of the cord may have important integrative tasks in the learning of motor performances, as alluded to above. In order that the central control can function properly, ascending pathways, bringing sensory information to the cord as well as to higher levels, must be intact.

As discussed in Chapter 2, several of the long descending fiber systems have been shown to influence the ascending transmission of sensory impulses from the cord, postsynaptically, by acting on "secondary" sensory neurons, or presynaptically, by acting on the terminations of dorsal root afferents. Here we are concerned with the effects of supraspinal structures on motor functions. It seems clear from many recent studies that most of these effects are exerted not directly on the motoneurons but on the spinal reflex mechanisms. Furthermore, the spinal reflex paths can be influenced by fibers from supraspinal structures via internuncials in the cord as well as by primary afferent depolarization (presynaptic inhibition of dorsal root afferents).

In any evaluation of supraspinal influences on the cord it should be recalled that the spinal reflexes are not fixed and stereotyped, but may be modified according to the stimulus site, the intensity of the stimulus, the concomitant application of other stimuli and other factors. Furthermore, there are numerous descending pathways which may act on the spinal reflex apparatus. Since all motor reflexes (except the stretch reflex) appear to have multisynaptic reflex arcs, it follows that the internuncial cells are of particular interest, the more so since in the cat (on which most experiments have been made) all supraspinal fiber systems (except possibly the lateral vestibulospinal tract) exert their action on the motoneurons via internuncial cells. In monkeys, and presumably also in man, the corticospinal tract can influence the motoneurons monosynaptically as well (whether this is true also for other descending systems is not known). Nevertheless, it appears likely that the spinal reflexes in man are organized essentially as in the cat, even though the spinal cord seems to lose some of its autonomy in the phylogenetic ascent.

As mentioned in an earlier section of this chapter it has been shown for several supraspinal fiber systems that they act on both α and γ neurons. In general these effects are parallel for a particular fiber system, and both (in the cat) are mediated via internuncials. Thus the corticospinal and rubrospinal tracts predominantly activate neurons supplying flexor muscles, and inhibit neurons supplying extensor muscles. Medullary reticulospinal fibers appear to have a similar action. On the other hand, vestibulospinal impulses have the opposite effect, facilitating (in part monosynaptically) extensor motoneurons. The pontine reticulospinal fibers appear to belong to the same group. In view of this the sites of termination of the descending fiber systems within the gray matter are of interest. As seen from Figs. 4-3, 4-7, and 4-8), the fiber systems giving facilitation of flexor motoneurons end in laminae V–VI and the dorsal part of VII, while those facilitating extensor motoneurons terminate more ventrally in lamina VII and the adjoining part of lamina VIII. These morphological observations suggest that different groups of interneurons may be involved in the supraspinal effects on flexor and extensor motoneurons.[31] This appears indeed to be the case since physiologically interneurons which are intercalated in reflex pathways excitatory to flexor and inhibitory to extensor motoneurons are found in laminae V–VII, chiefly laterally, while interneurons mediating the opposite effect are located in lamina VIII and adjacent parts of lamina VII (see Eccles, 1964). It is further of interest that the pathways giving excitation of flexor motoneurons are found more dorsally in the ventrolateral funiculus than those exciting extensor motoneurons (see Figs. 4-3, 4-7, and 4-8). The former pathways are considered to be phylogenetically younger than the latter (see Nyberg-Hansen, 1966a).

It appears from the above that in a very simplified fashion the supraspinal pathways may be subdivided into two groups, concerned in opposite actions on the motor apparatus (flexion and extension). While both act on motoneurons via interneurons, these appear to be different for the two groups. Some findings of Lawrence and Kuypers (1968b) support this view. Following recovery from a prior bilateral pyramidotomy in monkeys, lesions of the spinal cord were made. The lesions involved either the lateral or the ventral funiculus. In the latter the motor impairment concerned chiefly the axial and proximal extremity movements and there was a bias towards flexion of trunk and limbs. Lesions of the lateral funiculus gave rise to impairment of independent distal extremity and hand movements and an impaired capacity to flex the extended limb.

The anatomical and physiological observations mentioned above can give no more than a simple sketch of a very complex mechanism. The different supraspinal descending pathways within each of the two main groups are certainly not functionally equivalent in all respects. Furthermore, the intrinsic anatomical organization of the spinal cord, discussed in Chapters 2 and 3, is extremely complex and provides wide possibilities for interactions between several descending systems and a variety of spinal afferents. In recent years the effects of supraspinal systems on the reflex activity of the cord have been intensively studied, particularly by Lundberg and his collaborators. Some of the observations may be mentioned to give an idea of the complexity of the problems.

As mentioned previously, supraspinal structures may influence the reflex activity of the cord in two ways (leaving out of consideration here possible monosynaptic actions on motoneurons):

[31] It is of interest to notice (Nyberg-Hansen, 1966a) that in the cat the former groups of interneurons are situated closer to the motoneuron groups supplying the distal musculature, the latter closer to the groups for the proximal limb muscles and those of the trunk. The same appears to be the case in the monkey (Kuypers and Brinkman, 1970). Cf. also Chapter 3 on the somatotopic arrangement of motoneurons.

the descending fibers may act on interneurons which are intercalated in most reflex arcs, inhibiting or facilitating them, or the fibers may produce depolarization of afferent fibers and in this way reduce or block the reflex response that a sensory impulse would otherwise produce. Both types of action have been demonstrated (for reviews see Lundberg, 1966, 1975), and a very complex picture of the supraspinal control of reflexes is emerging. For example, the corticospinal tract has been shown to facilitate polysynaptic reflex arcs in which the afferents belong to Ia, Ib, cutaneous, and so-called flexor reflex afferents (Lundberg and Voorhoeve, 1962). This effect appears to be due to facilitation of interneurons of the segmental reflex arc. Likewise, the rubrospinal tract has a facilitatory action on interneurons of reflex pathways which mediate the Ib inhibition (Hongo, Jankowska, and Lundberg, 1965). In both cases the interneurons appear to be situated in the region of the gray matter where the two pathways end (see Fig. 4-3). Facilitatory actions have further been found from the rubrospinal tract on interneurons activated from group Ia muscle afferents and belonging to reflex arcs mediating inhibitory effects on flexor and extensor motoneurons (Hongo, Jankowska, and Lundberg, 1965). The Ia inhibitory pathway from extensors to flexor mononeurons (see Fig. 3-5) can be facilitated from the vestibulospinal tract through a monosynaptic excitatory action on the Ia inhibitory interneurons (Grillner, Hongo, and Lund, 1966a). However, interneurons may also be inhibited from supraspinal fiber systems (see Lundberg, 1966, 1967, 1975). Among recent relevant studies may be mentioned the papers of Illert, Lundberg, and Tanaka (1975, 1976), Illert and Tanaka (1978), and Illert, Lundberg, Padel, and Tanaka (1978). In recent years much attention has been devoted to the chemical aspects of transmission in the spinal cord (see Lundberg, 1966). However, these will not be considered here.

From a clinical point of view those aspects of the functional organization of the cord which are related to the supraspinal control of muscular tone and myotatic reflexes are of particular interest, since changes in these functions are common in lesions of the central nervous system. Increasing knowledge of the organization of the normal stretch reflex, discussed in Chapter 3, has provided valuable clues for the understanding of the phenomena of spasticity and rigidity. A consideration of these subjects will be postponed until the clinical conditions in which they occur have been dealt with.

Transverse lesions of the spinal cord and the Brown-Séquard syndrome. Spinal shock. In Chapter 2 the symptoms due to lesions of the sensory paths of the spinal cord were discussed. However, pathological processes affecting one of these fiber tracts exclusively are not very frequent. In most diseases affecting the cord the morbid changes will involve several ascending and descending tracts (whose fibers are even partly intermingled) and the gray matter of the cord. The lesions may be of several kinds or sizes, from complete transverse lesions in which all fiber tracts are interrupted to partial transverse lesions of half of the cord or less. Frequently the lesions will have an irregular distribution, affecting predominantly, for example, the gray matter. Possessing a knowledge of the anatomical features of the cord, the neurologist will be able to draw conclusions from the symptoms present concerning the approximate place of damage to the cord, its level in the cord, i.e., the segmental diagnosis, and its site in the transverse section of the cord. The number of possible combinations of lesions and consequently of symptoms is, of course, practically unlimited.

A typical clinical syndrome, the so-called *complete transverse syndrome,* is met with in cases in which an acute, complete interruption of the spinal cord has occurred. (For a detailed study, see Kuhn, 1950.) Usually this is caused by traumatic injury, gunshot wounds, stab wounds, or fracture dislocations of the spine (in other cases inflammatory processes, myelitis, may be the cause). Immediately following the injury a complete paralysis of all muscles which get their motor innervation from segments of the cord caudal to the lesion is observed; likewise a complete anesthesia is present in the same territory, all modalities of sensation being abolished, superficial as well as deep. The paralysis is, immediately after the injury, of the flaccid type: the tendon reflexes as well as the superficial, cutaneous reflexes are absent. Nearly always there is retention of urine (of practical importance!), and as a rule peristalsis is severely impaired and no emptying of the bowels takes place. Sweat secretion is abolished below the level of the lesion, and the blood pressure suffers a temporary drop. Erection is abolished.

If death does not ensue, this *intitial stage of spinal shock* gradually passes into what may be called the *"stage of reorganization"*: the spinal cord below the lesion resumes some of its functions, in spite of its lacking connections with the higher levels of the central nervous system. Usually the first signs of this are shown by the bladder and rectum. Occasionally a spontaneous emptying of the bladder occurs, eventually also of the rectum: *automatic reflex activity of bladder and rectum,* and sweat secretion reappear.[32] Somewhat later the muscles resume some of their tone, and simultaneously signs of reappearing *reflex activity of the muscles* will be noticed. Especially on painful stimuli or on cooling parts of the body below the lesion (e.g., when uncovering the patient) the so-called *reflexes of spinal automatism* appear, most frequently as a flexion of the legs in hip and knee joints, often accompanied by emptying of the bladder, and also of the rectum (Head and Riddoch's mass reflexes). These reflexes of spinal automatism can as a rule be elicited from any part of the body below the segment of lesion, but frequently not with the same ease from all regions. By and by the lower extremities become permanently fixed in a drawn-up position, with flexion at the hips and the knees and dorsiflexion in the ankle joint, due to a tendency to shortening of the flexor muscles. The result will be a *spastic paraplegia in flexion.* Later on the tendon reflexes reappear; the patellar and achilles reflexes can be elicited. "Mass extension" of one or both lower limbs may occur in response to proprioceptive stimuli, the stretch reflexes (myotatic reflexes) may be increased, and patellar and ankle clonus may appear. This "reorganization" usually requires several months. The same phenomena occur in animals following experimental transection of the cord, but in animals the "reorganization" occurs more rapidly, the more so the lower the position of the species on the evolutionary scale. It thus appears that the spinal reflex apparatus, which is responsible for these phenomena, loses more and more of its independence in phylogenetic ascent.

The occurrence of "spinal shock" following an acute transverse lesion of the cord cannot yet be adequately explained. It has been generally assumed that the shock is due to the disappearance of the normally present descending influences on the spinal gray matter, leaving its internuncial cells and the motoneurons at a lower level of excitability. From a comparison between the H-reflex[33] and the ankle jerk in patients with acute traumatic spinal cord lesions, Weaver, Landau,

---

[32] The symptoms referable to the autonomic system, including disturbances of bladder and rectum, are discussed more fully in Chapter 11.

[33] The H-reflex, described by P. Hoffman in 1918, has been used for studying certain features of the stretch reflexes in man. The H-reflex is elicited by electrical stimulation of the tibial nerve in the popliteal fossa, and the response is a contraction of the calf muscles (triceps surae). The central delay is so brief that the reflex may be assumed to have a monosynaptic reflex arc. The H-reflex is thus con-

and Higgins (1963) conclude that a depression of fusimotor function contributes to the general hyporeflexia in spinal shock, but it is not the only factor (see also Diamantopoulos and Zander Olsen, 1967). Illis (1967) has advanced a hypothesis based on experimental observations of boutons on cells in the spinal cord following interruption of afferents. In the first two days after the transection of afferents there is a general disorganization of the "synaptic zone" of the nerve cells, affecting all of the many thousand boutons. The boutons belonging to transected afferents will disintegrate, while the neighboring ones are not irreversibly changed and recover. The places on the neuronal surface occupied by the degenerating boutons are temporarily covered by glial cells, but ultimately appear to be again contacted by boutons, which are assumed to be formed by sprouting of intact afferents (see Chap. 1). The disorganization of the "synaptic zone," according to Illis (1967), may account for the reduced level of excitability of the spinal cord neurons in the first days, and the ensuing reorganization may explain the return of excitability, a process that requires some time. It appears that sprouts and terminals of dorsal root fiber axons in part take over vacated synaptic sites originally occupied by terminations of descending, now degenerated, fibers (cf. Chap. 1). Such a process has been suggested by some to contribute to the spasticity seen after lesions of descending pathways in the cord. (For a study of spinal shock in the monkey, see McCouch, Liu, and Chambers, 1966).

The development of the symptoms outlined above will not always appear in complete form. Particularly when urinary infections, leading to pyelonephritis, or decubitus ulcers (bed sores) with secondary infections and ensuing weakening of the organism, develop, the spastic paraplegia in flexion is usually replaced by a flaccid paralysis. The prognosis of a complete transverse lesion was formerly very bad. Following World War II great interest has been taken in improving the prognosis in paraplegic patients. It has been shown that if infections and the development of bed sores are prevented, it is possible by physiotherapy to rehabilitate many of these patients to a useful life. The therapy aims at enabling the patient to substitute for the functions of the paralyzed muscles by using those which are intact. In the first place the muscles of the shoulder girdles are utilized. It is indeed astonishing to see to what extent patients may regain mobility in this way.

The picture outlined above refers primarily to complete transections in the thoracic or lumbar region of the cord. Naturally the symptoms will not be wholly identical if other levels of the cord are injured. Thus an injury to the upper part of the cervical region may be instantly fatal, on account of respiratory paralysis due to involvement of the motor cells of the phrenic nerve (C_{3-4}). Even when the lesion is situated in the lower part of the cervical cord and is followed by paralysis of all four extremities, the phrenic nerve will frequently be affected secondarily (edema, circulatory disturbances). The symptomatology in lesions of the sacral region of the spinal cord is dominated by the autonomic disturbances and will be treated in the chapter on the autonomic system.

Above, only the acute, complete transverse lesion of the cord has been considered. If a complete transverse lesion develops gradually, the final result will be the same, but during its development certain clinical features appear rather regularly. In these cases the morbid changes may be of neoplastic type—an intraspinal, extramedullary or more seldom intramedullary tumor—or there may be inflammatory processes, myelitis. Diseases of the spine are another not infrequent

---

sidered to be an electrically induced equivalent to the ankle jerk, the difference being that in the latter the muscle spindles are stimulated by the stretching of the triceps surae, while in the H-reflex the spindle afferent fibers are stimulated directly.

cause of compression of the cord (metastases to the vertebrae from malignant tumors, spondylitis) or meningeal lesions may be responsible (arachnoiditis). The details of the symptomatology will depend on the site of the changes, not only its level in the cord but also from which side it takes its origin.

> An extramedullary tumor, for example, springing from a dorsal root, will usually start with radicular pain, accompanied by sensory changes (hyperesthesia) in the corresponding dermatome, but later on it will usually affect the adjacent roots, and a segmental sensory loss will appear. Pressure on the cord may then elicit symptoms, indicating damage to one or more of the long fiber tracts, for example sensory loss or a paresis of the spastic type below the segments of the lesion. This paresis, of course, will affect the muscles on the side of the lesion, and frequently at the start will be most conspicuous caudally, when the most superficial fibers of the descending tracts only are affected. This paresis is usually of the spastic type, with exaggerated tendon reflexes and extensor response of the plantar reflex, and weakened or abolished abdominal reflexes. The sensory loss which may occur on account of pressure on the dorsal funiculi will likewise be found on the side of the lesion. Sensory disturbances due to pressure on the anterolateral funiculus (cf. Chap. 2) will occur—or be most marked—on the contralateral side. The upper border of both motor and somatosensory disturbances may be found several segments lower than the lesion, owing to the segmental arrangement of fibers in the long tracts (the longest fibers are located most superficially).

The complete picture of a lesion comprising one lateral half of the cord is the so-called Brown-Séquard syndrome. This is characterized by a homolateral spastic paralysis below the level of the lesion, with extensor response of the plantar reflex, abolished abdominal reflexes, exaggerated patellar and achilles reflexes, abolished deep sensibility and sense of discrimination in the same regions, of the contralateral side of the body. In addition a segmental motor loss may also be found corresponding to the level and side of the lesion. The functions of the bladder, rectum, and genital organs are usually not interfered with (bilateral innervation).

> A complete Brown-Séquard syndrome is not frequently encountered, as might be anticipated. As a rule the lesion of the cord is more irregular and the symptoms consequently somewhat different. A feature which is usually found in incomplete lesions of the cord is the sparing of the sacral segments, already referred to.

Symptoms following lesions of the primary motor cortex and the internal capsule in man.   All efferent pathways from the primary motor cortex (Ms I and also neighboring areas, particularly Sm I) that are destined to act on the peripheral motor neurons—directly or via intercalated cell groups—descend in the internal capsule. Although this also contains other fibers which may be injured by a damage to the internal capsule but would escape involvement in a pure cortical lesion, the motor symptoms after lesions of the internal capsule or motor cortical areas are in the main similar. To avoid repetition, they will therefore be treated together. Affections of the corticofugal fibers from the cortical motor areas in man are far more frequently due to lesions of the internal capsule than to lesions of the cortex. The symptom complex resulting from lesions of this kind has in the past usually been referred to as the "pyramidal tract syndrome." The justification for this will be discussed later, following an account of the symptoms encountered in lesions of this kind. Before dealing with such "deficiency symptoms," it is appropriate to discuss some symptoms that may occur after lesions of the motor cortex and which are due to abnormal activity of its cells. The terms "irritative symptoms" or "irritative lesions" are commonly used in these instances.

Before the discovery that the cerebral cortex is electrically excitable and that stimulation of the motor cortex yields muscular contractions, the British neurologist Hughlings Jackson had prophesied the principal points in the organization of the primary motor area. His conclusions were based on close clinical examination of patients suffering from epileptic seizures. In some of these cases he had, like other clinicians, observed that clonic contractions may start locally in a circumscribed part of the body, for example, one foot, and then may be propagated in a regular sequence to other parts, for example, from the foot to the calf, further to the thigh, the trunk, and from here they may proceed to the shoulder, upper arm, forearm, hand and fingers, and finally to the muscles of the face. This type of convulsion, in which the patient frequently remains conscious for a relatively long period, is now commonly named a *Jacksonian fit*. As a result of present knowledge it is easy to explain these convulsions as being a result of an abnormal excitation which spreads in the cerebral cortex of the sensorimotor area, in the case referred to starting from the medial aspect of the hemisphere (due, for example, to a parasagittal meningeoma), and traveling downward in the primary motor area. The muscular contractions (mostly clonic twitchings) are the result of a stimulation of the cells in the corresponding site. *In general, a localized beginning of the convulsions in epileptic fits points to the presence of a pathological irritation somewhere in or near the primary motor area,* be it a scar, tumor, a foreign body, or an inflammatory process.[34]

The complex problems of the epilepsies will not be considered here, but some reference will be made to Jacksonian fits. These are a relatively rare type of focal attacks. Focal seizures, originating in somatosensory, auditory, visual, vestibular, olfactory, or other cortical areas, present other initial phenomena, depending on their sites of origin. In some patients having Jacksonian fits, the spreading of the abnormal irritation appears to be limited, and the convulsions may be confined, for example, to the fingers and hand or to the leg. Loss of consciousness is usually lacking, or the patient becomes unconscious only late in the seizure. If the convulsions have spread considerably, general convulsions are likely to end the seizure, accompanied by loss of consciousness. After a Jacksonian fit a paresis of the muscles which have been involved in the convulsions is frequent. This *postparoxysmal paresis* usually disappears in the course of some minutes or hours. If a paresis is present before the fit, it is usually more marked immediately afterward.

Certain particulars concerning focal motor seizures ought to be added. Jackson pointed out that these unilaterally beginning fits have a tendency to start at certain points, namely, the thumb or index finger, the angle of the mouth, and the big toe. This may be interpreted partly as a consequence of the larger area of "representation" of these parts in the motor cortex. When starting peripherally in an extremity, the convulsions are regularly propagated in a proximal direction in the limb, but if they spread to the other extremity they usually start proximally. The intensity of the convulsions at their onset shows a certain proportionality to the degree of spreading, e.g., faint contractions are more apt to be restricted than more violent ones. Some patients state that peripheral stimuli or emotional factors (frequently accompanied by hyperventilation) are apt to precipitate a convulsion. On the other hand, it is well known that some patients suffering from Jacksonian convulsions are able to prevent their development, e.g., by pressing the part in question firmly for a certain period, when they notice the start of an attack (for references see Walshe, 1943). These phenomena are probably due to alterations in cortical excitability which may occur on sensory stimulation.

[34] The classification of epileptic fits has been the subject of much debate. It was common to distinguish *symptomatic epilepsy* from *cryptogenetic, essential,* or *idiopathic epilepsy*. In the former the site of a lesion producing the seizures can be identified, while this is not so in the other type. According to the present international classification, distinction is made between *generalized epilepsies, partial or focal epilepsies,* and *nonclassifiable epilepsies*.

The occurrence of Jacksonian fits is an indication that a localized cerebral lesion exists. *An exact determination of the starting point of the convulsions will give information on where in the motor cortex the focus responsible for the attacks is likely to be found.* As a rule the convulsions start in the same muscles in the same patient. In numerous cases it has been verified by operation that a pathological change has been present at the place determined as the probable site of the lesion. It is important to realize that convulsions of the Jacksonian type may be the only clear-cut symptom of organic disease of the brain. The clinical neurological examination may be entirely negative, or the only positive finding may be, for example, a doubtful inversion of the plantar reflex (extensor plantar response) or a slight asymmetry of the tendon reflexes. This is easy to understand since the pathological process, giving rise to the abnormal irritation of the motor area, may be very small, e.g., a small depressed fracture or a dislocated bone spicule after a head injury.

Certain circumstances tend to minimize the diagnostic value of Jacksonian fits. Frequently the start and spread of the attacks is not sufficiently clearly observed by the patient or his relatives, and they are not able to give any information on this point. (Thus not infrequently there may even be doubt whether the convulsions start on the left or right side.) Methods aiming at provoking the EEG changes accompanying an attack may be used for diagnostic purposes (small doses of metrazol, flicker stimulation).

Far more frequent than the "irritative" symptoms described above are the *deficiency symptoms due to lesions of the motor areas and their efferent fibers.* A destructive lesion of the primary motor cortex will result in impairment of voluntary movements, particularly the discrete, skilled movements. On account of the somatotopic organization a circumscribed lesion may produce a *monoplegia*—that is, a paralysis (or a paresis) of one limb or part of it (see below). Thus a vascular disorder in the territory of the anterior cerebral artery may lead to a monoplegia affecting the contralateral toes and foot, and eventually also the leg, since the foci for these parts are found in the paracentral lobule supplied by this artery.

It might be expected that small lesions of the motor area would give rise to very circumscribed pareses, e.g., paretic flexion of the thumb. Such cases, are, however, very infrequently encountered, although they have been occasionally observed, particularly following small gunshot wounds in wartime (see, for example, Aring, 1944). Their infrequent occurrence is partly due to the rarity of such small injuries, but in addition the overlapping of foci, referred to previously, must be taken into consideration. Even if, for example, the center of the focus for flexion of the thumb is destroyed, there may be enough nerve cells of this focus left to maintain its function.

Identical monoplegias, in principle, may result from damage to the *internal capsule.* Since the corticofugal fibers in the cord and the brainstem lie rather closely packed in the internal capsule, *monoplegias* will be rare. As a rule the lesion of the internal capsule will more or less destroy all these fibers, and a *hemiplegia* will be the result: a paralysis or paresis of all finer voluntary movements of the contralateral leg and arm. In addition the contralateral lower part of the face and half of the tongue will frequently be involved. The most frequent causes of a capsular hemiplegia, as is well known, are hemorrhages, thromboses, and emboli in the territory of the middle cerebral artery.[35] Each of these conditions has its particular clinical features. These will, however, not be considered here. The consequences of an intracerebral bleeding are often referred to as *cerebral apoplexy*. The term *cerebral stroke* appears in general to be used as a common denominator for all three conditions.

It may be appropriate to consider somewhat more closely the *symptoms usually seen in cases of cerebral stroke*. The description below will refer to the changes seen after a rather complete interruption of the internal capsule or a fairly complete destruction of the motor cortex. It should be noted, however, that more often the tissue destruction (due to bleeding or infarction) is only partial and gives rise to symptoms that are less marked and also may be more restricted with regard to their distribution than described here. Immediately following the interruption of the cortifugal fibers, i.e., when the hemorrhage has occurred, the face, arm, and leg on the contralateral side of the body are toneless, there is a *flaccid hemiplegia*. As a rule the patient is unconscious for some time after the bleeding has taken place, and during this time, of course, all muscles are flaccid. Even when he regains consciousness, the paralysis of the affected parts is a flaccid one. The muscles are without tonus, and if an arm or a leg is elevated and then released, it drops passively. In addition to the paralysis and the abolished or very diminished muscular tone, all reflexes are abolished on the paralyzed side, the superficial as well as the deep. This *initial stage*, frequently called *shock stage*, is, however, transient if the patient recovers. The first evidence of recovery is shown by the reflexes. As early as 5 to 10 hours after the onset of the symptoms the plantar reflex can be elicited, but the *plantar reflex is now inverted*.[36] Somewhat later, tendon (myotatic) reflexes reappear, usually after 2 to 3 days or somewhat longer in the upper extremity. At first they are weak, but later they increase, and *some time after the hemorrhage the myotatic reflexes are as a rule clearly brisker on the paralyzed side than on the intact side*. Parallel with this increase in the tendon reflexes a *change in the muscular tone* takes place (both features are evidence of the fact that the spinal cord resumes its reflex function). The muscles present an *increased resistance to passive movements*.[37] As a rule, this *spasticity* and the hyperreflexia are more pronounced in the lower than the upper extremity. It is characteristic that the spasticity seen in capsular lesions in man is not equally distributed in flexors and extensors. In the *upper extremity* more resistance is felt when it is being extended passively, i.e., the *spasticity is more marked in the flexors,* and the arm has a tendency to remain in a flexed position. In the *lower extremity* it is different. The leg tends to be permanently extended since the *spasticity is greatest in the extensor muscles*.

[35] According to Alexander (1942a) this statement is too schematic. Some parts of the internal capsule, e.g., the anterior crus and the dorsal part of the posterior crus, are supplied from striate aterioles, which ultimately are derived from the middle cerebral artery in one third of all cases, whereas more frequently the anterior ones come from the anterior cerebral artery. The knee of the internal capsule is always supplied by small branches from the internal carotid. As a rule the anterior choroidal artery contributes to the supply of the posterior crus.

[36] The plantar reflex and the sign of Babinski will be discussed in a later section.

[37] However, in spite of this, the muscular tone as judged by palpation of the muscles may be lowered. Often the "clasp-knife reaction" is present. When a spastic muscle is progressively stretched the resistance increases up to a certain point at which it suddenly disappears and the muscle can be extended further without resistance. The basis of this phenomenon will be discussed later.

Other features of the same kind are also present. The arm tends not only to be flexed at the elbow, but also to be adducted at the shoulder, and pronation prevails in the hand, the fingers being flexed. In the hip there is a tendency toward adduction, in the foot and toes toward plantar flexion. This is clearly evident and easily recognized when one regards the position of the affected limbs in a patient who has suffered from a capsular hemiplegia.

Why the spasticity in these cases shows such a predilection in its distribution is not known. The position of the joints in the paralytic stage does not seem to be of particular importance. Thus if the leg of a patient suffering from capsular hemiplegia is kept in a flexed position in the early stages, the typical extension spasticity does not develop, but spasticity appears about equally in flexors and extensors. But as soon as the continuous flexor position is given up, the usual extensor overaction appears (Foerster, 1936b). Nor is there any exact correspondence between the strength of the different muscles and their liability to become spastic, even if, e.g. in the arm, the flexors are stronger than the extensors.

In a capsular lesion, as discussed above, other fibers than the corticofugal will of course be involved. Concomitant with the hemiplegia there will therefore commonly be some sensory loss on the same side as the paralysis, due to damage to somatosensory pathways having a relay in the thalamus (cf. Chap. 2). When the lesion is situated low in the capsule the fibers of the optic radiation may be affected, resulting in homonymous visual field defects (see Chap. 8). Aphasic disturbances (cf. Chap. 12) may also occur. A *pure motor hemiplegia* is relatively rare. Fisher and Curry (1965) studied 50 cases of this type, in 9 of which pathological anatomical examination could be performed. They concluded that as a rule a pure motor hemiplegia is the result of a small vessel thrombosis in the posterior part of the internal capsule. In other instances it may be due to a similar lesion in the basis of the pons. Other cases of pure motor hemiplegia have been described (Aleksic and George, 1973; Chokroverty and Rubino, 1975; Chokroverty, Rubino, and Haller, 1975; Levitt, Selkoe, Frankenfield, and Schoene, 1975; Leestma and Noronha, 1976; and others). On the basis of recordings of somastosensory cerebral-evoked responses after stimulation of the median nerve, Chokroverty and Rubino (1975) found that the hemiplegia may in fact not be purely motor. A pure motor hemiplegia is not always due to cerebrovascular disturbances. In the patient described by Levitt et al. (1975) it was caused by a glioma in the pons.

In most cases in which an initially complete paralysis of one side of the body is seen following a lesion of the internal capsule a *certain capacity for performing voluntary movements reappears after some time.* Conditions are essentially similar whether there has been a lesion of the primary motor area—perhaps more correctly the precentral gyrus—or a damage to the internal capsule. In some cases the paralysis affects more or less exclusively either the upper or the lower extremity. It is then referred to as a *monoplegia.*

In *crural monoplegia,* caused by a fairly restricted lesion of the internal capsule or by an affection of the paracentral lobule and the adjoining part of the uppermost parts of the precentral gyrus on the convexity, there appears, in the contralateral leg, a flaccid paralysis which soon is changed into a spastic paresis, with increased myotatic reflexes and inversion of the plantar reflex (sign of Babinski). The leg is kept in an extended position, with plantar flexion in the ankle joint. After some time a certain degree of *voluntary* movement reappears in the initially paralyzed leg. However, the movements which the patient can perform are only *crude, stereotyped, and massive.* Finely differentiated, delicate movements, such as movements of one toe only, are impossible. As discussed especially by Foerster (1936b), there are, strictly speaking, only two slow complex movements which the patient is able to perform voluntarily, one consisting primarily of flexion and called the *flexor synergy* by Foerster, the other an extension movement, the *extensor synergy*.

The extensor synergy in the leg consists of extension at the knee, adduction (more frequently an abduction) of the hip, plantar flexion of the ankle, and plantar flexion of the toes. These movements always occur in combination. If the patient intends to extend his leg at the knee, for example, there occurs concomitantly a plantar flexion of the foot and toes and, as a rule, also adduction of the hip. Isolated plantar flexion of the foot cannot be performed: it can only be combined with an extension of the knee. In the flexor synergy other movements take part; there occurs abduction of the hip, a flexion of hip and knee, dorsiflexion of the foot and toes, and usually inversion of the foot. As is the case with the extensor synergy, the different components of this movement cannot be performed individually, but are forcibly linked together. An attempt at flexing the knee is followed by a dorsiflexion of foot and toes and abduction in the hip.

In the *upper extremity* conditions are similar. Following a lesion of the arm area in the precentral gyrus all fine movements are lost; for example movements of a particular finger cannot be performed. After some time, however, two complex, slow, combined movements can be voluntarily executed. One is a *flexor synergy,* consisting of abduction and usually elevation of the shoulder, flexion in the elbow, pronation and volar flexion at the wrist, and a flexion (or in other cases an extension) of the fingers. Just as in the leg, the different components cannot be isolated, but always occur in combination. The other is the *extensor synergy,* consisting of adduction of the upper arm with lowering of the shoulder, extension of the elbow and wrist, and, as a rule, pronation of the hand (as in the flexor synergy). The fingers are extended or may be flexed.

As seen from the above, after a complete or almost complete destruction of the motor area or interruption of the internal capsule voluntary movements are not completely abolished. However, the complex synergic movements which remain are of a limited value to the patient. Particularly the movements of the hands are not very useful, mainly on account of the compulsory pronation of the hand which accompanies every attempt at flexing the arm. This makes it difficult for the patient to grasp objects and to use them properly (for example in taking a glass of water and drinking: the glass will be turned over and emptied before it reaches the patient's mouth, because the appropriate supination cannot be performed). In the lower extremity conditions are more fortunate. The flexor and extensor synergies will enable the patient to stand and walk, even if his steps are small and hampered by the existing spasticity.

Some additional observations on movements in hemiplegic patients may be mentioned (see Foerster, 1936b, for a complete account). As mentioned above, the capacity to perform discrete movements is opposed by the spasticity present. Since in the lower extremity this is most developed in the extensors and the plantar flexors, an isolated plantar flexion will be more easily performed than an isolated dorsiflexion of the foot (if the plantar flexion is not previously maximal). But a dorsiflexion is possible if some of the dorsal roots supplying the plantar flexors are sectioned. The spasticity, which is of reflex nature, is then diminished and does not prevent dorsiflexion of the foot. (A tenotomy of the Achilles tendon may also reveal that an isolated dorsiflexion is possible.)

If an ordinary hemiplegic patient tries to dorsiflex the affected foot, usually a complex, slow, combined movement occurs, the "flexor synergy" as described above. This is often called the tibialis phenomenon of Strümpell and is only one of several similar observations which can be made in patients with capsular lesions. These movements have been collectively termed "coordinated associated movements." Another phenomenon is the following: If a hemiplegic patient is asked to dorsiflex the nonparetic foot against resistance as strongly as possible, it is often noticed that the paretic foot also performs a correspondingly pure or nearly pure dorsiflexion. This is an example of the so-called "symmetric associated movements," which clinically have been described as signs of pyramidal tract lesions and often have been interpreted as reflexes. (As is well known, normal persons are usually able to perform a forcible dorsiflexion of one foot, without concomitant dorsiflexion of the other, or at least with only a minute movement.)

A satisfactory explanation of the "associated movements" described above can not as yet be given. It should be noted that the phenomena demonstrate that voluntary movements are not completely abolished, even if there is a severe damage of the internal capsule. Furthermore, it is of practical interest that the symmetric associated movements may be utilized in the rehabilitation of hemiplegic patients. It will be advantageous at the onset of the patient's physical exercises to let him try to perform the same movements in the intact extremity as in the paretic, and also at the very beginning, allow him to try to produce isolated movements of the paretic limb by maximal effort of the healthy one.

In capsular hemiplegias the same synergies are observed as after cortical ablations or lesions. However, in both cases the possibility of further improvement is not exhausted with the achievement of the combined movements, the synergies. *Some capacity for performing more differentiated delicate movements may reappear* at a later stage of the disease. However, this is frequently masked by the presence of a pronounced spasticity, which opposes the finer movements.[38] When the improvement takes place, the patient will be able, for example, to flex his foot without concomitant movements of the entire lower extremity.

In agreement with the results obtained from stimulation of the motor area the *capacity to regain discrete movements is not equal in different parts of the body*. Thus conditions are more favorable in the leg than in the arm, and in the arm the proximal parts are better situated than the distal ones. It is a common clinical experience that the capacity to perform discrete movements of the fingers is practically never completely regained after lesions of the internal capsule, or of the arm region of the motor area. (This is in accord with the fact that bilateral discrete movements of the fingers are practically never obtained on stimulation of the cortex.)

In consideration of the motor functions in capsular hemiplegias it is customary to assume that the impairment occurs only in the contralateral half of the body. This is probably an oversimplification and a consequence of the difficulties involved in recognizing minor defects of motor (and other) functions in ordinary clinical tests. After a rather mild stroke resulting in a left-sided hemiparesis (without recognizable sensory defects), the right-handed author observed (Brodal, 1973) that his handwriting was changed (uneven letters and lines, resembling the changes seen after cerebellar lesions). There was no discernible evidence of affection of the left cerebral hemisphere. The changes in handwriting (and subjective observations of some defects in other motor performances) are interpreted as evidence that *damage to one cerebral hemisphere will also influence functions of the other*. However, the changes (at least as concerns somatic functions) are rather

---

[38] The German neurosurgeon O. Foerster undertook transections of a series of lumbar dorsal roots as a treatment of spasticity in man. Although spasticity was markedly diminished or abolished, the subsequent disorders of sensations and trophic changes in the deafferented body parts were serious undesirable side effects. In recent years several neurosurgeons have used partial sections of dorsal roots and claim that this procedure is effective in spastic patients whose spasticity is not made worse by voluntary movements (Gros, Ouaknine, Vlahovitch, and Frerebeau, 1966; Fraioli and Guidetti, 1977; and others).

subtle and will therefore reveal themselves clearly only in very complex and specially learned manual skills, such as writing. (Presumably, the performance of the "intact" hand in a professional musician would be affected, but might not be considered surprising to the patient unless he had some neurological insight.) Only speculations can be made concerning the mechanism of this influence of one cerebral hemisphere on motor functions that are generally believed to be taken care of by the other hemisphere.

Interruption of commissural connections between the two hemispheres can hardly be of importance in most cases, since the damage of the internal capsule will usually be situated below the corpus callosum. An affection of the ipsilateral, uncrossed, corticospinal fibers might be imagined to play a role. It appears most likely, however, that the interruption of corticofugal fibers to the brainstem (descending in the internal capsule) is of major importance, especially of the corticopontine fibers, which make up the bulk of these fibers. Although the corticopontine pathway appears to be purely ipsilateral, the pontocerebellar projection is in part uncrossed (see Chap. 5). One cerebral hemisphere may thus influence both cerebellar halves. It is at present not possible to be more precise as to which cerebellar regions are particularly important for such functions as handwriting. It does not necessarily have to be the hemispheres with the most ample relations to the cerebral cortex. It should be recalled in this connection that the vermal region appears to be more important than the lateral parts for functions such as speech.

The assumption that the disturbances in handwriting described above are essentially due to some derangements of the cerebrocerebellar collaboration receives some support from the well-known fact that the movements of the affected limbs of a capsular hemiparetic patient are definitely atactic. This phenomenon generally receives little attention; the ataxia is usually considered to be a consequence of the affection of corticospinal connections and of the pareses and the spasticity. It is difficult to believe, however, that the interruption of the cerebrocerebellar circuits (via the pons and other relay stations in the brainstem) would not manifest itself by cerebellar symptoms. This problem might deserve closer study.

Mention was made above of the *restitution* that ordinarily occurs after a cerebral stroke. This restitution may be remarkably good. In Chapter 1 (p. 38 ff) some major factors were considered which may in part explain the recovery of function that may go on for long periods, often years, after the stroke. An apparently important process is "collateral sprouting"—that is, preserved afferent fibers give off (sprout) new terminals which establish synapses at synaptic sites denuded as a consequence of interruption of other afferents (see Chap. 1). "Unmasking" of hitherto inactive fiber projections appears to be another factor. Both "sprouting" and "unmasking" are obviously phenomena that indicate possibilities for changes in the patterns of impulse propagation. Although our knowledge about the "plasticity" of the nervous system (see Chap. 1) is still in its beginning, there is reason to believe that this plasticity is a general property of the central nervous system, and that it is a prerequisite for the capacity to learn (in general, be it motor patterns or pure intellectual capacities). Restitution after damage to the central nervous system may therefore in essence be likened to a learning process. Practical experience is in agreement with this.

As discussed in Chapter 1, several factors appear to determine the degree to which patients will regain physical and mental capacities that have been lost after a stroke. Among these factors are the severity and extent of destruction of brain tissue, the age of the patient, the "quality" of the damaged brain, and the retraining. Only the latter factor can be influenced after the stroke. It is appropriate to consider briefly some aspects of retraining, the ultimate goal of which is to establish new pathways, synapses, and neuronal circuits.

Most therapists who have experience with the rehabilitation of brain-damaged patients, maintain that it is essential that *exercises* which require efforts to overcome the functional defects after a stroke *have to be repeated over and over again* and during a period that often lasts several years. This is reminiscent of learning processes in general (cf. the necessity of daily, intensive practice to learn elaborate skills for musicians, ballet dancers, and others). In the case of a hemiplegia it is important that the patient receives physiotherapy and that it starts as early as his general condition permits. To perform exercises under the guidance of a physiotherapist for about half an hour each or every other day for a couple of months is, however, not sufficient to obtain optimal results.

It is often overlooked that *therapy or retraining should continue for years,* since improvement—albeit gradually more slowly progressing—may take place for 3–4 years or more (cf. again the learning of new capacities of healthy persons). In this connection it cannot be too strongly emphasized that *it is in essence up to the patient himself how far his retraining will bring him* (within the limits set by factors that are not amenable to environmental conditions). The main task of the physiotherapist is to teach the patient how best to practice, which exercises to do and so forth. There is little doubt that daily training for years is necessary if the patient wants to achieve optimal restitution. This explains the well-known fact that the patient's *motivation* for treatment is of decisive importance for the final result. Factual and honest information about what may be expected and encouragement are, therefore, essential in any rehabilitation program. It should not be overlooked that in the often difficult situation of rehabilitation after brain damage, psychological factors are of great importance (see Bach-y-Rita, 1980). Most aspects of the rehabilitation after brain injury are discussed in *Recovery of Function: Theoretical Considerations for Brain Injury Rehabilitation* (1980).

It is encouraging that in recent years much information has become available about the bases of restitution after brain damage, and that this knowledge appears to justify the therapeutic optimism we have witnessed in later years.

Symptoms in diseases of the basal ganglia and related nuclei. In this field we meet with an intricate complex of motor abnormalities, abnormal movements, and changes of muscular tone; and we are still far from comprehending clearly how and why the different symptoms appear in the various pathological conditions. Even though it is found that in some diseases there are lesions which are usually confined to one or two of the basal ganglia or associated nuclei, it is not permissible to conclude that this nucleus or these nuclei are "centers" regulating a particular phase of the motor functions. It is far more reasonable to interpret the facts in a general way as follows: The various lesions produce a disturbing effect on a normally existing finely adjusted collaboration and cooperation between various structures. If a lesion affects one of these, for example, the substantia nigra, this harmonious interaction will be disturbed in some way, yielding symptoms of one kind, while if another part is damaged, for example the pallidum, the ensuing disturbances will be of another type. Clearly this does not allow us to conclude that, for instance, the substantia nigra is a "center" for one or other of these functions. We shall return in a later section to attempts at explaining the symptomatology.

A brief survey of the principal symptoms following diseases of the basal ganglia is appropriate at this juncture. The symptoms are mainly of two types, disturbances of muscular tone and involuntary movements of different kinds, frequently called "hyperkinesias." Not unusually, however, symptoms of autonomic disturbances are also present, presumably due to concomitant lesions of the hypothalamus or of fiber connections in this region. Some of the pathological conditions form rather well-defined clinical entities, and will be described.[39]

A disease with fairly constant symptoms and fairly consistent localization of pathological changes is *paralysis agitans* or *Parkinson's disease*. This is characterized by three cardinal symptoms: rigidity, tremor, and akinesia. (For an account of the prevalence and natural history of Parkinson's disease, see Pollock and Hornabrook, 1966. For recent references, see Barbeau, 1976a.) It should be mentioned, however, that one or two of these symptoms may be absent, and that several other signs may inconsistently be present. In the opinion of Barbeau (1976b), Parkinson's disease is not a single entity, but rather a complex with some common features due to lesions involving the same part of the brain (see below).

The rigidity, which may be very marked, differs from the spasticity seen in lesions of the internal capsule in man. It affects all muscles approximately to the same degree, and in typical cases is of the cogwheel type. The voluntary movements suffer from a marked *retardation,* but muscular strength is well preserved. The myotatic reflexes are normal or slightly brisker than usual. There is often a *poverty in movements* (akinesia) which changes the patient's appearance. He appears rigid, the usual pendular movements of the arms in walking are reduced or lacking, and his physiognomy lacks the normal mimics (mask face). Speech is slow and monotonous, and the handwriting changes. The *tremor,* which may be absent, is usually described as a rhythmic tremor at rest, particularly prominent in the fingers, and comparable to the movements performed in pillrolling. This tremor is of quite another type than the coarse intentional tremor seen in cerebellar disease (cf. Chap. 5). The latter appears in voluntary movements, but the tremor of paralysis agitans may also sometimes be seen to increase in intentional movements, be these static or phasic.[40] It disappears during sleep, but is exaggerated on emotional tension. Frequently symptoms of autonomic disturbances are observed, such as an increased sebaceous secretion (greasy face), salivation, and vasomotor disturbances. Mental deterioration is not uncommon (see Pollock and Hornabrook, 1966).

The symptoms in paralysis agitans have been regarded by many as being due to a specific distribution of anatomical changes. *Changes are usually found in the substantia nigra* (particularly its pars compacta) and are frequently regarded as senile, degenerative. There is a loss of cells, and many of those that remain are abnormal, shrunken, and pale. Slight glial increase may be present. In some cases smaller changes are also found in the striatum, presumably caused by degeneration of the nigrostriatal fibers. Occasionally, less marked changes are found in other basal ganglia. Changes in the cortex have been described by some authors. In cases in which pronounced clinical symptoms have been found only unilaterally, the most clear-cut anatomical changes have usually been found in the contralateral substantia nigra. Cases may be observed clinically in which the symptoms of Parkinson's disease are limited at first to only one limb.

A clinical picture resembling that of paralysis agitans in all particulars, especially the cases without tremor, was formerly seen quite often and interpreted as

[39] Clinically a distinction is often made between two groups of syndromes. In the *hyperkinetic-dystonic syndromes* there is an excess of motor activity. In this group Jung and Hassler (1960) place the choreic, the ballistic, the athetoid, the dystonic, and the myotonic syndromes. In the *hypokinetic-rigid syndromes* there is a reduction in spontaneous motor manifestations, as in parkinsonism.

[40] Lance, Schwab, and Peterson (1963) discuss the differences between resting tremor and action tremor in Parkinson's disease.

representing the late stages of the *epidemic encephalitis,* or *encephalitis lethargica* (*v. Economo*). The condition was usually spoken of as *postencephalitic parkinsonism,* more correctly *chronic encephalitis,* since anatomical investigation often revealed clear-cut, inflammatory changes. As might be expected in a disease of inflammatory nature, the clinical symptoms, as well as the course and the distribution of the pathological changes, were more irregular than in the idiopathic Parkinson's disease. There are only a few suvivors from the epidemic of encephalitis lethargica in the 1920's, and parkinsonism with this etiology is becoming increasingly rare (Duvoisin et al., 1963).

> The pathological examination in the acute stages of epidemic encephalitis revealed inflammatory foci, with a predilection for affecting the gray matter, first and foremost the pallidum and the substantia nigra. Affection of the hypothalamus or the gray substance surrounding the aqueduct may explain the common occurrence of sleep disturbances and ocular palsies, respectively. In chronic encephalitis the pathological changes are found in the same regions as in the acute stages of the disease. An occasional flaring up of the inflammatory processes, which may be latent for long periods, explains the common fluctuations in the symptomatology, which may serve as useful guides in the differential diagnosis between paralysis agitans and chronic encephalitis. The scattered localization of the pathological foci also explains several of the concomitant symptoms which were frequent in chronic encephalitis, such as the oculogyric crises (probably due to damage to the supranuclear pathways to the eye muscle nuclei), or autonomic disturbances on account of hypothalamic foci. Changes in the cerebral cortex were assumed to be responsible for some of the changes in character and the mental deterioration which were not infrequent in chronic encephalitis, particularly when the acute stage occurred in infancy.

The dominant hereditary *Huntington's chorea* is characterized by involuntary muscular contractions, which usually begin at the age of forty and progress steadily. The choreatic movements may at first be limited to a small part of the body, the face, or an arm, but by and by they tend to involve all musculature of the body. The movements are jerky, irregular, uncoordinated, and changing. In more advanced cases the patient is never at rest. If the facial musculature is involved, speech and swallowing may suffer. There is no reduction of muscular power, but muscular tone is said to be diminished. This, however, is very difficult to judge on account of the incessant movements. In later stages mental changes appear, leading finally to dementia, but the mental changes do not develop parallel to the somatic alterations, and may start before the choreatic movements.[41]

The prominent pathological changes in this disease are found in the *striatum*. The nucleus caudatus and the putamen become atrophic, as can be seen in advanced cases even on macroscopical inspection of the brain. Microscopically the characteristic feature is a reduction or loss of the small nerve cells of the striatum, while the large cells are preserved or present only minor changes (see, however, Denny-Brown, 1962). There is an increase of the glial elements and a reduction of myelinated fibers in the striatum. Less marked changes may be found in the pallidum and the subthalamic nucleus, but most authors regard these as secondary to the affection of the striatum. Practically always, however, the *cerebral cortex* displays loss of cells, particularly in the frontal regions and the precentral gyrus and mostly in the deeper layers. These cortical changes are probably responsible for the mental deterioration.

[41] In rare instances Huntington's chorea occurs in children. The clinical picture then often differs from that in adults, but the pathological changes are chiefly the same (Jervis, 1963).

A localization of the pathological changes similar to that in Huntington's chorea is seen in *chorea infectiosa* or *chorea minor, Sydenham's chorea*. This disease, which occurs in children, most frequently girls, usually after a preceding infection ("rheumatic" or otherwise), shows involuntary movements very similar to those observed in Huntington's chorea. They are rapid, purposeless, and irregular. Since chorea minor is a benign disease, autopsy material is rare, and often other complicating diseases have been present which are responsible for the fatal outcome and which obscure the anatomical picture. As far as can be seen from the reports in the literature, the localization of the changes, which are mainly inflammatory, is first and foremost the striatum, the small cells of which suffer predominantly. The so-called *chorea gravidarum,* which is most frequently seen in women who have suffered from chorea minor in infancy, is also inadequately investigated with respect to its pathological anatomy, but it appears to present changes similar to those in chorea minor.

Some less usual types of involuntary movements will be briefly mentioned. *Athetoid movements* are usually very characteristic and easily recognized. They appear predominantly in the fingers and toes, and are as a rule less characteristic and less marked in the proximal muscles. The slow, bizarre movements of the fingers most frequently consist of a forcible extension or hyperestension of the metacarpophalangeal or interphalangeal joints, eventually combined with abduction movements of the fingers and extension of arm and hand. Like other types of involuntary movements, they increase during emotional stress.

Athetoid movements may be seen in various diseases. Thus the so-called *"infantile cerebral paralysis"* ("cerebral palsy"), particularly where it occurs as hemiplegia ("infantile hemiplegia"), may be accompanied by such movements. The anatomical basis of the disease is frequently an intracranial hemorrhage following birth trauma, or it may be a malformation or an encephalitis. In this syndrome the abnormal movements are not a cardinal symptom, but occur in addition to other signs of severe cerebral injury, such as hemiparesis or hemiplegia, severe reduction of intelligence, and frequently epileptiform convulsions.[42] Athetoid movements may also be observed in diseases of the basal ganglia that start in adults, particularly in *chronic encephalitis,* in which, occasionally, any type of involuntary movements may occur. The so-called *"athétose double"* described by O. Vogt is a rare disease in which athetoid movements occur bilaterally in early infancy without later progression. The anatomical changes underlying the athetosis in these cases are characteristic. On account of the anatomical picture the disease has been called *"status marmoratus of the striatum."* (For an account of the pathology see Alexander, 1942b.)

The term *torsion dystonia* has been used to describe a syndrome of abnormal involuntary movements irrespective of their cause (Marsden, 1976). Traditionally, the slow mass torsion movements of trunk and extremities which characterize the disease have been ascribed to pathological changes often found in the striatum. However, in cases of dystonia, changes can be found in many sites of the brain other than the striatum, and Zeman (1970) claimed that no pathologic–anatomic correlates of the clinical findings can be demonstrated in the brains of these patients. In most cases of *hemiballismus,* on the other hand, more or less isolated damage to the *subthalamic nucleus* has been found (see Whittier, 1947). In this

---

[42] In recent years great interest has been devoted to the study of this disease group. Modern treatment of affected children has given encouraging results. Interested readers will find information on all aspects of the condition in the journal *Developmental Medicine and Child Neurology.*

disorder, large-scale involuntary movements of arms, legs, and trunk occur contralateral to the brain lesion. Occasionally the abnormal movements are restricted to one arm or one leg (monoballism; for a review see Carpenter and Carpenter, 1951), a finding of interest in connection with some experimental evidence for the existence of a somatotopical organization within the subthalamic nucleus (Mettler and Stern, 1962). However, cases of hemiballism have been described with lesions in other locations and not in the subthalamic nucleus (Schwarz and Barrows, 1960). So-called *myoclonic twitchings* may be seen in chronic encephalitis in its acute stages. They consist of rhythmic, or more irregular, very rapid contractions of individual muscles or parts of muscles, which usually do not produce movements. The same symptom may also be found in the very rare *myoclonus-epilepsy,* in which "inclusion bodies" have been found in the nerve cells, particularly in the substantia nigra, often also in the striatum and dentate nucleus. Myoclonic twitchings in the muscles of the larynx, pharynx, and the soft palate have, probably erroneously, been attributed to lesions of the inferior olive and dentate nucleus which have been observed in these cases.

A type of involuntary movement which has been the subject of much debate is the co-called *tics:* simple involuntary twitchings of smaller muscle groups which occur repeatedly in a stereotyped manner at certain intervals. They may be suppressed by voluntary effort for a while and disappear during sleep. In most cases their cause remains unknown, but sometimes stereotyped movements of this type may be an initial symptom of some affection of the basal ganglia, such as Huntington's chorea. Even if it cannot be denied that in some cases tics may be purely "functional," having no organic basis, their occasional occurrence as an initial symptom and, likewise, the fact that they may remain as the only persistent symptoms in cases of, for example, chorea minor when the patient is otherwise free from symptoms, speaks in favor of their organic origin.

Correlations between clinical observations and experimental findings in animals. On account of the practical importance of disturbances of motor functions in man and their frequent occurrence it is no wonder that much work has been devoted to understanding their functional and structural basis. As seen from the preceding sections, there are certain clear differences between the motor disturbances following lesions of the central motor cortex and the internal capsule and those seen in cases in which the pathological changes are found in the basal ganglia. For a time this appeared to be explicable on the assumption that there are two separate supraspinal systems influencing motor functions, the pyramidal and the "extrapyramidal." As repeatedly alluded to in this chapter, this dualism can, however, no longer be upheld. For this reason, many of the interpretations and hypotheses set forth to explain results of experimental observations have turned out to be untenable and must be discarded, even if the observations themselves are essentially correct. (As a relevant example may be mentioned forced grasping, described above. Before the supplementary motor area was discovered the symptom was considered to be due to damage to the "extrapyramidal" cortical area 6.) Many observations made in the past, therefore, have to be reinterpreted in the light of present-day knowledge, Unfortunately, this is often not possible, because the observations made and the experimental conditions under which they were obtained have not been recorded in sufficient detail. Findings in animals cannot always be applied to man. It is therefore not yet possible to "explain" the disorders of motor functions occurring in diseases of the central nervous system in man. In the following, an attempt will be made to discuss certain features of such disorders in the light of recent studies, in animals as well as in man. For reasons given above, observations from the older literature will not be discussed at any length.

Decerebrate rigidity. Postural reflexes. Decerebrate rigidity is a condition which appears when the brain stem of an animal is transected above the vestibular nuclei. It is characterized primarily by an increased muscular tone in the extensors (antigravity muscles). The animal is able to stand with rigid legs, usually the tail is a little elevated and the chin is tilted somewhat upward. The myotatic reflexes (see Chap. 3) are strongly increased. Since Sherrington first observed this phenomenon in 1898, several investigators have studied it intensely. Rademaker performed sections of the brainstem at different levels and concluded that the presence or absence of the red nucleus was a deciding factor in the development of the rigidity. However, if the transection is made caudal to the level of the vestibular nuclei, the rigidity fails to develop and all muscles are flaccid. Furthermore, destruction of the lateral vestibular nucleus on one side nearly abolishes the rigidity on the same side, produced by a preceding more rostral transection of the brainstem. Descending impulses from the reticular formation of the brainstem appear to be, in part at least, responsible for the phenomenon of decrebate rigidity (see Brodal, Pompeiano, and Walberg, 1962, p. 109 ff., for some data). As has been mentioned, the lateral vestibular nucleus exerts a strong facilitatory action on ipsilateral extensor motoneurons (α as well as γ), and it appears that a main factor in the appearance of decerebrate rigidity is the following: the vestibular excitatory action on the extensor motoneurons lacks the normally opposing actions from other regions, such as the red nucleus, which facilitate flexor motoneurons. Since the spinal parts of the cerebellum (see Chap. 5) have an inhibitory effect on the lateral vestibular nucleus of Deiters, ablation of this part of the cerebellum will increase the decerebrate rigidity and even produce opisthotonus, while stimulation reduces it. The rigidity is, however, not dependent only on descending impulses to the motoneurons. Afferent impulses from the periphery are also of importance. This was clearly demonstrated by Sherrington, who found that when the dorsal roots supplying one of the extremities are sectioned, the rigidity disappears from this limb. It is the interruption of afferents from the muscle spindles which is the important factor. This type of decerebrate rigidity is, therefore, often spoken of as a "γ-rigidity" (Granit, 1955a), on the assumption that the main factor is an increased excitability of the γ neurons (which via the spindle afferents produce an excitation of the α neurons). However, when a so-called anemic decerebration is made (ligation of both carotids and the basilar artery) in which half the cerebellum and a considerable part of the pons are destroyed, the animal presents an intense rigidity, even when the dorsal roots are cut. In this instance, therefore, the α neurons are able to maintain an increased muscular tonus without support from the γ neurons and spindles (α rigidity). Even if it is possible that this distinction between α and γ rigidity may be too strict, and that most instances of α rigidity may include also a component of γ rigidity (see, for example, Batini, Moruzzi, and Pompeiano, 1957), there are obviously at least two different ways in which an increased excitability of the α motoneurons can occur, a fact of interest for clinical problems. The mechanisms involved in the production of decerebrate rigidity are, however, far more complex than appears from the above. Animals with decerebrate rigidity have proved very useful for the study of *postural reflexes*. This is a collective designation for a large number of reflexes which aim at preserving the normal posture of the body.[43] To these belong the tonic neck and labyrinthine reflexes.

The *tonic neck reflexes* are revealed as a change in the distribution of muscular tone when the position of the head relative to the body is altered.

During movements of the head impulses arise in the sensory epithelium of the labyrinth as well as in the proprioceptors of the neck. In order to elicit the tonic neck reflexes, therefore, the labyrinths must be destroyed. If, then, in a labyrinthectomized, decerebrate animal the head is rotated to one side, the increased extensor tonus following the decerebration is further augmented on the side to which the snout is pointing, while it diminishes on the other side. If the head is tilted to one side, the tonus increases on this side, but is reduced on the other. If the head is bent forward there ensues a flexion and reduced tone of the forelimbs but extension and increased tone of the hind limbs. If the head is bent backward, the opposite change appears: flexion of hindlimbs and extension of forelimbs. Since these reflexes disappear when the dorsal roots of the upper three cervical nerves are cut, it is clear that they are initiated by impulses from the neck. The afferent impulses which elicit the changes, arise in the receptors of the upper cervical joints (McCouch, Deering, and Ling, 1951).

If in a decerebrate animal the upper cervical dorsal roots are cut, or if the head is fixed so as to prevent any movement of it in relation to the body, the *tonic labyrinthine reflexes* can be elicited.

Changes of the animal's position in space are then followed by a regular alteration of tone distribution. In the supine position, with the snout elevated about 45°, the rigidity is maximal, but it is minimal if the animal is placed in the prone position with the snout tilted 45° below the horizontal axis. In intermediate positions the degree of the rigidity varies between the two extremes. These reflexes are elicited from the sensory epithelium of the utricle (and, possibly, the saccule) of the labyrinth (they disappear if the otoliths are detached from their proper place by centrifuging the animal) and they are maintained as long as the *position* is upheld. They are not to be confused with the vestibular reflexes which are provoked by *movements* in space and which are initiated from the semicircular canals (rotatory nystagmus, etc.; see Chap. 7). These tonic neck and labyrinthine reflexes are present in decerebrate animals in which the transection of the brainstem is made so far caudally that practically only the medulla oblongata is in connection with the spinal cord.

The same reflexes are present in animals in which, in addition, the mesencephalon and thalamus are in connection with the lower parts. In such animals several other postural reflexes can also be elicited.

One of them, which is dependent on the labyrinth, is the following: If a normal animal, or one in which the cerebral cortex has been removed (decortication), is blindfolded and then placed in various positions in space, it will always keep its head in the natural, horizontal position, whether the animal is placed on its side or in other positions. However, if both labyrinths are destroyed this reaction is abolished. This reflex righting of the head initiated from the labyrinth is called *the labyrinthine righting reflex,* and it serves to orientate the animal in space. When by means of this reflex the head has assumed its natural position, the rest of the body can be oriented

---

[43] For an exhaustive account of postural mechanisms see Roberts (1967).

after the head by means of other reflexes, initiated from the proprioceptors of the neck and of the trunk (neck righting reflexes, body righting reflexes acting upon the head, body righting reflexes acting upon the body). Other reflexes serving a similar purpose of orientation are dependent on the integrity of parts of the cortex. Thus the optic righting reflexes require an intact area striata, since they depend on vision. Hopping and placing reactions appear to be dependent on the motor cortex.

The postural reflexes are present in the normal human being and are essential for normal motor performances. In lesions of the brainstem the tonic neck and labyrinthine reflexes may appear in an exaggerated form. Furthermore, these reflexes may operate in influencing other reflexes, as mentioned previously with reference to forced grasping and the grasp reflex in infants (Pollack, 1960). They are, therefore, of clinical interest, the more so since some use may be made of these and other postural reflexes in physiotherapeutic exercises in some instances. In severe lesions of the brainstem decerebrate rigidity may be seen in man (see Chap. 6), resembling the same condition in animals. In these patients tonic neck and labyrinthine reflexes can often easily be elicited.

Spasticity and rigidity. The phenomena called spasticity and rigidity in man have been referred to previously. The condition called spasticity is regularly seen following lesions of the internal capsule and often in lesions involving the primary motor cortex,[44] while the term rigidity is used for the changes in muscular tone occurring in parkinsonism.[45]

Muscular tone was considered in Chapter 3. As mentioned there, muscular tone, as determined by the resistance offered by a muscle to being stretched, appears to be due to the inherent viscoelastic properties of the muscle and the tension set up by contraction. Although some reservations may have to be made as concerns man (see Chap. 3), experimental studies indicate that muscular tone is essentially a reflex phenomenon, being due to a continuous stream of impulses from the muscle spindles, activating the α motoneurons. These reflex mechanisms were considered in Chapter 3. In addition, impulses from other levels of the cord and from supraspinal regions influence muscular tone. When this is altered in a different way in capsular lesions than in parkinsonism, this must be due to different derangements of the normal supraspinal influences on the spinal reflex apparatus of the stretch reflex (including the motoneurons). No entirely satisfactory explanation can so far be given of these differences, but some information is available concerning the changes in the reflex mechanism occurring in the two conditions. The problems are far from solved, and no exhaustive treatment of the subject will be attempted.

Clinical spasticity in many ways resembles the situation in decerebrate rigidity in animals. As mentioned previously, sections of dorsal roots in patients with spastic hemiplegia abolishes spasticity (Foerster, 1911) as well as the increased stretch reflexes (in severe spasticity in man patellar and ankle clonus are common). This suggests that there must be an increased activity of the γ neurons in spasticity. Further support for this view was found in the fact that procaine infil-

---

[44] According to some authors lesions restricted to area 4 in man do not result in spasticity. However, such restricted lesions are rare.

[45] It is unfortunate that the experimentally produced condition in animals which in many respects appears to correspond to spasticity in man has been termed decerebrate *rigidity*. It should be kept in mind that this is quite different from the rigidity seen in parkinsonism.

tration around the nerves to spastic muscles eliminates spasticity without impairing voluntary power (Walshe, 1924; Rushworth, 1960). However, as mentioned previously, in a capsular hemiplegia muscle tone as judged by passive movements is regularly increased, while often little or no increase of tone is felt on palpation.

These findings have been explained, on the basis of experimental studies, on the assumption that there is an increase in the phasic stretch reflex (see Chap. 3), while the tonic stretch reflex is not changed or is less changed. On rapid movement the phasic reflex is brought into play, just as when a tendon jerk is elicited. The heightened excitability of the reflex explains that much slower movements will elicit it than is the case in normal persons. If in cases of mild spasticity stretching of a muscle is performed very slowly, there is no or very little increased resistance to passive movements. Thus, the speed of the movement is important.

As described in Chapter 3 (p. 156 ff), the innervation and function of the muscle spindles have turned out to be extremely complex. In view of this complexity it has become increasingly difficult to clarify the mechanisms underlying spasticity in man, the more so since in some respects the situation in man appears to be different from that in the experimental animals studied. As maintained by J. Jansen, Jr. (1962), the spastic states in man may in the end turn out to be a rather heterogeneous group with different degrees of release of the two intrafusal systems. Even though an essential general feature in spasticity appears to be a release of fusimotor innervation (see, however, below), nuclear bag and nuclear chain fibers may not be affected in parallel, and γ_1 and γ_2 motoneurons are not necessarily influenced together. Despite attempts of numerous students to analyze spasticity in man, there are still many open questions. Reference was made to some of these studies in connection with the description of the motor innervation and the muscle spindles in Chapter 3. It is beyond the scope of this text to discuss such questions thoroughly; only a few data will be mentioned.

Some authors who have attempted to analyze spasticity in man have made use of the H-reflex (see footnote 33), which is considered to correspond to the stretch reflex, except for the fact that the spindle afferents and not the spindles are stimulated. (According to Herman, 1970, and others, the two reflexes are not true counterparts of each other.) Landau and Clare (1964) concluded from studies utilizing this method that, although the motoneurons are hyperactive, "in spastic hemiplegias, there is no evidence which requires the presumption that fusimotor hypertonus is of etiological significance in pathological hypertonic states." (See also Diamantopoulos and Zander Olsen, 1967.) Support for this view was found in the observation that no hyperactivity of the fusimotor system accompanies hyperactive tendon jerks in the monkey (Meltzer, Hunt, and Landau, 1963). Increased excitability of the motoneurons in spasticity has been concluded to be present in several studies (Angel and Hofmann, 1963; and others). Herman (1970) maintains that the behavior of spastic muscle in man is mainly due to enhanced muscle afferent discharge. As referred to in Chapter 3, Hagbarth and his co-workers did not find clear evidence of increased spindle sensitivity in patients with spasticity. In this connection the findings of Gilman, Lieberman, and Marco (1974) in the monkey are of interest. After ablation of areas 4 and 6 they found spasticity but normal or reduced spindle activity.

The differing effects on spasticity of cooling of spastic muscles (positive in some cases, negative in others) have been taken to support the view that spasticity may not in all patients be due to hyperactivity of the fusimotor system. In some spastic patients the increased flow of spindle impulses may play no important role. It may be noted that in spastic hemiplegic patients the reciprocal Ia inhibition appears to be retained (see Yanagisawa, Tanaka, and Ito, 1976), in contrast to what is seen in paraplegia after transverse lesions of the cord.

The data above are only a few, selected from a large number of studies. On many points results opposing those mentioned have been reported. In general considerations of the problem of spasticity in man, however, the following should be emphasized: There are numerous descending fibers from several brainstem nuclei capable of influencing both α and γ neurons, and many of these nuclei will lose their innervation from the cerebral cortex after a capsular hemiplegia. These facts represent warnings against restricting attention to the *spinal* reflex mechanisms that may influence muscular tone and stretch reflexes in attempts to explain spasticity in man. The influences on the spinal mechanisms from supraspinal levels of the brain appear to be manifold and complex (as appears from studies of the actions on the spinal cord of the various descending fiber systems described earlier in this chapter). And, as mentioned above, an increased afferent input from muscle spindles and hyperactivity of fusimotor neurons does not appear to play an important role in all patients with spastic hemiplegias. A hyperexcitability of α neurons may be caused by influences from higher levels, without involving the γ neurons and muscle spindle activity.

It is noteworthy in this connection that, among the many brain nuclei that project to the spinal cord, the vestibular nuclei do not receive direct afferents from the cerebral cortex. Their activity will therefore not (or only to a minor extent) be affected in lesions of the internal capsule, and the strong facilitatory action exerted on extensor α and γ motoneurons by the lateral vestibular nucleus will lack the normally opposing influence of the corticospinal and corticorubrospinal pathways. This view fits in with the observation that there is an excess of extensor innervation in capsular lesions, and that spasticity affects the extensors.[46]

There are several similarities, but also differences, between decerebrate rigidity in animals and the spasticity seen in capsular hemiplegia in man. Many features in the organization of the nervous system appear to be different, anatomically and physiologically. It should be recalled, however, that some differences between the two conditions may be related to the fact that following decerebration even more supraspinal pathways are interrupted than in a capsular lesion, for example the tectospinal and interstitiospinal tracts.

The *"clasp-knife reaction,"* often seen in the spastic limbs in hemiplegia (see p. 255) and also found in decerebrate rigidity in animals, can be explained as an example of the so-called *lengthening reaction,* described by Sherrington. It is due to impulses arising in the Golgi organs in the muscle when this is stretched. As discussed in Chapter 3, the afferent impulses exert an inhibitory action on the agonists taking part in the stretch reflex, while they facilitate the antagonists (see Fig. 3-5). The Golgi organs appear, however, to have other functions than to prevent excessive tension of the muscle.

Several *therapeutic measures* have been suggested for the treatment of spasticity, a condition that can be very disabling. Mention has been made above of a method of *partial dorsal root section.* Other, nonsurgical procedures have also

---

[46] However, we are here again faced with the paradox that in man this is not true for the upper extremity, where the flexors are the spastic ones. One may indeed wonder whether man's assumption of the upright posture, in which the forelimbs are relieved of their body-supporting function, has been accompanied by some hitherto unrecognized changes in the organization of the supraspinal influence of the cord, and presumably reflected in the anatomical organization of the pathways involved. Recent research, however, has revealed interesting differences also in quadrupeds between nervous pathways related to forelimbs and hindlimbs.

been tried. Reference was made in Chapter 3 to the *tonic vibration reflex* (the muscle spindle afferents are activated by vibration of the muscle belly). In spastic pareses, vibration, applied to the antagonists of spastic muscle, often reduces the spasticity and may improve the execution of voluntary movements. Mention was likewise made in Chapter 3 of the beneficial effect in some patients of cooling spastic muscles (presumably due to reduction of the afferent input from muscle receptors).

As described in a preceding section, exaggerated stretch reflexes occur also following *transverse lesions of the cord*. However, there is no pure spasticity as in a capsular lesion or in decerebrate rigidity.[47] Flexion reflexes are also enhanced. Since in transverse lesions of the cord all descending pathways are interrupted, it is conceivable that the effects on the reflex mechanisms of the cord will differ from those in capsular hemiplegias and in decerebrate rigidity.

> In paraplegic patients there is a general excessive activity of motor units, as shown by Dimitrijević and Nathan (1967a, b) in extensive clinical and electromyographic studies. It is characteristic, in contrast to a spastic hemiplegia, that there is lack of normal reciprocal innervation. Certain muscles or groups of muscles are activated by every kind of stimulation (proprioceptive and exteroceptive). On eliciting stretch reflexes in a muscle there appears a recruiting of motor units after the stimulus has ceased, and this spreads to motoneurons of other muscles. The exaggerated stretch reflexes are taken to indicate that some of the spastic state is due to hypersensitivity of the spindles, an assumption which receives support from the observation that "chemical posterior rhizotomy" with phenol solution has been shown to have a beneficial effect on the spasms and to reduce the spontaneous activity which is usually present in most muscles in these patients.[48] With reference to physiological studies the increased excitability of the spinal cord is explained as a consequence of the loss of tonic inhibitory descending impulses from the brainstem, some of which act on interneurons intercalated in flexion reflex arcs.

The rigidity seen in parkinsonism differs in many respects from spasticity following lesions of the internal capsule. The resistance to passive movements involves flexors as well as extensors and is almost equal throughout the range of movement. This "plastic" rigidity is independent of the speed of the movement. The tendon (myotatic) reflexes are not appreciably altered, indicating that the phasic stretch reflex is not exaggerated in spasticity. However, since procaine infiltration of a muscle (presumably blocking the transmission in the γ efferents) abolishes rigidity in the muscle (Walshe, 1924; Rushworth, 1960), it has been concluded that rigidity is dependent on an increased sensitivity of the muscle spindles. An increased activity in the nuclear chain intrafusal system has been thought to augment the *static* responsiveness of the primary endings. Clinical observations (Schimazu et al., 1962) and other data (see Jansen, 1962) lend some support to the conclusion that exaggeration of the tonic properties of the stretch reflex is an important feature in parkinsonian rigidity. An increased descending

[47] However, in slowly developing damage to the cord due to diseases in the cord (intramedullary tumors, inflammations) or to pressure on the cord (extramedullary tumors, hematomas) there is often in the beginning a spastic mono-, para-, or hemiplegia, rather similar to that following a capsular lesion. No satisfactory explanation of this can be given. It may be presumed that first and foremost the dorsal parts of the lateral funiculus are affected.

[48] As mentioned earlier, Foerster used posterior rhizotomies in order to relieve exaggerated stretch reflexes in spasticity. Dimitrijević and Nathan (1967a) commented upon the fact that this operation has been so little used and advocate a more extensive therapeutic use of procedures which may cause a reduced inflow from the periphery to the spinal cord. See also footnote 38.

input to both α and static γ motoneurons has been suggested to be a central factor in the pathogenesis of parkinsonian rigidity (see Wallin, Hongell, and Hagbarth, 1973).

> During increasing stretch of a rigid muscle continuous lengthening reactions occur. This may be the explanation of the *"cogwheel phenomenon."* The frequency of cogwheeling is not consistently related to the rate of stretch, according to Lance, Schwab, and Peterson (1963), who suggest that the phenomenon is related to the resting or action tremor seen in some of these patients. (The tremor, often seen in parkinsonism, will be considered in a later section.)

The changes in the function of the motoneurons and muscle spindles in rigidity must ultimately be due to changed supraspinal influences on the reflex arc of the stretch reflex. However, since the nuclei which show pathological changes in parkinsonism (chiefly the globus pallidus and the substantia nigra) have only very sparse caudally directed efferents, they can scarcely influence the cord directly via descending pathways. We shall return to this problem when the functions of the basal ganglia are discussed.

"Pyramidal" and "extrapyramidal" cortical motor areas? In the discussion of the motor cortical areas the results of electrical stimulation of these areas were considered, while the results of ablations were only briefly mentioned, especially as concerns the primary motor cortex.

As mentioned previously, attempts were made in the 1930s, particularly by Fulton and his collaborators, to study the effects of lesions of area 4 in monkeys and chimpanzees, and to compare the resulting motor disturbances with those following lesions of the "premotor" area, and with combined lesions. The consequences of lesions of area 4 were considered to give information on the role of the pyramidal tract, since this was assumed at that time to arise only from area 4. Further, the results of isolated transection of the pyramidal tract were thought to correspond to those seen after extirpation of area 4. Although the observations made were correct, the interpretations of them are now chiefly of historical interest.

> In the monkey and chimpanzee a cortical lesion restricted to area 4 results in an impairment of finely coordinated movements on the contralateral side (Fulton and Kennard, 1934; and others, in the monkey; Fulton and Kennard, 1934; Tower, 1944; and others, in the chimpanzee). There is no spasticity (except for some transient involvement of the fingers) but a slight hypotonia and reduced tendon reflexes. In the chimpanzee an inverted plantar reflex (Babinski sign) occurs (Fulton and Keller, 1932). After some time the paretic muscles show some atrophy.
>
> A lesion confined to area 6 in the chimpanzee leads to a transitory weakness in the contralateral limbs. After about a week the gross motor performances are normal. But on closer analysis it is found that there is an impairment of the capacity to perform skilled movements, which are only awkwardly carried out. This is most evident when the animals have been trained, before the operation, to solve problems which require fine muscular adjustments (Jacobsen, 1934; Fulton, Jacobsen, and Kennard, 1932). The affected muscles do not develop spasticity, but all muscles present an approximately equal degree of increased resistance to passive movements. This is present throughout the range of the movements. The increased resistance is transient if the lesion is restricted to area 6. The tendon reflexes are moderately increased, but this feature disappears in the course of a week or more. The sign of Babinski is not present, but "forced grasping" is seen.
>
> Marion Hines (1937) described a "strip area" (area 4s) on the border between areas 4 and 6. On stimulation of this, movements elicited from area 4 were inhibited and contracted muscles relaxed. Later, other striplike areas similar to 4s (8s, 3s, 2s, 19s, and 24s) were described and collectively called "suppressor areas." Further research, however, brought forward evidence that led to abandonment of the notion of cortical "suppressor areas." Some of the phenomena seen after

stimulation of these areas appear to be results of so-called "spreading depression." (For early critical reviews, see Sloan and Jasper, 1950; Druckman, 1952; see also Kaada, 1960.) Travis (1955) found that ablation of the supplementary motor area in monkeys gives rise to spasticity.

As a result of the studies briefly reviewed above as well as others it was for some time generally held that lesions of the area 4 or of the pyramidal tract give rise to an impairment of discrete, coordinated voluntary movements and (in apes and man) the sign of Babinski but not to spasticity and increased tendon reflexes. The latter two phenomena were assumed to be due to damage to "extrapyramidal" cortical areas, especially to area 6 and area 4s, or their efferent projections. These conclusions rested on the assumption that the various cortical areas are functionally different.[49]

More recent studies have not found evidence of striking differences between ablations of area 4 and area 6 in monkeys. For example, Gilman, Lieberman, and Marco (1974) found that when these cortical areas were removed together the monkey showed a contralateral hemiparesis that in the beginning was hypotonic— like the paresis following pyramidotomy (see Gilman and Marco, 1971) but more marked. In both instances there is loss of contactual placing. However, in animals with combined area-4 and area-6 ablations the hypotonic paresis changes to a hypertonic one in the course of 4 to 6 weeks. The responses of muscle spindle primaries to passive extension, which were depressed in the hypotonic phase, reached—but did not exceed—control levels in the hypertonic phase. This indicates that the hypertonia is not due to a heightening of activity of fusimotor neurons.

As mentioned previously, in the primary motor region the trunk and proximal limb muscles are "represented" farthest from the central sulcus, and the effects of ablations of areas 6 and 4 on proximal and distal musculature, respectively, may therefore be related to the pattern of cortical representation, particularly since points giving rise to movements of the trunk and proximal extremities are found definitely in front of area 4. Different effects on proximal and distal muscles with regard to muscular tone and stretch reflexes may be dependent on differences in efferent projections from the various parts of the motor cortex. As has been described earlier, considerable and growing evidence shows that there are both anatomical and physiological differences between the components of the pyramidal tract derived from various cortical areas.

Further, the same areas give rise to efferents passing to various subcortical stations, more or less closely related to "motor functions," such as the red nucleus, the pontine nuclei, the reticular formation, and others. In view of these anatomical data it is to be expected that it is hardly possible to distinguish between cortical areas that are involved in each of the various facets of motor function or to outline, for example, a single cortical area whose destruction gives rise to spasticity (confer the results of Gilman, Lieberman, and Marco, 1974, described

---

[49] Attempts have been made by some (see for example, Hassler, 1966b) to indicate more specifically the motor functions of minor subdivisions of the main areas, described by students of cortical cytoarchitecture. Area 6 was subdivided by the Vogts and Foerster into areas 6a and 6b, area 6a again being divisible into 6aα and 6aβ. Area 4 of man was subdivided by von Bonin (1944) into subareas 4γ, 4a, and 4s. The studies and nomenclatures of the various investigators are not entirely consistent, however.

above). These and other experiences are in accord with the wealth of anatomical data showing that *it is not justified to distinguish between "pyramidal" and "extrapyramidal" cortical areas*. In view of the importance attributed in clinical neurology to the pyramidal tract it appears worthwhile to see to what extent available data necessitate modifications of the traditional concepts.

The "pyramidal tract syndrome"; a misnomer. The "pyramidal tract syndrome" has been mentioned several times in the preceding text. As described (p. 254 ff) it consists essentially of a loss of the capacity to perform coordinated skilled movements. This paresis is combined with spasticity of a particular distribution and exaggerated tendon reflexes. The cutaneous reflexes are abolished or diminished, and there is an inversion of the plantar reflex (sign of Babinski). The origin and course of the pyramidal fibers and the presence in the internal capsule of corticofugal fibers to many other destinations than the cord make it clear that *neither lesions of the cortex nor of the internal capsule can give rise to what may properly be called a "pyramidal tract syndrome."* This can occur only with lesions in the one situation where the pyramidal tract runs in isolation: the medullary pyramids. As we have seen, after isolated transection of the medullary pyramid, monkeys loose the capacity for discrete finger movements, whereas simpler hand and finger movements are little affected. Permanent hypotonia with weakened myotatic reflexes results, and the abdominal reflexes are abolished or weakened. Isolated destruction of the pyramids in man has been reported in a few instances (Brown and Fang, 1961; Meyer and Herndon, 1962; Chokroverty, Rubino, and Haller, 1975; Leestma and Noronha, 1976). However, a closer scrutiny shows that in none of these cases were the pathological changes of the medulla strictly limited to the pyramid, and a full-fledged pyramidal tract syndrome was not observed (the sign of Babinski was observed in all cases, however; see also footnote 52).

It has often been maintained that isolated damage of the pyramidal tract fibers in man might be achieved by a partial section of the cerebral peduncle, since the fibers of the corticospinal tract were assumed to occupy approximately the middle three-fifths of the peduncle (Fig. 4-2). As discussed above, this is probably an oversimplification, since it has been learned from experimental studies that in the cerebral peduncle (as in the internal capsule) corticospinal fibers run more or less intermingled with corticopontine fibers and efferent fibers from the cortex to nuclei in the brainstem. On the basis of the above assumption, however, some neurosurgeons undertook transections of the middle part or slightly more of the cerebral peduncle to relieve abnormal, involuntary movements (for references, see Bucy, Ladpli, and Ehrlich, 1966). The section is placed at the lowest level of the peduncle. Bucy (1957) and Bucy and Keplinger (1961) found that after the operation a considerable degree of useful finger movement is retained, and there is no spastic paralysis with hyperactive tendon reflexes, even if the latter may be somewhat increased. The Babinski sign is usually present.

Since it may be objected that the surgical interruption of the peduncle has not been as planned, it is of considerable interest that in one patient the lesion could be controlled at autopsy. Only some 17% of the fibers in the medullary pyramid appeared to be present some years after the operation (Bucy, Keplinger, and Siqueira, 1964). In order to gain more information on the subject, corresponding experimental studies were performed in monkeys, with anatomical control of the lesions (Bucy and Keplinger, 1961; Bucy, Ladpli, and Ehrlich, 1966). It turned out that monkeys in whom a complete degeneration of the pyramid was found at autopsy had been able to walk, climb, and jump, and when time was allowed for recovery, special tests showed that the animals had retained their ability to grasp and manipulate small objects even after bilateral pedunculotomies. More marked spasticity and hyperreflexia were seen only in animals in whom the damage involved the mesencephalic tegmentum. Walker and Richter (1966) in a corresponding study obtained largely corresponding results but on the whole found greater functional impairment. They emphasized particularly the lack of spontaneous movements of the paretic limbs.

It appears from the discussion above that reliable information on the consequences of interruption of the pyramidal tract in man is still lacking. However, it is clear that the interruption of a pyramid, and even the transaction of the corticospinal fibers together with some other corticofugal connections in a pedunculotomy, does not produce the clinical picture of the hemiplegia seen after a lesion of the internal capsule. In the latter case, many pathways from the cortex and descending in the internal capsule will be affected. There can then be little doubt that *the classical "pyramidal tract syndrome" does not follow lesions of the pyramidal tract and accordingly is a misnomer*. A *true* pyramidal tract syndrome in man can as yet not be characterized, but its principal features may be assumed to be impairment of discrete skilled movements of the fingers and usually the sign of Babinski (in man and in the chimpanzee). The marked differences between this and the classical pyramidal tract syndrome must be due to additional damage inflicted on other structures besides the pyramidal tract in cases showing the classical syndrome.

It follows from the discussion above that *the term "pyramidal tract syndrome" should be discarded*. The case of the "pyramidal tract syndrome" appears to be only one of many in which we must abandon the attempts to find a particular symptom or symptom constellation as the functional and clinical expression of injury to a particular structure. (The basal ganglia, to be considered below, furnish other examples.)

> From a practical point of view it may be disheartening to have to give up traditional views— and names. It is often convenient to have a designation of a symptom complex, but the name should not be such that it implies incorrect relations to definite structures. Since the symptom constellation of the "pyramidal tract syndrome" is by far most commonly seen in lesions of the internal capsule, another solution would be to speak of an "internal capsule syndrome." This has the advantage that it does not fix attention unduly on one component of the internal capsule only, but it suffers from the drawback that it may cover different symptoms according to the part of the capsule involved (for example, homonymous scotomas or hemianopsia when the posterior part, and aphasic disturbances when the anteriormost part, is included in the lesion). This again illustrates the difficulties mentioned above. It is to be feared that the ghost of the "pyramidal tract syndrome" will survive in clinical neurology for many years to come.

In the above discussion attention has been devoted mainly to the pyramidal tract as the pathway mediating cortical influences in voluntary movements. There can be no doubt, however, that other pathways are also involved. Even though we are not able to specify these pathways at present, it may be surmised from both anatomical and physiological findings that the corticorubrospinal route is particularly important in this respect (see the following section). The situation appears to be rather similar with regard to the other symptoms seen in the classical "pyramidal tract syndrome." Thus, increased tendon reflexes and muscular hypertonia are not necessarily signs of injury to the pyramidal tract. Since the sign of Babinski has almost always been considered the most decisive and reliable evidence of damage to the pyramidal tract, it is appropriate to consider it somewhat more closely. Likewise, some comments will be made on the weakened or abolished cutaneous reflexes belonging to the syndrome.

The sign of Babinski is "probably the most famous sign in clinical neurology" (Nathan and Smith, 1955b, p. 250). In clinical neurology it is often taken to be the most reliable sign of damage to the pyramidal tract. Normally a plantar flexion of the big toe is elicited when striking the sole of the foot with a pin or a fairly sharp object from the heel toward the toes. Often the other toes flex as well (and adduct). A dorsiflexion of the hallux following this stimulus is named an *inverted plantar* reflex and is considered a pathological phenomenon. Often there is at the same time a fanning of the toes. When this occurs there is a fully developed "sign of Babinski." Both variants were described by the French neurologist Babinski around the turn of the century and soon found their way into the test battery of the routine neurological examination. The clinically most decisive feature is the dorsiflexion of the great toe. (For a historical review, see van Gijn, 1977.) It is often said that noxious stimuli are the adequate stimuli. However, according to van Gijn (1975, 1976, 1977), who has undertaken extensive studies of the plantar reflex in man, this is not generally true, nor are brief electrical stimuli (used by some investigators) appropriate. Stroking of the sole of the foot is assumed to be effective because it generates spatial and temporal summation of afferent impulses.

Nathan and Smith (1955b) in a review of the development of views on the Babinski response since it was first described drew attention to the way hypotheses have been transformed gradually to a dictum: the presence of a Babinski response is evidence of a damage of the pyramidal tract. The reflex arc of the normal plantar reflex was considered to be long, involving the leg area of the primary motor region. Cases in which the sign of Babinski was found in lesions of this region have been described in man. In experimental studies in chimpanzees, Fulton and Keller (1932) found support for this view.

A perusal of the neurological literature from the first decades of the present century brings to light numerous observations which do not fit the theory. From the review of Nathan and Smith (1955b) as well as of those of Lassek (1944, 1945) it appears that the Babinski sign may be present when there is no damage to the pyramidal tract, and may have been absent in cases in which this tract was found affected in postmortem examinations. In a study of the spinal cord of 38 patients subjected to chordotomy Nathan and Smith (1955b) could confirm this lack of correlation between pyramidal tract injury and the sign of Babinski. Their conclusions were vehemently challenged by Walshe (1956), who also criticizes Lassek's work as well as conclusions reached by some physiologists concerning the mechanism of the plantar reflex and the sign of Babinski. Walshe emphasizes the dangers in attempting to explain clinical phenomena on the basis of findings in animals. In recent years clinical neurophysiological observations have been made which are of interest for the problem and may help to clarify the matter.

In these studies the activity in the muscles of the foot and leg have been recorded electromyographically. In order that exact measurements of the reflex time can be made some authors have used electrical stimulation by electrodes applied to the skin or deep tissues. These stimuli were assumed to be noxious. In such studies it is found that *in normal subjects* the reflex response varies according to the site of stimulation (Kugelberg, Eklund, and Grimby, 1960). For example, noxious stimulation of the ball of the toes elicits the general flexion reflex, including dorsiflexion of the great toe, while stimuli applied to the ball and hollow of the foot evoke the normal plantar reflex: plantar flexion of the toes and flexion at the ankle, knee, and hip.[50] These findings have been extended in further studies on normal subjects by Grimby (1963a), who emphasizes the existence of individual variations in the reflex pattern as the locus of stimulation is varied, and that changes in the subject's attention and expectancy may occasionally result in deviations from the basic pattern elicited from a particular site. The earliest responses have such short latencies (as little as 55 msec for the reflexes in the short hallux flexor and extensor, which are the main muscles activated and which show reciprocal innervation) that *only purely spinal reflex arcs can be involved.* (There are, however, also responses of latencies of more than 150 msec, which are generally elicited on weaker stimulation.)

The *pathological plantar response,* inversion of the plantar reflex or the sign of Babinski, has been held by clinical neurologists (see Walshe, 1956) to be an integral part of the general flexion reflex in man, homologous with the nociceptive flexion reflex studied extensively in spinal and decerebrate animals by Sherrington and others. In this a general flexion of a limb follows when a nociceptive stimulus is applied to it. It is a general experience in clinical neurological examinations that it may be extremely difficult to decide whether the reflex movement of the great toe is to be considered pathological or not. Van Gijn (1976, 1977) discusses the criteria that should be applied in the evaluation of equivocal responses. Electromyographic findings have shown that the dorsiflexion of the great toe (sign of Babinski) is due to contraction of the long extensor hallucis muscle (Landau and Clare, 1959; van Gijn, 1975). Examination of this muscle (by palpation or electromyography) when the plantar reflex is elicited may therefore help to distinguish a normal from a pathological reflex. It is important to note concomitant weak contractions of the anterior tibial muscle and usually also contractions of other (physiological) flexor muscles (cf. the Babinski response as part of the general flexion reflex) since, according to van Gijn, the long hallux extensor may be activated as an antagonist to the short hallux flexor, active in the normal plantar reflex.

> Clinical neurophysiological studies are in general agreement with the view that the Babinski sign is part of the general flexion reflex. It appears that the site of the stimulus and the receptive field are of importance for the resulting movement of the great toe.
>
> Grimby (1963b) found that the receptive field for the extensor hallux response (the ball of the foot) spreads to a larger area as a normal reflex pattern is converted into a pathological one. Individual variations in the receptive fields may be a factor in explaining why in some patients the pathological plantar reflex is most easily (or only) elicited on stimulation of the medial, whereas in others (and more often) on stimulation of the lateral aspect of the planta—a common clinical expe-

---

[50] It is important to be aware of the fact that dorsiflexions of the toes and ankle are physiologically flexion movements, even if the muscles producing these movements are called extensors. On the other hand, plantar flexions of the toes and ankle are in reality extensions. Thus dorsiflexion of the toes is part of a flexion reflex. Not infrequently an extensor response occurs in the other leg when a flexion reflex is elicited from the sole (see Brain and Wilkinson, 1959).

rience. Several of the conclusions of Kugelberg et al. (1960) and Grimby (1963a,) have been criticized by van Gijn (1977) based on the fact that that short electrical and mechanical stimuli are not equivalent in studies of the plantar reflex.

The normal plantar reflex appears to be a manifold phenomenon, different patterns of reflex responses having differently located receptive fields. The pathological response pattern (dorsiflexion of the great toe, etc.) appears to be a normal response to stimulation of the ball of the foot. Transitional forms to the normal response occur. Under pathological conditions the reflex pattern is less adaptive to the stimulus site.[51]

Both the normal and the pathological plantar reflexes are spinal (even though there may be some doubt as to the late responses). When a pathological response occurs, this is hardly due to interruption of a long reflex arc passing via the motor cortex. The reason for the disruption of the patterns of the plantar reflexes, resulting in the sign of Babinski, must presumably be sought in changes in the suprasegmental control of "the discriminating capacity of the reflex mechanism as regards the strength, modality and site of the sensory stimulus" (Kugelberg, Eklund, and Grimby, 1960). For some additional data see Grimby (1965). The pyramidal tract may well be concerned in this control (cf. the findings of Fisher and Curry, 1965). Van Gijn (1977, 1978) argued that the occurrence of the Babinski sign is related to damage of pyramidal tract fibers establishing monosynaptic contact with motoneurons. According to van Gijn (1977, p. 149), the sign is "closely linked with a disturbance of direct corticomotoneuronal connections which subserve differentiated movements of the foot and toes." In many hemiparetic patients he found that, when the sign of Babinski was present, the patient had difficulties in performing rapid, alternating movements of foot and toes, and there was some weakness of their muscles. On the basis of experiments in monkeys (described above) these findings are compatible with the assumption that the Babinski sign is a consequence of damage to the pyramidal tract. However, the frequent concomitant occurrence of increased myotatic reflexes or spasticity must be explained on the basis of damage to other descending fiber systems.[52]

It is still an open question whether the sign of Babinski is a reliable manifestation of an affection of the pyramidal tract. It should again be stressed that in most situations in which the pyramidal tract may be injured there will necessarily be damage to other fiber systems as well. This point appears to be disregarded by clinicians who attempt to defend a close relation between the pyramidal tract and

[51] Occasionally the dorsiflexion of the great toe is most easily elicited from the lateral side of the dorsum of the foot (Chaddock's sign). When in a hemiplegic patient presenting the sign of Babinski the peroneal nerve in the paretic leg is anesthetized, the Babinski response disappears since the dorsiflexors of the toes and ankle are paralyzed. However, the response to plantar stimulation is then a plantar flexion of the big toe. This seems compatible with the views discussed above on the organization of the plantar reflexes (see also Landau and Clare, 1959).

[52] Van Gijn (1977, 1978) cites the cases of "pure motor hemiplegia" described by Chokroverti, Rubino, and Haller (1975) and Leestma and Noronha (1976) as examples of pure pyramidal tract affections in man. In these patients there were, in addition to the sign of Babinski, increased myotatic reflexes. However, the pathological alterations were not restricted to the medullary pyramid. The two cases, therefore, can hardly be considered an illustration of the consequences of a pure pyramidal tract lesion in man. The finding (van Gijn, 1978) that the Babinski response and hyperreflexia (myotatic reflexes) regularly occur independently of each another does not strengthen the idea that both symptoms are due to damage of the same structure.

the sign of Babinski (see Walshe, 1956; Brain and Wilkinson, 1959). It appears indeed plausible that a derangement in spinal reflex patterns resulting in a Babinski sign may be caused by damage to several descending fiber systems. It is not even necessary to postulate that there is damage to one particular pathway. Even though little is known about this, also in man descending fiber systems other than the pyramidal tract exert influence on the spinal reflex apparatus (see for example, Lundberg, 1975). Presumably, therefore, lesser or partial damage to more than one tract may be effective. However, it should not be overlooked that a histologically demonstrable lesion cannot be expected to be present in all instances where a Babinski response is found. In the first place the microscopical changes may not be marked enough to permit recognition. Secondly, there is no doubt that functional disturbances of the nervous system may make the normal plantar response change into an extensor response. Thus a temporary Babinski response can be found in general anesthesia (see Grimby, Kugelberg, and Löfström, 1966) in hypoglycemia, following administration of various drugs, following an epileptic seizure, following long marches, during sleep, and in various general affections of the central nervous system (see Lassek, 1944, for references). The presence of a Babinski response should therefore not too hastily be interpreted as evidence of organic damage to the brain, and especially not to the pyramidal tract. The normal variations in the reflex response should be kept in mind.[53]

Weakening or absence of abdominal reflexes and the cremasteric reflex are usually considered parts of the "pyramidal tract syndrome." They belong to the cutaneous reflexes, whose receptors are situated in the skin. Kugelberg and Hagbarth (1958), using electromyographic recordings, concluded that the central latency was short enough to prove that the abdominal reflexes are mediated by a spinal reflex arc. (See also Teasdall and Magladery, 1959.) In addition to contraction of the abdominal muscles there occurs a reciprocal relaxation of the posterior trunk muscles. The reflexes are considered to belong to a spinal defense mechanism, ensuring that the resulting movement causes withdrawal from the stimulus. It is well known that the abdominal reflexes are subject to considerable incidental and individual variations. In further studies Hagbarth and Kugelberg (1958) analyzed factors influencing the reflex, such as habituation and sensitization, and concluded that such changes depend on "cerebral events." However, nothing can be said of pathways involved in this central influence. The abdominal reflexes have not been studied as intensively as the plantar reflexes.[54] However, there does not appear to be convincing evidence that weakening or abolishment of the abdominal reflexes, often seen in central nervous lesions, particularly of the internal capsule, can be ascribed to damage of the pyramidal tract. (The reflexes may obviously be affected segmentally in lesions of dorsal or ventral roots or in segmental lesions of the gray matter of the cord.)

The role of the pyramidal tract.   It is clear from the preceding account that the pyramidal tract is only one of several pathways leading from the cerebral

[53] In infants the plantar reflexes are usually said to be extensor until between the 9th and 24th month of life. In the beginning the receptive field extends to the abdomen or higher, but it gradually shrinks (see Brain and Wilkinson, 1959).

[54] The abdominal skin reflexes are present from birth in man, but their reflexogeneous zone is wider in infants than in later life (Harlem and Lønnum, 1957).

cortex, especially its sensorimotor regions, to the spinal cord. It is further obvious that the tract is not a homogeneous system, either anatomically or functionally. However, in one respect the pyramidal tract differs from all other corticospinal routes: it is the only fiber bundle which proceeds without synaptic interruption from the cortex to the cord. From a functional point of view this means that it enables the cortex to exert a more immediate control over the spinal mechanisms than does any other route. The relative increase in the size of the pyramidal tract in the phylogenetic scale of mammals makes it clear that this direct cortical control of the segmental mechanisms of the cord (as well as of corresponding mechanisms in the brainstem) acquires an increasing importance in higher mammals, reaching its peak in man. The same tendency is demonstrated by the fact that in the monkey and chimpanzee some fibers of the "motor" component of the tract establish direct synaptic contact with α motoneurons, while in the cat all contacts are made with interneurons. Presumably this monosynaptic component is even more prominent in man, but precise information is not available on this point. With regard to the role of the "motor" component of the pyramidal tract it should be recalled that its influence on flexor motoneurons is predominantly excitatory, while extensor motoneurons are inhibited.

Speculations concerning the particular functional role of the pyramidal tract have been especially directed to its importance in the initiation of skilled voluntary movements. Even if the pyramidal tract is not alone involved in mediating cortical impulses concerned in these, it may well play a special role (regardless of whether the patterning of such movements to a large extent may be a matter of the spinal cord; see p. 246). It has been pointed out by several students that the predominant flexor activities following stimulation of corticospinal fibers (and the same applies to the rubrospinal fibers) and of the cerebral cortex may be seen in relation to the function of the antigravity muscles. As expressed by Corazza, Fadiga, and Parmeggiani (1963, p. 360),

> . . . it does not seem too far-fetched to surmise that the functional meaning of the flexor prevalence of pyramidal excitatory drive (and of the extensor prevalence of the inhibitory one) may be that of counteracting, at the beginning of a "voluntary" pattern of movements, the extensor bias which under normal conditions must obtain as a result of myotatic reflexes from antigravity muscles.

Somewhat similar views have been expressed by others, for example Agnew, Preston, and Whitlock (1963). In this connection it is worth recalling that most of our skilled motor functions engage primarily flexor muscles.

As has been seen in previous sections of this chapter, extensive research from recent years on the anatomy and physiology of the corticospinal tract has brought much new information about particular aspects of this fiber system. To mention a few: motoneurons supplying proximal muscles are influenced by spatially extensive colonies of cortical cells whereas the colonies acting on distal muscles are more focal (Phillips, 1967), indicating a more specific organization of the corticospinal influence on distal muscles. Furthermore, distal motoneurons receive a stronger monosynaptic input than do proximal motoneurons.[55] Evarts (1965)

---

[55] The experimental study by Buxton and Goodman (1967) of the corticospinal projection in the raccoon is of interest in this connection. In this animal, which has a well-developed dexterity of the fingers, degeneration following motor cortical ablations extends into lamina IX in C_{7-8}, but not in other segments.

found slowly conducting pyramidal neurons to be tonically active in the absence of voluntary movements, whereas rapidly conducting neurons fired intensely only during the movement, supporting older conclusions concerning tonic and phasic properties of the pyramidal tract. The pyramidal tract fibers act on both α and γ motoneurons. It has been suggested by several authors that the influence on the γ system is a preparatory step for the activation of α fibers occurring in rapid, phasic movements. Afferent impulses from many sources and from different kinds of receptors interact in a very complex fashion and provide the sensory information for a necessary, properly adjusted, cortical influence on the motor apparatus.

Nevertheless, it is not yet possible to formulate more than rather general concepts of the functional importance of the pyramidal tract. It should also be recalled that our knowledge concerns almost exclusively the thick, rapidly conducting pyramidal tract fibers.[56] *It appears, however, that the following may safely be concluded:* The anatomical and functional peculiarities of *the pyramidal tract* strongly suggest that it *plays a particular role in the central control and initiation of skilled voluntary, discrete, and rapid movements of the hand and fingers.*[57] This role is of greater importance in primates than in other mammals and is especially great in man. This, however, should not induce us to overlook the fact that the corticospinal tract is scarcely alone in this function. Several indirect corticospinal routes are presumably involved. As far as the capacity to perform discrete movements is concerned, the corticorubrospinal pathway seems to be the most likely candidate. Since the time that this assumption was first ventured (Brodal, 1963, 1965a) further studies have provided supporting evidence. Reference was made earlier in this chapter to some striking similarities of organization between the corticospinal and the corticorubrospinal pathways, particularly the part of the former that arises in the anterior, "motor," part of the sensorimotor cortex.

Thus both pathways present a somatotopic localization, terminate in the same laminae of the spinal cord, and facilitate motoneurons to flexor muscles (γ as well as α). Some corticorubral fibers are collaterals of corticospinal axons, and both corticospinal and rubrospinal impulses have been found to excite monosynaptically the same propriospinal neurons in the cord (see Illert, Lundberg, and Tanaka, 1977) and to facilitate transmission in segmental reflex paths to motoneurons (for a brief review, see Jankowska, 1978). A considerable degree of integration appears to be possible between the corticospinal and the corticorubrospinal pathway, and the latter might be almost as well suited for inducing discrete movements as the corticospinal tract.[58] Other observations support the view.

[56] Phillips and Porter (1977, p. 406) concluded their thorough monograph of the pyramidal tract as follows: "Blinkered, however, by the limited resolving power of our present techniques, we remain ignorant of the activities and actions of the myriads of smaller PTN [pyramidal tract neurons] which furnish over 90 percent of the axons of the PT [pyramidal tract]."

[57] The facial and laryngeal movements should probably be included, but they have been studied relatively little.

[58] The main difference between them is the synaptic interruption of the former pathway in the red nucleus, where particularly cerebellar afferents may influence the corticospinal impulse transmission. From a functional point of view this input from the cerebellum is apparently important. For a discussion of the functional role of the red nucleus, see Massion (1967).

Lewis and Brindley (1965) who studied the responses to cortical stimulation in lightly anesthetized baboons found that a bilateral pyramidotomy had little or no effect on the character of the movements elicited or their somatotopic organization in the cortex. After the operation, however, there was a reduction in the number of movements that could be obtained at any one exploration of the cortex and an increased electrical threshold. Lewis and Brindley (1965) did not suggest a pathway for these impulses. In a corresponding study after pedunculotomy in the macaque (Walker and Richter, 1966) the responses were less clear and required strong stimuli. Lawrence and Kuypers (1968b) found that in monkeys, some time after a prior bilateral pyramidotomy, a unilateral interruption of the rubrospinal tract or destruction of the red nucleus had serious consequences for the capacity preserved after the first operation to perform finger and hand movements on the affected side. On cortical stimulation after unilateral or bilateral pyramidotomy in cats, Hongo and Jankowska (1967) obtained effects similar to those found on stimulation of the corticospinal tract (postsynaptic excitatory effects on motoneurons, mainly flexor; facilitation of spinal reflexes to motoneurons; and depolarization of presynaptic terminals of group Ib fibers). They recorded potentials in the rubrospinal tract after cortical stimulation. From observations of experimental animals after lesions of the spinal cord, they concluded that some effects are mediated via the corticorubrospinal pathway (facilitation of reflex paths and effects on primary afferents), while reticulospinal paths are involved in other effects.

In a recent study in the cat, Nieoullon and Gahéry (1978) obtained largely corresponding results. After bilateral pyramidotomy, cortical stimulation produced, as in normal animals, flexion movements and associated postural adjustments. The latencies of the responses were increased, especially in the forelimb (possibly indicating impulse passage via indirect pathways), but the thresholds for cortical stimulation were not changed.

The data presented above suggest that the corticorubrospinal pathway, at least in the cat and monkey, in many ways appears to function as a supplement to the corticospinal tract and to operate in close integration with it. Experiences with pedunculotomies are also of interest. As mentioned in a preceding section of this chapter, after a bilateral pedunculotomy, monkeys recover their ability to handle small objects (Bucy and Keplinger, 1961). In these operations the section is placed at the lower end of the cerebral peduncle, and most of the corticospinal fibers are presumably transected. However, many corticofugal fibers descending in the capsule with the pyramidal tract and entering the brainstem will not be interrupted. As pointed out by Bucy, it therefore seems a likely assumption that the preservation of these pathways may be essential for the monkey's remaining capacity to perform rather discrete voluntary movements following a pedunculotomy. It is not possible to decide which of these pathways is most important, but among them is the corticorubral tract.

Further studies are needed to clarify the role played by the different corticospinal routes from the cortex to the spinal cord which are involved in the production of movements on cortical stimulation when the pyramidal tract is interrupted. It does, however, seem rather likely that in the role played by the primary motor cortex in eliciting discrete movements, the corticospinal and other corticofugal pathways—especially the corticorubrospinal tract—collaborate closely. For perfect voluntary movements all pathways are probably needed. Lesion of only one of them leads to rather moderate disturbances.

The hypothesis presented above is based largely on studies in animals. Objections may be raised against applying it to man. However, as discussed previously, the assumption that man's rubrospinal tract is rudimentary may not be valid. If the hypothesis is tenable, one would expect that a lesion of the red nucleus in man would result in impairment of discrete voluntary movements. Such cases appear to be extremely rare. Usually other neighboring structures, not least the cerebral peduncle, are involved. However, Raymond and Raymond Cestan (1902), Halban and Infeld (1902), Marie and Guillain (1903), Gautier and Lereboullet (1927), and von Bogaert and Bertrand (1932) have described cases with anatomically verified circumscribed lesions of the red nucleus without involvement of the cerebral peduncle. Unfortunately, descriptions of the symptoms are far from complete, the cases are not identical in all respects, and the findings are difficult to evaluate. Clinicoanatomical studies of the corticorubrospinal pathways in man are needed to decide whether the hypothesis ventured here is tenable.

Studies on the function of the basal ganglia.   As described in a preceeding section, diseases of the basal ganglia and related nuclei may give rise to a variety of disturbances of motor functions. In the course of time a number of theories have been advanced to explain the relation between the site or sites of the pathological changes and the ensuing symptoms (see Jung and Hassler, 1960; Kaada, 1963; Denny-Brown, 1968). The earlier conceptions were mainly based on clinicopathological correlations. In later attempts at explanation the results of experimental studies have been taken into consideration as well, and the results of stereotactic operations on the basal ganglia and thalamus in man have given much useful information. The greatest contributions to our knowledge and understanding of the function of the basal ganglia in recent years, however, have come from extensive biochemical investigations (for some informative reviews see Barbeau, 1973, 1976c; Hornykiewicz, 1976). In spite of much work, it is, however, not yet possible to formulate clear concepts of the function of these parts of the brain, even though it may be safely concluded that they to a large extent must be involved in what is usually referred to as "integrative processes at a relatively high level," as must indeed be assumed from a study of the complex fiber connections. This complexity makes interpretations of physiological observations on these nuclei difficult.

There are several reasons which explain that our knowledge of the subject is still relatively scanty. It is difficult to make lesions which involve one of the nuclei completely without damaging others or by-passing fiber tracts. It is also quite possible that if parts of a nucleus are preserved they may be sufficient to maintain its functions. Observations of finer differences of movements and muscle tone in experimental animals are not easily made. In physiological investigations of the effect of lesions, careful anatomical control is far too often lacking. Here, as elsewhere, when an anatomical examination is performed, considerable changes may be found apart from those intentionally made. In attempts at stimulation of the basal ganglia it appears that spread of current to neighboring structures, particularly the internal capsule, has often occurred and has given rise to faulty conclusions. Variations in stimulus parameters, in anesthesia and species differences may explain some of the discrepancies in the literature. It is further evident that until quite recently attention has been focused almost exclusively on the effects of stimulation and lesions upon rather elementary motor functions, and other effects may easily have been overlooked. Finally, there is disagreement about the mutual relationships of the various types of movements seen in patients, for example as to whether two varieties are expressions of disturbances of a common mechanism or not.

In any present-day analysis it should be kept in mind that, as emphasized in a preceeding section, the basal ganglia and related nuclei give off only few fibers to lower levels. The structural organization of their fiber connections makes it clear that *most of the actions of the basal ganglia must be exerted on the cerebral cortex*. Their influence on motor functions must therefore to a considerable extent involve the cortex, as has indeed been suggested by neurosurgeons (see for example, Bucy, 1944b), who noted that following ablations of the precentral cortex in man pre-existing involuntary movements disappeared.[59] There is an increasing

[59] It is a common clinical experience that if a patient with parkinsonism has a cerebral stroke involving the internal capsule his tremor disappears on the side of the ensuing hemiplegia.

body of physiological evidence in support of this notion. With our increasing knowledge of the anatomy, physiology, pathology, and biochemistry of the basal ganglia and with the more sophisticated methods used for evaluating clinical symptoms, it has become clear that motor dysfunctions are only one of several— although often the most obvious—aspects of basal ganglia dysfunctions (Teuber, 1976).

In the following, only some of the many investigations of the function of the basal ganglia will be considered. It may be said at once that attempts to produce in animals symptoms resembling those seen in patients with diseases of these parts of the brain have not been too successful.

With regard to the effects of *lesions of the caudate nucleus, the putamen, and the globus pallidus* on motor performance, contradictory observations have been reported. Except for some tremor in cases with very extensive lesions (Mettler, 1942; Kennard, 1944; and others), no convincing changes in motor performance were seen in the monkey and the chimpanzee with lesions restricted to these nuclei. According to Mettler (1945), extensive bilateral lesions of the globus pallidus in monkeys result in marked poverty of movements. Chandler and Crosby (1975), on the other hand, reported a high incidence of abnormal movements in monkeys with bilateral radiofrequency lesions in the head of the caudate nucleus. Laursen (1963) concluded that in the cat there is no decisive evidence of changes in motor performance as a result of lesions confined to the corpus striatum. Harik and Morris (1973), on the other hand, reported that bilateral, small radiofrequency lesions restricted to the ventromedial part of the anterior caudate nuclei in the cat effectively prevent the increased locomotor activity induced by *l*-DOPA (see below). The discrepancy reported in the literature concerning the effects of lesions of the caudate nucleus can be partly explained as being due to the often concomitant damage to various parts of the cerebral cortex, particularly the frontal lobe.

In an extensive series of investigations in the cat, Villablanca and his colleagues tried to avoid concomitant damage to the frontal lobe when attempting to make bilateral caudate ablations (Villablanca, Marcus, and Olmstead, 1976a, b; Villablanca, Marcus, Olmstead, and Avery, 1976; Olmstead, Villablanca, Marcus, and Avery, 1976). The authors reported that after such lesions the animals were remarkably free of permanent gross neurological deficits. One of the principal symptoms exhibited was a stereotyped behavior characterized by a long-lasting tendency to approach or follow moving animals or objects. Further, the animals showed an impairment of contact-placing reactions proportional to the extent of the lesions. These effects were not seen following unilateral caudate lesions. The authors also reported a perseverative behavior and a marked slowness in the motor activity exhibited by acaudate cats. Such animals could not shift quickly from one motor response to another, and they were unable to perform two concurrent acts. This behavior is remarkably similar to the akinesia observed in many patients with lesions of the basal ganglia. On the basis of their extensive studies in acaudate cats, Villablanca and his collaborators concluded that control of *elementary* motor processes is not a primary function of the caudate nuclei, but that the nuclei are important for diverse aspects of higher sensorimotor integrations, such as modulation of multisensory inputs.

Some information has also been obtained in studies of single-unit recordings and of *stimulation of the basal ganglia,* even though, as pointed out by Laursen

(1963), there has presumably often been inadvertent stimulation of fibers descending in the internal capsule (corticospinal and others). In the awake monkey, single units in the caudate nucleus have been shown to increase their finding rate in relation to movements (Buser, Pouderoux, and Mereaux, 1974). Stimulation of the monkey's caudate nucleus is reported to result in contralateral limb flexion (Chandler and Crosby, 1975). This effect was abolished following radiofrequency lesions of the stimulated sites. In unanesthetized cats stimulation of the *caudate nucleus* results in head turning and circling movements toward the opposite side (Stevens, Kim, and MacLean, 1961; and others). It appears from the controlled studies of Laursen (see Laursen, 1963) that the effects are not due to spread of current. According to Forman and Ward (1957) the responses from the head of the caudate show a somatotopic pattern. However, on low-frequency electrical stimulation of the caudate (Akert and Andersson, 1951; Kaada, 1951), prolonged stimulation by injection of long-acting cholinergic agents in crystalline form (Stevens, Kim, and MacLean, 1961) or by injection of alumina cream (Spiegel and Szekely, 1961) the most conspicuous effect has been a cessation of spontaneous movements and a prolonged state of quietude. This effect is seen even if the frontal lobe or the "motor" cortex has been ablated (Spiegel and Szekely, 1961).

An interesting observation is that stimulation of the cat's caudate nucleus exerts an inhibitory influence on somatic afferent transmission in the cuneate nucleus (Jabbur, Harik, and Hush, 1976) and in the intralaminar thalamic nuclei (McKenzie, Gilbert, and Rogers, 1971; Rogers and McKenzie, 1973). These observations lend support to the conclusion, made in ablation studies, that an important function of the caudate nucleus is to modulate multisensory inputs (Villablanca, Marcus, and Olmstead, 1976a). According to Wood, Lake, Ziegler, and van Buren (1977), stimulation of the *human* caudate nucleus results in various psychic responses, including inappropriate smiling, confusion, feelings of anxiety, and slurred speech and, in addition, in a significant increase of noradrenaline in the lumbar cerebrospinal fluid. Too much emphasis must not be placed on these observations, however, since the stimulations were made in patients with various neurological disorders.

The *putamen* has been less extensively studied than the caudate, but Hassler and Dieckmann (1967), using the method employed by Akert and Andersson (1951), reported inhibition of movements (as well as an "arrest reaction" and an "empty gaze") following low-frequency stimulation in the awake cat. These authors briefly mentioned that similar effects were evoked from the pallidum. These results extend some early observations of Mettler, Ades, Lipman, and Culler (1939), who performed simultaneous stimulation of the striate body and the motor cortex in cats and monkeys. When movements were elicited by stimulation of the motor cortex, simultaneous stimulation of the caudate nucleus and the putamen reduced or inhibited the movements.

In recent years careful recordings of single units in the awake monkey's putamen and globus pallidus have been undertaken (DeLong, 1971, 1972, 1973; DeLong and Strick, 1974). These studies have revealed characteristic properties of neurons in the putamen and the two segments of the globus pallidus. The activity of these neurons increases *prior* to conditioned movements, indicating that these nuclei participate in the initiation of movements (DeLong, 1972). This does not, however, necessarily mean that the putamen discharges before the motor cortex or the cerebellum. Single units in the globus pallidus are very specific, since they are most often related to movements of only a single extremity and show different patterns of activity in relation to different movements studied (DeLong, 1971). With regard to the symptoms seen in patients with lesions of the basal ganglia, it is of particular interest that more than half of all movement-related units in the putamen discharge preferentially in relation to slow movements and less than 10% in relation to rapid movements. This observation is consistent with the hypothesis that a primary motor function of the basal ganglia is to generate slow (ramp) rather than rapid (ballistic) movements (DeLong, 1973).

It appears from the discussion above that lesions or stimulations of the caudate nucleus, putamen, or globus pallidus in cats and monkeys only occasionally result in symptoms resembling those seen when these parts of the brain are affected by disease in humans. The role played by these structures under pathological conditions in man are apparently far from simple. However, several observations strongly suggest that in addition to the pallidum the thalamus (especially VA and VL) is concerned in the appearance of tremor and rigidity in parkinsonism. We shall return to these problems later.

As mentioned above, the *substantia nigra* is practically always found changed in cases of parkinsonism. On the assumption that lesions of the substantia nigra may be responsible for rigidity and/or tremor seen in this disease, numerous authors have attempted to produce isolated lesions of the substantia nigra, particularly in monkeys. It is extremely difficult to obtain isolated lesions of this small nucleus, lying in close relation to several other structures, among these the cerebral peduncle. Some authors have reported that static tremor resembling that seen in parkinsonism follows lesions of the substantia nigra in monkeys. However, in subsequent studies it was established that the static tremor observed in these experiments is probably due to concomitant lesions of the ventral tegmentum, situated just dorsomedial to the substantia nigra (see below), and other authors (Carpenter and McMasters, 1964; Stern, 1966), succeeding in making more restricted lesions in the substantia nigra, failed to obtain tremor. In cases of partial or unilateral lesions usually no symptoms at all were observed. However, if bilateral and larger lesions are made, there ensues a *hypokinesia* with lack of spontaneous motor activity and a tendency to assume immobile postures (Stern, 1966). Additional damage to the globus pallidus increased this tendency. Caution should be exerted in applying conclusions drawn from such observations to conditions in man, but it is of interest that according to several clinical neurologists there is no constant relation between the appearance of hypokinesia and rigidity in parkinsonism, hypokinesia being a distinct clinical feature. From this point of view it is of interest that in several cases of parkinsonism no pathological changes have been found in the substantia nigra, while it has been found altered in the absence of symptoms of parkinsonism (for references see Mettler, 1964; Stern, 1966). Markham, Brown, and Rand (1966) in stereostatic lesions of the thalamus in parkinsonism obtained good effects on tremor and rigidity, but a present or developing akinesia was not influenced. The mechanisms by which the substantia nigra influences motor function and by which its lesion produces akinesia are not known. Several hypotheses have been suggested.

During the last decade a large number of investigators have attempted to reproduce several of the known symptoms of Parkinson's disease (and of other neurological disorders) by interfering with the mechanisms of production and release of known and putative neurotransmitters in the basal ganglia. Since dysfunction of the dopaminergic nigrostriatal projection probably accounts for at least some of the symptoms encountered in Parkinson's disease (see above), in a particularly large number of these studies, attempts have been made to interfere with dopamine in this projection, either by injection of dopamine agonists or antagonists in the striatum or by chemical destruction of the dopaminergic neurons by injecting 6-hydroxydopamine in the substantia nigra (see Ungerstedt, 1974; Iversen, 1974, for some references). These experiments leave little doubt about the importance played by the dopaminergic nigrostriatal pathway in stereotyped and locomotor behavior, but the exact mechanisms are still far from being resolved.

> One example will suffice to underline this point: It is well known that *l*-DOPA, a dopamine precursor, will not alleviate all the cardinal symptoms of Parkinson's disease to the same extent. The drug is most effective against the akinesia and secondarily against the rigidity. Tremor is the symptom that is most resistant to treatment with *l*-DOPA. Apomorphine which is known as a specific dopamine receptor agonist, is, on the other hand, most effective against tremor, less effective against rigidity, and only partially effective against akinesia (Barbeau, 1976c).

Considering the complexity of the connections and biochemistry of the basal ganglia, including the substantia nigra, it should come as no surprise that attempts to reproduce all symptoms of Parkinson's disease by manipulating mainly one fiber projection—the nigrostriatal—have met with little success. It appears relevant again to emphasize that not all dopaminergic nigral neurons project upon the striatum (see above). There is some experimental evidence that stereotyped behavior is related to the activity of striatal dopaminergic mechanisms, whereas increased locomotor activity is more related to the more medially projecting nigrofugal dopaminergic fibers (Kelly, 1975).[60] It should also be recalled that in patients with Parkinson's disease, changes have been reported in both the pars compacta and the pars reticulata of the substantia nigra. As mentioned above, neurons in the pars reticulata are probably not dopaminergic and have projections to brain structures other than the striatum, among them the thalamus.

In *Huntington's chorea* a major chemical abnormality appears to be a severe decrease in the levels of GABA and its synthesizing enzyme, GAD, in the basal ganglia, including the substantia nigra (Perry, Hansen, and Kloster, 1973; McGeer, McGeer, and Fibiger, 1973; Bird and Iversen, 1974). In addition, there is a less pronounced reduction of the acetylcholine-synthesizing enzyme, choline acetyltransferase (McGeer, McGeer, and Fibiger, 1973; Urquhart, Perry, Hansen, and Kennedy, 1975; Bird, 1976; McGeer and McGeer, 1976).

> It appears that the decrease of GAD, observed in the substantia nigra of patients with Huntington's chorea, is paralleled by a pronounced decrease of substance P, which has also been concluded to be involved in striatonigral functions (Kanazawa, Bird, O'Connell, and Powell, 1977). In recent years attempts have been made to reproduce the symptoms seen in Huntington's disease

---

[60] It is also interesting that the main result of degeneration of the dopaminergic nigrostriatal projection in the rat is a sensory inattention, and particularly a long-lasting inability to react to tactile stimuli (Ljungberg and Ungerstedt, 1976). These observations are compatible with earlier ablation and stimulation experiments of the caudate nucleus (see above).

by injecting kainic acid into the striatum in experimental animals (Coyle and Schwarcz, 1976). However, as emphasized by Schwarcz, Bennett, and Coyle (1977), this procedure provides a biochemical-histological model for Huntington's disease rather than a behavioral one, since no genuine choreatic movements could be elicited. Only asymmetric locomotor activity, reminiscent of that seen after injections of 6-hydroxydopamine in the substantia nigra, was observed.

Lesions of the subthalamic nucleus in man are regularly followed by hemiballismus. A corresponding dyskinesia has been produced experimentally in the monkey by sterotactic lesions of the nucleus, particularly by Carpenter and his collaborators (see Carpenter, 1961). In order to produce hyperkinesia the lesions must comprise at least 20% of the mass of the nucleus, and the globus pallidus and its efferent projections must not be damaged. The dyskinesia occurs in the limbs on the side contralateral to the lesion. There is some evidence for a somatotopic pattern within the subthalamic nucleus (Carpenter and Carpenter, 1951; Mettler and Stern, 1962; DeLong and Georgopoulos, 1979).

The production of dyskinesias of a choreiform-ballistic type by lesions of the subthalamic nucleus is the only instance where it has so far been possible, experimentally, to produce symptoms resembling those seen in patients with similarly placed affections. The observations made in monekys may therefore give information of the mechanism of dyskinesias of this type in man. From this point of view it is of interest that in animals in which subthalamic dyskinesia has been experimentally produced, this can be abolished by a subsequent lesion of the pallidum. Most effective are lesions of the medial segment, presumably because they interrupt a larger number of pallidofugal fibers (see Carpenter, 1961). Likewise, lesions of the lateral thalamic nuclear group are effective, while lesions of the substantia nigra are without effect (Strominger and Carpenter, 1965). The clinical experience that ablation of the precentral cortex alleviates or abolishes dyskinesias in the contralateral limbs in man, although leaving a hemiplegia, has been confirmed in monkeys having experimentally produced lesions of the subthalamic nucleus (Carpenter and Mettler, 1951). Furthermore, transections of the lateral funiculus of the spinal cord at the cervical level (interrupting, among other tracts, the corticospinal) abolish subthalamic dyskinesia, while lesions of the dorsal funiculus or the ventral part of the lateral and the ventral funiculus are without effect (Carpenter, 1961). From these observations it is concluded that the corticospinal tract is the chief pathway which ultimately transmits the impulses to the abnormal movements to the cord. The integrity of the pallidum and the lateral nuclear group of the thalamus appears to be necessary for the appearance of the subthalamic dyskinesias. It is assumed (see Carpenter, 1961; Carptenter and Strominger, 1967) that the impulses for the abnormal movements arise in these structures, and that the subthalamic nucleus normally exerts an inhibitory effect on the medial segment of the pallidum (via its efferent projection onto the latter, see Fig. 4-11). The occurrence of hemiballismus when the subthalamic nucleus is intact may be due to affections of afferent or efferent connections of the nucleus (Carpenter and Strominger, 1966). In a general way these views fit in with suggestions made concerning the role played by the basal ganglia, thalamus, and cortex in the appearance of parkinsonian tremor and rigidity. Modern neurophysiological investigations in animals and studies of patients subjected to surgical treatment for parkinsonism have given information of interest on this subject.

Surgical treatment of patients suffering from involuntary movements has a relatively short history (see for example, Meyers, 1958; Markham and Rand, 1963; Marion Smith, 1967; Tasker, 1976). As mentioned previously, some neurosurgeons observed good effects on tremor and other dyskinesias following ablations of the precentral cortex. Other procedures employed with success have been transection of the cerebral peduncle (discussed at some length above, p. 273) or transections of the dorsolateral funiculus in the spinal cord. In all these instances the abolition of the involuntary movements was followed by hemiparesis. As mentioned previously, these results supported the notion that corticospinal pathways were involved in the production of the abnormal movements. Attempts to destory parts of the basal ganglia by an open approach have been tried but carry condiderable risks. The introduction of stereotactic methods for placing lesions in the basal ganglia by Spiegel, Wycis, Marks, and Lee (1947) and others opened up new possibilities. Several variants of the technique have been developed in which the lesions are produced by such diverse means as alcohol injections, electrolysis, radiofrequency waves, ultrasound, placing of radioisotopes, or freezing. A number of nuclei have been the goal for the lesions, most frequently the globus pallidus and, somewhat later, the lateral thalamus, especially VA and VL. Some have attempted to destroy the centromedian nuclei or the substnatia nigra. Satisfactory relief has been reported following all of these procedures, although the results differ.

> There are several reasons for this, such as variations among patients selected for treatment as to age and stage of the disease and differences as to the size of the lesion achieved, but a main factor appears to be the difficulties inherent in the placing of the lesion precisely where it is intended without inadvertent damage to other structures. Even if certain landmarks, such as the foramen of Monroe and the anterior and posterior commissures can be found on ventriculography and can be used for comparisons with the many atlases of the human brain prepared for this purpose, the individual variations among human brains are too great to permit entirely reliable placements of the destruction in this way. In recent years the recording of potentials from microelectrodes inserted in the target area have made it possible to increase precision considerably (for example the VPL can be identified by the potential changes occurring in this on somatosensory stimulation, Albe-Fessard et al., 1963; Gaze et al., 1964). Other thalamic nuclei can also be identified (Albe-Fessard et al., 1967; Bertrand, Jasper, and Wong, 1967). Likewise, stimulation may give information, for example, of whether the electrode is situated in the internal capsule. Moderate cooling down to some + 10°C or somewhat less may produce a temporary inactivation of the target area (Le Beau, Dondey, and Albe-Fessard, 1962; see also Siegfried et al., 1962).

In spite of improvements in the methods for accurate placements of lesions in stereotactic surgery it is essential for an analysis of the results and for further progress to have precise information of the destructions actually made at operation. This information can only be obtained when careful postmortem studies are performed in patients who have been clinically thoroughly examined. Relatively few studies of this kind are available, and in most of them only the site of the lesion is described. In quite a number of cases in which good results have been obtained the internal capsule has been found to be involved in the lesion. This was the case in at least 12 of 15 patients whose brains were carefully examined by Marion Smith (1962) and in 15 of 17 operations on 12 patients studied by Nörholm and Thygstrup (1960) as well as in several others reported in the literature (see Marion Smith, 1967, for references). This is in keeping with other obser-

vations concerning the effects of cortical lesions and of lesions affecting corticospinal fibers, for example, in pedunculotomies. However, as mentioned
previously, the corticospinal fibers are situated in the posterior part of the internal
capsule, and they are therefore not very likely to be affected in the injury inflicted
on the internal capsule in stereotactic operations aiming at the pallidum or ventrolateral thalamus (see Fig. 4-9). Only in one case of several with lesions of the internal capsule did Marion Smith (1967) find degeneration of corticospinal fibers.
The favorable effects of lesions of the internal capsule (see for example, Gillingham, 1962) may therefore be due to interruption of some of the other fiber contingents passing in it. This may be important in view of the fact that lesions involving many different regions have been claimed to give satisfactory effects on
rigidity as well as tremor and other dyskinesias. We shall return to this question
below.

The steadily growing literature on the mechanisms and the treatment of dyskinesias and parkinsonian rigidity is rather confusing, and no attempt will be made
to discuss these subjects exhaustively.[61] Chiefly, points which are of relevance for
the anatomical basis will be considered. It will be practical to discuss separately
observations bearing on parkinsonian rigidity and on dyskinesias.

As to the mechanism underlying *parkinsonian rigidity* little is known for certain. It has apparently so far not been possible to produce a corresponding state
experimentally in animals by placing lesions in the brain or by stimulation. However, electrical stimulation in animals and man has given information of some interest.

Electrical stimulation of the VL and the pallidum influences the muscle spindles. Stern and
Ward (1960) found inhibition of the contralateral spindle discharge in cats. The integrity of the
motor cortex was necessary for this to occur. Langfitt et al. (1963), however, also in the cat, observed complex alterations in spontaneous activity of flexor and extensor γ neurons, including reciprocal and nonreciprocal inhibition and facilitation, and concluded that the cortex is not involved
in mediating the responses. According to Gilman and van der Meulen (1966), ablations of what
they refer to as "pyramidal" or "extrapyramidal" cortical areas in the monkey are followed by
alterations in the "tonic gamma as well as alpha mechanism." In man, Ohye et al. (1964) observed that high-frequency stimulation of the VL results in nonreciprocal increase of muscle tone
in the contralateral forelimb muscles, and Walter et al. (1963) report various alternations in the
patients' abnormal movements on stimulation of the globus pallidus and VL. The divergent results
obtained by these and other authors may perhaps in part be explained by differences in the experimental situations (anesthesia, etc.). Since there is evidence that especially the tonic stretch reflexes are increased in parkinsonism it is interesting that following stereotactic lesions of the
pallidum these are reduced while the phasic responses to stretch are not influenced (Shimazu et
al., 1962). Hassler (1966a) has observed that stimulation of his nucleus V.o.a. (apparently corresponding to VA) increases muscle tone in parkinsonian patients, while coagulation of this nucleus
abolishes rigidity as effectively as does coagulation of the pallidum. Hassler (1966a) further
reports that on stimulation of certain nuclei in the ventrobasal thalamus in conscious patients there
is an acceleration of contralateral movements and of speaking, while stimulation of other regions
has the opposite effect and results in an increase of muscle tone.

These and other observations indicate that the pallidum and thalamus (especially VL and VA) are able to influence the reflex apparatus of the cord (presumably chiefly the tonic stretch reflexes). The action appears to occur via the cortex

[61] For an account on various aspects of the problems see *Advances in Stereoencephalotomy*, III,
1967.

(chiefly the precentral) from which impulses are transmitted to the cord via corticofugal fibers descending in the internal capsule. These conclusions appear to be compatible with the results of stereotactic operations in man, since destruction of any of the above links may abolish the increased activity of the tonic stretch reflex in parkinsonism.

Dyskinesias of different kinds have been reported to be alleviated by surgical lesions of the pallidum or the ventrolateral thalamus (for example, dystonia musculorum deformans, Cooper, 1959, 1965b, 1976; hemiballismus, Martin and McCaul, 1959; intention tremor, Cooper, 1965b; Fox and Kurtzke, 1966; and others), but the bulk of observations concern *parkinsonian tremor*. This is essentially a tremor at rest and differs in several respects from the tremor in lesions of the cerebellum and from "physiological tremor."

> *Physiological tremor* is the name used for tremor found in many persons with no demonstrable disease. This shows great individual variations and has been found by several students to have a frequency of about 70/sec. It is influenced by a number of factors, such as the part of the body tested, the position of the extremity, the amount of work by the limb prior to testing, the weight of the extremity, the patient's age, the state of consciousness, emotional state and several other factors, among these intoxication. It appears to be accepted that physiological tremor "is due to a slight oscillation of the length servoloop which involves the complete stretch mechanism, including gamma efferents of the spinal segmental level" (Wachs and Boshes, 1961, p. 68).

According to Wachs and Boshes (1961) the tremor observed in parkinsonism shows a lower dominant frequency than the physiological tremor, and is less affected by the position of the extremities but more influenced by emotional and other endogenous factors. By employing a special recording device these authors sometimes found tremor to be present where it could not be seen clinically, for example on the nonaffected side in patients with unilateral symptoms. This corresponds to the pathological observations of Davison (1942). In patients with unilateral parkinsonism he often found bilateral changes in the pallidum and substantia nigra.

In recordings from the human thalamus during operations for Parkinson's disease a spontaneous rhythmic activity has been recorded when the patients are awake (Albe-Fessard, Arfel, and Guiot, 1963; Albe-Fessard, Guiot, Lamarre, and Arfel, 1966; Albe-Fessard et al., 1967; Crowell, Perret, Siegfried, and Villoz, 1968; Bates, 1969). The thalamic units fire in bursts at the frequency of the tremor, as further confirmed in simultaneous electromyographic recordings. It has been concluded that the thalamic rhythm is not due to afferent impulses resulting from the tremor movements, even though the central rhythm can be influenced from the periphery. These thalamic units were found in a region anterior to but partly overlapping the VPL, apparently in VL (but also in a more dorsoposterior region, the LP). Largely corresponding observations have been made by Jasper and Bertrand (1966) and Bertrand, Jasper, and Wong (1967).

Since the tremor is not abolished by deafferentation of the limbs in patients with Parkinson's disease (Pollock and Davis, 1930) or in monkeys with experimentally induced tremor (see below) (Ohye, Bouchard, Larochelle, Bédard, Boucher, Raphy, and Poirier, 1970), it has been suggested that the abnormal rhythmic bursts are generated centrally. Experimental evidence for this hypothesis has been provided by several authors (see Lamarre, Joffroy, Dumont, DeMontigny, Grou, and Lund, 1975). Whether the rhythmic activity arises locally in the thalamus or is triggered from other structures (from the cerebellum?) cannot be decided. It seems a likely assumption (Albe-Fessard et al., 1967), however, that in parkinsonism there occurs a release of a "rhythmic center" from an inhibition arising from a region that is destroyed in the disease.

Attempts to produce in animals an experimental tremor resembling that seen in parkinsonism have not been too successful.[62] However, Poirier and his collaborators have succeeded in producing a sustained postural Parkinson-like tremor in the contralateral limbs after unilateral lesions of the ventromedial part of the midbrain tegmentum in the monkey (Poirier, 1960; Poirier, Bouvier, Bédard, Boucher, Larochelle, Olivier, and Singh, 1969; Poirier, Sourkes, Bouvier, Boucher, and Carabin, 1966). According to these authors, the experimentally induced tremor is probably related to a combined damage of the nigrostriatal pathway and the rubro–olivo–cerebello–rubral loop (Poirier, Filion, Larochelle, and Péchadre, 1975). Other authors maintain that the Parkinson-like tremor seen in these animals is generated by a thalamocortical mechanism, while the olivocerebellar system is responsible for the faster "physiological" tremor observed in the animals (Lamarre, Joffroy, Dumont, DeMontigny, Grou, and Lund, 1975). Carpenter and McMasters (1964) suggest that the tremor that has been observed in monkeys after lesions of the ventromedial tegmentum is probably due to interruption of cerebellofugal fibers (Goldberger and Growden, 1971). As described in Chapter 5, lesions of the cerebellum involving the dentate nucleus or the brachium conjunctivum below the red nucleus result in coarse intentional tremor (as well as ataxia). Carpenter (1961) focuses attention on the fact that lesions of the ventrolateral thalamus abolish cerebellar tremor in monkeys (Carpenter, Glinsmann, and Fabrega, 1958), just as intention tremor has been abolished in man by similar lesions (Cooper, 1965b; Fox and Kurtzke, 1966; Tasker, 1976; and others). Experimental lesions of the pallidum appear to have some effect on cerebellar tremor and ataxia. It appears further from the experimental studies that impulses responsible for the appearance of cerebellar dyskinesias, including tremor, are transmitted to the cord by corticospinal connections, perhaps particularly the corticospinal tract. Thus, on many points there are marked similarities between the observations made in cases of cerebellar dyskinesias and those made in studies of subthalamic dyskinesias, described above. In both instances, as well as in parkinsonian tremor, it appears that impulses giving rise to dyskinesias are mediated via the ventrolateral thalamus-cortex-corticofugal fibers to the cord, and, probably except for cerebellar tremor, are transmitted to the cord by corticospinal connections, perhaps particularly the corticospinal tract. Thus, on many points there are marked similarities between the observations made in cases of cerebellar dyskinesias and those made in studies of subthalamic dyskinesias, described above. In both instances, as well as in parkinsonian tremor, it appears that impulses giving rise to dyskinesias are mediated via the ventrolateral thalamus-cortex-corticofugal fibers to the cord, and, probably except for cerebellar tremor, the pallidum should be included in these

---

[62] In any comparison between experimental observations and findings in man it should be recalled that most affections giving rise to dyskinesias in man develop slowly, and are not strictly parallel to an acute severe damage.

structures. Some authors advocate the importance of "nonspecific" thalamic nuclei as well.

Why tremors and dyskinesias result in diseases of the basal ganglia, the subthalmic nucleus, and the cerebellum is still not clearly understood. Even though many varieties of involuntary movements appear to be mediated via the same structures and pathways, it must be presumed that their basic mechanism is not identical. A number of theories have been set forth. These will not be discussed here. Suffice it to mention that in most attempts at explanation emphasis has been put on the presence of numerous closed loops within the connections of the basal ganglia and between these and other structures (see Fig. 4-11). These connections represent the basis for feedback mechanisms between various gray masses, and their importance was early stressed by Bucy (1944b). A closer scrutiny of the fiber connections reveals an abundance of mutual interrelationships which might be concerned. Any attmept to pick out destruction of links in a particular circuit as being responsible for one or another kind of dyskinesia will therefore necessarily represent a great oversimplification. Furthermore, the anatomical arrangement of the various connections thought to be involved is such that there is virtually no locus where only one fiber connection can be damaged in isolation. This has been discussed more fully by Marion Smith (1967), who draws attention to some salient points. The pallidofugal fibers running in the internal capsule have a very wide distribution both in the anteroposterior and craniocaudal planes. Furthermore, the pallidal efferents lie very close to the thalamic afferents from the cerebellum, and the dentatothalamic and pallidothalamic fibers end in the same nuclei of the thalamus, even though there are probably more pallidal efferents to VA and more cerebellar efferents to the VL. In view of these and other features in the fiber connections and considering that in stereotactic operations the lesions are certainly not always as circumscribed as planned and not placed precisely where intended, it is not surprising that destructions of almost any target (pallidum, ansa lenticularis, the thalamus (VL or VA), the internal capsule, and others) have been reported to alleviate different kinds of dyskinesias. Even in the reports of cases in which the anatomical site of the lesion has been mapped it is not possible so far to correlate a specific lesion with beneficial effect on a particular kind of dyskinesia (see for example the papers of Dierssen et al., 1962; Cooper, Bergmann, and Caracalos, 1963; Markham, Brown, and Rand, 1966; Beck and Bignami, 1968). It appears, however, that lesions of the pallidum are more effective in relieving rigidity, while thalamic lesions are superior in abolishing tremor. Bertrand and Martinez (1962) reported that, in a large material, tremor never responded satisfactorily to pallidal lesions, but results were very good in lesions of the ventrolateral thalamus. This and other observations may suggest that a disturbance in the impulses from the cerebellum to the thalamus may be of some importance in the production of tremor in parkinsonism and of other dyskinesias. Stereotactic destruction of the dentate nucleus in man has been tried in some cases of dyskinesias (see Chap. 5), but the results are not conclusive. However, whether the cerebellar impulses are of relevance or not, it is clear from the fiber connections that an appropriately placed lesion of the ventrolateral thalamus will interrupt pallidothalamic as well as dentatohalamic and rubrothalamic fibers, making thalamotomy preferable to pallidotomy as a method for alleviating dyskinesias. This indeed appears to be the experience of many neurosurgeons.

The considerations made above on the subject of the mechanisms and treatment of dyskinesias are rather elementary. A number of detailed physiological studies on particular aspects of the problem have not been considered. It is clear, however, that we are obviously still far from understanding the mechanisms underlying the various kinds of dyskinesias in man. However, this does not prevent the empirical search for regions whose destruction will alleviate particularly one or the other type of dyskinesia. In order to attain this goal "it is essential that more evidence is obtained by the rigorous follow-up of patients who have had stereotactic operations, ensuring that the actual location of the lesions is ascertained. Further, the pathological changes due to the underlying disease must be determined in these patients. Only when many such studies have been made will we be able to put these successful, though empirical, operations on a rational basis" (Marion Smith, 1967, p. 47).

5

The Cerebellum

DESPITE NUMEROUS anatomical and physiological studies devoted to the cerebellum we do not yet properly understand its function and its cooperation with other parts of the brain. Sherrington coined the term "the head ganglion of the proprioceptive system" for the cerebellum. However, subsequent research has shown that the cerebellum is not only related to "the proprioceptive system" but to activities in other functional spheres as well. Judging from its fiber connections, the cerebellum appears to be able to influence almost any other part of the brain. It might, therefore, be surmised that in general it coordinates and controls almost any function in which the nervous system is involved, in much the same way as it is known to regulate muscular activity. It may be essential for the full perfection of a number of bodily functions. The cerebellum is, however, not essential to life. In fact, individuals who are born without a cerebellum do not betray themselves in daily life by any obvious defects.

The wealth of new data on the anatomy and physiology of the cerebellum which has become available in the last decade makes it an impossible task today to give a simplified and yet meaningful account of the cerebellum, its connections, and its functions. The present chapter must therefore include a rather detailed presentation of new evidence. Even if much of the new information has improved our understanding of cerebellar organization, it has so far had modest consequences for clinical neurology. However, the rather detailed account of the cerebellum given here serves a particular purpose: It illustrates some general features of organization of the nervous system and exemplifies different technical approaches in its study.

Comparative anatomical aspects. Some comparative antomical data are valuable as a basis for understanding cerebellar function, since they reflect a functional subdivision within the cerebellum which is also demonstrated by other

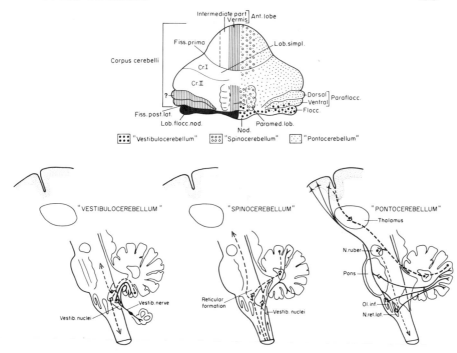

FIG. 5-1   Above, a simplified diagram of the mammalian cerebellum. In the left half the three main subdivisions, which can be recognized on a comparative anatomical basis, are seen. Black: Archicerebellum, flocculonodular lobe. Hatchings: palaeo-cerebellum, the vermis of the anterior lobe, the pyramis, uvula, and paraflocculus. White: neocerebellum. In the right half the main terminal areas of the vestibulo-cerebellar fibers are indicated by heavy dots, the main terminal areas of spino-cerebellar pathways by open rings, and the terminal areas of pontine afferents by small dots.

The three very simplified diagrams below illustrate that each of the main functional subdivisions, by way of its efferent projections, will influence first and fore-most the part of the nervous system from which it receives its main afferent input. See text. From Brodal (1972a).

methods of investigation. In the paired pimordia of the cerebellum two parts can be distinguished. One of these has an intimate relation to the matrix of the vestibular nuclei; the other develops immediately rostral to this. The former is present in a rather constant form in most vertebrates. In accordance with the designation applied to it by Larsell, who has studied the comparative anatomy of the cerebellum extensively, this part is commonly called the *flocculonodular lobe,* since it is made up of the flocculus and the nodulus (Fig. 5-1).

The other major part of the cerebellum is separated from the flocculonodular lobe by a fissure, the *fissura posterolateralis.* This fissure is the first to appear, phylogenetically as well as ontogenetically. The part of the cerebellum developing rostral to this fissure is called the *corpus cerebelli* (Fig. 5-1). In contrast to the flocculonodular lobe, the corpus cerebelli increases considerably in size in the phylogenetic ascent of vertebrates. However, this increase does not involve all parts of the corpus cerebelli to an equal extent. The most rostral part, called *lobus anterior* and separated from the rest by the so-called *fissura prima,* shows only moderate changes (the fissura prima is not, as previously commonly maintained, the oldest fissure of the cerebellum). The most caudal parts of the corpus cerebelli, the lobuli called the *pyramis* and *uvula,* are also relatively constant in most vertebrate species. It is the middle part of the corpus cerebelli which increases markedly in size. This holds true with respect to its vermal portions, and particularly its lateral parts. These are clearly developed only in mammals, and in monkeys, apes, and man their size is so large that they entirely overshadow the other parts of the cerebellum. These lateral parts correspond roughly to what is called the *cerebellar hemispheres.* In the increase of the lateral parts the lobus anterior also takes part.[1]

The lateral parts of the cerebellum and the middle portion of the vermis represent the phylogenetically youngest parts of the cerebellum, and are often collectively designated as the *neocerebellum* (white in Fig. 5-1, left half). In contradistinction to this, the other parts are sometimes grouped together as representing the paleocerebellum. However, the subdivision proposed by Larsell (1934, 1937) is to be preferred. According to this author the flocculonodular lobe is termed the *archicerebellum* (black in Fig. 5-1, left half), and the term *paleocerebellum* denotes the vermal part of the anterior lobe, the pyramis, uvula, and the paraflocculus (hatched in Fig. 5-1, to the left).

The subdivision of the cerebellum arrived at on the basis of comparative anatomical data, briefly outlined above, on the whole *corresponds to a subdivision made on the basis of the afferent cerebellar fiber connections* (see, however, below).

The archicerebellum is therefore sometimes referred to as the *vestibulocerebellum* (see Fig. 5-1), the paleocerebellum as the *spinocerebellum,* and the neocerebellum as the *pontocerebellum.* The basis for these designations, which will often be employed in this text, will be considered in a following section. Until fairly recently there has been some uncertainty concerning the homologies of some cerebellar lobules in man (especially in its posterior parts) with those of other mammals. Since most of our knowledge of the fiber connections and functional features of the various cerebellar subdivisions are derived from studies of animals, it is of interest that the main homologies appear now to be clarified, as shown in Fig. 5-2.

> The paraflocculus, especially, has been the subject of much dispute. According to comparative anatomical studies (Scholten, 1946; Jansen, 1950, 1954) the ventral part of the mammalian paraflocculus (the ventral paraflocculus) is homologous with the tonsilla of the human cerebellum, while the dorsal paraflocculus corresponds to the lobulus biventer. The paramedian lobule, on which much experimental information is available, corresponds to the lobulus gracilis. The paraflocculus, medially connected with the uvula and pyramis, is a peculiar part of the cerebellum, being enormous in aquatic mammals (see Jansen, 1950). Little is so far known of its function.

In the following account the various cerebellar lobules will be referred to as they are found in the cat and monkey. The names applied to corresponding lobules in the human cerebellum can be seen from the diagram of Fig. 5-2. Here are also

---

[1] For reviews of the comparative anatomy, embryology, and fiber connections of the cerebellum see Dow (1942a), Larsell (1945), Jansen and Brodal (1958), Brodal (1967a), Larsell (1967, 1970), and Larsell and Jansen (1972).

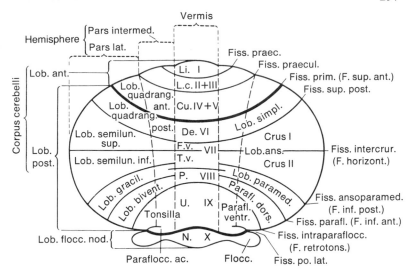

FIG. 5-2 Diagram of the subdivision of the human cerebellum, based on comparative studies of the mammalian cerebellum. In the left half of the diagram the classical names of the various lobules are shown, to be compared with the names used in mammals in general in the right half. On the left is further indicated the principal subdivision of the cerebellum in the flocculonodular lobe and corpus cerebelli, the latter again being subdivided into an anterior and posterior lobe. In the vermis the Roman numerals I to X suggested by Larsell (1952, and later) for the transverse foliation are shown. Finally, the longitudinal subdivision into vermis, intermediate zone, and lateral part (cf. text) is indicated. Abbreviations for vermal lobules: *Li.:* lingula; *L.c.:* Lobulus centralis; *Cu.:* culmen; *De.:* declive; *F.v.* and *T.v.:* folium and tuber vermis; *P.:* pyramis; *U.:* uvula; *N.:* nodulus. From Jansen and Brodal (1958).

included the Roman numerals for the various lobules, suggested by Larsell (1952, 1953) and now commonly used. For example, the pyramis and uvula are labeled VIII and IX, respectively, the nodulus being lobulus X. Corresponding hemispheral parts are referred to as VIIIH, etc. A detailed description of the human cerebellum can be found in the atlas of Angevine, Mancall, and Yakovlev (1961).

The cerebellum is connected with the brainstem by afferent and efferent fibers. They pass in three massive fiber bundles (Fig. 5-3), the *superior, middle, and inferior cerebellar peduncle* (or the brachium conjunctivum, the brachium pontis, and the corpus restiforme, respectively). The most massive, the middle cerebral peducle, contains only afferent fibers to the cerebellum. In man this peduncle has been estimated to contain some 20 million fibers (Tomasch, 1969). Almost all of them come from the pontine nuclei. The main pathway for the efferent cerebellar fibers is the superior peducle, carrying fibers passing from the cerebellar nuclei to the brainstem, the red nucleus, and the thalamus. In man it is calculated to contain about 0.8 million fibers (Heidary and Tomasch, 1969). In addition, it contains some cerebellar afferent fibers, most of them ascending from the spinal cord (the ventral spinocerebellar tract, see below). In the inferior cerebellar peduncle most of the fibers are afferent. Most of them (some 0.5 million in man) are derived from the inferior olive; others come from the spinal cord (the dorsal

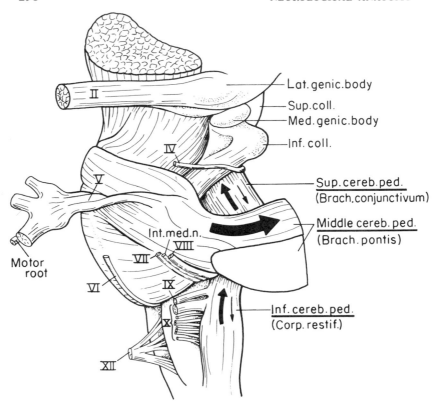

Lat. genic. body

Sup. coll.
Med. genic. body

Inf. coll.

Sup. cereb. ped.
(Brach. conjunctivum)

Middle cereb. ped.
(Brach. pontis)

Int. med. n.

Motor
root

Inf. cereb. ped.
(Corp. restif.)

FIG. 5-3 A drawing of the lower part of the human brainstem, seen from the left, to illustrate the gross features of its connections with the cerebellum (removed). Direction and thickness of arrows indicate direction of fibers and (roughly) their relative numbers in the three cerebellar peduncles. See text.

spinocerebellar tract). In addition, some efferent cerebellar fibers traverse the inferior peduncle, as will be considered below.

The fiber connections of the various parts of the cerebellum will be considered in a later section. Here, attention will be focused only on a general principle. The three main regions, the *vestibulocerebellum,* the *spinocerebellum,* and the *pontocerebellum,* distinguished on the basis of their afferents, by way of reciprocal connections exert their main influence on "corresponding" regions of the central nervous system, the vestibular nuclei, the spinal cord, and the cerebral cortex, respectively. The diagrams of Fig. 5-1 are very simplified illustrations of these reciprocal relations, which are, in fact, extremely complex and far less schematic than shown. Even in the simplified diagrams it can be seen, for example, that the spinal cord can be influenced via neural pathways from the vestibulocerebellum and pontocerebellum, as well as from the spinocerebellum.

The longitudinal zonal subdivision of the cerebellum. In recent years it has been shown that there exists within the cerebellar cortex a longitudinal pattern of organization in addition to the classical subdivision of the cerebellum in lobes and lobules. On the basis of a study of the projections from the cerebellar cortex to the cerebellar nuclei in the cat, rabbit, and monkey, Jansen and Brodal (1940, 1942) suggested a subdivision into three such zones, a medial zone (vermis) projecting to the fastigial nucleus, an intermediate zone (following the nomenclature of Hayashi's developmental studies, 1924) projecting to the nuclei interpositi, and a lateral zone supplying the lateral (dentate) nucleus (Fig. 5-4). This view received support from the physiological and anatomical studies of Chambers and Sprague (1955a, b). The longitudinal subdivision was later found to be far more detailed and precise. On the basis of studies of the ontogenetic development of the cerebellar cortex and nuclei in the rat and in whales, Korneliussen (1967, 1968, 1969) distinguished within the intermediate part two longitudinal zones, and within the medial part three subzones. The width of the zones differs among species and is correlated with the development of the corresponding cerebellar nucleus. Thus in *Cetacea,* where the nuclei interpositi are extremely large, the intermediate zone occupies most of the cerebellar cortex.

The most detailed mapping of the longitudinal subdivision of the cerebellum has been made by Voogd (1964, 1969). In transverse sections stained for myelin sheaths, alternating rather narrow strips of thin and thick fibers can be distinguished. On the basis of this, several longitudinal *compartments* can be recognized. To each compartment belongs a particular, more or less *longitudinal zone* in the cortex. The areas containing thin fibers (originally referred to as raphes; Voogd, 1964) form borders between the compartments and are not everywhere equally clear. Studies of the sites of termination of afferent fibers to the cortex and nuclei clearly show that many afferent fiber contingents are distributed in the cor-

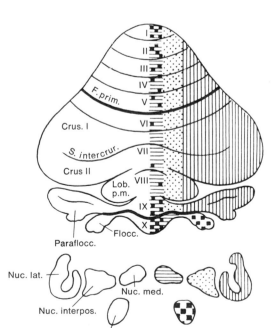

FIG. 5-4 A simplified diagram of the longitudinal zonal subdivision of the mammalian cerebellum based on the pattern in the corticonuclear projection. The cortical sites of origin of fibers to the various intracerebellar nuclei and the vestibular nuclei are indicated by symbols corresponding to those in the nuclei. From Jansen and Brodal (1958).

tex in strips. This is the case, for example, for the spinocerebellar tracts and the cuneocerebellar tract (see Voogd, 1969, for a survey). Each of these pathways terminates within more than one zone, an indication that they may carry functionally different categories of fibers (see below). The zonal pattern has been mapped in great detail, particularly as concerns the termination of climbing fibers from the inferior olive (see below and Fig. 5-18). The efferent fibers from the cerebellar cortex to the cerebellar nuclei appear to arise in cortical zones belonging to the same compartment as the nuclear region to which they project.

In the middle, and part of the intermediate portion of the cerebellum, the longitudinal pattern is rather purely sagittal. In the more lateral parts of the cerebellum the zones tend to make a lateral bend, making it difficult to identify them precisely. Figure 5-5 is a diagram of the zones in the cerebellum of the ferret. In principle, the same pattern has been found in several other mammals, including the monkey (see Voogd, 1969). As seen in Fig. 5-5, cortical zones A and B cover the vermis, while most of the intermediate part belongs to zones C_1, C_2, and C_3. The lateral parts belong to zones D_1 and D_2. As will appear from the following account, some of the zones can be further subdivided into subzones. The variations among animal species in the development of different parts of the cerebellum entail some differences of detail.[2]

Physiological studies have confirmed the correctness of the zonal longitudinal pattern in the cerebellum, and have given further information about differences between zones, particularly their afferent input (see below and Oscarsson, 1973). These studies and recent anatomical studies show that the longitudinal zonal pattern in the cerebellum is not as schematic as might appear from the above. Each zone is not necessarily present throughout the cerebellum in the adult animal (for

[2] It is possible, to some extent, to correlate Voogd's zones with the zones determined by Korneliussen.

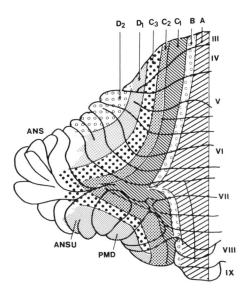

D_2 D_1 C_3 C_2 C_1 B A

FIG. 5-5 A diagram of the longitudinal zones in the cerebellar cortex of the ferret, according to Voogd. Note that the lateral zones are not arranged parasagitally. See text. From Voogd (1969).

example, zone B), and an afferent fiber system ending in a particular zone does not necessarily supply it throughout its extent, but may be restricted to part of it (for example, fiber systems carrying information from hindlimb or forelimb may supply different parts of a zone, in agreement with the well-known somatotopic pattern within some cerebellar areas; see below and Fig. 5-9). Examples will be given in the following.

The cerebellar cortex.   On account of its many deep sulci, the cerebellar cortex covers a very large surface area. If the human cerebellar cortex were imagined unfolded, the distance from its most anterior to its most posterior part would exceed one meter (Braitenberg and Atwood, 1958). Since most folia run transversely, the lateral extent of the cortex is only about one-seventh of the longitudinal.

In spite of some regional variations, the *cerebellar cortex is principally identically structured all over.*[3] (For details the reader is referred to other sources, for example Jansen and Brodal, 1958; Fox et al., 1967; Eccles, Ito, and Szentágothai, 1967; Mugnaini, 1972; Palay and Chan-Palay, 1974.) Electron microscopic studies have disclosed a wealth of details. In the following account only major points will be mentioned with reference to the diagram of Fig. 5-6.

The cerebellar cortex has three distinct layers, the *molecular layer,* the *Purkinje cell layer,* and the *granular layer*. The latter borders on the white matter. The most conspicuous cellular elements in the cortex are the *Purkinje cells*. The axons of the large flask-shaped, regularly arranged cell bodies are given off into the white matter, while richly branched dendritic trees extend outward into the molecular layer. It is peculiar that the dendritic tree is spread out in one plane only, perpendicular to the longitudinal axis of the folium, and that each tree appears to have its own territory. The dendritic branches are densely beset with spines, which are contacted by afferent fibers (see below).

> The number of dendritic spines (or thorns) on a single Purkinje cell is enormous. In the cat it has been calculated to be 80,000 (Palkovits, Magyar, and Szentágothai, 1974b), in the rat about 18,000 (Palay and Chan-Palay, 1974). According to electron microscopic studies (see Palay and Chan-Palay, 1974, for particulars), these thorns are contacted by parallel fibers (see below). Spines occur also on the proximal parts of the dendrites (which, before the era of the electron microscope, were often assumed to be smooth). The climbing fibers (see below) appear to establish contact with spines only.

The axons of the Purkinje cells, which pass either to the cerebellar nuclei or the vestibular nuclei, give off recurrent collaterals. These are largely oriented in the plane of the dendritic tree of the cell and may branch repeatedly. Some branches ascend to the level of the Purkinje cells and, as shown electron microscopically, may establish synaptic contact with somata and proximal dendrites of these (see Palay and Chan-Palay, 1974), while others establish synaptic contact with the *Golgi cells*.

[3] However, there are variations as to details. The flocculonodular lobe, for example, differs from other parts with regard to the types of mossy fibers present (Brodal and Drabløs, 1963) and with regard to the Golgi cells and myelinated fibers (see also Brodal, 1967a). The Purkinje cells and the granule cells are larger and more widely spaced in the flocculonodular lobe than in the corpus cerebelli (Lange, 1972). A new cell type, so-called "pale cells," has been claimed to exist in the granular layer (rat) and to be preferentially concentrated in the vestibulocerebellum (Altman and Bayer, 1977).

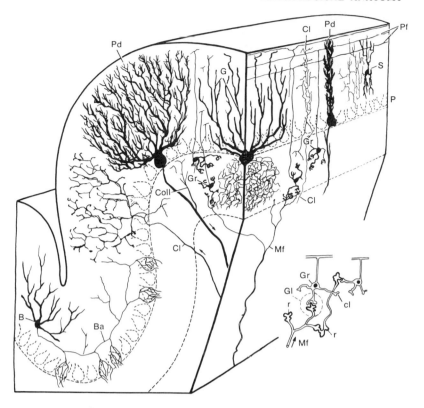

FIG. 5-6   Semidiagrammatic representation of part of a cerebellar folium to show the main elements of the cerebellar cortex and their topographical relationships and orientation. Note especially the arrangement of the Purkinje cell dendrites (*Pd*) and basket cell axons (*Ba*) in the transverse plane of the folium and the longitudinal arrangement of the parallel fibers (*Pf*). Other abbreviations: *B.*: basket cell; *Cl*: climbing fiber; *Coll*: recurrent collateral of Purkinje cell; *G*: Golgi cell; *Gr*: granule cell; *Mf*: mossy fiber; *P*: Purkinje cell; *Pd*: Purkinje cell dendrites; *Pf*: parallel fibers; *S*: stellate cell. Slightly altered from Jansen and Brodal (1958). The figure to the lower right illustrates diagrammatically the relation between the rosettes (*r*) of the mossy fibers (*Mf*) with the dendritic "claws" (*cl*) of the granule cells (*Gr*) in a glomerulus (*Gl*). Other elements of the glomerulus are not shown. Adapted from Hámori and Szentágothai (1966a).

The latter are found in the *granular layer*, which is made up chiefly of an enormous number of densely packed small cells, with very scanty cytoplasm, the *granular cells*. In Golgi preparations these are seen to be provided with about four to five short dendrites, radiating in various directions and ending with clawlike expansions (see Fig. 5-6). These are synaptically contacted by endings of mossy fibers (see below). The axons of the granule cells are characteristic. They ascend to the molecular layer where they bifurcate in a T-shaped manner, the two branches, called *parallel fibers*, always running along the longitudinal extent of the folia across the dendritic trees of the Purkinje cells.

Together the two branches of the parallel fibers are 1.5–3 mm long in the cat (Fox and Barnard, 1957), and even longer in man (Braitenberg and Atwood, 1958). Electron microscopic studies have shown that they establish synaptic contact with the dendritic spines of the Purkinje cells (in addition they contact stellate, basket, and Golgi cells). The number of parallel fibers passing through the dendritic tree of a single Purkinje cell has been estimated to be about 200,000 (Fox and Barnard, 1957) or even 400,000 (Palkovits, Magyar, and Szentágothai, 1971b) in the cat. It follows from this arrangement that a single parallel fiber may act on a large number of Purkinje cells (probably some 450) situated along a folium, and that each Purkinje cell is under the influence of a tremendous number of granule cells (see Fox, Siegesmund, and Dutta, 1964). For recent electron microscopic studies of the parallel fibers and their synaptic connections and for quantitative data, see Palay and Chan-Palay (1974).

The other element in the granular layer, the *Golgi cell* referred to above, in some respects resembles the Purkinje cell. The Golgi cells are large and have branching dendritic trees which largely extend outward into the molecular layer, but, unlike those of the Purkinje cells, these trees spread their branches in all directions (see Fig. 5-6). The dendrites are contacted by parallel fibers as well as other afferents (some dendrites remain in the granular layer and are contacted by mossy fibers). The axons of the Golgi cells are richly branched but do not leave the cerebellar cortex. They end in synaptic contact with dendrites of granule cells.

The *molecular layer* is dominated by fibers and contains relatively few nerve cells. Some of these are stellate cells, of which there are several varieties. A particular type is the so-called *basket cell*. These cells are situated just above the Purkinje cells. Their dendrites, like those of the Purkinje cells, are oriented in the transverse plane of the folium and receive collaterals of climbing fibers (see below). The characteristic feature of the basket cell is the arrangement of its axon. This passes for a considerable distance across the folium, immediately above the Purkinje cell bodies and at right angles gives off descending collaterals. These collateral branches surround the cell body of the Purkinje cell like a kind of basket (hence the name of the cells) and establish synaptic contact with it. The arrangement of the basket cells makes them capable of acting on a series of Purkinje cells arranged across the folium, in contrast to the parallel fibers, which provide for an activation of a series of Purkinje cells along the folium. These and other peculiar geometrical arrangements are of interest in analysis of the function of the cerebellar cortex.

The *afferent fibers entering the cerebellar cortex* are of two entirely different types (see Fig. 5-6). The *climbing fibers* are thin and pass through the granular layer. At the level of the Purkinje cells, each fiber divides into several branches, which closely follow and wind along the dendritic branches of a Purkinje cell. Synaptic contacts are established with the spines of these dendrites. While the main target of a climbing fiber is thus the dendrites of a Purkinje cell, some of its collaterals end on neighboring Purkinje cells, stellate cells, basket cells, and Golgi cells (Scheibel and Scheibel, 1954; Hámori and Szentágothai, 1966a). It appears from this anatomical arrangement that the climbing fibers must be able to exert a powerful synaptic action on the Purkinje cell, as has indeed been found to be the case (see below). The majority of the climbing fibers come from the inferior olive (Szentágothai and Rajkovits, 1959), but some may come from other nuclei in the brainstem.[4]

[4] It is often stated that there is a specific relation of one climbing fiber to one Purkinje cell. However while, anatomically, branching of climbing fibers has been seen only in or just beneath the cortex, and these branches supply two, three, or four Purkinje cells not far removed from each other (Scheibel and Scheibel, 1954; Fox et al., 1969), physiological investigations show that a single climbing fiber may branch and supply folia that are considerable distances apart (Faber and Murphy, 1969; Armstrong

Whether all climbing fibers are derived from the inferior olive is still not settled. According to the anatomical and physiological study of Batini, Corvisier, Destombes, Gioanni, and Everett (1976), nearly all, if not all, climbing fibers originate in the inferior olive (see, however, O'Leary et al., 1970, and a review by Brodal, 1972a). In this connection it is of interest that the number of nerve cells in both olives of man is about 1 million (Moatamed, 1966; Escobar, Sampedro, and Dow, 1968). This is only one-fifteenth of the calculated number of Purkinje cells (Braitenberg and Atwood, 1958). In the cat the inferior olives (both sides) contain some 121,000–145,000 cells (Escobar et al., 1968; Mlonyeni, 1973), while the number of Purkinje cells is given as 1.2–1.3 million (Palkovits, Magyar, and Szentágothai, 1971a) or 1.5 million (Mlonyeni, 1973).

The other type of afferent cerebellar fiber, the *mossy fiber,* differs in almost all respects from the climbing fiber. The mossy fibers are relatively thick and myelinated. Having entered the cerebellar cortex, they branch repeatedly. One fiber may supply two or even more folia. During their course they give off abundant collaterals, which, like the final branches, end in the granular layer with a cluster of small endings, forming what is often referred to as *rosettes.* These endings interdigitate with and are in synaptic contact with the clawlike dendritic terminations of the granule cell (see Fig. 5-6). The contacting elements belong to what is usually called a *cerebellar glomerulus.*

In cell-stained sections the glomeruli appear as empty spaces among the granule cells ("cerebellar islands"), since the nerve endings do not stain. There has been much diversity of opinion concerning the nature of the cerebellar glomeruli, but electron microscopic studies appear to have established their architecture (Gray, 1961; Hámori and Szentágothai, 1966b; Fox et al., 1967; Mugnaini, 1972; Palay and Chan-Palay, 1974; and others). The rosettes contain abundant, clear synaptic vesicles. Within a glomerulus a single mossy fiber may have synaptic contact with dendrites from many different granule cells. As referred to above, axons of Golgi cells likewise end in the glomerulus (see Fig. 1-5G). They are in synaptic contact with the granule cell dendrites. (And, vice versa, Golgi cell dendrites are contacted by mossy fiber terminals.) It is of interest that, while the influence of mossy fibers on granule cells is excitatory, the action of the Golgi cells is inhibitory. Since a single mossy fiber gives off collaterals to a large number of rather widely dispersed granule cells, and these give rise to the parallel fibers, it follows that an impulse entering the cerebellar cortex in a mossy fiber may act on a rather large cortical area. On account of the arrangement of the parallel fibers, this area will be long in the direction of the folium but relatively narrow in the transverse plane.

Most afferents to the cerebellar cortex, except those from the inferior olive, appear to end as mossy fibers. However, there may be certain differences between mossy fibers from various sources with regard to their degree of branching within the granular layer and presumably also in other respects (for some data see Brodal, 1967a).

The above account of the main features of the structure of the cerebellar cortex shows that this is indeed a very regularly constructed part of the brain. It is a likely assumption that this regularity and the geometrical patterns among its elements are reflected functionally also. The regularity, furthermore, facilitates functional studies. In recent years much work has been devoted to the analysis of the physiology of the cerebellar cortex, particularly by Eccles and his collaborators. From a series of detailed experimental studies there has emerged a picture of the mode of working of the cerebellar cortex that will be briefly considered below. (For reviews and references, see Eccles, 1966a; Eccles, Ito, and Szentágothai,

et al., 1971, 1973a; Cooke et al., 1972). The terminal points are situated within one of the longitudinal zones of the cerebellum (see below).

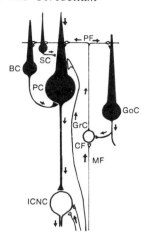

FIG. 5-7 Diagram of the most significant neuronal connections in the cerebellar cortex, according to physiological studies of Eccles and his collaborators. Cells and terminals shown in black are inhibitory. Abbreviations: *BC:* basket cell; *CF:* climbing fiber, *GoC:* Golgi cell; *GrC:* granule cell; *ICNC:* intracerebellar nuclei; *MF:* mossy fiber; *PC:* Purkinje cell; *SC:* stellate cell. See text. From Eccles (1966a).

1967.) The main concepts are summarized in Fig. 5-7, in which the actual anatomical arrangements are represented in a diagrammatic and simplified way.

One essential point is that the climbing fiber exerts a powerful excitatory action on the Purkinje cell. Stimulation of the inferior olive elicits excitatory monosynaptic responses in the Purkinje cells. The mossy fiber impulses have been found to excite the granule cells. The parallel fibers (i.e., the axons of the granule cells) excite the Purkinje cells (by way of their contact with its dendritic spines). Thus, both mossy and climbing fiber inputs to the cortex may excite the Purkinje cells. However, the situation is far more complex, particularly as concerns the mossy fibers. The granule cell axons (parallel fibers), in addition to exciting Purkinje cells, establish synaptic contacts with dendrites of basket and Golgi cells, both of which are inhibitory. The basket cell will thus inhibit the Purkinje cell; the Golgi cell will inhibit the granule cell and will then "counteract" the excitatory effects of mossy fiber impulses on the Purkinje cells. Since the inhibitory pathways to the Purkinje cells include one synapse more than the excitatory route (see Fig. 5-7), stimulation of parallel fibers gives rise first to an excitatory postsynaptic potential (EPSP) in the Purkinje cell, followed 1–2 msec later by an inhibitory potential (IPSP). This will tend to limit the area of cortex excited by an incoming volley. A final point deserves mention. It has been concluded (Ito and Yoshida, 1966) that the Purkinje cells have a purely inhibitory action on the cells with which they establish synaptic contact (cerebellar nuclei, vestibular nuclei). Not all authors are convinced, however, that this is so for all Purkinje cells.

The diagram of Fig. 5-7 shows only some of the many possible circuits which may be followed by the impulses set up in the cerebellar cortex on activation of climbing and mossy fibers. Obviously, the time of arrival of afferent impulses will be of importance for the resulting activity. It should, furthermore, be recalled that the mossy fibers ending in a particular region are derived from many sources, carrying information of different kinds and physiological significance. Conditions are certainly extremely complex, and far more so in the living being than in the experimental animal, where one or a few elements are studied separately. Nevertheless, the new data have given rise to interesting discussions of the working machinery of the cerebellar cortex and have prompted the construction of models of its principal mode of function, in which deductions based on the observations of the properties of the various elements and their geometrical arrangement have been combined. For an account of these subjects the reader is referred to Eccles (1966a) and Eccles, Ito, and Szentágothai (1967).

In recent years adrenergic and serotonergic fibers, coming chiefly from the nucleus locus coeruleus and the raphe nuclei, respectively (see Chap. 6), have been traced to the cerebellar cortex with the histofluorescence technique of Falck and Hillarp. Many studies have been devoted to these fibers. Their relation to the classical mossy and climbing fibers is not yet clear. Much still remains to be done before their structural and functional features are fully known. A review and a number of details can be found in the monograph of Chan-Palay (1977).

It may be mentioned that the relatively sparsely and diffusely distributed noradrenergic fibers appear to establish synaptic contact with Purkinje cell dendrites and thorns (Bloom, Hoffer, and Siggins, 1971). They have been concluded to have a direct inhibitory effect on the Purkinje cells (Hoffer, Siggins, and Bloom, 1971).

In a detailed electron microscopic study of fluorescent afferents to the lateral cerebellar nucleus in the rat, Chan-Palay (1973a) found two types, called CAT_1 and CAT_2 fibers, which she assumed correspond to noradrenergic and serotonergic afferents, respectively. The synaptic formations (en passage varicosities and terminals) of the former are characterized by containing chiefly large, dense-core vesicles; in the CAT_2 terminals most vesicles are small agranular ones. For further studies of the distribution of these newly discovered cerebellar afferents and their ultrastructure see, for example, Mugnaini and Dahl (1975); Chan-Palay (1975, 1977), and Landis and Bloom (1975).

The cerebellar nuclei.   In man there are four distinct cellular masses in the white matter of each half of the cerebellum (Fig. 5-8A). Most medial is the *fastigial nucleus;* then, more laterally, follow two small cell collections, the *nucleus globosus* and the *nucleus emboliformis,* and most laterally, deep in the hemisphere, the characteristic *dentate nucleus,* appearing in sections as a wrinkled band of gray matter, with a medioanteriorly directed hilus. In the rat, cat, monkey, and most mammals it is now generally accepted that four cerebellar nuclei can likewise be distinguished: *a nucleus medialis, a nucleus interpositus anterior, a nucleus interpositus posterior, and a nucleus lateralis* (Fig. 5-8B). The latter corresponds to the human dentate nucleus, the medial nucleus to the human fastigial nucleus, while the homology of the nuclei interpositi is not quite settled. (For an extensive account of the mammalian cerebellar nuclei, see Larsell and Jansen, 1972.) There are notable differences between animal species with regard to the size and configuration of the particular nuclei, as might indeed be expected in view of their different afferent and efferent connections (see below).

The cerebellar nuclei contain cells of different sizes and types, and, within each nucleus, minor regions can be distinguished on a cytoarchitectonic basis. This probably indicates that all parts of a particular nucleus are not uniform with regard to connections and functions. So far, however, relatively little is known about this subject, partly because of the difficulties of achieving isolated destructions of, or injections in, a particular cytoarchitectonic subdivision. Furthermore, even the transitions between the main nuclei are often indistinct. Differences in the fixing of borders and consequent differences in nomenclature may be responsible for some discrepant conclusions in the literature.

As concerns the animal most used in experimental studies, the cat, Flood and Jansen (1961) concluded from a cytoarchitectonic study that part of what has often, in this animal, been considered as belonging to the lateral nucleus actually belongs to the interpositus anterior. Courville and Brodal (1966) and Brodal and Courville (1973) concluded that the border between these two

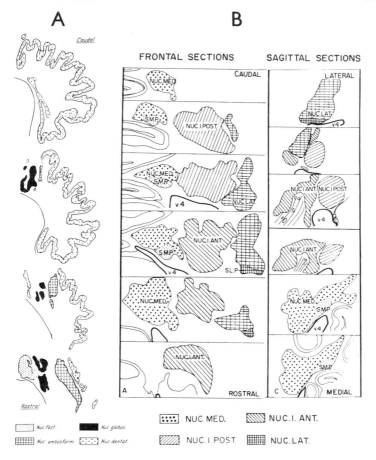

FIG. 5-8   Diagrams of the cerebellar nuclei in man (*A*) and in the cat (*B*). The drawings in *A*, selected from a series of frontal sections with unequal distances from caudal to rostral, are redrawn from Jansen and Brodal (1958). In *B*, series of equally spaced frontal and sagittal sections are reproduced from Flood and Jansen (1961). The small-celled parts of the medial and lateral cerebellar nucleus are indicated (*SMP* and *SLP*, respectively). See text.

nuclei should be placed even a little more laterally than suggested by Flood and Jansen (1961), preferring to speak of a *transition region* between the two nuclei. A final determination of this and other borders will probably require a complete and very detailed mapping of the afferent and efferent connections of the nuclei.

Food and Jansen (1961) have shown that, in the cat, two small nuclear regions consisting of only small cells can be distinguished, one in the ventral part of the fastigial nucleus and another in the ventrolateral part of the dentate nucleus, referred to as the subnucleus parvicellularis medialis and lateralis, respectively (see Fig. 5-8B). (These subdivisions may have their particular connections; see below.) A detailed study of the cerebellar nuclei in the macaque monkey has been published by Courville and Cooper (1970).

In man the dentate nucleus is enormous, both when compared with the other nuclei and when compared with its size in other species. In man the nucleus has been found to contain some 284,000 cells (Heidary and Tomasch, 1969; some earlier authors reported far higher values), while in the cat the total number of cells is found to be 46,000 (Palkovits, Mezey, Hamori, and Szentágothai, 1977).

The main *afferent fibers to the cerebellar nuclei* are the Purkinje cell axons. Other afferents have been found to come from the spinal cord, the inferior olive, the pontine nuclei, the precerebellar reticular nuclei, the red nucleus, and some other regions. Many of these fibers appear to be collaterals of axons that continue to the cerebellar cortex. *Almost all efferent fibers from the cerebellum are axons of cells in the cerebellar nuclei.* As mentioned above, these fibers leave the cerebellum to supply the vestibular nuclei, the red nucleus, and the thalamus; and in addition the reticular formations, the inferior olive, and other regions. In recent years numerous studies have been devoted to the intrinsic organization of the cerebellar nuclei and their function. Only some points will be mentioned here. Much information can be found in Chan-Palay's monograph (1977) on the dentate nucleus.

In principle, the organization of all four nuclei is said to be similar when studied in Golgi sections or electron miscroscopically (e.g., Matsushita and Iwahori, 1971a, c; Angaut and Sotelo, 1973; Sotelo and Angaut, 1973). When they enter the cerebellar nuclei, the axons of the Purkinje cells branch profusely, but each fiber supplies a roughly conical area, as described by Cajal (1909–11) in Golgi material and subsequently confirmed by others. In the cat the numerical ratio between Purkinje cells and nuclear neurons is found to be about 26 : 1, and it is calculated that on an average a single Purkinje cell axon may establish synapses (chiefly axodendritic) with 35 nuclear cells (Palkovits, Mezey, Hamori, and Szentágothai, 1977).

> When studied in Golgi preparations, both large and small neurons in the cerebellar nuclei appear to be of the "isodendritic" type, having relatively few, long, sparsely branching dendrites, many of them oriented dorsoventrally. Most cells are surrounded by a dense pericellular plexus, where fibers of different types converge. Matsushita and Iwahori (1971a, b, c, Golgi studies) and Angaut and Sotelo (1973, electron microscopically) distinguish three types of afferent fibers, while Chan-Palay (1977) describes six types in the rat. In the cat all types have been found to be in contact with the somata of large cells, while synapses on the somata of small cells are rare. Axodendritic synapses appear to occur on dendrites of all kinds of cells (for particulars of bouton types see Angaut and Sotelo, 1973; Chan-Palay, 1977; Hámori and Mezey, 1977). It appears likely that the different types of synapses found in the cerebellar nuclei are related to different kinds of afferents (see Chan-Palay, 1977).

The axons of the cells in the cerebellar nuclei give off recurrent collaterals to the nuclei. These collaterals are described as contacting small cells in the nuclei (Matsushita and Iwahori, 1971c). Whether some of the small neurons are pure internuncials (with axonal branches restricted to the nuclear territory) is not yet settled. Many of the small cells give off axons to distant regions.[5]

The somatotopic localization within the cerebellum.   Before considering the cerebellar fiber connections, it is of practical importance to mention a particular feature. Many of the fiber connections are arranged in a somatotopic pattern. The idea that a somatotopic localization might exist within the cerebellum was

---

[5] Numerous details concerning the fine structural organization of the cerebellar nuclei, not mentioned above, have been described. The extensive light and electron microscopic studies of Victoria Chan-Palay on the lateral nucleus of the rat, published in a series of papers in *Z. Anat. Entwickl.-Gesch.* in 1973 and included in her monograph (Chan-Palay, 1977) deserve special attention. Depending on the arrangement of the neurons, different zones can be distinguished within the nucleus, and ideas about the neuronal circuitry in the lateral (dentate) cerebellar nucleus in the rat have been formulated (for a brief account, see Chan-Palay, 1973b).

first set forth by the Dutch anatomist, Bolk, in 1906, on the basis of comparative anatomical investigations in mammals. Bolk concluded that there was a parallelism between the development of certain parts of the cerebellum and particular parts of the body with respect to their muscular masses. Thus, the extremities were "localized" to the large lateral parts of the cerebellum: the upper extremities to the anterior part (Bolk's crus I of his lobulus ansiformis), and the lower extremities to the posterior part of this lobule (crus II). Bolk's hypothesis was received with enthusiasm. However, when physiological experiments were performed to test it, neither stimulation nor extirpation yielded conclusive evidence in favor of the hypothesis, and anatomical studies brought forward data which were more or less incompatible with his view.

The first convincing demonstration that there is a somatotopic localization within certain parts of the cerebellum came with the works of Adrian (1943) and Snider and Stowell (1942, 1944), in which it was shown that stimulation (natural or electrical) of nerves from various parts of the body gives rise to action potentials in specific regions of the cerebellar cortex, depending on which part of the body is involved. Thus, as seen in Fig. 5-9, potentials arising in the hindlimb are found in the anterior part of the ipsilateral anterior lobe and in the caudalmost part of the paramedian lobule (bilaterally), while the forelimb is "represented" in the caudal part of the anterior lobe and rostrally in the paramedian lobule (bilaterally). Furthermore, face areas were found (see Fig. 5-9). It is to be noted that these responses were most easily obtained from skin receptors or cutaneous nerves. Finally, Snider and Stowell found that in the middle part of the vermis there is a zone (overlapping the face areas) where action potentials occur following acoustic and optic stimuli. Further studies demonstrated that the same somatotopic regions of the cerebellum could be activated by stimulation of the corresponding bodily areas in the sensorimotor region of the contralateral cerebral cortex (Adrian, 1943; Snider and Eldred, 1948, 1951, 1952; Hampson, 1949), and that stimulation of the acoustic and optic regions of the cerebral cortex gives rise to potentials in the audiovisual area of the cerebellum (Snider and Eldred, 1948; Hampson, 1949; Hampson, Harrison, and Woolsey, 1952). These findings have been confirmed and amplified by other workers, and further particulars concerning the somatotopic cerebellar localization have been discovered. For example, in the longitudinal

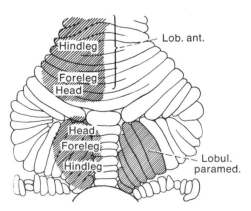

FIG. 5-9 Diagram of the surface of the cerebellar cortex of the monkey, showing (shaded) the areas in which discrete movements of hairs on the left side of the body give rise to action potentials. The somatotopic "representation" is indicated. In the middle part of the vermis and overlapping with the "head areas" is the audiovisual area. See text. From Snider (1950).

zones of the anterior lobe vermis, there is a mediolateral somatotopy with regard to spinal impulses mediated via the inferior olive (the lateral part of zone B is related to the hindlimb, the medial to the forelimb, and these are referred to as zones B_2 and B_1, respectively, see Fig. 5-21). Since a discussion of these problems presupposes a knowledge of the connections of the various parts of the cerebellum, we will return to them when the different connections are considered.

Recent studies in the rat (Shambes, Gibson, and Welker, 1978) with microelectrode recordings in the cerebellum of short-latency responses to natural stimulation of cutaneous mechanoreceptors indicate the presence of an extremely complex but, nevertheless, precise somatotopic "representation" of tactile sensibility in the crus I and II and paramedian lobule. Responses, chiefly ipsilateral, were found in columnar assemblies of granule cells, vertical to the pial surface. One particular receptive field may be represented in different small patches in the same lobule (multiple representation). In general, crus I receives projections from head and upper face, crus II from perioral and intraoral structures, and the paramedian lobule from the entire body. The largest patches are activated from functionally particularly important structures, such as the mystacial vibrissae and upper and lower lip. Corresponding precise tactile projections to some other cerebellar areas have been mapped, for example to the uvula (Joseph, Shambes, Gibson, and Welker, 1978).

Several anatomical and physiological studies from recent years (in part to be mentioned below) have provided strong indirect evidence of the presence of an extremely detailed topical precision in cerebellar organization. The findings mentioned above appear to represent the hitherto most convincing and direct demonstration of this precision. They show the importance of paying attention to details. However, the pathways along which observed tactile responses are conveyed remain to be ascertained.

Fiber connections of the cerebellum. Some general points.   The cerebellum is related by means of afferent and efferent connections to several other regions of the central nervous system. It receives afferent impulses from practically all kinds of receptors (proprioceptive, cutaneous, vestibular, visual, and others). These impulses are relayed to the cerebellum via some fairly direct and several indirect pathways. And, conversely, the cerebellum sends fibers to numerous other brain regions and accordingly is able to influence all, or almost all, functions. These connections, which are to a large extent reciprocal (see Fig. 5-1), have turned out to be extremely complicated. However, there are certain general features that should be noted. In the first place it is remarkable that the total number of efferent fibers from the cerebellum is far inferior to the number of afferents, having a ratio of approximately 1 : 40 (Heidary and Tomasch, 1969). Under the reasonable assumption that each fiber is used to its full capacity of information transfer, the marked discrepancy in the number of input and output fibers of the cerebellum provides an illustration of the efficiency of the operation of the cerebellar cortex. In order to transmit the outcome of this operation, the cerebellum requires less than 5% of the number of fibers necessary for transmitting the input signals used in the operation.

With regard to its *efferent connections,* the vast majority of efferents from the cerebellar cortex, the axons of the Purkinje cells, do not pass further than to the cerebellar nuclei. (Some of them supply the vestibular nuclei.) The cells of these nuclei give rise to fibers that, often via intercalated stations, convey the cerebellar messages to other regions (the spinal cord, the cerebral cortex, and others).

On the *afferent* side, some afferents mediating information from the vestibular apparatus pass directly to the cerebellum. Some afferent impulses coming from the body follow fairly simple routes, since some cells in the spinal cord give off fibers directly to the cerebellum (dorsal and ventral spinocerebellar tracts). Many of the afferent spinal pathways to the cerebellum, however, have additional relay stations in the brainstem. There relay stations (the inferior olive and several others) are also relays in pathways leading from other regions of the brain (the cerebral cortex, the superior colliculus, etc.) to the cerebellum. For this reason it is practical to describe each of these relay stations and the routes in which they are involved separately, and not to arrange the following account according to the various sites of origin of the afferent cerebellar fiber systems (from the spinal cord, the cerebral cortex). This is appropriate also because the relay stations are often the seat of convergence of impulses from various sources, and because many of them have two-way connections with the cerebellum. In these reciprocal connections, one link is generally quantitatively far more important than the other (see Fig. 5-3). After an account of the connections from the vestibular apparatus to the cerebellum (see also Chap. 7, section d) and a description of the "direct" spinocerebellar pathways (the dorsal, ventral, and rostral spinocerebellar tracts and the cuneocerebellar tract), the various precerebellar relay stations and their connections will be described. The most important of these relays are the pontine nuclei, the inferior olive, and the precerebellar reticular nuclei. In addition, several others have minor projections to the cerebellum (the perihypoglossal nuclei, the raphe nuclei, the nucleus locus coeruleus, the red nucleus, and others). The efferent cerebellar projections will be considered separately, but those influencing the relay stations mentioned will be referred to in passing.

The vestibulocerebellar pathways.   These pathways mediate impulses to the cerebellum from the vestibular apparatus, and consist in part of direct (primary) fibers—that is, axons of the vestibular ganglion cells. In part the fiber systems pass via relay stations.

Primary vestibular fibers were traced with the Marchi method by Dow (1936), in the cat and rat, to the flocculus, the nodulus, the ventral part of the uvula, and the fastigial nucleus. Following electrical stimulation of the statoacoustic nerve, cerebellar action potentials were recorded in the same regions (Dow, 1939). With the use of silver impregnation methods in the cat, fibers were also found to end in the ventral paraflocculus (see Fig. 5-10) and a few in the small-celled ventral part of the dentate nucleus, while termination in the fastigial nucleus could not be ascertained (Brodal and Høivik, 1964). It was established that the fibers end as mossy fibers.[6] Corresponding findings have been made in the monkey (Carpenter, Stein, and Peter, 1972). Recent studies suggest that a minor proportion of primary vestibular fibers supply the entire vermis. After HRP injections in various parts of the vermis, labeled cells were found in the vestibular gan-

[6] In a more recent study Korte and Mugnaini (1979) found the primary vestibular projection to the flocculus to be far more modest than that to the nodulus and uvula, and the termination in the paraflocculus could not be confirmed. Kotchabhakdi and Walberg (1978b) found no labeled cells in the vestibular nuclei following injections of HRP in the paraflocculus. The opinion that the "vestibulocerebellum" includes the ventral paraflocculus (Brodal and Høivik, 1964) may need revision, even though vestibular responses were recorded in it by Riva-Sanseverino and Urbano (1965).

glion (Kotchabhakdi and Walberg, 1978a). Potentials have been recorded from the entire vermis and areas of the intermediate part following electrical or natural stimulation of the vestibular apparatus (Precht, Volkind, and Blanks, 1977; see also Ferin, Gregorian, and Strata, 1971). These responses may, however, be mediated not only via primary fibers but indirectly via relay nuclei.

In addition to these primary fibers there is a *secondary vestibulocerebellar projection* via the vestibular nuclei. Certain regions of these receive primary vestibular fibers (see Chap. 7, section d). According to Dow (1936), the distribution of the secondary vestibular fibers within the cerebellum corresponds to that of the primary fibers, and they appear to end as mossy fibers (Carrea, Reissig, and Mettler, 1947). In studies with the modified Gudden method (Brodal and Torvik, 1957) it was found that the area of origin of secondary vestibulocerebellar fibers is restricted to certain regions of the medial and descending vestibular nucleus and to the group *x* (of Brodal and Pompeiano, 1957). In a systematic study with the HRP method, Kotchabhakdi and Walberg (1978b) confirmed these areas of origin and also demonstrated minor contributions from other parts of the vestibular complex.[7] They further found some labeled cells in the vestibular nuclei after HRP injections outside the flocculonodular lobe, for example in the anterior and posterior vermis (see also Precht, Volkind, and Blanks, 1977).

It appears from recent studies that the organization of the pathways mediating vestibular impulses to the cerebellum by primary and secondary vestibulocerebellar fibers is more complex that previously assumed. In addition to more direct pathways consisting of primary and secondary vestibulocerebellar fibers, there appear to be other, more or less indirect, routes, for example via nuclei in the brainstem, which receive secondary vestibular fibers and project onto the cerebellum, for example the lateral reticular nucleus, as suggested by Precht, Volkind, and Blanks (1977). There are still many unsolved questions.

Thus, while electrical single-shock stimulation of the vestibular nerve gives rise to mossy fiber potentials in the vestibulocerebellum (Precht and Llinas, 1969; Simpson, Precht, and Llinas, 1974), on natural stimulation of the horizontal canals some climbing fiber responses may occur in the nodulus and uvula (Ferin, Grigorian, and Strata, 1971; Precht, Simpson, and Llinas, 1976; Precht, Volkind, and Blanks, 1977). The mossy-fiber potentials are assumed to be due to transmission along primary (and possibly secondary) vestibulocerebellar fibers. Whether some primary (or secondary) vestibular fibers terminate as climbing fibers in mammals (as they appear to do in the frog according to physiological studies) has not been decided anatomically. However, some climbing-fiber responses are mediated via the inferior olive (Precht and Llinas, 1969). The functional significance of this dual (mossy fiber–climbing fiber) input is not yet clear. A dual input to the vestibulocerebellum has been found for proprioceptive and visual impulses as well.

The spinocerebellar tracts.   As mentioned above, this heading covers fiber connections that mediate impulses from the spinal cord more or less directly to the cerebellum. The cells of origin of these ascending tracts are acted upon by primary afferent dorsal root fibers. One often thinks only of the classical *dorsal* and *ventral*

---

[7] It is of relevance that some of the vestibular nuclear regions projecting onto the cerebellum do not appear to receive primary vestibular fibers (see Chap. 7, section d). The main afferents to the group *x*, for example, come from the spinal cord (see Brodal and Angaut, 1967; Wiksten, 1979b). The secondary vestibular fibers from this group, therefore, are apparently not of major importance in mediating vestibular impulses to the cerebellum but may be links in a spinocerebellar pathway.

spinocerebellar tracts as belonging to this group. However, it also comprises other routes: there is a recently discovered *rostral spinocerebellar tract* and a pathway via the external cuneate nucleus (which receives axons of primary sensory fibers ascending in the dorsal column and probably some from cells in the spinal cord; see Rustioni, 1977, and Chap. 2). This pathway is commonly called the *cuneocerebellar tract*. In addition, there is a recently discovered tract from the *central cervical nucleus* (see below). The indirect spinocerebellar pathways will be considered in connection with the precerebellar nuclei.

The *dorsal spinocerebellar tract* has been known for a long time to arise from the cells of Clarke's column, or the nucleus dorsalis (see Fig. 3-1), found in the base of the dorsal horn (in Rexed's lamina VII, see Fig. 2-3) at levels from Th_1 to L_2 in man (in the cat it extends to L_3–L_4). The rather thick and myelinated axons of the tract turn laterally into the ipsilateral lateral column, where they ascend dorsolaterally to the area of the lateral corticospinal tract (Fig. 2-3B), arranged in a segmental manner (Yoss, 1952). Via the restiform body, the fibers enter the cerebellum where they terminate in its "spinal regions" (see below).

The *column of Clarke* is slender rostrally and increases in volume caudally. This is related to its supply by afferent fibers, i.e., branches of dorsal root fibers. These branches may ascend ipsilaterally in the dorsal column for many segments before ending in the nucleus (Pass, 1933; Liu, 1956; Grant and Rexed, 1958; and others; Hogg, 1944, in man), and its caudal part is, therefore, charged with receiving the afferents not only from its own segments but also from the segments below (where the column is not present). The contribution of afferents from the upper thoracic levels is modest. The afferent fibers are distributed in a regular segmental somatotopic pattern (Szentágothai, 1961), the fibers from each segment occupying an area which winds spirally along the column. The primary fibers end with large boutons (Szentágothai and Albert, 1955), providing for a potentent synaptic action. Cells situated in the gray matter of the cord appear to give off axons to the column as well (Szentágothai, 1961). These findings have been confirmed and extended in an electron microscopic study by Réthely (1970).

In Marchi studies of the dorsal spinocerebellar tract, almost all authors found it to terminate chiefly ipsilaterally in the anterior lobe, covering the vermis and the region which is now referred to as the intermediate zone, and in the posterior vermis, chiefly the pyramis. This has been found also in human material (Brodal and Jansen, 1941; Marion Smith, 1961). Some authors traced fibers to the paramedian lobule. Although the presence of a somatotopic mode of termination of spinocerebellar tracts had been suggested on the basis of such material (Vachananda, 1959), it was only after using silver impregnation methods that Grant (1962a) succeeded in demonstrating that in the cat the *dorsal spinocerebellar fibers end only in the cerebellar hindlimb regions* (and part of the trunk area, which appears to be small and is difficult to outline) of the anterior lobe and the paramedian lobule, as shown in Fig. 5-10. Some fibers end in the caudal part of the pyramis (lobule VIII). Before reaching the cortex the fibers separate into a series of mediolaterally arranged bundles that terminate in longitudinally arranged zones. In several animal species Voogd (1969) distinguished six bundles in the anterior lobe, most of them supplying different strips in zone C (Fig. 5-5). The projection of dorsal spinocerebellar fibers to the hindlimb region of the anterior lobe, chiefly to its intermediate part, has been confirmed electrophysiologically (for a review, see Oscarsson, 1973). In agreement with the anatomical data (see Brodal and Grant, 1962), the responses elicited from the dorsal spinocerebellar tract are of the mossy-fiber type.

It has been known for some time that the dorsal spinocerebellar tract mediates impulses from proprioceptors (Grundfest and Campbell, 1942; and others). By recordings from single fibers in the dorsal spinocerebellar tract, Lundberg and his associates (see Lundberg and Oscarsson, 1960) have been able to distinguish *several functional components*. The tract conveys information from muscle spindles, tendon organs, pressure receptors in the hairless pads, and from touch and pressure receptors in hairy skin. Neurons that are monosynaptically excited from joint receptors are likewise found in the column of Clarke (Lindström and Takata, 1972). Some of these, however, appear to be situated outside the column (Kuno, Muñoz-Martinez, and Randic, 1973). Many of the fibers of the tract have small receptive fields. Thus some units activated from hairy skin may have receptive fields of only about 1 cm², or they may be activated from a single muscle (Lundberg and Oscarsson, 1960). The dorsal spinocerebellar tract obviously carries modality- and space-specific information from the lower trunk and the lower extremity.

It appears that some neurons of the column of Clarke are modality-specific, i.e., they are activated by one kind of receptor only, for example cutaneous, while, on other cells, information converges from different kinds of receptors (Lundberg and Oscarsson, 1960; Eccles, Oscarrson, and Willis, 1961). Further interesting details of the functioning of some of these cells have been brought forward by Jansen, Nicolaysen, and Rudjord (1966). (For a review of the functional organization of the dorsal spinocerebellar tract, see Oscarsson, 1973). The activity in the column of Clarke may be influenced from supraspinal levels (Holmqvist, Lundberg, and Oscarsson, 1960). Since descending fibers from supraspinal levels have not been found to end on the cells (see Nyberg-Hansen, 1966a) this action is presumably indirect, by way of cells in the dorsal horn.

The dorsal spinocerebellar tract in man appears to correspond rather closely to that in the cat and the monkey, as is learned from Marion Smith's (1957, 1961) careful studies of a number of human cases of cordotomy. However, a somatotopic cerebellar termination has not been found (Smith, 1961). Of practical consequence is a shifting of position of the tract as it ascends in the cord.

It follows from the course of the afferents to the column of Clarke that the dorsal spinocerebellar tract cannot be concerned in the mediation of impulses from the forelimb and only to a small extent in that from the upper part of the trunk. It is now generally agreed that *the cervical cord equivalent of the column of Clarke is the external (lateral or accessory) cuneate nucleus* (also called Monakow's nucleus), situated in the medulla lateral and a little rostral to the cuneate nucleus. It is histologically characterized by containing a fair number of large cells, in contrast to the main cuneate and gracile nuclei. The synaptic organization of the external cuneate nucleus has been studied electron microscopically by O'Neal and Westrum (1973). This nucleus appears to send most of its efferents to the cerebellum via the ipsilateral restiform body. Practically all its cells disappear following appropriate cerebellar lesions (Ferraro and Barrera, 1935a; Lafleur, De Lean, and Poirier, 1974, in the monkey; Brodal, 1941, in the cat). Its main afferents are branches of dorsal root fibers entering in the segments C_1–Th_{4-5}. The fibers from the nucleus to the cerebellum constitute the *cuneocerebellar tract*. This appears to be for the neck, forelimb, and upper trunk what the dorsal spinocerebellar tract is for the hindlimb and lower trunk. The organization of this projection is now fairly well known.

On the basis of retrograde cellular changes occurring in the nucleus following partial lesions of the cerebellum, it was found that the external cuneate nucleus projects onto the cerebellar vermis, particularly the anterior lobe (Brodal, 1941). With silver impregnation methods, Grant (1962b) succeeded in mapping more precisely the termination of the fibers in the cat. The *termination of the cuneocerebellar tract coincides with the cerebellar forelimb regions* as seen in Fig. 5-10—that is, the posterior part of the intermediate anterior lobe (chiefly lobule V of Larsell) and the anterior part of the paramedian lobule (compare Figs. 5-9 and 5-10). In addition, some fibers end in the posterior vermis. This cerebellar distribution of the cuneocerebellar fibers has been confirmed anatomically with the HRP method (Rinvik and Walberg, 1975) and physiologically with antidromic recording (except for the modest projection to the vermis; Cooke, Larson, Oscarsson, and

FIG. 5-10 Diagram of the cerebellar surface of the cat, showing the distribution of the fibers of the dorsal and ventral spinocerebellar tracts and of the fibers from the external cuneate nucleus as determined by Grant (1962a, 1962b). Note somatotopic pattern. The terminal areas of primary vestibular fibers as determined by Brodal and Høivik (1964) are also shown.

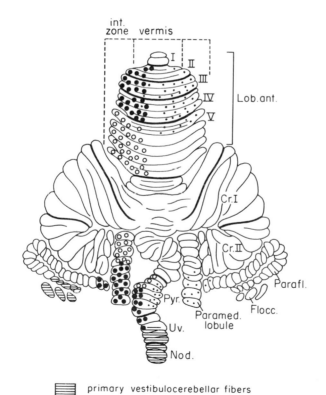

Sjölund, 1971a). Furthermore, the projection is topographically organized. In the cat, for example, cells in the caudal part of the external cuneate nucleus project to the caudal parts of lobule V and rostral parts of the paramedian lobule (Rinvik and Walberg, 1975). It appears from the studies of Cooke et al. (1971a) that most of the cells in the external cuneate nucleus have axons that divide and give off one branch to each of the projection areas, and a somatotopical pattern in the cuneo-cerebellar projection is confirmed. (Recently some cells of the external cuneate nucleus have been found to project to the thalamus; see Chap. 2.)

Likewise, the *afferents to the external cuneate nucleus,* dorsal root fibers from the cervical and upper thoracic nerves, end in a somatotopical, segmental pattern within the nucleus (Ferraro and Barrera, 1935b; Walker and Weaver, 1942; Liu, 1956; Shriver, Stein, and Carpenter, 1968). According to Liu (1956), in the cat the afferents are distributed in a rather complex spiral pattern. (A precise and careful electrophysiological mapping of the somatotopical pattern in the rat has been presented by Campbell, Parker, and Welker, 1974.) The patterns described are not entirely the same in all species, but on the whole it appears that they agree with the pattern in the cuneocerebellar projection in the same animal and indicate possibilities for a transmission of impulses with a high degree of spatial specificity via the external cuneate nucleus from the forelimb to the cerebellum.

As mentioned above, the spinal projection to the cerebellum via the cuneo-cerebellar tract functionally resembles that via the dorsal spinocerebellar tract. Both pathways have been found to transmit exteroceptive and proprioceptive information from various receptors (cutaneous, muscle afferents of group I and II, tendon organs).

It has repeatedly been found that muscle afferents of group I, and to a lesser extent of group II, activate external cuneate cells monosynaptically, as is the case for neurons in Clarke's column (for a review, see Oscarsson, 1973). However, although Campbell, Parker, and Welker (1974) found a precise musculotopic representation in the external cuneate nucleus of the rat, no evidence was obtained for tendon, joint, or cutaneous activation of its neurons. When units responding to cutaneous stimuli were encountered, they always turned out to be situated in the main cuneate nucleus. This has been noted by others as well (see Cooke, Larson, Oscarsson, and Sjölund, 1971b). These neurons are di- or polysynaptically activated by cutaneous stimuli.

It appears from the above (see also Oscarsson, 1973) that the cuneocerebellar tract consists of more than one component. According to physiological studies, some of its cells of origin are found in the main cuneate nucleus, and apparently also in the gracile nucleus (Gordon and Horrobin, 1967). Anatomical evidence of fibers to the cerebellum from the main cuneate nucleus was lacking until Rinvik and Walberg (1975) were able to demonstrate them with the HRP method. Their observations lend some support to the physiological conclusion that the pathways mediating proprioceptive impulses (from the external cuneate nucleus) terminate in the basal parts of the folia, whereas the exteroceptive pathways reach mainly the superficial parts of the folia (Cooke et al., 1971a). Both project to the same cerebellar areas.

The *ventral and rostral spinocerebellar tracts* in many ways differ from the two tracts considered above, although also they are direct pathways from the cord to the cerebellum. The *ventral spinocerebellar tract* (often called Gower's tract) was known to the classical anatomists. It arises from cells in the gray matter of the cord and ascends, segmentally arranged (Yoss, 1953; Marion Smith, 1957, in man), mainly contralaterally, in the lateral funiculus of the cord (see Fig. 2-3B), ventral to the dorsal spinocerebellar tract. Unlike this, however, the ventral tract does not enter the cerebellum in the restiform body, but ascends throughout the medulla and most of the pons. It then makes a bend over the trigeminal root fibers and turns dorsolaterally to enter the cerebellum in the superior cerebellar peduncle (see Fig. 5-32). According to Marion Smith (1957), in man, many fibers join the dorsal spinocerebellar tract.[8] In Marchi studies the fibers of the ventral spinocerebellar tract have been shown by a number of authors to end in the vermis of the anterior lobe, chiefly its anterior portions (for a review, see Jansen and Brodal, 1958). Some students have traced fibers to the posterior vermis and the paramedian lobule. Using silver impregnation methods Grant (1962a) could establish that in the cat the fibers terminate only within the physiologically identified "hindlimb regions" of the anterior lobe and paramedian lobule and in part of the posterior vermis, as shown in Fig. 5-10. On the whole the fibers are distributed somewhat more laterally than those of the dorsal spinocerebellar tract. Some contralateral terminations indicate that some fibers cross the midline in the cerebellum.

As one might expect from the site of termination, ventral spinocerebellar fibers have been conclusively demonstrated to come only from levels below the midthoracic cord. *The tract is thus concerned with transmission of impulses from the hindlimb and lower trunk only,* as confirmed physiologically (see Oscarsson, 1973). The fibers differ in calibers but are on the whole much thinner than those of the dorsal spinocerebellar tract (see Häggqvist, 1936). The cells of origin of the tract have been difficult to identify with anatomical methods. On the basis of studies of retrograde cellular changes following ablation of the cerebellum or transection of the brachium conjunctivum the tract has been found to come from lumbosacral segments of the cord. Cooper and Sherrington (1940) suggested that in the monkey the fibers arise from the so-called "spinal border cells," rather large cells situated superficially chiefly in the dorsolateral portion of the ventral horn. Similar results were obtained by Sprague (1953) in the monkey and by Ha and Liu (1968) in the cat, who in addition found retrograde changes in some cells in the neighboring regions. Recent studies with the HRP method (see Matsushita and Hosoya, 1979) seem to confirm this, as do physiological studies. With intracellular recordings in the cord, Hubbard and Oscarsson (1962) found cells fulfilling the criteria for cells of origin of the tract in what appears to be Rexed's (1952, 1954) laminae V–VII, chiefly VII, in lumbar segments 3–6 in the cat. Later physiological studies have further confirmed this (Burke, Lundberg, and Weight, 1971). Physiologically defined ventral spinocerebellar cells have even been made visible with iontophoretic intracellular injection of the fluorescent dye Procion Yellow (Jankowska and Lindström, 1970).

The main dorsal root afferents acting on the ventral spinocerebellar tract cells have been found to be flexor reflex afferents, which influence the cells polysynaptically.[9] However, some

---

[8] In the literature, mention is sometimes made of a so-called intermediate spinocerebellar tract. Different opinions have been held about this "tract" which is quantitatively modest (see Marion Smith, 1957; Jansen and Brodal, 1958, for some data).

[9] The flexor reflex afferents (often abbreviated to FRA) are defined as those myelinated afferents which, in the spinal animal, evoke the classical flexor reflex, i.e., they give excitation to ipsilateral flexor motoneurons and inhibition to ipsilateral extensor motoneurons (Eccles and Lundberg, 1959). This definition may, however, be too narrow.

cells receive monosynaptic excitation from tendon organ (Ib) or from muscle spindle (Ia) afferents. It is characteristic that the receptive fields of the flexor reflex afferents are large, and many comprise the entire ipsilateral hindlimb. It has been shown that the ventral spinocerebellar tract cells are markedly influenced from supraspinal levels—for example facilitated from the pyramidal tract and inhibited by reticulospinal paths (see Oscarsson, 1973, for a review). Vestibulospinal and rubrospinal fibers (Baldissera and Roberts, 1976; Baldissera and ten Bruggencate, 1976) have a monosynaptic excitatory effect on the neurons of the tract.

Since the ventral spinocerebellar tract thus apparently is related only to the hindlimb and posterior trunk, it is of great interest that a *forelimb equivalent* has been discovered by physiological methods (Oscarsson and Uddenberg, 1964). This tract is named the *rostral spinocerebellar tract*. Physiological studies show that it terminates in the "forelimb" regions of the cerebellum and that it appears functionally to correspond in most respects to the ventral spinocerebellar tract (see Oscarsson, 1973). The receptive fields of its cells of origin are ipsilateral and related to the forelimb. The rostral spinocerebellar tract fibers ascend uncrossed (whereas most of the ventral spinocerebellar tract fibers cross the midline in the spinal cord), and its fibers appear to pass partly in the brachium conjunctivum and partly in the restiform body. The cells of origin have so far not been anatomically identified. After injection of HRP in the cerebellum of rats, however, a particular group of labeled cells has been found by Matsushita and Hosoya (1979) in Rexed's lamina VII at levels C_4–C_8, which may give rise to the rostral spinocerebellar tract fibers (see also Petras and Cummings, 1977).

It appears that both the ventral and rostral spinocerebellar tracts are strongly activated polysynaptically be flexor reflex afferents and carry information that is highly integrated at the spinal level. The action of flexor reflex afferents on cells of the rostral tract appears to be mainly excitatory, whereas that on cells of the ventral tract is predominantly inhibitory. These two tracts obviously differ in many respects from the dorsal spinocerebellar and cuneocerebellar tracts. Whereas all four tracts appear to end as mossy fibers (see Miskolczy, 1931; Brodal and Grant, 1962; Grant, 1962a, b), there is physiological evidence that impulses mediated by single fibers in the dorsal spinocerebellar and cuneocerebellar tracts are distributed to much smaller cerebellar areas than are those from the two other tracts, indicating that the mossy fibers of the latter probably branch more amply (see Oscarsson, 1965).

A hitherto unknown spinocerebellar pathway was recently identified with the HRP method (Matsushita and Ikeda, 1975; Wiksten, 1975). It arises from the *central cervical nucleus*, a small cell group situated in Rexed's lamina VII of the first four cervical segments. Afferents to this nucleus have been described by many authors to come from the upper cervical dorsal roots, mainly C_1–C_4 (for references, see Wiksten, 1975). The efferents to the cerebellum appear to reach parts of the vermis, and particularly the most anterior lobule I (Wiksten, 1979a, b). There may well be some other, hitherto unrecognized, cell groups of the cord which send their fibers to the cerebellum (see Matsushita and Hosoya, 1979).

Pathways mediating spinal impulses to the cerebellar cortex have been considered above. The axons or collaterals of these fiber systems which end in the cerebellar nuclei will be considered in a later section. It may be appropriate to discuss briefly the *trigeminocerebellar* projection. Although it is doubtful that primary trigeminal fibers pass to the cerebellum in adult animals (see Woodburne, 1936; Larsell and Jansen, 1972), the presence of cerebellar afferents from the sensory trigeminal nucleus is established. Carpenter and Hanna (1961) followed fibers to the cerebellum from the *subnucleus interpolaris and oralis of the spinal fifth nucleus* in the cat (see Chap. 7), and Karamanlidis (1968) observed retrograde cellular changes in the same subdivisions after cerebellar lesions in the goat. Stewart and King (1963) and Roberts and Matzke (1974) found no cerebellar fibers after lesions of the subnucleus caudalis. The antidromic recordings of Darian-Smith and Phillips (1964), after stimulation of the cerebellum, do not contradict these findings. Most authors found the trigeminocerebellar fibers (passing in the restiform body) to end mainly in the ipsilateral parts of caudal lobule V and lobule VI, in good agreement with the site of the physiologically determined "face region" in the cerebellum (see Fig. 5-9).

Since the *mesencephalic trigeminal nucleus* appears to be concerned in the mediation of proprioceptive impulses from the face (see Chap. 7), it might be assumed that it gives off axons or collaterals to the cerebellum. Although such fibers have been described in normal material (Pearson, 1949a; 1949b, in man), attempts to demonstrate their site of termination experimentally have not given decisive results (Brodal and Fegersten Saugstad, 1965; Karamanlidis, 1968). Transection of the brachium conjunctivum, however, results in retrograde changes in the nucleus.

Tectocerebellar fibers have been described in early experimental studies (Ogawa and Mitomo, 1938) and in normal material, but their site of ending has not been established, and they appear to be scanty. For some data see Larsell and Jansen (1972).

The precerebellar nuclei and their connections. The term *precerebellar nuclei* refers to nuclei that give off most of their efferent fibers to the cerebellum. All of them receive afferents from many sources, among these a certain proportion from the cerebellum. They differ in many respects with regard to architecture and intrinsic structure and with regard to connections. Each of them receives its own characteristic set of afferents, and the relative number of fibers from different sources varies considerably. Further, they differ with regard to the pattern in their efferent projections to the cerebellum. It may be surmised from this that the precerebellar nuclei cannot be functionally equivalent, as appears also from physiological studies performed so far. However, it is obvious that under normal circumstances there must exist an intimate collaboration between the various precerebellar nuclei in their influence on the cerebellum.

The *pontine nuclei* (sometimes called the pontine gray) are the most massive of all precerebellar nuclei. They represent the most important relay for pathways from the cerebral cortex to the cerebellum.

The cells of the pontine nuclei surround the fibers of the cerebral peduncle with the corticospinal tract along the descending course through the pons (see Fig. 7-18). The cells are of different types and sizes, and on a cytoarchitechtonic basis several nuclei may be distinguished (Fig. 5-11). Thus the dorsolateral nucleus contains mostly large cells, while the ventral nucleus consists mainly of small cells. The nuclei are not sharply delimited from each other. Studies of the topography of the pontine nuclei have been performed in several animals, for example the opossum (King, Martin, and Biggert, 1968; Mihailoff and King, 1975), the rabbit

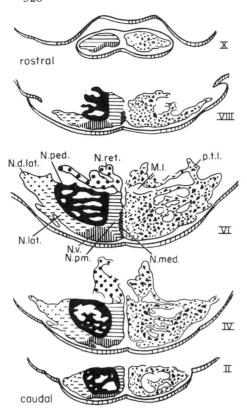

FIG. 5-11 Diagram of the pontine nuclei in the cat as seen at six equally spaced transverse levels. The dots in the right halves give an approximate picture of the distribution of cells of different sizes. In the left halves a commonly used subdivision of the pontine gray matter into different nuclei is shown. Abbreviations: *m.l.:* medial lemniscus; *N.d.lat.:* nucleus dorsolateralis; *N.lat:* nucleus lateralis; *N.med.:* nucleus medialis; *N.pm.:* nucleus paramedianus; *N.ped.:* nucleus peduncularis; *N.ret.:* nucleus reticularis tegmenti pontis; *N.v.:* nucleus ventralis; *p.t.l.:* processus tegmentosus lateralis. From Brodal and Jansen (1946).

(Brodal and Jansen, 1946; Messen and Olszewski, 1949), the cat (Brodal and Jansen, 1946), the monkey (Sunderland, 1940; Nyby and Jansen, 1951, Cooper and Fox, 1976), and in man (Olszewski and Baxter, 1954). In phylogenesis the pontine nuclei increase in proportion to the development of the cerebellar hemispheres and reach their peak of development in man, where there are some 20 million nerve cells on each side (Tomasch, 1969).

Immediately dorsal to the pontine nuclei proper is the *nucleus reticularis tegmenti pontis* of Bechterew (see Fig. 5-11), or nucleus papilioformis of Olszewski and Baxter (1954). Structurally, it resembles the reticular formation in many respects. Like the pontine nuclei it has its main efferent projection to the cerebellum. It will be considered below.

It appears from several studies, experimental as well as in human material, that all cells of the pontine nuclei send their axons to the cerebellum via the brachium pontis, mainly but not entirely the contralateral one. Most fibers are relatively thin (Szentágothai-Schimert, 1941b) and end as mossy fibers (Snider, 1936; Mettler and Lubin, 1942). Anatomical studies (see Jansen and Brodal, 1958; Larsell and Jansen, 1972, for reviews of the older literature) indicate that the cerebellar hemispheres and the paraflocculus are amply supplied with pontine fibers and that there is a somewhat smaller contribution to the vermis (in these studies distinction was not made between the vermis proper and the intermediate zone). Physiological studies (Dow, 1939; Jansen, Jr., 1957; and others) were in agreement with this. The connection with the hemispheres appears to be mainly contralateral, while the projection onto the vermis is bilateral. These findings have been confirmed in recent studies (see below), which in addition have demonstrated that some fibers pass to the flocculus. The nodulus appears to be the only part that does not receive pontine fibers; the cerebellar nuclei receive such fibers. In the following some particular features in the organization of the pontocerebellar projections will be considered.

In experimental studies of the retrograde cellular changes following lesions of the cerebellum in the cat and rabbit, Brodal and Jansen (1946) could determine that there is some degree of *localization in the pontocerebellar projection*. Voogd (1964) made largely corresponding findings after lesions of the pons.

It appears *a priori* extremely likely that there is a far greater degree of localization within the pontocerebellar projection than that described by the authors mentioned above. This suspicion was considerably strengthened when it was found that the afferents to the pons are distributed in a very specific pattern (see below). Until recently, however, adequate methods for studying details in the pontocerebellar projection were not available. The introduction of the HRP method (tracing of retrograde axonal transport of horseradish peroxidase, see Chap. 1) has dramatically changed the situation. With this method, principal points in the pontocerebellar projection have been elucidated in the cat and in the monkey, even though so far the projections to all parts of the cerebellum have not been studied, and many details remain to be clarified. It has been found that each of the cerebellar lobules studied (the paramedian lobule, Hoddevik, 1975; the middle lobules of the vermis, Hoddevik, Brodal, Kawamura, and Hashikawa, 1977; parts of the anterior lobe, P. Brodal and Walberg, 1977, in the cat; the flocculus and paraflocculus, Hoddevik, 1977; the uvula, Brodal and Hoddevik, 1978, in the rabbit) receives its pontine afferents from two or more different sites in the pons. After HRP injections in a small area of the cerebellum the labeled cells in the pons tend to be aggregated together as columns, most often rather longitudinally, even though these columns fuse at certain levels. Figure 5-12 shows an example. (The pontine projection to the cerebellar hemispheres in the rat appears to be organized in a corresponding pattern; Burne et al., 1978). Another general feature in the organization of the pontocerebellar projection in the cat has emerged: a particular column often projects to different parts of the cerebellar cortex. For example, a cell column in the paramedian pontine gray appears to project both to the paramedian lobule (Hoddevik, 1975) and to the intermediate part of the anterior lobe (P. Brodal and Walberg, 1977).[10] Finally, on the basis of the cerebellar projections, some of the columns can be concluded to show a somatotopic pattern.

[10] Whether this is due to branching of axons of pontine cells passing to different parts of a longitudinal zone, or whether the cells within a column differ in their zonal projection has so far not been determined. However, even if an injection of HRP is restricted to one of the longitudinal zones of the paramedian lobule, labeled cells are found in more than one of the columns projecting onto this lobule (Hoddevik and Walberg, 1979). This and other findings suggest that the pontine projection to the cerebellum is not organized in a zonal pattern, as, for example, the projection from the inferior olive (see below).

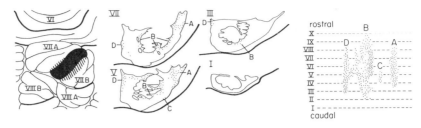

FIG. 5-12   An example of multiple projections from the pons upon a small cerebellar area. A small amount of horseradish peroxidase injected in lobule VII of the vermis resulted in labeled cells aggregated in the shape of four approximately longitudinally oriented columns in the pons (marked *A, B, C,* and *D*). To the left, the HRP-stained part of the cerebellar cortex; in the middle, the location of labeled cells (dots) as seen in four transverse sections of the pons; to the right, a diagram showing the relative longitudinal extent of the columns. Roman numerals indicate equidistant levels. From Hoddevik, Brodal, Kawamura, and Hashikawa (1977).

 After spatially restricted injections of radioactive leucine were made in the pons of the cat, fibers were traced to the cerebellum (see Kawamura and Hashikawa, 1975). Although this method does not give as precise information concerning the pontocerebellar localization as the HRP method, the results obtained so far are compatible with those referred to above. From such studies it appears that there is a projection to the cerebellar nuclei as well, possibly by collaterals from axons passing to the cortex. A pontine projection to the nucleus interpositus anterior was demonstrated with the HRP method by MacCrea, Bishop, and Kitai (1977). Their figure 6 indicates that this part of the projection is organized according to the same principles as that to the cortex. A projection to the dentate nucleus has been found with the same method (Chan-Palay, 1977).

 According to the recent HRP study of P. Brodal (1979), the pontocerebellar projection in the monkey is organized according to the same principles, even though there appear to be some differences. The columnar pattern of the projection area (i.e., the distribution of labeled cells) appears to be less clear, insofar as there is a higher degree of fusion among the columns. Furthermore, the areas projecting to some cerebellar regions appear to be rather different in the two species, and the projections to different cerebellar lobes and lobules appear to overlap less in the monkey than in the cat.[11] Figure 5-13 illustrates some main features of the pontocerebellar projection in the monkey.

 Even though many details remain to be determined, it appears from the recent studies that the *pontocerebellar projection is very precisely organized.* The pattern is such that impulses from different minor pontine regions may *converge* onto a particular cerebellar lobule. On the other hand, a *divergence* is possible as well, since a small region of the pons may send fibers to different cerebellar subdivisions. This organization must be considered in relation to the afferents to different pontine areas. As we shall see, the corticopontine projection appears to be organized according to similar principles.

 Although the pontine nuclei receive afferent inputs from many sources, the bulk comes from the cerebral cortex. Other afferents come from the superior and inferior colliculi, the cerebellum, and other sources.

[11] No convincing evidence of a zonal distribution of pontine fibers has been found in HRP studies or in studies of the degeneration in the cerebellum following pontine lesions (cf. footnote 10).

The *corticopontine projection,* as studied by classical anatomists (for a re-view, see Jansen and Brodal, 1958), was found to take its origin from all four lobes of the cerebrum. The fibers descend with those of the corticospinal tract in the internal capsule and the cerebral peduncle and terminate in the ipsilateral pons. In a March study in the monkey, Nyby and Jansen (1951) found that the various contingents have their preferential sites of termination (see also Sunderland, 1940). Thus, the fibers from the occipital and temporal lobes were found to end chiefly in the most lateral part of the pontine nuclei. The most massive corticopon-tine fiber contingent comes from the central region of the cerebral cortex, as ap-pears also from physiological studies. Recent studies have revealed functionally important details in the corticopontine projection, which has been most completely studied in the cat.

After making very small lesions in different parts of the cerebral cortex of the cat, P. Brodal identified the sites of termination of corticopontine fibers in silver impregnation studies. As seen from Fig. 5-14, in the cat the fibers from the first motor and the first somatosensory area (P. Brodal, 1968a) each end in two small pontine regions, whereas those from the second somatosensory area (P. Brodal, 1968b) have three different target areas. These terminal areas all have the shape of approximately longitudinally oriented columns. Furthermore, a *somatotopic pat-tern* is more or less clearly evident in all these columns. The terminal areas of

FIG. 5-13 A diagram showing some features of the pontocerebellar projection in the macaque monkey as determined on the basis of retrograde axonal transport of HRP. Fibers to major cerebellar subdivisions tend to arise in laminar or columnar groups of pontine cells. The general pattern resembles that of the terminal distribu-tion of corticopontine fibers shown in Fig. 5-15. From P. Brodal (1979).

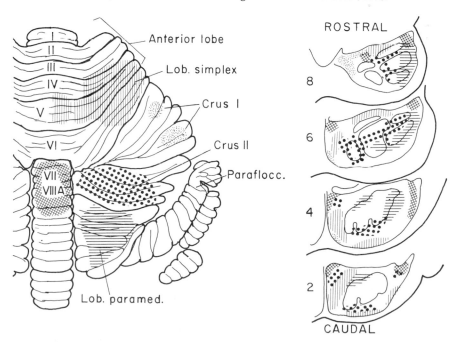

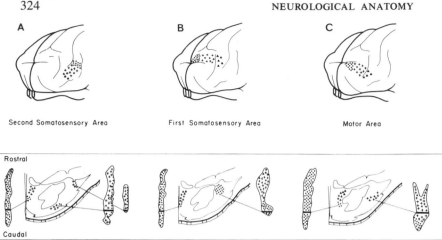

FIG. 5-14 Diagrams showing the projections from the motor and the first and second somatosensory cortical areas onto the pontine nuclei and the somatotopic pattern in these projections in the cat. The longitudinal pontine columns projected upon by the three cortical areas are represented on each side of a drawing of a transverse section through the lower part of the pons. Note different orientations of the somatotopic pattern. The first and second somatosensory areas have a medial column in common; otherwise the three areas project to different pontine areas. From P. Brodal (1968b).

pontine afferents from the proreate gyrus (P. Brodal, 1971a), the orbital gyrus (P. Brodal, 1971b), the visual cortex (P. Brodal, 1972a, b; see also Sanides, Fries, and Albus, 1978), and the auditory cortex (P. Brodal, 1972c) are arranged according to a similar principle, even though there are minor differences. In a study of efferent connections from the parietal lobe, Mizuno, Mochizuki, Akomoto, Matsushima, and Sasaki (1973) made corresponding observations. To some extent the fibers from different cortical regions end in identical pontine cellular columns. (This is the case, for example, for parts of the projections from the first and second somatosensory areas which share a terminal region in the medialmost column as seen in Fig. 5-14). Findings made in the opossum (Martin and King, 1968) indicate that the corticopontine projection in this animal is, in principle, arranged according to the same pattern. However, there are species differences, as indicated from studies in other animals, for example the rabbit (Abdel-Kader, 1968), the armadillo (Harting and Martin, 1970), and the tree shrew (Shriver and Noback, 1967). Of particular interest are the condition in the monkey.

In a recent study with silver impregnation methods, P. Brodal (1978b) has mapped the *corticopontine projection in the macaque monkey* in greater detail than was possible with the Marchi method used by Nyby and Jansen (1951). On many points the projection appears to be organized according to the same principles as in the cat. The heaviest projections come from the "motor" area 4, the first somatosensory area (Sm I, areas 3, 1, 2 of Brodmann) and area 5, and from parts of the visual areas (representing the peripheral visual field), while areas in front of and behind the central region contribute relatively little (Fig. 5-15B; see the original article for details). The fibers from a particular region of the cortex usually end within circumscribed, approximately longitudinally oriented columnar or more lamellar-shaped pontine areas. It appears that each cortical area projects to an area of the pontine nuclei which is at least partly separated, even though considerable overlap may occur between some projection areas (Fig. 5-15A). The projections from the first motor and somatosensory areas are somatotopically organized. There is a much clearer separation between the pontine projection fields of "sensory" and "motor" cortical areas in the monkey then in the cat, probably reflecting the corresponding clearer functional separation between these cortical areas in the monkey. Furthermore, whereas these projection fields have approximately the same location as in the cat, the projection fields of cortical areas 5 and 7 and the visual cortex are found chiefly lateral to the peduncle in the monkey; in the cat they also project to a medial pontine area. The "premotor" region (area 6) likewise projects differently in the cat and the monkey. It may be speculated that these differences are related to the functional role of the corresponding cortical regions in the two species. The patterns in the corticopontine projection from the motor

FIG. 5-15 Diagrams showing the main features in the corticopontine projection in the macaque monkey.
A: Termination of corticopontine fibers from various cortical regions as they appear in three transverse sections through the pons. The sites of termination of each contingent are not as clearly separated in particular columns as in the cat (Fig. 5-14), but tend to fuse to form more or less concentrically arranged lamellae. Some of these overlap, particularly in the dorsolateral part of the pons.
B: Approximate relative density of corticopontine fibers arising in various regions of the cortex. Densest projection comes from the precentral motor cortex
C: A simplified representation of the somatotopic pattern within the pontine projections in the macaque monkey from the primary motor cortex (triangles and horizontal lines) and the primary somatosensory cortex (squares and oblique lines). From P. Brodal (1978b).

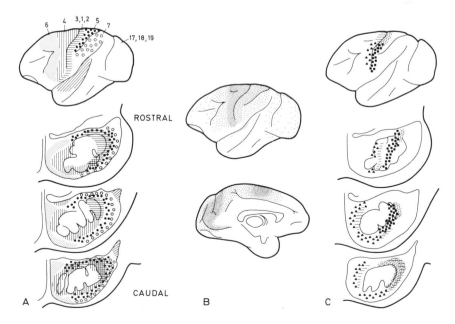

and primary sensory cortex in the squirrel monkey (Dhanarajan, Rüegg, and Wiesendanger, 1977), although in many respects similar to those in the macaque, appear to be somewhat different.

The *cells of origin of corticopontine fibers* have previously often been assumed to be the same as those giving rise to corticospinal fibers, and it has been assumed by many physiologists that the activation of pontine cells following cortical stimulation occurs by way of collaterals of corticospinal fibers. From an anatomical point of view it appears that this can be true to a limited extent only, since the number of fibers in the peduncle by far exceeds that of corticospinal fibers in the medullary pyramid (in man the ratio is about 20:1). The recent careful study of the distribution of retrogradely labeled cortical cells after injections of HRP in the cord, the pontine nuclei, the red nucleus, and the dorsal column nuclei in monkeys by Jones and Wise (1977) gives important information. Whereas all subcortical efferent projections arise from subgranular layers, particularly layer V, the cells giving rise to corticospinal fibers appear to be the largest ones and to have a laminar distribution different from that of cells projecting to other subcortical regions. Cells supplying axons to the pontine nuclei were found in Brodmann's areas 4, 3a, 3b, and 1, and in the second somatosensory area (Sm II). The authors concluded that *in general cells sending fibers to one site appear not to have collateral branches projecting to another site.* Corresponding results have been obtained by Humphrey and Rietz (1976) with regard to the corticospinal and rubrospinal projections.[12]

Fibers of the corticospinal tract give off collaterals to the pontine nuclei. However, as referred to above, most of the cortical activation of pontine cells must be assumed to occur via direct, separate, corticopontine fibers. Allen, Korn, Oshima, and Toyama (1975) concluded that about three-fourths of pontine neurons are activated by collaterals of pyramidal tract fibers, *inter alia* because they could be fired on stimulation of the medullary pyramid. However, these results can scarcely be taken as representative of the entire corticopontine projection. Rüegg, Séguin, and Wiesendanger (1977) found a value of 15%. It should be recalled that there are considerable variations among the pontine subdivisions both with regard to the density and sizes of their cells (see Fig. 5-11) and with regard to afferent and efferent connections, indicating functional differences between minor regions. It is not unlikely that cortical areas projecting to different sites in the pons may differ with regard to the proportion of direct and collateral innervation of pontine cells.

The *synaptic relationships* of the corticopontine fibers are incompletely known. In an electron microscopic study of the part of the pontine nuclei which receives fibers from the sensorimotor cortex in the cat, Holländer, Brodal, and Walberg (1969) found that after cortical lesions most of the degenerating corticopontine axons establish contact only with dendrites. (For a study of synaptic relationships in the monkey, see Cooper and Beal, 1978.) Interneurons have been claimed to exist on the basis of Golgi (Mihailoff and King, 1975; Cooper and Fox, 1976) and electron microscopic (Cooper and Beal, 1978) studies.

The corticopontine fibers exert a monosynaptic excitatory action on pontine neurons (see, for example, Allen, Korn, Oshima, and Toyama, 1975; Oka, Sasaki, Matsuda, Yasuda, and Mizuno, 1975; Allen, Oshima, and Toyama, 1977). Both fast and slow corticopontine fibers are involved (Allen, Korn, and Oshima,

---

[12] The cells of origin of corticopontine fibers from the visual areas in the cat have been concluded from physiological studies to be located in layer V (Albus and Donate-Oliver, 1977).

1975). It appears that pontine neurons giving off fast-conducting axons to the cerebellum receive a convergent input from both fast- and slow-conducting corticopontine fibers, while pontine neurons that have slow-conducting axons are innervated by slow corticopontine fibers (Allen, Korn, Oshima and Toyama, 1975).

As described above, in some pontine areas there is clear overlapping between afferents from different cerebral cortical areas in the cat (Fig. 5-14) and in the monkey (Fig. 5-15). Convergence of impulses from different cortical regions has also been observed physiologically, for example from the first and second somatosensory area in single neurons in the cat (Rüegg and Wiesendanger, 1975) and from somatosensory and motor areas in the squirrel monkey (Rüegg, Séguin, and Wiesendanger, 1977). On the other hand, there appears to be only slight convergence of impulses from the parietal and the motor cortex (Oka, Sasaki, Matsunda, Yasuda, and Mizuno, 1975).[13] This appears to be compatible with the anatomical data (see P. Brodal, 1978b; and Fig. 5-15A).

In their study in the squirrel monkey, Rüegg, Séguin, and Wiesendanger (1977) found that, of 69 antidromically identified pontine cells, 15 did not receive input from the motor and first sensorimotor cortex, 22 received input from one of these regions, and 18 from both—whereas 14 received input from multiple sites. They found that cells with a given cortical input appeared to be randomly distributed within the "transverse plane" of the pons (see also Oka et al., 1975). Precise identification of the sites of neurons recorded from would probably have made possible a correlation with their anatomical map of the pontine afferents from the cortical areas studied (Dhanarajan, Rüegg, and Wiesendanger, 1977), which supply only certain parts of the pons.

Despite such overlappings, research from the last few years shows that both links in the *corticopontocerebellar projection are rather precisely organized.* Just as each cortical efferent component has its particular distribution in the pons, each small cerebellar region receives afferents from circumscribed pontine regions. A somatotopic pattern can be identified in some of the corticopontine inputs and in some pontocerebellar projections. However, it is not yet possible to construct a map of the entire corticopontocerebellar projection either in the cat or in the monkey. Decisive correlations between sites of termination of corticopontine afferents and sites of origin of pontocerebellar fibers are hampered by the fact that often a particular afferent contingent supplies different parts of the pons, and a pontine region may project to different cerebellar lobules. The patterns are extremely complex. Nevertheless, on some points, fairly precise correlations between afferent and efferent connections are possible.

For example, in the cat the dorsolateral pontine nucleus projects to the middle vermis, particularly lobule VII (Hoddevik, Brodal, Kawamura, and Hashikawa, 1977), and is the site of termination of fibers from the acoustic cortex (P. Brodal, 1972c) and of some fibers from the visual cortex (see P. Brodal, 1972a, b). A medially situated column, projecting to the paramedian lobule (Hoddevik, 1975), coincides almost completely with one of the areas receiving cortical afferents from both the first and the second somatosensory cortex (see Fig. 5-14), and there appears to be a concordant somatotopic pattern within its afferent and its efferent connections. The same small pontine area also receives fibers from the orbital gyrus (P. Brodal, 1971b) and projects in a somatotopic pattern onto the intermediate part of the anterior lobe (P. Brodal and Walberg, 1977). For further correlations, see the original articles.

[13] According to Oka et al, impulses transmitted via the pons from the parietal association areas chiefly reach the lateral part of the cerebellum, those from the motor cortex mainly the intermediate part and the vermis.

It should be emphasized that in physiological studies of properties of pontine neurons, their synaptic relations, possible convergence of afferents, and so forth, it will in the future be essential to record precisely the sites within the pontine nuclei studied. In view of the complex and differentiated anatomical organization of this vast nuclear complex, it is extremely unlikely that findings made in one or another electrode penetration are generally valid for all pontine nuclear subdivisions.

Several studies of the physiology of pontine neurons and the influence from the cerebral cortex on them have been performed in recent years (only a few have been referred to above; see also Allen and Tsukuhara, 1974). For reasons mentioned above, and because there are still lacunae in the anatomical knowledge, correlations between anatomical and physiological observations are only possible to a limited extent.[14]

Even though the cerebral cortical input to the pontine nuclei is by far the most important (the ratio between pontine cells and corticopontine fibers being approximately 1:1.6), according to Tomasch (1969), there are several *other afferent systems to the pontine nuclei* which may influence the cerebrocerebellar transmission via the pontine nuclei, the most important possibly being afferents from the cerebellar nuclei.

The principle of multiple localized terminal areas found for the corticopontine projection appears to be valid for some, but not all, groups of pontine afferents. Those from the superior colliculus, the *tectopontine fibers,* have for a long time been known to have their terminal site in the dorsolateral pontine nucleus (see Chap. 7). There is some topical relation between the colliculus and the terminal area. The terminal area of some pontine afferents from the *inferior colliculus* coincides largely with that of fibers from the superior colliculus (Kawamura, 1975). Some fibers from the *ventral nucleus of the lateral geniculate body* appear to end in the paramedian pontine gray (Graybiel, 1974); others have been traced from the *pretectum* (Itoh, 1977). An interesting source of potine afferents is the cerebellum. *Cerebellopontine fibers,* which arise in the cerebellar nuclei, have been demonstrated by some authors (for example, Voogd, 1964). Most of them leave the cerebellum via the superior cerebellar peduncle and take off from the ascending brachium conjunctivum in its crossed descending limb (see section on the efferent connections of the cerebellar nuclei and Fig. 5-27). These cerebellopontine fibers arise in the dentate and the nucleus interpositus anterior. When studied with silver impregnation methods (Brodal, Destombes, Lacerda, and Angaut, 1972), they are found to terminate in three longitudinal cellular columns in the pons. Largely corresponding findings have been made in the opossum (Yuen, Dom, and Martin, 1974). While the nucleus interpositus posterior does not appear to contribute to the projection onto the pons (see also Angaut, 1970), some fibers from the fastigial nucleus, leaving via the hook bundle (see later), have been found in the opossum

---

[14] An interesting correlation may be mentioned. It has been suggested that the cerebellar hemispheres and the dentate nucleus are involved in the preprogramming of movements, while the intermediate part and the nuclei interpositi are concerned with updating of ongoing movements (see Allen and Tsukuhara, 1974). In view of the different role played by the motor cortex and the parietal association area in the initiation and performance of movements (see Chap. 4), the differences in the pontine projections from these two cortical territories and their further projection to the cerebellum (see also footnote 13) are of interest.

(Yuen, Dom, and Martin, 1974) and in the monkey (Batton, Jayaraman, Ruggiero, and Carpenter, 1977).[15]

> In addition to the afferents to the pons described above, there appear to be some minor con-
> tributions. For example, small contingents of *spinal afferents* have been described by some au-
> thors (Walberg and Brodal, 1953b; Kerr, 1966; Rüegg, Eldred, and Wiesendanger, 1978).
> In connection with the pontocerebellar projection, the *pontobulbar body* and the *arcuate
> nuclei* should be mentioned. Both of them appear to be related to the pontine nuclei (Essick,
> 1912). They are well developed only in man but show individual variations. It appears that they
> send their efferent fibers to the cerebellum and receive fibers from the cerebral cortex (see Larsell
> and Jansen, 1972, for an extensive account).

Even though afferents from the sources enumerated above may influence the activity of the pontine nuclei, these nuclei are first and foremost relay stations in the most important cerebrocerebellar pathway. With a few exceptions, some of which are mentioned above, the complex and detailed patterns of their afferent and efferent connections are so far not known in sufficient detail to permit reliable indications of relations between particular cerebral and cerebellar regions.[16] Functionally, the pontine nuclei appear to be particularly closely related to the execution of voluntary movements (see Chap. 4). Their most massive afferents come from cerebral cortical regions involved in this function, and the bulk of the pontine efferents pass to the cerebellar hemispheres, likewise important in this respect. Furthermore, their feedback input from the cerebellum comes preponderantly from the hemisphere and intermediate part. In consideration of the functional role of the pontine nuclei it should, however, be recalled that they may influence all parts of the cerebellum (probably with the exception of the nodulus) and thus may be concerned in the cerebellar control of many functions. Furthermore, there are other cerebrocerebellar routes, which presumably cooperate with the corticopontocerebellar pathway (see below and Fig. 5-25).

After the pontine nuclei, the inferior olive is the most important of the precerebellar relay nuclei. The axons of all its cells pass to the cerebellum and are generally considered to end as climbing fibers. During the last decade, the inferior olive has been intensively studied, physiologically as well as anatomically, and it appears at present to be one of the best known nuclei of the central nervous system. These studies have shown that the organization of the inferior olive is very precise and almost incredibly complex. Only the main points will be considered here. For more detailed accounts, the reader is referred to the original papers and to recent reviews of the physiology (Armstrong, 1974) and the anatomy (Brodal and Kawamura, 1980) of the olive and its connections.

The *inferior olive* is a folded gray mass in the medulla, dorsolateral to the pyramid (see Fig. 7-4). It consists of a principal olive and a dorsal and a medial accessory olive. The principal olive is a folded narrow cell band, in which a ven-

---

[15] According to the latter authors the fastigial fibers terminate in the dorsolateral pontine nucleus (which projects back to the vermis, particularly to lobule VII).

[16] It may be surmised that, to some extent, a particular pontine area or cell group is related preponderantly to one functional sphere. As discussed above, it is, for example, striking that in the cat the dorsolateral pontine nucleus (see Fig. 5-11), which receives some fibers from the visual and auditory cerebral cortices and its major afferent input from the superior and inferior colliculus, projects chiefly, or perhaps exclusively, to the middle vermis (particularly lobule VII), the chief "audiovisual" part of the cerebellum.

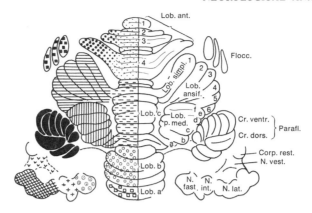

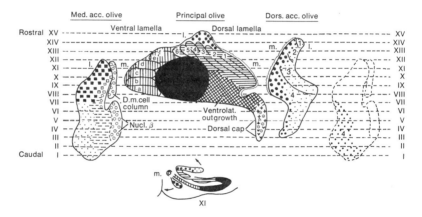

FIG. 5-16 A diagram of the olivocerebellar connections in the rabbit, on the basis of experimental investigations. Corresponding areas of the inferior olive and the cerebellum are labeled with identical symbols. The inferior olive is imagined unfolded in one plane, as indicated in the section through the inferior olive below. This diagram shows that all parts of the cerebellum and the cerebellar nuclei receive afferents from the olive, but the topographical patterns in the projection have turned out to be far more complex than shown. From Brodal (1940b).

tral and a dorsal lamella can be distinguished (see Fig. 5-16 and 17A). The axons of the cells in the olive cross the midline and enter the cerebellum in the contralateral restiform body. Henschen, in 1907, first suggested that the *olivocerebellar projection* in man presents a topical localization. This was confirmed by Holmes and Stewart (1908), who described the retrograde cell loss and cellular changes in the olive following destruction of isolated parts of the cerebellum in nine patients.

The first experimental study of the *olivocerebellar localization* (Brodal, 1940b) utilized the retrograde cellular changes in the inferior olive resulting from circumscribed lesions in various parts of the cerebellum in nearly newborn cats and rabbits (modified Gudden method, Brodal, 1940a). The olivocerebellar projection was mapped out in detail (Brodal, 1940b). Figure 5-16 represents a summary of the findings. It can be seen that the *olivocerebellar localization is astonishingly sharp, and that all parts of the cerebellar cortex and the cerebellar nuclei receive olivary fibers.* This has also been concluded from electrophysiological studies (Dow, 1939). Furthermore, each of the cerebellar lobules appears to receive its afferents from a particular region of the olivary complex.

With the development of the new anatomical methods based on retrograde and anterograde axonal transport (see Chap. 1) and with advanced neurophysiological techniques, it has been shown in recent years that the pattern in the olivocerebellar projection is far more complex than that found by Brodal (1940b). In the first place, *each cerebellar lobe or lobule receives afferents from more than one part of the olive.* This is most convincingly shown by the distribution of labeled cells after injections of horseradish peroxidase in various cerebellar lobules.[17] This has been found to be the case for all lobules of the cerebellum studied in the cat (Brodal, Walberg, and Hoddevik, 1975; Brodal, 1976; Hoddevik, Brodal and Walberg, 1976; Brodal and Walberg, 1977a; Kotchabhakdi, Walberg, and Brodal, 1978; Walberg, Kotchabhakdi, and Hoddevik, 1979), and in the rabbit (Hoddevik and Brodal, 1977). As an example, Fig. 5-17A shows that the paramedian lobule receives afferents not only from the ventral lamella of the principal olive, as found by Brodal (1940b, Fig. 5-16), but in addition from three other regions, one in the dorsal accessory, another in the medial accessory olive, and a third in the dorsal lamella. Figure 5-17B is a diagram of the entire projection to the paramedian lobule. A study of the entire olivocerebellar projection in the opossum (Linauts and Martin, 1978a) shows that it is in all essentials organized like the projection in the cat.[18] On the other hand, a *particular olivary region does not project to one lobule only, but to several.* (The olivary areas projecting onto the intermediate part of the anterior lobe and to the paramedian lobule even appear to be practically identical; see Brodal and Walberg, 1977a, b.) Correspondingly, different parts of the olive supply each of the cerebellar nuclei (see below).

The second important feature in the olivocerebellar projection is the demonstration that *each of the cerebellar longitudinal zones is supplied by fibers from a particular region of the inferior olive* (for example, the caudal part of the medial accessory olive projects to the middle zone throughout the vermis; see Fig. 5-19). This differential projection has been particularly clearly demonstrated in studies of the degeneration or labeling of fibers following lesions of, or injections of tritiated amino acids in, the inferior olive (Voogd, 1969; Courville, 1975; Groenewegen and Voogd, 1977; Groenewegen, Voogd, and Freedman, 1979; Kawamura and Hashikawa, 1979). Figure 5-18 shows the zonal pattern in the olivocerebellar projection as determined by Groenewegen, Voogd, and Freedman (1979). Zone C_2 (dots), for example, extends from the intermediate part of the anterior lobe through the ansiform and paramedian lobule (PMD) to the dorsal and ventral

---

[17] This was not recognized in the studies of retrograde cellular changes in the olive following partial cerebellar ablation (Brodal, 1940b), probably chiefly because the climbing fibers from the olive branch and give off collaterals to widely separated areas (within a particular cerebellar zone). Damage to one or a few of these collaterals is not sufficient to elicit identifiable retrograde changes or death of their cell of origin.

[18] Scattered observations on projections from the inferior olive to particular cerebellar lobules have been published, for example to the flocculus by Alley, Baker, and Simpson (1975).

paraflocculus (PFLD and PFLV) and occupies a lateralmost part of lobule IX, the uvula. All these cerebellar areas are supplied by fibers from the rostral half of the medial accessory olive.

The main points of this zonal pattern of the olivocerebellar projection agree with the results of the electrophysiological study of Armstrong, Harvey, and Schild (1974). They mapped the projection by recording antidromic responses in the olive after electrical stimulation of the cerebellar cortex. The sites recorded

FIG. 5-17 Diagram showing that several olivary regions project onto a particular cerebellar lobule.
A: The findings in a cat in which an injection of horseradish peroxidase covers about half of the left paramedian lobule. This results in the occurrence of labeled cells (dotted areas) in four subdivisions of the olive. Above to the left is shown the extent of the injection (black and hatched areas), to the right three representative transverse sections of the contralateral inferior olive. The total distribution of labeled cells can be seen in the diagram of the olive below, imagined unfolded in one plane (cf. Fig. 5-16, see also key at bottom of figure).
B: A diagram of the *total projection* of the inferior olive to the paramedian lobule. To the left, the paramedian lobule, to the right, the olive as imagined unfolded. Note topical correspondence between folia of the paramedian lobule and olivary areas (arrows). Abbreviations: *d.l.* and *v.l.*: dorsal and ventral lamella of principal olive; *l.*: lateral; *m.*: medial. Minor particular parts of the olive are indicated: *d.cap:* dorsal cap; *dm.c.col.*: dorsomedial cell column; *v.l.o.*: ventrolateral outgrowth. From Brodal, Walberg, and Hoddevik (1975).

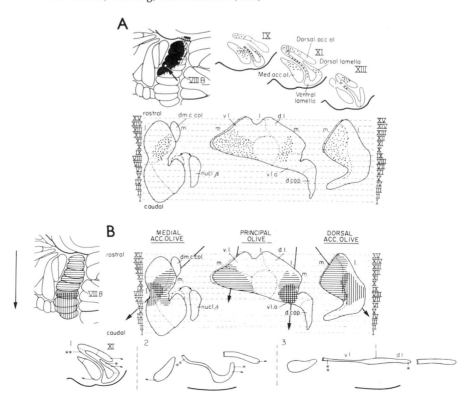

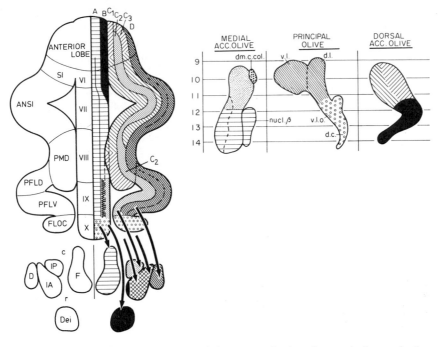

FIG. 5-18 Summarizing diagram of the pattern in the olivocerebellar projection as studied by tracing efferent fibers from the olive. As indicated by different symbols, particular parts of the inferior olivary complex project to different longitudinal zones. These extend more or less continuously through the cerebellum. The diagram further shows the general pattern in the efferent projections (arrows) from the various cerebellar cortical zones to the cerebellar nuclei and the lateral vestibular nucleus (of Deiters). The positions of the olivary subdivisions (right) have been rearranged to facilitate comparisons with the diagrams in Fig. 5-16, 17, 19, and 20. Some abbreviations: *D:* dentate nucleus; *Dei:* lateral vestibular nucleus; *F:* fastigial nucleus; *IA* and *IP:* nucleus interpositus anterior and posterior; *d.c.:* dorsal cap; *dm.c.col.:* dorsomedial cell column; *d.l.* and *v.l.:* dorsal and ventral lamellae; *v.l.o.:* ventrolateral outgrowth. From Groenewegen, Voogd, and Freedman (1979).

from in the olive were identified histologically.[19] When labeled cells occur in different subdivisions of the olive after injections in a particular lobule, this is due to the presence of more than one longitudinal zone in the lobule. After microinjections of HRP in the paramedian lobule, for example, labeling of cells is found in the particular olivary region which supplies the zone injected (Brodal and Walberg, 1977b; see also Walberg and Brodal, 1979).

At first sight it may appear difficult to combine the zonal distribution described above with the topical patterns found in HRP studies (see Fig. 5-17). However, on closer analysis, a very good agreement is found. *Within the projection from each olivary region to a particular cerebellar cortical zone, there is a*

[19] There are some discrepancies between their results and those obtained with anatomical methods with regard to details. The same is the case for the study of VanGilder and O'Leary (1970), who recorded responses in the cerebellum after electrical stimulation of the olive.

topographical pattern: that is, minor parts of such a region supply different lobules within the zone. In the diagram of Fig. 5-19, this is shown for the projections to zone A, which receives its afferents from the caudal half of the medial accessory olive. *A somatotopical pattern* is present in some zones, for example in zones C_1 and C_3. These are supplied by fibers from the rostral half of the dorsal accessory olive (see Fig. 5-18). As indicated by the hatch marks, the lateral and medial parts of this olivary region project to hindlimb and forelimb regions, respectively, in the anterior lobe and the paramedian lobule, in agreement with the classical somatotopic pattern (see Fig. 5-9). Corresponding topical relations between an olivary area and its zone of fiber terminations in the cerebellum are probably present within all zones. Even though there is some overlapping between the areas projecting to different parts within a zone, the patterns are evident.

The *olivocerebellar localization is extremely precise,* as shown by both ana-

FIG. 5-19   A diagram of the projection of the inferior olive to Voogd's zone A of the cerebellar vermis. To the left, a diagram of part of the cerebellar surface. Roman numerals refer to Larsell's lobules. All afferents to zone A, except those to the uvula (lobule IX), come from the caudal half of the medial accessory olive, shown to the right (reconstructed as seen from the ventromedial aspect, compare Figs. 5-17 and 5-18). Note the topical pattern: each lobule of the vermis (except the nodulus, lobule X) receives its afferents from a particular part of the total olivary area supplying zone A. Topical patterns can be found in the olivary projections to other cerebellar zones, arising from other parts of the olivary complex. Based on results of studies of Brodal and of Voogd and their collaborators. See text.

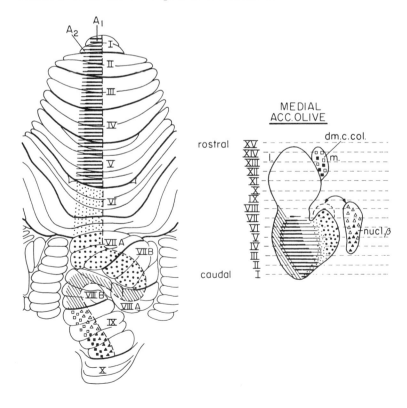

tomical and physiological methods. A particularly clear demonstration of the sharpness in the projection is seen in the fact that after microinjections of HRP (about 50 nl), a group of a few rather closely spaced labeled cells occurs in a particular part of the olive (Brodal and Walberg, 1977b; see also Beitz, 1976).

As mentioned above, the *inferior olive sends fibers* not only to the cerebellar cortex but *to the cerebellar nuclei* as well. (For some references to early authors see Matsushita and Ikeda, 1970a.) With recent methods the olivary projection to the cerebellar nuclei has been mapped in some detail, even though some points of uncertainty remain, in part because of the difficulties involved in obtaining HRP injections restricted to one of the cerebellar nuclei, or injections of tritiated amino acids to particular parts of the olive.

It appears from all studies performed so far that in the cat the olivary projections to the cerebellar nuclei, except for the fastigial nucleus, are preponderantly or purely contralateral.[20] (The conclusion of Matsushita and Ikeda (1970a) that the fibers project to the nuclei of both sides is scarcely tenable, since fibers crossing from the noninjured olive were transected in their lesions.) Each cerebellar nucleus appears to receive its afferents from a particular subdivision of the olive. It is generally held that olivary fibers to the cerebellar nuclei are collaterals of fibers to the cortex.

According to the autoradiographic and degeneration studies of Groenewegen and Voogd (1977) and Groenewegen et al. (1979), the dentate is supplied from the principal olive (their compartment D), the nucleus interpositus posterior from the rostral half of the medial accessory olive (compartment C_2), the interpositus anterior from the rostromedial part of the dorsal accessory olive (compartments C_1 and C_3), and the fastigial nucleus from the caudal part of the medial accessory olive (compartment A). On most points this is in agreement with studies of the distribution of labeled cells in the olive following injections of horseradish peroxidase into the cerebellar nuclei, for example as concerns the projection to the fastigial nucleus (Hoddevik, Brodal, and Walberg, 1976; Courville, Augustine, and Martel, 1977; Ruggiero, Batton, Jayaraman, and Carpenter, 1977) and the dentate nucleus (Beitz, 1976; Courville, Augustine, and Martel, 1977). In the olivodentate projection there is a precise topographical pattern (Beitz, 1976). However, according to HRP studies (Brodal, Walberg, and Hoddevik, 1975; Courville, Augustine, and Martel, 1977; Kitai, McCrea, Preston, and Bishop, 1977), the nuclei interpositus anterior and posterior both appear to receive afferents from the dorsal and the medial accessory olive and from largely overlapping regions.

As mentioned above, within some of the projections from a particular olivary region to a cerebellar lobule, a somatotopical pattern can be discerned. Before this problem is considered it is appropriate to review some data on the *afferent connections to the inferior olive*. In contrast to the pontine nuclei, which receive the overwhelming part of their afferent input from the cerebral cortex, the inferior olive shows a varied picture with regard to its afferents. The quantitatively most important contingents of afferents mediate spinal impulses, but in addition the olive receives afferents from the cerebral cortex, the red nucleus, the mesencephalic reticular formation, the superior colliculus, the pretectum, and, as recently shown, an important input from the cerebellar nuclei. Some of these afferent projections have been mapped in considerable detail. Each of the various afferent contingents has its particular sites of termination within the olive. Some

---

[20] In the monkey, according to Chan-Palay (1977), the projections are in part bilateral and also differ in other respects from those found in the cat.

of them supply extensive parts of the olive, while others are restricted to a small subdivision only. Furthermore, most parts of the olive receive afferents from several sources.

Of all afferent connections to the inferior olive those mediating *impulses from the limbs* are best known. Fibers arising in the spinal cord and ascending in the ventral funiculus of the cord, the *ventral or direct spino-olivary tract,* have been found by several authors to end in the medial and dorsal accessory olives only, in the cat (Brodal, Walberg, and Blackstad, 1950; Mizuno, 1966; Boesten and Voogd, 1975; Berkley and Hand, 1978a); the rabbit (Mizuno, 1966); and also in the hedgehog, pig, and opossum (for references, see Brodal and Kawamura, 1980). In addition, spinal inputs may influence the same olivary areas via a relay in the *dorsal column nuclei.* As first described by Gerebtzoff (1939) and recently confirmed by others (Ebbesson, 1968; Boesten and Voogd, 1975; Groenewegen, Boesten, and Voogd, 1975; Berkley and Hand, 1978a), the dorsal column nuclei send fibers to the contralateral inferior olive.

Furthermore, there is a somatotopic pattern within these projections, most clear-cut in the dorsal accessory olive (Fig. 5-20). The direct afferents from the *lumbar* cord end in the lateral parts of the medial and dorsal accessory olive; the *cervical* afferents terminate medially. As seen in Fig. 5-20, in the dorsal accessory olive, the pattern of termination of fibers from the dorsal column nuclei corresponds closely with that of the direct spinal afferents, insofar as fibers from the (hindlimb-related) gracile nucleus end laterally and those from the cuneate nucleus

FIG. 5-20   An example of the topical distribution of afferents from different sources within a particular olivary region. The dorsal accessory olive receives its main input from the spinal cord, via the dorsal column nuclei (*A*) and by direct spino-olivary fibers (*B*). Note the corresponding somatotopic pattern (hindlimb laterally, forelimb medially). The most medial region receives some trigeminal afferents (*A*). The sites of terminations of afferents from the cerebral cortex are shown in (*B*). Based on findings of several authors. See text.

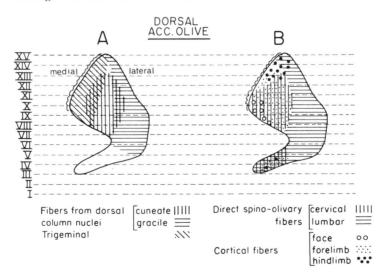

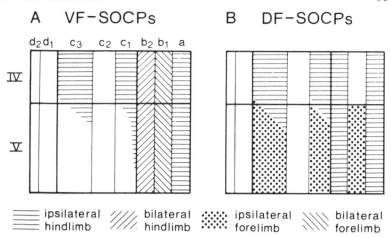

FIG. 5-21 Diagram showing the distribution of information relayed from the spinal cord to the cerebellum via the inferior olive by way of direct spino-olivary fibers (VF-SOCPs) or by way of fibers from the dorsal column nuclei (DF-SOCPs) as determined physiologically. IV and V refer to lobules IV and V of the anterior lobe. Note somatotopic patterns (longitudinal in zones a, b_1, and b_2, transverse in zones c_1 and c_3). See text. From Oscarsson and Sjölund (1977).

medially. In the medial accessory olive the correspondence is not as clear-cut.[21]

The somatotopic pattern in the distribution of spinal input to the inferior olive can be followed in the projections from the relevant parts of the olive to the cerebellum. Particularly important for an understanding of these problems have been the careful physiological studies of Oscarsson and his collaborators on the spino–olivo–cerebellar pathways. On the basis of differences in receptive fields, latency of the cerebellar responses to peripheral stimulations, and other criteria, they have distinguished several groups of pathways ascending through different funiculi of the spinal cord and mediating spinal impulses to the cerebellum via the inferior olive (for a review, see Oscarsson, 1973). All these pathways appear to terminate as climbing fibers, but they have different sites of termination in the cerebellum. For some of the ascending pathways, the presence of intercalated neurons (of unknown locations) makes correlations with anatomical data impossible.

Such correlations can, however, be made for Oscarsson's VF-SOCPs (ventral funiculus spino–olivo–cerebellar paths) and his DF-SOCPs (dorsal funiculus spino–olivo–cerebellar paths). The former correspond to the direct spino–olivo–cerebellar fibers ascending in the ventral funiculus and are studied in cats, where the cord is transected at the level of C_3, sparing only the ventral funiculus. The DF-SOCPs have a relay in the dorsal column nuclei, receiving fibers from the dorsal funiculi. Figure 5-21 is a diagram of the cerebellar regions in the anterior lobe,

[21] It is in keeping with the somatotopic pattern in the dorsal accessory olive that in its medialmost part, particularly rostrally, it receives a trigeminal input (Fig. 5-20A), as shown anatomically (Berkley and Hand, 1978a) and physiologically (Cook and Wiesendanger, 1976, in the rat). These fibers appear to come from the spinal trigeminal nucleus. Likewise, the somatotopical pattern in the projection from the cerebral cortex (Sousa-Pinto and Brodal, 1969) agrees fairly well (Fig. 5-20B).

where climbing fiber responses can be recorded when afferent inputs enter via these paths. The zones labeled a-d appears to correspond to Voogd's zones A-D. It is seen that *in the vermis (zones a, b_1, and b_2) there is a somatotopical pattern in the mediolateral direction*, zone a and zone b_2 receiving information from the hindlimb only, while b_1 is related to the forelimb.[22] In the intermediate part (zone c), the middle region (c_2), like zone d, does not receive spinal input via DF-SOCP or VF-SOCP, whereas *in zones c_1 and c_3 there is a rostrocaudal somatotopic pattern*, in agreement with the classical findings of Adrian (1943) and Snider and Stowell (1942, 1944) shown in Fig. 5-9. These findings are compatible with the patterns in the afferent and efferent projections of the various subdivisions of the inferior olive.

The projection to zone a (A) in the cerebellar cortex comes from the caudal half of the medial accessory olive (Fig. 5-19), and the main spinal input to this comes from lumbar levels of the cord, directly or via the nucleus gracilis. (The presence of some input from the cervical cord in this part of the olive may indicate that within zone a there may be a forelimb zone as well; see Brodal and Kawamura, 1980.) Zones b_1 and b_2 receive their afferents from the caudolateral part of the dorsal accessory olive (Fig. 5-18). This receives input from lumbar levels laterally and from cervical levels medially (Fig. 5-20); the lateral and medial regions appear to project to lateral and medial parts, respectively, of zone B (see Brodal and Kawamura, 1980). Finally, the afferents to zones c_1 and c_3 come from the rostral part of the dorsal accessory olive, where likewise spinal information (see Fig. 5-20) is relayed laterally from the hindlimb and medially from the forelimb. As seen from the diagram of Groenewegen, Voogd, and Freedman (1979) in Fig. 5-18, the medial part of this olivary area send its fibers to the posterior part of the anterior lobe (i.e., its forelimb region), while the lateral part of the area projects to the anterior part of the anterior lobe (i.e., its hindlimb region). This olivary area likewise projects to corresponding somatotopic parts of these zones in the paramedian lobule. The olivary areas projecting onto zones C_2 and D do not receive either direct spino-olivary fibers or fibers from the dorsal column nuclei.

It is obvious from the account above that there is remarkably good agreement between physiological and anatomical findings concerning the routes employed by information from the spinal cord via the inferior olive to certain longitudinal zones in the anterior lobe (a, b_1, b_2, c_1, and c_3). The same appears to be the case for the spinal input to the paramedian lobule via the olive. It is noteworthy that the olivary areas projecting to the intermediate part of the anterior lobe and the paramedian lobule appear to coincide. In physiological studies it has been shown by recording of antidromic responses that axons of cells in these olivary areas supply both lobules with collaterals, and that, more specifically, forelimb points in the anterior lobe correspond to forelimb points in the paramedian lobule, and so forth (Faber and Murphy, 1969; Armstrong, Harvey, and Schild, 1971, 1973a, 1973b; and others)

The DF-SOCP is found to be activated exclusively, and the VF-SOCP at least chiefly, by flexor reflex afferents. It appears from the studies of Oscarsson and his collaborators that three other groups of fibers, ascending in the cord and reaching the inferior olive via intercalated stations, may mediate spinal information to the anterior lobe, in part to zones other than those considered above (see Oscarsson, 1973).

Our knowledge of the *other, nonspinal afferents to the inferior olive* is far less complete, particularly as concerns physiological data. (Climbing-fiber re-

[22] In a recent study, zone b was further subdivided into five microzones (Andersson and Oscarsson, 1978).

sponses have been recorded in the cerebellar cortex by several authors after stimulation of brain regions that are, or are not, known to send direct fibers to the olive; see Armstrong, 1974, for a review.) In anatomical studies, olivary *afferents from the cerebral cortex* have been found to come chiefly from the motor cortex and to supply only parts of all three subdivisions of the olive (Walberg, 1956; Sousa-Pinto and Brodal, 1969; Berkley and Worden, 1978). These projections are somatotopically arranged (Sousa-Pinto and Brodal, 1969; see Fig. 5-20B). Other fibers come from area 6 (Sousa-Pinto, 1969). Following injections of HRP in the inferior olive, the origin of its afferents from the motor cortex has been studied by mapping retrogradely labeled cells in the cat (Bishop, McCrea, and Kitai, 1976), in the rat (Brown, Chan-Palay, and Palay, 1977), and in the monkey (Jones and Wise, 1977). The fibers to the olive are found to originate from medium-sized pyramidal cells in layer V.[23]

As judged from experimental silver impregnation studies, claims that projections to the olive come from the orbital gyrus and the parietal cortex appear to be doubtful (Mizuno, Sauerland, and Clemente, 1968; Mizuno, Mochizuki, Akimoto, Matsushima, and Sasaki, 1973). Cortico-olivary fibers have been described in animals other than the cat, such as the opossum, the tree shrew, the rat, and the armadillo (for references, see Brodal and Kawamura, 1980).

Physiologically, responses have been recorded in the inferior olive following electrical stimulation of the cerebral cortex, apparently coming chiefly from the motor cortex, in the cat (Armstrong and Harvey, 1966; Crill and Kennedy, 1967; Sedgwick and Williams, 1967; Crill, 1970). The points in the olive recorded from are usually not indicated.[24]

On the basis of the assumption that climbing-fiber responses recorded in the cerebellum are derived from olivocerebellar fibers, several authors have recorded such responses after stimulation of the cerebral cortex (Jansen, Jr., 1957; Provini, Redman, and Strata, 1968; Miller, Nezlina, and Oscarsson, 1969; Allen, Azzena, and Ohno, 1974; Rowe, 1977; Sasaki, Oka, Kawaguchi, Jinnai, and Yasuda, 1977; and others). Although the precise olivary relay has not been determined, it appears from these studies that stimulation of the forelimb and hindlimb areas of the sensorimotor cerebral cortex gives rise to climbing-fiber responses in the contralateral posterior (forelimb region) or anterior (hindlimb region), respectively, of the intermediate part of the anterior lobe.[25] In the vermis there appears to be a longitudinal somatotopic pattern (compare data on the transmission of spinal input to the vermis, considered above). It appears from the physiological data that in general, but not without exceptions, impulses from somatotopically corresponding regions in the periphery and in the cerebral cortex converge in the olive and activate the same olivary neurons. The reported findings can generally be correlated with anatomical knowledge of the afferent and efferent connections of

[23] According to Jones and Wise (1977), it appears that at least most of the fibers of the olive are not collaterals of corticospinal fibers.

[24] From Armstrong and Harvey's (1966) paper, however, some data can be gathered which agree with the anatomically determined sites of cortico-olivary fibers (see Sousa-Pinto and Brodal, 1969, p. 381).

[25] A corresponding cerebrocerebellar somatotopic relation is present for the mossy fiber responses as well (see, for example, Provini, Redman, and Strata, 1968), mediated via cortical fibers to the pontine nuclei or precerebellar reticular nuclei.

the inferior olive (for particulars see Brodal and Kawamura, 1980).[26] Climbing-fiber responses obtained on stimulation of the parietal cortex and recorded in approximately the same cerebellar areas as those from the sensorimotor cortex have a longer latency than the latter (Sasaki, Oka, Matsuda, Shimono, and Mizuno, 1975) and are assumed to be mediated via indirect cortico-olivary routes. This is in agreement with the absence of anatomical evidence of fibers to the olive from the parietal cortex.

Another important source of afferents to the inferior olive is the *red nucleus*. Rubro-olivary fibers have been described in the cat (Walberg, 1956; Himan and Carpenter, 1959; Edwards, 1972), in the monkey (Kuypers, Fleming, and Farin-holt, 1962; Poirier and Bouvier, 1966; Miller and Strominger, 1973; Courville and Otabe, 1974), and in the oppossum (Linauts and Martin, 1978b). The fibers have consistently been found to end in the dorsal lamella (which projects to lateral parts of the anterior lobe, the paramedian lobule, and the ansiform lobule).

Fibers from the *pretectal region* have been described in the cat (Itoh, 1977) and also in the rabbit (Mizuno, Mochizuki, Akimoto, and Matsushima, 1973) and monkey (Frankfurter, Weber, Royce, Strominger, and Harting, 1976).

This projection has been confirmed electron microscopically (Mizuno, Naka-mura, and Iwahori, 1974). It is of particular interest that most of the pretectal fibers end in a small division of the olive, the dorsal cap. This projects heavily to the flocculonodular lobe (see Fig. 5-18) and has been shown to be the olivary region involved in the transmission of visually evoked climbing-fiber responses in the flocculus, as will be considered in the following section.[27]

Fibers to the inferior olive have further been traced from the *superior colliculus,* as most recently shown in autoradiographic and HRP studies by Weber, Partlow, and Harting (1978; see also Kawamura, Brodal, and Hoddevik, 1974, their Fig. 2; Graham, 1977; Harting, 1977). These fibers appear to end chiefly in the medial part of the caudal medial accessory olive, projecting to lobule VII, belonging to the vermal visual area (see Fig. 5-19); they appear to be axons of small cells in layer IV of the superior colliculus (Weber, Partlow, and Harting, 1978). Several structures in the *mesencephalon* send fibers to the inferior olive. Their terminal areas appear to be rather extensive.

In addition to the ventral lamella, they include the rostral half of the medial accessory olive, which projects onto zone C$_2$ (see Fig. 5-18). Contributions have been traced from the mesen-cephalic reticular formation (Mabuchi and Kusama, 1970; Walberg, 1974),[28] the *nucleus of Darkschewitsch,* the *interstitial nucleus of Cajal,* and the *periaqueductal gray.* Some cells in these nuclei were labeled after injections of HRP in the inferior olive in the cat by Brown, Chan-Palay, and Palay (1977).[29] Alleged contributions from the basal ganglia have not been substantiated (see Oka and Jinnai, 1978).

[26] There is some discrepancy between anatomical and physiological findings concerning ipsi-lateral versus contralateral cerebrocerebellar relations. This may be due to the presence of indirect cor-tico-olivary connections, having a relay, for example, in the red nucleus or the reticular formation.

[27] Takeda and Maekawa (1976), after injecting HRP in the dorsal cap and its neighboring regions, found some labeled cells in the pretectal region and in the accessory oculomotor nuclei; but most cells were observed in the nucleus of the optic tract and the terminal nuclei of the optic tract.

[28] Autoradiographic studies in the opossum (Martin, Beattie, Hughes, Linauts, and Panneton, 1977) indicate that, in this species at least, the pontine and medullary reticular formation supply other olivary areas than does the mesencephalic reticular formation.

[29] These authors (extensive references) also found labeled cells in many other regions, for ex-ample the Edinger-Westphal nucleus, the perihypoglossal nuclei, parts of the vestibular nuclei, and the

A projection to the inferior olive from the *vestibular nuclei* appears now to be definitely established.

Climbing-fiber responses recorded in some parts of the cerebellum following vestibular stimulation have been assumed to be mediated via the olive. Anatomical evidence of fibers from the vestibular nuclei to the inferior olive has been scanty (see Brodal, 1974). Using methods of anterograde and retrograde transport of markers, Saint-Cyr and Courville (1979) found vestibulo-olivary fibers to arise from the medial and descending vestibular nuclei, the small groups x and z, and the nucleus prepositus hypoglossi. They end in the dorsomedial cell column and the nucleus β.

A final interesting afferent contingent to the olive comes from the contralateral *cerebellar nuclei*. Whereas a projection from the fastigial nucleus has not been found in the cat, rat, or monkey (Graybiel, Nauta, Lasek, and Nauta, 1973; Batton, Jayaraman, Ruggiero, and Carpenter, 1977; Brown, Chan-Palay, and Palay, 1977; Tolbert, Massopust, Murphy, and Young, 1977), it has been described in the opossum (Dom, King, and Martin, 1973). The main features in the projections from the nuclei interpositi and the dentate nucleus have recently been determined. Lesions or injections in one of these nuclei will, however, often involve part of another nucleus and make precise correlations difficult. The fibers have been claimed to arise from populations of small nuclear cells.

The findings are most conclusive concerning the projection from the *dentate nucleus*. Largely corresponding findings in HRP and autoradiographic studies in the cat (Beitz, 1976; Tolbert, Massopust, Murphy, and Young, 1977) indicate that the dentate projects exclusively to the principal olive in a topographical pattern. In the monkey, the pattern may be different (Chan-Palay, 1977).

The projections from the *nucleus interpositus anterior and posterior* have been concluded to end in the dorsal and medial accessory olive, respectively (Tolbert, Massopust, Murphy, and Young, 1977). It appears that the caudal half of the medial accessory olive does not receive afferents from the cerebellar nuclei (see, however, Berkley and Worden, 1978). Some findings indicate that there is some topical pattern in these projections (see Brodal and Kawamura, 1980, for a review).

The organization of the projection to the olive from the cerebellar nuclei appears to be, to a large extent but apparently not entirely, reciprocal to that of the olivonuclear projection, a feature of interest for functional considerations. This suggests that the activity in the inferior olive may be modulated by feedback loops from the cerebellar nuclei.

It is apparent from the discussion above that the inferior olive is a very complexly organized nucleus. The olivocerebellar projection (to the cortex and nuclei) shows an extremely precise pattern. Different olivary subdivisions project to particular longitudinal zones in the cerebellum, and within the projections to each of these zones there is a more or less clear topographical arrangement. Future studies may be expected to add further details to the map of Groeneweger, Voogd, and Freedman (1979) (Fig. 5-18). The afferents to the olive, however, are not arranged in a corresponding pattern. A particular afferent contingent most often ends in olivary areas that project to different cerebellar zones, and there are considerable differences between olivary areas concerning the sources of their afferents. Within the afferent fiber systems to the olive there thus appear to be ample possibilities for both *convergence and divergence* of impulses. Activity in an

---

lateral reticular nucleus; and also in the cerebral cortex, the red nucleus, and the pretectal complex, and other regions.

olivary afferent fiber system may, therefore, be imagined to influence many different parts of the cerebellum, and many sources of olivary afferents may contribute to the activation of a particular cerebellar region. These complex anatomical arrangements make physiological studies of the olive difficult, and represent a warning against simplified conceptions of relations between different cerebellar parts and the regions acting on them via the olive.

Physiological studies of the neurons of the inferior olive (for an extensive review, see Armstrong, 1974) describe them as having a tendency to slow, spontaneous rhythmic (synchronous) activity. Convergence of inputs from different afferent sources on the same neurons has repeatedly been observed. Reference has been made above to the physiological demonstration of branching of olivocerebellar climbing fibers within particular cerebellar zones. However, the synaptic "machinery" in the inferior olive is still imperfectly understood. Recent electron microscopic findings (see below), for example the demonstration of glomeruli and of gap junctions, show that it must be extremely complex. In agreement with the latter finding, electrotonic coupling between neurons has been found in the olive (Llinás, Baker, and Sotelo, 1974) and has been assumed to be instrumental in the production of synchronous firing of olivary neurons. It should be realized that the synaptic organization cannot *a priori* be assumed to be the same all over the olive. Our knowledge of this subject is still fragmentary, but there are structural differences between minor parts. (Differences between animal species likewise exist.)

> The olivary neurons, according to Golgi studies (see Scheibel and Scheibel, 1955), are of two main types. One type, provided with sparse, long unramified dendrites and considered to be more primitive, is not found in the principal olive but is abundant in the accessory olives. The afferents, when studied in Golgi sections (Scheibel and Scheibel, 1955), are morphologically of at least three types, which appear to occur with unequal frequencies within various olivary subdivisions (Scheibel, Scheibel, Walberg, and Brodal, 1956).

In electron microscopic studies most synapses observed are axodendritic; axosomatic synapses appear to be rare (Mizuno, Nakamura, and Iwahori, 1974; King, Martin, and Bowman, 1975; and others). Various types of synaptic vesicles have been found. Gap junctions occur (Sotelo, Llinás, and Baker, 1974; Rutherford and Gwyn, 1977; and others), and, recently, "synaptic clusters" or "synaptic glomeruli" have been described (Sotelo, Llinás, and Baker, 1974; King, 1976; and others). In the opossum, one of the afferent contingents to the clusters has been concluded from electron microscopic degeneration studies to be cerebelloolivary fibers (King, Andrezik, Falls, and Martin, 1976). Some terminals in the clusters have been found to contain dense-core vesicles, taken to suggest the presence of catecholaminergic transmitters.

> The distribution of catecholaminergic and indolaminergic fibers in the olive has been studied with the technique of Falck and Hillarp (see Chap. 1) in various animals (see, for example, Hoffman and Sladek, 1973; Sladek and Bowman, 1975; Wiklund, Björklund, and Sjögren, 1977). Likewise, the distribution of acetylcholinesterase has been studied (see Marani, Voogd, and Bokee, 1977). So far, correlations between these findings and the anatomical data have been of moderate success.

In contrast to all (?) other cerebellar afferent systems that terminate as mossy fibers, the inferior olive projects to the entire cerebellar cortex (and the cerebellar nuclei) via climbing fibers. The input to the cerebellum from the olive must, therefore, be assumed to play a particular and important role in the general mode of operation of all parts of the cerebellar cortex, regardless of which spheres of activity (vision, hearing, voluntary or reflex motor functions, etc.) a region is primarily concerned with. Numerous and conflicting hypotheses have been set forth on the "functional role" of the olive (see Armstrong, 1974). Some of them are based on observations of climbing-fiber responses in various situations, for example motor acts, while others are purely theoretical, but objections may be raised to all of them. They will not be discussed here.

An interesting observation, relating to the involvement of the olive in motor functions, may be mentioned. *Harmaline* (an alkaloid) has been known for a long time to induce a high-frequency tremor when injected into an experimental animal. The induced tremor is accompanied by 8–10/sec rhythmic bursts of activity of Purkinje cells due to a climbing-fiber input (see Llinás and Volkind, 1973). This synchronous rhythmic neuronal activity is concluded to arise in the inferior olive, and spreads (presumably via the cerebellum) to other regions, such as the vestibular nuclei and the spinal cord. It has been suggested that harmaline and other tremor-inducing drugs do not act directly on the olivary neurons but rather inhibit a tonic, possibly serotonergic, input to the olive (Headley, Lodge, and Duggan, 1976). In the experiments of Llinás and Volkind (1973), the rhythmic activity was particularly evident in the caudal medial accessory olive, and the Purkinje cell responses were found throughout the entire vermis. This finding is in agreement with the cerebellar projection from this part of the olive (Fig. 5-19).

It remains to consider *afferents to the cerebellum from the reticular formation*. As described in Chapter 6, three regions of the reticular formation, *the reticular tegmental pontine nucleus, the lateral reticular nucleus, and the paramedian reticular nucleus, can be considered as precerebellar nuclei*, since they give off most of their efferents to the cerebellum. So far, they have received less attention than the pons and the inferior olive. However, it is obvious that, like these, the reticular precerebellar nuclei are involved in influences exerted on the cerebellum from many sources. Each of them has its particular pattern of afferent and efferent connections. Their functional role must, therefore, be assumed to be different.

The *reticular tegmental nucleus* or reticulotegmental nucleus (nucleus reticularis tegmenti pontis of Bechterew, called the nucleus papillioformis by Olszewski and Baxter, 1954) was briefly mentioned above. It is situated dorsal to the pontine nuclei proper (see Fig. 5-11), separated from them by the medial lemniscus. The nucleus has an irregular configuration and contains cells of different sizes, many very large. Some cell strands connect it with the pontine nuclei. Its lateralmost part, often referred to as the processus tegmentosus lateralis (p.t.l. in Fig. 5-11), appears in some respects to be different from the rest of the nucleus.

The *projection to the cerebellum* from the reticular tegmental nucleus was established by early workers (see Brodal and Jansen, 1946, for some references). When the entire cerebellum is ablated in the cat, practically all cells of the nucleus disappear (Brodal and Jansen, 1946; see also Taber, Brodal, and Walberg, 1960). Brodal and Jansen (1946) found that the fibers to the cerebellum supply chiefly the vermis and probably adjacent lateral regions. Hoddevik (1978), in an HRP and degeneration study in the cat, confirmed that most efferents from the nucleus pass to vermal regions, particularly lobules VI and VII, and to the flocculus, but all parts of the cerebellar cortex appear to receive some fibers. The projection is bilateral but heavier contralaterally, and the fibers appear to end as mossy fibers. There is some degree of topical pattern in the projection. For example, the processus tegmentosus lateralis projects almost exclusively to lobule VII.[30] Fibers have been traced to the dentate nucleus (see Chan-Palay, 1977).

The *afferents to the nucleus reticularis tegmenti pontis* are fairly well known. The main inputs come from the cerebral cortex and from the cerebellar nuclei. Fibers have been traced to the nucleus from the *cerebral cortex* in the cat and monkey (Rossi and Brodal, 1956; Kuypers, 1958a, 1960; Kusama, Otani, and Kawana, 1966; Kuypers and Lawrence, 1967; Graybiel, Nauta, Lasek, and Nauta, 1973), in the chimpanzee (Kuypers, 1958b), and in man (Kuypers, 1958c). However, specific information on the sites of origin and termination of the cortical afferents is not given, except for Kusama and colleagues' statement that in the cat the fibers come from the sigmoid and coronal gyri. In an experimental study with the Nauta method this was confirmed (Brodal and Brodal, 1971), and some further details were determined.

According to Brodal and Brodal (1971), the main projection comes from the primary motor and sensory cortices (Ms I and Sm I, see Chap. 2), but there are also contributions from the second somatosensory cortex (Sm II), the proreate and orbital gyri, and from some other parts. The projection is bilateral, but mainly ipsilateral. The sites of termination of the cortical fibers are restricted to the ventral part of the reticular tegmental pontine nucleus. Within this area the various cortical afferent contingents have their preferential sites of termination, despite considerable overlapping. In the projections from the sensorimotor cortex there appears to be some degree of somatotopic pattern. For a recent study of the projection in the monkey, see P. Brodal (1980a).

The quantitatively most important afferent input to the nucleus reticularis tegmenti pontis comes from the *cerebellum*. The axons of this projection arise in cells of the cerebellar nuclei, and most of them ascend in the superior cerebellar peduncle. After having crossed the midline below the red nucleus, some fibers give off a caudally directed branch (see Fig. 5-27) or take a caudal course, forming the "crossed ventral descending limb" of the brachium conjunctivum (which supplies some other brainstem nuclei as well).

This pathway, noted by early anatomists, has been confirmed (Carpenter and Nova, 1960; Voogd, 1964; and others). Except for Voogd (1964) and Destombes (1971), authors have not given specific indications concerning the sites of termination, and there has been disagreement among authors concerning the precise origin of the fibers. After stereotactically placed lesions in the cerebellar nuclei in the cat, Brodal and Szikla (1972) found a massive projection from the dentate and the nuclei interpositi to cover a major central part of the contralateral nucleus reticularis tegmenti from rostral to caudal. Scattered fibers supply almost the entire nucleus. In a subsequent more detailed study (Brodal, Lacerda, Destombes, and Angaut, 1972), only the *lateral nucleus* (except its ventralmost part) and the *interpositus anterior* were found to give rise to this projection. In agreement with

---

[30] Mention of labeling of cells in the nucleus reticularis tegmenti following cerebellar injections of HRP has been made by others (see Hoddevik, 1978, for references). An HRP study of the projection in the monkey has recently appeared (P. Brodal, 1980b).

Angaut (1970), no afferents were found to arise in the *nucleus interpositus posterior*. Although there is considerable overlapping, there appears to be some degree of topical correlation in these projections. Chan-Palay (1977) describes a topical pattern in the projection from the dentate in the monkey.

A rather modest contribution from the *fastigial nucleus* does not pass into the brachium conjunctivum but takes a direct course (see section on the fastigial nucleus). According to Walberg, Pompeiano, Westrum, and Hauglie-Hanssen (1962) and Batton, Jayaraman, Ruggiero, and Carpenter (1977), these fibers end chiefly in the dorsal part of the nucleus. Afferents from the contralateral *vestibular nuclei* (especially the superior and lateral, Ladpli and Brodal, 1968) end chiefly in the ventral regions of the nucleus.[31] For a map of the terminations of afferents to the reticular tegmental nucleus, see Brodal and Brodal (1971).

In the cat, Kitai, Kocsis, and Kiyohara (1976) found that single-shock stimulation of a nucleus interpositus, the dentate nucleus, and the cerebral peduncle gives rise to monosynaptic excitatory potentials in the reticular tegmental nucleus, whereas the superior vestibular nucleus mediates a monosynaptic inhibition. In agreement with the anatomical data, they found an extensive convergence of inputs from the dentate and interpositus on single neurons, and also a convergence between the cerebral and cerebellar inputs. Responses to tilt (due to stimulation of the maculae of the labyrinth) have been found in the ventral part of the nucleus receiving afferents from the vestibular nuclei (Ghelarducci, Pompeiano, and Spyer, 1974).

The reticular tegmental pontine nucleus is one of several cerebrocerebellar pathways (see later and Fig. 5-25). In addition, and perhaps first and foremost, it must be considered a *link in a cerebello–reticulo–cerebellar feedback system*. However, when evaluating its role in this function, one should recall that most of its efferent fibers supply other areas of the cerebellum than those which project to the nucleus interpositus anterior and the dentate, the main sources of its afferents.

The second precerebellar reticular nucleus to be considered, the *lateral reticular nucleus* (or nucleus of the lateral funiculus), is situated just lateral to the inferior olive (Fig. 6-2). It is composed of cells of different sizes and types (Fig. 5-22B). In most mammals a ventrolateral parvicellular portion and a larger dorsomedial magnocellular portion can be distinguished (Walberg, 1952). (Often a subtrigeminal portion is also present, situated ventral to the spinal tract of the trigeminal nerve.) In recent years anatomical and physiological studies have considerably extended our knowledge of this nucleus. Its efferents pass to the cerebellum in the restiform body. Its main afferents come from the spinal cord, the cerebral cortex, and the red nucleus.

The *efferents to the cerebellum* appear to arise from all parts of the nucleus. After cerebellar ablations practically all cells of the lateral reticular nucleus show retrograde changes and finally disappear, as shown long ago (Blakeslee, Freiman, and Barrera, 1938; Brodal, 1943; see these papers for references to the early literature). More detailed information has been obtained with modern methods. According to the autoradiographic study of Künzle (1975a), most of the reticulocerebellar fibers pass in the ipsilateral restiform body, while some cross the midline in the medulla. The fibers end as mossy fibers, most densely in the an-

---

[31] There appear to be additional minor contingents of afferents to the tegmental reticular nucleus (see Brodal and Brodal, 1971, for some references). Fibers claimed to come from the red nucleus were, however, not found in the careful autoradiographic study of Edwards (1972). The same author (1975) described fibers from the mesencephalic reticular formation. Fibers have been described from the pretectum (Berman, 1977) and from the superior colliculus (Altman and Carpenter, 1961; Kawamura, Brodal, and Hoddevik, 1974; Graham, 1977; Harting, 1977).

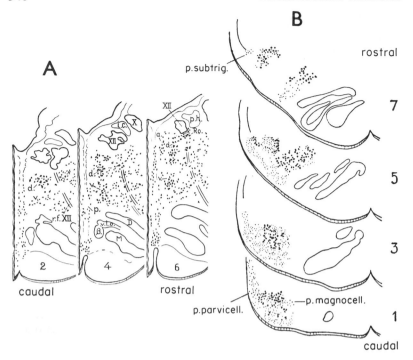

FIG. 5-22   Drawing showing main features in the topography and cytoarchitecture of the paramedian reticular nucleus (A) and the lateral reticular nucleus (B) in the cat, as seen in series of equally spaced transverse sections.
A: The three subdivisions of the paramedian reticular nucleus (*d.:* dorsal group, *v.:* ventral group; *a.:* accessory group) are not sharply circumscribed. They are all found medial to the emerging root fibers of the hypoglossal nerve. Dorsally the hypoglossal and perihypoglossal nuclei are seen (compare Fig. 5-24). From Brodal (1953).
B: The lateral reticular nucleus is situated lateral to the inferior olive. There is no distinct border between the parvicellular portion, which occupies the peripheral ventrolateral part of the nucleus, and the magnocellular portion. The small subtrigeminal portion is present most rostrally. From Walberg (1952).

terior lobe, its vermis, and its intermediate part; some end in the paramedian lobule and lobule VIII of the vermis. Matsushita and Ikeda (1976) obtained similar results in a degeneration study and also traced fibers to the nuclei fastigii and interpositi. In HRP studies P. Brodal (1975) demonstrated a topographical pattern in the projection to the anterior lobe and the paramedian lobule. The forelimb areas of both these cerebellar divisions (see Fig. 5-9) are supplied chiefly from the dorsomedial, magnocellular portion of the nucleus, the hindlimb areas from the ventrolateral, small-celled portion (Fig. 5-23A), but there is considerable overlapping. Smaller contributions to other cerebellar regions (some mentioned above) have been identified (Dietrichs and Walberg, 1979).

The ipsilateral projection to the nucleus interpositus anterior has been confirmed in an HRP study of McCrea, Bishop, and Kitai (1977), whereas Ruggiero, Batton, Jayaraman, and Carpenter (1977) did not find efferents to the fastigial nucleus. Chan-Palay (1977) describes a bilateral projection to the dentate in the rat and the monkey.

The results of recordings of antidromic responses in the lateral reticular nucleus of the cat following cerebellar cortical stimulation (Clendenin, Ekerot, Oscarsson, and Rosén, 1974a, b) are in general agreement with the anatomical findings (see P. Brodal, 1975, for particulars).[32]

The somatotopic pattern in the projection from the lateral reticular nucleus to the cerebellum fits in with what is known of the organization of its *main afferent input,* that from the *spinal cord.* Fibers ascending in the ventrolateral funiculus in company with spinothalamic fibers and fibers to other parts of the reticular formation were traced by early anatomists (for references see Brodal, 1949). In studies with silver impregnation methods (Brodal, 1949; Morin, Kennedy, and Gardner, 1966, in the cat; Mehler, Feferman, and Nauta, 1960, in the monkey) a segmental pattern of termination of the spinal afferents was found (Fig. 5-23B). This pattern has recently been mapped in detail in the cat by Künzle (1973) and confirmed in an HRP study by Corvaja, Grofová, Pompeiano, and Walberg (1977a). The projection is bilateral, with an ipsilateral preponderance. While the parvicellular (superficial) portion of the lateral reticular nucleus is supplied from lumbar levels, the magnocellular portion receives afferents chiefly from the cervical cord, the transition region from the thoracic cord.[33] Recordings of potentials mediated by spinal afferents (the bilateral ventral flexor reflex tract) by Clendenin, Ekerot, Oscarsson, and Rosén (1974b) are in general agreement, but show a more extensive overlap of spinal hindlimb and forelimb inputs than is apparent from the anatomical studies. As suggested by Künzle (1973) and Rosén and Scheid (1973b), this may be due to convergence of afferents from several spinal segments on the neurons giving off the ascending fibers.

The correspondence between the somatotopic pattern in the termination of spinal afferents in the lateral reticular nucleus and the pattern in its projection to the cerebellum explains that stimuli from forelimbs or hindlimbs, relayed in the lateral reticular nucleus, are recorded in somatotopically corresponding areas of the anterior lobe (Clendenin, Ekerot, Oscarsson, and Rosén, 1974b). Recent studies indicate that the role of the lateral reticular nucleus in the transmission of spinal impulses is more complex than appears from the discussion above.

Physiological studies have shown that the main input, described above, occurs via a bilateral ventral flexor reflex tract (bVFRT). The cells of origin of this pathway are activated polysynaptically by flexor reflex afferents from receptive fields that often include all four limbs (see Rosén and Scheid, 1973b). The ascending axons excite or inhibit cells of the lateral reticular nucleus monosynaptically. However, some cells in the nucleus are activated only from the *ipsilateral forelimb* (see Clendenin, Ekerot, and Oscarsson, 1974). They are concluded to receive their spinal input from a particular ascending component, called the iF (ipsilateral forelimb) tract. Cutaneous and high-threshold muscle afferents contribute to their excitation (see also Rosén and Scheid,

[32] Clendenin, Ekerot, Oscarsson, and Rosén (1974a) found some mediolateral differences in the cerebellum between the projections from the large-celled and small-celled portion of the nucleus; the latter projects preponderantly to the vermis (in agreement with Brodal, 1943). This may be related to the presence of a mediolateral somatotopic pattern in the vermis of the anterior lobe, as described above for the projection from the olive (see Fig. 5-21).

[33] According to the electron microscopic study of Mizuno and Nakamura (1973), most terminals appear to contact thin and medium-sized dendrites.

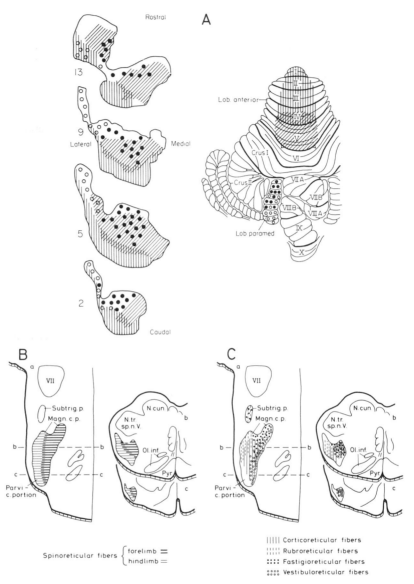

Rostral

A

Lob. anterior

Lateral   Medial

Crus I

Crus II

Lob. paramed

Caudal

B

a

VII

Subtrig. p.

Magn. c. p.

N. cun.

b

N. tr.
sp. n. V.

b

Ol. inf.

Pyr

c

Parvi-
c. portion

C

a

VII

Subtrig. p.

Magn. c. p.

N. cun.

b

N. tr.
sp. n. V.

b

Ol. inf.

Pyr

c

Parvi-
c. portion

Spinoreticular fibers { forelimb =
                        hindlimb =

‖‖‖ Corticoreticular fibers
┊┊┊┊ Rubroreticular fibers
∷∷∷ Fastigioreticular fibers
°°°° Vestibuloreticular fibers

FIG. 5-23   Simplified diagrams of the main connections of the lateral reticular nucleus in the cat.

Above (A) is shown the projection to the anterior lobe and the paramedian lobule as seen in four transverse sections of the nucleus. The projection is somatotopically organized. Cerebellar forelimb regions (oblique hatchings and black dots) receive their afferents mainly from the magnocellular portion, the hindlimb regions (vertical hatchings and open rings) from the parvicellular portion (cf. Fig. 5-22B). The subtrigeminal portion is not shown (cf. text). From P. Brodal (1975).

Below to the left (B) is a diagram of the sites of termination of spinal afferents to the lateral reticular nucleus shown in a horizontal (a) and two transverse (b and c) sections. Note somatotopic pattern. The parvicellular portion receives input from the hindlimb only.

Below to the right (C) is a similarly constructed diagram of the distribution of some nonspinal afferents. The parvicellular portion receives relatively few such afferents.

Diagrams below (from A. Brodal, 1972a) are based on findings of several authors (see text).

1973c). The iF tract appears to terminate in the dorsolateral part of the lateral reticular nucleus (i.e., its forelimb region). Supporting anatomical evidence has been brought foreward by Corvaya, Grofová, Pompeiano, and Walberg (1977a), who studied the origin of spinal neurons projecting to the lateral reticular nucleus after HRP injections in the nucleus. They confirmed that spinoreticular fibers arise bilaterally from all levels of the cord, most of them from cells in Rexed's lamina VIII and to some extent lamina IX. In addition, some rostrally projecting cells, located more dorsally in the gray matter and present only in the cervical segments, give rise to an ipsilateral tract to the lateral reticular nucleus.

In addition to its main afferent input from the spinal cord, the lateral reticular nucleus receives afferents from other sources (Fig. 5-23C). Thus the *cerebral cortex* was found by Kuypers (1958a, b, c) to give off fibers to the nucleus in the cat, the monkey, the chimpanzee, and in man; Walberg (1958), P. Brodal, Maršala, and A. Brodal (1967), and Künzle and Wiesendanger (1974) found the same situation in the cat. According to P. Brodal et al., most of the fibers come from the primary sensorimotor cortex (chiefly from the anterior sigmoid gyrus). Small contributions arise in other regions. The projection is preponderantly contralateral and appears to supply only the rostrodorsal part of the magnocellular portion (in rather good agreement with Kuypers, 1958a, b). A somatotopic pattern could not be ascertained.

Fibers from the *red nucleus* likewise end in the contralateral lateral reticular nucleus, as found with silver impregnation, autoradiographic, and electron microscopic methods (Walberg, 1958; Hinman and Carpenter, 1959; Courville, 1966a; Edwards, 1972; Mizuno, Mochizuki, Akimoto, Matsushima, and Nakamura, 1973). All authors agree that practically all fibers terminate rostrally in the dorsolateral regions of the magnocellular portion, that is, approximately the same region that receives afferents from the cerebral cortex.[34] Such fibers appear to establish chiefly axodendritic synapses (Mizuno et al., 1973). (Some fibers have been found to end in the subtrigeminal portion.) The rubroreticular projection appears to be somatotopically organized (Corvaja et al., 1977a), in agreement with the pattern in the corticorubral (Rinvik and Walberg, 1963) and the rubrospinal (Pompeiano and Brodal, 1957) projections.

Other afferents to the lateral reticular nucleus come from the *fastigial nucleus* (Cohen, Chambers, and Sprague, 1958; Walberg and Pompeiano, 1960; Batton, Jayaraman, Ruggiero, and Carpenter, 1977; Corvaja, Grofová, Pompeiano, and Walberg, 1977a) and end mainly in rostral regions of the magnocellular portion. Some fibers from the *ipsilateral lateral vestibular nucleus* appear to end chiefly medially and caudally (Ladpli and Brodal, 1968). Finally, the rostromedial part of the nucleus receives some afferents from the *superior colliculus* (Kawamura, Brodal, and Hoddevik, 1974). Afferents from other regions are probably present as well.[35]

In agreement with the anatomical data, neurons of the lateral reticular nucleus have been found to be monosynaptically activated from the cerebral cortex, the fastigial nucleus, and the red nucleus; they also respond to spinal inputs (see, for example, Bruckmoser, Hepp-Reymond, and Wiesendanger, 1970; Rosén and

---

[34] It is of interest that the cortical fibers to the red nucleus (Rinvik and Walberg, 1963) appear to come from the same cerebral region as do those to the lateral reticular nucleus. Responses recorded in the latter nucleus on cortical stimulation may, therefore, be mediated via at least two different routes (see Allen and Tsukuhara, 1974).

[35] Afferent input from fibers ascending in the dorsal funiculus has been described (see Clendenin, Ekerot, and Oscarsson, 1975), and there may be other spinal inputs via intercalated cell groups.

Scheid, 1973a; Kitai, de France, Hatada, and Kennedy, 1974). In agreement with anatomical data are also the findings of convergence of impulses from the spinal cord, the cerebral cortex, the red nucleus, and the fastigial nucleus (Bruckmoser, Hepp-Reymond, and Wiesendanger, 1970; Rosén and Scheid, 1973a; Kitai, de France, Hatada, and Kennedy, 1974; and others) and the responses to stimulation of the vestibular maculae (see Gherladucci, Pompeiano, and Spyer, 1974). It is to be expected from the anatomical distribution of afferents that most of the neurons found to show convergence of spinal input with other afferents would be located in the magnocellular portion.

It thus appears from the available data that the main spinal input to the lateral reticular nucleus supplies both its magno- and its parvicellular portion bilaterally and is somatotopically organized, as is its further projection mainly to "spinal" areas of the cerebellum (see Fig. 5-23). In addition, there is a particular ipsilateral forelimb projection to the magnocellular portion. It is noteworthy that most other afferent contingents appear to supply chiefly the rostrodorsal "forelimb" region of the nucleus (belonging to its magnocellular portion).[36] The afferent input to the nucleus may indicate an important difference between its dorsal (magnocellular) and ventral (parvicellular) portion and suggests that the latter is first and foremost related to the transmission of spinal information from the lower cord. The magnocellular portion, related to the forelimb, appears to be more complex and capable of a considerable degree of integration of input from the forelimb with impulses from several other sources. From physiological studies (see Oscarsson, 1973, for a review), the spinal afferents to the lateral reticular nucleus have been concluded to "carry information about activities in lower motor centers, rather than about external events" (Clendenin, Ekerot, Oscarsson, and Rosén, 1974c, p. 143). This information transfer is probably only one aspect of the functions of the nucleus.

The neurons giving off the spinoreticular axons are influenced from several supraspinal sources. Thus they are monosynaptically excited by the vestibulospinal tract (see Oscarsson, 1973). Such influences will thus be important for the activity in the spinal afferent projections to the lateral reticular nucleus. The pattern in the afferent and efferent connections of the lateral reticular nucleus with the cerebellum and their input from the vestibular nuclei (particularly from the "static" labyrinth) and other sources shows that the role of the spinocerebellar route via the lateral reticular nucleus must be very complex. It is presumably particularly related to motor functions. An example may be mentioned. After destruction of the lateral reticular nucleus in the cat, Corvaja, Grofová, Pompeiano, and Walberg (1977b) observed postural asymmetry, with ipsilateral hypertonia and contralateral hypotonia of the limb extensor muscles. In addition, there is a transient loss of the proprioceptive placing reaction in the ipsilateral limbs and a permanent defective tactile placing reflex. From an analysis of the effects of various surgical interventions, the authors conclude that the nucleus may functionally consist of two parts, one related to the bilateral somatotopical pathway, which supplies the entire nucleus, and another related to the ipsilateral forelimb afferents. The latter is assumed to be involved in the placing reaction, the former in the postural asymmetry. Possible routes involved in these mechanisms are discussed, indicating the magnitude of the complexities of the problems.

In addition to its role in mediating spinal impulses to the cerebellum, the lateral reticular nucleus must be considered as a link in a *cerebrocerebellar pathway*

---

[36] According to Rosén and Scheid (1973a), only stimulation of the cerebral *forelimb* area produces early excitatory responses in lateral reticular nucleus neurons. The sites of termination of cortical projections in the "forelimb" part of the nucleus are in agreement with this. Whether the afferents from the cortex come only or chiefly from its "forelimb areas" is, however, not clear.

(see Fig. 5-25). The cortical influence may occur not only directly via corticoreticular fibers, but also via a corticorubroreticular route. As considered above, only the magnocellular portion appears to serve as a cerebrocerebellar relay station. It influences mainly the intermediate part and the vermis of the anterior lobe and the paramedian lobule (probably in a somatotopic pattern). These regions are influenced from the cerebral cortex via the pontine nuclei as well. The relative importance of the two routes is unknown. Fibers from both relay stations end as mossy fibers, but little is to be learned about their cooperation from physiological studies (see, for example, Allen, Azzena, and Ohno, 1972).

The *subtrigeminal part* of the lateral reticular nucleus does not appear to receive afferents from the spinal cord. In the HRP studies of P. Brodal (1975) this part was found to project to the paramedian lobule and the anterior lobe, chiefly to its forelimb area, and to lobules VI and VII of the vermis. This suggests that the subtrigeminal part may be particularly related to transmission of impulses from the face. This assumption is supported by the findings of Darian-Smith and Phillips (1964). Cells in the dorsal part of the nucleus, apparently including the subtrigeminal part, were activated with short latencies from the face and could be antidromically activated from lobules V and VI ("face" area of the vermis). Clendenin, Ekerot, and Oscarsson (1975), who did not explore the subtrigeminal part, recorded responses to trigeminal stimulation throughout the main parts of the lateral reticular nucleus. In agreement with the pattern of the efferent projections of the nucleus (see above), trigeminal responses were recorded in different parts of the cerebellum, particularly in lobule V in the intermediate part of the anterior lobe. Künzle (1975a) and Matsushita and Ikeda (1976) found few or no fibers to lobules VI and VII probably because their lesions did not involve the subtrigeminal part.

The third precerebellar nucleus, the *paramedian reticular nucleus,* is a small but characteristic collection of neurons in the medial part of the reticular formation in the medulla, at the level of, and slightly above, the hypoglossal nucleus (Fig. 5-22A). It may be subdivided into three groups (dorsal, ventral, and accessory; see Brodal, 1953). The main efferent projection from this small nucleus goes to the cerebellum. Its afferent fibers are derived from various rostral levels and the spinal cord, but its connections are not completely known so far.

After decerebellation in the cat practically all cells in the paramedian reticular nucleus disappear retrogradely (Brodal, 1953). From studies of this kind it appears that the projection to the cerebellum goes chiefly to the anterior and posterior parts of the vermis (Brodal and Torvik, 1954). With the HRP method a wider distribution has been found, and it appears that the three subdivisions of the nucleus differ somewhat with regard to their cerebellar projection (Somana and Walberg, 1978).

Afferents have been traced with silver impregnation methods from the *spinal cord* in the cat (Brodal and Gogstad, 1957) and in the monkey (Mehler, Feferman, and Nauta, 1960) and also from the dorsal column nuclei (Brodal and Gogstad, 1957). There may be some from the *vestibular nuclei* (Ladpli and Brodal, 1968). Other fibers come from the *fastigial nucleus.* (Thomas, Kaufman, Sprague, and Chambers, 1956; Walberg, Pompeiano, Westrum, and Hauglie-Hanssen, 1962; Batton, Jayaraman, Ruggiero, and Carpenter, 1977). *Cerebral cortical afferents* were described by Brodal and Gogstad (1957) and Sousa-Pinto (1970a). The latter author found them to come chiefly from the first somatosensory area, particularly its "face region," and some from area 6.

The three small groups of the paramedian reticular nucleus are probably not quite identical functionally, since they appear to differ with regard to their projection to the cerebellum and since their afferent supply, for example from the cerebral cortex (see Sousa-Pinto, 1970a), is not the same. Particularly the dorsal group appears to be closely related to the perihypoglossal nuclei (see below). Further studies will possibly show that the various subdivisions of the paramedian reticular nucleus play particular functional roles.

Electrophysiological studies indicate that only a few cells of the paramedian reticular nucleus are influenced on peripheral nerve stimulation (Avanzino, Bradley, and Wolstencroft, 1975). Some cells are influenced from the maculae of the labyrinth (Gherladucci, Pompeiano, and Spyer, 1974). There is probably no direct input from the sinus nerve (Spyer and Wolstencroft, 1971). Most cells in this nucleus are excited by acetylcholine and serotonin and inhibited by noradrenaline (Avanzino et al., 1975). Duggan and Game (1975) report that about half of the cells could be antidromically activated from the cerebellum.

A further source of cerebellar afferents deserves particular consideration, namely some minor cell groups surrounding the hypoglossal nucleus. These cell groups, collectively referred to as the *perihypoglossal nuclei*, have attracted much interest in recent years (for a description, a consideration of nomenclature, and a review of earlier literature, see Brodal, 1952). They consist of three cell groups that are not clearly separated at all levels (Fig. 5-24).

The most caudally situated, the *nucleus intercalatus* of Staderini, contains mainly small cells. The *nucleus of Roller*, ventrolateral to the hypoglossal nucleus, contains many large cells. The largest of the nuclei is the *nucleus prepositus hypoglossi*, which extends rostral to the rostral end of the hypoglossal nucleus. It is more or less continuous laterally with the medial vestibular nucleus and the reticular formation and contains small, medium-sized, and large cells. (Some authors refer to its rostral part as the nucleus supragenualis nervi facialis). It should be noted that the rostral end of the nucleus prepositus fuses with the part of the reticular formation referred to as the PPRF, of importance for eye movements (see Chap 7). It appears that these perihypoglossal nuclei become increasingly differentiated in higher mammals, reaching their peak in man.

FIG. 5-24   A series of drawings of transverse sections of the dorsomedial part of the medulla in the cat, showing the position and the main cytoarchitectonic features of the perihypoglossal nuclei. Abbreviations: *a.* and *d.*: accessory and dorsal subdivisions of paramedian reticular nucleus (cf. Fig. 5-22A); *g.n. VII*: genu of facial nerve; *i.c.*: nucleus intercalatus; *p.h.*: nucleus praepositus hypoglossi; *Ro.*: nucleus of Roller; *VI, X,* and *XII*: cranial nerve nuclei. Rearranged from Brodal (1952).

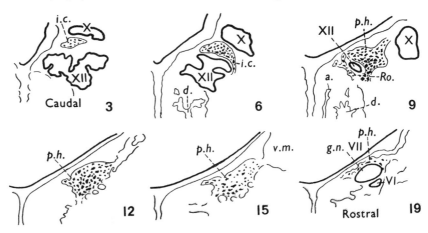

Although some early authors (for references see Brodal, 1952) assumed that the perihypoglossal nuclei send their efferents into the hypoglossal nerve, this has not been confirmed. Studies of the retrograde changes occurring after cerebellar lesions in the cat (Brodal, 1952) showed that a considerable proportion of the cells of these nuclei, both small and large, send their axons to the cerebellum. Further studies (Torvik and Brodal, 1954) indicated that most of the fibers terminate in the anterior lobe, the posterior vermis including the uvula, and the fastigial nucleus. A wider distribution could be shown with the HRP method. Thus Alley, Baker, and Simpson (1975) determined that, in addition, the flocculus and nodulus receive fibers from all three perihypoglossal nuclei, and Kotchabhakdi, Hoddevik, and Walberg (1978), in a systematic study, further found the projections to cover the entire vermis, including the visual area, the paraflocculus, and the nuclei interpositi. However, no afferents were found to the ansoparamedian lobule or the dentate nucleus. The projections are bilateral with an ipsilateral preponderance. Further, there appears to be a topographical pattern in the cerebellar projection from the perihypoglossal nuclei.

The projections to the fastigial nucleus and the nucleus interpositus anterior have been confirmed with the HRP method in the cat (Ruggiero, Batton, Jayaraman, and Carpenter, 1977; McCrea, Bishop, and Kitai, 1977, respectively) and likewise that to lobules VI and VII in the squirrel monkey (Frankfurter, Weber, and Harting, 1977).

In recent years neurons in the nucleus prepositus have been found to give off ascending fibers to the ocular motor nuclei, and it appears that they may be collaterals of axons passing to the cerebellum (see Baker and Berthoz, 1975). Labeled cells were found bilaterally throughout the nucleus prepositus after injection of HRP in one nucleus of the oculomotor nerve (Graybiel and Hartwieg, 1974; Gacek, 1977) and also after injection in the abducent nucleus (Maciewicz, Eagen, Kaneko, and Highstein, 1977). These and other findings have directed attention to the perihypoglossal nuclei as parts of the integrative mechanisms controlling eye movements. Knowledge of their afferents is essential for evaluating the function of the perihypoglossal nuclei. It has turned out that fibers from many sources impinge on these nuclei. Perhaps the most massive input comes from the cerebellum.

Lesions of the brainstem caudal to the mesencephalon in the cat were found to produce terminal degeneration in the perihypoglossal nuclei (Brodal, 1952); further, they appear to receive some spinal afferents (Brodal, 1952, in the cat; Mehler, Feferman, and Nauta, 1960, in the monkey). Subsequent studies indicated that most descending fibers are derived from the mesencephalon, particularly from the *nucleus interstitialis of Cajal* (Carpenter, Harbison, and Peter, 1970; Mabuchi and Kusama, 1970). A modest projection has been found from the *cerebral cortex*, particularly from the "face region" of the sensorimotor cortex (Sousa-Pinto, 1970a). Afferents from the *cerebellum* appear to be quantitatively most important. These are derived particularly from the nodulus and flocculus (Angaut and Brodal, 1957, see Fig. 5-31), and in addition there is a topically organized projection from the caudal part of the fastigial nucleus (Walberg, 1961; Voogd, 1964). In physiological studies Baker, Gresty, and Berthoz (1976) found that vestibular impulses reach the perihypoglossal nuclei, and recently Mergner, Pompeiano, and Corvaja (1977) concluded from an HRP study that the nucleus intercalatus receives afferents from the *descending and medial vestibular nuclei*. An interesting contingent of afferents from the *paramedian part of the pontine reticular formation* (PPRF) has been found in autoradiographic studies by Büttner-Ennever and Henn (1976) and Graybiel (1977). In evaluating the findings reviewed above, it should be recalled that many of the structures involved are very small. Lesions in such structures may interrupt bypassing fibers, and injections of marker solutions may easily spread to neighboring areas.

The small cell groups of the perihypoglossal nuclei are thus most intimately related to the cerebellum and structures in the mesencephalon which influence the extrinsic eye muscle. Some functional aspects will be considered in Chapter 7, section (g).

Among other, minor afferent projections to the cerebellum may be mentioned the recently established projections from the *raphe nuclei* and the *locus coeruleus,* which have attracted much interest (see Chap. 6). Afferents from the *sensory trigeminal nuclei* have been described above. Recently Kotchabhakdi and Walberg (1977) found some labeled large and medium-sized cells in the *motor cranial nerve nuclei* after injections of HRP in the cerebellum (anterior lobe, nodulus, flocculus, and fastigial nucleus). Whether these cerebellar afferents are collaterals of axons to the effector organ or belong to neurons supplying only the cerebellum could not be decided. Fibers from the *nucleus of the solitary tract* have been found (Somana and Walberg, 1979).

Many afferent cerebellar fiber systems give off fibers or collaterals to the cerebellar nuclei in addition to the cerebellar cortex (some have been referred to above). At present, only one afferent system is known which appears to end exclusively in the cerebellar nuclei, namely the *rubrocerebellar projection*. This and the other nuclear afferents will be considered more fully in the following section.

It appears from the preceding account of the afferent cerebellar connections that, like the spinal cord, the cerebral cortex disposes of several routes to the cerebellum. These may collectively be referred to as cerebrocerebellar pathways. Figure 5-25 is a simplified diagram of the main relays in these pathways. As described above, they are not identical with regard to the areas of the cerebellum on which they may act, and, from a quantitative point of view, there are great differences between them (the pontine nuclei being the most important by far). The relay nuclei differ further as concerns the possibilities of integration of cortical impulses with impulses from other regions within them. (For example, only few if any spinal afferents reach the pontine nuclei, whereas extensive integration of cortical and spinal inputs may occur in the lateral reticular nucleus.) Although physiological studies have demonstrated some functional properties of some of the cerebrocerebellar connections, little is known of how they cooperate in the cerebral influence on the cerebellum.

The corticonuclear projection and afferent connections to the cerebellar nuclei.   In the course of time, numerous studies have been made of the distribution of Purkinje cell axons (the only efferent fibers from the cerebellar cortex) within the cerebellar nuclei (for references to early studies see Jansen and Brodal, 1940). Many authors concluded from their experimental material that this corticonuclear projection must be organized in a very precise manner. With modern neuroanatomical methods this has generally been found to be correct, and other features have also been elucidated.

From studies with the Marchi methods (see Chap. 1) in the rabbit, cat, and monkey, Jansen and Brodal (1940, 1942) concluded, in agreement with some previously published data, that there is a definite *topographical pattern in the cor-*

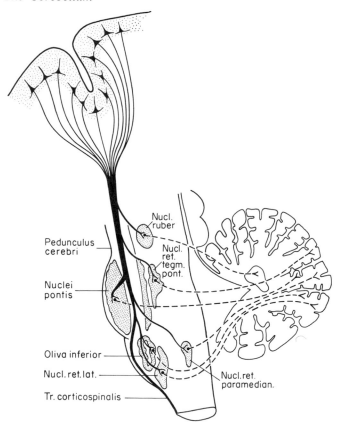

Pedunculus
cerebri

Nucl.
ruber

Nucl.
ret.
tegm.
pont.

Nuclei
pontis

Olivo inferior

Nucl. ret. lat.

Nucl. ret.
paramedian.

Tr. corticospinalis

FIG. 5-25 A simplified diagram of the main cerebrocerebellar routes consisting of two neurons. The first is located in the cerebral cortex, the second in various relay nuclei: the pontine nuclei, the red nucleus, the reticular tegmental pontine nucleus, the lateral reticular nucleus, the paramedian reticular nucleus, and the inferior olive. Note: The fibers to these nuclei are not all collaterals of the corticospinal tract and do not all have the same sites of origin (cf. text). From Brodal (1972a).

ticonuclear projection. Their diagram, reproduced in Fig. 5-26, indicates that the entire vermis projects to the nucleus fastigii; the intermediate part of the anterior lobe, the medial parts of crus I and crus II, and the paramedian lobule project to the nuclei interpositi; and the lateral parts of the hemisphere and the paraflocculus send their fibers to the lateral (dentate) nucleus. This subdivision into three zones is diagrammatically shown in Fig. 5-4.[37] Furthermore, rostral and caudal parts of each zone generally project to rostral and caudal parts of the corresponding cerebellar nucleus respectively.

With the introduction of the silver impregnation methods (see Chap. 1) greater precision could be achieved, and numerous studies of the corticonuclear projection were performed in several animal species (Eager, 1963a, 1966; Goodman, Hallett, and Welch, 1963; Walberg and Jansen, 1964; Voogd, 1964; van

[37] In this figure the vermal projection to the vestibular nuclei, to be considered below, is included.

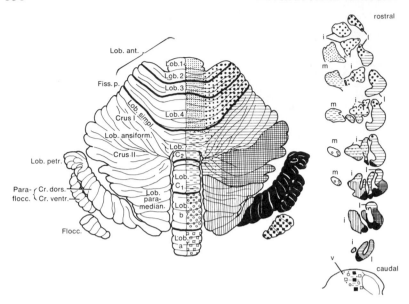

FIG. 5-26 A diagram of the corticonuclear projection in the cerebellum of the monkey based on experimental investigations. Corresponding areas of the cortex and the nuclei labeled with identical symbols. *m:* nucleus medialis or fastigii; *i:* nucleus interpositus; *l:* nucleus lateralis or dentatus; *v:* vestibular nuclei. The projection of the flocculonodular lobe after Dow (1936, 1938a). From Jansen and Brodal (1942).

Rossum, 1969; Brodal and Courville, 1973; Courville, Diakiw, and Brodal, 1973; Sreesai, 1974; Haines, 1975b, 1976, 1977a; Courville and Diakiw, 1976; Haines, Culberson, and Martin, 1976; Haines and Whitworth, 1978). The distribution of degenerating fine fibers and terminals within the cerebellar nuclei has been mapped after small cortical lesions were produced. The results of early Marchi studies have been confirmed and extended. The zonal subdivision has been found to be less schematic than indicated in Fig. 5-4. Almost all students found the corticonuclear projection to be strictly ipsilateral.

With regard to the *corticonuclear projection from the vermis,* practically all recent authors agree that this is restricted to the fastigial nucleus (see Courville and Diakiw, 1976, in the cat; Haines, 1975b, 1976, in the galago, a prosimian primate). The projection is ipsilateral, probably except for a narrow region in the midline (Haines, 1976).[38] Furthermore, the projection is topographically organized, with a sequence "such that each lobule projects to the portion of the nucleus which is at the shortest distance" (Courville and Diakiw, 1976, p. 17). Even though there is some overlapping of terminals from adjacent lobules, the pattern is

[38] Fibers from the nodulus and uvula have been found by some authors to end in the small-celled ventral part of the dentate nucleus, which receives primary vestibular and floccular afferents as well. As will be discussed below, this nuclear region may be a station in a vestibulo-ocular pathway. Physiologically, a projection to the trochlear nuclei from the fastigial nucleus has been described (Hirai, Uchino, and Watanabe, 1977), suggesting a similar route for cerebellar influences on eye movement (see Haines, 1977b).

distinct.[39] The fastigial afferents appear to come from a medial zone of the vermis only, corresponding to Voogd's longitudinal zone A, while a lateral strip corresponds to zone B (Haines, 1976; Haines and Rubertone, 1977; and others) and gives rise to vermal fibers to the lateral vestibular nucleus (Deiters), as maintained by Voogd (1969) and van Rossum (1969). (The cerebellovestibular projection will be considered in a later section of this chapter.)

Knowledge of most other regions of the cerebellum is less extensive. However, it appears from the extensive studies with silver impregnation methods in the cat and rabbit by Voogd (1964) and van Rossum (1969) that the principle of projections of rostral and caudal parts of the cerebellar region to rostral and caudal parts, respectively, of its target nucleus is valid. The topical arrangement is not purely rostrocaudal, however, but conforms to the principle formulated by Courville and Diakiw (1976, quoted above) concerning the projection onto the fastigial nucleus. The question of whether each cerebellar nucleus is supplied only from cortical regions belonging to a particular longitudinal zone, as maintained by Voogd (1964) and van Rossum (1969), has not been finally settled. According to Voogd (1969), the nucleus interpositus anterior belongs to his compartment C_1 and C_3, the interpositus posterior to his compartment C_2, and the lateral nucleus to compartment D (see Fig. 5-18). It appears, however, that the situation may be less schematic.

In the cat the projection from the *paramedian lobule* was found by Courville, Diakiw, and Brodal (1973) to go to three separate regions: an area in the ventral part of the lateral cerebellar nucleus, another in the nucleus interpositus anterior, and a third in the interpositus posterior (see also Sreesai, 1974, in the opossum). Within all three projections there is a somatotopical pattern in the cat, with rostral parts of the lobule (i.e., its face and forelimb region) projecting more laterally than the caudal parts (hindlimb region, refer to Fig. 5-9). Zones C_1–C_3, C_2, and D are all present in the paramedian lobule (see Fig. 5-18). Since even a rather small cortical lesion in this lobule may involve more than one zone, the findings of Courville, Diakiw, and Brodal (1973) do not exclude the possibility that there is a zonal pattern in the corticonuclear projection from the paramedian lobule. Studies of the corticonuclear projection from *crus II* (Brodal and Courville, 1973) showed a topically arranged projection to the dentate nucleus and to both nuclei interpositi. Again evidence of a zonal projection could not be found. Thus lesions of the lateralmost part (belonging to zone D) of crus II gave rise to degeneration not only in the dentate (or lateral) nucleus but in the nuclei interpositi as well.

The corticonuclear projection from the *paraflocculus* has recently been studied in the tree shrew by Haines and Whitworth (1978). In agreement with some previous authors, they found fibers from this lobule to supply restricted regions of the lateral nucleus and the interpositus posterior. The dorsal and ventral paraflocculus have their particular sites of termination, with some overlapping. The dual terminations are explained by the presence of zone D and C_2 in the paraflocculus (an assumption supported by studies of the olivocerebellar projection). In a study of the corticonuclear projections of the *flocculonodular lobe* (projecting chiefly to the vestibular

[39] The results of an autoradiographic study in the rat by Armstrong and Schild (1978a) largely agree with those described for the cat. However, the topical pattern in the corticonuclear projection appears to be less sharp, and, like some authors studying the cat, Armstrong and Schild described some fibers from the vermis as passing to the nuclei interpositi.

When very small cortical lesions are made in silver impregnation studies, a rather discrete patch of degeneration is usually found in the nuclei. It should be recalled, however, that in anatomical studies of this kind, it is almost impossible to demonstrate the topographical relations precisely, since the degeneration after a minimal lesion of the cortex may be too scanty to be identified. Furthermore, the anatomical spread of fibers from a small part of the cortex is likely to exceed the central area from which potentials may be picked up. (Cf. the principle of lateral inhibition, referred to in Chap. 2.)

nuclei, see below) in the galago (a prosimian primate) Haines (1977a) confirmed some previous studies (Angaut and Brodal, 1967; van Rossum, 1969) showing that the ventrolateral (small-celled) part of the dentate nucleus receives fibers from the nodulus and from the flocculus.[40]

Even though many details remain to be clarified, there appears to be little doubt that the cerebellar corticonuclear projection is very precisely organized. As mentioned above, the Purkinje cell axons have been found to have a monosynaptic inhibitory action on the cerebellar nuclear neurons (see Eccles, Ito, and Szentágothai, 1967.[41] This made it almost a logical necessity that there must be *other afferent inputs* that have an excitatory action, and after stimulation of many afferent fiber systems supplying the cerebellar cortex, various investigators observed such actions (see Eccles, Ito, and Szentágothai, 1967, for references).[42] Many authors who found degenerating fragments in the nuclei in anatomical studies following lesions of various afferent cerebellar fiber systems were hesitant to interpret them as true terminals. In recent years, however, evidence has been brought forward that there are indeed several afferent inputs to the nuclei apart from the Purkinje cell axons. As might be expected, the various afferent pathways do not supply all cerebellar nuclei, but each appears to have its particular terminal territory.

Primary vestibular afferents have been traced to the small-celled part of the lateral (dentate) nucleus (Brodal and Høivik, 1964; Carpenter, Stein, and Peter, 1972). *Secondary vestibular fibers* have been shown to supply the fastigial nucleus (Brodal and Torvik, 1957; Carpenter, Bard, and Alling, 1959). The sites within the vestibular nuclei, determined in the studies above as giving rise to these fibers are not exactly identical with those found in an HRP study by Ruggiero, Batton, Jayaraman, and Carpenter (1977).

Spinal afferents have recently been described in the cat (Matsushita and Ikeda, 1970b) and in the rabbit and the rat (Matsushita and Ueyama, 1973a) to end chiefly in the fastigial nucleus, including its parvicellular part, but also partly in the nuclei interpositi. Most of the fibers appear to pass with the ventral spinocerebellar tract, and a small number with the dorsal spinocerebellar tract.

Different *brainstem nuclei* send fibers to the cerebellar nuclei. An important source appears to be the *inferior olive,* which supplies all cerebellar nuclei in a topical pattern (see above on projections from the olive). Fibers from the *lateral reticular nucleus* have been traced to the ipsilateral fastigial nucleus and the nuclei interpositi (Matsushita and Ikeda, 1976). The projection to the fastigial nucleus could, however, not be definitely confirmed in the HRP study of Ruggiero, Bat-

[40] In general, the results of an autoradiographic study of the corticonuclear projection in the rat (Armstrong and Schild, 1978b) agree with the findings described above.

[41] The putative transmitter is γ-aminobutyric acid (GABA) as determined from several studies (see, for example, Fonnum and Walberg, 1973). It has been maintained by some that inhibitory synapses are provided with elliptical (flattened) vesicles, excitatory with round vesicles (see Chap. 1). In an electron microscopic study of autoradiographically identified Purkinje cell terminals, Walberg, Holländer, and Grofová (1976) found, however, that round vesicles occur in various numbers in many of the terminals.

[42] Among demonstrations of monosynaptic excitatory potentials in the cerebellar nuclei following stimulation of sources giving off efferents to these nuclei may be mentioned the nucleus interpositus anterior from spinal afferents (Eccles, Rosén, Scheid, and Táboriková, 1972), from the lateral reticular nucleus (the latter confirmed by McCrea, Bishop, and Kitai, 1977), and from the pontine nuclei (Tsukahara and Bando, 1970).

ton, Jayaraman, and Carpenter (1977), whereas that to the interpositus anterior was confirmed with the same method by McCrea, Bishop, and Kitai (1977). The latter authors found projections from the *nucleus prepositus hypoglossi* to the fastigial nucleus. The *tegmental reticular nucleus* appears to project particularly to the dentate nucleus (see McCrea, Bishop, and Kitai, 1977). A projection from the *pontine nuclei* has likewise been found (see Kawamura and Hashikawa, 1975). For an HRP study of afferents to the dentate nucleus in the rat, see Eller and Chan-Palay (1976).

While all the afferent fiber contingents listed above (and there may be more!) supply particular areas of the cerebellar cortex in addition to one or more of the cerebellar nuclei (probably by way of collaterals), the situation is different with regard to the *red nucleus*. Its fibers to the cerebellum appear not to reach the cortex, but to be restricted to the nuclei, particularly the nucleus interpositus anterior (see Fig. 5-25).

A *rubrocerebellar projection* was concluded to exist by early authors (see Courville and Brodal, 1966, for references). Its presence was confirmed in the cat. After cerebellar ablations involving the cerebellar nuclei in the cat, retrograde cellular changes occur in the red nucleus, chiefly in its caudal third (Brodal and Gogstad, 1954). Both small and large cells contribute to the projection.

This projection has also been demonstrated following lesions of the red nucleus (Hinman and Carpenter, 1959; Courville and Brodal, 1966). After crossing the midline, the fibers descend in the superior cerebellar peduncle to the contralateral cerebellar half. The distribution of degeneration after stereotactically placed lesions in the ''forelimb'' or ''hindlimb'' part of the red nucleus (see Chap. 4 and Fig. 4-6) suggests the presence of a somatotopic pattern in the projection, which appears to supply only the *nucleus interpositus anterior* (Courville and Brodal, 1966). From a comparison of the numbers of changed cells in the red nucleus after transection of the spinal cord and cerebellar ablations, it appears that the fibers to the cerebellum may be branches or collaterals of descending axons to the spinal cord (Brodal and Gogstad, 1954).

The rubrocerebellar projection appears in many respects to be reciprocal to the cerebellorubral projection, which has been more extensively studied (see below). It has attracted considerable interest from a functional point of view as a particular link in the collaboration between the cerebral cortex and the cerebellum.

It can be seen from the discussion above that *practically all information transmitted to the cerebellar cortex may also influence one or more of the cerebellar nuclei*. Future studies may disclose an even more precise pattern in the afferent supply of the cerebellar nuclei than is known at present. It appears from the available data that the afferent input to each nucleus (or even part of this) is generally much the same as that reaching the part of the cerebellar cortex which projects onto it. A further indication of a close functional, and precise topical, relation between the cerebellar nuclei and the cerebellar cortex has recently been discovered. As will be described in the following section, some axons or collaterals of nuclear efferents supply the cerebellar cortex; that is, there is both a nucleocortical and a corticonuclear projection.

Efferent connections of the cerebellar nuclei. The efferent fibers from the cerebellar nuclei are distributed to numerous other parts of the central nervous system, indicating that the cerebellum controls a variety of functions, as has been shown physiologically. However, the four main nuclei differ with regard to the targets of their efferent projections; to some extent even minor parts of a single nucleus show differences in this respect. There is little reason to doubt that further studies will reveal a far greater specificity within the efferent nuclear projections than is known at present. An argument in favor of this assumption is the highly ordered pattern within the most important afferent contingent, the Purkinje cell axons.

In the course of time, numerous studies of the efferent cerebellar projections have been performed (for a review of the older literature, see, for example, Jansen and Brodal, 1958; Larsell and Jansen, 1972). Some of the fibers pass to the vestibular nuclei. These will be considered separately. Here the *nonvestibular projections* will be discussed.

As referred to in a preceding section, most of the efferents leave the cerebellum via the superior cerebellar peduncle (Fig. 5-3), and a minor proportion in the restiform body and the so-called juxtarestiform body, medial to the latter. In general, the fibers from the dentate and interpositus nuclei ascend in the ipsilateral superior peduncle, forming the brachium conjunctivum. (Afferent, cerebellopetal fibers in the superior peduncle belong to the ventral spinocerebellar tract and the rubrocerebellar tract, and some come from the mesencephalic nucleus of the trigeminal nerve.) Some fibers from the contralateral fastigial nucleus (passing in the uncinate fasciculus; see below) likewise enter the brachium conjunctivum. Other fastigial efferents enter the ipsilateral restiform body. It appears from Flood and Jansen's (1966) studies with the modified Gudden method (Brodal, 1940a) that the great majority of the nuclear cells, both large and small, disappear or are heavily changed after transection of the three cerebellar peduncles in the kitten; that is, they give off efferent fibers. (None of these, however, appear to pass in the middle cerebellar peduncle, according to these and other authors.) Whether there may be true internuncials (Golgi type II cells) in the nuclei could, however, not be definitely decided.

The *efferent fibers from the fastigial nucleus* have been studied by many authors (Thomas, Kaufman, Sprague, and Chambers, 1956; Carpenter, Brittin, and Pines, 1958; Cohen, Chambers, and Sprague, 1958; Walberg, Pompeiano, Westrum, and Hauglie-Hanssen, 1962; Walberg, Pompeiano, Brodal, and Jansen, 1962; Angaut and Bowsher, 1970; Martin, King, and Dom, 1974; Batton, Jayaraman, Ruggiero, and Carpenter, 1977; Fukushima et al., 1977; Matsushita and Hosoya, 1978; and others). Only some main points will be summarized. The fibers from the rostral part of the nucleus make up almost half of all fastigiobulbar fibers (Flood and Jansen, 1966). They leave the cerebellum in the ipsilateral restiform body. The rest, coming from the caudal part, cross the midline in the cerebellum, in part thereby (unfortunately for experimental studies) passing through the opposite fastigial nucleus. They form the *hook bundle of Russell* (uncinate fasciculus), which bends around the superior cerebellar peduncle. The terminal distributions of the two components differ to some extent, indicating functional differences between the anterior and posterior parts of the fastigial nucleus.

The main targets of efferents from the fastigial nucleus are the main reticular formation and the vestibular nuclei (see below and Fig. 5-30). After lesions of the fastigial nucleus in the cat many degenerating fibers are distributed throughout the *medullary and pontine reticular formation,* chiefly contralaterally, without any pattern of localization (Walberg, Pompeiano, Westrum, and Hauglie-Hanssen, 1962; Voogd, 1964). Their main sites of termination appear to be the nuclei reticularis gigantocellularis and pontis caudalis (see Chap. 6), as confirmed autoradiographically in the monkey (Batton, Jayaraman, Ruggiero, and Carpenter, 1977). Most of the crossing fastigioreticular fibers, passing in the hook bundle, appear to arise from the caudal part of the nucleus, while most of the direct, ipsilateral fibers come from the rostral part—without sharp border, however (Jansen and Jansen, 1955; Walberg, Pompeiano, Westrum, and Hauglie-Hanssen, 1962; Voogd, 1964; Flood and Jansen, 1966; and others). In concordance with the anatomical data, the results of stimulation or destruction of the rostral and caudal parts of the fastigial nucleus are different, as shown in a series of studies by Moruzzi and Pompeiano and collaborators (see Pompeiano, 1967, for a review).

Thus a unilateral lesion of the rostral part of the nucleus fastigii in the decerebrate cat gives rise to an ipsilateral hypotonia, whereas destruction of the caudal part of the fastigial nucleus results in atonia in the contralateral limbs (crossed fastigial atonia, Moruzzi and Pompeiano, 1956. Furthermore, the effects from the medial and lateral parts of the rostral half of the nucleus differ, stimulation of the former giving augmentation, of the latter inhibition of postural tonus (Batini and Pompeiano, 1958). In the caudal part there are corresponding differences between medial and lateral regions, but the augmentatory and inhibitory regions are found in the reverse order (Pompeiano, 1962). An anatomical counterpart to these observations of mediolateral differences has not yet been found, but they tally with other findings which indicate that the longitudinal zones of the cerebellum (in this case the vermis proper) can be further subdivided. The fastigiofugal fibers appear to have primarily a monosynaptic excitatory action on their target neurons in the reticular formation (Ito, Uno, Mano, and Kawai, 1970) as on the vestibular nuclei.

The terminal regions of the fastigioreticular fibers cover parts of the reticular formation which give off both descending and ascending fibers, indicating that the fastigial nucleus (and the cerebellar vermis) via the reticular formation may influence the spinal cord and "higher levels" of the brain (see Chap. 6). However, such regions appear to receive a modest number of direct fastigial efferents, leaving in the hook bundle and the ascending brachium conjunctivum.

According to the autoradiographic study of the monkey by Batton, Jayaraman, Ruggiero, and Carpenter (1977) and other studies (see Batton et al., 1977, and Angaut and Bowsher, 1970, for references), these fibers come only from the caudal parts of the fastigial nucleus and are entirely crossed.[43] Terminations have been found by many authors in the *superior colliculus* (its deeper layers, see Chap. 7) and the *nucleus of the posterior commissure*. While several authors have described terminations in *thalamic nuclei,* there is some disagreement concerning details. According to Batton et al. (1977), the major terminal region is the contralateral nucleus ventralis posterior lateralis (VPL, see Chap. 2 and Fig. 2-14), particularly rostrally; some fibers are found in the rostral part of the nucleus ventralis lateralis (VL). Other authors indicate the thalamic sites of termination somewhat differently. Angaut and Bowsher (1970) found terminations in what they refer to as the VM (ventralis medialis) in the cat, in addition to fibers to the medial part of the VL (see Fig. 5-28). Kievit and Kuypers (1972) found a more widespread termination in the VL in the monkey than in the cat. Species differences and prob-

---

[43] The bilateral projections found by some authors in degeneration studies are probably due to damage to crossing fibers in the cerebellum.

lems in the nomenclature and delimitations of the thalamic nuclei may explain some of the differences in the findings reported in the literature. The question of the precise sites of termination of fastigiothalamic fibers is of interest for the evaluation of a possible influence of the fastigial nucleus on the motor cortex (which receives thalamic fibers chiefly from the VL; see below and Chaps. 2 and 4).

> Some efferent ascending fastigial fibers have been claimed to end in other regions as well, such as the hypothalamus, the septal nuclei, and some cerebral cortical regions, for example the orbital gyrus (Harper and Heath, 1973, 1974, silver impregnation methods). The pathway to the septal nuclei has been described as being cholinergic (Paul, Heath, and Ellison, 1973).

The presence of fibers passing from the fastigial nucleus directly to the *spinal cord* was advocated by Thomas, Kaufman, Sprague, and Chambers (1956) but has been doubted by many. After injections of horseradish peroxidase in the spinal cord in cats, however, Fukushima et al. (1977) and Matsushita and Hosoya (1978) found labeled cells in the contralateral fastigial nucleus. The modest tract appears to supply chiefly the cervical segments. The physiological studies of Wilson, Uchino, Susswein, and Fukushima (1977) and the autoradiographic studies of Batton, Jayaraman, Ruggiero, and Carpenter (1977) are in agreement with this (see also Ware and Mufson, 1979, in the tree shrew).

Other minor contingents of fastigial efferents have been mentioned above as passing to the *pontine nuclei proper,* the *nucleus reticularis tegmenti pontis,* the *lateral reticular nucleus,* the *paramedian reticular nucleus,* and the *perihypoglossal nuclei* (whereas fibers to the inferior olive have not been found).

The *efferent fibers from the nuclei interpositi and the lateral (dentate) nucleus* all appear to leave the cerebellum in the ipsilateral superior cerebellar peduncle. Within this, the fiber groups arising in different parts of these nuclei have their particular place, as studied by numerous authors (for an exhaustive account see Larsell and Jansen, 1972). The fibers from the interpositus posterior are found medially, those from the lateral nucleus most laterally.

A considerable number of the fibers of the brachium conjunctivum end in *the thalamus,* many in *the red nucleus,* and some have been traced to other destinations, particularly *some nuclei in the brainstem.* (According to McCrea, Bishop, and Kitai, 1978, there are at least three morphologically distinguishable populations of projection neurons in the nuclei interpositi.) After crossing the midline, the ascending fibers pass through the red nucleus, and many of them give off a descending branch (crossed ventral descending limb of the brachium conjunctivum). It has been disputed whether some of the ascending cerebellar efferents end in the red nucleus, or whether the latter is supplied only by collaterals of fibers continuing further rostrally, and whether fibers supplying the red nucleus have their separate sites of origin within the complex of the cerebellar nuclei. Studies of the distribution of degeneration after transections of the brachium conjunctivum (for example, the recent study of Miller and Strominger, 1977, in the monkey), although they supply information on the total area of termination, do not answer these questions; but recent studies using other approaches have given interesting information. There are still some points of dispute, in part due to problems of identification and delimitation of the thalamic nuclei and of the nucleus interpositus anterior and the dentate (see section on the cerebellar nuclei above), but

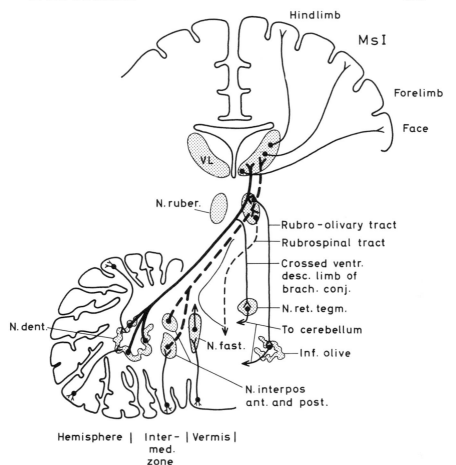

FIG. 5-27 A very simplified diagram of the main ascending connections from the cerebellar nuclei to the VL of the thalamus and the red nucleus. (The contribution from the fastigial nucleus is not shown.) Note the crossed descending limb of the brachium conjunctivum and its ending in the nucleus reticularis tegmenti pontis (and other regions), establishing a direct feedback route to the cerebellum. Of other, indirect feedback pathways, the rubral projection to the inferior olive is shown. The main efferent output from the red nucleus goes to the spinal cord (rubrospinal tract). The main projection from the VL to the cerebral cortex is somatotopically organized. For particulars see text.

the main features in the efferent projections from the lateral (dentate) nucleus and the nuclei interpositi appear fairly well established. To avoid repetitions, it is deemed practical to discuss them according to their sites of terminations and not according to their sites of origin. Figure 5-27 shows a simplified diagram of the cerebellar projections to the red nucleus and the thalamus.

Starting with the *cerebellar projection to the red nucleus,* it appears that this is a rather precisely organized pathway.[44] Studies of the degeneration following

---

[44] This pathway appears to be almost completely reciprocal to the rubrocerebellar tract, considered above.

stereotactic lesions of the cerebellar nuclei in the cat (Courville, 1966c) indicate that the *nucleus interpositus anterior projects in a topical pattern to the caudal two-thirds of the red nucleus*. The termination of fibers from the nucleus interpositus anterior in the caudal two-thirds, or a little more, of the red nucleus has been confirmed in the monkey (Flumerfelt, Otabe, and Courville, 1973) and in the rat (Caughell and Flumerfelt, 1977) and is apparently valid for the opossum as well (King, Dom, Conner, and Martin, 1973). A correlation of the sites of termination of these fibers with the somatotopic pattern in the red nucleus (Pompeiano and Brodal, 1957; see Chap. 4) suggests that the rostral and caudal parts of the interpositus anterior are related to the hindlimb and forelimb, respectively. A somatotopic pattern appears to be present in the monkey (Flumerfelt, Otabe, and Courville, 1973). Physiological studies tend to confirm it (see, for example, Allen, Gilbert, Marini, Schultz, and Yin, 1977).

> The somatotopical pattern, deduced on the basis of fiber connections, agrees with the results of stimulation of the interpositus complex in the cat (Pompeiano, 1959). From both Courville's and Pompeiano's studies, it appears that a distinction can further be made between medial and lateral parts of the nucleus interpositus anterior (see Courville, 1966c, and Pompeiano, 1967, for particulars).

In addition to the major contribution of afferents from the nucleus interpositus anterior, *the red nucleus receives afferents from the lateral (dentate) nucleus* (see Flumerfelt, Otabe, and Courville, 1973) and some from *the interpositus posterior* (Angaut, 1970). The fibers from the lateral nucleus are found to end in the rostralmost part of the red nucleus (Fig. 5-27). The scanty contribution from the interpositus posterior appears to be restricted to the medialmost part of the red nucleus (Angaut, 1970). Electrophysiological studies have confirmed that fibers from the dentate supply only the rostral part of the nucleus (Condé and Angaut, 1970).

> Interestingly, the terminals from the dentate nucleus appear to be rather fine, as judged from their appearance in silver-impregnated sections when they undergo degeneration, while those from the interpositus anterior are much coarser (see Flumerfelt, Otabe, and Courville, 1973; Caughell and Flumerfelt, 1977). The difference in distribution of fibers from the two nuclei may be related to the efferent projections of the red nucleus (see Chap. 4). Most of the rubrospinal efferents arise in the caudal two-thirds (containing many large neurons), whereas the rubro-olivary tract appears to arise only from the rostral small-celled part of the nucleus (see section on the inferior olive). This suggests that the red nucleus is a link in two more or less separate pathways: one from the intermediate part of the cerebellar cortex via the nucleus interpositus anterior and the caudal two-thirds of the red nucleus to the spinal cord, the other from the lateral parts of the cerebellum via the lateral cerebellar nucleus and the rostral third of the red nucleus to the inferior olive.

As mentioned in Chapter 4, in electron microscopic studies the rubral afferents from the nucleus interpositus have been found to establish synaptic contact preferentially with somata and proximal dendrites of the neurons and to mediate monosynaptic excitatory potentials. In recent years, several physiological studies have been devoted to the interpositorubral projection and its role in the performance of movements.

Whether some neurons in the red nucleus give off axons or collaterals which join the cerebellar nuclear fibers that pass to the thalamus, has been a subject of controversy. Decisive results have been difficult to obtain. Lesions of the red nucleus will inevitably interrupt direct cerebellothalamic fibers. After lesions of the red nucleus in monkeys, decerebellated more than one year previously (allowing time for degeneration of the cerebellothalamic fibers), no degenerated fibers could be traced to the thalamus by Hopkins and Lawrence (1975), nor did Edwards (1972), after tritiated leucine injection in the red nucleus of the cat, find reliable evidence of a rubrothalamic projection.

The *thalamic terminations of cerebellofugal fibers from the nuclei interpositi and lateralis* have been described by early authors in various animals (for an extensive review, see Larsell and Jansen, 1972). After transections of the brachium conjunctivum or lesions of the cerebellar nuclei, fiber terminations were consistently found in the VL (ventral lateral thalamic nucleus, see Fig. 2-14). In addition, various other sites of terminations were reported. In recent studies the main terminal region has likewise been found to be the VL, in the cat (Angaut, 1969a, 1970, 1973; Rinvik and Grofová, 1974), the rat (Chan-Palay, 1977), the hedgehog (Earle and Matzke, 1974), the opossum (Martin, King, and Dom, 1974) and the monkey (Chan-Palay, 1977; Miller and Strominger, 1977). Most authors found the terminal region of the fibers to extend into the caudal part of the VA (ventralis anterior) and into the adjoining part of VPL (ventralis posterior lateralis). There are some discrepancies among studies concerning details, caused, at least in part, by uncertainties in the delimitation of the thalamic nuclei (see Chap. 2). The same is the case with regard to cerebellar projections to the "unspecific" thalamic nuclei. The nucleus centralis lateralis, the nucleus paracentralis, the nucleus centrum medianum, and other groups, such as the thalamic reticular nucleus, have been claimed by one or more authors to receive fibers ascending in the brachium conjunctivum. A postulated projection to the basal ganglia is denied by most authors. We shall be concerned chiefly with the projections to the "specific" thalamic nuclei (relay nuclei) in the following discussion.

As might be expected, the efferent fibers from the two interpositi and the dentate do not have identical sites of terminations, even if they overlap to some extent, However, the literature concerning this problem is relatively sparse. The thalamic projection from the *nucleus interpositus posterior* is modest (Voogd, 1964; Angaut, 1970; and others). In the cat (Angaut, 1970), most fibers end in the lateral part of the VL, but some end laterally in the VA and the medialmost region of VPL (as seen in Fig. 5-28). The projection was found to be topically organized: the rostral and caudal parts of the interpositus posterior project to ventral and dorsal parts, respectively, of the VL. When considered in relation to the thalamocortical projection (see Chap. 2), this suggests that there may be a somatotopical pattern in the nucleus interpositus posterior, with the forelimb represented rostrally and the hindlimb caudally.

Figure 5-28 shows, on the basis of studies of degeneration of fibers after destruction of individual cerebellar nuclei (Angaut, 1969a, 1970, 1973), that in the cat the projection of the *nucleus interpositus anterior* is concentrated in two dorsoventrally oriented laminar zones, which fuse caudally and rostrally, where they extend into the VA. (The modest, medially situated contribution of the fastigial nucleus was described above.) The projections from the two nuclei interpositi overlap more or less with the relatively extensive area of fibers from the *dentate*. Within the latter projection there appears to be a topical pattern (not shown in the figure). Recordings of the sites of monosynaptic potentials evoked in the VL on

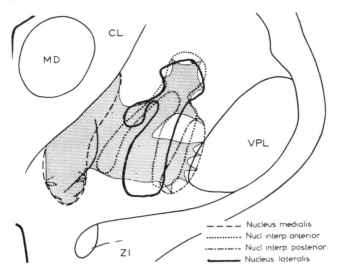

FIG. 5-28   A simplified diagram of the terminal distribution in the VL of fibers from the four cerebellar nuclei in the cat (symbols below). The stippled area represents a homunculus of the projections onto the cerebral cortex. Note: face represented medially and below, distal forelimb ventrolaterally. *CL:* nucleus centralis lateralis; *MD:* nucleus medialis dorsalis; *VPL:* nucleus ventralis lateralis posterior; *ZI:* zona inzertia. From Angaut (1973).

stimulation of the cerebellar nuclei (Rispal-Padel and Grangetto, 1977) are in general agreement with the studies of Angaut (see Fig. 5-28). Responses to stimulation of the nucleus lateralis were found chiefly dorsally and medially, those from the interpositus more ventrally and laterally. As in the anatomical studies, considerable overlap is present.

> The thalamic projection from the *dentate nucleus* in the rat and the monkey has recently been studied in great detail autoradiographically by Chan-Palay (1977). After injections in different small parts of the dentate nucleus in the monkey the distribution of labeled terminals in the thalamus differs, suggesting that ''there is a precise representation of each area of the dentate in these thalamic nuclei'' (loc. cit. p., 341).[45]

In experimental electron microscopic studies (Rinvik and Grofová, 1974), the cerebellothalamic fibers to the VL and VA have been found to end with large boutons that establish synaptic contact with dendrites of neurons projecting to the cerebral cortex and of local circuit neurons. The fibers appear to end in clusters. In agreement with the anatomical findings, stimulation of the nuclei interpositi and lateralis gives rise to short-latency, monosynaptic activation in the VL (Angaut, Guilbaud, and Reymond, 1968; Condé and Angaut, 1970; Uno, Yoshida, and Hirota, 1970; and others), and to polysynaptic potentials.

It should be recalled that the cerebellar afferents are only one of many afferent fiber contingents that supply the VL (and VA). Recent anatomical and physiological studies have given considerable insight into the synaptic organiza-

[45] This pattern appears to be such that rostral and caudal dentate areas correspond to lateral and medial areas, respectively, in the thalamic field of termination, while lateral dentate areas and areas around the hilus of the nucleus correspond to dorsal and ventral thalamic regions, respectively.

tion and operation of these thalamic nuclei. These problems have been discussed in Chapter 2. Here we are concerned with the cerebellar afferents to the thalamus.

The pattern of distribution of cerebellar fibers within the thalamus is of particular interest for an understanding of the cooperation between the cerebellum and the cerebrum in motor activities, since the "specific" thalamic nuclei receiving cerebellar fibers project onto the cerebral cortex. As described in Chapter 2, the VL and probably the adjoining part of VA project to the motor cortex (Ms I) in a somatotopic pattern. This was shown by early anatomists (Walker, 1934; see also Chow and Pribram, 1956). Figure 5-29 shows Walker's classical diagram. This pattern has been confirmed in man by recordings of action potentials in the precentral gyrus after stimulation of the VL (Uno et al., 1967). Later studies have given more detailed information. According to the study of Strick (1973) in the cat, the VL can be divided into two architectonically different parts, one dorsal and one ventral. Lesions in the dorsal part of the VL give rise to degeneration in the anterior part of the motor cortex, which on stimulation elicits movements of the proximal musculature, while the ventral part of VL sends fibers to the cortex governing distal musculature and vibrissae (chiefly area 4). The caudal part of the VA was found to project onto the cortical hindlimb region and to evoke hindlimb movements on stimulation (suggesting that it should be considered a part of the VL). The ventral part of the VL appears to be more discretely somatotopically organized than the dorsal part.

The preciseness in the thalamocortical projection has been verified in physiological studies as well (Rispal-Padel and Massion, 1970; Strick, 1973; and others). Stimulation of a small area of the VL gives rise to contractions of contralateral limb muscles; with microstimulations even a single muscle may contract (Asanuma and Hunsperger, 1975). The available anatomical data suggest that an equally precisely organized pattern is present in the cerebellothalmic projection (see particularly Chan-Palay, 1977). Furthermore, after stimulation of the nuclei interpositi and lateralis, responses are recorded in particular regions of the motor cortex (Rispal-Padel and Latreille, 1974), and contractions of various limb muscles occur on stimulation of different sites within the cerebellar nuclei (Asanuma and Hunsperger, 1975). These and other observations suggest that there must be a top-

FIG. 5-29 A diagram of the thalamocortical projection to the central region of the cerebral cortex in the monkey, showing the somatotopic pattern in the projection. Corresponding areas of the thalamus and cortex are identically labeled. Redrawn from Walker (1934).

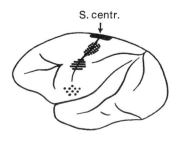

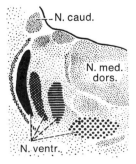

ical organization throughout the entire cerebellothalamocortical projection. This does not necessarily mean that there is a point-to-point arrangement throughout this route. The situation appears to be far more complex.

> While neurons in a rather restricted field of the VL were found to be activated with short (monosynaptic) latencies on stimulation of cerebellar nuclei, cells responding with longer latencies could be found in more extensive regions (Rispal-Padel and Grangetto, 1977). Furthermore, neurons situated at a considerable distance from each other could often be activated from the same cerebellar site, and some cells responded to stimulation of both the dentate and the interpositus posterior. This indicates that the terminals of the cerebellar fibers branch widely, as appears also from anatomical studies (see Rinvik and Grofová, 1974). On the other hand, the same cortical point may be activated from different sites in the VL, in agreement with anatomical demonstrations of a wide terminal branching of thalamocortical fibers from the VL (see especially Asanuma, Fernandez, Scheibel, and Scheibel, 1974). However, often "neurons excited from the same cerebellar site but separated in the VL excite the same cortical point" (Rispal-Padel and Grangetto, 1977, p. 119). This divergence–convergence pattern would secure a precise cerebellocortical impulse transmission and "concentrate the cerebellar effects on a small number of muscles involved in the movements of a joint" (Rispal-Padel and Grangetto, 1977, p. 120). In addition to this, there is a mechanism by which impulses from a cerebellar point may spread to wider cortical areas by branching of thalamocortical fibers. According to Rispal-Padel and Grangetto, these routes may be concerned in the realization of precise movements that are limited to a single joint but need support from other parts of the limb.

In view of the general principles considered above, exact knowledge of the correlations between the sites of origin of thalamocortical fibers and the sites of termination of cerebellothalamic fibers would be of interest.[46] However, satisfactory correlations are not yet possible. It appears to be established that there is a rather definite somatotopic pattern in the thalamocortical projection of the VL and that certain regions of this project to cortical areas influencing particular limbs and proximal and distal parts of these. It is scarcely tenable, however, as attempted by some authors, to correlate these data with the general view that the interpositus nuclei are concerned in the control of axial and proximal limb muscles, while the dentate is chiefly concerned in the control of finer movements, particularly influencing the distal parts of the limbs. The position of the sites of termination of cerebellar afferents from the various cerebellar nuclei (see Fig. 5-28) and physiological studies (see Rispal-Padel and Grangetto, 1977) indicate that the area of the VL receiving interpositus afferents must influence cortical areas concerned with movements of both distal and proximal parts of the limbs. And, *vice versa,* projections from the dentate supply areas related to proximal and distal muscles of the extremities.

It has been convincingly shown, particularly by the experiments of Brooks and his collaborators, with transient cooling of the dentate in monkeys (see Chap. 4) that the dentate nucleus is essential for the coordination of skilled movements. The arrangement in the cerebellothalamocortical projections is a strong indication that the interpositi, and even the fastigial nucleus, must be operative as well in the cerebellar control of cortical regions presumed to be especially involved in the production of voluntary (skilled, discrete, purposive) movements. So far, little attention appears to have been devoted to the particular functional importance of the

---

[46] Studies of this subject would probably enable us to elucidate somatotopic patterns in the cerebellar nuclei, about which little is known so far (see Brodal and Courville, 1973; Courville, Diakiw, and Brodal, 1973; Asanuma and Hunsperger, 1975, for some data).

inputs to the VL from the fastigial and interpositus nuclei, even though some data are available (for example, the activation of the same VL units from the interpositus posterior and the dentate, referred to above). It should again be recalled that the VL is much more than a simple relay station between the dentate and the motor cortex!

Some of the *nonthalamic projections of the nuclei interpositi and lateralis* have been considered in previous sections of the present chapter, for example the projections to the *inferior olive* (p. 341) and the *pontine nuclei* (p. 328). The relatively massive bundle to the *reticular tegmental nucleus* (p. 344) is made up of descending collaterals of the ascending brachium conjunctivum (crossed descending limb of the brachium conjunctivum). A relatively small contingent supplies some of the *raphe nuclei* (see Chap. 6). Some other sites of termination, e.g., parts of the *reticular formation* (see Miller and Strominger, 1977, for a list) have been described.[47] The functional role of the projections from the cerebellar nuclei to the *"unspecific thalamic" nuclei* is rather enigmatic. They will not be considered here. However, two other interesting efferent projections of the cerebellar nuclei will be briefly considered.

It has for a long time been regarded as established that all efferent fibers from the cerebellar nuclei pass to extracerebellar regions. Early anatomical (Carrea, Reissig, and Mettler, 1947) and physiological reports of a connection from the nuclei to the cerebellar cortex have not been considered decisive. However, in the last few years it has been shown anatomically by means of modern fiber tracing methods (see Chap. 1) as well as physiologically by recording of antidromic responses that *cells in the cerebellar nuclei project to the cerebellar cortex* in the cat (Tolbert, Bantli, and Bloedel, 1976; Gould and Graybiel, 1976), in the tree shrew (Haines, 1977b), and in the monkey (Tolbert, Bantli, and Bloedel, 1977; Chan-Palay, 1977). From physiological experiments, the fibers ending in the cerebellar cortex (presumably as mossy fibers) have been concluded to be collaterals of efferent nuclear fibers passing to the thalamus (Tolbcrt, Bantli, and Bloedel, 1978) or ending in the inferior olive (Ban and Ohno, 1977; Tolbert et al., 1978). The nucleocortical projection appears to be topographically organized and to arise from relatively small cells. The existence of these nucleocortical fibers is of great interest, since it suggests a feedback control of the cerebellar cortex via the first synaptic station in its efferent projections (in addition to feedback systems involving precerebellar nuclei).

It remains to consider briefly a projection passing directly to the *oculomotor nucleus* and the accessory oculomotor nuclei described in the monkey by Carpenter and Strominger (1964) and confirmed autoradiographically by Miller and Strominger (1977) and Chan-Palay (1977) and in the opossum by Martin, King, and Dom (1974) in a degeneration study. These fibers are held to arise from a ventral area in the caudal pole of the dentate (Chan-Palay, 1977) which contains large and small cells. Whether this region corresponds to the small-celled part of the lateral nucleus of the cat, described by Flood and Jansen (1961, see Fig. 5-8B), is not clear (see, for example, Courville and Cooper, 1970; Haines, 1978). However,

---

[47] The lateral reticular nucleus, the paramedian reticular nucleus, and the perihypoglossal nuclei appear to receive cerebellar afferents from the fastigial nucleus only and not from the other cerebellar nuclei.

this region has been shown to receive primary vestibular fibers in the cat (Brodal and Høivik, 1964) and in the monkey (Carpenter, Stein, and Peter, 1972); fibers from the flocculus in the cat (Voogd, 1964; Angaut and Brodal, 1967); from the flocculus, nodulus, and uvula in the galago (a prosimian primate, Haines, 1977a); and from the ventral part of the uvula in the rabbit (van Rossum, 1969). It appears likely, therefore, that the dentato-oculomotor fibers are a link in a vestibulo-ocular pathway, in addition to other routes (see Chap. 7), as suggested by Haines, among others (see Haines, 1977b).[48]

Cerebellovestibular connections. As briefly referred to above, the vestibular nuclei receive afferent fibers from the cerebellar nuclei as well as from the cerebellar cortex. However, these fibers originate only in certain parts of the cortex and the nuclei. The cerebellovestibular connections of both kinds are known in some detail and have turned out to be organized according to rather complex patterns. It will be practical to consider the two components separately.

Cerebellar corticovestibular fibers arise from certain parts of the cerebellar cortex: *the anterior and posterior parts of the vermis and the vestibulocerebellum* (the flocculonodular lobe and adjoining regions of the uvula and probably the ventral paraflocculus). The former projection appears to be organized in a somewhat simpler pattern than the latter and to be quantitatively the most important. *Direct vermal cerebellovestibular fibers* have been described by several authors who chiefly used the Marchi method, and by some who used silver impregnation methods. Walberg and Jansen (1961) could thereby establish two features of considerable interest. In the first place, most of the fibers from the anterior-lobe vermis end in the ipsilateral lateral vestibular nucleus, while a few end in the adjoining parts of the descending nucleus. Secondly, the projection onto the lateral nucleus is somatotopically organized (Fig. 5-30), fibers from the caudal part of the anterior-lobe vermis ending in the "forelimb region" of the nucleus. Largely corresponding findings have been made by other authors in the cat (Voogd, 1964), the rabbit (van Rossum, 1969), the opossum (Sreesai, 1974), and the galago (Haines, 1976). These cortical cerebellovestibular fibers appear to arise only from the lateral parts of the vermis, Voogd's zone B (Voogd, 1964; van Rossum, 1969; Haines, 1976). In agreement with suggestions made in silver impregnation studies, the experimental electron microscopic study of Mugnaini and Walberg (1967) has shown that the anterior-lobe efferents to the nucleus of Deiters establish synaptic contact with cells of all sizes and with both cell somata and proximal and distal dendrites. The somatotopical pattern found anatomically is in agreement with the physiological findings of Pompeiano and Cotti (1959).

Although lobules VI and VII of the vermis do not appear to project to the vestibular nuclei, as indicated in Fig. 5-30 (see particularly van Rossum, 1969), the posterior part of the vermis (lobule VIII and the rostral part of the uvula) send fibers to the lateral vestibular nucleus. A somatotopic pattern could not be found in this part of the projection by Walberg and Jansen (1961). Haines (1975a) found

---

[48] Autoradiographically, fibers have been traced from the fastigial nucleus to the accessory oculomotor nuclei (Batton, Jayaraman, Ruggiero, and Carpenter, 1977). Gacek (1977) found labeled cells in the nucleus interpositus (anterior?) after injections of HRP in the oculomotor nucleus, and there may be fibers from the fastigial nucleus to the trochlear nucleus (see footnote 38).

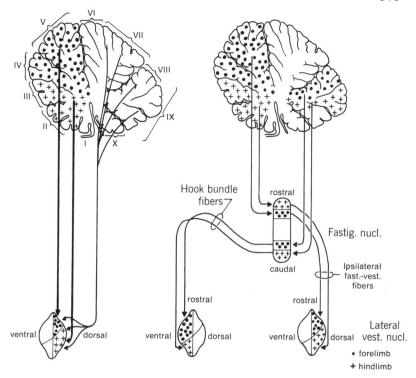

FIG. 5-30 Diagram illustrating major features in the projection from the cerebellar cortex onto the nucleus of Deiters (to the left) and (to the right) in the projections from the cerebellar cortex onto the fastigial nucleus and from this to the lateral vestibular nuclei. Note that the direct cerebellovestibular fibers and the projection from the rostral part of the fastigial nucleus end in the dorsal half of the ipsilateral lateral vestibular nucleus, while the fibers from the caudal part of the fastigial nucleus via the hook bundle supply the ventral half of the contralateral lateral vestibular nucleus. Within each of these projections there is a somatotopic localization. (See text). Slightly altered from Brodal, Pompeiano, and Walberg (1962).

a more detailed pattern in the projection from the posterior-lobe vermis in the galago.

In addition to the *direct route* considered above, the same parts of the cerebellar vermis are able to influence the vestibular nuclei *indirectly,* via *a relay in the fastigial nucleus.* As previously described in this chapter, the vermis of the anterior lobe gives off an orderly arranged projection to the fastigial nucleus. In general, rostral and caudal parts of the vermis send fibers to rostral and caudal parts, respectively, of the fastigial nucleus. Several authors have studied the fastigiovestibular projection (see Brodal, Pompeiano, and Walberg, 1962, Brodal, 1974, for references). Even though the direct and indirect vermal cerebellovestibular pathways differ in many respects, on some points there are remarkable similarities.[49] In a detailed study of the fastigiovestibular projection in the cat after lesions

[49] In both anatomical and physiological studies, the course of the crossing fastigial efferents and the course of direct vermal corticovestibular fibers through the rostrolateral part of the fastigial nucleus represent important sources of error in the interpretation of results.

of the fastigial nucleus, Walberg, Pompeiano, Brodal, and Jansen (1962; see also Voogd, 1964) found that ipsilateral fibers, arising chiefly in the rostral part of the nucleus, supply all four main vestibular nuclei, as do the crossed fibers coming mainly from the caudal part. However, fibers from the two components tend to supply different parts of each of the nuclei, with some overlapping (except in the superior vestibular nucleus, where both end preferentially in the peripheral regions; see Chap. 7). Of particular interest are the projections onto the nucleus of Deiters (the lateral vestibular nucleus) whose spinal projection shows a somatotopic pattern (see Chap. 7). The ipsilateral fibers from the rostral part of the fastigial nucleus supply the dorsal half of the lateral nucleus (Fig. 5-30) in a topical pattern.[50] When correlated with the pattern in the projection from the anterior-lobe vermis, it appears that the anterior lobe has at its disposal two somatotopically organized pathways to the lateral vestibular nucleus, one direct and one involving the fastigial nucleus.

> The relative roles of the direct and indirect routes from the vermis to the lateral vestibular nucleus are not yet entirely clear (see Pompeiano, 1974, for an exhaustive review). They are scarcely identical. Since the axons of the Purkinje cells have an inhibitory action, the vermis, via its direct fibers, will inhibit the neurons of Deiters' nucleus, while the fastigiovestibular fibers have an excitatory action. The fastigial neurons may be excited by afferents of noncerebellar origin, presumably chiefly collaterals of fibers to the cortex, which influence the Purkinje cells as well. The situation is further complicated by the recent demonstration of projections from the cerebellar nuclei to the cerebellar cortex. It has generally been assumed with regard to the fastigial nucleus that its "excitatory input provides an amorphous background excitation that is given somatotopic form by the sculpturing action of the more specifically organized Purkyně cell discharges" (Eccles et al., 1974, p. 115).

The patterns in the direct and indirect vestibular projections from the posterior vermis are less definitely known than those from the anterior-lobe vermis, but physiological studies (see Pompeiano, 1974, for a review) favor a somatotopic pattern in the direct pathways. The largely different terminal regions in the vestibular complex of fibers from the rostral and caudal parts of the fastigial nucleus indicate that the two parts play different functional roles. Physiological studies likewise indicate such differences. Furthermore, there appear to be anatomical and physiological differences between lateral and medial regions of the fastigial nucleus.

> The diagrams of Fig. 5-30 are greatly simplified to bring out some essential points in the projections to the lateral vestibular nucleus. The distinction between a rostral and a caudal part of the fastigial nucleus and their projections is not as sharp as indicated. Crossed projections do not come exclusively from the caudal part, but to some extent from rostral parts, and some ipsilateral fibers arise in the caudal part. The somatotopic pattern in the fastigial nucleus is less sharp than shown, as is also apparent from electrophysiological studies (for example, from studies of the input to the fastigial nucleus in response to cutaneous stimulation of the forelimb and hindlimb by Eccles, Rantucci, Sabah, and Táboříková, 1974).

The cerebellovestibular projection from the vestibulocerebellum was investigated in the rat, cat, and monkey by Dow (1936, 1938a) in Marchi preparations. Voogd (1964) and van Rossum (1969) described further differences in the sites of termination of fibers from various parts of the vestibulocerebellum. The patterns in

---

[50] An autoradiographic study of the distribution of fastigiovestibular fibers in the monkey (Batton, Jayaraman, Ruggiero, and Carpenter, 1977) has given somewhat different results.

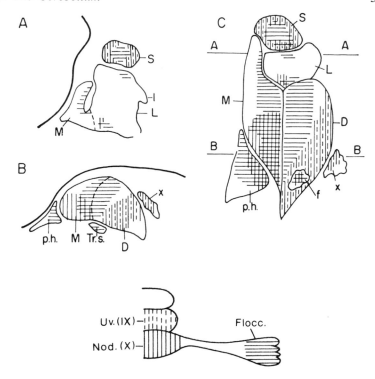

FIG. 5-31 Diagram showing the sites of terminations in the vestibular nuclei of fibers from the flocculus, nodulus, and uvula in the cat (see key below). A and B, from transverse sections at two levels indicated in C, which is from a horizontal section. Differences in densities of terminations are not shown. Note that contingents of fibers from the vermal and hemispheral parts of the vestibulocerebellum (nodulus-uvula and flocculus, respectively) largely have their particular fields of termination. Some abbreviations: *D, L, M,* and *S:* descending (spinal), lateral, medial, and superior vestibular nucleus, respectively; *f* and *x:* small cell groups of the vestibular complex; *p.h.:* nucleus prepositus hypoglossi; *Tr.s.:* tractus solitarius. See text. Redrawn from Angaut and Brodal (1967).

these projections were studied in more detail in the cat with the Nauta method by Angaut and Brodal (1967) and in the galago with the Fink and Heimer method by Haines (1977a). Apart from minor differences, which may reflect variations among species, the two studies are in agreement. The fibers are distributed ipsilaterally. In keeping with what has been found for other afferents to the vestibular nuclei (see section on the vestibular nerve, Chap. 7 and Brodal, 1974), the afferents from the flocculus, nodulus, ventral part of the uvula each have their particular sites of termination within the vestibular nuclei, even though there is some overlap, as can be seen in Fig. 5-31. All authors agree that the lateral vestibular nucleus receives only a few fibers from the vestibulocerebellum.

There is considerable overlap between the nodular and the uvular projection areas, while the flocculus to a large extent supplies other regions (see also Fig. 13 in Haines, 1977a). This may be seen as an indication that the flocculus and nodulus-uvula play different functional roles, as appears from physiological studies. Some of the minor cell groups of the vestibular nuclear complex (see Chap. 7) likewise receive fibers. A correlation between the sites of ending of cerebellovestibular fibers and other afferents to the vestibular nuclei is of interest for functional correlations (see Brodal, 1974). For example, in the superior vestibular nucleus the central area that receives fibers from the flocculus is also the main site of termination of primary vestibular afferents, whereas the peripheral part, receiving fibers from the nodulus-uvula, is supplied by afferents from the fastigial nucleus.

It can be seen from the above that the cerebellovestibular pathways from the vestibular parts of the cerebellum differ in many respects from those from the vermis. The anatomy of these pathways makes it clear that the cerebellar influences on the vestibular nuclei must be very complex. Greatly simplified, however, it may be said that the vermal areas projecting onto the vestibular nuclei (particularly the direct projection onto the nucleus of Deiters) generally exert their main influence on the spinal cord and are related first and foremost to the control of postural mechanisms, in agreement with the important spinal input to these parts of the cerebellum. The efferents from the vestibulocerebellum, on the other hand, supply mainly other parts of the vestibular nuclei which are related to the vestibular influences on oculomotor mechanisms. However, the distinction is not sharp. Furthermore, as mentioned above, the vermal and hemispheral parts of the vestibulocerebellum differ both in their projections and functionally.

The preceding accounts of the many afferent and efferent cerebellar fiber connections, although far from complete, convey an impression of the complexity in their organization. Each of the fiber contingents and projections appears to have its particular pattern of organization. Anatomical interrelations between various "systems" are far more abundant and intricate than was believed even a few years ago. An attempt to include all fiber connections in a single diagram would give a rather confusing picture. Figure 5-32 intends to show only some main features.

Before attempting to present some of the principal points from current views on the function of the cerebellum, it is appropriate to describe the symptoms most often met with in cerebellar affections in man.

Cerebellar symptoms.   The classical study of cerebellar symptoms is still the paper by Gordon Holmes (1939). In recent years relatively few authors have been interested in the subject (see, however, Wyke, 1947; Brown, 1949; Wartenberg, 1954). It has been extensively reviewed by Dow and Moruzzi (1958) and Dow (1969). Leaving for a later section the question whether particular symptoms can be related to specific parts of the cerebellum, the symptoms most commonly observed in cerebellar lesions will be briefly described.

Hypotonia. The tonus of the muscles is reduced; this can usually be ascertained either by palpating the muscles or by examining their resistance to passive movements. When hypotonia is present, the tendon (myotatic) reflexes are usually less brisk than normal, and they may be pendulous. In a unilateral lesion these changes occur on the side of the lesion.

A certain degree of *asthenia* and increased fatigability of the muscles is commonly present together with the hypotonia, and is regarded by some (see Holmes, 1939) as one of the fundamental disturbances in cerebellar lesions.

Other symptoms may be ascribed to a lack of cooperation between the muscles in voluntary contractions, affecting the rate, regularity, and force of the movements. Symptoms of this kind are somewhat difficult to analyze, and have been given different names by various neurologists. However, it appears doubtful that they really are more than different aspects of the same fundamental disturbance. Among such symptoms may be mentioned:

Asynergia. This is defined as a lack of capacity to adjust correctly the impulses of innervation in the various muscles participating in a movement. The different stages of a compound movement are performed as isolated, successive

FIG. 5-32 A simplified diagram of some of the principal cerebellar connections. The lateral reticular nucleus and its connections are not included, nor are the spino-olivary and certain other connections shown. Modified and redrawn from Rasmussen (1932).

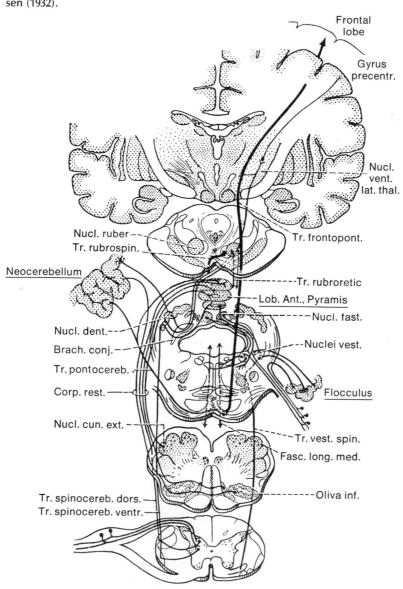

movements, and not as a single synchronous, harmonic performance (*décomposition des mouvements*). The excursions of the movements may be excessive or too small, *dysmetria,* frequently then followed in the next performance by a compensation, leading to an exaggeration in the opposite sense. The impairment of performing rapid alternate movements, such as pronation and supination of the hand, *dysdiadochokinesis,* is probably also a feature of asynergia.

Tremor. This occurs in voluntary movements, but as a rule not when the extremity is at rest. This "cerebellar tremor" is coarse and arrhythmic and of a different type than the tremor seen, for example, in paralysis agitans.

Spontaneous past pointing. When the patient is told to point in a certain direction with his eyes closed, e.g., to the finger of the examiner, he points to the side, below, or above the right point. Most frequently the deviation is outward.

Deviation of gait may be observed when the patient is made to walk with his eyes closed; usually the deviation occurs in the same direction as the past pointing. When taking some steps alternately forward and backward, the phenomenon of compass gait may become evident.

Unsteadiness of gait, staggering (titubatio), may be accompanied by a tendency to fall to one particular side, or it may occur without any definite directional preponderance.

Nystagmus and other ocular disturbances may occur in cerebellar lesions. They will be considered below when the cerebellar relations to visual functions are discussed (see also Chap. 7, the section on the vestibular nerve).

Speech disturbances are seen, particularly with rapidly developing and severe lesions. There may be defects in the proper cooperation of articulation, phonation, and respiration in speech, resulting in slowness, explosiveness, and so-called scanning speech (for particulars, see Zentay, 1937).

The function of the cerebellum as inferred from experimental studies. In the course of time, a number of studies have been performed to elucidate the function of the cerebellum. For a critical review of earlier literature, the monograph of Dow and Moruzzi (1958) should be consulted. As it gradually became clear that various parts of the cerebellum differ with regard to their connections, experiments were performed to study the function of the various subdivisions specifically, and considerable progress has been made. However, the course of the afferent and efferent cerebellar fibers is such that in experiments with ablations or stimulations of various parts of the cerebellum, there is considerable risk of stimulating or destroying connections which are not immediately relevant to the cerebellar subdivision under study. The geography of the cerebellar connections often makes this unavoidable, but the fact should be kept in mind when the functional changes observed are interpreted. Furthermore, it is to be expected that the differences in development of the cerebellar subdivisions among animal species will be reflected in functional studies.

In the preceding sections of this chapter, mention has been made of a number of findings concerning the origin and terminations of cerebellar connections and of physiological studies of recordings of afferent impulses to the cerebellum and of results of stimulation or destruction of particular structures. These new data bear evidence of the extremely complex organization of the cerebellum and are still difficult to fit into a coherent picture. Many of these data are as yet of little immediate relevance to the diagnosis of cerebellar diseases in man, mainly for two reasons: (1) The lesions of the cerebellum met with in patients are usually not restricted to one minor region which appears to have its particular functional role (for example the rostromedial part of the fastigial nucleus). (2) The methods available for the diagnosis of cerebellar disorders in man are generally still rather crude. In the following, emphasis will be put on the results of ablations and, to a lesser extent, stimulations of different parts of the cerebellum. At the outset, it will be useful to recapitulate briefly some general points.

The uniform structure of the cerebellar cortex indicates that the basic "machinery" of the cortex is the same in all parts of the cerebellum. The functional roles of the climbing fibers and mossy fibers are quite different. If, as is generally believed, the inferior olive is the sole source of climbing fibers, its role in the activation of the cerebellum will be different from that of all other afferent fiber systems (which exert their influence by way of mossy fibers). The cerebellum receives information from most, if not all, sensory modalities. It appears that each sensory modality, and also impulses from other sources, are mediated both via mossy fiber systems and via climbing fibers (via the olive). This "dual input" (see Strata, 1975, for a short review) appears to be a basic pattern, but its functional significance is still in the realm of speculations.

Because of the uniform structure (and mode of operation) of all regions of the cerebellar cortex, the influence of its different parts on particular bodily "functions" will depend on the afferent input and the targets of the outgoing messages (or commands) of each part. In these respects there are marked differences between various cerebellar regions. The cerebellar nuclei must play a far more important role in the working of the cerebellum than previously believed (when they were thought of as functioning solely as relay stations in efferent cerebellar pathways). Not only do they receive afferent information, as does the cortex, but in addition they may influence the latter by nucleocortical projections. The principle of feedback demonstrated in these connections appears to be generally valid for the cerebellum as a whole. It is a general rule that regions which receive impulses from the cerebellum (cortex or nuclei) are able to act in turn on the cerebellum by means of direct or more indirect routes. It should be recalled that these feedback systems are often not completely reciprocal.[51] In considerations of the functional role of such feedback systems, it is therefore essential to take into account topical features.

The fiber connections of the cerebellum and physiological studies show that it is involved in many spheres of activity, not only of the skeletal musculature. Thus it influences visceral functions, and some areas are particularly related to visual or acoustic functions. These aspects will be briefly considered below. It will be practical to take as a starting point the traditional subdivision of the cerebellum into three main regions (see Fig. 5-1), loosely referred to as the "vestibulocerebellum," the "spinocerebellum," and the "pontocerebellum," even though the distinction is not sharp and there are ample possibilities for cooperation between the three regions.

Thus, on the afferent side, pontocerebellar fibers end in considerable numbers in the cortex of the "spinocerebellum." Spinal afferents may act on the "vestibulocerebellum" by way of endings in certain parts of the vestibular nuclei, such as the group *x*, which projects onto the flocculus and nodulus-uvula. On the efferent side the overlap is equally evident, as witnessed for example, by the fibers from the vermis and fastigial nucleus to the vestibular nuclei, the pathways to the thalamus from the cerebellar nuclei. Possibilities for direct interactions between the main subdivisions, however, appear to be scanty since the *cerebellar cortical association fibers* are sparse

---

[51] For example, all parts of the lateral reticular nucleus project to the cerebellum, but only its magnocellular part appears to receive afferents from the cerebellum.

and most of them interconnect neighboring folia (Jansen, 1933; Eager, 1963b). According to Frezik (1963) the association fibers are recurrent collaterals of Purkinje cell axons.

The *vestibulocerebellum* (consisting of the flocculus, nodulus, caudal part of the uvula, and probably the ventral paraflocculus, see Fig. 5-1) is difficult to approach, particularly the flocculus. If a *lesion* in the monkey is restricted to the *nodulus* (Dow, 1938c), or to the nodulus and the adjoining parts of the uvula (Botterell and Fulton, 1938), disturbances of equilibrium appear, but none of the other cerebellar symptoms are observed. The animal sways when standing and walks unsteadily. However, if the trunk is supported, as when the animal is sitting in a corner of its cage, movements of the extremities are performed without ataxia. There are no signs of asynergia or tremor, and there is no hypotonia of the muscles. If the same operation is performed in animals whose labyrinths have been destroyed bilaterally, no additional symptoms appear following the ablation of the nodulus. These results were confirmed by Carrea and Mettler (1947). In the cat, Fernández and Frederickson (1964) observed disequilibrium without ataxia to follow lesions of the nodulus (and the adjoining part of the uvula), as well as positional nystagmus (nystagmus influenced by the position of the animal) and prolonged vestibular reactions to rotatory and caloric tests (see section on the vestibular nerve, Chap. 7). Their findings following attempts to destroy the flocculus were inconclusive, while Carrea and Mettler (1947) noted disturbances in the monkey resembling those following lesions of the nodulus. We shall return to the ocular signs which follow ablation or stimulation of the vestibulocerebellum when we discuss the cerebellar relations to visual functions.

The *"spinocerebellum,"* when defined as the region receiving spinal impulses, covers most of the vermis and intermediate zone of the cerebellum (see Fig. 5-1). These parts of the cerebellum are chiefly concerned in the control of posture. Studies from recent years have shown that the "spinocerebellum" is not an entity, but that several minor regions are to be distinguished, as indicated from data mentioned in previous sections. Among the first important demonstrations of this were the studies of Chambers and Sprague (1955a, b) which showed that the longitudinal subdivision distinguished in the corpus cerebelli on the basis of anatomical findings (see Fig. 5-4) has its functional counterpart. These authors undertook an analysis of the functional disturbances following ablation of the vermis proper, the intermediate zone, and the lateral parts in the cat. They concluded that the vermal zone "controls the posture, tone, locomotion, and equilibrium of the entire body"; the intermediate zone "controls the postural placing and hopping reflexes, tone, and individual movements of the ipsilateral limbs"; the lateral zone in the cat "does not appear to be involved in the regulation of posture or progression," but its function "is limited at most to a regulation of placing and hopping reflexes and volitional movements of the ipsilateral limbs." Further studies have brought additional data which support these conclusions, but in addition have demonstrated differences between rostral and caudal parts within the vermis and intermediate zone, as is evident from what has been described in previous sections. It was found relatively early that *the anterior lobe exerts an inhibitory influence on postural tone.* Most of these experiments have been performed on decerebrate cats, in which the presence of an extensor rigidity provides a useful background for testing a reduction in extensor tone. Following ablation (or temporary cooling) of the anterior lobe in a decerebrate cat there is an increase of decerebrate rigidity, while stimulation reduces it. Later studies have shown that this effect is strictly a property of the anterior-lobe vermis, while the intermediate zone has an opposite effect. However, in total ablations of the anterior lobe the latter effect is not evident, presumably because the vermal action is the strongest. The situation is similar with regard to the posterior part of the spinocerebellum, the pyramis, the rostral part of the uvula, and the paramedian lobule, but on the whole these regions appear to be less potent than the corresponding zones in the anterior lobe. For all these subdivisions *the effect following localized ablations or stimulations has been shown to be somatotopically localized.*

In the light of many detailed studies it seems possible to explain the main features of these changes as follows (see Pompeiano, 1967): The Purkinje cells of the anterior and posterior cerebellar vermis exert an (at least predominantly) inhibitory action on the cells of Deiters's nucleus, which have a monosynaptically excitatory action on extensor motoneurons (in addition, however, the vermis acts on the nucleus of Deiters via the fastigial nucleus). The loss of this inhibition increases the extensor tone. The intermediate part of the anterior lobe normally inhibits the nucleus interpositus, which has been shown to have a monosynaptic excitatory action on the cells of the red nucleus, which again, via the rubrospinal tract, activate flexor motoneurons. In both instances α as well as γ neurons are acted upon.

As mentioned in Chapters 3 and 4, under normal circumstances α and γ motoneurons are activated together. The rigidity seen following decerebration in experimental animals appears to be mainly a "gamma rigidity" due to increased γ activity. When the anterior lobe is removed or inactivated by cooling, the ensuing rigidity is found to be due mainly to hyperactivity of the α motoneurons and some depression of fusimotor activity (Granit, Holmgren, and Merton, 1955; see also Granit, 1970).

The simplified scheme of the increases of postural tone following removal of the anterior lobe given above does not take into account the projections from the fastigial nucleus onto facilitatory as well as inhibitory regions of the reticular formation, which act on the motoneurons (cf. Chaps. 4 and 6). The actions of these connections are not known in sufficient detail, but among other things the fact that the influences of the vermis proper as well as the intermediate zone are somatotopically organized indicates that the action via the lateral vestibular nucleus and the red nucleus are important in the cerebellar control of muscle tone. The simplified presentation given above, furthermore, does not take into account the many detailed physiological observations, particularly of Moruzzi and Pompeiano and their collaborators, partially described previously, which show that minor parts of the fastigial nucleus and the nucleus interpositus have different actions. Furthermore, symptoms following lesions of the nuclei do not mirror those seen after lesions of the cortex projecting onto the particular nucleus.

Postural reflexes, some of which have been referred to in Chapter 4, have been found to be influenced by lesions of the anterior lobe. Another of these reflexes may be mentioned here, the so-called positive supporting reflex. This consists of an extension of the extremity in response to a light pressure on the pad. It is in keeping with the anatomy of the spinocerebellar pathways and of its efferent connections that the cerebellum may influence postural reflexes. According to Chambers and Sprague (1955a, b) the intermediate zone is particularly involved in these functions, a point of interest in view of the termination of the dorsal spinocerebellar and cuneocerebellar tracts chiefly in this zone (Fig. 5-10). That the tactile placing reactions are likewise influenced fits in with the demonstration that tactile stimuli give rise to action potentials in the spinocerebellum. In general it appears that lesions of the spinocerebellum, especially the intermediate zone, result in exaggeration of the postural reflexes, particularly the "positive supporting reflexes," while loss or reduction of some other postural reflexes has been described. From the study of Corvaja, Grofová, Pompeiano, and Walberg (1977b) referred to above, it appears that the lateral reticular nucleus may be a particularly important station in the neuronal circuits that control postural tonus and placing reactions.

As we have seen, the main efferent projections from the vermis and intermediate zone of the cerebellum pass to the fastigial and interpositus nuclei, respectively. While these cerebellar regions and nuclei thus may act on spinal mechanisms by way of efferents to the vestibular nuclei, the reticular formation, and the red nucleus, it should not be overlooked that they may in addition influence "higher levels" of the brain, not least the thalamus (especially the VL, projecting onto the primary motor cortex). This indicates that these parts of the cerebellum may influence the corticospinal tract and other corticofugal connections as well, and suggests that they are not without relation to what is generally spoken of as voluntary movements (see Chap. 4 and below).

Concerning the *functional role of the middle part of the vermis and intermediate zone,* it has been mentioned above that they receive impulses from the face and also optic and acoustic impulses. It is appropriate at this junction to consider the role of the cerebellum in visual and acoustic functions, even though other parts of the cerebellum are involved as well.

Cerebellar relations to visual functions. Our knowledge of the relation of the cerebellum to visual functions, including movements of the eyes, has increased considerably in recent years. Cerebellar responses to visual stimuli and to stimulation of proprioceptors in the extrinsic ocular muscles have been repeatedly demonstrated, as have cerebellar influences on ocular movements. Certain cerebellar regions appear to be particularly closely linked to these functions, especially *vermal lobules VI and VII* (perhaps VIII) and the *flocculus and nodulus.* The pathways engaged in these phenomena are not completely known, as is apparent from the discussion above of the cerebellar fiber connections. An attempt to summarize the most relevant points of our knowledge concerning the oculocerebellar relations may, however, be of interest.

Following the initial demonstration by Snider and Stowell (1942, 1944) referred to above, that optic impulses (from both the retina and the visual cortex) are transmitted to the middle part of the vermis, several studies have confirmed this. The area in question appears to be vermal lobules VI, VII, and VIII, particularly lobule VII, but the area is not sharply circumscribed (for references and a review, see Fadiga and Pupilli, 1964), and visually evoked potentials have been recorded, under certain experimental conditions, even from the hemispheres (see Mortimer, 1975).

The *visual information reaching the cerebellar vermis* may follow different routes. As described above, the most important pathway appears to be that from deeper layers of the superior colliculus to the dorsolateral pontine nucleus and from this to the vermis (see, for example, Kawamura and Brodal, 1973, and Hoddevik, Brodal, Kawamura, and Hashikawa, 1977). In addition, this pontine area receives some fibers from the visual cortex (P. Brodal, 1972a, b; Kawamura and Brodal, 1973). Some impulses from the visual cortex and from the ventral lateral geniculate may be transmitted to the vermal visual area by some other pontine cell groups (see Hoddevik et al., 1977), as may the visual responses recorded outside the vermal visual area. It is in agreement with the anatomical data that, after visual stimulation, the mossy-fiber input to these lobules is more widespread and more consistent than the climbing-fiber input (Buchtel, Iosif,

Marchesi, Provini, and Strata, 1972). The latter is assumed to be mediated via the inferior olive. Of the olivary areas projecting to the visual area of the vermis (Hoddevik, Brodal, and Walberg, 1976), only the sparsely projecting nucleus β receives an afferent input related to visual functions. Other pathways for visual impulses to cerebellar lobules VI–VIII may be involved as well. Among them a route via the lateral reticular nucleus may be important, since this nucleus receives a fair amount of afferents from the superior colliculus and the pretectum.

The *visually evoked potentials recorded in the nodulus and flocculus* (Maekawa and Simpson, 1973; Simpson and Alley, 1974; and others) have been found to be predominantly climbing-fiber responses and to be direction-selective. In addition, mossy-fiber responses have been observed in a circumscribed part of the flocculus (Maekawa and Takeda, 1975). As mentioned above, the part of the inferior olive projecting onto the flocculus and nodulus, chiefly the dorsal cap of Kooy (Alley, Baker, and Simpson, 1975; Hoddevik and Brodal, 1977; see Fig. 5–18), receives afferents from the pretectum and from the terminal nucleus of the accessory optic tract, and thus appears to be a link in the transmission of the climbing-fiber responses to the vestibulocerebellum. A possible relay station for the mossy-fiber impulses may be a restricted area of the pons which gives off some fibers to the flocculus (Hoddevik, 1977). Some fibers, probably mediating visual impulses, have been traced to these pontine regions.

The cerebellum has further been shown to receive information from the *extrinsic ocular muscles,* probably from their stretch receptors (Fuchs and Kornhuber, 1969; Baker, Precht, and Llinás, 1972; Batini, Buisseret, and Kado, 1974; Schwarz and Tomlinson, 1977). Climbing- and mossy-fiber responses have been recorded mainly from lobules VI and VII and have been found to converge on the same Purkinje cells (which also can be activated from trigeminal proprioceptors; Batini et al., 1974). The pathways used in the transmission of these proprioceptive impulses are incompletely known, even though such impulses have been recorded from the mesencephalic trigeminal nucleus (see Chap. 7, section g). The long latency observed for the potentials in the cerebellum may suggest a multisynaptic pathway. The subnuclei oralis and interpolaris of the spinal trigeminal nucleus may be considered, since they send fibers to the cerebellum, as mentioned above.

The cerebellar areas receiving visual information or information from the extrinsic ocular muscles have been found to *influence the oculomotor nuclei and eye movements.* However, there are still many open questions, not least concerning the collaboration between the cerebellar areas that may influence different "systems" of ocular movements. Only some data from the extremely complex field will be mentioned below. (For reviews see, for example, Hoyt and Daroff, 1971; Cohen and Highstein, 1972; Precht, 1975; *Control of Gaze by Brain Stem Neurons,* 1977).

Particularly well established is the production of ipsilateral, conjugate horizontal eye movements on stimulation of *lobules VI–VII* of the vermis and adjoining regions. From studies in alert monkeys (Ron and Robinson, 1973) it appears that this area contains a "map" of saccadic movements in different directions. Saccadic movements can likewise be obtained from the fastigial nucleus (which receives fibers from the vermis). Whether disturbances in saccadic eye movements follow experimental lesions of the cerebellum in animals appears not to be settled yet, but potentials associated with saccadic eye movements have been recorded in vermal lobules VI and VII (Wolfe, 1971; and others). In man, dysmetria of saccadic ocular movements ("overshoot dysmetria") has been found in cerebellar affections (see Selhorst, Stark, Ochs, and Hoyt, 1976), and it appears that it is due to dysfunction of part of the vermis.

More marked cerebellar influences of oculomotor functions appear to be exerted by the *vestibulocerebellum.* As briefly mentioned above, *positional nystagmus* is observed following *lesions* of the nodulus. This has been found in several animal species and man (Fernández and Fredrickson, 1964; Grant, Aschan, and Ekvall, 1964; Grand, 1971; and others). In *stimulation experiments* in rabbits Aschan, Ekvall, and Grant (1964), using electronystagmography, found that stimulation laterally in the nodulus resulted in positional nystagmus. In alert monkeys, Ron and Robinson (1973) observed nystagmus on stimulation of the flocculus, nodulus, and ventral uvula (quick phase directed contralaterally on weak stimulation).[52]

   Spontaneous nystagmus has repeatedly been observed after unilateral lesions of the vestibulocerebellum. Evoked caloric and optokinetic nystagmus is likewise affected, and the capacity to perform smooth-pursuit eye movements suffers. Many other varieties of abnormalities in vestibulo-ocular reflexes may occur with lesions of the vestibulocerebellum (for accounts, see Hoyt and Daroff, 1971; Dichgans and Jung, 1975). It is characteristic that many of the disturbances of ocular movements observed after damage to the cerebellum are transient both in animals and man.

Many theoretical attempts have been made to explain how the cerebellum influences eye movements. It appears that most researchers favor the view that the cerebellar regions concerned are not integrative parts of the basic mechanism for the different "systems" of eye movements (saccades, smooth-pursuit movements, vergence movements, and so forth.) and that the cerebellum is not essential for eye movements (see, for example, Cohen and Highstein, 1972). It is more likely that the cerebellum is responsible for a final precise control of other regions concerned in movements of the eyes, be they elicited on vestibular stimulation (vestibulo-ocular reflexes), by ocular stimuli (visual and proprioceptive), or in other ways (see Chap. 7).[53] This is compatible with known anatomical data.

   Not only do the relevant cerebellar areas receive the necessary information for this task (see above), but, as described in other sections of this chapter, via more-or-less direct routes they also project to brain regions involved in eye movements (the vestibular nuclei, the nucleus prepositus hypoglossi, the PPRF of the reticular formation, and to some extent indirectly even to the nuclei of the ocular muscles and the accessory oculomotor nuclei). Since the connections of the visual vermal area and the vestibulocerebellum (and even of the flocculus and nodulus) differ on important points, the suggestion set forth by some students that these cerebellar regions may control different "systems" of eye movements appears to be rather likely, but little is settled concerning this problem. It may be mentioned in this context that cerebellar stimulation (of the fastigial nucleus) has been found to influence the parasympathetic oculomotor neurons that innervate the intrinsic ocular muscles and thus to influence the light reflex and accommodation (see Hultborn, Mori, and Tsukahara, 1978; Hosoba, Bando, and Tsukahara, 1978).

An *acoustic area in the cerebellum* was described by Snider and Stowell (1942, 1944). After acoustic stimuli and stimulation of the auditory cortex they recorded action potentials in the middle part of the vermis, in an area approximately coinciding with the vermal visual area. This has been fully confirmed in later research, even though acoustic responses have been found outside this region as well (see Fadiga and Pupilli, 1964, for a review). In unanesthetized monkeys mossy- as well as climbing-fiber responses to acoustic stimuli were found in rather wide cerebellar regions outside the classical acoustic area and in the nuclei interpositi and the dentate nucleus (Mortimer, 1975).

---

   [52] Positional nystagmus must be considered as being due to an interference with ocular reflexes elicited from the "static labyrinth" (the maculae equipped with otoliths, see Chap. 7). Many primary vestibular afferents from the utricular macula appear to end in the descending and medial vestibular nuclei (see Chap. 7), which project to the flocculus and nodulus, as discussed above in the present chapter. Positional nystagmus may also be seen after lesions of the vestibular organ, particularly of the utricle.
   [53] The inhibitory action of the Purkinje-cell axons is probably an important factor in such control mechanisms. Loss of this inhibition may possibly explain some oculomotor disturbances seen after unilateral cerebellar lesions.

The properties of single units responding to acoustic stimuli of different kinds have been explored (see, for example, Altman, Bechterev, Radionova, Shmigidina, and Syka, 1976). Mossy- and climbing-fiber responses have been studied. From the description of the fiber connections given above, it seems likely that the mossy-fiber responses may be mediated via a pathway from the inferior colliculus to the pons, particularly the dorsolateral pontine nucleus (Kawamura, 1975), which projects heavily onto vermal lobules VI–VIII (Hoddevik et al., 1977). This pontine region also receives fibers from the auditory cortex (P. Brodal, 1972c). A possible pathway for the climbing-fiber responses is less clear. Connections from the inferior colliculus to the inferior olive have apparently not been demonstrated. It may be of relevance that the inferior colliculus projects to the superior colliculus (see Chap. 7 and Fig. 7-31).

In view of the close topical correspondence between vermal visual and acoustic areas, it is of interest to note the very similar projection onto the pontine nuclei from the superior and the inferior colliculus. This strongly suggests that an extensive integration of visual and acoustic messages occurs in the dorsolateral pontine nucleus, which, furthermore, receives fibers from the acoustic cortex. The projection of the inferior onto the superior colliculus reflects an integration of this kind in the latter as well, the more so since fibers from the cortical acoustic area end in the superior colliculus (its deeper layers).[54] It appears likely that the acoustic input to the cerebellum contributes to the sensory basis of cerebellar motor control. After lesions of the middle part of the vermis, cats show reduced startle responses to loud noises (Chambers and Sprague, 1955b; and others).

Cerebellar influences on visceral (autonomic functions) have been described by many authors. Cardiovascular functions (for a review see Calaresu, Faiers, and Mogenson, 1975), for example the carotid sinus reflex (Moruzzi, 1940; and later authors), blood pressure, and heart rate are influenced on electrical stimulation of the cerebellum; likewise, pupillary constriction ensues (Moruzzi, 1950; and later authors). The micturition reflex was found by Chambers (1947) to be influenced from points in the rostral part of the fastigial nucleus. Intestinal and gastric motility and the defecation reflex are other functions that may be influenced from the cerebellum (see Martner, 1975).

The effects on autonomic functions have been observed chiefly after stimulation of the middle part of the vermis and the posterior part of the anterior-lobe vermis, and not least of the fastigial nucleus, particularly its rostral part (see Martner, 1975, for a recent study). The fiber connections concerned in the cerebellar control of visceral functions are largely unknown. Ultimately they must act on the preganglionic neurons in the sympathetic and parasympathetic system (see Chap. 11).[55]

The cerebellar influences on these neurons must occur chiefly by way of different multisynaptic routes. Projections from the fastigial nucleus to the sympathetic intermediolateral column have apparently not been found. Fibers from the fastigial nucleus pass to the nucleus parasolitarius, sit-

---

[54] There are minor differences between the sites of termination of fibers from the superior and inferior colliculus in the dorsolateral pontine nucleus (see Kawamura, 1975). When evaluated in connection with the projection of the latter onto the cerebellum, it appears that the classical visual and acoustic areas in the cerebellar vermis do not coincide completely, as is also apparent from physiological studies.

[55] This has been confirmed in electrophysiological studies by, for example, Nisimaru and Yamamoto (1977), who recorded depressed sympathetic discharges in the renal nerve of rabbits after stimulation of the vermis, particularly lobule VII.

uated closely lateral to the nucleus of the solitary tract (Walberg, Pompeiano, Brodal, and Jansen, 1962; Batton, Jayaraman, Ruggiero, and Carpenter, 1977). This pathway may possibly mediate an influence on the dorsal motor vagus nucleus. Ascending fibers from the cerebellar nuclei to the accessory oculomotor nuclei may be a link mediating an influence on the Edinger-Westphal nucleus.

It is of interest that visceral afferent impulses have been recorded in the cerebellum, mainly in the areas found to influence autonomic functions, by Rubia (1970), for example, after stimulation of the splanchnic nerve in the cat.

Cerebellar influences on speech mechanisms were referred to above. Gordon Holmes (1939) suggested on the basis of clinical evidence that the vermis and the paramedian regions are the cerebellar areas that are particularly involved in control of speech. Rather similar conclusions have been drawn by others (see Brown, Darley, and Aronson, 1970).

> Modern methods (electromyographic studies of laryngeal muscles, precise recordings of frequency and intensity of phonation) have made it possible to study various aspects of speech function in the monkey after lesions of the cerebellum. It appears from such studies that the cerebellum exerts control of laryngeal and respiratory mechanisms involved in speech and that lesions of the cerebellum produce a disruption of the coordination of these mechanisms. It is possible that different parts of the cerebellum (anterior lobe, posterior vermis, paramedian lobules, cerebellar nuclei) play different roles in this coordination, but little is known about this (for a recent study see Larson, Sutton, and Lindeman, 1978).

The *"pontocerebellum"* remains to be considered. On account of the more extensive development of the cerebellar hemispheres in man than in monkeys (and man's greater capacity to perform precise and complex voluntary movements), symptoms caused by lesions of the neocerebellum are more marked in man than in monkeys. We shall return briefly to some findings in man in the following section. These are largely compatible with the conclusions drawn from experiments in monkeys.

In cats and dogs the symptoms following ablations of the ansoparamedian lobule (making up most of the "pontocerebellum") are not marked (see Bremer, 1935b; Chambers and Sprague, 1955a, b; Dow and Moruzzi, 1958), but of the same type as seen in monkeys. Ablations of the hemisphere in monkeys result in disturbances which are different from those following lesions of the spinocerebellum (Botterell and Fulton, 1938; Carrea and Mettler, 1947). Hypotonia and clumsiness develop, as well as atactic movements of the extremities (asynergia). These symptoms are always seen on the side on which the lesion is performed. In the upper extremity, ataxia is particularly evident when the animal uses its hand for grasping or seizing an object; in the lower extremity it is mainly revealed when the animal is walking. If the dentate nucleus is damaged in addition, these symptoms are more enduring and a tremor also occurs. The tremor is not evident unless the animal performs voluntary movements. In monkeys these symptoms disappear in the course of some three weeks, but in apes they persist much longer (Botterell and Fulton, 1938). The tremor following a lesion of the dentate nucleus or a transection of the brachium conjunctivum (Walker and Botterell, 1937) is a coarse, irregular tremor, which is increased at the termination of the movements. It may be designated a typical intentional tremor. These results have generally been confirmed in later studies in monkeys.

However, in these experimental studies, and also in clinical studies of patients with cerebral affections, disturbances of voluntary movements of the limbs with dysmetria and tremor are frequently seen in affections of the intermediate region of the cerebellum—not only when the lateral part is damaged. This is indeed what might be expected, since the coordination of voluntary movements requires a precisely organized activity of postural mechanisms. To understand this, it will suffice to recall how practically every voluntary movement will alter the equilibrial status of the whole body and therefore will necessitate several adjustments in the muscular apparatus of the rest of the body. The fiber connections of the various main subdivisions of the cerebellum bear witness that there are possibilities for such collaboration between them. It appears to be an undue oversimplification to consider different parts of the cerebellum as being concerned in postural mechanisms and voluntary movements, respectively.

Reference was made above to the disturbances in *muscular tone* following extirpation of the anterior lobe (hypertonia) and following isolated lesions of the fastigial nucleus (hypotonia). When the entire cerebellum is ablated, hypotonia (decreased resistance to stretch) of limb muscles occurs. This is less marked in the cat than in the monkey (but is a marked feature in most cerebellar affections in man). The neocerebellum thus exerts an influence on muscular tone. It has been assumed that this hypotonia is due to disruption of the interaction between the dentate nucleus and the cerebral cortex. The synapses in all links of the projection dentate nucleus–thalamus–cerebral cortex are excitatory (see, for example, Massion, 1973). This view receives some support from the findings of Gilman (1969). Decerebellation in monkeys produces a decrease in the tonic sensitivity of muscle spindle afferents, presumably because of a decrease in the tonic discharge of fusimotor neurons, activated via the cerebellocorticopyramidal route. In agreement with this view it has been found that during the first weeks following lesions of areas 4 and 6, there is a decrease in the monkey's spindle responses to stretch (Gilman, Lieberman, and Marco, 1974). This is also the case following pyramidotomy (Gilman, Marco, and Ebel, 1971). In the motor cortex the corticospinal neurons acting on α and γ motoneurons are closely linked (see Chap. 4). An α–γ linkage, demonstrated in the action of the cerebellar anterior lobe (see above), is probably operative in the effects exerted on muscular tone by the neocerebellum as well (see Granit, 1977).

According to Bantli and Bloedel's (1976) studies in the monkey, the action of the dentate on the spinal cord, responsible for cerebellar hypotonia, may be partly mediated via pathways other than the cerebral cortex, since ablation of the sensorimotor cortex does not eliminate excitability changes in α motoneurons following stimulation of the dentate. The anatomical routes for these effects are conjectural.

Hypotonia following cerebellar lesions is sometimes considered a symptom separate from disturbances of movements. This is scarcely justified. Even though hypotonia can be observed in a resting extremity, all movements are accompanied by adjustments in muscular tone. It is unlikely that the mechanisms controlling muscular tone and movements can be separated.

In recent years refined experimental methods have been used to study the role played by the cerebellum in the coordination of voluntary movements. Animals, most often monkeys, have been trained to perform various types of movements or sequences of movements, which are recorded cinematographically or by other methods. Temporary cooling of the cerebellar nuclei, most often the dentate, by a stereotaxically inserted cooling probe results in a transient elimination of the activity of the nucleus. From studies of this kind it has been learned that cooling of the dentate affects the range, rate, and force with which the monkey executes a movement and results in disturbances of movements in the ipsilateral limbs resembling those seen after lesions of the cerebellum in man (Brooks, Kozlovskaya, Atkin, Horvath, and Uno, 1973; Conrad and Brooks, 1974; and others).

Further information on the role of the cerebellum in voluntary movements has been obtained by recording potentials in the cerebellar nuclei before, during, and after limb movements in trained animals. Many studies of this kind have appeared in recent years. Only a few data will be mentioned.

> The neuronal activity of the cerebellar nuclei appears not to be identical for all types of movement (fast or "ballistic," slow or "ramp," abruptly arrested, and others), even though many units in the dentate discharge in relation to both slow and rapid movements. Many units start changing their activity before the movement begins (see Thach, 1970a) and therefore could not have been brought into action by feedback from the movement. (Purkinje cells discharge as well; Thach, 1970b.) A dentate unit may be involved in only part of a motor sequence. Some units show "burst" activity, others fire more continuously, and still others appear to be silent (Grimm and Rushmer, 1974). As discussed in Chapter 4, the temporal relations between the activity of cells in the dentate nucleus and the "upper motor neurons" in the motor cortex have attracted much interest. It is particularly noteworthy that some dentate neurons have been found to be activated before any activity in the motor cortex can be discovered (see Chap. 4).

Some observations on the *nuclei interpositi* indicate important similarities but also differences between these and the dentate concerning their involvement in the performance of voluntary movements (Thach, 1970a; Uno, Kozlovskaya, and Brooks, 1973, in the monkey; Burton and Onoda, 1977, in the cat). Errors of range, rate, and force were most prominent in dentate cooling (Brooks, Kozlovskaya, Atkin, Horvath, and Uno, 1973). It appears that the interpositi are less involved in the accurate execution of voluntary movements than the dentate.

While it is obvious, not least for anatomical reasons, that the dentate nucleus influences first and foremost the cerebral cortex via the thalamus, and thus may be especially important for the execution of voluntary movements, it follows from its efferent projections that it may influence other brain regions as well. Conversely, the nuclei interpositi are not devoid of an influence on the cortex (see Figs. 5-27 and 5-28), even though their main outflow goes to other regions and appears to be particularly important for the control of postural mechanisms. In addition to the observations mentioned above (Thach; Brooks and collaborators; Burton and Onoda), other data illustrate such dual functions. For example, on electrical microstimulation of the dentate which permitted localized activation, Schultz, Montgomery, and Marini (1976) most often obtained in the forelimb of monkeys a stereotyped flexion movement. This was present after removal of the contralateral motor cortex, indicating that the movements represent a postural mechanism that is not mediated via the thalamocortical route.

On the basis of studies like those referred to above and numerous others, attempts have been made to interpret the role of the cerebellum in the execution and control of movements. Generally speaking, the dentate nucleus has been concluded to be involved in organizing the temporal relationship of motoneurons innervating agonistic and antagonistic muscles (cf. dysdiadokinesis in lesions) and to be concerned in the control mechanisms that regulate the force and rate of voluntary movements (cf. dysmetria in lesions). The lateral cerebellum has been considered to "program" movement parameters, such as force, duration, and direction, and to be essential in the learning of motor skills.

The cerebral cortical influence on the cerebellum must be considered in evaluations of the role played by the cerebellum in voluntary movements. It is generally assumed that the immediate commands to motoneurons engaged in voluntary movements arise in the cerebral cortex (see Chap. 4). The cooperation between it and the cerebellum is extremely complex and cannot be reduced to involve only the cerebellar hemispheres and the dentate nucleus. It has been suggested that when commands to motoneurons involved in a particular movement are issued from the cerebral cortex, leaving it via corticospinal (and presumably other corticofugal) fibers, this information is simultaneously transmitted to the cerebellum via corticopontine fibers (only some of these are collaterals of corticospinal fibers) and cerebral projections to the precerebellar nuclei.[56] The cerebellar nuclei, particularly the dentate, react back on the cortex, and are capable of modifying and adjusting its activity during an ongoing movement. Information from extero- and proprioceptors and other afferent inputs inform the cerebellum, as well as the cerebral cortex, about the movment.[57] In analyses of the cerebrocerebellar cooperation the numerous feedback and "closed-loop" connections of the cerebellum with other regions have to be taken into account, as must also its relations with other regions known to influence motor functions (for example the basal ganglia). Present-day knowledge of the complexity of the interconnections of the cerebellum with other regions of the brain, as well as physiological studies, show that we are still far from a complete understanding of how the cerebellum controls movement. Many interesting hypothetical schemes have been set forth, and attempts at computer modeling of the cerebellum have been made. For recent reviews, see for example Eccles (1973), Massion (1973), Allen and Tsukuhara (1974), Szentágothai and Arbib (1974), Brooks (1975), and Chan-Palay (1977). These subjects are beyond the scope of the present text. As stated by Eccles (1973, p. 143), when discussing this subject, "Much of this theory is speculative, going beyond what has been scientifically demonstrated." Eccles is further certainly right when he states ". . . there have been many attempts at computer modeling of the cerebellum. Unfortunately this work has not led to any further understanding of the cerebellum. It fails, I think, because it is premature. We do not yet have sufficient 'hard data' as a basis for computer modeling." This is true in both the anatomical and the physiological sphere.

Despite several attempts to demonstrate a somatotopic pattern within the cerebellar hemisphere, as suspected by Bolk, convincing evidence in favor of this view has so far not been produced. Early authors who made positive findings probably made lesions that involved the intermediate zone, as pointed out by Chambers and Sprague (1955b). However, it is *a priori* extremely

---

[56] The observation that some dentate neurons may start firing before the cortical pyramidal tract neurons may be explained if it is assumed that the cortical association areas are concerned in the initiation of voluntary movements and are a primary source for activation of corticospinal neurons (chiefly from the motor region) and corticopontine neurons arising in the parietal association area (cf. Chap. 4).

[57] The cerebellum has been thought to function as an *error detector* during the execution of movements. Via the dentato–thalamo–cortical pathway, it may adjust the further output of motor signals to the motoneurons and might, through information from other sources, be able to correct errors of intended movements before the final discharge of impulses to the motoneurons occurs.

likely that a somatotopic pattern exists in the cortex of the lateral part of the cerebellum projecting to the dentate nucleus, since the latter (see Chan-Palay, 1977) and its thalamic projection appear to have such a pattern (as described above). Some differences observed between the effects of cooling lateral and medial parts of the interpositus complex may reflect a somatotopic pattern, but precise mapping will require more refined physiological methods and accurate correlated anatomical investigations. See also p. 310 and Shambes, Gibson, and Welker (1978).

On the basis of experimental studies it has been common to distinguish *three rather characteristic cerebellar syndromes,* each of them representing the changes that follow a lesion of one of the three main subdivisions of the cerebellum.

1. *The flocculonodular syndrome.* This occurs in lesions of the nodulus (and probably also the flocculus) and the adjoining part of the uvula. It is characterized by difficulties in maintaining equilibrium (sometimes called "trunk ataxia," but in fact not really ataxia). There is no ataxia of the extremities if the body is supported, no tremor, and no hypotonia. Positional nystagmus may occur.

2. *The lobus anterior syndrome.* This is distinguished by an increase of postural reflexes and decerebrate rigidity (when the anterior lobe is ablated). Whether other symptoms may also occur in this case is not yet definitely settled. The changes are somatotopically localized, affecting the hindlimb with anterior lesions, the forelimb with posterior lesions.

3. *The neocerebellar syndrome.* Here homolateral hypotonia and atactic movements, asynergic and clumsy, appear, and when the dentate nucleus is involved tremor also develops. The tremor is an irregular, coarse, intentional tremor. There is no somatotopical pattern.

It should be emphasized that this is a very crude scheme, which is apparent from the data mentioned above, particularly concerning the anatomical possibility for cooperation between the three "main" cerebellar subdivisions. Nevertheless, these three syndromes are of some practical value as a guide to the topical diagnosis of cerebellar affections in man.

Cerebellar affections in man.   There is a vast literature on symptoms of cerebellar diseases in man (see Dow and Moruzzi, 1958, and Dow, 1969, for references). In recent years advances in methods have made possible more precise studies than previously. On the whole, observations are compatible with the results of experimental studies. The dominance in man of symptoms due to involvement of the cerebellar hemispheres explains why most studies have been devoted to cerebellar hypotonia and ataxic disturbances.

In man a clinical picture corresponding to the *flocculonodular syndrome* may often be met with in a special type of glioma, the so-called *medulloblastoma,* which occurs most often in the cerebellum in children between five and ten years of age. With these tumors there is usually an initial period of general symptoms of increased intracranial pressure (headache, vomiting, etc.). When cerebellar symptoms appear, they are at first limited to unsteadiness of gait and standing (Bailey and Cushing, 1925). As a rule, however, there is no incoordination of the extremities when the child is lying in its bed. These tumors practically always arise in the vermis ("midcerebellar tumor"), and in the most posterior part of it, the nodulus, according to Ostertag (1936). They have been assumed to arise from undifferentiated cells at this place, and Raaf and Kernohan (1944) have verified the common occurrence of groups of undifferentiated neuroblasts at the attachment of the posterior medullary velum to the nodulus in human embryos and infants. As will be seen, this origin and symptomatology fit in well with the findings in experimental animals, the clinical picture in the early stages corresponding to the flocculonodular syndrome as described above. When the tumor grows, eventually the cerebellar hemispheres will also be damaged and neocerebellar symptoms appear in addition (apart from symptoms due to increased intracranial pressure, among which should be remembered enlargement of the head in these cases, since the tumors affect children).

In experimental animals positional nystagmus has been seen following unilateral lesions of the nodulus (see p. 381 and p. 388). Positional nystagmus may be observed in cerebellar lesions in man, but it is often impossible to decide whether it is due to the cerebellar lesion or to involvement of vestibular nuclei or pathways (see Nylén, 1950). According to Grand (1971), positional nystagmus may be an early sign in medulloblastoma. An exhaustive review of the many abnormalities of ocular function which may occur in cerebellar affections in man has been given by Dichgans and Jung (1975).

A clinical picture which corresponds to the experimentally produced *lobus anterior syndrome* has not yet been clearly defined in man. However, there is reason to believe that in cases in which the postural reflexes are exaggerated there is an involvement of the anterior lobe (see Brown, 1949). The type of cerebellar atrophy named after Marie, Foix, and Alajouanine (1922), their *"atrophie cérébelleuse tardive à predominance cortical,"* affects primarily the anterior lobe. These patients usually walk with a wide base and stagger to both sides, while they have little or no symptoms in the upper extremities. The myotatic reflexes in the legs may be increased. The atrophy appears to begin in the anterior part of the anterior lobe. Victor, Adams, and Mancall (1959) published an exhaustive clinical and pathological analysis of cerebellar cortical degeneration in alcoholic patients. This disease appears to have a localization and a symptomatology similar to those of the atrophy of Marie, Foix, and Alajouanine, and the authors consider it likely that the affection which occurs first and is most marked in the legs may be explained by the predominant and early involvement of the anterior parts of the anterior lobe. According to Nyberg-Hansen and Horn (1972), even in cases in which mainly the anterior lobe is affected, adequate evidence of increase of muscle tone and postural reflexes has not been presented, and occurrence of a complete anterior-lobe syndrome does not appear to have been observed in man.

As referred to above, in lesions of the spinocerebellum, particularly of parts caudal to the anterior lobe, *disturbances of speech* may occur. Certain types of defects in *oculomotor control,* for example saccadic overshoot dysmetria (Selhorst, Stark, Ochs, and Hoyt, 1976) likewise appear to point to an affection of approximately the same part of the vermis.

In cerebellar diseases in man, symptoms belonging to the *neocerebellar syndrome* as defined in animals are most commonly found. In unilateral lesions the symptoms occur on the side of the lesion.[58] If this is restricted to the hemisphere,

---

[58] In an extensive study of clinical signs in a group of 282 patients affected with cerebellar tumors, Amici, Avanzini, and Pacini (1976) indicated that bilateral symptoms are more common than generally assumed. Hypotonia was observed also in medially situated tumors. Since the extension of

hypotonia, asynergia, dysdiadochokinesis, and (if the dentate nucleus is involved) intentional tremor usually occur. It appears probable that past pointing and deviation of gait, both to the affected side, belong to the same group of symptoms, since they have been observed in pure hemispheral lesions, e.g., in war injuries (Holmes, 1917). In a lesion restricted to one hemisphere the ensuing ataxia on the homolateral side occurs in the lower as well as in the upper extremity, but the characteristic titubatio, the staggering gait, is not present unless other parts of the cerebellum are also damaged (directly or secondarily by pressure, edema, or circulatory disturbances). Detailed descriptions of the symptomatology and of tests for cerebellar functions have been given by Holmes (1939), Dow and Moruzzi (1958), and Dow (1969).

On most points recent studies have confirmed the careful observations presented in the classical papers of Gordon Holmes (1939, and earlier) and his conclusions. In many of these studies electromyographic recordings have been used. Most work has been devoted to an analysis of voluntary movements. For example, Marsden et al. (1977) found a breakdown in the normal timing of activation of agonist and antagonist muscles at the onset of fast ballistic movements and a failure of long-latency compensating load reflexes during slow ramp movements (see also Terzuolo, Soechting, and Viviani, 1973; Hallett, Shahani, and Young, 1975b). According to Grimby and Hannertz (1975), the recruitment order of high- and low-frequency motor units (see Chap. 3) is unstable in patients with cerebellar ataxia.

It should be recalled that occlusion of the superior cerebellar artery is likely to affect the fibers in the brachium conjunctivum with resultant cerebellar symptoms, especially ipsilateral incoordination of voluntary movements. If the posterior inferior cerebellar artery is occluded, cerebellar symptoms are likewise usually present, in combination with signs of involvement of the medulla, e.g., in Wallenberg's syndrome.

> Different opinions have been held as to whether all movement disorders commonly seen in cerebellar affections in man can be referred to disturbance of a single mechanism, or whether they are expressions of derangements of particular mechanisms. Growdon, Chambers, and Liu (1967) concluded that ataxia and tremor "represent a single neurologic disturbance," while others hold different opinions. The mechanisms controlling rapid movements have been concluded to be separate from those for slow, smooth movements (see Hallett, Shahani, and Young, 1975a). According to these authors (Hallett, Shahani, and Young, 1975b), dysmetria may occur without dysdiadochokinesis. Some authors are inclined to consider hypotonia and ataxia to be caused by disturbance of different mechanisms. These still controversial questions will not be considered here.

Surgical removal of a cerebellar hemisphere, or, more particularly, destruction of a dentate nucleus, has been attempted for the relief of certain neurological disorders, most often in hyperkinesias or for the alleviation of muscular spasticity in cases of "cerebral palsy" (see, for example, Nashold and Slaughter, 1969; Siegfried, Esslen, Gretener, Ketz, and Perret, 1970; Nádvorník, Šramka, Lisý, and Svička, 1972). Some authors report good immediate results, but the rationale of dentatectomy in man is not clear.

---

the tumors was judged at operation only, it is possible that these discrepancies are partly due to secondary changes in the vicinity of the tumor or to pressure effects. The monograph should be consulted for particular data.

Electrical stimulation of the cerebellum in animals has long been known to elicit potential changes in the motor and sensory cortex (as well as postural changes). In man EEG changes may be recorded in the contralateral cerebral hemisphere on cerebellar stimulation (Snider and Wetzel, 1965; and later authors). In patients with cerebellar affections EEG changes are often observed.[59] When it was found that stimulation of the cerebellum may shorten or stop seizure discharges provoked experimentally in the cerebrum of animals (Cooke and Snider, 1955; and others), attempts were made, particularly by I. S. Cooper (see *The Cerebellum, Epilepsy and Behavior*, 1974, and *Cerebellar Stimulation in Man*, 1978) to treat patients suffering from intractable epilepsy with electrical stimulation of the cerebellum by means of permanently implanted electrodes. These can be made to discharge by the patient.

It is assumed that the cerebellar stimulation inhibits the abnormal activity in the cortex and hippocampus. Some believe that this is caused by excitation of Purkinje cells, while others believe that an antidromic effect on reticular formation cells is responsible. However, the latter explanation seems very unlikely since most, if not all, of the reticular nuclei which project onto the cerebellum do not give off ascending fibers. On the other hand, the Purkinje cells have often been found reduced or even absent in places where the electrode has been applied, but such changes may also be seen in epileptics not treated surgically.

Continuous cerebellar stimulation has been attempted in the treatment of cerebral palsy and has been claimed to give good results. Cerebellar stimulation in man has further been reported to alter responses to pain and to influence behavior, as observed in animals. (For a recent review, see Grabow, Ebersold, Albers, and Schima, 1974.) The effectiveness of cerebellar stimulation as a therapeutic procedure is still under judgment. Variable results have been obtained. Factors such as position and types of electrodes and stimulation variables appear to play a role in the effects obtained. Marsden et al. (1977) report that they were unable to detect any effect of cerebellar cortical stimulation on the performance of usual neurological tests for cerebellar function.

The diagnosis of cerebellar disease. It appears from the preceding account that an analysis of the cerebellar symptoms in man in some instances will yield information concerning the part of the cerebellum affected. It should be realized, however, that the pure cerebellar syndromes, as they have been defined experimentally in animals, will not very frequently be encountered in man. There are several reasons for this. The pathological processes affecting the cerebellum will frequently have reached a considerable magnitude before they are diagnosed or can be diagnosed, and by that time they are apt to affect more than one of the functionally different subdivisions of the cerebellum. Because of their size the hemispheres will as a rule soon be involved, even if the process starts in one of the other subdivisions. Furthermore, even if the lesion itself is confined to one of the subdivisions, the others may be secondarily affected by pressure, dislocation, or obliteration of vessels with ensuing circulatory disturbances. It is a common experience that pressure effects are particularly apt to occur in diseases in the infratentorial part of the cranial cavity, where the possibilities of unhindered expansion

[59] For a recent extensive study and references to the literature see Amici, Avanzini, and Pacini (1976).

are limited. Such pressure effects may be due to a vascular incident or more frequently to a tumor. For the same reason symptoms due to mechanical pressure may appear from structures outside the cerebellum itself, e.g., the medulla or pons. Such symptoms include signs of pyramidal tract injury without direct involvement of the pyramidal tract, or symptoms of vestibular disturbances or cranial nerve affections. On the other hand, extracerebellar lesions not infrequently give rise to cerebellar symptoms on account of their pressing on the cerebellum. Particularly important in this connection are the tumors of the cerebellopontine angle, most frequently the *acoustic nerve neurinomas* (cf. Chap. 7).

It is to be expected that cerebellar symptoms may also appear when the large afferent and efferent pathways to and from the cerebellum are injured. This has been observed both clinically and experimentally in animals. Thus transection of the superior cerebellar peduncle, as has been mentioned, gives rise to a neocerebellar syndrome on the homolateral side, and following transection of the restiform body, which, *inter alia,* carries the vestibular fibers to the cerebellum, symptoms appear which are reminiscent of the flocculonodular syndrome.

The latter facts raise the question as to whether the symptoms usually designated as cerebellar are specific to lesions of the cerebellum. With several of them this is clearly not the case. Thus hypotonia and weakened tendon reflexes are also met with in lesions of the dorsal roots, and unsteadiness of gait and standing is seen in lesions of the vestibular apparatus. In this case past pointing may also occur (cf. Chap. 7). Disturbances of coordination will be present in lesions of the dorsal columns of the cord or of the dorsal roots, although typical dysmetria and asynergia are as a rule not conspicuous. In hemipareses following cerebral vascular disturbances (thrombosis, emboli, bleedings) the patient, in addition to exhibiting symptoms caused by the interruption of corticofugal pathways (see Chap. 4), has difficulties in coordination of the affected limbs. Such disturbances are usually thought to be consequences of the paresis and spasticity of the affected muscles. However, as discussed in Chapter 4, vascular lesions of the internal capsule will interrupt several cortical efferents terminating in the precerebellar relay nuclei (inferior olive, lateral reticular nucleus, tegmental pontine reticular nucleus, pontine nuclei, and others), and the cerebral influence on the cerebellum will consequently be disturbed. It appears extremely likely that some of the defects in motor functions in these patients are due to derangements of the cerebellar control of movements. As concerns elaborate motor tasks, for example writing, slight cerebellar symptoms may even be recognized in the ipsilateral limbs (Brodal, 1973). This is probably to be explained by the fact that the cerebrocerebellar connections are not strictly ipsilateral, but to some extent bilateral. No wonder, then, that the diagnosis of a cerebellar disease may be made without any primary affection of the cerebellum being present. As an example may be mentioned *internal hydrocephalus,* in which the dilatation of the ventricles, particularly the fourth, may produce cerebellar symptoms.[60] Not infrequently *tumors of the frontal lobe* may be accom-

---

[60] In cases in which the hydrocephalus develops very early in fetal life it may be accompanied by, and is probably the cause of, a defective development of the cerebellar vermis (see Brodal, 1945). The anomaly is often referred to as the Dandy-Walker syndrome and believed to be due to a congenital atresia of the foramina of Magendie and Luschka. However, these may be found patent in such cases (Brodal and Hauglie-Hanssen, 1959; Portugal and Brock, 1962). This rare developmental anomaly has

panied by cerebellar symptoms. In this case the cerebellar symptoms occur contralateral to the side of the tumor. The chronological development of the symptoms may give a clue to the correct diagnosis, since frontal-lobe tumors commonly start with mental symptoms. When these appear in cerebellar lesions they usually occur late, and are due to the effects of the increased intracranial pressure. As a rule ventriculography, angiography and computer tomography will be able to yield decisive information. The reason for the appearance of cerebellar symptoms in frontal-lobe affections is not quite clear, but it appears most reasonable that they are due to an affection of the corticopontine tract.

The few features mentioned will suffice to make clear that the diagnosis of a cerebellar lesion is not always an easy task, particularly not without resorting to special diagnostic measures. The symptoms are sometimes very difficult to evaluate and may be very slight (cf. below), and may be produced by lesions of structures other than the cerebellum. In the etiological diagnosis the case history must be taken into account just as in other cases, and a knowledge of the chronological development of the different symptoms may also give hints concerning the topographical diagnosis. This, for instance, applies to the so-called *heredo-ataxias,* an ill-defined group of slowly progressing degenerative changes, which may affect the cerebellum or its afferent and efferent tracts and also other systems such as the pyramidal tract.

A peculiar feature which contributes to making the diagnosis of cerebellar lesions difficult should finally be emphasized, namely, the *remarkable degree of compensation which occurs in lesions of the cerebellum.* This compensation may be practically complete, with the consequence that some time after the initial lesion only traces are left of the original cerebellar symptoms. As a general rule it can be stated that the more rapidly a pathological process develops in the cerebellum, and the more widespread the damage, the more clear-cut are the symptoms. In slowly progressing affections the symptoms may be very scanty and feeble, because they are masked by the compensation which takes place. The compensation is particularly marked in children and young individuals. In cerebellar defects arising in fetal life or early childhood all signs of cerebellar malfunction may be lacking, but nevertheless autopsy may reveal the absence of a cerebellar hemisphere, or other grave defects. It is assumed that the compensation in cerebellar lesions is due to other systems taking over the function of the cerebellum or that intact parts of the cerebellum are able to cope with the functions of the damaged or deficient parts as well. However, nothing definite is known of the mechanism of compensation, in spite of its great importance in clinical neurology.

---

also been observed as a hereditary condition in mice (Bonnevie, 1943; Bonnevie and Brodal, 1946; see also Portugal and Brock, 1962).

6

The Reticular Formation and Some Related Nuclei

THE RETICULAR formation of the brainstem was recognized as a separate part of the central nervous system by the classical neuronanatomists. By the turn of the century some of its connections were known in broad outline, and speculations on its functions were ventured. However, neither anatomists and physiologists nor clinicians devoted much attention to it during the first half of this century. Following the publication of the now classical paper by Moruzzi and Magoun, "Brain stem reticular formation and activation of the E.E.G.," in 1949, this situation was radically altered, chiefly because it appeared that the reticular formation is involved in the maintenance of the enigmatic function which we call consciousness. In the following years further physiological investigations confirmed and extended the original observations of Moruzzi and Magoun (for an early review, see Rossi and Zanchetti, 1957). Since we shall often refer to these problems in the following, a brief outline is appropriate.

The reticular formation and the "ascending activating system." Following high-frequency electrical stimulation of the brainstem in chloralosane-anesthetized cats, the synchronized discharges of high-voltage slow waves in the electroencephalogram (EEG) are replaced by low-voltage fast activity, closely resembling the corresponding changes observed in the human EEG on transition from a relaxed or drowsy state to attention or alertness. These changes are generally referred to as *desynchronization, activation,* or *the arousal reaction*. They occur diffusely over the entire cortex, are assumed to be mediated via the "diffuse thalamic system" (mentioned in Chap. 2 and to be considered later in the present chapter). These changes can be obtained by stimulation of the medial bulbar reticular formation, pontile and midbrain tegmentum, and dorsal hypothalamus and subthalamus. The same changes can be produced by natural or electrical stimulation of spinal nerves, the trigeminal, the splanchnic, vagus, acoustic, and some other cranial nerves (for references see Rossi and Zanchetti, 1957). Impulses entering the reticular formation by way of these channels are assumed to produce a state of activity in the reticular formation, where potentials can be recorded on such stimulation. Electrical stimulation or strychninization of the cerebral cortex likewise gives rise to potentials in the reticular formation, and may also produce EEG arousal in sleeping *"encéphale isolé"* cats (cats in which the spinal cord is separated from the rest of the brain). Stimulation of the fastigial nucleus influences the electrocortical activity via the reticular formation. Following transection of the "specific sensory paths" in the brainstem (the medial lemniscus and the spinothalamic tract), activation of the EEG can still be obtained on stimulation of nerves or of the reticular formation. Lesions in the central parts of the brainstem, however, result in behavioral somnolence and electrocortical synchrony, and activation cannot be obtained with stimuli which are effective when the central part of the brainstem is intact. Corresponding results were obtained in monkeys (French, von Amerongen, and Magoun, 1952).

From these and other observations it was concluded that there is an "ascending reticular system" in the brainstem. The concept was further elaborated in subsequent studies. The system was found to be tonically active, its level of activity depending on the amount of afferent impulses from a number of sources as well as on humoral agents, such as adrenaline and carbon dioxide. The activity of the system is reflected in the EEG and was assumed to determine the level of consciousness in all its variations from complete alertness and attention to drowsiness and sleep. The system was assumed to be entirely diffusely organized, in contrast to the "specific" sensory "system." The structural basis of the ascending activating system was assumed to be the reticular formation of the brainstem and the "nonspecific" thalamic nuclei, including ascending connections from the former to the latter and further projections from the thalamus to the cortex. Physiological observations on the ascending conduction led to the postulate that the ascending pathways were composed of series of short-axoned cells. Furthermore, it was concluded that activation of the system occurred primarily by "collateral spread" from the specific sensory pathways. The reticular formation was assumed to be diffusely organized, since potentials could be led off from large territories following stimulation of afferents from a particular source, and since impulses from several sources were found to reach the same region. Its capacity for integration was especially stressed.

While the existence of an "activating system" in the brainstem has been definitely established, and its discovery has greatly improved our understanding of important functions of the brain, some of the deductions made by the early workers in the field have had to be modified, especially their assumption concerning the anatomical substrate of the system. Moruzzi and Magoun (1949) clearly stated that on this point their conclusions were tentative, since little information on necessary anatomical details was available. However, in subsequent studies the reservations made by the original workers—as usually happens—have often been disregarded.[1]

---

[1] Moruzzi (1963) points to the fact that the hypothesis made possible an understanding and correlation of physiological, pharmacological, and clinical data which had so far appeared to be unrelated, and adds: "It was probably because of this success that the distinction between direct experimental

It is unfortunate that the authors included the word "reticular" in the designation. This is a morphological term and refers to the netlike appearance of the central part of the brainstem (recticular formation). It is clear from the physiological findings that the structures mediating the activation extend rostrally beyond the reticular formation as anatomically defined (see below). Even if there is a nucleus in the thalamus which, on account of its structure, was named the reticular nucleus by the old anatomists, and even if this were involved in the mechanism of activation, this does not justify the use of the word "reticular" as an epithet to the term "ascending activating system," and even less to speak of parts of the thalamus as belonging to the reticular formation. Much confusion has arisen on account of this lack of clarity in designations. To use the term "reticular system" as synonymous with the "activating system," as is often done, serves only to confuse the issue. *It should be made perfectly clear that the "activating system" is a functional concept, the "reticular formation" a morphological one, and it has been obvious for many years that these do not correspond.* For this reason, when referring to Moruzzi and Magoun's concept in the present text, the word "reticular" will not be used. To retain the term "ascending" might likewise be misleading since it is known that regions related to the ascending activation have a corresponding action on the spinal cord. When speaking of the "activating system," it is further advisable to delete the words "of the brainstem." This designation may be confusing since the term *brainstem* is defined in different ways.. In agreement with common usage *it will here be employed in its gross macroscopical sense, comprising the medulla oblongata, the pons, and the mesencephalon.* After an account of the structure and fiber connections of the reticular formation we shall return to questions of its functions and of the activating system.

Anatomy of the reticular formation.   Even if certain regions of the brain have a so-called reticular structure and therefore may be referred to as a "reticular formation," for example in the spinal cord, we shall here deal only with the reticular formation of the brainstem. According to the delimitation mentioned above, this will be the reticular formation of the medulla, pons, and mesencephalon. These regions represent a phylogenetically old part of the brain. It was named by the old anatomists, and is generally taken to comprise those areas of the brainstem which are characterized structurally by being made up of diffuse aggregations of cells of different types and sizes, separated by a wealth of fibers traveling in all directions. Circumscribed cell groups, such as the red nucleus, the superior olive, or the cranial nerve nuclei are not included. Thus the chief criterion for considering a cellular area of the brainstem as part of the reticular formation is its structure. From this point of view some fairly well circumscribed nuclei, such as the lateral reticular nucleus (nucleus of the lateral funiculus) and the nucleus reticularis tegmenti pontis of Bechterew, when studied in Nissl sections, rightly belong to the reticular formation, although they are usually omitted when one speaks of the reticular formation in a general way. This may be justified because these nuclei (as well as the paramedian reticular nucleus), in contrast to the rest of

---

proof and suggestive evidence, a discrimination which should always remain clear in our minds, was frequently forgotten." (Moruzzi, 1963, p. 235.)

the reticular formation, project onto the cerebellum, as described in Chapter 5, and thus differ functionally from the remaining part.[2] On the other hand, there are some minor cell groups, for example those of the raphe, which despite their reticular structure have been referred to under particular names and, therefore, are often not included. Since these nuclei resemble other parts of the reticular formation with regard to their connections, it appears likely that they are functionally closely related to it. Occasionally it may be a matter of taste whether a cell group is considered part of the reticular formation or not.[3] In the following account *the term "reticular formation" will be employed as a general denominator for those areas of the brainstem* (as defined above) *which have a reticular structure, with the exception of the three cerebellar projecting nuclei mentioned above.* Nuclei having special traditional names will, however, be referred to under those names. Broadly speaking the region to be considered will cover the central areas of the brainstem. Peripherally, each half of the reticular formation borders on the long ascending and descending fiber bundles traversing the brainstem (medial longitudinal fasciculus, medial lemniscus, spinothalamic tract, etc.) and on particular nuclei, which in certain places intrude somewhat into its territory.

Although some early neuroanatomists had described certain regions of the reticular formation as having their particular structural features, it was generally held that it was a rather diffuse mass of cells. However, in systematic studies of the cytoarchitectonics of the brainstem of the rabbit and man, Meessen and Olszewski (1949) and Olszewski and Baxter (1954), respectively, were able to distinguish a number of cell groups within the reticular formation.[4] The main groups which they recognized can be found in other mammals as well, for example in the cat (Brodal, 1957; Taber, 1961), the rat (Valverde, 1962), and the guinea pig (Petrovicky, 1966). There are certain species differences, and the borders between the groups are not equally clear everywhere. Furthermore, the borders between the reticular formation and many nuclei, for example the vestibular and sensory trigeminal nuclei, are in places rather indefinite. The main point is, however, *that there are obvious architectonic differences between relatively small areas of the reticular formation.* This makes it appear likely that there are other differences as well between these areas, with regard to fiber connections and function, as has indeed been found to be the case (see below). Figure 6-1 shows a drawing of a section from the human reticular formation, Fig. 6-2 a series of drawings from the cat's brainstem. The latter gives an impression of the general arrangement. A prominent feature is that the large cells of the reticular formation are restricted to its medial part, approximately the medial two-thirds, where they are intermingled with small and medium-sized cells. In the lateral third there are only small cells. Furthermore, large cells are found particularly at certain levels (see Fig. 6-4). One

[2] In Golgi sections the cerebellar projecting nuclei are found to differ from the reticular formation proper with regard to the dendritic patterns of their cells (Leontovich and Zhukova, 1963).

[3] Because of this, Olszewski (1954) has suggested that the epithet "reticular" should be omitted altogether when speaking of cell groups in the reticular formation, the more so since it can be subdivided into a number of fairly well circumscribed cellular areas which can be referred to as nuclei. A discussion of the criteria for distinguishing the reticular formation as a particular part of the brain is given by Ramón-Moliner and Nauta (1966), who stress the dendritic patterns of the nerve cells as an important feature.

[4] Some of the findings of Olszewski and Baxter (1954) in man have been criticized by Feremutsch and Simma (1959). See also Koikegami (1957).

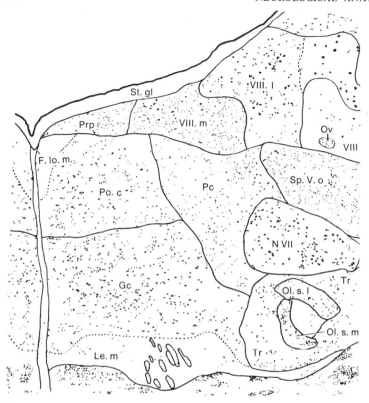

FIG. 6-1 Semischematic representation of part of a Nissl-stained section through the human brainstem at a low pontine level. The facial nucleus (N. VII), the rostral part of the spinal trigeminal nucleus (Sp. V.o), parts of the vestibular nuclei (VIII.l and VIII.m), the nucleus praepositus (Prp), the medial longitudinal fasciculus (F.lo.m), and the medial lemniscus (Le.m) surround the reticular formation. Within this three subdivisions can be separated at this level, the nuclei gigantocellularis (Gc), pontis caudalis (Po.c), and parvicellularis (Pc). From Olszewski and Baxter (1954).

of the large-celled nuclei is the nucleus reticularis gigantocellularis (R. gc.) in the medulla. Rostral to this is another, the nucleus reticularis pontis caudalis (R.p.c.). The latter passes rather gradually into the nucleus reticularis pontis oralis (not seen in Fig. 6-2), which lacks giant cells.

Studies of the reticular formation with the Golgi method give additional information about the cells present and their dendritic and axonal patterns. From such studies (Scheibel and Scheibel, 1958; Leontovich and Zhukova, 1963) it has been concluded that there are in the reticular formation no typical Golgi II cells (cells with short axons branching profusely close to the perikarya and considered as prototypes of internuncial or association cells). All axons "appear to project at least some distance rostral and/or caudal" (Scheibel and Scheibel, 1958, p. 37). In the rat, Valverde (1961b) found an exceedingly small number of Golgi II cells in the lateral parvicellular region. It is further remarkable that all long axons appear to give off several collaterals along their course and that collaterals from the same axon may vary with regard to their pattern of branching. The dendrites are usually long and radiating, characterizing the cells as being of the so-called iso-dendritic type (Ramón-Moliner and Nauta, 1966; and others). It is characteristic that most of the dendrites of reticular cells are spread out in a plane perpendicular to the long axis of the brainstem (Fig. 6-3). These and other features of interest for functional correlations will be considered later.

Along the midline of the brainstem there are collections of nerve cells, together referred to as the *raphe nuclei*. They may be subdivided into a number of minor units, which in part are characterized by differences in cytoarchitecture (see Taber, Brodal, and Walberg, 1960). With regard to cell types and in several other respects these nuclei resemble the reticular formation. In recent years these nuclei have attracted much interest. They will be considered more fully later in this chapter. A particular, diffusely outlined region in the dorsal part of the reticular formation, referred to as the *PPRF* (paramedian pontine reticular formation) and concluded to be related to the control of horizontal gaze will be treated in Chapter 7 (section g). Some authors prefer to use the expression "the reticular core of the brainstem" as a common denominator for the reticular formation (as outlined above) and some other cell groups (see Scheibel, 1980, for a discussion).

FIG. 6-2   A cytoarchitectonic map of the reticular formation of the cat. In a series of transverse sections are plotted the various cell groups and their composition of small, medium-sized, and large cells. Some abbreviations: *Coe.:* nucleus sub-coeruleus; *F.l.m.:* medial longitudinal fasciculus; *N.c.e.:* external cuneate nucleus; *N.f.c.:* nucleus cuneatus; *N.f.g.:* nucleus gracilis; *N.r.:* red nucleus; *N.r.t.:* nucleus reticularis tegmenti pontis; *N.tr.sp.V:* spinal nucleus of trigeminal nerve; *P.:* pontine nuclei; *P.g.:* periaqueductal gray; *R.gc.:* nucleus reticularis gigantocellularis; *R.l.:* lateral reticular nucleus (nucleus of lateral funiculus); *R.mes.:* reticular formation of the mesencephalon; *R.n.:* nuclei of the raphe; *R.p.c.:* nucleus reticularis pontis caudalis; *R.pc.:* nucleus reticularis parvicellularis; *R.v.:* nucleus reticularis ventralis. From Brodal (1957).

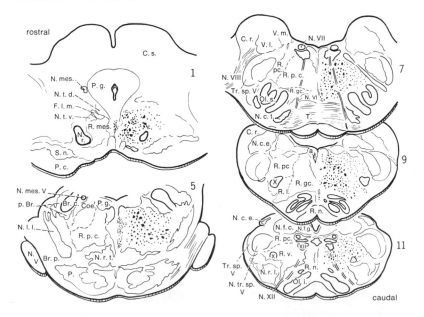

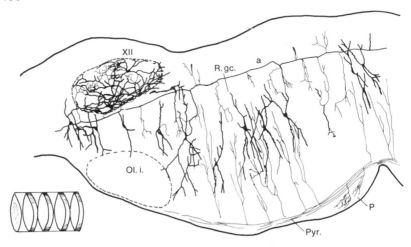

FIG. 6-3 A drawing of a Golgi-impregnated parasagittal section through the lower brainstem of a young rat, illustrating the orientation of the dendrites of cells in the reticular formation in planes perpendicular to the long axis of the brainstem. Collaterals of axons descending in the pyramid (*Pyr.*) take off in the same plane as do also the collaterals of a long axon (*a*) of a reticular cell. (Note different organization of the dendrites in the hypoglossal nucleus, XII.) *Ol.i.:* inferior olive; *P:* pons; *R.gc.:* nucleus reticularis gigantocellularis. The inset at the lower left shows how the reticular formation may be considered a series of neuropil segments (cf. text). From Scheibel and Scheibel (1958).

Some authors have studied the *neurochemistry of the reticular formation*. In the medullary and pontine reticular formation some neurons have been found which are excited by acetylcholine while others are inhibited (Salmoiraghi and Steiner, 1963; Bradley, Dhawan, and Wolstencroft, 1964; and others). Other cells are influenced by l-noradrenaline (Bradley and Wolstencroft, 1962), and most neurons respond to 5-hydroxytryptamine (5-HT). Using fluorescence methods for identifying neurons containing catecholamines and 5-HT, Dahlström and Fuxe (1964) have mapped their distribution in the brainstem. 5-HT (serotonin) cells are numerous in the raphe nuclei, whereas in other regions, for example the nucleus locus coeruleus, most neurons contain noradrenaline (see below). In many articles dealing with neurochemistry, however, the region studied is unfortunately not identified precisely, a fact which makes correlations with observations from other fields difficult. In neurochemical studies careful anatomical identification of the cells studied is essential.

Fiber connections of the reticular formation. These link the reticular formation, directly or indirectly, with many other parts of the central nervous system. The *efferent fibers* pass chiefly to the spinal cord, to the thalamus, and to other nuclei within and above the brainstem. The efferent projections may be studied by making lesions in the reticular formation and tracing the ensuing descending or ascending degeneration of fibers. With this approach there is considerable risk of damaging bypassing fibers and thus getting misleading results. Or one may look for retrograde cellular changes in the reticular formation following interruption of its efferent fibers. In this way the sites of origin of the fibers may be established accurately, but the sites of termination cannot be determined precisely. If lesions are made in regions which are known to give off efferents, more unequivocal results can be obtained in studies of anterograde fiber degeneration. In recent years some authors have studied some of the connections by the HRP and autoradiography methods (see Chap. 1).

Fibers passing from the reticular formation to the spinal cord were established by the old anatomists using the Marchi method. Present knowledge of these tracts, which is fairly complete, was reviewed in Chapter 4 (p. 205 ff). As will be recalled, reticulospinal fibers come from small as well as large cells, chiefly in two regions in the medial two-thirds of the reticular formation, one in the pons, the other in the medulla (Fig. 6-4). Fibers from the medulla descend crossed as well as uncrossed, those from the pons chiefly uncrossed. In the cord the fibers do not appear to end in synaptic contact with motoneurons, but end in laminae VI–VIII, the pontine fibers terminating more ventrally than the medullary (see Fig. 4-8). No somatotopical pattern was found within the areas of origin (Torvik and Brodal, 1957).

Contrary to what has often been assumed, particularly in the early physiological studies of the ascending activating system, *there is an abundant projection of long ascending fibers from the reticular formation*. Many studies have been devoted to these projections. Like the long descending fibers, the ascending ones do not arise in equal numbers from all parts of the reticular formation, but have their preferential sites of origin. Following lesions rostral to the mesencephalon in newborn kittens, a mapping of the ensuing retrograde cellular changes (Brodal and Rossi, 1955) shows that, although scattered fibers arise from almost all parts of the medial two-thirds of the reticular formation, the majority come from two regions, one in the medulla, the other in the lower pons-upper medulla (Fig. 6-4). Both regions give off crossed as well as uncrossed ascending fibers. However, in the caudal region giant cells do not take part in the rostral projection, indicating that the giant cells here act only on the cord.

It appears that at least one-third of all cells in the regions mentioned have long axons ascending beyond the mesencephalon. In addition there are ascending fibers from the mesencephalic reticular formation (Nauta and Kuypers, 1958). The presence of a considerable number of long ascending fibers from the medial part of the reticular formation has further been confirmed in studies with the Nauta method (Nauta and Kuypers, 1958) as well as in Golgi studies (Scheibel and Scheibel, 1958; Valverde, 1961a). Such studies have further indicated that there are also cells in the lateral small-celled third of the reticular formation which give off ascending axons, but these are relatively short. In a recent study with antidromic identification of rostrally projecting cells, Fuller (1975) could not localize such cells from levels at or caudal to the inferior olive, probably because the rostrally projecting cells at this level are small (see Fig. 6-4).

As to the *terminations of the long ascending fibers* from the reticular formation, there are still open questions. In general, fibers arising in the mesencephalon pass further than those coming from more caudal regions. There appears to be agreement among authors using the Golgi method (Scheibel and Scheibel, 1958; Valverde, 1961a) and the Nauta method following lesions of the reticular formation (Nauta and Kuypers, 1958) that a fair number of medullary and pontine ascending fibers end in some of the "nonspecific" thalamic nuclei, as will be discussed in a later section (see Fig. 6-12). Fibers from the mesencephalic reticular formation (chiefly its medial parts) have been traced to the hypothalamus, the

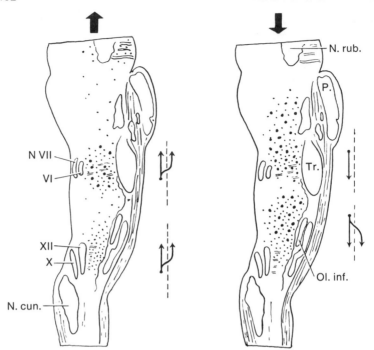

FIG. 6-4 The distribution of cells of the reticular formation sending long axons to the spinal cord (right) and of cells having long axons ascending beyond the mesencephalon (left) is mapped in parasagittal sections of the cat's brainstem. Large dots indicate giant cells. Note that, although both types of fibers take origin from practically the entire longitudinal extent of the reticular formation, they both have their maximal regions of origin. In spite of some overlapping, these maximal regions are different. The arrows to the right of each drawing indicate that the pontine reticulospinal fibers descend homolaterally, while all other contingents are crossed as well as uncrossed. From Brodal (1956).

preoptic area, and the medial septal nucleus. Some fibers pass even to the caudate and lentiform nuclei. These mesencephalic fibers do not pass through the thalamus but below it, a point of some interest that will be considered later. The findings above were largely confirmed in a study by Lynch, Smith, and Robertson using silver impregnation methods (1973). Large lesions of the reticular formation were made in cats (see also Bowsher, 1975). Fibers passing beyond the subthalamus were not found following lesions caudal to the pontine reticular formation, while with more rostral lesions some fibers could be traced, for example to the caudate nucleus and even the frontal cortex.

On the basis of histochemical studies it has been concluded (Shute and Lewis, 1967) that at least some of the widely distributed ascending fibers from the reticular formation are cholinergic. Studies with the Falck-Hillarp fluorescence method suggest that there are two abundant noradrenergic rostral projections from the brainstem reticular formation (see below; Fuxe, Hökfelt, and Ungerstedt, 1970). The terminal distribution of this system appears to correspond on many points to that determined with silver impregnation methods. The two histochemically studied systems have been claimed to be part of the ascending activating system.

Two features in the anatomy of the long ascending and descending projections deserve particular emphasis. In the first place it was learned from Golgi studies (Scheibel and Scheibel, 1958; Valverde, 1961a) that a considerable number of the cells give off an axon that dichotomizes and has a long ascending, as well as a long descending, branch (see Fig. 6-6). In the mouse and rat one branch of such cells has been traced to the cord, the other to the thalamus (Fig. 6-5). Thus, the same cell will act in a rostral as well as a caudal direction. From a quantitative analysis of the retrograde cellular changes following lesions of the spinal cord or lesions above the mesencephalon in the cat it has been conservatively estimated that in the rostral region of origin (see Fig. 6-4) more than half of the cells must be of this type (Torvik and Brodal, 1957). Such cells have also been identified physiologically (Magni and Willis, 1963). The other noteworthy feature is that the rostrally projecting cells are on the whole situated more caudally than those projecting to the cord, although there is some overlapping of rostrally and caudally projecting regions. Since the long axons give off collaterals along their course, this must mean that impulses passing to the cord may act on cells projecting rostrally, and vice versa, as shown in Fig. 6-6. The two morphological features mentioned indicate that *there must be a very close integration between the influences exerted by the reticular formation in the rostral and caudal direction.*

In addition to acting on the spinal cord and/or the thalamus and other rostral structures, the reticular formation may influence nuclei in the brainstem by collaterals of ascending or descending axons, or by way of the apparently few cells which do not have long ascending or descending axons. Scheibel and Scheibel

FIG. 6-5 Drawing of a sagittal Golgi section from the brainstem of a 2-day-old rat, showing a single large cell in the magnocellular nucleus. It emits an axon which bifurcates into an ascending and a descending branch. The latter gives off collaterals to the adjacent reticular formation (*N.gc.*, nucleus reticularis gigantocellularis), to the nucleus gracilis (*N.g.*), and to the ventral horn in the spinal cord. The ascending branch gives off collaterals to the reticular formation and the periaqueductal gray (*P.g.*); it then appears to supply several thalamic nuclei (*Pf. & Pc.*; parafascicularis and paracentralis; *Re:* reuniens, and others), the hypothalamus (*H*), and the so-called zona incerta (*Z*). *MD:* dorsomedial thalamic nucleus. From Scheibel and Scheibel (1958).

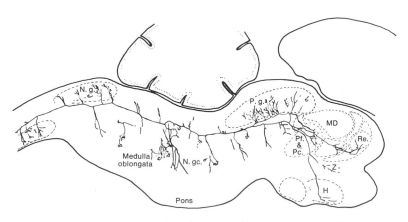

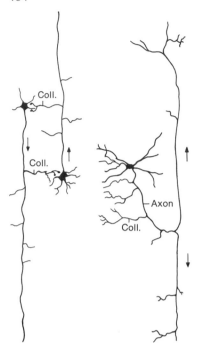

FIG. 6-6 Diagram illustrating two anatomical features which make possible a close correlation of caudally and rostrally directed influences of the reticular formation. To the right a simplified drawing of a typical cell in the medial reticular formation. Its axon gives off a long ascending and a long descending branch provided with collaterals (cf. Fig. 6-3). To the left a diagram showing how cells giving off ascending and descending axons, respectively, may influence each other by way of collaterals. From Brodal (1964).

(1958) traced collaterals to all cranial nerve nuclei in the brainstem, sensory and motor. Collaterals appear to reach the *dorsal column nuclei* as well. Physiologically, cells projecting to those nuclei have been identified in the nucleus reticularis gigantocellularis (Sotgiu and Margnelli, 1976) and following HRP injections in the cuneate nucleus (Sotgiu and Marini, 1977). According to Odutola's (1977) Golgi studies, they occur in some other reticular nuclei as well. The projection of the reticular formation onto the *vestibular nuclei,* seen by several authors in Golgi preparation, was found to be rather modest in an experimental study in the cat (Hoddevik, Brodal, and Walberg, 1975). It takes origin mainly from the magnocellular parts of the reticular formation and supplies all four main vestibular nuclei. Fibers from the mesencephalic reticular formation do not appear to pass to the vestibular nuclei (see also Pompeiano and Walberg, 1957; Edwards, 1975). Physiological studies have shown that the reticular formation influences the activity of neurons in the vestibular nuclei (for a review see Pompeiano, 1972c). Some *reticuloreticular connections* have been demonstrated experimentally with the tritiated leucine method, for example commissural connections (Walberg, 1974b) and fibers from the mesencephalic to the pontine and medullary reticular formation (see Edwards, 1975).

Afferent connections to the reticular formation come from many sources. Most impressive are the *afferent connections from the spinal cord.* Contrary to what was inferred in early physiological studies, only some of the spinal activation of the reticular formation appears to take place via collaterals of secondary sensory fibers. Thus several anatomical studies have failed to demonstrate collaterals to the reticular formation of the fibers of the medial lemniscus (Matzke, 1951, in the cat;

Bowsher, 1958; Nauta and Kuypers, 1958, in the monkey; Valverde, 1961a, in the rat). Supporting evidence comes from physiological studies (see for example, Morillo and Baylor, 1963). Nor do collaterals from the spinothalamic tract appear to be of great importance since this tract is poorly developed (except perhaps in man, see Chap. 2). However, there is *a massive influx of direct spinoreticular fibers* which ascend in the ventrolateral funiculus intermingled with the spinothalamic ones. In the brainstem the fibers take off in a dorsomedial direction, and, as stated above, their terminal branches and collaterals run approximately perpendicular to the long axis of the brainstem, in the same plane as most dendrites of the cells (see Fig. 6-3). The termination of many fibers in the reticular formation following transection of the ventrolateral funiculus has been demonstrated by early neuroanatomists using the Marchi method, and later with silver impregnation methods in the cat (Rossi and Brodal, 1957; Anderson and Berry, 1959), in the sheep (Rao, Breazile, and Kitchell, 1969), in the monkey (Mehler, Feferman, and Nauta, 1960), and in man (Bowsher, 1957, 1962; Mehler, 1962). In most studies the entire ventrolateral funiculus was transected more-or-less completely. Kerr (1975) has shown in the monkey that some of the spinoreticular fibers ascend in the ventral funiculus, and suggests that they may be functionally different from those ascending in the lateral part of the ventrolateral funiculus. Attempts to determine the sites of spinoreticular neurons by means of antidromic recordings have given somewhat conflicting results (see Albe-Fessard, Levante, and Lamour, 1974; Fields, Clanton, and Anderson, 1977; Maunz, Pitts, and Peterson, 1978). There may be differences between the cat and the monkey.

It should be noted that although some spinoreticular fibers appear to reach almost all parts of the reticular formation bilaterally, the majority are distributed to its medial two-thirds and, furthermore, to particular levels of this. In the cat (Rossi and Brodal, 1957), there is one maximal area of termination in the medulla (corresponding approximately to the nucleus reticularis gigantocellularis) and another in the pons (within the nuclei reticularis pontis caudalis and oralis). These two regions correspond approximately to those giving off long ascending afferents (see Fig. 6-4), indicating that there is a direct route from the spinal cord to the thalamus via the reticular formation.[5] From Bowsher's (1962) study of the brains of seven patients subjected to chordotomy or medullary tractotomy (one case), it appears that in man there is, in addition, a third maximal terminal region at the transition from the pons to the mesencephalon.[6] In animals as well as in man there appears to be no somatotopical pattern in the distribution of the spinoreticular fibers.

The courses followed by *sensory impulses in the cranial nerves entering the reticular formation* are not completely known. A small number of primary sensory fibers have been traced to it after lesions of the vagal, glossopharyngeal, and trigeminal nerves (see Torvik, 1956b; Clarke and Bowsher, 1962, in the rat; Kerr, 1962, in the cat). This has been confirmed in man for the first two nerves (Kunc

[5] Electron microscopically, degenerating boutons of spinoreticular fibers have been found, which contact oligo- and polydendritic neurons in the medullary reticular formation of the cat (Bowsher and Westman, 1970).

[6] In the cat, as well as in the monkey and man, quite heavy degeneration is also found in a particular nucleus dorsal in the upper pons, the nucleus subcoeruleus (Coe. in Fig. 6-2; Scoe. in Fig. 6-10).

and Maršala, 1962). However, such fibers appear to be relatively scanty and to have a restricted distribution. *Secondary sensory fibers,* however, have been shown in Golgi studies to give off collaterals to the reticular formation (see Brodal, 1957; and Rossi and Zanchetti, 1957, for reviews), and some of these connections have been studied experimentally, making it possible to decide their areas of distribution more specifically. This is true for the nucleus of the spinal trigeminal tract (see Carpenter and Hanna, 1961; Stewart and King, 1963), and the vestibular nuclei (see Brodal, 1972e, 1974, for reviews). It has been shown (Ladpli and Brodal, 1968) that (secondary) *vestibuloreticular fibers* are organized according to a rather specific pattern. They arise from all four main vestibular nuclei (cf. Chap. 7, section (d)), but the distribution of the fibers from each of these nuclei is not identical. Their main areas of termination are in the nucleus reticularis gigantocellularis and reticularis pontis caudalis, within the territories that give off long descending and ascending fibers. A particular region of the reticular formation, receiving secondary vestibular fibers and referred to as the PPRF, has been found to be a neural substrate for horizontal eye movements (see Chap. 7). This region appears to receive fibers from the superior colliculus as well (Kawamura, Brodal, and Hoddevik, 1974). Collaterals of the ascending acoustic pathways have been found to end in the reticular formation (see Chap. 9). Optic impulses presumably reach it via optic nerve fibers to the superior colliculus (see Chap. 7), from which tectoreticular fibers originate. The route followed by olfactory fibers is less clear, but fiber connections that may presumably be utilized have been described (see Guillery, 1956, 1957; Nauta, 1956, 1958).

As described in the preceding chapter, *fibers from the cerebellum,* particularly the fastigial nucleus, are distributed throughout the medial two-thirds of especially the medullary reticular formation (see Walberg, Pompeiano, Westrum, and Hauglie-Hanssen, 1962). These fibers supply both rostrally and caudally projecting areas.

Among *fibers to the reticular formation from "higher levels"* there appear to be some from the *lateral hypothalamus* (see Nauta, 1958, and Chap. 11) and from the *pallidum* (see Johnson and Clemente, 1959; Nauta and Mehler, 1966), both groups ending chiefly in the mesencephalon. *Tectoreticular fibers* were mentioned above. These fibers have been demonstrated in various animal species. In the cat they arise from the deep layers of the *superior colliculus* (Altman and Carpenter, 1961; Kawamura, Brodal, and Hoddevik, 1974), which is also the case in the opossum (Rafols and Matzke, 1970). See also Chapter 7.

> Most authors found the projection to consist of two components, a crossed one to the medial parts of the medullary and pontine and an uncrossed one to the mesencephalic reticular formation. From silver impregnation studies of the terminal distribution of these fibers in the cat (Kawamura, Brodal, and Hoddevik, 1974) it has been learned that the fibers supply preferentially certain regions. In the pons and medulla two distinct maxima of termination coincide approximately with the main sites of endings of corticoreticular fibers (see below). Many neurons, including some identified as giving rise to reticulospinal fibers, are monosynaptically excited on stimulation of the superior colliculus (Udo and Mano, 1970; Peterson, Anderson, and Filion, 1974).

More impressive than the tectoreticular are the *corticoreticular fibers,* which presumably mediate the short latency responses recorded in the reticular formation following stimulation of the cerebral cortex (2–6 msec according to Hugelin,

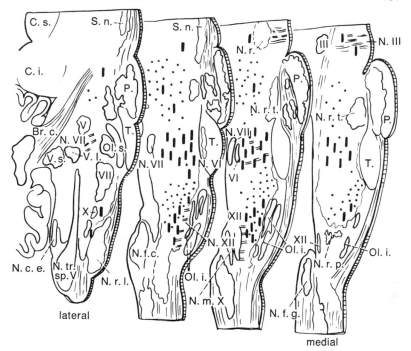

FIG. 6-7 In parasagittal sections through the cat's brainstem are plotted the regions of the medial two-thirds of the reticular formation which receive the maximum of fibers from the cerebral cortex (dots) and the regions from which the majority of long ascending fibers take origin (vertical bars). Note overlap in the region dorsal to the middle part of the inferior olive. Some abbreviations: *Br.c*: brachium conjunctivum; *N.f.c.* and *N.f.g.*: nucleus cuneatus and nucleus gracilis; *N. VII*: facial nerve; *V, VI, VII, X,* and *XII*: motor nuclei of cranial nerves.

Bonvallet, and Dell, 1953; for further data see Brodal, 1957; Rossi and Zanchetti, 1957). The existence of these fibers was established by early neuroanatomical workers. They come chiefly from the "sensorimotor" cortex, with contributions from other regions, and descend with the corticospinal fibers, leaving these during their course in the brainstem.[7] (Some of the corticoreticular fibers may be collaterals of corticospinal fibers.) When the terminal distribution of the corticoreticular fibers is studied with silver impregnation methods, it is learned that, like so many other afferent connections, they do not supply the reticular formation in equal density throughout. In the cat (see Fig. 6-7) there are two chief terminal regions (Rossi and Brodal, 1956a), a caudal one which coincides approximately with the nucleus gigantocellularis, and a rostral one (covering the nucleus reticularis pontis caudalis and the caudal part of the nucleus reticularis pontis rostra-

---

[7] Following injections of horseradish peroxidase in the medullary reticular formation in the cat, labeled cells were found only in front of the cruciate sulcus, and mainly in area 6 (Berrevoets and Kuypers, 1975). According to this study, the site of origin of corticoreticular fibers appears to a large extent to be different from that of cortical fibers to the spinal cord and dorsal column nuclei. (The latter are found chiefly more posteriorly, in areas 3 and 4.) The findings suggest that fibers to these different destinations are in part derived from different sets of cortical neurons. See also Wise and Jones (1977).

lis). In subsequent studies in the cat, the monkey, and man (Kuypers, 1958a, b, c, 1960) and in the rat (Valverde, 1962, 1966), largely corresponding findings have been made, and evidence has been obtained that the distribution may be even more specific. Thus most fibers to the medial magnocellular reticular formation appear to come from cortical regions situated more frontally than those which supply the lateral parvicellular reticular formation with its more modest projection (Kuypers, 1960, in the monkey; Valverde, 1962, in the rat). There are also direct corticofugal fibers to the mesencephalic reticular formation (Szentágothai and Rajkovits, 1958; Pearce, 1960; Valverde, 1962).[8]

It is of some interest for functional considerations that, although most of the corticoreticular fibers end in regions of the reticular formation which give off reticulospinal fibers, their terminal area in the medulla covers part of the region which gives off long ascending fibers (Fig. 6-7). We shall return to this and other anatomical features of the organization of the reticular formation later.

In addition to the afferents to the reticular formation described above, the new methods for tracing fiber connections have demonstrated the presence of some hitherto unknown or doubtful input systems. For example, fibers have been found to come from the substantia nigra (pars reticulata) and to supply mesencephalic, pontine, and medullary levels of the reticular formation (Rinvik, Grofová, and Ottersen, 1976; Jayaraman, Batton, and Carpenter, 1977). Other afferents come from the raphe nuclei (see below).

Organization of the reticular formation.   It can be seen from the preceding account that what is known of the fiber connections of the reticular formation supports the notion that it is not diffusely organized. Maximal sites of endings of some main afferents, as well as sites of origin of long efferent fibers, can be indicated fairly precisely, even if these areas of termination or origin do not correspond exactly to particular nuclear groups. Furthermore, there are noteworthy differences between the medial two-thirds and the lateral one-third of the reticular formation. In the latter part there are only small cells, and their ascending or descending axons are relatively short. The medial two-thirds alone, containing many large cells, give rise to long ascending and descending fibers in considerable numbers. It appears, therefore, that what we may call *effector functions of the reticular formation* are mainly mediated by its medial two-thirds, while the lateral one-third appears to act to a large extent on the former, by medially directed axons of its cells. It seems a likely assumption that particularly the lateral part of the reticular formation functions as what is often referred to loosely as an area of association. Within the medial two-thirds there is, furthermore, a segregation, although incomplete, between levels that are equipped to exert their main action on the spinal cord and levels that may act preponderantly on more rostral parts of the brain (see Fig. 6-4). These features in the anatomical organization of the reticular formation argue strongly against the assumption that the reticular formation is functionally a diffusely organized entity.

Studies of cytoarchitectonics and fiber connections, however, give us only

---

[8] Several physiological studies (see Rossi and Zanchetti, 1957, and later in the present chapter) have demonstrated that following stimulation of various cortical areas, potentials can be recorded in the reticular formation.

part of the picture. Important additional information is derived from studies of Golgi preparations. As mentioned previously, several of the observations made in Golgi studies are in complete accord with those obtained in experimental investigations. However, there are features that can only be studied in Golgi material, such as dendritic patterns and arborizations of axons. As referred to previously, the typical cell of the reticular formation has fairly long dendrites, and in Golgi preparations it is striking to see how the "dendritic fields" of cells overlap. Furthermore, it can be seen that collaterals of fibers belonging to different afferent systems overlap extensively. As emphasized by Scheibel and Scheibel (1958), this occurs predominantly in the medial parts of the reticular formation, from which the long ascending and descending projections arise. This overlap of axonal branches "makes it difficult to see how any specificity of input can be maintained" (Scheibel and Scheibel, 1958, p. 34).

There are other features which likewise leave one with the impression that the reticular formation is diffusely organized. In many places, dendrites from cells in the reticular formation extend into neighboring nuclei and vice versa (see Ramón-Moliner and Nauta, 1966). It appears that cells in all cranial nerve nuclei give off axons or collaterals to the reticular formation, just as cells in the reticular formation send collaterals or axons to all these nuclei (Scheibel and Scheibel, 1958). The absence of cells of the Golgi II type was mentioned above (most cells have axons which project for some way along the length of the brainstem). Scheibel and Scheibel (1958) have presented an interesting discussion of the intrinsic organization of the reticular formation, and reach the conclusion (loc. cit. p. 42) that "Rather, it can be said that the fine structure of the brain stem core is such that given the proper physiologic conditions an impulse pattern can probably describe any conceivable path within the reticular formation, so extensive is the interconnectivity of the elements." They draw attention, however, to an interesting feature which is of relevance if one attempts to correlate the conclusions made on the basis of Golgi sections with the results of experimental studies. As stated above, the majority of dendrites of reticular cells are oriented in the transverse plane in the brainstem (see Fig. 6-3). The collaterals entering the reticular formation run in the same plane, taking off from the parent fibers at approximately right angles (Fig. 6-3). The reticular formation may thus be described as being composed of a series of segments, as illustrated in the inset of Fig. 6-3. However, even if the fundamental pattern of dendrites and axons is thus fairly uniform throughout the reticular formation, and each segment appears to be principally organized in the same manner, the segments may well differ functionally. In the first place, the cells in different segments do not send their axons to the same destinations (at some levels, for example, most axons are descending, at others ascending, see Fig. 6-4). Secondly, the afferents do not come from the same source at all levels. This is so far best known as concerns the spinal afferents, which supply chiefly certain "segments." It seems a likely assumption that collaterals coming from different afferent fiber systems end predominantly at particular levels, but this problem has so far not been extensively studied.

A feature of some interest emerges from recent detailed studies on the distribution of reticular formation afferents from various sources. The preferential distribution of the cortical afferents of two regions of the reticular formation, one potine and one medullary, was described above (see Fig. 6-7). The same regions have been found to be the main recipients of afferents from the cerebellum (the fastigial nucleus, see Walberg, Pompeiano, Westrum and Hauglie-Hanssen, 1962), the superior colliculus (Kawamura, Brodal, and Hoddevik, 1974), and the vestibular nuclei (Ladpli and Brodal, 1968). The larger part of these terminal regions coincides with the sites of origin of reticulospinal fibers (Fig. 6-4). There is far less overlap with the regions giving off long ascending axons and with the terminal distribution of spinoreticular fibers. Cortical, tectal, cerebellar, and vestibular inputs would accordingly be assumed to influence first and foremost, but not only, the reticular control of the spinal cord.

Physiological studies have brought forward a great deal of data in good accord with the anatomical findings summarized above. Like the latter, they make it clear that the reticular formation is not a functionally homogeneous unit.

The afferent input to the reticular formation is often said to be diffuse, since sensory information from various sources appears to reach the entire reticular formation. This, however, is not strictly correct. Scheibel, Scheibel, Mollica, and Moruzzi (1955) performed microelectrode recordings of the responses of neurons in the medullary reticular formation to stimulation of the cerebellum, the cerebral cortex, peripheral nerves (proprioceptive and exteroceptive), the vagus nerve, and to acoustic stimuli. Most units could be influenced from several sources, although in different combinations. Units responding to acoustic stimuli are easily found in the mesencephalon (Amassian and de Vito, 1954). Convergence of somatosensory, acoustic, and vestibular impulses on the same units have been observed (Duensing and Schaefer, 1957b). It was concluded by Scheibel, Scheibel, Mollica, and Moruzzi (1955) that even if there is widespread convergence of afferent impulses on units of the reticular formation, this convergence is not unlimited. Patterns of convergence vary markedly among individual units. Further evidence on the minute functional organization of the reticular formation comes from studies in which the neurons recorded from have been identified.

In the medial medullary and pontine reticular formation, neurons projecting to the spinal cord (see Fig. 6-4) have been identified electrophysiologically. About half of them were excited by somatic stimuli from all parts of the body (Wolstencroft, 1964; Magni and Willis, 1964b). For some neurons stimuli applied to different parts of the body were not equally effective (Wolstencroft, 1964). Some neurons were excited from one part of the body and inhibited from others. Magni and Willis (1964a) found that most reticular neurons which project to the spinal cord were excited on cortical stimulation, as were reticular neurons with axons projecting both rostrally and caudally, while neurons with long ascending axons were only occasionally influenced. The cortical excitatory action on reticulospinal neurons appears to a large extent to occur monosynaptically (Peterson, Anderson, and Filion, 1974; Pilyavsky, 1975). Reticulospinal neurons appear to be subject to monosynaptic convergence of cortical, tectal, and cutaneous inputs (Peterson et al., 1974) and to be monosynaptically excited by vestibular impulses as well (Peterson, Filion, Felpel, and Abzug, 1975). This extensive convergence of impulses from many sources is in agreement with the anatomical data on the sites of termination of afferent fiber groups mentioned above. However, even if there is a widespread convergence of excitatory actions from many areas of the cerebral cortex upon individual reticular neurons, the pattern of convergence varies among the neurons. With regard to units responding to stimulation of spinal nerves, Pompeiano and Swett (1962a, 1962b) found that a majority of them are located in the nucleus reticularis gigantocellularis (where many spinoreticular fibers terminate). Not all kinds of spinal impulses are equally effective in influencing the reticular formation. Single unit studies (see, for example, Wolstencroft, 1964; Pompeiano and Swett, 1962a, 1963; Magni and Willis, 1964b) indicate that cutaneous impulses and group II muscle afferents are effective, while proprioceptive impulses mediated by group I fibers do not influence reticular neurons. Some of the units studied and identified as reticulospinal neurons have been found to respond to nociceptive cutaneous stimuli (Wolstencroft, 1964). Obviously the interplay of impulses entering the reticular formation from various sources is extremely complex, and involves excitatory and inhibitory actions. On stimulation of a particular source of afferents a particular neuron may show excitation, followed by inhibition (see, for example, Magni and Willis, 1964b), or from the same source some neurons may be excited and others inhibited, as found for reticulospinal neurons (Wolstencroft, 1964). It appears that reticular neurons may periodically alter their responsiveness to incoming stimuli (Scheibel and Scheibel, 1965), indicating a flexibility in the impulse passages within the reticular formation.

Most of the units studied in the medial part of the pontine and medullary reticular formation appear to be active "spontaneously," that is in the absence of any intentional stimulation of sensory receptors or of central structures, in anesthetized as well as unanesthetized animals. Further, some neurons may fire spontaneously, according to different patterns. This activity has often been found to be rhythmical. For particulars and references, see the review by Pompeiano (1973). For a recent physiological study of the spinoreticular neurons and references to earlier work on this subject, see Fields, Clanton, and Anderson (1977).

The above data are compatible with the view of the organization of the reticular formation derived from anatomical studies, but further studies are needed, in which a precise mapping of the distribution of units responding to particular kinds of sensory stimulation is done. A number of microelectrode studies, in addition to those mentioned, have been performed in recent years, but it is scarcely possible as yet to put the results together into a coherent picture of the modes of operation of the reticular formation. Although a knowledge of the function of the reticular formation at the cellular level is highly desirable, other findings are of more immmediate relevance from a clinical point of view. Before discussing this subject, it is appropriate to consider three small brain regions which are rather closely related to the reticular formation in a restricted sense, namely the raphe nuclei, the periaqueductal (central) gray, and the nucleus locus coeruleus.

The raphe nuclei. As briefly referred to on p. 399, this nuclear complex consists of several minor cell groups that can be distinguished on the basis of their cytoarchitecture. Because of the recent interest in these nuclei, they will be considered in some detail. Together the raphe nuclei form a rather narrow, more or less continuous collection of cells which extends along the midline of the brainstem (raphe = seam) from the caudal end of the medulla to the rostral mesencephalon. The various subgroups fuse with each other at some levels and are not everywhere sharply delimited from the reticular formation. Figure 6-8 shows their main topography in the cat in horizontal sections and as plotted diagrammatically on a midsagittal section of the brainstem. It can be seen that altogether eight nuclei are distinguished, named from caudal to rostral, the nuclei raphe obscurus, raphe pallidus, raphe magnus, raphe pontis, nucleus centralis superior, nucleus raphe dorsalis, and nuclei linearis intermedius and linearis rostralis. This subdivision (Taber, Brodal, and Walberg, 1960), except for minor points, is in accord with those made by previous authors in various animals and man (see Taber et al., 1960, for particulars and references).

With regard to its anatomical organization and connections, the raphe complex shows several similarities with the main reticular formation (see Taber, Brodal, and Walberg, 1960). The assumption that the various raphe nuclei are to

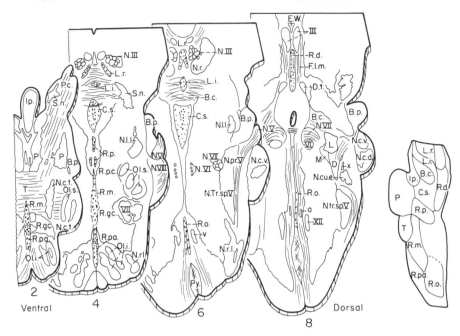

FIG. 6-8 The raphe nuclei of the cat, as seen in four equally spaced horizontal sections through the brainstem, to the right as projected on a midsagittal section. The eight nuclei distinguished differ with regard to cell types and cell density (roughly indicated by differences in size and density of dots in the drawings). Some abbreviations: *B.c.:* brachium conjunctivum; *C.s.:* nucleus centralis superior; *Ip.:* nucleus interpeduncularis; *L.i.* and *L.r.:* nucleus linearis intermedius and rostralis; *P:* pons; *R.d.:* nucleus raphe dorsalis; *R.m.:* nucleus raphe magnus; *R.o.:* nucleus raphe obscurus; *R.p.:* nucleus raphe pontis; *R.pa.:* nucleus raphe pallidus; *T:* trapezoid body. From Taber, Brodal, and Walberg (1960).

some extent particular units receives support from anatomical and functional studies. However, because the raphe nuclei are so small, it is almost impossible to study their connections and functions by stimulating, destroying, or injecting only one of them. Involvement of more than one particular nuclear group or of adjoining structures has been the rule in all studies of this kind performed so far. Approaches in which damage to the nuclei is avoided, for example studies of their efferents by means of the HRP method, carry less risk of misinterpretation.

The demonstration that the raphe nuclei (particularly the nucleus raphe dorsalis and centralis superior) appear to be the principal parts of the brain which harbor serotonin-containing (5-HT) neurons (see, for example, Dahlström and Fuxe, 1964; Björklund, Falck, and Stenevi, 1971) prompted numerous investigations of these nuclei, in normal material and experimentally.[9]

The *efferent connections* of the raphe nuclei have been most extensively studied (see Fig. 6-9). In an investigation of the efferents from the raphe nuclei by means of the modified Gudden method (Brodal, 1940a) in the cat, Brodal, Taber,

[9] Hubbard and di Carlo (1974b) describe eight serotonin raphe groups in the squirrel monkey that appear to correspond to those outlined cytoarchitectonically.

and Walberg (1960) found clear-cut retrograde cellular changes in certain of the raphe nuclei following transections of the spinal cord, transections of the brain-stem at the mesencephalic level, or lesions of the cerebellum. Long ascending fibers make up the efferent contingent that is most important quantitatively. They arise from all raphe nuclei, but only to a little extent from the nuclei raphe pontis and magnus. Smaller contingents to the cerebellum and the spinal cord come mainly from the nucleus raphe pontis and magnus, respectively.

Studies of the retrograde cellular changes in a nucleus give information about the origin of efferent fibers (although negative findings are not conclusive), but tell little about their exact terminal distribution. With silver impregnation methods and by autoradiography, it has been confirmed that the *ascending efferent connections* are abundant and extremely widespread in the rat (Conrad, Leonard, and Pfaff, 1974) and in the cat (Taber, Foote, and Hobson, 1976; Bobillier et al., 1976; and others). It has been concluded that the fibers supply several cell groups in the mesencephalon (for example, the periaqueductal gray), nuclei in the hypothala-mus, intralaminar and other thalamic nuclei, the cerebellum, parts of the amygda-loid nucleus and of the hippocampal formation, the septum, the caudate and pu-tamen, and even the cerebral cortex, especially the frontal cortex.

With the Falck-Hillarp method, serotonergic fibers have been traced rostrally in a distribution resembling that found with other methods (see, for example, Fuxe, Hökfelt, and Ungerstedt, 1970), and serotonin is assumed to be the transmitter at the endings of these fibers. While all the efferents described may not be serotonergic, it appears that such fibers arise particularly from cer-tain parts of the raphe, among them the nucleus raphe dorsalis and the centralis superior (particu-larly the former) and that they pass to the forebrain (Lorens and Guldberg, 1974). Biochemical analyses of the target organs support the data on different efferent connections from these nuclei. A large literature has grown up on the serotonin pathways and their functional role. (See, for ex-ample, Wang and Aghajanian, 1977, on the action on the amygdala.)

A particular target of raphe efferents, the cerebral ventricles, deserves men-tion.

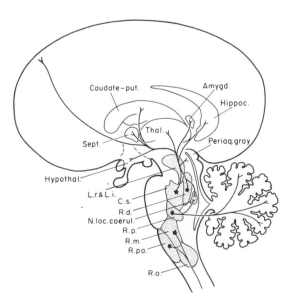

FIG. 6-9 A highly simplified diagram of the main efferent connections of the raphe nu-clei. While many connections appear to come from more than one raphe nucleus, only the main nucleus of origin is indi-cated for each fiber contingent. For abbreviations of the raphe nuclei, see legend to Fig. 6-8 (cf. text).

Brightman and Palay's (1963) description of supraependymal fine varicose nerve fibers with terminals in the cerebral ventricles has been confirmed in several electron microscopic studies (see, for example, Westergaard, 1972). The fibers are found among the cilia and microvilli on the surface of the ependymal cells and are particularly well demonstrated in scanning electron micrographs (Noack, Dumitrescu, and Schweichel, 1972). According to Richards, Lorez, and Tranzer (1973), these varicose fibers are serotonergic, as determined with the Falck-Hillarp technique (see Chap. 1). Following lesions of the rostral raphe nuclei, they show degenerative changes and a loss in their serotonin content (Aghajanian and Gallager, 1975). The function of these axonal terminals of efferent serotonergic neurons of the raphe nuclei remains conjectural (synaptic action on ependymal cells? liberation of substances that may act on the choroid plexus or the neurohypophysis via the median eminence? receptors registering the composition of the cerebrospinal fluid?).

The *descending efferents* of the raphe nuclei are less abundant than the ascending ones. In autoradiographic studies they have been traced, for example, to the reticular formation of the pons and the medulla, to cranial nerve nuclei, to the cerebellum, and some to the spinal cord (Taber, Foote, and Hobson, 1976; Bobillier et al., 1976; the latter list more than 100 nuclei that are claimed to receive afferents from the raphe nuclei).

It has been shown in these and other studies that the fibers from the various raphe nuclei to some extent have their preferential sites of termination. The most precise information about this subject and about the sites of origin of fibers within the raphe complex may be gathered from studies using HRP injections in the assumed target nuclei. In this way it has been confirmed that the caudate and putamen receive raphe fibers chiefly from the dorsal raphe nucleus (Nauta, Pritz, and Lasek, 1974; Miller et al., 1975), that fibers to the hippocampus appear to come mainly from the centralis superior and the dorsal raphe nucleus (Segal and Landis, 1974a; Pasquier and Reinoso-Suarez, 1977), and that fibers to the locus coeruleus come mainly from the nucleus raphe pontis and dorsalis (Sakai et al., 1977). Fibers to cerebellar lobules VI and VII, noted by Shinnar, Maciewicz, and Shofer (1975), and fibers to the crus II, were found by Taber, Hoddevik, and Walberg (1977) to arise chiefly from the nuclei raphe pontis and obscurus, in agreement with the results of studies of retrograde cellular changes (Brodal, Taber, and Walberg, 1960). Recent anatomical and electrophysiological studies (Basbaum, Clanton, and Fields, 1976; Anderson, Basbaum, and Fields, 1977) indicate that axons of spinal projecting raphe neurons descend in the dorsolateral funiculus and end in laminae I, II, and V. They appear to arise mainly in the nucleus raphe magnus and pallidus. An increasing body of information appears to indicate that descending raphe fibers end in the intermediolateral cell column and exert a monosynaptic inhibitory action on preganglionic sympathetic neurons (for a recent report see Cabot, Wild, and Cohen (1979), in the pigeon).

The distribution of *afferents* within the raphe complex likewise appears to occur in a particular pattern, but so far rather little is known. In a study with silver impregnation methods, Brodal, Walberg, and Taber (1960) found afferents from the spinal cord, the cerebellum, and the cerebral cortex to be relatively scanty and to supply only four of the nuclei (the pattern of distribution differs somewhat among the three contingents studied). The nucleus raphe magnus, a main source for efferents to the cord, appears to be the most important target for all of them. It may be significant that the nucleus raphe pontis, projecting to the cerebellum (see above), appears to receive its main afferent input from the cerebellum, which also supplies the nucleus raphe magnus (Brodal, Walberg, and Taber, 1960; Miller and Strominger, 1977). Further, it may be noted that the nucleus raphe dorsalis is almost free from direct fibers from the cord, cerebral cortex, or cerebellum. The major afferent input to the raphe nuclei probably comes from other sources than those mentioned, and afferents from several other regions have been described, including the septum, the lateral preoptic region, the lateral hypothalamus, the habenula, and the prefrontal cortex, but their specific sites of ending are not always indicated (for references, see Mosko, Haubrich, and Jacobs, 1977; Aghajanian and Wang, 1977).

However, Edwards (1975) autoradiographically traced fibers from the mesencephalic reticular formation to the nuclei raphe magnus and dorsalis, and Usunoff, Hassler, Wagner, and Bak (1974) found fibers from the caudate to end in the nucleus raphe magnus. Chu and Bloom (1974; see also Olson and Fuxe, 1971) traced adrenergic fibers from the nucleus locus coeruleus to the dorsal raphe nucleus, and to the raphe pointis, magnus, and centralis superior. Following HRP injections in the dorsal and median raphe nuclei in the rat, Aghajanian and Wang (1977) found some differences in the afferent projections to the two nuclei. For example, fibers from the nucleus of the solitary tract end in the dorsal nucleus only. The projection from the habenula appears to be particularly massive to both nuclei. HRP studies indicate that the raphe dorsalis receives abundant fibers from the nucleus centralis superior (Mosko, Haubrich, and Jacobs, 1977). Likewise, there are ample connections between these raphe nuclei and the reticular formation and some minor nuclei in the brainstem.

Scheibel, Tomiyasu, and Scheibel (1975) have recently described some structural peculiarities of the raphe nuclei which are likely to be functionally important. In extensive Golgi studies of brains from mice, rats, cats, monkeys, and man, they found a particularly close relation between the nervous elements and the vessels in the nucleus raphe pontis and nucleus linearis rostralis. The vessels (the afferent ones coming from the basilar artery) ascend in the ventrodorsal direction on both sides of the midline. The nerve cells are assembled in dense aggregates along the vessels and are in close contact with the vessel wall (studied electron microscopically). The neurons give off dense bundles of fine dendrites, arranged dorsoventrally and closely applied to one or more underlying blood vessel. The dendrites are essentially spineless but have frequent elongated nodules along their course. The functional importance of this relation between nerve cells and vessels is still conjectural (see Scheibel et al., 1975). One possibility might be that the raphe neurons in these nuclei act as chemosensors for substances circulating in the blood and thus may be involved, for example, in the sleep–waking rhythm.

Despite marked differences between its minor units, the complex of the raphe nuclei appears to be to a large degree an entity. It may obviously influence extensive regions of the brain by its widespread efferent projections. The simplified diagram of Fig. 6-9 illustrates this. Attempts have been made (mainly in the rat) to study the importance of the raphe nuclei, and particularly the ascending serotonergic system, by making lesions in the nuclei and analyzing subsequent behavioral changes. It appears that such lesions may produce an array of disturbances, such as increase of locomotor activity, insomnia, hyperreactivity, and aggression, but that the effects of lesions of the various nuclei are not identical, as might indeed be expected.

Thus, lesions of the centralis superior (usually called the median raphe nucleus in the rat) have more marked effects than lesions of the nucleus raphe dorsalis (see, for example, Jacobs,

Wise, and Taylor, 1974; Srebro and Lorens, 1975). It should be recalled in evaluating the results of such behavioral studies, that "most of the behavioral effects of lesions in the raphe nuclei have been observed following other central nervous system lesions" (Srebro and Lorens, 1975, p. 321). Some authors include the raphe nuclei in the enigmatic "limbic system" (see Chap. 10). Among the roles attributed to the raphe nuclei is their involvement in the mechanisms of sleep, as will be briefly considered later in this chapter. The raphe nuclei, particularly the nucleus raphe magnus, have further been claimed to be involved in pain mechanisms. Stimulation results in analgesia, possibly through inhibition of spinothalamic or other neurons with nociceptive input (see, for example, Beall et al., 1976; Fields, Basbaum, Clanton, and Anderson, 1977; Guilbaud, Oliveras, Giesler, and Besson, 1977). The mechanism appears to be much the same as in the case of analgesia produced by stimulation of the periaqueductal gray (see below). Likewise, much interest has recently been devoted to the role of serotonin and the raphe nuclei in morphine analgesia (see, for example, Basbaum, Clanton, and Fields, 1976).

The nucleus locus coeruleus. It is appropriate to consider briefly in this chapter this small brain region, which in recent years has attracted much interest. This particular cell group was noted by the early neuroanatomists because of the heavy pigmentation of its cells (often called the *nucleus pigmentosus pontis*). It appears to be present in all mammalian species studied (Russell, 1955).

It is found near the floor of the rostral part of the fourth ventricle (see Fig. 6-10), begins caudally slightly rostral to the principal trigeminal nucleus, and extends rostrally ventral to the mesencephalic trigeminal nucleus (some intermingling of cells). The nucleus (see Russell, 1955; Olszewski and Baxter, 1954, in man) consists mainly of medium-sized neurons, with rather coarse particles of melanin granules in the cytoplasm. In addition, there are smaller, usually nonpigmented, neurons. (The designation *nucleus subcoeruleus* refers to a rather diffuse cell group ventrolateral to the nucleus coeruleus.)

Knowledge of the fiber connections of the nucleus coeruleus was very limited until quite recently, when it was found that the nucleus belongs to regions of the brainstem which are very rich in noradrenaline. In their 1964 study, Dahlström and Fuxe described altogether 12 such cell groups in the rat (named A 1–A 12) containing noradrenaline (norepinephrine), as witnessed by green fluorescence with the Falck-Hillarp method. One of these, group A 6, was found to correspond approximately to the nucleus locus coeruleus (refer to Fig. 1-21). This nucleus is rich in noradrenergic neurons in the cat (Chu and Bloom, 1974; Jones and Moore, 1974), and in monkeys as well (Hubbard and di Carlo, 1973, 1974a; Garver and Sladek, 1975; German and Bowden, 1975). The same appears to be the case in man (Farley and Hornykiewiez, 1977). However, it should be noted that the location of noradrenergic neurons does not correspond precisely to the territory of the nucleus coeruleus in either the cat or the monkey but extends, for example, into the nucleus subcoeruleus and regions close to the brachium conjunctivum. Even if the groups distinguished in general appear to be similar in different animals, there are clear species differences (see, for example, Hubbard and di Carlo, 1974a). In some species, at least, a larger dorsal and a smaller ventral part can be distinguished (Swanson, 1976).

The *efferent connections* of the nucleus locus coeruleus have been found to be extremely widespread. From studies using the histofluorescence method (tracing of fibers in normal material and after lesions of the nucleus or its efferents) and from correlations with chemical analyses, two main ascending noradrenergic

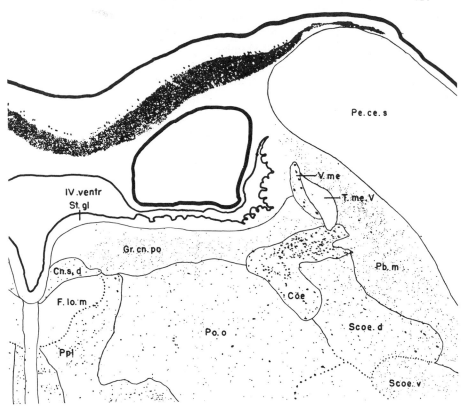

FIG. 6-10 Semischematic representation of part of a Nissl-stained transverse section of the human brainstem illustrating the topography of the nucleus locus coeruleus (*Coe*). Other abbreviations: *Cn.s.d:* nucleus centralis superior, subnucleus dorsalis; *F.lo.m:* fasciculus longitudinalis medialis; *Gr.cn.po:* griseum centrale pontis; Pbm: nucleus parabrachialis medialis; *Pe.ce.s:* pedunculus cerebelli superior; *Po.o:* nucleus pontis centralis oralis; *Ppl:* nucleus papillioformis (nucleus reticularis tegmenti); *Scoe.d* and *Scoe.v:* nucleus subcoeruleus, subnucleus dorsalis and ventralis; *St.gl:* stratum gliosum subependymale; *T.me.V:* tractus nervi trigemini mesencephalicus; *V.me:* nucleus nervi trigemini mesencephalicus; *IV. ventr.:* fourth ventricle. From Olszewski and Baxter (1954).

pathways have been identified. A dorsal one preferentially innervates the *entire cerebral cortex, the hippocampal formation, and the amygdala* (see Fuxe, Hökfelt, and Ungerstedt, 1970; Maeda and Shimizu, 1972; and others) and gives off abundant collaterals to the *thalamic relay nuclei.* A ventral or intermediate pathway supplies the *hypothalamus.* Other fibers pass via the superior cerebellar peduncle to the *cerebellum* (Olson and Fuxe, 1971; and others),[10] and a caudal projection has been traced to the *lower brainstem,* including the reticular forma-

[10] The fibers to the cerebellum appear to be relatively sparse but diffusely distributed. According to electron microscopic studies following intracisternal injections of tritiated noradrenaline, the fibers contact dendrites of Purkinje cells (Bloom, Hoffer, and Siggins, 1971). For some other data see Chapter 5.

tion (Olson and Fuxe, 1972). It has been concluded from the findings that a single noradrenergic neuron can supply, via its branches, the entire cerebral cortex (Olson and Fuxe, 1971; Maeda and Shimizu, 1972). The terminal branches are extremely fine and densely beset with varicosities. It appears further that the various efferent components in part have different preferential areas of origin within the nucleus coeruleus and the adjoining areas.

Following these studies of the connections of the nucleus locus coeruleus with the Falck-Hillarp method, attempts were made to trace them with conventional anatomical methods. The findings made in autoradiographic studies of the efferent fibers in the rat (Pickel, Segal, and Bloom, 1974; German and Bowden, 1975) largely agree with the results of histofluorescence studies. Detailed accounts have recently been given by Jones and Moore (1977), and by Bowden, German, and Poynter (1978) for the monkey. Using modifications of the Fink and Heimer method, Shimizu, Ohnishi, Tohyama, and Maeda (1974) confirmed the presence of fibers from the nucleus locus coeruleus to the cerebral cortex, chiefly the frontal cortex. However, following HRP injections in different parts of the cerebral cortex in the cat, Llamas, Reinoso-Suárez, and Martinez-Moreno (1975) found labeled cells in the nucleus locus coeruleus only with injections in the gyrus proreus. With the same method, however, Gatter and Powell (1977) found the nucleus locus coeruleus in the monkey to project bilaterally to extensive parts of the cerebral cortex. Further, the projection to the hippocampus has been confirmed with the HRP method (Pasquier and Reinoso-Suarez, 1977), and cells in the nucleus locus coeruleus and in the nucleus subcoeruleus were found to project to the *lumbar spinal cord* in the cat and the monkey (Hancock and Foggerousse, 1976; see also Kuypers and Maisky, 1975). Fibers from the nucleus coeruleus to the cord appear to descend in the ipsilateral ventral funiculus (Pickel, Segal, and Bloom, 1974; see also Nygren and Olson, 1977). Other fibers enter some of the raphe nuclei (Olson and Fuxe, 1971; Chu and Bloom, 1974).

Relatively little is known about *afferents* to the nucleus locus coeruleus. It may be mentioned that in an electron microscopic study in the rabbit, Mizuno and Nakamura (1970) found evidence of a projection from the *hypothalamus* (which contains some noradrenergic neurons), and that Domesick (1969) traced fibers from the *cingulate gyrus* in the rat. Fibers from the *cerebellum* have been described (Snider, 1975). In a recent study using the HRP and autoradiographic methods, Sakai et al. (1977) described afferents from the *raphe nuclei*, particularly the raphe pontis and dorsalis, and from the *substantia nigra* and other regions mentioned above (see also Cedarbaum and Aghajanian, 1978). From HRP and autoradiographic studies, Hopkins and Holstege (1978) concluded that the nucleus locus coeruleus receives afferents from the *nucleus amygdalae* (see also Chap. 10).

The nucleus locus coeruleus and neighboring regions thus appear, much in the same way as the raphe nuclei, to give rise to extremely widespread efferent projections, terminating in structures almost all over the central nervous system. In view of the fairly restricted number of cells in the nucleus—altogether about 1400 in the rat (Swanson, 1976)—it must be concluded that the axons branch profusely. This has recently been confirmed in Golgi studies by Scheibel and Scheibel (1977), as can be seen in Fig. 6-11. It appears that this system is able to mediate a general, rather diffuse noradrenergically mediated influence and, as stated by Swanson (1976), is "involved in a bewildering variety of behavioral, physiological and neuroendocrine functions." It has been hypothesized that it may be an essential part of the activating system and may be concerned, for example, in cortical arousal, in the induction of paradoxical sleep (see Jouvet, 1972), in

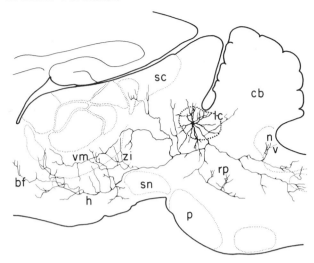

FIG. 6-11 In Golgi preparations the widespread distribution of axonal collaterals of cells in the nucleus locus coeruleus can be visualized, as seen in this drawing based on two adjacent sections from a 14-day mouse (rapid Golgi variant). *bf:* basal fore-brain; *cb:* cerebellum; *h:* hypothalamus; *lc:* locus coeruleus; *nv:* vestibular nuclei; *p:* pons; *rp:* nucleus raphe pontis; *sn:* substantia nigra; *sc:* superior colliculus; *vm:* ventromedial thalamus; *zi:* zona incerta. From Scheibel and Scheibel (1977).

mediating pressor responses to peripheral vessels via its descending projections to the cord (see Ward and Gunn, 1976), and in other phenomena.

The periaqueductal gray. It is appropriate at this junction to consider briefly the gray matter surrounding the cerebral aqueduct, the *periaqueductal gray*. This area, in spite of being easily distinguished in Nissl-stained sections of the brain, has offered great problems for both anatomists and physiologists. It consists of rather densely packed, mainly small, cells and has been subdivided into minor areas. Hamilton and Skultety (1970) and Hamilton (1973a) distinguished three areas arranged concentrically around the aqueduct (the medial, lateral, and dorsal nuclei), in agreement with the subdivision suggested for the human brain by Olszewski and Baxter (1954). A number of *efferent connections* of the periaqueductal gray have been described (see, for example, Nauta, 1958; Chi, 1970; Hamilton and Skultety, 1970; Hamilton, 1973b). In studies using silver impregnation methods, in which attempts were made to destroy a particular subdivision only, most previously described connections were confirmed, but the three nuclei were found to differ with regard to their projections (Hamilton, 1973b).

The medial nucleus was found to project ventrally to the tegmentum in a radiating pattern and mainly in a rostral direction to the diffusely outlined field of Forel and the ventral tegmental area. Fibers from the dorsal nucleus were traced to the ipsilateral pretectal area and the lateral habenular nucleus, while the lateral nucleus supplies the posterior hypothalamus and several thalamic nuclei. Fibers have further been traced to the inferior olive (Walberg, 1956, 1974a). Although it appears so far to have been little studied, it is to be expected that the three nuclei differ as well with regard to their *afferent connections*. These have been described to come from different sources, for example the cingulate (Domesick, 1969) and prefrontal cortex (Leonard, 1969), the hippocampus, possibly the septal nuclei, the lateral hypothalamus, the habenula, and interpeduncular nucleus (Nauta, 1958). Afferents from the spinal cord and the reticular formation have likewise been described. In a recent HRP study Grofová, Ottersen, and Rinvik (1978) found afferents to come from other regions as well, among them the substantia nigra, the zona incerta, and the reticular thalamic nucleus, in addition to a heavy projection from the ventromedial hypothalamic nucleus.

The periaqueductal gray thus seems to be a small region which is linked, in many instances by two-way connections, with several regions of the brain. Physiologically, the periaqueductal gray has been claimed to be involved in many functions, for example rage reactions, feeding responses, influence on bladder tonus, and pain.

It has repeatedly been shown that electrical stimulation of the periaqueductal gray (to a lesser extent also of other brain regions) abolishes responsiveness to noxious stimuli (see, for example, Mayer and Liebeskind, 1974). It is generally assumed that the effect is due to an inhibitory action on pain-transmitting spinal neurons. There appear to be no fibers from the periaqueductal gray passing directly to the spinal cord. Since stimulation of the raphe nuclei, particularly the raphe magnus (see above), produces analgesia, it has been suggested that these nuclei serve as a relay between the periaqueductal gray and the spinal neurons. However, the periaqueductal gray does not appear to send fibers to the raphe nuclei, but the latter may, nevertheless, be activated by the periaqueductal gray via relays in the reticular formation. The experimental data mentioned above have prompted attempts to alleviate pain in man by electrical stimulation of the periaqueductal gray (see later).

The periaqueductal gray, like the raphe nuclei, has been assumed to be involved in morphine-induced analgesia (see Lewis and Gebhart, 1977, for a recent critical study). Morphine and other opiates appear to act by binding to particular receptors on nerve cells. It is to be noted that opiate receptors have been identified in numerous parts of the brain other than the periaqueductal gray and the raphe (see, for example, Atweh and Kuhar, 1977). Even though there are similarities between the analgesia produced by brain stimulation and that evoked by morphine, much is still conjectural concerning both mechanisms. (See also Chap. 2.)

Functional aspects of the reticular formation.   A number of functions are known to be influenced from the reticular formation. In fact, it appears that this, like the cerebellum (see Chap. 5), may be of importance for almost all "functions" controlled by the nervous system, as might indeed be expected when one considers its interconnections with other parts of the brain. Only some aspects of the functional role of the reticular formation will be considered in this text. Owing to the many problems connected with the studies and concepts of the ascending activating system, this will be dealt with separately in a following section. Here our concern will be mainly with the "descending" actions of the reticular formation, i.e., its influence on spinal cord mechanisms, somatic and visceral.

It has been known for a long time that the brainstem contains regions that are essential for functions such as muscle tone and reflexes, respiration, and several autonomic functions, particularly cardiovascular control. In keeping with general views on the organization of the central nervous system some 40 years ago, it was assumed that the brainstem harbors particular "centers" engaged in the control of such functions. From experiments with stimulation or destruction of particular parts of the reticular formation, centers for inhibition or facilitation of motor activity, depression or elevation of blood pressure, and inspiration or expiration, were outlined. To a great extent these pairs of centers appeared to cover identical regions. For example, there was a rather close spatial correspondence of muscular inhibitory, vascular depressor, and inspiratory regions (see, for example, Text-fig. 17 in Brodal, 1957).

Advances in neurophysiological techniques and changing views of the organization of the nervous system (see Chap. 1) have made it clear that the idea of pairs of mutually antagonistic "centers" for these functions is an undue oversimplification. Furthermore, many other regions of the brain are known to influence the same functions, even though many of the effects are mediated via structures in the brainstem, particularly the reticular formation. Considering further that changes in muscular activity, in blood pressure and cardiac rhythm, in respiration, and in other functions all occur in an integrated pattern in almost every situation, ranging from changes in the position of the body to emotional reactions, it is no wonder that on stimulation of the same point one may observe changes in muscle tone, blood pressure, and respiration simultaneously. This was observed long ago by Bach (1952), who drew attention to the complexity in the integrated control of functions influenced from the reticular formation. A priori one must assume that this integration requires abundant and complexly organized interconnections between many parts of the brain, among which cooperation may even change from one moment to the next and may be influenced by humoral factors. This cooperation does not exclude the possibility that certain areas of the brain are more specifically concerned in some "functions" than in others, as indeed appears to be the case, as was learned from several physiological attempts to outline such particular brain regions. Basically, all cooperation between brain regions depends on their interconnections, direct and indirect. Our knowledge of the anatomical organization of neuronal networks in the brain is, however, still fragmentary.

The problems of the *control of muscular functions, respiration, and the cardiovascular system by the reticular formation* and other parts of the brain have turned out to be immensely complex. Only a few points that permit some correlations with anatomical data will be briefly considered here.

The beginning of a more systematic approach to the problem was made by Magoun and his collaborators, starting with the papers of Magoun and Rhines (1946) and Rhines and Magoun (1946). It was shown that *electrical stimulation of the reticular formation could alter all kinds of motor activity.* Stimulation of the ventromedial part of the medullary reticular formation (in cats and monkeys) inhibited myotatic reflexes and muscular tone (the rigid limbs in decerebrate animals became flaccid). Likewise, movements evoked by cerebral cortical stimulation were inhibited. Opposite, facilitatory effects were obtained on stimulation of the reticular formation lateral to the inhibitory region and at more rostral levels, in the pons, mesencephalon, and even from midline structures in the thalamus, the central gray, and the subthalamus and hypothalamus. The effects were assumed to be mediated via a final link of reticulospinal projections.

According to Magoun and his collaborators the two regions of the brain exert a generalized inhibitory or facilitatory effect, respectively. Later studies (for references see Rossi and Zanchetti, 1957) have, however, shown that most of the responses obtained on stimulation of the lower reticular formation are reciprocally organized. As was mentioned in Chapter 4, α as well as γ

motoneurons can be influenced—facilitated or inhibited—from the reticular formation. The complexity is well illustrated by findings such as those of Lindblom and Ottosen (1956), that some reflex activity can be inhibited and some facilitated from the same bulbar point. Rossi and Zanchetti (1957) early suggested that the reciprocity depends chiefly on properties of the spinal cord and not on the intrinsic organization of the reticular formation.

From an anatomical point of view it is of interest that the "inhibitory region" of Magoun coincides fairly well with the medullary region that gives rise to reticulospinal fibers. On stimulation of this region, inhibitory postsynaptic potentials can be recorded in flexor and extensor motoneurons (see, for example, Jankowska, Lund, Lundberg, and Pompeiano, 1968). These inhibitory effects may, at least in part, be mediated via direct reticulospinal fibers, contacting interneurons in the cord. As to the facilitatory effects, the situation is more complex. The area of the brainstem found to yield facilitation extends far outside the regions from which reticulospinal fibers originate, although it includes part of the pontine region (see Chap. 4). It appears likely, therefore, that stimulation of reticular neurons projecting directly onto the cord can only to a slight extent be concerned in the facilitatory effects, and that most of these must occur by way of fibers originating in the facilitatory areas and terminating directly or via internuncials on reticulospinal neurons or on other caudally projecting neurons of the brainstem.

In recent years several physiological studies have been devoted to the influence of the reticular formation on the cord, on α and γ motoneurons, on spinal reflexes elicited by various kinds of afferent impulses (see Pompeiano, 1973, for a review), and on other functions.

It is usual to distinguish between dorsal and ventral reticulospinal systems in these studies. However, these "systems" do not appear to correspond to anatomically determined reticulospinal tracts, but to include other descending pathways as well (medial longitudinal fasciculus, descending fibers from the raphe nuclei and others). To a considerable extent the caudal propagation of impulses within the cord appears to involve propriospinal neurons. Despite numerous observations on details of reticular formation influence on the motor and reflex activities of the cord, it does not appear possible to distinguish particular regions within it as being either inhibitory or facilitatory (cf. the general comments above). It should further be recalled that, in modifying motor activity, the reticular formation collaborates with several other supraspinal structures, such as the vestibular nuclei, the red nucleus, the superior colliculus, the fastigial nucleus, and the cerebral cortex, all of which may act on the spinal cord. Some of these structures act directly on the reticular formation as well. Furthermore, by way of its projections, for example onto the vestibular nuclei, and via connections leading to the cerebral cortex, the reticular formation may act on the spinal motor apparatus indirectly. The role of these mechanisms in the intact animal is not yet clear.

Mention was made in Chapter 2 of the *effects exerted by the reticular formation on the central transmission of sensory impulses* entering in the spinal nerves; in Chapter 7 corresponding observations concerning the cranial nerves will be considered. The pathways involved are presumably reticulospinal projections and reticular fibers entering the cranial nerve nuclei, respectively.

Respiratory movements and cardiovascular functions are influenced from the reticular formation. The physiology of these mechanisms will not be discussed here, but some relevant anatomical data will be briefly considered. Mainly on the basis of stimulation experiments (Pitts, Magoun, and Ranson, 1939, and others), an *inspiratory area* or "center" was found to be located in the medulla, and to be almost coextensive with the muscular inhibitory region. An *expiratory region* was concluded to be situated rostral and lateral to the inspiratory region, located largely outside the spinal projection areas of the reticular formation. (In addition an "apneustic center" and a "pneumotaxic center" have been distinguished. The latter is said by some to be situated in the area ventral to the brachium conjunctivum, covering the so-called nucleus parabrachialis medialis, and is assumed to exert its action via the more elementary inspiratory and expiratory "centers.")

The original concepts about the function of centers controlling respiratory movements have, as might be expected, turned out to be far too simple. Refined neurophysiological techniques, among which recordings of single units have proved significant, have made this clear. Thus respiratory-related neurons are of different types (see, for example, Hukuhara, 1974; Vibert, Bertrand, Denavit-Saubié, and Hugelin, 1976a). Not all of them are simply inspiratory or expiratory.

The respiratory-related neurons (like other neurons of the reticular formation) appear to be rather heterogeneous, for example with regard to their response to afferent impulses from the mesencephalon and the spinal cord and afferent vagal impulses (see Hukuhara, 1974, for some references). Some inspiratory neurons, for example, are inhibited on stimulation of baroreceptors, signaling intra-arterial pressure, as shown by intracellular recordings (Richter and Seller, 1975).

Studies devoted to determination of the location of respiratory-related neurons in the brainstem have not given entirely concordant results.

Merill (1970), in agreement with some previous authors, found respiratory-related neurons to be concentrated within a longitudinal cell column, located laterally in the medulla, dorsal to the nucleus ambiguus (corresponding to the so-called nucleus retroambigualis) and extending from spinal cord segment C_1 to about 4 mm above the obex. In spite of some overlapping, the caudal part of the group contains mainly expiratory neurons, the rostral part inspiratory neurons. (Some respiratory-related neurons were found also in the nucleus ambiguus, reflecting the activation of laryngeal and pharyngeal muscles in respiration.) Hukuhara (1974) plots the site of different kinds of respiratory-related neurons as being mainly in the lateral part of the reticular formation at pontine levels. A wider spatial distribution of respiratory-related neurons was found by Vibert, Bertrand, Denavit-Saubié, and Hugelin (1976a, 1976b), who recorded responses from many thousands of neurons and mapped their localization anatomically. Respiratory-related neurons were found at all levels of the entire bulbopontine reticular formation. Even though most of the inspiratory neurons are located dorsolaterally in the caudal third of the total area and most of the expiratory neurons rostrally and ventromedially, the two kinds of neurons have different spatial distributions and are aggregated in uninterrupted columns without much overlapping. In addition, respiratory-related neurons have been found in several nuclei outside the reticular formation—for example, in the nucleus ambiguus, as mentioned above, and in the trigeminal complex (Bertrand, Hugelin, and Vibert, 1973).

In spite of some discrepancies as to details, recent studies leave little doubt that reticular neurons related to respiration (rhythmic respiration and respiratory reflexes) are widely distributed within the brainstem. Their activity may be influenced by afferent impulses from many sources and may change functionally in different situations. It is of interest that some respiratory-related reticular neurons have been found to send their axons rostrally to the mesencephalon. Hukuhara (1974) found 12% of the neurons to be of this kind and 15% to project to the spinal cord. Merill (1970) found that most respiratory-related neurons could be activated antidromically from the spinal cord, indicating that they have descending axons. The recent physiological findings are in general agreement with data on the anatomical organization of the reticular formation and its connections, even if detailed correlations are so far not possible. (Hukuhara, 1974, has suggested an interesting scheme for a possible neuronal organization of the central respiratory mechanisms in the reticular formation, which includes the role played by interneurons.)

With regard to the *alleged "centers" in the brainstem for the control of cardiovascular functions,* early authors such as Wang and Ranson (1939) and Alexander (1946) distinguished a depressor and a pressor region in the medulla. On electrical stimulation, fall or rise, respectively, in arterial blood pressure and heart rate, could be obtained. The depressor area was located to the medial and caudal parts of the medulla; the pressor area was found at more rostral levels and more laterally, extending into the pons.

In the following years the central regulation of cardiovascular functions was extensively studied. It has subsequently become clear that the original concepts need considerable revision. The brain mechanisms controlling cardiovascular function appear to be even more complex than those related to respiration.[11] They exert their action both on the heart and on the peripheral circulation and serve to regulate tissue blood flow by means of several local mechanisms.

> As is in part known from clinical experience, and as amply demonstrated in experimental physiology, cardiovascular functions can be influenced by inputs from a number of peripheral sources (see Korner, 1971): arterial baroreceptors such as the carotid sinus, arterial chemoreceptors, cardiac mechanoreceptors, lung inflation receptors, receptors in the trigeminal area and in the skin in other regions of the body, and receptors in muscle and in the gastrointestinal and urinary tracts. Changes in cardiovascular functions can further be elicited from "higher" regions of the brain, such as the cerebral cortex (particularly the orbitofrontal), the hypothalamus (see Chap. 11), and the cerebellum.

Studies of the central neural cardiovascular regulation meet with considerable technical and conceptual problems, as discussed by Calaresu, Faiers, and Mogenson (1975). On electrical stimulation, alterations of stimulus parameters may affect the type of response elicited. For example, the frequency or intensity of stimulation may determine whether the blood pressure rises or falls. Stimulation of a particular region may affect neighboring structures by direct spread of current or via activation of neuronal interconnections. The use and type of anesthetics may influence the results. The criteria used for the identification of "cardiovascular" neurons are essential. It has been amply demonstrated that, following stimulation of baroreceptors and chemoreceptors of the arterial system, evoked potentials can be recorded in several different regions of the central nervous system. However, this does not necessarily mean that all these structures are important in the central regulation of cardiovascular functions. Only a few data will be considered in the following discussion, with emphasis on possible correlations with anatomical knowledge of the structures involved.

---

[11] For some recent surveys, see Korner, 1971; Smith, 1974; *Central Rhythmic and Regulation,* 1974; *Central Organization of the Autonomic Nervous System,* 1975; Calaresu, Faiers, and Mogenson, 1975.

In single-unit studies, Langhorst and Werz (1974) classified neurons of the reticular formation as "cardiovascular neurons" when they showed spontaneous activity related to cardiac rhythm; spontaneous rhythmical changes of tonic activity parallel to blood pressure waves; and changes in activity during experimentally induced changes in blood pressure. On the basis of these criteria, Langhorst and Werz (1974) did extensive studies of the behavior of "cardiovascular neurons." It turned out, as has been described by others, that most neurons of this kind may be influenced by inputs from many different sources. Different combinations of afferent influences could be observed, changing from one neuron to the other. (There appears to be a particularly close relationship between these neurons and those controlling respiration.) This type of neuron, which the authors call class 2 of cardiovascular neurons, "appears to be intermingled with other ascending and descending reticular elements," and to have a complex pattern of integrative functions. Other neurons (class 1) appear to be specifically related to afferent cardiovascular impulses. These neurons are found first and foremost in the nucleus of the solitary tract and its vicinity, but also in the nucleus ambiguus and in the dorsal motor nucleus of the vagus. (As mentioned in Chap. 7, there is increasing evidence that many or all visceromotor fibers to the heart originate from the nucleus ambiguus.) Physiologically, short-latency baroreceptor responses have been recorded from the nucleus of the solitary tract and its immediate vicinity (see Spyer, 1975, for some references). Practically all afferent vagal-glossopharyngeal fibers end in this nucleus (see Chap. 7). The two other nuclei mentioned receive afferents from the nucleus of the solitary tract and give off efferent fibers to the heart (see Chap. 7). Neurons found at these three sites can, therefore, be considered to be specifically related to cardiovascular regulation, particularly to reflex changes (for example bradycardia) resulting from stimulation of aortic and other baro- and chemoreceptors and to be links in a direct vagoinhibitory reflex pathway (see also Spyer, 1975; Thomas and Calaresu, 1974).

More-or-less convincing attempts have been made to locate cardiovascular neurons to other nuclei in the brainstem (for example, the paramedian reticular nucleus, the raphe nuclei, the nucleus reticularis gigantocellularis, the perihypoglossal nuclei, the lateral reticular nucleus, and other reticular nuclei; see Henry and Calaresu, 1974a). By way of its efferent projections (see Chap. 7), the nucleus of the solitary tract may influence these. The claim by some physiologists that some of these nuclei may be activated monosynaptically from the carotid sinus nerve is still debated.[12] Most often, only long-latency responses have been recorded from them, suggesting a polysynaptic afferent route. On electrical stimulation of some of the nuclei listed above, cardiac slowing and fall in blood pressure have been observed (paramedian reticular nucleus, raphe nuclei, and the gigantocellular reticular nucleus) while stimulation of others (the perihypoglossal nuclei) has been found to give rise to cardiac acceleration and arterial hypertension. The efferent pathways from the "controlling" regions are made up of descending connections that influence the neurons of the sympathetic intermediolateral cell column or the parasympathetic preganglionic neurons, particularly those belonging to the vagus. With regard to the connections to the spinal cord, electrophysiological studies indicate that some of the nuclei mentioned, for example the gigantocellular reticular nucleus and the lateral reticular nucleus, provide an excitatory input to sympathetic cardiovascular neurons, while the paramedian and the raphe nuclei provide an inhibitory input to the same neurons. (For some references, see Calaresu, Faiers, and Mogenson, 1975.) Anatomically, projections to the cord have been demonstrated for some of these nuclei, for example the gigantocellular reticular nucleus and the raphe nuclei, while spinal projections from other nuclei are not firmly established. The parasympathetic cardioinhibitory effects elicitable from some of the nuclei (see above) are probably mediated via the nucleus ambiguus (and the dorsal motor nucleus of the vagus?) but are not known in detail.

Attempts have been made to determine the site in the cord of the descending fibers that influence cardiovascular (and other autonomic) functions by stimulation of discrete parts of the white columns in the cord or by partial transections com-

---

[12] Spyer and Wolstencroft (1971) were unable to confirm this for the paramedian reticular nucleus. Nor did Spyer (1975) find evidence for such input to the medial reticular formation, the classical "depressor region."

bined with stimulation of regions in the brainstem (Wang and Ranson, 1939; Kerr and Alexander, 1964; Illert and Gabriel, 1972; Henry and Calaresu, 1974b; and others). Most authors distinguish sympathicoexcitatory and sympathicoinhibitory pathways, but there is disagreement concerning their positions in the cord.

According to some authors, for example Henry and Calaresu (1974b), cardioexcitatory (sympathicoexcitatory) fibers descend in the dorsolateral funiculus, while inhibitory fibers pass in this as well as in the ventral funiculus. These authors further concluded that sympathetic preganglionic neurons receive inputs from more than one excitatory and one inhibitory structure in the brainstem (Henry and Calaresu, 1974a) and that, accordingly, some integration of supraspinal cardiovascular reflexes occurs in the spinal cord. The largest number of cardioacceleratory neurons was found in the second thoracic segment of the intermediolateral cell column (Henry and Calaresu, 1972). There appears to be neither anatomical nor physiological evidence for a monosynaptic contact between the descending fibers and the sympathetic neurons in the intermediolateral cell column.

Conclusions drawn from physiological studies about pathways of descending fibers from the reticular formation must, however, be evaluated with caution. The presence of abundant propriospinal fibers, involvement of indirect pathways, temporary tissue changes in acute experiments, stimulus parameters, and other factors might influence the results obtained. Meticulous anatomical control, for example of the extent of a lesion throughout its levels, is necessary. Since neurons related to autonomic functions have been identified at both pontine and medullary levels and at different sites of the reticular formation and adjacent nuclei, it appears likely that descending fibers mediating effects in any of the autonomic spheres are widely distributed within the cord. The recent findings concerning the functional properties and the distribution of such reticular neurons are difficult to reconcile with the idea that, for example, cardioinhibitory and cardioexcitatory fibers pass in different segregated parts of a particular funiculus. Furthermore, it is doubtful whether a particular reticular neuron is concerned in the control of one function only (for example, cardiovascular, respiratory, or somatic muscular). A differential effect of activity in a reticulospinal fiber may, nevertheless, be imagined to be possible, for example by way of different afferents ending on the internuncials interposed in the pathway from the "center" to the visceral effector cells.

In view of the many difficulties involved in physiological studies of central cardiovascular control, referred to above, it is no wonder that understanding of these mechanisms is still fragmentary. Not the least of these is the phenomenon of the integration of the influence of the hypothalamus, the cerebral cortex, and the cerebellum (see Chaps. 11, 12 and 5, respectively) with those of the postulated "centers" in the brainstem. Even though some pathways are known from those supraspinal areas to the medullary regions discussed in this chapter, it appears (Calaresu, Faiers, and Mogenson, 1975, p. 24) "that the central neural structures involved in cardiovascular regulation do not function simply in series, with the higher neural structures modulating the activity of the lower, and phylogenetically older, neural centers." Hilton (1975, p. 218) states emphatically: "It seems timely to cease talking and writing about a special medullary vasomotor centre." (Corresponding statements may be made concerning the role of the reticular formation in the control of other functions, such as respiratory, somatomotor, micturition, and digestion.)

Hypotheses on the role played by various areas of the brain in cardiovascular regulation, constructed on the basis of physiological studies, will have to await anatomical verification of postulated interconnections. Tentative "wiring diagrams" of the connections of sites in the brain involved in cardiovascular regulation, on the basis of present-day knowledge (see, for example, Calaresu, Faiers, and Mogenson, 1975), are extremely complex (as might indeed have been expected) and in part hypothetical.

The ascending activating system and the "nonspecific" thalamic nuclei. The original views on the organization and function of the ascending activating system were summarized in the introduction to this chapter. As alluded to, continued research has necessitated revisions on some points and, in particular, has brought forward a quantity of data demonstrating that matters are far more complex than originally assumed. Since the early 1950s a countless number of papers have appeared on this subject, and there has been a continuous change in views. What was originally conceived of as one particular "system," concerned first and foremost with the maintenance of consciousness, has turned out to be a very differentiated complex, anatomically as well as functionally. It is becoming more and more obvious that it is untenable to consider an ascending activating system as separate from other parts and functions of the brain.

> The latter circumstance, as well as the misuse of the designation "reticular formation," may have contributed to the tendency to widen the concept of the reticular formation almost without limits, as exemplified by the contents of the Henry Ford Hospital Symposium "The Reticular Formation of the Brain" (1958). The situation was properly characterized at that symposium by the late Sir Geoffrey Jefferson (1958, p. 729) when he remarked: "I confess that three of four years ago I thought that I understood the concept of the reticular formation, but now I find that it has turned into a system which, like a big flourishing and expanding business, has bought up all its competitors." Alluding to another feature in the development he added: "It would not be too absurd to say that wherever any really interesting fun was going on in brain research, that part was immediately claimed as part of the reticular formation."

Under these circumstances and because of the complexities of the problems and the overwhelming literature on the subject, it is impossible to give a comprehensive and synthesized presentation of the ascending activating system. In the following, some recent data will be emphasized, and an attempt will be made to draw attention to those aspects which are of particular interest from a clinical point of view. A number of specialized physiological and other observations will not be considered at all. It may be practical to start with some of the findings made in the early 1950s and then pass rapidly to the most recent information. Although the "nonspecific" thalamic nuclei are not covered by the heading of the present chapter, it is appropriate to consider them in this connection. They were briefly referred to in the general discussion of the thalamus in Chapter 2 (p. 99).

It was mentioned in the introduction to this chapter that the cortical desynchronization (activation) occurring on high-frequency electrical stimulation of the brainstem was assumed to be mediated via the "nonspecific" thalamic nuclei. This assumption was based on several lines of evidence. Thus stimulation of "nonspecific" thalamic nuclei (Dempsey and Morison, 1942; Morison and Dempsey, 1942) gives rise to widespread long-latency recruiting responses in the cerebral cortex, while stimulation of "specific" thalamic nuclei results in localized short-latency responses, the distribution being determined by the cortical projection of the nucleus stimulated. It was well known at that time that there is a parallelism between the EEG (electroencephalographic) records and the state of consciousness. Both in man and in animals the alpha rhythm in the EEG (waves of relatively high voltage and a frequency of 8–12/sec) changes to a low-voltage fast activity when the animal or man passes from a state of relaxation or drowsiness to an alert, attentive state, for example in response to an external stimulus. As mentioned in the introduction, the former pattern, generally referred to as *synchronization* (on the assumption that it represents the synchronous activity of many cortical neurons) was broken and replaced by widespread *desynchronization,* often referred to as EEG *arousal* or *activation.* On the other hand, when an animal passes from wakefulness to sleep, the low-voltage fast activity in the cerebral cortex gradually changes to high-voltage slow rhythms with so-called spindle bursts. It was further shown that electrical stimulation of the "nonspecific" thalamic nuclei could give rise to both EEG patterns, depending on the frequency of stimulation. Low-frequency stimulation of the midline thalamic region produces inattention, drowsiness, and sleep, accompanied by slow waves and spindle bursts; high-frequency stimulation arouses a sleeping animal or alerts a waking animal, and there is desynchronization of electrocortical activity.[13]

The essential part of the ascending activating system is located in the brainstem, as noted in Bremer's (1935a) early observation that an animal which had its brainstem transected at the mesencephalic level presented a permanently synchronized EEG and behavioral sleep. The fact that lesions in the brainstem result in coma (Lindsley et al., 1950, in the cat; French, von Amerongen, and Magoun, 1952, in the monkey; and others), and that stimulation of the reticular formation may alter the cortical responses following low-frequency thalamic stimulation further supported this view (for references, see Rossi and Zanchetti, 1957). There appears to be a tonic activity in the ascending activating system which is upheld by afferent stimuli. However, different kinds of stimuli are not equally effective in influencing it. Acoustic impulses appear to be more potent than optic for example, and among impulses entering via the trigeminal and spinal nerves, those from pain receptors are particularly effective. Electrical stimulation of certain parts of the cerebral cortex may give electrocortical desynchronization by acting on the reticular formaton (see Rossi and Zanchetti, 1957, for a review). Some general anesthetics, such as pentobarbital, have been shown to block the ascending transmission in the activating system while leaving transmission in the "specific" sensory pathways relatively unchanged. Adrenaline and CO_2 increase the activity of the system.

In the studies on the ascending activating system, early attention was focused on the "nonspecific" thalamic nuclei. Attempts were made to specify the nuclei involved and their mutual connections and relations by studying the physiological effects of lesions in various regions of the thalamus and neighboring structures. Differing opinions were expressed and hypotheses were set forth to explain the

---

[13] Electrical stimulation with a 3-per-sec stimulus was shown to produce a regular wave-and-spike complex, as seen in petit mal epilepsy (Jasper and Droogleever-Fortuyn, 1947). The problems of epilepsy and the mechanisms underlying the appearance of different kinds of epileptic seizures will not be considered in this text (see the monograph of Penfield and Jasper, 1954, and *Basic Mechanisms of the Epilepsies,* 1969).

phenomena.[14] This is no wonder, considering the structural complexities of the thalamus, the difficulties involved in producing small and precisely located lesions, the risk of interrupting fibers of passage, and other technical difficulties in the experimental work. These studies will not be reviewed here. Suffice it to mention that until some years ago the most generally accepted view of the mechanism of EEG and behavioral arousal (see, for example, Hanbery, Ajmone-Marsan, and Dilworth, 1954) appeared to be the following: by reticulothalamic connections the reticular formation of the brainstem influences the midline and intralaminar nuclei. These project onto the thalamic reticular nucleus. Projections from this to the cortex are responsible for the widespread cortical distribution of the EEG desynchronization. However, it has already been emphasized in the pioneer study of Moruzzi and Magoun (1949) that mechanisms other than the diffuse thalamic projection may be involved in the cortical activation following stimulation of the reticular formation. Rather early, some authors maintained that the ventral anterior thalamic nucleus (VA) may be concerned.

The problems are still far from being solved, but later research has clarified some points. Some findings will be presented below in a rather simplified way. It is appropriate to start with a survey of present anatomical knowledge of the "nonspecific" thalamic nuclei.

According to their location, three groups of *"unspecific" or "nonspecific"* thalamic nuclei (sometimes called reticular, diffuse, or recruiting nuclei) may be distinguished: *the intralaminar nuclei, the midline nuclei,* and *the reticular thalamic nucleus* (see Fig. 2-14). Within the former two groups, a parcelling into a number of minor nuclei is possible (see Fig. 6-12). They will not be considered separately here, except for the centromedian nucleus (CM).[15] The difference in architecture of the small groups and our knowledge of their connections make it a likely assumption that they are not functionally identical.

As referred to previously, *fibers arising in the medullary, pontine, and mesencephalic reticular formation have been traced to several of the "nonspecific" nuclei* (as well as to the hypothalamus, the septal area, and the caudate and lentiform nuclei). However, some "specific" thalamic nuclei also receive ascending fibers from the reticular formation (Nauta and Kuypers, 1958; Scheibel and Scheibel, 1958, 1967). In a corresponding manner, some spinothalamic fibers have been traced to "nonspecific" thalamic nuclei, although authors are not in complete accord as concerns particulars in the distribution (see Gerebtzoff, 1939; Getz, 1952; Anderson and Berry, 1959; Mehler, Feferman, and Nauta, 1960;

---

[14] A vast number of studies have been undertaken to clarify the neural events occurring in the thalamus and the cortex during synchronized and desynchronized activity. This complex subject is outside the scope of the present text. It appears that there are still many unsolved problems and differences of opinion among authors.

[15] As will be seen from what follows, it is scarcely possible to maintain a clear distinction between "specific" and "nonspecific" thalamic nuclei. Some authors include the nucleus ventralis anterior (VA), the nucleus ventralis medialis (VM), and the nucleus suprageniculatus (SG) among the "nonspecific" thalamic nuclei. According to Ajmone Marsan (1965), the following groups have so far not been properly placed in either category: the nuclei limitans, submedius, parafascicularis, paraventricularis anterior, parataenialis, and subparafascicularis. On the basis of an analysis of its afferent fiber connections, Jones and Powell (1971) suggested that the posterior group of thalamic nuclei (see Chap. 2) should be included with the intralaminar nuclei.

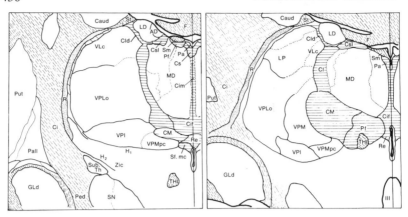

FIG. 6-12 Drawings of transverse sections through the thalamus of the macaque monkey, showing the arrangement and subdivisions of the "nonspecific" thalamic nuclei. The reticular nucleus is indicated by dots, the intralaminar and midline nuclei by horizontal hatchings. The section to the left is placed most rostrally. Some abbreviations: *Ci:* internal capsule; *Cif:* nucleus centralis inferior; *Cim:* nucleus centralis intermedialis; *Cl:* nucleus centralis lateralis; *CM:* centromedian nucleus; *Cs:* nucleus centralis superior; *Csl:* nucleus centralis superior lateralis; *GLd:* nucleus geniculatus dorsalis; *LP:* nucleus lateralis posterior: *MD:* dorsomedial nucleus; *Pa:* nucleus paraventricularis; *Pf:* nucleus parafascicularis; *R:* nucleus reticularis; *Re:* nucleus reuniens; *Sf.mc:* nucleus subfascicularis pars magnocellularis; *Sm:* stria medullaris; *SN:* substantia nigra; *St:* stria terminalis; *THl:* tractus habenulointerpeduncularis; *VLc:* nucleus ventralis lateralis, pars caudalis; *VPI:* nucleus ventralis inferior; *VPLo:* nucleus ventralis posterior lateralis, pars oralis; *VPM:* nucleus ventralis posterior medialis; *VPMpc:* nucleus ventralis posterior medialis, pars parvocellularis. From Olszewski (1952).

Bowsher, 1961; Scheibel and Scheibel, 1967; Rao, Breazile, and Kitchell, 1969; and others). Ascending afferents have further been described to come from various other regions, for example from the periaqueductal gray to midline and intralaminar nuclei (Chi, 1970), and from the raphe nuclei (Conrad, Leonhard, and Pfaff, 1974). In addition, efferent cerebellar fibers have been traced to these nuclei (see Chap. 5). The projections from the cerebral cortex are considered in Chapter 2. As concerns the *centromedian nucleus,* some authors have described terminations of reticulothalamic fibers in this, while others are inclined to interpret the degenerating fibers found in this nucleus as fibers of passage. Mehler (1966a) emphatically denies their existence. These divergencies may in part be due to semantic problems.[16] In physiological studies, responses (although of relatively long latencies—10 to 20 msec) have been recorded in the parafascicular-centromedian complex following stimulation of peripheral nerves (see for example, Albe-Fessard and Bowsher, 1965). This pathway appears to have a relay in the reticularis gigantocellularis in the medulla (Bowsher, Mallart, Petit, and Albe-Fessard, 1968). Responses have further been recorded in the centromedian nucleus follow-

---

[16] As reviewed by Mehler (1966b), the nucleus centrum medianum of primates consists of a small-celled ventrolateral region and a large-celled dorsomedial region. There is evidence from phylogenetic studies that only the former should be considered the CM, while the latter corresponds to the parafascicular nucleus, which is relatively better developed in nonprimate species.

ing stimulation of the mesencephalic (Dila, 1971) or mesencephalic and rostral pontine reticular formation (Robertson, Lynch, and Thompson, 1973).

The *efferent connections of the intralaminar and midline nuclei* are difficult to clarify with the methods of anterograde degeneration, since the lesions must be small and there is considerable risk of interrupting fibers of passage. Nauta and Whitlock (1954) were careful in interpreting their results in a study of this kind. Among the conclusions they felt could safely be made were the following: there are longitudinal interconnections between the various nuclei, predominantly directed rostrally, but there are also transversely running associational connections, and there are interconnections with the adjacent "specific" nuclei. The centromedian nucleus was concluded to send fibers to all other "nonspecific" nuclei, including some contralateral connections and a projection to the rostral part of the reticular nucleus. Furthermore, the centromedian has a fairly massive projection to the ventral nuclear complex of the thalamus, including the ventralis anterior (VA). Fibers could further be traced from the centromedian to the putamen and in lesser numbers to the globus pallidus, caudate nucleus, and claustrum. Some of these connections were also described by Kaelber and Mitchell (1967). Using the method of retrograde cell changes, Cowan and Powell (1955) in the rabbit, and Powell and Cowan in the monkey (1956), found evidence for a topically organized projection of the centromedian and parafascicular nuclei to the putamen, but projection of intralaminar nuclei onto the head of the caudate nucleus. As to projections to the cerebral cortex (a point of particular relevance for physiological interpretations), Nauta and Whitlock (1954) denied a cortical projection from the centromedian nucleus, but concluded that the rostrally situated complex of midline and intralaminar nuclei project to phylogenetically older parts of the cortex, among these the prepiriform cortex, the "limbic" cortex (see Chap. 10), and the entorhinal area. Some of these projections were also found by Powell and Cowan, 1956; see also Murray, 1966.

In view of the sources of error inherent in experimental studies of the fiber connections of the "nonspecific" thalamic nuclei, the extensive studies of Scheibel and Scheibel with the Golgi method were highly welcome (Scheibel and Scheibel, 1966c, 1966d, 1966e, 1967). Their observations are based on a large number of preparations, chiefly of brains of mice, rats, kittens, and some other animals, and in the main agree with the conclusions drawn by Nauta and Whitlock (1954). As concerns the midline and intramedullary nuclei (Scheibel and Scheibel, 1967) the following may be noted. The neurons of these nuclei resemble those of the reticular formation of the brainstem (see p. 398), and their richly branching axons enter neighboring "specific" and "nonspecific" nuclei. No Golgi II type cells appear to be present. From the rostral part of the nuclei axons enter the medial portions of the VA and the rostral part of the reticular nucleus, and some axons could be traced to the orbitofrontal cortex. Of particular interest is the observation that there is a considerable contingent of fibers which takes off from the rostrally projecting bundle from the "nonspecific" nuclei and turns back toward the mesencephalon (Fig. 6-13).

With the advent of new neuroanatomical fiber tracing techniques, further information has been obtained. Findings made previously by other methods have been confirmed on several points. For example, Nauta and Whitlock (1954) were

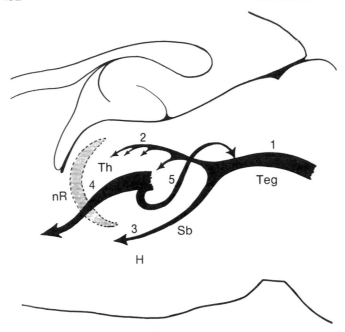

FIG. 6-13 Schematic drawing of some main features in the fiber projections assumed to be related to the ascending activating system. Fibers from the reticular
formation of the brainstem (1) ascend through the tegmentum (*Teg*). Caudal to the
thalamus they bifurcate into a dorsal leaf (2) which is lost in the intralaminar (*Th*)
and dorsomedial thalamic fields and a ventral leaf (3). This runs ventral and lateral
through the subthalamus (*Sb*) and hypothalamus (*H*) and thereby swings ventral to
the thalamic reticular nucleus (*nR*). Axons of cells in the "nonspecific" thalamic
projections have a rostral component (4) which perforates the reticular nucleus and
continues rostrally in the inferior thalamic peduncle. Another component (5) turns
caudally and runs back to the tegmental level (cf. text). From Scheibel and Scheibel
(1967).

right when they maintained that the afferents to the cortex, described as "nonspecific" afferents according to their mode of termination (see Chap. 12), do not
arise only from "nonspecific" but also from "specific" thalamic nuclei. Among
other data may be mentioned the results of studies using the HRP method. Jones
and Leavitt (1974) studied the occurrence of labeled cells in the thalamus following injections in the striatum and cerebral cortex in the rat, cat, and monkey. The
intralaminar nuclei, including the centromedian and parafascicular nucleus, were
found to project densely to the striatum, but sparsely and diffusely to the cerebral
cortex.[17] Divac, LaVail, Rakic, and Winston (1977) described a projection to
Brodmann's area 7, and Macchi et al. (1977) a projection from the centromedian

[17] The faint labeling of cells in the intralaminar nuclei following injections in the cerebral cortex
(Jones and Leavitt, 1974) may be due to an extensive branching or collateralization of the axons passing to the cortex. This would explain why cells in a restricted area of an intralaminar nucleus could be
labeled from wide areas of the cortex, while cells in the thalamic relay nuclei could be heavily labeled
following injections in their particular cortical projection area, in agreement with the spatial restriction
of their terminals in the cortex (see Chap. 2). Another possibility would be that the fibers to the cortex
are collaterals of axons ending in the striatum.

to the "sensory and motor" cortical areas. In agreement with the findings of Cowan and Powell (1955) and Powell and Cowan (1954a, 1956), more anterior parts of the intralaminar nuclei project onto more anterior parts of the striatum; more posterior parts, including (in primates) the nucleus centromedian, to more posterior parts. Jones (1975c) found some efferent fibers from intralaminar nuclei to pass to the reticular thalamic nucleus.

The *reticular thalamic nucleus* (Fig. 6-12) has played an important role in attempts to explain the mechanisms involved in the elicitation of cortical activation or desynchronization from the thalamus. As discussed previously, it was for a long period generally assumed to be the final link in the cortical projections from the "nonspecific" nuclei. (Stimulation of the reticular nucleus gives rise to diffuse recruiting responses in the cortex, see Jasper, 1949; and others.) Following localized cortical lesions topically distributed cell changes (anterior part of the cortex–anterior part of the reticular nucleus, etc.) have been found in the reticular nucleus by Rose and Woolsey (1949a) and others. However, the possibility that the changes found in the nucleus are not retrograde but transneuronal was emphasized by Rose (1952) and others. In experiments in the rabbit with silver impregnation methods Carman, Cowan, and Powell (1964b) found that the reticular nucleus receives a topically arranged projection from the entire cerebral cortex. This might explain the cell changes observed in the reticular nucleus following cortical lesions. In the Golgi studies of Scheibel and Scheibel (1966d), only a few axons of the multipolar cells of the nucleus were found to proceed in a rostral direction, and these could not be traced further than to the striatum. More than 90% of the cells have *caudally* directed axons, some of which could be traced to the mesencephalon. The axons give off abundant collaterals which supply both "specific" and "nonspecific" thalamic nuclei. It is characteristic of the reticular nucleus that the dendrites of its cells are largely oriented in the plane of the flattened nucleus. Because of its position the reticular nucleus is traversed by virtually all fibers interconnecting the thalamus and the cerebral cortex. When these fibers penetrate the dendritic neuropil of the reticular nucleus, they give off collaterals at right angles which appear to establish synaptic contacts with the dendrites (characterized by having long and filamented spines) of the cells in the nucleus.

Recent experimental studies with modern methods based on axonal transport have confirmed the Golgi studies of Scheibel and Scheibel (1966d) concerning the connections of the reticular thalamic nucleus, perhaps better called the thalamic reticular complex (Jones, 1975c). Following injections of tritiated amino acids in different thalamic nuclei or different parts of the cortex, Jones (1975c), in a careful study in the monkey, cat, and rat, confirmed that both corticothalamic and thalamocortical fibers give rise to dense terminal arborizations within the reticular nucleus when they traverse it. Reciprocally interconnected parts of the cortex and thalamus (see Chap. 2) establish contact with identical parts of the reticular nucleus, even if there is some overlapping between the fields of termination from different areas. Cells in the reticular complex further give off axons to the thalamic nucleus related to them (see also Minderhoud, 1971). However, no evidence was found of efferent fibers passing to the cerebral cortex. Efferent projections in the caudal direction were found to be sparse. However, with other methods, the Golgi findings of a caudal projection (Scheibel and Scheibel, 1966d) have been confirmed. In a study with the HRP method, Grofová, Ottersen, and Rinvik (1978) have been able to demonstrate that a large number of cells throughout the reticular thalamic nucleus send their axons caudally, at least as far as the caudal end of the lower mesencephalic tegmentum. A clear-cut projection to the superior colliculus is likewise present.

It appears from these studies that the *thalamic reticular nucleus cannot be considered a final relay in the projection of "nonspecific" thalamic nuclei upon the cortex.* It appears, as stated by Scheibel and Scheibel (1966d, 1967), that the reticular thalamic nucleus presumably has another function: to integrate thalamo-cortical and corticothalamic impulses (entering it by collaterals of the perforating fibers) and to act back on other "nonspecific" and "specific" thalamic nuclei and the mesencephalon and to modify their patterns of activity.

What then is the final link in the "nonspecific" thalamocortical projection? Several physiological studies indicate that the ventral anterior thalamic nucleus (VA) plays an essential role (Starzl and Whitlock, 1952; Hanbery, Ajmone-Marsan, and Dilworth, 1954; Weinberger, Velasco, and Lindsley, 1965, see also below). This nucleus, as referred to above, receives fibers from many "nonspecific" thalamic nuclei (in addition to fibers from the pallidum) and projects to the cortex, although its projection area is not quite certain (see Chap. 4). Its efferents to the cortex penetrate the anterior part of the reticular nucleus. This might explain the cortical responses obtained on stimulation of the anterior part of the reticular nucleus, but, as pointed out by Scheibel and Scheibel and others, there may be additional possibilities for an impulse transmission from "nonspecific" thalamic nuclei to the cortex. The associational connections between "nonspecific" and "specific" thalamic nuclei may be involved.[18]

The diagram of Scheibel and Scheibel shown in Fig. 6-13 summarizes their anatomical observations (see legend). It is of interest that there appear to be two routes for rostrally projecting impulses from the reticular formation. One (2 in the figure) goes via midline and intralaminar nuclei, from which efferent fibers pass to the VA. From this, fibers (penetrating the reticular nucleus, 4 in the figure) continue rostrally to the cortex. Another ventral or caudal route of ascending projections from the reticular formation passes outside the thalamus, through the subthalamus and hypothalamus. (The connections of these regions with the cortex are not fully clarified, but the hypothalamus sends fibers to the dorsomedial nucleus (MD) which projects onto the frontal lobe, see Chap. 11.) The two routes described by Scheibel and Scheibel (Fig. 6-13) have been confirmed in physiological studies in the cat (Robertson, Lynch, and Thompson, 1973; Fuller, 1975).

At present it does not seem possible to reach definite conclusions as to the role played by the various components of the diffuse thalamic system in its action on the cerebral cortex and to correlate physiological and anatomical observations in a fruitful manner. However, some relevant observations may be mentioned. Extending observations made by previous authors, Skinner and Lindsley (1967) found that lesions, or temporary block by local freezing, of the ventral anterior nucleus, the inferior thalamic peduncle (where the rostrally projecting fibers pass),

---

[18] According to Hassler (1964) the cortical projection areas of the relevant thalamic nuclei correspond to the cortical regions from which recruiting responses are most easily obtained.

or the orbitofrontal cortex reduce or abolish the recruiting responses. However, such rostrally placed lesions do not affect the desynchronization in the EEG following high-frequency thalamic stimulation, as lesions of the mesencephalic reticular formation do (see also Weinberger, Velasco, and Lindsley, 1965). These observations may be correlated with the demonstration of a descending projection from the "nonspecific" nuclei to the mesencephalon (5 in Fig. 6-13). It appears that separate paths are involved in the production of cortical desynchronization (the ventral route) and the recruitment phenomena (the dorsal route). The suggestion that the ventral ascending route (3 in Fig. 6-13) may be concerned with cortical desynchronization may further be correlated with the observation that most of the ascending fibers to the subthalamus, hypothalamus, and septum come from the mesencephalic reticular formation (see p. 401), since this has been shown to be the most potent part of the brainstem in producing cortical desynchronization.

New anatomical and physiological data have necessitated changes in our views on the ascending activating system. Its "machinery" is far more complicated than originally assumed. No attempt will be made to discuss these difficult problems here. It may be appropriate to underline, however, that, anatomically, a clear distinction between "specific" and "nonspecific" thalamic nuclei cannot be made. Physiological findings tend to support this (see Andersen and Andersson, 1968). There are obviously intimate functional relations between the striatum and certain parts of the thalamus, and, among cortical regions, the orbital cortex appears to be of particular interest. Since the reticular formation exerts a potent influence on the thalamus, it is of great interest that an increasing body of evidence indicates the presence in the reticular formation of regions which do not activate the cortex and are not correlated with behavioral arousal but which have the opposite effect, causing electrocortical synchronization and behavioral sleep.

Sleep and consciousness. Before proceeding, a few comments on the subjects of sleep and consciousness are appropriate. Sleep is generally considered a particular state of consciousness. It has so far not been possible to give a satisfactory *definition of consciousness,* in spite of a vast literature in which the problems of consciousness have been approached from philosophical, psychological, and biological points of view.[19] Any discussion of consciousness inevitably leads to another problem, the "mind-body relationship." This probably is—and will forever be—insoluble. As Kuhlenbeck (1957, p. 93) phrases it: ". . . any attempt to explain how nervous impulses can be translated into a mental experience is an impossible attempt." Hebb (1954) has expressed a similar opinion. Nevertheless, from a practical point of view a few features of consciousness may be mentioned. In the first place it is important to distinguish between the "level" of consciousness and its "content." As to the latter, we cannot imagine a state of consciousness without "content." Kubie (1954, p. 447) has stressed: ". . . *consciousness of something* exists; but *consciousness* is merely an abstraction." This aspect of the problem of consciousness can only be studied in man, who is able to com-

[19] An exhaustive and critical account of relevant problems has been given by Kuhlenbeck (1957). Other enlightening papers can be found in *Brain Mechanisms and Consciousness* (1954) and in *Consciousness and the Brain* (1976). See also Kuhlenbeck (1959).

municate his perceptions and thoughts to the investigator.[20] The problem of factors which influence the *level* of consciousness, however, may be approached in animal studies. The aim of such studies has been expressed by Kubie (1954, p. 446) as follows: "The psychologist and psychiatrist must challenge the neurophysiologist with the necessity of undertaking to explain not 'Consciousness' but the multiple phenomenology of varied conscious states."

As to the levels of consciousness, it is common to undertake a gradation in attention, alertness, relaxed mood, drowsiness, sleep, stupor, and coma, but clearcut criteria for these various states are difficult to establish since there are fleeting transitions. (An attempt to define characteristics of various degrees of reduced consciousness in human beings has been made by Mollaret and Goulon, 1959, and Jouvet, 1964). A final aspect of the problem is to define in structural terms those parts of the nervous system which are involved. Anatomical studies cannot by themselves give information of this kind. The task for the morphologist will be restricted to making an attempt to answer the following question: To what extent are anatomical data compatible with conclusions drawn from physiological and clinical studies as to the structural organization of those regions of the brain which are believed to bear some relation to consciousness?

As discussed elsewhere (Brodal, 1964), this is not an easy question to answer, for several reasons. In the first place, our knowledge of the anatomy as well as the physiology of the brain is still too fragmentary to permit a close correlation between findings made in the two fields of research. Furthermore, the approaches of the anatomist and the physiologist are different. The physiologist of necessity concentrates upon a particular region or function at a time, since it is impossible to record everything that goes on in the brain during an experiment, and he may be tempted to oversimplify the matter. The anatomist, being familiar with the innumerable pathways for impulse transmission which are available, is apt to be skeptical about simple hypotheses of functional organization and parcellations of the brain, and may be inclined to "overcomplicate" the issue.

For some time it was generally held that it is the activity in the ascending activating system which determines the level of consciousness. The functional importance of the system would be to act as an "energizer" to exert a facilitatory influence on other parts of the nervous system.[21] It would be concerned mainly with maintaining, in the cortex, an optimal level of activity which could be modulated by the specific sensory impulses. The impulses entering the reticular formation from different sources were assumed to merge and lose their specificity within this. A cessation or reduction of the amount of these impulses would reduce the activity in the system and eventually produce sleep. This, of course, is a rather simple concept.

At the symposium on "Brain Mechanisms and Consciousness" in 1953 (printed 1954) Lord Adrian appropriately raised the question whether one should regard the ascending activating system "as just coming in to wake us up in the morning and to send us to sleep at night, to do nonspecific activation," or whether it has "something to do with the direction of attention, with the actual work of the conscious brain." At the same meeting Moruzzi in the final discussion pointed

[20] Some relevant data will be discussed in Chapter 12.
[21] The same factors which produce an activation with desynchronization in the EEG and behavioral arousal will also act on the spinal mechanism, increasing muscular tone and myotatic reflexes, heart rate, and blood pressure. Furthermore, an increased excretion of ACTH from the hypophysis occurs.

to the fact that research so far had been concerned with the studies of the extremes of the functions of the reticular formation only, and he argued that the intermediate ranges of activity deserve attention as well. Only in this way could one hope to disclose a more detailed and particular functional fractionation.

As has been seen, anatomical data strongly suggest that the reticular formation is not an entity. Physiological data (some of which have been mentioned) have been forthcoming and point in the same direction. It appears that the reticular formation is concerned not only in activation, but also in "deactivation," and that the two opposing functions are related to different regions. Recent observations on sleep mechanisms are particularly illuminating.

The *"mechanism" of sleep* has attracted much interest in recent years, and several symposia have been devoted to the subject.[22] As referred to above, during quiet sleep the EEG shows synchronization (different phases are distinguished; see Kleitman, 1963). From the early 1950s it was realized that during quiet sleep there are intermittent phases of desynchronization in the EEG. Since these are combined with rapid eye movements, one often speaks of REM sleep (or paradoxical or desynchronized sleep). These sleep phases are generally accompanied by dreams, and in addition to the rapid eye movements there are other bodily changes: a reduction or abolition of muscular tone with occasional twitches, weakening of spinal reflexes, and changes in blood pressure and heart rate. These phases of sleep are often referred to also as deep sleep, since the thresholds for sensory stimuli which produce behavioral arousal appear to be increased.

According to the concept of the ascending activating system outlined above, sleep would be a passive process, due to a reduced input of afferent impulses to a tonically active excitatory mechanism. This view contrasts with several previous observations which indicated that sleep may be an *active* process, due to an active inhibition originating from regions often referred to as "sleep centers."

In clinicopathological studies of lethargic encephalitis, von Economo (1929) found that lesions affecting the posterior hypothalamus and rostral mesencephalon were usually associated with hypersomnia, while lesions in the anterior hypothalamus often resulted in insomnia. The latter region would then be a "sleep center," the former a "waking center." In a review of the literature, Akert (1965) reports further cases in which lesions producing hypersomnia involved the medial thalamus, subthalamus, and even the pons and medulla. In his pioneer studies with implanted electrodes in cats, Hess (1944) could induce "natural" sleep by low-frequency stimulation of points in the medial thalamus, the caudate nucleus, and the preoptic and supraoptic parts of the hypothalamus.[23] (High-frequency stimulation of the same points gave an awakening reaction.) Lesions of the anterior hypothalamus in rats (Nauta, 1946) result in an almost complete insomnia and hyperactivity which eventually is fatal. Subsequent studies have confirmed that areas in the anterior hypothalamus and neighboring regions are of importance in the induction of sleep (see for example Hernández Peón and Chávez Ibarra, 1963; Clemente and Sterman, 1963), and a general activation and behavioral arousal have been obtained from regions assumed to belong to the "waking center," in agreement with the findings of Moruzzi and Magoun (1949) and others.

[22] See, for example, the Ciba Symposium *The Nature of Sleep*, 1961, and *Sleep Mechanisms*, 1965. See also Moruzzi, 1972; Jouvet, 1972; *Advances in Sleep Research*, Vol. 1, 1974; Vol. 2, 1976.

[23] It should be noted that on stimulation of the sleep-inducing regions, other inhibitory actions occur, such as inhibition of muscular activity, myotatic reflexes, respiration, and lowering of blood pressure. It appears that the active region comprises also the subcallosal gyrus and some neighboring parts of the cortex (see Kaada, 1951, 1960).

The observations referred to above make it clear that *the theory of sleep as being a purely passive process is not satisfactory*. Certain mechanisms in the brain appear to play an active role in inducing sleep. Certain regions of the reticular formation and other regions of the brain are involved in these processes. Only a few of the relevant observations will be mentioned. For an exhaustive account and an illuminating discussion of the neurophysiological mechanisms involved in regulation of sleep the reader should consult Moruzzi's review (1972).

In experiments in cats it was found that the effects of transections of the brainstem at the midpontine and rostropontine level differ. In the first case the animal is alert for most of the time and shows an activated desynchronized EEG (Batini et al., 1959). These observations lead to the conclusion that there are synchronizing structures in the lower brainstem, which are tonically active. In the following years these results were confirmed and extended. After low-frequency stimulation of the medulla in the region of the solitary tract (Magnes, Moruzzi, and Pompeiano, 1961), a widespread bilateral cortical synchronization was obtained (in Bremer's *encéphale isolé* preparations, with a transection between the medulla and spinal cord). The responsible cells do not appear to belong to the nucleus of the solitary tract. However, synchronizing responses could be obtained also from other regions of the medullary reticular formation, thus from the nucleus reticularis ventralis (see Fig. 6-2). Signs of behavioral sleep were found to accompany EEG synchronization on low-frequency stimulations (in unanesthetized animals) in several regions of the reticular formation (Favale, Loeb, Rossi, and Sacco, 1961). Continued studies by Rossi and his collaborators and others have brought further information of synchronizing structures in the brainstem (see Moruzzi, 1972). On the basis of evidence of electrical stimulation and of lesions in various parts of the brainstem, the conclusion was reached that there is a region medial in the lower pons which facilitates cortical EEG synchronization. This region seems to coincide with the pontine region that gives off abundant long ascending fibers (see Fig. 6-4, and Fig. 12 in Rossi, (1965) as determined anatomically (Brodal and Rossi, 1955). Further support was derived from studies on injections of small doses of barbiturates in the vertebral artery in *encéphale isolé* cats when the basilar artery was clamped at the midpontine level (Magni, Moruzzi, Rossi, and Zanchetti, 1959). This caused an immediate suppression of EEG synchronization. This is interpreted as being due to a temporary elimination of caudal synchronizing structures, while the desynchronizing region in the mesencephalon is unaffected by the anesthetic. Synchronization reappeared when the action of the barbiturate had vanished. Local cooling of the region in the floor of the fourth ventricle at the medullary level is followed by activation of the EEG and behavioral arousal, while opposite results are obtained on cooling of the upper (pontine) floor of the fourth ventricle (Berlucchi, Maffei, Moruzzi, and Strata, 1964). Since destruction of the raphe nuclei (particularly the rostral ones) results in insomnia (lasting for some days), the raphe nuclei have been considered a deactivating region (see below).

These and other findings suggest that there are at least two regions in the brainstem which exert a synchronizing effect on the electrocorticogram, and, when activated, have a sleep-inducing effect. Two of them appear to cover parts of the reticular formation.

The discovery that there exists in the reticular formation a dual system, one part of it having a deactivating (EEG synchronizing) and the other an activating (EEG desynchronizing) effect, has been an important one and has provided a stimulus to further research and comprehension. Thus, studies focusing on how the opposing regions are brought into play by different kinds of stimuli were initiated: the regions seem to differ in this respect.

According to Pompeiano and Swett (1962a, 1962b), EEG synchronization, and often behavioral sleep, may be induced by low-rate stimulation of low-threshold cutaneous fibers, belonging to group II (while high-rate stimulation of these, as well as of group III afferents, induces arousal). The synchronizing impulses were shown to influence especially the nucleus reticularis gigantocellularis, and appeared to ascend in the spinoreticular tract (Pompeiano and Swett, 1963). Vagal afferent impulses may induce synchronization (but may also give rise to desynchronization, see Chase et al., 1967). Synchronization of the EEG, obtainable by stimulation of the orbitofrontal cortex, may be mediated via the medullary reticular formation (see Dell, Bonvallet, and Hugelin, 1961), which, as shown in Fig. 6-7, receives cortical fibers.

As seen from the above, in the brainstem, synchronizing structures are to be sought in its lower parts and the desynchronizing, activating structures, in the mesencephalon. Several findings suggest that the deactivation is an active process, probably related to inhibition of the activating "system." Most of the conclusions drawn concerning a relation between sleep mechanisms and the reticular formation are based on results of electrical stimulation in the cat. In spite of criticisms that may be raised against this approach, it appears that the results should be considered valid (Moruzzi, 1972, p. 111 ff.). However, other regions of the brain likewise influence the sleep–waking cycle, and the role of the brainstem regions involved in this mechanism may "simply be one aspect of the control of all cerebral activities, during conscious and unconscious life" (Moruzzi, 1972, p. 48). While the cerebral cortex appears to play a minor role, the hypothalamus is definitely involved in the regulation of the sleep–waking cycle, as referred to above. Bilateral lesions in the anterior hypothalamus-preoptic region give rise to insomnia, while lesions of the posterior hypothalamus give rise to profound somnolence or lethargy. (Concerning fiber connections between the hypothalamus and the reticular formation of the brainstem, see Chap. 11.) The stimulation of basal forebrain areas, like the lateral preoptic area (see for example, Clemente and Sterman, 1963; Hernández-Peón and Chávez-Ibana, 1963) has a sleep-inducing effect. The nucleus fastigii of the cerebellum (projecting onto the reticular formation, see Chap. 5) is also of some relevance. Lesions of its rostrolateral part give rise to EEG synchronization, and lesions of its rostromedial part to EEG desynchronization (Fadiga et al., 1968, see also Gianazzo et al., 1969) and to changes in the sleep–waking cycle. Stimulation experiments have given concordant results.

In recent years much interest has been devoted to *chemical aspects of the sleep–waking cycle,* and a number of speculations have been set forth. They are based on findings such as the insomnia produced by lesions of the rostral part of the serotonergic raphe nuclei and the concomitant reduction in the content of serotonin (5-HTP) in the cerebral cortex and on numerous other pieces of more or less indirect evidence. According to the "monoamine theory of sleep" (see Jouvet, 1972) the serotonergic brainstem neurons, particularly in the rostral raphe nuclei, are responsible for behavioral and EEG changes of synchronized (slow-wave) sleep, while catecholamine and acetylcholine "systems" are involved in the maintenance of tonic behavioral and EEG arousal. Among these "systems" the ascending noradrenergic projections from the nucleus coeruleus have been assumed to be an integral component of the ascending reticular activating system. Morgane and Stern (1974) have recently reviewed the "chemical anatomy" of brain structures and connections assumed to be involved in the sleep–waking process.

A large component of these "chemical" studies of sleep mechanisms consists of chemical analyses of the amount and distribution of the relevant transmitter substances following destruction of loci or connections assumed to be involved, and of pharmacological manipulations that have been shown to alter the amount and duration of EEG changes. Numerous conflicting observations have been made, and many assumptions will certainly have to be rejected in the future.

> For example, the conclusion that noradrenergic projections from the nucleus locus coeruleus are essential for wakefulness and EEG activation appears doubtful. Destruction of the locus coeruleus in cats by radiofrequency lesions (involving, in addition, some neighboring structures) did not alter the amount of EEG activation on wakefulness in the cat (Jones, Harper, and Halaris, 1977), although there was a marked depletion of noradrenaline in the paleo- and neocortex.

Attempts to correlate the findings in "chemical sleep research" with what is known about the organization of the brain from modern anatomical and electrophysiological observations are singularly deficient and so far appear to have been largely unsuccessful. Among the different phases of sleep the *REM, paradoxical, or desynchronized sleep episodes* (see above, p. 437) have attracted much interest. They always occur on a background of slow-wave (synchronized) sleep and are accompanied by various phasic events. Some data indicate that the neuronal substrate for REM sleep is not entirely identical with regions that are tonically active in slow-wave, desynchronized, sleep.

> Destruction of the region of the nucleus reticularis pontis oralis and immediately caudal, neighboring areas has been found to abolish both EEG and somatic manifestations of paradoxical sleep. When a lesion of the raphe nucleus involves its rostral part but not the caudal, the result is a permanent arousal during the first days, with episodes of paradoxical sleep still appearing (see Jouvet, 1972). The nucleus locus coeruleus and its noradrenergic pathways have likewise been thought to be involved in the production of paradoxical sleep by some investigators. However, in a recent study (Jones, Harper, and Halaris, 1977) no support was found for this claim. It appears to be established that the vestibular nuclei are important for the appearance of some of the somatic phenomena of deep sleep, as shown by Pompeiano and Morrison in a series of studies. Complete destruction of the vestibular nuclei on both sides does not prevent the appearance of the synchronized or desynchronized phases of sleep. However, the bursts of rapid eye movements and the related phasic inhibition of monosynaptic reflexes are abolished when the medial and descending vestibular nuclei are destroyed bilaterally (see Pompeiano and Morrison, 1966). While lesions of the nucleus fastigii influence the sleep–waking cycle, cerebellar lesions do not appear to affect the episodes of REM sleep (Gianazzo et al., 1969; see, however, Marchesi and Strata, 1970).

Investigations of functions of the brain in the sleep–waking cycle are technically difficult and crowded with pitfalls in interpretation. (Thus electrocortical activation does not always parallel behavioral arousal.) As is apparent from the few data mentioned above, there is more or less convincing evidence that certain regions of the brainstem, as well as of the hypothalamus and the basal forebrain, are involved in the production of the various phases of sleep, and that different kinds of transmitters are operative. However, only to some extent is it possible to associate these regions with definite anatomical structures. The ample interconnections between these structures, and connections with neighboring regions suggest that under normal circumstances there must occur an intimate collaboration between neurons scattered over extensive areas of the brain. This is evident from physiological studies as well. Further, the same sites of the brain (both in the brainstem and in the hypothalamus) are involved in several other functions besides influencing sleep—and thus the level of consciousness (cardiovascular, somatic, and other phenomena).

The presence of activating and deactivating "systems" does not by itself explain the natural sleep–waking rhythm, even though it must be assumed that the alternation between sleep and wakefulness implies a reciprocal organization between the two "systems." These are probably triggered and their responses integrated in some way (a phenomenon probably related to the circadian rhythm). As briefly formulated by Moruzzi (1972, p. 118), "The existence of a cycle is likely to be due to the slow accumulation and dissipation of chemical products within well defined groups of neurons." *Where* rhythmicity arises in the normal animal is yet not clear. It may well be a property of extensive parts of the brain. According to Moruzzi (1972), the hypothalamus is a particularly likely candidate.

The recent studies on sleep are of immediate interest for the problems related to the level of consciousness in general. Rossi (1964, p. 187) has suggested that: "The regulation of the level of consciousness under physiological conditions is achieved by the interplay of two opposite, competing neural mechanisms: a facilitatory, activating arousing system and an inhibitory, deactivating, sleep-producing system, with the most important structures involved lying in the brainstem." This means that a decrease in the level of consciousness may depend not only on a reduced activity of the activating structures, but may also occur when the deactivating influences are brought into action (representing a passive and an active reduction of the level of consciousness, respectively). The results of the sleep–waking studies show that parts of the brain other than the reticular formation harbor activating and deactivating mechanisms and are involved in influencing the level of consciousness. Little is definitely known about how the cooperation between these regions occurs. These problems are extremely complex.

It is not yet quite clear whether conditions in man are exactly similar to those in the cat. As referred to previously, there are some anatomical differences between the reticular formation in the two species. For example, the spinoreticular fibers appear to extend further rostrally in man and to have three maximal sites of termination. Studies on the behavioral and EEG effects of injecting quickly acting barbiturates into the vertebral artery in human patients are of relevance. The injection produces a temporary "anesthesia" in the areas supplied by the vertebral arteries. Its distribution within the brainstem can be judged approximately from examination of function of the cranial nerves. Following such injections the patient is alert (he is, for example, able to push a button) and his EEG is desynchronized (Alemà, Perria, Rosadini, Rossi, and Zattoni, 1966). If the injection was made with the patient relaxed and with a mild EEG synchronization, as occasionally was possible, there was an immediately clear-cut desynchronization. These findings were interpreted as evidence that the desynchronizing-activating part of the reticular formation in man is situated at the mesencephalic level. This is compatible with the assumption that deactivating (EEG-synchronizing) structures are present in the lower part of the brainstem (as in the cat) but does not prove this. The consequences of brainstem lesions in man are in agreement with the above conception, insofar as lesions of the brainstem involving the mesencephalon are regularly followed by EEG changes of synchronization and loss of consciousness in man, indicating involvement of the activating system (see also the following section).

Even though there are still many open questions, it appears that in man, as in experimental animals, the reticular formation is important in controlling the level of consciousness. That it has a dual influence seems indeed likely. However, it follows from what has been discussed above, that it would be *entirely misleading to consider the reticular formation the "seat of consciousness."*

In recent years much attention has been paid to the role of the reticular formation and the ascending activating system in processes such as sensory perception and discrimination, learning, adaptation, habituation, and modification of behavior.[24] These functions are all closely related to consciousness, and a number of experimental studies have been done in this field, the more important being behavioral studies of animals. The use of implanted electrodes permits observations in the unanesthetized state, and radio control may be used for stimulation and recording (telemetry). (For an illuminating study of this kind see Delgado, 1963.) These studies leave no doubt that the reticular formation cooperates with many other parts of the brain in influencing functions such as alertness, attention, perception, and discrimination. Changes in many of these functions have, furthermore, been observed following lesions or stimulation of many other parts of the brain, such as the entorhinal area, the nucleus amygdalae, the caudate nucleus, and the hypothalamus. The problems are extremely complex, and it is probably right to say that no clear conception can as yet be formulated. The only safe conclusion which can be drawn from all these observations is that *a number of brain regions are of importance for consciousness,* even for its simplest aspect, its level—a conclusion in agreement with clinical observations.

It ought finally to be emphasized, as will have appeared from several data mentioned in this chapter, that *it is no longer possible to maintain a distinction between "specific" and "unspecific" systems.* In an experimental study, Sprague et al. (1963), following bilateral lesions of the rostral midbrain without significant involvement of the reticular formation, found marked effects on functions which are generally thought to be influenced from the reticular formation, such as affective behavior and attention. They conclude (loc. cit. p. 226) that "there is probably a continuous gradient in the functional organization of ascending systems between the most highly localized lemniscal path and the most diffuse of the reticular paths. Anatomically, there is intermingling of the two. Thus the best placed lesion can achieve no perfect separation of specific and reticular systems."

Clinical aspects.   Even if much remains to be learned before we can say that we "understand" the reticular formation, the many *recent experimental studies have cast light on phenomena which are common in daily life.* The alerting effect of a sudden noise, a light flash, or a painful stimulus is well known, and may be explained by the influence of such stimuli on the ascending activating system. It is likewise a common experience that increased attention and alertness are accompanied by an increased heart rate and often also other autonomic phenomena. This is easily explained by the general, ascending and descending, actions of the activating system (cf. axons with dichotomizing, ascending and descending branches in the reticular formation). On the other hand, monotonous and usually weak stimuli, such as gentle stroking of the skin, slow rhythmic movements (rock-

---

[24] See, for example: *Neurological Basis of Behaviour,* 1958; *The Physiological Basis of Mental Activity,* 1963.

ing chair!), monotonous sounds and quiet music, are well known to have a relaxing and often sleep-inducing effect. Physiotherapy, especially massage and exercises aiming at muscular relaxation, have a beneficial effect in favoring reduction of mental as well as physical tension. It seems a likely assumption that all these procedures owe at least part of their effect to an influence on the synchronizing parts of the reticular formation. The cerebral influence on the reticular formation can, however, scarcely be overrated. Our general alertness is influenced by words we hear, scenes we see, and processes which require consciousness and interpretation of perceptions and which certainly are dependent on cortical activity (e.g., the awakening effect on a drowsy audience of a humorous or otherwise engaging remark by the lecturer). More impressive even is the common experience that mental imagery and daydreaming may have a general alerting effect as marked as that following any external stimulus. It appears very likely that the corticoreticular projections are involved in these processes. These connections, presumably in collaboration with others, appear further to be important in the inhibition and facilitation of sensory impulses, which make it possible for us to concentrate upon a particular sensory stimulus and neglect all others. (The most illuminating illustration is perhaps what everybody will have experienced at a lively cocktail party: among a number of loud voices one is able to pick out a single voice and perceive this, without paying attention to all the others.)

Turning to more direct *clinical aspects* it may be noted that there are good reasons to believe that several *anesthetic agents* exert at least part of their action by reducing the activity of the activating (ascending and descending) systems (see Domino, 1958). It appears further that the actions of some *psychopharmaca* are, in part, due to their influence on the reticular formation (see Domino, 1958). These subjects are beyond the scope of the present text. It should be stressed, however, that studies in these fields are complicated by the fact that all these substances, as well as neurohumors, scarcely affect the reticular formation in isolation.

When turning to the symptoms following pathological changes in this part of the brain, we meet with a similar complexity and corresponding difficulties in interpreting results. It is certainly also true for the human brain that the reticular formation cooperates with many other parts in influencing functions such as alertness, attention, perception, and discrimination. It is no wonder, therefore, that disturbances in such functions may be seen in humans with lesions affecting various parts of the brain, often the cerebral cortex. Clinical observations may serve as a warning against putting too much emphasis on the reticular formation in these complex functions. It appears that there has been a tendency to overrate the role of the reticular formation and to underrate the importance of the cerebral cortex, as maintained by Sir Geoffrey Jefferson (1958).

A *pathological process affecting the reticular formation* will probably never be restricted to the reticular formation alone. A tumor, a vascular disturbance, or an inflammatory process is likely to involve other structures as well, with resulting signs and symptoms from cranial nerves, long ascending and descending pathways, and various nuclei. These symptoms have been considered in other chapters. Here our concern will be with symptoms which may be seen in affections of the reticular formation proper and its related nuclei.

In *poliomyelitis* changes are often found in the reticular formation (Barnhart, Rhines, McCarter, and Magoun, 1948; Bodian, 1949, and others). Pathological anatomical studies have shown that the changes may extend throughout the brainstem. Whether these changes are responsible for the "spasms" often occurring in the acute stages of poliomyelitis has been debated. In a study of 80 cases, Baker, Matzke, and Brown (1950) point to the frequent affection of the area dorsal to the inferior olive.

As has been seen in the account of the raphe nuclei, the nucleus locus coeruleus, and the periaqueductal gray, animal experiments have permitted some tentative conclusions about their functional role. Most of our knowledge in this field has so far been of little relevance for clinical medicine. However, the information obtained in animal experiments mentioned above (p. 420), concerning the abolition of responses to noxious stimuli on electrical stimulation of the periaqueductal gray, has been utilized in clinical practice (Hosobuchi, Adams, and Linchitz, 1977; Richardson and Akil, 1977a, b). It has been found that pain in clinical disorders and normal pain perception in humans can be blocked by electrical stimulation of the periaqueductal gray. By means of permanently implanted electrodes, the patient is able to produce stimulation in this part of the brain to relieve his pain. (For a brief account of the technical aspects, see Hosobuchi, Adams, and Linchitz, 1977). In many cases (most often in patients suffering from pain due to a carcinoma or metastases from this) satisfactory pain relief has been reported. Postmortem anatomical control of some cases indicates that the effective site is as expected, while failures have been noted when the electrode placement was outside the target area.

From a clinical point of view the *possible relation of the reticular formation to consciousness* has attracted considerable interest. The role which may be played by a deranged function of the reticular formation in conditions such as diabetic, hepatic, or uremic coma and in disturbances of consciousness in general asphyxia appears to be little known. It seems likely that the general metabolic and humoral changes in these conditions affect widespread areas of the brain. More information has been obtained from studies of relatively local damage to the brainstem in cases of tumors or vascular affections. It appears to be the general experience of nurosurgeons that tumors involving the mesencephalon and diencephalon are generally followed by loss of consciousness which often may last for months (Cairns, 1952; French, 1952; M. Jefferson, 1952, and others). The EEG may show synchronization. According to Cairns (1952) the most common disturbance in these cases is hypersomnia. This condition may also occur in tumors below the floor of the third ventricle, arising from remnants of the craniopharyngeal pouch, and if the tumor is cystic and is emptied by aspiration, the stupor may promptly disappear (Cairns, 1952). In other instances of tumors in the upper mesencephalon-diencephalon, decerebrate rigidity may be present together with loss of consciousness. This may also occur in occlusion of the basilar artery, which will affect the brainstem (and in which possible pressure effects may be excluded). Disturbances of consciousness, from slight confusion to deep coma, are an early and constant feature in the clinical picture (Kubik and Adams, 1946). Occasionally "akinetic mutism" has been seen in lower brainstem lesions (see Cairns, 1952; Cravioto et al., 1960).

It appears, from a relatively large number of observations on patients with tumors of the upper brainstem and diencephalon, that these regions are essential for the maintenance of what is sometimes referred to as "crude consciousness." The clinical findings as well as the accompanying EEG changes are compatible with the view that these brain regions are concerned in a general activation of the brain in man, as in experimental animals.

The importance of the lower brainstem for consciousness is less clear. On account of the role played by the medulla and lower pons in the regulation of respiration and cardiovascular functions, lesions of the lower brainstem are apt to be rapidly fatal. Unconsciousness in these cases is usually accompanied by disturbances in breathing, lowering of blood pressure, and signs of involvement of other structures in the brainstem. According to Cairns (1952), the loss of consciousness is usually sudden in onset and deep, and it may be that the cerebral anoxia resulting from the deranged cardiovascular and respiratory functions is the cause of the coma. However, some cases have been reported in the literature in which patients with lesions of the pons and medulla have survived for longer periods, and in which the EEG has shown a desynchronized pattern, as in the waking state. Behaviorally, however, in most instances the patients have shown a reduced level of consciousness or have been in stupor or coma (see, for example, Loeb and Poggio, 1953; Lundervold, Hauge, and Løken, 1956; Kaada, Harkmark, and Stokke, 1961; Chatrian, White, and Shaw, 1964; Marquardsen and Harvald, 1964; also Aléma et al., 1966). Observations of this kind, as well as the results of intravertebral injection of barbiturates, referred to above, seem at least to indicate that "in man the neurons having a tonic activating function are not located in the whole brainstem, but only in its rostralmost part (i.e., rostral midbrain) and diencephalon" (Rossi, 1965, p. 275). The presence of a synchronizing region in the lower brainstem has so far not been proved, and the relation between loss of consciousness with lesions in this region and the accompanying changes in the EEG is not clear.[25]

From a clinical point of view it is important to recall that *disturbances of consciousness,* as found in lesions of the brainstem and diencephalon, *may be caused by processes outside the brainstem when this is subjected to compression or traction or there are secondary vascular disturbances.* Cairns (1952) in discussing this subject draws attention to the frequent occurrence of compression of the brainstem against the tentorium when a tumor, particularly in the temporal lobe or in the upper cerebellar vermis, produces a displacement of the brain. Release of these brain herniations by operation may often be followed by immediate return of consciousness. The disturbances of consciousness occurring with intense rise of intracranial pressure are probably due to tentorial herniation or tonsillar herniation (downward protrusion of the cerebellar tonsilla into the foramen magnum). In such cases, as well as with local lesions of the brainstem, intermittent attacks of unconsciousness may occur, most likely due to transitory circulatory disturbances (ischemia). The attacks may be accompanied by tonic fits.

There is abundant clinical evidence that *consciousness may be disturbed and*

---

[25] Gastaut (1954, p. 279) early drew attention to the fact that the changes in EEG and in the level of consciousness "are not directly dependent on each other and may evolve independently."

reduced in man in affections of the brain which do not involve the brainstem. In some patients who had shown prolonged altered states of consciousness for months or years, following severe head trauma, no focal lesions were found in the brainstem at autopsy, but there was a diffuse degeneration or often a necrosis of the white matter in the cerebrum, with degeneration of descending tracts (Jouvet and Jouvet, 1963; Kristiansen, 1964; and others). In traumatic head injuries, disturbances of consciousness are commonly seen. However, in a great proportion of such cases which take a fatal course no evidence of structural damage of the brainstem is found at autopsy (Tandon and Kristiansen, 1966).The structural basis of the symptoms in cases of blunt head injury is often difficult to evaluate, since secondary brain changes commonly develop, but obviously disturbances of consciousness are by themselves no reliable criterion of damage to the brainstem in such cases. Some other regions appear to bear a rather close relation to disturbances of consciousness. G. Jefferson (1958, 1960) emphasized the frequent occurrence of hypersomnia passing into coma in cases of bleeding from aneurysms of the anterior cerebral and anterior communicating arteries. He speaks of the region around the lamina terminalis (in front of the third ventricle) as an "anterior critical point" in contrast to the "posterior critical point" at the level of the tentorium. Reference was made above to the frequent occurrence of hypersomnia in cases of tumors below the floor of the third ventricle. That lesions around the hypothalamus may affect the ascending activating pathways (by pressure or bleeding) can scarcely be denied, but more often it appears that this is not the case and does not explain the changes in consciousness. Disturbances of "crude" consciousness may be seen, although less often, in lesions in many other parts of the brain (see Kristiansen, 1964). In some of these the alterations in the "contents" of consciousness are marked. While it is a general observation by neurosurgeons that quite extensive parts of the cerebral cortex may be ablated without overt signs of a reduced level of consciousness, certain aspects of consciousness, taken in its broader sense (for example those related to perceptual processes), may be affected by cortical lesions. This is particularly well illustrated by lesions of the temporoparietal cortex (see Chap. 12) and in epileptic seizures in such cases, often appearing as so-called psychical seizures (see Penfield and Jasper, 1954). These may occur as an aura to a larger epileptic attack. Little is known of the mechanisms underlying these and other varieties of disturbances of consciousness, and these subjects will not be discussed here.

In closed head injuries unconsciousness is a frequent and important symptom. The depth and duration of unconsciousness may vary. Many attempts have been made to explain the pathogenesis of cerebral concussion. The frequent occurrence of transitory signs of disturbances in cranial nerve functions, in coordination, and in spinal reflexes (Kristiansen, 1949) suggests that there are reversible changes in the brainstem in many of the patients. The mechanism of these changes has so far not been clarified. The transient disturbances of consciousness which occasionally occur during vertebral angiography (see Hauge, 1954) are presumably due to a temporary ischemia of the brainstem.

As will be apparent from the few data considered in this section, there appears to be good evidence that in man, as in animals, the activity of the reticular formation is important in regulating the *level* of consciousness. However, in this function it certainly collaborates with other parts of the brain, such as the thalamus and the cerebral cortex. Concerning those parts of the brain which are related to the many other, more subtle, aspects of consciousness, little can be said except that certain cortical regions are necessary for those aspects of consciousness which are related to acoustic, visual, or other sensory perceptions. It may be wise to be very careful in drawing conclusions about particular anatomical structures related to these complex functions, and Penfield's (1958, p. 232) words, "There is no room or place where consciousness dwells," should be borne in mind.

7

The Cranial Nerves

The cranial nerves in general. The structure of the spinal cord is in general much the same throughout its length, and the plan of organization of its cell groups and fibers can be quite clearly recognized. The brainstem, continuing the spinal cord in the skull, at first glance shows only a slight resemblance to the cord. However, a closer analysis reveals many common principles of organization. Present knowledge of these principles is based mainly on comparative anatomical investigations, for the fundamental pattern of organization is more clearly discernible in animals low in the evolutionary scale. Among the anatomists who have contributed most to this subject, C. Judson Herrick should be especially mentioned.

During evolution the cranial part of the body assumes a steadily increasing importance, due primarily to the development of the special sense organs. This is reflected in the structure of the nervous system. The dominant influence of the cranial parts of the body reaches its peak with the development of a cerebral cortex, which, especially in the higher mammals and man, not only in regard to its mass but also to its structural complexity, is an organ of overwhelming importance for the function of the entire nervous system. It follows from this that the primitive pattern of the brainstem is most easily discerned and most completely preserved in its caudal part, the medulla oblongata.

In the myelencephalon and metencephalon, the *sulcus limitans* is present in man also in the fully developed brain. It is clearly seen in the floor of the 4th ventricle. In accordance with its functional significance in the development of the spinal cord, it indicates also in the brainstem the borderline between the zone of efferent nuclei in the floor plate and the zone of terminal nuclei of the afferent fibers in the alar plate, which is here bent laterally. In embryonic development the cells giving origin to efferent and afferent nuclei are arranged in longitudinal columns, which later become partly broken up into distinct cell groups or nuclei. However, most of them retain approximately their original place, and even in the adult, therefore, nuclei derived from the different primary columns are found arranged in a columnar manner.

Whereas in the spinal cord only four functionally separate categories of fibers are present—somatic efferent, visceral efferent, visceral afferent, and somatic afferent—the development in the cranial part of the body of the special senses and the gill apparatus with the branchial arches is accompanied by a corresponding complexity of fiber categories and nuclei.

However, the composition of the different cranial nerves is not identical, some having only efferent fibers, others only afferent, whereas some are mixed.

The arrangement is presented diagrammatically in Fig. 7-1. The sulcus limitans marks the boundary between efferent and afferent nuclear zones. Within these zones a further differentiation is possible. The medialmost column of efferent nuclei corresponds closely to the groups of ventral horn cells and is embryologically continuous with them. Like them the efferent fibers innervate striated muscle, derived embryologically from the myotomes of the somites. In the head, muscle of myotomic origin, according to most observers, is represented only by the extraocular muscles and the intrinsic muscles of the tongue. Accordingly such *somatic efferent nuclei* are the motor nuclei of the IIIrd, IVth, VIth, and XIIth cranial nerves.

The cells of these nuclei are morphologically of the same type as the ventral horn cells, and the nuclei are found immediately ventral to the floor of the 4th ventricle and the aqueduct.

Lateral to the somatic efferent nuclei in the basal plate of the brainstem are

FIG. 7-1   A diagram of the nuclear columns in the brainstem, illustrating the type of structures supplied by the different categories and the nerves containing fibers from the different nuclear columns. The accessory nerve is omitted. Altered from Strong and Elwyn (1943).

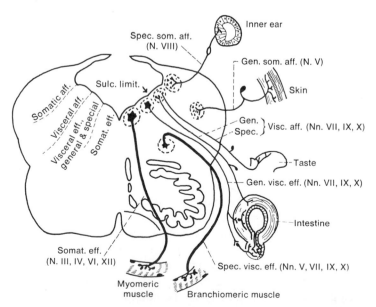

found the two categories of *visceral efferent nuclei,* forming two separate columns. The most medial of these, which is displaced somewhat in a ventral direction, is an interrupted column, the fibers of which pass to striated muscles. These muscles are developed from the mesoderm of the branchial arches, and comprise those from the first branchial arch, the masticatory muscles (+ the anterior belly of the digastric and the mylohyoid muscles); those from the second branchial arch, the facial, mimetic muscles (+ the posterior belly of the digastric and the stylohyoid muscles); and the striated muscles developed from the third and fourth branchial arches, namely those of the pharynx and larynx. The nuclei are the motor nuclei of the Vth and VIIth cranial nerves, and one of the vagus-glossopharyneal nuclei, the nucleus ambiguus. Probably the nucleus of the XIth nerve belongs to this group. These nuclei and the fibers they give off are called *special visceral efferent.*

Just as there is a morphological similarity between the striated muscle innervated by these nuclei and the musculature supplied by the nuclei of the somatic efferent group, so their nerve cells also are of the same type. On account of these similarities the name "lateral or special somatic efferent nuclei and fibers" is sometimes applied to them.

Medial to the sulcus limitans are found the nuclei belonging to the *general visceral efferent* category, consisting of the nucleus of Edinger-Westphal of the IIIrd nerve, the superior and inferior salivatory nuclei of the VIIth and IXth nerve, respectively, and the dorsal motor nucleus of the vagus. The cells of these nuclei morphologically resemble those in the sympathetic intermediolateral cell column in the cord and functionally represent the cranial division of the parasympathetic giving rise to preganglionic fibers, which end in autonomic ganglia. The postganglionic neurons supply smooth, unstriped muscle in the organs innervated by the nerves mentioned and glands (e.g., the lacrimal and salivary glands).

Immediately lateral to the sulcus limitans is found a sensory column receiving *visceral afferent* fibers. This consists of only one nucleus, the nucleus of the tractus solitarius, extending throughout the entire length of the medulla oblongata. The afferent fibers ending in it pass through the VIIth (intermedius), IXth, and Xth nerves, have their cells in the ganglia of these nerves, and before entering the nucleus descend as the tractus solitarius. Usually a distinction is made between two types of visceral afferent fibers, and there is some comparative data indicating a corresponding subdivision of the nucleus in certain forms (see Barnard, 1936). Some of the fibers convey impulses of taste and are called *special visceral afferent.* (It may be justifiable to place the olfactory fibers also in this group.) Other fibers convey general impulses from the viscera and are termed *general visceral afferent.*

The *somatic afferent nuclei* are found in the lateralmost part of the alar plate, and here a subdivision can also be made. Fibers conveying impulses of superficial (and probably also deep) sensibility from the face travel through the trigeminal nerve, have their cells in the semilunar ganglion, and end in the principal or superior sensory nucleus of the Vth nerve and in the nucleus of the spinal tract of the trigeminal nerve, after having passed for some distance caudally in the medulla oblongata. The homology of the trigeminal nuclei with the dorsal horns reveals itself morphologically in the continuity of the nucleus of the spinal tract and the gelatinous substance of the dorsal horns. This is the *general somatic afferent* component. In the vagus, glossopharyngeal, and intermediate nerves some fibers of this type are also present.

The *special somatic afferent* nerves are the vestibular and cochlear, conveying impulses from the sensory apparatus in the inner ear, the former proprioceptive, the latter exteroceptive. Their nuclei are found in the extreme lateral and dorsolateral parts of the medulla oblongata.

It will be seen (see Fig. 7-2) that whereas nearly all of the efferent nuclear columns are broken up into distinct parts, belonging to the different nerves containing fibers of the kind in question, this is not the case with the afferent nuclei. This is particularly evident in the general somatic afferent nuclei of the trigeminal, which forms a continuous cell mass in receipt of fibers from the Vth, VIIth, IXth, and Xth nerves, and in the nucleus of the tractus solitarius, which receives the visceral afferent fibers of the VIIth, IXth, and Xth nerves. This presumably has a functional significance. Kappers has interpreted this as an expression of a general law in morphogenesis: fibers transmitting the same sensory modality and usually stimulated simultaneously tend to take the same course within the central nervous system and end in the same nuclear complex.

In a general consideration of the cranial nerves it should be remembered that they are not only links in the sensory and effector mechanisms related to the conscious acts subserved by the cerebral cortex. Equally significant is the role they

FIG. 7-2 Diagram indicating the columnar arrangement of the cranial nerve nuclei. The nuclei belonging to the same categories are indicated by identical symbols.

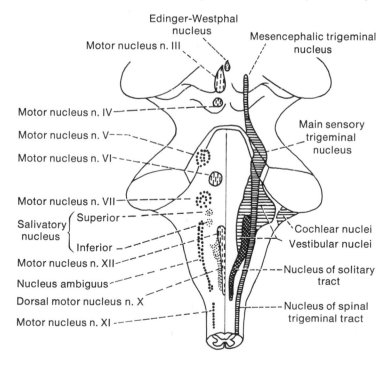

play in many of the vitally important reflexes, evoked from and partly effected through the mechanisms in the head and its organs. Some of these will be touched upon when dealing with the appropriate cranial nerves.

Regarding the cranial nerves as a whole, it will be apparent that they differ to a considerable degree in their physiological properties, as well as in their anatomical features. From a clinical viewpoint, not all of them are of equal importance. Some of them are highly specific sensory nerves forming links in rather well-defined functional systems. This applies especially to the IInd cranial nerve, the optic (which strictly speaking is not a cranial nerve at all), the cochlear division of the VIIIth nerve, and the olfactory (Ist) nerve. It has on this account been deemed appropriate to treat these cranial nerves in special chapters devoted to the mecha-

FIG. 7-3 Diagrams illustrating some principal features in the organization of cranial nerve nuclei.

A: The neurons in an *afferent (sensory) cranial nerve nucleus* (for example the trigeminal nucleus) send their main axons centrally to the thalamus (from which a new set of neurons projects to the cerebral cortex). The ascending fibers from the nucleus give off collaterals (1) to the reticular formation (R.F.) and to motor cranial nerve nuclei (2), which in addition may be acted upon by afferent impulses via the R.F. (3). (Routes 2 and 1–3 represent reflex arcs.) The sensory nucleus may be influenced both from the R.F. (4) and from the cerebral cortex (5).

B: The neurons of an *efferent (motor) cranial nerve nucleus* (for example, the nucleus ambiguus) send their axons to the periphery. Its cells are acted upon by collaterals of ascending sensory fibers (2) and by axons (3) from cells in the reticular formation (R.F.), descending fibers from the cerebral cortex (5), as well as from other sources (4). Cells in the reticular formation sending axons (3) to the motor nucleus may further be influenced from collaterals of ascending fibers (1). In man some cortical fibers (5') appear to establish monosynaptic contact with the motoneurons.

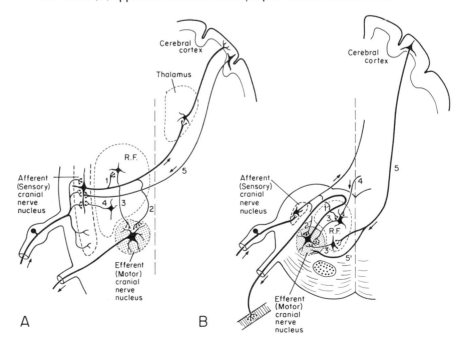

nisms of vision, hearing, and smell, respectively. In the following description, therefore, these cranial nerves are omitted. In connection with the description of the other nerves some special problems relating to more than one nerve are more comprehensively treated with only one of them, for example taste together with the VIIth cranial nerve. The peripheral course of the nerves will be treated only briefly. For a more complete account the reader is referred to the author's book on the cranial nerves (Brodal, 1965b).

The cranial nerve nuclei are acted upon by fibers from many sources. The diagrams of Fig. 7-3 illustrate the main features of the fiber connections commonly seen in afferent (A) and efferent (B) cranial nerve nuclei (see legend). It should be noted that axons or axonal collaterals of cells in the reticular formation appear to pass to virtually all cranial nerve nuclei (see Scheibel and Scheibel, 1958). The ascending axons from secondary sensory neurons in the afferent, sensory, cranial nerve nuclei likewise give off collaterals, particularly to the reticular formation. The neurons of the efferent cranial nerve nuclei send their axons out of the brainstem, but primary afferent fibers have been traced only to some of them, indicating that at least one neuron must be intercalated in most reflex arcs of the brainstem. Fibers descending from "higher levels," among them the cerebral cortex, pass to both motor and sensory nuclei. The latter connections indicate possibilities for a control of the central transmission of sensory impulses entering the cranial nerves, as is the case for afferent impulses entering the spinal cord (see Chap. 2). Concerning the descending fibers, those from the "motor cortex" to the somatic and special visceral efferent nuclei are of particular interest. It appears that in man (Kuypers, 1958c), as in the monkey (Kuypers, 1958b), some corticobulbar fibers end in synaptic contact with neurons in the motor cranial nerve nuclei, while in the cat such fibers can be traced only to their vicinity (Walberg, 1957b; Kuypers, 1958a; Szentágothai and Rajkovits, 1958; and others), where they probably end on "internuncial" cells. This, as will be seen, corresponds to the patterns of ending of cortical fibers in the spinal cord.

(a) THE HYPOGLOSSAL NERVE

Anatomy. The fibers of the hypoglossal nerve spring from the hypoglossal nucleus. This belongs to the somatic efferent group of nuclei and is made up of cells of the same type as the motoneurons in the cord. In many animals and embryologically in man the nucleus is continuous with the ventral horn, thus indicating its functional relationship, but after birth in man it becomes separated from this. The nucleus, forming a longitudinal cell column (see Fig. 7-2), is found close beneath the floor of the 4th ventricle in the trigonum nervi hypoglossi, and extends rostrally to the striae medullares. Caudally it reaches the caudalmost part of the medulla, where it is found immediately ventral to the central canal (Fig. 7-4). The small nuclei situated in its immediate neighborhood (the nucleus intercalatus of Staderini, the nucleus of Roller, and the nucleus prepositus hypoglossi, collectively referred to as the perihypoglossal nuclei) do not send fibers into the hypoglossal nerve but have other projections (see Chap. 5). The hypoglossal nucleus consists of several distinct cell groups. By a study of retrograde changes in the cells after extirpation of various tongue muscles it has been shown that each

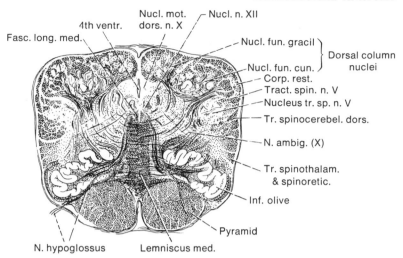

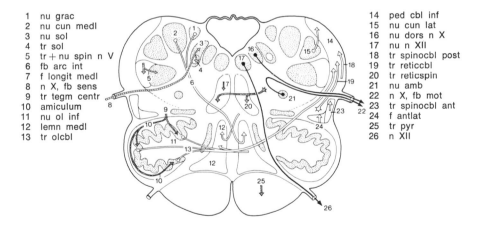

FIG. 7-4 Above, a drawing of a transverse section through the caudal part of the medulla oblongata in man. Weigert's myelin sheath stain. The fibers are clearly visible, the nuclear groups evident as light areas. (Approx. × 4.)

Below, a schematic representation of nuclei and fiber tracts of the medulla oblongata (taken from Niewenhuys, Voogd, and van Huijzen, 1978) at approximately the same level as the Weigert section above. Explanations on the sides of the figure. (Some of the names are not used in the text.)

group supplies certain of the tongue muscles (Barnard, 1940, and earlier authors). Thus a localization pattern is revealed in the structure of the nucleus (cf. similar findings in the ventral horns and other motor nuclei).

The axons from the hypoglossal nucleus emanate as bundles which course in a ventral and slightly lateral direction, traverse the reticular formation of the medulla oblongata, pierce the medial parts of the inferior olive and leave the brainstem in the lateral ventral sulcus, between the pyramid and the olivary eminence (see Figs. 2-11 and 7-7).

The nucleus receives impulses from axons derived from several sources. Fibers of the corticobulbar tract, originating in the central area of the cerebral cortex, are presumably involved when voluntary movements of the tongue are performed. In man some fibers appear to end on cells of the nucleus (see Kuypers, 1958c). It appears that most of the corticobulbar fibers acting on the hypoglossal nucleus in man are crossed, since cortical lesions or lesions of the internal capsule give rise only to contralateral changes in the tongue. Furthermore, fibers reach the nucleus from the reticular formation and possibly also from the sensory trigeminal nuclei and the nucleus of the tractus solitarius. These fibers are involved in the actions of reflex sucking, swallowing, chewing, etc.

The fibers of the hypoglossal nerve, which are mainly of the larger and medium-sized myelinated type, leave the medulla as 10 to 15 slender rootlets, which soon coalesce and pass through the hypoglossal canal in the occipital bone, surrounded by a plexus of veins.

Turning lateral to the nodose ganglion of the vagus, with which it is usually connected by connective tissue, the nerve descends in a caudally arched course, passes downward and ventrally to reach the tongue near its base, and distributes its fibers to the muscles of the tongue. The hypoglossal fibers supply not only the intrinsic muscles of the tongue, but also the styloglossus, hyoglossus, genioglossus, and geniohyoid muscles (e.g., Pearson, 1939). The fibers terminate with typical motor end plates, *each nerve being limited to the homolateral half of the tongue.* The motor units are small.

From the hypoglossal nerve proper a peculiar closed arch of fiber bundles (the so-called descending hypoglossal ramus) courses downward, and continues in the *ansa hypoglossi.* This, however, is not formed by fibers from the hypoglossal nucleus, but by fibers from the anterior rami of the upper two (or three) cervical nerves. Section of the descending ramus provokes no changes in the hypoglossal nucleus. Some of these fibers join the nerve in the beginning of its extracerebral

FIG. 7-5   A diagram of the hypoglossal nerve and its connections with spinal nerve fibers, supplying the infrahyoid muscles via the ansa hypoglossi (see text).

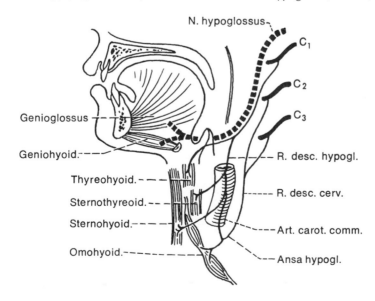

course but soon leave it again. The fibers of the ansa are distributed to the infra-hyoid muscles (sternohyoid, omohyoid, sternothyreoid, and thyreohyoid muscles) which are thus supplied by motor fibers from the upper cervical segments (Fig. 7-5) in the ansa. Clearly a proximal lesion of the hypoglossal trunk will thus not affect the infrahyoid muscles.

From a practical point of view, knowledge of the motor innervation of the tongue as outlined above is sufficient to interpret disturbances in its function. However, some additional data are of interest for an understanding of the motor functioning of the tongue.[1] It might be surmised that an organ such as the tongue, employed in chewing and swallowing and in the rather delicate and differentiated movements of speech, should be equipped with a fairly elaborate proprioceptive innervation. A few receptors of a special type ("flower spray") at the origin of the genioglossus muscle have been found in the rabbit by Weddell, Harpman, Lambley, and Young (1940), and Cooper (1953) has demonstrated typical muscle spindles in the human tongue. The possibility exists that there might also be, in the tongue, other types of sensory endings subserving proprio-ception. The presence of sensory fibers in the hypoglossal nerve has been claimed by several au-thors, mainly on account of the presence in some animals of a small ganglion on the course of the nerve ("Froriep's ganglion"). Boyd (1941) in a special study of this matter in rabbits found no trace of sensory cells in the hypoglossal nerve. Zimny, Sobusiak, and Matlosz (1970/71) de-scribed afferent nerve fibers in the hypoglossal nerve of the cat and the dog. Some physiological studies indicate that sensory impulses may travel centrally in the hypoglossal nerve and may act on motor cranial nerve nuclei (for a recent study and references, see Tanaka, 1975). However, when the lingual nerves in man are anesthetized, proprioceptive impressions from the anterior two-thirds of the tongue are abolished (Weddell et al., 1940).

The hypoglossal nerve conveys, however, not only somatic motor fibers to the lingual musculature, but also sympathetic fibers, functionally vasoconstrictors. From anatomical (see Weddell et al.) as well as physiological data it is concluded that these fibers partly join the nerve through an anastomosis with the superior cer-vical ganglion or the carotid plexus. Unilateral extirpation of the cervical sympa-thetic chain is followed by degeneration of the fine nerve fibers in the hypoglossal nerve and on the vessels of the tongue on the same side.

Symptomatology. *The trunk of the hypoglossal nerve may be damaged* where it passes through the hypoglossal canal or more distally. The cause may be a tumor, fracture of the occipital bone, or traumatic wounds. As a rule lesions of this type are unilateral. The resulting changes are essentially the same as in lesions of the peripheral motor neurons of the cord. The muscles of half of the tongue on the side of the lesion are paretic or paralytic. This becomes conspicuous when the tongue is protruded. The intact genioglossus muscle (assisted by the geniohyoid) will pull the base of the tongue forward, whereas this will not occur on the paralyzed side. Consequently the *tongue will deviate to the side of the lesion* when it is protruded. When kept in the mouth a deviation will not be apparent. If the pa-ralysis occurs acutely there will at first usually be some difficulty in moving the food within the mouth, and the patient will be unable to press the tip of his tongue against the cheek on the opposite side, but in unilateral lesions swallowing is not severely interfered with and any eventual dysarthria is moderate. *Atrophy* of the paralyzed half of the tongue occurs rapidly, the surface of the tongue becomes

[1] The sensory functions, especially the gustatory mechanisms, are treated in connection with the Vth and the VIIth cranial nerves.

wrinkled,[2] and on palpation this side of the tongue is found to be *flaccid*. Fasciculations are usually seen, especially when the tongue is protruded; these appear to be of the same nature as those observed in lesions of the peripheral motor neurons of the cord (see Chap. 3).

The symptoms outlined above will also appear in unilateral *lesions of the hypoglossal nucleus*. Nuclear lesions, however, because of the close proximity of the right and left nucleus, are most frequently bilateral. They may be caused, for example, by *poliomyelitis,* when the brain stem is involved (bulbar types). Usually in such cases other motor cranial nerves are also affected and the disease often takes a fatal course. More characteristic is the involvement of the hypoglossal nuclei in *chronic bulbar palsy,* where the motor neurons of the cranial nerves, especially the lower, fall victim to a slowly progressing degeneration. This clinical picture will be more fully discussed in connection with the vagus-glossopharyngeal complex. Not infrequently the first symptom of the disease is fasciculation of the tongue musculature.

Vascular lesions in the medulla oblongata may also affect the hypoglossal nuclei, but more frequently they injure the fibers in their course in the medulla. A peripheral hypoglossal paresis usually occurs in vascular disorders of the anterior spinal artery or the vertebral artery, regularly combined with a complete or incomplete hemiplegia on the other side on account of concomitant damage to long descending fiber tracts.

Because of the crossed path of the corticobulbar fibers to the hypoglossal nuclei, and the usually strictly homolateral peripheral innervation, a *lesion of the corticobulbar fibers above the decussation* will result in a *paralysis of the contralateral half of the tongue*. Thus the common capsular hemiplegias are as a rule combined with a paralysis of the tongue on the hemiplegic side. As this is an "upper motor neuron paralysis," lingual atrophy and fasciculations will be lacking. As a rule there is therefore no difficulty in distinguishing the two types of lingual paralysis. Bilateral affection of the tongue muscles of the "upper motor neuron type" is seen especially in cerebral atherosclerosis where both sides of the brain are usually affected. The clinical picture of this *pseudobulbar palsy* will be referred to later. As always when both sides of the tongue are involved, the disturbances tend to be severe.

(b) THE ACCESSORY NERVE

Anatomy.   The XIth cranial nerve, the accessory, is usually regarded as a purely efferent nerve, supplying the sternocleidomastoid and trapezius muscles with motor fibers. It is customary to describe the accessory nerve as having *two parts, a spinal and a cranial*. These fuse in the jugular foramen to form a single nerve trunk which again gives rise to *an internal ramus,* joining the vagus, and *an external ramus,* representing the accessory nerve proper (Fig. 7-6). It is generally agreed that the fibers of the cranial portion, which take their origin mainly from the nucleus ambiguus, are those leaving the trunk as the internal ramus. Therefore

---

[2] This atrophy of the muscles is not to be confused with the atrophy of the mucous membrane of the tongue, with disappearance of the papillae, seen, for example, in pernicious anemia.

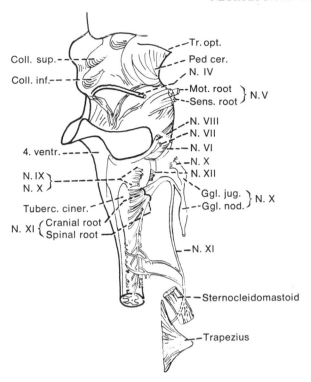

FIG. 7-6 Drawing of the caudal part of the brainstem with the rootlets of cranial nerves IV to XII, illustrating particularly the relations of the accessory nerve (see text).

this part may be looked upon as an aberrant vagus fasciculus, and will not be considered more closely.

The efferent fibers of the external ramus arise from the *accessory nucleus,* a column of cells of the somatic efferent type, extending from the 2nd to the 5th-6th cervical segment of the cord. Caudally it is present as a group of cells in the dorsolateral part of the ventral horn, while more cranially it attains a certain degree of independence and approaches the caudal end of the nucleus ambiguus[3] (Pearson, 1938). The fibers pass dorsolaterally and leave the cord as bundles in the lateral funiculus, dorsal to the ligamentum denticulatum, between the dorsal and ventral roots (Fig. 7-6). They bend in a rostral direction and unite to form the spinal part of the nerve which passes through the jugular foramen. Having given off the internal ramus, the remainder descends as the external ramus, bending slightly laterally and dorsally, and reaches the inner aspect of the sternocleidomastoid, pierces it, and then swings from the posterior border of this muscle to the trapezius, where it ends.

During its *extracranial course* the accessory nerve is joined by fibers derived

---

[3] This fact together with some other findings are taken by some authors to support the view that the accessory nerve belongs to the special visceral efferent group and that the sternocleidomastoid and trapezius muscles are derived from branchial mesoderm.

from the 3rd and 4th upper cervical ventral rami. On the basis of clinical findings in cases of pure accessory nerve lesions these spinal nerve fibers have been assumed by some neurologists to be concerned chiefly in the innervation of the caudal part of the trapezius, its middle and cranial parts as well as the sternocleidomastoid being supplied predominantly by the accessory nerve.

Besides the motor fibers the accessory nerve also contains some *afferent fibers*. Some of these pass centrally with the accessory nerve, which in its intracranial portion regularly contains some unipolar ganglion cells, in man (Pearson, 1938) as well as in the sheep, cat, and monkey. These ganglion cells undergo retrograde changes after section of the accessory nerve (Windle, 1931). In addition, sensory fibers are found in the spinal nerves joining it. Anatomical and physiological experiments in rabbits by Yee, Harrison, and Corbin (1939) indicate that the fibers are proprioceptive.

Symptomatology. *Lesions of the accessory nerve* are not frequently encountered. The nerve may be damaged by traumatic injury, tumors at the base of the skull, lesions during operations, or by fractures. There ensues, in a complete lesion, a loss of motor power in the sternocleidomastoid and the trapezius, less pronounced in its lower part (see above). If the lesion is situated distal to the entrance of the nerve into the sternocleidomastoid, as is more common, the trapezius alone will be affected. The paralytic muscles are atrophic and flaccid. The characteristic contour of the sternocleidomastoid in the neck is lost, and likewise the anterior upper border of the trapezius does not reveal itself as distinctly as usual. The atrophy of the sternocleidomastoid is particularly evident when the head is turned to the other side and the chin is elevated. A unilateral paralysis of this muscle as a rule does not inconvenience the patient very much. Bilateral lesions (such as are characteristic of the myotonic type of progressive muscular dystrophy, dystrophia myotonica, which are due, however, to primary muscle changes and not to lesions of the nerve) disable the patient considerably, e.g., when trying to lift his head from the pillow.

Complete paralysis of the *trapezius,* especially if combined with lesions of the upper cervical nerves, is a serious handicap. On account of its widespread origin from the occiput to the 12th dorsal vertebra and the convergence of its fibers to their insertion on the lateral part of the clavicle, the acromion, and the whole of the scapular spine, the muscle is an important assistant to the serratus anterior in elevation of the arm. Elevation of the arm is performed with less power than usual, and frequently cannot be executed to its full extent, mainly because the rotation of the scapula around its anteroposterior axis is reduced. The outward rotation of the arm also suffers when the scapula cannot be brought as close to the spine as normally. The scapula is frequently displaced a little in a lateral direction and presents some degree of winging (scapula alata), as in paralysis of the serratus anterior.

In hemiplegias a contralateral paresis of the muscles supplied by the accessory nerve is a common finding, due to the practically total crossing of the corticobulbar fibers. This paresis, as will be understood, is not accompanied by marked atrophy.

(c) THE VAGUS AND GLOSSOPHARYNGEAL NERVES

Anatomy. On account of the close anatomical and functional relationship of the Xth and IXth cranial nerves they will be treated together. Both contain special and general visceral efferent, general somatic afferent, and special and general visceral afferent fibers. The special visceral efferent fibers of the vagus and glossopharyngeal spring from the nucleus ambiguus, and the general visceral efferent from the dorsal motor nucleus of the vagus and from the so-called inferior salivatory nucleus, respectively. The *nucleus ambiguus,* composed of cells of the same type as the hypoglossal nucleus, forms a longitudinal column in the medulla oblongata (Fig. 7-2). From its original embryonic position it is displaced ventrolaterally and is found in the reticular formation medial to the spinal tract of the Vth nerve and its nucleus (Figs. 7-4 and 7-7). The fibers leaving the nucleus pass backward and then bend sharply forward and laterally to reach the surface of the medulla behind the inferior olive (Fig. 7-7). The nucleus ambiguus is the common special visceral efferent nucleus to nerves IX and X. Most of its fibers pass in the vagus. Some of these make their way through the cranial portion of the accessory nerve to join the vagus again through the internal ramus. A smaller contingent of the fibers runs in the glossopharyngeal nerve. These fibers in the IXth and Xth nerve supply the striated muscles of the pharynx and larynx and the upper part of the esophagus.

> The nucleus ambiguus may be subdivided into subgroups with somewhat different cytoarchitecture, in man as well as in animals. Following sections of peripheral branches of the vagus and glossopharyngeal nerves, several authors (Getz and Sirnes, 1949; Szabo and Dussardier, 1964; Lawn, 1966; and others) have studied the ensuing retrograde cellular changes in the nucleus and established a localization within it: the fibers of the glossopharyngeal nerve arise in the rostral regions, the fibers to the laryngeal muscles chiefly caudally. The cricothyroid muscle appears to be innervated from a particular cell group (Lawn, 1966, in the rabbit). According to Szentágothai (1943a) the other laryngeal muscles likewise are innervated from fairly well circumscribed regions of the nucleus.

The general visceral efferent *dorsal motor nucleus of the vagus* is found lateral to the hypoglossal nucleus, as a longitudinal column, under the floor of the 4th ventricle (Figs. 7-2, 7-4, and 7-7). Its fibers course ventrally to join those from the nucleus ambiguus. The corresponding nucleus of the glossopharyngeal nerve is the *inferior salivatory nucleus,* which, however, is represented by relatively scattered cells and does not form a proper nucleus. These nuclei give rise to preganglionic fibers in the parasympathetic system, and supply glandular structures and smooth muscle. They will be considered in more detail in Chapter 11 on the autonomic system.

Most of the *afferent fibers in the vagus and glossopharyngeal* are regarded as being of the *visceral type,* conveying impulses of general and special visceral sensations. The latter, the taste fibers, terminate in the same nucleus as the general visceral afferent fibers, the nucleus of the solitary tract. The fibers have their perikarya in ganglia of the corresponding nerves (the nodose and petrous ganglion, respectively) and, having entered the medulla, bend in a caudal direction, thus forming a conspicuous fiber bundle, the *solitary tract,* situated lateral to the dorsal motor nucleus of the vagus nerve and somewhat deeper (Fig. 7-2 and 7-7). This

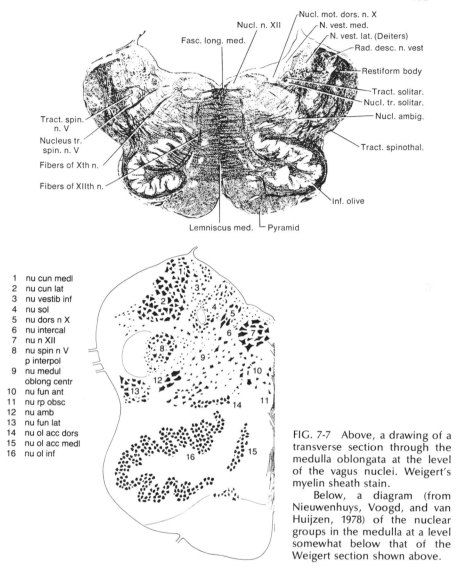

1 nu cun medl
2 nu cun lat
3 nu vestib inf
4 nu sol
5 nu dors n X
6 nu intercal
7 nu n XII
8 nu spin n V
   p interpol
9 nu medul
   oblong centr
10 nu fun ant
11 nu rp obsc
12 nu amb
13 nu fun lat
14 nu ol acc dors
15 nu ol acc medl
16 nu ol inf

FIG. 7-7 Above, a drawing of a transverse section through the medulla oblongata at the level of the vagus nuclei. Weigert's myelin sheath stain.

Below, a diagram (from Nieuwenhuys, Voogd, and van Huijzen, 1978) of the nuclear groups in the medulla at a level somewhat below that of the Weigert section shown above.

bundle extends throughout the length of the medulla oblongata, and its fibers and their collaterals end in a fairly circumscribed cell column, lateral to the tract, forming the *nucleus of the tractus solitarius*. It is generally agreed that the vagal-glossopharyngeal fibers end in the caudal part of the nucleus. Some fibers pass to the spinal trigeminal nucleus and the reticular formation (see Culberson and Kimmel, 1972, for references and a recent study). Some visceral afferent fibers of the intermediate nerve join the solitary tract and terminate in the nucleus. The intermediate, glossopharyngeal, and vagal fibers end in particular rostrocaudally arranged, although overlapping, regions of the nucleus of the solitary tract (see Torvik, 1956b; Kerr, 1962). A similar arrangement is found in man (Schwartz, Roulhac, Lam, and O'Leary, 1951). The nucleus of the solitary tract is made up of small nerve cells, considerably smaller than those in the dorsal motor nucleus of the vagus. Since it is almost impossible to destroy the nucleus without concomitant damage to neighboring structures, anatomical tracing of its efferent connections are difficult. In a careful study, Morest (1967) was unable to trace fibers ascending rostrally in the brainstem, which he suspected might transmit impulses directly to the thalamus for example. However, neurons of the nucleus of the solitary tract (particularly in its caudal part, receiving vagal and glossopharyngeal fibers) give off axons or collaterals to many other regions, such as the nucleus ambiguus and the dorsal motor nucleus of the vagus (see also Cottle and Calaresu, 1975; Norgren, 1978).

In a recent, careful study using autoradiographic and HRP methods in the rat, Ricardo and Koh (1978) have shown that there are, in fact, *ascending connections from the nucleus of the solitary tract* to several forebrain structures. Among these are the paraventricular, the dorsomedial, and the arcuate hypothalamic nuclei; the medial preoptic area; and the central amygdaloid nucleus. These connections arise chiefly from the caudal part of the nucleus of the tractus solitarius (while its rostral part appears to be related particularly to the relay of gustatory impulses).[4] Thus, visceral sensory impulses relayed through the nucleus of the solitary tract may reach areas concerned in the control of visceral and autonomic functions via direct and indirect pathways (see Chap. 11). In this connection it should be mentioned that direct fiber connections from the hypothalamus to the preganglionic autonomic efferent neurons have recently been demonstrated (Saper, Loewy, Swanson, and Cowan, 1976). The functional interplay of direct and indirect reciprocal connections between the hypothalamus and the medullary autonomic nuclei is so far little known (for references to anatomic data and functional correlations, see Richardo and Koh, 1978).

In an HRP study, Aghajanian and Wang (1977) found the nucleus of the solitary tract to project to the dorsal raphe nucleus (see Chap. 6). The major efferent projections of the nucleus of the solitary tract appear to supply the dorsolateral reticular formation. Some fibers have been traced to the cerebellum (Somana and Walberg, 1979). Experiments have shown other fibers, particularly from the caudal parts of the nucleus that contain some larger cells, to pass to the spinal cord as a *solitariospinal tract* (see Torvik, 1957b; Norgren, 1978). Its origin has recently been demonstrated following injections of HRP in the spinal cord of the cat (Kuypers and Maisky, 1975; Peterson and Coulter, 1977). The latter authors traced the tract to lumbar levels. For a recent study, see Loewy and Burton (1978). The solitariospinal tract probably represents an efferent link in the arc of some visceral reflexes, the afferent link of which is formed by visceral afferent fibers.

> The projections mentioned above make it clear that the nucleus of the solitary tract cannot be considered a pure relay station in visceral afferent pathways to higher levels. That a considerable degree of integration can take place in the nucleus is made clear by the fact that fibers from the cerebral cortex (Brodal, Szabo, and Torvik, 1956; and others), the spinal cord (Rossi and Brodal, 1956b), the vestibular part of the cerebellum (Angaut and Brodal, 1967), and other sources end in it.

[4] For particulars see section (e) of this chapter. Recently, olfactory responses have been recorded in the gustatory part of the nucleus of the solitary tract (see footnote 1 in Chap. 10).

The dorsolateral part of the reticular formation has been found to be involved in a number of reflex functions, such as swallowing, vomiting, and cardiovascular and rspiratory regulation. The vagal afferents of relevance terminating in the nucleus of the solitary tract, for example those from the carotid sinus (Biscoe and Sampson, 1970; Miura, 1975; and others), are probably concerned not only in medullary reflex arcs, but may also influence higher levels. Thus cerebral cortical responses have been recorded following stimulation of vagal afferents in the superior laryngeal nerve (Car, 1971), and taste impulses reach the cortex (see section (e) in this chapter). For accounts of relays in the pathways for deglutition and taste, see Car, Jean, and Roman (1975) and Norgren and Pfaffmann (1975), respectively.

General somatic afferent fibers are found in the vagus, and are involved in the innervation of the skin in the concha of the ear. These fibers enter the trigeminal nuclei, which represent the terminal station for all general somatic afferent fibers from the face. Some fibers from the glossopharyngeal nerve (as well as some from the intermediate nerve) likewise end in the trigeminal nuclei, in animals as well as in man. This will be considered in more detail in section (f) of this chapter.

The various nuclei treated above are acted upon by fibers from various sources. *Fibers from the corticobulbar tract,* presumably via internuncial cells, act on the nucleus ambiguus. The fibers are partly crossed, partly uncrossed, and are involved in the voluntary movements of the striated muscles supplied by the IXth and Xth nerves (pharynx and larynx, see below). In addition, collaterals from afferent visceral and somatic fibers and axons and collaterals from visceral and somatic sensory nuclei (see Stewart and King, 1963; Morest, 1967) mediate the participation of the pharyngeal and laryngeal muscles in reflex acts, such as coughing, swallowing, and vomiting. Further, descending fibers from subcortical levels influence the effector nuclei, presumably largely via the reticular formation.

The *roots of the vagus and glossopharyngeal nerves* leave the medulla in a shallow sulcus at the dorsal border of the inferior olive (Figs. 7-6 and 7-7). The upper, fewer bundles unite to form the glossopharyngeal, whereas the lower and more numerous bundles fuse to form the vagus nerve. Both leave the skull through the jugular foramen and have two ganglia, each composed of typical (pseudo-) unipolar cells. The IXth nerve ganglia are the superior and petrous, the Xth nerve ganglia the jugular and nodose. Between its ganglia the vagus is joined by the internal ramus of the accessory nerve.

The *peripheral distribution* of the nerves may be briefly outlined as follows: the *glossopharyngeal* descends on the lateral side of the pharynx, bending gradually forward, ultimately penetrating the muscular coat of the pharynx to reach the base of the tongue, the posterior third of which it supplies with sensory fibers; it also supplies the adjacent parts of the mucous membrane, the tonsillar region and posterior palatal arch, and the soft palate. The nerve sends a motor branch to the stylopharyngeus muscle, and partakes with the vagus in the formation of the pharyngeal plexus, in which sympathetic fibers from the cervical sympathetic trunk are also present.

There is some diversity of opinion concerning the relative importance of the vagus and glossopharyngeal in the innervation of the pharynx and the soft palate. In cats, section of the vagus but not of the glossopharyngeal produces a unilateral paralysis of the pharynx (Sjöberg, 1943), and some clinicians have made corresponding findings in human cases of operative sectioning of the IXth and Xth nerves (e.g., Fay, 1927). Sensory loss in the pharynx following intracranial section of the IXth nerve has been described, e.g., by Lewis and Dandy (1930), Reichert (1934), and Bohm and Strang (1962). Foerster maintained that the glossopharyngeal alone is concerned in the innervation of the pharynx, whereas the vagus supplies the soft palate with motor fibers (together with the facial, which innervates the levator palati, and the trigeminal nerve, which innervates the tensor palati).

The glossopharyngeal, besides containing special visceral afferent fibers for taste from the posterior third of the tongue, as already mentioned, also has general visceral efferents to the glands in the region it supplies. In addition, secretory fibers to the parotid gland pass through the small tympanic nerve and the lesser superficial petrosal nerve to the otic ganglion, where the postganglionic fibers have their perikarya. Worthy of mention is a special sensory branch from the glossopharyngeal, the carotid branch, which has constantly been shown to innervate the *carotid sinus* (Boyd, 1937), the dilatation at the beginning of the internal carotid, playing an important role as a receptor mechanism for the regulation of blood pressure.

The peripheral course of the vagus. The trunk of the vagus or pneumogastric nerve descends in the sheath common to the internal carotid artery and internal jugular vein in the neck, passes, on the right side anterior to the subclavian artery, on the left anterior to the aortic arch, then on both sides behind the root of the lung. Ultimately the left nerve is displaced ventrally onto the anterior surface of the esophagus, the right being found on the posterior surface. Correspondingly the left nerve breaks up into a plexus on the ventral side of the stomach, the right on the posterior. From these plexuses the terminal fibers of the nerve are distributed to the viscera in the upper abdomen (see Chap. 11).

The following branches deserve special mention:

The *auricular ramus* leaves the nerve between the two ganglia, frequently anastomoses with the glossopharyngeal, penetrates the mastoid process, and ultimately innervates the skin in the concha of the auricle. This branch is a purely general somatic sensory nerve.

The *pharyngeal rami* join the pharyngeal plexus, referred to above.

The *superior laryngeal nerve,* on the wall of the middle pharyngeal constrictor, gives off a predominantly motor *external ramus* which ends in the circothyroid muscles, and an *internal ramus.* The latter pierces, the thyreohyoid membrane and provides the larynx with sensory fibers.

The inferior laryngeal or recurrent nerve is different on the two sides, the right bending upward posteriorly under the right subclavian artery, the left under the aortic arch. Ascending in the tracheoesophageal sulcus, it finally divides into an anterior and a posterior ramus which supply all the laryngeal muscles, except the circothyroid, with motor fibers.

Apart from these more definite branches the vagus gives off the *superior cardiac rami,* usually below the superior laryngeal nerve. They follow the carotids downward to the aorta and partake in the formation of the *cardiac plexus,* which is also supplied by the *inferior cardiac rami* from the recurrent nerve. (The superior, middle, and inferior cardiac *nerves* from the corresponding ganglia of the sympathetic trunk form another component of the cardiac plexus.) From the thoracic part of the vagus emanate the pericardial, bronchial, and esophageal rami, which form nervous plexuses with sympathetic fibers on the organs mentioned and will be discussed in connection with the abdominal branches of the vagus in the chapter on the autonomic system.

It is evident from the description above that the *vagus is preponderantly a visceral nerve,* the somatic afferent constituent being only composed of the fibers in the auricular ramus. The special visceral efferent component (which, functionally at least, behaves as a somatic element) is concerned only in the innervation of the pharynx and larynx. This fact is revealed also in the fiber types present in the different parts of the nerve. The auricular branch is composed mainly of thicker myelinated fibers, and at least most, if not all, of them have their cells in the jugular ganglion. After section of this ramus, 69–77% of these cells show retograde changes (DuBois and Foley, 1937). Most of the cells of the visceral afferent fibers, representing a far larger number, are thus located in the larger nodose ganglion. Below this the nerve contains predominantly (67–77%) unmyelinated and some small myelinated fibers (Foley and DuBois, 1937). The larger fibers present in this part of the nerve appear to be concerned in the motor innervation of the larynx, and are found in the recurrent nerve, a lesser portion also in the superior laryngeal, supplying the cricothyroid. However, some of the larger fibers do not undergo degeneration after section of the vagus cranial to its ganglia, and are thus afferent and commonly believed to be proprioceptive. Some of the thick, afferent fibers are probably derived from muscle spindles, which are present in all human laryngeal muscles (Paulsen, 1958; Lucas Keene, 1961; Grim, 1967).[5] In the posterior cricoarytenoid muscle, for example, there are on an average 5.4 spindles (Grim, 1967). Other types of sensory endings have been described (Lucas Keene, 1961; Rossi and Cortesina, 1965b). The total number of spindle afferents is, however, relatively modest. In the monkey (Brocklehurst and Edgeworth, 1940), only about 3% of the fibers in the recurrent nerve appear to be afferent against some 30% in the superior laryngeal nerve. In man, Ogura and Lam (1953) found the latter nerve to contain some 15,000 fibers, some 30% being thick and myelinated, and from electrical stimulation of it, they concluded that it mediates sensations of pain and touch. The ample sensory innervation of the larynx by way of the superior laryngeal nerve reflects its importance in the coughing reflex.[6] The cooperation of the laryngeal muscles in phonation is obviously a very complex matter.

Analyses of the muscular actions of the human larynx by electromyography have clearly demonstrated this complexity (see Faaborg-Andersen, 1957; Kotby and Haugen, 1970a, b). They have shown that there is an intimate collaboration between the various intrinsic laryngeal muscles. For example, the posterior cricoarytenoid muscle appears not to be as essential for abduction of the vocal cords as previously believed (Kotby and Haugen, 1970b). The infrahyoid muscles (extrinsic laryngeal muscles), for example, the sternothyroid, likewise take part. The small motor units (less than 30 muscle fibers per nerve fiber; English and Blevins, 1969) display a resting activity considered postural.

Activity of the laryngeal muscles is necessary for vocalization, but this is only one component of the complex mechanisms involved in speech (for some data on man, see Chap. 12). In monkeys, eight different types of vocalization can be elicited on electrical stimulation of various cortical and brainstem regions (Jürgens and Ploog, 1970). Among the latter regions, the mesencephalic periaqueductal gray seems to be particularly important. A cortical laryngeal area responsible for vocal cord adduction and vocalization has been physiologically identified in the precentral cortex, in part of area 6. Within this, different laryngeal muscles, for example the cricothyroid, the thyreoarytenoid, and the extrinsic laryngeal muscles, have separate, although overlapping, "representations" (Hast, Fischer, Wetzel, and Thompson, 1974), as is the case for

---

[5] The somata of afferent fibers carrying sensory impulses from the larynx appear to occupy a particular region of the nodose ganglion (Mei, 1970).

[6] It is worthy of notice that sensory impulses from the laryngeal and pharyngeal mucosa are consciously perceived, contrasting with impulses from lower parts of the respiratory and digestive system. Yet the visceral origin of these parts is equally clear. It may be asked whether the usual classification of sensory impulses is appropriate, as it makes no distinction between the consciously and subconsciously perceived stimuli, a functionally important point.

the innervation of skeletal muscle (see Chap. 4). However, this cortical laryngeal area has efferent projections to several other areas, cortical and subcortical, a subject recently studied by tracing of the connections with tritiated leucine (Jürgens, 1976). Among the cortical areas acted upon is the supplementary motor area, which in man has been found to be a speech area (see Chap. 12). Interestingly, fibers were traced to the nucleus of the solitary tract, but not to the ambiguus nucleus, which innervates the laryngeal muscles. The neural mechanisms underlying vocalization are extremely complex and as yet far from fully understood. Many different parts of the brain are obviously involved (see Jürgens, 1976, for some data and references).

Symptomatology. The symptoms caused by involvement of the visceral, especially the visceral efferent, components of the IXth and Xth nerves are treated in the chapter on the autonomic system, and will be mentioned here only to complete the descriptions of the clinical pictures. Our concern here is mainly with the somatic functions of these nerves.

A lesion of the supranuclear tracts to the efferent vagus and glossopharyngeal nuclei will, if unilateral, give no clear symptoms. This follows from the bilateral distribution of the corticobulbar and other descending tracts. If bilateral, however, such lesions will be severe (see below).

On account of the close spatial relationship between the two nerves and their nuclei, it follows that symptoms referable to both of them will be more frequently encountered than isolated lesions. However, isolated lesions may occur. The chief *symptoms in lesions of the glossopharyngeal nerve* are those due to interruption of its sensory fibers. They comprise loss of sensibility in the posterior third of the tongue, the palatal arches (especially the posterior), the tonsillar region, the velum, and in most cases at least, the pharynx (see above). Taste is lost in the posterior third of the tongue. This has been verified in operative sections of the nerve in man (e.g., Lewis and Dandy, 1930; Reichert, 1934; and others). The motor impairment, according to most authors, will consist of a homolateral paresis of the pharynx. When the pharynx is made to contract, e.g., in phonation, the raphe of the pharynx will be drawn over to the intact side (*mouvement de rideau*) owing to the effect of the constrictors. The gagging reflex cannot be elicited from the affected side. The causes of glossopharyngeal lesions are the same as those producing affections of the vagus nerve and will be referred to below.

The glossopharyngeal may also be the site of irritative lesions, manifesting themselves usually as the so-called *glossopharyngeal neuralgia*. As in trigeminal neuralgia this may be caused by local affections of the nerve, for example tumors, but in most cases the etiology remains obscure. The main feature of the affection is the pain, which usually occurs in paroxysms, starting at the base of the tongue, the tonsil, or in the region of the palatal arches, irradiating eventually to the ear region. "Trigger points" may be present in the areas mentioned, and paroxysms are often elicited on chewing or swallowing. This condition is not frequent, but it is important to distinguish it from trigeminal neuralgia.

In a recent study Bohm and Strang (1962) distinguish two types of glossopharyngeal neuralgia, the otitic and the oropharyngeal, and point to some clinical features of these conditions which differ from those seen in trigeminal neuralgia. (Occasionally pain is felt in the trigeminal area, presumably due to overlapping of glossopharyngeal and trigeminal fibers in the spinal trigeminal

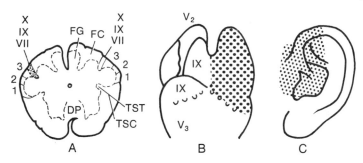

FIG. 7-8 *A:* Diagram showing the position of the fibers from the intermediate, glossopharyngeal, and vagus nerves (VII, IX, X) in the spinal trigeminal tract; 1, 2, and 3 refer to positions of the fibers from the first, second, and third branch of the trigeminal nerve, respectively, in the spinal trigeminal tract. *B:* The area of analgesia in the pharynx and palatal arches following a selective tractotomy of the stippled area in *A*. From Kunc (1965). *C:* The analgesic area in the auricle following a medullary tractotomy. From Brodal (1947a).

nucleus.) Relief may be obtained by intracranial section of glossopharyngeal nerve (see Bohm and Strang, 1962, for a recent account). However, since the somatic afferent fibers in the nerve join the spinal trigeminal tract, a medullary tractotomy (section of the spinal tract of the trigeminal nerve; see section (f) of this chapter) may also be employed. Since the glossopharyngeal fibers occupy a restricted area, dorsalmost within the spinal tract (Torvik, 1956b; Kunc and Maršala, 1962; Kerr, 1962; Rhoton et al., 1966), as shown in Fig. 7-8, it is possible to destroy this part selectively and thus to spare the fibers from the trigeminal nerve, as shown by Kunc (1965, 1970).

An isolated *lesion of one vagus nerve* will, according to most observers, result in a unilateral paresis or paralysis of the soft palate, which in phonation will not be elevated fully and when at rest will frequently hang somewhat lower on the paralyzed side. In contraction of the velum the uvula will deviate to the unaffected side. In bilateral lesions, particularly, the speech will have a nasal character, and on swallowing, food and especially fluids will enter the nasal cavity as the occlusion of the nasopharynx from the oropharynx, necessary for proper deglutition, will be imperfect. The interruption of the motor fibers to the larynx will produce a unilateral vocal cord palsy. This will also be the case if the damage is restricted to the recurrent laryngeal nerve. Because of its practical significance, recurrent laryngeal nerve palsy will be described more fully.

The *symptoms following unilateral recurrent laryngeal nerve affections are usually transitory.* When the one vocal cord loses its innervation, the intact vocal cord will usually, after some time, adapt itself to the altered conditions, and the initial hoarseness and occasional coughing attacks will disappear. *Therefore, if a unilateral recurrent nerve or vagus nerve affection is suspected, a laryngoscopic examination should always be performed, even if there are no symptoms from the larynx.* In paralysis of the recurrent laryngeal nerve due to peripheral injury it is a peculiar feature that the abductor muscles of the larynx are affected first. This circumstance was first described by Semon and consequently is usually referred to as

Semon's law. A satisfactory explanation has not been found as yet.[7] Normally, the posterior cricoarytenoid muscles which rotate the vocal process of the arytenoid cartilages laterally will produce a widening of the rima glottidis and consequently act as abductors. These muscles are supplied by the posterior ramus of the recurrent nerve. The adductors, most important of which is the lateral cricoarytenoid muscle, innervated by the interior ramus, will act as antagonists and narrow the rima. Consequently the predilection of the posterior ramus to be involved first in affections of the nerve will produce an abductor palsy, and the cord will, by the unopposed action of the antagonists, assume a position in the median line.

Apart from a transitory hoarseness an abductor palsy will usually give rise to no symptoms.[8] If the anterior ramus and the adductor muscles are also affected, the vocal cord will assume the cadaveric position, midway between adduction and abduction. This is usually compensated after some time, as the vocal cord of the intact side can be brought to approach the paralyzed one.

In *bilateral palsies of the recurrent laryngeal nerve,* far more seldom met with, symptoms are always present. A bilateral abductor paralysis will produce an extreme narrowing of the rima, and, especially if it occurs rapidly, it may necessitate immediate tracheotomy, if death resulting from asphyxia is to be prevented. If less pronounced, there will be a marked stridor on inspiration and some degree of asphyxia. In bilateral complete lesions of both nerves, the rima is widened, and as the cords cannot be brought into contact, the voice is lost (aphonia). Whispering, however, is possible.

Partial or total paralysis of the recurrent laryngeal nerve is more frequent on the left side than on the right (the left nerve being longer) and in about two-thirds of all cases is due to local processes affecting the nerve. Of such cases may be mentioned *aneurysms of the aortic arch* or *the subclavian artery, enlargement of the tracheobronchial lymph nodes* in tuberculosis, Hodgkin's disease or diseases of the blood (leukemia), and *tumors of the mediastinum.* Recurrent nerve palsy is also a complication encountered in some cases of *thyroidectomy,* where the nerve may be affected by ligature of the inferior thyroid artery, which comes into close contact with the nerve. Isolated paralysis—so-called idiopathic—of the nerve occurs in many cases (one-third to one-quarter of all) where no definite cause can be ascertained. In some instances it is assumed to be due to a "mononeuritis" of the type far more frequently seen in the facial nerve.[9]

It should be emphasized, as follows from the electromyographic studies mentioned above (see p. 465), that the common interpretation of abductor and adductor paralyses as simply being caused by the affection of the posterior or lateral cricoarytenoid muscles, respectively, is too simple. Electromyography may be of diagnostic use in vocal cord palsies. For example, not a few unilateral vocal cord palsies, assumed to be due to lesions of the nerve, are caused by ankylosis of the

---

[7] It has been assumed that the fibers to the abductor muscles occupy a particular part of the nerve. However, according to Sunderland and Swaney (1952) the fibers to the various muscles are intermingled in the recurrent nerve.

[8] On inspiration there normally occurs a moderate widening of the rima glottidis, due to contraction of the abductors. On severe physical exertion a unilateral abductor palsy may therefore cause some dyspnea.

[9] The occurrence of hysterical aphonia should be borne in mind.

cricoarytenoid joints (Weddell, Feinstein, and Pattle, 1944; Faaborg-Andersen, 1957; Kotby and Haugen, 1970c).

It goes without saying that recurrent laryngeal nerve palsies will be a frequent symptom also in lesions of the vagus nerve superior to the origin of the former. In these instances, however, other symptoms usually accompany the vocal cord palsy. Thus the affection of the fibers in the superior laryngeal nerve will produce a partial anesthesia of the larynx, and a unilateral paralysis of the velum will be found.

Other symptoms which may be observed are due to the involvement of the parasympathetic fibers, but in unilateral lesions, as a rule, these are negligible. However, it is worthy of notice that isolated paralysis of the laryngeal muscles may also be caused by diseases affecting the medulla oblongata, for example poliomyelitis. The explanation is found in the arrangement of the motor cells of the nucleus ambiguus, referred to above.

On the whole, unilateral lesions of the extracranial part of the vagus are not common, apart from the recurrent nerve affections already referred to. Tumors, aneurysms of the internal carotid, enlargement of the deep cervical lymph nodes, and traumatic injuries may be mentioned as causes. Unilateral affections of the vagus and glossopharyngeal nuclei may be found, as already mentioned above, in poliomyelitis; furthermore, they are also found in intramedullary vascular disturbances, especially those affecting the posterior inferior cerebellar artery (Wallenberg's syndrome).

It will appear from what has been said that unilateral affections of the vagus and glossopharyngeal nerves give no alarming symptoms. Not so with *bilateral lesions* which, as a rule, prove fatal after some time. It is obvious that such bilateral affections will most frequently be due to diseases of the medulla oblongata. Before turning to these, however, a condition of practical importance should be mentioned, in which the noxious agent probably attacks the peripheral nerve fibers bilaterally, i.e., the common *postdiphtheritic velum palsy.* Diphtheria toxin may damage nerve fibers nearly everywhere in the organism, but, perhaps owing to local factors, the fibers to the velum are most frequently involved (see Fisher and Adams, 1956). As a consequence, some weeks after the onset of diphtheria the patient's speech will be nasal and difficulties of deglutition will occur (cf. above).

Bilateral lesions of the glossopharyngeal and vagus nerves of central origin as a rule will affect all parts of the nerves, although partial affections of the nuclei may be encountered. Of practical importance are chronic bulbar palsy and *poliomyelitis.* In the latter disease the motor cells of the medulla oblongata may be involved in the same manner as those of the spinal cord.

As the distribution, however, is irregular and fortuitous, all transitions may occur from a selective affection of one small nucleus or part of a nucleus to an involvement of nearly all motor nuclei of the brainstem. In cases of this type death usually ensues rapidly, due frequently to the damage to the vagus and glossopharyngeal nuclei. Briefly mentioned, the symptoms referable to the vagus and glossopharyngeal in these cases will be a paralysis of the velum, with regurgitation into the nose on swallowing. Swallowing itself will also be severely interfered with on account of the paralysis of the laryngeal muscles. If the patient does not succumb to a respiratory paralysis, due to concomitant injury to the nucleus of the phrenic nerve, an aspiration pneumonia will regularly result from the lack of coughing reflexes.

In *chronic bulbar palsy* the motor neurons of the brainstem are affected by a slowly progressing degeneration, parallel to that seen in progressive spinal muscular atrophy. The symptoms referable to the vagus and glossopharyngeal nerves are those of most prognostic importance in this disease, but as a rule nearly all motor nuclei in the pons and medulla oblongata are successively affected in a somewhat varying sequence, during the course of three or four years from the beginning of the disease until death. Usually a *dysarthria* is the first symptom, due mainly to lesions of the XIIth and VIIth nuclei. The tongue, the seat of atrophy with fasciculations, cannot be protruded (cf. section (a) of this chapter), the mouth cannot be properly closed, and there is frequently dribbling of saliva. The face becomes motionless and acquires an ''empty'' expression. The dysarthria will be more prominent when the vagus is included, since then the voice grows gradually weaker until complete aphonia occurs, and disturbances in swallowing appear as described above. Finally an aspiration pneumonia ends the sad condition of the patient. It should be remembered that symptoms referable to concomitant degeneration of the motoneurons of the spinal cord are not uncommon.

Somewhat similar clinical manifestations are found in *pseudobulbar palsy*. In this disease, however, the peripheral motor neurons of the bulb are not affected, but the damage responsible for the symptoms is located in the fibers of the corticobulbar tract, usually produced by cerebral atherosclerosis and consequently appearing bilaterally. An important differential diagnostic clue is the lack of muscle atrophy in this disease, most easily ascertained in the tongue. Another peculiarity of pseudobulbar palsy is the frequent occurrence of convulsive laughter and weeping (cf. section (e) in this chapter). *Myasthenia gravis,* which also shows a predilection for the muscles of the head, must be borne in mind in the differential diagnosis. Here the morbid alterations are related to the motor end plates, and apart from the frequent diurnal variations in the paresis and the rapid tiring of the muscles, a prostigmine test will aid in the diagnosis.

In unilateral *lesions of the cerebral hemispheres* (tumors, bleedings), symptoms caused by disturbance of the supranuclear innervation of the glossopharyngeal and vagus nerves are usually absent or negligible. Slight *dysphagia* (difficulty in swallowing) may, however, occur. In rare instances dysphagia may be the presenting or the only symptom. It appears that the symptom is related to an affection of the lower part of the precentral gyrus (see Meadows, 1973).

As a rule *the visceral symptoms* following unilateral lesions of the vagus and glossopharyngeal are negligible. In bilateral lesions, especially if they develop acutely, an acceleration of the heart rate may occur, frequently leading to death. This may be seen in the terminal stages of chronic bulbar palsy. The influence of the vagus on the heart is seen also in some other types of the heredodegenerative diseases, e.g., Friedreich's ataxia, in which electrocardiographic examinations have revealed disturbances thought to be caused by affection of the vagus nuclei.

(d) THE VESTIBULAR DIVISION OF THE VIIITH CRANIAL NERVE

The old anatomists called *the VIIIth cranial nerve* the acoustic, since it supplied the internal ear. However, the so-called acoustic nerve is really made up of two functionally different nerves, *the cochlear nerve* and *the vestibular nerve*. The cochlear nerve transmits exteroceptive impulses perceived as hearing, from the organ of Corti. The vestibular nerve transmits proprioceptive impulses from the saccule and utricle of the vestibulum and from the semicircular ducts, which are important for the maintenance of equilibrium and for orientation in space. The impulses conveyed by the vestibular nerve are only to a limited extent consciously perceived, and, therefore, escaped recognition for a far longer period than did the acoustic function of the cochlear division.

The epithelium of the internal ear develops in early embryonic life as a groove of the ectoderm, and later becomes separated from this to form a vesicle. Certain areas of the epithelium later become differentiated into sensory epithelium. The cochlear duct as well as the semicircular ducts and the saccule and utricle are all formed from the original primary otic vesicle. Phylogenetically the vestibular apparatus appeared long before the cochlear. All vertebrates possess some sort of vestibular organ, already fairly well developed in sharks, whereas an organ of hearing makes its appearance first in amphibians.

From a comparative anatomical point of view the inner ear of vertebrates can be subdivided into a *pars superior* and a *pars inferior*. Structural features and details in the nervous connections support this conception. Other sensory areas than those usually described in man are present in some vertebrates. The pars superior is assumed to be predominantly concerned in the propriocep-tive, "equilibratory" functions of the inner ear, the pars inferior with the exteroceptive function of hearing. To the sensory areas of the pars superior belong the cristae of the three semicircular ducts and the macula of the utricle (and a varying "papilla neglecta"). The pars inferior includes the sensory area of the saccule, the cochlear duct or its forerunner, and another small area. Com-parative anatomical studies by Weston (1939) of the relative size of the different sensory areas have given interesting results. The pars superior is rather constant in phylogenesis, but presents variations in the relative development of the macula utriculi and the cristae of the ducts. The pars inferior shows a progressive development, which particularly concerns the cochlear duct. In man the cochlear sensory area constitutes more than half of the sensory areas of the pars inferior, and a quarter of the entire sensory areas of the inner ear. There is still some uncertainty concerning the function of the saccule (see Kornhuber, 1966, for references), and it will not be considered in the following description.

The cochlear division of the VIIIth cranial nerve will be discussed in Chapter 9. The impressions mediated by the vestibular part of the nerve are transmitted through various pathways to influence ultimately chiefly the motor apparatus, by means of reflex connections of several kinds. The vestibular apparatus is func-tionally particularly closely related to the spinal cord and the cerebellum. This has in part been considered in Chapters 4 and 5. In addition, vestibular impulses influ-ence the oculomotor apparatus. In the present section some principal features of the vestibular nerve and its connections will be dealt with, and some anatomical and clinical features not referred to at other places will be described. It should be realized that the vestibular nerve and its connections have widespread influences on other parts of the nervous system.

The vestibular receptors. The vestibular nerve and nuclei. *The ves-tibular nerve* enters the brainstem (Fig. 7-6) at the lower border of the pons, on the lateral aspect, behind the facial nerve and separated from this only by the tiny intermediate nerve and the cochlear nerve. Together with the nerves mentioned it can be followed in a peripheral direction into the internal acoustic meatus, where

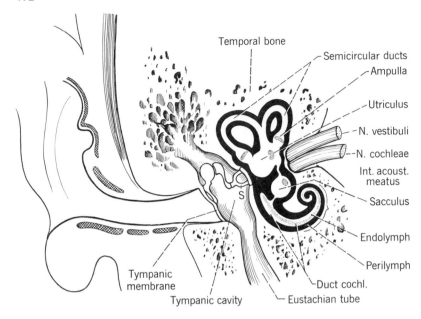

FIG. 7-9 Diagram of the internal ear. The areas containing sensory epithelium are stippled.

it subdivides into smaller branches, passing to different portions of the labyrinth. Where they pierce the bottom of the internal meatus the vestibular nerve fibers have their perikarya. These, like those of the cochlear nerve, are bipolar ganglion cells of the primitive type and are collectively designated *the vestibular ganglion*. Whereas the central processes of these cells form the vestibular nerve, the distal ones are very short and end in relation to the sensory epithelium of the labyrinth.

It will be appropriate here to recall briefly the *principal features of the structure of the labyrinth* (see Fig. 7-9). The parts of the labyrinth generally assumed to be concerned with *vestibular function* are the two sac-like divisions, the *saccule* and the *utricle*, and the *three semicircular ducts*, which are in open communication with the lumen of the utricle. The whole thin-walled membranous labyrinth, filled with endolymph, is enclosed within the osseous labyrinth, separated from its periosteum by the perilymph. The petrous bone in which these structures are embedded is particularly compact. At five places distinct areas of sensory epithelium are present. In the widened ampulla of each semicircular duct there is a transversely projecting crest, consisting of supporting and sensory cells. The hairs of the latter (the hair cells) are covered by a gelatinous mass, the cupula. The sensory epithelium of the utricle is found in a small area of this only, measuring 3 × 2 mm, and its plane corresponds approximately to that of the base of the skull. The corresponding macula of the saccule has the same size, but its plane is vertical, perpendicular to that of the macula of the utricle. Both these maculae consist of hair cells and supporting cells and are covered by a gelatinous mass, into which the hairs of the sensory cells dip. In this mass small prisms consisting mainly of calcium carbonate are found. These so-called otoliths induce different stimuli to the hairs in various positions of the head, according to the action of gravity on them. The hair cells of the ampullary crests, however, are stimulated by movements of the endolymph in the semicircular canals producing a deflexion of the cupula. Thus the latter will react to movements of the head, the former to alterations in its position. The semicircular ducts (horizontal or lateral, superior, and posterior) are arranged at right angles to each other, and should thus be able to react to movements in all directions. Other relevant data will be mentioned later.

The part of the membranous labyrinth concerned in *hearing* is the *cochlear duct,* which is found as 2½ spiral turns within the bony cochlea. It is in open communication near its blindly ending base with the saccule, and its apical coil ends blindly at the apex of the cochlea (Fig. 7-9). The duct is triangular in cross section, attached broadly to the outer walls of the bony cochlear canal, and is separated from the perilymph by the thin vestibular membrane above, and the more solid basilar membrane below. This extends from the bony spiral lamina to the outer wall of the bony duct. On the basilar membrane the sensory apparatus, the organ of Corti, is found. Its sensory hair cells are surrounded by supporting cells. The hair cells are arranged in a single inner row along the duct, and three outer rows. The number of hair cells has been estimated to be about 3,500 and 20,000 in the inner and outer rows, respectively. Between the two rows and their adjoining supporting cells the tunnel of Corti is found, traversed by the nerve fibers passing to the outer hair cells. On their free surface the hair cells are in contact with the tectorial membrane.

The majority of the central processes of the cells of the vestibular ganglion, forming together the vestibular nerve, divide into an ascending and a descending branch when they reach the medulla oblongata. Before dividing they pass between the restiform body and the spinal tract of the trigeminal. The branches are distributed to the vestibular nuclei (Figs. 7-2, 7-7, 7-13, and 7-15).

Within the *vestibular nuclear complex* it is customary to distinguish four large or main nuclei: the *superior* (or nucleus of Bechterew), the *lateral* (or nucleus of Deiters), the *medial* (triangular or nucleus of Schwalbe), and the *descending* (inferior or spinal). The entire vestibular nuclear complex occupies a rather large region just below the floor of the 4th ventricle (see Figs. 7-2, 7-7). On closer analysis it is learned that in addition to the four main nuclei there are several minor cell groups. These have been mapped in detail in the cat (Brodal and Pompeiano, 1957), and most of them can be identified in man as well (Sadjadpour and Brodal, 1968). These groups differ with regard to cytoarchitecture, and there are, furthermore, regional architectonic variations within each of the main nuclei (see Fig. 7-14). For example, in the central region of the superior nucleus the cells are on the whole larger and somewhat more densely packed than in the peripheral region. The lateral vestibular nucleus of Deiters is characterized by containing a fair number of large and giant (Deiters') cells; these are (in the cat) more numerous and on the whole larger in the caudodorsal part of the nucleus than in the rostroventral part. The assumption that these and several other architectonic features reflect functional differences is strongly supported by experimental studies of the fiber connections of the vestibular nuclei and by physiological observations, as will appear from the following. The anatomical and functional organization of the vestibular nuclei has been shown to be extremely complex and far more differentiated than was assumed a few decades ago. (For a review, see Brodal, Pompeiano, and Walberg, 1962; for more recent data, see also Brodal, 1964, 1974.) Most of the more recent information has been obtained from experimental anatomical and physiological studies in the cat. One point of particular interest is the fact (mentioned in Chap. 4) that the nucleus of Deiters presents a somatotopic organization (Pompeiano and Brodal, 1957a) as shown in Figs. 4-7 and 7-10.

As referred to above, the vestibular nerve, when entering the vestibular nuclei, divides into an ascending and a descending branch because most (if not all) primary vestibular fibers divide in a dichotomous fashion. Many of the myelinated parent fibers, most of which are rather coarse, enter the lateral vestibular nucleus, and from here the ascending fibers pass to the superior nucleus (and the cerebel-

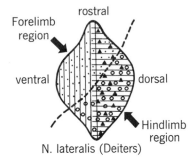

FIG. 7-10 A diagram of the lateral vestibular nucleus of Deiters in the cat in a sagittal view, showing the somatotopic pattern within the nucleus and the approximate distribution within it of some of its main afferents to particular territories. Note restriction of primary vestibular fibers to the "forelimb region." From Brodal (1967b).

Fibers from:
Labyrinth
Spinal cord
Cerebellar vermis
N. fast. rostral
N. fast. caudal

lum); the descending ones pass to the descending and medial vestibular nuclei. Contrary to what has been generally believed, the *primary vestibular fibers do not supply the entire territory of the vestibular nuclear complex*. Within each of the main nuclei there are quite extensive regions which are free from vestibular afferents (Walberg, Bowsher, and Brodal, 1958; Stein and Carpenter, 1967; Gacek, 1969). For example, only the rostroventral part of the nucleus of Deiters (its forelimb region) receives primary vestibular fibers (see Fig. 7-10 and Fig. 1-9A). This has been confirmed physiologically (Wilson et al., 1966). (In the superior nucleus such fibers are restricted to its central regions, and only some of the small groups receive primary fibers.) Obviously, the regions which are supplied by primary vestibular fibers must be under a more direct influence of vestibular impulses than the rest of the nuclear complex.[10]

Since the different parts of the vestibular organ are not functionally identical (see also below), it would be of interest to know the central distribution of the fibers from the various receptor regions of the labyrinth. In experimental studies following small lesions of the vestibular ganglion, the termination of degenerating fibers has been studied with silver impregnation methods. Fibers from the different receptor areas have their particular terminal areas in the monkey (Stein and Carpenter, 1967) and in the cat (Gacek, 1969), even though terminal areas overlap somewhat.

It appears from these studies (for a review, see Brodal, 1974) for example, that fibers from the ampullae of the semicircular ducts mainly supply the superior vestibular nucleus and the rostral part of the medial nucleus, sending some fibers to the descending nucleus. In the superior nucleus the terminal areas of the three cristae appear to differ to some extent (Gacek, 1969). Fibers from the utricular macula appear to end mainly in the lateral vestibular nucleus (rostroven-

[10] It may be argued that the term "vestibular nuclei" should be restricted to those parts of the complex which receive primary vestibular fibers. However, for practical reasons, it is advisable to retain the current terminology.

trally, cp. above, its ''forelimb region''); some fibers end in the descending and medial, but none in the superior nucleus. This is in agreement with physiological studies, since units responding to stimulation of the horizontal semicircular duct appear to be located mainly in the medial and superior nuclei (Shimazu and Precht, 1965). Units responding to tilting have been found in the ventral part of Deiters' nucleus (Peterson, 1970; see also Sans, Raymond, and Marty (1972). In a recent study with the Fink and Heimer silver impregnation method, Korte (1979) traced some primary vestibular afferents to the reticular formation close to the abducent nucleus and to the rostral pole of the external cuneate nucleus. The findings mentioned above are of considerable functional interest, since at least some efferent projections from the vestibular nuclei arise largely from specific parts of the complex (see below).

In addition to the afferent fibers coming from various subdivisions of the labyrinth, an *efferent component* has been demonstrated in the vestibular nerve (Gacek, 1960). The number of efferent fibers is small. According to Rossi and Cortesina (1965a) the fibers arise in three small cell groups, one of them in the lateral vestibular nucleus. More definite results can be obtained by means of the retrograde axonal transport of horseradish peroxidase. In studies with this method, labeled cells have been found, bilaterally, in a small region of the reticular formation, lateral to the abducent nucleus, ventromedial to the ventral part of Deiters' nucleus (Gacek and Lyon, 1974; Warr, 1975). Peripherally the fibers have been traced to all parts of the vestibular sensory epithelia (see also below). These fibers presumably mediate a central control of vestibular receptors. How this occurs is debated (for a discussion, see Klinke and Galley, 1974; Precht, 1974a).

The *vestibular receptors,* as mentioned above, are the hair cells in the utricular and saccular maculae and in the cristae of the semicircular ducts. Although light microscope studies, particularly by Polyak and Lorente de Nó, had indicated that there are two types of receptor cells in the sensory epithelia and that there are differences in the nerve endings establishing contact with them, electron microscope studies were needed to clarify these problems. Such studies have also revealed other morphological features which are of functional interest. For particulars the reader is referred to Wersäll (1956), Engström and Wersäll (1958), Spoendlin (1964), Wersäll, Gleisner, and Lundquist (1967), and Lindeman (1969).

The two types of sensory cells found within the vestibular sensory epithelia and separated by supporting cells, differ in several respects (Fig. 7-11A). The type I cell is bottle-shaped, while the type II cell is slender. The sensory cells do not reach the basal membrane. Both kinds of cells contain several mitochondria and are provided with sensory hairs. On electron microscopic examination it appears that the sensory hairs are stereocilia (40–110 per cell; Spoendlin, 1964), arranged in a regular manner according to their length. In addition, however, each cell is equipped with a peripherally situated kinocilium, having the characteristics of those found in other regions. When the orientation of the hairs is examined it becomes apparent that there is a regular organization (see Flock, 1964; Spoenlin, 1964), a spatial polarization (see Fig. 7-11C), suggesting that the sensibility of the epithelium is not the same to forces acting upon it from all directions. It appears that deviation of the sensory hairs toward the site of the kinocilium (see Fig. 7-11B) increases the firing rate of its afferent nerve fiber, while deviation in the opposite direction reduces the firing rate (see Flock and Duvall, 1965). The cells of type I and II show clear-cut differences with regard to their contacts with the afferent nerve fibers (the distal process of the vestibular ganglion cells). The type I cells are almost completely surrounded by chalyces of nerve terminals, while the slender type II cells are in contact with smaller bouton-like nerve endings at their bottom (Fig. 7-11A). In addition to these endings there occur small boutons which, in contrast to the two other kinds of endings, are provided with synaptic vesicles. These endings are assumed to belong to the

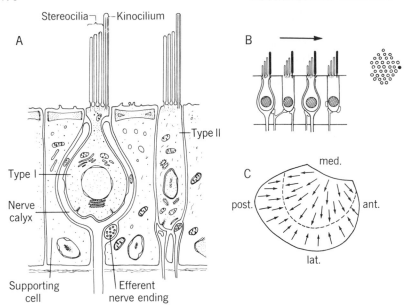

FIG. 7-11 *A:* Schematic drawing of the two types of hair cells in the vestibular epithelia of mammals. Note the different kinds of nerve endings in contact with the cells, the symmetric arrangement of the bundles of sensory hairs, many stereocilia with a single kinocilium in the periphery. From Wersäll, Gleisner, and Lundquist (1967). *B:* A diagram showing the orientation of the sensory hairs, the direction of displacement of the hairs which produces excitation (arrow), and (to the right) a view of the upper surface of a hair cell. *C:* A diagrammatic surface view of the human utricular macula, indicating (arrows) the direction of polarization of the sensory cells in different regions. Figures *B* and *C:* courtesy of Dr. H. Lindeman, Jr.

efferent vestibular fibers, an assumption which is supported by the observation that they degenerate following transection of the vestibular nerve.

These morphological observations indicate that the sensory epithelia of the vestibular apparatus are very complexly organized. This is substantiated by neurophysiological studies. Most reliable information is obtained in studies in which the receptors are stimulated naturally, by rotatory movements of the animal or tilting of the head. Recordings of action potentials in single fibers in the vestibular nerve have shown that each duct is maximally sensitive to movements in its own plane (Löwenstein and Sand, 1940). Furthermore, such studies have revealed functional differences between the cells in a crista or a macula. Units have been found in the cristae which react with an increase in their impulse frequency at the onset of rotation in one direction, for example to the right, while they are silent at the onset of a rotation to the left. During rest they show a slight resting discharge. A few units respond to rotation in both directions, either by excitation or inhibition (Löwenstein and Sand, 1940; Gernandt, 1949, and others). These data may be related to the polarization of the receptor cells described above. In the case referred to above it is assumed that the potentials recorded arise from stimulation of cells in the cristae. However, strictly speaking, it is not the rotation as such which is the adequate stimulus; it is more correct to speak of *angular acceleration and deceleration* as stimuli. This is exemplified by the following: When a rotation starts, a particular unit increases its discharge. In the course of some time, usually about 20 sec, it gradually reverts to its slow resting discharge frequency, and when the rotation stops, its discharge ceases entirely. This can be explained in the following way: a rotation of the head in the plane of a semicircular duct will, on account of the inertia of the endolymph, produce a deviation of the cupula with the sensory hairs. Thus the cell is stimulated. After the rotation has continued for some time the fluid will follow the movement of the head, the cupula resumes its resting position,

FIG. 7-12 Scanning electron micrographs of the vestibular receptor apparatus.
A shows the bundles of hairs on sensory cells of the macula utriculi in the chinchilla. Note the orientation of the stereocila in the same direction. The surface of one hair cell is indicated. The supporting cells are provided with microvilli. × 6700. B shows otoliths (statoconia) of different sizes covering the macula utriculi in the cat. × 1400. Compare text and Fig. 7-11. From Lindeman (1973).

and there is no stimulus. When the rotation stops, the cupula again deviates, but now in the opposite direction. This does not stimulate the particular sensory cell, but silences it. The deviations of the cupula were first observed directly by Steinhausen (1931). The cupula behaves like a highly damped torsion pendulum (see Lowenstein, 1966). Scanning electron microscopy (Fig. 7-12) shows the otoliths and the stereocilia of the sensory cells and their orientation particularly clearly.

Angular acceleration or deceleration as described above *appears to be the adequate stimuli for the canals. The maculae* appear to be chiefly *position recorders,* which signal the position of the head in space. They are sometimes referred to as the static labyrinth, while the semicircular ducts represent the kinetic labyrinth. (Some data indicate that the maculae may also be vibration receptors.) On tilting the head the otolithic macula of the utricle, on account of its weight, will exert a pull on the sensory hairs, making them respond. These receptors adapt very slowly and most of them continue to fire if the same position of the head is upheld. The spatial polarization within the macula, referred to above (see Fig. 7-11B), gives a rational explanation of the fact that changes of the position of the head in different directions may stimulate hair cells. In one position cells in one area of the macula will be excited, in another position the pull exerted by the otolithic membrane will stimulate cells in another region of the macula. This agrees with results of single-unit analyses of afferent fibers, in which some units respond to an increase in tilting in one direction, others to tilting in other directions.

The above account is a simplified presentation of a very complicated mechanism which is still not completely understood. The semicurcular ducts may be characterized as "bidirectional angular accelerometers," while the utricular macula is more of a "position recorder." However, linear acceleration may influence the semicircular canals to some extent, while it appears doubtful whether angular acceleration affects the macula (see Lowenstein, 1966). The data referred to above make it appear likely that the central parts of the vestibular pathways are also complexly organized. This, indeed, has turned out to be true. Some main points will be considered in the following.

Fiber connections of the vestibular nuclei. In a general way it may be said that the vestibular nuclei are connected by afferent and efferent fibers with the following regions: *the spinal cord, the cerebellum, certain nuclei in the brainstem, and the reticular formation.* However, the afferent and efferent links in each of these reciprocal pathways with other structures are not quantitatively equal, nor are they related to the same subdivisions of the nuclear complex, a further indication of its intricate organization. Figure 7-13 shows a very simplified diagram of the principal pattern of the connections of the vestibular nuclei. Some particular features can be seen in Fig. 7-15.

There are two well-known efferent pathways by which the vestibular nuclei may influence the activity of the *spinal cord.* As described previously (p. 201), the *lateral vestibulospinal tract* (see Fig. 4-7) arises only from the ipsilateral nucleus of Deiters and from small and large cells; it is somatotopically organized, can be traced to sacral levels of the cord, and terminates chiefly in Rexed's laminae VII to VIII. It exerts a facilitatory action on extensor motoneurons (α and γ). It appears that the lateral vestibular nucleus in man is likewise somatotopically

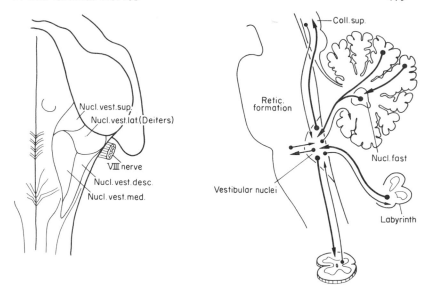

FIG. 7-13 To the left, a diagram of the four main vestibular nuclei in man as projected on the dorsal surface of the brainstem (cf. text). To the right, a simplified diagram of the main connections of the vestibular nuclei. Note that the vestibular nuclei have reciprocal connections with the spinal cord, the labyrinth, the cerebellum, the reticular formation, and the mesencephalon at the level of the superior colliculus (oculomotor and adjacent nuclear groups, see text). The connections indicated by heavy lines are more massive than those shown as thin lines. It can be seen that the major afferent input to the vestibular nuclei comes from the labyrinth and the cerebellum, while their main output goes to the spinal cord and the mesencephalon (mainly the motor eye nuclei).

organized (Løken and Brodal, 1969). The other (*medial*) *vestibulospinal pathway* comes chiefly from the medial vestibular nucleus (see Figs. 4-7 and 7-15B), descends crossed and uncrossed only to upper thoracic segments, and ends in the same laminae of the spinal gray as does the lateral tract.[11] In contrast to these conspicuous vestibulospinal pathways, there is only a modest direct pathway conducting in the opposite direction, from the cord to the vestibular nuclei. The *spinovestibular fibers,* which may to some extent represent collaterals of dorsal spinocerebellar tract fibers, end in a restricted number in the most caudal parts of the descending, medial, and lateral vestibular nuclei. Most of them terminate in the small group x (of Brodal and Pompeiano, 1957) in the cat (Pompeiano and Brodal, 1957b; Brodal and Angaut, 1967) and probably in man (Bowsher, 1962). These nuclear regions appear to receive few or no primary vestibular fibers.[12]

---

[11] Using the horseradish peroxidase method (see Chap. 1), Peterson and Coulter (1977) recently suggested that there is in the cat a third vestibulospinal projection, originating in the caudal poles of the medial and descending vestibular nuclei and the cell group *f* of Brodal and Pompeiano (1957). It could be traced to the lumbar enlargement.

[12] The group x (see Fig. 7-14) and the other regions of the vestibular complex receiving direct spinal afferents send fibers to the cerebellum (Brodal and Torvik, 1957), presumably mainly the vestibulocerebellum (see Chap. 5). These groups may thus be relays in a supplementary, usually neglected, spinovestibulocerebellar pathway. Nothing is known physiologically about this. It is of interest here to refer to another small part of the vestibular complex, the group z (Fig. 7-14), outlined on the

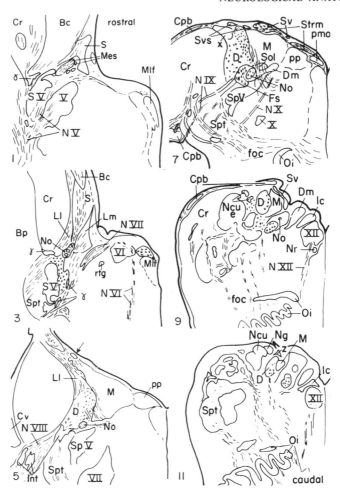

FIG. 7-14 A series of transverse sections of the human brainstem taken with equal intervals from rostral (no. 1) to caudal (no. 11) to illustrate main features of the topography of the vestibular nuclei and their surroundings. Some abbreviations: *Bc:* brachium conjunctivum, *Bp:* brachium pontis; *Cr:* corpus restiforme; *Cv:* ventral cochlear nucleus; *D:* descending (inferior) vestibular nucleus; *Dm:* dorsal motor vagus nucleus; *x, z:* small groups of the vestibular nuclei (Brodal and Pompeiano, 1957); *Ic:* nucleus intercalatus (Staderini); *Int:* interstitial vestibular nucleus of Cajal; *Ll* and *Lm:* parts of the lateral vestibular nucleus (Deiters); *M:* medial vestibular nucleus; *mes:* mesencephalic nucleus and tract of trigeminal nerves; *Mlf:* medial longitudinal fasciculus; *Ncu, Ncue,* and *Ng:* cuneate, external cuneate, and gracile nucleus, respectively; *NV, NVI, NVII, NVIII, NX,* and *NXII:* cranial nerves; *pp:* nucleus prepositus nervi hypoglossi; *S:* superior vestibular nucleus; *SV* and *Spv:* main sensory nucleus and spinal nucleus of the trigeminal nerve; *V, VI, VII, X,* and *XII:* motor cranial nerve nuclei. From Sadjadpour and Brodal (1968).

basis of cytoarchitectonics by Brodal and Pompeiano (1957). Like the group *x*, it receives spinal afferents but not primary vestibular fibers. It has been shown to be a relay nucleus for the transmission of group I muscle afferents via the thalamus to the cerebral cortex (Landgren and Silfvenius, 1969, 1971). On the other hand, the small group *y*, which according to Gacek (1969) receives primary vestibular afferents from the sacculus, has been concluded from physiological (Highstein, 1971) and anatomical (Graybiel and Hartwieg, 1974) studies to project onto the oculomotor nucleus.

Physiological studies (Wilson, Kato, Thomas, and Peterson, 1966) have shown that spinal impulses may reach the vestibular nuclei indirectly via other routes, probably chiefly the reticular formation (spinoreticular and reticulovestibular fibers). According to Fredrickson, Schwarz, and Kornhuber (1965) and others, joint movements are particularly effective in eliciting responses. Another indirect route passes through the cerebellum. (See Pompeiano, 1972b, for a discussion of the complex problems of spinal influences on the vestibular nuclei.)

The relations of the vestibular nuclei with the *cerebellum* have been considered in Chapter 5. Suffice it here to recapitulate some points. A small number of fibers from the vestibular nuclei pass to the cerebellum and appear to end in its "vestibular part." They are derived from restricted parts of the medial and descending vestibular nuclei (Brodal and Torvik, 1957), which appear not to receive primary vestibular fibers (see Chap. 5, however). Nevertheless, vestibular impulses may reach the cerebellum directly via *primary* vestibulocerebellar fibers (see Fig. 5-10) which pass to the nodulus, the ventral part of the uvula, the flocculus (Dow, 1936), and the ventral paraflocculus (Brodal and Høivik, 1964). In the monkey the same appears to be the case (Carpenter, Stein, and Peter, 1972), while Korte and Mugnaini (1979) in a recent study in the cat did not confirm the projection to the paraflocculus (see Chap. 5). Thus, except for its vestibular part, the cerebellum does not appear to be much influenced from the vestibular apparatus, but it exerts a marked influence on the vestibular nuclei. As described in Chapter 5, the "vestibulocerebellum" sends only a relatively modest number of fibers back to certain regions within the vestibular complex, but there are two other more massive routes by which the cerebellum may influence the vestibular nuclei. Both these pathways arise in parts of the vermis, one being direct, the other having a synaptic relay in the fastigial nucleus (see Fig. 7-13). Of particular interest is the observation that within the projections onto the lateral vestibular nucleus of Deiters, both pathways present a somatotopical localization, as illustrated in Fig. 5-30, from which further details may be seen. By way of these fibers the "spinal parts" of the cerebellum (which appear to receive few vestibular impulses) may act in a somatotopically localized manner on the motor apparatus of the spinal cord.

The connections of the vestibular nuclei with the *reticular formation* when studied in Golgi material (see Scheibel and Scheibel, 1958) show that axons of cells in the reticular formation appear to reach all nuclei in the vestibular complex and that the latter give off fibers or collaterals to widespread regions of the reticular formation. In experimental studies the projections can be studied in greater detail. With regard to the projection onto the reticular formation (see Chap. 6), the vestibular nuclei are not identical (see Ladpli and Brodal, 1968). For example, only the nucleus of Deiters gives off fibers to the lateral reticular nucleus (which projects back to the cerebellum). Reticulovestibular fibers come mainly from the medullary and caudal pontine reticular formation (Hoddevik, Brodal, and Walberg, 1975; see also Graybiel, 1977; Pompeiano, Mergner, and Corvaja, 1978). Physiological studies are in agreement with the antomical findings (Remmel, Skinner, and Pola, 1977). By way of these connections the vestibular nuclei may cooperate with the reticular formation and, for example, influence the activity of the spinal cord and certain parts of the thalamus (see Chap. 6), since vestibular efferents pass to regions which give rise to long descending and ascending fibers.

The vestibuloreticular connections are presumably involved in vomiting and cardiovascular reactions observed on vestibular irritation (see below).

Connections of the vestibular nuclei with the *higher brainstem*, particularly with the nuclei of the nerves of the ocular muscles, have attracted much interest. Such fibers pass in the *fasciculus longitudinalis medialis*. This is a rather complicated bundle in spite of its course as a distinct fiber tract. It extends from the region of the oculomotor nucleus down the brainstem and continues to the spinal cord. It is phylogenetically very old, and in ontogenesis it stands out very clearly on account of its early myelinization. In the brainstem it is situated close to the median line beneath the aqueduct and 4th ventricle (cf. Figs. 7-4, 7-7, 7-18, 7-22), and is closely related to the nuclei of the eye muscles. In the caudal part of the medulla oblongata it bends ventrally and continues in the ventral funiculus of the spinal cord close to the ventral median fissure. In the cord it is usually called the *sulcomarginal fasciculus*. In addition to fibers from the vestibular nuclei the medial longitudinal fasciculus contains other fibers. Some of these are descending and come from the interstitial nucleus of Cajal in the mesencephalon situated close to the oculomotor nucleus (Fig. 7-28). Some of them end in a restricted part of the medial vestibular nucleus (Pompeiano and Walberg, 1957), while others descend to the cord as the *interstitiospinal tract*, mentioned in Chapter 4 (see Fig. 4-1).

The bulk of fibers in the medial longitudinal fasciculus come from the ves-

FIG. 7-15 Simplified diagrams of main features in the organization of primary vestibular fibers to the vestibular nuclei and of the ascending (A) and descending (B) projections from these nuclei. D, L, M, and S: descending, lateral, medial, and superior vestibular nucleus, respectively. For particulars see text.

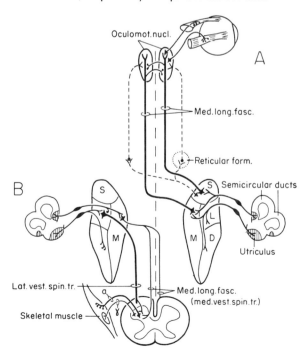

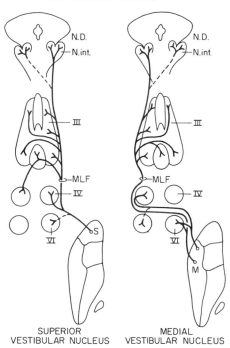

SUPERIOR
VESTIBULAR NUCLEUS

MEDIAL
VESTIBULAR NUCLEUS

FIG. 7-16 Diagram illustrating the main points in the ascending projections from the vestibular nuclei. Note that their origin is restricted to the superior nucleus (*S*), which gives off an ipsilateral projection, and the rostral part of the medial nucleus (*M*), whose fibers ascend in the contralateral medial longitudinal fasciculus (*MLF*). Both components to a large extent supply most groups of oculomotor nuclei (except IV) bilaterally by collaterals crossing the midline. Some fibers ascend to the nucleus of Darkschewitsch (*N.D.*) and the interstitial nucleus of Cajal (*N. int.*). From Tarlov (1970).

tibular nuclei. A smaller proportion passes to the cord as the medial vestibulospinal tract (Figs. 4-6, 7-15B), referred to above. While these descending fibers are derived from the medial vestibular nucleus, possibly also the descending, conflicting opinions have been held concerning the origin of the ascending fibers (Fig. 7-15A) (for reviews see Brodal and Pompeiano, 1958; Brodal, Pompeiano, and Walberg, 1962; Tarlov, 1969; Brodal, 1974). A component arising in the superior vestibular nucleus has long been recognized as coursing only homolaterally (see McMasters, Weiss, and Carpenter, 1966). The origin and termination of the ascending vestibular fibers have recently been studied in great detail by Tarlov (1970) and Gacek (1971) by tracing degenerating fibers following minute lesions in individual vestibular nuclei. The main points of their results agree. Figure 7-16 shows a summary of Tarlov's (1970) findings.

In addition to confirming the exclusive origin of ipsilaterally ascending fibers from the superior nucleus, both authors found the crossed ones to be derived from the rostral part of the medial vestibular nucleus.[13] However, both crossed and

[13] Gacek (1971) also found ascending fibers to the IIIrd nerve nucleus from the ventral part of the ipsilateral lateral nucleus of Deiters. These fibers ascend lateral to, and separated from, the medial longitudinal fasciculus. The origin of this projection (Graybiel and Hartwieg, 1974), and of a corresponding projection to the abducent nucleus (Maciewicz, Eagen, Kaneko, and Highstein, 1977) has been confirmed in HRP studies. Some fibers to these nuclei appear to arise in the descending nucleus (cf. Brodal and Pompeiano, 1958 and in the group y of Brodal and Pompeiano (1957), which appears to receive its afferents from the sacculus (Gacek, 1969). These fibers and those from the nucleus prepositus hypoglossi (see Chap. 5 and section (g) in this chapter) appear to ascend in or close to the superior cerebellar peduncle and not in the medial longitudinal fasciculus. The same appears to be the case for some fibers arising in the most dorsal part of the superior vestibular nucleus (Yamamoto, Shimoyama, and Highstein, 1978).

uncrossed ascending fiber components supply structures on both sides of the brainstem by means of branches (collaterals) across the midline. Most ascending vestibular fibers terminate in the nuclei of the extrinsic eye muscles (III, IV, and VI). The pattern of distribution is fairly specific insofar as certain cell groups which innervate particular extraocular muscles (see section g in this chapter) receive their afferents, at least predominantly, from certain areas within the superior and rostral part of the medial vestibular nuclei (see Tarlov, 1970). From a correlation with what is known of the endings of fibers from the cristae and maculae in the vestibular nuclei (see above), it appears that impulses from the cristae are transmitted chiefly via the superior and medial nucleus, while those from the maculae are relayed in part in the rostral region of the medial nucleus and in part in the rostroventral region of the lateral nucleus (see Gacek, 1971). Numerous physiological studies have been devoted to analyses of the ascending vestibulo-oculomotor connections.

Natural stimulation of a semicircular duct induces eye movements (nystagmus, see below) in a particular direction, and stimulation of a nerve from a duct results in conjugate deviation of the eyes in the plane of the duct (Fluur, 1959; Cohen, Suzuki, and Bender, 1964; and others). As will be discussed in the section on the oculomotor nerves, conjugate eye movements require a precise cooperation of several muscles of both eyes. Furthermore, the various extrinsic eye muscles appear to be supplied from particular cell groups in the oculomotor nuclei. This strongly suggests that there must be an extremely precise correlation between the sites of ending of primary vestibular fibers from individual receptor organs in the nuclei and the distribution to the oculomotor nuclei of the fibers arising from the particular terminal areas in the vestibular nuclei. This indeed appears to be the case.

On the basis of anatomical and physiological studies, Szentágothai (1952a, 1964b) and others concluded that the vestibulo-oculomotor fibers in the medial longitudinal fasciculus are links in *elementary three-neuron reflex arcs* (primary sensory neuron–vestibular neuron–peripheral motor neuron). In addition, however, the vestibular nuclei may influence the nuclei of the ocular muscles via *indirect routes,* and it seems clear that these are involved as well when conjugate eye movements occur following vestibular stimulation. These indirect vestibulo-ocular reflex arcs (via the cerebellum, the nucleus prepositus hypoglossi, and the part of the reticular formation referred to as PPRF) are considered in Chapter 5 and in section g of this chapter. For a recent discussion of the various indirect vestibulo-ocular reflex arcs and their influence (excitatory or inhibitory actions and other aspects) on the eye motor nuclei, see the extensive review of Cohen (1974; see also Precht, 1978). Some data on the direct three-neuron vestibulo-ocular reflex pathways will be briefly mentioned here.

In an extensive study in the monkey, Uemura and Cohen (1973) found that lesions of various parts of the vestibular nuclear complex have different effects on vestibulo-ocular reflexes, as well as on posture. There seems to be agreement that the fibers ascending from the superior vestibular nucleus mediate only inhibitory actions on the eye muscle nuclei, while those from the superior part of the medial nucleus mediate both excitatory and inhibitory effects (see Cohen, 1974, for references).

With recent neurophysiological techniques, far more complex patterns have been discerned. Impulses from the semicircular canals as well as from the maculae influence the motor ocular nuclei via the direct vestibulo-ocular route. Altogether six direct, short-latency, excitatory, and six short-latency, inhibitory pathways from the two labyrinths to the two eyes have been distinguished (see Precht, 1978). Excitatory and inhibitory pathways cooperate and secure the appropriate reciprocal innervation of a pair of extrinsic eye muscles. (For example, stimulation of a horizontal canal excites the motoneurons supplying the contralateral lateral rectus and the ipsilateral medial rectus but inhibits the motoneurons passing to the ipsilateral lateral rectus and contralateral medial rectus, as might be expected from the eye movements to the contralateral side following stimulation of a horizontal canal.) Furthermore, on the basis of anatomical and physiological observations, it has to some extent been possible to locate the nuclear region within the vestibular complex which serves as a relay between the afferents from a particular subdivision of the labyrinth and the motoneurons of a particular muscle. For an instructive table see Precht (1978, p. 170), likewise for references and discussions of particular neurophysiological features.

The three-neuron, disynaptic short-latency, direct vestibulo-ocular reflex arc is obviously finely differentiated and a potent mediator of vestibular influences on the ocular muscles. The indirect vestibulo-ocular pathways are all polysynaptic, and their mechanism is not yet completely clarified.

Some of the ascending fibers in the medial longitudinal fasciculus proceed beyond the oculomotor nucleus and end in the small neighboring nuclei, the nucleus of Darkschewitsch, the nucleus of the posterior commissure, and the interstitial nucleus of Cajal (see McMasters, Weiss, and Carpenter, 1960; Tarlov, 1969, 1970; Gacek, 1971). According to a recent autoradiographic study in the monkey by Büttner-Ennever and Lang (1978), only few of the ascending fibers from the vestibular nuclei supply the nucleus of Darkschewitsch, while the interstitial nucleus of Cajal receives many, suggesting that the latter is far more directly involved in ocular movement control than the nucleus of Darkschewitsch. In addition, vestibular fibers were found to end in the so-called rostral nucleus of the medial longitudinal fasciculus, assumed to be particularly important for vertical gaze (see section g of this chapter).

Since vestibular stimulation is known to give rise to consciously perceived sensations it has been assumed for a long time that there are *pathways from the vestibular nuclei to the cerebral cortex.* Several workers have recorded action potentials from the cerebral cortex in animals following natural or electrical stimulation of the vestibular receptors or the vestibular nerve, but there has been some disagreement concerning the regions of the cortex receiving the vestibular impulses as well as concerning the pathways followed and the laterality of the projections. Early workers (see Brodal, Pompeiano, and Walberg, 1962, for references) recorded responses in a cortical area close to the cortical acoustic area (see Chap. 9), mainly lateral and rostral to this. Some findings suggest that this projection may be rather specifically organized. Thus, according to Andersson and Gernandt (1954), stimulation of individual branches of the vestibular nerve gives responses in limited portions of the total "vestibular cortical area," and Massopust and Daigle (1960) found the responses elicited from the medial and descending nuclei to differ spatially in the cortex.

In recent years interest has been focused on another cortical vestibular area. In the monkey it occupies a small region in the face subdivision of Sm I belonging to Brodmann's area 3. (For a recent study, see Ødkvist, Schwarz, Fredrickson,

and Hassler, 1974.) This area has also been found in the rabbit and cat (see Ødkvist, Liedgren, Larsby, and Jerlvall, 1975) and shows a marked convergence of
vestibular and peripheral somatosensory impulses, from group IA muscle afferents
and skin and joint afferents. Vestibular responses have finally been recorded from
the motor cortex (see Boisacq-Schepens and Hanus, 1972). It seems a likely assumption that the cortical vestibular areas are of importance for the conscious appreciation of spatial orientation (see Fredrickson et al., 1966; Kornhuber, 1966),
and that the influence of vestibular input on pyramidal tract cells (Boisacq-
Schepens and Hanus, 1972) may play a role in the performance of movements.

Little is still known about the pathways mediating vestibulocortical impulses. (For some references to earlier studies see Hassler, 1956; Szentágothai, 1964b). Sans, Raymond, and Marty
(1970) recorded vestibular responses in two regions of the thalamus of the cat (in addition to
responses in the two cortical vestibular areas mentioned above), one in the VPL (nucl. ventr. post.
lat.), another in the VL (nucl. ventr. lat.). The distribution of labeled cells in the thalamus following injections of HRP in the two cortical vestibular areas in the cat (Liedgren, Kristensson,
Larsby, and Ødkvist, 1976) is in partial agreement with these findings. In the monkey, a thalamic
focus responding to natural vestibular stimuli or to vestibular nerve stimulation has been located in
the VPI (nucl. ventr. inf., see Chap. 2) between the VPM (nucl. ventr. post. med.) and the VPL
(Deecke, Schwarz, and Fredrickson, 1974; Büttner and Henn, 1976).

Anatomically, fibers passing from the vestibular nuclei to thalamic nuclei,
particularly the VPM, have been described in the monkey (Carpenter and Strominger, 1965) and in the cat (Carpenter and Hanna, 1962; and others). However, in
a careful experimental study in macaques, baboons, and a chimpanzee, Tarlov
(1969) found no evidence of such fibers, whereas Raymond, Sans, and Marty
(1974) recently described in the cat a rather sparse contralateral projection to the
thalamus, particularly from the medial and superior vestibular nuclei. The terminal
region is found in the ventrocaudal part of the thalamus. The region is not sharply
delimited, but appears to cover parts of the VL adjoining the VPM. Autoradiographically, in the monkey, Lang, Büttner-Ennever, and Büttner (1978)
traced vestibular fibers bilaterally to the oral part of VPL and some to the VPI.
Even if there are still some discrepancies between authors concerning details,
there is no doubt that the vestibulocortical pathway has a thalamic relay. It appears
further that this relay receives information from nonvestibular sensory afferents as
well, for example from muscle afferents (see Chap. 2).

Even though a direct vestibulothalamocortical pathway exists, this is scarcely
the only route used by vestibular impulses to the cortex, even if little is known
about alternative pathways (via the cerebellum?, via the reticular formation?). The
possibility that there may be an ascending route from the nucleus of Darkschewitsch and the interstitial nucleus (which receive abundant vestibular nuclear
fibers) has received relatively little attention. Isolated lesions of these small nuclei
are difficult to achieve. However, following injections of horseradish peroxidase
in the sensorimotor cortex of the cat, labeled cells were found in the nucleus of
Darkschewitsch by Avendaño (1976). The complexity of the interplay between the
vestibular nuclei and the cerebral cortex is further emphasized by the observation
that, following stimulation of cortical vestibular areas, single-unit responses have
been found in some of the vestibular nuclei (Gildenberg and Hassler, 1971). The
pathway is assumed to be multisynaptic. Only guesses can be made concerning the
cells and fibers involved.

In addition to the connections of the vestibular nuclei considered above, the presence of *commissural connections* between the nuclei on the two sides (see Brodal, 1972c, for a review) is of great functional importance. In an experimental anatomical study in the cat, Ladpli and Brodal (1968) found that the four main vestibular nuclei differ markedly with regard to their commissural connections. For example, while the two lateral nuclei are only sparsely interconnected, the superior and descending nuclei have ample projections to the opposite side. In addition to supplying their partner on the other side, they give off commissural fibers to the three other nuclei as well. Largely corresponding findings were made by Tarlov (1969) in the monkey. Precise information on the origin of the commissural fibers in the cat was obtained with the HRP method by Pompeiano, Mergner, and Corvaja (1978) whose report should be consulted for particulars. Several detailed electrophysiological studies from recent years are in agreement with the anatomical data and have brought forward valuable information about the collaboration between the two vestibular nuclear complexes (for a review, see Shimazu, 1972). They have also shown, in agreement with anatomical findings, that there are possibilities for cooperation via indirect connections, presumably via the reticular formation.

It is apparent even from the incomplete account given here that anatomically the vestibular nuclei are organized in a very complex manner. Even within each of the four main nuclei a differentiation is possible, not only with regard to architecture but also as concerns the sites of termination of afferent fibers and the sites of origin of efferent projections. This is shown for the nucleus of Deiters in Fig. 7-10, but the patterns within the other nuclei are at least as complex (see Brodal, 1972b, 1974, for reviews). Some observations which further illustrate the complexity may be mentioned.

For example, the afferents from the cerebellar cortex appear to contact preferentially large and giant cells in the nucleus of Deiters (Walberg and Jansen, 1964), while those from the fastigial nucleus first and foremost contact small cells (Walberg, Pompeiano, Brodal, and Jansen, 1962). When studied in the electron microscope, the speed of degeneration and the appearance of degenerating boutons of the primary vestibular fibers can be seen to differ from those of the direct corticocerebellovestibular fibers (see Mugnaini, Walberg, and Brodal, 1967; and Mugnaini and Walberg, 1967, respectively). Electron microscopic studies of the lateral vestibular nuclei in the cat (Mugnaini, Walberg, and Hauglie-Hanssen, 1967) and rat (Sotelo and Palay, 1970) have shown the presence of several types of synapses. In the rat, "gap junctions," assumed to represent sites of electronic synapses (see Chap. 1), have been found.

The question of anatomical possibilities for interaction between the various units of the vestibular complex is of great functional interest. In studies with the Golgi method, only scattered cells which can be considered true internuncials have been found (Hauglie-Hanssen, 1968). Physiological studies have demonstrated the existence of a close collaboration between the vestibular nuclei of the two sides. Like the other fiber connections, the commissural connections are arranged in a specific manner (Ladpli and Brodal, 1968). To some extent this is true for the reticulovestibular projection (Hoddevik, Brodal, and Walberg, 1975). An illuminating example of specificity of connections is found in the different termination of fibers from various parts of the vestibulocerebellum in the vestibular nuclei (Angaut and Brodal, 1967; see Fig. 5-31).

Precise neurophysiological studies with exact identification of the sites recorded from have given much information on the functional organization of the vestibular nuclei, confirming and extending the anatomical findings. These studies likewise bear witness to an extremely intricate organization of this nuclear complex: For example, various functional types of neurons may be distinguished (see Duensing and Schaefer, 1958) which have in part different locations in the nuclei.

Among those responding to horizontal rotation, the type I neurons increase their discharge on ipsilateral angular acceleration and decrease it on contralateral acceleration, while type II neurons behave in the opposite way (Duensing and Schaefer, 1958). Whether these units are influenced by different kinds of receptor cells in the cristae is unknown. The type I neurons again are of two kinds, tonic and kinetic, according to their response characteristics to horizontal angular acceleration (Shimazu and Precht, 1965). It appears that kinetic neurons are excited monosynaptically, the tonic ones multisynaptically (Precht and Shimazu, 1965). Some of the type II neurons are only excited from the contralateral labyrinth (Shimazu and Precht, 1966); these are located in regions which receive commissural fibers (Ladpli and Brodal, 1968). The picture is further complicated by the fact that the same neurons in the nuclei may be influenced from different vestibular receptors (Duensing and Schaefer, 1959). This is compatible with what is known of the distribution of primary vestibular fibers in the nuclei. Physiological data on some of the other afferent and efferent connections of the vestibular nuclei have been referred to elsewhere in this text (see Chaps. 4 and 5). For extensive reviews of the physiology of the vestibular nuclei see Precht (1974b) and Wilson and Jones (1979).

Functional aspects. Nystagmus. The disturbances resulting from lesions of the vestibular nerve and the labyrinth have particularly attracted the interest of otologists. Increasingly accurate methods of investigation have drawn attention to some special features which are of diagnostic value. However, a satisfactory correlation of the phenomena observed clinically with the new data on the structural and functional organization of the vestibular receptors and their central connections is not yet possible. No attempt will be made here to treat this subject exhaustively, but some main features will be considered.[14]

It is clear from the preceding account that one may with some reservations distinguish between a ''static'' labyrinth, represented by the utricle (perhaps the saccule), and a ''kinetic'' labyrinth, consisting of the three semicircular ducts. The former signals the position of the head in space, the latter records angular movements of the head in space. Impulses from the utricle influence chiefly the distribution of muscular tone in various parts of the body, while the semicircular ducts appear to be concerned chiefly in adjusting the position of the eyes so that visual orientation is secured during movements.

Animal experiments which illustrate some features in the function of the *''tonic labyrinth''* have been referred to previously. In Chapter 4 mention was made of decerebrate rigidity (p. 265) as well as of the tonic labyrinthine reflexes, and the role of the cerebellum has been discussed in Chapter 5. The role played by the labyrinth in ocular movements will be considered again in connection with the oculomotor nerves.

It is important to be aware of the fact that under normal circumstances the vestibular impulses interact with impulses from other sources which reach the vestibular nuclei (see Gernandt and Gilman, 1960). The nuclei further collaborate with other parts of the brain. Analysis of this cooperation is, however, difficult. As examples illustrating such collaboration may be mentioned that anesthesia of

---

[14] An extensive review has been published by Kornhuber (1966). Here the reader will find a critical evaluation of clinical methods used in the examination of vestibular function, of their anatomicophysiological basis, and of their clinical value. Several clinical and theoretical aspects of the vestibular mechanisms have been treated recently in symposia: *Neurological Aspects of Auditory and Vestibular Disorders,* 1964; *International Vestibular Symposium,* 1964; *Second Symposium of the Role of the Vestibular Organs in Space Exploration,* 1966; *Myotatic, Kinesthetic and Vestibular Mechanisms,* 1967; *Handbook of Sensory Physiology,* Vol. 6, 1974; *The Vestibular System,* 1975.

the upper three cervical dorsal roots in monkeys results in severe disorientation, imbalance, and incoordination, resembling the symptoms appearing in labyrinthectomized animals (Cohen, 1961). This, in addition, demonstrates the great functional role of the tonic neck reflexes considered in Chapter 4. Since the vestibular nuclei have ample projections to the reticular formation (units in the reticular formation responding to vestibular stimulation which produce nystagmus have been recorded; see Duensing and Schaefer 1957a), it obviously cooperates with the vestibular nuclei when they influence the spinal cord as well as with the oculomotor nuclei.

Some further points of relevance for clinical neurology deserve mention. It is obvious from experimental studies that the results of destruction of the labyrinth or transection of the vestibular nerve are not identical to results of destruction of the vestibular nuclei. Nor are the results of stimulation of the nerve and the nuclei the same (see Brodal, Pompeiano, and Walberg, 1962, for some data). It appears that in man the vestibular apparatus is less important than in many animals. Thus, persons with no demonstrable labyrinthine function preserved manage quite well in daily life and may, for example, drive a car. In the absence of information from the labyrinth they rely on visual orientation and on proprioceptive information. It is only when vision is excluded and they have to walk on uneven ground or on a soft mattress that they lose their equilibrium (Martin, 1967).

In view of the complex organization of the vestibular mechanisms it is no wonder that *methods of studying vestibular function in man* are still imperfect, and that diagnostic conclusions often have to be tentative. Attempts have been made to study clinically the influence of the labyrinthine impulses on the spinal motor apparatus. The *stepping test* (see Peitersen, 1965) may give some information in cases of unilateral lesions, but the results must be evaluated with great caution. More objective methods of recording differences in the postural innervation on the two sides of the body have been suggested (Henriksson, Johansson, and Østlund, 1967), but their value is not yet clear. Far better and more precise methods are available for the study of the vestibular influences on the oculomotor apparatus. These are based on the examination of *nystagmus*. On account of its clinical importance it will be appropriate to consider nystagmus and the methods used for its study in some detail. The literature on nystagmus is enormous. While the nystagmus occurring in affections of the labyrinth and in part in affections of the vestibular nuclei is fairly well understood, much less is known of the basis for nystagmus in affections of several other parts of the brain.

The name nystagmus is used for a particular kind of conjugate movement of the eyes, which may be elicited in normal subjects on stimulation of the vestibular apparatus or by a particular kind of optic stimulation (optokinetic nystagmus, see section g). We are here concerned with *vestibular nystagmus*. This depends essentially on the stimulation of the sensory hairs of the cristae of the semicircular ducts, due to the deflection of the cupula which occurs on angular accelerations and decelerations. As referred to in a preceding section of this chapter, when the head is rotated, at the start the inertia of the endolymph in the semicircular ducts will deflect the cupula in one direction. If the rotation is continued in the same direction, the fluid will, after some time, obtain the same velocity of rotation as the canal, and irritation ceases. If then the movement stops short, the inertia of the endolymph will make it continue its flow, with a renewed stimulation of the cupula, but this time in an opposite direction. The correctness of this view, first set forth by Bárány on the basis of clinical studies, has been confirmed, as we have seen, in experimental studies. Since the semicircular canals are situated at right angles to each other, rotation around any axis will induce some change in the endolymph. The stimulation of the vestibular receptors gives rise to potentials transmitted to the vestibular nuclei, and from these, impulses are sent to various cell groups of the oculomotor nuclei, initiating conjugate movements of the eyes, as discussed above.

The nystagmus elicited from the labyrinth consists essentially in rhythmic conjugate movements of the eyes, the movement being slow in one direction, rapid in the other. In clinical terminology the nystagmus is named according to its fast component, but this is merely for convenience, the slow movement really being the active phase, the rapid being a reflex return to the starting position. In slighter degrees a pathological nystagmus appears only if the patient looks in the direction of the fast component; in the severest forms it is present also when he looks the other way. Among many *variants of pathological nystagmus, positional nystagmus* may be mentioned. This is present only in a certain position of the head, for example when the patient is lying on his right side. Most commonly a horizontal nystagmus is observed, but it may also be vertical, oblique, or rotary. The latter type most often occurs in combination with one of the other types.[15]

Nystagmus is always pathological if it occurs spontaneously, but it may be induced by different methods in normal individuals (induced nystagmus). This procedure is much used as a test of the function of the labyrinth. Most employed are the *postrotatory* and the *caloric nystagmus.*[16] When an individual is made to rotate around his axis, the movements set up in the semicircular canals are those described above. (Clinically usually 10 rotations within 20 sec are used.) During the rotation the impulses try to keep the axis of the eyes fixed at a given point. The eyes will lag behind in the rotatory movement, but when they finally lose sight of the object they perform a quick movement to the original position, another point in the environment is fixed and the process repeats itself. Thus during rotation the quick component will be in the direction of the rotation. When this is stopped, the inertia of the endolymph will produce a stimulation of opposite order to the ampullary crista (of the duct whose plane coincides with that of the rotation). Consequently the postrotatory nystagmus will be one with the quick component opposite to the direction of the rotation. The postrotatory nystagmus normally lasts for a definite time after rotation has stopped.

In the caloric tests movements in the endolymph are produced by irrigation of the external auditory meatus with hot or cold water. To avoid misinterpretations, tests with both hot and cold water should be employed. With the head tilted in different directions the various canals are brought into the most favorable position. For the horizontal canal, which responds most easily to caloric stimulation, this is

---

[15] Positional nystagmus is by many assumed to be due to damage to otolith organs or dysfunction in otolith–ocular reflex arcs (see Cohen, 1974). In this connection it is of interest that responses to tilting of the head have been recorded in the nodulus of the cat (Marini, Provini, and Rosina, 1975), and that positional nystagmus has been observed following lesions or stimulation of the nodulus (see Chap. 5).

[16] In addition the otogalvanic test may be employed. In this, however, the vestibular nuclei may react to the current, even if the peripheral vestibular apparatus is destroyed.

with the head inclined 60° backward, when the canal is in a vertical position. The nystagmus appears after a short latency and lasts ½ to 2 min, depending on conditions such as the temperature of the water and the duration of the stimulus. The effect of cold water is contrary to that of hot. The caloric test has the advantage over the rotatory because it permits the examination of one ear at a time. A reduction in vestibular function will betray itself in abnormal results in these tests. A nystagmus lasting a shorter time than normally, or shorter when elicited from one ear than from the other, thus indicates some reduced function of the corresponding labyrinth (Examination of nystagmus is usually performed with the patient carrying strong convex lenses to exclude the fixation reflex, see section g.)

A number of methods have been evolved in order to secure a more precise recording of nystagmus movements. Graphic registrations can be made in several ways (see Kornhuber, 1966). Most used of these methods appears to be *electronystagmography*, in which the retinocorneal potential and its changes on movements of the eyes is recorded by electrodes fixed to the skin around the eyes (see Aschan, Bergstedt, and Stahle, 1956; Aschan, 1964). This has the advantage that it may be done with the eyes closed (of interest in some conditions) and in darkness. The graphic registrations permit precise comparison of the results of examinations performed on different occasions. In present-day otoneurology a number of tests for nystagmus are employed, and many variants can be distinguished. To some extent particular patterns of change in the experimentally induced nystagmus may give diagnostic information.

The nystagmus movements elicited in the caloric and postrotatory tests in healthy individuals are, of course, not normal in a restricted sense, since they occur in response to stimuli which are not usually acting on the labyrinth. However, they disclose some principal features of the vestibular control of eye movements. The tests also reveal that the semicircular duct impulses are not only concerned with the adjustment of the gaze in movements of the head, but also influence the muscular status of the body as a whole. Thus the phenomena of *past pointing* which follows both tests mentioned above must be interpreted in this manner. When the individual is asked to touch a certain point with his eyes closed, the arm will be seen to deviate in the opposite direction to the nystagmus. Furthermore, he will have a tendency to fall in this direction.[17] Experimental studies in cats and monkeys (see Suzuki and Cohen, 1964) have shown that following stimulation of single ampullary nerves (by implanted electrodes) there occur ipsilateral forelimb extension and contralateral forelimb flexion as well as some other motor phenomena. This supports the interpretation of past pointing as being a result of stimulation of semicircular ducts.

Nystagmus may be produced in healthy persons in other ways than by employing the tests described above. Thus a *positional nystagmus* may be induced in humans (as well as in rabbits) by alcoholic intoxication (see Aschan, 1958; Kornhuber, 1966).[18] Certain drugs have been shown to influence the labyrinth (see, for

[17] It will be seen that the direction of the past pointing and tendency of falling (e.g., in Romberg's test) are in accord with the fact that the slow component of the nystagmus is the active phase. All phenomena tend to direct the patient to the same side.

[18] It appears that this is due, in part at least, to an action on the labyrinth (Aschan, Bergstedt, and Goldberg, 1964), but in addition there may be a depressing effect on the normally occurring inhibitory action of the nodulus on the vestibular nuclei (Aschan, Ekvall, and Grant, 1964) as discussed in Chapter 5.

example, Jongkees and Philipszoon, 1960), a problem of interest in the evaluation of motion sickness drugs. It is well known that experimental as well as spontaneous nystagmus is reduced when the individual is tired and increases when he is excited. Changes of this kind may occur rather rapidly. In examination of nystagmus (as in most clinical tests), it should further be recalled that psychological factors, for example distraction, may influence the results in tests of caloric nystagmus. It is well known that psychological factors may influence the vertigo of seasickness, and may precipitate attacks of Ménière's disease (for some references, see Gildenberg and Hassler, 1971). It seems a likely assumption that the connections between the cortical vestibular areas and the vestibular nuclei described above are in some way or other concerned in these phenomena. During sleep nystagmus ceases.

The labyrinths of the two sides collaborate normally, although there is evidence that each labyrinth is concerned primarily in the reactions involving adaptations in one direction. Thus extirpatation of a labyrinth in monkeys is followed by nystagmus, with the quick component toward the normal side and rotation of head and neck to the same side (see, for example, Dow, 1938b). Following labyrinthectomy in man, symptoms of the same kind but of somewhat different order have been observed (Cawthorne, Fitzgerald, and Hallpike, 1942). In man the effects are less, and less enduring, than in monkeys. The chimpanzee appears to hold an intermediate position.

Symptomatology.   Disturbances in vestibular function play a more important role in otology than in neurology, even if they are also frequently encountered in patients suffering from diseases of the nervous system. But their diagnostic value in the latter cases is rather limited. In this text the vestibular symptoms provoked by disease of the internal ear will not be considered, nor will the particular differences in symptomatology which in some cases allow the otologist to decide which part of the labyrinth is affected. It will suffice to point out a few characteristic features. In *acute vestibular affections* (labyrinthitis, bleedings, etc.) the vestibular symptoms, such as nystagmus, past pointing, tendency to fall, and vertigo, are frequently accompanied by symptoms pointing to an involvement of the autonomic system. Nausea and vomiting, lowering of blood pressure, tachycardia, and excessive perspiration may occur in the beginning. Following the examination by rotary or caloric tests the same symptoms may be provoked, but usually to a more moderate degree. There is good reason to believe that these autonomic effects are mediated by reflex connections from the vestibular nuclei to the visceral efferent nuclei of the lower cranial nerves, particularly the vagus.

The most frequently observed symptom of vestibular origin is without doubt *nystagmus*. When this is due to an affection of the labyrinth or the vestibular nerve it is called "peripheral"; when it it produced by lesions within the central nervous system it is referred to as "central." When examining nystagmus, however, some care has to be taken. In perfectly healthy persons, particularly if they are fatigued, slight *nystagmoid jerks* may result from gazing markedly in a lateral direction. Irregular movements of the eyes, of an oscillating type and lacking as a rule the characteristic rhythmic appearance, are common in persons with severely reduced vision, particularly in congenital amblyopia. This *ocular nystagmus* is regarded as being due to a defective power of fixation, and the "miner's nystagmus," which is usually also oscillating, is perhaps of the same nature. Thus, these types of nystagmus must not be ascribed to vestibular lesions. The cerebellar nystagmus will be referred to below.

The *peripheral nystagmus* is described as being usually of the combined horizontal-rotary type, with its quick component toward the nonaffected side. It is usually accompanied by vertigo, tendency to fall, past pointing toward the affected side (cf. above), and autonomic symptoms. It is transitory, a fact explained by assuming that the vestibular nuclei are able to compensate for the altered distribution of impulses which results when one labyrinth ceases to function. Apart from appearing in diseases of the labyrinth, these symptoms may occur in *affections of the vestibular nerve*. The processes affecting this will be the same as those mentioned as causes of cochlear nerve lesions (cf. Chap. 9). Thus in an *acoustic nerve tumor,* vertigo may occur, and examination with vestibular tests may reveal signs of reduced vestibular function.[19] Lesions of the vestibular and cochlear nerves may give rise to symptoms of Ménière's disease, where attacks of nausea, vertigo, nystagmus, and falling occur and the hearing is reduced following each attack, until the patient is finally completely deaf on the affected side. Abnormal vessels pressing on the VIIIth nerve may result in tinnitus and vertigo, but this is rare.

Occasionally, most often in the course of a febrile illness, patients complain of episodes of vertigo. These may be associated with objective signs of vestibular dysfunction, such as nystagmus. The condition has been assumed to be due to an affection of the vestibular nerve or the labyrinth and is usually referred to as "vestibular neuronitis," "neurolabyrinthitis," or "epidemic vertigo," since epidemic occurrence has been noted (see Pedersen, 1959). It appears that the pathogenesis is not the same in all cases (see Pedersen, 1959; Harrison, 1962).

"Central" nystagmus differs in certain respects from the peripheral which demonstrates all the symptoms known to be due to disturbed function of the vestibular apparatus. Vertigo is most commonly absent or of slight degree, and the nystagmus is said to be usually to the side of the lesion, but no definite rule can be seen to prevail. The nystagmus in affections of the central nervous system is attributed to lesions involving the vestibular nuclei or the secondary vestibular tracts. It is most commonly observed in *disseminated sclerosis,* in which the nystagmus in combination with a characteristic speech disturbance and intentional tremor forms the classical triad of Charcot. The nystagmus in disseminated sclerosis differs from the peripheral type also in its tendency to persist for years, although remissions may occur as with the other symptoms in this disease. On account of the multiplicity of foci in disseminated sclerosis it has not been possible to decide which structure has to be damaged in order for nystagmus to result. It appears probable that the symptom may follow lesions of several localizations, since nystagmus is a rather constant and frequently an early symptom. On the

---

[19] Hearing loss, due to affection of the cochlear nerve, may be an early symptom in these tumors (often referred to as cerebellopontine angle tumors), likewise, facial weakness (facial nerve), more seldom facial numbness (trigeminal nerve). According to Baloh, Konrad, Dirks, and Honrubia (1976) a battery of different audiometric and vestibulo-ocular tests may allow conclusions concerning tumor size, location, and type.

basis of clinical experience the medial longitudinal fasciculus has frequently been held to be responsible. However, experimental studies by Bender and Weinstein (1944) in monkeys did not give much support to this contention. Following bilateral small lesions to the medial part of the fascicle they observed nystagmus only on lateral gaze. It appears that lesions of the medial longitudinal fasciculus may give rise to so-called *internuclear ophthalmoplegia* (see Carpenter, 1964; Carpenter and Stominger, 1965; Harrington et al., 1966; Ross and deMyer, 1966). Here a paresis of the medial recti muscles of the eye is a prominent feature, associated with particular types of nystagmus.[20] Lesions of the vestibular nuclei may be followed by alterations in induced nystagmus, as shown for example by Shanzer and Bender (1959) in the monkey, and by Carmichael, Dix, and Hallpike (1965) in man. Nystagmus may also appear in *other lesions of the brainstem* (syringobulbia, vascular disorders, tumors) but usually in combination with symptoms of widespread damage. Finally it may occur in conditions of increased intracranial pressure, presumably on account of affection of the vestibular nuclei.

In cerebral lesions so-called *directional preponderance* of nystagmus is often seen. Attempts have been made to utilize these and other disturbances of nystagmus in the focal diagnosis of brain lesions. Fitzgerald and Hallpike (1942) found a directional preponderance of caloric nystagmus to the side of the lesion only when this was situated in the temporal lobe, but not in any of the other lobes. However, in spite of much work and much speculation, little is yet known of the mechanism of the observed changes. In view of the complex collaboration between certain parts of the cerebral cortex, the brainstem, and the vestibular nuclei in the control of ocular movements, this is not surprising. For some observations and interpretations the reader is referred to Carmichael, Dix, and Hallpike (1954, 1965).

Much discussion has been devoted to the question of the existence of a *cerebellar nystagmus* (see Chap. 5). The "nystagmus" seen in many cases of lesions to the cerebellum is not always a true nystagmus movement, but may rather be interpreted as ataxia of the eye muscles. However, in view of the relations between the vestibular apparatus and the eye muscles, the occurrence of a true cerebellar nystagmus can scarcely be denied. It should be remembered that cerebellar lesions frequently affect the brainstem directly (by pressure) or secondarily (by vascular disturbances), and a typical nystagmus might thus be caused by damage to the vestibular nuclei. In cerebellar lesions in man, positional nystagmus appears to be common, and positional nystagmus has been observed following cerebellar lesions in different animal species (see Cohen and Highstein, 1972, for references). As mentioned in Chapter 5, the nodulus is of particular interest in this connection.

[20] Most often there is paralysis of both medial rectus muscles on attempted conjugate lateral gaze, without other evidence of third-nerve paralysis. The syndrome appears to be almost invariably due to multiple sclerosis and is generally said to be caused by an affection of the medial longitudinal fasciculus. In view of the complexity of the central nervous mechanisms that control gaze (see section g in this chapter), this explanation is probably too simple. Many theories have been set forth to explain the disturbances of eye movements in internuclear ophthalmoplegia. For a recent study, see Pola and Robinson (1976).

In so-called *progressive supranuclear paresis* (see Hoyt and Daroff, 1971, for some data) there are, in addition to a supranuclear gaze paresis, usually bradykinesia, rigidity, and other symptoms that resemble those seen in parkinsonism. The disease is therefore often misdiagnosed, but clinical and pathological findings indicate that it is a particular disease entity.

It may be mentioned that in severe damage to the brainstem sometimes a condition resembling decerebrate rigidity in animals may be seen, presumably when the vestibular nuclei are released from impulses from higher levels. In such cases the tonic labyrinthine reflexes may be demonstrated in man. "Vestibular sensations" have been occasionally reported in man following cortical stimulation close to the primary acoustic area (Penfield and Jasper, 1954).

It will be seen from the account in this chapter that although the vestibular apparatus is of considerable importance in the normal functioning of the organism and is extremely precisely organized, the inferences of diagnostic value for the neurologist which can be drawn from lesions of these structures are limited.

(e) THE INTERMEDIOFACIAL NERVE

Anatomy. The VIIth cranial nerve, the facial or, better, the intermediofacial nerve, is a mixed nerve. Quantitatively the motor (special visceral efferent) fibers supplying the facial mimetic musculature form the most important part, the *facial nerve strictly speaking;* but in addition a minor portion made up of visceral and somatic afferent and general visceral efferent fibers joins it. This minor part is called the *nervus intermedius* and is commonly spoken of as the sensory root of the facial. It will be appropriate to consider the motor root first.

The *fibers of the motor root* spring from the *motor facial nucleus,* which belongs to the special visceral efferent nuclear group (Figs. 7-1 and 7-2). It is found just rostral to the nucleus ambiguus in the caudal part of the pons as a column 4 mm in length, in the lateral part of the reticular formation, dorsal to the superior olive and medial to the nucleus of the spinal tract of the Vth nerve. Its cells are of the usual motor type found in the ventral horns and the somatic efferent cranial nerve nuclei. The axons of these cells do not pass ventrally at once but form a loop (Figs. 7-17 and 7-18). The first part of this is directed dorsomedially, where the fibers approach the floor of the 4th venticle and then ascend immediately dorsal to the abducens nucleus. In a transverse section at this level of the pons this

FIG. 7-17   Diagram of the pons illustrating, *inter alia,* the intrapontine course of the facial nerve.

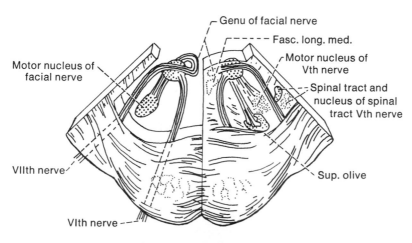

Genu of facial nerve

Fasc. long. med.

Motor nucleus of facial nerve

Motor nucleus of Vth nerve

Spinal tract and nucleus of spinal tract Vth nerve

VIIth nerve

Sup. olive

VIth nerve

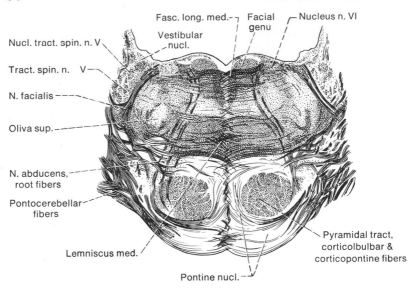

FIG. 7-18 Drawing of a section through the caudal part of the pons in man. Weigert's myelin sheath stain.

part of the nerve stands out clearly as a compact bundle (Fig. 7-18 to the right). At the rostral part of the abducens nucleus it bends over its dorsal surface (forming the genu of the facial nerve) and then passes directly ventrolaterally and somewhat caudally to its point of exit at the caudal border of the pons on the lateral aspect of the brain stem. Together with the intermedius and the VIIIth cranial nerve it enters the internal auditory meatus, passes through this to its bottom, covered by the meninges in a sheath common to these nerves. The facial and intermediate nerves then enter the facial canal, which in its first part is directed laterally, then at the geniculate ganglion performs a sharp bend in a dorsolateral direction, and after a short course in this direction ultimately proceeds in a caudal direction to leave the skull through the stylomastoid foramen. In the last part of the dorsolateral and the first part of the caudal course the facial canal lies close to the tympanic cavity, separated from it only by a thin bony lamella, a fact of practical importance. The fibers belonging to the intermedius are given off at various places to be considered below.

After leaving the stylomastoid foramen the facial nerve pierces the substance of the parotid gland and is here split into several branches which spread out in a fanlike manner to reach all the superficial, mimetic muscles of the head. These muscles are derived from the second branchial arch, the nerve of which is the facial. Apart from innervating the mimetic (e.g., orbicularis oris and oculi, buccinator, zygomaticus, frontalis, occipitalis, etc.) muscles, the facial nerve also gives off motor fibers to the stapedius muscle in the middle ear and extracranially to the stylohyoid, the posterior belly of the digastric muscle, and the platysma. In man, the facial nucleus is the largest among the motor cranial nerve nuclei and contains some 7,000 to 10,500 nerve cells (Tomasch, 1963).

In view of the finely differentiated movements performed by the facial musculature, determining the extremely variable expression of the face, it is no wonder that the facial nucleus presents a subdivision into several cellular groups, each of which is concerned in the innervation of certain facial muscles. This question has been studied rather extensively in animals by recording the distribution of retrograde changes in the different parts of the nucleus after section of individual branches of the nerve. The results of various authors using this approach agree fairly well (see Vraa-Jensen, 1942; Courville, 1966b). Some human cases have also been reported.[21] The facial nucleus is more differentiated in man than in other mammals, a fact in harmony with the importance of the facial muscles in speech and emotional facial expression. The oral muscles particularly show a considerable development both in regard to their number and the finer differentiation of the individual muscles. These are supplied by a lateral cell group which is particularly large in man. The frontalis, corrugator supercilii, and orbicularis oculi muscles are innervated by a clearly delimited part of the nucleus, the pars intermedia. The anatomical independence of this part is of interest in view of the clinical findings in supranuclear lesions of the facial nerve (cf. below).

It is in keeping with the almost infinite variations in facial expression in man that the motor units in the mimetic muscles are small—in the platysma 25 mucles fibers per nerve fiber (Feinstein et al., 1954). The total number of fibers in the human facial nerve is some 7,000, and most of them are relatively thick—7–10 μm (van Buskirk, 1945). Relatively little is know about *afferent innervation of the facial muscles*. One would assume, a priori, that the mimetic muscles are amply provided with proprioceptors. However, the number of muscle spindles in the mimetic muscles appears to be small. Voss (1956) found six in the human stylohyoid muscle. The perikarya of the γ fibers to the spindles have not yet been identified (see Courville, 1966b, for a discussion), nor is it known along which route the afferents from the muscle spindles and other possible proprioceptors reach the brainstem. According to Bruesch (1944), in the cat a small number of afferent fibers (140) is found in the branches of the facial nerve to the mimetic muscles. Some clinical studies (for example, Huber, 1930; Carmichael and Wollard, 1933) indicate that proprioceptive sensations are preserved evven if the trigeminal nerve is anesthetized. It has been suggested that the proprioceptive fibers from the mimetic muscles have their perikarya in the mesencephalic trigeminal nucleus (Kimmel, 1941; Pearson, 1945).

The motor facial nucleus is played upon by impulses from different sources. The facial musculature is involved in several *reflexes*, which are initiated by optic and acoustic stimuli, as well as by sensory impulses from the face and mouth. Fibers entering the nucleus or ending in its vicinity have been traced from the superior colliculi, an optic reflex center, from the superior olive (acoustic impulses) as well as from the sensory trigeminal nuclei and the nucleus of the solitary tract. Reflexes mediated by these fibers are, for example, blinking and eventually closing the eyes in response to strong illumination, closing of the eye on touching the cornea (corneal reflex, blink reflex),[22] contraction or relaxation of ·the stapedius

---

[21] A full review of earlier literature is found in the monograph by Vraa-Jensen (1942). Recording of antidromic potentials in the facial nucleus of the cat following stimulation of some of its branches (Kitai, Tanaka, Tsukahara, and Yu, 1972) has given results that largely agree with those of Courville (1966b).

[22] In electromyographic studies of the corneal reflex in man Magladery and Teasdall (1961) found its reflex latency to be relatively long; 40 to 64 msec on the side ipsilateral to the stimulation. This strongly suggests that the reflex arc is multisynaptic. In experimental animal studies, the blink reflex has been found to consist of two successive phasic contractions of the palpebral part of the orbicularis oculi muscle. The first response appears to be mediated by a three-neuron arc via the sensory trigeminal and the facial nucleus (latency about 7.5 msec). The second response has a latency of some

muscle in response to sounds of varying intensity (acoustic middle ear or stapedius reflex),[23] and chewing and sucking on introduction of food into the mouth. Several authors have traced afferents to the nucleus from the spinal cord (see Kerr, 1975).

> The projection is bilateral, arises mainly from cervical levels of the cord, and appears to supply mainly the medial part of the nucleus. According to the HRP study of Tanaka, Takeuchi, and Nakano (1978), most cells of origin are located in Rexed's laminae VII–VIII and V. The action is mainly monosynaptic and excitatory.

In addition, the motor facial nucleus is influenced by descending fibers from higher levels, including some from the "motor" cortex. In agreement with the anatomical finding that in the cat these fibers do not end in the facial nucleus (Kuypers, 1958a; Walberg, 1957), electrophysiological studies show that the cells of the facial nucleus are di- or polysynaptically excited on stimulation of the cerebral peduncle (Tanaka, 1976). In man, some of the corticobulbar fibers may end on cells of the nucleus (Kuypers, 1958c). The corticobulbar tract presumably mediates the voluntary movements of the face. From clinical findings it appears that these fibers are crossed, except some of those going to the nuclear division supplying the muscles above and around the eye (see above), since lesions of the corticobulbar tracts, e.g., in hemiplegia, result in paralysis of the lower part of the facial muscles only. Some support for this view has been obtained in an anatomical study of human brains by Kuypers (1958c). Among afferents to the facial nucleus from other "higher" regions, the presence of fibers from the red nucleus has been definitely established (Courville, 1966a; Edwards, 1972, in the cat; Miller and Strominger, 1973, in the monkey; Mizuno et al., 1973, in the rabbit). This projection is crossed and the fibers end only in certain subdivisions (not in those supplying the perioral muscles). In addition, there appears to be a bilateral projection from the mesencephalic reticular formation (Edwards, 1975). Other pathways, for example directly, or more likely indirectly, from the globus pallidus, may exist but are so far little known. Such connections have been assumed to be involved in emotional facial movements (see below).

The composition of the *intermediate nerve* is more complex than that of the motor facial. Since it is a small nerve and the contingents belonging to the different fiber types are relatively scanty and the peripheral course complicated, it is no wonder that it has been difficult to establish the details of its anatomy, some of which are of some practical relevance. Most of our present-day knowledge has been obtained from experimentation with animals. Some of the principal features

---

15 msec and must be multisynaptic. According to the recent study of Hiraoka and Shimamura (1977), this reflex arc probably passes via the reticular formation. In patients with cerebral lesions, some authors found the corneal (blink) reflex to be slightly increased, while others found it diminished (for example Oliver, 1952). According to Ross (1972), the absence of a contralateral corneal reflex may result from deep cerebral lesions, particularly of the parietal lobe. These reflex changes in man do not necessarily indicate that the corneal reflex has a cerebral reflex arc. They may also be explained by interference with pathways descending to relays in a brainstem reflex arc.

[23] Examinations of the acoustic middle ear reflex may be useful in clinical otology. The pathways for this reflex are not yet entirely clear (see, for example, Borg, 1973b). The location of the facial nucleus neurons innervating the stapedius muscle has been disputed. In an HRP study. Lyon (1978) found most of the perikarya to be present in the region between the rostral end of the facial nucleus and the caudal end of the lateral superior olivary nucleus. This may indicate that the stapedius motoneurons are activated by first- or second-order accoustic afferents (see Chap. 9).

will be mentioned below. Concerning the general visceral efferent fibers which supply the lacrimal, the submaxillary and the sublingual glands and the glands of the nasal and part of the oral cavity, and which convey preganglionic parasympathetic (secretory) fibers to them, the reader is referred to the chapter on the autonomic system. Here our concern is with the *sensory components* of the nervus intermedius.

The sensory fibers of the intermediate nerve have their perikarya in the small *geniculate ganglion,* referred to above. The cells are of the usual pseudounipolar type. A considerable number of the sensory fibers are special visceral afferent, conveying taste inpulses from the anterior two-thirds of the tongue. In addition the nerve also contains other sensory fibers.[24] The branches from the intermediate nerve are firstly the *greater superficial petrosal nerve* passing to the sphenopalatine ganglion, then the *chorda tympani* leaving the nerve in the tympanic cavity and anastomosing with the lingual nerve to reach the tongue. But not all afferent fibers are distributed to these two branches. Some, as referred to above, follow the facial nerve in its further course. In cats, according to Foley and DuBois (1943), about 33 to 34% of the sensory fibers, which are mainly small and myelinated, pass with the great superficial petrosal nerve; 45 to 55% with the chorda tympani; and 12 to 15% proceed further with the facial nerve. Whereas at least a considerable number of the sensory fibers in the two above-mentioned branches are gustatory (to be treated in more detail below), those passing with the facial nerve proper cannot be so, but must be general sensory afferents. From comparative anatomical data (see Kappers, Huber, and Crosby, 1936) as well as from experimental findings (Foley and DuBois, 1943; Bruesch, 1944) it is apparent that most of these fibers are distributed to the external ear, partaking in the sensory cutaneous innervation of the concha of the auricle and sometimes an area behind the ear. (These regions are also supplied by the auricular ramus of the vagus nerve.) This has been confirmed in physiological studies (Iwata, Kitai, and Olson, 1972). In the cat, afferent impulses appear to pass in the facial nerve following stimulation of its branches (see Kitai, Tanaka, Tsukuhara, and Yu, 1972).

A *cutaneous sensory branch of the facial nerve* is present also in man (Larsell and Fenton, 1928; Pearson, 1945). As early as 1907 Ramsay Hunt had postulated the existence of such fibers based on a study of the symptomatology of herpes zoster oticus and a certain type of neuralgia which he called geniculate neuralgia (see below). These fibers must be regarded as somatic sensory and are probably among those found to end in the trigeminal nuclei (Woodburne, 1936; Kimmel, 1941; Torvik, 1956b; Kunc and Maršala, 1962; Kerr, 1962; Rhoton, 1968).

Facial reflexes (contractions of facial muscles in response to sensory stimuli of the face) are largely based on an afferent input from branches of the trigeminal nerve to the spinal trigeminal nucleus (see for example Iwata, Kitai, and Olson, 1972). It appears, however, that afferents in the peripheral facial nerve (having their perikarya in the geniculate ganglion) may be involved as well, according to the studies in man by Willer and Lamour (1977).

---

[24] According to the physiological study of Boudreau et al. (1971), different neuronal populations can be distinguished in the geniculate ganglion in the cat. Some units discharge in response to movement of hairs in the external ear; others respond to mechanical or chemical stimulation of the tongue, see below.

As in the facial (intermediate) nerve, some fibers in the glossopharyngeal and vagus nerves can be traced to the spinal trigeminal nucleus. All three nerves, contrary to what is often stated, thus contain a *somatic afferent fiber component*. In the case of the vagus these fibers (as well as some glossopharyngeal fibers) are concerned in the innervation of the areas, already mentioned, in the external ear, whereas most of the fibers of the IXth nerve must be those which supply the posterior third of the tongue, the palatal arches, and the pharyngeal wall. This has been learned from cases of medullary tractotomy (operative sectioning of the spinal tract of the trigeminal nerve, cf. p. 519), since this surgical procedure is followed by analgesia in the areas mentioned as well as in the trigeminal field proper (Brodal, 1947a; Falconer, 1949, and others). The practical consequences of this distribution of the somatic afferent fibers in the facial intermediate nerve will be mentioned below.

The pathways for taste.   The *sense of taste* from the anterior two-thirds of the tongue is the only sensory quality conveyed through the intermediate nerve which remains for our consideration. Because of the practical importance of taste sensations for the topical diagnosis of a lesion of the facial nerve, the question of the gustatory pathways will be considered in some detail (cf. Fig. 7-19).

Diversity of opinion has prevailed with regard to the part played by the facial nerve. Animal experiments have their obvious shortcomings in studies of taste. In human studies it is worth remembering that the examination of taste, like the examination of other sensory functions, is dependent on the subjective interpretation, and mental power of the patient and his willingness to cooperate, and that several sources of error are involved in the technical side of the examination. Lastly, the number of taste buds on the anterior two-thirds of the tongue is not very large, and their number decreases progressively with age.[25] Consequently, the acuity of taste shows considerable individual variation in normal persons. In man, the most important nerve concerned in the mediation of taste is the glossopharyngeal, which supplies the vallate papillae.

Although most investigators agree that operative interruption of the trigeminal nerve does not affect taste (e.g., Lewis and Dandy, 1930; Schwartz and Weddell, 1938), some reports to the contrary are found, but in many cases the reduction of gustatory acuity has been found to be slight or transient. The more enduring reduction, or loss in some instances, is reasonably explained as Rowbotham (1939) expressed it, "The taste impulses lack the background of common sensation." A third possibility is that the geniculate ganglion or the greater superficial petrosal nerve may be damaged in the operation. The nerve is mentioned in this connection because in some instances it appears to be concerned in transmitting taste sensation.

In most persons interference with the chorda tympani, for example in mastoidectomy, has been observed to lead to loss of taste in the anterior two-thirds of the tongue, in accordance with the current conception of the course of the taste fibers. However, Schwartz and Weddell (1938) have reported cases in which the chorda tympani was destroyed without ensuing gustatory disturbances. This indicates that in some persons taste fibers from the anterior two-thirds of the tongue

[25] For some data on the morphology and physiology of the taste buds and gustatory receptors see *Taste and Smell in Vertebrates,* 1970.

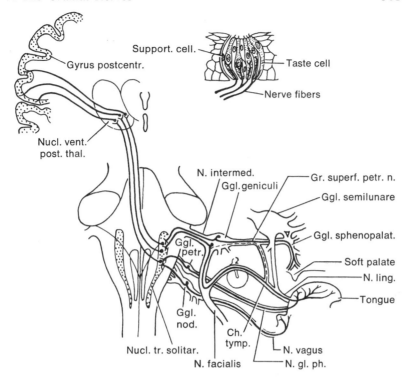

FIG. 7-19 Diagram of the course of the gustatory fibers. Heavy lines indicate the usual paths, stippling the alternative route taken by fibers from the anterior two-thirds of the tongue in certain cases. Above to the right a diagram of a taste bud.

take another course. This pathway appears to be through the greater superficial petrosal nerve, since the same authors have observed some cases in which this nerve was damaged but the chorda remained intact, with resulting loss of taste in the anterior two-thirds of the tongue. The fibers are believed to follow the chorda for a short distance and then to leave it through a small anastomotic branch to the otic ganglion. From here they are assumed to pass by another anastomosis to the greater superficial petrosal nerve and then centralward as usual. This variation is apt to cause confusion in diagnostic reasoning in some cases of facial nerve affections. In Fig. 7-19 the usual course of the gustatory fibers of the intermedius is indicated by heavy lines, the alternative by dots.

It is generally agreed that the glossopharyngeal nerve is concerned in the transmission of taste impulses from the posterior third of the tongue. Several cases of intracranial section of this nerve have given conclusive evidence of this (e.g., Fay; 1927; Lewis and Dandy, 1930; Reichert, 1934). Likewise, taste sensations from the soft palate and palatal arches, where some taste buds are found, presumably pass by the glossopharyngeal, but others may pass via the sphenopalatine ganglion (cf. Fig. 7-19). It is furthermore generally assumed, mainly on an anatomical basis, that the vagus carries some special visceral afferent taste fibers from the extreme dorsal part of the tongue and the superior surface of the epiglottis. These regions are, of course, very difficult to examine with regard to taste, and their taste buds disappear in early infancy.

All taste fibers, those of the intermediate as well as the glossopharyngeal and the vagus nerve, are *distributed to the nucleus of the solitary tract* and, as anatomical evidence shows, its rostral part (cf. section c on the glossopharyngeal-vagus). The central course of the secondary gustatory fibers has been difficult to identify (see Gerebtzoff, 1939, for an early study). It has been known for some time that gustatory impulses are transmitted to the thalamus and ultimately reach the cerebral cortex. The afferents to the thalamus, however, do not appear to arise from cells in the nucleus of the solitary tract (see section c of this chapter), but from a region referred to as the *pontine taste area*. This area consists of cells situated ventral and dorsal to and within the brachium conjunctivum, and the area appears to bear some relation to the parabrachial nucleus. Cells in this region respond to gustatory stimuli (Norgren and Pfaffman, 1975). According to Norgren and Leonard (1973; see also Ricardo and Koh, 1978), this region receives fibers from the nucleus of the solitary tract and projects (bilaterally) to the thalamus. Other fibers have been traced to the hypothalamus (see Norgren and Leonard, 1973, Ricardo and Koh, 1978) and must be assumed to influence autonomic functions, particularly those related to feeding. Responses (excitatory and inhibitory) have been recorded in the hypothalamus of the rat following application of gustatory stimuli (Norgren, 1970). (Other projections appear to pass to the ventral telencephalon; see Norgren, 1974; and also Ricardo and Koh, 1978.)

Previously, *the cortical area for taste* was generally assumed to be closely related to the cortical olfactory regions, but recent research has made it clear that the cortical taste area is located in close relation to the focus for somatic sensibility of the tongue.

It is customary to distinguish four elementary taste qualities: salt, sour, bitter, and sweet. Tests of gustatory sensibility are made by applying substances of different kinds to small spots on the tongue. Usually solutions of sodium chloride, citric acid, quinine hydrochloride, or cane sugar are used as test substances. The subjective sensations experienced bear no clear relation to the chemical composition of the test substance. In experimental animals, recordings of action potentials in fibers of the nerves conveying taste impulses can be made. Likewise, potentials may be picked up in the central nervous system by the insertion of electrodes.

From studies of this kind it has been found, in agreement with anatomical data on the central distribution of fibers in the nerves conveying sensations of taste, that the rostral part of the nucleus of the solitary tract is the first relay in the taste pathways (Makous et al., 1963; and others). Most units found respond to the application of taste stimuli of more than one of the elementary qualities, and often they respond to other kinds of stimulation as well. Following gustatory stimuli, action potentials occur in the thalamus, more precisely in the medialmost part of the VPM. According to Blomquist, Benjamin, and Emmers (1962), Emmers, Benjamin, and Blomquist (1962), and Norgren and Wolf (1975), in the monkey and in the rat, the thalamic "taste region" is largely separate from and more medially situated than the region activated by other stimuli of the tongue. Destruction involving this part of the thalamus is followed by impairment of taste (Patton, Ruch, and Walker, 1944; and others). From the VPM, thalamocortical fibers transmit the impulses to the cerebral cortex, where action potentials can likewise be recorded on gustatory stimulation. Most authors found the "cortical taste region" to be situated below and a little in front of the "face area" of the sensorimotor cortex. In contrast to Cohen et al. (1957) in the cat, Benjamin and his co-workers found that in the rat this area is separate from the nongustatory lingual area, and shows little overlap. Norgren and Wolf (1975) came to the same conclusion in a combined physiological-anatomical study in the rat (see also Ganchrow and Erikson, 1972, ample references). Furthermore, the taste areas for the intermediate and glossopharyngeal nerve are separate in the rat, as they are in the squirrel monkey (see Benjamin, 1962). Ablation of this area in the rat abolishes taste discrimination (Benjamin and Akert, 1959), as does destruction of the appropriate thalamic region. (However, in monkeys, corresponding cortical ablations did not have this result.) There are obvious but still poorly understood species differences in the central mechanisms of taste.[26]

Experiences in man are in agreement with the results of anatomical studies. Thus cases have been described of contralateral impairment of taste following tumors involving the VPM of the thalamus (Adler, 1934). On stimulation of the anterior portion of the second sensorimotor cortex where it approaches the insula, subjective sensations of taste have been recorded (see Penfield and Jasper, 1954), and occasionally a gustatory aura heralds an epileptic convulsion starting from this part of the cortex (Shenkin and Lewey, 1944; Penfield and Kristiansen, 1951). Finally, lesions of this part of the cortex have been shown to be accompanied by impairment of taste in the contralateral half of the tongue (Bornstein, 1940).

Symptomatology. *Lesions of the central pathways of the somatic and visceral afferent components of the intermediate nerve* are not frequently recognized, partly because the ensuing symptoms are usually insignificant and are obscured by more prominent signs. However, from what has been stated above, it is clear that disturbances of taste may be observed when sought in certain cortical and more deeply situated lesions of the brain, and such cases have been described. Loss of taste (ageusia) may occasionally be seen following head trauma (Sumner, 1967; Schechter and Henkin, 1974). Lesions of the central parts of the *efferent* facial mechanism, however, are not uncommon, and lesions of the peripheral intermediofacial nerve belong to the most common of the affections which the neurologist encounters. It is readily understood that according to the site of the lesion, the symptoms will be somewhat different and will depend on whether the intermediate nerve is also affected. It is appropriate to treat the pure motor facial paralysis first, and then supplement the account of this with a description of the features that accompany it when the intermediate nerve is also involved, and finally discuss some symptoms ascribable to an affection of the intermediate nerve alone.

The most frequent type of *peripheral facial nerve palsy* is that usually spoken of as being "rheumatic," setting in acutely sometimes in the course of an infection of the upper air passages or in connection with exposure to drafts or cold. Little is known concerning the true etiology of this condition, the so-called *Bell's palsy*. A viral etiology has been suggested (see Leibowitz, 1969). The assumption that the nerve is compressed due to edema in the facial canal has prompted operative intervention with opening of the canal. This has been reported to give good results in many cases, but the results are difficult to evaluate on account of the frequent occurrence of spontaneous recovery. There appears, however, to be ample room for the nerve in the canal, since about 50% of the space in the canal is occupied by loose connective tissue and vessels (Sunderland and Cossar, 1953). According to Langworth and Taverner (1963), the prognosis of a facial palsy can be reliably established in the first few days by studying the conduction time in the

---

[26] In recent years considerable attention has been devoted to this subject. The interested reader will find much information in the monographs *Olfaction and Taste* (Vol. V, 1975, and earlier), which carry reports from international symposia. Zotterman (1975) has given a short review of some aspects of the physiology of taste in man.

nerve branches. This may thus aid in a decision on whether operative decompression is likely to be of benefit.[27]

If the paralysis, usually unilateral, *is complete,* which is the rule, all movements of the mimetic facial muscles are abolished, and the face is asymmetrical at rest as well as in motion. The patient cannot frown or raise his eyebrow on the affected side, he cannot close his eye on account of the paralysis of the orbicularis oculi, and the palpebral fissure is widened. On attempting to close his eye the eyeball will be seen to deviate upward and slightly outward (Bell's phenomenon), due to a relaxation of the inferior rectus and contraction of the superior rectus. This also occurs normally when the eyes are closed but becomes more distinctly visible when the closure of the eye is impossible. The blinking reflex (corneal reflex) is abolished, and this is of practical relevance since it necessarily leads to a drying of the cornea with resulting ulceration. Therefore one of the most important therapeutic measures of Bell's palsy is the protection of the eye from drying. The most efficient therapy is to perform a tarsorrhaphy. A patient with a peripheral facial palsy cannot properly close his mouth, and the affected oral angle is usually lower than the other, sometimes with dribbling of saliva. On speaking, laughing, weeping, etc., the affected side does not partake in the movements. As the paralysis is of the peripheral type, the flaccidity of the mimetic muscles will betray itself in a more-or-less complete loss of the habitual wrinkles, which are due to the insertion of several of the facial muscles in the skin, and the affected side of the face will on this account appear smoother than the other. The paralysis and lack of tone of the buccinator muscle will frequently reveal itself on eating as a tendency to accumulation of food within the cheek. On account of the paralysis of the buccinator it is, for example, difficult for the patient to blow out a candle. After some time the muscles will fall the victim of atrophy. Fasciculation may be seen and can be recorded electromyographically. The diagnosis of a complete peripheral facial palsy is easy. If the palsy is bilateral, difficulties arise in articulation and partly in eating.

In *incomplete lesions* the distribution of the paresis can be variable and the testing of power is important (power of orbicularis oris on closing the mouth, of the frontalis on raising the eyebrow, the power of keeping the eyes closed when the eyelids are forcibly drawn apart, etc.) as in slight affections movements on inspection may appear to be normal, and a slight asymmetry of the face is a normal occurrence. Furthermore, the determination of the location of the paretic muscles is important in order to distinguish a peripheral from a central facial palsy. In the latter the upper part of the face is usually not affected. Electromyography may be of value in diagnosis and prognosis (Weddell, Feinstein, and Pattle, 1944; Taverner, 1955; see also Langworth and Taverner, 1963; Buchthal, 1965). A depression of the corneal reflex due to interference with the efferent fibers in the facial nerve may signal the disease. A facial palsy usually progresses slowly, often preceded by signs of damage to the VIIIth nerve.

A feature peculiar to peripheral facial paresis is the frequent occurrence of *contractures of the muscles* when the paralysis is complete and of somewhat longer duration. On inspection, the contracted side appears at first to be normal;

---

[27] For a review of clinical aspects of facial palsy, see Taverner (1969).

only when the examiner tests for movements is the actual state revealed. Electromyographic studies suggest that the contracture is due to abnormal activity of some residual motor units (Taverner, 1955).

Less frequently than these "rheumatic" palsies, a peripheral facial palsy is seen in *affections of the middle ear*. In uncomplicated otitis and in matoiditis the nerve may be affected as it passes in the facial canal in the wall of the tympanic cavity. The nerve may be affected with edema and circulatory disturbances, or it may be directly involved in the inflammation when the bone covering it is encroached upon by the disease process in the middle ear. The prognosis will vary according to the pathological alterations of the nerve. A facial palsy in the course of an acute otitis media usually recovers spontaneously. If the palsy occurs during a chronic middle ear infection, operation is often necessary.

Among other causes of a peripheral facial palsy may be mentioned *traumatic injury* to the nerve in fractures of the base of the skull involving the temporal bone. In *tumors of the parotid gland* an eventual paralysis will usually not be complete, at least not in the beginning. *Operations* aimed at removing or incising submaxillary lymph nodes may cause a partial paralysis affecting the muscles of the lower lip. Another cause of partial facial palsy, although not frequently encountered now, is *leprosy*. In both the maculoanesthetic and the tuberous forms of leprosy, partial affections of the branches of the facial nerve are a common and early symptom (Monrad-Krohn, 1927).

In its *intracranial course* the facial nerve may also be injured, as in diseases of the meninges, e.g., syphilis. Another cause is the affection of the nerve by a *neurinoma of the VIIIth nerve*. Here a facial palsy is one of the symptoms which frequently signals the disease. Usually the paresis progresses slowly, accompanied or preceded by signs of damage to the VIIIth nerve (cf. the latter). Facial palsy, often bilateral, may be the first manifestation of *polyneuritis*.

The *motor nucleus of the facial nerve* and its fibers within the pons may be affected by pontile hemorrhages or other *vascular changes*, in which case there will be motor and frequently also sensory symptoms (resulting in, for example, a "crossed hemiplegia"). The facial nucleus is not infrequently involved in *poliomyelitis*. In some cases the facial nucleus alone is affected, or, if other structures are involved, only symptoms due to the damage of this nucleus are manifest as a unilateral peripheral facial palsy (Wallgren, 1929; Schumacher, 1940). In other instances the facial nucleus is affected together with other cranial nerve nuclei (bulbopontile form of poliomyelitis). The frequent affection of the facial nucleus in *chronic bulbar palsy* has already been referred to on p. 470. (Facial palsy is also common in pseudobulbar palsy, but then of the central type.) *Congenital facial palsies*, usually bilateral, are occasionally found, and are due to an aplasia or hypoplasia of the nucleus (see, for example, Henderson, 1939). Similar aplasias and congenital paralyses may occur concomitantly in other motor cranial nerves, particularly the abducens.

Before leaving the peripheral motor facial nerve, reference should be made to the frequent occurrence of *involuntary facial movements*. These may appear as a facial hemispasm, consisting of irregularly occurring contractions of the muscles on one side, usually increased by emotional stimuli but absent during sleep. The etiology of this phenomenon is obscure. It may occur following a facial palsy, but usually no preceding disease can be ascertained. It ought to be remembered, however, that involuntary movements of this type without accompanying paresis may be an early sign of an acoustic nerve tumor or a cerebellar glioma (personal observation).

A supranuclear paralysis of the facial nerve is a frequent feature of *capsular hemiplegia*. The damage at the internal capsule also interrupts the corticobulbar fibers destined for the facial nucleus. As repeatedly mentioned, this central paresis is characterized by being limited to the lower part of the face. Thus a hemiplegic patient can close his eyes and frown, but is somewhat hampered in speech. This central paresis is not followed by marked atrophy of the facial muscles, and is situated on the side opposite the lesion, on the same side as the other paralyses which may be present. This ensues from the crossing of the corticobulbar and corticospinal fibers. A similar paresis or paralysis of the facial muscles may occur in *lesions of the face region of the precentral cortex*. In these instances it is not so frequently accompanied by total hemiplegia, as is easily explained by the larger space occupied by the cortical motor region compared with the narrow zone taken up by the corticobulbar and corticospinal fibers in the internal capsule.

Reference was made above to the *emotional involuntary innervation of the facial muscles*. It is well known that in paralysis agitans and postencephalitic parkinsonism the patient is able to show his teeth, whistle, frown, etc., i.e., there is no facial palsy, but his emotions are not reflected in his mimics, and he usually has a stiff, masklike facial expression (poker face). On the other hand, a hemiplegic patient, suffering from a damage of the corticobulbar fibers and having a complete central paresis of the lower part of the face, may be able to smile a spontaneous smile, for instance when he enjoys a joke. Monrad-Krohn (1939) has drawn attention to the fact that this spontaneous smile on the paralyzed side is frequently exaggerated, occurring earlier and exceeding that which appears on the nonparalyzed side. This *dissociation between the voluntary and emotional facial innervation* can so far not be adequately explained. It appears a likely assumption, however, that the fibers mediating the impulses to the emotional innervation do not descend in the internal capsule. Possibly hypothalamic and pallidal efferents are involved when the patient smiles emotionally, while a "social" smile requires that the corticobulbar projections be intact. Monrad-Krohn is inclined to interpret these phenomena as due to lack of inhibition by the cortical impulses, which in normal individuals suppress some of the involuntary emotional impulses. In the same manner one can explain the exaggerated emotional expressions frequently met with in the form of weeping or laughing following trifling, normally inadequate stimuli, in cases of capsular hemiplegias. The same explanation may be applied to the attacks of spasmodic laughter or weeping so frequent in pseudobulbar palsy. Here on both sides the corticobulbar fibers to the efferent cranial nerve nuclei are as a rule affected, and presumably the lack of cortical control is still more complete than in the hemiplegias.[28]

Turning now to *symptoms referable to the peripheral part of the intermediate nerve,* it will be clear from the anatomical description that they may vary. Apart from deficiency symptoms, irritative symptoms also occur, and are discussed below.

A lesion of the facial nerve anywhere from its origin at the lower border of

---

[28] Some aspects of emotional expression have been discussed by Brown (1967).

the pons to the tympanic cavity will be apt to affect the intermedius also. The affection of its general visceral efferent, parasympathetic fibers will usually not be recognized, unless particularly sought for by special methods used to detect some diminution of secretion of saliva and tears. Interference with the gustatory fibers may also escape the patient's attention, but it is more easily ascertained, and is of a certain importance. As a general rule it may be said that a *lesion of the facial nerve anywhere between the pons and the departure of the chorda tympani will be followed by loss of the sense of taste in the anterior two-thirds of the homolateral half of the tongue.*[29] A lesion situated distal to the departure of the chorda tympani will not have this consequence. However, this statement requires the qualification, evident from the anatomical data, that in some instances the nerve may be damaged between the geniculate ganglion and the chorda tympani, and yet no gustatory loss is present. This will appear in those cases where the taste fibers pass by way of the greater superficial petrosal nerve. But it is safe to state that, when loss of taste in the anterior two-thirds of the tongue accompanies a facial palsy, the lesion must be situated central to the departure of the greater superficial petrosal nerve and probably central to the chorda tympani. Further clues cannot be obtained from an analysis of taste.[30] Another finding may be of some use, namely the occurrence of *hyperacusis,* when the fibers to the stapedius muscle are involved. This phenomenon then, when present, points to the location of the lesion as being central to the pyramidal eminence in the tympanic cavity. Lesions situated proximal to the tympanic cavity will frequently be accompanied by symptoms due to involvement of the VIIIth nerve (tinnitus, reduced hearing, or vestibular disturbances).

In later stages of a peripheral facial palsy occasionally the *syndrome of crocodile tears* occurs. On eating or following other stimuli which normally produce secretion of saliva, lacrimation appears in the eye on the affected side. This is seen in lesions of the facial nerve central to the geniculate ganglion and is explained as being due to erroneous regeneration of nerve fibers (see Taverner, 1955). Fibers of the intermediate nerve which had originally supplied salivatory glands on regeneration enter Schwann sheaths which belong to degenerated fibers having innervated the lacrimal gland. Another possibility is the establishment of "artificial synapses," by means of ephaptic transmission between neighboring nerve fibers (see Sadjadpour, 1975).

Deficiency symptoms due to affection of the general sensory fibers, the cutaneous and alleged proprioceptive fibers in the intermediate nerve, will scarcely be detected clinically. However, irritative phenomena referable to this part of the nerve deserve brief mention. The occurrence of the so-called *geniculate neuralgia* of Ramsay Hunt has already been alluded to. This admittedly is not very frequent, although there is reason to believe that it is often not correctly diagnosed. It may appear following a herpes zoster oticus which, as suggested already by Ramsay Hunt (1907, 1937), is due to an inflammation of the geniculate ganglion, the blis-

[29] This ought to be emphasized since several neurological textbooks still contain the statement that only lesions between the geniculate ganglion and the chorda tympani will cause gustatory loss. Since it has been definitely proved by section of the intermediofacial nerve in man (see, for example, Lewis and Dandy, 1930) that the gustatory fibers enter the brain with the facial, this can obviously not be true.

[30] Of course an affection of the taste fibers to the anterior two-thirds of the tongue may be due to a lesion of the chorda peripheral to the tympanic cavity, or where it has joined the lingual nerve. In the latter case the concomitant anesthesia of the same regions of the tongue will settle the matter.

ters occurring in the cutaneous field of the nerve, the concha, and eventually in a limited area behind the ear.[31] The herpetic eruption may be followed in some days by a facial palsy, or this may appear simultaneously. In other instances the neuralgia sets in without any preceding herpetic eruption. Attacks of pain, frequently very severe and localized to the concha and also felt deep in the ear (involvement probably of a tympanic branch of the intermediate), occur in fits. They may be provoked on touching the external ear or sometimes on swallowing or yawning, and frequently irradiate forward deep in the face. This irradiation is explained as being due to the involvement of fibers of the greater superficial petrosal nerve, passing, *inter alia,* to the posterior parts of the nasal and oral cavities. (The irradiating pain marks a transition to the so-called Sluder's neuralgia of the sphenopalatine ganglion.) Neuralgias of the geniculate type have been observed to disappear in certain cases when the intermediate nerve has been cut, and they may also be attacked by tractotomy as referred to above.

> Probably due to an affection of the sensory fibers of the intermediate nerve are the pains in the ear region, which not infrequently accompany an ordinary "rheumatic" *peripheral facial palsy,* especially in its beginning. Likewise it may happen that patients affected with a *neurinoma of the VIIIth nerve* tell a story of their sufferings starting with a pain situated deep in the ear. This symptom has been observed to be the initial single symptom in several cases, and the possibility of a tumor is worth remembering when no otitis or other symptoms of disease of the ear can be ascertained in cases of otalgia. It appears probable that the pain is due to affection of the intermediate nerve, lying closer to the acoustic than the facial.

(f) THE TRIGEMINAL NERVE

The nerve and its nuclei.   The Vth cranial nerve, the trigeminal, is a mixed nerve supplying mainly the masticatory muscles with motor fibers and the skin of the face, the conjunctiva, and a large part of the mucous membrane lining the nasal and oral cavities with sensory fibers. Apart from its first division, the ophthalmic nerve, it is the nerve of the 1st branchial arch, as the facial is that of the 2nd, the glossopharyngeal that of the 3rd, and the vagus the nerve of the 4th and following arches.

The *sensory part of the nerve* is far larger than the motor. At the exit of the nerve on the lateral aspect of the pons (see Fig. 7-6) the two parts can clearly be distinguished as a larger, lateral, or sensory root, *portio major,* and a smaller, medial, motor root, *portio minor.* The latter joins the third division of the sensory root. The sensory fibers will be considered first.

The pseudounipolar perikarya of the fibers of the sensory trigeminal root are found in the large *semilunar or Gasserian ganglion,* situated on the cerebral surface of the petrous bone near its apex in the middle cerebral fossa. Peripherally this gives off three branches, the three principal divisions of the nerve, *ophthalmic, maxillary,* and *mandibular* nerves, leaving the cranial cavity through the superior orbital fissure, the foramen rotundum, and the foramen ovale, respectively.[32] The further peripheral course will not be considered in detail. The

---

[31] In addition some vesicles may appear on the soft palate, explained by an affection of sensory intermedius fibers passing via the sphenopalatine ganglion.

[32] Allen's (1924) original observation that there is a topical pattern in the semilunar ganglion has been confirmed both physiologically and anatomically, recently with the HRP method (Arvidsson, 1975; see this article for references). The ganglion cells belonging to the mandibular, maxillary, and ophthalmic nerve are orderly situated from posterolateral to anteromedial (in accordance with the sites of entry of the three nerves in the ganglion).

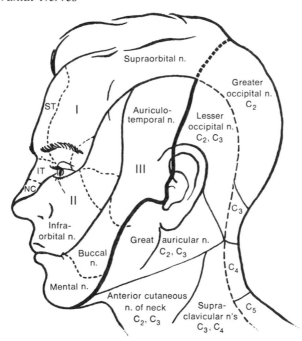

Supraorbital n.

Greater
occipital n.
C_2

ST

I

Auriculo-
temporal n.

Lesser
occipital n.
C_2, C_3

IT

III

NC

II

C_3

Infra-
orbital n.

Great auricular n.
C_2, C_3

Buccal
n.

C_4

Mental n.

Anterior cutaneous
n. of neck
C_2, C_3

Supra-
clavicular n's
C_3, C_4

C_5

FIG. 7-20   The cutaneous distribution of the trigeminal nerve and its branches, with the adjacent cervical nerves. Modified from Haymaker and Woodhall (1945).

cutaneous distribution of the different branches will be evident from Fig. 7-20. These sensory trigeminal fibers are *somatic afferent*, and convey impulses of exteroceptive cutaneous sensations. The overlap between the areas supplied by the three divisions of the trigeminal nerve is generally said to be slight (in contrast to the overlapping of the dermatomes), likewise the borders of the cutaneous fields of the trigeminal to those of the spinal nerves. Experimental studies in the monkey by Denny-Brown and Yanagisawa (1973) cast doubt on this view.

The cutaneous distribution of the trigeminal nerve has been ascertained by dissections of the nerve and from findings in clinical cases. With the advance of neurosurgery it has been possible to check these findings when section of the nerve or extirpation of the semilunar ganglion has been performed. These findings correspond well with the anatomical data. It should be noted that the lower border of the mandibular nerve territory is found several centimeters above the lower margin of the mandible, and that whereas the trigeminal as a rule does not supply the concha of the auricle (innervated by the VIIth, the Xth, and possibly the IXth nerves), it is usually concerned in the cutaneous innervation of the anterior wall of the external auditory meatus and the anterior part of the tympanic membrane (Cushing, 1904). It is frequently overlooked that the trigeminal nerve innervates the mucous membranes of the nasal and oral cavities and the maxillary and frontal sinuses. The dura is innervated by meningeal branches from the trigeminal divisions, except the dura in the infratentorial portion of the posterior cranial fossa, which receives its sensory innervation mainly from the vagus.

Just as there are differences in size of the cells in the trigeminal ganglion, so the sensory trigeminal fibers are found to be of widely varying diameter, and partly differ as regards their terminal nuclei. The *sensory nuclei of the trigeminal nerve* consist of three different parts, together extending through the brainstem from the 2nd cervical segment of the cord and upward through the mesencephalon. In a caudorostral direction these nuclei are the *nucleus of the spinal tract, the main (or principal) sensory nucleus,* and the *mesencephalic sensory nucleus* (Figs. 7-2, 7-21B). The fibers of the sensory root, on entering the pons, traverse its basilar portions, coursing dorsomedially, in the direction of the main sensory nucleus. Many of the fibers then dichotomize into ascending and descending branches. Several of the latter are very long and descend as a distinct bundle, the *spinal tract of the trigeminal nerve* (Figs. 7-2, 7-4, 7-7, 7-18, and 7-21), to the caudal end of the medulla oblongata where it fuses with the dorsolateral tract of Lissauer in the spinal cord.[33] As the tract descends, collaterals and terminal fibers are given off to a long small-celled nucleus, lying immediately medial to the tract, the *nucleus of the spinal tract,* which is continuous with the gelatinous substance of the dorsal horn. In the medulla oblongata the tract and its nucleus are situated beneath the surface, its upper part producing the elevation called the tuberculum cinereum (Fig. 7-6). The higher parts of it are covered by the fibers of the brachium pontis (Figs. 7-6 and 7-18). At its rostral end the nucleus of the spinal tract is continuous with the *main sensory nucleus.* The latter nucleus is phylogenetically younger than the nucleus of the spinal tract, but is well developed in most mammals and man. This fact, as well as the knowledge of the central connections of the two nuclei, the distal fusing of the nucleus of the spinal tract with the dorsal horn, and the difference in their afferent fibers and their functions (cf. below), makes it *plausible to regard the main sensory nucleus as being homologous to the nuclei of the dorsal funiculi of the cord, and the nucleus of the spinal tract as being homologous to the dorsalmost laminae of the dorsal horn.*

The details of the distribution of afferent fibers to these two trigeminal nuclei are of practical as well as theoretical interest. From correlated clinical and pathological observations it has, for some 50 years, generally been assumed that the main nucleus is concerned primarily in the transmission of tactile sensibility of the face, whereas the spinal tract and its nucleus have to do with pain and thermal sensibility. Experimental findings, to be considered below, show, however, that the situation is less schematic.

The *third sensory trigeminal nucleus,* the so-called *mesencephalic nucleus,* presents several peculiarities. It extends as a slender column of cells from the rostral end of the main sensory nucleus to the superior colliculus, occupying a position somewhat lateral to the upper part of the 4th ventricle and the aqueduct (Figs. 7-2, 7-21, and 7-23). It is made up chiefly of pseudounipolar cells resembling those of the semilunar and other ganglia. These cells, for this and other reasons, are generally assumed to be sensory. Their alleged origin in the neural crest is, however, disputed. The cell column is accompanied by a fine tract of fibers, the *mesencephalic root* of the trigeminal nerve. Most of these fibers are derived from the cells of the mesencephalic nucleus and descend. These descending fibers,

---

[33] The designation may be confusing, since it is common to use the term "tract" (tractus) for fiber bundles connecting nuclei within the central nervous system, while the spinal trigeminal tract (as well as the solitary tract) is composed of the central processes of cells having their perikarya outside the central nervous system.

many of which are of the large myelinated type, have been traced in degeneration experiments to the portio minor and to the branches of the mandibular nerve supplying the masseter, temporalis, and pterygoid muscles, i.e., the muscles of mastication, and have therefore been assumed to mediate proprioceptive impulses. This assumption has gained support from recent studies to be considered below.

The *motor nucleus of the trigeminal nerve* is the uppermost of the special visceral efferent column nuclei (see Fig. 7-2). It is composed of cells of the usual "motor" type and is situated in the middle of the pons, closely medial to the main sensory nucleus (see Fig. 7-21B). Its fibers, as already mentioned, all leave the pons as the portio minor, join the mandibular nerve and supply the masticatory muscles proper: the masseter, temporalis, and external and internal pterygoid muscles. In addition motor fibers are given off to the tensor tympani, the tensor palati, the mylohyoid, and the anterior belly of the digastric muscle. Like the other motor nuclei in the brainstem, the masticatory nucleus consists of several cell groups, and the particular muscles receive their motor innervation from fairly well circumscribed regions of the nucleus (see Szentágothai, 1949). A somewhat different pattern was found in HRP studies by Mizuno, Konishi, and Sato (1975). For some physiological data, see Landgren and Olsson (1976). The motoneurons for the tensor tympani were found collected in a ventrolateral group in the motor trigeminal nucleus by Borg (1973a; see however, Lyon, 1975).

The motor trigeminal nucleus is acted upon by fibers from higher levels, among them *corticobulbar fibers* (see Fig. 7-3A). In the cat these fibers do not end in the motor nucleus (Walberg, 1957b). They appear to activate motoneurons via internuncials, probably in adjacent parts of the reticular formation. In the monkey (Kuypers and Lawrence, 1967), and probably in man, there are direct corticomotoneuronal connections from the precentral gyrus to the motor trigeminal nucleus. There appear not to be afferents from the red nucleus (Edwards, 1972). From physiological experiments it appears that the hypothalamus and the amygdala may influence the jaw reflexes (see Landgren and Olsson, 1977, for some data). Which fiber connections are involved is not known precisely.

The main input to the motor trigeminal nucleus comes via the *sensory branches of the trigeminal nerve*. However, the primary sensory afferents do not appear to end on cells in the motor nucleus according to anatomical (Kerr, 1961; and others) and physiological studies (see below). Impulses from *other sensory cranial nerves* likewise influence the motor trigeminal nucleus via internuncials. Fibers from the mesencephalic trigeminal nucleus, however, end in direct synaptic contact with the cells of the motor nucleus (see below).

From a clinical point of view the trigeminal nerve and its central connections are of great importance. Research from recent years has greatly increased our knowledge of these subjects. Since many of the observations made, particularly on the sensory trigeminal nuclei, are of relevance to clinical problems, some main points in their anatomical and functional organization will be considered. The complexity is far greater than was previously assumed.

Organization and connections of the sensory trigeminal nuclei. As shown by Olszewski (1950), the nucleus of the spinal tract may be subdivided into three architectonically different portions, called the nucleus caudalis, interpolaris,

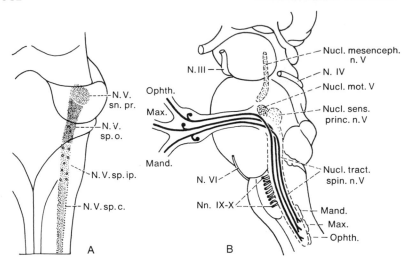

FIG. 7-21 *A:* Diagram showing the subdivision of the trigeminal sensory nuclei. The upper region *(N.V.sn.pr.)* is the main sensory nucleus. Below this follow the three subdivisions of the spinal nucleus: the oralis *(N.S.sp.o.)*, the interpolaris *(N.S.sp.ip.)*, and the caudalis *(N.V.sp.c.)*. The latter continues caudally into the dorsal horn. From Olszewski (1950). *B:* A diagram showing the topical arrangement within the spinal trigeminal tract of the fibers belonging to the main divisions of the trigeminal nerve. The fibers terminate in the nuclei according to the same pattern. The successive rostrocaudal terminations of the three groups shown in the diagram is disputed (see text).

and oralis, respectively (Fig. 21A). These subdivisions can be identified in a number of animal species and in man (Olszewski and Baxter, 1954). Structurally, the nucleus caudalis closely resembles the dorsal horn (being made up of a peripheral marginal zone, an intermediate gelatinous part, and a more massive deep part, the subnucleus magnocellularis).[34] Within the other nuclei, likewise, different subgroups may be distinguished. These features indicate the presence of functional differences between minor parts of the nucleus of the spinal tract.

On entering the brainstem almost half of the root fibers of the trigeminal nerve divide into an ascending and a descending branch (Windle, 1926), the former ending in the main sensory nucleus, the latter in the nucleus of the spinal tract. During their course they give off numerous collaterals to the nuclei. The fibers differ in calibers, and a considerable proportion are thin. In the spinal trigeminal tract in man approximately 90% have a diameter of less than 4μm (Sjöqvist, 1938). At caudal levels thin fibers appear to be relatively more numerous than more rostrally. It is of some practical interest that the fibers belonging

[34] Cytologically, the marginal layer (or zone) is narrow. Many of its cells are rather large (Waldeyer's marginal cells). The gelatinous layer is broader and is characterized by a dense network of mainly unmyelinated fibers and relatively scanty small cells. The largest, magnocellular layer contains neurons of different sizes and shapes. As will be seen below, these layers differ with regard to their connections and their content of projection and internuncial cells. Physiological differences have been observed as well. When compared with the dorsal horn (see Chap. 2), the marginal layer corresponds to lamina I of Rexed, the gelatinous to laminae II and III, and the magnocellular layer probably to lamina IV. The medially adjacent area of the reticular formation has been assumed by some to correspond to laminae V and VI.

to the three main branches of the nerve are arranged in a definite order within the tract, the ophthalmic being found most ventrally, the mandibular most dorsally (see Fig. 7-8A and 7-21B). This has been repeatedly demonstrated anatomically in the rat (Torvik, 1956b; Grant and Arvidsson, 1975), the cat, and the monkey (Kerr, 1963a; and others). Physiological studies are in general agreement. It has further been held, chiefly on the basis of studies of clinical cases of vascular disturbances of the posterior inferior cerebellar artery (Smyth, 1939) and electrophysiological studies (Harrison and Corbin, 1942; McKinley and Magoun, 1942), that the ophthalmic fibers extend most caudally, while the maxillary and mandibular fibers end successively more rostrally. Thus the ophthalmic division, in the nucleus as well as in its cutaneous distribution, would be in series with the sensory 2nd cervical nerve. However, Torvik (1956b) and Kerr (1963a) found no clear evidence of a successive pattern in the distribution. It is of interest that a fair number of trigeminal fibers descend as far as the upper part of the C_2 segment of the cord and even to C_4 in the rat (Torvik, 1956b). Furthermore, the termination in the main nucleus as well as in the divisions of the spinal nucleus shows the same dorsoventral somatotopical order as found in the spinal tract (Torvik, 1956b; Kerr, 1963a). This agrees with physiological observations (Kruger and Michel, 1962a; Eisenman et al., 1963; see also Darian-Smith, 1966). There appears to be little overlap between the terminal territories of the three branches in the nuclei. Mainly on the basis of clinical evidence (first apparently by Déjérine in 1914), another kind of topical representation in the trigeminal sensory nuclei has been advocated. The essence of this so-called ''onion-skin'' pattern is that facial sensory regions near the midline, around the mouth and nose, are ''represented'' rostrally in the caudal nucleus, while axons from successively more lateral regions of the face supply more caudal parts of the nucleus. We shall return to these questions below.

In addition to the primary sensory fibers the main and spinal trigeminal nuclei have *other afferent connections*. In the first place, as referred to in previous sections, some fibers in the *intermediate, glossopharyngeal, and vagus nerves* join the spinal tract, where they occupy the dorsalmost part (see Fig. 7-8A). These fibers can be traced caudally to the level of C_2 in the cat (Kerr, 1962) or even C_3 in the monkey (Rhoton, O'Leary, and Ferguson, 1966). Furthermore, there are fibers in the spinal tract from *the spinal cord* (Torvik, 1956b; Rossi and Brodal, 1956b; Mehler, Feferman, and Nauta, 1960; Bowsher, 1962). *Dorsal root fibers of the upper cervical nerves* end in the nucleus of the spinal trigeminal tract (see Kerr, 1961). Finally, the main nucleus as well as all subdivisions of the spinal nucleus receive fibers from the *cerebral cortex* (Brodal, Szabo, and Torvik, 1956; Kuypers, 1960; Kawana, 1969; and others). Most of these fibers come from the sensorimotor cortex, Sm I and Sm II, particularly the face regions.

Wold and Brodal (1973) confirmed this and clarified some details. The projections appears to be quantitatively rather modest and predominantly contralateral (from Sm II purely contralateral). Although all subdivisions receive some corticotrigeminal fibers, most of those from the sensorimotor regions end in the nucleus principalis and the adjoining rostral part of the pars oralis of the spinal nucleus. Furthermore (Wold and Brodal, 1974, in the cat; see also Mizuno, Sauerland, and Clemente, 1968), there are small contingents of fibers from the orbital and proreate gyrus as well, but these end preferentially in the nucleus interpolaris, which receives only a few fibers from the sensorimotor cortex. (The nucleus interpolaris projects to the cerebellum; see Chap. 5.) The corticotrigeminal fibers enable the cerebral cortex to influence the central transmission of primary sensory impulses entering it (see below).[35]

Among other afferents to the sensory trigeminal nuclei may be mentioned fibers from *the red nucleus* (Miller and Strominger, 1973, in the monkey). In an autoradiographic study in the cat, Edwards (1972) found the rubral fibers to supply chiefly the main sensory nucleus and the pars interpolaris of the spinal trigeminal nucleus (while no fibers were traced to the motor trigeminal nucleus).

Reference should also be made to afferent fibers from the *reticular formation.* Fibers from many different sources thus converge on the main and spinal trigeminal nuclei, suggesting that they are more than pure relay stations. The convergence of trigeminal, intermediate, glossopharyngeal, vagal, and spinal afferents in the spinal nucleus is of some clinical relevance (see below). The fine structure and synaptology of the sensory trigeminal nuclei and correlations with physiological observations will be considered in the following section, dealing with functional aspects of the trigeminal nuclei (p. 519 ff).

The ascending connections from the main and spinal trigeminal nuclei have been the subject of much controversy.[36] In studies of human cases the lesions have not been circumscribed, and in experimental studies stererotactic placing of lesions has been done only in recent years. However, the main features appear now to be clarified in animals, and it is reasonable to assume that conditions are, in principle, similar in man. Using the modified Gudden method (see Chap. 1), Torvik (1957a) could establish in the cat that *practically all cells of the main sensory nucleus give off ascending fibers passing to the thalamus.* A larger, crossed projection arises in the ventral two-thirds of the nucleus, a smaller uncrossed component from the dorsomedial third. This has been confirmed in retrograde studies in the cat (Smith, 1975) and in the dog and pig (Michail and Karamanlidis, 1970).[37] Studies of degenerating fibers following lesions of the main sensory nucleus are in general agreement and further demonstrate that the two components of the trigeminothalamic projections differ with regard to their termination. The fibers of the major *ventral crossed tract* ascend with the medial lemniscus, where they (in the medulla) are located dorsomedially (see Fig. 2-11). These fibers, often referred to as the "trigeminal lemniscus" have rather consistently been found to be distributed to the contralateral VPM of the thalamus in the monkey (Walker, 1939; Smith, 1975) and in the cat (Mizuno, 1970) and to a corresponding area of the ventrobasal thalamic complex in the rat (Smith, 1973). In addition, some fibers end in the ventral part of the zone incerta. The *uncrossed dorsal trigeminothalamic tract,* according to several investigations (Walker, 1939; Carpenter,

---

[35] The termination of corticotrigeminal fibers in the sensory trigeminal nuclei has been confirmed in electron microscopic studies (Gobel, 1971). In addition, some fibers end in the adjoining reticular formation (particularly in the region between the main sensory and the motor nucleus), as described by several authors (Torvik, 1956b; Valverde, 1962; Kuypers and Tuerk, 1964; Kuypers and Lawrence, 1967; Smith, 1975).

[36] For an exhaustive historical review, see Smith (1975).

[37] There appear to be marked species differences among animals as concerns the relative development of the dorsal and ventral trigeminothalamic tracts and the related thalamic regions. For example, the dorsal trigeminothalamic tract appears not to be present in the rat (Smith, 1973), while the ventral tract was not found in the goat (Karamanlidis and Voogd, 1970). According to an HRP study by Karamanlidis, Michaloudi, Mangana, and Saigal (1978) in the rabbit all subdivisions of the trigeminal sensory nucleus give off only crossed connections.

1957; Karamanlidis and Voogd, 1970; Mizuno, 1970; Smith, 1975; and others), likewise ends in the VPM of the thalamus but, according to Smith (1975), it ends in a dorsomedial part of this that is not supplied by the crossed ventral tract.[38]

The presence and course of *ascending fibers from the spinal nucleus* have been debated. While Carpenter and Hanna (1961) described some fibers ascending from the interpolaris and oralis parts, others observed ascending fibers from the subnucleus caudalis (Stewart and King, 1963; Roberts and Matzke, 1971; Tiwari and King, 1974). Most of these appear to be ipsilateral, and a relatively small number have been traced to the thalamus, particularly to the VPM, and the intralaminar nuclei (see Carpenter and Hanna, 1961; Stewart and King, 1963). It has been noted by several authors that the ascending fibers from the spinal trigeminal nucleus ascend in the reticular formation.

From an experimental study in monkeys following small radiofrequency lesions of the nucleus caudalis, Tiwari and King (1974) concluded that the fibers traced to the VPM and the intralaminar thalamic nuclei originate from the region of the reticular formation situated immediately medial to the subnucleus caudalis. Following HRP injections in the thalamus, retrogradely labeled cells were found in this region and in addition in the marginal layer of the contralateral nucleus caudalis in the cat (Hockfield and Gobel, 1978; see also Albe-Fessard, Boivie, Grant, and Levante, 1975, in the monkey). Virtually no labeled neurons were found in the gelatinous and magnocellular layers.[39]

Recording of antidromic responses in the nucleus caudalis following electrical stimulation of the VPM in the monkey (Price, Dubner, and Hu, 1976) has given largely corresponding results. Antidromically activated cells were found in the marginal zone and in the reticular formation area medioventrally to the magnocellular layer of the caudal nucleus, but also in the magnocellular layer itself. Many of these thalamotrigeminal neurons were responsive to sensory trigeminal input (some only to noxious stimuli, others to tactile stimuli only). Likewise, axons of cells in the main trigeminal nucleus, mediating tactile stimuli and having restricted receptive fields, have been identified as ending in the VPM (see Darian-Smith, 1966, 1973, for particulars and references).[40]

A fair number of ascending fibers from the main and some fibers from the spinal trigeminal nucleus thus appear to end in the VPM of the thalamus, the face region of the thalamic nucleus ventralis posterior (cf. Chap. 2). As mentioned above, there appears to be some degree of topical arrangement within this projection, fibers from the mandibular division of the trigeminal nuclei ending most dorsomedially.

The contribution of thalamic afferents from the reticular formation adjacent to the trigeminal nuclei is worthy of notice. Many details in the trigeminothalamic projections remain to be worked out. Physiological studies show that a clear dis-

[38] The different sites of origin of the two trigeminothalamic tracts appear to indicate that the dorsal tract is mainly related to the mandibular division of the trigeminal nerve, the ventral tract to the two other divisions. Correspondingly, within the terminal area in the thalamus, the part relaying information from face areas supplied by the mandibular nerve is found most medially (for some data, see Smith, 1975).

[39] Some cells of the marginal layer of the caudal nucleus were found to project to the spinal cord by Burton and Loewy (1976; HRP studies in the cat).

[40] In more posterior parts of the thalamus (the PO region of Poggio and Mountcastle, 1960), some units responding to tactile stimuli applied to the face have also been found, but these units have quite extensive receptive fields and are less specific, some of them responding to acoustic stimuli.

tinction between parts of the nuclei concerned in the transmission of tactile, thermal, and nociceptive stimuli is not possible (see also below).

The final links in the ascending pathways for facial sensation are parts of the
thalamocortical projections. As described in Chapter 2, the VPM projects to the
lower part of the primary sensory cortical region (Sm I) to a rather extensive face
area (see, for example, Figs. 2-18, 2-19, and 2-22). Likewise, there is a face area
in the rostral part of the second somatosensory region (Sm II), and in a third sensory region, S III (Darian-Smith, Isbister, Mok, and Yokota, 1966). The latter is
found in the cat immediately caudal to the part of Sm I related to the head and has
been concluded to lie within Brodmann's area 5. According to physiological studies, in all three areas most units responding to tactile stimuli are specific as to
mode and place, and all areas present a somatotopic pattern.

The thalamic relay for the sensory projection to the cortical somatosensory
face region in Sm I has long been known to be the VPM, while the site of origin
of thalamic fibers to the two other areas has been more difficult to determine. It
appears now to be settled that the VPM projects to the face area of Sm II as well.
Following stimulation of the VPM, Guillery, Adrian, Woolsey, and Rose (1966)
recorded evoked responses in topographically related parts of both Sm I and Sm
II. According to the experimental anatomical study (silver impregnation method)
of Jones and Powell (1969c) in the cat, the VPM projects in a somatotopical pattern
both to Sm II and Sm I. In both areas the fibers end chiefly in cortical layer IV,
but they are less abundant in Sm II than in Sm I. (Fibers were not traced from the
VPM to the area S III.) It appears from recordings of antidromic potentials that
about half of the cortical projecting thalamic neurons send branches to Sm I only,
most of the others to both Sm I and Sm II, while a small proportion projects to Sm
II only. The thalamic relay to the third sensory area (S III) is still debated (see
Chap. 2).

The topical precision in the central representation of the face is remarkable.
Some findings concerning the cortical input from the vibrissae in some rodents
give a particularly clear demonstration of this and will be briefly mentioned. They
further furnish an example of the principle of the organization of the cerebral cortex in functional vertical columns, a problem to which we will return in Chapter
12.

In tangential sections of the cerebral cortex of the mouse, Woolsey and Van
der Loos (1970) described a part of the head region of area Sm I as the barrel
field. This name refers to a characteristic arrangement of neurons: a roughly
cylindric hollow is surrounded by a wall consisting mainly of perikarya, arranged
perpendicularly to the surface of the cortex in its layer IV (see Fig. 12-7A and B).
In tangential sections of the cortex the wall of a barrel appears as a ring. The posteromedial part of the cortical barrel field forms a subfield of barrels of greater
size, arranged in five rows in a constant number (Fig. 12-7C). This number corresponds to the number of the animal's mystacial vibrissae. On the basis of quantitative data and careful studies, the authors put forward the hypothesis that each barrel is the correlate of a single controlateral sinus hair, and each of the larger
barrels in the posteromedial barrel field is related to one mystacial vibrissa.

Further studies have strongly supported the hypothesis of Woolsey and Van der Loos (1970).
Barrels related to vibrissae have been found in the primary sensory cortex of the rat (Welker and

Woolsey, 1974; Welker, 1976) and some other species. The projection is extremely precise.[41] Neurons in a particular small focus (presumably a barrel) are each excited by the same vibrissa and only that vibrissa (Welker, 1971). (It appears that the vibrissa receptors are capable of coding mechanical stimuli in considerable detail; see below.)

The relay stations in the pathways to the cerebral cortex have been studied. Exploring the *thalamic region* that responds to movements of the vibrissae, Waite (1973) found that most cells respond to movement of one vibrissa only, and that there is a precise somatotopic pattern in the thalamic representation of the vibrissae. After physiological identification of thalamic points responsive to light stimuli of vibrissae, Donaldson, Hand, and Morrison (1975) made minute lesions in these points. In silver impregnation studies the ensuing degeneration was confined to one or a few barrels in the cortex. Presumably, the projection of primary afferents of the vibrissae to the sensory trigeminal nuclei (particularly the main nucleus) is likewise very precise. Each vibrissa appears to have its own fiber to the nuclei (Zucker and Welker, 1969). Observations on the situation in the nuclei are in part conflicting, but Kruger and Michel (1962a) found that most cells in the main nucleus respond to input from one vibrissa receptor.

The recent studies of the sensory information from the vibrissae in the rat and mouse, briefly discussed above, are of general interest. In the first place, the sharpness of the topical pattern in the transmission lines from the receptors to the cerebral cortex is remarkable. (This does not mean that in all relays there is a strict one-to-one relation between an afferent fiber and a projection cell.) Second, the findings show the preponderant termination of "specific" sensory afferents in layer IV of the cortex. Further, in addition to enabling the animal to *localize* mechanical stimuli in the exploration of its environment, the system gives essential information about *stimulus characteristics*. Zucker and Welker (1969, p. 154) concluded from a very careful physiological study that "when a rat encounters objects which deflect its mobile or stationary vibrissae, a relatively large proportion of first-order somatic sensory neurons is capable of coding the following aspects of mechanical stimuli: peripheral location, deflection direction, onset, termination, amplitude, velocity, duration, repetition rate, and temporal pattern." The areas related to the vibrissae, so important in the life of these animals, occupy a major part of the entire thalamic and cortical regions concerned with sensory information from the face.

This is an example of a general rule: it is the functional importance and not the size that determines the representation of a receptor field in the central nervous system. Other examples are found in the optic system (Chap. 8), the acoustic system (Chap. 9), and the somatosensory pathways from the body (Chap. 2). On the "motor" side (Chaps. 3 and 4) and in the cerebellum (Chap. 5) the same is true.

The occurrence of barrels is of particular relevance to the columnar organization of the unilateral cortex, which will be discussed in Chapter 12.

As mentioned in the preceding section of this chapter, sensory information from the mouth is relayed to the cerebral cortex in part via the trigeminal nuclei (from the mucuous membranes of the nasal and oral cavities) and in part via the nucleus of the solitary tract (gustatory impulses). The "representation" of taste in the thalamus is found in a particular region most medially in the VPM, and the cortical taste field is not identical with the areas receiving somatosensory information from the face, including its mucous membranes. Integration of gustatory and

---

[41] For a study of afferent and efferent connections of the posteromedial barrel field in the mouse, see White and DeAmicis (1977).

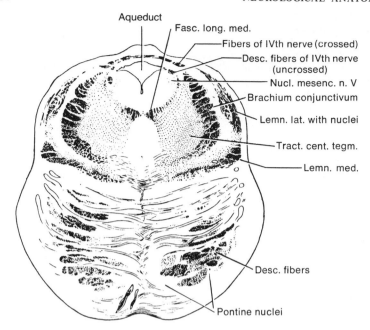

FIG. 7-22 Drawing of a tranverse section through the rostral part of the pons in a
20-day-old infant. Weigert's myelin sheath stain. A large proportion of the fibers
from the cerebral peduncle and the pontocerebellar fibers are still unmyelinated
and therefore not clearly visible.

mucuous membrane information from the mouth presumably takes place chiefly at
the cortical level.

The *mesencephalic trigeminal nucleus* differs in many respects from the two
other sensory trigeminal nuclei. Some trigeminal afferents from the mandibular
division appear to end in its caudal part (Torvik, 1956b), but their functional role
is not clear. There appears to be reliable evidence for concluding that the mesencephalic nucleus is concerned in the mediation of proprioceptive impulses. Thus
short-latency action potentials have been recorded from it on stretching of the extrinsic ocular muscles (see section (g) of the present chapter) as well as on stretching of the muscles of mastication (Corbin and Harrison, 1940; Cooper, Daniel,
and Whitteridge, 1953a). In agreement with these physiological observations, the
peripheral, largely myelinated (sensory) fibers from the cells of the nucleus can,
when degenerating, be followed into the muscular branches of the mandibular
nerve (Szentágothai, 1948; and others) and appear to supply the muscle spindles
which have been identified in the muscles of mastication (see Cooper, 1960).
Since the fibers of the mesencephalic root give off collaterals to the motor trigeminal nucleus (Szentágothai, 1948) a two-neuron reflex arc is formed for proprioceptive impulses from the masticatory muscles, analogous to the condition existing in the spinal cord (afferent fibers from the muscle spindles establishing
monosynaptic contact with motoneurons). This appears to be the basis for the *jaw
reflex* tested in clinical examinations. This reflex appears to be monosynaptic, as

has been concluded from animal studies.[42] For a study in man, see Godaux and Desmedt (1975a).

Some of the peripheral mesencephalic fibers, however, appear to run in sensory branches of the trigeminal nerve such as the alveolar nerves (Corbin, 1940) and are assumed to mediate sensations of pressure from the teeth and the peridontium. Action potentials can be led off from these nerves when pressure is applied to the teeth in animals (Pfaffman, 1939; and others) and in man (Johansson and Olsson, 1976) and also from the mesencephalic nucleus (Jerge, 1963a). It therefore appears likely that these fibers may be concerned in a mechanism which controls the force of the bite. The termination of some mesencephalic fibers in the cerebellum (see Chap. 5) fits in with the view of its function outlined above.[43]

Some functional aspects. Reference was made in the preceding section to some data of functional interest. Below other features of more immediate clinical relevance will be considered. Further, some data of the fine structure and synaptology of the trigeminal nuclei and their relation to functional observations will be dealt with. As mentioned above, it has been generally held that the caudal part of the spinal trigeminal nucleus is concerned chiefly in the mediation of impulses arising on stimulation of pain and temperature receptors, while the main nucleus has been related to tactile sensation. (For an account of the receptors involved, see Chap. 2.) This view is based on several lines of evidence, not least on observations of clinical material. It received strong support when Sjöqvist (1938) introduced his method of *medullary trigeminal tractotomy*. Since the majority of the fibers in the caudal part of the spinal trigeminal tract are thin, and pain sensations were assumed to be mediated by thin and in part unmyelinated fibers, Sjöqvist undertook to transect the spinal tract at the level of the obex in patients suffering from major trigeminal neuralgia. Following this operation the patients were usually relieved of their pain attacks, and there was a loss of pain and temperature sensibility in the skin and mucous membranes of the face. The great advantage of this operative procedure, compared with those performed on the nerve or the ganglion, is that corneal sensibility is not abolished. Thus the corneal reflex is not lost following the operation, and the complication of corneal ulceration is avoided. Further experiences (see, for example, Grant and Weinberger, 1941; Olivecrona, 1942; and others) indicated that the section of the spinal tract may be performed at a more caudal level than originally proposed by Sjöqvist, thereby minimizing the risk of complications, such as ataxia (on account of lesions of the restiform body) and unilateral vocal cord palsy (damage to the vagus). This indicated that the very caudal part of the nucleus and its continuation in the dorsal horn at the levels of C_1 and C_2 were the regions receiving the fibers mediating sensations of pain. It soon became clear, however, that following the operation there was usually also some loss of tactile sensation (see, for example, Olivecrona, 1942; Weinberger and Grant, 1942; Falconer, 1949). In view of the large number of primary afferent trigeminal fibers which give off one branch to the main nucleus, another to the

---

[42] A jaw-opening reflex, where the digastricus muscle is involved, appears to be disynaptic.

[43] It should be recalled that the mesencephalic trigeminal nucleus is not a nucleus in the proper sense, but rather is to be compared with the sensory ganglia.

spinal nucleus, it would indeed not be strange if the separation of sensory modalities within the nucleus is not as distinct as was formerly believed. This has been confirmed in neurophysiological studies (see below). There is no clear segregation within the sensory trigeminal nucleus with regard to areas concerned in the transmission of tactile and pain impulses.[44] However, it seems equally clear from clinical experience that the caudal part of the spinal nucleus and the upper cervical segments of the dorsal horn are especially important in the conduction of pain from the face. Findings from various fields of research demonstrate that the mechanism of pain transmission from the face is not a simple process.

Before considering these problems further, it is appropriate to mention some relevant points concerning the finer anatomical organization of the sensory trigeminal nuclei. An important feature is the presence of *interconnections between the different subdivisions of the sensory trigeminal nuclei,* providing morphological evidence for close cooperation between them.

> When the sensory root of the trigeminal nerve is transected, the ensuing degeneration of the afferent fibers in the spinal tract does not involve its deeper, inner part. The chiefly fine (unmyelinated and small myelinated) fibers present here are preserved. According to Gobel and Purvis's (1972) studies (normal preparations, Golgi material, and electron microscopy), some of these fibers course within the subdivisions of the spinal nucleus as well. They consist of both ascending and descending axons. Most of them have their cell body in the spinal nucleus; some are found as groups of small neurons in the bundles of the spinal tract. These axons appear to generate a large proportion of the synaptic endings in the spinal nucleus.

From experimental studies with silver impregnation methods it appears that most of these internuclear fibers are ascending. They have been found to take origin in the nucleus oralis and interpolaris (Carpenter and Hanna, 1961) and the nucleus caudalis (Stewart and King, 1963; Robert and Matzke, 1971; Tiwari and King, 1974). Some of the fibers from the nucleus of the spinal tract end in the principal sensory nucleus.

Possibilities for a *collaboration between the sensory trigeminal nuclei and the reticular formation* are evident by the presence of projections from the nuclei to the adjacent areas of the reticular formation. Furthermore, some primary trigeminal fibers pass through the nuclei to end in the same areas of the reticular formation. One of these areas has been considered to correspond to the so-called *nucleus supratrigeminalis* of Lorente de Nó (see Åström, 1953), another to the so-called *nucleus intertrigeminalis,* and a third to *an area medial to the subnucleus caudalis.* However, the borders of these regions are indistinct, as appears from anatomical and physiological studies.

> The *nucleus supratrigeminalis* is a diffusely delimited small area of loosely arranged cells, situated dorsolateral to the motor trigeminal nucleus and dorsomedial to the main sensory trigeminal nucleus. The *nucleus intertrigeminalis* is a rather similarly built area, found just ventral to the two trigeminal nuclei mentioned. Several authors have traced primary sensory trigeminal fibers to one or more of these three reticular regions (Torvik, 1956b; Kerr, 1961).
>
> These regions further appear to receive fibers from the spinal nucleus (Stewart and King, 1963; Tiwari and King, 1974; and others) and from the main nucleus (see Smith, 1973, 1975). Finally, they receive projections from the cerebral cortex (see footnote 35) and from the spinal cord (Torvik, 1956b; and others) and give off ascending fibers to the thalamus (Tiwari and King, 1974;

---

[44] It is of interest that in trigeminal neuralgia, pain paroxysms are often easily triggered by weak cutaneous stimulation while painful stimuli are far less effective.

Hockfield and Gobel, 1978). According to Jerge (1936b) the supratrigeminal nucleus receives proprioceptive impulses from the jaws.

With regard to their afferent and efferent connections the three small regions of the reticular formation discussed above thus bear a marked resemblance to the sensory trigeminal nuclei, to which they are topographically closely related. Physiological investigations (see below) support the contention that they should be considered to *form integral parts of the trigeminal nuclear complex*. They may function as interneuronal systems between trigeminal sensory and cranial motor nerve nuclei, but they also appear to be important for the central transmission of trigeminal input (see below).

"Commissural" connections—fibers crossing the midline—have been found to arise in the sensory trigeminal nuclei and particularly from cells in the trigeminus-related regions of the reticular formation (Carpenter and Hanna, 1961; Mizuno, 1970; Tiwari and King, 1970; Smith, 1973, 1975; and others). Some authors traced such fibers to the contralateral motor trigeminal nucleus.

Studies of the *minute structure of the sensory trigeminal nuclei and their synaptic organization* have revealed a far-reaching complexity. In recent years, electron microscopic studies in particular have disclosed many interesting features. Only some data will be mentioned.

Golgi studies have confirmed the classical observations of Cajal that most cells have dendrites that extend beyond the nuclear subdivision in which they are situated. Within each of the subdivisions different cell types occur, and each subdivision appears to have its peculiarities.

The *nucleus caudalis* has been most extensively studied so far. In its gelatinous zone, for example, Gobel (1975a) distinguished three types of cells with different dendritic arborizations in Golgi material: cylinderlike, pyramidal-shaped, and spherical (cells of the first type may have two axons; Gobel, 1975b). Cells of these types are all considered to function as internucials between primary afferents and second-order sensory neurons of the nucleus caudalis.

As might be expected from light microscopic studies, with the electron microscope primary trigeminal sensory fibers have been found to establish synaptic contacts with neurons in all three layers of the nucleus caudalis (Kerr, 1970b, 1971), although least in the gelatinous layer. They end predominantly on dendritic spines and fine dendritic shafts (see Kerr, 1970b, 1971; Gobel and Brinck, 1977). Most terminal boutons present in the nuclei, however, do not appear to belong to primary trigeminal afferents but to axons of internuncial cells.

Axoaxonic synapses have been repeatedly observed in the caudal nucleus (Kerr, 1970a; Gobel, 1974). From electron microscopic degeneration studies it appears that most often the boutons of primary afferents are postsynaptic to boutons derived from axons of other (presumably internuncial) cells (Kerr, 1970b, 1971). Glomerular formations (see Chap. 1, Fig. 1-5G) are frequent in the subnucleus caudalis (its gelatinous zone; see Gobel, 1974). In these structures an usually large central ending (bouton) is surrounded by smaller axonal endings and small dendrites. The complex is more or less completely encapsulated by astrocytes. At least most of the central endings appear to belong to primary sensory fibers (Gobel, 1971, 1974). Axoaxonic and dendrodendritic synapses occur in the glomeruli (for particulars, see Gobel, 1974).

As concerns its fine structure, the *main sensory nucleus* has been less extensively studied than the nucleus caudalis. An exhaustive study has been published by Gobel and Dubner (1969). This nucleus harbors numerous glomeruli, and, within these, axoaxonic contacts are frequent. Following section of the trigeminal root most of the central endings in the glomeruli degenerate (Gobel, 1971). A small number of primary trigeminal afferents end on dendrites outside the glomeruli. The light microscopic finding that fibers from the cerebral cortex end in the main sensory nucleus has been confirmed electron microscopically. However, their number appears to be modest. The fibers appear to give off small endings, which are widely dispersed along the dendritic trees of neurons in the nucleus and do not appear to enter the glomeruli or to form axoaxonic synapses (Gobel, 1971).

Several other observations on the fine structure and the cytological organization of the trigeminal sensory nuclei have been made, for example concerning types of vesicles present in the terminals, details of synaptic contact, and patterns of axonal arborization. The emerging picture is one of extreme complexity, and it is still not possible to interpret all data and to correlate them with the equally complex information that has been brought forward in recent years concerning the functional aspects of the sensory trigeminal nuclei.

Numerous *physiological studies of the trigeminal nuclei* have been performed. On the whole, they tally with anatomical observation. Only some data with relevance to the latter will be mentioned. For particulars concerning functional properties of the neurons and interpretations, the original articles should be consulted. (For a review see Darian-Smith, 1973. See also *Pain in the Trigeminal Region,* 1977.)

Units responding to *tactile stimuli* (for example, bending of hairs or light touch) or to electrical stimulation of branches of the trigeminal nerve have been found in the main nucleus and in all subdivisions of the spinal nucleus (see Darian-Smith, 1973, for references; Rowe and Sessle, 1972; Mosso and Kruger, 1973; Khayyat, Yu, and King, 1975).[45] Most mechanoreceptive cells in the main sensory nucleus and the nucleus oralis have a restricted receptive field and adapt rapidly, and many appear to be modality specific. Functionally different cell types can be distinguished. Most of the mechanoreceptive neurons appear to project to the thalamus. Units responding to tactile stimuli in the face have further been found in the reticular formation medial to the nucleus of the spinal tract. They most often have wide receptive fields, and many of them are excited by auditory and visual stimuli as well (for references see Darian-Smith, 1973).

The recording of responses to *noxious stimuli* has encountered greater difficulties than studies of tactile sensory qualities. Under the assumption that tooth pulp afferents ending in the sensory trigeminal nuclei are concerned in pain transmission, the responses to electrical stimulation of tooth pulps have often been used to study and to locate trigeminal units responding to pain. (Whether all tooth pulp afferents are related to pain transmission may, however, be doubted.) It appears from several studies that the role played by the trigeminal nuclei in the transmission of nociceptive impulses is very complex.

Many authors have found neurons responding to noxious stimuli in the main sensory (and oralis) nucleus (Eisenman, Landgren, and Novin, 1963; and others). The input to this may be even more important than the input to the spinal nucleus (see Sessle and Greenwood, 1976). It appears from recent studies that there may be a dual representation of nociceptive units in the nucleus caudalis. Such neurons have repeatedly been found in the ventromedial aspects of the nucleus caudalis (that is, its deeper magnocellular layer) and, in addition, particularly in the medially adjoining area of the reticular formation referred to above (see, for example, Mosso and Kruger, 1972, 1973; Nord and Ross, 1973; Nord and Young, 1975).[46] The precise site of the

---

[45] Responses to electrical stimulation of the glossopharyngeal and vagus nerves have likewise been found in the main sensory nucleus (for example, see Sessle and Greenwood, 1976), in agreement with the anatomical data on the distribution of afferents in these nerves (see above).

[46] In a recent detailed study five cell types (three of them responsive to noxious stimulation) could be physiologically distinguished in the nucleus caudalis (see Price, Dubner, and Hu, 1976).

nociceptive cells in the nucleus caudalis has been debated. In rather good agreement with some previous authors, Price, Dubner, and Hu (1976) and Shigenaga, Sakai, and Okada (1976) located cells responding to noxious stimulation in the marginal and magnocellular layer of the caudal nucleus.[47] According to the latter authors, most of these cells are activated monosynaptically and often are "pulp-specific" (see also Mosso and Kruger, 1973). The nociceptive cells in the reticular formation medial to the nucleus, on the other hand, are supposed to be activated from pulp afferents by a route that involves more synapses. They appear to be specifically activated by pulp stimulation and cutaneous impulses and possess wide receptive fields. They differ also in other respects from the nociceptive cells in the nucleus caudalis.

Information concerning the trigeminal nuclei and mediation of *responses to thermal stimuli* is relatively sparse.

According to Rowe and Sessle (1972), units responding to thermal changes are present both in the nucleus oralis and in the nucleus caudalis (see also Price, Dubner, and Hu, 1976). All such units studied were also mechanosensitive. It is considered likely that this may be explained by the presence of bimodal primary afferents and not by a convergence of separate fibers from thermoreceptors and mechanoreceptors on the same nuclear neurons. Poulos and Molt (1977) found specific thermoreceptive neurons to be clustered in a somatotopic pattern in the marginal zone of the caudal nucleus. (Neurons in the thalamus responding to thermal stimuli have been found to be mechanosensitive as well.) For some studies on trigeminal temperature mechanisms, see *Oral-Facial Sensory and Motor Mechanisms* (1971).

The afferent connections of the sensory trigeminal nuclei attest that the activity in the nuclei must be subject to alterations from several sources. Hernández-Peón and Hagbarth (1955) first demonstrated that stimulation of the sensorimotor cortical regions and the reticular formation may inhibit the central transmission of trigeminal impulses through the nucleus. The former phenomenon is mediated via corticotrigeminal fibers. In agreement with anatomical data (see p. 513) antidromic responses have been recorded in cortical neurons following stimulation of the main sensory nucleus and the subnucleus caudalis of the spinal nucleus (Dubner and Sessle, 1971). Both first and second somatosensory areas are concerned, chiefly their "face regions" (see Darian-Smith, 1966; Dubner, 1967. Inhibitory effects appear to be predominant, and some of these effects may be due to presynaptic inhibition (Darian-Smith, 1965; Stewart, Scibetta, and King, 1967). However, excitatory effects are also seen (Darian-Smith and Yokota, 1966; Wiesendanger and Felix, 1969). Projection neurons and interneurons are affected. The situation is thus similar to that in other somatosensory pathways (see Chap. 2), where the transmission of sensory impulses is subject to central control.[48] Since the corticotrigeminal fibers end in all subdivisions of the nuclei, this effect presumably concerns all sensory modalities. As referred to above, however, the corticotrigeminal supply of the different nuclear subdivisions varies with regard to quantity and in other respects. The functional role of the cortical input, therefore, can scarcely be expected to be identical for all nuclear subdivisions.

It is apparent from the above that, despite much information from recent years, there are still many unsolved problems concerning the function of the sensory trigeminal nuclei, not least concerning their role in the transmission of facial pain, a problem of relevance to trigeminal neuralgia (to be considered below). To

[47] According to the anatomical study of Westrum, Canfield, and Black (1976), the tooth pulp afferents end chiefly in the interpolaris.

[48] It should be recalled that influences on the central transmission of sensory information may also occur via corticothalamic fibers (see Chap. 2).

sum up some points: It is not possible to indicate particular regions of the tri-geminal sensory nuclei that relate purely to transmission of tactile sensation, tem-perature, or pain. Furthermore, some areas of the reticular formation adjoining the sensory nuclei appear to be integral parts of the mechanism related to sensory in-formation from the face. There is anatomical and physiological evidence of an in-timate collaboration (by way of internuncial fibers) between different parts of the sensory nuclei and between these and the reticular formation. Primary trigeminal afferents end on relay cells (projecting to the thalamus) and on internuncial cells. The presence of axoaxonic and dendrodendritic contacts and of glomerular forma-tions in the nuclei is evidence that complex synaptic interactions must occur, as is evident also from physiological studies, but much remains to be resolved before satisfactory correlations between these two sets of data can be made.

In agreement with clinical studies, it appears to be generally accepted that pure *tactile and other mechanical stimuli* from the face are relayed in the main sensory nucleus (and probably the adjoining part of the pars rostralis of the spinal nucleus). Even if the main nucleus harbors pain-receptive units, most of its units respond to mechanical stimuli, have small receptive fields, and adapt rapidly. Many units are stimulus specific. The entire pathway from the mechanoreceptors to the cortex appears to be very precisely topographically organized (compare the observations on the mystacial vibrissae in the rat and mouse, mentioned on p. 516). The system appears to be well suited to serve discriminative tactile percep-tion.

The situation is far more complex when we turn to the *nucleus of the spinal tract and its functional importance*. Presumably, each of its subdivisions (oralis, interpolaris, and caudalis) and the zones in the nucleus caudalis have their particu-lar role to play. There is considerable evidence that the interpolaris is particularly related to the cerebellum. The main problem concerning the rest of the spinal nucleus is its relation to transmission of painful sensations. Problems related to the neural substrates for transmission of pain and their functional organization have been briefly discussed in Chapter 2. Evidently, much is still hypothetical. The same is the case concerning pain from the face mediated via the trigeminal nuclei.

It was originally assumed that the facial analgesia following Sjöqvist's (1938) medullary tractotomy could be explained as a consequence of the transection of primary afferent pain-mediating trigeminal fibers in the spinal tract. Later studies, in particular experimental physiological ones (some referred to above), have shown that this is too simple an explanation. (Both pain and tactile impulses are mediated via the nucleus caudalis and via the main sensory nucleus, and nocicep-tive neurons in both nuclei project to the thalamus.)

Numerous physiological studies of the properties of trigeminal nuclear cells and their synaptic relations (receptive fields, receptor specificity, synaptic and presynaptic excitability changes, and other characteristics) have been performed. Attempts to correlate these observations with knowledge of the ultrastructure of the nuclei have prompted hypotheses about the role of the trigeminal nuclei in pain transmission. It is not possible to discuss these problems in any detail here.

Only a few points will be mentioned. In the first place, it is clear that even if cells in the nucleus caudalis project to the thalamus, the nucleus functions as more than a pure relay center in the transmission of pain impulses. It has been shown that it influences (probably by way of internuclear association fibers) the transmission of sensory impulses through the main sensory nucleus (see Denny-Brown and Yanagisawa, 1973; Greenwood and Sessle, 1976). The two nuclei appear to cooperate closely. The caudal nucleus further influences oral facial reflex activities (these have been found to be depressed following trigeminal tractotomy). It appears that the marginal layer may be more directly related to specific nociceptive input than the magnocellular layer and the adjoining reticular formation. Under normal circumstances there appears to be close cooperation between tactile and noxious stimuli in pain transmission and between cells in different layers of the caudal nucleus. More particularly, it appears that "normal facial pain sensibility requires the transmission of spatially summating inhibitory and facilitatory effects from broad facial fields to a polyneuronal network via the descending tract of the trigeminal nerve" (Nord and Young, 1975; see also Denny-Brown and Yanagisawa, 1973). The gate theory of Melzack and Wall (1965; see Chap. 2) has been applied to the problems of pain transmission in the trigeminal nuclei.[49] So far no entirely satisfactory theory of the functional mechanism of pain transmission from the face has been presented.[50] Some points of interest, originating from clinical experience with medullary tractotomy, will be considered below.

Symptomatology. Lesions of the *peripheral branches* of the trigeminal nerve may be caused by fractured bones of the face or skull or by tumors. The infraorbital nerve is frequently damaged in maxillary fractures. Pain is usually absent in such cases, but the ensuing sensory loss in the face and oral and nasal cavities as a rule easily permits the determination of the branches affected, the more so as the extent of overlapping between the areas of the three divisions and their separate branches is small. The peripheral distribution is seen in Fig. 7-20, and a detailed description is superfluous.

If the *mandibular nerve* is damaged, e.g., by a skull fracture passing through the foramen ovale, *the motor fibers will usually also be affected,* and there results a homolateral *paresis or paralysis of the masticatory muscles,* The paralyses of the anterior belly of the digastric and the mylohyoid which usually accompany it are of little practical importance, although they may easily be ascertained when looked for. Likewise, the eventually concomitant paralyses of the tensor tympani and tensor palati muscles are of little importance. The paralysis of the masticatory muscles is easily recognized. The masseter and temporalis muscles are flaccid, as the paralysis is one affecting the peripheral motor neuron, and their flaccidity and lack of contraction are felt and seen when the patient is asked to bite. After some time, the atrophy of these muscles can be seen. If only a partial lesion of the nerve

[49] On the basis of anatomical and physiological studies it appears that there are clear similarities between the trigeminal nuclei and the spinal cord as concerns the transmission of somatosensory information. The main sensory trigeminal nucleus (and possibly the nucleus oralis) appears to correspond to the dorsal column nuclei (gracilis and cuneatus) and to belong to the "lemniscal system." The nucleus caudalis of the spinal nucleus corresponds to the dorsal horn (see also footnote 34) and is part of the "spinothalamic system." The nucleus interpolaris has been assumed by some authors to correspond to the external cuneate nucleus (they both project to the cerebellum).

[50] It has recently been reported that stimulation of the raphe nuclei produces long-lasting inhibition of the responses of neurons in the nucleus oralis and caudalis to excitation by peripheral inputs (see *Pain in the Trigeminal Region,* 1977). This is of some interest in relation to the analgesic effects observed following stimulation of the raphe nuclei (see Chap. 6).

is present, the reduced power on biting can be felt clearly when the closing of the opened mouth is resisted by the examiner. The paralysis of the external pterygoid reveals itself when the patient opens his mouth. This muscle normally pulls the mandible forward on opening the mouth, as it is attached to the neck of the mandible and reaches it from the anterior aspect. Because of this a *deviation of the jaw to the paralyzed side* will appear, since the paralyzed external pterygoid does not participate in the protrusion of the mandible. The patient will chew on the normal side only. In rare cases the masticatory muscles will be affected on both sides. The lower jaw then droops, but the mouth can be closed with reduced power by means of the facial muscles. Chewing is impossible, and swallowing is difficult.

The lesions of the trigeminal nerve referred to above, affecting peripheral branches of one or more divisions, are usually not accompanied by pain. However, such lesions are far less frequently observed than those in which *facial pain* is the outstanding or only complaint of the patient. The facial pain may occur in several forms and may be due to a variety of causes. Some of the more important will be mentioned.

First, on account of its frequency, the typical or *major trigeminal neuralgia* should be remembered. It manifests itself in a characteristic clinical picture. Paroxysms of pain, localized to the peripheral area of one or more of the three divisions of the nerve, and lasting for some seconds only, occur with varying frequency. The pain is severe, even excruciating, commonly described as stabbing, cutting, grinding, or tearing. Between the paroxysms there is no pains. As a rule "trigger zones" are present, most commonly near the eye, the nose, or on the alveolar margins. The slightest stimulus at these points induces a paroxysm. Frequently chewing, swallowing, washing the face, or even the slightest tactile stimulus provokes a paroxysm. Signs of autonomic irritation frequently accompany the paroxysms of pain. Thus lacrimation, conjunctival injection, salivation, and flushing on the painful side of the face are not uncommon. In the typical cases, which occur predominantly in the latter half of life, apart from the pain *no neurological disturbances are present*. There are, for example, no trigeminal areas with altered sensibility.

The etiology of this peculiar and distressing disease is not clear. There appears to be no neuritis of the nerve or the semilunar ganglion. Some authors assume that the symptoms are due to a central process; others believe that changes in the nerve or at its entry into the brainstem are responsible.

On the basis of neurophysiological considerations of observations of 50 patients with trigeminal neuralgia, Kugelberg and Lindblom (1959) reached the conclusion that the mechanism responsible for the paroxysmal pain is situated centrally: probably in the brainstem in structures related to the spinal trigeminal nucleus. This conclusion supports the view of List and Williams (1957) that the paroxysms are due to a "pathological multineuronal reflex in the trigeminal systems of the brainstem." These views are, however, challenged by Kerr (1963b) who points to certain features of the disease which are difficult to reconcile with the assumption of a central origin, such as the rare occurrence of concomitant neuralgia of the glossopharyngeal or intermediate nerve, the predominant involvement of the third or second divisions, and the age and sex distribution. Kerr draws attention to certain features in the anatomical relation between the semilunar ganglion and the internal carotid artery (Fig. 7-25) and certain changes occurring in this region with age, which he has found in autopsy studies. Carney (1967) reports favorable results of correcting asymmetric protrusion of the jaw, and assumes that traction on the third division in such instances is responsible for trigeminal neuralgia. Some authors are inclined to believe that ephaptic transmission between axons with destroyed sheaths due to long-standing pressure is an essential factor. For reviews on the etiology of trigeminal neuralgia see *Trigeminal Neuralgia* (1970). See also *Pain in the Trigeminal Region* (1977).

In spite of insufficient knowledge of the etiology of trigeminal neuralgia, treatment on an empirical basis has given valuable results. Modern neurosurgery is able to cope with practically all severe cases in which drug treatment fails. The introduction of Tegretol (carbamazepine), an anticonvulsant related to imipramine (Blom, 1963), has substantially reduced the number of patients in need of surgical treatment. For some references see *Trigeminal Neuralgia* (1970).

The different measures at the surgeon's disposal consist of interruption of the trigeminal pathways, for example by alcohol injections[51] in the maxillary or mandibular nerves at the foramen rotundum or ovale or into the semilunar ganglion, or section of the trigeminal nerve between the pons and the ganglion in various technical procedures. The most common method is still the retrogasserian root section introduced by Spiller and Frazier (1901). In most of these cases an ensuing anesthesia of the cornea is a serious danger to the eye of the patient, since keratitis is apt to develop unless special care is taken. In the medullary or trigeminal tractotomy of Sjöqvist this is avoided. Taarnhøj (1952) introduced a method of "decompression" of the ganglion by splitting the dural envelope of the semilunar ganglion, while Shelden et al. (1955) attempted compression of the nerve. It seems a likely assumption that the effect of the latter two procedures depends on destruction of ganglion cells. Experimental support for this view has been produced by Baker and Kerr (1963), who studied the degeneration resulting from compression of the semilunar ganglion in the cat.

The trigeminal secondary neurons have been attacked by so-called mesencephalic tractotomy (Walker, 1942b). The latter operation and medullary tractotomy may to a certain extent be regarded as an indirect outcome of anatomical knowledge, as practical applications of theoretical achievements. For accounts of trigeminal neuralgia and its treatment see Stookey and Ransohoff (1955) and *Trigeminal Neuralgia* (1970). In recent years percutaneous, elective, partial coagulation of the semilunar ganglion has been employed by some surgeons (see Schürmann, Butz, and Brock, 1972). In this way and with graded heating by means of a radiofrequency current, loss of pain with preservation of some touch sensation has been achieved (see White and Sweet, 1969). It appears that with this procedure it is possible to destroy only small myelinated and unmyelinated (A and C) fibers concerned in pain conduction.

Clinical experience in the treatment of trigeminal neuralgia with the medullary tractotomy of Sjöqvist (1938) has given some information of both theoretical and practical interest and will therefore be briefly considered. The usually concomitant reduction of tactile (particularly two-point discrimination) sensibility is understandable considering present knowledge of the representation of tactile- and pain-responsive units in the nucleus caudalis and their cooperation. Medullary tractotomy is apparently now not used as much as in former years. It appears to have fallen into disrepute because of the rather frequent occurrence of concomitant

---

[51] Phenol injections have been recommended (see Jefferson, 1963).

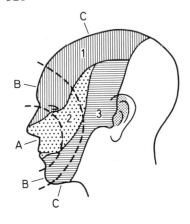
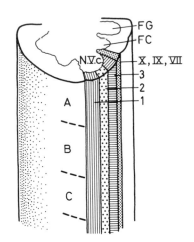

FIG. 7-23   Diagrams to explain how the pattern of segmental distribution in the spinal trigeminal tract (and its nucleus) of fibers from the three divisions of the trigeminal nerve (1, 2, and 3) can be reconciled with the onion-skin pattern of anesthesia (particularly analgesia and thermanesthesia), often seen in patients in whom parts of the spinal trigeminal tract and its nucleus are damaged.

To the right a diagram of the segmental termination of fibers from the trigeminal divisions, 1, 2, and 3, from ventral to dorsal, and the termination of fibers from cranial nerves VII, IX, and X most dorsally (cf. Fig. 7-8). Within each of the cutaneous territories of the trigeminal divisions (1, 2, 3), fibers from the middle (anterior, perioral, and perinasal) facial regions (zone A) are assumed to terminate most rostrally in the nucleus of the spinal tract, while fibers from zones B and C end at successively lower levels. A process affecting the caudal part of the nucleus of the spinal tract (for example, in syringomyelia) will result in sensory defects in zone C only, but within the distributions of all divisions of the nerve. A similar pattern is presumably present in the trigeminal innervation of the oral and nasal cavities. FC and FG: nuclei of cuneate and gracile funiculus, respectively; N.V.c: nucleus of spinal trigeminal tract.

damage to the vagus nerve or the restiform body. According to Kunc (see Kunc, 1970), however, if a tractotomy is made below the obex, there is little risk of producing inadvertent damage. The appropriate level (see below) of the incision is calculated from its distance from the uppermost filament of the second sensory dorsal root. The distance from this to the rostral end of the nucleus caudalis is about 15 mm. Furthermore, if proper care is taken in placing the incision of the spinal trigeminal tract, it is possible to achieve analgesia more or less entirely limited to a particular branch of the trigeminal nerve. This is explicable considering the anatomical segregation of descending fibers from the three trigeminal nerve divisions in the tract, described earlier in this chapter (see Figs. 7-8 and 7-23). (The appropriate site of the incision is determined on the basis of local stimulation of the tract during operation.) Isolated transection of the dorsalmost part, carrying the somatic sensory fibers from cranial nerves VII, IX, and X, has been reported to give relief of glossopharyngeal neuralgia (Kunc, 1965, 1970).[52] The clinical

[52] The suggestion that trigeminal tractotomy might prove to be useful in the treatment of glossopharyngeal or geniculate neuralgia was first set forth on the basis of anatomical considerations of the results of examinations of the analgesia in the mouth, throat, and external ear in patients treated with tractotomy for trigeminal neuralgia (Brodal, 1947a; see also Falconer, 1949).

studies of the distribution of analgesia following selective tractotomies are thus in complete accord with anatomical knowledge.

Examinations of the distribution of the analgesia in cases of medullary tractotomy have further given valuable information concerning another aspect of sensory representation in the spinal tract and its nucleus, particularly the nucleus caudalis. As mentioned above, in patients with affections of the trigeminal nuclei, the distribution of analgesia and anesthesia does often not conform to the pattern of distribution of the three trigeminal branches. For example, in syringomyelia, analgesia may be absent in the region around the nose and mouth. As referred to above (p. 513), these clinical observations suggest an "onion-skin" pattern of representation in the trigeminal spinal nucleus. These observations have been difficult to reconcile with the topical pattern of fibers from the trigeminal branches in the spinal tract and some have thought clinical and anatomical data to be incompatible. However, this appears not to be so. According to Kunc, who has extensive experience with medullary tractotomies, the "onion-skin" pattern reflects a rostrocaudal sequence of representation in the terminations in the caudal nucleus. Within each of the three main branches the fibers from the anterior (perioral) regions of the face end most rostrally, those from the most posterior (lateral) regions caudally. Only with transections of the spinal tract at levels at the rostral end of the caudal nucleus will the analgesia include the entire trigeminal cutaneous area, whereas with tractotomies at lower levels the anterior regions (zone A, or A and part of B in Fig. 7-23) will be spared. Experimental studies (see Yokota and Nishikawa, 1977) support the clinical findings. Figure 7-23 shows a diagram of how the two patterns of distribution of the trigeminal nerve can be explained on an anatomical basis. Pains referred by the patient to one or more of the trigeminal divisions may be caused by a variety of diseases in the face. These facial pains are frequently termed *symptomatic trigeminal neuralgia.* This name is misleading since true neuralgic paroxysmal pain is rare in these conditions. Usually there is a more continuous aching that is sometimes intensified in paroxysms. Facial pain of this type may be caused by any process which is apt to irritate the nerve directly. The distribution will commonly, at least in the beginning, be limited to the division or branch involved, but has a tendency to irradiate in the course of time to other divisions.

The mechanism of this phenomenon is not entirely clear. Since fibers from all divisions of the nerve extend to the caudalmost part of the nucleus caudalis and the dorsal horn at C_1 and C_2, irradiation of pain may perhaps be explained in the same way as suggested by Kerr for so-called *atypical facial neuralgia.* Here pain may be felt outside the trigeminal area, including the neck, on the same side. Kerr (1962) points to the convergence of afferents of the trigeminal, intermediate, glossopharyngeal-vagal and spinal nerves in the nucleus caudalis and C_1–C_2 (see Kerr, 1961), and cites clinical evidence in favor of this explanation. Units responding to volleys in both the trigeminal and the first and second dorsal roots have, furthermore, been found in the dorsal horn at C_1 and C_2 in the cat (Kerr and Olafson, 1961). Some dorsal root fibers from the upper cervical nerves ascend to reach the nucleus caudalis, as shown anatomically and physiologically. It has been suggested by several students that the frequently observed diffuse spreading of pain in the trigeminal region to the neck and vice versa can be explained by the common territory of termination of sensory afferents from the face and upper cervical segments.

Infections of the nasal sinuses are perhaps the most common of the morbid conditions which may be the causal factor of pains in the trigeminal area. The intraorbital nerve, and especially the superior alveolar branches, will easily be af-

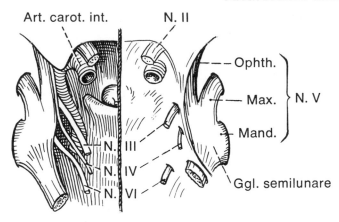

FIG. 7-24 Diagram of dissection of the cavernous sinus region, displaying the course of the IInd to VIth cranial nerves. In the right half of the diagram the dura is retained, in the left it is removed. (Cf. text.)

fected by edema and hyperemia or may even be the seat of real inflammation in sinusitis of the maxillary sinus, as only tiny bone lamellae or even only the mucous membrane cover the nerves.

Other conditions capable of producing trigeminal pain are tumors of the mouth, tongue, nasal sinuses, or in the neck. Dental infections are responsible less frequently than usually assumed. Facial pain of a trigeminal distribution may also be caused by impacted wisdom teeth or diseases of the temporomandibular joint, and likewise diseases of the eye, especially iritis and glaucoma, are sometimes accompanied by pain irradiating in one of the major divisions. These pains are usually assumed to be of the type called "referred pain" (cf. Chap. 11).

Before leaving the topic of trigeminal neuralgia, the occurrence of glossopharyngeal and intermediate nerve neuralgia, referred to previously, should be remembered, as they may be confused with true trigeminal neuralgia. Finally, the so-called *atypical trigeminal neuralgia* should be mentioned. This designation is usually applied to conditions in which facial pain is the outstanding symptom, frequently the only symptom, and in which no local affections of the face and skull can be held responsible for the symptom. The etiology is unknown. The main difference from the typical forms is that the pain is more continuous, lasting for hours or even days, but frequently acute exacerbations occur in addition. These paroxysmal pains, as well as the more persistent ones, are usually less clearly confined to one or more divisions of the nerve.

Let us now consider the *intracranial division* of the trigeminal nerve (Fig. 7-24). Apart from its occasional involvement in *diseases of the meninges,* such as arachnoiditis and syphilitic infection, some special instances may be particularly mentioned. *Herpes zoster,* a virus infection, occurs also in the trigeminal ganglion. It manifests itself pathologically as an inflammation of the ganglion with concomitant necrosis of ganglion cells, localized leptomeningitis, and less severe changes within the root and in the gray matter where the root enters. This will result in typical herpetic eruptions, localized to the entire trigeminal area or the area innervated by one of its divisions or an even smaller area. Facial herpes is most commonly seen in the region of the ophthalmic nerve (herpes zoster ophthalmicus) in which case the cornea is also frequently involved. The inflammatory changes, obviously representing a highly abnormal irritation of the ganglion cells and fibers, must be assumed to be the cause of the facial pain which is usually premonitory to the eruption of the vesicles. The appearance of the latter is explained by some as being due to antidromic impulses in the sensory fibers (cf. Chap. 11). Occasionally the inflammation of the motor root or the motor nucleus may cause an accompanying paresis of the masticatory muscles, or in ophthalmic herpes a paralysis or paresis of one of the motor ocular nerves may occur, an indication of the spread of the infection within the gray matter of the brainstem. When the inflammation recedes, several ganglion cells are as a rule lost and the ensuing scar formation in the ganglion will represent an abnormal irritant to the remaining cells. This explains the occurrence of pains, frequently of neuralgic type (postherpetic neuralgia), which often last for years after a herpes zoster infection, and also the anesthesia of the scars, which usually persists in the patches where the eruption was located. The fibers of the corresponding twigs and branches must be assumed to have been destroyed.

Involvement of the trigeminal nerve may occur in cases of *aneurysms of the internal carotid,* particularly in the so-called infraclinoid carotid aneurysms (Jefferson, 1938). Owing to the intimate relationships of the carotid to the oculomotor, trochlear, and abducens nerves, symptoms from these will usually also be present. The more posteriorly the aneurysm is seated, the greater is the tendency to affect all three divisions of the trigeminal nerve (cf. Fig. 7-25). Apart from the ocular palsies, which occur especially in the IIIrd and IVth nerves, trigeminal pain may be an early symptom, usually starting in the first division and hence being located to the forehead and orbit. The first sign of an affection of the first division is often a weakening of the corneal reflex. When the conductive capacity of the fibers is lost, a hypoesthesia or anesthesia will appear.

With increasing growth of the aneurysm, pressure may be exerted on several other structures, e.g., the optic nerve with resulting visual impairment, or the hypothalamic region with autonomic symptoms (see, for example, Nyquist, Refsum, and Torkildsen, 1939). The symptoms referred to above will in some instances develop gradually, while in other cases they become manifest only after rupture of the aneurysm, when this is not immediately fatal. It is a peculiar feature that the facial pains as well as the ocular palsies not infrequently appear intermittently. Probably this must

FIG. 7-25 *A:* Lateral view of the right internal carotid artery and related nerves in the cavernous sinus. *B:* The alterations seen in cases of aneurysm of the carotid artery (posterior aneurysm of the infraclinoid type). (See text.) From Jefferson (1938).

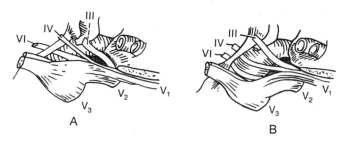

be attributed to changes taking place in the aneurysm. The ensuing symptoms in this case may present themselves as an intermittent painful ocular palsy or "migraine ophthalmoplègique." Following each attack a certain residual paresis is frequent. Ophthalmoplegic migraine may, however, occur in the absence of an obvious structural abnormality (Friedman, Harter, and Merritt, 1962).

Another condition apt to affect the trigeminal nerve is *inflammation of the pneumatic cells in the apex of the petrous temporal bone,* a not uncommon complication of suppurative otitis. This so-called Gradenigo's syndrome reveals itself as pain in the trigeminal area, usually in the first division, accompanied by a paralysis of the external rectus muscle, due to affection of the abducent nerve. The fact that these nerves are especially apt to suffer will be evident from their close relationship with the apex (cf. Fig. 7-24). Among other possibilities of injury to the intracranial part of the trigeminal may be mentioned *acoustic nerve tumors.* As a rule, however, the trigeminal pain in this instance will first appear late in the course of the disease when the tumor has reached a considerable size. Occasionally an acoustic nerve tumor or a meningeoma in the posterior cranial fossa may elicit symptoms from the trigeminal nerve (on the ipsilateral or contralateral side or bilaterally) due to displacement of the brainstem with distortion of the trigeminal sensory root. The involvement of the trigeminal nerve will be a false localizing sign. For a recent report, see O'Connel (1978).

Lesions of the trigeminal nuclei are met with particularly in two morbid conditions: *syringobulbia* and *vascular disturbances* of the medulla oblongata and pons, particularly of the posterior inferior cerebellar artery.[53] In both instances the lesion will also involve other structures. The main symptoms due to lesions of the trigeminal nuclei and their secondary tracts will be evident from a knowledge of the anatomical data and have been mentioned in connection with the account of the sensory system. It should, however, be pointed out that because of the imperfect knowledge and the varying distribution and size of the lesions, an exact anatomical diagnosis cannot always be reached in these cases. Lesions of the sensory trigeminal pathways in the thalamus and the cortical area of the face are discussed in Chapter 2.

(g) THE ABDUCENT, TROCHLEAR, AND OCULOMOTOR NERVES

The three cranial nerves concerned in the innervation of the ocular muscles are conveniently treated collectively on account of their close anatomical and functional relationship.

The nerves and their nuclei. Morphologically as well as functionally, there are common features among these three nerves, the VIth, IVth, and IIIrd, which are usually regarded as being purely efferent. Apart from the parasympathetic division of the oculomotor nucleus, supplying the intrinsic muscles of the eye, the other nuclei belong to the somatic efferent nuclear column (see Fig. 7-2) and are composed mainly of large polygonal cells like the motoneurons in the cord.

---

[53] In the latter case, which gives rise to Wallenberg's syndrome, analgesia in the trigeminal region may be more pronounced than the thermanesthesia and, particularly, than loss of tactile sensibility, indicating that the lesion of the sensory trigeminal nuclear complex involves mainly its caudal part.

Concerning particulars of the three nuclei and the peripheral distribution of their fibers, the following data may be noted.

The nucleus of the abducent nerve has already been referred to in connection with the description of the facial nerve. It is situated beneath the floor of the fourth ventricle, covered by the facial fibers, and separated from the median plane by the medial longitudinal fasciculus (See Figs. 7-17 and 7-18).

The emerging fibers course forward through the pons, pass immediately lateral to the pyramid, and leave the brainstem at the lower border of the pons (cf. Fig. 7-6). The nerve then pierces the dura, traverses the cavernous sinus in its lateral part (Fig. 7-24), enters the orbit through the superior orbital fissure, and ends in the lateral rectus muscle of the eyeball, supplying it with motor fibers. In the sinus it is joined by some sympathetic fibers from the carotid plexus.

The nucleus of the trochlear nerve is found in the mesencephalon at the level of the inferior colliculus, a little ventral to the aqueduct (see Fig. 7-2) near the midline. Immediately ventral to it is the medial longitudinal fasciculus. The trochlear nerve is the only cranial nerve emerging on the dorsal aspect of the brainstem. Leaving the cells of the nucleus, the axons descend somewhat and then cross the median line dorsal to the aqueduct (see Fig. 7-22). The nerve emerges slightly above and lateral to the anterior medullary velum, then swings along the upper border of the pons around the cerebral peduncle (cf. Fig. 7-6), and pierces the dura in the anterior attaching fold of the tentorium (Fig. 7-24). After having traversed the cavernous sinus, where it is situated near the lateral wall, it enters the orbit through the superior orbital fissure and innervates the *superior oblique muscle* (Fig. 7-26). It will be seen that the right trochlear nucleus innervates the left superior oblique muscle and vice versa.

The nucleus of the oculomotor nerve, situated at the level of the superior colliculus, is more complex. Like the trochlear nucleus it is situated near the midline, ventral to the aqueduct, and has on its lateral and ventral side the medial longitudinal fasciculus (Fig. 4-5). The axons course ventral in the mesencephalon in several bundles, which leave the brainstem in the interpeduncular fossa, medial

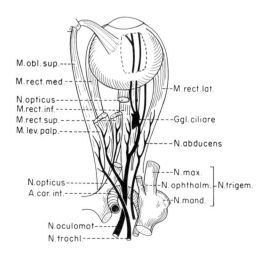

M.obl.sup.---
M.rect.med.--
N.opticus----
M.rect.inf.----
M.rect.sup.-----
M.lev.palp.------
                    --M.rect.lat.
                    ----Ggl.ciliare
                    -----N.abducens
N.opticus------
A.car.int.-----
                    ---N.max.
                    -N.ophthalm. ⎱N.trigem.
                    -N.mand. ⎰
N.oculomot----
N.trochl------

FIG. 7-26 The right eyeball and its muscles seen from above. The levator palpebrae and the superior rectus muscle and the optic nerve have been transected, and the inferior oblique muscle is not shown. Lateral to the optic nerve the ciliary ganglion is shown with its preganglionic branches from the oculomotor nerve and its postganglionic fibers to the intrinsic eye muscles (visible in a window of the eyeball). Modified from Rauber-Kopsch.

to the cerebral peduncle. The bundles partly penetrate the red nucleus (Fig. 4-5); the most lateral of them even penetrate the medial part of the cerebral peduncle. The several bundles then unite to form the oculomotor nerve, which courses forward for a distance before piercing the dura and entering the cavernous sinus. Its superior position here can be seen from Figs. 7-24 and 7-25. Then bending somewhat laterally and downward, the nerve enters the orbit in the medial part of the superior orbital fissure. In the orbit (Fig. 7-26) a superior ramus supplies the *superior rectus muscle* and the *levator palpebrae superioris,* and an inferior ramus splits into branches to the *inferior rectus,* the *inferior oblique,* and the *medial rectus muscles.* Furthermore, branches are given off to the *ciliary ganglion,* where the preganglionic parasympathetic fibers to the pupillary sphincter and the ciliary muscle (cf. below) terminate. In the cavernous sinus the oculomotor and the trochlear nerves are joined by some sympathetic fibers.

Unlike the abducent and trochlear nerves, the oculomotor is concerned in the innervation of more than one of the extrinsic eye muscles, and, furthermore, has a contingent of parasympathetic, visceral efferent fibers. Several groups and sub-groups have been distinguished (for references see, for example, Pearson, 1944; Warwick, 1953). Broadly speaking, the nucleus is made up of a larger *lateral nucleus* situated a short distance from the midline, a *median nucleus,* which is un-paired and extends in front of the lateral, and the *visceral efferent nucleus of Edinger-Westphal* (see Fig. 7-2), paired and interposed between the other parts. The latter is composed of small, pyriform cells of the preganglionic autonomic type, whereas the lateral and part of the median nucleus are composed chiefly of cells of the somatic motor type, as are the nuclei of the abducent and trochlear nerves.[54] It might be expected that definite regions of the somatic efferent part of the oculomotor nucleus should be related to the different extrinsic eye muscles in analogy with the situation in, for example, the facial nucleus. This has turned out to be the case.

Several investigators have studied this problem. Some investigators have used physiological methods, for example the recording of antidromic potentials in cells of the oculomotor nucleus following stimulation of the nerves to the different muscles (Bienfang, 1968; Naito, Tanimura, Taga, and Hosoya, 1974), whereas others studied the problem anatomically. Thus Warwick (1953) analyzed the retrograde changes occurring after extirpation of the individual extrinsic eye muscles in adult monkeys, while Tarlov and Tarlov (1971) used the modified Gudden method (Brodal, 1940a) in kittens for the same purpose, and Gacek (1974), likewise in kittens, used the HRP method (see Chap. 1) to identify cells supplying the different muscles. These studies are in general agreement concerning the main points and show that the neurons supplying a particular muscle are aggregated as groups within the oculomotor complex. Furthermore, there is agreement that the groups supplying the medial rectus, the inferior rectus, and the inferior oblique are located ipsilaterally; that the group supplying the superior rectus is located contralaterally; and that the levator palpebrae is supplied bilaterally from a caudal central group (caudal central nucleus).

As to details concerning the findings in the cat (Tarlov and Tarlov, 1971; Gacek, 1974; Naito et al., 1974) there are some minor discrepancies, presumably on account of differences in methods. However, it appears that there are some real differences between the patterns of oculo-motor innervation in the cat and the monkey. Species differences of this kind may be related to different degrees of binocular vision. Since one muscle may be active in different movements (cf.

[54] As will be discussed below, the morphologically outlined nucleus of Edinger-Westphal does not correspond precisely to the collection of visceral efferent neurons which, via the ciliary ganglion, act on the ciliary muscle and the sphincter of the iris.

below), it might furthermore be surmised that localization in the ocular nuclei will to some extent be in terms of movements rather than of muscles, as pointed out by Le Gros Clark (1926). Human cases of nuclear paresis or paralysis of one extrinsic eye muscle only are rare, and even more rarely are histological studies in such cases. Little is known, therefore, concerning a possible localization within the oculomotor nucleus in man (for references see Warwick, 1953).

The cells in the oculomotor, trochlear, and abducent nuclei are of different sizes, and not all are of the motoneuron type. In retrograde studies or HRP studies, all cells of a nuclear group have never been found to be affected or labeled. This may indicate that some neurons are true interneurons (Tarlov and Tarlov, 1971; Gacek, 1974) in the sense that their axons do not leave the territory of the nucleus. However, it might be assumed that the unlabeled cells in the studies mentioned represent "interneurons" between different oculomotor nuclei. (Such neurons should be called internuclear neurons rather than interneurons.) This assumption received support when it was found that retrogradely labeled cells occur in the oculomotor nucleus following injections of HRP in the abducent nucleus and that antidromic responses can be recorded in the former on stimulation of the latter (Maciewicz, Kaneko, Highstein, and Baker, 1975). This projection from the oculomotor to the abducent nucleus has been confirmed autoradiographically (Graybiel, 1977). The presence of interneurons (or better internuclear neurons) in the abducent nucleus, projecting to the contralateral oculomotor complex, has been demonstrated with the same methods (Graybiel and Hartwieg, 1974; Baker and Highstein, 1975). According to the autoradiographic study of Bienfang (1978), these fibers end in that part of the oculomotor nucleus which supplies the (contralateral) medial rectus muscle (the main agonist to the lateral rectus). It has been suggested that fibers interconnecting various oculomotor nuclei are collaterals of axons of motoneurons. Results with a technique of double retrograde tracing (Steiger and Büttner-Ennever, 1978) indicate, however, that internuclear neurons and motoneurons are two different kinds of cells. It may be mentioned that from HRP studies some neurons in the oculomotor nuclei (III, IV, and VI) have been concluded to project to the cerebellum, particularly to the vestibulocerebellum (Kotchabhakdi and Walberg, 1977).[55]

Thus, in principle, each extrinsic eye muscle receives its motor innervation from a particular group of the oculomotor nuclei, even if the pattern may not be rigid (see Gacek, 1974). The presence of neurons interconnecting the oculomotor and abducent nuclei is only one of many examples of arrangements that makes possible the finely adjusted cooperation of the various muscles in movements of the eyes.

The efferent fibers from the oculomotor nucleus are partly crossed and partly uncrossed, but exact knowledge of the proportions of crossed and uncrossed fibers from the different parts of the nucleus in man is lacking. In the monkey, according to Warwick (1953), the medial and inferior recti and the inferior oblique are supplied with uncrossed fibers only; the superior rectus receives only crossed fibers; while the levator palpebrae superioris has a bilateral innervation, a point of interest since as rule both eyelids are lifted simultaneously.

[55] Much information on the connections and interconnections of the oculomotor nuclei can be found in *Control of Gaze by Brain Stem Neurons* (1977).

Eye movements. The ocular muscles are brought into play both reflexly and voluntarily. The voluntary innervation presumably takes place through cortical fibers, originating from the cortical eye field (cf. below), passing through the internal capsule and cerebral peduncle. However, most of the movements performed by the eyes are not strictly voluntary. Thus those concerned in fixation, i.e., orienting the ocular axes to a given point, occur reflexly. Further, vestibular impulses (see section (d) of this chapter), impulses from the neck muscles, and auditory and other stimuli may elicit reflex movements of the eyes. The movements of the eyes are of different kinds. Some can be characterized as *slow movements* (pursuit movements, when the eyes follow a moving object, and convergence movements on focusing on a near object). The slow "active" phase of nystagmus likewise is of this type. Other movements are *fast* (the saccadic movements, which bring the fovea of the eye to the image of the target, and the quick phase of nystagmus). Finally, the extrinsic eye muscles must be able to hold the eyes in position during periods of fixation.

In recent years, owing to progress in research methods, a wealth of information on the underlying mechanism of the various kinds of eye movements has been brought forward. Recordings from eye muscles, from oculomotor neurons, and from sites in the brainstem active during different kinds of eye movements have been made. In spite of a great number of physiological studies, there are still unsolved questions, particularly concerning the mechanisms of central control of eye movements. For surveys of this field and references, the reader should consult special texts.[56] Here, mainly anatomical data of relevance for the subject will be considered. It will be appropriate to begin with some comments on the extrinsic eye muscles. The *mechanical features* of the action of the individual eye muscles can be understood in principle when their position relative to the optical axis of the eye is visualized.

> The origin of the lateral and medial rectus at the sides of the optic nerve and their insertion onto the lateral and medial aspect of the eyeball, anterior to its equator, make them act practically as a pure abductor and adductor, respectively. With the superior and inferior rectus the situation is more complicated, since their longitudinal axes diverge from the optical axis of the eye at rest. Only when the eye is abducted some 25° are these two axes parallel, and consequently the muscles in question will pull the eye purely upward or downward only in this position. With the eye axis directed straight forward, the superior and inferior recti will have an additional effect of adduction. The superior and inferior oblique muscles, with their insertion tendons reaching the eyeball from a medial and anterior direction, will move the eye around its transverse axis when it is adducted, the maximal effect being reached in an adduction of some 50°. In this position the superior oblique has only the effect of lowering the visual axis, whereas the inferior oblique will be a pure elevator. In intermediate positions the superior oblique collaborates with the inferior rectus, the inferior oblique with the superior rectus, the oblique muscles compensating by a component of abduction,[57] the adductor tendency of the recti. Thus it is safe to state that in most of the movements of the eye (except in pure lateral and medial movements) an oblique and a rectus muscle collaborate. The components in the action of the individual muscles are represented diagrammatically in Fig. 7-27.

[56] See, for example, *The Control of Eye Movements* (1971), *The Oculomotor System and Brain Functions* (1973), Cohen (1974), *Basic Mechanisms of Ocular Mobility and their Clinical Implications* (1975), *Control of Gaze by Brain Stem Neurons* (1977), Raphan and Cohen (1978).
　　　[57] The abductor component of the oblique muscles results from their being inserted behind the equator of the eye.

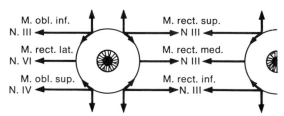

M. obl. inf.	M. rect. sup.
N. III	N III
M. rect. lat.	M. rect. med.
N. VI	N. III
M. obl. sup.	M. rect. inf.
N. IV	N. III

FIG. 7-27 Starling's diagram of the action of the extrinsic eye muscles.

The extrinsic eye muscles and their innervation present a number of interesting problems related to the capacity of these muscles to produce extremely finely coordinated conjugate movements of the eyes. Several anatomical features are known which cast some light on this phenomenon. Thus the muscles are amply innervated. In man, close to the brainstem the oculomotor nerve has some 24,000 fibers, the trochlear and the abducent nerves about 3,400 and 6,600, respectively (Björkman and Wohfart, 1936). Furthermore, the muscle fibers are thin and the motor units are small; that is, each nerve cell supplies only a few muscle fibers, probably about six (see Whitteridge, 1960).

Recent studies (for reviews see Peachey, 1971; Alvarado and Van Horn, 1975; Bach-y-Rita, 1975) have shown that the muscle fibers of the extrinsic eye muscles can be divided into five types, differing with regard to size, content of mitochondria, and in other respects. Two of these types appear to be multiply innervated, twitch fibers. The latter appear to be "phasic" and involved in rapid eye movements (as in saccadic movements and the fast phase of nystagmus), while the three former types are "tonic." They are active in maintaining fixation and are presumably responsible for highly precise, slow following movements, and for the slow phase in nystagmus.

Electromyographic investigations of the human extrinsic eye muscles (Björk and Kugelberg, 1953a, b; see also Teasdall and Sears, 1960) illuminate the finely coordinated activity of the extrinsic eye muscles referred to above. They show that there is activity, revealed as motor unit potentials, in the eye muscles in all positions of the eyes, except on maximal contraction of antagonists (for example, in the lateral rectus when the eye bulb is turned maximally inward). On looking straight ahead, all four recti muscles are active, and the activity in the individual muscles varies according to the position of the bulb. Electromyography may be of some value in the diagnosis of ocular muscle palsies (for a review, see Jampolsky, 1970). Single-unit recordings can also be made (see Collins, 1975).

The nervous factors implied in the ocular movements are many, and for details reference must be made to special textbooks. Some principal points only will be mentioned. First, a contraction of one of the eye muscles is always followed by a relaxation of its antagonist (Sherrington's law of reciprocal innervation). This is clearly seen also in ocular palsies. For example, in a complete pure abducens paralysis, where the paralyzed eye cannot be abducted actively, a slight abduction is nevertheless observed when the gaze is directed to the paralyzed side, due partly to a relaxation of the medial rectus (see later discussion). Secondly, there is always some degree of differential tone of the eye muscles, which tends to keep the axes of the two eyes directed to the same point in space. This compulsory tendency of fusion is temporarily abolished when vision is excluded from one of the eyes. In nearly all normal persons a slight squint then occurs (heterophoria). This is clear evidence that as a rule mechanical factors alone are not sufficient to secure a correct position of the eyeballs. For the same reason a squint is prone to develop when the vision of one eye is so reduced that corresponding visual impressions from both eyes cannot be fused to one image. Heterophoria should be borne in mind as a factor which may make the diagnosis of ocular palsies uncertain in some cases.

Another point which is of a certain practical interest in evaluating changes in ocular movements should also be mentioned. Normally there is a tendency for convergence of the eyeballs with the gaze directed downward, and inversely a tendency of divergence when the gaze is directed upward. There are, however, other phenomena of a similar nature which occur normally, and which frequently make the exact diagnosis of an ocular muscle palsy a more difficult matter than should perhaps be expected from their rather schematic mechanical arrangement. This will not be discussed further in this text, but it is important not to overlook functional factors when anomalies of ocular movements are to be analyzed.

The discussion above is an extremely simplified account of a very complex problem. Many new data from recent years illustrate the intricacy of the mechanisms involved in movements of the eyes. Before discussing anatomical information on the central control of eye movements, it is advantageous to consider the pathways along which stimuli of various origins may reach the nuclei of the extrinsic eye muscles.

Afferent connections of the oculomotor nuclei. The proprioceptive innervation of the extrinsic eye muscles. In order to effect the delicately adjusted and coordinated conjugate movements of the eyes, the motor nuclei supplying the extrinsic eye muscles must receive information of many kinds, first and foremost from the retinae and from the muscles themselves. In addition, impulses from other sources, such as the vestibular apparatus (adjustments of the position of the eyes to changes in posture and movements of the head), the cerebral cortex (voluntary influences on movements of the eyes), and the cerebellum are necessary. Our knowledge of where and how the integration of these impulses occurs is still incomplete. Only some of the afferent pathways concerned have been traced directly to the oculomotor nuclei. It should be noted that *retinal afferents* are not among these. Nor do proprioceptive impulses from the extrinsic ocular muscles reach these nuclei. Further, claims of fibers from the *cerebral cortex,* particularly from the frontal eye field, area 8, passing directly to the nuclei of the extrinsic eye muscles have not been confirmed in recent studies with silver impregnation (Kuypers and Lawrence, 1967; Astruc, 1971) and autoradiographic methods (Künzle and Akert, 1977), nor have fibers been traced from the *superior colliculus (see Harting, 1977) or from the pretectal nuclei* (Tigges and O'Steen, 1974; Berman, 1977) to the nuclei of the extrinsic ocular muscles.

However, in a recent experimental study in the cat, Edwards and Henkel (1978) found fibers from the superior colliculus to end in areas in the central (periaqueductal) gray directly overlying the oculomotor complex and containing dendrites of oculomotor neurons. Tectal afferents were further traced to cell groups in close relation to the trochlear and abducent nuclei. Some appear to pass directly to the abducent nucleus, which in addition receives some afferents from the region of the periaqueductal gray that receives fibers from the superior colliculus. It appears from these studies that the superior colliculus may be able to influence some of the ocular motoneurons more directly than was previously assumed.

Among afferent fiber connections that pass directly to the oculomotor, the abducent, and the trochlear nuclei, those from the *vestibular nuclei* have been most extensively studied. As described in section (d) of this chapter, fibers from different parts of the vestibular nuclei appear to supply particular regions of the nuclei of the extrinsic eye muscles in a rather specific pattern. These connections are first and foremost links in direct vestibulo-ocular reflex arcs. Other direct fibers to the oculomotor nuclei have been traced from the *cerebellum* (see Chap. 5). Recently, anatomical as well as physiological evidence has been brought forward describing a projection from the *perihypoglossal nuclei,* particularly from the nucleus prepositus, to the oculomotor nuclei (see Chap. 5). Fibers from the *reticular formation* have been seen in Golgi material to give off collaterals to the oculomotor nuclei (Scheibel and Scheibel, 1958; Szentágothai, 1964b), and in experimental anatomical studies fibers have been traced to the abducent nucleus (Büttner-Ennever and Henn, 1976; Graybiel, 1977; and others). Further, in physiological studies monosynaptic potentials have been recorded in the oculomotor nucleus following stimulation of the potine reticular formation (Highstein, Cohen, and Matsunami, 1974; see also below). It is well known that impulses from the *neck* influence eye movements. However, while spinal afferent fibers supply parts of the vestibular nuclei and via these impulses may be transmitted to the oculomotor nuclei (see section (d) of this chapter), the existence of fibers ascending directly from the spinal cord to the oculomotor nuclei is uncertain. For a recent HRP study of the afferent connections of the oculomotor nucleus in the monkey, see Steiger and Büttner-Ennever (1979).

We will consider in some detail the afferent information to the oculomotor nuclei provided by *proprioceptive impulses from the extrinsic ocular muscles.* Very elaborate proprioceptive mechanisms are obviously necessary for the performance of the extremely finely graded eye movements. It might therefore be assumed a priori that the eye muscles are endowed with proprioceptors. The occurrence of muscle spindles in the extrinsic eye muscles was described in different species and man by early workers (for a review, see Cooper and Daniel, 1949), but only during the last 30 years have we obtained precise information on the sensory organs of these muscles. Modern electrophysiological methods have further made it possible to study their proprioceptive function in great detail (for a recent review, see Bach-y-Rita, 1975). Many problems related to the proprioceptive innervation of the extraocular muscles are, however, still far from solved. Complicating factors in such studies are the differences among animal species with regard to the proprioceptive end organs. Although all animals have some kind of proprioceptive receptors, muscle spindles have been demonstrated only in some, for example in pig, sheep, goat, and man, but not in cat, dog, and rabbit. In human ocular muscles Daniel (1946), by using very thick sections of silver-stained material, found an abundance of spiral nerve endings, which in several turns encircle single muscle fibers. These were assumed to be sensory. In further studies in which the muscles were cut serially, Cooper and Daniel (1949) and Merrillees, Sunderland, and Hayhow (1950) could demonstrate typical muscle spindles in all human extraocular muscles.

From 22 to 71 spindles are found in one muscle. The spindles are located only in certain regions, mainly near the proximal ends of the muscles, and have more delicate capsules and thinner intrafusal fibers than those in skeletal muscles. Small motor end plates can be seen at the poles of the spindles (Cooper, Daniel, and Whitteridge, 1955), representing an innervation of γ fibers. In addition to muscle spindles the extraocular muscles of man contain other kinds of nerve endings which are apparently sensory, for example some that resemble the Golgi tendon organs. The γ neurons supplying the extrinsic ocular muscles remain to be identified, but the efferent innervation of the spindles provides morphologic evidence that they are under central control.

According to the physiological studies of Cooper, Daniel, and Whitteridge (1951) and others, it appears that some of the proprioceptive impulses from the ocular muscles reach the brainstem via the ocular motor nerves. It appears from several studies (see, for example, Batini, Buisserat, and Buisserat-Delmas, 1975) that most of the impulses pass from the ocular motor nerves to the ophthalmic branch of the trigeminal nerve by way of anastomoses in the orbit or in the cavernous sinus. Clusters of ganglion cells have been observed in the oculomotor nerve in man and in some animals (Bortolami et al., 1977).

Physiologically static and dynamic fusimotor innervation (see Chap. 3) has been found in spindles in the extrinsic ocular muscles (in the pig), and responses from primary and secondary endings and responses that appear to be derived from Golgi tendon organs have been recorded (see Bach-y-Rita, 1975, for a survey and particulars).[58] After entering the brainstem the proprioceptive impulses appear to follow different routes.

On stretching the muscles, investigators recorded action potentials with short latencies from the mesencephalic trigeminal nucleus along its entire length in the goat (Cooper, Daniel, and Whitteridge, 1953a, b) and in the cat (Fillenz, 1955; Jerge, 1963a), whereas Alvarado-Mallart et al. (1975) found responses only in a restricted part in the cat. (Following HRP injections in the lateral rectus muscle of the cat, the latter authors found some labeled cells in the same area of the mesencephalic nucleus.) The potentials recorded apparently are of the same kind as those recorded in this nucleus following stretching of the muscles of mastication. Some fibers from the mesencephalic nucleus can be traced to the nuclei of the extrinsic eye muscles (Pearson, 1949a; and others). It is possible, therefore, that there exists a two-neuron reflex arc for a stretch reflex in these muscles (neuron of the mesencephalic nucleus–periperal motor neuron), as for the muscles of mastication, but physiological studies do not appear to support this assumption (see Schwarz and Tomlinson, 1977, for some references).

Other authors, particularly Manni and his collaborators, have recorded responses to stretch of extrinsic ocular muscles in the trigeminal ganglion (in the lamb, pig, and cat). They even found a topical representation of different muscles in the ganglion (Manni and Pettorossi, 1976). From combined electrophysiological and anatomical experiments they concluded that the central processes of the ganglion cells receiving impulses from the extrinsic ocular muscles end mainly in the pars oralis of the spinal trigeminal nucleus (Manni, Palmieri, and Marini, 1972). From the pars oralis the impulses are propagated ipsilaterally in a rostral direction, in part via the medial lemniscus, to the VPL and VPM of the thalamus (Manni, Palmieri, and Marini, 1974). Proprioceptive impulses passing via the trigeminal

[58] It appears that in man proprioceptive impulses from the extrinsic ocular muscules are not, as often assumed, of importance for the "position sense" of the eye. If the conjunctiva is anesthetized, the subject is unable to detect passive movements of the eye if visual clues are excluded (Brindley and Merton, 1960).

ganglion probably must follow a rather circumvential route before they can influence the extrinsic eye muscles.

The fact that potentials can be recorded in the vermal visual area of the cerebellum of the cat
following stretching of extrinsic eye muscles, and both as mossy and climbing fiber responses,
agrees with observations that the cerebellum plays a role in the control of the extrinsic eye
muscles (see Chap. 5). Following stimulation of extraocular proprioceptors, responses have further been recorded in the superior colliculus (Abrahams and Rose, 1975). Their long latency
suggests that they do not follow a direct route.

**The accessory oculomotor nuclei, the pretectal region, and the superior
colliculus.** It is evident from the discussion above that the integration of messages from various sources that influence the movements of the eye cannot take
place in the oculomotor nuclei and must be assumed to occur elsewhere. Three
cellular regions, situated in the vicinity of the oculomotor nucleus and related to
visual functions, are of particular interest: the so-called *accessory oculomotor nuclei,* the *pretectal region,* and the *superior colliculus.* Our knowledge of their connections and functions are, however, still far from complete. Experimental studies
particularly of some of them are extremely difficult because the cell groups are
small and situated closely together.

The *accessory oculomotor nuclei* (see Carpenter and Peter, 1970/71) consist
of three small cell groups situated close to, and partly within, the periaqueductal
gray (see Fig. 7-28). The *nucleus of the posterior commissure* is found most dorsally and is connected with its partner on the other side by fibers of the posterior
commissure, which also carries fibers of other origins. The small *nucleus of
Darkschewitsch* is partly embedded in the periaqueductal gray. The most ventrally
situated, the *interstitial nucleus of Cajal,* is closely related to the medial longitudinal fascicle, whose fibers in part pass through it. *These nuclei receive afferent
fibers from several regions, but not from the retina* (cf. Fig. 7-31).

Fibers arising in the vestibular nuclei and ascending in the medial longitudinal fasciculus end
in the interstitial nucleus of Cajal (section (d) of this chapter). Otherwise, little is known of the afferent connections of the accessory oculomotor nuclei. A few fibers to the interstitial nucleus appear to come from the frontal visual area 8 (Astruc, 1971; see, however, Künzle and Akert,
1977), and Kuypers and Lawrence (1967) describe fibers from the precentral and the postcentral
region (in the monkey) to the nucleus of Darkschewitsch. Further, these nuclei appear to receive
some afferents from the superior colliculus (Altman and Carpenter, 1961), from the cerebellar
nuclei, and from the nucleus prepositus hypoglossi (see Chap. 5). From experiments in which
lesions of the retina were induced, fibers appear not to have been traced to these small nuclei, but
some nuclei of the pretectal region, receiving retinal afferents, have been concluded to send fibers
to the nucleus of Darkschewitsch and the interstitial nucleus (Berman, 1977; Benevento, Rezak,
and Santos-Anderson, 1977). Some spinal fibers, ascending in the ventral funiculus, were traced
in monkeys to the nucleus of Darkschewitsch by Kerr (1975).

In physiological experiments in which the interstitial nucleus is assumed to be
stimulated, rotation of the head, ocular torsion, and deviation of the eyes have
been observed. This is compatible with what is known of the efferent connections
of this small nucleus.

It gives off fibers to the *spinal cord,* (the interstitiospinal tract, see Chap. 4),
the *vestibular nuclei* (see section (d) of this chapter), and the *perihypoglossal
nuclei* and the dorsal part of the *paramedian reticular formation,* as recently

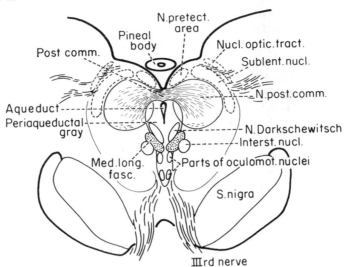

FIG. 7-28   Outline drawing of a section through the monkey brainstem at the level of the posterior commissure (immediately rostral to the superior colliculus), showing the position of the accessory oculomotor nuclei and the pretectal region. Three of the subnuclei of the latter are outlined with broken lines. The accessory oculomotor nuclei (interstitial nucleus of Cajal, nucleus of Darkschewitsch, and the nucleus of the posterior commissure) are indicated. The medial longitudinal fascicle is shown dotted. See text. Slightly altered from Carpenter and Pierson (1973).

shown by Carpenter, Harbison, and Peter (1970).[59] These authors further studied (silver impregnation method) the projections from the accessory to the *proper oculomotor nuclei*. It appears from their study that the interstitial nucleus gives off fibers which, after crossing in the posterior commissure, bilaterally but differentially supply all except one of the cell groups of the oculomotor nuclei, as well as the trochlear nucleus.

Isolated lesions of the nucleus of Darkschewitsch were not achieved, and conclusions about its efferent connections could not be made. Nor could it be decided whether the nucleus of the posterior commissure projects to the oculomotor nuclei. In an HRP study, Graybiel and Hartwieg (1974) confirmed the projection from the interstitial nucleus and found evidence suggesting a projection from the nucleus of the posterior commissure to the third nerve nucleus.

As far as can be surmised from the anatomical data, the accessory oculomotor nuclei may serve particularly to integrate vestibular impulses with impulses from the cerebral cortex and the superior colliculus, while input from the retina may contribute only via intercalated stations, first and foremost the pretectal region, which receives afferents from the retina.

The *pretectal region*, situated at the level of the posterior commissure and just rostral to the superior colliculus, is a cellular area that can be subdivided into subnuclei (Authors differ somewhat in distinguishing the nomenclature and sub-

[59] Loewy and Saper (1978) recently reported finding labeled cells in the nucleus of Darkschewitsch and the interstitial nucleus of Cajal following injections of HRP in different regions of the medulla. Although this must be interpreted to show a descending projection from these nuclei, their sites of termination cannot be definitely ascertained from these experiments.

divisions; see Carpenter and Peter, 1970; Scalia, 1972; Kanaseki and Sprague, 1974; Benevento, Rezak, and Santos-Anderson, 1977.) Three of these nuclear groups can be seen in Fig. 7-28. Particularly in recent years, much new anatomical information about the intricate fiber connections of the pretectal region has appeared. It has been possible to obtain some idea of the general pattern of its organization. Only main points will be considered here (compare Fig. 7-31).

When traced autoradiographically in the monkey, fibers from the *retina* have been found by several authors (Hendrickson, Wilson, and Toyne, 1970; Tigges and O'Steen, 1974; Pierson and Carpenter, 1974; Benevento, Rezak, and Santos-Anderson, 1977) to end mainly contralaterally in the sublentiform nucleus (see Fig. 7-31) and in a small region called the pretectal olivary nucleus (not seen in Fig. 7-28, but closely associated with the sublentiform nucleus). Some authors found fibers to the nucleus of the pretectal area as well. In the cat a somewhat different projection was found by Berman (1977). In addition to retinal afferents, the pretectal nuclei receive fibers from several other regions, among them the cerebral cortex and the superior colliculus (see Fig. 7-31). It should be noted that *no evidence has been found for fibers from the pretectal nuclei to any of the somatic motor nuclei of cranial nerves III, IV, and VI.*

The fibers from the *cerebral cortex* appear to come chiefly from the visual areas in the occipital lobe (Kuypers and Lawrence, 1967; Harting and Noback, 1971; Kawamura, Sprague, and Niimi, 1974). These, as well as the afferents from the *superior colliculus* (Tarlov and Moore, 1966; Rafols and Matzke, 1970; Berman, 1977), appear according to Benevento and his collaborators (see Benevento et al., 1977) to terminate in those pretectal nuclei that receive afferents from the retina. Further, fibers have been traced from the frontal eye field (Astruc, 1971; Künzle and Akert, 1977) and from the lateral geniculate body (from the ventral geniculate nucleus, in primates called the pregeniculate nucleus, receiving retinal fibers; see, for example, Harting and Noback, 1971; Swanson, Cowan, and Jones, 1974; Edwards, Rosenquist and Palmer, 1974; Ribak and Peters, 1975; Berman, 1977), possibly from the posterior regions of the thalamus (see Carpenter and Pierson, 1973), and from the pontine reticular formation (Graybiel, 1977).

The *efferent* connections of the pretectal nuclear groups are to some extent reciprocal to their afferents. In autoradiographic studies in the cat (Berman, 1977) and in the monkey (Benevento, Rezak, and Santos-Anderson, 1977), fibers were traced to the superior colliculus, the ventral lateral geniculate nucleus, the accessory oculomotor nuclei, and the pulvinar (see also Carpenter and Pierson, 1973), as well as to other regions less immediately related to visual functions, such as the reticular thalamic nucleus and parts of the intralaminar thalamic nuclei, the reticular formation of the brainstem, the hypothalamus, and some other regions. Most of these connections are described by Itoh (1977) in degeneration studies following electrolytic lesions in the pretectal region in the cat. In addition, he describes fibers to the reticular formation, pons, and inferior olive. There are interconnections between the different pretectal nuclei.

The pretectal region has been considered to be an area involved in the reflex arc of the pupillary light reflex. A question of particular interest is, therefore, whether *it projects onto the visceral motor nucleus of Edinger-Westphal.* In monkeys the groups of the oculomotor nucleus assumed to provide the visceral innervation of the eye have been determined on the basis of retrograde cellular changes following extirpation of the ciliary ganglion (Warwick, 1954; Pierson and Carpenter, 1974). Fibers to these visceral nuclear groups were traced by Carpenter and Pierson (1973) from the olivary pretectal nucleus, but they apparently do not supply the entire territory of the visceral cell groups. According to Benevento, Rezak, and Santos-Anderson (1977), the fibers from the pretectal region passing to the visceral oculomotor nucleus in the monkey come from the sublentiform as well as the olivary nucleus (that is the pretectal nuclei that receive retinal afferents). Some of the fibers cross in the posterior commissure and supply the contralateral visceral nucleus.

These recent findings contribute to an understanding of the light and accommodation reflexes, to which we shall return later (see Fig. 7-31). Furthermore, it is interesting, as pointed out by Benevento, Rezak, and Santos-Anderson (1977), that there appear to be in the pretectum (as in the superior colliculus; see below) certain differences in connections between nuclear groups which are, or are not, closely related to visual functions (for particulars the original article should be consulted).

The *superior colliculus,* the third region mentioned above as an area of interest for the integration of impulses influencing the oculomotor nuclei, has been extensively studied in recent years. Anatomical and numerous physiological studies have produced evidence that it is a brain region quite differentiated and related to various aspects of visual function. It is possible here to consider only some salient points concerning the organization and function of the superior colliculus. Recent references will be mentioned primarily.

When studied anatomically, the superior colliculus has a laminar structure, with alternating fiber-rich and cell-rich laminae. It appears to be more important for visual functions in lower than in higher vertebrate species.

> In lower vertebrates the superior colliculus appears to be the most highly differentiated visual structure. When the cerebral cortex develops in the phylogenetic ascent, the striate area of the cortex eventually becomes the main end station for the visual impulses (via a relay station in the lateral geniculate body). The fibers that terminate in the superior colliculus gradually become reduced in number. This has for a long time been considered an example of the "migration" of functions to higher levels in the central nervous system, usually referred to as *"corticalization"* or *"encephalization."* However, recent data suggest that the problem is less simple (see Ingle and Sprague, 1975). Nevertheless, it is important to be aware of species differences when experimental physiological results are to be compared with clinical findings in human beings, and it is necessary to be cautious with regard to conclusions concerning homologies.

The many afferent and efferent fiber connections of the superior colliculus have been found to be related more or less clearly to particular layers (see below). In mammals four main layers have often been distinguished. Most superficial is layer I, the *stratum zonale,* consisting of thin fibers derived mainly from the occipital cortex. Then follows the *stratum cinereum* (II) or outer gray layer, consisting of mainly small cells, giving off their axons to or through the deeper layers. The *stratum opticum,* layer III, is a fiber-rich layer containing many optic tract fibers. The remaining layers, often referred to collectively as the *stratum lemnisci,* can be further subdivided into fiber- and cell-rich layers. Kanaseki and Sprague (1974) who made a detailed study of the lamination of the cat's superior colliculus (Fig. 7-29; see their article for particulars) subdivide the stratum lemnisci into a lamina griseum intermedium (IV), a lamina album intermedium (V), a lamina griseum profundum (VI), and a lamina album profundum (VIII). In general, the superficial layers receive their main input from the retina and visual cortex, while the layers below these receive a variety of other afferent connections (see below) and give rise to most of the efferents of the superior colliculus (see Fig. 7-31).

The functionally most important input to the superior colliculus is probably

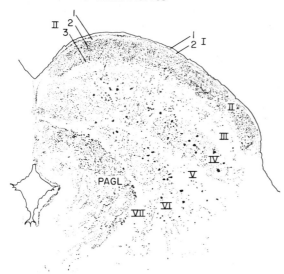

FIG. 7-29 Projection drawing of the superior colliculus in the cat, showing its lamination as seen in a Nissl-stained section. *I:* stratum zonale; *II:* stratum griseum superficiale; *III:* stratum opticum; *IV:* stratum griseum intermedium; *V:* stratum album intermedium; *VI:* stratum griseum profundum; *VII:* stratum album profundum; *PAGL:* lateral nucleus of periaqueductal gray. Layers I and II can be further subdivided (1, 2, 3). From Kanaseki and Sprague (1974).

that from the *retina*. Some of the optic nerve fibers do not follow the course of the majority through the optic tract to the lateral geniculate body (see Chap. 8), but leave the tract as it enters the latter. These fibers, situated on the medial aspect of the optic tract (most of them having crossed in the chiasma), pass through the superior quadrigeminal brachium to the superior colliculus on the same side, but some also cross and reach the contralateral superior colliculus. The fibers enter in the stratum opticum and, to a lesser extent, in the stratum zonale. Early experimental anatomical studies demonstrated a strictly localized arrangement in the projection of the retina on the superior colliculus in several lower vertebrates as well as in the rat (Lashley, 1934), the rabbit (Brouwer, Zeeman, and Houwer, 1923; Putnam and Putnam, 1926), and the opossum Bodian, 1937). In higher mammals the number of optic fibers passing to the superior colliculus becomes progressively reduced, and a clear-cut localization has been more difficult to determine. From recent studies it has been learned, however, that a *retinotopical projection* is present in the cat (Laties and Sprague, 1966; Kanaseki and Sprague, 1974; Graybiel, 1975a; Harting and Guillery, 1976) and in the monkey (Wilson and Toyne, 1970; Tigges and O'Steen, 1974; Hubel, LeVay, and Wiesel, 1975).

It appears from these studies that the lower half of the retina (that is, upper half of the visual field) projects to the medial side of the superior colliculus, and the upper half of the retina to the lateral side of the colliculus. The horizontal meridian runs from anterolateral to posteromedial (Fig. 7-30). The central part of the visual field, represented rostrolaterally in the colliculus, occupies a relatively larger territory than the peripheral areas. Further, homonymous halves of the visual fields of both eyes are represented in the contralateral superior colliculus (some fibers do not cross in the optic chiasm; compare the course of retinogeniculate fibers, Chap. 8). In Julia Apter's pioneer study (1945) a corresponding topical pattern in the retinocollicular projection was shown electrophysiologically, and has later been confirmed by others, for example Feldon, Feldon, and Kruger (1970); Goldberg and Wurtz (1972).

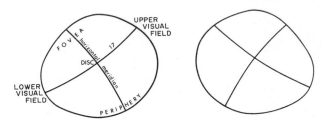

FIG. 7-30 Diagram of the representation of various parts of the visual field in the superior colliculus, as determined in an autoradiographic study in the monkey by Hubel, LeVay, and Wiesel (1975).

In autoradiographic studies, following injection of tritiated amino acids in the eye, further details have been demonstrated, for example an anatomical (bandlike) segregation of the terminals from the two eyes, in the monkey (Hubel, LeVay, and Wiesel, 1975) and in the cat (Graybiel, 1975b). The fibers terminate mainly on dendrites in the superficial part of the outer gray layer (lamina II_1), as confirmed also electron microscopically (Lund, 1972; Valverde, 1973; Sterling, 1973; Hubel, LeVay, and Wiesel, 1975). Golgi studies in the mouse (Valverde, 1973) and the rat (Langer and Lund, 1974) show that the superficial layers, receiving the retinal afferents, contain many different cell types.

Information originating in the retina may reach the superior colliculus not only directly but also indirectly, via the cortical visual areas (see Chap. 8) by way of *corticotectal fibers*. Numerous studies have been devoted to these fibers (see Altman, 1962, for references to earlier literature). In recent years it has been found that not only the classical cortical occipital visual areas (see Chap. 8) and the frontal eye field (area 8) but also other cortical areas send fibers to the superior colliculus. Furthermore, some subcortical regions that receive visual input project onto the superior colliculus.

The *occipital corticotectal projections* (Fig. 7-31) have been shown in several studies to arise from areas 18 and 19 as well as 17 (Altman, 1962; Garey, Jones, and Powell, 1968; Kawamura, Sprague, and Niimi, 1974). Most authors consider these projections to be strictly ipsilateral (see, however, Powell, 1976). The projections from these areas (entering the colliculus via the brachium quadrigeminum superius) are *topically organized* in agreement with the pattern of the retinocollicular projection (Fig. 7-30), as shown anatomically (Garey, Jones, and Powell, 1968; Harting and Noback, 1971; Kawamura, Sprague, and Niimi, 1974; and others) and physiologically (McIlwain, 1973; and others). HRP studies indicate that the cells of origin are pyramidal cells in layer V of the cortex of areas 17, 18, and 19 (Holländer, 1974a; Magalhães-Castro, Saraiva, and Magalhães-Castro, 1975). Physiologically, the fibers have likewise been found to come from layer V (Palmer and Rosenquist, 1974). Their axons appear to end on cells chiefly in the superficial gray layer of the colliculus, but deeper than the retinal afferents (see Ingle and Sprague, 1975).

In addition to fibers from the occipital visual areas 17, 18, and 19, the superior colliculus receives fibers from the ipsilateral *"frontal eye field,"* corresponding to area 8 of Brodmann, from which conjugate eye movements can be elicited. Although such fibers were not identified in studies with the Marchi method, they have been found with silver impregnation methods in the monkey (Astruc, 1971) and confirmed autoradiographically (Künzle and Akert, 1977).

Of *other cortical regions* found to send fibers to the superior colliculus may be mentioned the so-called *Clare-Bishop area* (situated in the lateral suprasylvian gyrus in the cat) which receives an optic input (see Chap. 8). Fibers have further been traced from some "nonvisual" regions, such as the *primary sensorimotor cortex* (Kuypers and Lawrence, 1967; Garey, Jones, and Powell, 1968; Price and Webster, 1972), the *temporal and parietal cortex* (Kuypers and Lawrence, 1967), the *prefrontal cortex* outside the "frontal visual field" (Goldman and Nauta, 1976), and parts of the *acoustic cortex* (Garey, Jones, and Powell, 1968; Paula-Barbosa and Sousa-Pinto, 1973). According to Wise and Jones (1977), in the rat the corticotectal fibers from the first somato sensory area (Sm I) are derived from large cells in cortical layer V. The projection is somatotopically organized.

As appears from the discussion above, several nonvisual cortical areas may influence the superior colliculus. Likewise, many of the other regions projecting onto the superior colliculus are not, or only remotely, related to visual functions.

Afferents from the *spinal cord* (spinotectal fibers) were described by the early

FIG. 7-31 Simplified diagram of the main afferent and efferent connections of the superior colliculus (left side of figure) and of the pretectal nuclei (right side). The various nuclei of the pretectum are not indicated (see text). Efferent connections from a region are shown by heavy lines, its afferent connections by thin lines. Note that the layers of the superior colliculus differ with regard to the origin of their afferents and the destination of efferents as further described in the text. The connection shown as a broken line from the pretectal nuclei to the visceral (Edinger-Westphal) nucleus is considered to be part of the elementary reflex arc for the light reflex. Many known connections are not included. To simplify the drawing, all connections except the tectospinal tract are shown as being ipsilateral. For particulars see text. Concerning layers of the superior colliculus see Fig. 7-29. *D:* nucleus of Darkschewitsch; *Int.:* interstitial nucleus of Cajal.

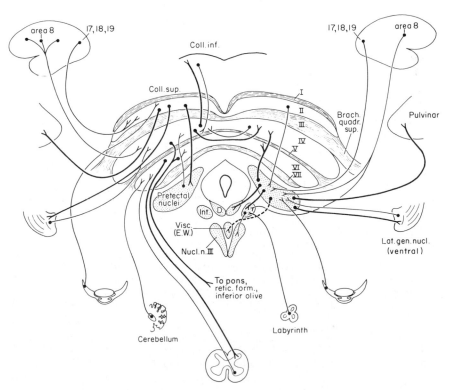

authors referred to by Mehler, Feferman, and Nauta (1960), who found the termination of such afferents in the monkey to be restricted to the deep layers of the colliculus.

Spinotectal fibers were noted in man by Bowsher (1957). On the basis of experimental studies (silver impregnation method) in the rat, Antonetty and Webster (1975) found the spinotectal fibers to end in a transverse somatotopical pattern (sacral segments caudally, cervical segments rostrally) in the deep layers in the caudal half of the contralateral superior colliculus. After injections of HRP in the superior colliculus in cats, Baleydier and Mauguiere (1978) found labeled cells in the contralateral cervical spinal cord and doral column nuclei (confirming some previous findings) and in the contralateral spinal and main trigeminal nuclei. The somatotopic pattern found in the termination of spinal afferents (Antonetty and Webster, 1975) agrees with that determined autoradiographically in the collicular projection from Sm I in the rat. The cortical face region projects to a large rostral part (Wise and Jones, 1977), as also found physiologically (see also below).

Fibers from the *inferior colliculus* which receive acoustic information (see Chap. 9) end in the deep layers of the superior colliculus (Moore and Goldberg, 1963, 1966; Powell and Hatton, 1969). The *pretectal region*, receiving fibers from the superior colliculus, projects back onto its deeper and intermediate layers according to autoradiographic studies (Benevento, Rezak, and Santos-Anderson, 1977; Berman, 1977; Grofová, Ottersen, and Rinvik, 1978). The intermediate layers further receive fibers—mainly ipsilateral and relatively sparse—from the *ventral lateral geniculate nucleus* (Edwards, Rosenquist, and Palmer, 1974; Graybiel, 1974; Swanson, Cowan, and Jones, 1974; Ribak and Peters, 1975). Fibers from the *cerebellum* (the fastigial nucleus) have been found to end in the deepest layers of the superior colliculus (Angaut, 1969; Angaut and Bowsher, 1970; Batton, Jayaraman, Ruggiero, and Carpenter, 1977), suggesting a cerebellar control of the movements elicitable from the superior colliculus (see Chap. 5). Other afferents have likewise been described, for example from the *substantia nigra,* its pars reticulata (Rinvik, Grofová, and Ottersen, 1976; Jayaraman, Batton, and Carpenter, 1977), and the *periaqueductal gray* (Hamilton and Skultety, 1970).[60] A study of the sources of afferents from the mesencephalon and diencephalon (Grofová, Ottersen, and Rinvik, 1978), with the method of retrograde axonal transport of horseradish peroxidase, confirmed most of those reported by other authors and in addition disclosed other afferents, including fibers from the *zona incerta,* the *hypothalamus,* and the *reticular thalamic nucleus.*

The superior colliculus gives off *efferent fibers* both in the caudal and in the rostral direction (Fig. 7-31). *Descending* efferents pass to some nuclei in the brainstem and to the spinal cord. The latter fibers constitute the *tectospinal tract* (see Chap. 4), which is crossed and terminates mainly in Rexed's laminae VI and VII in the upper cervical segments. It is generally assumed that the tectospinal tract is concerned in the head turning that can be elicited from the superior colliculus.[61]

[60] According to autoradiographic studies in the cat (Graybiel, 1978), the fibers from the substantia nigra terminate in the intermediate layers of the colliculus in a pattern of bands oriented roughly longitudinally. A corresponding pattern has been found in the termination of retinal afferents as well (Graybiel, 1975a). It is likely that the nigrotectal projection is a link in pathways by which the corpus striatum (see Chap. 4) may influence eye movements. Disturbances of visuomotor functions may be seen in disease of the basal ganglia, for example parkinsonism.

[61] About 50% of antidromically identified cells of origin of the tectospinal tract, located mainly within intermediate and deep layers of the superior colliculus, were found to receive a convergent input from extraocular muscles, neck afferents, and the retina (Abrahams and Rose, 1975).

Tectal efferents further pass to the pontine nuclei and the reticular formation of the medulla, pons, and mesencephalon, and some pass to the inferior olive. The *tectopontine tract* was known to the classical neuroanatomists and has been demonstrated in many animal species, for example in the cat (Altman and Carpenter, 1961), the opossum (Rafols and Matzke, 1970), and the monkey (Harting, 1977). It descends ipsilaterally and terminates in the dorsolateral pontine nucleus.

From a detailed analysis of its terminal area in the cat (Kawamura and Brodal, 1973), it appears that there is some degree of topical relation between the colliculus and its pontine projection area (see also Hashikawa and Kawamura, 1977), and that the latter coincides almost completely with one of the pontine areas that project onto the vermal visual area of the cerebellum (Hoddevik, Brodal, Kawamura, and Hashikawa, 1977). As discussed in Chapter 5, the tectopontocerebellar pathway is probably the main route for the influences exerted on the cerebellum by visual impulses.

The descending fibers from the superior colliculus to the *reticular formation of the brainstem* have been found in various species to be ipsilateral to the mesencephalic reticular formation, and crossed to the medullary and pontine reticular formation.

According to Kawamura, Brodal, and Hoddevik (1974; see also Graham, 1977; see these reports for references) and as described in Chapter 6, the maximal sites of termination of tectoreticular fibers in the pontine and medullary reticular formation appear to coincide approximately with those areas that give off the majority of reticulospinal fibers. In the monkey the fibers were traced only to the pontine reticular formation (Harting, 1977). While these data indicate a second, indirect route for influences of the superior colliculus on the spinal cord, the tectoreticular fibers may have other functions as well. Thus, some fibers end in those reticular nuclei, particularly the nucleus reticularis tegmenti pontis, which project onto the cerebellum (see Chap. 5). Reticular projections to the motor eye nuclei, for example from the region referred to as the PPRF (see later in this chapter), may be involved when movements of the eyes are elicited from the superior colliculus. In a physiological study with recording of potentials in motoneurons of the abducens nucleus following stimulation of the superior colliculus, Precht, Schwindt, and Magherini (1974) found that the ipsilateral abducens neurons were inhibited and contralateral excited (in agreement with the direction of eye movements evoked). The pathway involved is polysynaptic, and from lesion experiments it appears that it passes via the reticular formation.

The superior colliculus finally gives off some descending fibers to the *inferior olive*. In recent study Weber, Partlow, and Harting (1978) found the terminal area of these fibers to be a region that projects onto the vermal visual area of the cerebellum (see Chap. 5). From autoradiographic studies (Harting, 1977, in the monkey; Graham, 1977, in the cat) fibers from the superior colliculus have further been concluded to end in some other nuclei in the brainstem, such as the *cuneiform nucleus* and parabigeminal nucleus, *the raphe nuclei,* and the *substantia nigra*.

Commissural connections between the two superior colliculi have been mentioned by many authors. According to the autoradiographic and HRP studies of Edwards (1977; see also Magalhães-Castro et al., 1978) in the cat, they are rather specifically organized and interconnect mainly the stratum griseum intermedium in the rostral part of the colliculi concerned with central vision). Fibers further pass to the *inferior colliculus*. Of particular interest are projections to the *pretectal nuclei* (Benevento and Fallon, 1975; Berman, 1977). According to Benevento, Rezak, and Santos-Anderson (1977), the fibers end particularly in those tectal nuclei that receive retinal afferents.

In addition to the efferent projections described above, the superior colliculus gives off numerous *ascending connections*. To a large extent these are reciprocal to afferents to the colliculus from more rostrally situated structures. Fibers have been traced to different subdivisions of the thalamus, for example the *pulvinar* in the cat (Altman and Carpenter, 1961; Graham, 1977) and in the monkey (Mathers, 1971).

In autoradiographic studies in the monkey the same has been described (Benevento and Fallon, 1975; Benevento and Rezak, 1976), and it appears that the rather heavy projection is restricted to the inferior pulvinar. Other thalamic nuclei that have been claimed to receive tectal efferents are the dorsomedial nucleus, the intralaminar nuclei, the reticular nucleus, the dorsal and ventral lateral geniculate nucleus, and some other cellular groups Altman and Carpenter, 1961; Benevento and Fallon, 1975).

As referred to above, the lamination of the superior colliculus is related to the sites of ending of afferent fibers (in part mentioned above) and to the destination of axons of its cells. The main features in these patterns have been clarified. Since the data are of functional interest, they will be briefly summarized (see Fig. 7-31).

The *retinal afferents* end in the superficial part of the superficial stratum griseum (II_1 in Fig. 7-29), while the cortical fibers from the *occipital visual areas* 17, 18, and 19 are distributed to the deeper parts of this layer (II_{2-3}). Still deeper, in part in the stratum griseum intermedium (IV), end the fibers from other cortical areas, including those from the *frontal eye field,* area 8, and the *acoustic cortex,* and some from the *ventral lateral geniculate nucleus.* The main input to the deeper layers, however, is nonvisual and noncortical. This applies to the fibers from the *pretectal region* and the *substantia nigra,* which end mainly in the intermediate gray layer (IV), while the *spinotectal afferents* and the *fibers from the cerebellum* end mainly in the deepest gray layer (VI).

In a corresponding manner the *efferents* have their particular sites of origin. These data have been derived from lesion experiments, autoradiographic tracing of efferent connections, and, most reliably, by mapping the cells of origin by means of the retrograde axonal transport of horseradish peroxidase. The *tectospinal* fibers appear to come only from deep layers (Nyberg-Hansen, 1964b; Harting, 1977; see also Abrahams and Rose, 1975), apparently chiefly from large cells (Kuypers and Maisky, 1975; Weber, Partlow, and Harting, 1978). The tectopontine fibers originate, at least mainly, in deep and intermediate layers (Altman and Carpenter, 1961; Kawamura and Brodal, 1973; Harting, 1977; Hashikawa and Kawamura, 1977), chiefly from small cells. Likewise, the cells giving rise to *tectoreticular* (Altman and Carpenter, 1961; Kawamura, Brodal, and Hoddevik, 1974; Harting, 1977) and *tecto-olivary* fibers come from layers deeper than the stratum opticum (particularly the stratum griseum intermedium), and the majority of them are small (Kawamura and Hashikawa, 1978; Weber, Partlow, Harting, 1978).

Information is less complete concerning the sites of origin of ascending fibers from the superior colliculus. Many of them appear to arise from the superficial layers. According to Benevento and Fallon (1975), this is the case in the monkey for the fibers passing to the dorsal and ventral lateral geniculate nucleus, the inferior pulvinar and the dorsomedial thalamic nucleus, and to the pretectal nuclei (see also Berman, 1977, in the cat), while the deeper layers give origin to several other projections, for example to the medial and oral pulvinar, the intralaminar thalamic nuclei, and several other cellular regions, such as the suprageniculate nucleus.

The lengthy but, nevertheless, incomplete account of the fiber connections of the superior colliculus given above indicates that this small part of the brain cooperates with many other regions and that it is not related to visual functions only. A wealth of other information has been brought forward in recent years and has made possible an improved understanding of the role of the superior colliculus. In general, there is good agreement between anatomical, physiological, and behavioral data. Some points of interest will be considered below.

Following electrical stimulation of various parts of the superior colliculus in the cat and monkey, several authors have recorded conjugate eye movements in various directions (for example, Hyde and Eliasson, 1957, and, among recent studies, Robinson, 1972; Straschill and Rieger, 1973; Stryker and Schiller, 1975; Stein, Goldberg, and Clamann 1976; Roucoux and Crommelinck, 1976). As described above, the retinocollicular projection is topographically organized (Fig. 7-30). On stimulation of the colliculus, both eyes are turned conjugately so that the center of the gaze is brought to the point in the visual field which corresponds to the point stimulated in the colliculus, as first shown by Julia Apter (1946) in the cat. The "motor" collicular map worked out from such experiments corresponds to the "sensory" map of the retinocollicular projection as determined anatomically and physiologically. In unrestrained animals, in which the colliculus can be stimulated by means of implanted electrodes, the conjugate eye movements are always accompanied by a corresponding turning of the head and body. This integrated eye-head-body movement is considered an *"orienting response"* and has been referred to as the *"visual grasp reflex"* or *"foveation reflex."* The superior colliculus thus is important for locating and tracking targets in space. Corresponding phenomena can be elicited on stimulation of the frontal and occipital cortical eye fields.(See below. The corticotectal projections, as mentioned above, are retinotopically organized.)

Several physiological studies have been performed to elucidate the functional "machinery" of the superior colliculus, by recordings from single units, stimulation experiments, studies of ablations, and other procedures, as recently reviewed by Ingle and Sprague (1975). It appears that on certain points there are differences between cats and monkeys (depending possibly in part on the wider range of ocular movements in the monkey than in the cat?). Only the main data will be considered here.

Electrophysiological recordings of single units have shown that units in various layers differ in the responses to visual stimuli. As mentioned above, the retinocollicular fibers end chiefly in the superficial part of the stratum griseum superficiale (layer II_1), while the corticotectal fibers from areas 17, 18, and 19 terminate in deeper parts of this layer (II_2 and II_3, see Fig. 7-31). In agreement with this, in the *superficial layers* (lamina II), units have been found that are movement-selective; that is, they respond only to moving visual stimuli. Most of the units show direction selectivity: they respond to a stimulus moving in a particular direction. Most units are binocular. It appears that the retinal input to these collicular units may occur not only directly by retinocollicular fibers, but also via the lateral geniculate body–visual cortex. Furthermore, the activity of cells in the superficial layers is influenced from the occipital visual cortex (areas 17 and 18), from cells in its lamina V. These cells are mainly binocularly activated direction-selective cells of the complex type of Hubel and Wiesel, as described in Chapter 8 (Palmer and Rosenquist, 1974). (It has been suggested that the route for visual impulses via the cortical areas is responsible for the binocular responses of collicular cells, since the retinocollicular projection is mainly crossed.)

Single-unit studies of cells in the deeper layers (beneath the stratum opticum, layer III) show that these cells have other properties than those in the superficial layers. Fewer cells are responsive to visual stimuli. Many units are multimodal and respond to acoustic and tactile as well as to visual stimuli. (Acoustic stimuli are effective in eliciting orienting reactions.) According to Stein, Magalhães-Castro, and Kruger (1976), cells responsive to tactile stimuli are present only in and below the intermediate layers. It is of interest that responses to trigeminal stimulation are found in nearly the entire rostral part of the colliculus. These findings are in agreement with anatomical data reviewed above (see Fig. 7-31). The visual input to these ccells appears to be mediated via cells in the superficial layers. Cells in intermediate layers are less responsive to visual stimuli than cells in the superficial layer, and discharge with a greater latency, but before eye movements occur. The stimulus thresholds necessary to produce eye movements and to activate spinal motoneurons are considerably lower from deep layers than from superficial layers, in agreement with what is known of the origin of efferent fibers from the colliculus (see above and Fig. 7-31). Unit activity in the superior colliculus appears to be related to rapid eye movements (saccades and rapid phase of nystagmus) and not to slow movements.

It appears that the main task of the superficial cell layers of the superior colliculus is to serve as sensory analyzers of the visual environment. The deeper layers are particularly important for integrating visual afferent input with other kinds of information and for mediating appropriate movements of eyes, head, and body in response to such stimuli. These layers may collectively be referred to as forming a sensorimotor integration "center" or an optomotor region.

A main function of the superior colliculus is thus the orientation of eyes, head, and body toward visual and acoustic stimuli in the environment (the "visual grasp reflex").[62] Following ablation of the superior colliculus, an animal shows deficits in orientation to visual and acoustic cues. It is well known that what we refer to as "attention" or "motivation" is of importance as to whether an orienting response to a given stimulus will occur. This is a clear indication that other regions of the brain are involved, presumably first and foremost the cortical visual areas, which project to the colliculus (and in turn, partly via circumvential pathways, receive information from the colliculus). It has repeatedly been shown that the responses of single collicular units to visual stimuli can be influenced from the cortex, and it appears that the cortex plays a dominant role in controlling the flow of information through the deep layers of the colliculus (see Ingle and Sprague, 1975, for a recent survey). It appears further that the cortex is necessary for the initiation of the orienting response in active or "voluntary" searching movements when no visual (or acoustic) stimuli elicit the response.

Pathways through which the superior colliculus may influence spinal motoneurons (that is, produce head and body movements) in the visual orienting response are known in part (tectospinal and tectoreticulospinal routes). The influence of the superior colliculus on the oculomotor neurons of cranial nerves III, IV, and VI appears to be more complexly arranged. Although some collicular fibers pass to the pretectal nuclei, and there appear to be fibers from these to the accessory oculomotor nuclei (see above and Fig. 7-31), neither the colliculus nor the pretectal nuclei have been convincingly shown to project directly to the somatic oculomotor nuclei. (It appears doubtful that lesions of the pretectum alone result in deficits in visual orientation.) Obviously, the conjugate ocular movements that can be elicited require an extremely precise pattern of integrated and coordinated activity of a number of extrinsic eye muscles. If the superior colliculus (its deeper layers) were the final station in transmitting a highly differentiated pattern of excitation and inhibition to these muscles, one might have expected a richer and far more elaborate system of fiber connections from this to the oculomotor nuclei

---

[62] In the "visual grasp reflex," turning of the eyes and head occurs in a highly integrated pattern. Turning of the head starts some 20–40 msec later than the eye movements. The turning of the head requires an adjustment of the ongoing activity of the eye muscles (a compensatory eye movement). It appears that vestibular impulses, set up on movements of the head, are essential for this compensation (see Bizzi, 1974, for a popular presentation).

than that found so far. The question therefore arises: Is the integration pattern of afferent impulses found in the deeper layers of the colliculus (from which conjugate movements are elicited on stimulation) only a first, relatively crudely organized part of a more extensive integrative machinery? While a final answer can scarcely be given as yet, it is of interest to consider this question in relation to what is known of other regions of the brainstem which on stimulation may give rise to conjugate eye movements.

This necessitates consideration of the so-called "gaze centers" located in the brainstem. This problem is extremely complex. A large number of physiological and clinical studies have been concerned with these "centers," and many important discoveries have been made. Only some of the main points will be considered below, with particular emphasis on recent relevant anatomical findings.

"Gaze centers." As indicated from the discussion above, many parts of the brain must be involved in the performance and control of eye movements and may be referred to as "gaze centers." In the light of our present knowledge of the organization of the nervous system, it is clear that a "center" for a particular function, in this case for conjugate movements of the eyes, cannot be considered to be a circumscribed and well-defined region of the brain, being alone responsible for production of the function. Rather, the region may be thought of as a particularly important "nodal point" in a network of connections involved in the integration of ocular movements.

A "gaze center" for horizontal eye movements has been postulated to exist in the vicinity of the abducent nucleus. Clinical and experimental observations (Carpenter, McMasters, and Hanna, 1963; Carpenter and McMasters, 1963; Carpenter and Strominger, 1965; Cohen, Komatsuzaki, and Bender, 1968; Goebel Komatsuzaki, Bender, and Cohen, 1971; Keller, 1974; and others) show that lesions of this area result in paralysis of conjugate gaze to the ipsilateral side and other ocular symptoms (while a lesion of the abducent nerve produces paralysis of lateral movement of the ipsilateral eye only). On electrical stimulation of this region conjugate horizontal deviation of the eyes to the ipsilateral side occurs (position of fixation and slow and rapid movements may be influenced; Cohen and Komatsuzaki, 1972; and others). Gross potential changes in this region precede every rapid eye movement by 10–20 msec. Many neurons change their activity preferentially before the onset of horizontal eye movements. The latency of eye deviations induced from this part of the reticular formation is short, and it has been concluded that there are no more than one or two synapses between the neurons in the brainstem and the oculomotor neurons (see Cohen, 1974, for a review).

The area of interest comprises a part of the reticular formation close to the abducent nucleus, at levels between this and the trochlear nucleus. It is usually referred to as the *paramedian pontine reticular formation (PPRF)* (see Fig. 7-32). Anatomically it is not clearly outlined but appears to belong to the nucleus reticularis pontis caudalis and oralis, apparently mainly the former. The PPRF has recently attracted much interest. Several studies of the behavior of its cells have prompted further anatomical and physiological studies.

As referred to above, like stimulation of the PPRF, stimulation of a horizon-

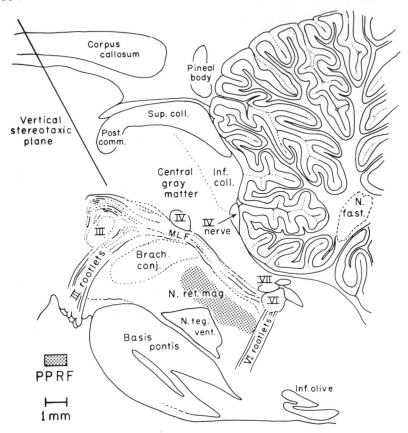

FIG. 7-32 Sagittal section of the brainstem of the monkey 0.75 mm from the midline, showing the position of the part of the reticular formation referred to as the
paramedian pontine reticular formation (*PPRF*, crosshatched) concerned in the control of conjugate lateral eye movements. From Goebel, Komatsuzaki, Bender, and
Cohen (1971).

tal canal of the vestibular apparatus produces conjugate deviation of the eyes to
the contralateral side; in other words, the contralateral lateral rectus and the ipsilateral medial rectus are brought into action (at least primarily). While the basic
reflex arc for this (primary vestibular afferents–axons of neurons in the vestibular
nuclei to the oculomotor nuclei–fibers from these to the extrinsic eye muscles) is
fairly well known (see section d of this chapter), other pathways, passing outside
the medial longitudinal fasciculus, are also involved in the mediation of these
reflexes. The PPRF, which receives fibers from the vestibular nuclei (see below)
appears to be an important station in these reflex arcs. It has been suggested that
the PPRF is the immediate premotor structure that initiates conjugate eye movements to the ipsilateral side, that is, it excites the ipsilateral abducent nucleus and
motoneurons in the oculomotor nucleus which supply the contralateral medial rectus. This view receives support from recent findings. Thus fibers from the PPRF to
the ipsilateral abducent nucleus have been demonstrated with the HRP method

(Kaneko et al., 1975) and autoradiographically (Büttner-Ennever and Henn, 1976; Graybiel, 1977). According to Kaneko et al. (1975), this connection is monosynaptic. Highstein, Maekawa, Steinacker, and Cohen (1976) recorded monosynaptic excitatory potentials in the ipsilateral, and disynaptic inhibitory potentials in the contralateral, abducent nucleus. Likewise (probably monosynaptic), activation of oculomotor neurons has been observed following stimulation of the PPRF (Highstein, Cohen, and Matsunami, 1974). However, the activation was not restricted to cells in a particular subgroup of the nucleus. Fibers from the PPRF have been traced autoradiographically to the ipsilateral oculomotor complex (Büttner-Ennever and Henn, 1976; see also Graybiel, 1977). According to Büttner-Ennever, Miles, and Henn (1975), these fibers are found in the immediate surroundings of the ipsilateral group of neurons which innervate the medial rectus (inactive during PPRF stimulation). A sparser projection was found to the same group on the contralateral side. It appears that the areas of the PPRF supplying the abducent and the oculomotor nucleus may be different (Büttner-Ennever and Henn, 1976; Graybiel, 1977), and that fibers to the former are more abundant than those to the latter.[63]

As might be expected, the *efferents of the PPRF* are not restricted to the oculomotor nuclei. In autoradiographic studies Büttner-Ennever and Henn (1976) and Graybiel (1977) traced fibers from this part of the reticular formation to the *nucleus prepositus hypoglossi* and to parts of the *vestibular complex*. Ascending fibers pass through and perhaps end in the *nucleus interstitialis* of Cajal and the *nucleus of Darkschewitsch* and appear to supply many regions, such as the mesencephalic reticular formation, the periaqueductal gray, the substantia nigra, some intralaminar thalamic nuclei, the hypothalamus, and the mammillary body. Of particular interest are projections traced to the *pretectal area* and to the *cerebellum* (Büttner-Ennever and Henn, 1976). It may be noted that all the ascending connections appear to follow a paramedian route, outside the medial longitudinal fasciculus.

The data mentioned above lend some support to the view that the PPRF may serve as a premotor structure for the initiation of conjugate eye movements.[64] It is of interest in this connection that internuclear nuerons connect the oculomotor and abducent nuclei (see p. 535). The presence of neurons in the abducent nucleus which project to the contralateral oculomotor nucleus is especially noteworthy.

When evaluating the possible role of the PPRF as a "gaze center," it is necessary to consider its *afferents*. As described in Chapter 6, a large number of afferent fiber systems have been traced to the reticular formation. While all of them do not appear to supply the region of the PPRF, the latter also receives fibers from the *vestibular nuclei* (Ladpli and Brodal, 1968) and some from the *cerebellum*

[63] As mentioned in the general discussion of eye movements (p. 536), it is an undue oversimplification to assume that only the lateral and medial rectus muscles are involved in horizontal eye movements. To quote Henn and Cohen (1976, p. 323): "Thus for each eye movement a very specific ratio of innervation has to be established between motoneurons of every eye muscle." In this connection it is interesting that according to autoradiographic studies the projection from the PPRF, although massive to the nuclear region of the medial rectus, involves other parts of the oculomotor complex as well (see Graybiel, 1977). In addition, the fact that dentrites of oculomotor neurons extend considerably outside the cellular nuclear area (Cajal, 1909–1911) is of relevance.

[64] It appears that the PPRF is engaged in the generation of slow movements as well as fast eye movements, and that different neurons are concerned in the two functions.

(fastigial nucleus, Walberg, Pompeiano, Westrum, and Hauglie-Hanssen, 1962). Visual stimuli may influence the PPRF via the *superior colliculus,* which gives off fibers to the PPRF (Kawamura, Brodal, and Hoddevik, 1974). Fibers traced from the pretectal nuclei to the reticular formation appear to end more ventrally (Itoh, 1977; Graybiel, 1977) and might be assumed to be less important. Acoustic stimuli may reach the PPRF via the *inferior colliculus,* which projects to the superior colliculus. A modest number of *spinal afferents* appear to end in the PPRF, and it receives fibers from other parts of the reticular formation. Finally, there may be afferents from the perihypoglossal nuclei, recently found to be related to visual functions (see below).

In agreement with anatomical data, Peterson (1977) reports monosynaptic responses of units in the PPRF to stimulation of the cerebral cortex and the superior colliculus, and disynaptic responses from the labyrinth. Responses to somatic stimuli and to stimulation of the fastigial nuclei have likewise been observed. Peterson (1977) further found that most reticulospinal neurons acting chiefly on neck motoneurons are located in a dorsal region of the reticular formation, which in part appears to correspond to the PPRF. This suggests that the PPRF may be involved in the head turning that accompanies shift of gaze. A large number of physiological studies have been made of the responses of units in the PPRF to stimuli of different kinds, of their relation to types of ocular movements, and so forth. (For some relevant papers, see *Control of Gaze by Brain Stem Neurons,* 1977.) The field still abounds with unsolved problems, but the fact that this part of the brainstem is essential in the performance of horizontal conjugate movements appears to be established.

It seems, from the critical study of Graybiel (1977) whose paper should be consulted for details, that the PPRF is not a unit. As mentioned above, the caudal part is particularly closely linked to the abducent nucleus and the perihypoglossal nuclei, while the rostral part appears to be the most important site of origins of ascending connections. These and other differences suggest a functional differentiation within the PPRF with regard to its influence of eye–head coordination. Some physiological data (for references, see Graybiel, 1977) are compatible with this view.

The existence of a *"gaze center" for vertical and rotatory eye movements* has been the subject of much debate. According to Cohen (1974, p. 515), "The organization of central pathways for rotatory and vertical eye movements is still unclear." Vertical eye movements are usually normal after small lesions of the pons. Units related to oblique eye movements have, however, been observed in the pontine reticular formation.

A small cell collection, referred to as the *rostral interstitial nucleus of the medial longitudinal fasciculus* has recently been shown to be particularly closely linked with vertical eye movements. It is found ventral to the nucleus of Darkschewitsch and the interstitial nucleus. According to Büttner-Ennever (1977), it receives afferents from the PPRF (see Büttner-Ennever and Henn, 1976) and the vestibular nuclei, and it projects to the oculomotor nucleus. Units in this region alter their activity in relation to vertical eye movements and start discharging before the movement (Büttner, Büttner-Ennever, and Henn, 1977, in the monkey). There must, of course, be close collaboration between regions that influence, at least preferentially, horizontal and those that influence vertical eye movements. Büttner-Ennever and Büttner (1978) found that the rostral interstitial nucleus of the medial longitudinal fasciculus in the monkey not only receives afferents from, but also gives off efferents to, the PPRF. This is clear morphological evidence of a collaboration. Büttner, Büttner-Ennever, and Henn (1977) suggest that the rostral part of the PPRF is particularly important in this respect, and that the coding of vertical eye movements may originally occur in the PPRF. This is compatible with the observation that large bilateral lesions in the PPRF cause defects in vertical as well as horizontal eye movements, whereas small lesions result in horizontal gaze paralysis only.

In the last few years evidence has been accumulating that another small region of the brainstem is concerned in the control of eye movements, namely the *perihypoglossal nuclei,* particularly their largest component, the *nucleus prepositus hypoglossi.* These nuclei have been described (refer to Fig. 5-24) and their main connections reviewed in Chapter 5. It may be recalled that they give off abundant efferents to the cerebellum and project upon the ocular motor nuclei. R. Baker and his collaborators have provided electrophysiological information concerning the function of the nucleus prepositus hypoglossi. (For a recent review, see Baker, 1977.)

Details concerning the behavior of neurons, i.e., whether they are monosynaptically or polysynaptically activated, whether they exhibit excitatory or inhibitory effects on stimulation of afferents, and so forth, will not be systematically considered here. Suffice it to mention a few points. Baker and Berthoz (1975) identified neurons of the nucleus prepositus in the cat by antidromic stimulation from the cerebellum (flocculus or vermis) or the oculomotor complex. It appears from their findings that fibers to the cerebellum and to the oculomotor complex may be branches of the same initial axons. Such prepositus neurons were found to be influenced by ipsilateral and contralateral stimulation of the vestibular nerve and on natural stimulation of horizontal semicircular canals (Blanks, Volkind, Precht, and Baker, 1977). The responses were interpreted as being disynaptic, indicating a relay, probably in the vestibular nuclei. Recordings of single units in the nucleus prepositus of the cat during eye movements (Baker, Gresty, and Berthoz, 1975) disclosed that there are both position-sensitive and velocity-sensitive neurons, and, further, that they are involved in vertical as well as horizontal eye movements. They may show a burst of activity preceding saccadic eye movements by up to 20 msec. Cells responsive to visual stimuli show direction specificity and have binocular receptive fields (Gresty and Baker, 1976). Units responsive to neck movements were likewise found. On stimulation of the nucleus prepositus a modest monosynaptic excitatory action has been recorded in trochlear and oculomotor neurons (Baker, Berthoz, and Delgado-Garcia, 1977). There is evidence that there may be a functional differentiation between parts of the nucleus.

The physiological evidence reported above is in accord with anatomical knowledge of the nucleus prepositus hypoglossi. It appears to be a region capable of integrating visual, vestibular, cerebellar, and spinal impulses and of influencing conjugate movements of the eyes by acting directly on the motoneurons that supply the extrinsic ocular muscles. Detailed electrophysiological studies show that the "intrinsic machinery" of the nucleus is extremely complex (for a recent survey, see Baker, 1977). Its important two-way connections with the cerebellum suggest a particularly close functional relation. The special role played by the nucleus prepositus in the control of integrated oculomotor functions is, however, still enigmatic.

The discovery that the nucleus prepositus is a brain region intimately related to the control of eye movements adds a third brain region to those already known to be essential for this function (the superior colliculus with the pretectal nuclei and the accessory oculomotor nuclei and the PPRF area in the lower brainstem). Each of them appears to be a "nodal point" in a very complex "system." Obviously, they must all cooperate with each other in a very precise manner, and with the cerebral cortex and other parts of the brain as well. However, despite many observations of details, little, is known of how this cooperation takes place. The available anatomical and physiological evidence suggests that much of the control of eye movements exerted by the superior colliculus occurs via the PPRF (and perhaps the nucleus prepositus). Both have a more immediate influence on the oculomotor effector cells than the superior colliculus. However, the cortical influence on eye movements may be exerted first and foremost via cortical connections to the superior colliculus. Finally, it should be recalled that the cerebellum, particularly the visual area of the vermis and the vestibulocerebellum, influences eye movements, as considered in Chapter 5. The extreme complexity in the interconnections of the various regions involved in eye movements makes it clear that their cooperation must be very elaborate. As stated by Graybiel (1977, p. 59), "The oculomotor system is among the most mystifying with respect to its central nervous organization." For a review of recent data illustrating this, see *Control of Gaze by Brain Stem Neurons* (1977).

The voluntary eye movements and the cortical eye fields. It has already been mentioned that most of our eye movements are not voluntarily initiated. But some of the reflex ocular movements involve reflex arcs reaching the cerebral cortex, and it is appropriate to consider these reflexes in connection with the cortical eye fields. Reference was made above to the so-called *frontal eye field*, situated anterior to the precentral face region. It appears to correspond approximately to Brodmann's area 8 (see Fig. 12-2) and has been assumed to play a role in the voluntary innervation of the extrinsic eye muscles similar to the role played by the precentral cortex in the innervation of the other striated muscles of the body. From this region ocular movements have been elicited experimentally in monkeys, apes, and man by a number of investigators. On the whole their findings are in good accord, but there has been some difference of opinion with regard to the exact relation of cytoarchitectonic areas to the fields responding with ocular movements on stimulation. Smith (1944a) reached the conclusion that, in man, eye movements are probably elicited not only from area 8 but also from portions of areas 6 and 9 (in front of area 8). Studies in monkeys (see Wagman, 1964) are in agreement with this. The threshold to stimulation is lowest in area 8.

The most constantly observed response on electrical stimulation of the frontal eye field is a conjugate deviation of the eyes to the contralateral side. This may be accompanied by a turning of the head in the same direction. However, movements in other directions may be elicited from the frontal eye field as well; for example vertical conjugate movements of the eyes, and occasionally dilatation of the pupils, occur (for a review see Crosby, 1953). The dominant response in all instances, however, is the lateral conjugate deviation of the eyes, a finding which is probably related to the fact that under normal circumstances lateral movements are functionally most important. In studies of single units in the frontal eye field of unanesthetized monkeys, cells have been found to fire during voluntary eye movements. Other interesting particulars have been observed as well (see Bizzi and Schiller, 1970).

The pathways of the impulses from the frontal eye field to the nuclei of the ocular nerves are not fully known. In Marchi preparations from the monkey (Levin, 1936) degenerated fibers have been followed from this cortical region through the internal capsule and the pes pedunculi. In the former they are found posteriorly in the anterior crus near the genu; in the latter they are located in the medial part of the peduncle (cf. Figs. 4-2 and 4-9). The frontal corticofugal fibers have so far not been conclusively traced to the motor nuclei of the ocular nerves themselves, either in silver impregnation studies (Astruc, 1971) or in autoradiographic studies (Künzle and Akert, 1977). The accessory oculomotor nuclei (the nucleus of Darkschewitsch and others) likewise appear to lack afferent projections from area 8. However, area 8 has been found to project to the superior colliculus and to the pretectal area.[65] As mentioned above, these corticotectal fibers end chiefly in layer IV of the superior colliculus, deeper than its retinal afferents and the fibers from the occipital visual areas. Further, it has been shown electrophysiologically that the cortex influences the activity in these layers of the superior colliculus.

The fibers from the frontal eye field obviously must influence the motor nuclei of the nerves to the extrinsic ocular muscles *indirectly*. As discussed above, it is not unlikely that the action takes place via the PPRF (see, for example, Szentágothai, 1943b), but this is not the only possibility.

Even if the ultimate pathways of the voluntary impulses initiating movements of the eyes are not finally settled, there seems to be no doubt that the cerebral cortex, especially the frontal eye field, is essential in initiating voluntary movements of the eyes, as is indeed supported by clinical findings, to be considered below. However, the frontal eye field is scarcely to be considered the ''starting'' point for voluntary eye movements. Like the precentral cortex, initiating voluntary movements of limb muscles, it is presumably ''played upon'' from other regions, not least the cortical visual and association areas. Furthermore, it should be realized that these eye movements, like all other voluntary movements, are influenced by reflex impulses. In the case of the eyes, particularly optic and probably also proprioceptive impulses from the eye muscles are important. Furthermore, impulses from the labyrinth, from the neck muscles, and from the cerebellum play a role.

Generally turning of the eyes is associated with turning of the head in the same direction. Physiological recordings of single units in area 8 in unanesthetized, unrestrained monkeys (Bizzi and Schiller, 1970) indicate that some neurons are exclusively related to head movements, while two other types signal eye position and eye movements, respectively. The latter discharge only after the initiation of the eye movement. This suggests that the frontal eye field may not be a ''motor'' area in the strict sense.

Voluntary influences on eye movements appear to have only indirect relations to the other cortical fields from which conjugate movements of the eyes can be

[65] These are the major corticofugal projections from area 8. In addition, fibers are given off to the caudate and putamen, to the reticular thalamic nucleus, part of the nucleus medialis dorsalis and some of the unspecific thalamic nuclei, and to the pontine nuclei and some other nuclei (see Astruc, 1971; Künzle and Akert, 1977).

It is of some interest that the thalamic afferents to the frontal eye field are not derived from the VL, which projects, at least mainly, onto the precentral gyrus. They come, at least chiefly, from the so-called paralamellar portion of the MD (see Fig. 2-14), according to Walker (1940b) and Scollo-Lavizzari and Akert (1963). Correspondingly, it appears that area 8 projects in part back to the dorsomedial nucleus (Rinvik, 1968c; Astruc, 1971; Künzle and Akert, 1977).

elicited on stimulation. It appears that the occipital visual areas mainly function as "reflex centers." (The possible function of the other cortical areas found to project to the superior colliculus, mentioned above, will not be considered here.) As already mentioned, *movements of the eyes can be elicited on stimulation of the occipital cortex,* from the striate area 17, as well as from areas 18 and 19 (see Fig. 12-2). The resulting movement is essentially a conjugate deviation of both eyes to the contralateral side. (On stimulation of area 17 in man the patient usually has a visual sensation; see Penfield and Jasper, 1954.) However, according to Crosby and Henderson (1948), stimulation of the upper part of the striate area evokes downward movements of the eyes, of the lower part upward movements. This pattern agrees with the "representation" of the retina in the striate area (see Chap. 8). The basis of production of conjugate movements of the eyes must be sought in corticofugal fibers from these areas. At least most of them pass to the superior colliculus and pretectal region. As described above (p. 546), these corticocollicular projections are retinotopically organized; that is, a focus in the cortex "representing" a particular spot in the retina (and visual field) projects to the area in the colliculus which "represents" the same spot.

The occipital eye fields are not concerned in the strictly voluntary movements of the eyes. If they are destroyed, the patient is still able to look, for example, to the right when asked to do so, whereas when the frontal eye field (in this case the left one) is damaged, this is not possible. On the other hand, reflex movements are not affected in lesions of the frontal eye field. This becomes evident when the patient follows a slowly moving object with his eyes. Then the eyes may deviate to a normal extent to the side to which the patient is unable to direct the gaze on voluntary effort. Relatively little is so far known about details of the role played by the occipital eye fields in eye movements. It may not be appropriate to consider these functions as being reflexes at all. However, for the sake of simplicity, in the following this designation will be used. Before considering the visual reflexes that depend on the integrity of the occipital cortex, some more elementary reflexes will be dealt with.

Subcortical optic reflexes and reflex arcs.   Vision is the most important of the senses in man. It is therefore not surprising that the ocular reflexes are numerous and functionally important. In order to understand them and, still more, the diagnostic value of disturbances of these reflexes, a knowledge of the anatomical substratum, i.e., the reflex arcs, is indispensable. However, on many points there are still gaps in our knowledge.

Reflex ocular movements, as has been mentioned, are initiated from different sources: by visual, vestibular, and acoustic impulses and from the muscles of the neck. The *efferent link* in the various reflex paths is made up of the efferent neurons from the nuclei of the eye muscles (somatic and visceral), and the *afferent* of fibers in the pathways of the optic, vestibular, acoustic, and ascending spinal fiber tracts. The reflexes may be somewhat arbitrarily subdivided into two groups. One group is concerned in the regulation of the position of the eyes and ensures their appropriate movements, when the position of the object being looked at or of the head is altered. The other is concerned in the finer adjustment of the optic apparatus necessary for visual perception (accommodation and pupillary changes). In many cases, however, reflexes of both types occur simultaneously.

The light reflex, as is well known, consists in constriction of the pupil on illumination of the eye. The efferent link in the reflex arc is represented by visceral efferent fibers in the oculomotor nerve coming from the Edinger-Westphal nuclei (however, see below). These preganglionic parasympathetic fibers pass to the ciliary ganglion, and from this the postganglionic fibers reach the sphincter of the pupil via the short ciliary nerves (see Fig. 11-2). *The afferent link* in the reflex arc is represented by optic nerve fibers that pass via the superior quadrigeminal brachium. Some of them end in the superior colliculus, others in the pretectal region (see Fig. 7-31). How the afferent impulses reach the visceral efferent motor neurons has been a controversial question. Since pupillary constriction occurs in both eyes if light is thrown into one eye (consensual reaction), afferent optic impulses must reach the visceral nucleus of both sides. Ranson and Magoun (1933) found that destruction of the superior colliculus does not abolish the light reflex, whereas the reflex disappears when the pretectal area is damaged, making it probable that axons of cells in this area transmit the impulses to the visceral nucleus (see, for example, Hare, Magoun, and Ranson, 1935). The pretectal region has, therefore, generally been held to be a "center" for the light reflex. This conception appears to be confirmed by the results of recent anatomical and physiological studies, which have provided some valuable new information.

Among these new data is the discovery (referred to above) that the classical Edinger-Westphal nucleus coincides only in part with the cells of the oculomotor nuclear complex that give rise to visceral efferent (preganglionic) fibers to the ciliary ganglion.

According to Carpenter and Peter (1970/71), in the monkey the visceral nuclei consist of a slender paired anterior median nucleus (nucleus of Perlia) which is continuous with the medial part of a dorsal visceral cell column. Retrograde changes following extirpation of the ciliary ganglion are seen only in these parts of the oculomotor complex (see also Warwick, 1954). (Only part of the Edinger-Westphal nucleus is included in this "visceral" region.) The visceral efferent fibers emerge in the ipsilateral oculomotor nerve. In physiological studies, largely corresponding locations of visceral efferent oculomotor neurons have been found. It appears further that different populations of the visceral cell groups are related to pupilloconstriction and accommodation. Some of them supply (via postganglionic cells in the ciliary ganglion) the sphincter of the iris, while others are concerned in the mediation of accommodation (acting finally on the ciliary muscle). However, there is disagreement among authors concerning details (see Jampel and Mindel, 1967; Sillito and Zbrozyna, 1970).[66] In a recent study utilizing HRP injections in the ciliary ganglion of the cat, Loewy, Saper, and Yamodis (1978) located visceral efferent neurons throughout the Edinger-Wesphal nucleus, and, in addition, in neighboring areas of the periaqueductal gray and the ventral tegmental area. They further concluded from other experiments that some neurons of the Edinger-Westphal nucleus project to lower levels, such as the spinal cord, the dorsal column nuclei, the spinal trigeminal nucleus, and the inferior olive (Loewy and Saper, 1978).

Concerning the light reflex arc, some new evidence is of interest. Most authors assume the primary afferent link to be retinal afferents to the pretectal nuclei. A possible pathway via the ventral lateral geniculate nucleus (via retinal fibers to this and secondary fibers to the pretectal nuclei) has been discussed. As

[66] According to Sunderland and Hughes (1946) the majority of the constrictor fibers for the pupil in the proximal part of the oculomotor nerve are found in its upper segment. This has been confirmed by Kerr and Hollowell (1964). On the basis of experimental findings in the dog and the monkey, these authors stressed that the oculomotor nerve is the only supratentorial structure which on compression gives rise to pupillary dilatation (mydriasis). Compression of the brainstem results in miosis. These observations may be of diagnostic value. The authors further found that the fibers mediating accommodation run together with the pupillomotor fibers in the third nerve.

described above, retinal fibers terminate predominantly (or exclusively?) in those pretectal nuclei (the olivary pretectal nucleus and the sublentiform nucleus) that have been found to send fibers to the visceral efferent cell groups in the oculomotor complex. The *basic arc for the light reflex* (see Fig. 7-31) thus appears to consist of the following neuronal links: fibers from the retina to the pretectal nuclei– fibers from these to the visceral oculomotor neurons–axons of these to the ciliary ganglion–postganglionic fibers to the sphincter of the iris. It is of relevance that the retinopretectal afferents and the pretecto-oculomotor (visceral) fibers are both crossed and uncrossed. This explains why both pupils constrict when only one eye is exposed to light—the so-called *consensual reaction* or *consensual light reflex*— even though the efferents from the visceral nuclei are ipsilateral.

It is difficult to draw conclusions concerning the reflex ''center'' for pupillary constriction from studies of the effects of lesions of the pretectal nuclei, in part because of the small size of the nuclei and in part because of their ample interconnections and connections with other regions. Unilateral lesions that fairly selectively destroyed parts of one of the pretectal nuclei in the monkey (Carpenter and Pierson, 1973) did not impair the pupillary light reflex. As suggested by Benevento, Rezak, and Santos-Anderson (1977), these negative results may be explained by the bilaterality of the retinopretectal and pretectovisceromotor connections.

It follows from the discussion above that a *lesion of the oculomotor nerve or the ciliary ganglion* will prevent the impulses from the visceral oculomotor nucleus from reaching the homolateral pupillary sphincter; that is, the light reflex will be abolished on the side of the lesion, while the consensual reaction will be preserved (the afferent impulses traveling via the optic nerve are distributed to both sides of the brainstem). A unilaterally absent light reflex due to an optic nerve lesion can therefore be distinguished from one due to a lesion of the oculomotor. In the case of an *optic nerve lesion,* illumination of the nonaffected eye causes pupillary constriction in both eyes (direct and consensual reaction to light), but illumination of the affected eye is not followed by pupillary changes in either eye. If the lesion of the optic nerve is incomplete, the reaction may be present, but is reduced in magnitude and rapidity (cf. also Chap. 8, on the diagnostic value of pupillary changes in lesions of the optic tract).

As judged from the afferent connections of the pretectal nuclei (see above and Fig. 7-31), the basic (subcortical) reflex arc for the light reflex considered above must be subject to influences from many different sources, particularly from the superior colliculus (its superficial layers) and from different parts of the cortex. It is not surprising, therefore, that pupillary changes have been observed following electrical stimulation of the superior colliculus and of various regions of the cortex (see, for example, Barris, 1936; Jampel, 1960), but little is known of the details of the intrinsic connections involved. The fact that the size of the pupil is influenced by impulses from sources other than the eye should be borne in mind when pupillary changes, so frequently met with in neurological patients, are to be evaluated.

Dilatation of the pupil can be seen on electrical stimulation of many regions of the brain, for example the cerebral cortex, the amygdala, the hypothalamus, the subthalamic nucleus, the septal area, and the periaqueductal gray (for references see Loewy, Araujo, and Kerr, 1973). Whether all these regions send fibers to the cells in the sympathetic intermediolateral cell column (the ciliospinal center) from which the preganglionic fibers arise, is questionable. Most of the fiber systems involved may, however, infuence the ciliospinal center via relays (in the mesencephalic reticular formation, the superior colliculi, or others; see, for example, Enoch and Kerr, 1967a, b). Recently, evidence has been brought forward of a direct projection from the hypothalamus to the intermediolateral cell column (see Chap. 11).

It is generally said that the size of the pupil is determined by a balance between the sympathetic and parasympathetic system, and that constriction of the pupil is due to excitation of the parasympathetic, dilatation to excitation of the sympathetic fibers.[67] Although this in principle appears to be correct, the situation is less simple than was originally assumed.

As shown by Langworthy and Ortega (1943) and others, the pupillary dilatation following stimulation of the sympathetic appears to be due, at least mainly, to the constriction of the vessels of the iris, which are provided with sympathetic vasoconstrictor nerves (see Fig. 11-9).

Several findings have been made which tend to show that the varying size of the pupil under different circumstances is first and foremost an expression of a varying tonic influence by the parasympathetic fibers of the oculomotor nerve, and that the sympathetic plays a subordinate role in this respect. Thus Kuntz and Richins (1946) as well as other authors have shown that the reflex dilatation of the pupil which follows painful stimuli is not abolished in cats and dogs if the sympathetic impulses to the eye are eliminated, e.g., after section of the cervical sympathetic trunk. This reflex had previously been commonly regarded as mediated through the sympathetic fibers to the dilator of the pupil. According to recent findings, however, the pupillary dilatation appears to be due to an inhibition of the parasympathetic oculomotor neurons supplying the sphincter of the iris. Loewy, Araujo, and Kerr (1973) identified two ascending pathways in the brainstem which on stimulation produce pupillary dilatation, even if both sympathetic trunks are transected. The ascending fibers appear to end in the Edinger-Westphal nucleus. Kerr (1975) traced some of these fibers which ascend in the lateral funiculus from the spinal cord. It is assumed that these fibers may be concerned in the pupillary dilatation occurring in response to painful stimuli. A corresponding mechanism appears to explain the pupillary dilatation occurring on stimulation of the frontal eye fields. Pupillary dilatation can still be elicited from the frontal cortex if the cervical sympathetic is removed.

There seems to be agreement that the pupillary dilatation following peripheral, particularly painful, stimulation is first and foremost due to inhibition of the parasympathetic innervation by the oculomotor nerve. The same applies to the dilatation occurring in psychical excitement. However, there is no doubt that active stimulation of the sympathetic is also able to produce a pupillary dilatation and that, conversely, damage to the sympathetic supply to the eye is followed by reduction of the size of the pupil (cf. also Chap. 11). It is probable that maximal dilatation of the pupil requires both parasympathetic inhibition and sympathetic stimulation, and some findings seem to indicate that conditions are not wholly identical in different animal species.

Apart from impulses from the frontal eye fields and the peripheral impulses, both of which may elicit pupillary dilatation, other cortical impulses may produce *pupillary constriction*. This effect is obtained on electrical stimulation of the occipital cortex, and, it appears, most constantly and most easily from area 19 both in man and in animals (Barris, 1936). This area appears to be that sending most corticofugal fibers to the midbrain, and the occurrence of pupillary constriction

---

[67] Although this subject should logically be treated in the chapter on the autonomic system, it is considered here for practical reasons. Further reference to the sympathetic parasympathetic "balance" is made in Chapter 11.

from stimulation of it is in complete harmony with the function of the occipital eye fields (areas 17, 18, and 19) known to be concerned in the cortically integrated optic reflexes, among which are those of fixation and accommodation. On account of their practical importance these reflexes will be dealt with in some detail.

Cortical optic reflexes. Fixation. Accommodation.   The cortical optic reflexes in man are far more important and elaborate than in any animal, and, on account of this, inferences from animal experiments in this field can only be drawn with the utmost care. The reflex direction of the eye toward an object attracting attention appears to be at least predominantly subserved by reflex arcs involving the cerebral cortex. In the following we shall consider chiefly the *fixation reflex in man*. Most of our information concerning this subject has been obtained from studies of patients suffering from lesions of the central nervous system, and it is reasonable to postpone the detailed description until the symptomatology is considered. At this point it is sufficient to draw attention to some of the data revealed by clinical studies. This rather complicated subject may well be approached by making reference to a phenomenon familiar to everyone, namely the so-called *railway nystagmus*. As is well known, when a person is sitting in a moving train, looking out of the window, his eyes will fix some object in the landscape and keep the image of it in the center of the visual field until it disappears at the border of the window. Then the eyes make a quick movement in the direction in which the train is going and fix another object. It requires a considerable effort to avoid this fixation under these circumstances, a fact which clearly betrays the fact that a tendency to involuntary fixation is present. The same nystagmus may be observed when a person is looking at a series of moving objects. When experimentally induced by making a patient look, for example, at a horizontally revolving screen with alternate light and dark vertical stripes, the phenomenon is referred to as *optokinetic nystagmus*. This may be *subcortical,* mediated, it appears, via the superior colliculi, or *cortical*. The former may be elicited even if the cerebral cortex is destroyed, while the latter is often said to be dependent on the integrity of the occipital cortex. It seems to be generally accepted that in man optokinetic nystagmus is mainly cortical (see for example, Kornhuber, 1966). The optokinetic nystagmus requires the attention of the individual. It seems, however, that it may be present even if the striate area is destroyed and there is hemianopsia (see below). The testing of optokinetic nystagmus can often give information of diagnostic value.

The reason for mentioning optokinetic nystagmus at this point is that it so clearly betrays *the strong tendency to fixation* which obtains in normal persons and, furthermore, that this is *reflex in nature*. The fixation aims essentially at maintaining the position of the eyes so that the image of the object looked at is kept on the fovea of both eyes. The fixation reflex displayed in normal persons requires far weaker stimuli than those necessary for eliciting optokinetic nystagmus. The reflex arc involved has its afferent link in the visual pathways, which ultimately reach the cortical visual areas. The efferent link must be presumed to be corticofugal connections to the superior colliculus, pretectal region, and ultimately to the eye muscle nuclei.[68] Under normal circumstances, it is necessary for the optic impulses to be consciously perceived if they are to elicit the fixation reflex, and attention and interest for particular objects in the visual field, as is well known, determine which objects will elicit the reflex. It is, for example, brought into action when a person is reading. The importance of the occipital cortex in the fixation reflex is not entirely clear. It appears that disorders of fixation do not always follow lesions of the occipital cortex (cf. below). A "center" for a cortical fixation reflex can scarcely be precisely located. It would probably include extensive parts of the occipitoparietal cortex, in keeping with the importance of attention for its occurrence. It may be possibly superimposed on the basic subcortical mechanism (cf. subcortical optokinetic nystagmus).

In the monkey, conjugate movements elicited on stimulation of the striate area occur in different directions according to the place of the electrode within the area (Walker and Weaver, Jr., 1940; Crosby and Henderson, 1948; and others), and the resulting change in position of the eyes in general conforms with what might be expected if they should serve the purpose of fixation. On the basis of these findings and other data it is tempting to assume that there exists an organization of the occipital cortex as concerns its role in the fixation reflex similar to that present in the superior colliculus. However, the activity of the occipital reflex center apparently is subject to alterations by impulses from the rest of the cortex, presumably the most important being derived from the adjoining areas and being concerned in the role played by the subject's attention and interest.

Evidently the cortical fixation reflex is an extremely precisely working mechanism, which ultimately influences the differential tone and contraction of the external muscles of both eyes. It is a prerequisite for the appropriate perception of the retinal images that the fixation reflex functions satisfactorily since it ensures that the image of the object looked at falls on corresponding parts of the retinae and particularly on the fovea in both eyes. The fixation reflex should, however, not be conceived of as a rigidly organized mechanism, as some of the facts mentioned might suggest. On the contrary, it acts in an extremely finely graded, functionally plastic manner.

The extreme delicacy displayed in the process of fixation is still more astonishing when we consider that fixation is usually maintained even if the head or the head and body are moving. Also in these instances, as is well known, we are able to keep the eyes fixed on an object of interest. To secure this, the cortical fixation reflex is supported by *vestibular reflexes*. These have been discussed in section (d) of this chapter. Impulses from the labyrinth are able to act on the different ocular muscles in an extremely precise manner. Many details of these reflex arcs are as yet obscure, but their delicacy will be realized when it is remembered, for example, that the slightest tilting of the head is accompanied by a rotation of the eyeballs around their longitudinal axes (performed by an appropriate interaction between the two oblique and the superior and inferior recti muscles of both eyes). In a similar way *proprioceptive impulses* from the muscles of the neck are concerned in the fixation reflex. These impulses are probably transmitted through the medial lemniscus.

After this account of the fixation reflex, some words must be devoted to another optic reflex which is also dependent on the occipital cortex, namely the *accommodation reflex*. Accommodation, as will be known, consists in an adaptation of the visual apparatus of the eye for near vision, and is accomplished by an increased curvature of the lens.[69] The smooth muscles of the ciliary body are

[68] The mechanisms underlying the eye movements in fixation of an object, requiring a varying degree of convergence of the axis of the two eyes, are imperfectly known; for a review see Bishop (1973). It may be mentioned that Hubel and Wiesel (1970) have found cortical cells in area 18 of the monkey that respond selectively to binocular convergence (binocular depth cells).

[69] The mechanism appears to be as follows: When the ciliary muscle is made to contract, the pull exerted on the lens by its suspensory ligament (the zonula) will decrease. On account of its elastic properties the lens will then increase its curvature. With advancing age the elasticity of the lens

supplied by the oculomotor nerve by parasympathetic fibers having a relay in the ciliary ganglion. Stimulation of the oculomotor nerve leads to accommodation, i.e., increased curvature of the lens.

Normally accommodation occurs only when a near object is looked at, and thus, if both eyes are utilized, is accompanied by fixation, consisting essentially of converging movements of the eyeballs. Exactly how the accommodation reflex is integrated is not known, but it appears to be established that the reflex involves the occipital cortex. It is a familiar experience that the act of accommodation requires the attention of the individual and the desire to get a clear image of a near object. Although this process may require some volitional effort during states of fatigue, the frontal eye fields appear not to be directly concerned in it. Probably the efferent link in the reflex path is made up of corticofugal fibers to the pretectal region, from which the appropriate part of the oculomotor nucleus is innervated. The afferent link in the reflex path obviously must be represented by fibers in the optic nerve and geniculocalcarine projection.

It should be remembered that, *under normal circumstances, accommodation is always accompanied by pupillary constriction.*[70] It is reasonable to regard this as a feature favorable to acute visual perception, since the narrowing of the pupil will counteract the chromatic and spherical aberrations of the lens. The assumption that the accommodation and accompanying pupillary constriction are initiated from the occipital cortex is in harmony with the fact, mentioned previously, that pupillary constriction may be elicited from the occipital cortex.

Some other reflexes acting upon the eyes have already been mentioned, such as the *vestibular reflexes* and the *proprioceptive reflexes from the neck muscles.* The reflex pupillary dilatation observed on *peripheral stimuli,* particularly of painful nature, has also been described. Other reflexes are of less importance. The turning of the eyes and the head in the direction of a sudden noise clearly is an *acoustic reflex,* mediated probably through the acoustic impulses which reach the superior colliculus (cf. also Chap. 9). The closing of the eyes in response to strong illumination has already been mentioned.

Symptoms following lesions of the abducent, trochlear, and oculomotor nerves.   Among all lesions affecting the oculomotor apparatus including its reflex arcs, those affecting the nerves themselves are by far the most common and practically important. The causes may be of various kinds, a fact which is easily understood in view of the long and partly intracranial course of the nerves.

A lesion occurring acutely and abolishing the conductive capacity of the

---

decreases and the power of accommodation is gradually reduced and finally completely lost, usually at an age of about 60.

    [70] Although the phenomena of accommodation, pupillary constriction, and convergence normally are closely linked and occur in combination, it has been shown by testing these functions with prisms or lenses in front of the eyes that each of them may occur separately. As referred to above, there is evidence that different parts of the visceral oculomotor nucleus are concerned in pupillary constriction and in accommodation. On the basis of clinical evidence, Nathan and Turner (1942) suggested that there are two efferent pathways for pupillary contraction, one serving the light reflex (discussed above) and another serving the accommodation–convergence synkinesia. They suggest that possibly the latter efferent pathway has a relay not in the ciliary ganglion but in the so-called episcleral postsynaptic neurons found in the episcleral tissue and sending axons to the ciliary muscle.

fibers (with or without anatomical interruption of their continuity) is regularly followed by symptoms which are of the same type regardless of whether the oculomotor, trochlear, or abducent nerve is affected. There occur *strabismus, diplopia,
vertigo,* and an *altered posture of the head*. Diplopia and vertigo usually disappear
after some time, even if the paralysis is enduring, because the patient learns to
suppress the image of the paralyzed eye. Thus he will no longer be disturbed by
dissimilar retinal images, and since probably these are responsible for the vertigo,
this will disappear. Diplopia and vertigo occur mainly when the patient is looking
in the direction of the action of the paralyzed muscle or muscles. In order to avoid
diplopia the patient soon learns to hold the head habitually in a position which
does not require the cooperation of the paralyzed muscle for binocular vision. In
these paralytic forms of strabismus, the secondary deviation of the eyes is greater
than the primary (i.e., the squint is more pronounced when the paralyzed eye is
used for fixation than when the sound eye is utilized). This is a differential diagnostic feature which distinguishes these paralyses from cases of concomitant strabismus (developed, for example, on the basis of amblyopia of one eye) where the
primary and secondary squint angles are identical.

Some reference should be made to the paralyses of the individual nerves.
Since the abducent and trochlear nerves each supply one muscle only, the resultant
symptoms are simpler than those following lesions of the oculomotor nerve.

In a pure *abducent nerve palsy* the affected eye cannot be abducted actively. However, when
the patient is asked to look to the affected side, some abduction movement of the eye on this side
is nevertheless observed, but this is due mainly to relaxation of the internal rectus, which would
normally occur under this circumstances, and in addition the two intact oblique muscles have a
slight abductor effect (cf. Fig. 7-27). The head is usually turned to the paralyzed side and is kept
so habitually in the beginning, as diplopia will be avoided with the head in this position. Abducent
paralysis is the ocular palsy most frequently observed and in about 25% of all cases occurs
without concomitant signs of injury to the trochlear or oculomotor nerves.

An isolated *paralysis of the trochlear nerve* is not infrequently encountered. According to
Burger, Kalvin, and Smith (1970), who examined 35 cases, the three most common causes are
head injury, vascular disease, and diabetes. The abnormal position of the affected eye will be
most conspicuous when the patient is looking downward and medially with the affected eye, since
the superior oblique muscle then should exert its maximal effect as a pure depressor of the visual
axis (cf. Fig. 7-27). However, on looking downward in other directions, as well as in abduction
and inward rotation, some deviation will occur, since the muscle takes part in these movements
(cf. Fig. 7-27). The head is usually kept tilted a little forward and sometimes also rotated to the
other side in an attempt to compensate for the lacking action of the superior oblique, thus avoiding
diplopia.

The *symptoms following lesions of the oculomotor nerve* are more complex, since it innervates several extrinsic muscles with different actions and in addition supplies the sphincter of the
iris and the ciliary muscle with motor fibers. In a *complete IIIrd cranial nerve palsy* there will be a
marked ptosis (due to the paralysis of the levator palpebrae). However, the patient usually tries to
compensate for this by a contraction of the frontalis muscle (innervated by the facial nerve), visible as wrinkles on his forehead on the affected side. If the lid is elevated passively the eye is
found to be in an abducted and somewhat depressed position, due to the combined action of the
intact external rectus and superior oblique muscles. The pupil will be found to be larger than on
the normal side due to the absence of innervation of the sphincter, and accommodation is abolished. The direct and consensual light reflexes are lost in the affected eye. Usually it appears that
the eyeball is protruding somewhat (exophthalmos). This is explained as being due to the lack of
retraction by the paralyzed muscles.

Incomplete oculomotor nerve lesions are, however, not infrequently found. They may mark the beginning of a later complete paralysis or remain as incomplete, when the lesion responsible is not of a progressive nature. Their distribution will depend on the parts of the nerve affected, and if due to pathological processes in the orbit may be limited to the superior or inferior branch of the nerve with resulting paralyses of the muscles supplied by them only (cf. above). Such cases are, however, not common. Partial oculomotor palsies are more often observed in cases in which the whole nerve has been the subject of damage, but for some cause or other not always clearly understood, only some of the fibers are injured. Most commonly such partial affections reveal themselves as a ptosis, due to paralysis of the levator palpebrae superioris. The ptosis may even be incomplete, or if complete, unaccompanied by other signs of damage to the oculomotor nerve. Two particular types of incomplete oculomotor palsies are the so-called *external* and *internal oculomotor palsies*. In the former the extrinsic muscles to the eye only are paralyzed, while in the latter these are not affected, but there is paralysis of the pupillary sphincter and paralysis of accommodation. These types are, however, more frequently observed in cases in which the lesion is nuclear and does not involve the nerve fibers but their nuclei of origin.

By careful examination of the ocular movements, of the pupillary changes, and of accommodation, it will usually be possible to decide which muscles and nerves or parts of nerves are suffering in a particular case. However, this is by no means an easy task in many instances, as has already been pointed out in the description of ocular movements. Apart from the factors mentioned which also come into play normally, the secondary contracture which usually occurs after some time in the antagonists of the paralyzed muscles should be remembered. For details, textbooks of ophthalmology should be consulted. Here only some points of particular interest to the neurologist will be mentioned. One of these concerns the *ptosis*. When evaluating the degree of a ptosis the compensatory frontalis contraction should be taken into account. Furthermore, the possibility that the ptosis may be due to a lesion of the sympathetic (cf. Chap. 11) should be remembered. If paralyses of extrinsic eye muscles are not concomitantly present, the condition of the pupil will frequently settle the question. If the ptosis is due to an oculomotor nerve lesion, the pupil will be larger than that of the normal side. In this case the light reflex of the affected eye will also be abolished or weaker than the other (when elicited from both eyes). A ptosis due to a lesion of the sympathetic innervation of the eye leading to a paralysis of the smooth tarsal muscle will be accompanied by a smaller pupil than that in the other eye, and the light reflex will be preserved. This miosis, however, is most clearly seen in the early stages (cf. Chap. 11) and the reduced size of the pupil will make the light reflex less conspicuous than in the intact eye. Ptosis may occasionally be seen following a hemiparesis due to a cerebral stroke (Caplan, 1974).

Lesions of the abducent, trochlear, and oculomotor nerves in their peripheral course, as is clear from their anatomical features, may be produced by processes of various sites and types. *Orbital disease,* most commonly *tumors* (although these are not very frequent), may damage one or more of the nerves within the orbit. Protrusion of the eyeball is usually present in these cases.[71] More important from the neurologist's point of view are *intracranial disturbances*. The course of the three nerves in the cavernous sinus makes them liable to be damaged in *aneurysms of the internal carotid artery* as described in the section on the trigeminal nerve (see Fig. 7-25). The common affection of the abducent nerve in

[71] In cranial arteritis ("arteritis temporalis") pareses or paralyses of extrinsic ocular muscles may occur. The cause appears to be inflammatory involvement of the small arterial branches to the eye muscles, producing ischemia of these (Barricks, Traviesa, Glaser, and Levy, 1977).

inflammation of the cells in the apex of the petrous bone has also been pointed to. It may be added that a *thrombosis of the cavernous sinus* will affect these nerves. The intimate and extensive relations of the nerves to the meninges make them liable to be affected in disease of these, and ocular palsies are among the most common signs of involvement of cranial nerves in meningeal disease. In *epidemic meningitis,* as well as in other types of *acute(purulent) meningitis,* ocular palsies are, therefore, common symptoms. Chronic meningeal diseases are also frequently accompanied by ocular palsies. They are particularly common in *syphilitic meningitis.* In *tabes dorsalis* a partial oculomotor paresis, betraying itself first and foremost by ptosis and an internal ophthalmoplegia, is one of the most constant symptoms. The pupillary changes, usually present in the form of the so-called Argyll Robertson pupil, will be referred to below.

Apart from these processes and *tuberculous meningitis,* other diseases of the meninges will commonly give symptoms from the ocular nerves. Without going into details as concerns their mode of action, which in most instances is probably by pressure, the following diseases may be mentioned: *subarachnoid hemorrhage, subdural hematoma* (in the latter the changes may give a valuable clue to which side is affected), and *sarcomatosis and other tumor metastases of the meninges.* In this connection should also be mentioned *meningeomas,* which when arising from the tuberculum sellae, the alae parvae of the sphenoid bone, or the olfactory grove, may cause ocular palsies, although usually only late, when other more conspicuous symptoms have been apparent for some time (visual defects, olfactory loss). Other *intracranial tumors* may occasionally also give rise to ocular palsies. Thus tumors of the temporal lobe, if they are situated medially, may affect the nerves, and also neoplasms of the semilunar ganglion will usually soon damage particularly the abducent nerve. In all cases in which extracerebral but intracranial lesions are responsible the ocular palsies are frequently accompanied by other cranial nerve palsies.

The ocular palsies are, however, not always produced by a direct action on the nerves. Thus tumors, particularly in the temporal lobe, or an epidural or an acute subdural or intracerebral hematoma, may press the medial part of the temporal lobe with the uncus through the tentorial notch ("tentorial herniation"). The cerebral peduncle is displaced to the opposite side and caudally and the *oculomotor nerve* is stretched (Sunderland and Bradley, 1953). The oculomotor nerve of the other side may also be affected in this way, but as a rule later. The posterior cerebral artery may also be dislocated and compressed, with a resultant infarction of the medial aspect of the occipital lobe.

The diagnostic value of ocular palsies, occurring in intracranial tumors, should, however, not be overestimated. This applies particularly to abducent palsy. An abducent palsy is often seen in cases in which the intracranial pressure is increased when the pathological process, particularly a tumor, does not affect the nerve directly. It is often assumed that the paralysis is due to the stretching of the long slender nerve that occurs when the brain is dislocated. Anatomical studies (Sunderland, 1948) make it probable that it may be due to traction on the nerve exerted by the anterior cerebellar artery, which as a rule runs just below it. (Possibly a similar mechanism is responsible for the abducent nerve palsy which occasionally occurs following spinal anesthesia.) Van Allen (1967) presented evidence that the paralysis, which is often transient and recurring, is due in part to ischemia of the brainstem. Among causes of *local damage* to the ocular nerves, *cranial fractures* should be mentioned; even *closed head injuries* are not uncommonly accompanied by ocular nerve palsies. These may be due to lesions of the nerves or in other cases to intracerebral lesions of the nuclei and are frequently transitory.[72]

In most of the instances mentioned above a clear-cut local cause is responsible for the paralyses of the ocular nerves. However, it happens that an ocular palsy develops in the course of a few days, and in some days or weeks usually disappears again completely or leaves only traces of the original symptoms. This may occur several times, with intervals of months or years. Even the most thorough examination may fail to reveal any obvious cause of the paralysis, which may affect one or more but rarely several of the oculomotor nerves. Such cases of *recurring ocular palsies* are usually designated as being "rheumatic" for lack of precise knowledge of their etiology. That the ocular palsies in aneurysms of the internal carotid artery may occur intermittently, under the picture of a *migraine ophthalmoplégique,* has been mentioned in discussing the trigeminal nerve. Congenital oculomotor palsy, assumed to be due to lesion of the nerve, has been described (see Victor, 1977).

Although not dependent on lesions of the peripheral nerves, the pareses of the ocular muscles observed in *myasthenia gravis* should be mentioned, since they are not infrequently misinterpreted. Particularly common is a ptosis, usually one of the initial symptoms in this disease. The diurnal variations, the fatigability of the muscles, the occurrence of other pareses of the same type, and finally the result of the prostigmine test should give the correct diagnosis, if the possibility of this disease is borne in mind. That the abnormal fatigability of the muscles is first revealed in the ocular muscles appears to be due to their being almost constantly innervated when the person is awake, as demonstrated electromyographically (Björk and Kugelberg, 1953b).

Chronic progressive ophthalmoplegia was formerly often assumed to be due to a progressive degeneration of nerve cells in the nuclei of the ocular nerves. However, several studies indicate that the primary pathological process usually occurs in the muscles and resembles that seen in muscular dystrophies. This appears from anatomical (Kiloh and Nevin, 1951; Nicolaissen and Brodal, 1959; and others) and from electromyographic studies (Teasdall and Sears, 1960; Lees and Liversedge, 1962). The intrinsic (smooth) muscles of the eye are not affected, but dystrophic changes have often been found in skeletal muscles.

The concept of a primary myopathic origin of the disease has been challenged by Drachman et al. (1969). From a study of 50 of their own patients and an analysis of 226 cases described in the literature, Danta, Hilton, and Lynch (1975) concluded that chronic progressive external ophthalmoplegia represents a number of different degenerative disorders. Their group of patients was clinically, genetically, and histologically heterogeneous.

[72] Paralyses or pareses of ocular muscles may occur in affections of the brainstem at many levels, usually accompanied by symptoms due to involvement of various other structures. A bewildering variety of syndromes has been described as being characteristic of lesions of particular regions. These syndromes are often named after one or more authors (for example, Foville's, Villaret's, Benedikt's, Bertolotti-Garcin's, Magendie-Hertwig's, and Parinaud's syndromes). Only a few are mentioned in this text. Most often they do not occur in their pure form.

Symptoms in intracerebral lesions of the oculomotor mechanisms. Apart from lesions of the nuclei of the IIIrd, IVth, and VIth cranial nerves, lesions at other places in the central nervous system may interfere with the proper innervation of the ocular muscles, as will be realized when the widespread connections of these nuclei are recalled. In these instances the symptoms frequently occur bilaterally. A perplexing variety of disturbances of ocular function may be observed: paralysis of conjugate gaze in one or more directions, forcible deviation of the eyes in one direction, so-called ocular motor apraxia, disturbances of fixation, various disturbances of the normal pattern of vestibular induced or optokinetic nystagmus, and many others. There may be isolated defects in (the slow) pursuit movements or in (the rapid) saccadic movements (for a succinct review, see Hoyt and Daroff, 1971). While some of these disturbances are seen preferentially in affections of particular parts of the brain, and thus may have diagnostic value, little is known of the underlying mechanisms that are disturbed. Considering the immense complexity in the interconnections among the many regions more or less directly involved in the control of ocular movements, this is indeed not surprising. In spite of much clinical research of ocular disturbances occurring in lesions of the brain and many experimental anatomical studies, the field of neuro-ophthalmology abounds with unsolved problems and hypotheses. Only some of the main points will be mentioned below. Much interesting and relevant data can be found in *The Control of Eye Movements* (1971), *The Oculomotor System and Brain Functions* (1973), Cohen (1974), *Basic Mechanisms of Ocular Motility and their Clinical Implications* (1975), *Control of Gaze by Brain Stem Neurons* (1977).

To start with the *nuclei,* these may be quite selectively affected in rare instances. *Congenital aplasias* or hypoplasias of the nuclei in question may occur and will appear as a congenital paralysis of the respective muscles. Most commonly a *congenital ptosis* is observed. The pupillary reactions are usually not affected.

Inflammatory diseases of the central nervous system may also affect the oculomotor nuclei. Of practical relevance are, in particular, acute poliomyelitis and epidemic (lethargic) encephalitis. The occasional occurrence of isolated cranial nerve palsies in *poliomyelitis* has been referred to previously. The process may be limited to the nuclei of one or more eye muscles. The author has observed a case in which only the one abducent nerve was paralyzed. If the general symptoms of the disease are vague or are overlooked, such cases are apt to be misinterpreted. The same may be the case in *epidemic encephalitis,* in which ocular palsies, revealing themselves as ptosis, strabismus, convergence palsies, paralysis of accommodation, or pupillary disturbances, are among the most common symptoms in the acute stages. This is explained by the predilection of the pathological process to attack the gray matter around the aqueduct, and the nature of the process explains that the ocular palsies as a rule are transitory and changing in intensity. Other signs of interference with the oculomotor mechanisms may also be found in this disease, such as involuntary, usually conjugate, movements of the eyes, so-called oculogyric crises. These, however, are scarcely due to lesions of the nuclei themselves but more likely to interference with supranuclear connections (cf. below).

If the nuclei of the eye muscles are involved in *vascular disturbances* (embolism, thrombosis, or hemorrhage) or *tumors* of the brainstem paralyses or pareses of the appropriate muscles will, of course, occur. Since these lesions as a rule are of a larger extent, there will be other symptoms as well. Thus in a unilateral mesencephalic vascular accident the oculomotor palsy on the side of the lesion will frequently be accompanied by a hemiplegia of the other half of the body including the face, since the pyramidal tract is situated not far from the oculomotor nucleus (so-called Weber's syndrome). Other structures in the vicinity (see Fig. 4-5) may also be involved, e.g., the medial lemniscus and the spinothalamic tract, with a resultant contralateral hemianesthesia. Or the red nucleus and the fibers of the brachium conjunctivum may be damaged, producing in addition to the oculomotor palsy, ataxia and other cerebellar symptoms. The different syndromes observed in these and other instances have been designated by the names of those first to describe them, but there is no need to burden the mind with these designations. Occasionally cases have been observed in which partial loss of oculomotor function has been noted, such as isolated loss of convergence and accommodation. Present knowledge in this field is, however, still too scanty to be utilized practically, and in addition such cases are very rare. If the *abducent nucleus* is affected by vascular accidents or tumors a homolateral facial palsy will regularly appear since the facial fibers bend around the nucleus (cf. Figs. 7-17 and 7-18).

Lesions of the nuclei of the abducent, trochlear, and oculomotor nerves, or of the nerves themselves, may thus be the cause of ocular palsies. However, ocular movements can be interfered with also by lesions sparing these structures, and it is on this account important not to limit the diagnostic considerations to them, when ocular disturbances are observed. On the whole the symptoms in lesions of the nerves or the nuclei are fairly simple, consisting of more or less widespread and more or less severe paralyses. The *symptoms observed in lesions of other parts of the nervous oculomotor structures* are more complex, more difficult to analyze clinically, and far less fully understood as concerns their mechanism. This is no wonder, since the neuronal connections subserving ocular reflexes, as has been seen, are imperfectly known. *These disturbances are never limited to a single muscle only and practically always affect muscles of both eyes.*

The *nystagmus*, observed in lesions of the vestibular nuclei and tracts, has been considered in section d of this chapter. It remains to deal particularly with the disturbances of conjugate ocular movements and some pupillary changes that cannot be explained as being due to a pure nerve or nuclear lesion.

As described above, several brain regions appear to be engaged in the production of conjugate eye movements. In agreement with this, *in man disturbances of conjugate eye movements may occur following lesions of the cerebral cortex (first and foremost the frontal and occipital), in lesions interrupting corticofugal pathways, lesions of the superior colliculus, lesions of the mesencephalon ("vertical gaze center"), lesions of the reticular formation of the pontine level ("horizontal gaze center"), and lesions of the brainstem affecting the medial longitudinal fasciculus.* Lesions of the *superior cerebellar peduncle,* interrupting fibers ascending to the oculomotor nuclei, for example from the nucleus prepositus hypoglossi, may be expected to cause disturbances of eye movements, as may lesions of certain parts of the *basal ganglia*. Finally lesions of the *cerebellum* and of cerebellar connections should be remembered (see Chap. 5).

Starting with lesions of the *superior colliculus,* in certain cases in man, particularly in pineal tumors, paralysis of gaze in an upward direction is commonly observed, followed frequently in later stages by a paralysis of downward gaze in addition, and eventually also of horizontal gaze.[73] Some authors are of the opinion that affections restricted to the superior colliculus in man do not result in paralysis of gaze.

As mentioned above, recent experimental studies indicate that a small region in the mesencephalon, ventral to the nucleus of Darkschewitsch (the rostral interstitial nucleus of the medial longitudinal fasciculus), bears a particular relation to the premotor mechanism of vertical conjugate eye movements. This lends support to the view that there is a particular "vertical gaze center" in the mesencephalon. In *lesions of the mesencephalon* (see, for example, Cogan, 1974), paralysis of downward gaze has often been found, sometimes without paralysis of upward gaze (Jacobs, Anderson, and Bender, 1973). Some studies of the pathological changes in such cases appear to be compatible with data from recent physiological research, as discussed by Christoff (1974), who further suggests that the lesions responsible for upward- and downward-gaze paralysis may be differently situated.

Oculogyric crises were seen in some cases of *epidemic encephalitis*. They may further occur in parkinsonism. The "crises" most frequently reveal themselves in a forcible involuntary upward movement of both eyes, lasting for seconds or minutes. Several hypotheses have been set forth to explain the anatomical connections, lesions of which will result in such symptoms, but definite evidence in favor of one or another is still lacking (for some references see Hoyt and Daroff, 1971). In parkinsonism it may be surmised that affections of nigrotectal connections are involved.

Whether the disturbances of conjugate eye movements seen in lesions of the superior colliculus in man (for example, pineal tumors) are due to an influence on this "center" or are consequences of damage to the mechanisms of the superior colliculus itself is still an open question.

As described above, the existence of a pontine gaze "center" appears to be established. It appears to be represented by a part of the reticular formation near the abducent nucleus, referred to as the PPRF (see Fig. 7-32). From this area conjugate lateral eye movements can be elicited. Paralysis of conjugate lateral gaze is often found in *pontine lesions* in man. In these patients the eyes cannot be moved conjugately to the side of the lesion. Reflex lateral movements of the eyes are also abolished (when tested, for example, with visual stimuli as in optokinetic nystagmus). As a rule there is paralysis of the abducent nerve on the same side, but the internal rectus of the other side, which is also concerned in the lateral conjugate deviation, is not paralyzed, as shown, for example, by the preserved capacity of convergence. In addition, a homolateral facial palsy and also a contralateral hemiplegia are frequently present. Some diagnostic value in differentiating between pontine and cortical paralyses of conjugate lateral gaze can be attributed to these accompanying symptoms, as will be described below. In rare cases bilateral pa-

[73] In a study of 47 cases of pineal tumor Müller and Wohlfart (1947) found paralysis of upward conjugate movements in 17 cases; in 2 there was paralysis of upward and downward gaze.

ralyses of conjugate gaze occur. These are usually due to midline pontine lesions.[74]

Since conjugate movements, as has been mentioned, can be elicited in man as well as in animals from the frontal eye field and the occipital eye fields, it is to be expected that damage to these cortical regions or to their efferent fiber tracts will be followed by ocular symptoms. This, indeed, is the case. The lateral conjugate deviation of the eyes to the contralateral side which occurs in epileptic seizures originating in these regions is commonly explained as being due to an abnormal *irritation*.

Considering first the *frontal eye field,* a *destructive* lesion of this will be followed by a paralysis of conjugate ocular movements to the contralateral side. This has also been observed in animals. In man it is most frequently found in the ordinary vascular accidents which damage the corticofugal fibers in the internal capsule and thus will interrupt the pathway for the voluntary impulses to the eye muscle nuclei. If asked to look to the side opposite the lesion, the patient is unable to do so. The paralysis, however, is not enduring, and if the lesion is a slowly developing one, the symptom may not occur at all. Some sort of compensation, presumably from the eye field of the other side, seems to occur. With regard to lateral movements each side of the brain appears usually to be concerned predominantly in the innervations necessary for the eyes to be directed to the opposite side, and when one eye field is destroyed the normal balance is for some time disturbed. At the very onset of a lesion of the type mentioned, it may be seen that the patient's eyes are forcibly directed toward the other side ("the patient looks away from his focus"). This presumably is due to an abnormal stimulation of the damaged fibers, and thus to a certain extent is comparable to the phenomena of ocular deviation seen in epileptic seizures.

Even if voluntary eye movements, particularly in the lateral direction, are completely paralyzed, this does not mean that the eyes cannot be moved in this direction. This is evident from the following fact, which also shows that the eye muscles themselves are not paralyzed. If the patient is asked to look at an object which is placed in front of him, and the object is then slowly moved toward the side to which voluntary movements are not possible, the eyes can be seen to follow the object as in a normal person. As maintained, for example, by Holmes (1938) the *fixation reflex* is responsible for this. When the object is moved rapidly or irregularly, the patient is no longer capable of following it, since then the image will be removed from the fovea from which the strongest fixation reflex is elicited. That extrafoveal impulses, however, are also able to elicit the fixation reflex, and to bring the image perceived extrafoveally in the focus, is revealed by the fact that such patients are able to read slowly. It may occasionally also be observed that the fixation reflex is exaggerated when normal volitional control is lacking, making it

---

[74] Fisher (1967) pointed out that episodic deviation of one or both eyes may occur in transient ischemic attacks associated with cerebrovascular disease in the territory of the basal and vertebral arteries. In the same article Fisher reports a number of neuro-ophthalmological observations that may be of diagnostic value. Thus pupillary changes are often present. Small (pinpoint) pupils are usually seen in severe damage to the pons (hemorrhage or infarction). This has been explained as being due to interruption of the descending pupillodilator pathway without impairment of the parasympathetic innervation of the eye. On close examination the light reflex can be seen to be preserved. When the lesion involves the mesencephalon and comprises the third nerve nucleus, maximally dilated pupils will occur. The different sizes of the pupils in these instances may be of some value in the differential diagnosis.

difficult for the patient to divert his gaze from the object which he has fixed, and patients may be observed to use several kinds of tricks in order to break the "spasmodic fixation," such as blinking, suddenly turning the head, etc. (Holmes, 1938). This phenomenon is a parallel to other conditions in which the loss of voluntary control allows the usually subordinate reflex mechanisms to become exaggerated (cf., for example, pseudobulbar paralysis). In spite of many studies the mechanisms underlying spasms of visual fixation are not clear (for a recent study, see Denny-Brown, 1977).

> The occurrence of paralysis of lateral gaze or loss of conjugate lateral movements of the eyes in pontine lesions was mentioned above. Certain features usually enable a differential diagnosis to be made. In a case in which the paralysis of lateral gaze is due to a lesion in the cortex or the subcortical matter, usually fibers from the "motor" area will also be involved. Thus there will be a hemiplegia on the side opposite the lesion, i.e., on the side to which the patient is unable to look voluntarily, since the corticofugal fibers are damaged before they cross. Particularly common will be a facial palsy contralateral to the side of the lesion, since the facial fibers are found anterior to the fibers to the body in the internal capsule. This facial palsy, then, will be of the central type, with sparing of the supraorbital facial muscles. In pontine lesions, on the other hand, the facial palsy will be of the peripheral type, involving as a rule all facial muscles, since the peripheral fibers or the nucleus itself are damaged in the pons, and, furthermore, it will be present on the side of the lesion (but also in this case on the side to which the patient is unable to look). An accompanying hemiplegia will be contralateral to the lesion. Finally, the fixation reflex will be preserved in lesions of the frontal eye field, as described above, whereas in pontine lesions the eyes cannot be moved to the contralateral side even when a slowly moving object is fixed. Optokinetic nystagmus is lost in the latter case in the direction of the paralysis but preserved in lesions of the frontal eyefield (the subcortical mechanism, presumably involving the superior colliculus, is not seriously disturbed).

In *pseudobulbar palsy* an impairment of voluntary eye movements may be observed, with well-preserved reflex movements. The mechanism of the symptom is the same as that just described, i.e., damage to the corticofugal frontal fibers.

Lesions of the occipital cortex or of the underlying white matter may interrupt the corticofugal pathways concerned in the optic reflexes which are dependent on the occipital cortex (fixation, accommodation, and fusion of the two retinal images). However, in such instances very frequently the optic pathways or the striate, peristriate, or parastriate areas (as concerns symptoms see Chap. 8) will be injured in addition, and since these reflexes are dependent on visual impulses reaching the cortex, they might be expected to be lost when the visual pathways or cortical visual areas are damaged. However, in lesions of the occipital lobe, disorders of fixation are not always found. Gaze paralysis is generally incomplete and transitory. Optokinetic nystagmus may be decreased to the side opposite the lesion. Experience following occipital lobectomy in man is in agreement with other clinical experience in indicating that marked disturbances of oculomotor functions occur only when the affection of the occipital lobe and the visual cortex involves the parietal lobe in addition (particularly the supramarginal and angular gyri). In rare cases the corticomesencephalic fibers may be injured without damage to the afferent link in the reflex arc. Then the lesion is most frequently found in the posterior part of the thalamus or the pulvinar, sparing the lateral geniculate body and the optic radiation (cf. Chap. 8). In the rare cases of this type which have been analyzed, it has been found that the eyes can be voluntarily moved in all directions, but it is not possible for the patient to keep the eyes fixed on any particular object. This is most prominent when the patient is to fix an object on the side opposite the lesion. He also experiences difficulties in obtaining clear vision if he is moving, e.g., in walking or driving a car. As Holmes (1938) has pointed out, this is not due only to diplopia or failure of fusion of the two images, since the difficulty is still present if one eye is closed. In the extremely rare cases of bilateral lesions of this type the disturbances are still more prominent.

The power of accommodation will also suffer in the latter cases. Even if there are no visual defects, and visual acuity is good, the patient experiences difficulties in obtaining a clear image when accommodation is required. That the capacity of fusing the images of the two retinae may also be reduced is evidenced by the fact that the patients may prefer using one eye only, e.g., in reading.

A comparison of the defects following lesions of the frontal with those following lesions of the occipital cortical eye fields and their efferent connections gives clear evidence that neither functions appropriately without the aid of the other. Normally an intimate collaboration exists, and it is reasonable to assume that in those activities of the cortical optic reflex apparatus in which mental factors, such as attention, play a role, other cortical areas are also involved.

In addition to the disturbances mentioned above and the visual field defects, described in Chapter 8, affections of the occipital lobe, particularly when they invade the parietal lobe may result in derangements of what is referred to as "psychovisual functions." Conditions such as visual agnosia, alexia, metamorphopsia, defects in visual memory, and others, may occur (for a review, see Gassel, 1969). Some of these disturbances will be briefly touched upon in Chapter 12.

Some aspects of pupillary function. Argyll Robertson's sign. The size of the pupil under normal conditions, as has been seen, is dependent on several factors. Ultimately it is influenced by parasympathetic impulses to the pupillary sphincter and to a lesser extent by sympathetic impulses, acting certainly on the vessels of the iris and possibly also on a functionally less important dilator muscle. The pupillary size may be varied by factors which act through these nervous pathways, such as changes in illumination, peripheral sensations, particularly painful ones, emotional factors, and lastly impulses reaching it in the act of accommodation.[75] It has already been mentioned that in accommodation there occur normally a concomitant pupillary constriction and a convergence of the eye axes. These three phenomena appear to be functionally intimately linked together, a fact of which it is important to be aware, e.g., in the ordinary testing for pupillary reactions. As is well known, a light reflex may be falsely assumed to be present when the patient is tested for this with a small lamp close to his eyes, since he then will be inclined to fix on the lamp, i.e., accommodate.

In pathological cases of damage to the oculomotor nerve, pupillary reflex constriction to light and accommodation and convergence are usually affected to a corresponding degree. Occasionally, however, a dissociation between these phenomena may be observed. Thus the *postdiphtheritic paralysis of accommodation* is not as a rule followed by a corresponding diminution of the light reflex.[76] The

[75] The normal variations in pupillary size according to age should be borne in mind, particularly the normally small pupils in old age and the commonly wide pupils in young adults and children.

[76] As mentioned earlier in this text, there is suggestive evidence that different cells in the visceral oculomotor nucleus are concerned in the production of pupillary constriction and accommodation (contraction of the sphincter of the iris and the ciliary muscle, respectively). It is even possible that the vis-

most important of these instances, however, is the symptom complex called *Argyll Robertson's sign,* described in 1869. This is found first and foremost in syphilis of the central nervous system, and particularly in tabes dorsalis, where it is stated to be present in some 90% of the patients or even more, according to the duration of the disease. Its presence is, therefore, rightly taken to indicate the probability of a syphilitic infection of the nervous system, even if the symptom may occur occasionally also in other diseases, such as encephalitis, disseminated sclerosis, syringobulbia, tumors of the pineal body (see Müller and Wohlfart, 1947), the superior colliculus, or the 3rd ventricle, and some other rare conditions. The Argyll Robertson pupil is a small pupil, which does not react to light but reacts well on accommodation. In the nonsyphilitic diseases mentioned the pupil is not always small, and thus is not a true Argyll Robertson pupil. It should also be made clear that a prerequisite for designating the pupillary anomalies mentioned as an Argyll Robertson pupil is that vision in the eye in question must not be too greatly reduced, since this clearly will impair the light reflex. Sometimes, in otherwise normal persons, a pupil is found which reacts to light only very slowly, but the pupil is usually larger than the other and on accommodation it also contracts slowly. This pseudo-Argyll Robertson pupil may be misinterpreted, particularly if the knee-jerks are absent. The combination of such "myotonic" pupils and absent patellar reflexes is usually designated Adie's syndrome, but is a harmless condition which bears no relation to syphilis (Adie, 1932). Severe neuronal degeneration in the ciliary ganglion has been assumed by some to explain the pupillary anomalies in Adie's syndrome. (See Harriman and Garland, 1968.)

On account of its considerable clinical importance, the Argyll Robertson pupil has attracted a lively interest among neurologists, but in spite of numerous attempts at explaining the underlying mechanism, it is not clearly understood. Some authors have interpreted the pupillary changes as being due to basilar meningeal changes, while others have looked for changes in the central gray matter which might be responsible for the symptom, but none of their hypotheses overcome the difficulty of explaining the unilateral Argyll Robertson pupil. Langworthy and Ortega (1943) emphasized particularly the common occurrence of local changes in the iris in syphilis, consisting, *inter alia,* of atrophy of the stroma and of the vessels. The Argyll Robertson pupil is frequently irregular in outline, which may be due to the local changes or to partial damage of the fibers, those supplying one or more sectors of the iris only being affected. The probability that local changes are the cause of the irregularity is brought out by the fact that the pupils remain irregular after death and by the not uncommon occurrence of a unilateral Argyll Robertson pupil (and of cases where the changes are far more advanced in one eye than in the other; see Apter, 1954). These data are incompatible with the theory of a central nervous system pathogenesis of the Argyll Robertson pupil and suggest that a peripheral injury, perhaps changes in the iris itself, may produce the characteristics of the true Argyll Robertson pupil. This hypothesis is of considerable interest, but it does not satisfactorily explain the preserved reaction on accommodation.

---

ceral oculomotor fibers involved in constriction of the pupil in response to light and in accommodation are not the same.

8

The Optic System*

THE EXAMINATION of the optic system is of great importance in clinical neurology since the findings allow the neurologist to draw conclusions of value with regard to the localization of a lesion and frequently with regard to its nature. The fact that the optic pathways extend from the eyeball to the posterior pole of the hemisphere explains that lesions in various parts of the brain may be followed by symptoms due to injury of the optic system, and that symptoms of this type are relatively frequent. In order to evaluate the symptoms correctly, a knowledge of the anatomical structures is indispensable.

The pathways of the visual impulses in general and the partial crossing of the optic nerve fibers. The receptor organ for visual impressions, *the retina,* is really an evaginated part of the hemisphere. It develops very early in fetal life as the optic vesicle in open communication with the first cerebral vesicle. This origin is reflected in the structure of the retina, which mainly consists of nerve cells. One layer of these, the *bipolar cells,* send their dendrites outward in the direction of the pigmented layer of the retina (formed by the outer wall of the optic vesicle). The dendrites enter into synaptic connection with the *visual receptors, the rod and cone cells* (cf. Fig. 8-1, below to the left). The axons of the bipolar cells, conducting in a central direction, end in synaptic contact with the *ganglion cells,* and the long axons from these transmit the visual impulses as far as the lateral geniculate body (Fig. 8-1). They pass first in the inner, fibrillar layer of the retina, then pierce the wall of the eyeball at the papilla (optic disk) and collect to form the optic nerve.[1] After a partial crossing in the optic chiasma the fibers continue in the

*Revised by Eric Rinvik.
[1] According to the investigations of Arey and Gore (1942) in the dog, all ganglion cells of the retina appear to send their axons into the optic nerve. The existence of internuncial ganglion cells could not be sustained.

optic tracts to end ultimately in the *lateral geniculate body,* which is the largest and best known relay nucleus in the visual system. It is appropriate to emphasize, however, that a substantial number of retinal fibers terminate in the *superior colliculus* and in some *pretectal nuclei.* These fibers form the basis of optic reflex arcs and are treated in Chapter 7.[2] In addition to the retinogeniculate, retinocollicular, and retinopretectal projections, fibers from the retina terminate in the so-called *nucleus of the accessory optic tract* in the midbrain tegmentum (Hayhow, 1959, cat; Giolli, 1961, rabbit; Giolli, 1963; Hendrickson, Wilson, and Toyne, 1970; Pasik, Pasik, and Hamori, 1973, monkey). The realization of the existence of the accessory optic tract dates back over a hundred years (Gudden, 1870, 1881), but only in recent years has it attracted renewed interest (for a review see Marg, 1973). In particular the experiments carried out by Pasik and Pasik (for references see Pasik, Pasik, and Hamori, 1973) in the monkey appear to indicate that the accessory optic system is necessary for a basic residual visual function in the absence of the striate cortex. The further course of the optic impulses from the lateral geniculate body is by means of the axons of its cells, which collect to form the *optic radiation,* passing through the white matter of the hemisphere to reach the cerebral cortex of the occipital lobe, more precisely the *striate area,* a cytoarchitecturally characteristic region on both sides of the calcarine fissure.

There can be little doubt that vision is the most important of the special senses in man. A morphological expression of this may be seen in the fact that the total amount of fibers in the optic nerve of man (about one million; Kupfer, Chumbley, and Downer, 1967) constitutes as much as 38% of the total number of afferent and efferent nerve fibers of all cranial nerves together (Bruesch and Arey, 1942).

From the point of view of practical neurology it is fortunate that the optic fibers are not intermingled irregularly in their course. On the contrary, a detailed orderly arrangement of the fibers and their cells of origin is characteristic of the optic system. The partial decussation of the optic nerve fibers in the optic chiasma has been known for a long time. The fibers originating from ganglion cells in the nasal half of the retina cross in the chiasma to the tract of the other side, whereas the temporal fibers continue in the homolateral optic tract. As a result, *all visual stimuli which impinge upon homonymous halves of the retinae* (e.g., both right halves) *are ultimately transmitted to the lateral geniculate body on the same side and finally to the homolateral striate area.* All light waves coming, for example, from the left will fall on the right halves of both retinae, namely, the temporal half of the right retina, the nasal of the left (cf. diagram in Fig. 8-1). The fibers from the right eye pass without crossing to the right optic tract, those from the nasal half of the left eye cross to continue in the right optic tract. During their further course the two groups of fibers will both reach the right lateral geniculate body and from this the efferent fibers pass to the right striate area. Consequently, *the right striate area is concerned in the perception of objects situated to the left of the vertical median line in the visual fields.* This condition corresponds to the fact that the right cerebral hemisphere is concerned in the motor and sensory activities

---

[2] The retinal fibers that terminate in the hypothalamus will be described in Chapter 11.

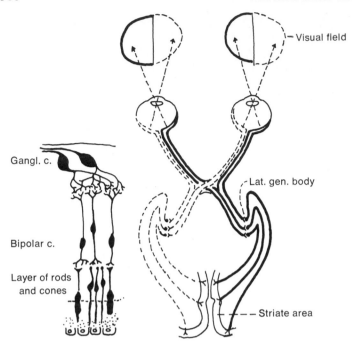

FIG. 8-1   Simplified diagram of the basic optic pathways, illustrating the partial decussation of the optic fibers in the chiasma. Below to the left, a figure indicating the arrangement of the cells of the retina.

of the left half of the body. The significance of this arrangement for symptomatology will be considered below.

Before proceeding, it is appropriate to quote Hubel and Wiesel (1965, p. 229): "To understand vision in physiological terms represents a formidable problem for the biologist. It amounts to learning how the nervous system handles incoming messages so that form, color, movement, and depth can be perceived and interpreted." In recent years considerable advances have been made toward this goal. These studies have revealed an overwhelming complexity in the structural and functional organization of the receptor organ, the retina, as well as the parts of the central nervous system related to vision. Electron microscopy has clarified a number of structural details in the retina of importance for functional interpretations (see Stell, 1972, for a review), and physiological studies have given much information on the function of the receptors and the other elements in the retina. However, even if the retina is the starting point for all visual impulses, and its organization, therefore, is fundamental in determining the processes occurring in the central parts of the visual system, the subject is of relatively remote interest for clinical neurology. For this reason it will only be referred to very briefly in the following account, which will deal primarily with the anatomical organization of the visual pathways. Physiological data will be correlated with anatomical and clinical observations.[3]

[3] Various aspects of the physiology of vision are considered in a report of a symposium in Freiburg, 1960: *Neurophysiologie und Psychophysik des visuellen Systems,* 1961 and in *Handbook of Sensory Physiology,* vol. VII, part 3, 1973.

Localization within the optic system. The localized arrangement within the optic system goes much farther than to the semidecussation of the fibers in the chiasma. It has been established that *there is a very definite point-to-point locali-zation of each small part of the retina throughout the entire optic system.* The light impulses which reach the various minute parts of the retina are transmitted through clearly localized paths to the striate area. With a certain simplification it is permissible to say that in the striate area is formed an "image" which is a true copy of the image formed in the retina. Because of this Henschen designated the striate area as "the cortical retina."

Henschen (about 1890) was the first to claim the existence of a detailed localization within the optic system. He drew his conclusions from an extensive study of human cases. Since his first papers appeared in a period when most of the leading neurologists were strongly opposed to the idea of cerebral localization, his views were vigorously challenged. Gradually, and particularly since 1920, after experimental investigations of the optic system had been made, his opinion has proved to be correct.

Since the fibers of the optic nerve have their cell bodies in the retina, a de-struction of the retina will be followed by a degeneration of the optic nerve fibers from the injured parts of the retina. These degenerating fibers can be traced by the Marchi method, since they are myelinated. By a comparison of which fibers degenerate when different parts of the retina are destroyed, the pattern of localiza-tion can be mapped out. It has been demonstrated that the fibers originating from the various retinal quadrants occupy different parts of the optic nerve and optic tract. In the nerve the position of the fibers from the four quadrants corresponds approximately to their mutual relations in the retina. Fibers from the upper quad-rants are found superiorly, from the lower quadrants inferiorly, whereas the cen-tral fibers from the macular region occupy a central position (Brouwer and Zee-man in monkeys, 1926). In the optic chiasma the upper retinal fibers cross dorsally, the lower ventrally. When the partial decussation occurs in the chiasma, crossed and uncrossed fibers are intermingled and there is a certain rearrangement, the fibers assuming an arrangement corresponding to their relative end stations in the lateral geniculate body. Therefore, in the optic tract the crossed and uncrossed fibers from the homonymous halves of the central, macular regions of the retinae are found to occupy a large area of the cross section dorsolaterally, those from the upper retinal quadrants are found medially. The same pattern is present in the lat-eral geniculate body, as will be seen from Fig. 8-2. *The macular fibers end in an extensive area, centrally and posteriorly, in the lateral geniculate body; the fibers from the peripheral parts of the retina terminate more anteriorly,* and further-more, the fibers from the upper quadrants end medially, while those from the lower quadrants end laterally.[4] Within the macular area the same distinction be-tween fibers from lower and upper parts of the macula can be found.

A more detailed mapping out of the retinogeniculate projection has not been possible by the Marchi method. Experiments utilizing other methods have shown that the localization is actually very distinct. Thus Le Gros Clark and Penman (1934) succeeded in demonstrating this very clearly in monkeys by means of the transneuronal degeneration method described in Chapter 1. When the ganglion

[4] For a study of the macular projection in the lateral geniculate body of man, see Kupfer (1962).

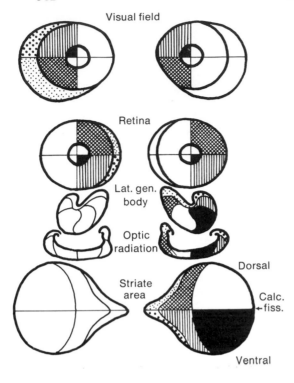

FIG. 8-2   Diagram of the projection of the different parts of the retina in the lateral geniculate body, the optic radiation, and to the striate area. Redrawn from Polyak (1932).

cells of the retina are damaged, the cells of the lateral geniculate body to which their axons are distributed degenerate. Figure 1-15 demonstrates the circumscribed changes in the lateral geniculate body of a monkey following a lesion of a restricted part of the retina. The sharp boundaries of the changes give evidence that the localization must be very accurate. Another feature, which is seen in the photograph and should also be given attention, is the fact that the cellular degeneration affects only three of the six layers which compose the lateral geniculate body (in monkeys and man). The explanation of this peculiar fact is that the intact areas of the layers between the changed ones are the terminal areas for fibers from corresponding parts of the intact retina, i.e., from the region of the retina which corresponds to the damaged area in the operated eye. If one eye is extirpated, the cells will disappear in one set (three out of the six present) of cellular layers in the homolateral lateral geniculate body, but in the other set in the contralateral body, namely the layers which are intact on the side of the lesion. In other words, *the crossed and uncrossed retinal fibers end in alternating layers of the lateral geniculate body, but in such a manner that fibers from corresponding parts of the two retinae end in neighboring parts of the different layers* (see Fig. 8-3).

The very precise retinotopical projection onto the lateral geniculate nucleus and the distribution of the crossed and uncrossed optic fibers to alternating layers of the lateral geniculate body have also been verified by studies with silver impregnation methods after section of the optic nerve in the monkey (Glees and Le Gros Clark, 1941) and in the cat (Guillery, 1970) and after lesions of the retina in the cat (Hayhow, 1958; Laties and Sprague, 1966; Stone and Hansen, 1966;

Garey and Powell, 1968). In an autoradiographic study of the retinogeniculate projections in the cat and the fox, Hickey and Guillery (1974) similarly found that there is no significant binocular overlap between the retinal projections to alternate geniculate layers. In a detailed study of the retinogeniculate projection in the monkey, using the technique of retrograde axonal transport of horseradish peroxidase, Bunt, Hendrickson, Lund, Lund, and Fuchs (1975) have shown that virtually all retinal ganglion cells project to the dorsal lateral geniculate body. Their study has also corroborated the precise organization of the retinogeniculate projections described by earlier authors using degeneration techniques (Polyak, 1957). When the fibers enter the lateral geniculate body, crossed and uncrossed fibers are, however, not yet segregated.

A localization of the same sharpness as in the retinogeniculate projection is also present in the next link of the system, the projection of the lateral geniculate body on the striate area. This projection has been mapped out in detail in animal experiments. If a small lesion is made in the striate area of a monkey and the animal is killed some weeks later, the cells in the homolateral lateral geniculate body which send their fibers to the damaged part will be affected by retrograde degeneration. Figure 8-4 is a photograph from the lateral geniculate body of a monkey in which three small lesions had previously been made in the striate area (Polyak, 1933). Corresponding to the lesions, three small patches of cell loss are found in the lateral geniculate body. On the basis of investigations of this kind the organization of the geniculocortical projection has been determined as it is represented diagrammatically in Fig. 8-2.

In the diagram the projection of the retina on the lateral geniculate body is seen as described above. In the further course the localization is maintained, and it

FIG. 8-3 Diagram to explain the anatomical basis for the impulse propagation from corresponding parts of the two retinae to the same region in the striate area. (See text.) Redrawn from Le Gros Clark (1941a).

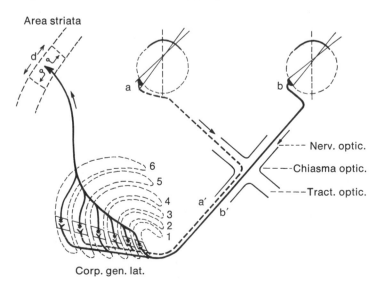

Area striata

Nerv. optic.

Chiasma optic.

Tract. optic.

Corp. gen. lat.

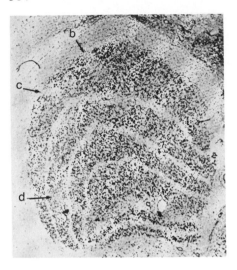

FIG. 8-4 Photograph from a section through the lateral geniculate body of a monkey. In the three small areas, labeled *b*, *c*, and *d*, the cells have disappeared because of three small lesions which were made in the striate area some weeks before death. From Polyak (1933).

will be seen that *the upper retinal quadrants are represented in the striate area superior or dorsal to the calcarine fissure, the lower quadrants below or ventral to the fissure*. This principle applies to the peripheral parts of the retina as well as to the central, macular region. It will also be seen that *the macular representation comprises a large posterior part of the striate area*, and this area is relatively much larger in proportion to the entire striate area than is the macular region in proportion to the whole retina. Thus *the macula must be more amply represented in the striate area than the peripheral parts of the retina*, a finding in complete harmony with the fact that the macula is the part of the retina in which the visual acuity is highest and in which the sensory cells are most densely packed. Anterior to the macular area the homonymous peripheral retinal quadrants are represented, the upper above, the lower below the calcarine fissure. The extreme anterior part of the striate area is occupied by the representation of the most peripheral nasal parts of the retina, which correspond to the extreme temporal crescent of the visual field where vision is monocular. In man conditions appear to be principally identical with those studied so closely in monkeys. A knowledge of this pattern of localization is of practical value when the visual field defects in lesions of the striate area are to be interpreted, as will be described below.

If the minute localization described above is to have any functional meaning, it must be presupposed that a similar clear-cut localization must be present in the first link of the optic pathways, the conduction within the retina itself from the rod and cone cells via the bipolar to the ganglion cells. This has indeed been demonstrated to be the case. In painstaking investigations of normal preparations from the retina of monkey and man, Polyak (1936 and 1941) has found that each bipolar cell of the retina is in synaptic connection with only a few rods and cones in the macular region, even to some extent only with a single cone cell. The axons of the bipolar cells are similarly not distributed to large numbers of ganglion cells, but only a few—in the macular region partly to only one or two ganglion cells. Owing to this arrangement, the possibility exists that impulses can be conveyed in a central direction by separate paths from a very small number of sensory cells (cones).[5]

[5] However, the cone cells in the fovea, which exhibit a one-to-one relationship to bipolar and ganglion cells, are also in synaptic relation to other types of bipolars (Polyak, 1941, 1957). The exten-

The lateral geniculate body. In the simplified presentation given above, the lateral geniculate body has been treated as if it were a pure relay station in the transmission of impulses from the retina to the striate area. Actually, the lateral geniculate body consists of two subdivisions, a larger dorsal nucleus and a smaller ventral one. The latter can again be divided into several cytoarchitectonically distinct subgroups (Jordan and Holländer, 1972). Although both divisions of the lateral geniculate receive retinal afferents (Sanides, 1975), only the dorsal nucleus appears to convey visual information to the cerebral cortex. Autoradiographic studies (Edwards, Rosenquist, and Palmer, 1974; Swanson, Cowan, and Jones, 1974; Ribak and Peters, 1975) have shown that the ventral nucleus of the lateral geniculate body projects upon several subcortical nuclei, such as the zona incerta, the pretectum, the superior colliculus, the lateral terminal nucleus of the accessory optic tract, and the suprachiasmatic nucleus. It is remarkable that the projections from the retina and from the ventral nucleus of the lateral geniculate body to several of the above-mentioned nuclei appear to overlap (Swanson, Cowan, and Jones, 1974). The functional significance of the ventral nucleus of the lateral geniculate body and its fiber projections is still poorly understood. Therefore, whenever reference is made to the lateral geniculate body in the following, its dorsal division is implied unless otherwise stated. The main task of the lateral geniculate body would be to make possible a fusion of the impulses from corresponding parts of the two retinae (see Fig. 8-3). Fibers from corresponding points (a and b in Fig. 8-3) of each of the two retinae end in small areas within three of the six laminae of the lateral geniculate body. The patches supplied from corresponding points of both retinae are arranged in a regular manner, representing together approximately a column extending through all layers of the lateral geniculate body. Cells in all six laminae within this column project onto a limited area (d) of the striate cortex, as shown in the rabbit (Rose and Malis, 1965b), in the cat (Garey and Powell, 1967), and in the monkey (Winfield, Gatter, and Powell, 1975). Impulses from corresponding points of the two retinae will therefore meet only when they reach the striate area. This has been confirmed in physiological studies, since in the striate area a majority of the cells respond to stimuli of both eyes (see below), while such cells are very rarely found in the lateral geniculate body (Erulkar and Fillenz, 1960; Hubel and Wiesel, 1961; and others).

However, it appears from recent research that the lateral geniculate body is more than a pure relay station, and its projection onto the cortex is not as simple as presented above. Although most of the new data concerning this subject are as yet scarcely of practical diagnostic value, some deserve to be mentioned.

As referred to above, in man and monkeys six layers of cells, separated by fibers, can be distinguished in the main dorsal part of the lateral geniculate body (Figs. 8-3 and 8-4).[6] They are usually numbered 1 to 6, beginning ventrally.

---

sive use of the electron microscope during the last 15 years has greatly increased our understanding of the immensely complex structure of the primate retina (Dowling and Boycott, 1966; Boycott and Dowling, 1969; Dowling, 1970; Stell, 1972; and others) and particularly of the role played by the horizontal and the amacrine cells in its organization (Dowling and Boycott, 1966; Boycott and Kolb, 1973; Kolb, 1974; Fisher and Boycott, 1974; Famiglietti and Kolb, 1975).

[6]The lateral geniculate body of the monkey is rather similar to that of man. In the cat's lateral geniculate body only three distinct layers can be seen in normal Nissl- or myelin-stained material. How-

Layers 1 and 2 are composed of large cells, the others of smaller ones. In the peripheral part (corresponding to the termination of fibers from the periphery of the retina, where rods only are present), some of the layers (3 and 5, 4 and 6) fuse (not shown in Fig. 8-3) so that there are only four laminae, and here the proportion of large to small cells is much greater than in the central regions.

In this context it is of interest that in the monkey only the largest retinal ganglion cells appear to project to the magnocellular layers of the lateral geniculate body (Bunt, Hendrickson, Lund, Lund, and Fuchs, 1975). However, all ganglion cells are labeled after injections of horseradish peroxidase in the parvocellular layers. This observation seems to support the suggestion that large ganglion cells have large, fast-conducting axons that terminate on large cells of the lateral geniculate body (Stone and Fukuda, 1974; Boycott and Wässle, 1974). Furthermore, recent studies using the horseradish peroxidase technique in the cat (Gilbert and Kelly, 1975; Vanegas, Holländer, and Distel, 1977; Holländer and Vanegas, 1977) support earlier observations by Garey and Powell (1967) that only large geniculate cells project upon visual area 18, whereas the striate cortex (area 17) receives afferents from geniculate cells of all sizes. Together these morphological findings support the physiological evidence for two distinct pathways for impulse conduction from the retina through the lateral geniculate body to the visual cortices (Stone and Hoffmann, 1971; Hoffmann, Stone, and Sherman, 1972; Stone and Dreher, 1973; Stone, 1972).

The total number of cells in the human lateral geniculate body has been estimated to be about 1 million (Sullivan et al., 1958; Kupfer, Chumbley, and Downer, 1967), a figure which corresponds approximately to the number of fibers in the human optic nerve. This gives a 1:1 ratio of afferent optic fibers and cells in the lateral geniculate body, but does not mean that one fiber establishes contact with one cell only. Numerous experimental electron microscopic studies in various species (for references, see below) have shown that the optic nerve fibers establish synaptic contact with the dendrites of geniculate cells.

Authors working with the Golgi method (O'Leary, 1940; Bishop, 1953; Guillery, 1966; Szentágothai, 1973a) have described several types of cells in the lateral geniculate body on the basis of differences in cell size, shape, dendritic arborization, and axonal ramification. The dendrites of the geniculate neurons are largely distributed within their particular laminae, but some cell types have dendrites that extend into neighboring laminae (Guillery, 1966; Tömböl, 1969; Famiglietti, 1970; Szentágothai, 1973a). A substantial number of short-axon cells of the Golgi II type have been described in Golgi studies in the lateral geniculate body (O'Leary, 1940; Bishop, 1953; Guillery, 1966; Szentágothai, 1973a). In fact, some authors have described two different kinds of Golgi II type cells in this nucleus (Tömböl, 1969; Famiglietti and Peters, 1972; Pasik, Pasik, Hámori, and Szentágothai, 1973). In an investigation of the geniculocortical projection in the cat using the horseradish peroxidase technique (Laemle, 1975) five morphologically defined cell types were distinguished in the lateral geniculate body. Four of these cell types were shown to have long axons that project directly and topographically to the striate cortex.

The *synaptic organization* of the lateral geniculate body has been extensively studied electron microscopically by several authors in normal and experimental material of the cat (Szentágothai,

---

ever, a recent autoradiographic study of the cat's retinogeniculate pathway has revealed six layers (Hickey and Guillery, 1974), and the main principles of organization of this nucleus appear to be the same as in the monkey and man (see Meikle and Sprague, 1964, for an account of the visual pathways in cats).

1963; Peters and Palay, 1966; Szentágothai, Hamori and Tömböl, 1966; Guillery, 1969a, 1969b; Jones and Powell, 1969d; Guillery and Scott, 1971; Famiglietti and Peters, 1972), in the monkey (Colonnier and Guillery, 1964; Pecci-Saveedra and Vacarezza, 1968; Guillery and Colonnier, 1970; Le Vay, 1971; Wong-Riley, 1972a, 1972b, 1972c; Pasik, Pasik, Hamori, and Szentágothai, 1973), in the rat (McMahan, 1967; Lieberman and Webster, 1974), and in the mouse (Rafols and Valverde, 1973).

The synaptic organization of the lateral geniculate body appears to be basically similar in the species studied. A conspicuous feature is the so-called *synaptic glomerulus* (Szentágothai, 1963) centered around the optic nerve terminal and resembling the cerebellar glomeruli (cf. Fig. 1-5G). Optic nerve fibers end on dendrites of Golgi II cells as well as on dendrites of thalamocortical relay cells. Other axonal endings in these glomeruli appear to belong to Golgi II cells and still others to afferents from the occipital cortex. Dendrites of Golgi II cells are prominent in the glomerular organization and contribute to its complexity by the fact that they contain synaptic vesicles and establish dendrodendritic synapses of the chemical type (Famiglietti, 1970; Pasik, Pasik, Hamori, and Szentágothai, 1973; Wong-Riley, 1972a) such as have been described in other thalamic relay nuclei (Ralston and Herman, 1969, ventrobasal complex; Morest, 1971, medial geniculate body; Rinvik, 1972, ventral lateral nucleus of the thalamus). Obviously, the existence of a large number of dendrodendritic synapses in the lateral geniculate body adds to the difficulties in interpreting the physiological events that take place at the cellular level during transmission of information from the retina to the visual cortex. So far the function of dendrodendritic synapses appears to be established only in the olfactory bulb (Rall, Shepherd, Reese, and Brightman, 1966), although general hypotheses for their possible function have been suggested by some authors (Ralston, 1971a; Schmitt, Dev, and Smith, 1976). With few exceptions, however (Purpura, 1972; Hamori, Pasik, Pasik, and Szentágothai, 1974), the existence of dendrodendritic synapses in the lateral geniculate nucleus appears to have been largely neglected in the interpretation of physiological observations. The extensive literature on the physiology of the lateral geniculate body bears witness to the complex organization of this nucleus (for a review see Freund, 1973).

It appears from the discussion above that the activity of the lateral geniculate body can be influenced from the occipital cortex, which presumably may exert a central control of the transmission of optic impulses.[7] Interaction of corticofugal and retinal impulses has been demonstrated in the lateral geniculate body (Hubel and Wiesel, 1961; Widén and Ajmone Marsan, 1961; see Freund, 1973, for a review), and stimulation of the reticular formation has been claimed to influence it. Presynaptic cortical inhibition has been reported to occur. It is still not possible to obtain a completely satisfactory correlation between the detailed neurophysiological and neuroanatomical data concerning the organization of the lateral geniculate body. (For a discussion of some points see Szentágothai, Hamori, and Tömböl, 1966; Szentágothai, 1973a). Both kinds of research demonstrate, however, that the lateral geniculate body is far from being a simple relay station in the visual pathways. Some other recent data on the lateral geniculate body will be considered in the account of the striate area.

The optic radiation. The optic radiation, consisting of the axons of the cells of the lateral geniculate body and ending in the striate area, takes a peculiar course in man, important to the neurologist. After leaving the lateral geniculate body, the fibers pass for a short distance laterally, and to some extent anteriorly,

---

[7] A central influence on the transmission of impulses from the retina to the brain has been physiologically demonstrated (Granit, 1955b; Dodt, 1956) and is assumed to occur via efferent fibers in the optic nerve. Such fibers and their origin have been conclusively demonstrated in birds (Cowan and Powell, 1963; La Vail and La Vail, 1972). In electron microscopic studies degeneration of terminal boutons has been found in the retina of the cat and monkey following lesions of the optic tract or nerve (Brooke, Downer, and Powell, 1965).

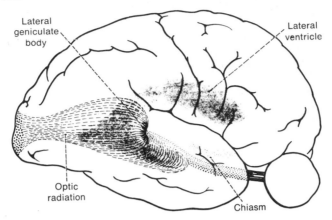

FIG. 8-5 Diagrammatic representation of the course of the fibers of the optic radiation. The geniculocalcarine fibers are seen to swing from the geniculate body (not
visible) around the lateral ventricle to reach the striate area. From Sanford and Bair
(1939) after Cushing (1922).

in the most posterior part of the internal capsule, the so-called pars retrolen-
ticularis. In this way the fibers come to occupy a position anterior to the lateral
ventricle. Some of them are found anterior to the transition between the temporal
horn and the central part (cella media) of the ventricle, while others, the lower
fibers of the bundle, are situated anterior to the anterior and upper limit of the temporal horn (see Fig. 8-5). In their further course the fibers are found on the lateral
surface of the temporal and occipital horns of the lateral ventricle, to finally reach
the striate area. In this part of their course they are found immediately beneath the
lateral wall of the ventricle, in the so-called external sagittal stratum (Pfeifer,
1925, myelogenetic studies). The original view, that the optic radiation fibers took
a straight course to the striate area, has been disproved, by both anatomical and
clinico-pathological findings. The clinical importance of the complicated course of
the optic radiation will be discussed below. It is of interest for the interpretation of
the symptoms in lesions of the occipital and temporal lobes.

The fibers in the optic radiation are arranged in a localized manner. The
fibers ending in the upper part of the striate area, for example, are found in the
upper part of the optic radiation; the macular fibers from the central part of the lateral geniculate body occupy the largest central part of the radiation. Even if there
appears to be some degree of intermingling of fibers from different parts of the
geniculate body in the radiation, the principle of localization appears to be valid.
This is learned from anatomical studies in monkey (Polyak, 1957) and in human
material (Van Buren and Borke, 1972) and confirmed in clinical observations (see
especially Spalding, 1952a). Localized lesions of the optic radiation may result in
well-delimited homonymous visual field defects (scotomas), as will be discussed
below.

The striate area. The striate area (Brodmann's area 17), receiving the
fibers of the optic radiation, is in man found almost entirely on the medial surface

of the occipital lobe (see Figs. 10-3, and 12-2). In lower mammals it is located more or less exclusively on the convexity of the cortex, and in monkeys a certain part, mainly the macular area, is still on the lateral surface. The area striata received its name on account of a light strip visible even in fresh preparations of this part of the cortex. This line of Gennari or Vicq d'Azyr consists of a layer of myelinated fibers, which appear whitish within the gray cortex. In myelin sheath preparations it can easily be recognized as a dark line in the cortex (Fig. 8-6). It is limited to the striate area. With regard to cytoarchitecture the striate area is also characteristic (Fig. 8-7c; see, however, Billings-Gagliardi, Chan-Palay, and Palay, 1974). The inner and outer granular layers (cf. Chap. 12) are well developed. The inner granular layer, IV, is particularly thick, and can be subdivided into sublayers, IVa, b, and c. Layer IVb corresponds to the line of Gennari and in Nissl preparations appears poor in cells. The third cortical layer, the pyramidal layer, also contains predominantly granule cells. A Vth layer cannot be clearly identified. The ample occurrence of granule cells and the high degree of development of the granular cortex so characteristic of the striate area are found to a lesser extent also in other cortical areas mainly subserving receptory functions (cf. Chap. 12).

Surrounding the striate area (area 17 after Brodmann) two other cortical areas are found which are related to the visual functions of the brain. Brodmann's area 18 (area parastriata) immediately surrounds the striate area (cf. Fig. 12-2) and is sharply delimited from it. It contains abundant granular cells, but in addition some pyramidal cells are present, mainly in the IIIrd and Vth layers. Area 19 (area peristriata) has a structure somewhat similar to that of area 18 and surrounds it just as area 18 surrounds area 17. The peristriate area, however, is larger than the parastriate, and occupies a considerable portion of the convexity of the occipital lobe.

When the cortex of the occipital lobe is electrically stimulated in conscious human beings, subjective optic sensations are usually elicited (see Brindley,

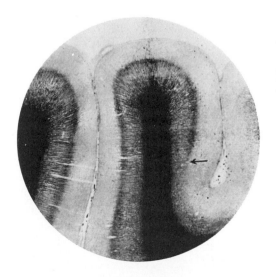

FIG. 8-6 A photograph of a myelin-stained section through the striate area of man. The abrupt transition from the striate area (area 17) with the line of Gennari to the surrounding area 18 is clearly seen (arrow).

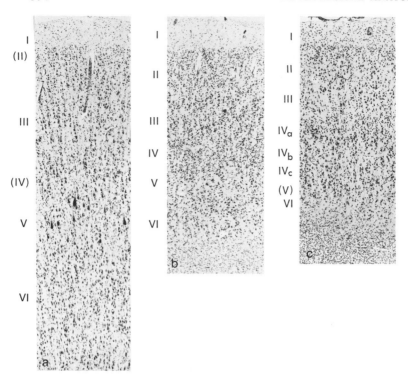

FIG. 8-7 Photomicrographs from three different cortical areas of the human brain (Nissl stain). The various layers are shown, the figures in parentheses indicating that the corresponding layer cannot be clearly identified (× 134). (a) From the area 4. A Betz cell is visible in layer V, and layers II and IV are indistinct. (b) Area 3. The granular layers (II and IV) are well developed. (c) Striate area 17. Typical granular cortex. Layer IV subdivided by the line of Gennari at the level of IVb. Since the sections are not quite perpendicular to the surface, the relative thickness of the cortex of the three areas cannot be judged from the photographs. Note columnar arrangement of cells.

1973). They differ somewhat according to the area stimulated; simple optic "hallucinations" are usually produced; the patient says that he sees sparks and flashes of light, sometimes in the form of the flicker scotomas occurring in migraine. These visual impressions are commonly observed to arise from different parts of the visual field, according to which part of the striate area is stimulated, and the localization corresponds to the known projection of the visual fields on the striate area. Stimulation of the anterior parts of the area, on which the peripheral parts of the retina are projected, gives rise to "hallucinations" which start in the periphery of the visual field and usually move toward the center, thus imitating a common type of visual prodromata in migraine. They start from the left side when the right striate area is stimulated and vice versa, and from the lower part of the visual field if the superior (supracalcarine) portion of the striate area is the site of stimulation. When the macular region, the posterior part of the striate area, is stimulated, the visual sensations are referred to the center of the visual field (Foerster, 1929).

The visual sensations following stimulation of the area 18, and particularly area 19, were by Foerster (1929) reported to be more complex, although simple visual sensations are not uncommon. The patient may inform the investigator that he sees animals, men, figures, or objects of various kinds. According to Penfield and Jasper (1954) simple visual sensations are the rule. (The designation of these optic sensations as hallucinations is unfortunate, since the patient as a rule is fully aware that the things he sees do not really exist.) From the visual areas movements of the eyes are elicited on electrical stimulation, as has been discussed in the preceding chapter.

From the available anatomical and physiological data, as well as from the findings in clinical cases, to be treated later, it must be concluded that the striate area 17 is the primary visual cortex, to which the visual impressions are relayed (cf. the "cortical retina" of Henschen). The two other areas, particularly area 19, appear to be essential for the interpretation of the visual impressions and their integration with other sensory impressions. Furthermore, the visual areas are concerned in cortical optic reflexes, which are of supreme importance for normal visual function in securing the fixation of the object to which interest is attached. The symptoms following lesions of the occipital cortex have given evidence in accord with the views mentioned, as will be seen below. Information obtained in meticulous studies in animals in recent years is on the whole compatible with clinical observations and has considerably enhanced our understanding of the visual functions of the brain. Particularly illuminating have been the studies of Hubel and Wiesel.

In a series of extensive and carefully controlled experiments Hubel and Wiesel studied the responses of neurons in the cat's and the monkey's visual cortex following the application of light stimuli to the retina. A unit in the visual cortex is characterized on the basis of its *receptive field,* defined as the region of the retina (or visual field) over which one can influence the firing of the cell. It turned out that in contrast to the lateral geniculate body, where there are two types of cells, the visual cortex contains a large number of functionally different cell types. One type is called "simple," others are "complex" or even "hypercomplex."

The responses of the cortical units were recorded with microelectrodes in anesthetized monkeys following exposure of the retina to white light stimuli of different shapes, such as stripes of light, edges between dark and light fields, etc., stationary or moving. In the lateral geniculate body the receptive fields are approximately circular, having a center which excites the cell (on-response) and a periphery which inhibits it (off-response), or vice versa. The simple cortical receptive fields likewise have excitatory and inhibitory subdivisions, but in contrast to those in the lateral geniculate they are linear with a parallel arrangement of excitatory and inhibitory areas. These cells thus respond to a stripe of light, and for each cell a particular orientation of the stripe in the visual field is the most effective (Hubel and Wiesel, 1962, 1968, 1974a, b). In fact, often a slight change in the orientation of the illuminating line will make the stimulus ineffective for the particular cell. Cells have been registered which respond specifically to a vertical, a horizontal, or any intermediate orientation of the illuminating line. The difference in the shape of the receptive fields of the geniculate and striate area neurons can be explained (Hubel and Wiesel, 1962) by assuming that there are a number of cells in the lateral geniculate body which have receptive fields with their circular "on-centers" arranged along a straight line in the retina, and that the axons of all these cells converge on a particular cortical cell and excite it. The cells with simple receptive fields react only to stimuli applied to a particular region of the retina, in agreement with the topical projection onto the striate area.

The "complex" cells respond to an appropriately oriented stimulus regardless of where in the visual field it is positioned. Some of these units respond well to contrasting edges between dark and light but poorly or not at all to stripes of light. Many other variants have been encountered. It is assumed that these "complex" units receive axons from several "simple" cells in the visual cortex. Thus it appears that the latter cells represent an early stage in cortical integration, the "complex" cells a later stage. These findings are illuminating when attempts are made to explain certain features in visual perception. For example, a hierarchy of "complex" cells will make it possible to explain how a form can be recognized regardless of its position in the visual field. For further interesting considerations the original articles should be consulted.

The explanations given by Hubel and Wiesel (1962, 1968, 1974a, b) receive further support from data obtained by the authors on the location of the units. Cells with common orientation of the axis of their receptive field are not scattered at random through the cortex but tend to be grouped together, arranged in vertical columns which extend throughout the cortex, just as is the case for the modality-specific units in the first somatosensory cortex (see Chap. 2). However, within a single column both "simple" and "complex" units occur.

The physiological observations made by Hubel and Wiesel appear to find substantial support in much of what is known of the anatomical organization of the striate area and its afferent connections.[8] Here only some main points will be considered. In light microscopic studies a large number of investigators have confirmed that, following lesions of the lateral geniculate body, lamina IV of the striate cortex is the main area of termination of *degenerating axons* (Polyak, 1927; Wilson and Cragg, 1967; Valverde, 1968, mouse; Colonnier and Rossignol, 1969; Hubel and Wiesel, 1969a, 1972; Garey and Powell, 1971; Rossignol and Colonnier, 1971, cat; Lund, 1973, monkey).[9] Autoradiographic studies (Specht and Grafstein, 1973, mouse; Rosenquist, Edwards, and Palmer, 1974, cat; Ribak and Peters, 1975, rat) have corroborated the degeneration studies. Similarly, electron microscopic studies have confirmed that most degenerating geniculocortical boutons are found in layer IV, although occasional degenerating boutons are seen in all cortical layers, particularly the deep part of layer III and in layers I and VI (Garey and Powell, 1971; Peters and Saldanha, 1976; Winfield and Powell, 1976). A great majority of the geniculocortical boutons contact dendritic spines. It is interesting that the "simple" cells of Hubel and Wiesel are particularly abundant in layer IV. In an elegant combined physiological and morphological investigation Kelly and van Essen (1974) have shown that most "simple" cells in the striate area of the cat are stellate cells and located in layer IV, whereas "complex" and "hypercomplex" cells are pyramidal neurons concentrated in layers superficial to or deeper than layer IV.

Two major types of functional columns have been identified in the striate cortex, "ocular dominance" and "orientation" columns. The cells in layer IV respond to light presented to one or the other eye only (monocular). However, cells lying in superficial or deeper layers, although binocularly driven, will respond more favorably to light from that eye which drives cells in layer IV within a narrow column through the striate cortex perpendicular to the pial surface. The ocular dominance column is 250–500 μm wide, as shown in a series of elegant experimental and normal studies (Hubel and Wiesel, 1969a, 1972, 1974a, b; Wiesel, Hubel, and Lam, 1974; LeVay, Hubel, and Wiesel, 1975). The total width of an adjoining pair of right and left ocular dominance columns is very similar to the width of an orientation column that encompasses a complete cycle of orientations through 180 degrees. The dimension of the "ocular dominance" and "orientation" columns appears to find support in a series of meticulous experimental studies on the intrinsic fiber connections of the visual cortices in the cat and monkey (Fisken, Garey, and Powell, 1975). Their report should be consulted for a com-

---

[8] The reader is referred to the comprehensive review by Szentágothai (1973b) for a detailed account of the synaptology of the striate cortex.

[9] Within lamina IV sublayer IVB, containing the stria of Gennari, is relatively free of degenerating geniculocortical axons. Hubel and Wiesel (1972) have shown in the monkey that the dorsal laminae of the lateral geniculate body project to sublayers IVA and IVC whereas the ventral laminae project upon the junction of sublayers IVB and IVC as a single band.

prehensive and critical evaluation of possible correlations between the physiological observations of Hubel and Wiesel—particularly with regard to the localization of "complex" cells—and present knowledge of the intrinsic morphological organization of the striate cortex.

As mentioned above, there is clinical (and experimental) evidence that areas outside the striate, especially areas 18 and 19, are important for visual perceptions of a more complex order. Recent experimental studies have provided interesting information confirming this. It has been known for some time that on stimulation of the retina with light, action potentials can be recorded outside the striate cortex in areas often referred to as visual areas II and III (Hubel and Wiesel, 1965). It appears that in the cat these areas correspond to the areas which Otsuka and Hassler (1962) and Sanides and Hoffman (1969), on a cytoarchitectonic basis, have homologized with areas 18 and 19 in primates. A third extrastriate cortical area related to visual function has been described in the cat's suprasylvian gyrus (Clare and Bishop, 1954; Hubel and Wiesel, 1969b). This so-called Clare-Bishop area appears to correspond to a visual area in the primate's superior temproal sulcus (for references see Zeki, 1974).

The pathways followed by visual impulses to these cortical areas have been disputed, but recent anatomical studies have given some results that are compatible with the physiological observations. Optic nerve fibers have been traced to the ventral part of the lateral geniculate body (see Sanides, 1975, for references), to the superior colliculus and the pretectal region (see Singleton and Peele, 1965; Laties and Sprague, 1966; Garey and Powell, 1968; Sprague, 1975, for recent studies and references). Contrary to most authors, who deny the presence of optic fibers to the pulvinar, Berman and Jones (1977) in an autoradiographic study in the cat found evidence of a substantial bilateral retinopulvinar projection. Recent radiographic studies (Edwards, Rosenquist, and Palmer, 1974; Swanson, Cowan, and Jones, 1974; Ribak and Peters, 1975) have shown that, although the ventral lateral geniculate body does not project to the visual cortex, it has connections with several subcortical structures related to visual functions, such as the superior colliculus and the pretectal area. From these structures (see Chap. 7) ascending projections have been described to the lateral posterior nucleus of the thalamus and to the pulvinar—i.e., to thalamic nuclei which in turn may convey visual impulses to the peristriate cortical areas (for recent studies and references see Graybiel, 1972; Jones, 1974; Kawamura, 1974; Lin, Wagor, and Kaas, 1974; Benevento and Fallon, 1975; Maciewicz, 1975; Winfield, Gatter, and Powell, 1975; Partlow, Colonnier, and Szabo, 1977; Berson and Graybiel, 1978). Furthermore, the peristriate areas may be activated by optic impulses via area 17, since the latter sends association fibers to areas 18 and 19 and to the cortex in the superior temporal sulcus, as recently confirmed in an autoradiographic study in the monkey by Martinez-Millán and Holländer (1975). These authors were not able to demonstrate commissural projections from the striate cortex, but such fibers have been described in the monkey from the part of area 17 bordering on area 18 (Zeki, 1971).

Furthermore, strong morphological evidence for the close relations of the peristriate cortical areas to visual function is the fact that the geniculocortical projections pass not only to area 17 but to areas 18 and 19 as well. This has been shown in anterograde degeneration studies by several authors (Glickstein, King, Miller, and Berkley, 1967; Wilson and Cragg, 1967; Colonnier and Rossignol, 1969; Burrows and Hayhow, 1971; Garey and Powell, 1971; Heath and Jones, 1971; Rossignol and Colonnier, 1971) and confirmed in studies of retrograde cellular changes in the lateral geniculate nucleus after lesions of the visual areas (Rose and Malis, 1965a; Garey and Powell, 1967; Niimi and Sprague, 1970; Burrows and Hayhow, 1971). Recent autoradiographic studies in the cat (Rosenquist, Edwards, and Palmer, 1974) and studies using the retrograde axonal transport of horseradish peroxidase in the cat (Gilbert and Kelly, 1975; Maciewicz, 1975; Holländer and Vanegas, 1977) and in the monkey (Wong-Riley, 1976b) have generally confirmed the experimental degeneration studies cited above, but have in addition disclosed further details in the projections from the lateral geniculate body to the visual cortices. Thus, Holländer and Vanegas (1977) have qualitatively and quantitatively confirmed the observations of Garey and Powell (1967) that only the large cells in the lateral geniculate body project to area 18, whereas geniculate cells of all sizes project to area 17. There is, furthermore, good morphological evidence (Garey and Powell, 1967; Vanegas, Holländer, and Distel, 1978; Holländer and Vanegas, 1977) that large geniculate cells send dichotomizing branches to areas 17 and 18. These anatomical observations have their physiological counterpart in the experiments of Stone and Hoffman (1971), Hoffmann, Stone, and Sherman (1972), and Stone and Dreher (1973), which have shown that the presumably largest, fast-conducting relay cells in the lateral geniculate body (so-called Y-cells) project to both areas 17 and 18 by means of a branching axon, and that the presumably smaller, slower-conducting relay cells (so-called X-cells) project only to area 17.

Only some main points of the complex organization of the visual pathways to the cerebral cortex have been considered above, and the original articles should be consulted for further details. One cannot, however, leave the subject without emphasizing that it has been known for a long time that the visual cortical areas, and area 17 in particular, project heavily back upon the lateral geniculate body and other subcortical visual relay centers, such as the superior colliculus, pretectum (see Chap. 7), and certain thalamic nuclei (for references and recent details see Guillery, 1967; Niimi, Kawamura, and Ishimaru, 1971; Campos-Ortega and Hayhow, 1972; Holländer, 1974b; Kawamura, Sprague, and Niimi, 1974; Holländer and Martinez-Millán, 1975; Lund, Lund, Hendrickson, Bunt, and Fuchs, 1975; Macchi and Rinvik, 1976). Here, again, only a few points will be mentioned to underline the complexity of the organization of the corticofugal projection. In an autoradiographic study in the monkey, Holländer (1974b) has shown that the terminal fields of the projections from area 17 extend through all of the layers of the lateral geniculate body. In addition to the projection to the lateral geniculate body, each small injection of tritiated leucine within area 17 resulted in labeling in several other thalamic nuclei, notably the posterior nucleus of the thalamus, the inferior and lateral pulvinar, the reticular nucleus of the thalamus, and in the pregeniculate nucleus. In most of the projections the localization of the terminal field is clearly dependent on the site of injection in area 17, indicating a retinotopic organization. The studies of Lund, Lund, Hendrickson, Bunt, and Fuchs (1975), using the horseradish peroxidase technique in the monkey, have contributed further to our understanding of these projections. These authors have shown that the striatal cells projecting to the lateral geniculate body are small to medium-sized pyramidal neurons located in layer VI, whereas pyramidal cells of all sizes in layer V B project to the superior colliculus and to the inferior pulvinar. On the other hand, the autoradiographic studies of Martinez-Millán and Holländer (1975) in the monkey indicate that the striatal association projections to areas 18 and 19 arise from layers superficial to lamina IV. It thus appears from the detailed Golgi, electron microscopical, autoradiographic, and horseradish peroxidase studies that have been carried out in the last few years that there are marked differences between the different cortical layers in area 17 both with regard to intrinsic organization and with regard to afferent and efferent fiber projections. As pointed out by Lund et al. (1975), some of these data are at present difficult to reconcile with the observations by Hubel and Wiesel (1962, 1965, 1974b) of physiological similarity in terms of orientation specificity for all cells within a narrow column from pia to white matter. Undoubtedly, much more experimental work remains to be done before one can fully understand how the visual cortex is organized, both morphologically and functionally.

Symptoms following lesions of the optic system. Lesions of the optic chiasma. From a knowledge of the general anatomical features of the optic system it is easy to understand the main symptoms following lesions of its different parts. Most important for the focal diagnosis are the defects that will appear in the visual fields.[10]

Defects in the visual fields occur in diseases affecting the retina and the optic nerve. However, these will not be considered extensively. From a neurological point of view the so-called "axial neuritis" is of interest, revealing itself in a *central scotoma,* frequently at first only for colors, since it is a rather frequent symptom in *disseminated sclerosis* (retrobulbar neuritis). The visual field defects in *glaucoma* should also be recalled. They are due to a compression of the nerve fibers at the margin of the papilla when this becomes excavated. A complete lesion of the optic nerve will, of course, produce complete blindness of the corresponding eye. In incomplete lesions the visual field defects will be partial, their localization being determined by the arrangement of the fibers as described previously.[11]

In *lesions of the optic chiasma* different types of visual field defects may occur. The most characteristic is a *bitemporal hemianopsia,* which occurs when the crossing fibers are damaged, whereas the laterally situated uncrossed fibers escape destruction (Fig. 8-8B). The light impinging upon the nasal halves of the two retinae will not be perceived, i.e., the *temporal halves of both visual fields are blind.* This type of lesion of the chiasma is met with most typically in *hypophyseal* or *pituitary tumors.* When the tumor expands it will exert a pressure on the

[10] The occurrence of papilledema or choked disks in cases of increased intracranial pressure will not be considered in this connection. Although it may in some cases aid in the topographic diagnosis (being, for example, usually pronounced and appearing early in tumors of the infratentorial type), its main importance is as a general symptom of increased intracranial pressure.

[11] For an exhaustive account of the symptomatology in lesions of the optic system, see Cogan, 1966; Walsh and Hoyt, 1969.

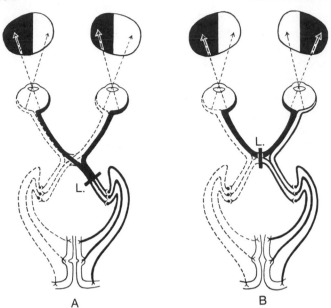

A   B

FIG. 8-8   A diagram of the defects in the visual field (black) following: *A*, a lesion of the optic tract (homonymous hemianopsia); and *B*, a lesion of the optic chiasma (bitemporal hemianopsia). *L*, lesion.

chiasma from below, and consequently the lowermost fibers of the chiasma, namely those derived from the lower regions of both nasal halves of the retinae, will be damaged first.[12] Consequently the visual defect usually starts as an *upper bitemporal quadrantanopsia*. As a rule this starts from the periphery, but sometimes the first defect appears somewhat nearer to the center. (Therefore it is important to test for visual defects in a radial direction from the periphery to the center, and not to content oneself with making sure that the outer limits of the visual fields are intact.) In the very beginning the defect is only for colors, a finding quite commonly made also in other types of visual defects. Visual acuity commonly remains good for a long time, an expression of the anatomical fact that the macular fibers are not attacked until later, and since these tumors are followed by increased intracranial pressure only late in the course of the disease, the fundi of the eyes will remain normal for a long time, the more so since the atrophy of the fibers occurs slowly. It is not uncommon for the bitemporal defects in the visual field to be the first symptom to make the patient seek medical advice.

Since the pituitary is separated from the cranial cavity proper by the firm diaphragma sellae the tumor must reach some size before it is able to dislocate the diaphragm and exert a pressure on the optic chiasma. Two other factors, which explain that the bitemporal hemianopsia does not

[12] In a recent study of the gross anatomy of the human optic chiasma O'Connell (1973) brings forward good evidence that the visual field defect in bitemporal hemianopsia is due to stretching of the crossing fibers in the chiasma and not to compression. The visual-field defects seen in lesions of the optic chiasma are not always those that would be expected on the basis of the main features of the course of fibers from different parts of the retina described here. This would be explained on the basis of finer details in the pattern of the fibers in the chiasma, as discussed by Walsh and Hoyt (1969).

occur as early as might be expected, should also be borne in mind. First, the chiasma is not rigidly fixed, but is capable of a considerable amount of dislocation. Second, the position of the chiasma varies somewhat. It may be found resting on the sulcus chiasmatis, but usually lies more posteriorly, and in some cases extremely so. According to this varying position of the chiasma the length of the intracranial part of the optic nerve will vary from a very short distance to a very long one. Apart from irregularities in the symptomatology due to these conditions, it is obvious that it is possible for the tumor to expand not exactly in the median plane. The visual defects of the two eyes, therefore, are frequently not exactly alike, and additional abberations may be caused by concomitant pressure on one of the optic nerves or optic tracts.

Although bitemporal defects of the visual fields are common in tumors of the hypophysis they may also be observed in other instances. Thus similar defects are seen in cases of *craniopharyngeomas* ("adamantinomas," hypophyseal duct tumors) which arise from remnants of the embryonal Rathke's pouch. Since this tumor, as well as those mentioned below, arises supratentorially, it is apt to exert pressure also on the third ventricle and the hypothalamus, resulting in corresponding symptoms (cf. Chap. 11). The more uncommon *gliomas* arising from the chiasma may cause bitemporal hemianopsia, usually of a more irregular type, and combined with signs of lesions of the optic nerve or tracts as well, since the tumors are apt to infiltrate these structures. The so-called *suprasellar meningeomas,* most frequently arising from the sulcus chiasmatis or the dorsum of the sella, should also be mentioned. They will as a rule dislocate the chiasma in an upward, posterior direction, and the visual defect is commonly not purely bitemporal. This is even more the case with the visual defects produced by other tumors in this region, such as *meningeomas from the olfactory groove, gliomas of the frontal lobe,* or tumors in relation to the interpeduncular fossa. However, the examination of the visual fields will usually give valuable diagnostic information.

Far less common than bitemporal hemianopsias are the *binasal hemianopsias,* occurring when pressure is exerted against the lateral sides of the optic chiasma. In this case the uncrossed retinal fibers, arising from the temporal half of each retina, will be damaged, and, consequently, visual impulses set up by objects in the nasal halves of both visual fields will not be perceived. The most frequent cause of a binasal hemianopsia is an *aneurysm of the internal carotid artery.* If a binasal hemianopsia is to develop, there must be aneurysms on both sides. Aneurysms may also occur in other arteries of the circulus arteriosus and may cause various types of defects of the visual fields. For a recent account of binasal hemianopsia, see O'Connell and Du Boulay (1973). See also O'Connell (1973).

Of other pathological conditions which less frequently give rise to visual field defects of the chiasmatic type may be mentioned *syphilitic and tuberculous basal meningitis* (direct pressure or impairment of vascular supply) and *localized arachnoiditis* in the cisterna chiasmatis. *Skull fractures* involving the base of the skull may occasionally affect the optic chiasma. In all these instances, however, the defects of the visual fields are rarely purely bitemporal.

Symptoms in lesions of the optic tract. The bitemporal and binasal hemianopsias seen in lesions of the optic chiasma are *heteronymous.* In contrast to this, lesions affecting the optic pathways posterior to the chiasma will always be *homonymous,* i.e., they will affect the same half of the visual fields of both eyes, the right or the left according to whether the lesion is in the left or the right side of the brain. This is a consequence of the partial crossing of the fibers in the chiasma and the intermingling of crossed and uncrossed fibers in the optic tract. The differences observed in the visual defects in lesions of the various parts of the pathways posterior to the chiasma are not always sufficiently marked to allow a detailed diagnosis of the site of the lesion. Generally, a complete hemianopsia is more likely to develop when the lesion affects a part where the fibers are tightly packed, as for instance the optic tract; whereas, when the lesion affects a structure in which the fibers are more widely spread, such as the optic radiation or the striate area, partial defects are more likely to develop in the form of partial homonymous field defects, quadrant scotomas, or smaller scotomas. However, the definite localization throughout the optic system explains that even small, circumscribed lesions may produce rather sharply delimited scotomas, which will be present in corresponding places in the visual fields of both eyes.

A *lesion of the optic tract,* by a tumor or a traumatic injury, as a rule will result in a *complete homonymous hemianopsia.* An interruption of the right optic tract will cut off impulses originating in the right halves of both retinae, and consequently the left halves of the visual fields are blind (Fig. 8-8A). The limit between the intact and the blind halves of the visual fields goes through the point of fixation, and as a rule is completely vertical.

> However, a lesion situated more posteriorly may also produce a complete homonymous hemianopsia. A distinguishing feature of some value is the difference observed when testing for the so-called *hemianopic pupillary reflex.* When this is present, the pupils react only when light is thrown onto the intact halves of the retinae, but not when the blind halves are illuminated. This indicates a lesion of the optic tract. The explanation of this is to be found in the course of the retinal fibers concerned in the light reflex (cf. Chap. 7). These leave the optic tract to reach the superior colliculus and pretectal region. A lesion of the optic tract will also interrupt these fibers, but they will not be affected when the optic radiation or the striate area is injured. However, unless special precautions are taken, it is difficult to make sure that the light will fall only on one half of the retina. Special apparatuses have been designed for this purpose, but on the whole the practical value of the hemianopic pupillary reflex is moderate. More important as a distinguishing feature between optic tract lesions and lesions of the radiation or visual cortex is the fact that, since the tract is rather small, an optic tract lesion most commonly will be accompanied by evidence of damage to neighboring structures in addition, such as the cerebral peduncle.

Apart from tumors which may produce a lesion of the optic tract, basal meningitis should be recalled.

Lesions of the lateral geniculate body will result in a complete, or nearly complete, homonymous hemianopsia, if they affect the entire nucleus, or most of it. As a rule other symptoms from neighboring structures will be present.

Symptoms in lesions of the optic radiation and the striate area.   A *complete homonymous hemianopsia* will occur in lesions of the *optic radiation* or the *striate area* if all fibers of the radiation are interrupted or the entire area is destroyed. However, owing to the larger extent of these structures, incomplete lesions and *incomplete hemianopsias are far more common than in lesions of the optic tract.* Many authors have stressed that in lesions of the optic radiation and of the striate area, the macular region is frequently spared, in contrast to the condition with lesions of the optic tract. However, this is not an entirely reliable criterion; central vision may occasionally be preserved also in lesions of the tract, and in posteriorly situated lesions the hemianopsia may be complete, without sparing of the macula.

There has been much discussion concerning this "macular sparing" frequently observed in these lesions. Some authors have maintained that the macula is represented bilaterally in the striate area, in contrast to the rest of the retina. Most of the adherents to this view assume that some fibers join the optic radiation on the other side, reaching this through the posterior part of the corpus callosum. There would thus be a second crossing of some of the fibers in the optic system. This of course would explain the macular sparing. This view is, however, based on anatomical investigations of cases of lesions in the temporal and occipital lobes, and it is practically impossible to be sure in such cases that *all* fibers have been interrupted or that the *entire* striate area has been functionless. In monkeys in which the entire striate area has been destroyed, all cells of the lateral geniculate body disintegrate, as was mentioned previously, a fact which demonstrates that crossing fibers of the type assumed to be present in man cannot exist in these animals (monkeys and apes). It appears reasonable that conditions should be essentially similar in man. In investigations aiming especially at tracing the fibers of the optic radiation in man, crossing fibers have not been ascertained (Putnam, 1926). Human cases have also been described in which a total ablation of the occipital lobe was not followed by any macular sparing (e.g., Halstead, Walker, and Bucy, 1940). Following unilateral occipital lobectomy no cellular degeneration has been observed in the contralateral lateral geniculate body (German and Brody, 1940). It should be remembered that it is extremely difficult to ascertain a moderate sparing of macular vision, and that some investigators are inclined to assume that a certain degree of shifting of the fixation point occurs normally. A factor of some importance in this connection is that the macular representation in man occupies a very large proportion of the striate area, and probably even extends anteriorly in the depth of the calcarine fissure. The macular sparing might therefore be explained on this basis as being due to preservation of these regions in some cases. As Putnam and Liebman (1942) emphasized in a review of this problem, what is most urgently needed is a minute anatomical control of relevant cases which have been examined with exact methods during life. Even if such studies have not yet been made, Spalding's (1952b) analysis of 72 cases of traumatic injuries of the striate cortex leave little doubt that in man there is no bilateral representation of the macula and that, when "macular sparing" occurs in lesions of the striate area, this is due to incomplete involvement of the macular representation. (See also Kupfer, 1962).

However, irrespective of the theoretical explanations which may be given, the frequent occurrence of macular sparing in lesions of the striate area has some diagnostic value in distinguishing hemisphere hemianopsias from optic tract hemianopsias. Still greater diagnostic importance may be attributed to the less complete types of homonymous hemianopsias. If, for instance, part of the upper or lower homonymous quadrants is spared, this is a more or less definite indication that a lesion of the tract is not responsible, but that the lesion has to be looked for in more posterior parts of the brain. Purely quadrantic homonymous anopsias point even more to a lesion of the hemisphere, and they are most commonly encountered in lesions of the striate area. The arrangement of the anatomical projection explains that a partial lesion of the striate area will result in partial defects of the homonymous visual fields, varying according to the situation of the lesion. For example, a destruction of the striate area below the calcarine fissure will reveal itself in an upper homonymous quadrantic anopsia contralateral to the side of the lesion. Circumscribed central (homonymous) scotomas will result from lesions of the posterior part of the striate area. Purely peripheral scotomas, homonymous and as a rule congruent in the two eyes, will develop in lesions of the anterior part of the striate area. The different types of such small homonymous scotomas have been observed mainly in traumatic lesions of the occipital lobe, particularly in wartime (Holmes and Lister, 1916), but they may occasionally be due to vascular lesions. Details concerning the visual defects in lesions of the striate area and the optic radiation can be found in the papers of Spalding (1952a, b), van Buren and Baldwin (1958), and Falconer and Wilson (1958).

From the anatomical features described previously it will be clear that *diseases affecting the temporal lobe can produce visual field defects of the same type as lesions of the occipital lobe*. This is due to the peculiar course of the optic radiation, and is a fact worth remembering for the clinical neurologist. A tumor of the temporal lobe, even if it is situated only some centimeters behind the temporal pole, will be likely to destroy parts of the optic radiation, particularly the lower fibers (cf. Fig. 8-5). Consequently the visual defect in these cases usually starts as an upper homonymous quadrantic anopsia (see, for example, Falconer and Wilson, 1958). It might be expected that the optical irritative phenomena which may occur should give a clue to the differential diagnosis between temporal and occipital lobe lesions, since electrical stimulation of the striate area gives rise to simple visual sensations only, whereas more complex "hallucinations" are elicited from areas 18 and 19. However, such phenomena of irritation are not very common, and furthermore it should be remembered that the striate area is not a very large region. A tumor involving the striate area will therefore frequently also affect the surrounding areas and the subjective sensations which may occur will be of a mixed type. Other symptoms may give more information (cf. Chap. 9), and ventriculography usually settles the question.

When the defects in the visual fields are due to a tumor, they will gradually increase, and if edema or circulatory disturbances develop, a sudden deterioration may occur. If the cause is a vascular accident or a traumatic lesion, however, a gradual improvement usually takes place. This is due to the fact that fibers which are at first only damaged by edema or pressure due to hemorrhage, thus abolishing their conducting capacity but not destroying them, eventually regain their function. In this restitution it is a regular feature that the perception of white is first regained, and only later does the perception of colors reappear. Usually also the definitive scotomas for colors are more extensive than is that for white. As mentioned above, the opposite development is common when the scotomas start. In these cases color vision is affected first and may be severely imparied before any defect can be ascertained by testing the visual field with white objects. The examination of the visual field for colors may therefore give valuable information in cases in which there is reason to suspect a visual field defect.[13]

As mentioned above, it is commonly assumed that the interpretation and the comprehension of visual impressions take place mainly in the areas surrounding the striate area, collaborating in this function with other areas, mainly in the parietal and probably also the temporal lobe. This conception is founded mainly on the phenomena which have been observed in more extensive lesions of the occipital lobe where it borders on the other lobes. In such cases various types of defective visual functions occur, even if the striate area is intact and there is no visual field defect. For example, such patients may be unable to recognize persons whom they know very well, or they may have lost the capacity to read, although vision is

[13] The facts mentioned most probably can be explained by the assumption that the perception of colors is a very complex process, which requires intact pathways, whereas the perception of white may be possible also when a slight damage to the optic pathways is present.

good, or other types of disturbances may occur. These symptoms usually occur only with lesions in the left hemisphere in right-handed persons. They are grouped among the so-called psychosomatic disturbances, which are to be considered in Chapter 12. It should be emphasized that disturbances of the type mentioned very frequently occur in association with other signs of impaired psychosomatic functions.

In recent years recordings of evoked potentials have made it clear that impulses arising on visual stimulation may affect numerous other parts of the brain than the areas formerly believed to receive optic information. Among these regions are the temporal and parietal association areas (see Chap. 12) but also many other regions, for example parts of the amygdala (see Chap. 10).

In focal epilepsy originating in the occipital lobe, especially the striate area, the patient may have a *visual aura,* seeing light, often moving, white, or colored; or there may be dimming of vision. The phenomena occur in the opposite visual fields. If the discharge is localized to areas 18 or 19, similar sensations may occur, but the movement is more apt to be twinkling or pulsating (Penfield and Jasper, 1954). These auras resemble the *flicker scotomas* occurring in some patients suffering from migraine, usually at the start of an attack. Most often a semicircular flickering light sensation, sometimes in color, moves from the periphery of the visual field toward the center, indicating a spread of cortical irritation progressing from the anterior part of the striate cortex toward its macular part.

Finally, mention should be made of transient *visual disturbances following vertebral angiography.* There may be blindness lasting for some days with complete or incomplete recovery. Visual ''hallucinations,'' visual agnosia, and various other changes in visual function and mental changes may also occur (see, for example, Hauge, 1954; Silverman et al., 1961). Obviously the visual symptoms are due to transitory circulatory disturbances in the territory of the posterior cerebral artery which supplies the striate area.

In evaluating visual disturbances it should be borne in mind that psychological factors are of importance. Recent research has brought forward interesting data concerning these subjects. Since they have, as yet, a moderate practical importance for the neurologist, they will not be considered here. The interested reader may be referred to Gassel and Williams (1963).

9

The Auditory System[*]

THE MAIN subjects of this chapter will be the innervation of the auditory receptor organ, the cochlear nerve, and the central auditory pathways. The cochlear duct with the organ of Corti will be treated only briefly.[1]

The organ of Corti. The auditory receptors are the hair cells of the cochlear duct. In man this is 35 mm long and spirals 2¾ turns within the bony cochlea. Near its basal end the basal coil of the duct is in open communication with the saccule. The apical coil terminates blindly at the apex of the cochlea (see Fig. 7-9). The duct is triangular in cross section (Fig. 9-1). It is attached broadly to the bony walls of the cochlear canal and is separated from the perilymph by the thin vestibular membrane toward the scala tympani. The latter membrane extends from the bony spiral lamina to the outer wall of the bony canal and supports the organ of Corti.

Like the vestibular apparatus, the organ of Corti contains two morphologically different types of sensory cells (Fig. 9-1). The *inner hair cells* are bottle-shaped with a central nucleus and organelles scattered throughout the cytoplasm (see, among others, Smith, 1975, 1978). The *outer hair cells* are more specialized. Their basally situated nucleus divides the cytoplasm into a small infranuclear

---

*Revised by Kirsten Kjelsberg Osen.

[1] For reviews covering various aspects of the auditory mechanisms the reader is referred to Whitfield (1967) and the following symposia and reviews: *Neural Mechanisms of the Auditory and Vestibular Systems*, 1960; *Neurological Aspects of Auditory and Vestibular Disorders*, 1964; *Sensory Hearing Processes and Disorders*, 1967; *Hearing Mechanisms in Vertebrates*, 1968; *Contributions to Sensory Physiology*, Vol. 4, 1970; *Sensorineural Hearing Loss*, 1970; *Physiology of the Auditory System. A Workshop*, 1971; *Neuroanatomy of the Auditory System. Report on Workshop*, 1973; *Basic Mechanisms in Hearing*, 1973; *Otophysiology. Advances in Oto-Rhino-Laryngology, Vol. 20*, 1973; *Handbook of Sensory Physiology*, Vol. V/1, 1974; *The Nervous System, Vol. 3, Human Communication and its Disorders*, 1975; and *Evoked Electrical Activity in the Auditory Nervous System*, 1978.

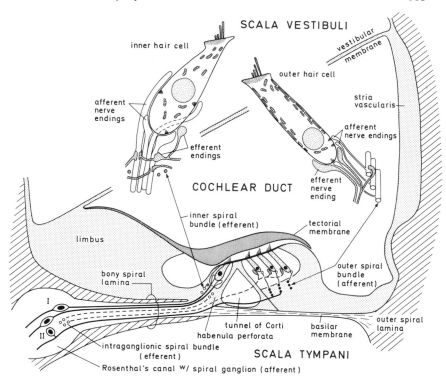

FIG. 9-1 Diagrammatic cross section through the cochlear duct with the organ of Corti and closeup views of an inner and an outer hair cell. Afferent nerve fibers are indicated by fully drawn lines and by dots where transected; terminal swellings are undotted. The efferent nerve fibers are stippled and shown as open circles when transected. Their terminal swellings are finely dotted. The intraganglionic and inner spiral bundles are composed of efferent fibers; the outer spiral bundle of afferent fibers. Note the inverse relationship of afferent and efferent nerve terminals on inner and outer hair cells. See text. Redrawn and simplified after Smith (1975).

and a large supranuclear portion. The latter contains an organized system of sublemmal cisternae of smooth endoplasmic reticulum with adjacent mitochondria. At their basal end, both types of cells establish synapses with nerve fibers (discussed later). At their free end the cells are provided with sensory hairs, stereocilia. On the outer hair cells, these are arranged in three or more rows, forming a W pattern (Fig. 9-2B). On the inner hair cells the stereocilia form almost straight rows arranged parallel to the cochlear duct (Fig. 9-2A). In contrast to the situation in the vestibular organ, kinocilia are present only during fetal or very early postnatal life (Lindemann, Ades, Bredberg, and Engström, 1971).

In man there are about 3,500 inner and 15,000 outer hair cells (Guild, 1932; Bredberg, 1968). The inner hair cells are present in a single row, whereas the outer cells are arranged in three parallel rows (Fig. 9-2A). The inner and outer hair cells with their supporting cells are separated by a canal, the tunnel of Corti, which is traversed by nerve fibers passing to the outer hair cells (Fig. 9-1). For scanning electron microscopy of the organ of Corti see, for example, *Inner Ear Studies* (1972).

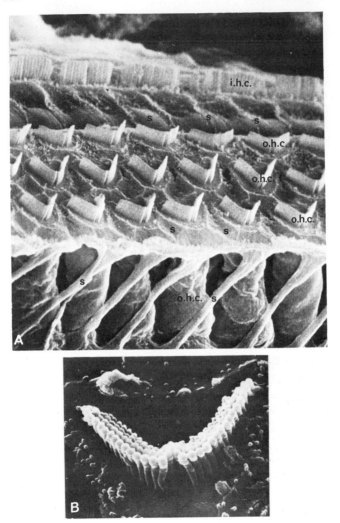

FIG. 9-2   Scanning electron micrographs of the organ of Corti of the guinea pig.
A: The single row of inner hair cells (*i.h.c.*) with sensory hairs arranged in a straight
line and the three rows of outer hair cells (*o.h.c.*) with a W pattern of sensory hairs.
Abbreviation: *s:* supporting cells. *B:* The surface of one outer hair cell with four par-
allel rows of sensory hairs in a typical W pattern. Courtesy of H. Engstrøm and B.
Engstrøm.

The mechanism producing stimulation of the cochlear sensory cells has been difficult to clarify. At present the following theory appears to be generally accepted. The stereocilia of the sensory cells appear to be in contact with the jellylike tectorial membrane, which, via the limbus is
supported by the bony spiral lamina. By oscillations of the basilar membrane that occur on auditory stimulation, the nonoscillating tectorial membrane presumably causes a tilting of the stiff
stereocilia, whereby the hair cells are stimulated. How the basilar membrane oscillates, and how
this motion results in nerve fiber potentials are problems not fully understood. The final solution
probably has to await the results of intracellular recordings of the stimulated cochlea. (For a
review see, for example, Fex, 1974.)

As has been known for some time, there is a spatial representation of tonal frequencies, so-called *tonotopical organization,* in the organ of Corti. The cochlear nerve fibers show a corresponding specificity for tonal frequencies, defined by their characteristic frequency. Because of the relevance of these phenomena to the organization of the central auditory pathway, they will be briefly considered. Originally two alternative theories of frequency discrimination were presented, the *place theory* of Helmholtz (1863) and the *frequency theory* of Rutherford (1886). The place theory assumes a spatial distribution of activity along the cochlear duct as a function of frequency. The frequency theory, on the other hand, assumes a direct representation of stimulus frequency by the frequency of the nerve impulses. Recent investigations have shown that with certain limitations both theories are actually valid, and a third theory, the *volley theory* (Wever, 1964), has been proposed as a combination of the two.

According to Helmholtz' *place theory,* the basal end of the organ of Corti is stimulated by high-frequency tones, the apical end by low frequencies. On auditory stimulation traveling waves are set up in the basilar membrane, transmitted to it by vibrations of the tympanic membrane, the middle ear ossicles, and the inner ear fluid (Békésy, 1960, and later). The point of maximal excursion of the basilar membrane moves progressively toward the apex as the frequency of the stimulating tone decreases.[2] By near-threshold stimulation with pure tones, action potentials are probably generated only by those sensory cells that are located over the maximally oscillating part of the membrane. This assumption is supported by recordings from the afferent fibers in the nerve central to the cochlea (Kiang, 1965; Whitfield, 1967).

A spatial frequency representation in the organ of Corti is also supported by the specific frequency hearing loss that occurs with localized lesions of hair cells (Held and Kleinknecht, 1927; Schuknecht, 1960) and by the occurrence of localized hair cell loss following excessive tonal stimulation in animals (Bredberg, Ades, and Engström, 1972). Postmortem studies of cochleas from patients with a specific frequency hearing loss also have shown localized defects in the organ of Corti and in corresponding afferent fibers (Guild, Crowe, Bunch, and Polvogt, 1931; Crowe, Guild, and Polvogt, 1934; Bredberg, 1968). In man the frequency range of the cochlea is from 16 to 20,000 cycles/sec, with the lowest threshold to both stimulation and noise damage in the higher frequency zone.

A particular problem is to explain how the frequency specificity of the cochlear nerve fibers is possible in view of the apparently widespread oscillation of the basilar membrane. Recordings from the cochlear nerve during high-intensity tonal stimulation show that a single fiber responds to a fairly broad band of frequencies. With near-threshold stimulation the response band is, in contrast, extremely narrow. A "response area" is defined as the area on a graph of "stimulus intensity" versus "stimulus frequency" which contains all intensities and frequencies capable of producing a detectable response. The outline of the response area is called the *"tuning curve,"* and the frequency to which the unit is most sensitive (the tip of the curve) is called the *characteristic frequency* (CF) or best frequency (Kiang, 1965). Depending on the width of the curve, the unit is characterized as sharply or broadly tuned. In attempts to explain the sharp tuning of the

[2] The afferent fibers from the apical coil are slightly thicker than those from the basal coil (see Fig. 9-6, cat) (Arnesen and Osen, 1978). Possibly this arrangement serves to compensate for part of the time differences in the wave traveling in the basilar membrane.

cochlear nerve fibers, various theories have been proposed. Local interaction between nervous elements and sensory cells and inhibitory feedback mechanisms from the brainstem have been suggested to be involved.

While the place theory explains the spatial representation of tonal frequencies in the receptor organ and the central auditory pathways, the *frequency theory* is of importance for the understanding of low-frequency hearing. By stimulation with low tonal frequencies the firing of the activated cochlear nerve fibers is locked to the phase of the stimulating tone. This phenomenon is of great importance for detection of binaural phase or time differences and, thereby, for localization of sound in space, as explained later.

The innervation of the auditory receptors. The cochlea is supplied by three sets of nerve fibers: (1) afferent fibers of the cochlear nerve, (2) efferent fibers (of the olivocochlear bundle), and (3) sympathetic fibers. Only the two former sets are in direct contact with the hair cells. In the following each set of fibers will be described separately.

The *cochlear nerve* enters the brainstem in company with the vestibular nerve at the lower border of the pons, slightly lateral to the facial and intermediate nerves (Figs. 5-3 and 7-6). These nerves all pass through the internal acoustic meatus, where the facial nerve is situated above, the cochlear nerve below, and the vestibular nerve is interprosed between them (see Fig. 7-9). At the bottom of the meatus the cochlear nerve, when followed in a peripheral direction, bends anteriorly into the central modiolar part of the cochlea, whose apex also points anteriorly. In the modiolus, fascicles of fibers radiate through small bony channels toward the *cochlear or spiral ganglion,* which forms a continuous row of ganglion cells along the inner curvature of the cochlear spiral (Figs. 9-1, 9-6). The ganglion is situated in a spiraled canal (Rosenthal's canal) at the base of the bony spiral lamina.

The cochlear nerve is composed of the central processes, or axons, of the bipolar spiral ganglion cells. This nerve is, therefore, by definition, not a peripheral nerve, but rather a sensory nerve root. The axons are organized in a strict cochleotopic pattern (Lorente de Nó, 1933a; Sando, 1965; Arnesen and Osen, 1978). In correspondence with their sites of origin in the cochlea, the cochlear nerve fibers are sequentially arranged, forming a roll with the basal coil fibers situated outside the apical ones (Fig. 9-6). The peripheral processes or dendrites of the spiral ganglion cells, on the other hand, take a straight radial course through the bony spiral lamina toward the organ of Corti. Where the dendrites leave the bone, at the habenula perforata (Fig. 9-1), they lose their myelin sheaths. The fibers of the cochlear nerve, like those of the vestibular nerve, belong to the special somatic afferent category (Fig. 7-1).

Our knowledge of the distribution of the afferent fibers to the rows of cochlear hair cells, originally described by Polyak (1927a), Lorente de Nó (1933a), and Fernandez (1951), has been considerably increased recently, especially by Spoendlin's (1972, 1973) work using the cat. As described by him, the fiber-to-cell ratio is drastically different for inner an outer hair cells (Fig. 9-3). Whereas the majority of afferent fibers innervates the relatively small population of inner hair cells, only a minority of fibers supplies the far more numerous outer hair

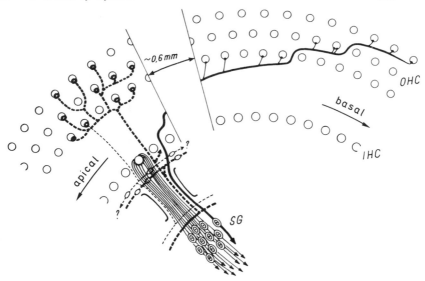

FIG. 9-3 Schematic outline of the fiber distribution in the organ of Corti. Fully drawn thick lines: afferent fibers to the outer hair cells (*OHC*); full thin lines: afferent fibers to the inner hair cells (*IHC*); broken thick lines: efferent nerve fibers from the crossed olivocochlear bundle; broken thin lines: efferent nerve fibers from the uncrossed olivocochlear bundle. Compare Fig. 9-1. Abbreviation: *SG:* spiral ganglion. From Spoendlin (1970).

cells. The size and number of afferent nerve terminals on the two types of hair cells also show considerable differences (Figs. 9-1, 9-3).

In man there are about 32,000 myelinated fibers in the cochlear nerve (Rasmussen, 1940a). In the cat the number is about 50,000 (Gacek and Rasmussen, 1961). About 90–95% of these supply inner hair cells (Spoendlin, 1972). Each fiber takes a short radial course from the habenula perforata toward an inner hair cell on which it terminates unbranched with a large synapse (Figs. 9-1, 9-3). About 20 fibers converge on a single inner hair cell. Only 5–10% of the afferent fibers supply the outer hair cells. The outer hair cell afferents, in contrast to those of the inner hair cells, spiral in the basal direction for about 0.6 mm in the outer spiral bundle before branching and forming one or two small synapses with each of several cells (Spoendlin, 1970). In spite of the branching of the outer hair cell afferents, each cell is provided with considerably fewer and smaller synapses than are the inner cells. According to Spoendlin the inner and outer hair cell afferents belong to different types of spiral ganglion cells, defined as type I and type II (see Fig. 9-1). The existence of two different types of cochlear nerve fibers, however, still remains to be proved.

The description of the cochlear nerve given here is based mainly on findings in the cat. The recent progress in the clinical application of cochlear prostheses, however, has increased the need for a similarly thorough analysis of the cochlear nerve and the cochlear ganglion in man.

Besides the afferent innervation the auditory receptor cells receive a rich *efferent nerve supply* from the superior olivary complex in the brainstem. These fibers reach the cochlea via the *olivocochlear bundle,* first described by Rasmussen (1946, 1953) in the cat and the opossum. Later it was identified in several other species, including man (Gacek, 1961; Nomura and Schuknecht, 1965; Ishii, Murakami, and Balogh, 1967; Moore and Osen, 1979). The olivocochlear fibers are probably cholinergic (for a pharmacological review see Guth, Norris, and Bobbin, 1976). The various components of the system are easily recognized histochemically by means of their strong acetylcholinesterase activity.

By acetylcholinesterase staining of the brainstem combined with HRP injections in the cochlea (Warr, 1975), the olivocochlear fibers have been found to originate in two distinct types of superior olivary neurons, defined as large and small acetylcholinesterase-positive cells (Osen and Roth, 1969). The large cells are situated in the periolivary nuclei (see below). Their axons, probably myelinated (see below), cross the midline in the floor of the fourth ventricle (Fig. 9.-5). They leave the brainstem in the vestibular nerve root and pass over to the cochlear nerve in the vestibulocochlear anastomosis, which bridges the two nerves in the internal acoustic meatus. The fibers can be followed to the outer hair cells (Rasmussen, 1960; Smith and Rasmussen, 1963), where they form large axosomatic synapses that far outnumber the afferent ones (Figs. 9-1, 9-3).

> The existence of two ultrastructurally different types of nerve endings on the outer hair cells, vesiculated and nonvesiculated ones, was first described by Spoendlin (1956, 1960) and later by Engström (1958, 1967), Engström and Wersäll (1958), Smith and Sjöstrand (1961), and others. Degeneration studies following lesions of the crossed olivocochlear fibers in the floor of the fourth ventricle have identified the vesiculated endings as belonging to the efferent fibers (Iurato, 1962; Kimura and Wersäll, 1962; Spoendlin and Gacek, 1963; Smith and Rasmussen, 1963, 1965). The nonvesiculated endings, on the other hand, belong to the afferent system.

In contrast to the large olivocochlear neurons, the small acetylcholinesterase-positive cells of the cat are located at the margin of the lateral superior olive. Their tiny axons, probably uncrossed and unmyelinated (see below), join the crossed olivocochlear fibers in the ipsilateral vestibular nerve root (Rasmussen, 1960). The distribution of these fibers within the organ of Corti is unclear, but most likely they are among the efferent fibers that spiral underneath the inner hair cells in the inner spiral bundle, making *en passage* synapses with the inner hair cell afferents (Figs. 9-1, 9-3) (Smith, 1961; Iurato, 1974; Iurato et al., 1978).

> In accordance with the presence of two types of olivocochlear neurons, the olivocochlear bundle contains two types of axons: relatively large (2–3μm) myelinated axons and small unmyelinated ones (Rasmussen, 1960; Terayama, Yamamoto, and Sakamoto, 1969; for further references, see Arnesen and Osen, 1980). The ratio of myelinated to unmyelinated axons—two to three (Arnesen and Osen, 1980)—is about the same as the ratio of large to small cells (Warr, 1975). The two types of axons and the two types of cells probably are mutually interrelated. On the basis of autoradiographic studies in kittens, Warr (1978) has postulated the existence of two olivocochlear pathways, a *large-cell pathway,* which is mainly crossed and supplies outer hair cells, and a *small-cell pathway,* which is mainly uncrossed and supplies inner hair cells. Although this conclusion would fit many other observations, more studies are required to confirm this interesting hypothesis. One of the functional implications of the theory would be that the two pathways may differ with regard to conduction velocity and hence latency.

In spite of intensive investigations, the role of the olivocochlear system in audition remains unclear. Electrical stimulation of the crossed olivocochlear fibers in the floor of the fourth ventricle is found to reduce the auditory nerve activity aroused by sound stimulation (Galambos, 1956; Desmedt, 1962; Fex, 1967). Since the perikarya of the olivocochlear fibers are probably influenced both by ascending and descending auditory fibers (see later), they may possibly be involved in feedback mechanisms that serve to suppress unwanted auditory signals. On the other hand, a role in the sharpening of the frequency resolving power of the inner ear has also been discussed (Capps and Ades, 1968; see, however, Agarashi, Cranford, Nakai, and Alford, 1979). (For review of the olivocochlear bundle see Engström, Ades, and Andersson, 1966; Klinke and Galley, 1974; Iurato, 1974.)

Although it has been disputed (Ross, 1969), there can be little doubt that the olivocochlear bundle is a direct efferent pathway that is not to be regarded as part of the visceral nervous system. Besides these fibers, however, the cochlea receives *noradrenergic fibers* from ganglia of the sympathetic trunk. These fibers represent two different systems that may have different functions—that is, a vessel-dependent and a vessel-independent system (Spoendlin and Lichtensteiger, 1966, 1967; Ross, 1973; Densert, 1974; and others). The functional significance of these fiber systems is unclear, but their existence implies that the possibility of hearing disturbances due to imbalance in the sympathetic nervous system cannot be rejected.

> The *vessel-dependent system* originates in the stellate ganglion. The fibers course along the vertebral and labyrinthine arteries and seem to terminate in relation to the cochlear vesels. The *vessel-independent system* originates in the superior cervical ganglion. The fibers reach the cochlea via the vagus and facial nerves and terminate in the cochlea central to the hair cells (the bony spiral lamina and the limbus).

The ascending pathway in general. In the following the general principles of organization of the ascending auditory pathway will be considered. The various auditory centers are treated separately, as we follow the route of auditory impulses from the organ of Corti to the cerebral cortex (see Fig. 9-4). Parallel to the ascending projection is a descending pathway, which is treated separately in a subsequent section (Fig. 9-5). The present account is based largely on experimental findings, but, where appropriate, reference will be made to observations in man. For a recent review of the ascending system see Harrison and Howe (1974a).

The ascending pathway, transmitting the signals from the auditory receptors, can be traced to the cerebral cortex. The pathway is interrupted at several levels, notably the cochlear nuclei, the superior olive, the inferior colliculus, and the medial geniculate body. These cellular assemblies do not only serve as relays in the central propagation of auditory information leading to conscious perception. Each of them appears to have its particular role in influencing the central transmission of auditory information and in mediating an interaction of this with other kinds of information. Each of the stations in the auditory pathway is subject to descending influences from higher levels, not least the cerebral cortex. Further, they are important for a variety of reflexes that may be elicited on auditory stimulation.

The afferent cochlear nerve fibers pass to the cochlear complex, where they bifurcate into ascending and descending branches. In the nuclear complex, as in the nerve, the cochlear fibers are arranged in a strict cochleotopic pattern that is fundamental for the spatial representation of tonal frequencies. By means of collateral branching and specific types of synaptic swellings (see below), the primary afferents supply various categories of second-order neurons which occupy specific regions of the nuclear complex. These neurons in turn have different central projections, either via the *trapezoid* body, which courses ventral to the inferior cerebellar peduncle, or via the *intermediate* or *dorsal acoustic striae*, which bend above the peduncle (Figs. 9-4, 9-5). In this way, *the unichannel system of the cochlear nerve is converted into a multichannel system* with different trajectories

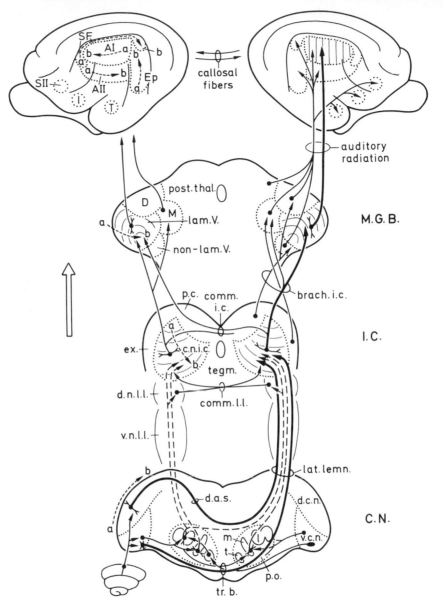

FIG. 9-4 Diagram of the ascending auditory pathway of the cat, including the auditory centers at the medullary, mesencephalic, diencephalic, and cortical levels. Thick, full lines indicate the auditory core projection to the contralateral primary auditory cortex. Other fibers are drawn as thin lines; the tertiary auditory fibers are stippled. The laminar organization of nuclei in the core projection is indicated by parallel lines, and the tonotopical organization by stippled arrows, pointing from apical (*a*) to basal (*b*) areas of cochlear representation. See text. For explanation of the auditory cortical areas, see Fig. 9-9. Abbreviations: *brach.i.c.:* brachium of the inferior colliculus; *comm.i.c.:* commissure of the inferior colliculus; *comm.1.1.:* commissure of the lateral lemniscus; *C.N.:* cochlear nuclear complex; *c.n.i.c.:* central nucleus of the inferior colliculus; *D:* dorsal division of the medial geniculate body; *d.a.s.:* dorsal acoustic stria; *d.c.n.:* dorsal cochlear nucleus; *d.n.l.l.:* dorsal nucleus of the lateral lemniscus; *ex.:* external nucleus of the inferior colliculus; *i.a.s.:* intermediate acoustic stria; *I.C.:* inferior colliculus; *1:* lateral superior olive;

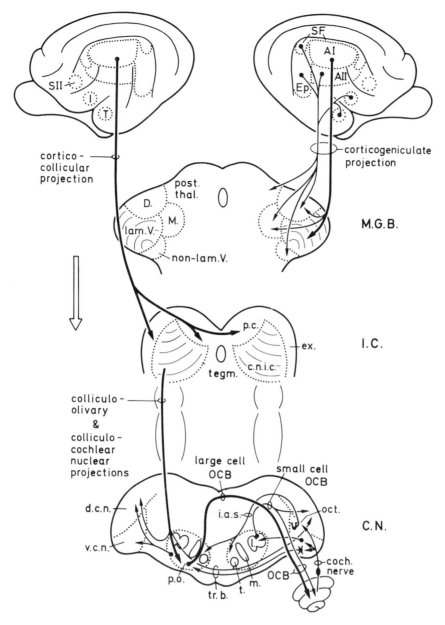

FIG. 9-5 Diagram of the descending auditory pathway of the cat. See text. For abbreviations see Fig. 9-4.

---

lam. V.: laminated portion of the ventral division of the medial geniculate body; *lat. lemn.:* lateral lemniscus; *M:* medial division of the medial geniculate body; *m:* medial superior olive; *M.G.B.:* medial geniculate body; *non-lam.V.:* nonlaminated nuclei of the ventral division of the medial geniculate body; *OCB:* olivocochlear bundle; *oct.:* octopus cell area; *p.c.:* pericentral nucleus of the inferior colliculus; *p.o.:* periolivary nuclei; *post. thal.:* posterior thalamic group; *t:* nucleus of the trapezoid body; *tegm.:* midbrain tegmentum; *tr.b.:* trapezoid body; *v.c.n.:* ventral cochlear nucleus; *v.n.l.l.:* ventral nucleus of the lateral lemniscus.

through the brainstem. During their central course, the channels arising in the cochlear nuclei all converge toward the opposite *lateral lemniscus* and, through this, toward the *central nucleus of the inferior colliculus*. Some of the channels (made up of axons of the pyramidal cells and multipolar cells in the cochlear nuclei) ascend directly to the inferior colliculus; others (arising in the spherical cells, globular cells, and octopus cells) are interrupted by synapses in various subnuclei of the *superior olivary complex*. In these nuclei, the signals from one ear interact with signals from the other ear, a point of importance,—for, among other things, localization of sound in space (see later). During their ascending course, most axons probably give off collaterals to other auditory nuclei besides their main target. A multitude of differently processed signals ultimately reach the central nucleus of the inferior colliculus, which in many respects has a key position in the ascending auditory pathway.

From the inferior colliculus, the ascending pathway continues rostrally in the *brachium of the inferior colliculus* to the *medial geniculate body* of the thalamus. Via the *auditory radiation,* the impulses finally reach the *auditory cortex* in the temporal lobe (Fig. 9-4). At the inferior colliculus and above, the ascending auditory pathway may be divided into a *core projection* and a *belt projection*. The final target of the core projection is the auditory *koniocortex* (AI in Fig. 9-4), which is the type of cortex present where direct sensory systems terminate (cf. Chap. 12). The target of the belt projection, on the other hand, is primarily the auditory cortical areas surrounding the koniocortex. These areas were collectively referred to by Rose (1949) as the auditory *cortical belt areas,* hence the term belt projection. As pointed out by Casseday, Diamond, and Harting (1976), the division of the ascending auditory pathway into core and belt portions seems to be characteristic of all mammals and thus is deemed essential in comparative studies. Although in certain respects it may represent an oversimplification, a distinction between core and belt structures will be emphasized in this account, when feasible.

The *auditory core projection* includes the central nucleus of the inferior colliculus, the laminated portion of the ventral division of the medial geniculate body, and the primary auditory cortex (the pathway is marked by thick lines in Fig. 9-4). The two nuclei are characterized by a laminated structure that, by anatomical and electrophysiological methods, is found to establish the basis for a spatial frequency representation, which is also preserved at the cortical level.

The *auditory belt projection* includes the pericentral region of the inferior colliculus and the dorsal midbrain tegmentum, both of which project to the nonlaminated portions of the medial geniculate body (the pathway is marked by thin lines in Fig. 9-4). These in turn project primarily to the auditory cortical belt areas. The core and belt projections are interconnected by fibers to be described later. In contrast to the core projection, which is exclusively auditory, the belt projection also receives input from visual and somesthetic pathways. These inputs may possibly be of importance for, among other things, the interaction of various sensory modalities in motor functions, such as speech. An interesting aspect of this interaction may be the possible benefit of sensory stimulation, other than auditory, in speech training of deaf children.

The ascending auditory pathway is characterized, at nearly all levels, by well-developed commissural connections (Fig. 9-4). Besides the crossing over in the trapezoid body and the intermediate and dorsal acoustic striae, auditory fibers also traverse the midline in the medullary and pontine reticular formation, the commissure of the lateral lemniscus, the commissure of the inferior colliculus, and the corpus callosum. After unilateral lesions at subcortical levels, therefore, the auditory cortex in both hemispheres may still be activated by auditory stimulation. The contralateral cortical representation of each ear, however, seems to predominate, possibly as a consequence of the ampler projection of second-order neurons to the inferior colliculus on the contralateral than on the ipsilateral side.

> According to fiber degeneration studies in the chimpanzee (Strominger, Nelson, and Dougherty, 1977), there may be more second-order fibers reaching the inferior colliculus on the ipsilateral side in primates. In these species, moreover, there may be a larger contingent of ascending fibers bypassing the inferior colliculus to terminate directly in the medial geniculate body.

The cochlear nuclei. In mammals, the cochlear nuclear complex is found lateral to the inferior cerebellar peduncle in the floor of the lateral recess of the fourth ventricle. In the human brainstem (Fig. 7-2) the complex is situated so far laterally that it is usually cut off with the cerebellum in routine postmortem studies. The complex consists of a ventral and a dorsal nucleus (Fig. 9-6). The *ventral nucleus* is divided by the entering cochlear nerve root into an *anteroventral* and a *posteroventral nucleus*. The *dorsal nucleus* is situated dorsolateral to the latter. As mentioned above, the cochlear nerve fibers bifurcate within the ventral nucleus into ascending and descending branches, like primary afferents in general. The branches are arranged in a regular cochleotopic order that forms the basis for the tonotopical organization of the system (Lorente de Nó, 1933a; Lewy and Kobrak, 1936; Rasmussen, Gacek, McCrane, and Baker, 1960; and others). In all three subnuclei the branches from the basal part of the organ of Corti, containing the basal (high-frequency) fibers, are located dorsally, whereas the apical (low-frequency) fibers are found ventrally (Fig. 9-6, cat). These anatomical observations agree completely with Rose, Galambos, and Hughes' (1959) mapping of characteristic frequencies within the nuclear complex (Fig. 9-7). In the human dorsal nucleus, however, the frequency axis may differ, as suggested by the special course of the descending cochlear branches (see Fig. 9-6, man).

As appears from various cytoarchitectonic studies (Lorente de Nó, 1933b, 1976; Harrison and Irving, 1965, 1966 a, b; Osen, 1969; Brawer, Morest, and Kane, 1974), the cochlear nuclei are composed of several classes of neurons which occur more or less intermingled. The cells are either interneurons or principal neurons that give rise to the various second-order auditory channels mentioned above. The cytology of the cochlear nuclei, thus, is a key to understanding the auditory pathway. In fact, it was not until attention was paid to the *cell types* rather than to the traditional subnuclei of the cochlear nuclear complex that any substantial progress in the elucidation of the brainstem auditory centers was achieved.

In the Nissl study by Osen (1969), nine different types of neurons were defined. Most of these are also distinguishable by their electrophysiological properties and their afferent and efferent fiber connections (see below). With slight modifications, the Nissl picture is also compatible with findings in Golgi material

(Brawer, Morest, and Kane, 1974; Lorente de Nó, 1933b, 1976). As described by Osen (1969), the areas of distribution of the particular cell types within the nuclear complex do not in all places correspond to the borderlines of the three subnuclei referred to above.

FIG. 9-6 Diagram of the cochlear nuclei in man and in the cat, in lateral views. In the *cat* the cochlear nerve fibers can be followed centrally from their perikarya in the spiral ganglion through their bifurcation in the cochlear nerve root and to the sites of termination of their branches. The ascending branches terminate in the area of spherical cells (*sph.*), the descending branches in the dorsal cochlear nucleus (*d.c.n.*). Note the gradual increase in thickness of the fibers from the basal to the apical coil and the spatial representation of the cochlea in the nerve and the nuclei. In the cross section of the cat cochlear nerve, to the lower left, the cochleotopic organization is indicated by different symbols (key below) and by an arrow pointing from low- (*a:* apical) to high- (*b:* basal) frequency regions. In *man* the distribution of fibers appears to be the same as in the cat, except for the orientation of the descending branches. On entering the dorsal nucleus, these fibers take a straight course instead of making a 180° turn, as they do in the cat. The relative size of the various subdivisions of the ventral nucleus differs in the two species. The areas of octopus cells (*oct.*) and spherical cells (*sph.*) are relatively small in man, while the central region (*cent.*) and the peripheral cap of small cells (*cap*) are well developed. A granule cell layer (*gran.*) is present only in the cat. Abbreviations: *a.v.e.n.:* anteroventral cochlear nucleus; *p.v.e.n.:* posteroventral cochlear nucleus. CAT: from Arnesen and Osen (1978). MAN: from Moore and Osen (1979).

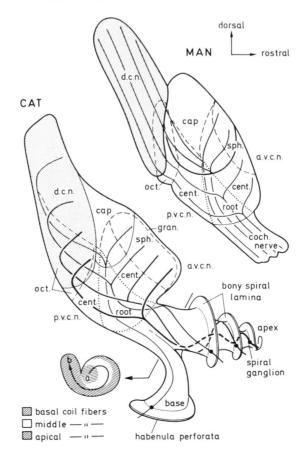

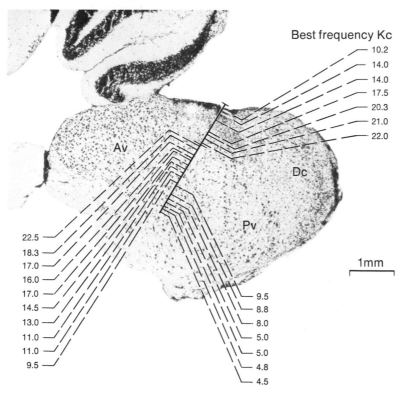

FIG. 9-7 Photomicrograph of a sagittal section through the cochlear nuclei in the cat, showing an example of the spatial representation of the cochlea as determined by mapping of characteristic (best) frequencies. See text. *Av, Pv,* and *Dc:* anteroventral, posteroventral, and dorsal cochlear nucleus, respectively. From Rose, Galambos, and Hughes (1959).

Most of our present knowledge concerning the cochlear nuclear complex refers to observations in the cat. The situation in this species will, therefore, be treated first, with reference to Figs. 9-6, 9-7, and 9-8. The first figure, for comparison, also contains a diagram of the human cochlear nuclear complex, which will be dealt with briefly at the end of this section.

In the *cat* the *ventral cochlear nucleus* may be divided, on cytological grounds, into three portions (Fig. 9-6): a rostral *area of spherical cells* (sph.), a caudal *area of octopus cells* (oct.), and an intervening *central region* (cent.) (Osen, 1969). The latter contains a mixture of *multipolar cells, globular cells,* and *small cells.* Dorsolateral to the ventral nucleus proper is a *cap of small cells* (cap). In the cat, the ventral nucleus with the cap is, moreover, covered by a *layer of granule cells* (gran.).

In subprimate mammals the *dorsal cochlear nucleus* is stratified (not illustrated here) with a superficial molecular layer and four deeper cellular layers distinguishable by their cyto- and fiber-architecture (Lorente de Nó, 1933b). The main projection neurons are the radially oriented, bipolar *pyramidal cells,* which may have a position in the dorsal cochlear nucleus similar to that of the Purkinje cells in the cerebellar cortex.

The pyramidal cells of the dorsal cochlear nucleus in subprimates are unique among second-order sensory neurons because they are influenced, via the granule cells, by a neuronal circuitry that has many features in common with that of the cerebellar cortex (Mugnaini, Osen, Dahl, Friedrich, and Korte, 1980). The cochlear granule cell dendrites are engaged in glomerular synaptic arrays centered on a mossy-fiber terminal that may originate in higher auditory centers (McDonald and Rasmussen, 1971; Kane, 1977a) or possibly within the cochlear nuclear complex. The granule cell axons form parallel fibers in the molecular layer of the dorsal nucleus, establishing synapses *en passage* with the pyramidal cell apical dendrites (Mugnaini, Warr, and Osen, 1980). On their basal dendrites, the pyramidal cells receive primary afferents and fibers descending from, among others, the inferior colliculus. The dorsal cochlear nucleus may possibly play a role in the auditory system, which is analogous in some respects to that of the cerebellar nodulus in the vestibular system. So far, however, nothing is known about the functional significance of this interesting nucleus in audition.

A simplified wiring diagram of the main cell types of the cochlear nuclear complex, on the basis of findings in the cat and the rat, is shown in Fig. 9-8. The cochlear nerve fibers provide different types of synaptic contacts with the various types of cochlear nuclear neurons (see, among others, Lorente de Nó, 1933b, 1976; Lenn and Reese, 1966; Osen, 1970; Brawer and Morest, 1975; Kane, 1974; Tolbert and Morest, 1978). The electrophysiological properties of the cells, as revealed by extracellular recordings on auditory stimulation (Godfrey, Kiang, and Norris, 1975a, b; van Gisbergen, Grashuis, Johannesma, and Vendrik, 1975; Bourk, 1976), probably reflect these structural peculiarities.

The spherical cells and the globular cells which receive a restricted number of large, axosomatic endings (bulbs of Held) are characterized as "primary-like" owing to the similarity of their firing pattern to that of the primary afferents. Other cell types, like the multipolar cells, the octopus cells, and the pyramidal cells, which receive a large number of small boutons on their soma and dendrites, have more complicated firing patterns. This is partly due to a larger convergence of primary afferents on individual neurons, and partly to interaction of intrinsic and descending fiber systems (not illustrated).

The cell types of the cochlear nuclei also differ in their central projection as described later and illustrated in Figs. 9-4, 9-5, and 9-8. Here it is sufficient to mention that the spherical cells, the globular cells, and the octopus cells have their main target in the superior olivary complex (Harrison and Irving, 1964, 1966a, b; Warr, 1966, 1969; and others), whereas the pyramidal cells and the multipolar cells project as far rostral as to the contralateral inferior colliculus (Osen, 1972; Adams, 1977, 1979; and others).

In *man* the cochlear nuclear complex is of about the same size as in the cat (Fig. 9-6). The volume ratio of the dorsal to the ventral nucleus, though, is slightly larger in man (1 to 2.8; Hall, 1964) than in the cat (1 to 4; Kiang, Godfrey, Norris, and Moxon, 1975). This is probably due to the somewhat smaller size of the spherical and octopus cell areas in man (Moore and Osen, 1979), a fact that accords well with the relative smallness of the superior olivary complex in the human brainstem (Moore and Moore, 1971). In the human ventral nucleus the cap of small cells is relatively large, but, in contrast to the situation in the cat, there is not a vestige of a granule cell layer. The human dorsal cochlear nucleus differs from that of the cat by a different orientation of the descending cochlear branches (see above) and by not being stratified. The pyramidal cells have lost their typical orientation, and only a vestige of a molecular layer is found. In man, thus, the dorsal cochlear nucleus cannot be regarded as a structural analogue of the cerebellar cortex. Since the function of the dorsal nucleus in subprimates is unknown, there is no easy explanation for the different pattern in man. The change, which seems to be the result of a gradual disappearance of cochlear granule cells in primates (Moore, 1980), possibly means that the processing of auditory information is, to a larger extent, taken care of by higher auditory centers. The predominance in the ventral nucleus of neurons like the multipolar cells and the small cap cells, which have a direct collicular connection (Adams, 1977, 1979), may be part of the same trend.

The superior olivary complex and the nuclei of the lateral lemniscus. The superior olivary complex is situated in the trapezoid body, which in the human brain is found in the caudal part of the pons, immediately dorsal to the pontine gray. The complex shows considerable variation among species (Irving and Harrison, 1967; Feldman and Harrison, 1971; Harrison, 1974). In the cat, it consists of three well-defined nuclei: the *medial superior olive*, the *lateral superior olive*, and the *nucleus of the trapezoid body* (m.s.o., l.s.o., and n.t.b., respectively, in Fig. 9-8). These nuclei are surrounded by a sphere of scattered neurons, here collectively referred to as the *periolivary nuclei*. The lateral superior olive and the nucleus of the trapezoid body are mutually related. Both are poorly developed in the human brain (Moore and Moore, 1971). The medial superior olive, however, is about the same size in the cat and in man.

In transverse sections of the brainstem, the *medial superior olive* appears as a

FIG. 9-8 A simplified wiring diagram of the cochlear nuclei and the superior olivary complex, on the basis of observations in the cat. The cochlear nuclear neurons are supplied from the cochlear nerve by different types of synaptic endings. Each cell type has its specific projection onto higher auditory brainstem centers. As shown in the diagram, where only main targets are indicated, the spherical and globular cell axons are responsible for the binaural input to the medial superior olive (*m.s.o.*) and the lateral superior olive (*l.s.o.*), the latter nucleus receiving its contralateral input via the nucleus of the trapezoid body (*n.t.b.*). Abbreviations: *i.c.*: inferior colliculus; *p.o.*: periolivary nuclei.

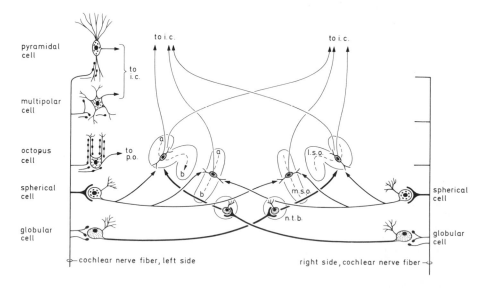

parasagittal row of transversely oriented bipolar neurons. Each neuron is influenced bilaterally by means of axons arising in the spherical cells of the cochlear nucleus. The laterally extending dendrites are in synaptic contact with afferents from the ipsilateral side, and the medially extending dendrites with fibers from the opposite side (Stotler, 1953; and others) (Fig. 9-8). The nucleus responds chiefly to low-frequency tones and its neurons are apparently sensitive to interaural time or phase differences (Goldberg and Brown, 1968; Moushegian, Rupert, and Langford, 1967).[3] The medial superior olive is presumably involved in spatial localization of sounds, especially of low frequencies. It projects ipsilaterally to the central nucleus of the inferior colliculus (Stotler, 1953; Elverland, 1978; Adams, 1979) (Fig. 9-4).

The *nucleus of the trapezoid body* is situated ventromedially in the superior olivary complex. Its principal neurons resemble the globular cells of the cochlear nuclei. As illustrated in Fig. 9-8, the nucleus is supplied from the globular cells of the contralateral side by means of thick axons, which terminate in large calyces in a one-to-one relationship (Harrison and Irving, 1964, 1966a; Lenn and Reese, 1966; Jean-Baptiste and Morest, 1975). The cells project to the lateral superior olive on the same side (Rasmussen, 1967; van Noort, 1969; Elverland, 1978).

The *lateral superior olive* of the cat forms an S-shaped row of bipolar neurons. Like the medial superior olive, it is influenced from both ears (Fig. 9-8). The ipsilateral input comes directly from the spherical cells (Warr, 1966; and others). The contralateral input is indirect from the globular cells via the nucleus of the trapezoid body (see above). In contrast to the situation in medial superior olive, all audible frequencies are represented. Mapping of characteristic frequencies (Tsuchitani and Boudreau, 1966) and fiber degeneration studies (personal observations) have found a strict tonotopical order, with low frequencies represented laterally and high frequencies medially. The neurons seem to be sensitive to interaural intensity differences (Boudreau and Tsuchitani, 1973). The lateral superior olive, therefore, may also be involved in localization of sound in space, but, in contrast to the medial superior olive, it may operate best for high-frequency tones.[4] The lateral superior olive projects bilaterally to the central nucleus of the inferior colliculus (Stotler, 1953; Adams, 1979) and possibly to the dorsal nucleus of the lateral lemniscus (Elverland, 1978) (Fig. 9-4).

The *periolivary nuclei* contain many different cell types, some of which may be cholinergic, while others are apparently not. These nuclei are probably to a large extent involved in descending pathways, and they will therefore be treated

---

[3] When a sound reaches the ears from one side of the body, it will hit the ipsilateral ear slightly earlier than the opposite side and in a different phase. While the tympanic membrane moves inward on one side, it may move outward on the other side. This difference is interpreted by the central nervous system as a displacement of the sound toward the ear in which the phase is leading. By its phase-locked, primary-like firing, the bilateral spherical cell input to the medial superior olive may be ideal for interpretation of interaural phase differences.

[4] According to Tsuchitani (1977), the activity of the neurons in the lateral superior olive results from an interaction between the ipsilateral excitatory and the contralateral inhibitory input. The delay of the contralateral input across the midline is possibly kept at a minimum by the position of the globular cells close to the cochlear nerve root, the large size of their axons, and the calyciform synapses in the nucleus of the trapezoid body, which are among the largest in the mammalian brain. The globular cells, like the spherical cells, have a primary-like firing pattern, which probably is able to replicate the cochlear nerve activity.

together with these in a subsequent section. Here it is sufficient to mention that the periolivary neurons, in contrast to those of the three main nuclei of the superior olivary complex, receive descending fibers from the inferior colliculus and, probably, collateral input from several types of cochlear nuclear neurons, including the octopus cells (Warr, 1969).

The *nuclei of the lateral lemniscus* are groups of cells situated largely within the fiber bundles of the lateral lemniscus. Usually two nuclei can be distinguished, a ventral and a dorsal. They appear to be stations in the ascending auditory pathway, but precise knowledge is scanty. Nothing is known about the function of these nuclei, which together constitute a substantial amount of auditory neurons.

The *ventral nucleus of the lateral lemniscus* is probably supplied by collaterals of lemniscal fibers from various sources, including the cochlear nuclei on the contralateral side and the superior olivary complex on both sides (Warr, 1969; and others). It projects to the central nucleus of the ipsilateral inferior colliculus (Adams, 1979). The *dorsal nucleus of the lateral lemniscus*, in contrast, may not be supplied from the cochlear nuclei. According to autoradiographic studies, it may receive fibers from the lateral superior olive on both sides (Elverland, 1978). It projects bilaterally to the central nucleus of the inferior colliculus, with fibers crossing in the commissure of the lateral lemniscus (Fig. 9-4).

The inferior colliculus. This important auditory center is part of the midbrain tectum (Figs. 7-6, 9-4, 9-5). Ventrally, it borders on the dorsal midbrain tegmentum, to which it is anatomically and functionally related. Dorsally, the two colliculi are interconnected by the *commissure of the inferior colliculus*. The inferior colliculus is a relay in the auditory pathway between fibers of the lateral lemniscus and those of the *brachium of the inferior colliculus*, which leads to the medial geniculate body. Like auditory structures in general, both the lateral lemniscus and the brachium are superficially located in the brainstem. The functional role of the inferior colliculus differs in many respects from that of the superior colliculus in the visual system. The inferior colliculus, thus, has a key position in the auditory pathway as an obligatory relay station for all (or nearly all) ascending and descending auditory fibers reaching the midbrain level (Woollard and Harpman, 1940; Barnes, Magoun, and Ranson, 1943; Moore and Goldberg, 1963, 1966; Goldberg and Moore, 1967; van Noorth, 1969; Casseday, Diamond, and Harting, 1976; Strominger, Nelson, and Dougherty, 1977).

The inferior colliculus, as a whole, *receives fibers from* the dorsal and ventral cochlear nuclei, the superior olive, the nuclei of the lateral lemniscus, and the auditory cortex. It *gives off fibers to* the medial geniculate body, the midbrain tegmentum, the periaqueductal gray, the periolivary nuclei, the dorsal cochlear nucleus, and the dorsolateral pontine nucleus. As indicated in Figs. 9-4 and 9-5, the connections are either ipsilateral, contralateral, or bilateral. As a rule, however, the projection ascending *to* the inferior colliculus is larger from the contralateral than from the ipsilateral ear. The pathway ascending *from* the collicle, on the other hand, is larger on the ipsilateral side, thus favoring the ultimate representation of the contralateral ear in the auditory cortex.

Whereas the fibers that reach the inferior colliculus from the lower auditory centers all terminate in the central nucleus (see below), every part of the collicle partakes in the ascending projection to the medial geniculate body. The various subdivisions of the collicle, nevertheless, have different targets in the medial geniculate body. As mentioned in a previous section, at the midbrain level and above, the auditory pathway may be separated into a core and a belt projection. The *core projection* arises in the central nucleus of the inferior colliculus (see below). The *belt projection,* on the other hand, takes its origin in the peripheral gray matter surrounding the central nucleus. As will be considered later, this peripheral gray matter also differs from the central nucleus by being the target of fibers descending from the auditory cortex.

The cells of the *central nucleus of the inferior colliculus* are activated first and foremost by fibers of the lateral lemniscus. This nucleus is characterized by a laminar arrangement of disk-shaped neurons oriented in parallel with the lemniscal afferents. This has been shown by Golgi studies in several species including the cat (Morest, 1964a; Rockel and Jones, 1973), the monkey (FitzPatrick, 1975; Fullerton, 1975), and man (Geneic and Morest, 1971). Physiological mapping of characteristic frequencies has demonstrated the existence of isofrequency lamellae corresponding to the fibrodendritic lamination (Rose, Greenwood, Goldberg, and Hind, 1963; Merzenich and Reid, 1974; FitzPatrick, 1975). As a microelectrode is advanced from superficial to deep portions of the central nucleus, a regular progression of characteristic frequencies from lower to higher is encountered (indicated by the arrow from a to b in the left c.n.i.c. in Fig. 9-4).

The *principal efferent projection of the central nucleus is to the laminated portion of the ventral division* of the medial geniculate body (Moore and Goldberg, 1963, 1966; Casseday, Diamond, and Harting, 1976; Oliver and Hall, 1978a; Kudo and Niimi, 1978), which in turn projects to the primary auditory cortex (Fig. 9-4, thick lines). It is compatible with what is known of the anatomical organization of the central nucleus that this projection must be of crucial importance, not only for frequency discrimination but also for other auditory functions, as will be discussed later.

The lemniscal input to the central nucleus of the inferior colliculus originates in a variety of lower auditory centers, including the dorsal and ventral cochlear nuclei, the medial and lateral superior olives, and the ventral and dorsal nuclei of the lateral lemniscus (Stotler, 1953; Osen, 1972; Adams, 1979; Elverland, 1978). In the recent study of Adams (1979), as many as 24 different types of cells in the lower auditory centers were labeled after HRP injections in the inferior colliculus. With the possible exception of the nuclei of the lateral lemniscus, these centers are all tonotopically organized, as are also their projections to the central nucleus. Although both ears are represented in both colliculi, the main input is from the contralateral side, as illustrated by the ampler projection of second-order fibers from the contralateral cochlear nuclei.

The projection of the central nucleus to the ventral division of the medial geniculate body is bilateral but larger on the ipsilateral side, thus favoring the ultimate representation of the contralateral ear in the auditory cortex. The central nucleus also provides fibers to the medial division of the medial geniculate body and the midbrain tegmentum bilaterally, and the peripheral gray matter of the inferior colliculus ipsilaterally. The contralateral projection crosses the midline in the intercollicular commissure.

The *peripheral gray matter of the inferior colliculus* has been subdivided differently by various authors (for a review see Harrison and Howe, 1974a). There seems to be general agreement, though, to consider the field on the lateral side of the central nucleus a separate nucleus, termed the lateral zone or the *external nucleus* (ex. in Fig. 9-4). According to Morest (1964a), this nucleus is part of the midbrain tegmentum. The rest of the peripheral gray matter will be referred to here as the *pericentral nucleus* (as done by Casseday, Diamond, and Harting, 1976). This nucleus, which in this account also includes the dorsomedial portion of the central nucleus, was described by Geneic and Morest (1971) as a five-layered cortical structure.

As indicated above, the external and pericentral nuclei differ considerably from the central nucleus in their fiber connections. With the rest of the midbrain tegmentum, they receive auditory input from the central nucleus via fibers in the brachium of the inferior colliculus (Rasmussen, 1961; Moore and Goldberg, 1963, 1966; van Noort, 1969; Casseday, Diamond, and Harting, 1976). They also receive fibers of the spinothalamic tract and the medial lemniscus transmitting impulses of somesthetic origin (van Noort, 1969; Schroeder and Jane, 1971; Oliver and Hall, 1978a) and fibers from the auditory cortex, to be described later in the section on the descending pathway. These nuclei project fibers to the nonlaminated portions of the medial geniculate body as part of the auditory belt projection. (For more detailed information regarding these connections see the HRP studies by Oliver and Hall, 1978 a, b.) The pericentral and external nuclei moreover project fibers to the *deep layers of the superior colliculus* (HRP study by Edwards, Ginsburgh, Henkel, and Stein, 1979). These layers also receive input from, among others, the auditory cortex (to be described later) and several brainstem centers related to vision and somesthetic senses (Edwards, Ginsburgh, Henkel, and Stein, 1979). Among their many efferent projections are the tectoreticular (Kawamura, Brodal, and Hoddevik, 1974) and *tectopontine* tracts (cf. Chaps. 4 and 5 and Fig. 7-31). The latter, in addition, receives a contingent of fibers directly from the inferior colliculus (see below). The tectopontine tract, via a relay in the dorsolateral pontine nucleus (Kawamura and Brodal, 1973), constitutes one of many pathways leading teleceptive impulses from the midbrain to the cerebellum, including the audiovisual area in the vermis (Hoddevik, Brodal, Kawamura, and Hashikawa, 1977). The output from the vermis and the fastigial nucleus is complex, but may include direct nuclear fibers to the superior colliculus (Edwards, Ginsburgh, Henkel, and Stein, 1979) and to the nonlaminated portions of the medial geniculate body (Carpenter, 1960). The latter also receives direct fibers from the superior colliculus (Altman and Carpenter, 1961; Oliver and Hall, 1978a), and, as mentioned above, the inferior colliculus. The inferior and superior colliculi are thus involved in, among other things, neuronal circuits including the cerebellum. The functional significance of these circuits is poorly understood. It may be reasonable, however, to suggest a role in auditory (and visually) guided motor functions, like the startle response on the one extreme and speech on the other (see Fagida and Pupilli, 1964; Mortimer, 1975).

Although a contribution of the inferior colliculus to the tectopontine tract has been disputed, recent fiber degeneration studies have demonstrated a definite sparse projection from all parts of the inferior colliculus to the dorsolateral pontine nucleus (Kawamura and Brodal, 1973; Kawamura, 1975; Aitkin and Boyd, 1978). A relation of the inferior colliculus to the *autonomic nervous system* has also been suggested, since the collicle has a projection to the periaqueductal gray (Moore and Goldberg, 1963; Matano and Ban, 1967). The periaqueductal gray and the midbrain tegmentum may, however, also be involved in the production of species-specific calls as demonstrated both in primates and in other species (see Jürgens and Pratt, 1979). The *descending projections* of the inferior colliculus to the superior olivary complex and the dorsal cochlear nucleus are considered with the descending pathways.

The medial geniculate body. This nuclear complex is situated on the caudal, subpial aspect of the thalamus to which it belongs (Fig. 2-14). The medial geniculate body is found medial to the lateral geniculate body in man, ventral to it in cats. Like its namesake in the visual system, the medial geniculate body is a relay in the ascending auditory pathway, intercalated between the fibers of the brachium of the inferior colliculus and those of the *auditory radiation* (Fig. 9-4). The latter passes to the auditory cortex via the posterior portion of the internal capsule (Pfeifer, 1920; Bremer and Dow, 1939; Woollard and Harpman, 1939; and many others). The laminated portion of the ventral division of the medial geniculate body (see below) projects to the auditory koniocortex, as part of the auditory *core projection*. The other nonlaminated portions provide fibers to the auditory cortical belt areas, as part of the auditory *belt projection*. Exactly in parallel to the ascending projection there is a recurrent corticothalamic projection (to be described later). In contrast to the more caudal and rostral auditory centers, the medial geniculate bodies are not joined by commissural connections. The geniculocortical projection is also entirely ipsilateral.

The medial geniculate body is composed of several subdivisions that differ in cytoarchitecture and fiber connections. The subdivisions are not easily distinguished in ordinary histological sections, nor is there general agreement so far regarding the parcellation of the complex and its delimitation against the posterior thalamic group (for reviews see Morest, 1964b; Harrison and Howe, 1974a). Among the alternative subdivisions, the one proposed by Morest (1964b, 1965a), on the basis of Golgi studies in the cat, has proved the most advantageous for further morphological and electrophysiological studies. (For further subdivisions along the same line, see Oliver and Hall, 1975, 1978a.)

Morest (1964b) divided the medial geniculate body into three main divisions—medial, dorsal, and ventral (M.,D., lam. V., and non-lam. V. in Fig. 9-4).[5] Of these, the dorsal and ventral divisions were further divided in three subnuclei each. The *ventral division* is composed of a major ventral nucleus and two minor nuclei. The ventral nucleus differs from all other subdivisions of the medial geniculate body by a distinct laminar organization. It will therefore be referred to here as the *laminated portion of the ventral division* (lam. V. in Figs. 9-4 and 9-5). The lamination is evident from the dendritic pattern of the principal (tufted) neurons and the orientation of the afferent fibers.[6] Mapping of characteristic frequencies in the cat (Galambos and Rose, 1952; Aitkin and Webster, 1972) and the monkey (Gross, Lifschitz, and Anderson, 1974) suggests that, as in the central nucleus of the inferior colliculus, the lamination is related to a spatial representation of the cochlea; the apical (low-frequency) part of the cochlea is represented in the more lateral laminae, the basal (high-frequency) part in the more medial ones (indicated by arrow, connecting a and b in Fig. 9-4).

The neurons of the laminated portion are activated by fibers arising in the central nucleus of the inferior colliculus. Except for the reciprocal corticothalamic fibers (see below), this region receives few, if any, fibers of other origins. It is

---

[5] The ventral and dorsal divisions together correspond roughly to the principal division of Rioch (1929), and the medial division to his magnocellular division.

[6] For descriptions of the fine structure of the laminae, the reader should consult, among others, Morest, 1971, 1975; Majorossy and Kiss, 1976 a,b.

therefore predominantly an auditory structure. As mentioned above, it projects to the auditory koniocortex (A I) as the final link in the auditory core projection. Recent studies based on fiber degeneration and axonal transport methods have proved, beyond doubt, that this projection is organized in a strict tonotopical order, as considered more closely in the next section.

The *nonlaminated portions of the medial geniculate body,* consisting of the medial division, the dorsal division, and the nonlaminated nuclei of the ventral division, constitute an intricate jungle of subdivisions and fiber connections, whose functional significance is poorly understood. Any detailed description of these structures is beyond the scope of this account, and they will therefore be considered only briefly and collectively. (For details, see, among others, Casseday, Diamond, and Harting, 1976; Winer, Diamond, and Raczkowski, 1977; Oliver and Hall, 1978 a,b; Niimi and Matsuoka, 1979.)

Among the nonlaminated portions, the *medial division* contains the largest neurons and is therefore often referred to as the magnocellular division. This division receives a certain amount of fibers from the central nucleus of the inferior colliculus. Except for these fibers, all the nonlaminated portions receive auditory input from the belt regions of the inferior colliculus—that is, the pericentral nucleus and/or the external nucleus with the rest of the midbrain tegmentum (thin lines in Fig. 9-4). In addition to the auditory input, the medial and dorsal divisions are also influenced by other sensory modalities through fibers (or collaterals) from the spinothalamic tract, the medial lemniscus (Nauta and Kuypers, 1958; Lund and Webster, 1967 a, b; Oliver and Hall, 1978b), and the brachium of the superior colliculus (Altman and Carpenter, 1961; Morest, 1965b). As mentioned in a previous section, the belt areas of the medial geniculate body also receive fibers from the fastigial nucleus (Carpenter, 1960) which is related to the vestibulocerebellum and the audiovisual area in the cerebellar vermis. In agreement with these anatomical data, there is electrophysiological evidence for convergence of somatic, vestibular, and auditory inputs on single neurons (Blum, Abraham, and Gilman, 1979). Although the functional significance of this interaction remains to be clarified, its net effect must necessarily be conveyed to the auditory cortex that is the target of the medial geniculate body.

As mentioned above, the nonlaminated portions send fibers primarily to the auditory cortical belt areas (thin lines in Fig. 9-4). These fibers form the final link in the auditory belt projection. The medial division probably supplies all of the auditory cortical areas, including the koniocortex and the secondary somesthetic area. The other portions may have a more restricted cortical representation, but as a rule they all supply more than a single area.

The auditory cortex. The auditory region of the cerebral cortex has been identified in several species by studies of cytoarchitecture, thalamic connections, and auditory-evoked potentials recorded under deep barbiturate anesthesia. In man and monkey, the auditory cortex occupies a region on the upper bank of the temporal lobe, buried in the Sylvian fissure. In the cat, it is located on the lateral side of the hemisphere below the suprasylvian sulcus (Fig. 9-9). Unfortunately, detailed comparisons of neocortical structures are hampered by considerable variations between species. In all mammals studied so far, however, a central konio-

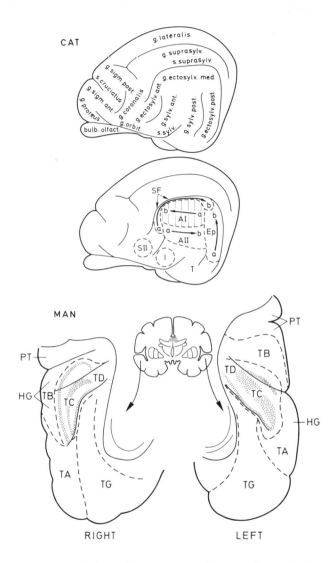

FIG. 9-9 Diagrams of the auditory cortex in the cat and man. In the upper drawing the main gyri and sulci on the lateral aspect of the hemisphere of the cat are shown and labeled according to Hassler and Muhs-Clement (1964). The second drawing shows the auditory cortical areas of the cat, as defined by Rose (1949). These are the primary auditory area (*A I*), the secondary auditory area (*A II*), the posterior ectosylvian area (*Ep*), and the suprasylvian fringe sector (*SF*), with the probable addition of the insular area (*I*), the secondary somatosensory area (*S II*), and the temporal area (*T*). The spatial representation of tonal frequencies within each area is indicated by arrows pointing from low- (*a*) to high- (*b*) frequency regions. Vertical lines in *A I* indicate isofrequency strips. Below are shown the right and left supratemporal planes of the human brain (simplified from Fig. 5 of von Economo and Horn, 1930). The auditory cortex is composed of two fields, *TD* (area supratemporalis intercalata) and *TC* (area supratemporalis transversa). On the right side these occupy a double transverse temporal (*HG:* Heschl's) gyrus; on the left they correspond to a single Heschl's gyrus and a part of the more caudally situated temporal plane (*PT*). Areas *TD* and *TC* are composed of markedly granular (dotted) and less heavily granular (blank) subareas. Note distinct right–left asymmetries, with a larger temporal plane on the left side. Abbreviations: *TA:* area temporalis superior; *TB:* area supratemporalis magnocellularis; *TG:* area temporopolaris.

cortex (see Chap. 12), with a complete and orderly representation of audible frequencies, can be distinguished. This cortex is surrounded by belt areas with a reversed and often less accurate frequency representation (Woolsey, 1960). In the following, the auditory cortex of the cat, the most commonly used experimental animal, will be considered. The less well-known situations in the monkey and man will be commented on more briefly.

Before the comprehensive cytoarchitectonic study of Rose (1949), considerable uncertainty prevailed regarding the accurate location and subdivision of the auditory cortex (for review, see Rose, 1949). Although some modifications of Rose's scheme have been introduced on the basis of later physiological and anatomical studies, particularly those using modern methods for the study of axonal transport (Merzenich et al., 1977 and later), his cytoarchitectonic map of the cat auditory cortex will be used in this account.

As concluded by Rose, the *auditory cortex of the cat* consists of a small-celled and highly cellular koniocortex, the first auditory area, surrounded by a belt of auditory cortical fields. The cytoarchitecture of these fields shows transitions toward the adjacent cortical regions. The *first auditory area (A I)* occupies the anterior portion of the middle ectosylvian gyrus below the suprasylvian sulcus (Fig. 9-9). Rose defined three belt areas, the *suprasylvian fringe sector (SF)*, which occupies the lower bank of the suprasylvian sulcus, the *second auditory area (A II)*, which is adjacent to the lower border of A I in a parainsular position, and the *posterior ectosylvian area* (Ep), which is situated caudal to A I and A II in the posterior ectosylvian gyrus. Later, as a result of electrophysiological recordings, also the *insular area (I)*, the *temporal area (T)*, and a part (A III) of the *second somatosensory area (S II)* have been included in the auditory cortical belt areas of the cat (Woolsey, 1960, 1964).

All of the auditory cortical areas receive fibers from the medial geniculate body. In the foregoing sections, the subdivision of the ascending auditory pathway into a core and a belt projection has repeatedly been touched upon. It should be noticed, however, that in their final, thalamocortical link, the two projections are not completely separate. Whereas *A I is the* target of the core projection from the laminated portion of the medial geniculate body, this area also receives fibers from certain of the nonlaminated portions of the medial geniculate body (as indicated in Fig. 9-4). Owing to various convergent and divergent connections within the geniculocortical belt projection, *all auditory cortical areas, including A I, are supplied from more than one portion of the medial geniculate body.* The existence of principal differences in the alignment of the core and belt projections, considered likely already on the basis of the classical ablation studies, is supported by recent anatomical studies.

The ablation experiments were based on retrograde cell degeneration in the thalamus following removal of various portions of the auditory cortex (Rose and Woolsey, 1949b; Walker and Fulton, 1938; Butler, Diamond, and Neff, 1957; Diamond, Chow, and Neff, 1958; Locke, 1961; and others). Whereas lesions restricted to A I were regularly followed by cell degeneration in the medial geniculate body, extensive lesions of several cortical belt areas were required to obtain a similar effect. The reason for this was assumed to be a combination of divergence and convergence in the corticothalamic belt projection, as confirmed in recent studies on the basis of fiber degeneration and axonal transport methods (Sousa-Pinto, 1973; Casseday, Diamond, and Harting, 1976; Winer, Diamond, and Raczkowski, 1977; Niimi and Matsuoka, 1979). It should be noticed, however, that in Oliver and Hall's (1978b) HRP study in the tree shrew, less overlap in the belt projection was found. This discrepancy may be due to different principles of parcellation of the medial geniculate body. The more complex subdivision employed by Oliver and Hall (1978b) may possibly give a higher resolution of the pathway.

The primary auditory cortex (A I) in the cat is characterized by a strict, *spatial representation of tonal frequencies.* Anatomical and physiological evidence (see below) speaks in favor of an organization of A I in vertical isofrequency strips oriented across the longitudinal axis of the brain, with high frequencies represented anteriorly and low frequencies posteriorly (arrow from a to b in Fig. 9-9). In SF and Ep a reversed and in A II a much less accurate tonotopy is found (Woolsey, 1960, 1971).

A spatial representation of the cochlea within the auditory cortex was indicated by Woolsey and Walzl in 1942. By electrical stimulation of nerve fibers at the free edge of the bony spiral lamina in surgically exposed cat cochleas (see Fig. 9-6), evoked potentials could be recorded in the middle ectosylvian gyrus. When the basal coil was stimulated, the potentials were located anteriorly in the gyrus; when the apical coil was stimulated, the potentials were found posteriorly. An orderly, tonotopical organization was also demonstrated by Tunturi (1950) in the dog. After local application of strychnine to the cortex of *deeply anesthetized* animals, tones of different frequencies produced action potentials in narrow isofrequency strips oriented across the anteroposterior axis of the middle ectosylvian gyrus. Recent mappings of characteristic frequencies in anesthetized animals have given similar results (Merzenich and Brugge, 1973; Merzenich, Knight, and Roth, 1975; Oliver, Merzenich, Roth, Hall, and Kaas, 1976; Merzenich, Kaas, and Roth, 1976). Like the situation in the lower auditory centers, the higher-octave bands have a relatively larger surface representation in the primary auditory cortex than the lower octaves, corresponding to the higher innervation density in the basal portion of the cochlea (Spoendlin, 1972). As a curiosity, it may be mentioned that in echolocating bats, an extraordinarily large proportion of the cortical surface is devoted to the specific frequency of the Doppler-shifted echo used in locating the prey (Suga and O'Neill, 1978).

The tonotopical organization of the primary auditory cortex, as revealed by electrophysiological studies in anesthetized animals, is presumably based on a strict topographical organization of the afferent projection from the laminated portion of the medial geniculate body, as appears from recent anatomical studies with axonal transport methods.

After minute injections of HRP in A I, narrow laminae of labeled cells are found in the laminated portion of the medial geniculate body (Winer, Diamond, and Raczkowski, 1977; Oliver and Hall, 1978b; Niimi and Matsuoka, 1979). After injections in the anterior (high-frequency) region of A I, the labeled cells are found in the medial (high-frequency) laminae of the ventral division. After injections in the caudal (low-frequency) region of AI, however, the labeled cells occur in the lateral (low-frequency) laminae.

Recent electrophysiological recordings from *unanesthetized* animals, however, have raised some doubt about the validity of the concept of the "tonotopical organization" of the auditory cortex (Whitfield, 1967; Goldstein, Abeles, Daly, and McIntosh, 1970; Clopton, Winfield, and Flammino, 1974; Goldstein and Abeles, 1975). The neuronal coding of tonal frequency in the cortical neurons is apparently a very complex function, which is influenced not only by the thalamocortical core projection but also by other afferent fibers and intrinsic neuronal circuits, which may possibly be more active in alert animals.

There may also be a *spatial representation of sound intensity* within the primary auditory cortex of the dog, as first proposed by Tunturi (1952). For sound stimulation of the ipsilateral ear, he found that the intensity required to excite the cortex was less in the dorsal than in the ventral region of A I, thus indicating a dorsoventral threshold gradient along the isofrequency strips. For stimulation of the contralateral ear, the threshold was equally low all over the field. As a curiosity, it may be mentioned that in the echolocating bats there appear to be special sectors of the cortical area that are related to intensity and frequency discrimination, respectively (Suga and O'Neill, 1978).

The *association and commissural fibers* of the auditory cortex appear to be organized in much the same way as those in the visual and somesthetic cortex. According to fiber degeneration studies in the cat (Diamond, Jones, and Powell, 1968a), the total *commissural projection of the auditory cortex* is confined to the homotypic area in the contralateral hemisphere. Slightly modified, it may be stated that each of the cortical areas is reciprocally connected with its homologue over the midline. The ipsilateral association fibers will be dealt with below, in the description of the auditory cortex in the monkey, in which these connections are more elaborately developed. For a silver impregnation study of the association fibers in the cat, see Diamond, Jones, and Powell (1968b). The *efferent, descending projection* of the auditory cortex to the medial geniculate body and the midbrain in the cat will be described in a subsequent section in conjunction with the descending pathway.

In the *monkey,* as in the cat, the auditory cortex consists of a central koniocortex surrounded by at least four different belt areas, which are named differently in various species and by various authors (Pandya and Sanides, 1973; Merzenich and Brugge, 1973; Imig, Ruggero, Kitzes, Javel, and Brugge, 1977).[7] The koniocortex of the monkey consists of two cytoarchitectonically different areas, a *medial auditory koniocortex (Kam)* and a *lateral auditory koniocortex (Kalt)*. These are adjacent to each other and, apparently, share the same tonotopical pattern, with a complete and orderly representation of all audible frequencies, like A I in the cat. However, the frequency axis is reversed, with low frequencies represented rostrolaterally and high frequencies caudomedially (Bailey, von Bonin, Garol, and McCulloch, 1943; Merzenich and Brugge, 1973; Imig, Ruggero, Kitzes, Javel, and Brugge, 1977). In the belt areas there is a reversed and less accurate tonotopical pattern, as in the corresponding areas in the cat.

Concerning the fiber connections of the auditory cortex in the monkey, it may be noticed that the commissural connections, the reciprocal connections with the medial geniculate body, and the descending projection to the inferior colliculus seem to be organized according to the same principles as in subprimates (Polyak, 1932; Le Gros Clark, 1936b; Walker, 1938b; Walker and Fulton, 1938; Ades and Felder, 1942; Akert, Woolsey, Diamond, and Neff, 1959; Forbes and Moskowitz, 1974; Mesulam and Pandya, 1973; Pandya and Sanides, 1973; Thompson, 1978). Corresponding to the more differentiated neocortex in primates, there is a more elaborate system of *cortical association fibers* than in cats. As in the visual and somesthetic systems, Jones and Powell (1970a), on the basis of Nauta studies, have

---

[7] The *homologies* of the auditory cortical areas in the monkey are discussed by Pandya and Sanides (1973). During the growth of the temporal lobe in primates, the entire auditory cortical field apparently becomes rotated and hidden from view on the lower bank of the Sylvian fissure. This is probably the explanation of the reversal of the frequency axis of the koniocortex, as compared with lower mammals. The medial belt area in the monkey is interposed between the koniocortex and the insula and is probably the homologue of the feline A II, which is also in a parainsular position. The superior temporal gyrus, or part of it, may be the homologue of the posterior ectosylvian gyrus.

described a stepwise, outward progression of connections from the primary auditory cortex, each step representing reciprocal connections. The *first step* is represented by the projection from the primary auditory cortex to the cortical belt areas and to area 8a in the frontal lobe. *Step two* consists of fibers from the belt area to the adjacent area 22 in the superior temporal gyrus and from 8a to the adjacent and more frontally located area 9. *Step three* is a further progression of projections from the supratemporal gyrus to the supratemporal sulcus and from area 9 to areas 10 and 12 close to the frontal pole of the brain (see Fig. 12-2). The latter regions, especially the cortex in the supratemporal sulcus, is a multisensory region in which auditory information is integrated with visual and somesthetic inputs (cp. Chap. 12).

The auditory cortex of man is even less well known than that of the monkey. As in the monkey, it is situated on the supratemporal plane, buried in the Sylvian fissure. The extent of the auditory cortex and the borders between the koniocortex and the cortical belt areas remain to be clarified. According to the comprehensive cytoarchitectonic study of von Economo and Horn (1930), the auditory field of the human cerebral cortex occupies the transverse temporal gyrus (Heschl's gyrus) and a variable amount of the caudally situated planum temporale—that is, a field corresponding approximately to Brodmann's areas 41 and 42 (see Fig. 12-2). Von Economo and Horn described two auditory areas, a medial area (TD, area supratemporalis intercalata) and a lateral area (TC, area supratemporalis transversa) (Fig. 9-9). The homologies of the two areas are unclear, however, and it is not known whether they include both the koniocortex and the cortical belt areas.

> From the description by von Economo and Horn (1930), the structure of the two auditory areas TD and TC seems to be very much the same. Both areas appear to consist of several more or less distinct granular subregions, which in the diagrams may traverse the borderlines between the two main areas. In general, a distinct granular region is found in the center of the auditory field. The surrounding more or less distinct granular zones tend to form alternating bands in parallel with the long axis of Heschl's gyrus. In most brains, the central area with typical granular cortex is largest on the left side. In Harrison and Howe's (1974a) view, the acoustic field of von Economo and Horn includes both koniocortical and parakoniocortical areas, the latter corresponding to cortical belt areas. The variations in cytoarchitecture within the auditory fields is interpreted as an intermingling of koniocortical and parakoniocortical areas as described also for certain primates (see Harrison and Howe, 1974a). On the other hand, a relation to binaural interaction columns (see below) may also be considered.

In any case, the acoustic field of von Economo and Horn corresponds closely to the part of the supratemporal plane which, in studies of surgically exposed human brains, is found to respond with short-latency potentials to auditory stimulation (Celesia, Broughton, Rasmussen, and Branch, 1968; Celesia, 1976).

> In the studies of Celesia and co-workers in fully anesthetized patients, no auditory-evoked potentials could be recorded outside the auditory field of von Economo and Horn. In alert patients, on the other hand, potentials with longer latencies were recorded from the superior temporal gyrus and the frontal and parietal operculum. In view of the anatomical studies of the auditory association fibers in the monkey referred to above, these regions may possibly represent auditory association areas. According to Arezzo, Pickoff, and Vaughan (1975), however, an unspecific "volume-conduction" of potentials originally generated in the auditory cortex on the supratemporal plane also has to be taken into account.

In *man auditory sensations* can be produced by electrical stimulation of Heschl's gyrus (Penfield and Jasper, 1954). The sensations are usually described as ringing, humming, clicking, buzzing, and so forth and most often are referred to the opposite ear. A tonotopical localization has not been demonstrated in this way in man. (More complex auditory sensations may occur on stimulation of other parts of the temporal lobe; see Chap. 12.) Pfeifer (1936), from a study of human cases of temporal lobe lesions, however, was led to conclude that if there exists any tonotopical localization in the primary auditory cortex of man, the highest frequencies must be "represented" medially, those of lowest frequency laterally. This is in full agreement with findings in the monkey and chimpanzee referred to above. On the whole, the available evidence suggests that there is a tonotopical pattern in the geniculocortical projection also in man, but detailed information is lacking.

von Economo and Horn were among the first to point out the pronounced individual variation and the considerable right–left asymmetries in the size and convolution of the supratemporal plane. The Heschl's gyrus may be solitary or double (HG in Fig. 9-9). According to von Economo and Horn, the gyrus is usually solitary and longer on the left side. In general, the planum temporale, caudal to the Heschl's gyrus, is larger on the left than on the right side (PT in Fig. 9-9) (see also Witelson and Pallie, 1973). In some brains the situation is reversed. These asymmetries, which are visible even in computerized axial tomography (Galaburda, LeMay, Kemper, and Geschwind, 1978), speak in favor of a functional differentiation between the auditory cortex in the right and the left hemisphere. (For further information see Chap. 12.) A functional difference between the auditory functions of the two hemispheres is apparent also from *dichotic listening experiments*. In these experiments subjects may have to respond to various kinds of tests delivered randomly to either ear through earphones coupled to a dual-channel tape recorder. The scores for the two ears indicate the relative role of the contralateral hemisphere in the auditory function tested. Usually the right ear (left hemisphere) shows a significantly better score for verbal tests, while the left ear (right hemisphere) is superior in recognition of music (Kimura, 1961, 1964; Bryden, 1963; Gordon, 1970).

The *columnar organization* of the sensory cortices has attracted considerable attention in recent years (see Chaps. 8 and 12). There is some experimental evidence of a columnar organization also in the auditory cortex. According to the electrophysiological study of Merzenich, Knight, and Roth (1975), the isofrequency strips of the primary auditory cortex probably represent *isofrequency columns* with the same characteristic frequency for all units throughout the entire depth of the cortex. *Binaural interaction columns* have also been described, exhibiting either binaural summation or inhibition (Imig and Adrian, 1977; Middlebrooks, Dykes, and Merzenich, 1978). The summation and inhibition columns occupy alternating bands oriented across the isofrequency strips or columns. Within the binaural summation columns, there is a mosaic of minor *ipsi- or contralateral dominance columns*.

The descending auditory pathway. Parallel to the ascending pathway there is a stepwise, descending projection from higher to lower auditory centers, starting with the auditory cortex at the one extreme and ending at the cochlear hair cells on the other. The relays of this projection are probably influenced by ascending fibers, establishing feedback loops of varied sizes and complexities. Despite

some physiological evidence of inhibitory and facilitatory actions, the role of these
loops in audition is still poorly understood. A simplified diagram of the descending pathway of the cat is presented in Fig. 9-5. (For reviews, see Rasmussen,
1960, 1964, 1967; Harrison and Howe, 1974b.)

The auditory cortex gives off three descending tracts, one terminating in the
thalamus, the other in the midbrain, and the third in the pons. As illustrated in
Fig. 9-5, the medial geniculate body, although influenced from the auditory cortex, is not a link in the descending cortical projection to the lower auditory centers. In this respect, the position of the medial geniculate body as a relay in the auditory system is principally different from that of, for example, the inferior
colliculus. The descending tracts originate in different layers of the auditory cortex, as demonstrated by HRP studies in the hamster, the cat, and the monkey
(Ravizza, Straw, and Long, 1976; Jones, Casseday, and Diamond, 1976; Thompson, 1978). Physiological evidence speaks in favor of inhibitory actions (Massopust and Ordy, 1962; Amato, LaGrutta, and Enida, 1969).

As in the case of other corticothalamic connections to specific thalamic nuclei, the *corticogeniculate projection* from the auditory cortex is reciprocal to the
geniculocortical projection. Each auditory cortical area projects to the sources of
its afferent fibers in the medial geniculate body. Thus, short feedback loops are established. This has been demonstrated in several anatomical studies (Cajal, 1911;
Diamond, Jones, and Powell, 1969; Forbes and Moskowitz, 1974; Pontes, Reis,
and Sousa-Pinto, 1975), including studies using combined HRP and autoradiographic methods (Oliver and Hall, 1978b).

The *corticocollicular projection,* as mentioned above, bypasses the medial
geniculate body on its way to the inferior colliculus (see Fig. 9-5). As shown by
fiber degeneration and axonal transport methods in the cat and the monkey, the
corticocollicular projection, in contrast to the corticogeniculate, arises largely
from the primary auditory cortex, while the cortical belt areas are less involved. It
terminates bilaterally in the pericentral nucleus of the inferior colliculus but apparently not in the central nucleus (Massopust and Ordy, 1962; Rasmussen, 1964;
van Noort, 1969; Diamond, Jones, and Powell, 1969; Rockel and Jones, 1973;
Jones, Casseday, and Diamond, 1976; Oliver and Hall, 1978a, b; Thompson,
1978).[8] In view of its ascending projection, the pericentral nucleus of the inferior
colliculus may, among other things, be involved in cortico–colliculo–geniculo–
cortical loops, involving both core and belt areas of the auditory system.

The auditory cortex also gives off fibers to the *superior colliculus* (see Chap.
7) (not illustrated here). These fibers seem to arise largely from the auditory belt
areas (Paula-Barbosa and Sousa-Pinto, 1973). As mentioned in a previous section,
the inferior and superior colliculi both contribute fibers to the tectopontine tract
that terminates in the dorsolateral pontine nucleus. This nucleus, which is a relay
to the audiovisual area in the cerebellar vermis (Hoddevik, Brodal, Kawamura,
and Hashikawa, 1977), also receives fibers directly from the auditory cortex as
part of the *corticopontine* projection (P. Brodal, 1972c). Impulses from the auditory cortex, thus, are obviously transmitted to the cerebellum by pathways of differ-

---

[8] According to some authors (FitzPatrick and Imig, 1976; FitzPatrick, Diamond, and Raczkowski, 1978), the cortical fibers also supply the central nucleus. The difference of opinion may possibly be due to different definitions of the borders of this nucleus.

ent complexities, some of which also involve the inferior and superior colliculi. The possible role of these connections in audition is touched upon elsewhere.

The second link in the main descending auditory pathway consists of fibers from the inferior colliculus. The interrelationship of the ascending and descending pathways in the inferior colliculus is still unclear, as is also the exact location of neurons giving rise to the descending fibers. The fibers, which descend in the lateral lemniscus, constitute a colliculo-olivary and a colliculo-cochleonuclear projection (illustrated collectively in Fig. 9-5) (Rasmussen, 1955, 1960; van Noort, 1969; Casseday, Diamond, and Harting, 1976; Kane, 1977a; Oliver and Hall, 1978a). The latter may not be present in rodents (Borg, 1973a).

The *colliculo-olivary projection* terminates in the periolivary nuclei, preferentially ipsilaterally, and in the regions giving rise to the *crossed olivocochlear bundle* (see above). As indicated in Fig. 9-5, these regions are also provided with fibers from the cochlear nuclei, especially from the area of octopus cells (Warr, 1969; and others). Although not yet proved at the ultrastructural level, the cells of origin of the crossed olivocochlear bundle may thus be influenced by both ascending and descending fiber tracts. There is less evidence of a connection of the collicular fibers to the *uncrossed olivocochlear bundle,* which originates from cells at the margin of the lateral superior olive. These cells, instead, may share the cochlear nuclear input to the latter nucleus.The olivocochlear fibers pass dorsally through the reticular formation to the floor of the fourth ventricle. Several fibers cross the midline between the facial genua. Crossed and uncrossed fibers gather in the vestibular nerve root. At the site where this bypasses the cochlear nuclear complex, fibers (or collaterals) are given off for the latter. These terminate largely in the granule cell domain of the cochlear nuclei (Rasmussen, 1960, 1964; Osen and Roth, 1969), possibly as mossy fibers (McDonald and Rasmussen, 1971; Mugnaini, Osen, Dahl, Friedrich, and Korte, 1980). In man, the latter fibers are apparently missing, in accordance with the absence of granule cells in the human cochlear nuclear complex (Moore and Osen, 1979).

The *colliculo-cochleonuclear projection* is bilateral. The fibers pass via the trapezoid body to the deep, cellular layers of the dorsal cochlear nucleus, where they terminate (Fig. 9-5) (van Noort, 1969). A *periolivary-cochleonuclear projection* also provides a substantial number of fibers that pass via the trapezoid body and possibly the dorsal and intermediate acoustic striae to the ventral and dorsal cochlear nuclei, both ipsi- and contralaterally (Rasmussen, 1967; Borg, 1973a; Adams and Warr, 1976; Kane, 1976, 1977b; Kane and Finn, 1977; Elverland, 1977). These fibers, which apparently originate in acetylcholinesterase-negative periolivary cells, probably establish short cochlear nuclear–periolivary–cochlear nuclear loops, which may be influenced by fibers descending from the inferior colliculus. In the cochlear nuclei, both inhibitory and fascilitatory actions have been demonstrated on electrical stimulation of the superior olivary complex (Whitfield, 1967). The cochlear nuclei, on their side, do not provide fibers to the cochlea, which receives its entire efferent nerve supply directly from the superior olivary complex, as described in a previous section. Possible connections of the cochlear nuclei and the superior olivary complex to the facial and motor trigeminal nuclei, as part of middle ear reflex arcs (Borg, 1973b), are discussed in the next section.

Functional aspects.   In animals many *audiomotor reflexes* are evidently of vital importance, for example the auditory-evoked startle response. Such reflexes include, among many other things, directioning of the pinnae, the eyes, and the head toward the source of a sound. More local reflexes, also important in man, are concerned with the changes in the tone of the stapedius and tensor tympani muscles, by means of which the position and tension of the middle ear ossicles are influenced. A reflex contraction of these muscles occurs on stimulation of the ear with sounds of high intensity. From extratympanic manometry and electromyography in surgically exposed human ears (for references, see Djupesland, 1980), it appears that loud acoustic stimuli usually lead to reflex contraction of the stapedius muscle alone. The tensor tympani muscle contracts only when the sound is of such a high intensity and presented in such a way that it gives rise to a startle defense reaction, including contraction of the facial muscles.

The middle ear reflexes probably protect the organ of Corti against excessive stimulation, which is known to have a deleterious effect on the sensory epithelia. The reflexes, which are more efficient in the low-frequency range of sounds, probably also serve to filter out disturbing noise from the head region itself, arising, for example, from the articulation of the jaw during speech and chewing. The reflexes are subject to central control, as revealed by several observations. For example, the muscles contract in advance of the sounds produced in vocalization, and the reflexes may be influenced by previous experience with loud noise (Carmel and Starr, 1963).

The stapedius and tensor tympani muscles are innervated from the facial and motor trigeminal nuclei, respectively (see Chap. 7), but the central circuitries of the middle ear reflexes are not known in detail. From combined physiological and anatomical fiber degeneration studies in the rabbit, the existence of two principally different arcs for each reflex has been proposed (Borg, 1973b).

One reflex arc is concluded to be oligosynaptic and restricted to the lower brainstem; the other is multisynaptic and also involves higher auditory centers. The *oligosynaptic* circuit has a first synapse in the ventral cochlear nucleus. Fibers from this lead via the trapezoid body to the medial portion of the superior olivary complex bilaterally. This region, in turn, may provide fibers to the facial and motor trigeminal nuclei. The facial nucleus may also receive fibers directly from the ventral nucleus. In this way, an oligosynaptic reflex arc with three or four synapses is formed. In human experiments, the latency of this reflex is about 10 msec (Djupesland, 1980). Owing to the partial crossing in the brainstem, stimulation of one ear normally elicits contraction of the stapedius muscle in both ears. Like the pupillary light reflex, therefore, the crossed and uncrossed stapedius reflex has become a useful tool in examination for brainstem disorders (Brask, 1978).

The *multisynaptic* reflex arc, in contrast to the oligosynaptic circuit, has longer latencies, is sensitive to barbiturate anesthesia, and therefore probably involves higher auditory centers. The circuitry for this reflex is not known. The inferior colliculus may be involved. This has usually been considered an auditory reflex center. As mentioned previously, however, the pericentral and external nuclei of the inferior colliculus, with the midbrain tegmentum and possibly the deep superior colliculus, are more likely candidates for this function than is the central nucleus, which is part of the auditory core projection.

Although in man the audiomotor reflexes may be of less vital importance than in many animals, it is clear that the *analytic aspect of hearing has achieved a higher differentiation*. The analytical powers of the auditory apparatus, as is well known, are considerable, allowing us to discriminate between tones of only slightly different pitch, and, in addition, probably by means of an analysis of the simultaneous overtones which determine the timbre of the sound, to discriminate between the sounds of different musical instruments and different voices. Furthermore, we are able to select from a multitude of simultaneous auditory stimuli precisely those in which we are interested at the moment (cf. Chap. 6), and to determine where in our surroundings a sound comes from.

Little is known of how the brain handles the more complex kinds of auditory information, for example speech. However, experimental studies in animals have given valuable clues to our understanding of the more elementary features of auditory perception, even though much is still disputed. Four elementary functions have been fairly extensively studied: the abilities to recognize frequency of sound, intensity of sound, pattern of sound, and the source of sound (its localization in space).

Much of our present knowledge in this field is derived from behavioral studies combined with uni- or bilateral ablations of the auditory cortex. Normally, the cortex is possibly involved in the final refinement of all auditory perception. In animal experiments, it seems to be *essential,* however, only for the more complex processing that involves, among other things, short-time memory. According to Neff (1960), discrimination after cortical ablation probably requires that *new* neuronal units be regularly excited when the positive stimulus is presented. This would explain why frequency and intensity discrimination, which are, in part, based on a spatial organization of subcortical structures, are less influenced by cortical ablation than discrimination of temporal patterns, which, to a larger extent, may depend on short-time memory. Some data on these subjects will be mentioned.

The capacity to discriminate between frequencies of sound has been touched upon repeatedly in the preceding account, and reference has been made to the existence of a tonotopical localization within the auditory system. At the level of the cochlear nuclei and peripheral to these, the entire cross section of the ascending pathway is clearly tonotopically organized. At higher levels, however, this appears to be the case mainly for the structures belonging to the core projection. Although the primary auditory cortex is tonotopically organized, it is not essential for frequency discrimination. Thus, in the cat, the sensitivity to sound and the ability to respond to changes in frequencies are unimpaired following bilateral ablation of the auditory cortex (Butler, Diamond, and Neff, 1957; Goldberg and Neff, 1961; and others). Since these lesions cause profound retrograde degeneration of the medial geniculate body, the perception of tonal frequencies in the ablated animals probably takes place at or below the level of the inferior colliculus. The impairment of frequency discrimination following subsequent transection of the brachium of the inferior colliculus, as described by Goldberg and Neff (1961), may possibly be ascribed to inactivation of the midbrain tegmentum, which may play a role in the attention to sound, rather than to further inactivation of the medial geniculate body, as suggested by those authors.

In the primary auditory cortex, Whitfield and Evans (1965) found numerous units that were sensitive to frequency-modulated tones—that is, tones with rising or falling frequencies. Units were also found that were frequency oriented; that is, they respond to a rising tone but not a falling tone, or vice versa, in the same frequency range. (This resembles the situation in the visual cortex, where units responding to lines of different orientation have been found; see Chap. 8.) It seems a likely assumption that units of this kind may play a role in speech recognition in man. In the cat, however, the auditory cortex is not essential for discrimination of frequency-modulated tones (Kelly and Whitfield, 1971).

The discrimination of intensity of sound appears to depend in part on the number of hair cells and neurons stimulated and in part on the rates at which they fire. Experimental studies indicate that the cortex is not necessary for simple intensity discrimination, and it may be that some discrimination of this kind may be mediated even by structures below the collicular level (see Whitfield, 1967). As mentioned previously, however, electrophysiological experiments in the dog (Tunturi, 1952) and the bat (Suga and O'Neill, 1978) have revealed spatial threshold gradients within the primary auditory cortex which may possibly be related to intensity discrimination.

Discrimination of temporal auditory patterns most probably represents an important feature of human speech perception. In auditory pattern discrimination tests, the animals are trained to respond to sequences of different auditory signals, for example rows of shorter and longer tone pips. Following bilateral ablation of the auditory cortex in the cat, the learning of auditory temporal patterns is completely abolished (Diamond and Neff, 1957; Kelly, 1973; and others). Even the smallest remnant of the auditory cortex is, however, sufficient to mask the ablation effect. The critical relation of pattern discrimination to the auditory cortex is explained by the temporal feature of the pattern, which necessitates a certain degree of short-time memory for comparison of the various elements of the test.

Localization of sound in space is functionally important in man, as in animals. As everyone knows, our capacity to localize the source of sound in our surroundings is remarkably precise. Much work has been devoted to the mechanisms underlying this capacity. Several factors appear to be involved. First and foremost, localization is dependent on binaural audition. The ability is greatly reduced by deafness in one ear. When a sound reaches the ear from one side of the body, it will hit the ipsilateral ear with a greater intensity than the contralateral ear, and also a little later. Interaural time and intensity differences, therefore, are important clues for localization of sound in space (see Whitfield, 1967).

Structures related to this function probably include the spherical and globular cells of the ventral cochlear nucleus, the medial and lateral superior olives, and the nucleus of the trapezoid body, as described in a previous section. In accordance with this assumption, lesions of the trapezoid body in the cat greatly impair the localization ability (Masterton, Jane, and Diamond, 1967; Moore, Casseday, and Neff, 1974). The medial and lateral superior olives have a common target for their efferent projections in the central nucleus of the inferior colliculus (Adams, 1979). In the central nucleus of the colliculus and in the auditory cortex, many units are found to be sensitive either to interaural time differences or interaural intensity differences (Rose, Gross, Geisler, and Hind, 1966; Aitkin, Blake, Fryman, and Bock, 1972; Brugge and Merzenich, 1973). The higher auditory centers, therefore, may play a role in a more integrated aspect of sound localization. As expressed by Masterton, Jane, and Diamond (1967, p. 359), "The superior olives analyse the binaural time disparity; auditory cortex integrates sounds into an organized auditory space."

As to the role of the cerebral cortex in sound localization, the results of various studies are somewhat contradictory. According to some reports (Neff, Fisher, Diamond, and Yela, 1956; Masterton, Jane, and Diamond, 1968), bilateral ablation of the auditory cortex abolishes the capacity to localize sound. According to Ravizza and Diamond (1974), however, the performance differs depending on the test stimulus used.

From their comparative ablation studies of the hedgehog and the bushbaby, Ravizza and Diamond (1974) concluded that two different mechanisms of localization may be employed, depending on the duration of the stimulus. The one, more "primitive" form, involves localization or "tracking" of a continuous sound source. The other, more "refined" form, represents localization of a very brief sound. The latter task probably requires a combination of perceiving the locus, remembering it, and moving toward it. Like other abilities dependent on short-time memory, the refined form is abolished following cortical ablation, whereas the primitive form remains intact. In this connection, it may be pertinent to mention that the *deep layers of the superior colliculus* contain units that respond both to auditory and to visual stimuli (see Chap. 7) and are activated only when the stimulus is moved through a particular region of space (Gordon, 1973). Possibly, the tracking of auditory and of visual stimuli are analogous functions that may both be controlled from the optic tectum. The connection of the superior colliculus to the audiovisual cerebellar centers (see above) may speak in favor of this suggestion.

Relatively little is known of the *function of the central auditory pathways and nuclei in man*. It is reasonable to assume that, in principle, conditions are as in the cat. However, the difference in development of certain parts, such as various cell types of the cochlear nuclei and the main subdivisions of the superior olivary complex, may indicate certain quantitative differences. Comprehensive studies of patients with lesions of the auditory pathway unfortunately are relatively few and not entirely consistent. As in animal experiments, much of the controversy concerns the effect of cortical damage on the ability to localize sound in space. Walsh (1957) found that lesions destroying the auditory cortex on one side did not affect the capacity to localize sound in the horizontal plane. On the other hand, Sanchez-Longo and Forster (1958), in patients with unilateral temporal lobe damage, found impaired accuracy of localization of sounds, especially on the side contralateral to the lesion.

Jerger, Lovering, and Wertz (1972) reported an interesting case. A 62-year-old man had bilateral temporal infarctions, which occurred in two steps. The first, left-sided attack was followed after 16 months by a right-sided one. Bilateral damage of the auditory cortex was confirmed by autopsy 5 months after the last attack. After the first (left-sided) attack, the patient experienced slight, transient disturbances of speech discrimination. After the second (right-sided) attack, speech comprehension was completely abolished. The sensitivity to sound, as measured by pure tone audiometry 3 months after the last attack, was nearly normal. The patient described his difficulty as "I can hear you talking, but I can't translate it" (Jerger, Lovering, and Wertz, 1972, p. 524). The ability to localize sound was also normal. The symptoms in this patient corresponded closely to the ablation studies in the cat, as mentioned above.

The effect of a cortical damage on speech and pattern discrimination is strikingly similar. A complete loss of speech perception, just like a loss of pattern discrimination, seems to require a complete, bilateral lesion of the auditory cortex. To cite Jerger, Lovering, and Wertz (1972, p. 533), "Here we see a fundamental point of sharp contrast between unilateral and bilateral disorders. Whereas the unilateral deficit in speech understanding is subtle and can be elicited only by creating a rather difficult listening task, the bilateral deficit is profound and may be demonstrated even under the easiest of listening conditions." As suggested by Galaburda, LeMay, Kemper, and Geschwind (1978), however, a unilateral lesion of the dominant hemisphere in an individual with a prominent left–right asymmetry of the temporal cortex may possibly be more deleterious to speech perception than a lesion of the nondominant hemisphere. As to the effect of subcortical lesions, gross hearing defects are seldom found, probably as a result of the extensive, binaural interaction at several levels of the auditory pathway (Penfield and Erickson, 1941).

Symptoms in lesions of the auditory system.[9]   Disturbances in the function of the auditory system in man are, in the overwhelming majority of cases, due to pathological processes affecting the sensory apparatus itself. Since, therefore, these disorders belong to the domain of otology, they will not be treated here. From a neurological point of view, it is important to remember the frequency with which diseases of the ear may lead to impaired hearing, and to be careful in attributing too much diagnostic importance to reduced hearing before the presence of a local process in the ear is excluded. In the case of diseases of the middle ear this is, as a rule, not difficult, but with affections of the inner ear it may be impossible in some cases to be sure whether the damage is in the cochlea or in the cochlear nerve, especially since disturbances from the vestibular apparatus or nerve are frequently present at the same time in both types of lesions. Modern methods of analysis of hearing and vestibular functions may, however, enable the otologist, with whom collaboration is recommended in such cases, to obtain valuable information.

Omitting here the symptoms observed in labyrinthitis, otosclerosis, Ménière's disease, professional deafness, and other conditions that affect in some degree the sensorineuronal auditory apparatus, we shall confine ourselves to the symptoms resulting from affections of the intracranial parts of the auditory system.

Of most practical importance from a neurological point of view are the *symptoms following injury to the cochlear nerve*. Among the diseases affecting this nerve, there is one that is far more common than the others and at the same time more important, since surgical therapy is possible, namely the so-called *acoustic tumor, neurinoma, or Schwannoma of the VIIIth nerve*. Taking its origin from the Schwann cells of the sheath, this tumor, when its slow growth has progressed sufficiently, will exert pressure not only on the cochlear but also on the vestibular, facial, and intermediate nerves. The symptoms resulting from the damage of the latter nerves are discussed in Chapter 7. It is clear that the deeper in the meatus the tumor starts its growth, the earlier it is apt to produce symptoms, due to compression of the nerves and circulatory disturbances. The symptoms from the cochlear nerve are usually at first subjective, abnormal sensations of hearing of various kinds, collectively termed *tinnitus* and interpreted as due to irritation of the fibers. Eventually the fibers lose their conductive capacity, and an impairment of hearing follows. The tinnitus may occur periodically but is usually absent when all fibers are interrupted, and the patient is entirely deaf on the affected side. On account of the slow growth of the tumor, the patient may have forgotten the initial tinnitus when he seeks medical advice many years later. The unilateral deafness, however, will give persistent difficulties in, among other things, localization of sound in space, which may be an important handicap.

---

[9] For a recent presentation of the sensorineuronal auditory pathology, see Dublin (1976).

It is noteworthy that discrimination of speech in patients with an acoustic tumor often is more impaired than would be expected from pure tone audiometry. The explanation for this may possibly be that the compression affects one part of the nerve more than another, thus causing sigificant informational gaps (Bredberg, 1968; Dublin, 1976). The cochleotopic organization of the nerve, as shown in Fig. 9-6, may favor the occurrence of restricted lesions in the speech frequency range by pressure from outside. A partial damage of the cochlear nerve is also followed by a threshold change and decay of the stapedius reflex (Djupesland, 1980). With complete destruction of the nerve, this reflex cannot be elicited at all from the affected ear. This is in contrast to a pure cochlear deafness, in which the stapedius reflex, which even normally requires high-intensity stimulation, is still present.

Since the distance from the internal auditory meatus to the medulla oblongata is very short, a tumor of even a moderate size will soon be able to exert a pressure on the neighboring structures. This will be exerted both laterally and medially. Medially, the medulla oblongata will as a rule be pushed aside, without serious symptoms in the beginning. Laterally, the continuous pressure on the internal auditory meatus will result in its being slowly widened, and this may be readily recognized in radiograms taken in the appropriate projection. This widening of the meatus is a very valuable localizing symptom in acoustic tumors. If the tumor originally starts in the depths of the meatus, it will cause the same changes, since it will as a rule press itself toward the medial opening of the bony canal. When the tumor grows, usually pressure effects on the brain eventually become apparent. Before the occurrence of more obvious symptoms, an influence of the tumor on the brainstem may be revealed by examinations of the crossed and uncrossed stapedius reflex and by the auditory-evoked brainstem response.

Like other neoplasms in the posterior cranial fossa, the acoustic tumor will soon give rise to an increased intracranial pressure, betraying itself in headache, vomiting, choked discs, and the other symptoms frequently present when the intracranial pressure is raised. These symptoms may appear earlier than any clear-cut sign of local damage to the central nervous structures, but when such symptoms occur, and occasionally they may appear early, they point most conspicuously to involvement of the homolateral cerebellar hemisphere. Homolateral ataxia, tremor, and hypotonia appear and may make it difficult to decide if the tumor is a primary cerebellar neoplasm or an acoustic tumor. As a rule, however, the acoustic tumors grow very slowly, and occasionally they may reach an astonishing size before they are diagnosed. The medulla oblongata and pons appear to be able to accommodate themselves to a certain extent to the narrowed space without impairment of their functions.[10]

Far less common than acoustic tumors are other pathological processes affecting the cochlear nerve. Among such conditions may be mentioned *meningeomas of the cerebellopontine angle, syphilitic meningitis, arachnoiditis,* and very infrequently aneurysms of the anterior inferior cerebellar artery. A *cerebellar tumor* in a more advanced stage may give symptoms very similar to those of an acoustic

---

[10] It is generally held that the first symptoms in acoustic Schwannomas are those indicating involvement of the eighth cranial nerve. Especially when the tumor arises medially, atypical features in the clinical picture, indicating an early affection of the brainstem, are not uncommon (see Sheppard and Wadia, 1956).

nerve tumor, but the development of the symptoms will be different. On account of the nearness of the nerve to the posterior inferior cerebellar artery, changes in this may also lead to cochlear nerve symptoms (see, for example, Stibbe, 1939).

Lesions of the ascending acoustic tracts may occur in all sorts of brainstem lesions, particularly tumors or vascular disorders. A slightly reduced capacity of hearing may be part of the symptom complex in some of these cases. Other disturbances, for example what may be referred to as a peculiar type of acoustic dysesthesia may occur. In general, however, on account of the binaural representation in all auditory centers above the level of the cochlear nuclei, restricted lesions of the auditory pathway will give few, if any, symptoms. The diagnostic value of the stapedius reflex and the auditory-evoked brainstem response in brainstem disorders is mentioned above.

> Although a selective, high-frequency hearing loss is usually caused by lesions of the cochlea, it may also be caused by damage to the brain. It is seen characteristically as a consequence of anoxic brain lesions such as erythroblastosis, in which a specific cell loss is found in the dorsal, high-frequency zones of the ventral cochlear nucleus (Dublin, 1976). It appears that the high-frequency, second-order neurons exhibit a selective vulnerability similar to that of the receptors in the basal turn of the cochlea. Interestingly, after experimental lesions of the cochlear nerve, the Wallerian degeneration also occurs first in the high-frequency cochlear nerve fibers and their terminals in the dorsal zones of the cochlear nuclei (Arnesen, Osen, and Mugnaini, 1978). The reason for this is not known, but a higher metabolic rate in the high-frequency elements has been suggested to be a causal factor.

Lesions of the medial geniculate body or *of the auditory cortex* will not lead to complete deafness unless they are bilateral. In some cases of temporal lobe tumors a slight impairment of hearing has been observed, and, as referred to above, there may be defects in the capacity to localize sounds. It seems a likely assumption that some impairment of other integrative aspects of hearing will occur in unilateral lesions of the auditory area, but little appears so far to have been delineated in this field. Studies of this subject require the use of elaborate testing methods, and the conclusions which can be drawn are restricted because the cortical lesions usually extend outside the auditory cortex. For a discussion of the importance of the cortex in the elementary features of auditory perception, see the preceding section.

Subjective acoustic sensations, usually in the form of tinnitus, may occur in temporal lobe lesions involving the auditory area (particularly in tumors). More infrequently, more complex phenomena result from irritation of the temporal cortex, such as hearing of voices. From the experiences obtained in electrical stimulation of the cortex, it appears that these complex acoustic "hallucinations" are elicited from regions outside the auditory cortex (see Chap. 12). Epileptic fits, when originating from the temporal lobe, may be preceded by an acoustic aura (frequently combined with vertigo), presumably the expression of irritation of auditory areas. If a temporal lobe lesion occurs in the left hemisphere of a right-handed person, aphasic disturbances, primarily as concerns interpretation of auditory symbols, often occur. According to Penfield and Erickson (1941) the middle and inferior temporal gyri can, however, be removed without producing aphasic disorders. (The aphasic disorders are briefly described in Chap. 12.)

It should be stressed that the symptoms from the auditory areas are not the only ones which may be found in lesions of the temporal lobe (see Chap. 11). On the whole, this is a part of the brain which gives few local symptoms. Particularly if the nondominant hemisphere is affected, auditory disturbances may be entirely absent. Of considerable practical importance is the occurrence of defects in the visual field, which are frequently found on account of the course of the optic radiation (cf. Chap. 8). The so-called uncinate attacks, ascribed commonly to a lesion of the uncus and adjacent parts of the gyrus hippocampi (cf. Chap 10), may occur particularly if the tumor is localized medially. Now and then vestibular disturbances, mainly dizziness, are seen. This fact has been interpreted as indicating the involvement of the cortical area concerned in the reception of vestibular impulses (cf. Chap. 7).

10

The Olfactory Pathways
The Amygdala
The Hippocampus
The "Limbic System"

FROM A clinical point of view the importance of the olfactory system is slight, just as the sense of smell is of relatively minor importance in the normal life of civilized man. However, certain diagnostic information may be obtained from investigations of the sense of smell in clinical cases, particularly when more elaborate test methods are employed. A knowledge of fundamental features in the anatomy of the olfactory connections is, therefore, useful for the neurologist, but no attempts will be made to treat the subject exhaustively. In mammals only part of the amygdaloid nucleus appears to be related to olfaction, and the hippocampus probably has no such relation at all, but for practical purposes both these parts of the brain will be considered here. Likewise the septum and the so-called "limbic system" will be briefly discussed.

The olfactory receptors. The olfactory epithelium is found in the nasal cavity in a limited area on the upper posterior part of the nasal septum and the opposite part of the lateral wall, where it partly covers the superior concha. This part of the nasal mucous membrane has a yellow tint. The slender sensory cells of the olfactory epithelium are kept together by supporting cells. The sensory hairs are situated on a protrusion from the free surface of the sensory cells, an "olfactory rod." This carries 6 to 20 fine hairs which in the electron microscope show the structure of cilia that protrude into the overlying mucus. Electron microscopic studies have demonstrated a number of other details, such as the presence of true tight junctions between the apices of sensory and supporting cells and presumably synaptic contacts between the former. (For particulars see, for example, de Lorenzo, 1963, 1970; Reese and Brightman, 1970.) The supporting cells are provided with microvilli. The olfactory sensory epithelium differs from other types of sensory cells in the organism (for example hair cells of the inner ear or the rod and cone cells of the retina). From the basal end of each olfactory cell a fine threadlike process transmits the impulses in a central direction. This morphological feature shows that the olfactory epithelium represents the primitive type of sensory cell in the vertebrate phylum, and supports the view that the sense of smell is phylogenetically the oldest and most primitive of all senses. The unmyelinated proximal fibers of the olfactory cells are comparable to axons.

The extent of the olfactory area in the nasal cavity varies greatly in different animals, according to their development of the sense of smell. (In the rabbit there are about 50 million receptors on each side, distributed over about 4.5 cm^2; Allison and Warwick, 1949.) As is well known, humans are able to distinguish a great variety of odors. The odors generally used in tests for the sense of olfaction are arbitrarily selected and do not represent any principal or fundamental types. There is some degree of histological differentiation between the olfactory receptor cells. However, attempts to distinguish particular types on a morphological basis have been unsuccessful in the light microscope (Le Gros Clark, 1956) as well as in the electron microscope (see de Lorenzo, 1963).

Several theories have been suggested to explain how olfactory stimuli can be discriminated, but a final solution has not been achieved. For some articles on the sense of smell and its physiology, the reader should consult the symposium reports *Olfaction and Taste* (1975), *Taste and Smell in Vertebrates* (1970), and the monograph of Douek (1974).

Amoore and his collaborators (see Amoore and Venstrom, 1967) have produced evidence that there is a strong positive correlation between molecular size and shape and the odor quality of different odorous substances. Substances giving rise to similar subjective olfactory sensations resemble each other with regard to molecular configuration. The interesting suggestion has been ventured that the correlation between odor and molecular configuration ultimately depends on a correspondence between the geometric shape of the molecules and ultramicroscopic slots or hollows in the surface of the receptors. (For a popular presentation of the theory, see Amoore, Johnston, Jr., and Rubin, 1964.) In continued studies (see Amoore, 1970, 1975, for brief reports) 31 "primary" odors have been distinguished. Some persons are "odor-blind" to a particular smell (about 2% of the population, for example, is unable to smell isobutyric acid and related medium-chain fatty acids). They suffer from what is called "specific anosmia" and have proved useful as test subjects in studies of olfaction in man. Four primary odors, "sweaty," "spermous," "fishy," and "musky," have been described as related to particular human secretions and particular biological functions. Numerous studies have shown that odorous substances play a great role as so-called *pheromones,* eliciting particular behavior patterns (sexual and other) in invertebrates as well as in mammals.[1]

[1] Responses to odors (not irritating) have been recorded from branches of the trigeminal nerve innervating the nasal mucosa. These fibers end in parts of the sensory trigeminal nuclei. Recently, responses to odors (apparently mediated via the ethmoid nerve) have been found in most of the taste-responsive cells in the nucleus of the solitary tract (Van Buskirk and Erickson, 1977). A corresponding convergence of olfactory, gustatory, and oral-tactile impulses has been described in othe brain regions as well, for example the prefrontal cerebral cortex (for references see Van Buskirk and Erickson, 1977). These observations appear to illustrate the general principle that stimuli from several kinds of receptors related to a particular "function," in this case ingestion, tend to converge in the central nervous system.

The extremely thin fibers from the olfactory epithelium (about 0.2 μm in diameter) unite to form slender bundles which penetrate the cribriform plate of the ethmoid bone to enter the brain in the olfactory bulb. The collection of these olfactory fiber bundles represents the 1st cranial nerve, the *olfactory nerve*.

Some comparative anatomical data.   In fishes and amphibians the telencephalon is dominated by afferent fibers carrying olfactory impulses to its pallial part, which betrays no features of a cortexlike structure as seen, for example, in mammals. From comparative anatomical studies it is inferred that the three main regions which can be distinguished within the pallium in amphibians correspond to definite parts which can be recognized in the pallium of higher vertebrates. The medial of these represents the *archicortex* or the anlage of the hippocampus, the lateral the *paleocortex* or the piriform area. Between these is interposed a socalled *dorsal area,* which is considered to correspond to the dorsal cortex or *neopallium,* the *neocortex* of mammals (cf. Fig. 10-1, 1). In reptiles these three divisions

FIG. 10-1   Diagram (after Herrick, 1933) to illustrate the development of the olfactory parts of the brain. In amphibians (drawing 1) a dorsal area is found between the hippocampal area (archipallium) and the piriform area (palaeopallium). In reptiles the dorsal area has the appearance of a cortex (drawing 2); in mammals this develops progressively as the neocortex or neopallium (drawing 3 from opossum) and in man (drawing 4) entirely overshadows the piriform cortex and the hippocampal (archipallial) cortex. Compare text.

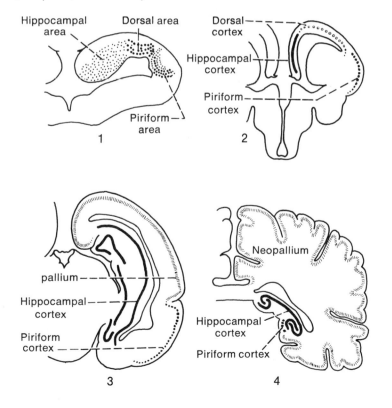

become more distinct, although the cortical structure is of a very primitive type (Fig. 10-1, 2). It is only in mammals that the dorsal cortex undergoes a marked development, and increases progressively in the phylogenetic ascent, to reach its peak of development in man, in whom it forms the bulk of the entire pallium (Fig. 10-1, 3 and 4). The distribution of olfactory impulses during this development becomes restricted to the paleo- and archicortex, which on the whole do not undergo further development in higher mammals after they have reached a high degree of differentiation in lower mammals.[2] The archicortex becomes folded, and by the development of the hippocampal fissure will bulge into the medial wall of the lateral ventricle as the *hippocampus*.

With the development of the neopallium or neocortex, the two more primitive areas are finally pushed medially, and in man are found entirely on the medial aspect of the hemisphere, as illustrated in Fig. 10-1, 4. The growth of the neocortex is also responsible for a change in shape of the other areas, since when the occipital pole of the hemisphere is developed, the paleo- and archicortex are drawn posteriorly, and eventually, later on, with the development of a temporal lobe, again anteriorly and ventrally. On account of this the representatives of the archicortex and paleo-cortex in man and most mammals are found as nearly circular structures, extending backward roughly from the region in front of the interventricular foramen, then curving downward and forward, to reach finally the base of the brain below their starting-point. According to this diagrammatic presentation, the archicortex, represented mainly by the hippocampus and the dentate gyrus, should be present along this entire line. However, when the corpus callosum appears in mammals owing to the development of abundant interhemispheral association fibers between the neocortex of the two hemispheres, the part of the archicortex situated at this place becomes much reduced. It is generally held that its homologue here is the induseum griseum. The subcallosal gyrus, beneath the genu of the corpus callosum (cf. Fig. 10-4), is regarded (by some authors) as being the representative of the most anterior part of the hippocampus. The paleopallium develops posteriorly in mammals to form the piriform lobe, in man represented by most of the hippocampal gyrus, and the rest of it is regarded as being present in the form of the cingular gyrus and the retrosplenial cortex, connecting the cingulate with the hippocampal gyrus. However, many authors are of the opinion that the cingulate cortex is really a transitional cortex between the paleo- and neocortex (see below on the "limbic system"). Within the paleo- and archicortex several cytoarchitectonic areas can be distinguished. At this point it is sufficient to draw attention to area 28 of Brodmann in the hippocampal gyrus (Fig. 10-4), the entorhinal area.

The description given above is highly diagrammatic, but it will serve to give an impression of the development of the structures mentioned in phylogenesis. It explains some peculiar features of the fiber connections in mammals and man, such as the long, curved course of the fornix and the stria terminalis. Above all, the comparative anatomical data clearly reveal how the parts of the brain (the paleo- and archicortex), which in lower vertebrates represent the entire hemi-

---

[2] When the structure of the cerebral cortex is considered, the neocortex is also designated "isocortex" or "homogenetic cortex," whereas the paleo- and archicortex together are named "allocortex" or "heterogenetic cortex." This differs in some respects from the isocortex (cf. Chap. 12).

sphere, become entirely overshadowed in mammals by the neocortex. The paleo-
and archicortical parts of the brain, regarded in lower animals as being first and
foremost concerned in olfactory functions, in mammals, and particularly higher
mammals, have taken over other important functions.

The term *rhinencephalon* is often used, but unfortunately not always in the
same sense. On the basis of phylogenetic studies the term is commonly taken to
cover the pallial archi- and paleocortex, parts of the basal areas of the telencephalon, such as the septal areas, the olfactory tubercle, tract, and bulb. However, the term has a physiological connotation. An analysis of fiber connections
makes clear that in mammals we are not entitled to consider many parts of the so-
called rhinencephalon as related especially to smell, as the name would imply. In
the author's opinion it is unfortunate to use a term of which it can rightly be said:
"From a historical point of view the reader is at present caught up in a morass of
terminology, which has led to an almost individual approach to the terminology of
that portion of the brain now referred to as the rhinencephalon" (White, Jr.,
1965b, p. 28).[3]

The olfactory bulb.   The centripetal fibers from the olfactory epithelium
end in the paired olfactory bulbs, which are parts of the brain, originally evaginated from the telencephalon. From the bulb, situated at the cranial surface of the
ethmoid cribriform plate, begin the fiber connections transmitting the olfactory impulses in a central direction. These connections are not yet fully known. In the following discussion only the main features will be considered.

The *olfactory bulb* has much the same structure in all vertebrates (see Allison, 1953b). Broadly speaking, the afferent fibers from the olfactory sensory cells
interlock in the so-called *glomeruli* with the dendrites of second-order neurons
(Fig. 10-2, see also Fig. 1-22), the *mitral* and *tufted* cells. The former are provided with many fairly coarse secondary dendrites and send their axons into the
olfactory tract. Somewhat similar are the tufted cells, which likewise give off
many dendrites, some of them to the glomeruli, and appear to project centrally, at
least in part.[4] On the basis of Golgi sections, some of the axons of tufted cells
have been claimed to be intrinsic to the bulb (see Valverde, 1965). Finally, there
are large numbers of *granule* cells, situated below the so-called *external plexiform
layer*. The granule cells send numerous dendrites into this layer, where they establish dendrodendritic contacts with the long dendrites of mitral and tufted cells.
Scattered so-called periglomerular cells occur in the vicinity of the glomeruli.

The finer structure of the olfactory bulb has turned out to be rather complex.
Fundamental features were clarified by Cajal (1909–1911). It has been studied
with the light microscope (for example, Allison and Warwick, 1949; Allison,
1953a; Valverde, 1965; Lohman and Mentink, 1969; Scheibel and Scheibel, 1975a;

[3] An impression of the diverse views on the "rhinencephalon" can be gained from a discussion
published as a final chapter in *The Rhinencephalon and Related Structures*, 1963.

[4] All authors appear to agree that the axons of the mitral cells pass centrally in the olfactory tract.
Whether the same is true of the tufted cells has been debated, but has been found to be the case in
studies with silver impregnation methods (see, for example, Lohman and Mentink, 1969). A recent
HRP study of Haberly and Price (1977) confirms this (see their article for references) and indicates that
the axons of tufted cells reach mainly the rostral terminal stations of the olfactory tract fibers (see also
Skeen and Hall, 1977). Neither mitral nor tufted cells appear to send axons across the midline in the
anterior commissure.

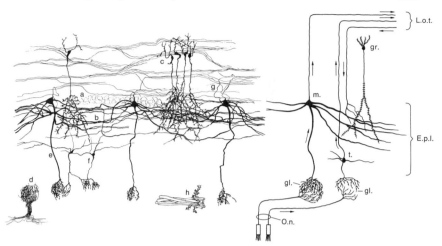

FIG. 10-2 To the right a simplified diagram of the arrangement of the main neuronal elements in the olfactory bulb as described in the text (periglomerular cells and some other cell types are not shown). The fibers from the olfactory receptors entering in the olfactory nerve (*O.n.*) terminate in the glomeruli (*gl.*), where they have contact with dendrites of mitral (*m.*) and tufted (*t.*) cells. Note the broad external plexiform layer (*E.p.l.*) where dendrites of mitral and tufted cells establish dendrodendritic contacts with granule cells (*gr.*). Afferent fibers (*Af.*), entering in the lateral olfactory tract (*L.o.t.*), appear to establish axodendritic contacts with dendrites of granule cells.

To the left, a drawing of a Golgi section from the olfactory bulb of the rat, taken from Scheibel and Scheibel (1975a). *a:* layer of mitral cells; *b:* external plexiform layer containing masses of mitral-cells secondary dendrites; *c:* granule cells; *d:* olfactory glomerulus; *e:* primary dendrites of mitral cells; *f:* tufted cells; *g:* masses of centripetal axons of mitral cells; *h:* enlargement showing details of a dendrite-bundle complex and a single spine bearing granule cell dendrite.

and others) and more recently with the electron microscope (Andres, 1965; Rall, Shepherd, Reese, and Brightman, 1966; Reese and Brightman, 1970; Price and Powell, 1970c, Pinching and Powell, 1971; White, 1973; and others) in normal and experimental material and in various animal species. Attempts to clarify the *synaptic arrangements* have attracted particular interest.

The synapses of the central processes (axons) of the olfactory receptor cells on the dendrites in the glomeruli have been found by several authors to be of the asymmetric type with round vesicles and are generally considered to be excitatory. The dendrodendritic synapses referred to above between mitral and granule cell dendrites are reciprocally presynaptic to each other.[5] The mitral-to-granule cell synapses are of the asymmetric type with spherical vesicles, the granule-to-mitral cell synapses are of the symmetric type with flattened vesicles. The dendrites of the granule cells are densely beset with spines. (Dendrodendritic contacts occur also in the glomeruli.) It appears that the granule cells are devoid of an axon in the

---

[5] In the external plexiform layer the heavy and long secondary dendrites of the mitral cells are aggregated in bundles (Scheibel and Scheibel, 1975a). Such dendrite bundles occur in other regions of the central nervous system as well, such as the spinal cord, reticular formation, and cerebral cortex. They probably serve a particular, so far hypothetical, function. For some views see Scheibel and Scheibel (1975b).

classical sense (cf. amacrine cells in the retina) and, therefore, must exert their action on other cells via dendrites only. These and other anatomical and physiological findings (see Shepherd, 1972; Shepherd, Getchell, and Kauer, 1975, for reviews) have been interpreted as showing that the mitral cells via their long dendrites excite the granule cells, while the latter in turn inhibit the mitral cells. The granule cells appear to be essential in the inhibition of activity of the olfactory bulb which can be elicitecd from central structures that send fibers to the bulb (see below). Even if all details in the synaptic organization of the olfactory bulb are not yet known, recent studies of this part of the brain have been essential in prompting the formulation of certain general principles of central nervous system activity, among others the importance of what has come to be referred to as *local circuits* and their organization (see Chap. 1 and Fig. 1-22).[6]

Studies of the histochemistry of the olfactory bulb have been performed, not least concerning transmitter histochemistry. For example, noradrenergic terminals appear to be present in the granular layer. For a recent study and references see Halász, Ljungdahl, and Hökfelt (1978). Correlations with knowledge of the synaptology have been attempted, but much is hypothetical.

In view of the large number of receptor cells it is striking that the numbers of glomeruli and mitral cells are modest, being 1,900 and 48,000, respectively, in the rabbit (Allison and Warwick, 1949). This shows that there is an extensive convergence of impulses from the receptors in the bulb (about 1,000 receptors per mitral cell). In spite of this the discriminative capacity of the sense of olfaction is considerable and presumably depends in part on processes in the olfactory bulb, since ablations of extensive parts of the cortex and of central olfactory nuclei (see below) do not seriously impair simple olfactory discrimination (Allen, 1941). Physiological (Adrian, 1950a) and anatomical studies (Le Gros Clark, 1951) and depth electrode recordings in man (Sem-Jacobsen et al., 1956) indicate some degree of topical localization in the bulb, particular zones being related to special regions of the epithelium and to different odors (see also Le Gros Clark, 1957; Døving, 1967). These early studies have recently been confirmed. The topographical pattern in the projection from the olfactory mucosa to the olfactory bulb in the rabbit has been described by Land (1973). After exposing rats to a single odor for weeks or months, Døving and Pinching (1973) observed transneuronal degeneration of tufted and mitral cells in a restricted part of the bulb. With different odorants (altogether 44 were tested) Pinching and Døving (1974) found a specific pattern of degeneration for each odor, demonstrating a topological representation of different odors in the olfactory bulb. It is of interest that in spite of clear transneuronal changes in the cells of the bulb, no pathological alterations were found in the olfactory nerve fibers (electron microscopy). This suggests that changes in the activity of afferent fibers may influence cells postsynaptically without morphological changes in their terminals (cf. similar findings in the visual system, Chap. 8).

Connections of the olfactory bulb. The olfactory bulb is the starting point for pathways transmitting olfactory impulses to other parts of the brain. Its main afferents are, as described, primary olfactory fibers, the central processes of the olfactory sensory cells. However, the olfactory bulb also receives afferents from certain regions of the brain, to a large extent those to which it projects. These afferent connections of the olfactory bulb will be considered below, after a presentation of the main features of its efferent projections.

[6] Numerous studies have been devoted to the physiology of the olfactory bulb, but this subject will not be considered here. For a brief review of the physiology of olfaction see Døving (1967). Other articles may be found in the symposia reports on *Olfaction and Taste,* in *Taste and Smell in Vertebrates* (1970); Shepherd (1972); Douek (1974); Shepherd, Getchell, and Kauer (1975).

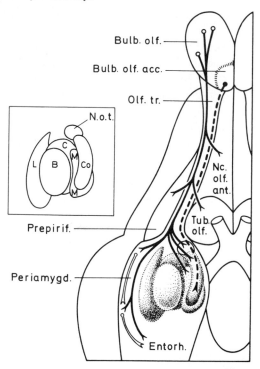

FIG. 10-3   A diagram showing the course and sites of termination of efferent fibers
from the main olfactory bulb (uninterrupted line) and from the accessory olfactory
bulb (interrupted line). Projections from the periamygdaloid cortex to the entorhinal area (white) are also shown. The inset diagram above is a key to the identification of subdivisions of the amygdaloid complex (see Fig. 10-5). *N.o.t.:* nucleus
of lateral olfactory tract. Note: olfactory fibers terminate in this and the cortical (and
medial) amygdaloid nucleus. Compare text. From Lammers (1972).

The *central connections from the olfactory bulb,* the axons of the mitral and
tufted cells, leave the bulb as the *olfactory tract* (Figs. 10-3 and 10-4). Posteriorly
this is flattened and the fiber bundles are found relatively separated as the *lateral
and medial olfactory striae* (Fig. 10-4). Where the two striae diverge they form a
triangular area, the *olfactory trigone.* The striae contribute fibers to several nuclear
structures. After lesions of the olfactory bulb or the olfactory tract, the central
connections have been studied in several animal species with the Marchi method,
with various silver impregnation methods, and electron microscopically. In recent
years autoradiographic tracing of efferents has been carried out. The HRP method
(see Chap. 1) has likewise given important information. The results agree on most
points.

The *distribution of the centrally coursing fibers differs to some extent among
animals* (see Allison, 1953b). In mammals the fibers have been traced with degeneration methods in the opossum (Scalia and Winans, 1975), the rat (Ban and Zyo,
1962; White, Jr., 1965a; Powell, Cowan, and Raisman, 1965; Heimer, 1968;
Scalia and Winans, 1975), the rat (Ban and Zyo, 1962; White, Jr., 1965a; Powell,
Cowan, and Raisman, 1965; Heimer, 1968; Scalia and Winans, 1975), the guinea

pig (Lohman, 1963), the rabbit (Le Gros Clark and Meyer, 1947; Scalia, 1966; Scalia and Winans, 1975), the cat (Fox and Schmitz, 1943; Mascitti and Ortega, 1966), the tree shrew (Skeen and Hall, 1977), and the monkey (Meyer and Allison, 1949; Turner, Gupta, and Mishkin, 1978). Autoradiographic studies have been performed in the rat (Price, 1973), the rabbit (Broadwell, 1975a), the tree shrew (Skeen and Hall, 1977), and the monkey (Turner, Gupta, and Mishkin, 1978).

Despite the minor discrepancies between authors, in part presumably caused by the use of different methods, and not considering some minor end stations, it may as a general rule be stated that in mammals fibers from the olfactory bulb are distributed to the *anterior olfactory nucleus,* the *olfactory tubercle,* to some subdivisions of the *amygdaloid nucleus,* to the cortex of the *piriform lobe,* the *septal nuclei,* and some to the *hypothalamus.* Below some data on these terminations and the terminal stations will be mentioned (see Fig. 10-3).

The *anterior olfactory nucleus,* situated at the base of the olfactory tract, has consistently been found to receive afferents from the olfactory bulb. In Golgi and in anterograde and retrograde experimental studies it has been found to project to its partner on the other side (for example, Brodal, 1948; White, 1965a; Powell, Cowan, and Raisman, 1965; Valverde, 1965; Ferrer, 1969; Price and Powell, 1971).

> This was confirmed by Broadwell (1975b) in autoradiographic and HRP studies of the efferent connections of the anterior olfactory nucleus in the rabbit. Largely in agreement with data obtained by previous authors, Broadwell (1975b) mapped efferents passing back from the anterior olfactory nucleus to the olfactory bulb. Other fibers pass to the olfactory tubercle and the prepiriform cortex, and minor contingents to some other regions, for example the hypothalamus.

It appears that in general the anterior olfactory nucleus projects to structures that also receive afferent input from the olfactory bulb, and that it represents a relay in an indirect route for olfactory impulses to certain areas, complementary to the direct pathway from the olfactory bulb. For a recent physiological study of the anterior olfactory nucleus and some references see Daval and Levetau (1974).

The *olfactory tubercle* (apparently corresponding to the *anterior perforated space* in human anatomy and situated immediately behind the olfactory trigone, as seen in Figs. 10-3 and 10-4) has been shown to receive fibers from the olfactory bulb and from the anterior olfactory nucleus (Heimer, 1968; Price and Powell, 1971; Broadwell, 1975b; Skeen and Hall, 1977; and others).

> The olfactory tubercle is fairly large in microsmatic animals and of considerable size even in the anosmatic whales. In the monkey its supply by olfactory afferents is sparse. However, Van Hoesen, Mesulam, and Haaxma (1976) found considerable afferent fiber contingents from the temporal lobe in the monkey (areas 20, 21, and 35; see Fig. 12-2). It appears that in primates the olfactory tubercle bears little relation to smell and can scarcely be considered a "secondary" olfactory area. The authors stress the extensive connections with other parts of the cerebral cortex (see Chap. 12) of those temporal lobe areas that project to the olfactory tubercle. However, little is known about functional aspects.

The fibers from the olfactory bulb to the *amygdaloid nucleus* (to be considered below) have been found to be restricted to its corticomedial regions, the cortical and medial nucleus, as determined experimentally in silver impregnation studies (Le Gros Clark and Meyer, 1947; Powell, Cowan, and Raisman, 1965;

Scalia, 1966, in the rabbit; Meyer and Allison, 1949, in the monkey; Adey, 1953, in a marsupial; and others). The results of autoradiographic studies (Price, 1973, in the cat; Broadwell, 1975a, in the rabbit; Skeen and Hall, 1977, in the tree shrew; Turner, Gupta, and Mishkin, 1978, in the monkey) are in general agreement but appear to indicate even more clearly than previous studies that the cortical amygdaloid nucleus is the main receiving nucleus. A projection to the central nucleus, claimed by some, has been confirmed by Lohman (1963); Cowan, Raisman, and Powell (1965); Turner, Gupta, and Mishkin (1978); and others.

Even though there appears to be general agreement among authors concerning main points, valid comparisons are difficult because of the use of different experimental animal species, methods, and nomenclature. From a functional point of view it is of interest that the olfactory fibers ending in the amygdaloid nucleus supply first and foremost the subdivisions of the complex which project to the hypothalamus, especially to the ventromedial nucleus (see Chap. 11). There appears thus to be a rather direct, oliogosynaptic route between olfactory receptors and the "feeding center" in the hypothalamus.

However, *direct olfactohypothalamic* routes are available as well, since fibers from the olfactory bulb have been found in different animals to pass to the hypothalamus, particularly its anterior and lateral regions (Powell, Cowan, and Raisman, 1963; Scott and Leonard, 1971). These findings are of interest in connection with observations of the role played by olfactory stimuli in reproductive and other kinds of behavior patterns that are influenced from the hypothalamus (cf. Chap. 11).

Fibers from the olfactory bulb have rather consistently been traced to the *piriform lobe*. As described above, this belongs to the paleocortex. There has been much dispute concerning the subdivision of the piriform lobe cortex into different cytoarchitectural areas, and there are considerable species differences. It is common to distinguish *three main areas* from rostral to caudal, *prepiriform, piriform* (or *periamygdaloid*), and *entorhinal* (see for example, Valverde, 1965). The entorhinal area (Brodmann's area 28) is the largest in man, where it occupies a large anterior part of the hippocampal gyrus (see Fig. 10-4). The prepiriform and periamygdaloid areas are found on the upper anterior aspect of the uncus.

> There are marked species differences with regard to development and relative size of the various parts of the piriform lobe. The prepiriform area is often described as reaching for a considerable distance in the rostral direction along the rhinal fissure and the lateral olfactory tract, extending almost to the olfactory bulb. This extension rostrally beyond the macroscopically identified piriform lobe explains the name "prepiriform" for this area. The periamygdaloid area receives its name from its location on the surface of the amygdaloid nucleus. (In Fig. 10-4 a distinction between the prepiriform and the periamygdaloid area is not indicated.) It appears that in man the periamygdaloid area is represented in the gyrus semilunaris, and the prepiriform are in the anterior part of the gyrus ambiens, both of which are hardly visible gyri close to the uncus (cf. Fig. 11 in the atlas of Niewenhuys, Voogd, and van Huijzen, 1978). The three areas of the piriform lobe, particularly the entorhinal area, have been subdivided on a cytoarchitectonic basis into several subareas (Lorente de Nó, 1933c; Rose, 1935). The entorhinal area appears to correspond to what Cajal (1909–1911) called "écorce temporale supérieure ou postérieur." For a description of cell types in the four-layered cortex of the piriform lobe, see Valverde (1965).

Most early authors were of the opinion that the sites of termination of olfactory fibers in the piriform lobe did not extend beyond the prepiriform and peri-

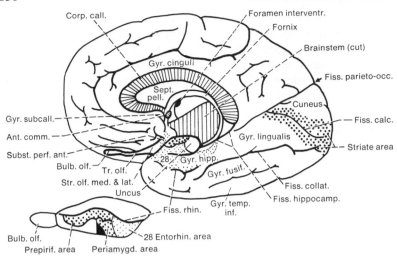

FIG. 10-4   A drawing of the medial aspect of a right human hemisphere, showing some of the olfactory structures. The primary olfactory cortex, indicated by coarse stippling (like the striate area 17), covers parts of two small gyri which appear to correspond to the prepiriform and periamygdaloid areas (see text). An extension of the primary olfactory cortex into the anterior part of the entorhinal area (fine stippling) as determined in animals, is not indicated. Lower left, a diagram of the basal surface of the rabbit's brain for comparison. The primary olfactory area is indicated according to Rose (1935), and symbols correspond to those in the drawing of the human brain.

amygdaloid areas. However, with the use of modern methods it has been shown in many animal species that the distribution of the olfactory bulb fibers in the piriform lobe covers most of the entorhinal area (Brodmann's area 28) except its most medioposterior part (White, 1965a; Heimer, 1968; Price, 1973; Broadwell, 1975a; Skeen and Hall, 1977; Scalia and Winans, 1975). That the olfactory bulb fibers terminate in a large part of the entorhinal area is of interest, since the entorhinal area is the origin of a major afferent input to the hippocampus. This observation indicates that the pathway mediating olfactory impulses to the hippocampus is "one synapse shorter" than was previously held. We shall return to the connections of the entorhinal area below.

Fibers from the olfactory bulb to the *hippocampus* proper have never been convincingly demonstrated.

However, in some animals olfactory fibers have been followed into what is generally referred to as the subcallosal part of the *hippocampal rudiment* (also called taenia tecta), for example in radiographic studies of Scalia and Winans (1975, opposum), Broadwell (1975a, rabbit), Skeen and Hall (1977, tree shrew), in agreement with earlier silver impregnation studies of others. This projection appears to be absent in the microsmatic monkey (see Turner, Gupta, and Mishkin, 1978).

Physiological studies of the distribution of evoked potentials after olfactory stimuli and electrical stimulation of the olfactory bulb are in general agreement with anatomical data (for a recent study see, for example, Dennis and Kerr, 1975).

Responses with short latencies have been recorded in all regions found in anatomical studies to receive fibers from the olfactory bulb. It is particularly noteworthy that the lateral part of the entorhinal area is one of these (see also below). Responses to olfactory stimulation have been recorded in other regions as well (see, for example, Motokizawa, 1974), even as far caudally as in the mesencephalic reticular formation. This must be assumed to occur via fibers descending from structures receiving fibers from the olfactory bulb (see also footnote 1).

The question may be raised: What should be considered *the primary olfactory cortex* (corresponding to the primary sensory cortices of other sensory qualities)? Different opinions have been held. It appears that most authors would restrict the term to cover those regions of the piriform lobe which receive direct fibers from the olfactory bulb and the anterior olfactory nucleus.[7] Accordingly, the primary olfactory cortex consists of the prepiriform and the periamygdaloid areas and, as learned from recent studies, a considerable lateral part of the entorhinal area as well. This primary olfactory cortex is relatively large in some mammals, such as the rabbit, but in man it occupies only a small area on the anterior end of the hippocampal gyrus and uncus, as shown in Fig. 10-4, where an attempt has been made to indicate approximately the extent of this area on the basis of cytoarchitectonic and experimental studies.[8] (The probable extension of the primary olfactory cortex into the entorhinal area also in man is not shown in the diagram.) In contrast to the primary sensory areas for vision, hearing, taste, and somatic sensibility, the primary olfactory area is found in the allocortex (cf. Chap. 12). This part of the cortex lacks the typical granular appearance characteristic of the other primary sensory areas. It differs from them also in another respect: its afferent, sensory fibers do not reach it from the interior of the brain after a relay in the thalamus, but arrive from the surface.

It may be surmised that the relatively small primary olfactory region in man is first and foremost concerned in the conscious perception of olfactory stimuli, as appears from studies in man (see below). The other terminal regions are presumably related chiefly to reflex activities and behavioral reactions elicited in response to olfactory stimuli. However, second-order connections from the primary olfactory cortex and the amygdaloid nucleus must be taken into account as well (see below).

The central connections of the olfactory bulb, receiving input from receptors in the olfactory mucosa, have been considered above. However, in addition to the main olfactory bulb many animals possess an *accessory olfactory bulb*. This receives its main afferents from the vomeronasal organ (of Jacobsohn), structurally very similar to the olfactory mucosa and believed to function as an olfactory receptor. The vomeronasal organ is found in most macrosmatic mammals, but is absent or rudimentary in monkey and man. It is of some interest that the central distributions of the efferents from the main and the accessory olfactory bulb (Fig. 10-3) appear to be to some extent different. For example, they supply different parts of the corticomedial amygdaloid nucleus and of the piriform cortex (Scalia and Winans, 1975; see also Broadwell, 1975a, and Skeen and Hall, 1977). These and several other data strongly suggest that there exist dual olfactory systems, which are dissimilar both anatomically and functionally. For some data and discussion see Raisman (1972, 1975), Scalia and Winans (1975), Broadwell (1975a), and Skeen and Hall (1977). Like the olfactory bulb, the accessory olfactory bulb receives afferents from some of the regions

---

[7] Others, for example Price (1973), include all end stations of afferent fibers from the olfactory bulb, for example the anterior olfactory nucleus and the cortical amygdaloid nucleus.

[8] A discussion of these cortical areas may be found in the papers of Brodal (1947b) and Meyer and Allison (1949). See also Allison (1954).

to which it projects, for example the amygdaloid nucleus (see Raisman, 1975; de Olmos, Hardy, and Heimer, 1978).

In addition to fibers from the olfactory receptors, the olfactory bulb receives afferents from the contralateral bulb (see below) and from the brain, as maintained by Cajal (1909–11). The presence of *centrifugal fibers in the lateral olfactory tract* (see Fig. 10-2) has been demonstrated experimentally in various animal species by Cragg (1962), Powell and Cowan (1963), Powell, Cowan, and Raisman (1965), Heimer (1968), Price and Powell (1970a), and others.

On the basis of anterograde degeneration studies the fibers have been concluded to come from the prepiriform cortex and from cell groups referred to as the "horizontal nucleus of the diagonal band" and some other regions. After injection of HRP in the olfactory bulb, retrogradely labeled cells were found not only in the two regions mentioned but also in the nucleus of the olfactory tract and the olfactory tubercle (Broadwell, 1975b, in the rabbit; Dennis and Kerr, 1976, in the cat; de Olmos, Hardy, and Heimer, 1978, in the rat). Minor contingents have been found in other regions as well, such as the lateral hypothalamus and the cortical amygdaloid nucleus. The results of physiological mappings by Dennis and Kerr (1975) in the ferret are in general agreement.

In silver impregnation studies some afferent fibers to the olfactory bulb have been described to pass as far as the glomeruli. In an experimental electron microscope study in the rat, however, Price and Powell (1970b) found the terminals of the centrifugal fibers to establish asymmetrical synapses with the granule cells (for particulars see their paper).

As to the functional role of the centrifugal fibers, Kerr and Hagbarth (1955) early demonstrated an inhibitory influence on the activity of the olfactory bulb on stimulation of various regions of the brain, such as the prepiriform cortex. Recent neurophysiological studies in the rabbit likewise indicate that the centrifugal fibers mainly exert a depressive influence on the olfactory bulb and, furthermore, that the granule cell acts as an inhibitory interneuron (Nakashima, Mori, and Takagi, 1978).

As seen from the discussion above, there appear to be feedback connections to the olfactory bulb from most of the regions to which it projects. It is of further interest to note that the principle of central control of afferent impulses is valid for the sense of olfaction, as it is for other sensory qualities.

Commissural connections, passing in the anterior commissure, have been held to interconnect the olfactory bulbs. However, it appears that at least most of these fibers pass from the anterior olfactory nucleus on one side to the opposite olfactory bulb (see Broadwell, 1975b). (The anterior commissure carriers other fibers as well, see below.)

Other "olfactory" connections. The fibers from the nasal mucosa to the olfactory bulb (the axons of the olfactory receptor cells) represent primary olfactory fibers in a restricted sense. The axons of the mitral and tufted cells, passing centrally in the olfactory tract and described above, are strictly speaking, all secondary olfactory fibers. Their regions of termination—the piriform cortex, the amygdala, and other structures considered above—influence other brain regions by way of efferent connections. These pathways scarcely transmit uncontaminated olfactory information, since their regions of origin receive fibers from many sources. Nevertheless, for practical reasons it may be permissible to refer to these pathways (secondary or tertiary) as olfactory connections.

These connections and their functions have been extensively studied, not least on account of the great interest devoted to the "limbic system" (to be considered later). Here we shall survey the cellular regions concerned and their fiber connections. It is practical to consider them with reference to the particular relevant regions of gray matter, the most important of these being the primary olfactory cortex, the amygdaloid nucleus, the hippocampus, the septum, the entorhinal area, and the cingulate gyrus. Only main features will be discussed; for details the original articles should be consulted.

Among the regions mentioned, the *primary olfactory cortex* is the one in whose afferent input the olfactory impulses appear to be most dominant. From a functional point of view, knowledge of the *efferent projections from the primary olfactory cortex* is of some interest. Efferents have been traced from the primary olfactory cortex (prepiriform and periamygdaloid areas and lateral part of the entorhinal area 28) to several other nuclei. After lesions restricted to the piriform cortex in the rabbit, Powell, Cowan, and Raisman (1965), using the Nauta method, traced degenerating fibers to the *entorhinal area, the basolateral amygdaloid nucleus, the hypothalamus,* and to the *dorsomedial thalamic nucleus (MD).* Recent research has on the whole confirmed the results of Powell, Cowan, and Raisman (1965), and some additional sites of fiber termination have been reported.

Mention was made above of fibers passing to the *olfactory bulb* and the anterior olfactory nucleus and establishing feedback connections between these and the olfactory cortex. There are association connections from the anterior parts of the piriform lobe to the *entorhinal area* (to be considered below), probably chiefly to its lateral part (see Beckstead, 1978). (There is an important projection from the entorhinal area to the hippocampus.) The fibers from the olfactory cortex to the *amygdaloid complex* (see below on the amygdala) appear to terminate mainly in the basolateral nucleus. (This gives rise to efferent fibers passing to several regions.) With silver impregnation methods, fibers from the olfactory cortex have been traced to the *hypothalamus,* particularly its lateral and anterior regions (Cragg, 1961; Powell, Cowan, and Raisman, 1965; Scott and Leonard, 1971). However, this projection has not been convincingly demonstrated electron microscopically (Scott and Chafin, 1975).[9] The projection from the primary olfactory cortex to the *dorsomedial thalamic nucleus* (MD, its medial magnocellular part) has been studied experimentally with the light and electron microscope by Heimer (1972).[10] The fibers appear to pass in the stria medullaris (see also Scott and Leonard, 1971). (Via the MD the olfactory cortex may influence the prefrontal granular cortex, see Chap. 12.) By way of fibers to the habenula, the piriform cortex may act on the midbrain (see below).

It can be seen from the description above that via the olfactory cortex impulses arising on olfactory stimulation may influence many brain regions that do not receive direct afferents from the olfactory bulb (for example, the MD of the thalamus and some of the nuclei in the amygdaloid complex). The hypothalamus is remarkable in receiving in addition to fibers from the olfactory cortex also direct afferents from the olfactory bulb and fibers from the olfactory fiber receiving part of the amygdala. These multiple roads to the hypothalamus are presumably an expression of the well-known influence exerted by olfactory stimuli on autonomic functions and behavioral reactions.

The other regions mentioned above (p. 648) are less dominated by the olfactory input than is the primary olfactory cortex. However, olfactory impulses have

[9] It is noteworthy that fibers from the amygdaloid complex to the hypothalamus (its anterior regions and the ventromedial nucleus) have been found to arise mainly from the part of the complex which receives input from the olfactory bulb (the corticomedial nuclear group).

[10] From physiological investigations it appears that, in agreement with anatomical findings, responses in the MD (magnocellular part) to olfactory stimuli are relayed via the piriform cortex (Benjamin and Jackson, 1974, in the monkey).

been shown to reach them and appear to influence their function. This has perhaps been most clearly shown for the amygdaloid nucleus.

The amygdaloid nucleus, or *nucleus amygdalae,* received its name from its resemblance to an almond. Its position was mentioned previously. The nucleus has been described in a series of mammals (for references to older literature, see Brodal, 1947c; for more recent ones, see, for example, Hall, 1972). In man it is found immediately underneath the uncus (see Fig. 10-4). In all animal species studied the amygdaloid nucleus can be subdivided into a number of cell groups or subnuclei, differing in architecture. The general pattern is similar in all mammals, even though there are differences in details. In a simplified manner a distinction can be made between a *corticomedial* and a *basolateral group of nuclei.* (In addition there are some minor cell groups and transitional areas which will not be considered here.) In the phylogenesis of mammals the basolateral group (containing the basal and lateral nuclei) increases progressively and is especially well developed in man (see Crosby and Humphrey, 1941; Allison, 1954; Humphrey, 1972). In Fig. 10-5 this is clearly seen; and likewise it is apparent that the medial, central and cortical nuclei are small in man. These differences in size are related to the connections and functions of the two main subdivisions.

There has been considerable disagreement among authors concerning details in the subdivisions of the amygdaloid nucleus. It is common to refer to the entire nucleus as the *amygdaloid complex,* since particular nuclei do not always have distinct borders in cell-stained sections and a nucleus may often be composed of minor regions, differing cytoarchitectonically (see stippled lines in Fig. 10-5B, in-

FIG. 10-5   Diagram of the amygdaloid complex in the rat (A) and in man (B and C), showing its various nuclei (subdivisions of individual nuclei are indicated by broken lines). In C the position of the entire complex in man is shown in black, in B the various nuclei are indicated. Redrawn from Brodal (1947c, A) and Humphrey (1972, B and C). Abbreviations: *bas.:* nucleus basalis; *centr.:* nucleus centralis; *cort.:* nucleus corticalis; *i.:* intercalated masses; *lat.:* nucleus lateralis; *med.:* nucleus medialis; *ol.:* nucleus of lateral olfactory tract; *pr.:* piriform cortex; *tr.a.:* amygdalopiriform transition area. Note the large size of the basal and lateral nuclei in man as compared with the small central, medial, and cortical nuclei.

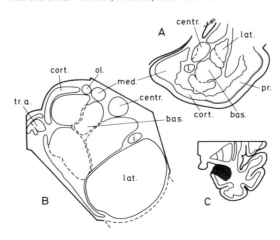

dicating nuclear subdivisions). In Golgi sections differences are likewise apparent (see Hall, 1972).

Studies of the *chemoarchitecture* of the amygdaloid complex have been undertaken in various animal species, and, likewise, studies of the distribution of putative transmitters. The maps arrived at in such studies cannot always be correlated with the cytoarchitectonic pattern, for example with regard to the intensity of staining for cholinesterase (see Hall and Geneser Jensen, 1971). However, cholinergic fibers appear to supply particularly the basolateral nuclei (Ben-Ari, Ziegmond, Shute, and Lewis, 1977). Most of the nuclei appear to have a high concentration of serotonin (see for example Saavedra, Brownstein, and Palkovits, 1974), and of noradrenaline (Ben-Ari, Ziegmond, and Moore (1975), while the central nucleus appears to be rich in dopamine. Regional variations are further seen when the amygdala is stained using procedures considered to demonstrate the presence of zinc (for example, the silver-sulfide method of Timm). For a brief review, see Hall (1972).[11]

The *fiber connections of the amygdaloid complex* are multifarious. The afferents come from several different brain regions. However, afferents from the different sources do not end indiscriminately throughout the complex. In general, a particular afferent contingent supplies only one or a few nuclei. This, as well as a similar arrangement of the origin of amygdaloid efferents, bears witness that the amygdala is to be considered a complex of anatomically and functionally different units. The finer patterns of the connections have been difficult to decipher. However, studies using recent anatomical methods have provided many new data illustrating the functional heterogeneity of the complex.

A fairly complete list of *afferents to the amygdaloid complex* would include fibers from the olfactory bulb and anterior olfactory nucleus, the primary olfactory cortex, the anterior cingulate gyrus, the "prefrontal" granular cortex, the temporal neocortex, the dorsomedial thalamic nucleus (MD) and other regions of the thalamus, the hypothalamus, the raphe nuclei, the nucleus of the solitary tract, and some other regions of the brainstem (Fig. 10-6A).

The terminal region of fibers from the *olfactory bulb* and the anterior olfactory nucleus, as mentioned previously, appears to be the corticomedial nucleus, particularly the cortical. Possibly, some fibers end in the central nucleus. The *primary olfactory cortex* has been found to project to the basolateral group (Powell, Cowan, and Raisman, 1965; and others). This group may thus be influenced indirectly by olfactory impulses. In the monkey a projection from the *anterior cingulate gyrus* (area 24) to the basal nucleus has been described by Pandya, van Hoesen, and Domesick (1973). Likewise, the basolateral group is the only or the main terminal station of projections from parts of the *neocortex*. Thus in the monkey fibers were traced chiefly to the basal nucleus from the *"prefrontal" granular cortex* (see Chap. 12) by Leichnetz and Astruc (1976) and from the *temporal neocortex* by Whitlock and Nauta (1956) and Jones and Powell (1970a)

[11] The amygdala is one of the brain areas richest in enkephalins. Enkephalins are pentapeptides shown to be endogenous ligands for the cerebral opiate receptors. They occur in particularly large amounts in the central nucleus. They have been demonstrated in the amygdalofugal component of the stria terminalis, but not as yet in amygdalopetal pathways.

The opiate receptors in the amygdala seem to be involved in the mediation of the cataleptic and epileptic effects of opiate administration rather than the analgesic effects. Recent studies also suggest that the amygdala, particularly its central nucleus, plays an important role in the morphine withdrawal syndrome. For some references see, for example, Sar et al. (1978) and Calvino, Lagowska, and Ben-Ari (1979).

using silver impregnation methods. Electron microscopic studies of the temporoamygdaloid projections (see Hall, 1972) have confirmed this. According to the autoradiographic study of Herzog and van Hoesen (1976) the temporal cortex projects in a topographical pattern on the amygdala, particularly to the basal and lateral nuclei. Physiologically, however, the projection has been found to be diffuse (see, for example, Prelević, Burnham, and Gloor, 1976). There appear to be afferents from the temporal allocortex as well, for example from the *entorhinal area* and the subiculum (see Veening, 1978a).

The projection from the *thalamus,* formerly believed to arise from the dorsomedial thalamic nucleus (MD) only, has recently been found to have a far wider origin. Some nuclei near the midline and other thalamic nuclei contribute. Some of these afferent components have been found to be rather diffusely distributed, while others appear to supply a particular nucleus only.

A projection from the *dorsomedial thalamic nucleus* (MD) to the amygdala, claimed to be present on the basis of the findings in normal material, was first demonstrated experimentally by Nauta (1962). Other studies, most recently the autoradiographic one by Krettek and Price (1977a), confirmed this in the rat. The terminal area was found to be the basolateral nucleus. Ottersen and Ben-Ari (1978a, 1979), however, after HRP injections in this amygdaloid nucleus, found only a few labeled cells in the MD in the rat and cat. This discrepancy may be due to methodological factors, and the negative findings do not disprove the presence of a mediodorsoamygdaloid projection. Species differences may be concerned as well.

Studies with the HRP method have given evidence of a far more complex and extensive thalamoamygdaloid projection. Thus, the lateral amygdaloid nucleus in the rat and cat receives afferents from several of the "unspecific" small midline and *intralaminar nuclei* of the thalamus (Ottersen and Ben-Ari, 1978a; see also Veening, 1978b). Continued studies by Ottersen and Ben-Ari (1979) have further elaborated this subject. Their data indicate, for example, that the parafascicular thalamic nucleus projects to the central amygdaloid nucleus only. The central and medial nuclei likewise are the end stations of fibers from the "taste area" in the ventromedial thalamus (see Chap. 7). The suggested presence of a projection from the *medial geniculate body* to the amygdala has likewise been confirmed (see Ottersen and Ben-Ari, 1979). These fibers appear to end in the centrocorticomedial part of the amygdala. The *pulvinar* has been found in autoradiographic studies to project to the lateral amygdaloid nucleus in the monkey (Jones and Burton, 1976a). However, this could not be confirmed in HRP studies in the rat (Veening, 1978b; Ottersen and Ben-Ari, 1979).

In addition to afferents from cortex and thalamus, the amygdala receives ascending afferents from several nuclei of the *brainstem.* These have recently been studied by mapping labeled cells after injections of small amounts of HRP in different parts of the amygdaloid complex.

It appears that the *central amygdaloid nucleus* receives fibers from the *dorsal raphe nucleus,* the *nucleus locus coeruleus,* and the *parabrachial nucleus* at the pontine level and from the *substantia nigra* and some other areas in the mesencephalon (Veening, 1978b; Ottersen and Ben-Ari, 1978b, in the rat). The central nucleus also receives afferents from the *nucleus of the solitary tract,* as shown by Ricardo and Koh (1978), and from the *hypothalamus* (Conrad and Pfaff, 1976b; Veening, 1978b). According to the HRP and electrophysiological studies of Renaud and Hopkins (1977), the distribution of the afferents from the hypothalamus may extend beyond the central nucleus and, interestingly, most of them appear to come from the ventromedial hypothalamic nucleus (which is a main target for amygdalohypothalamic fibers; see below).

Just as the amygdaloid complex as a whole receives afferents from many sources, it has ample *efferent projections* (Fig. 10-6B). Many of these projections

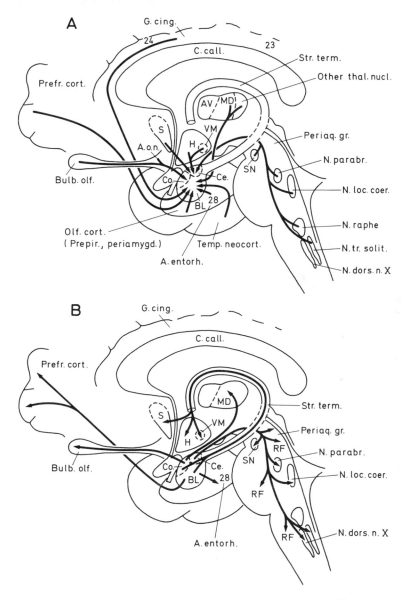

FIG. 10-6   Very simplified diagrams of the main afferent (A) and efferent (B) connections of the amygdaloid complex. To some extent sites of origin and endings of fibers within particular nuclei of the complex are shown. Note the many reciprocal connections. Not all connections are shown (for example, in *B*, the fibers to the subiculum). See text for particulars. Abbreviations: *A.o.n.:* anterior olfactory nucleus; *AV:* anteroventral thalamic nucleus; *BL:* basolateral amygdaloid nucleus; *Ce.:* central amygdaloid nucleus; *Co.:* cortical amygdaloid nucleus; *H:* hypothalamus; *MD:* dorsomedial thalamic nucleus; *RF:* reticular formation; *S:* septum; *SN:* substantia nigra; *VM:* ventromedial hypothalamic nucleus.

arise from particular nuclei of the complex. To a considerable extent these connections are reciprocal to the afferents to the amygdala.

The efferent fibers from the amygdala are generally described as taking two routes, one called the *dorsal amygdalofugal pathway, passing in the stria terminalis,* and the other the *ventral amygdalofugal pathway.*[12] It is generally agreed that the stria terminalis (Fig. 10-6) carries a fair number of efferent fibers from the amygdala. The stria terminalis is a macroscopically visible fiber bundle that pursues a curved course along the medial surface of the caudate nucleus in the lateral ventricle. Some fibers descend anterior, others posterior to the anterior commissure and are distributed to, for example, the preoptic area, the hypothalamus, and some minor nuclear groups. The ventral amygdalofugal pathway is a common denominator for fibers from the amygdala which proceed along the base of the brain. In an early study in the monkey, Nauta (1961) stressed the considerable volume of this pathway. Some of its fibers run medial to the olfactory tubercle and neighboring regions or pass to the thalamus. Others take a caudal course to the mesencephalon, while still others are assumed to pass anteriorly in association bundles of the hemisphere.

> One might perhaps have expected that the two amygdalofugal pathways would be different with regard to both the origin of their fibers within the amygdaloid complex and their sites of termination. This question has been difficult to answer. It appears that in general there is no clear pattern, insofar as both pathways to some extent carry fibers from the same parts of the amygdaloid complex to the same targets. However, it is often stated that the ventral amygdalofugal pathway consists mainly of fibers from the basal and lateral nuclei, while most fibers from the corticomedial group pass in the stria terminalis.

By way of summary, it may be stated that one or both *amygdalofugal pathways supply the following regions:* the *hypothalamus,* the *thalamus,* parts of the *neocortex,* parts of the *allocortex,* regions in the *brainstem,* and the *olfactory bulb. Commissural connections* between the nuclei of the two sides pass in the anterior commissure.[13] In the rat they appear to come chiefly from the medial and cortical amygdaloid nuclei (Brodal, 1948; de Olmos, 1972).

This incomplete summary gives an impression of the complexity in the efferent projections of the amygdaloid nuclei. It appears likely that the individual nuclei have more or less specific targets for their projections. This question has been studied by numerous authors, but the reports in the literature have been rather confusing (see, for example, Ban and Onukai, 1959; Nauta, 1961, 1962; Hall, 1963; Valverde, 1963; Cowan, Raisman, and Powell, 1965; Heimer and Nauta, 1969). With the advent of the recent anatomical methods based on axonal transport (anterograde and retrograde, see Chap. 1), much new insight has been gained (see below), but many points are still unclear. To a large extent this is presumably due to difficulties in producing lesions, or achieving injections of markers, restricted to one nucleus. Further, fibers from one subnucleus pass through others, fibers coming from the piriform cortex pass through the amygdaloid complex, and there are interamygdaloid connections (see de Olmos, 1972).

---

[12] These names are not quite appropriate, since both pathways also contain afferent fibers to the amygdala.

[13] The anterior commissure is a complex fiber system. In addition to the amygdaloid component and the fibers interconnecting the olfactory bulbs, mentioned before, it contains commissural fibers between the piriform cortex, the entorhinal area, the temporal cortex, and some other regions.

These features are sources of error in physiological studies of the amygdala as well. Nevertheless, the origin of some of the efferent connections appears to be rather definitely established.

The fibers to the *hypothalamus* and the *preoptic area* make up a fair proportion of the stria terminalis. They have been found to terminate particularly in the ventromedial hypothalamic nucleus (Heimer and Nauta, 1969; Leonard and Scott, 1971; McBride and Sutin, 1977), and appear to come, at least mainly, from the cortical amygdaloid nucleus (Leonard and Scott, 1971; McBride and Sutin, 1977), the main recipient of afferents from the olfactory bulb. In the rat, electrical stimulation of the amygdala exerts a prominent excitatory influence on many neurons of the ventromedial hypothalamic nucleus (Renaud, 1976). (The fibers to the preoptic area appear to originate in a more extensive part of the amygdala.) Fibers to the septum have been described.

As referred to above, various *thalamic nuclei* project *to* the amygdala. The projection in the reverse direction appears to be far more modest. Fibers to the MD (its medial part) have been found in the monkey (Fox, 1949; Nauta, 1961; Pribram, Chow, and Semmes, 1953) and in the rat (Krettek and Price, 1974, 1977b, autoradiography). According to the HRP study of Siegel, Fukushima, Meibach, Burke, Edinger, and Weiner (1977), the fibers arise from most parts of the amygdaloid complex with the exception of the central and lateral nuclei, in partial agreement with the findings of Krettek and Price (1977b).

The *neocortex* of the frontal lobe, more precisely the "prefrontal" granular cortex (see Chap. 12), may be influenced from the amygdala via the dorsomedial thalamic nucleus (MD), since its medial, magnocellular division gives off an efferent projection to this cortex, particularly the orbitofrontal cortex (see Chap. 12). However, it has recently been shown that the amygdaloid nucleus also has possibilities for influencing the "prefrontal" cortex *directly*—that is monosynaptically. Krettek and Price (1974) traced such ipsilateral fibers autoradiographically in the rat and cat to restricted areas of the cortex on the medial and convex surfaces of the frontal lobe.

Studies using the HRP method in the monkey indicate, as suggested by the studies above, that the direct amygdaloprefrontal fibers arise chiefly from the basal nucleus (Jacobson and Trojanowski, 1975a). In the cat, Llamas, Avendaño, and Reinoso-Suárez (1977) made largely corresponding findings. In addition, they advocated the presence of amygdaloid projections to the motor and "premotor" cortex. (These may be of interest in the explanation of motor effects elicitable from the amygdala.)

The amygdala further has efferent projections to various parts of the *allocortex*. Thus the presence of fibers to the *entorhinal area* (particularly its lateral part) and the *perirhinal cortex* has been confirmed both autoradiographically (Krettek and Price, 1974, 1977b, 1977c) and with the HRP method (Beckstead, 1978). The origin of these fibers was found to be the lateral amygdaloid nucleus, which gives off some fibers to the *subiculum* as well (Krettek and Price, 1977c). Nauta (1961) advocated the existence of a projection to the *cingulum,* but this has apparently not been confirmed. A projection from the basolateral nucleus to the *caudate nucleus* has recently been described in an HRP study in the cat (Royce, 1978b).

In contrast to most of the projections described above, fibers destined for the *olfactory bulb* (see de Olmos, Hardy, and Heimer, 1978), like those to the hypothalamus, appear to arise chiefly from the cortical nucleus (receiving the bulk of amygdala afferents from the bulb; compare Figs. 10-6A and B). The same appears to be the case for fibers to the *accessory olfactory bulb*.

Recently, *descending fibers* from the amygdala to the brainstem, concluded from physiological studies by Gloor (1955) to be present, have been demonstrated anatomically. After HRP injections at different levels of the brainstem, Hopkins (1975) found labeled cells in the amygdala, mainly in the central nucleus (which, as noted above, is the main terminal station in the amygdala for fibers ascending in the brainstem).

From continued HRP and autoradiographic studies Hopkins and Holstege (1978) concluded that amygdalofugal fibers arising from the central nucleus, terminate in the substantia nigra and periaqueductal gray, the reticular formation of the mesencephalon, pons, and medulla. Furthermore, fibers were traced to the parabrachial nucleus, the nucleus of the solitary tract, and the dorsal motor nucleus of the vagus.

Many features in the connections of the amygdaloid nucleus are still far from clear. However, as has been seen from the description above, to some extent the individual amygdaloid nuclei have connections with particular other parts of the brain (presumably an indication of functional differences). Like so many other structures the amygdaloid complex has reciprocal connections with the areas to which it sends fibers. Among the connections some authors stress the reciprocal amygdalohypothalamic relationships, while others have put special emphasis on the circuit amygdala–dorsomedial thalamic nucleus–orbitofrontotemporal cortex–amygdala. Recent findings have focused attention on the reciprocal amygdala–brainstem relations (apparently involving mainly the central nucleus). These and other circuits have played a role in the attempts to interpret the manifold effects observed following stimulations or lesions of the amygdaloid nuclei. The various theories and hypotheses set forth will not be considered here. Brief mention will, however, be made of some of the many physiological observations (for a review, see Gloor, 1960). Many relevant papers and references will be found in *The Neurobiology of the Amygdala* (1972).

Functions of the amygdaloid nucleus.   In general it may be stated that the amygdala is involved in a variety of responses in the autonomic, endocrine, and somatomotor spheres related to motivation and emotions. Correlations between anatomical and physiological observations are possible only to some extent. In agreement with anatomical findings sensory information of different kinds has been shown to reach the amygdala. However, precise data on correlations between receiving areas (nuclei) of the amygdala and types of sensory input is scanty. In physiological studies potentials with relatively short latency have been recorded from the basolateral group of the amygdaloid nucleus after stimulation of various kinds of sensory afferents—somatosensory, auditory, and visual (Machne and Segundo, 1956; Wendt and Albe-Fessard, 1962; and others). The pathways concerned are not known. However, the recently demonstrated afferents to the amygdala from the reticular formation and various other nuclei in the brainstem may be candidates (even though most of these appear to end in the central nucleus). Responses to visual stimuli may reach the amygdala via the inferior temporal cortex (cf. Chap. 12). Responses to olfactory stimuli or electrical stimulation of the olfactory bulb have repeatedly been found in the amygdala in animals (see Gloor, 1960). However, responses with short latencies do not occur only in regions that are concluded to belong to the cortical nucleus, but in other amygdaloid nuclei as well. Cain and Bindra (1972) reported responses in the rat in the basal and medial as well as in the cortical nucleus. Thus, there is no complete agreement with the results of anatomical studies. [14]

Physiological recordings and anatomical tracing of fiber connections show that within the amygdaloid complex there must occur an extensive integration of impulses from many sources. In view of this and judging from the distribution of the efferents from the amygdala, it is no wonder that after electrical *stimulation* of the amygdaloid nucleus in unanesthetized animals a variety of *responses in the motor and vegetative spheres* have been recorded: arrest of spontaneous ongoing movements ("arrest reaction"); inhibition or facilitation of spinal reflexes and cortically evoked movements; contraversive turning movements of head and eyes; complex rhythmic movements related to eating, such as swallowing, licking, and chewing; changes in respiration and cardiovascular functions; inhibition or activation of gastric motility and secretion; micturition and defecation; uterine contractions; pupillary dilatation; piloerection, etc. Many of these responses appear to be merely components in more complex *behavioral reactions* which can be elicited in unanaesthetized animals by stimulation in or near the amygdaloid nucleus. Kaada (1951) and Kaada, Andersen, and Jansen, Jr. (1954), as well as others, in cats, noted the appearance of searching movements to the contralateral side with facial expressions of attention with some bewilderment and anxiety, sometimes fear or anger. This has been called the "attention response" or the "orienting response." It is accompanied by EEG desynchronization (Ursin and Kaada, 1960). In continued studies it has been shown that the attention response always precedes certain behavior patterns indicating emotional changes which may be obtained on stimulation of the amygdala. These reactions may be of two kinds, generally referred to as fear and anger, respectively. In the "fear response" the searching movements become more rapid, there are anxious glancing movements, the animal becomes restless and finally runs away and hides. In the "anger response" there is growling and hissing and piloerection. The rage or anger seems to be directed toward something imaginary. The former response may perhaps more adequately be referred to as the "flight response," the latter as the "defense reaction" (see Kaada, 1967, 1972).

The question whether different neural structures are related to the various kinds of responses has been studied in cats. Kaada, Andersen, and Jansen, Jr. (1954) found that the autonomic responses and chewing, sniffing, licking, as well as tonic and clonic movements were elicited mainly on stimulation of the corticomedial group of nuclei, which receives olfactory fibers. The affective responses and the attention response occurred mainly on stimulation of the basolateral group. In continued studies Ursin and Kaada (1960) and Ursin (1965)

---

[14] According to Halgren, Rausch, and Crandall (1977), in man, units responding to odors do not occur in the basolateral amygdaloid nuclei. It appears likely a priori, that the relation of the amygdala to smell is scarcely very close, at least not in man. This is suggested by the modest size of the main target within the amygdala of the olfactory bulb fibers, the cortical nucleus, in relation to the entire amygdaloid complex, particularly in man (see Fig. 10-5). Further, it is remarkable that an amygdaloid complex and even a cortical nucleus are present in anosmatic animals (see Addison, 1915).

outlined topographically separate zones as being more particularly related to either fear (flight) or anger (defense) responses. Other authors found little evidence of a separation of this kind. However, as stated by Kaada (1972, p. 259) with reference to the amygdala in general, "A denial of the existence of any functional representation within the amygdaloid complex finds no support in the wealth of experimental data available." It may be added that present-day knowledge of the architecture and fiber connections of the amygdala is not compatible with such a denial either. The *neuronal mechanisms* involved in the responses to stimulation of the amygdaloid complex are rather imperfectly known, even though some studies have been undertaken. For example, single units in the ventromedial nucleus of the hypothalamus may be monosynaptically excited from the amygdaloid nucleus (see Renaud, 1976), but other responses may occur as well. For some data see articles in *The Neurobiology of the Amygdala* (1972) dealing with the neurophysiology of this nuclear complex. See also Isaacson (1974).

The results of isolated *destructions of the amygdaloid nucleus* or parts of it are even less clear-cut than the observations in stimulation experiments (see Gloor, 1960; Kaada, 1972). Most often an increased tameness has been reported to follow destruction of the amygdala (which has to be bilateral). Ursin (1965) found a specific reduction of the flight response without loss of defense behavior following lesions restricted to the "flight zones." However, after lesions of the amygdaloid complex, aggressive behavior has been observed as well. According to Kaada (1972), this is probably caused by removal of areas that exert inhibitory influences on aggression.

In addition to the behavioral effects of stimulation or destruction of different parts of the amygdala referred to above, amygdaloid influences on *secretion of hormones*, mediated by an action on the hypothalamus (see Chap. 11), have been investigated. Changes in secretion of gonadotrophic hormone, adrenocorticotrophin, thyreotrophin, and vasopressin, and possibly others, have been observed. This is indeed what might be expected, since hormonal changes are obviously involved in many facets of the behavioral reactions elicitable from the amygdala. Some studies of the *transmitters* active in the amygdaloid nuclei and of their relation to observed behavioral changes have been undertaken. These studies are in part closely related to histochemical investigations of cholinergic and catecholaminergic neurons and fibers. As mentioned above, electrophysiological studies have been undertaken of single units and synaptic actions. A discussion of these aspects of amygdaloid function is beyond the scope of the present text. For some relevant articles see *The Neurobiology of the Amygdala* (1972). *Observations in man* which may elucidate the function of the amygdala will be briefly considered later in connection with the symptoms of damage to the temporal lobe.

It appears from the observations mentioned above that the *amygdala is functionally related to emotional experiences and behavioral reactions. However, it is not the only region concerned in these functions.* There is convincing evidence that other regions are also involved, especially the hypothalamus (see Chap. 11). Lesions of the septum (see below) have been reported to result in aggressive behavior; bilateral ablation of area 24 is said to produce increased tameness and reduced aggressiveness; and the mesencephalic gray matter may on stimulation give rage reactions (see Gloor, 1960; Delgado, 1964; Kaada, 1967, 1972, for references). For a review of the brain regions related to aggressive behavior see Clemente and Chase (1973). All these regions of the brain are more or less directly interconnected by fibers, making it understandable that they are functionally related. It has been surmised that the amygdaloid nucleus is more of an integrative center for these functions than the others (see, for example, Fernandez de Molina and Hunsperger, 1962). A review of the effects of amygdalectomy on social affective behavior in monkeys has been given by Kling (1972).

When evaluating the experimental results described above, it should be recalled that the behavioral reactions elicited from the amygdala are, in fact, extremely complex processes. Like other behavioral actions they develop over a certain time and consist of sequences of reactions. As emphasized by Delgado (1964, p. 436) in a review: "Electrical stimulation of the brain is a rather crude procedure, and to explain the finesse, co-ordination and drive of many of the evoked reactions, it is necessary to assume the activation of physiological mechanisms." Many behavioral reactions appearing on stimulation of the amygdala and other parts of the brain show considerable flexibility in response to changes in the external circumstances (just as is the case with corresponding reactions in normal animals). This has become especially clear in studies where electrically induced behavioral reactions are produced by implanted electrodes set in action by remote control. The animal can then be watched when free to move under fairly normal circumstances, as discussed by Delgado (1964). This author points to the need for precise criteria for definition and classification of behavior and stresses the importance of considering individual variations among animals, anatomically as well as physiologically. A serious difficulty is further presented by the restricted possibilities of performing stimulation of discrete anatomical units without spread of current to neighboring structures or stimulation of passing fibers. Finally, a precise anatomical identification of the stimulus site is necessary.

The considerations above apply also to the studies of behavioral changes occurring on stimulation of the amygdaloid nuclei. When different reactions (flight or aggressive behavior) can be elicited preferentially from different regions, this does not necessarily mean that there are separate neural substrates. Much work has been devoted especially to identifying structures related to aggressive behavior.[15] Many theories have been set forth. When considered from the anatomical point of view, theories assuming complex actions to be mediated by extensive collaboration of many parts of the brain appear, a priori, most plausible.[16] In fact, all the reactions that can be elicited from the amygdaloid nucleus can be obtained also from other regions of the brain, although not all of them from the same region. Of these regions the hypothalamus will be considered in Chapter 11. Some other regions of relevance will be dealt with below.

The septal nuclei. The *septal region* develops from the telencephalon (septum telencephali). It consists of a sheet of gray matter, traversed by many fibers, arranged in the vertical plane of the medial wall of the anterior horn of the lateral ventricle, chiefly in front of the anterior commissure. It shows some variations in different mammals. The septal nuclei are well developed in lower mammals, and

---

[15] For some data see *Aggression and Defense* (1967) and Clemente and Chase (1973).

[16] Delgado (1964) has suggested a working hypothesis of fragmental organization of behavior: "Behavioral categories are formed by a series of fragments which have anatomical and functional bases inside the cerebral organization. Single behavioral fragments form part of different behavioral categories and may have a different functional meaning" (loc. cit., p. 437).

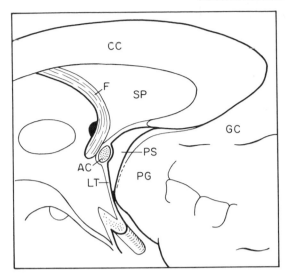

FIG. 10-7   Drawing of the medial aspect of the human cerebrum, showing the position of the septal region. Its upper part forms the nerve-cell-free septum pellucidum (*SP*). Its lower part, the precommissural septum (*PS*), is found in the medial wall of the anterior horn of the lateral ventricle and bulges slightly in the medial direction in front of the lamina terminalis (*LT*). *AC:* anterior commissure; *CC:* corpus callosum; *F:* fornix; *GC:* gyrus cinguli; *PG:* parolfactory area. Slightly altered from Nauta and Haymaker (1969).

according to Stephan and Andy (1962) there is a good correlation between the relative size of these nuclei and the hippocampus. There has been some uncertainty as to the homologue of the septal nuclei in man. Most authors are of the opinion that in man (Fig. 10-7) the upper part of the septal region forms the thin septum pellucidum, free of nerve cells. The lower part (the "precommissural" septum) is usually subdivided in a lateral and a medial septal nucleus. Dorsal to the septum is the fornix. Readers interested in more detailed and precise information on the septum should consult the exhaustive account of Stephan (1975).

> Closely topographically related to the septal nuclei are several minor cell groups or fiber bundles such as the diagonal band of Broca and its nucleus, the septohippocampal nucleus, the septofimbrial nucleus, and the nucleus accumbens. (For a brief survey, see Nauta and Haymaker, 1969; Stephan, 1975). In a detailed study of the human septal nuclei Andy and Stephan (1968) distinguished several small cell groups, including some of those listed above, as part of the septal region (for example, the bed nucleus of the stria terminalis and the bed nucleus of the anterior commissure).

In recent years much interest has been devoted to the septal nuclei, since they seem to influence various behavioral patterns and autonomic functions. (For an exhaustive account see *The Septal Nuclei,* 1976.) Studies of their fiber connections are difficult, and conflicting data can be found in the literature. Their main connections appear to be established with the hippocampus, and are reciprocal (see Figs. 10-8 and 10-11). It appears that the lateral septal nucleus is the main region receiving afferents and that the medial septal nucleus gives rise to most of the septal efferents. Some data on the septal connections will be mentioned.[17]

Among the *afferent connections* of the septal nuclei the projections from the *hippocampal formation* are well established (see also below). They have been found to be topographically organized (Raisman, 1966a; Raisman, Cowan, and Powell, 1966; Segal and Landis, 1974b; Siegel, Edinger, and Ohgami, 1974; Siegel, Ohgami, and Edinger, 1975; Poletti and Creswell, 1977; and others). From autoradiographic studies Swanson and Cowan (1976, 1977) concluded that all afferents from the hippocampus proper end in the lateral septal nucleus. The fibers pass in the precommissural component of the fornix. They appear to exert an excitatory, probably monosynaptic, action on its cells (DeFrance, Kitai, and Shimono, 1973). In addition, the septum has been found by many authors to receive fibers from the *subiculum* (Chronister, Sikes, and White, 1976; Meibach and Siegel, 1977a; Swanson and Cowan, 1977; Harkmark, Köhler, and Srebro, 1977; and others). Other septal afferents come from areas that do not belong to the hippocampal formation.

[17] In the following account no attempt is made to describe particular features in the septal connections, for example with regard to connections of minor subdivisions, or to discuss diverging results. Mainly recent studies will be mentioned. Reference to others will be found in these and in the articles of Raisman (1966a), Swanson and Cowan (1976), and Swanson (1978). See also *Functions of the Septo-Hippocampal System* (1978).

FIG. 10-8   A simplified diagram of the main afferent and efferent connections of the septal nuclei. Some projections are not included. See text for particulars. *M:* mammillary body. For other abbreviations see legend to Fig. 10-6.

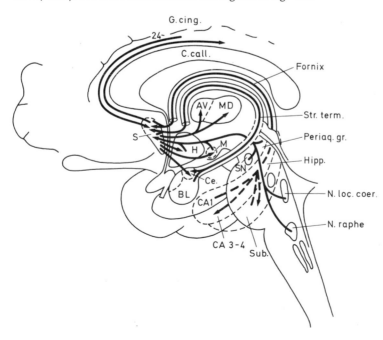

Thus some ascending fibers reach the septal nuclei from the *periaqueductal gray,* the *nucleus locus coeruleus,* the *raphe nuclei* (for references, see Chap. 6), and the *ventral tegmental areas.* The last three regions give off catecholaminergic fibers. (Their neurons appear to be noradrenergic, serotoninergic, and dopaminergic, respectively.) Afferents from the *fastigial nucleus* have been reported (Paul, Heath, and Ellison, 1973; Harper and Heath, 1973). Fibers from the amygdala seem to terminate only in the bed nucleus of the stria terminalis (which some consider part of the septum). A projection from the *hypothalamus* has been described (see, for example, Segal and Landis, 1974b; Conrad and Pfaff, 1976b). In an HRP study in the rat, Srebro, Köhler, and Harkmark (personal communication), in addition to confirming the projection from the lateral hypothalamus and the hippocampus (CA 1), found labeling of cells in the *mammillary body,* the *substantia nigra,* and the *ventral tegmental area.* The only part of the cerebral cortex for which there is some evidence of a projection to the septum is the *gyrus cinguli.* Autoradiographically, Powell and Robinson (1975) traced cinguloseptal fibers in the monkey; silver impregnation studies in the cat (Powell, Akagi, and Hatton, 1974) were less conclusive.[18]

To a large extent the *efferent connections* of the septal nuclei are reciprocal to the afferents. The most important one appears to be the projection back to the *hippocampus* (Daitz and Powell, 1954; Cragg and Hamlyn, 1957; Cragg, 1965; DeVito and White, 1966; Raisman, Cowan, and Powell, 1965; Raisman, 1966a; Mosko, Lynch, and Cotman, 1973; Mellgren and Srebro, 1973; Segal and Landis, 1974b; Meibach and Siegel, 1977a). Most reliable information on the origin of the septohippocampal fibers can be expected from studies of the septum after HRP injections in the hippocampus (see Segal and Landis, 1974a; Meibach and Siegel, 1977a). It appears that the fibers arise mainly in the medial septal nucleus (and the diagonal band of Broca). Most authors have concluded that the septohippocampal projection supplies all fields of the hippocampus (while CA 1, together with the subiculum, is the main origin of the fibers from the hippocampal region to the septum). The septohippocampal fibers appear to be distributed rather diffusely throughout the length of the hippocampus. They are cholinergic (Lewis and Shute, 1967; Mellgren and Srebro, 1973; and others). Acetylcholinesterase activity has been found in their terminals. This seems to be the case also in man (Mellgren, Harkmark, and Srebro, 1977).

Other efferents from the septal nuclei appear to pass to the *hypothalamus* according to anatomical (Meibach and Siegel, 1977a) and physiological (Culberson and Bach, 1973) findings. Fibers have further been traced to the *thalamus.* This projection is said to be more marked than that to the hypothalamus. It has been described by various authors as supplying several thalamic nuclei, among them the anteromedial, anteroventral, and dorsomedial (for recent accounts, see Powell, 1973; Meibach and Siegel, 1977a). There are fibers to the medial nucleus of the *mammillary body* (see Simmons and Powell, 1972; Swanson and Cowan, 1976). A projection to the *cingulate gyrus* has been advocated in the monkey on the basis of retrograde cellular changes (in the lateral septal nucleus) after lesions of the gyrus by Kemper, Wright, and Locke (1972). Finally, fibers passing to the *amygdala* (particularly the central and medial nuclei) appear to be present (see Swanson and Cowan, 1976), and fibers have been traced to the *habenula* (medial nucleus) after lesions of the septal region (see, for example, Raisman, 1966a) and in HRP studies (Herkenham and Nauta, 1977).

[18] The architecture of the lateral and medial septal nuclei differs both in Nissl-stained and in Golgi-impregnated sections (see Raisman, 1969a). Many details of their finer organization remain to be clarified. In experimental electron microscopic studies in the rat, Raisman (1969a) found differences in synaptic arrangements and terminals between fibers coming from the hippocampus and fibers assumed to arise in the hypothalamus (degenerating as a consequence of transection of the rostral part of the medial forebrain bundle). The hippocampal afferents terminate mainly with axodendritic boutons. For particulars the original article should be consulted. See also Chapter 1 and Fig. 1-23 concerning reinnervation of the septum.

It appears from the description above that the septal nuclei are first and foremost mutually interconnected with the hippocampus, in a topographical pattern, while to a lesser extent they are able to act on and to be influenced from other regions, such as the amygdala and the hypothalamus (see Fig. 10-8). The complexity in their organization is far from being understood. According to Golgi studies, dendrites of cells in one group usually extend into neighboring areas and may thus be acted upon by afferents ending outside the particular nucleus. Conclusions about the functions of the septal nuclei should be made with great caution. Commenting upon our insufficient knowledge of the morphological complexity of the septal region, Swanson and Cowan (1976) appropriately concluded their review as follows: "Functionally, it [the septal region] is likely to prove at least as complex, and until we know more about the types of information it receives from such structures as the amygdala, hippocampus, and brainstem, we shall continue to be frustrated in our efforts to assign a function to the system as a whole or to any of its anatomical subdivisions."

Several functions have been found to be influenced from the septum (by stimulations or lesions). The name "septal syndrome" has been coined as a common denominator for the changes seen after destructions of the septal nuclei. In general, this syndrome is said to be characterized by behavioral overreaction to most environmental stimuli. Behavioral changes occur in many spheres, including sexual and reproductive behavior, feeding and drinking, and rage reactions. Particular interest has been devoted to the reduction of aggressive behavior following lesions of the septum (see below). As is the case for behavior changes related to the amygdala, analyses of hormones (corticotrophins and others) have been performed and hormonal changes have been found to accompany the behavioral changes elicitable from the septum. Several single-unit studies of septal neurons have likewise been undertaken. See *The Septal Nuclei* (1976) and *Functions of the Septo-Hippocampal System* (1978) for recent accounts on the septal nuclei and their functions.

In view of the many interconnections between the septal nuclei, the hippocampus, the amygdala, and the hypothalamus it is no wonder that lesions of any one of them give rise to a large number of rather similar symptoms, even if there are functional differences between these brain regions. Another illustration of their close relationship is the fact that, as stated by Grossman (1976, p. 361) in his review of functions of the septum: "Just about every behavior and/or psychological function which has been investigated to date has been shown to be affected in some way by septal lesions, . . . ".

The cingulate gyrus. The mammillary body. The stria medullaris. Many regions of the brain have intimate fiber connections with—and presumably functional relations to—the septum and the amygdala (described above) and the hippocampus (see below). Even though these regions and their connections are referred to when other relevant regions of the brain are treated, it might be appropriate to consider each of them separately. They have all received much attention in connection with discussions of the "limbic system."

The *cingulate gyrus* has been considered a station in the so-called "circuit of Papez" (see below). The gyrus is situated above the corpus callosum (Figs. 10-4

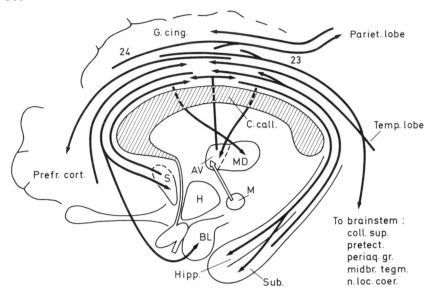

FIG. 10-9 A simplified diagram of some main connections of the cingulate gyrus (see also Figs. 10-6A, 10-8, 12-4, 12-9). The mammillothalamic bundle is shown (white arrow). *M:* mammillary body. For other abbreviations see legend to Fig. 10-6.

and 10-9) and harbors Brodmann's areas 24, 23, and some others (see Fig. 12-2). It is often considered to be an alloneocortical transitional region.

The *cingulate gyrus receives afferents from many sources* (Fig. 10-9). An important input comes from the anterior thalamic nucleus. This nucleus (see Fig. 2-14) is generally subdivided into an anteromedial (AM), an anterodorsal (AD), and an anteroventral (AV) nucleus. In man, the latter is best developed; it occupies the bulk of the entire nucleus and produces the anterior tubercle of the thalamus. The nucleus projects in a topically arranged pattern upon the cingulate gyrus.

Fibers from various subdivisions of the anterior nucleus appear to supply cortical areas 32, 23, and 24 and retrosplenial areas 29 and 30 (Fig. 12-2). The AV appears to project chiefly to the anterior areas of the cingulate gyrus. The fibers leave the thalamus in the inferior thalamic peduncle and reach the cortex via the anterior limb of the internal capsule. This projection has been studied experimentally in various animals by mapping retrograde cellular changes in the thalamus after cortical lesions (Le Gros Clark and Boggon, 1933; Lashley, 1941; Bodian, 1942; Rose and Woolsey, 1948; Powell and Cowan, 1954; and others). Studies of brains from patients subjected to prefrontal leucotomy (see Chap. 12) have been undertaken by Meyer (see Meyer and Beck, 1954) and by Yakolev and his collaborators. The latter have also made experimental studies in monkeys (see, for example, Yakovlev, Locke, Koskoff, and Patton, 1960; Locke, Angevine, and Yakovlev, 1964).

Recent studies indicate that the thalamic projection onto the cingulate gyrus is more complex than was previously assumed. Studies using silver impregnation techniques for demonstration of degenerating fibers (Domesick, 1969; Niimi, 1978) and studies using the HRP method (Niimi, Niimi, and Okada, 1978) have shown, for example, that the AV (in the rat and cat) projects to all areas of the cingulate gyrus and to the presubiculum. Most of these areas receive afferents from the other subdivisions of the anterior thalamic nucleus as well (and to some extent from the dorsomedial thalamic nucleus).[19] However, even though different thalamic nuclei may project to the same cortical field of the cingulate gyrus, fibers from the various thalamic subdivisions differ with regard to their laminar distribution of terminals within the cortical field. The projections appear to be topographically organized (for particulars, see the articles cited above).

By way of the anterior nucleus of the thalamus the cingulate gyrus may be influenced indirectly from other regions. A main, afferent route to the anterior thalamic nucleus, particularly the AV, is the macroscopically visible *mammillothalamic bundle* (white arrows in Figs. 10-9 and 10-11), arising in the mammillary body. We shall return to this and its connections below. Here it is sufficient to mention that, according to recent studies (see below), afferents to the mammillary body reaching it in the fornix all derive from the subiculum (see Fig. 10-11). The cingulate gyrus may thus be influenced from the hippocampal formation. This *indirect hippocampocingulate pathway* (via the thalamus) is supplemented by a *direct* one by fibers that pass from the hippocampus and the subiculum to the cingulate gyrus (see below and Figs. 10-9 and 10-11). The fibers course in the *cingulum,* a fiber bundle running along the cingulate gyrus. (The bundle also contains shorter fibers connecting anterior and posterior parts of the cingulate gyrus.) It may finally be mentioned that there appears to be a projection to the cingulate gyrus from the lateral *septal nucleus* (Kemper, Wright, and Locke, 1972).

Of particular interest is the rather recent recognition that the gyrus cinguli receives a fair amount of information from *neocortical areas.* As described in Chapter 12, the parietal, the temporal, and the granular "prefrontal" cortex all have been found to send fibers to the cingulate gyrus (see Figs. 10-9, 12-4, and 12-9). To some extent efferent fibers from the cingulate gyrus enable it to act back on the regions from which it receives its afferents.

The *cortex of the cingulate gyrus gives off efferent fibers to several other regions.* All details are not clear, but, summarizing, it may be said that with different anatomical methods, cingulofugal fibers have been found to supply at least the following structures (see Fig. 10-9): the *hippocampal* formation, the *amygdala,* the *septum,* parts of the *thalamus,* and parts of the *neocortex* (the "prefrontal" granular cortex and the association cortex in the parietal lobe). Finally, there appear to be projections descending in the *brainstem* (superior colliculus, pretectal area, periaqueductal gray, midbrain tegmentum).

The efferent connections of the cingulate gyrus to the *hippocampal formation* pass in the posterior direction in the cingulum bundle. They were observed by early neuroanatomists and have been studied with silver impregnation methods in the rat (White, Jr., 1959), the rabbit (Adey, 1951; Cragg and Hamlyn, 1959), the macaque (Adey and Meyer, 1952a), and the squirrel monkey (DeVito and White, Jr., 1966). Some of the fibers are described as ending in the *hippocampus proper.* In the rat, most of them end in the *subiculum* and *parasubiculum,* from which there are connections to the hippocampus. Electrophysiological studies in the cat and monkey have given concordant results (White, Jr., Nelson, and Foltz, 1960). The efferents to the amygdala, according to Pandya, van Hoesen, and Domesick (1973), come from the anterior part of the cingulate gyrus (area 24) and end chiefly in the basal amygdaloid nucleus (Fig. 10-9). The efferents to the

[19] The cortical projections and other connections of the MD will be considered in Chap. 12.

thalamus do not appear to be clearly reciprocal to the thalamocingulate projection mentioned above. As determined experimentally by Domesick (1969, rat) and Siegel, Troiano, and Royce (1973, cat), the posterior granular region projects to the AV of the thalamus, the anterior region chiefly to the MD (see Fig. 10-9). According to the same authors, the anterior as well as the posterior parts of the cingulate gyrus project to structures in the brainstem, among them the *superior colliculus,* the *pretectal area,* the *periaqueductal gray,* the *midbrain tegmentum,* the *nucleus locus coeruleus,* and parts of the *pontine gray,* while fibers to the hypothalamus do not appear to exist. Fibers to the *septal nuclei* have been doubted but were described in the monkey by Powell and Robinson (1975). Finally, it should be noted, as described in more detail in Chapter 12, that in HRP studies in the monkey the cingulate cortex has been found to give rise to efferents to neocortical regions, such as the *parietal lobe* (Mesulam, Van Hoesen, Pandya, and Geschwind, 1977) and the *"prefrontal"* granular cortex (Jacobson and Trojanowski, 1977a).

It has been common to consider the cingulate gyrus a part of the "limbic system." The recent demonstrations of the presence of two-way connections between the cingulate gyrus and neocortical regions and of its efferents to several brainstem nuclei indicate that the cingulate gyrus has important functions far beyond being a link in the so-called "circuit of Papez" (see below).

After stimulation of the cingulate gyrus, behavioral changes of many kinds have been observed, and, as mentioned above, many of them resemble the effects that can be elicited from the amygdala and the septum. Particular interest has been devoted to the influence of the cingulate cortex on kinds of behavior referred to as aggressive. Bilateral ablation of the anterior cingulate cortex has been reported by several authors to result in an increased tameness and "social indifference" (Smith, 1944b; Ward, 1948; Glees, Cole, Whitty, and Cairns, 1950; and others), while others found only small changes in social behavior (see Myers, Swett, and Miller, 1973). Some authors observed a temporary increase in aggressiveness (Mirsky, Rosvold, and Pribram, 1957). These experimental studies have prompted surgical procedures directed at the cingulum as a treatment for certain psychic disorders. We shall return briefly to this subject in Chapter 12, in connection with considerations of "psychosurgery."

Reference was made above to the mammillothalamic bundle (white arrow in Figs. 10-9 and 10-11), made up of fibers from the mammillary body to the anterior thalamic nucleus, particularly the anteroventral nucleus (AV). The *mammillary body,* the most posterior part of the hypothalamus (see Fig. 11-10), is usually subdivided into a lateral (large-celled) and a medial (small-celled) nucleus. In man the latter is by far the largest (in addition, a small intermediate nucleus is present). The two main mammillary nuclei differ with regard to their fiber connections.

The most important contingent of *afferent fibers to the mammillary body* is made up of fibers from the *hippocampal formation,* descending in the postcommissural fornix (Fig. 10-11). It was formerly generally held that these fibers came from the hippocampus proper. However, recent studies have made it clear that they all come from the *subiculum* (see below and Fig. 10-11). The fibers end, at least chiefly, in the medial mammillary nucleus (Simpson, 1952b, in the monkey; see also Swanson and Cowan, 1977, in the rat), from which the bulk of fibers in the mallillothalamic bundle arises. Another important afferent fiber contingent comes from the *septum* (see above), chiefly from the medial septal nucleus. It ends in the medial mammillary nucleus (see Swanson and Cowan, 1976). Still other, *ascending afferents* reach the mammillary body in the so-called *mammillary peduncle,* which arises from some *small cell groups in the mesencephalon,* the ventral and dorsal tegmental nucleus. (The latter is situated closely ventral to the aqueduct at the level of the inferior colliculus.) This has been concluded from several studies (Guillery, 1956; Cowan, Guillery, and Powell, 1964; and others). In addition to the afferents mentioned above, the mammillary nuclei receive some others, apparently quantitatively less important.

The size of the (macroscopically visible) mammillothalamic bundle bears witness that this must be an important route for *efferent fibers from the mammillary body.* These fibers must influence the *anterior thalamic nucleus* which in turn, as described above, projects to the cingulate gyrus in a specifically arranged pattern. The mammillothalamic projection likewise appears to be very specifically organized according to a topographical pattern.

This appears from silver impregnation studies and from studies of retrograde cellular changes (see Powell and Cowan, 1954b; Fry et al., 1963; Fry and Cowan, 1972). A recent autoradiographic study (Cruce, 1975) is largely in agreement. It appears that both the medial and the lateral mammillary nuclei project to the anterior thalamic nucleus, but the input to the AV appears to come chiefly from the medial mammillary nucleus (innervated from the subiculum; see above). In combined autoradiographic and electron microscopic studies, Dekker and Kuypers (1976) have shown the presence of differences in synaptic relations between three kinds of afferents to the AV in the rat. The afferents from the mammillary body appear to end with large terminals on proximal dendrites.

Efferents from the mammillary body appear further to pass to the septal nuclei (see above, and Fig. 10-8).[20] Numerous efferents from the mammillary body form a *descending* projection to the brainstem. These fibers are usually referred to as the *mammillotegmental tract.* The fibers originate, together with those of the mammillothalamic tract, as branches of common axons and have been traced in the caudal direction to the dorsal tegmental nucleus of Gudden. In an autoradiographic study (Cruce, 1977), the descending fibers were found to end in the *ventral* and the *dorsal tegmental nucleus.* Some fibers pass to the *pontine gray,* and the *nucleus reticularis tegmenti pontis.*

Knowledge of the *functional role of the mammillary bodies* is still fragmentary. Attempts to relate them to memory appear to rest on loose ground (see below). According to Rosenstock, Field, and Greene (1977), lesions of the mammillary body result in defects in spatial discrimination. The mammillary bodies obviously act in collaboration with many other parts of the brain. Two conclusions may presumably be made on the basis of the connections of the mammillary nuclei. In the first place, as seen from the discussion above, the mammillary body, by way of the mammillotegmental tract, may influence the activity in the upper brainstem. This descending tract and the largely reciprocal ascending mammillary peduncle may be part of a so-called "limbic system–midbrain circuit." The fact that the efferent ascending and descending fibers appear to be branches of the same axons (see Fry and Cowan, 1972) is a morphological indication of a correlated action of the mammillary body in the rostral and caudal direction (compare

---

[20] Some efferents from the vicinity of the mammillary body have been traced back through the fornix (Fig. 10-11) to the hippocampus and the subiculum, and some efferents have been described as ending in the subthalamic nucleus.

cells in the reticular formation, Chap. 6, Fig. 6-6). Another chief function of the mammillary body must be mediation of subicular influences on the cingulate gyrus (via the anterior nucleus of the thalamus).

The mammillary body forms a well-defined relay in the so-called *"circuit of Papez,"* suggested by Papez (1937) to be an essential part of the structural basis of emotions. The circuit was pictured as consisting of the following links: fibers from the hippocampus to the mammillary body–mammillothalamic tract–fibers from the AV of the thalamus to the cingulate gyrus–cingulohippocampal fibers.

McLean (1975, p. 181), referring to his considerations on the circuit of Papez in a paper from 1949, described the view on the circuit as follows: "It was suggested that the hippocampal formation combines information of internal and external origin into affective feelings that find further elaboration and expression through connections with the amygdala, septum, striatum, and hypothalamus, as well as through reentry paths to the limbic lobe via the mammillothalamic tract and thalamic projections to the cingulate gyrus (the so-called Papez circuit)."

The demonstration in recent years of numerous hitherto unknown connections necessitates thorough revisions of our thinking about areas of the brain which may be related to emotions. There is no sound biological basis for selecting a few brain regions as being involved in these complex functions, and today the theory of Papez is of historical interest only.

The *stria medullaris* and the *habenula* have not been considered in the preceding text. A brief account of these structures is therefore appropriate. The *stria medullaris* is a fiber bundle passing in the anteroposterior direction along either side of the midline on the dorsomedial aspect of the thalamus. Posteriorly, this stria medullaris (thalami) broadens and ends in a swelling, the *habenula*. Within this cell mass a lateral and a medial nucleus can be distinguished. The two habenulae are interconnected by the *habenular commissure*. From the habenula a conspicuous fiber bundle courses caudoventrally and can be followed to the *interpeduncular nucleus* as the *habenulointerpeduncular tract* or the *fasciculus retroflexus*. Most fibers in the stria medullaris run anteroposteriorly and end in the habenular nuclei.

The origin of these fibers is not clear in all details. Herkenham and Nauta (1977; see this article for references), after injections of HRP in the habenula in the rat, found numerous retrogradely labeled cells in the *septum* and the *lateral hypothalamus* and smaller numbers in other nuclei, such as the nucleus of the diagonal band and the lateral preoptic area. They assumed that the stria terminalis also probably carries fibers, described by others, that are axons of cells located in olfactory structures and that do not end in the habenula, but bypass it.

Other afferents to the habenula ascend from the brainstem. Herkenham and Nauta (1977) in their HRP study found many labeled cells in the *interpeduncular nucleus* and also some in the *mesencephalic raphe*, the *periaqueductal gray*, and the *ventral tegmental area*. There are some differences between the medial and the lateral habenular nuclei with regard to the sources of their afferents.

Efferents from the habenula have been traced chiefly in the caudal direction. Among these are some fibers of the stria medullaris that only pass through the habenula, in part after crossing in the commissure. The descending fibers pass in the fasciculus retroflexus and end largely in the *interpeduncular nucleus*. Other fibers have been traced to the *periaqueductal gray* and to the *mesencephalic reticular formation*. To a large extent these connections are thus reciprocal to the ascending connections to the habenula.

Some efferent fibers from the habenular nuclei have been found to course anteriorly in the stria medullaris. It appears that the contribution of the medial and lateral nuclei of the habenula to its efferent projections are different.

The functional role of the stria terminalis–habenula "system" is not clear. Judging from its fiber connections, however, it must serve to interconnect structures in the rostral brainstem with telencephalic structures, such as the septum and the hypothalamus. It appears to belong to what Nauta has called the "limbic system–midbrain circuit."

The hippocampus (sometimes called Ammon's horn) got its name from its resemblance to a sea horse. In man it extends with its main part along the floor of the temporal horn of the lateral ventricle and can be considered a gyrus of the allocortex, folded inward and covering the hippocampal fissure. Accordingly, the surface facing the ventricle is the deepest layer, and consists of myelinated fibers which collect on its surface as the *alveus,* covered by ependyma. Most of these fibers are efferent and unite to form the *fornix.*[21] While the delimitation of the hippocampus as a gross anatomical structure is fairly clear, its relations to neighboring regions is less obvious. It is common to speak of a "hippocampal formation" or "hippocampal region," including in addition to the hippocampus itself, the dentate gyrus and the subiculum (see Fig. 10-10). The *dentate gyrus* accompanies the hippocampus as a narrow band of cortex (for a description see textbooks on neuroanatomy). The cortical areas medial to the dentate gyrus are collectively referred to as the *subiculum* (more specifically, a distinction may be made between prosubiculum, subiculum, presubiculum, and parasubiculum; see Fig. 10-10). The subiculum occupies part of the hippocampal gyrus (now usually referred to as the parahippocampal gyrus), and is contiguous with the area entorhinalis (Brodmann's area 28; see Figs. 10-3, 10-10, and 12-2) belonging to the paleopallium. As mentioned above, the entorhinal area makes up the large posterior part of the piriform cortex. (The entorhinal area is included in the "hippocampal region.") The two fornices are dorsally connected by fibers crossing the midline underneath the corpus callosum. These (largely commissural) fibers are called the *hippocampal commissure or psalterium.*

The general principles of organization of the hippocampal formation are simpler than in the neocortex, and for this reason the hippocampal formation has been considered a favorable site for studying general anatomical as well as physiological features of cortical gray matter and for anatomophysiological correlations. A great many studies performed in recent years have made it clear that the organization of the hippocampal formation is after all rather complex.[22] There are architectonic differences between minor parts. Experimental anatomical mapping of afferents has shown that they are arranged in a very specific manner, fibers from one

[21] As is apparent from the sketch of the development of the brain given in the beginning of this chapter, the hippocampus originally extended in an arch to a region in front of the interventricular foramen. Its most anterior end in man is generally said to be represented by the subcallosal gyrus, from which rudimentary gray masses, the induseum griseum, continue dorsally over the corpus allosum.

[22] The literature on the hippocampal formation and the hippocampus is overwhelming. For recent comprehensive surveys see, for example, Haug (1974); Stephan (1975); *The Hippocampus,* Vols. 1 and 2 (1975); *Functions of the Septo-Hippocampal System* (1978); see also *The Septal Nuclei* (1976); O'Keefe and Nadel (1978).

source ending in restricted laminae within particular architectonic zones. These differences on many points are reflected in histochemical differences. By correlating light and electron microscopic and electrophysiological studies, it has been possible in some instances to determine whether a particular afferent fiber system has excitatory or inhibitory functions. Most of these investigations have been made in animals, especially in the rat. In the following discussion our main concern will be with the hippocampus proper; the other parts of hippocampal formation, except the entorhinal area, will not be considered separately.[23]

In *man* the hippocampus is by far the largest part of the hippocampal formation. The principles of structural organization of the hippocampus appear to be largely the same in most mammals. It is possible here to give only a sketchy outline of a very complicated pattern of organization. The principal data will be discussed with reference to Fig. 10-10, which is an extremely simplified diagram.

The most impressive cellular elements of the hippocampus are the large *pyramidal cells*. They make up the *pyramidal layer* (stratum pyramidale) and have their base directed toward the ventricular surface of the hippocampus with the alveus. Basal dendrites extend in this direction and belong to the *stratum oriens*. From the other end of the pyramids, slender apical dendrites extend inward. On account of their regular arrangement, the layer harboring the apical dendrites is called the *stratum radiatum*. Distally the apical dendrites branch repeatedly and spread out in the innermost layer, the *stratum lacunosum-moleculare*. The axons of the pyramidal cells give off recurrent collaterals, many of which ascend to and pervade the stratum radiatum. Other collaterals are given off to the stratum oriens. While the regular arrangement and shape of the pyramidal cells determine the overall pattern of the hippocampus, it should be recalled that these cells are not the only ones present. In all layers there are cells of other types with short axons which take various courses. Among these are the so-called *basket cells*, found in the stratum oriens. They give off axons which ramify between the pyramidal cells, and appear to be the only fibers which end on the perikarya of these cells. The basket and other cells are concerned in the intrinsic activity of the hippocampus.[24] The *efferent fibers* leaving the hippocampus in the fornix are axons of pyramidal cells. *Afferent fibers* to the hippocampus enter it from the entorhinal area (medial and lateral perforant path; see Fig. 10-10). In addition, some afferents pass in the fornix (see below). In the *dentate gyrus* (Fig. 10-10), there is a dense layer of *granule cells*. These have ramifying dendrites extending into a so-called *molecular layer* where the perforant paths and other afferents enter. The axons of the granule cells of the dentate gyrus are the so-called *mossy fibers*. These pursue a course along the hippocampal pyramidal cells and make synaptic contacts with the basal parts of their apical dendrites.

[23] A consideration of the hippocampus is well suited to give an idea of the intricate organization of this part of the brain. The hippocampus is selected as an example because so far it is best known. It should be emphasized, however, that there is evidence from recent studies of ample and specifically arranged interconnections between the different parts of the hippocampal formation. Even though the patterns in these connections are known in part only, what is known strongly suggests the presence of functional differences between minor regions. It is beyond the scope of the present text to discuss more than a few salient points concerning these subjects.

[24] In exhaustive studies with the Golgi method, Amaral (1978) distinguished more than 20 different cell types in fields CA 3 and CA 4. Some have not been described previously.

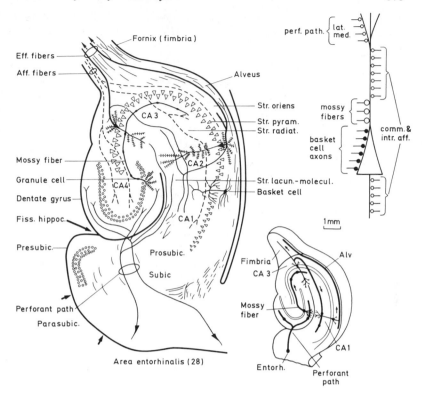

FIG. 10-10   Schematic drawings to illustrate some main features in the organization of the hippocampal formation. The transverse section (medial to the left) shows the sequential arrangement of areas: dentate gyrus, areas CA 4, CA 3, CA 2, and CA 1 of the hippocampus proper, subiculum (with prosubiculum, presubiculum, and parasubiculum) and entorhinal area. The main cellular elements and layers in the dentate gyrus and the hippocampus are shown. Compare text.
Inset, above right, illustrates the specific distribution on the pyramidal cells of afferents from various sources.
Inset, below right (from Andersen, Bliss, and Skrede, 1971), illustrates a main path of transmission of impulses entering from the entorhinal area in fibers of the perforant path. Compare text.

On the basis of architectonic differences, the hippocampus may be subdivided into different fields running along its length. They are usually referred to as fields CA 1, CA 2, CA 3, and CA 4. The latter is the field closest to—and in part fusing with—the dentate gyrus; CA 1 is the field adjacent to the subiculum. The features in the anatomical organization of the hippocampus described above were largely clarified by Cajal (1909–11) and supplemented by extensive Golgi studies by Lorente de Nó (1934). More recent experimental anatomical (in part electron microscopic), histochemical, and electrophysiological studies have brought forward further details and necessitated revisions of older concepts. Before discussing the finer organization of the hippocampus, it is appropriate to give an orientation of its main afferent and efferent connections. For recent detailed accounts of the fiber connections of the hippocampal formation, the reader should consult the articles of Blackstad (1977) and Swanson (1978).

All authors agree that among the *afferent connections to the hippocampus* the projection from the *entorhinal area* (area 28, Figs. 10-4 and 10-10) is, at least quantitatively, the most important one. Some notes on the entorhinal area and its main connections deserve mention.

The *entorhinal area* (area 28) is not homogeneous, either cytoarchitecturally, or with regard to fiber connections. The entorhinal area is considered to be greatly expanded in man. It is common to distinguish a medial and a lateral part (areas 28a and 28b; see for example Blackstad, 1956). Many students have described a finer parcellation, particularly of the lateral part (see, for example, Haug, 1976; Krettek and Price, 1977c). The differences between the two parts are apparent, not least with regard to their afferent connections.

The *lateral part* has been found to be the main recipient of the fibers from the *olfactory bulb* (described above) and of those from the *prepiriform and periamygdaloid cortex* (Cragg, 1961; Powell, Cowan, and Raisman, 1965; Price, 1973; Van Hoesen and Pandya, 1975a; Krettek and Price, 1977c; and others). There are fibers (Fig. 12-4) from *neocortical temporal areas* (see Van Hoesen and Pandya, 1975a)[25] and from the *prefrontal granular cortex* (Van Hoesen, Pandya, and Butters, 1975; Leichnetz and Astruc, 1976). Other afferents to the lateral entorhinal area come from the amygdala (from its basolateral nuclei), as shown by Krettek and Price (1977c; see also de Olmos, 1972; Beckstead, 1978; Veening, 1978a), and from certain parts of the *hippocampal formation* (see Fig. 10-11). Some other sources of afferents have been described as well, for example the *medial septal nucleus, the dorsal raphe nucleus,* and the *nucleus locus coeruleus* (see Segal, 1977; Beckstead, 1978).

The various afferent contingents have been shown to terminate in different layers of the entorhinal cortex. For example, the fibers from the olfactory bulb and the prepiriform cortex project to layers IA and IB, respectively, and those from the lateral amygdaloid nucleus to layer III (see Krettek and Price, 1977c, for further data and particulars).

The medial part of the entorhinal area receives its afferents in part from other regions than does the lateral. Projections have been described from the *presubiculum* (Shipley, 1975; Segal, 1977; Beckstead, 1978; Köhler, Shipley, Srebro, and Harkmark, 1978) and from the *hippocampal field CA 3* (Hjorth-Simonsen, 1971). There are additional afferents, in part corresponding to those reaching the lateral entorhinal area. Subcortical afferents have been described in HRP studies from the *nucleus locus coeruleus* and the *raphe nuclei,* and further from the *septum* and from parts of the *thalamus* (see Segal, 1977).

On the basis of their afferent connections, the two parts of the entorhinal area must be assumed to be functionally different. Of the two, only the lateral part receives afferents mediating olfactory impulses (more or less indirectly). Its numerous other afferents indicate, however, that the lateral part of the entorhinal area is the seat of convergence and integration of impulses of olfactory origin with others.

The functional dissimilarity of lateral and medial parts of the entorhinal area is seen not only in their afferent but also in their *efferent projections*. As mentioned above, the vast majority of the fibers enter the *hippocampus*. They pass in the lateral and medial perforant paths (Fig. 10-10), coming from the lateral and medial part of the entorhinal area, respectively.[26] In addition, the entorhinal area

[25] Between the entorhinal area 28 and the neighboring region, called the perirhinal area (area 35), there is a transitional region, often referred to as the prorhinal cortex. See Van Hoesen and Pandya (1975a) for some data. According to these authors, projections from the neocortex of the temporal lobe pass to area 35. By connections from this to the entorhinal area, the latter may be influenced from the temporal neocortex indirectly as well as directly.

[26] On the basis of Golgi preparations, Cajal and Lorente de Nó and others described, in addition to a perforant path, a so-called *alvear path* from (the medial part of) the entorhinal cortex to the hippocampus. The existence of this path has not been confirmed in experimental studies (see Hjorth-Simonsen, 1972; Van Hoesen and Pandya, 1975b). See, however, Steward (1976).

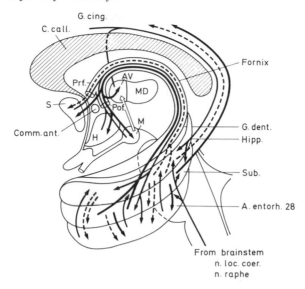

FIG. 10-11 Simplified diagram of the main afferent and efferent connections of the hippocampus proper and of some of the connections of the subiculum. The diagram shows the main fiber components of the fornix. Note that the efferent fibers from the hippocampus proper (broken lines), passing in the precommissural fornix (*prf.*), supply the septal nuclei only. Efferent fibers from the subiculum (thick lines) are distributed to most end stations of the fornix via the postcommissural fornix (*pof.*). Some afferents to the hippocampus are shown (thin lines). For particulars see text. For some abbreviations see legend to Fig. 10-6.

sends fibers to some other regions, apparently chiefly neighboring parts of the allocortex (see, for example, Van Hoesen and Pandya, 1975b), but little is known about this.

Let us now *return to the hippocampus proper* (with the dentate gyrus) and its afferent connections. Its main input comes, as mentioned above, from the entorhinal area (28). Below we shall restrict ourselves to some main points. The *medial* and the *lateral perforant paths* (Fig. 10-10), coming from the medial and lateral parts of the *entorhinal area,* respectively, have been found to supply the entire length of the hippocampus and the dentate gyrus (from the temporal tip to the subsplenial portion). There appears to be a spatial topographical pattern within these connections, and the termination of fibers of the two components has a distinctly different laminar distribution (see Fig. 10-10) as shown by Hjorth-Simonsen (1972) in degeneration studies and autoradiographically by Steward (1976).

It appears from these and from earlier studies (see, for example, Blackstad, 1958; Raisman, Cowan, and Powell, 1965, in the rat; Adey and Meyer, 1952b, in the monkey) that the entorhinal–hippocampal projection is specifically organized. That the entorhinal area markedly influences hippocampal activity was found rather early in physiological studies (see Andersen, Holmqvist, and Voorhoeve, 1966).

Among *other afferents* that influence the activity of the hippocampus, much interest has in recent years been devoted to the *septohippocampal connections.*[27]

[27] For reports from a recent Ciba Symposium, see *Functions of the Septo-Hippocampal System* (1978).

Compared with the afferent input to the hippocampus from the entorhinal area, the fibers from the septal nuclei (Fig. 10-11) appear to be relatively modest in number. As mentioned above in the account on the septum (pp. 663 ff.), it appears that most of the fibers are derived from the *medial* septal nucleus and that they supply all fields of the hippocampus (and the subiculum) in a rather diffuse way. The fibers have been shown to be cholinergic.

Other afferents traced to the hippocampus in various studies have recently been confirmed in HRP studies. Thus there are fibers from the *hypothalamus* (chiefly from cell groups in the vicinity of the mammillary body), as found by Segal and Landis (1974a) and Pasquier and Reinoso-Suarez (1976). According to Segal (1979), the hypothalamus exerts a strong inhibitory influence on the hippocampus. Fibers to the hippocampus from the *anterior thalamic nucleus* have been described (see Swanson, 1978, for references). Much interest has further been devoted to hippocampal afferents from the *raphe nuclei* (dorsal and medial) and the *nucleus locus coeruleus* (Segal and Landis, 1974a; Pickel, Segal, and Bloom, 1974; Moore and Halaris, 1975).

> The connections from the raphe and the nucleus coeruleus have been investigated with the fluorescence method of Falck and Hillarp (see Chap. 1). *Noradrenergic* fibers (coming from the nucleus locus coeruleus) have been traced to the hippocampus and to the dentate gyrus. A *dopaminergic* input to the hippocampal formation, particularly to the entorhinal area, appears to be present. It probably arises in the ventral tegmental area. *Serotonergic* fibers, arising in the (dorsal and median) raphe nuclei, have likewise been traced to the hippocampus. (For some references to relevant studies, see Blackstad, 1977; Swanson, 1978).

The account above shows that the activity of the hippocampus may be influenced by direct fiber connections from various regions. In addition to the major input from the entorhinal area, and a presumably important one from the septum, contributions are made by the hypothalamus (particularly areas close to the mammillary body), from the anterior thalamic nucleus, and from some brainstem nuclei harboring monoaminergic neurons. To a considerable extent the various afferent fiber contingents have their particular sites of termination within the hippocampus and the dentate gyrus. These details, obviously important for a proper analysis, cannot be fully considered in the present sketchy presentation. (Some data on the terminations and synaptology of the afferents will be mentioned below.)

To some extent the hippocampus is able to act back on the regions that influence it. There are great differences, however. The reciprocal connections with the septum are conspicuous, whereas the situation is different with regard to the entorhinal area. It is generally agreed that the *efferent projection from the hippocampus* is made up of the axons of its pyramidal cells. There seems to be general agreement that most efferents pass in the *fornix*. This massive fiber bundle pursues a characteristic forward course beneath the corpus callosum. Immediately in front of the interventricular foramen (Figs. 10-4 and 10-7), the bundle bends downward and splits into minor components reaching different regions. Some of the fornix fibers descend anterior to the anterior commissure (precommissural fornix); others descend posterior to it (postcommissural fornix). In callosal mammals a minor contingent of hippocampal fibers passes on the dorsal surface of the corpus callosum in the dorsal longitudinal striae of Lancisi (dorsal fornix). Some of these fibers break through the corpus callosum to join fibers from the main fornix. The presence of fornix fibers crossing the midline in the hippocampal commissure was mentioned above. In addition to hippocampal efferents, the fornix contains some fibers coursing in the opposite direction.

The fornix contains some 1,200,000 fibers in man (Powell, Guillery, and Cowan, 1957) and some 500,000 in the monkey (Daitz and Powell, 1954). Following lesions of the hippocampus, the course and sites of termination of the fornix fibers were mapped experimentally by several authors with silver impregnation methods in the rat, cat, and monkey (Simpson, 1925b; Guillery, 1956; Nauta, 1956, 1958; Raisman, Cowan, and Powell, 1966; Poletti and Creswell, 1977; and others).

The findings of these authors agree in all esentials. Fibers of the postcommissural fornix were found to be distributed to the *mammillary body* and the *anterior thalamic nucleus,* but some fibers continue caudally and were traced to the rostral part of the *periaqueductal gray matter of the mesencephalon* (see Nauta, 1958) and the *pontine nuclei* (Cragg and Hamlyn, 1959). Some of the precommissural fornix fibers likewise were found to reach the mammillary body and anterior thalamic nucleus, while others were distributed to the *preoptic region,* the *lateral hypothalamus,* and the *nuclei of the septum.*

Until 1975 it was generally accepted that the efferent fibers in the fornix originate from the pyramidal cells in the hippocampus proper.[28] With the advent of the modern tracing methods it became possible to study the origin of the fornix fibers more reliably than before. The hippocampus cannot be approached without damage being inflicted on the cerebral cortex or other parts of the hippocampal formation. The fiber degeneration following lesions of the hippocampus might, therefore, be the consequence of damage to fibers of passage from non-hippocampal structures. With the autoradiographic tracing method this source of error is avoided, since the radioactive markers (tritiated amino acids) are taken up by cell somata only. In the course of 1977 Swanson and Cowan and Meibach and Siegel published complete reports of their autoradiographic studies of the efferent projections of the hippocampal formation in the rat. In the main their results are concordant. (Figure 10-11 shows some main features of the connections of the hippocampus proper and of the fibers passing in the fornix.) These authors (see also Chronister, Sikes, and White, 1976) demonstrated a far more complex pattern in the efferent projections from the hippocampal formation than was previously believed to be present. Each component of the hippocampal formation appears to have its own pattern of efferent connections. The most exciting finding of both groups of workers was the discovery that *most of the fibers of the fornix are not,* as believed for some 100 years, *derived from the hippocampus proper, but from the subiculum.* The *subiculum* (and the prosubiculum) was found to give origin to all fibers distributed to the *hypothalamus* and many of the fibers to the *septum.* Most of these fibers pass in the *postcommissural fornix,* and some in the *precommissural fornix.* The *mammillary body* is thus supplied by fornix fibers from the subiculum. The fornix fibers originating in the *hippocampus proper* (from fields

---

[28] It was rather surprising, however, that the pyramidal cells of the hippocampus did not show retrograde changes following transection of the fornix (see Daitz and Powell, 1954; McLardy, 1955a). This might be explained by the presence of abundant recurrent collaterals, which could be assumed to protect the parent cells from suffering retrograde cellular changes.

CA 1–3) all appear to take their course through the precommissural fornix and end in the *septum* (lateral nucleus).[29] There is a topographical pattern within these connections.

While the projection to the septum (via the precommissural fornix) appears to represent the main outflow from the hippocampus proper, this has in addition some other efferent projections.[30] Several such projections, suggested on the basis of lesion and degeneration experiments, could be confirmed in autoradiographic studies. Thus fibers from the hippocampus proper pass to the *cingulate gyrus,* and to the *entorhinal area* and the *subiculum* (see, for example, Hjorth-Simonsen, 1971, 1973; Swanson and Cowan, 1977). (The subiculum also gives off fibers to the cingulate gyrus and the entorhinal area.) Fornix fibers, assumed to arise in the hippocampus, were traced with silver impregnation methods to the *anterior thalamic* nucleus, mainly the anteroventral nucleus. According to autoradiographic studies (Swanson and Cowan, 1977; Chronister, Sikes, and White, 1976; Sikes, Chronister, and White, 1977), however, these fibers come from the subiculum. Likewise, efferent fibers from the hippocampal formation (non-mammillary) to the *hypothalamus,* formerly believed to come from the hippocampus proper, have been found with autoradiography to come from the subiculum (Swanson and Cowan, 1977). This projection, sometimes referred to as the medial corticohypothalamic tract, appears to end chiefly in the ventromedial hypothalamic nucleus.

In the preceding account of fiber connections of the hippocampal formation, reference has occasionally been made to the presence of regional structural differences between minor parts. Such *areal parcellations,* originally based on traditional cytoarchitectonic studies, have been supplemented by investigations with other methods, for example studies of differences in the distribution of afferents and in their patterns of ending. Further, histochemical studies of the distribution of enzymes and metals (see, for example, Haug, 1974, 1976), and studies of the distribution of putative transmitters (see Storm-Mathisen, 1978) have been undertaken. It is likely that such methods will permit finer, and perhaps more meaningful, parcellations than can be obtained in classical studies of cell morphology and hodology only. These problems will, however, not be considered here.

A few observations may be mentioned. In the hippocampus acetylcholinesterase is distributed according to a specific pattern. It is particularly abundant in the stratum oriens and the pyramidal layer. A peculiar finding is the selective occurrence of heavy metals in certain regions, particularly in the layer of mossy fiber. Heavy metals, probably especially zinc, appear to be present only in the boutons of the mossy fibers and not in the postsynaptic excrescences of the basal parts of the apical dendrites. The physiological significance of this and many other findings is so far not known.

Before leaving the morphology of the hippocampal formation we must mention its commissural connections and present a sketchy outline of some features of

[29] Some fibers have been traced to the anterior olfactory nucleus, the bed nucleus of the stria terminalis, and some other cell groups (see Swanson and Cowan, 1977).

[30] The unexpected finding that the hippocampus proper (via the precommissural fornix) supplies only the septum raises a puzzling question. Numerical estimates (see Swanson and Cowan, 1977) indicate that the number of fibers passing to the septum represents a relatively modest proportion of the axons of all pyramidal hippocampal cells. Possible explanations may be ventured, but the question is still open.

the intrinsic organization of the hippocampus proper. *Commissural connections* between the two hippocampal formations are numerous. Tracings with silver impregnation methods (Blackstad, 1956; and others) indicate that the *hippocampal commissure* contains partly connections between corresponding regions of the two hippocampal formations and partly heterotopic connections. For example, some fibers from the entorhinal area end in certain parts of the contralateral hippocampus proper. Autoradiographic (Gottlieb and Cowan, 1973; Fricke and Cowan, 1978) and HRP studies (Segal and Landis, 1974a) have brought forward further data. Considering the parcellation of the hippocampal formation and the differences between minor parts with regard to, for example, afferent connections, it is no wonder that the pattern of the commissural connections is very intricate. It is still imperfectly known.

The *finer organization and the synaptology of the hippocampus proper* have been extensively studied. The principal points in the organization of the hippocampus proper were briefly referred to above and are shown in Fig. 10-10. The apparent structural simplicity of the hippocampus has made it a favorite subject for anatomical and physiological studies of fundamental features in neuronal interrelationships, such as morphological and functional properties of synapses. A feature of considerable physiological interest was noted by early workers such as Cajal and Lorente de Nó: the various contingents of afferents that establish synaptic contact with the pyramidal cells have their preferential site of termination on the dendrites or the cell body. The inset above in Fig. 10-10 is a simplified diagram of this arrangement, which has been confirmed by later authors. This pattern facilitates electron microscopic studies of the morphology of synapses of afferents from different sources. (For some early studies see Blackstad and Kjaerheim, 1961; Westrum and Blackstad, 1962; Hamlyn, 1962, 1963; Blackstad and Flood, 1963; Andersen, Blackstad, and Lømo, 1966; and the review by Blackstad, 1967).

It can be seen in Fig. 10-10 (upper inset) that the important *afferents to the hippocampus* from the *entorhinal area* (via the perforant path) end in the stratum lacunosum-moleculare. The fibers from the lateral and medial entorhinal area are distributed in different layers. They establish contact with the most distal ramifications of pyramidal cell apical dendrites (Blackstad, 1958; Raisman, Cowan, and Powell, 1965; Hjorth-Simonsen, 1972; Hjorth-Simonsen and Jeune, 1972). This has been confirmed electron microscopically (Nafstad, 1967). Entorhinal fibers further end on the outer portion of the dendritic tree of the granule cells of the dentate gyrus. Electron microscopic studies show that the fibers establish synaptic contact with dendritic spines (for a recent study see Matthews, Cotman, and Lynch, 1976). Little is known about the terminations of afferents from the *septum*. Raisman, Cowan, and Powell (1966) found them to end in the stratum oriens and stratum radiatum. An electron microscopic study has been published (Rose, Hattori, and Fibiger, 1976). *Commissural fibers* have likewise been found to be distributed chiefly to the stratum oriens and radiatum (see Gottlieb and Cowan, 1973). According to Blackstad (1956), this is the case particularly for the homotopic commissural connections. The heterotopic commissural afferents from the entorhinal area (like the ipsilateral entorhinal-hippocampal fibers) end in the stratum lacunosum-moleculare. The *mossy fibers* (the axons of the granule cells) end on the initial part of the pyramidal dendrites with giant boutons, surrounding large and ramified spines (Blackstad and Kjaerheim, 1961; Hamlyn, 1962). The axons of the *basket cells* appear to be the only afferents contacting the soma of the pyramidal cells. This occurs by way of densely packed boutons (Blackstad and Flood, 1963). The recurrent collaterals of the pyramidal cell axons end on both basal and apical dendrites of pyramidal cells. For a study on—and references to— intrinsic connections of the hippocampus, see Hjorth-Simonsen (1973).

Electrophysiological studies of the synaptic events and transmission of impulses in the hippocampus are in good agreement with the anatomical data described above. A particular structural feature in the organization of the hippocampus has been important for the success of electrophysiological findings and should be noted. The regular picture seen in transverse sections (Fig. 10-10) reflects an organization of the hippocampus in *lamellae,* running approximately transversely to the longitudinal extent of the hippocampus. This is learned from anatomical (see Blackstad, Brink, Hem, and Jeune, 1970) and physiological studies (see Andersen, Bliss, and Skrede, 1971).[31] (In addition, there are elements interconnecting the lamellae.) When inserted in the appropriate direction, the tip of a recording or stimulating electrode can, therefore, be moved succesively from one layer to the next within a lamella, and layers and cellular elements under study can be identified.

One *fundamental route of impulse transmission* appears to be as follows (Fig. 10-10, lower inset): cells of the entorhinal area give off axons that pass through the subiculum and, via the perforant path, establish contact with dendrites of granule cells in the dentate gyrus. The granule cell axons (the mossy fibers) contact apical dendrites of pyramidal cells in field CA 3, whose recurrent collaterals establish synaptic contact with apical dendrites of pyramidal cells in field CA 1. This important afferent pathway appears to be principally excitatory in function. All commissural connections appear to have an excitatory action on the pyramidal cells (Andersen, 1960; Andersen, Blackstad, and Lømo, 1966; and others). On the other hand, most authors hold that the basket cell (see Fig. 10-10) is inhibitory, regardless of from which source it is excited (Andersen, Eccles, and Løyning, 1964a, b; and others). Recurrent collaterals of pyramidal cell axons, ending on the apical dendrites, have been concluded to be excitatory (Andersen and Lømo, 1965; and others). For reviews of the electrophysiology of the hippocampus and references see Andersen (1975) and Lopes de Silva and Arnold (1978). The impulse transmission within the dentate gyrus and the hippocampus proper (the best investigated parts of the hippocampal formation) is still far from completely understood.

Correlations of electrophysiological observations of properties of cells and of synaptic transmission with effects of stimulation or destruction of the hippocampus are still in the realm of hypotheses. Nevertheless, it is appropriate to consider briefly some data from this field of research. Many of the results described in the literature may not have been caused by interference with the activity of the hippocampus proper, even though they have often been interpreted in that way.

[31] This situation has been used in a recently developed physiological in vitro technique (see Skrede and Westgaard, 1971). A slice of tissue (some 300–500 μm thick), cut perpendicularly to the surface, is removed from the hippocampus immediately after the animal has been killed (without anesthesia). With proper care a slice may be kept viable for several hours when placed in a constant-flow incubation chamber. The various layers are visible to the naked eye. Microelectrodes can under visual guidance (dissecting microscope) be placed in the various elements found in the slice, and their electrical behavior can be studied. The results obtained by this "slice method" are in general agreement with those of other studies, even though the interruption of some, normally present, connections and other factors will entail certain differences. For a paper presenting results obtained with this method see Schwartzkroin (1975).

Function of the hippocampus. The characteristic gross appearance of the hippocampus and its considerable size in man may in part explain why the question of the function of the hippocampus always has attracted much interest among workers in various fields of neurological research. In attempting to understand the function of the hippocampus, one previously had to rely mainly on inferences from comparative anatomical findings. In recent years more direct approaches to the question have been made, possibly owing to progress in physiological methods of study.

It will appear from the account given above of the connections of the hippocampus that it must obviously collaborate very closely with many other regions of the brain. Furthermore, connections between different subdivisions of the hippocampal formation bear witness to a collaboration between these subdivisions. Finally, it follows that each of these cannot be a functional unit. The recent demonstrations of many, hitherto unknown, connections will make it necessary to reevaluate and reintegrate many physiological findings made on the hippocampus and the hippocampal formation.

A case in point is the new insight into the composition of the fornix, showing that most of its fibers arise in the subiculum and not in the hippocampus proper. Obviously, the effects of stimulations or transections of the fornix cannot be interpreted as being related to the hippocampus proper only, as had frequently been thought in the past.

The anatomical complexity of the hippocampal formation makes it clear that it will be difficult—and perhaps impossible—ever to define satisfactorily the "function" of the hippocampus or of any other part of the hippocampal formation. Let us emphasize at the outset that it would certainly be misleading to consider the hippocampus a "center" for any particular function. The best that can be achieved will probably be to identify certain functions in which the hippocampus is concerned to a greater or lesser extent. The technical difficulties involved in stimulating or ablating the hippocampus or parts of it without affecting other regions are considerable. The account given below will be restricted to some observations assumed to elucidate the "function" of the hippocampus proper.

The hippocampus, as mentioned previously, is generally included among the regions of the brain collectively referred to as the "rhinencephalon." For a long time it was accepted that the hippocampus was related to olfactory functions, although it was known that there is no relation between the size of the hippocampus and the development of the sense of smell in various animals. The whales, which are anosmatic or nearly anosmatic, possess a hippocampus (see, for example, Addison, 1915; Ries and Langworthy, 1937), and in some human brains in which the olfactory bulbs and tracts have been absent, the hippocampus (as well as the fornix, mammillary body, and cingulate gyrus) has been found to be well developed (see, for example, Stewart, 1939). Microsmatic man has the largest hippocampus of all animals. From a review of the data available in the literature it was clear as early as in 1947 that *any relation between the hippocampus and the sense of smell must be very remote* (see Brodal, 1947b; Allison, 1953b). As mentioned previously, the hippocampus does not receive olfactory fibers from the bulb (with the exception of the so-called anterior continuation of the hippocampus, which is rudimentary). Impulses of olfactory origin may, however, reach the hippocampus indirectly, particularly via the entorhinal area. This, as described, projects to the hippocampus and the dentate gyrus. It has recently been shown that some fibers from the olfactory bulb reach the anterior part of the entorhinal area (its lateral part), and there are ample projections from the prepiriform and periamygdaloid areas to the entorhinal area (see Fig. 10-3). Although some authors have reported action potentials in the hippocampus following stimulation of the olfactory bulb in animals (Cragg, 1960; and others), these responses appear to be weak and to be mediated via multisynaptic pathways. While the role of the hippocampus in olfactory function thus appears to be modest, there is evidence of its involvement in other functions.

In view of the many sources of afferents to the hippocampus from other parts of the hippocampal formation, and not least the many (indirect) connections from the neocortex, it is no wonder that in electrophysiological studies action potentials have been reported in the hippocampus following peripheral stimulation (visual, acoustic, gustatory, somatosensory, and other kinds of stimuli). However, these responses are complex, labile, and easily modified by various factors. Stimulation of certain regions of the cortex and various subcortical structures may be followed by responses in the hippocampus. Conversely, stimulation of the hippocampus has been found to give rise to potentials in many other regions of the brain, in agreement with its multifarious efferent connections.

Attempts to elucidate the function of the hippocampus by studies of the *effects of stimulations or lesions of the hippocampus* have not permitted decisive conclusions. Many authors have investigated behavioral changes of different kinds. Mainly on the basis of recordings of motor activity, changes have been reported in the fields of sexual, reproductive, and other types of behavior. Visceral and endocrine functions have been examined. Motor deficits do not appear to occur.

The hippocampus has been thought to be concerned in such processes as attention and alertness. For example, Kaada, Jansen, Jr., and Andersen (1953) found in unanesthetized cats that on stimulation of the hippocampus the animals performed quick glancing or searching movements to the contralateral side. Their facial expression indicated ''attention'' associated with some bewilderment and anxiety. Their reaction to external stimuli was decreased, and their attention appeared to be intensely fixed on ''something'' in the environment which they seemed to experience. Corresponding observations have been made by other authors and have also been described as betraying ''orientation,'' ''hallucinations,'' ''defensive behavior,'' and ''arrest.'' It has been assumed that the motor reactions involved in such behavior patterns are initiated by emotional and psychic processes.

Numerous hypotheses have been set forth to explain the function of the hippocampus.[32] In a simplified way one may subdivide them into two main types, as

---

[32] Behavioral responses, similar to those mentioned above, have been noted also following lesions or stimulations of other parts of the brain, especially regions considered to belong to the ''rhinencephalon'' (taken in various senses). As mentioned previously, when evaluating the findings reported, the complex fiber connections and the difficulties involved in producing isolated lesions or stimulations of the hippocampus should be kept in mind. Green (1964, p. 591) is ''tempted to discount many of the results of rhinencephalic lesions and stimulation on the score that they represent spread of activity to other more relevant regions.''

done by Black (1975, p. 158): "One type of theory suggests that a major function of the hippocampus is to modulate motor control systems directly. The second type of theory suggests that the major function of the hippocampus concerns some nonmotor behavioral or psychological process." (Some authors have concluded that a main function of the hippocampus is to exert a nonspecific inhibition of emotional reactivity in general.)

Numerous attempts have been made to decide if the *electrical activity of the hippocampus* can be correlated with particular behavior patterns. The *hippocampal EEG* (electroencephalogram) is characterized by the occurrence of the so-called ϑ (theta) waves (Green and Arduini, 1954). These are rhythmic sinusoidal waves of 4 to 7/sec.[33] The spontaneous activity of the hippocampus appears to bear a relation to the various states of consciousness. In situations in which the cerebral neocortex becomes "desynchronized" or "activated" (see Chap. 6), the hippocampus shows "synchronization" and presents theta waves. When the cortex shows synchronization, the hippocampal activity is desynchronized. The functional role of this reciprocal relationship is not clearly understood. Both structures may be influenced from the reticular formation. As described above, there are fiber connections that may mediate such influences. It has been maintained that the septum may be especially important with regard to hippocampal activity, and that it serves as a "pacemaker" for the hippocampus. The hippocampus has a very low seizure threshold, and the relation of hippocampal activity to epileptic seizures has attracted considerable interest.

The theta waves and the electrical activity of the hippocampus have been studied in various behavioral situations. The theta waves have been maintained to be related not only to attention and alertness but also to numerous other phenomena.

> Black (1975, p. 131) lists the following: processing of informational input, general motivational changes, reactions of frustration to nonreward and to punishment, learning and memory, low-intensity nonspecific motivational responses involved in approach behavior, the control of species-typical acts, voluntary phasic skeletal activity, and motor processes produced by brainstem reticular activity. In recent years much attention has been devoted to the idea that the hippocampus plays a role in spatial discrimination. In rats, enduring deficits in performance in tests of spatial discrimination have been found after destruction of any of the main extrinsic connections of the hippocampus (Olton, Walker, and Gage, 1978). The anatomical relations of the hippocampus to parietotemporal regions of the cortex, concerned in visual and other kinds of discrimination (see Chap. 12), are compatible with the assumption that one of the main tasks of the hippocampus is to function as a cognitive map (see O'Keefe and Nadel, 1978).[34]

In view of recent anatomical and some electrophysiological studies on the hippocampal formation there is little doubt that many observations made in the past and interpreted as being due to stimulation of—or damage to—the hippocampus proper may have resulted from interference with other parts of the hippocampal formation. New interpretations of the old findings are needed. Not least

---

[33] There are some differences among animal species with regard to the frequency of the theta waves. These waves are also referred to as RSA (rhythmic slow activity). Two other wave forms, LIA (large-amplitude irregular activity) and SIA (small-amplitude irregular activity), have been distinguished as well.

[34] For further information on the function of the hippocampus and on the hypotheses mentioned the reader should consult *The Hippocampus* (1975) and the monograph by O'Keefe and Nadel (1978) that contains abundant references to the literature on the hippocampus.

should it be emphasized that the connections between certain parts of the hippocampal formation and the neocortex are far more abundant than formerly believed. Again, the subiculum appears to be an important case in point. As mentioned in one of the recent studies on the subiculum (Rosene and Van Hoesen, 1977), these findings may contribute to an explanation of some paradoxical clinical observations in affections of the temporal lobe (concerning, for example, spread of epileptic seizures and memory disturbances).

What is known about the hippocampus supports the assumption that it cannot be particularly closely related to any special and relatively simple function. Apparently it performs functions of the kind that are generally called "integrative." One of these, for a long time assumed to be related to the hippocampus, is *memory,* particularly "recent memory." "Recent memory" is often and better described as "short-term memory." The term refers to the capacity to *store new information.* Even if this faculty is impaired, the capacity to *retrieve information stored in the past* may be intact. There is a close relation between the processes of learning and of remembering.[35]

> In general, *loss of recent memory* is said to be present when the patient is unable to establish lasting new memories, even if he can remember relatively small amounts of information for seconds or minutes *if he is not distracted* (see Douglas, 1967). This defect prevents learning, and as time passes gives rise to a retrograde amnesia of increasing length. Barbizet (1963) describes this as follows (p. 128): "Such patients who no longer fix the present live constantly in a past which preceded the onset of their illness."

The belief that the hippocampus is involved in memory functions is based chiefly on experience with patients who have either had destructive diseases of the temporal lobe or in whom parts of it have been removed surgically. In the course of the last few years our views on this question have changed markedly. Since the item is of some clinical relevance, it will be briefly surveyed.

Among the early observations that may be mentioned are those of Glees and Griffith (1952), who reported loss of recent memory and progressive dementia in a patient whose brain at autopsy showed extensive bilateral destruction of the hippocampus and gyrus hippocampi. Other cases of loss of recent memory following bilateral lesions or ablations of the hippocampus with more or less concomitant involvement of neighboring structures have been reported (Scoville and Milner, 1957; Penfield and Milner, 1958; Victor, Angevine, Mancall, and Fisher, 1961; Drachman and Arbit, 1966; and others). In most cases studied there has been no loss of old memories or deterioration in personality or general intelligence. The degree of memory deficit was concluded by Scoville and Milner (1957) to vary in proportion to the extent of removal. Some experimental findings were interpreted as supporting this notion (Stepien, Cordeau, and Rasmussen, 1960; Drachman and Ommaya, 1964, in the monkey; and others). Penfield and Milner (1958) found that in man unilateral partial temporal lobectomy, including the amygdaloid nucleus, the hippocampus, and the hippocampal gyrus with the uncus, did not cause any serious psychological impairment if the other temporal lobe was functioning

---

[35] There is a vast literature on memory and learning. In most of these publications the question of a relation of the hippocampus to memory is considered. A few references to recent publications may be mentioned: Douglas (1967); *Biology of Memory* (1970); Mark (1974); Isaacson (1975b); Izquierdo (1975); *Short-Term Memory* (1975); *The Structure of Human Memory* (1976); Horel (1978).

normally. Autopsy was carried out on only a few of the patients, and convincing evidence has never been produced that damage to the hippocampus is responsible for the loss of recent memory that occurs following bilateral removal of the medial part of the temporal lobe (see Horel, 1978, for references). While some authors have reported loss of recent memory after bilateral destruction of the fornix in man (see Heilman and Sypert, 1977), others did not notice this (for a recent study, see Woolsey and Nelson, 1975). Congenital maldevelopment of the hippocampus with absence of the fornix may occur without defects in memory (Nathan and Smith, 1950). There appear to be no reports of cases where the fornix has been interrupted in isolation, without damage to other regions, and in which there has been defects in memory (see Horel, 1978, for references).

However, the assumption of a relation between the hippocampus and recent memory was strengthened by other data. Thus emphasis was put on the observation that in patients with Korsakoff's psychosis or Wernicke's encephalopathy, the mammillary bodies are often affected. Since the mammillary body was believed to receive its fornix fibers from the hippocampus, it appeared reasonable that one of the characteristics of these diseases is loss of recent memory.

Korsakoff's syndrome or psychosis has also been called the "amnestic confabulatory syndrome." In addition to a conspicuous characteristic loss of recent memory (for example, the patient does not remember having seen the doctor half an hour ago, what he has had for dinner, and so forth), there is often a tendency to confusion and fabrication and diminution in spontaneity and initiative. The syndrome is most commonly seen in chronic alcoholism (where deficient supply of thiamine plays an etiological role). When postmortem studies have been performed, most often the mammillary bodies have been found to be affected with vascular, inflammatory, or degenerative changes. The Korsakoff syndrome may occur as an element in the so-called *Wernicke's polioencephalitis hemorrhagica superior* (in which acute ocular palsies, nystagmus, and ataxia of gait usually are present). The syndrome may occur following head injuries and in cases of *tumors at the base of the brain* involving the floor of the third ventricle (see Williams and Pennybaker, 1954). Whitty and Lewin (1960) noted a transient Korsakoff's syndrome in eight patients after bilateral ablation of the anterior part of the cingulate gyrus.

In the light of present-day anatomical knowledge it is important to note that even if transection of the fornix had been found to influence memory functions, this would carry little weight as an argument for a relation of the hippocampus to memory, since most fornix fibers are derived from the subiculum, as repeatedly mentioned above. Nor is the presence of pathological changes in the mammillary body in cases of Korsakoff's psychosis significant, since the fornix fibers that supply the mammillary body do not arise in the hippocampus proper. It is interesting that this view receives support from the critical and exhaustive study by Victor, Adams, and Collins (1971), undertaken when it was still generally believed that the hippocampus projected to the mammillary body. These authors studied 245 patients with Wernicke's disease and Korsakoff's psychosis (with postmortem examinations of 82 cases). In addition to a frequent affection of the mammillary bodies, they often found symmetrical pathological changes in other brain regions, such as other parts of the hypothalamus, some thalamic nuclei, the periaqueductal gray, the vestibular nuclei, and the anterior lobe of the cerebellum. According to Victor, Adams, and Collins (1971) the amnesic defect is not related to an affection of the mammillary bodies but to involvement of parts of the diencephalon, particularly the dorsomedial thalamic nuclei.

It may be mentioned that in some experimental animal studies, changes that may be interpreted as defects in short-term memory have been observed in lesions of the thalamus. Furthermore, memory disturbances have been observed in some patients who have had lesions of the thalamus, especially in the region of the dorsomedial nucleus, or who have been treated surgically with thalamotomies. (For references, see Horel, 1978. The dorsomedial thalamic nucleus will be considered in Chap. 12. A diagram of its afferent connection is shown in Fig. 12-8.)

On the basis of recent anatomical insight, a relation between the mammillary body and memory is unlikely. Critical evaluation of available observations in man and animals gives the same result. The role of the hippocampus is more difficult to decide. It appears that in all human cases described, lesions of the hippocampus found to lead to defects in recent memory have included other brain regions as well, especially medial parts of the temporal lobe (covering the subiculum and the entorhinal area) and more or less of the neocortex. The question may be asked: Are the memory disturbances due to affection of non-hippocampal parts of the temporal lobe? Horel (1978), who discusses these relations in an extensive study, produces good arguments for an affirmative answer to this question.

According to Horel (1978), there is no evidence (in animals or man) that the amygdaloid nucleus or the entorhinal area is of relevance. However, Horel and collaborators (see Horel, 1978) have found that in the monkey if a transection is made of what they refer to as "the temporal stem," severe deficits appear in learning and the ability to retain learned tasks. The term "temporal stem" (or "albal stalk") is used to describe the white matter lateral to the temporal horn of the lateral ventricle and medial to the three temporal gyri. "It is the white matter that contains connections of afferents and efferents of temporal cortex and amygdala, but carries no connections of the hippocampus" (Horel, 1978, p. 405). This white matter appears to be more or less injured in medial temporal lobectomies, and Horel (1978) suggests that interruption of fibers passing in the "temporal stem" is responsible for the amnesic disturbances that occur after these operations, regardless of whether there has been concomitant encroachment on the hippocampus. Attention is drawn to the many connections of the temporal lobe, not least those with the orbitofrontal cortex and the dorsomedial thalamic nucleus (MD, projecting onto the orbitofrontal cortex and receiving afferents from many sources, among them the inferior temporal cortex; see Chap. 12 and Fig. 12-8B). The regular occurrence of pathological changes in the MD in Korsakoff's psychosis (Victor, Adams, and Collins, 1971) may indicate that the connections with this nucleus are of particular importance. Furthermore, it appears that the temporal lobe cortex is important for functions that may be essential for memory, for example, cognitive and perceptual faculties (cf. Klüver-Bucy syndrome, to be considered below; see also Chap. 12 on the temporal lobe). No final conclusion is as yet possible about the structures in the temporal lobe whose damage results in loss of recent memory. However, at present evidence for a close relation of the hippocampus to memory functions is rather weak.

Innumerable psychological studies have been devoted to human memory functions. Distinctions have been made between several categories of memory, and various hypotheses have been formulated about the underlying mechanisms. Theoretical conceptions of human memory appear to have undergone radical changes in recent years. There appears to be a growing skepticism about theories that consider certain brain regions specifically related to memory. As expressed by O'Keefe and Nadel (1978, p. 373) "it appears that there are different types of memory, relating perhaps to different kinds of information, and that these are localized in many, possibly most, neural systems." There may be "memory areas, each responsible for a different form of information storage" (loc. cit. p. 374). As touched upon in other chapters, more or less overt disturbances of one or another aspect of memory functions may be seen after damage to almost any part of the brain.[36] Perhaps the most important common factor is a general disturbance of brain function. There are interesting features in common between the search for "the site of memory" and the search for "the site of consciousness," discussed in Chapter 6.

At the end of this account of the hippocampus it may be appropriate to quote some words of Weiskrantz in his Chairman's introduction to the symposium on *Functions of the Septo-Hippocampal System* (1978, p. 1). Weiskrantz characterizes our present state of knowledge as follows: "The striking aspect of the hippocampus is the anatomical elegance of its structure, revealed in detail in the past few years. In contrast there is really appalling ignorance about what this elegance means."

The limbic system. Reference has been made to this term previously. A perusal of neurological, especially neurophysiological, periodicals from the last twenty years shows that "limbic system" occurs in the headings of an increasing number of articles.[37] The contents of the term seem to be steadily expanding, but a generally accepted definition of the "limbic system" appears never to have been given.[38]

As far as can be judged from the literature, the concept of a "limbic system" originated from the term "limbic lobe," introduced in the nomenclature mainly on the basis of comparative anatomical studies. Unfortunately, there is no unanimously accepted definition of the "limbic lobe." The name is often credited to Broca, who in 1878 spoke of *"le grand lobe limbique."* It was conceived of as a ring of gray matter bordering the hemisphere against the central parts of the brain and arranged in a circular manner around the interventricular foramen. The distinction between the "limbic lobe" and "rhinencephalon" is far from clear. (For a historical review see White, Jr., 1965b.) Most authors include in the limbic lobe the following structures: the cingulate gyrus and the induseum griseum, the hippocampus and dentate gyrus, the subiculum, presubiculum, parasubiculum, the

---

[36] Amnesia (often transitory) may occur when the posterior cerebral artery is occluded (see Beneson, Marsden, and Meadows, 1974, for a recent study). That the amnesia is due to anoxia of the hippocampus appears doubtful.

[37] In addition there are several reviews, monographs, and proceedings of symposia dealing with the subject. To mention a few: *Structure and Function of the Limbic System* (1967); *Limbic System Mechanics and Autonomic Function* (1972); Isaacson (1974); *Limbic and Autonomic Nervous Systems Research* (1974); and *Neural Integration of Physiological Mechanisms and Behavior* (1975). The anatomy of the "limbic system" has been surveyed more specifically by Hall (1975). See also Hamilton (1976) on the system in the rat.

[38] The "limbic system" is usually described in loose and general terms, like the following statement, given by Mesulam, Van Hoesen, Pandya, and Geschwind (1977, p. 407): "The 'limbic system' is a construct which is commonly used to denote a set of interconnected structures which have intimate, albeit multisynaptic, connections with the hypothalamus and parts of the mesencephalon."

entorhinal area, the prepiriform cortex, the septum, the olfactory tubercle, the medial and cortical amygdaloid nucleus, and some minor gray masses. Other authors extend the term to cover other structures in addition, such as the subcallosal gyrus, the posterior orbital cortex in the frontal lobe, the anterior insula, and the temporal polar region, in part on the basis of cytoarchitectonic resemblances to the allocortex. Even if the various regions listed above have in common an early appearance in phylogenesis, their finer structural organization differs considerably. *It is difficult to see that the lumping together of these different regions under one anatomical heading, "the limbic lobe," serves any purpose* (except to illustrate the truth of Goethe's words, quoted in the Preface!).

It is even less justifiable to speak of a "limbic system." This concept is even more diffuse than that covered by the term "limbic lobe." The "limbic system" appears to be a phrase used collectively for functions attributed to the "limbic lobe," under the tacit assumption that those parts of the brain which have been given a particular name by the old anatomists form part of one functional "system."[39] Even though stimulation or ablation of various of these regions has in part given corresponding results, it is becoming increasingly evident, as Kaada (1960, p. 1346) emphasized, that: "Recent anatomical and physiological research rather tends to fractionate the 'limbic system' into several units with quite different projections and functional significance." On the other hand, as research progresses it becomes increasingly difficult to separate functionally different regions of the brain. The borderlines between "functional systems" become more and more diffuse. This is in complete accord with the outcome of studies of the fiber connections, and is especially evident from behavioral studies undertaken with modern methods. The "limbic system" appears to be on its way to including all brain regions and functions. As this process continues, the value of the term as a useful concept is correspondingly reduced.

Functionally, the "limbic system" is generally said to be concerned with visceral processes, particularly those associated with the emotional status of the organism. (The term "visceral brain" has been used almost synonymously with the "limbic system.")

For reasons given above it is the author's opinion that *the use of the terms "limbic lobe" and "limbic system" should be abandoned.* For the same reasons no attempt will be made to discuss these subjects here, the more so since correlations of functional observations and anatomical data can only be made to a limited extent. Some of the observations made on particular brain regions belonging to the "system" have been considered in the present chapter; reference to some others has been made occasionally at other places in the text. Several of the general considerations made when discussing data from the hippocampus and the amygdala may be applied to other brain regions which are included as part of the structural basis of the "limbic system."

---

[39] Increased knowledge of the fiber connections of the brain reveals numerous examples. Only one of them will be quoted. According to Isaacson (1975b, p. 331), "the limbic system is a system only because each of its components has relatively direct connections with the hypothalamus." On this basis several nuclei of the brainstem, among them the raphe nuclei, the nucleus locus coeruleus, the nucleus of the solitary tract, and the dorsal motor vagus nucleus and further the intermediolateral cell column in the spinal cord, should be considered parts of the "limbic system," since they all have connections—even *direct* ones—with the hypothalamus (see Chap. 11).

Some functional aspects of the central "olfactory" structures. The regions of the brain considered in this chapter, although all are often listed as parts of the "rhinencephalon" or "limbic system," differ in important respects anatomically and functionally. They are not all related to smell. Several functional aspects have been considered in the preceding sections. Some additional data, particularly with reference to the sense of olfaction, will be mentioned.

In fishes and amphibians the sense of smell is of supreme importance. It is essential for informing the animal of approaching enemies as well as in guiding it to its food and mates. With the increasing differentiation of the other senses in the vertebrate phylum, smell loses this supremacy. However, even in man, the capacity to discriminate between smells of different types is a factor of some importance in his reaction to food and drink as well as in his sexual behavior. The widespread use of perfumes in civilized and primitive communities points to the importance of smell for man, and the connoisseurs of perfumes have noted many peculiar relationships between special odors and their psychological effect.

From a clinical point of view the olfactory reflexes are of minor interest. The cortical aspect of olfactory function, however, is of some importance. Animal experiments have given some information on this point. Decorticate cats (i.e., cats in which the entire neocortex has been removed and only the thalamus and striatum are left in connection with the paleo- and archicortex) suffer no marked reduction in their olfactory capacity. They are still in possession of the usual feeding reflexes (Dusser de Barenne, 1933). Even with rather extensive damage to the paleo- and archicortex, highly complicated feeding reflexes are present (Bard and McK. Rioch, 1937). It appears that most of the olfactory reflexes are mediated through subcortical structures. According to Allen's classical studies, conditioned positive and negative reactions established to agreeable and disagreeable smells, respectively, are not abolished after bilateral extirpation of the hippocampus (Allen, 1940), but the correct differential response is lost when the piriform cortex is also ablated, although the elementary reflexes persist (Allen, 1941). These findings tend to show that the piriform lobe is essential for the discrimination of smells of different kinds, a finding in harmony with the anatomical features.

On the whole, the relatively sparse data on the human brain are in agreement with experimental observations. In conscious human beings stimulation of the olfactory bulb is followed by olfactory sensations, whereas electrical stimulation of the hippocampus yields no response at all (Penfield and Erickson, 1941). Olfactory sensations occurring in the so-called uncinate fits, to be considered below, appear to be associated with lesions of the uncus and the hippocampal gyrus, but not with lesions of the hippocampus. In patients with epileptic fits a marked cell loss is frequent in the hippocampus, but there is no decisive evidence that these changes are responsible for the disturbances of smell which occasionally may occur in these patients as well as in others.[40]

The examination of the sense of smell. Before the symptoms following lesions of the olfactory structures of the brain in man are considered, it is appro-

---

[40] The cell loss is usually found only in part of the hippocampus (Sommer's sector). There has been much discussion of whether these changes are of etiological significance for the epileptic manifestations or whether they are secondary, due to vascular changes occurring during the seizures (see Meyer, 1957).

priate to consider the examination of the sense of smell. Just as in examination of other sensory functions the cooperation of the patient is essential, and the results, therefore, are dependent on his intellectual level and to a certain extent also his particular familiarity with various orders. It is, therefore, of importance in testing to choose odors which are known by most persons if comparisons have to be made, since it is extremely difficult to describe a particular smell without reference to some object or circumstance commonly associated with it. The patient should not be allowed to recognize the test substances, for example by sight or by feel prior to its application. The odor should be applied to one nostril at a time, and it should always be made certain that local affections of the nasal cavity which interfere with the perception are not present. It is well known that a common cold reduces the olfactory acuity, and this reduction has been shown by Elsberg, Brewer, and Levy (1935) to be present in a decreasing degree for several weeks. The selection of test odors should be made with due consideration of the fact that certain odors have a marked capacity to irritate the muccous membrane and, therefore, are as much trigeminal as olfactory irritants. Pinching (1977) strongly recommends the use of "pure" olfactory stimulants, such as musks and floral odors, instead of the conventionally used ones (camphor, peppermint, and others). The use of floral odors considerably increased the sensitivity of clinical olfactory testing.

In the routine examination smell is usually tested simply by letting the patient inhale vapors of different odor separately by each nostril. This, however, is a rather crude method and as a rule permits only the statement that smell appears normal, or that there is some degree of hyposmia or that anosmia is present. Several methods have been suggested to obtain a more accurate test. A method worked out by Elsberg and collaborators (Elsberg and Levy, 1935), allows a quantitative determination of olfactory acuity.

> Graded quantities of odorous air are injected into the nostrils (monorhinal or birhinal), and the minimal quantity necessary for recognition of the smell is determined. By this "blast injection method" the variable factor of the force of inspiration in sniffing is avoided (the patient holds his breath during the injection), and the quantity of air acting on the olfactory epithelium can be changed at will. The values for the individual odors appear to be approximately the same in different individuals. By applying a continuous stream of odorous air for a certain time, the olfactory fatigue can be determined ("stream injection method"). The method has been used to advantage by others, and its diagnostic value further examined. It is particularly important to be aware of the physiological variations of the threshold values (see Zilstorff-Pedersen, 1964).

If the olfactory epithelium is subjected to a certain smell for a time, as is well known, the individual ceases to perceive it. This fatigue appears to be a central process, since it is increased in central lesions which do not affect the olfactory bulb or tracts (Elsberg, 1935; Elsberg and Spotnitz, 1942). With lesions in the latter parts there is no increased fatigability, but the olfactory acuity is reduced, larger amounts of the odorous substances being required to elicit olfactory impressions. The reduction of olfactory acuity does not necessarily affect the perception of all types of olfactory stimuli, but may be limited to certain qualities of smell.[41]

----

[41] The "fatigue" referred to above may in part be due to adaptation. According to Adrian (1950b), this depends on changes in the olfactory bulb and not in the receptors. The subject is briefly discussed in *Taste and Smell in Vertebrates* (1970).

Several other interesting features have also been brought out by these investigations. For example, the olfactory fatigue seen in central lesions is homolateral to the side of the lesion. This tends to show that the bulk of afferent impulses from the olfactory bulb are transmitted to the homolateral half of the brain, and this is in good accord with knowledge of the olfactory pathways.

Olfactory disturbances in lesions of the nervous system. Apart from signs of loss of function of the olfactory structures, irritative symptoms may also occur, but it appears that symptoms of this type have as yet only been observed in the so-called uncinate fits. The more important deficiency symptoms will be treated first.

A *destruction* or a damage leading to interruption of conductive capacity of *the olfactory bulb or tract* may be caused by various processes. *Cranial fractures,* involving the cribriform lamina of the ethmoid bone are not uncommonly responsible for bilateral or unilateral anosmia or hyposmia. The fine olfactory fibers are easily damaged in their passage through the bone by the accompanying bleeding or the ensuing scar formation. It is worth emphasizing, however, that also in cases where no fracture is ascertained, a *closed head injury* may be followed by transient or permanent impairment of smell. The anatomical causes may be traction of the nerve and fibers, or a secondary damage due to bleeding, or even undiagnosed fractures. For an extensive account of post-traumatic anosmia see Sumner (1964).

The olfactory bulb and tract will frequently be affected by *tumors* in the anterior cranial fossa. Most consistently this will occur in the *meningeomas arising from the olfactory groove,* in which case unilateral hyposmia or anosmia will be the initial symptom, frequently, however, not noticed by the patient. It is only later on, when the growing tumor causes more alarming symptoms, such as visual disturbances with scotomas and optic atrophy, or when mental changes develop on account of involvement of the frontal lobe, that the disease is discovered. In this stage the anosmia will frequently be bilateral, since the tumor will affect both olfactory lobes or tracts. Other tumors that not infrequently are accompanied by impairment of the sense of smell are the *suprasellar meningeomas,* or those arising from the ala parva of the sphenoid bone. These may give unilateral or bilateral olfactory loss according to their site and size. *Osteomas* from the frontal bone (orbital lamina) may also encroach upon the olfactory bulb or tract. Likewise *diseases of the frontal lobe, gliomas or abscesses,* will sooner or later exert a pressure on these structures. Since the destruction of the frontal lobe may be considerable before clear-cut symptoms appear, the affection of smell may be a valuable guide in the diagnosis. Finally *pituitary tumors* should be remembered, although they affect the olfactory bulb or tract only when they transcend the sella, and as a rule bilaterally. Occasionally *aneurysms of the circle of Willis* may affect the olfactory bulb or tract. In all cases mentioned above there will be a reduction of olfactory acuity, but no increased fatigability, according to Elsberg (1936) (except when the processes affect also the interior of the frontal lobe).

It is not known whether lesions of the anterior perforated space and the adjacent regions are followed by particular olfactory symptoms, but as a rule lesions of these structures will be associated with damage of the olfactory tract. The impairment of smell found in some cases of *increased intracranial pressure* without evidence of local damage to the olfactory structures should be mentioned, although it is difficult to explain satisfactorily. The *hysterical anosmia* should be borne in mind as a source of error. In this case, as a rule, also odors that have a marked trigeminal component are said not to be recognized, whereas a patient with an organic anosmia usually will perceive the trigeminal component.

From an analysis of the anatomical and physiological data available, it might be assumed that lesions of the temporal lobe, and more particularly the gyrus hippocampi, would result in defects in olfactory discrimination and interpretation. However, cases sufficiently clear-cut to decide these matters appear not to have been observed.

The most decisive evidence that the cortex of the hippocampal gyrus is concerned in perception or even interpretation of olfactory impressions appears to come from the observation of olfactory "hallucinations" in some patients with lesions of this region. After the original description by Hughlings Jackson and Beevor in 1890 of a lesion limited to the uncus (see Fig. 10-4) and adjoining parts of the amygdaloid nucleus giving rise to a peculiar disturbance called by Jackson "uncinate fits," a considerable number of similar cases has been observed. The olfactory sensations which the patients experience in these fits are usually of a disagreeable character, and may be accompanied by movements of the lips and tongue (smacking, etc.) and quite frequently also by a so-called "dreamy state," which, however, may occur without olfactory disturbances in cases of epilepsy. The uncinate fits are regarded as representing a variety of epilepsy; and it appears reasonable to consider the olfactory sensations in the same manner as the other sensory phenomena occurring in epileptic seizures elicited from other sensory areas (visual, acoustic, and sensory) as due to irritation of the olfactory cortex. An uncinate fit may be produced by stimulation of the uncus (Penfield and Kristiansen, 1951).

The "dreamy state" is characterized by the patient having a feeling that things seem unreal as in a dream. He appears to lose contact with his environment. Frequently he tells afterward that during the attack he had the feeling that he had experienced the same situation before, although this cannot have been the case. These psychic phenomena are probably due to the spread of the irritation to other areas of the cerebral cortex (the "highest levels" of Jackson). An uncinate attack may be followed by a generalized seizure, in which case the uncinate attack represents an aura.

Although uncinate fits are most often seen with lesions of the uncus and hippocampal gyrus, they may be observed also in cases where these structures are not directly damaged, but are affected by pressure, for example by tumors in other parts of the temporal lobe.

Symptoms in lesions of the temporal lobe.   Some of these have been mentioned previously, for example homonymous anopsias due to affection of the optic radiation (Chap. 8), the occurrence of an acoustic aura in epileptic fits originating from the superior temporal gyrus and aphasic symptoms, and certain disturbances in auditory functions in affections of this gyrus (Chap. 9). (The disturbances in speech mechanisms will be briefly discussed in Chap. 12). The uncinate fits were mentioned above. The bewildering occurrence of ipsilateral hemiparesis in cases of tumors in the temporal lobe should be recalled. When the tumor reaches a certain size, it will push the brainstem to the other side, and the contralateral cerebral peduncle will be compressed against the edge of the tentorium. (Since this takes place above the crossing of the descending fibers in the peduncle, the pareses will appear on the side of the lesion.)

Certain disturbances are usually observed only in *bilateral lesions of the temporal lobe*. This is the case with loss of recent memory described above. Bilateral, more or less symmetrical lesions of the temporal lobe are rare in man. Studies of the consequences of bilateral temporal lobe ablations in experimental animals have, however, been performed by several authors, following the description by Klüver and Bucy (1937, and later) of what is often called, after them, the Klüver-Bucy syndrome.

The Klüver-Bucy syndrome comprises a number of symptoms. There is *visual agnosia:* the animal appears to have lost the ability to recognize and detect the meaning of objects on the basis of visual criteria alone. The animal shows *"oral tendencies,"* a tendency to examine all objects by mouth. There is *"hypermetamorphosis,"* a tendency to pay attention to every visual stimulus. There are changes in *emotional behavior,* an absence of emotional responses. There is a striking increase in *sexual activities* and in the diversity of sexual manifestations. Finally there are conspicuous changes in *dietary habits.* Similar symptoms have been observed in man following bilateral temporal lobectomies (Terzian and Ore, 1955). Since in these lesions the hippocampus, the amygdala, and the hippocampal gyrus, as well as extensive parts of the neocortex, are removed, it might be surmised that the various symptoms are consequences of the ablation of particular structures. However, as stated by Klüver (1958, p. 177): "Intensive researches along behavioural, clinical, anatomical, and electrophysiological lines during the last twenty years have not settled the question of the particular anatomical structures related to each of the symptoms; they have also not settled the question whether the polysymptomatic picture results from a few fundamental alterations in behaviour or even from only one basic behaviour disturbance." It may perhaps be assumed that the involvement of the amygdala is particularly important in producing the changes in emotional behavior. The results of bilateral stereotaxic amygdalatomy in man (Narabayashi et al., 1963; and others) seem to lend some support to this view. However, following ablations in the monkey of the neocortical parts only of the temporal lobe (sparing the amygdala, gyrus hippocampi, and hippocampus) there develops a partial Klüver-Bucy syndrome, including emotional behavior changes, as shown by Akert, Gruesen, Woolsey, and Meyer (1961).

In a recent study of the Klüver-Bucy syndrome in the monkey, Horel, Keating, and Misantone (1975; see their paper for references to earlier studies) found that destructions restricted to either the temporal neocortex or the amygdala produced so-called oral behavior, decreased emotionality, and increased reaction to stimuli. The visual defects in the syndrome, however, are assumed to be probable consequences of a lesion of the temporal neocortex, particularly the middle and inferior temporal gyrus (see Chap. 12).

It is noteworthy that studies in man likewise indicate that it is scarcely possible to relate the various symptoms following temporal lobectomy to particular structures in the temporal lobe. Of interest for this question are the results of *hippocampal stimulation in man.* Such studies may also be expected to give some information of the role of the hippocampus in mental activities. Penfield and many later students noted that after electrical stimulation of the temporal lobe in conscious man, the patient may experience complex hallucinations, fear or anxiety, déjà vu sensations, alimentary sensations, and amnesia. Little is known, however, of whether the various mental phenomena are due to stimulation of particular structures in the temporal lobe (the hippocampus, amygdala, temporal cortex). Of relevance to this question is the recent study of Halgren, Walter, Cherlow, and Crandall (1978).

The authors carefully recorded the results of stimulations of the medial temporal lobe in 36 patients (with psychomotor epilepsy). Various mental phenomena, classified into 20 categories, were reported, such as unformed visual or auditory hallucinations, memory-like hallucinations, visceral sensations from heart or digestive organs, somesthetic sensations, emotions, and amnesia.

Only about 1/13th of all stimulations resulted in mental responses. Olfactory sensations or sensations of anger were not reported. Some electrode sites were in the hippocampus, others in the amygdala, and still others in the entorhinal area or other regions.

The studies of Halgren, Walter, Cherlow, and Crandall (1978) indicate that the types of mental responses to stimulation of the temporal lobe show little correlation with the anatomical site stimulated and are rather, as the authors phrase it, related to "patient-specific variables including, specifically, their personality." Whatever the mechanism underlying these phenomena, this conclusion appears reasonable. It is compatible with the presence of numerous interconnections between the various neuronal assemblies in the temporal lobe and with the view that complex nervous phenomena such as those considered here, can scarcely be taken care of by a minor cell assembly only, but require the cooperation of rather extensive parts of the brain (even others than structures found in the temporal lobe).

"Temporal lobe epilepsy" has attracted much interest in recent years. Epileptic seizures originating in the temporal lobe are relatively common. There are often special kinds of disturbances in consciousness, occurring as an initial symptom or during or after the seizure. Reference has been made to the occurrence of acoustic sensations as an aura (Chap. 9) and to olfactory sensations in the "uncinate fits." In other cases complex subjective psychic experiences are described as an aura, and it is common that during the attack the patient is unresponsive, and there is some degree of loss of understanding, even though motor control and reception of sensory stimuli usually are preserved. The patient may be able to carry out rather elaborate acts during the attack, but following it he has a complete amnesia for what has happened. These types of epileptic seizures are often called "psychomotor attacks," or "automatisms."[42] The accompanying EEG changes are often rather characteristic, and changes may be present between the overt attacks. Operative treatment by partial temporal lobectomy has given favorable results in many cases in which medical treatment fails (see Penfield and Jasper, 1954). With these operations usually the amygdaloid nucleus, the uncus, the anterior part of the hippocampus, and more or less of the three temporal gyri are involved. As mentioned above (footnote 40), a cell loss in the so-called Sommer's sector of the hippocampus is common in brains from epileptic patients. In material removed at operation the pathological changes have most often been found to be of the atrophic type, so-called Ammon's horn's sclerosis. (Some authors prefer the term "mesial temporal sclerosis," since the changes often involve the amygdala and other regions in the neighborhood of the hippocampus). The next most frequent pathological change assumed to play a role in the pathogenesis of the epileptic seizures appears to be a scar resulting from traumatic injury (see Corellis, 1970). According to Engel, Driver, and Falconer (1975), there is good correlation between the pathology and electroencephalographic findings. The origin or focus eliciting the attacks has been assumed by many to be in the amygdaloid nucleus (see, for example, Falconer, Serafetinides, and Corsellis, 1964), whereas others consider changes in the hippocampus to be responsible in most cases (see, for example, Malamud, 1966). A succinct review of the place of surgery in the treatment of epilepsy has been given by Rasmussen (1976).

---

[42] Behavioral abnormalities between the attacks are found much more frequently in patients with psychomotor epilepsy than in any of the other types. It has been suggested that this may be due to a continuous state of subictal irritation by the focus (see Gloor, 1960).

As mentioned above, in view of the ample interconnections and functional interrelations between the different structures contained in the temporal lobe, it is not surprising that approximately identical disturbances may result from an abnormal activity starting in almost any of these structures. The initial symptoms may in some instances point to the site of origin (for example, olfactory sensations to the uncus). The occurrence of complex psychomotor disturbances, especially in epilepsy originating from the temporal lobe, is in agreement with other observations which indicate that this part of the brain is important for some rather complex functions, such as comprehension of verbal symbols (see Chap. 12).

The experimental demonstration that the amygdala bears some relation to emotional reactions and behavioral changes of different types has prompted attempts to utilize this knowledge in the therapy of certain mental disturbances. Some observations in patients were in agreement with results of animal experiments. Thus, stimulation of the amygdala in conscious patients (undertaken as a diagnostic procedure in the treatment of epilepsy of temporal lobe origin, "psychomotor epilepsy") had often been observed to give rise to a sensation of fear, as may occur at the start of an epileptic seizure of this type (see Penfield and Jasper, 1954).[43] With the advent of stereotactic neurosurgery possibilities for new "psychosurgical" approaches were opened. One of these was the introduction of so-called amygdalectomy, destruction of the amygdaloid nucleus, which by some (rather uncritically!) has been claimed to represent "the highest center of behavior."

This operation has been performed on some epileptic patients who suffer from psychomotor seizures (assumed to be elicited from the temporal lobe) with accompanying emotional and behavioral disturbances. Some authors extend the indications to comprise also nonepileptic patients with episodes of violent and aggressive behavior.[44] Good results have been reported by some, whereas others disagree. In an early paper, in which the patients were described as being improved, they were said to "become obedient, calm and even trainable and educable"! However, long-time follow-up studies are scanty. Intellectual deficits are said not to follow the operation.

Most authors maintain that better results are obtained with destruction of the medial than the lateral parts of the amygdala. This is what might be expected from experimental anatomical and physiological data on the amygdala.

As discussed above, the corticomedial group is the main receiving region of olfactory and probably other, "visceral" impulses and appears to be the major source of efferent amygdaloid fibers to the hypothalamus and the mesencephalon. On the other hand, the afferent connections with neocortical regions and the MD of the thalamus appear to concern mainly nuclei in the basolateral group. A close relation of the corticomedial group to emotion has likewise been adduced from the finding of a large monoaminergic concentration in this part of the nucleus, and from electrophysiological and behavioral studies. On rather loose evidence it is often said that the amygdala exerts its effect on behavioral mechanisms via the hypothalamus.

For some general comments on so-called "psychosurgery," see Chapter 12.

[43] It has been suggested that in psychomotor epilepsy the amygdala is in a state of relatively greater excitability in patients showing explosive and aggressive behavior than in those who do not show these behavioral features. Studies on potentials evoked by olfactory stimuli appear to support this view (see Andy and Jurko, 1977).

[44] Amygdalectomy and posterior hypothalamotomy (see Chap. 11) have been claimed to be the most successful operations in the field of so-called "sedative neurosurgery."

11

The Autonomic Nervous
System
The Hypothalamus

THE AUTONOMIC (vegetative or visceral) system is a part of the nervous system
which has acquired a large and steadily growing importance in neurology as well
as in several other clinical disciplines. Therefore it deserves considerable interest
and a rather thorough description. Below, the most important anatomical features
will be considered, as well as some of the more fundamental physiological observations, and finally some clinical findings in disturbances of the autonomic system
will be dealt with.

Definition. Characteristics of the autonomic nervous system. Originally
the concept autonomic or vegetative system comprised only structures found outside the central nervous system, i.e., the sympathetic trunk and the large nervous
plexuses in the body cavities. Later the rami communicantes were discovered, and
it was ascertained that certain cells of the brainstem and spinal cord sent their
axons to the autonomic system. The delimitation of the latter then became less
clear, and this was even more the case after the discovery of integrative autonomic
centers in the diencephalon. Finally, the discovery that the cerebral cortex also influences the functions attributed to the autonomic system apparently makes any attempt at subdivision into cerebrospinal (somatic) and autonomic (visceral) systems
futile. Only more general sweeping definitions of the autonomic system are possible, like that given by Greving (1928): ''As autonomic nervous system the entire
mass of those nerve cells and fibers is designated which are concerned in the innervation of the internal organs, in so far as these are made up of smooth muscles
or belong to glandular organs. They subserve the regulation of processes which

usually are not under voluntary influence.'' From a morphological point of view a feature characteristic of the autonomic nervous system should be emphasized: *The efferent conducting pathway from the central nervous system to the innervated organs is always constituted by two succeeding neurons, one preganglionic* with its perikaryon in the central nervous system, and the other *postganglionic* with its perikaryon outside this. The nerve cells within the central nervous system which are concerned with autonomic functions as a rule differ also morphologically from those subserving the somatic processes, and frequently are found arranged as definite cell groups or nuclei.

Commonly the autonomic nervous system is subdivided into two major parts, the *sympathetic* or orthosympathetic and the *parasympathetic*. This distinction, introduced by Langley on the basis of physiological studies, applies mainly to the efferent components of the autonomic system. Whether a corresponding segregation can be made as concerns the afferent components is doubtful. It is common to speak of visceral afferent fibers and impulses only, thus avoiding the difficulty.

The preganglionic efferent neurons of the sympathetic nervous system in man have their perikarya in the spinal cord, more precisely in all the thoracic and the uppermost two lumbar segments. The preganglionic fibers in the parasympathetic system spring from perikarya situated in certain divisions of the brain stem and in the 3rd and 4th sacral segments of the cord. Synonymous designations for the sympathetic and parasympathetic system therefore are the *thoracolumbar* and *craniosacral* systems.

Another morphological difference between the two divisions of the autonomic system is seen in the position of the postganglionic perikarya. In the sympathetic system these are found in the vertebral and prevertebral ganglia. Those belonging to the parasympathetic are situated either in or on the walls of the organ they supply or in close proximity to it. From the anatomical arrangement of the sympathetic and parasympathetic it is also to be inferred that the former must be involved predominantly as a whole in diffuse reactions affecting the entire organism, whereas in the latter, the structural organization permits more restricted, localized effects. With certain qualifications this holds good.

In many instances these two systems exert an antagonistic influence: where one has a depressing, inhibitory effect on function, the other is stimulating, increasing, accelerating. Certain drugs act preferentially on one of the systems (thus atropine ''paralyses'' the parasympathetic, pilocarpine stimulates it). Furthermore, the two systems appear to be related to different hormones, noradrenaline being the specific sympathetic hormone, acetylcholine the parasympathetic. Dominated by these facts, conceptions of the autonomic system have shown a tendency to rigid schematization. However, advancing knowledge has made clear that great care must be taken in schematization, in this instance as in most others. The alleged antagonism between the sympathetic and the parasympathetic has especially proved to be far less absolute than originally assumed.[1] Both anatomical and physiological data opposing such a conception are available and a distinction is especially difficult as regards the higher levels of the autonomic centers. It is important to realize that the two divisions of the autonomic system in most instances

[1] Particularly as concerns the eye and the bladder the sympathetic innervation appears to play a minor role. The varying parasympathetic tone almost alone determines the autonomic status.

collaborate, in spite of their partly antagonistic effect on various functions. Both of them take part in the intricate regulation of visceral functions, ensuring a proper adjustment of the functioning of the various organs. This influence is not limited only to adjusting them in regard to each other, but in regard to somatic functions as well. Thus, to take a single example, vasomotor changes, as is well known, are involved in the functional levels of activity of the striated skeletal (somatic) musculature. Correspondingly, the morphological delineation of structures subserving visceral and somatic functions becomes extremely difficult at the higher levels of the central nervous system, where the highest integrative "centers" for the visceral functions cannot be separated from structures involved in somatic integrative functions. Bearing these qualifications in mind, the customary subdivision of the autonomic system into a sympathetic and parasympathetic division may be upheld, and in the following description it will be utilized.

The autonomic preganglionic neurons in the spinal cord and brainstem. *The perikarya of the preganglionic sympathetic neurons,* as already mentioned, are found in all the thoracic and the two upper lumbar segments of the spinal cord.[2] Some authors assume that some of the cervical segments contribute also, but this appears to be a rare exception according to Wrete (1930) and Pick and Sheehan (1946). However, Wiesman, Jones, and Randall (1966) have adduced evidence that in the dog and in man preganglionic sympathetic fibers arise in the lower cervical segments. The cells of origin form a distinct group in the lateral horn of the gray matter of the cord, the *intermediolateral cell column* (Fig. 11-1).[3] This nucleus is composed of rather small nerve cells, most of them considerably smaller than the motoneurons in the ventral horns. They are often gathered in small groups and vary in form, being either spindle-shaped or polygonal. The nucleus is relatively large, the nucleolus distinct. The rather scanty cytoplasm contains only finely dispersed tigroid granules. In this respect, also, these cells are clearly different from the motoneurons (cf. Fig. 11-1, a and b). With Golgi methods Réthely (1972) observed that the dendrites of the cells are strictly longitudinally oriented, as are also the fine axons which enter the intermediolateral cell column from the lateral funiculus. With the electron microscope he could show that the axons largely "climb" along the dendrites and establish synaptic contacts by means of three different kinds of terminals. The great majority of these are of the type which contains dense-core vesicles and are presumably noradrenergic. The neurons themselves, however, are rich in acetylcholinesterase (Navaratnam and Lewis, 1970).

That the cells of the intermediolateral cell column virtually send their axons through the ventral roots of the spinal nerves has been shown with the Golgi method in several animals (Bok, 1922; Poljak, 1924; Réthely, 1972). Recent as well as earlier studies of retrograde cellular changes following thoracic or lumbar sympathectomy (see Petras and Cummings, 1972, in the

---

[2] In other mammals upper and lower borders are not always identical with those in man. Confusion has arisen on this point because the number of thoracic, lumbar, and sacral segments of the cord varies in different species, a fact which has not always been considered when delimiting the sympathetic spinal outflow.
    [3] Some authors distinguish other cell groups in the cord as being of visceral nature, e.g., an intermediomedial cell group, placed close to the central canal. Laruelle (1937) traced this column throughout the cord, and assumed it to belong to the parasympathetic system. See also below.

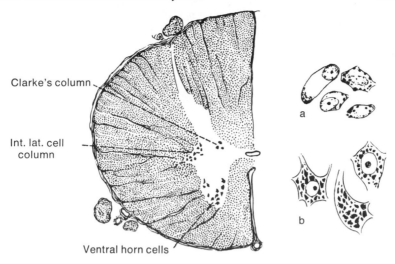

Clarke's column

Int. lat. cell
column

Ventral horn cells

a

b

FIG. 11-1 Drawing of a Nissl-stained preparation from the thoracic spinal cord of man. To the right, representative cells from the same section drawn with higher magnification. Note the differences between the somatic motoneurons from the ventral horn (b) and the visceral efferent cells from the intermediolateral cell column (a).

monkey) and studies of labeling of cells in the cord following HRP injections in sympathetic ganglia (Chung, Chung and Wurster, 1975) have confirmed that the great majority of preganglionic sympathetic neurons are localized in the intermediolateral cell column, but smaller contributions come from some medially situated cell groups as well. For a description of these cell groups in the dog, see Petras and Faden (1978).

It is of some interest that the intermediolateral cell column, in the cat extending from $C_8–Th_1$ to L_4, contains a larger number of cells in the male than in the female, on an average 55,500 and 45,700, respectively, and that the most densely populated levels of the column are $T_1–T_2$ and $L_3–L_4$ (Henry and Calaresu, 1972). According to Petras and Cummings (1972) the total number of preganglionic visceral efferent neurons in the cord exceeds that of somatic motor neurons.

The axons from the sympathetic preganglionic neurons leave the cord in the ventral roots and then pass via the ventral rami of the spinal nerves and the white communicant rami (to be described below) to the sympathetic trunk (Figs. 11-3, 11-4, 11-5).

Utilizing the retrograde transport of horseradish peroxidase following injections in single paravertebral ganglia, Faden and Petras (1978) found in the dog that preganglionic sympathetic cells were labeled in as many as six segments rostral or caudal to the injected ganglion. This indicates, in agreement with some physiological studies, that the preganglionic sympathetic fibers leaving the cord in one ventral root are derived from many segments and that some of these axons course longitudinally for some distance before leaving the cord. It further appears that preganglionic axons projecting to thoracic ganglia arise ipsilaterally, those to lumbar ganglia bilaterally.

The *preganglionic fibers of the craniosacral, parasympathetic system* take origin from several nuclei in the brainstem and spinal cord. The preganglionic fibers in the *sacral division* come from cells of the same type as those in the intermediolateral column, located as a corresponding column in the (2nd), 3rd, and 4th segments. The preganglionic fibers in the *cranial division* of the parasympathetic

join some of the cranial nerves, constituting their general visceral efferent components (see Chap. 7). The perikarya of these fibers are groups of cells situated more
or less in the proximity of the somatic efferent nuclei of the nerves. Visceral
efferent fibers of the parasympathetic type are present in the IIIrd, VIIth, IXth,
and Xth cranial nerves (cf. Fig. 11-2). The largest mass of them join the *vagus
nerve*. They spring from the *dorsal motor nucleus* of the vagus, situated immediately lateral to the hypoglossal nucleus, under the floor of the 4th ventricle
(Figs. 7-2, 7-4, 7-7). Transection of the vagus in the neck causes the cells of the
nucleus to undergo retrograde changes. The same is the case with the cells of the
nucleus ambiguus, the other effector vagus nucleus. As described in Chapter 7,
the cells in the nucleus ambiguus are of the same type as the motoneurons,
whereas those of the dorsal motor nucleus are similar to the cells in the intermediolateral cell column of the cord, suggesting that the nucleus ambiguus must
be concerned with the innervation of the striated muscles of the larynx, pharynx,
and esophagus supplied by the vagus, and the dorsal motor nucleus with the innervation of glands, smooth muscles, and the heart. Transection of the vagus
below the diaphragm, where it supplies no striated muscles, leads to retrograde
changes only in the dorsal motor nucleus. This has been taken to support the
notion above. However, recent evidence suggests that the innervation of the heart
may come from the nucleus ambiguus (see below).

On closer examination it is found that there are differences in the cell types in
different parts of the dorsal motor nucleus in animals as well as in man (Olszewski
and Baxter, 1954). Several authors have attempted to decide whether various
regions of the nucleus are related to the innervation of particular organs, by studying the retrograde cullular changes in the nucleus following transections of
branches of the vagus nerve. While Getz and Sirnes (1949) in the rabbit and
Mitchell and Warwick (1955) in the monkey found evidence that the fibers to the
lungs come largely from other regions than those to the heart and esophagus, other
authors (for example Szabo and Dussardier, 1964, in the sheep) found only
meager evidence in favor of a somatotopic localization, as did also authors working with the HRP method, for example Yamamoto et al. (1977), who injected
HRP in different parts of the wall of the stomach of the cat.

The notion that the dorsal motor vagal nucleus is the source of all visceral efferent fibers in
the vagus was challenged by Szentágothai (1952b), who failed to find degenerating fibers in the
nerve following lesions of the nucleus. Kerr (1969) reported that visceromotor effects on stimulation of the cervical vagus are not altered significantly if the dorsal motor vagal nucleus has been
destroyed several weeks previously (leaving time for efferent fibers to degenerate). He therefore
suggested that the nucleus may possibly give rise only to secretomotor fibers, and that the neurons
responsible for vagal visceromotor activity are found in an area between the nucleus ambiguus and
the spinal trigeminal nucleus. Several physiological studies support the view that cardiomotor
neurons are found in and close to the nucleus ambiguus (see, for example, Thomas and Calaresu,
1974; Spyer, 1975). Geis and Wurster (1980) after HRP injections in the pericardium of cats
found retrogradely labeled cells chiefly in the nucleus ambiguus and a smaller number in the
dorsal motor vagal nucleus. With the same method injections in the stomach give labeling in the
dorsal motor nucleus only (Ellison and Clark, 1975). In the vagal rootlets leaving the brainstem,
cardioinhibitory fibers are found in the caudal rootlets only, whereas motor fibers acting on the
esophagus and bronchi and duodenum occur in all rootlets (Kerr, Hendler, and Bowron, 1970).
The assumption that the cardiomotor vagal fibers are derived from the nucleus ambiguus is, however, weakened by the negative findings of Todo et al. (1977). Following injection of HRP in the

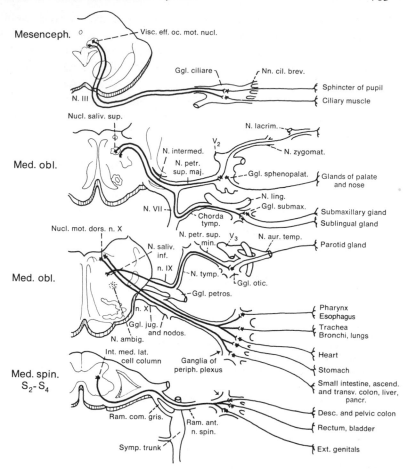

FIG. 11-2 Diagram of the craniosacral division of the autonomic nervous system. Preganglionic fibers indicated by heavy lines, postganglionic with light lines. The preganglionic fibers, their endings in the ganglia, and the distribution of the postganglionic fibers are indicated. Modified from Rasmussen (1932).

sinoatrial and atrioventricular regions of the heart in cats, these authors did not find labeled cells in the ambiguus but in the dorsal motor nucleus (and in parts of the nucleus of the solitary tract).

It is noteworthy that, according to the HRP studies of Satomi, Yamamoto, Ise, and Takatama (1978), some cells of the dorsal motor nucleus of the vagus innervate the descending colon and the upper rectum, which receive most of their preganglionic parasympathetic fibers from the sacral segments.

The visceral efferent (parasympathetic) nuclei related to the *glossopharyngeal and facial* (intermedius) *nerves* are termed the *inferior* and *superior salivatory nucleus,* respectively (Figs. 7-2 and 11-2), their most important function being the innervation of the salivary glands. They are usually described as forming two small groups of cells of the visceral efferent type, situated in the rostral prolongation of the dorsal motor nucleus of the vagus. The findings in these nuclei following transections of the nerves mentioned have been equivocal. Taking advantage of the modified Gudden method, Torvik (1957c) has been able to show in the cat that the visceral efferent fibers in the glossopharyngeal and vagus nerves come from cells lying rather scattered in certain regions of the reticular formation, which are not collected as clearly delimited nuclei. This has recently been confirmed with the HRP method by Hiura (1977).

The visceral efferent (parasympathetic) fibers, belonging to the *oculomotor* nerve and supplying the sphincter pupillae and the ciliary muscle, come from cells of the same type as those composing the other (general) visceral efferent nuclei and are grouped with the somatic efferent nuclei innervating the extrinsic ocular muscles, ventral to the aqueduct of the mesencephalon at the level of the superior colliculus, as mentioned in Chapter 7. These cells form the so-called *Edinger-Westphal nucleus,* situated on both sides of the median plane, dorsomedial to the somatic efferent cell groups (Figs. 7-2 and 11-2). From this nucleus preganglionic fibers follow the oculomotor nerve to the ciliary ganglion. The position of the different parasympathetic nuclei is indicated in the diagram of Fig. 11-2 (cf. also Chap. 7).

Terminal nuclei of the afferent autonomic fibers.   The efferent nuclei of the sympathetic and parasympathetic fibers differ only in their localization, not in their cell type. As concerns the *afferent fibers* of the two divisions of the autonomic system, with one exception it has hitherto not been possible to identify particular cell groups as being especially concerned in the transmission of visceral afferent impulses. The exception is the *nucleus of the solitary tract* (Figs. 7-2, 7-7, and 7-18), described in Chapter 7, forming the terminal station of the visceral afferent fibers of the facial (intermedius), glossopharyngeal, and vagus nerves. As mentioned previously, it receives some trigeminal afferents as well.[4] It is concerned chiefly in the transmission of visceral afferent impressions from the mouth, pharynx, lungs, heart, esophagus, and upper part of the digestive tube with its glands in the abdomen. (In addition, it transmits gustatory impulses; see Chap. 7.) These afferent impulses are important mainly in visceral reflexes, but it appears that the afferent vagus fibers especially are also involved in the transmission of certain diffuse sensations, such as nausea and others. Contrary to the afferent fibers in the sympathetic system, they appear not to be concerned with the conduction of pain. The afferent *sympathetic* fibers terminate on cells in the dorsal horn and intermediate zone of the spinal cord gray matter, but their precise synaptic relationships are not known. Our knowledge concerning the visceral afferent impulses is still scanty and, as referred to above, it is doubtful if it is appropriate to distinguish between "sympathetic" and "parasympathetic" afferent impulses. The formerly widely accepted fundamental difference between somatic and visceral afferent impulses also appears to be less well founded than formerly believed.

The peripheral parts of the autonomic nervous system.   A knowledge of the peripheral parts of the autonomic system[5] is of considerable practical impor-

[4] Other connections of the nucleus of the solitary tract have been considered in Chapter 7, section c.
[5] Exhaustive accounts of this subject can be found in the monographs of Mitchell (1953) and Pick (1970).

tance, first and foremost for the neurosurgeon. Starting with the *thoracolumbar, sympathetic system,* it has already been mentioned that the axons from the cells of the intermediolateral cell column leave the cord in the ventral roots. However, they soon leave these again and enter the sympathetic trunk (Figs. 11-3, 11-4) via the so-called *white rami communicantes.* The sympathetic trunk consists of a paired chain of cell aggregations, situated on the ventrolateral side of the vertebral column, forming the sympathetic, *vertebral ganglia,* and being interconnected by longitudinally arranged bundles of nerve fibers. Transverse connections are also present, in man only below the level of the 5th lumbar vertebra (Pick and Sheehan, 1946). The sympathetic trunk is connected with the spinal nerves by white and *gray rami communicantes.*

Originally a primordium of a sympathetic ganglion is laid down on each side for each segment of the spinal cord. During development this original pattern is partly lost. Two or more adjacent primordia may fuse, or one may partake in the formation of two ganglia. On account of these and other variations in development, there are usually only three ganglia in the cervical sympathetic trunk: the superior, middle (which may be absent), and inferior cervical ganglion. The inferior is furthermore frequently fused with the 1st and 2nd thoracic ganglia to form the so-called stellate ganglion. The *superior cervical ganglion* is large, elongated, and placed on the ventral aspect of the transverse processes of the upper two cervical vertebrae in proximity to the nodose ganglion of the vagus. When present, the *middle cervical ganglion* is found at the level of the 6th cervical vertebra. *The inferior ganglion,* or the *stellate ganglion,* is as a rule situated behind the subclavian artery. Between the middle and inferior ganglia an intermediate ganglion is commonly present. From this a loop of fibers, forming the *ansa subclavia,* goes around the subclavian artery, and unites the intermediate ganglion with the inferior. There is usually one ganglion for each segment in the thoracic part of the trunk, the first, however, being usually fused in the *stellate ganglion.* In the lumbar part there may be four or five ganglia, in the sacral part usually four are present, and lastly there is frequently an unpaired coccygeal ganglion. It is of practical revelance that individual variations exist in the number, size, location, and extent of fusion of the sympathetic vertebral ganglia.[6] It is also of considerable interest *that sympathetic ganglion cells* are not confined to the ganglia proper, but that such cells *are constantly found in larger or smaller numbers along the gray rami communicantes as intermediate ganglia,* especially in the cervical and lumbar regions (Wrete, 1935).[7] This is explained by the development, the ganglia arising from cells migrating along the dorsal and ventral roots (Kuntz, 1920). Since the axons from the cells of the intermediolateral column are myelinated, the rami conveying them from the spinal nerve to the sympathetic trunk are whitish, and generally are called *white rami communicantes.* Corresponding to the extension of the intermediolateral cell column in the cord, the spinal autonomic outflow and the occurrence of white rami is limited to the segments Th_1-L_2, inclusive

---

[6] Sheehan and Pick (1943) and Pick and Sheehan (1946) have, from studies in monkey and man, stressed the importance of numbering the ganglia not according to their place relative to the level of the vertebral column but according to their connections with the spinal nerves, the gray rami (where separately present) being the most reliable criterion.

[7] For an exhaustive account of the occurrence of intermediate ganglia in man and their clinical importance the monograph of Monro (1959) should be consulted.

(Sheehan, 1941). Some of the preganglionic fibers in the white rami end in the ganglion of the corresponding segment, but other fibers traverse the ganglion without interruption (cf. Figs. 11-4 and 11-5). Some of these preganglionic fibers ascend or descend in the sympathetic trunk, to end in ganglia situated farther cranially or caudally, where they end in synaptic contact with the postganglionic neurons. Other preganglionic fibers which pass through the vertebral ganglia con-

FIG. 11-3 Drawing of part of the right sympathetic trunk (black), showing some ganglia and their connection with the spinal nerves by communicant rami and the greater and lesser splanchnic nerves. Redrawn from Spalteholz.

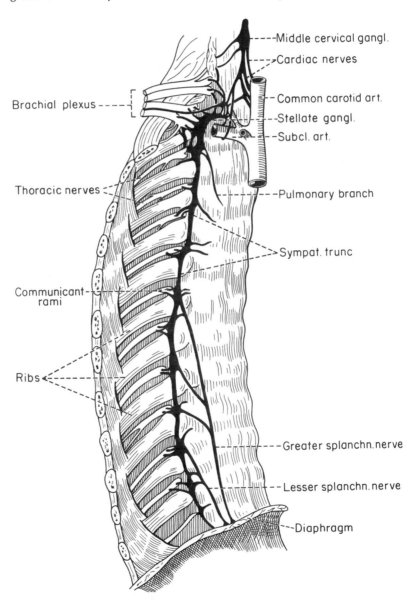

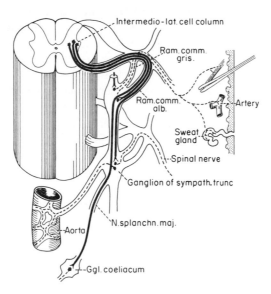

FIG. 11-4 Diagram of the arrangement of the efferent fibers in the sympathetic nervous system. *Preganglionic* sympathetic fibers (solid lines) from cells in the intermediolateral cell column leave the spinal cord through ventral roots and enter the sympathetic trunk via white communicant rami to end in one or more sympathetic ganglia. Some preganglionic fibers continue in the splanchnic nerves. Most *postganglionic* fibers (broken lines) from cells in the ganglia reenter the spinal nerve via gray communicant rami and follow the nerve peripherally, for example to the skin, supplying glands and smooth piloarrector muscles or muscles in small vessels. Other postganglionic fibers pass to large vessels, for example the aorta.

tinue in branches which pass from the sympathetic trunk to the *prevertebral ganglia* related to viscera. The most important of these latter branches are the *greater and lesser splanchnic nerves.* (It should be noted that these are made up of axons of preganglionic neurons, and correspond to white rami communicantes.) The former arises from branches from the 5th to 6th to the 9th to 10th thoracic ganglia (individual variations are usual), the latter from the (9th) 10th to 11th thoracic ganglia. Both pierce the diaphragm and enter the celiac plexus around the celiac artery, where most of the fibers of the greater splanchnic nerve establish connections with the postganglionic cells of the *celiac ganglion.* The celiac plexus extends along the branches from the abdominal aorta, forming several other plexuses with other ganglia, most conspicuous of which are the *superior and inferior mesenteric ganglia.* The fibers of the lesser splanchnic nerve, following the greater, are mainly distributed to the *renal ganglion,* and partly go to the adrenal medulla. In addition some fibers from the 12th thoracic segment, usually designated the least splanchnic nerve, participate in the innervation of the adrenal. The prevertebral ganglia considered above thus correspond to the vertebral ganglia of the sympathetic system. Both types of ganglia are the terminal station for preganglionic thoracolumbar fibers.

In the vertebral and prevertebral ganglia the postganglionic cells of the sympathetic system are found, sending their axons to the organs to be supplied. The postganglionic fibers, being unmyelinated, may behave in two different ways. They may return to the spinal nerves as gray rami communicantes or they may join the arteries and follow these as plexuses of nerve fibers. Gray rami are, in contradistinction to the white, found along the entire sympathetic trunk, connecting the ganglia with the corresponding spinal nerves. Postganglionic sympathetic fibers (mediating vasoconstriction, sweat secretion, and piloarrection) therefore are present in the spinal nerves.[8]

The fibers joining the arteries follow these as plexuses along their course. Thus several postganglionic fibers from the superior cervical ganglion accompany the internal carotid and its branches to supply the structures of the head with sympathetic fibers. Some of these fibers later join several of the cranial nerves (cf. Fig. 11-5). From the stellate ganglion fibers join the subclavian artery to supply the upper extremity (in addition to fibers joining the inferior cervical nerves via gray rami communicantes). Other branches from the cervical ganglia are the *superior, middle, and inferior cardiac nerves,* arising from the corresponding cervical ganglia where the postganglionic neurons are situated. These nerves take part in the formation of the *cardiac plexus,* in combination with postganglionic sympathetic fibers from the upper 4–5 thoracic ganglia and parasympathetic vagus fibers.

From the celiac ganglion and the associated abdominal ganglia, postganglionic sympathetic fibers accompany the branches of the abdominal aorta, as the hepatic, splenic, renal, phrenic and other plexuses. The fibers extend also along the ovarian and testicular arteries, and along the iliac artery and its branches, here being supplemented by fibers from the lumbar and sacral vertebral ganglia, especially in the formation of the (superior) *hypogastric plexus* from which sympathetic fibers are distributed to most of the pelvic organs. The postganglionic fibers to the intestine arise in the superior mesenteric ganglion for the small intestine and the ascending and transverse colon, in the inferior mesenteric ganglion for the descending and sigmoid colon and rectum. The fibers to the lower extremity are partly derived from plexuses along the arteries (which later join the nerves, cf. below), partly from gray rami from the lumbar and sacral sympathetic ganglia joining the corresponding spinal nerves.

It should be emphasized that, in addition to sympathetic fibers, the plexuses of the visceral organs also contain parasympathetic fibers, intermingling with the former in the plexuses (cf. below).

In spite of the restriction of the intermediolateral column to only a part of the spinal cord (Th_1–L_2), all regions of the body are supplied with sympathetic fibers.

---

[8] The distinction between white and gray rami has proved to be too schematic. Sheehan and Pick (1943) have emphasized that typical white rami are composed of myelinated fibers of a diameter up to $15\mu m$ whereas the gray rami are of two types, one containing almost entirely small unmyelinated fibers, the other, in addition, many small myelinated fibers (under $3\mu m$ in diameter). A few heavier myelinated fibers may also be present in the gray rami. A fourth type of ramus is of a mixed type, one part containing fibers of the type found in the white rami, the rest resembling the gray rami. The medullated fibers in the gray rami are assumed to be at least partly afferent fibers. These anatomical findings (Kuntz and Farnsworth, 1931) are of some practical relevance, since they yield support to the contention of surgeons that white and gray rami cannot be differentiated during operations. The occurrence of mixed rami explains why in some instances only one ramus is found, where theoretically two should be present. On the basis of electron microscope studies in the cat, Coggeshal and Galbraith (1978) have calculated the numbers of myelinated, unmyelinated, and sensory fibers in the midthoracic white and gray communicant rami.

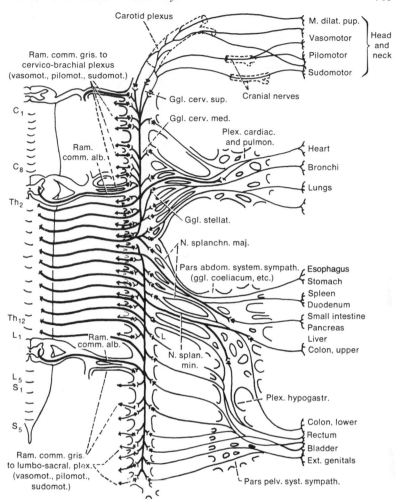

FIG. 11-5 Diagram of the sympathetic division of the autonomic nervous system. Same principle used as in Fig. 11-2. Modified from Rasmussen (1932).

This is made possible by the arrangement of the preganglionic sympathetic fibers, some of them traversing the sympathetic ganglia without interruption, to ascend or descend in the sympathetic trunk. Thus many of the fibers originating in the upper thoracic segments end in the superior cervical ganglion, and the sacral vertebral ganglia receive their preganglionic fibers from the lower thoracic and upper two lumbar segments of the cord. This arrangement is diagrammatically represented in Fig. 11-5. The sympathetic innervation of the extremities is not included, but is evident from Figs. 11-6 and 11-7 and is represented also in the table on p. 765. Concerning the segmental sympathetic innervation, see below.

Turning to the *peripheral parts of the parasympathetic system,* it has already been mentioned that the preganglionic fibers of the cranial division join several of the cranial nerves. Contrary to the arrangement in the sympathetic system, the

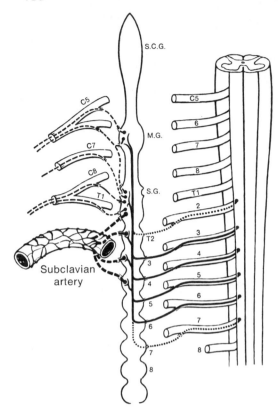

FIG. 11-6 Diagram of the origin and course of the sympathetic fibers to the upper extremity. The dotted preganglionic fibers are somewhat equivocal. Postganglionic fibers are indicated by broken lines. *S.C.G.:* superior cervical ganglion; *M.G.:* middle cervical ganglion; *S.G.:* stellate ganglion. From Haymaker and Woodhall (1945) after Foerster.

preganglionic fibers of the parasympathetic system can be followed to the organs to be supplied or close to them, where the fibers then end in synaptic contact with the postganglionic cells, forming discrete ganglia or more scattered ganglionic plexuses in the walls of the organs.

The *visceral efferent fibers in the oculomotor nerve* accompany the somatic fibers of the nerve to the orbit, where they separate from them and enter the *ciliary ganglion,* behind the eyeball (see Fig. 11-2). From here the postganglionic fibers reach the eye as the short ciliary nerves, and terminate in the ciliary muscle and the sphincter of the iris.

The *visceral efferent preganglionic fibers following the facial nerve,* as part of the intermedius, originate from the so-called superior salivatory nucleus (see above) and pursue a complicated course (Fig. 11-2). Some of them leave the nerve in the facial canal, and traverse the tympanic cavity as the chorda tympani which after having left the cranium joins the lingual nerve. In its terminal branches to the *sublingual and submandibular glands* several nerve cells are present, representing the perikarya of the postganglionic neurons and passing to the glandular cells. These cell aggregates are usually termed the *submandibular ganglion.* Other visceral efferent fibers in the intermedius leave the nerve at the geniculate ganglion as the great superficial petrosal nerve, pass through the sphenoid bone (pterygoid canal) to the *sphenopalatine ganglion* in the sphenopalatine fossa. The postganglionic fibers from this ganglion partly pass to the maxillary nerve, and through an anastomosis with the lacrimal nerve reach the lacrimal gland (Fig. 11-2). Other fibers are distributed to the nasal cavity (sphenopalatine nerves) and the oral cavity through the descending palatine nerves.

The *visceral efferent preganglionic fibers of the glossopharyngeal nerve,* arising from the "inferior salivatory nucleus," leave the nerve at the petrous ganglion in the petrous fossa when the nerve has just left the jugular foramen (Fig. 11-2). Forming the tympanic nerve the fibers pierce the floor of the tympanic cavity, ascend on the medial wall, and penetrate the thin roof of the cavity to appear as the lesser superficial petrosal nerve intracranially. This then again leaves the cranial cavity and ends in the *otic ganglion,* medial to the exit of the third division of the trigeminal nerve in the foramen ovale. The postganglionic fibers from this ganglion pass through the auriculotemporal nerve to the *parotid gland.*

FIG. 11-7 Diagram of the sympathetic supply of the lower extremity following the same principle as in Fig. 11-6. The parasympathetic fibers to the pelvic organs are included on the right side. From Haymaker and Woodhall (1945) after Foerster.

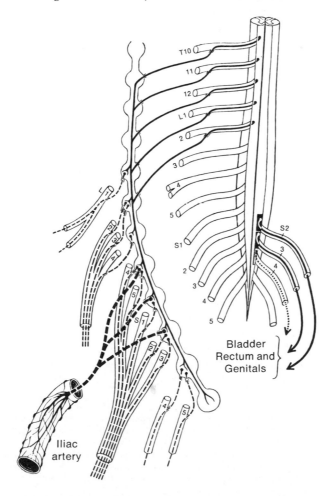

By far the most important contingent of preganglionic parasympathetic cranial fibers, however, is contained in the *vagus nerve*. The fibers arise from the dorsal motor nucleus (as mentioned above, those to the heart possibly from the nucleus ambiguus). These visceral efferent fibers accompany several of the rami of the nerve (see Chap. 7 concerning these). They end in ganglia and plexuses containing ganglion cells in or near the innervated organs (see Fig. 11-2). Thus fibers in the *pharyngeal rami* end in relation to ganglion cells in the pharyngeal plexus, from which the glands of the pharynx are innervated. Other fibers are distributed in a similar manner to the *esophagus* (secretory fibers and motor fibers to the smooth muscles). The *heart* receives parasympathetic fibers through the superior, middle, and inferior cardiac branches, which terminate on cells in the cardiac plexus, particularly well developed around the beginning of the aorta and pulmonary artery. This plexus continues also along the coronary arteries.[9] In addition to parasympathetic fibers, the plexus mentioned above contain sympathetic fibers, as do also the tracheal, bronchial, and pulmonary plexuses, supplied by branches from the vagus. Having pierced the diaphragm the left vagus is distributed mainly on the anterior surface of the stomach, the right on the posterior, and they then take part in the formation of the celiac plexus. Other vagus fibers from the celiac plexus accompany the sympathetic fibers in the plexuses surrounding the arteries and provide parasympathetic fibers to the duodenum, pancreas, small intestine, and large intestine to the left flexure. Fibers to the liver and gall bladder, participating in the hepatic plexus, are derived mainly from the right vagus, and end in relation to postganglionic cells in the hepatic plexus in the porta hepatis and in the wall of the gall bladder and cystic and common bile ducts (Alexander, 1940). In the wall of the stomach a gastric plexus is formed, consisting of fibers and postganglionic cells with short axons. One part of the plexus is found between the muscular layers (plexus myentericus), another in the submucosa (plexus submucosus). Sympathetic fibers are also present in these plexuses which appear to possess a certain degree of local autonomy. Plexuses of a similar type are found also in the other parts of the digestive tube in the abdomen. Collectively they are frequently termed *"enteric plexuses."* The postganglionic fibers of these plexuses are in general secretory and motor.[10]

The *preganglionic fibers belonging to the sacral division of the parasympathetic system* (cf. Fig. 11-2 and Fig. 11-7 to the right) leave the cord in the ventral roots from the 3rd and 4th sacral nerves (occasionally also S_2 and S_5). They follow the nerves forming the pudendal plexus, and leave them as the pelvic nerves (nervi erigentes) on each side of the rectum. Following transection of the pelvic nerve in the cat, Oliver, Bradley, and Fletcher (1969a) found retrograde changes in most of the cells in the intermediolateral cell column in S_2 and S_3, and in some of the cells in the intermediomedial cell column. HRP studies appear to be in agreement. Thus, Sato, Mizuno, and Konishi (1978) found labeling of these

---

[9] The innervation of the heart is treated more fully later in this chapter.

[10] According to Kuntz (1938, 1940) there are afferent fibers from these plexuses to the celiac and mesenteric ganglia. Following interruption of all centrally derived afferents to the ganglia, some fibers with intact terminal arborizations are still present in them. The presence of such axons from the enteric plexuses to the prevertebral ganglia might explain the fact that visceral reflex activity, for example, of the stomach, is still possible after the celiac ganglion is separated from the central nervous system. If this is so it will be unnecessary to regard these functions as due to axon reflexes.

cells following injection of HRP in the wall of the urinary bladder in cats. Peripherally the preganglionic parasympathetic fibers come in contact with the hypogastric plexus and are intermingled with sympathetic fibers. They synapse with postganglionic neurons situated within the walls of the bladder, rectum, and genitals; the descending and sigmoid colon also receive their parasympathetic innervation from the sacral division.

In spite of the restricted origin of the parasympathetic preganglionic neurons, it is seen that at least the majority of the *visceral* organs are supplied by parasympathetic fibers as well as sympathetic. In their peripheral course the fibers of the two divisions of the autonomic system are intermingled, but they retain their functional independence. The principal points of autonomic innervation are summarized in the table on p. 765 in which some functional aspects are also included.

The minute structure of the autonomic ganglia. The nerve cells composing the autonomic ganglia differ in certain respects from those in the central nervous system. Several cell types may be distinguished and there are also structural differences between the various ganglia.[11] In spite of this, however, there are enough common features to allow us to regard these cells as belonging to a fairly uniform group. Reliable criteria which permit a distinction between cells and fibers belonging to the sympathetic and parasympathetic have not been found.

The *autonomic ganglion cells* of the ganglia of the sympathetic trunk and in the peripheral ganglia are usually multipolar and are characterized by numerous and long dendrites (Fig. 11-8). The tigroid substance is finely granular and pigment inclusions are common. The cells, like the spinal ganglion cells, are always surrounded by a capsule with flattened cells, homologous to the glial cells, lying in close contact with the nerve cells (sheath cells). The numerous dendrites may arborize abundantly within the capsule or they may penetrate it. On account of the numerous and long dendrites it is often difficult to identify cell processes which can be designated as axons, but under fortunate circumstances such may be seen. The axon is frequently very thin, and may either spring from the perikaryon or from one of the dendrites. The axons from the postganglionic cells are usually unmyelinated.

Between the ganglion cells an extremely intricate feltwork of fine fibers is present. These fibers are partly dendrites which have pierced the capsules, partly axons and collaterals from preganglionic fibers terminating in the ganglion, and, finally, some of them are fibers passing through the ganglion without interruption. The terminal parts of the preganglionic fibers are frequently unmyelinated or covered with a very delicate myelin investment only, and on this account they cannot, or can only with difficulty, be distinguished microscopically from the dendrites present.

The mode of termination of the axons is highly variable. They may penetrate the capsule surrounding the ganglion cells and end in an intracapsular plexus, or

[11] Numerous authors have attempted classifications of autonomic ganglion cells according to size, arrangement of Nissl granules, dendritic arborizations, or other criteria (for some references see Hillarp, 1960; Gabella, 1976). Even if there are some physiological indications that the cells in a ganglion may differ functionally, correlations with morphological differences between cells are so far not possible. The ganglia in man appear to be built like those in animals (Kirsche, 1958; see, however, Pick, 1970).

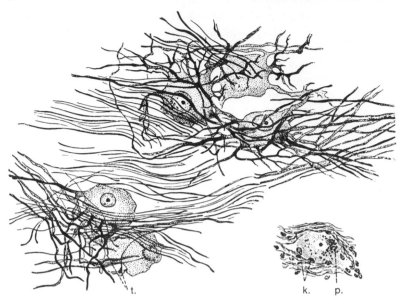

FIG. 11-8   Drawing of cell types from autonomic ganglia. Lower right, a cell from
the celiac ganglion in man (autopsy material, modified Cajal method with contrast
Safranin staining). The other cells are from a lumbar sympathetic ganglion of man
(Bodian's method on autopsy material). *t:* terminal ramifications; *k:* sheath cells; *p:*
argentaffine pigment (see text). Courtesy of J. Cammermeyer.

they may end outside the capsules interwoven between dendrites which have pene-
trated the latter. The axons as well as the dendrites frequently pass in several
spirals around the cells, and their course within the ganglia is frequently tortuous
and of considerable length. The fibers end with different kinds of terminal formations (ordinary end bulbs, chalyces, pericellular arborizations).

The *nerve bundles interconnecting the autonomic cell groups* are sometimes
composed of predominantly myelinated fibers, as, for example, the white rami
communicantes and the splanchnic nerves. Sometimes the majority of the fibers
are unmyelinated, as is the case with the gray rami communicantes and the plexuses around the arteries. As a general rule it may be said that preganglionic fibers
tend to be myelinated, postganglionic not, but there are exceptions. In the vertebral ganglia thicker myelinated fibers take their course through them.

In view of the complicated pictures presented by the autonomic ganglia, it is
no wonder that opinions have differed concerning their interpretation. *Electron microscopic studies* of the *structure of the autonomic ganglion cells* have largely
confirmed conclusions drawn from light microscopic work, and in addition demonstrated some details (see, for example, Pick, 1970).

On the basis of light microscopic studies some early authors had maintained
that the nerve cells form a true syncytium. According to this view the neuron
theory should not be valid for the peripheral parts of the autonomic system. However, even before the advent of electron microscopy most authors (Kunz, 1938,
and several later workers) were convinced that the preganglionic fibers that do not
bypass the ganglia are in true synaptic contact with its cells, as Langley was led to
conclude from his physiological experiments. This has been fully confirmed in

electron microscopic studies (Taxi, 1965; Grillo, 1966; and others). Numerous studies devoted to the structural organization of the autonomic ganglia have clarified further details. Only a few will be mentioned here (for a recent review, see Gabella, 1976). Thus different kinds of nerve endings have been identified electron microscopically, and on most points there is rather good correspondence with light microscopic pictures, as appears, for example, from Taxi's (1965) extensive study. Axosomatic and axodendritic synapses have been seen on the ganglion cells, but in most ganglia studied the great majority are axodendritic. There appear to be variations among ganglia and between species in this respect. In spite of variations with regard to the fine structure of synaptic contacts the synaptic structures in the autonomic ganglia in general correspond to those in the central nervous system (Szentágothai, 1964c; Grillo, 1966; and others).

The *preganglionic autonomic fibers* entering the ganglia, for example the ciliary ganglion or the superior cervical ganglion, can be stained using procedures demonstrating the presence of cholinesterases (see for example Szentágothai, Donhoffer, and Rajkovits, 1954; Taxi, 1965), and it appears that acetylcholine is present in the fibers and their synaptic endings. Most of these contain small agranular synaptic vesicles, assumed to be engaged in cholinergic transmission. However, in sympathetic ganglia, many terminals contain in addition a varying number of granular vesicles (dense-core vesicles), small or large. Particularly the former appear to be related to monoamines. (It has been suggested that these endings may belong, not to preganglionic fibers, but to axons of cells in the ganglion.)

Confirming earlier, more indirect evidence, histochemical studies of the *postganglionic neurons* and their usually unmyelinated axons have made it clear that there are important differences between ganglia belonging to the sympathetic and the parasympathetic system. In the vertebral and prevertebral *sympathetic ganglia* most of the cells are noradrenergic as determined with the Falck-Hillarp fluorescence method (see Chap. 1), as are their axons and terminal branches in the organs they supply. However, a minor proportion of the ganglion cells and their axons are cholinergic when studied histochemically. These cholinergic sympathetic fibers have been shown to supply the sweat glands and to act as vasodilators in skeletal muscle and the tongue. However, in the *parasympathetic ganglia,* for example the ciliary ganglion, all neurons appear to be cholinergic. Whether the same is valid for all parasympathetic ganglia is not clear.

On the whole, recent findings are thus in agreement with conclusions reached in earlier physiological and pharmacological studies which indicate that as a general rule, preganglionic and parasympathetic postganglionic neurons are cholinergic, sympathetic postganglionic neurons adrenergic. However, there are still many unsolved problems related to the structural and functional organization of the autonomic ganglia. Thus the synaptic arrangements within the ganglia are far from being fully known. There is reason to believe that the autonomic ganglia are not simple relays, serving only the transmission of impulses from preganglionic to postganglionic neurons (see Hillarp, 1960, for some data). A question of particular interest is whether there are internuncial cells in the ganglia.

Many sympathetic ganglia contain so-called *chromaffin cells,* i.e., cells which can be made visible with bichromate salts. (For an extensive review, see Gabella, 1976.) They are small and have a few short processes. Electron microscopically they have been shown to contain large granular vesicles and appear to be rich in catecholamines. Their functional role is not clear. The suggestion that they may function as interneurons has so far not been finally confirmed.

In addition to the *differences* considered above *between the sympathetic and the parasympathetic system, a further difference is noteworthy: The ratio of preganglionic to postganglionic neurons is not the same.* Counts of preganglionic fibers entering an autonomic ganglion and of its cells show that in the parasympathetic system the ratio is low. In the ciliary ganglion of the cat, Wolf (1941) found the ratio between preganglionic fibers and ganglion cells to be only 1:2. In the sympathetic system the ratio is much higher. Billingsley and Ranson (1918) counted more than 120,000 cells in the superior cervical ganglion of the cat, whereas the sympathetic trunk immediately below the ganglion contains only 3,800 myelinated fibers, presumably preganglionic. This gives a ratio of about 32 cells to each preganglionic fiber. Other values have been reported (see Gabella, 1976, for a review). It appears that the number of ganglion cells in a particular ganglion shows species differences, and increases with body weight, as does also the ratio above. In the superior cervical ganglia from three humans Ebbesson (1968b) found from 760,000 to 1 million cells and a preganglionic–postganglionic ratio of 1:63 to 1:196. In human lumbar sympathetic ganglia Webber (1958) found ratios ranging from 1:28 to 1:183. (The figures calculated in such counts will depend on the types of fibers included.) These anatomical observations are in accord with a number of physiological findings which show that, in general, the parasympathetic has considerable possibilities for influencing local functions, while the sympathetic by way of its more diffuse distribution tends to have widespread and general bodily effects (for example on the cutaneous vessels). It should further be noted that anatomically the parasympathetic division is rather clearly fractionated into minor separate cell groups concerned with the regulation of particular functions or organs. This is perhaps most clearly seen in the cranial part, where, for example, in the oculomotor nucleus particular cell groups are delegated to the control of the intrinsic ocular muscles only (for further differentiations, see Chap. 7, section g). Likewise the various salivary glands and the lacrimal gland are supplied by particular cell groups. Segregations of this kind appear to be less marked in the intermediolateral cell column, but are not entirely absent.

> While many preganglionic neurons in the cranial division of the parasympathetic system send their axons to rather well delimited groups of postganglionic neurons (the ciliary ganglion, submandibular ganglion etc.), in other instances, particularly in the sacral division, the postganglionic cells are diffusely distributed in the walls of the target organ, for example in the myenteric and submucosal plexuses in the intestine, the cardiac plexus, and others. The number of nerve cells in these ganglia appears to be considerable, but the values found by different authors show great discrepancies (see Gabella, 1976, for a list) presumably because different techniques were used. (In the duodenum of the cat, for example, the number of ganglion cells per cm² was found by one author to be 49,081, by another 12,170.) The minute organization of such ganglia has been particularly difficult to clarify, in part because of the wealth of fibers present in the nerve plexuses where the ganglion cells occur. While the ganglion cells are not adrenergic, noradrenergic fibers are abundant in these ganglia. Many of these fibers disappear following extrinsic denervation of the ganglionic collection, and appear thus to be sympathetic fibers, passing through the ganglion. Other endings persist, however, and appear to be derived from neurons in the ganglion, possibly projecting to other ganglion cells.

The peripheral autonomic fibers. Neuroeffector mechanisms.   All organs innervated by the autonomic system contain plexuses of very fine nerve fibers.

The details of the organization of the plexuses and of the relations between the nerve fibers and the effector structures (smooth muscles and glandular cells) have been extremely difficult to clarify.

On the basis of studies of silver-impregnated sections, many histologists— Boeke, Støhr, and others—strongly advocated that the terminal nerve fibers form an anastomosing network of extremely thin fibers, a syncytial "terminal" or "preterminal reticulum" or "sympathetic ground plexus." The terminal fibers of the plexus were held to be continuous with the cytoplasm of the effector cells. The cells found in between the fibers of the plexuses, generally called "interstitial cells," have been the subject of much debate. They have been claimed to be connective tissue elements, supporting cells of the Schwann cell type, or nervous elements interposed between the terminal fibers and the effector cells. When electron microscopic studies became possible several of the controversies were solved.

The fibers in the peripheral autonomic plexuses have been confirmed to be extremely thin, as small as some 200 Å. (This may explain why they have been misinterpreted in the light microscope as "neurofibrils.") The unmyelinated fibers are covered by the cytoplasm of cells of the Schwann cell type, sometimes called lemmocytes. Within the "cisterns" of these cells there are usually a number of thin axons, a fact which may be of relevance for the conduction of impulses, since depolarization of one axon may influence other axons running in the same cistern (see Ruska and Ruska, 1961). Convincing evidence of a syncytial connection between the lemmocytes has not been obtained. In addition to the lemmocytes the autonomic plexuses contain some other interstitial cells. These differ structurally from the lemmocytes and are most likely connective tissue elements (Richardson, 1958; Taxi, 1965; Rogers and Burnstock, 1966). Anastomoses between individual nerve fibers have not been observed, even when the fibers run in interlocking bundles, together making up the plexus seen in the light microscope. Since the diameters of the postganglionic fibers are very small (0.3–1.3 μm) their conduction velocity is slow, 0.7–2.3 m/sec. With the Falck-Hillarp technique terminal fine branches in the plexuses in several organs have been found to contain noradrenaline, indicating that they are derived from postganglionic sympathetic neurons (see Norberg, 1967, for an extensive review). There are quantitative differences between organs with regard to terminals of noradrenergic fibers (Fillenz and Pollard, 1976). Other fibers in the plexuses appear to be cholinergic (parasympathetic) when studied histochemically or as deduced from the type of synaptic vesicles in their terminals.

> There are considerable differences among organs (as well as among species) as concerns the proportion of the two types of terminals. For example, in the iris of the guinea pig the finest terminal fibers are cholinergic; only about 15% have been found to be noradrenergic (Nishida and Sears, 1969). The latter are thought to be possibly inhibitory. In the alimentary tract of the same animal the proportion of noradrenergic fibers is much greater in the muscular layer of the sphincter regions than elsewhere (Furness and Costa, 1973; and others. For data on other organs, see Gabella, 1976.)

The question of *how impulses in the peripheral fibers of the autonomic system activate the effector elements* (smooth muscle and glandular cells) in the autonomically innervated organs has been far more difficult to solve than the corresponding question for the somatic nervous system (the innervation of extrafusal and intrafusal muscle fibers, see Chap. 3). There are still many open questions, and there are divergent views on some points. Noradrenergic terminals have often been found to establish synaptic contact with neurons (postganglionic parasympathetic) in the plexus of the intestine, and it has been suggested that this arrangement may be the basis of the sympathetic inhibition of intestinal motility (Norberg, 1967). Other fine fibers in the plexuses have been found to end in relation to effectors.

The view that there is a cytoplasmic continuity between the peripheral autonomic nerve fibers and the effector cells has, however, had to be discarded. Again electron microscopic observations have settled an old dispute. Such studies are extremely difficult to perform, and all organs receiving an autonomic innervation have not yet been examined. Most studies have been concerned with the innervation of smooth muscles (see for example Richardson, 1962, 1964; Taxi, 1965). Some authors have studied glands (for some references, see Pick, 1970; Gabella, 1976). It appears that the situation is not exactly the same in all organs (Taxi, 1965). Most favorable for study are regions with a rich supply of autonomic innervation, for example the vas deferens and the iris (see Fig. 11-9A). In the vas deferens of the rat (Richardson, 1962; see also Dixon and Gosling, 1972)

FIG. 11-9 *A:* The iris of the rat, 2 days following partial sympathetic denervation (Falck-Hillarp method). Since the great majority of adrenergic fibers have degenerated, the course of a single remaining fiber can be identified by its fluorescence. Note profuse branching of terminal fibers. The varicosities are not too clearly seen at this magnification. From Olson and Malmfors (1970).

  B: Two varicosities of a postganglionic fiber forming close contact relationships with two muscle cells. Over half the surface of each varicosity (*V*) is covered by muscle (*m*) and separated from it by less than 200 Å. Synaptic vesicles are seen in both varicosities and the narrow part (*I*) of the axon between them. From Bennett (1972).

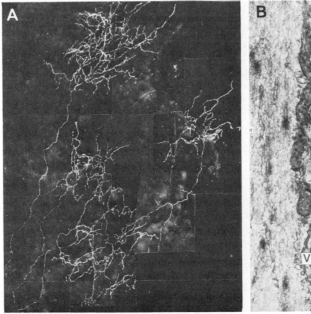
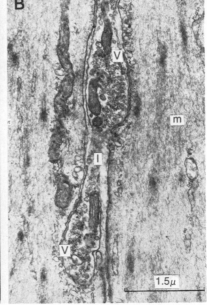

the fine, unmyelinated axons pass without a Schwann cell covering between the muscle fibers. At their ends and along their course, the nerve fibers show small swellings in apposition to the muscle fiber membrane. In some organs, for example the vas deferens, there is close contact between the swellings and the muscle fiber (Fig. 11-9B). In other organs, such as the uterus and the gastrointestinal tract, there is a distance of some 1000–3000 Å (see Bennett, 1972). The swellings or varicosities contain vesicles and mitochondria and in several respects resemble terminal boutons and boutons en passage in the central nervous system. The vesicles are of different kinds.

Some are small (300–600 Å) and may be clear or contain a granular core. Others are large (600–1500 Å) and always contain a granular core. One kind of terminal or varicosity contains small agranular and granular vesicles; in another type there are only large granular vesicles (which may, however, occur in the other type as well). Largely corresponding findings have been made in many other organs (for references see Grillo, 1966; Bennett, 1972; Gabella, 1976).

The electron microscopic findings permit the conclusion that in autonomically innervated organs there are structures that may be considered neuroeffector junctions, in agreement with the conclusion of Burnstock and Holman (1961) from physiological studies.[12] It is generally assumed that the different types of synaptic vesicles can be related to different kinds of transmitters. While the nature of the large, dense-core vesicles is still disputed, it appears that the small, clear vesicles are found in terminals of the cholinergic type; the small, dense-core vesicles are found in noradrenergic synapses. For example, in the iris of the rat, the sphincter, which has a rich parasympathetic innervation, has nerve endings containing vesicles of the agranular type, while granular vesicles are found in the dilator muscle, supplied by the sympathetic (Richardson, 1962). However, the common occurrence of different types of vesicles in the same terminal indicates that the situation is less schematic than appears from this finding. In spite of extensive research with refined methods, a definite correlation between a particular transmitter and a particular type of vesicle in the terminals has not yet been established (for some data, see Geffen and Livett, 1971; Bloom, 1972).

While there seems to be general agreement that the *autonomic effectors are set in action by chemical transmission*, the mode of action of the transmitters on the target cells has been the subject of much discussion. The transmitter appears to be transported along the axon and at the synaptic site to be bound to a receptor substance and involved in complex enzymatic reactions. In the adrenergic systems α- and β-receptors (possibly more) can be distinguished. The α-receptors are concerned in certain functions, such as vasoconstriction, intestinal relaxation, and dilatation of the pupil, while other effects, such as vasodilatation (especially in muscle), increase in force and contraction of the heart, and bronchial relaxation, are mediated via β-receptors.

It is of practical interest that certain chemical substances and drugs act specifically on either α- or β-receptors and may block the synaptic transmission. β-blockers, for example, may counteract the sympathetic, noradrenergic effects on the heart. These blocking agents have been assumed to act by competing with the natural transmitter at the receptor sites. They have turned out to be

---

[12] Other possible ways of activation of the effector organs by autonomic fibers have been discussed as well (see von Euler, 1959). For an extensive account of the anatomy and physiology of neuromuscular autonomic innervation, see Bennett (1972).

effective therapeutic agents in several diseases where disorders of the sympathetic nervous system occur (for some data see Johnson and Spalding, 1974).

As has been repeatedly mentioned, the transmitter at parasympathetic and some sympathetic postganglionic terminals (as well as at all preganglionic terminals) appears to be acetylcholine. The action of acetylcholine at the peripheral endings of parasympathetic fibers has long been known to be inhibited by atropine, which acts as a cholinergic blocker, while anticholinesterases have the opposite effect. However, the cholinergic synapses of the preganglionic fibers are apparently not influenced by atropine.

Atropine has been utilized clinically in several instances, e.g., in ophthalmology to dilate the pupil (homatropine). In bronchial asthma, relaxation of the bronchial muscles and reduced secretion result, while in the stomach, peristalsis and secretion of gastric juice are reduced. As previously mentioned, sweat secretion is diminished or abolished by atropine, since the sympathetic innervation of the sweat glands is cholinergic.

Pilocarpine has the opposite effect. In general, it stimulates cholinergically innervated organs and thus produces secretion of salivary, lacrimal, sweat glands, and gastrointestinal glands—and causes contraction of the pupil. The effect is similar to the action of *eserine, physostigmine,* and other anticholinesterases, which act by inhibiting acetylcholinesterase. (These act on motor end plates as well and are useful in, for example, myasthenia gravis.)

Today, several drugs and substances are known that may influence the synaptic transmission in the autonomic system.[13] Much information on this subject, obtained in animal studies, has been made use of in the treatment of diseases in man. In the clinical use of ganglion blockers or other substances acting on the autonomic system, it is important to be aware that *the effects are usually general and will influence transmission at all synapses of a particular kind.* Accordingly, one often sees undesirable side effects, more or less predictable on the basis of knowledge of the autonomic nervous system, such as diarrhea and bradycardia when patients with myasthenia gravis are treated with anticholinesterases.

Increased knowledge of the role and the chemistry of the transmitter substances has furthered the understanding of the effects of certain drugs on autonomically controlled functions. A rational basis has been found for many well-known observations. On the other hand, progress in neurochemistry and neuropharmacology has shown that these problems are extremely complex and the mechanisms involved very intricate.

Sensitization of autonomically denervated organs. Regeneration in the autonomic system. It has been known for some time that when the sympathetic innervation of an organ is interrupted by section of the postganglionic fibers or extirpation of the ganglion in which the postganglionic cells are located, the effector organ, in a week or two, usually shows an increased susceptibility to the action of noradrenaline and adrenaline, whether this be excitatory or inhibitory to the organ in question. For example, following extirpation of the stellate ganglion in Raynaud's disease the immediate effect is a vasodilatation in the head, neck, and arm on the operated side. During the following days, however, this becomes less conspicuous, and a week or so after the operation, a tendency to vasoconstriction is clearly evident on occasions in which an extra output of adrenaline (and noradren-

[13] For a recent survey, see Carrier's textbook (1972).

aline) occurs, for example under emotional stress or exposure to cold. It appears that the effector organ acquires an increased susceptibility to adrenaline, commonly referred to as *sensitization* or "supersensitivity." This phenomenon has been observed not only in adrenergically innervated organs, but in cholinergically innervated organs as well (it applies to the pupillary contraction and to sweat secretion in response to acetylcholine). The *postganglionic* neurons can also become sensitized after section of the preganglionic fibers, although clinically the phenomena of sensitization are less marked after preganglionic than after postganglionic ramisection.

The supersensitivity following autonomic denervation appears to be an example of Cannon's "law of denervation." When one neuron in a series of efferent neurons is destroyed, there develops an increased irritability to chemical agents in the isolated structures, most marked in the directly denervated part. Several attempts have been made to explain this sensitization, but no generally accepted theory has been set forth (for some data see Hillarp, 1960). The phenomenon is of some clinical interest, not least in surgery on the sympathetic nervous system, although its importance has been questioned by some authors (see, for example, Monro, 1959). Alternative explanations for the recurrence of autonomic functions after surgery have been set forth.[14] As described in an earlier section of this chapter, sympathetic postganglionic neurons are regularly present outside the vertebral (and prevertebral) ganglia proper, in the sympathetic trunk and the communicant rami. If these ganglion cells are not removed, for example in a sympathectomy, they may be imagined to be responsible for the recurrence of symptoms. It has further been suggested that regeneration may be responsible. Transected fibers from ganglion cells not removed may regenerate and reinnervate the target, or collaterals of axons from intact cells may be thought to grow out to take over the innervation of the removed fibers. Much speculation and hypothetization has been devoted to this question in the past, since convincing data have been difficult to produce (see Pick, 1970, for some early references). Furthermore, it has been pointed out that humoral factors, for example bradykinin, may play a role, particularly with regard to vascular changes. Recent experimental research, anatomical as well as physiological, has confirmed previous data showing that autonomic neurons possess great regenerative capacities.

Most experiments have been done on the superior cervical ganglion, following transection of the cervical sympathetic trunk. In the cat and rabbit, Butson (1950) observed functional recovery following such transections within 2 to 4 weeks (as measured by the disappearance of the operatively produced miosis, ptosis, and vasodilatation of the face). Following transection of the cervical sympathetic trunk the great majority of synaptic endings in the ganglion have been found to degenerate and finally disappear, while the (postganglionic) neurons in the ganglion appear to show little change. Electron microscopically, the denervated, vacated, synaptic sites can be seen to persist for a long time. However, when the presynaptic (preganglionic) fibers in the trunk are allowed to regenerate, synapses begin to reappear after 1 month, and after 2 months their numbers

[14] Experimentally, a sympathectomy can be achieved by chemical and immunological means as well (see Thoenen, 1972, for a review). For example, following prolonged administration of guanethidine, the majority of nerve cells in the sympathetic ganglia are destroyed (Burnstock et al., 1971; Angeletti, Levi-Montalcini, and Caramia, 1972; and others). Following treatment with nerve growth factor (NGF) antiserum, the results appear to differ widely among ganglia (for some references see Thoenen, 1972; Gabella, 1976).

reach normal levels, as determined by Raisman et al. (1974). Studies of some enzyme activities likewise suggest that the cells function normally.[15] Quantitative electron microscopic studies of the number of fibers in the sympathetic trunk of the rat in such experiments show that in the first period there is a great excess of fibers, while after 6 months the values are within the normal range (Bray and Aguayo, 1974). This supports the findings of earlier authors that a vigorous collateral sprouting of preganglionic axons occurs following their transection. The surplus of collaterals is gradually removed.

This active regenerative capacity of *preganglionic* sympathetic neurons is of relevance for the evaluation of the results of surgical preganglionic sympathectomy (i.e., interruption of these fibers by section of the sympathetic trunk or of the white communicant rami). Less is known as concerns the regenerative capacities of the *postganglionic* neurons, which are removed by extirpation of autonomic ganglia. There appear to be variations between organs and species.

Afferent fibers in the autonomic system.   Above, mainly the efferent part of the autonomic system has been dealt with, and only occasionally have afferent fibers been mentioned. These are less well known than the efferent, in spite of their great practical significance. The efferent visceral impulses normally play their most important role in the various visceral reflexes responsible for the preservation of the ''internal equilibrium'' of the body and the mutual adjustment of the different visceral functions. It should be emphasized that visceral reflexes are not only initiated by visceral afferent impulses, but by somatic impulses as well, even including those from the special sense organs, e.g., vomiting following irritation of the vestibular apparatus. A review of somatosympathetic reflexes has been given by Sato and Schmidt (1973). The visceral afferent fibers which are present in some of the cranial nerves (which might be designated parasympathetic afferents, although objections can be made to this terminology) appear to have their prime importance as afferent links in visceral reflexes. The vagus nerve does not appear to carry pain sensations. Certain visceral afferent impressions mediated through the cranial nerves, however, also reach consciousness, e.g., taste sensations. These afferent visceral fibers have their perikarya in the sensory ganglia of the cranial nerves in question, those of the vagus predominantly in the nodose ganglion, those of the glossopharyngeal in the petrous ganglion, and those of the intermediate in the geniculate ganglion. These fibers are treated more fully in Chapter 7 on the cranial nerves.

Other visceral afferent fibers enter the spinal cord, follow the efferent sympathetic fibers peripherally, and apparently have the same segmental distribution. They are distributed to the viscera and the blood vessels, where their terminal ramifications act as receptors. Most of these present themselves morphologically as free nerve endings, but other more specialized endings are also found. In the mesentery there are, for example, Pacinian corpuscles. It is now generally agreed that these visceral afferent fibers, some of which mediate pain (see below), like

[15] The new synapses can be shown to belong to regenerating presynaptic fibers because they disappear after surgical section of the regenerated sympathetic trunk (Raisman et al., 1974). Thus the situation is not quite parallel to that in the septal nuclei (described in Chap. 1), since in that case the regenerating synapses are established by another fiber system than that to which the original terminals belonged. In the superior cervical ganglion the new synapses are formed by the fibers which originally supplied the vacated synaptic sites. Whether the detailed pattern in the innervation corresponds to the original remains an open question.

the somatic afferents, have their perikarya in the spinal ganglia. The ganglia contain labeled cells following injections of HRP in the organs from which the fibers are derived (Ellison and Clark, 1975). The afferent fibers reach the spinal nerve through the white rami communicantes. In the cervical, lower lumbar, and sacral regions some of them pass through the gray rami to the sympathetic trunk and ascend or descend in this to the thoracic and upper lumbar segments. Few experimental studies are available. It may be mentioned that in the guinea pig, Elfvin and Dalsgaard (1977), using the HRP method, found afferents from the inferior mesenteric ganglion to have their perikarya chiefly in the ganglia of L_2 and L_3. The visceral afferent fibers from the lower colon, bladder, prostate, and uterine cervix appear to enter the cord in the 2nd to 4th sacral nerve, as is corroborated by surgical observations.

Relatively few experimental studies have been made of the *visceral receptors* (for a recent review, see Leek, 1977). Almost nothing is known of their structure. Some are located in the mucosa and react to chemical stimuli. Mechanoreceptors occur mainly in the muscular layers. In a recent study of the mechanoreceptors of the gastrointestinal tract in the cat, Ranieri, Mei, and Crousillat (1973) distinguished two types, slowly adapting ones and receptors of the on–off type. They are activated by intestinal contractions and by artificial stimuli (distension, etc.). In the peritoneum, particularly the mesentery, fibers coming from the Pacinian corpuscles are thicker and more rapidly conducting than the others. Among the afferents from the gastrointestinal tract, passing in the splanchnic nerves, about 40% are nonmyelinated, and most of the myelinated ones are thin, 1–4 μm (Ranieri, Crousillat, and Mei, 1975). Recordings from dorsal root ganglia show that most of the splanchnic afferents in the cat have their perikarya in thoracic ganglia Th_9–Th_{11}. Most intestinal mechanoreceptors appear to function as "tension receptors"—that is, they give information about the volume or the degree of filling of a viscus, and are important for the mechanisms of moving or expelling of intestinal contents.

The central termination of the visceral afferents is insufficiently known. They probably end in synaptic contact with cells in the dorsal horn of the spinal cord. However, visceral afferent fibers do not appear to end on the visceral efferent cells in the intermediolateral cell column, but on interneurons. Following transections of dorsal roots, degenerating terminals are not found on cells in the intermediolateral cell column which show retrograde changes following a preceding thoracic or abdominal sympathectomy (Petras and Cummings, 1972). This and other findings suggest that the visceral reflex arcs are polysynaptic. Visceral impulses further ascend in the anterolateral funiculus of the cord, and some may ultimately reach the level of consciousness. For some data on physiological studies of ascending pathways and parts of the central nervous system (brainstem, thalamus, hypothalamus, cerebral cortex) related to visceral afferent impulses, see Newman (1974).

Some general aspects. Before continuing the account of the anatomy of the autonomic system, it is appropriate to consider some aspects of its physiology. This is a vast field, in which knowledge has increased considerably in recent years. Only some general features will be dealt with here. A few historical remarks may be of interest.

The introduction by Langley of his nicotine method is undoubtedly one of the events which most decisively furthered study in the physiology of the autonomic system. In several now classical papers, dating from about 1890 and later, Langley and his collaborators showed that the application of nicotine (in a solution of a certain concentration) to an autonomic ganglion after a transient facilitation blocked the transmission in the preganglionic–postganglionic synapses, whereas the preganglionic fibers passing through the ganglion were not affected. In other words, nicotine influenced the synapses but not the fibers. In this manner Langley was able to show that stimulation of the uppermost thoracic ventral roots produced dilatation of the pupil when nicotine was painted on the stellate or middle cervical ganglion, but not when it was applied to the superior cervical ganglion. From this the conclusion was drawn that the preganglionic fibers to the eye pass without interruption to the latter ganglion, where the postganglionic neurons begin. In a similar manner the position of the synapses in other efferent autonomic pathways was determined. Furthermore, Langley showed that stimulation of the preganglionic sympathetic fibers led to vasoconstriction, piloarrection, and sweat secretion, and that the different body regions are supplied with sympathetic efferents from distinct segments of the spinal cord.

A further great step was made with the discovery that certain chemical substances mediate the synaptic transmission in the autonomic nervous system, as was later shown to be the case also in the central nervous system. In 1921 O. Loewi observed that on stimulation of the vagus nerve in the frog, a substance was liberated in the heart, which when transferred to another animal had the same physiological effect as vagal stimulation. This substance had properties similar to acetylcholine. It is now generally agreed that this vagus substance is *acetylocholine*. Stimulation of the sympathetic was followed by the liberation of a substance very similar to *adrenaline* (Cannon and collaborators) and named *sympathin*. Both substances are set free at the terminal points of the corresponding postganglionic fibers. Sir Henry Dale on this account proposed a distinction between *cholinergic and adrenergic postganglionic fibers,* a nomenclature often used.

While confirming the fundamental principles established by the pioneers in the field, later research has shown the existence of more specific patterns of functional organization within the autonomic nervous system. Thus, concerning the transmitter substance, the *sympathetic transmitter,* originally called sympathin, has been shown to be, not adrenaline, but *noradrenaline,* as demonstrated especially by von Euler and his collaborators. (Adrenaline is liberated from the adrenal medulla on stimulation of the splanchnic nerve.) As mentioned above, there appears to be some evidence that the two autonomic transmitters, noradrenaline and acetylcholine, are related to different kinds of synaptic vesicles. The presence in the sympathetic system of some cholinergic postganglionic fibers that supply the sweat glands. [16] and muscular arteries are examples showing that the two anatomically distinguished divisions, the sympathetic and the parasympathetic, should not be referred to as the adrenergic and the cholinergic system, respectively. More examples may be found in the future.

In spite of these and other details, clarified by recent research, it still seems permissible to uphold the traditional view that *in general the two divisions of the autonomic system have antagonistic functions:* where stimulation of the parasympathetic results in an increased activity (motility, secretion), stimulation of the sympathetic has the opposite effects, and vice versa. However, this is not an absolute rule. The principle is valid for the innervation of the stomach and the gut, where parasympathetic stimulation increases peristalsis and secretion, while sympathetic stimulation has an inhibitory action. In the salivary glands parasympathetic stimulation gives rise to a copious, mucous secretion, while the effect of sympathetic stimulation on secretion appears to be disputed.[17] Under normal

[16] The cholinergic nature of the postganglionic fibers to the sweat glands explains the long-known fact that the sweat glands are susceptible to atropine and pilocarpine (causing inhibition and stimulation, respectively) like parasympathetic fibers in other instances.

[17] The parasympathetic fibers appear to end on the mucous cells of these glands only. With the fluorescence method a rich plexus of noradrenergic fibers was found around the serious acini of the submandibular gland, while there were no such fibers in the predominantly mucous sublingual gland (Norberg and Olson, 1965; see also Cowley and Shackleford, 1970).

circumstances the two divisions operate in an integrated manner, as seen in the nervous control of the heart.

In spite of the cooperation between the two divisions, there are certain noteworthy differences, for example with regard to the ratios of preganglionic to postganglionic fibers. As considered above, the proportion of preganglionic to postganglionic fibers is far lower in the parasympathetic division than in the sympathetic, indicating a greater possibility for divergence of impulses in the latter division. This tallies with many physiological findings which show that, in general, the parasympathetic has considerable possibilities for influencing local functions, while the sympathetic, by way of its more diffuse distribution, tends to have widespread and general bodily effects (for example on the cutaneous vessels), even if there are many exceptions to this.

From a clinical point of view it is important to be aware of the fact that disturbances of autonomic functions are common but often overlooked, in numerous affections of the nervous system (for an exhaustive review, see Johnson and Spalding, 1974). A careful examination of autonomic functions may give valuable diagnostic information (see also below).

The term *"vegetative tonus"* was formerly sometimes used in clinical language to characterize the conditions within the autonomic system as a whole. This "tonus" is thought to vary from individual to individual and also in the same person at different times. Eppinger and Hess suggested that one could distinguish between vagotonic and sympathicotonic individuals according to whether they present signs of relative dominance of the parasympathetic or sympathetic system, respectively. However, in most persons signs of sympathetic preponderance are coupled with other signs betraying a parasympathetic overaction. Frequently, only one organ reveals a clear-cut parasympathetic or sympathetic dominance (local vagotonia or sympathicotonia). Cannon proposed a more general biological mode of regarding the functions of the autonomic system: Broadly speaking, the parasympathetic functions are protective and serve to maintain and restore the bodily reserves. Thus the parasympathetic protects the retina from excessive light, promotes the secretion of the digestive tract, and is concerned in the emptying of the bladder and rectum. The sympathetic, on the other hand, is mainly brought into play in emergency states, when the bodily reserves are drawn upon, such as physical exertion, fear, rage, and other stress situations. In these the adrenal medulla is activated by way of the splanchnic nerves, and the consequent outflow of adrenaline is important and explains some of the functional changes. In stress situations the sympathetic effects of acceleration of the heart rate, dilatation of the bronchi, elevation of the blood sugar level, but inhibition of gastrointestinal secretion and motility, are all appropriate. It is in keeping with Cannon's views that animals deprived of their sympathetic trunks may live for years in the sheltered conditions of the laboratory, but when exposed to their usual environmental conditions their resistance is reduced and their power of adaptation is impaired. Cannon's original views have, however, had to be modified. The autonomic nervous system is more specifically organized than was assumed in his time. Numerous experimental studies have shown, for example, that changes in autonomic functions mediated by the sympathetic system often are localized and not as general as previously assumed. Further, in heavy stress or emergency states in man it is well known that unfavorable autonomic reactions may occur, such as slowing of the heart, with loss of consciousness and eventually death, loss of control of the sphincters of bladder and rectum, and other reactions, phenomena which are not compatible with the idea of a general sympathetic activation in fear and stress. It should be recalled that the more easily observed phenomena in stress reactions are only the outward manifestations of numerous extremely complex changes in the organism. These also involve the central nervous system, which has the task of coordinating and controlling the various components of the total reactions. In these functions there are close interrelationships between the nervous and endocrine controlling systems.[18]

[18] The roles of noradrenaline and adrenaline (and other biogenic amines such as serotonin) in influencing emotional states have recently attracted great interest because it appears that many drugs used in the psychiatric treatment of patients with affective disorders cause changes in the metabolism and function of various biogenic amines, peripherally and centrally.

Central levels of autonomic regulation. The hypothalamus.   In the pre-ceding chapters it has been seen that autonomic functions can be influenced from many regions of the brain, among these various parts of the cerebral cortex, the hippocampus, the entorhinal area, parts of the thalamus, basal ganglia, the re-ticular formation, and the cerebellum. Many of these actions appear to occur by way of the *hypothalamus*. This receives direct or indirect connections from many of the regions mentioned, and appears to be the main part of the brain concerned in the integration of functions in the autonomic sphere. Its influence on such func-tions is made possible by efferent fibers from the hypothalamus, which are the first links in pathways leading to the preganglionic autonomic neurons located in the brainstem and spinal cord. In addition, the hypothalamus has intimate nervous and vascular relations to the hypophysis which explain much of its influence on the endocrine system. Increasing knowledge of these relations has led to the emergence of a special field of medicine: *neuroendocrinology,* which is steadily expanding.

While the hypothalamus must be considered as probably the "highest" level of the brain concerned in the integration of autonomic functions, for many of them it is not essential. Some regions in the brainstem are able to take care of a fair degree of integration, as exemplified by the fact that cardiovascular and respira-tory regulations function rather satisfactorily in animals and humans even follow-ing complete interruption of the brainstem above the pontine region. These and other observations demonstrate the presence of integrative "centers" for these functions in the brainstem. As discussed in Chapter 6, certain regions in the medullary, pontine, and mesencephalic reticular formation have been outlined as being concerned in these regulations. Other "centers" have likewise been de-scribed, for example a "vomiting center" and a "center for micturition." As we have seen, there are direct and indirect pathways from these regions of the re-ticular formation to the spinal cord, which mediate the effects (even if it appears that reticulospinal fibers do not establish synaptic contact with the cells of the in-termediolateral cell column). Under natural circumstances changes in respiration or cardiovascular function scarcely ever occur in isolation, but as parts of a series of functional changes occurring in response to a variety of stimuli. For example, changes in cardiovascular functions accompany such diverse processes as diges-tion, temperature regulation, muscular exercise, sexual function, and others. The integration of the various changes occurring in the autonomically innervated organs under such circumstances appears to be the task first and foremost of the hypothalamus, as will be discussed later.

In recent years an enormous number of studies have been devoted to the structure and function of the hypothalamus, not least its relations to the hypo-physis and its influence on endocrine organs. The following account will mainly be concerned with anatomical data on the hypothalamus and its connections with other parts of the brain and with the hypophysis.

Quantitatively *the hypothalamus* represents only a trifling part of the whole brain, consisting of those parts of the thin-walled 3rd ventricle which are found below the hypothalamic sulcus (Fig. 11-10). The borders of the hypothalamus toward neighboring regions of the brain are rather diffuse, particularly laterally and caudally. (For an account, see Nauta and Haymaker, 1969.) Many of these regions are interconnected with the hypothalamus (for example, the septal region, compare Chap. 10). In spite of its smallness the hypothalamus presents no large-scale variations in different vertebrate species, a fact in harmony with the presence within it of centers regulating the autonomic functions which are more or less uniform in different vertebrates.[19]

In Fig. 11-10 a diagram is reproduced, based on the studies of Le Gros Clark (1938). Only a simplified presentation of this intricate region will be attempted here. It can be seen that posterior to the optic chiasma emerges the infundibulum, to which the hypophysis is attached. Posteriorly the infundibulum is continuous with a slightly bulging region in the floor of the 3rd ventricle, the *tuber cinereum*. Then the floor increases in thickness at the transition to the mesencephalon below the aqueduct. Here the *mammillary bodies* are found in the interpeduncular fossa, one on each side of the median plane.

A microscopic study reveals that the lateral walls and the floor of the 3rd ventricle are made up of gray matter with some minor fiber bundles. Of the larger fiber bundles the *fornix* should be mentioned, passing from the mammillary body forwards and upwards close beneath the ventricular walls, and revealing itself as a curved elevation anterior to the interventricular foramen. The gray matter consists of small nerve cells which extend close to the ependyma lining the ventricle. This *substantia grisea centralis* or central gray matter contains several groups of cells presenting themselves more or less distinctly as separate nuclei. Some of these will be mentioned.

The *supraoptic nucleus* (Fig. 11-10) is one of them. The name refers to its location, above the beginning of the optic tract on both sides. At its most anterior end it is situated above and lateral to the optic chiasma. The cells of the supraoptic nucleus are considerably larger than those composing the central gray matter, being piriform to round in shape, and on account of the arrangement of the tigroid granules and a frequently somewhat peripherally situated nucleus they resemble cells undergoing retrograde changes. Cells of a similar type are found in another hypothalamic nucleus, the *paraventricular nucleus,* approximately forming a plate immediately beneath the ependyma. In the cells of this nucleus, as well as in those of the supraoptic nucleus, a colloid substance is present (see below). These two nuclei together with some smaller nuclei make up an *anterior group*. In man each supraoptic nucleus contains on an average 75,500 cells, the paraventricular 55,000 (Morton, 1969).

A short distance posterior to the supraoptic nucleus, several small nuclei, usually collectively designated the *nucleus tuberis*, are found (Fig. 11-10). They are situated in the tuber cinereum near its surface, and consist of rather small, multipolar cells. These nuclei are regarded as characteristic primate nuclei by most observers (Ingram, 1940).

Structurally clearly separated from the tuberal nuclei are other nuclear groups exhibiting many features in common. They have been designated by various names. Le Gros Clark (1938) has called them the *dorsomedial, ventromedial,* and *lateral hypothalamic nuclei*. Situated above the tuberal nuclei and composed of

[19] Among recent monographs dealing with the hypothalamus may be mentioned *The Hypothalamus,* 1969; *The Hypothalamus,* 1970; *Integrative Hypothalamic Activity,* 1974; *The Hypothalamus,* 1978.

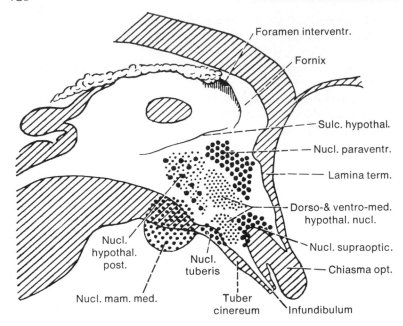

FIG. 11-10   Diagram of the ventricular surface of the hypothalamus in man, showing the position and extent of the most conspicuous hypothalamic nuclei (see text). Slightly modified from Le Gros Clark, Beattie, Riddoch, and Dott (1938).

small and medium-sized nerve cells, they form, together with the tuberal nuclei, a *middle nuclear group*. The lateral nucleus is especially well developed in anthropoids and man.

A *posterior group* is made up of the nuclei of the mammillary body and a nucleus situated immediately above this. In the mammillary body is found a larger, medial small-celled *nucleus mammillaris medialis* and a smaller large-celled *lateral nucleus*. The small-celled part is well developed in anthropoids and man. The other nucleus in the posterior group is the *posterior hypothalamic nucleus*, composed of scattered large cells between a majority of small ones. Apart from the nuclei mentioned, several other nuclear groups are present in the hypothalamus, which will not be considered here.

Other authors have arrived at somewhat different conclusions concerning the subdivisions of the human hypothalamus. Feremutsch (1955) stressed the individual variations in the architecture and was reluctant to consider all the particular cell groups indicated by Le Gros Clark as nuclei. Other authors distinguish a greater number of nuclei. For a discussion of the problems involved in cytoarchitectonic studies of the hypothalamus and references to the literature, the reader should consult the review of Christ (1969) which also contains an atlas of the human hypothalamus (see also Nauta and Haymaker, 1969). Golgi studies of the hypothalamus of the rat (Szentágothai, Flerkó, Mess, and Halász, 1968) confirm the existence within the hypothalamus of more or less well circumscribed cell groups within a more diffuse conglomerate of cells. These groups differ in part with regard to axonal and dendritic patterns and terminal arborization of afferent fibers. For particulars the reader is referred to the monograph of these authors, which also contains some electron microscopic observations. The authors stressed the multifarious interconnections within the hypothalamus, and emphasized certain general differences between the medial and lateral areas. Concerning the former they concluded (loc. cit., p. 56): "Thus a nervous network is established in which, besides the impulses leaving the hypothalamus through the main axons, excitation can spread from a given focus in any direction and can establish an infinite number of closed self-re-exciting chains." The lateral hypothalamic area appears to have fewer local interconnections. The large neurons present here all appear to project to distant structures. On the other hand, most afferent hypothalamic connections appear to end in this area. This is traversed by fiber bundles of the phylogenetically ancient medial forebrain bundle.

Some electrophysiological studies have been carried out to elucidate some of the complex interconnections between different hypothalamic nuclei visible in Golgi preparations. From studies with antidromic recordings from hypothalamic nerve cells it appears, for example, that neurons in the arcuate,[20] ventromedial, and dorsomedial nuclei and the periventricular area project to the preoptic and anterior hypothalamic areas; some of these also project to the paraventricular nucleus (Harris and Sanghera, 1974; Makara and Hodács, 1975). Some neurons in these regions can be antidromically activated from the median eminence as well. Whether a single neuron gives off branches to both end stations is disputed.

In recent years many studies have been devoted to the "chemical anatomy" of the hypothalamus. The hypothalamus appears to be the region of the brain which contains the highest concentration of noradrenaline. Other putative transmitters, such as acetylcholine, serotonin, dopamine, and histamine, have likewise been found. However, the various substances are not equally distributed.

For example, the highest concentration of noradrenaline was determined by Palkovits, Brownstein, Saavedra, and Axelrod (1974) to be in the paraventricular and dorsomedial nuclei and the retrochiasmatic area. Dopamine, although present in all nuclei, was found to be relatively highly concentrated in the same nuclei. Serotonin concentrations appear to be highest in the nuclei of the basal and posterior hypothalamus (Saavedra, Palkovits, Brownstein, and Axelrod, 1974). Acetylcholine, as judged from the presence of acetylcholinesterase, is found in the cells of several nuclei but is virtually absent in others, for example the ventromedial nucleus (see Shute and Lewis, 1966). For a recent review of the histochemistry of the hypothalamus, see Pilgrim (1974).

Little is so far known about whether the transmitters found in the various hypothalamic nuclei are present in their neurons or in the terminals of their afferent fibers. It appears, for example, that at least most of the noradrenaline in the dorsomedial and some other nuclei is contained in terminals of afferent noradrenergic fibers, coming, at least in part, from the nucleus locus coeruleus (Fuxe, 1965; and others). Dopamine concentration, on the other hand, does not change markedly following total hypothalamic deafferentation (Palkovits, Fekete, Makara, and Herman, 1977), suggesting that most of the dopaminergic neurons are intrahypothalamic. Extremely complex maps of the noradrenergic and cholinergic connections of the hypothalamic nuclei have been presented. They will not be considered in the following discussion.

The available anatomical information indicates that even if certain cell groups

[20] The arcuate nucleus is not shown in Fig. 11-10. It is a small-celled group (sometimes referred to as the infundibular nucleus) in the most ventral part of the third ventricle near the entrance to the infundibular recess. It extends into the median eminence (see below).

stand out as particular units within the hypothalamus, in much of its territory it is difficult or impossible to draw borders between cell groups. This lack of delineation, as well as the ample interconnections, should be kept in mind in attempts to relate certain regions of the hypothalamus to particular functions. We will return to this question later. For the interpretation of functional observations, a knowledge of the connections of the hypothalamus with other parts of the central nervous system is important. In the following discussion the connections of the hypothalamus with other parts of the central nervous system and its relations to the hypophysis will be considered separately.

Fiber connections of the hypothalamus with other regions of the central nervous system.   Because the hypothalamus is so small and because it is composed of numerous minute anatomical "units," studies of its connections are extremely difficult.[21] For this reason precise knowledge of the subject is scanty. A considerable proportion of the efferent and afferent fibers of the hypothalamus course in the medial forebrain bundle.

The *medial forebrain bundle* is an old designation of a diffusely outlined system of fibers. Its anterior beginning can be discerned in front of the level of the anterior commissure. From here it continues parasagittally, laterally to, and in part infiltrating, the hypothalamus, and is lost posteriorly in the paramedian zone of the mesencephalon. Essentially it appears as a fiber system which links basal forebrain structures with the brainstem in both directions. Experimental studies of the various components of the medial forebrain bundle meet with considerable difficulties. It appears, however, from such studies as well as from Golgi studies (see Valverde, 1965) that most of the fibers that make up the medial forebrain bundle are short, and serve to interconnect adjacent neighboring basal structures, among them the septal nuclei and the hypothalamus. Figure 11-11 shows a diagram of the main afferent and efferent connections of the hypothalamus.

The *afferent connections of the hypothalamus* provide evidence that impulses from a number of other regions converge on this part of the brain, as has also been concluded from physiological studies. Information from sensory receptors of different kinds reaches the hypothalamus by more or less direct routes. *Retinal fibers* to the hypothalamus have been a matter of dispute for decades, but their existence appears now to be definitely confirmed.

Studies of normal material and experimental studies using the Marchi or silver impregnation methods were not decisive. In recent years autoradiographic investigations have demonstrated retinohypothalamic fibers in various mammals (O'Steen and Vaughan, 1968; Moore and Lenn, 1972; Hendrickson, Wagoner, and Cowan, 1972; Moore, 1973). These fibers are crossed as well as uncrossed and have so far been convincingly traced only to the suprachiasmatic nucleus (not shown in Fig. 11-10, but present close to the midline above the chiasma). Electron microscopically, degenerating boutons have been found in the suprachiasmatic nucleus following lesions of the retina in the cat (Hendrickson, Wagoner, and Cowan, 1972) and in the rat (Moore and Lenn, 1972). Sousa-Pinto (1970b) described endings in the arcuate nucleus in the rat. In electrophysiological studies some cells in the suprachiasmatic nuclei were found to receive visual input (Sawaki, 1977).

---

[21] In several anatomical as well as physiological respects the mammillary body differs from the rest of the hypothalamus—being, for example, apparently less closely related to autonomic and endocrine functions. It was considered in Chapter 10 and will be only briefly referred to here.

The termination of retinal fibers in the hypothalamus is of interest for the understanding of the effect of light on hypothalamic functions, such as its influence on the estrus cycle in many mammals and on circadian behavioral patterns.

Some olfactory fibers coming, it appears, largely from the anterior olfactory nucleus, have been claimed to pass directly to the hypothalamus. They appear to be scanty and to end in its lateral regions (see Scott and Leonard, 1971). Olfactory stimuli have been found to excite cells in this area. In addition, as described in Chapter 10, the hypothalamus may be influenced indirectly by olfactory impulses, by way of olfactory projections onto the amygdaloid nucleus and the piriform cortex, both of which give off fibers to the hypothalamus.

In physiological studies potentials have also been recorded in the hypothalamus following auditory and cutaneous stimuli and following stimulation of the medial lemniscus (see Cross and Silver, 1966, for references). The anatomical bases for some of these impulse transmissions are not entirely clear. Direct fibers from *the spinal cord* or from the dorsal column nuclei have been debated. Kerr (1975) recently found some fibers ascending in the ventral funiculus to end in the posteromedial hypothalamus. It appears likely that sensory (visceral and somatic) impulses from the head and body may influence the hypothalamus through nuclear groups which receive such information from ascending pathways. Thus hypothalamic afferents have been traced from the *reticular formation* (described in Chap. 6) and may perhaps be involved. Other ascending connections come from the *nucleus locus coeruleus,* the *raphe nuclei,* and the *periaqueductal gray* (see Chap. 6). For a recent study see Sakumoto et al. (1978). Some of these fibers have been traced to the median eminence (Palkovits, Léránth, Záborszky, and Brownstein, 1977). Influence from the midbrain is further possible by way of the connection in the *mammillary peduncle* (see Chap. 10) from the dorsal and deep tegmental nuclei to the mammillary body and to the lateral hypothalamus. There appear to be some fibers from the cerebellum (see Chap. 5). Of particular interest is the recent demonstration by Ricardo and Koh (1978) of direct projections from the *nucleus of the solitary tract* to the paraventricular, dorsomedial, and arcuate hypothalamic nuclei (see Chap. 7, section c).

Several "higher" regions of the brain send fibers to the hypothalamus. There are direct fibers from the *hippocampal formation* to the hypothalamus, the most massive being the projection to the mammillary body. Secondary connections from the hippocampal formation may be established via the *septum.* This area and its connections have been considered in Chapter 10, as have the connections of the *gyrus cinguli,* which may influence the hypothalamus indirectly via the septum (Fig. 10-8). Other hypothalamic afferents come from the *piriform cortex* and the *amygdala* (see Chap. 10). Fibers from the basal ganglia have been claimed to exist but have so far not been convincingly demonstrated. There appear to be some afferents from the thalamus (for some references, see Nauta and Haymaker, 1969; see, however, Siegel, Edinger, and Troiano, 1973). The presence of hypothalamic afferents from the *neocortex* has been a matter of dispute. In contrast to some previous investigators, Lundberg (1960) found no evidence for such fibers in an extensive experimental study in the rabbit, nor did Szentágothai et al. (1968) in the cat. In the monkey, however, Nauta (1962), in agreement with some previous authors, traced fibers from the orbitofrontal cortex to the hypothalamus, particularly its lateral regions. Indirect neocortical pathways to the hypothalamus, having a relay in the dorsomedial thalamic nucleus (MD, see Chap. 12), have been described, for example a projection from the orbitofrontal cortex to the MD (Nauta, 1962).

Although far from complete, the survey above clearly demonstrates that the hypothalamus must be influenced—directly or indirectly—from almost all other parts of the brain. It may be conjectured that the various afferents have more or less specific sites of termination within the hypothalamus. Little is known concerning this, however (see Szentágothai et al., 1968, for some data). In a corresponding manner the *efferent connections of the hypothalamus* proceed to many other parts of the brain. Leaving out of consideration the fibers passing to the hypophysis, to be considered in the following section, a summary of our relatively scanty knowledge of the subject is given below (for reviews see Szentágothai et al., 1968; Raisman, 1966b; Nauta and Haymaker, 1969).

To some extent the efferent hypothalamic fiber connections are reciprocal to the afferent ones (Fig. 11-11). Among "ascending" projections the mammillothalamic tract passing to the *anterior thalamic nucleus* (see Chap. 10) is the most massive. Some hypothalamic fibers appear to end in the *amygdala;* others have been traced to the *septum* (see Chap. 10). Fibers passing to the *hippocampus* have been described (Guillery, 1957). According to HRP studies they appear to

FIG. 11-11   Simplified diagram of the main afferent and efferent connections of the hypothalamus. See text for particulars. The connections of the mammillary body and the hypothalamohypophysial projections are not included. Abbreviations: *BL, Ce.,* and *Co.:* basolateral, central, and cortical amygdaloid nucleus, respectively: *H:* hypothalamus; *M:* mammillary body; *MD:* dorsomedial thalamic nucleus; *RF:* reticular formation; *S:* septum; *SN:* substantia nigra; *VM:* ventromedial hypothalamic nucleus.

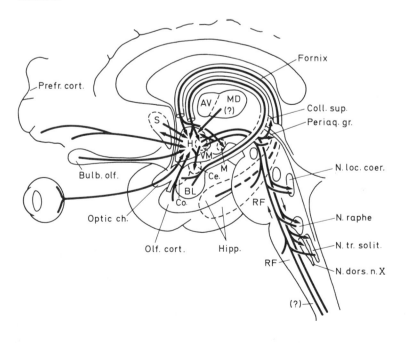

originate from cell groups adjacent to the mammillary body (Pasquier and Reinoso-Suarez, 1976). Using the same method, Yoshii, Fujii, and Mizokami (1978) described a restricted hypothalamic projection to the pulvinar. Of great interest are the *pathways by which the hypothalamus may be imagined to act on the preganglionic sympathetic and parasympathetic neurons.* Until the introduction of modern neuroanatomical tracer methods it was generally believed that this influence occurred via multisynaptic pathways, with relays in various rostral brain nuclei, since direct connections had not been convincingly demonstrated. Following small lesions in the hypothalamus and using silver impregnation methods, several authors (see Guillery, 1957; Nauta, 1958; Wolf and Sutin, 1966; Enoch and Kerr, 1967b, and Szentágothai et al., 1968) had traced fibers to the *mesencephalic periventricular gray* substance. These fibers have recently been found with the HRP method to arise especially in the ventromedial nucleus (Grofová, Ottersen, and Rinvik, 1978). In addition, fibers have been found to supply the *pretectal area* and the *superior colliculus,* the *dorsal and ventral tegmental nuclei of Gudden* (see Chap. 10), the *raphe nuclei,* and the *nucleus locus coeruleus* (see Chap. 6). A final caudally directed pathway is made up of fibers of the mammillotegmental tract, passing from the mammillary body to the *midbrain reticular formation* (see Chap. 10).

Research from the last few years has established that there are indeed *pathways that may mediate hypothalamic messages directly to the preganglionic visceral nuclei.* Following injections of HRP in the spinal cord of the cat, Kuypers and Maisky (1975) found labeled cells in the dorsal hypothalamic area. In a study in the rat, cat, and monkey, Saper, Loewy, Swanson, and Cowan (1976) located hypothalamic neurons projecting to the cord, more specifically from the paraventricular nuclei, lateral and dorsal hypothalamic areas, and lateral part of the dorsomedial nucleus, as well as the caudal part of the hypothalamus. The origin of the direct hypothalamospinal projections is rather widespread. The projections are chiefly uncrossed. Largely corresponding findings have been made in the rat by Hancock (1976; see also Hosoya and Matsushita, 1979).

Information on the terminal distribution of these projections has been obtained in studies with tracing of injected tritiated amino acids. While confirming and extending previous knowledge of hypothalamic fibers to various sites in the brainstem and mapping some intrahypothalamic connections (Conrad and Pfaff, 1976a, 1976b), such studies have also demonstrated hypothalamic fibers to the *dorsal motor nucleus of the vagus,* the *nucleus of the solitary tract,* and some to the *nucleus ambiguus* (Saper, Loewy, Swanson, and Cowan, 1976). It is of particular interest that abundant fibers were traced to the *intermediolateral cell column* in the cord. Most of the fibers descend uncrossed.

The demonstration of direct pathways from the hypothalamus to the preganglionic sympathetic and parasympathetic neurons does not exclude the possibility that the hypothalamic influence may occur via indirect pathways as well. Nor is it known whether the hypothalamic neurons involved also project to other regions. The synaptic relationships between the hypothalamic fibers and the visceral efferent neurons are likewise not yet clear and will require electron microscopic investigations.

Several physiological studies have been undertaken in order to determine the site within the brain and the spinal cord of the descending fibers influencing the autonomic preganglionic neurons. Most often conclusions have been drawn from the effects of partial lesions of the cord on autonomic effects elicited from the hypothalamus or from medullary or pontine regions that influence autonomic functions (the latter regions have been briefly considered in Chap. 6). Other authors have used stimulation of the white matter of the cord or the brainstem. To mention a few examples: Kerr and Alexander (1964; see also Enoch and Kerr, 1967a, b) concluded that the spinal pathways for the vasomotor pressor responses are situated along most of the surface of the lateral funiculus, while those for vesicomotor and pilomotor responses are found in the dorsal part of the same area.[22] In the cervical cord the pupillodilator pathway (passing to the ciliospinal center in C_8-TH_1, see below) was located underneath the surface of the ventral part of the lateral funiculus (Kerr and Brown, 1946; see also Loewy, Araujo, and Kerr, 1973). The results of a number of physiological studies on the descending course of fibers influencing autonomic functions are highly equivocal (see, for example, the review of Calaresu, Faiers, and Mogenson, 1975, on central cardiovascular regulation). Correlations with anatomical data are difficult, and the subject will not be considered further here.

In the preceding summary of the fiber connections of the hypothalamus, most details about particular sites of origin or ending of the various fiber contingents have been omitted. On these points knowledge is still fragmentary and difficult to integrate. Apparently there may be marked differences between minor cell groups with regard to their pattern of connections. Only a few areas or nuclei have been extensively studied. One of the hypothalamic nuclei, the *ventromedial*, appears to have attracted particular interest, possibly because of its alleged relation to specific functions such as feeding and affective behavior. The main source of afferents to this nucleus appears to be the amygdala (see Chap. 10 and Leonard and Scott, 1971, for references).

Golgi studies (see Millhouse, 1973) show that most cells of this nucleus have long and unramified dendrites with a number of spines. The dendrites extend beyond the confines of the nucleus. Many axons ramify extensively within the nucleus itself, others can be traced to other hypothalamic nuclei. Electron microscopically it has been found that the fibers from the amygdala end preponderantly upon dendritic spines (Field, 1972). Potentials have been recorded in the ventromedial nucleus following electrical stimulation of the corticomedial group of the amygdala and, in addition, of the septum and the dorsal hippocampus (Culberson and Bach, 1973). The amygdala appears to exert a prominent monosynaptic influence on the activity of many of the cells of the ventromedial nucleus (Renaud, 1978b), but they respond to stimulation of other intra- and extrahypothalamic sources of afferents as well (Renaud and Martin, 1975). Among hypothalamic efferents found to arise in the ventromedial nucleus may be mentioned fibers to the periaqueductal gray (Grofová, Ottersen, and Rinvik, 1978).

Hypothalamohypophysial relationships. It is an old clinical experience that lesions of the central nervous system, especially of the hypothalamus, may be accompanied by changes in functions which are controlled by the endocrine organs. Of these, only the adrenal medulla receives an inflow of autonomic (sympathetic) fibers of any importance. This explains that the stress-induced increase in adrenal hormone release occurs almost immediately. Most of the other nervous system influences on the endocrine organs evolve much less rapidly. Many of

---

[22] It is worthy of notice that the fibers mediating vesicopressor and vasopressor responses appear to pass together in the hypothalamus and midbrain (Enoch and Kerr, 1967b) since the former finally influence the parasympathetic—and the latter the sympathetic—division of the peripheral autonomic system.

them are now known to be due to the action of hormones from the anterior lobe of the hypophysis on other endocrine organs, such as the gonads, the adrenal cortex, and the thyroid. In recent years the major points in the mechanisms involved in these phenomena have been clarified. It has been shown that the hypothalamus sends abundant nerve fibers to the posterior lobe of the hypophysis, and that there is a vascular link between the hypophysial stalk and the anterior lobe, the hypophysial portal system, which brings the anterior lobe under the control of the hypothalamus. In the following description the main anatomical data will be reviewed and some functional observations, related to structure, will be mentioned.[23] Some differences exist between man and various animals in the structure and topography of the hypophysis and the hypothalamus (see Holmes and Ball, 1974; Daniel and Prichard, 1975). For a recent review of knowledge on the hypophysis see *The Pituitary. A Current Review, 1977.*

Cajal and other early authors described nerve fibers passing from the hypothalamus to the hypophysis in the hypophysial stalk. Knowledge of these connections was subsequently considerably extended. The fibers are unmyelinated and numerous: 40,000 in the monkey, some 100,000 in man (Rasmussen, 1940b). Following transection of the stalk in animals, retrograde cellular changes and cell loss were observed to a marked degree in the supraoptic and paraventricular nuclei (see Fig. 11-10), the more so the more proximal the section is made. In man likewise a massive cell loss is seen following lesions of the infundibulum (Rasmussen, 1940b). This has been confirmed in studies of brains from a large number of patients who had been subjected to section of the hypophysial stalk or to hypophysectomy for therapeutic purposes (Morton, 1970; see especially Daniel and Prichard, 1975). After one year 85% of the cells disappear (Morton, 1970. Similar values have been found in the dog and rat by Olivecrona, 1957.) This means that at least most of the cells of the supraoptic and paraventricular nuclei send their

---

[23] The hypophysis or pituitary gland consists of an anterior lobe or adenohypophysis and a posterior lobe or neurohypophysis. The former is developed embryologically from the epithelium of the primitive pharynx as Rathke's pouch, the latter from the neural tube. The two main lobes may be further subdivided, the most commonly used subdivision being the following: the bulk of the anterior lobe is formed by its pars distalis; dorsally follow the pars tuberalis rostrally and the pars intermedia caudally. The neurohypophysis consists of the median eminence of the tuber cinereum, the infundibular stem caudal to it, and the infundibular process or neural lobe.

The epithelial cells of the adrenohypophysis are of different kinds, and are surrounded by an ample network of sinusoid capillaries. On the basis of their appearance in hematoxylin-eosin stained sections it is common to distinguish between acidophilic (or eosinophilic), basophilic, and neutrophilic (or chromophobe) cells (according to Rasmussen, 1938, about 40, 10, and 50%, respectively). A further subdivision into cell types is possible. There appears to be increasing evidence that there are as many cell types as hormones in the anterior lobe, each type being related to one hormone (see Purves, 1966). Different kinds of basophilic cells are related to the manufacture of the follicle-stimulating (FSH), the luteinizing (LH), and the thyreotrophic (TH) hormone. Two kinds of acidophilic cells produce the growth hormone or somatotrophin (STH) and the lactogenic hormone or prolactin, respectively, while the chromophobe cells are concerned in the production of the adrenocorticotrophic hormone (ACTH). For an exhaustive treatment of the hypophysis see the three-volume monograph: *The Pituitary Gland* (1966). For some recent data on the ultra-structure of the adenohypophysis see *The Anterior Pituitary* (1975).

The main cellular elements in the *posterior lobe* are the so-called pituicytes. At one time they were believed to be the secretory elements of the neurophypophysis. Presumably they represent a particular kind of glial cell, but their functional role is not yet clear. Electron microscopically, they have been found to be contacted by terminals provided with neurosecretory granules. These may belong to fibers from the supraoptic-paraventricular nuclei. In addition there are terminals of other types (see, for example, Nakai, 1970).

axons to the hypophysis, a conclusion that has recently been confirmed using the
HRP method. Following injection of HRP in the hypophysis of the rat almost all
cells in these two nuclei are labeled (Sherlock, Field, and Raisman, 1975). In addition some labeled magnocellular neurosecretory neurons were found in certain
other (small) cell groups. The *efferent fibers of the supraoptic and paraventricular
nuclei* are collectively referred to as the *supraopticohypophysial tract*.

In addition, there are other hypothalamohypophysial fibers, arising in other
nuclei of the hypothalamus, collectively often referred to as the *tuberal nuclei*.
These consist of several smaller cell groups. Extending into the infundibulum is a
nucleus infundibularis or arcuatus (not shown in Fig. 11-10; see footnote 20). This
region of the hypothalamus is often referred to as the *hypophysiotrophic area*.[24]
Fibers from this pass in the *tuberohypophysial* (or *tuberoinfundibular*) *tract* to the
hypophysis. Convincing cell loss has not been observed in these nuclei following
transection of the hypophysial stalk, and the precise origin of the fibers of the tract
has been difficult to determine (see below).

In contrast to the fibers in the supraopticohypophysial tract, which terminate
in the neural lobe or infundibular process of the neurohypophysis, the tuberoinfundibular tract does not pass further than to the median eminence. It should be noted
that *hypothalamohypophysial fibers have never convincingly been traced to the
pars distalis of the anterior lobe.* They are restricted to the posterior lobe (the median eminence, the infundibular stem, and the neural lobe) with a few entering the
pars intermedia. It is now well established that *the two hypothalamohypophysial
connections described above are functionally different.* The fibers to the neural
lobe are concerned in the secretion of the "posterior lobe hormones," vasopressin
and oxytocin; those to the median eminence act via the portal vessels on the hormone production of the anterior lobe. The main anatomical features of the two
"systems" will be considered below. In spite of differences between them they
have essential features in common. Thus they both act by way of a process called
neurosecretion. Cells that are true nerve cells, with axons, dendrites, Nissl bodies,
and other organelles, and which may be activated via axonal terminals of processes of other nerve cells, produce substances that are transmitted along the
axons and are liberated at their endings. This was first demonstrated for the cells
in the supraoptic and paraventricular nuclei.[25] The historical development of our

---

[24] The designation indicates that this hypothalamic area is first and foremost concerned in influences exerted on the hypophysis, particularly the anterior lobe (see below). The hypophysiotrophic area
cannot be precisely outlined and includes, or at least encroaches on, several hypothalamic nuclei—according to some authors, the suprachiasmatic, paraventricular, periventricular, anterior hypothalamic,
and premammillary nuclei and the medial half of the ventromedial nucleus. (see Fig. 11-14.)

[25] Neurosecretory cells are found also in other locations, in vertebrates as well as invertebrates.
For some data see Ortmann (1960). The concept of neurosecretion, when it was first set forth appeared
to distinguish a particular type of neurons from others. When it was later shown that conventional
neurons produce transmitter substances that are transported along their axons and released at the synaptic sites, it became clear that there are scarcely principal differences between neurosecretory and conventional neurons. For a discussion of this subject see Smith (1971) and B. Scharrer (1976).

The problems of production, transport, and release of neurosecretory and transmitter substances
have been extensively studied in recent years. To a large extent biochemical studies have been combined with electron microscopic ones, to identify the morphological substratum of the processes going
on. It is beyond the scope of the present text to consider these problems. Interested readers will find
several relevant data in *A Discussion on Subcellular and Macromolecular Aspects of Synaptic Transmission* (1971). See also *The Hypothalamus* (1978), *The Pituitary. A Current Review* (1977).

knowledge of the neurosecretory function of these nuclei is of interest and will be briefly considered below. (The situation in the tuberoinfundibular system will be referred to later.) The beginning was made when E. Scharrer in a series of studies from the 1930s described the regular occurrence of what he termed ''colloid droplets.'' They are of various sizes and can be found in the cytoplasm as well as between the cells. Such droplets are found in a number of mammals studied, including man, and also in lower vertebrates (for a comprehensive review see Scharrer and Scharrer, 1954). Scharrer's assumption that these droplets are secretion products, and that the cells are neurosecretory, has been amply confirmed. Using a special method (Gomori's hematoxylin-phloxin stain), Bargmann (1949, and later, see Bargmann, 1954) succeeded in staining the droplets within the cells and could further show that they can be traced along the axons of the hypothalamohypophysial fibers down to the neurohypophysis. In some instances stained droplets were found in the capillaries in the neurohypophysis (Hanström, 1952; and others). Following transection of the hypophysial stalk the stainable substance is found to accumulate above the cut, while it disappears distal to the cut (Hild and Zetler, 1953), and in studies, with the phase-contrast microscope, of tissue cultures from the neurohypophysis, Hild (1954) found evidence that the droplets move along the axons. The relation between the endings of these neurosecretory fibers and the capillaries in the neurohypophysis, where one would assume the substance to enter the blood, has been more difficult to clarify. In silver-impregnated sections from the human neurohypophysis Hagen (1949–50) found that the fibers terminate with small swellings on the capillaries. In the neurohypophysis of the opossum Bodian (1951) observed anatomical features which were favorable for the study of its organization. The neural lobe of the opossum is characterized by a regular subdivision into small lobules. A bundle of nerve fibers enters each of these, and the fibers radiate toward the periphery where they end as palisades of small bulblike swellings against the vascular connective tissue septa surrounding the lobules. The swellings as well as the fibers contain droplets which can be stained with the Gomori method. Following the discovery of the characteristic pattern in the neurohypophysis of the opossum, Bodian (1951) was able to show that in principle the same pattern is present also in man and monkey, although it is less clear.

The neurosecretory material can be stained with other methods than the Gomori stain (see Ortmann, 1960; Holmes and Ball, 1974). The so-called *Herring bodies*, rather large, often irregularly shaped, granular or homogeneous bodies visible in ordinary histological sections from the neurohypophysis and noted by the classical anatomists, appear to be particularly large accumulations of neurosecretory material present as swellings at the end of nerve fibers.[26]

From electron microscopic studies it appears that the neurosecretory material consists of aggregates of granules, 500 to 2,000 Å in diameter (Bargmann and Knoop, 1957; Palay, 1957; Bodian, 1963, 1966a; and others). In addition the palisade terminals contain smaller vesicles of about 300 Å in diameter, resembling the synaptic vesicles found elsewhere in nerve endings. It appears that at least two types of terminals can be distinguished on the basis of differences in their neurosecretory vesicles (see Holmes and Ball, 1974), in agreement with the dual origin of

[26] Herring bodies have been found in all mammalian neurohypophyses examined.

fibers from the supraoptic and paraventricular nucleus. Electron microscopic studies have further clarified the relations between the palisade terminals and the capillaries in the neural lobe. Bodian (1963), in the opossum, found that the terminals are not separated from the blood vessels by a glial sheath as elsewhere in the brain. The terminals end on the perivascular collagen layer, separated from this only by a basement membrane. Occasionally neurosecretory granules are seen in the collagen layer. Palay (1955, 1957) observed small pores in the capillary endothelium in the neurohypophysis (fenestrated capillaries), a finding which has later been confirmed in many studies.

On the basis of the findings reviewed above, as well as others, it appears that the colloid droplets containing the neurosecretory substance are formed in the neurosecretory cells of the nucleus supraopticus and paraventricularis, and that they reach the neurohypophysis by being transported along the fibers of the hypothalamohypophysial tracts. They are finally given off to the blood in the neurohypophysis (see Fig. 11-12). The droplets contain the posterior lobe hormones or precursors to them.[27]

Early investigators observed in man as well as in experimental animals (Fisher, Ingram, Hare, and Ranson, 1935) that diabetes inspidus follows both interruption of the hypophysial stalk and destruction of the posterior lobe. In both instances the antidiuretic hormone production fails. In the hypothalamus the hormones were found in extracts from the supraoptic and paraventricular nuclei but not in extracts from other parts of the hypothalamus (Hild and Zetler, 1951; and others; see Heller, 1966). In his pioneer studies with implanted electrodes and remote stimulation in rabbits, Harris (1947) demonstrated that stimulation in the region of the supraoptic nucleus or the supraopticohypophysial tract in the course of a few minutes resulted in a reduced diuresis; in animals in estrus it resulted in increased uterine contractions. Confirmatory findings were made by many authors. Numerous studies have been made of the supraoptic and paraventricular nuclei in animals subjected to thirst for some days. This results in a depletion of stainable neurosecretory substance in the hypothalamus (Hild and Zetler, 1953) as does also the ingestion of massive doses of sodium chloride (Ortmann, 1951). In both cases, excessive demands are made on the manufacturing of vasopressin. These and other studies in which cytological details such as changes in nuclear and nucleolar size have been examined support the general view of the neurosecretory hypothalamohypophysial system outlined above.

In recent years a vast literature has grown up on the electron microscopical, electrophysiological, and biochemical aspects of neurosecretory mechanisms. Most recently, immunohistochemical methods and radioimmunoassays have turned out to be useful tools. A wealth of information has been brought forth. It

[27] It seems to be generally agreed that in mammals there are only two such hormones: *oxytocin*, acting chiefly on the smooth muscles of the uterus and the myoepithelial cells of the mammary gland (milk-ejecting factor) and in addition playing a role in parturition and sperm transport in the male and female, and *vasopressin* (corresponding to the former vasopressin and adiuretin) acting particularly on smooth muscles in the small arteries and the intestinal tract and on the epithelia of the renal tubules. (There appear to be two kinds of vasopressin, arginine and lysine vasopressin.) The precursors may be the hormones linked with so-called neurophysins (sulfur-rich "carrier" proteins) which are synthesized in the same cells, but the question is not yet clear (see Pickering, 1976, for a brief review). According to Defendini and Zimmerman (1978), vasopressin is associated intracellularly with one neurophysin "carrier" protein and oxytocin with a different one.

may be mentioned here that the transport of hormones appears to proceed rather fast, about 3 mm/hour, and that their liberation appears to take place by exocytosis; i.e., the vesicles containing the hormones are emptied into the interceullular space. There is evidence that some of the posterior lobe hormones (or precursors) enter the portal vessels (see below) and thus may influence the anterior lobe as well. (This is in agreement with the anatomical finding that some fibers of the supraopticohypophysial tract end in the median eminence; see below.) The role of release of some hormones into the cerebrospinal fluid remains unclear. For exhaustive accounts of the fine structure of the posterior lobe see Palay (1957), and Holmes and Ball (1974). For some other references see Hayward (1975) and Pickering (1976).

There is good evidence that of the two distinct nuclei in the anterior hypothalamus, the supraoptic nucleus is related especially to vasopressin (adiuretin), the paraventricular to oxytocin.[28] Experimental lesions of the latter nucleus result in a marked decrease in oxytocin in the posterior lobe (Olivecrona, 1957; see also Heller, 1966). This is in keeping with the observation that a differential release of the two hormones may occur under certain conditions; for example, the baby's sucking releases preponderantly oxytocin, dehydration chiefly vasopressin.

Some additional observations on the regulation of vasopressin production and release deserve mention. A number of studies have shown that there is in the hypothalamus a region responsible for the regulation of water intake. Following the observations of Andersson (see Andersson, 1957) that drinking can be induced in goats by electrical stimulation of the anterior hypothalamus, several studies have been devoted to the presence of a "thirst center" in the hypothalamus. Fitzsimons (1966) concluded that while the hypothalamus is obviously very important in regulating water intake, it is probably part of a more extensive system concerned in regulating food and water intake. Various stimuli (exertion, changes in emotional state, alterations of blood volume, and others) are known to cause increased drinking, but the mechanism by which these stimuli act is not entirely clear. Under several circumstances an increase in the osmotic pressure of the body fluids appears to be the effective stimulus.

Verney (1947) demonstrated that the osmolarity of the blood in the carotids influences the release of antidiuretic hormone (vasopressin) and created the term *osmoreceptors* for elements in the hypothalamus which are sensitive to osmotic changes. According to Verney the osmoreceptors are found in the anterior part of the hypothalamus, and it was suggested that they are cells situated in or close to the supraoptic nucleus. This assumption received support from studies of different kinds, reviewed by Joynt (1966). Vincent, Arnaud, and Nicolescu-Catargi (1972) recorded the responses of cells in the supraoptic nucleus and its vicinity to antidromic stimulation from the posterior hypophysis and to intracarotid osmotic stimuli in monkeys. The authors concluded that the osmoreceptive cells of the nucleus are not those which project to the posterior lobe. The osmoreceptive cells

---

[28] Recent studies indicate that vasopressin occurs not exclusively in the supraoptic nucleus but in the paraventricular and suprachiasmatic nuclei as well (see George and Jacobwitz, 1975); compare the findings in the HRP study of Sherlock, Field, and Raisman (1975) mentioned above. According to immunohistochemical studies (Defendini and Zimmerman, 1978), the two posterior lobe hormones are present in both the paraventricular and the supraoptic nucleus in all species studied.

were found on the border and in the immediate vicinity of the supraoptic nucleus. It appears from this, as well as some other findings, that there is one or several synapses between the osmoreceptors and the neurosecretory neurons.

Anatomical data suggest that the cells of the supraoptic nucleus must be especially well suited to the role of osmoreceptors. The supraoptic nucleus and the paraventricular nucleus, especially the former, have an extremely ample blood supply. The capillary density of the supraoptic nucleus exceeds that of any other part of the brain (Finley, 1940; and others). In sections in which the capillaries are stained, the two nuclei can be discerned with the naked eye. Furthermore, each cell is surrounded by a network of capillaries without a glial sheath between the capillary and the cell. Occasionally a capillary has even been seen to pass through the cytoplasm of a cell (Scharrer and Gaupp, 1933, in man). Finally it is of interest that the blood supply to the anterior hypothalamus comes directly from the internal carotid or its branches and the circle of Willis (see Dawson, 1958; Daniel, 1966). The situation appears to be essentially the same in man and most other mammals studied.

The cells in the supraoptic nucleus are obviously in an especially favorable situation for reacting to changes in the osmolarity of the blood, increasing osmolarity leading to increased production of antidiuretic hormone (vasopressin).[29] The ways in which the cells are acted upon when diuresis is influenced, for example in emotional changes, are not known. According to Léránth, Záborszky, Marton, and Palkovits (1975) only about one-third of the axon terminals present in the nucleus originate outside the nucleus. In Golgi sections, the sparse network of afferents found appears to reach it from above (Szentágothai et al., 1968). On the basis of an electron microscopic study of terminal degeneration in the supraoptic nucleus in the rat, Záborszky, Léránth, Makara, and Palkovits (1975) confirmed some data from light microscopic studies and attempted a quantitative analysis of different afferent contingents. They concluded that the most massive contingent comes from the brainstem, others from the medial basal hypothalamus, the amygdala, the septum, the hippocampus, the olfactory tubercle, and probably some rostral cortical regions. It thus appears that the activity of the supraoptic nucleus may be influenced from many brain regions. The authors suggested that the afferents from the brainstem may be monoaminergic fibers that mediate a brainstem influence on vasopressin release. Some of the other afferents may be involved when stimulation of certain structures induces changes in vasopressin secretion.

Observations in man are consistent with the results of experimental studies. Diabetes insipidus[30] is met with in tumors which affect the hypothalamus (frequently combined with other hypothalamic symptoms), and sometimes in chronic epidemic encephalitis. Traumatic injuries to the hypothalamus may be followed by diabetes insipidus, but in some cases no explanation for its appearance can be found. Cases of diabetes insipidus have been described in which pathological changes have been found in the anterior part of the hypothalamus, and there appears to be no doubt that also in man diabetes insipidus may be produced by purely hypothalamic as well as by purely hypophysial lesions.

As mentioned above, the other "posterior lobe hormone," *oxytocin*, chemically closely related to vasopressin, appears to be produced chiefly in the paraven-

---

[29] For some data on the neurohypophysis, see Farrel, Fabre, and Rauschkolb (1968), Holmes and Ball (1974), Hayward (1975), *The Pituitary. A Current Review* (1977), *The Hypothalamus* (1978).

[30] The polyuria is the primary feature of the disease, due to deficient reabsorption of water in the renal tubules; the polydipsia is a consequence of the increased elimination of water.

tricular nucleus. Its physiological effects on the uterine muscles and on the mammary gland have been extensively studied. Some data concerning the latter action may serve to illustrate both mechanisms. As mentioned previously, oxytocin causes the (noninnervated) ectodermal myoepithelial cells surrounding the alveoli of the mammary gland to contract, resulting in emptying of the breast.[31] This is generally referred to as milk let-down. Electrical stimulation of the paraventricular nucleus produces milk let-down, but not when the hypophysial stalk has been transected. A nervous and hormonal reflex is responsible for the milk let-down which occurs in the mammary gland during sucking of the infant. The afferent part of the reflex arc is nervous, the first link being sensory fibers from the nipple and areola passing to the spinal cord in the intercostal nerves. Along routes which are not fully known the nervous impulses finally act on the paraventricular nucleus, with resulting release of oxytocin. The phenomenon that milk let-down can be influenced from "higher" levels of the nervous system has been demonstrated experimentally, and is well known from daily life. During the period of lactation most mothers observe that milk let-down occurs not only when the baby is sucking. The very thought of putting the baby to the breast may produce some milk let-down.[32] In what way the imagination in this case results in activations of the paraventricular nucleus is not known, but it must occur via nervous connections which ultimately lead to the hypothalamus.

> It is of theoretical as well as practical interest that following transection of the hypophysial stalk or hypophysectomy the transected fibers of the supraopticohypophysial tract sprout new terminals. These establish new neurohemal contacts with regenerating vessels, so that a "miniature neural lobe" is formed. This regeneration is associated with recovery of antidiuretic function (for recent studies and references see Raisman, 1973b. See also Daniel and Prichard, 1975). The paraventricular nucleus, like the supraoptic, appears to be less purely related to the neurohypophysis than was originally believed. There is some evidence that some of the descending hypothalamic fibers (described above) originate from the paraventricular nucleus (see Swanson, 1977; Hosoya and Matsushita, 1979).

To sum up: The two hormones formerly attributed to the posterior lobe of the hypophysis are not produced there but in the supraoptic and paraventricular nuclei. They are transported by axoplasmic flow in the axons of the cells in the nuclei to their terminals in the neural lobe, and from these they pass through the fenestrated walls of the capillaries into the bloodstream. Vasopressin and oxytocin appear to be octapeptides. Many problems still remain unsolved.

The *hypothalamic influence on the anterior lobe occurs via a somewhat different route and involves a vascular link* (*the hypophyseal portal system*). As mentioned above, another important link in this route is made up of *nerve fibers from the tuberal region to the neurohypophysis*. In contrast to the fibers of the supraopticohypophysial tract, *these fibers can be traced only to the median eminence and the infundibular stem*. They are collectively named the tuberohypophysial tract or,

---

[31] The *secretion* of milk is a complex process in which several anterior lobe hormones are concerned, first and foremost the lactogenic hormone, prolactin, produced presumably by the acidophilic cells.

[32] The concomitant activation of the uterine musculature during sucking illustrates the other main action of oxytocin. It may be imagined that this has a beneficial effect on postpartum involution of the uterus (and serves as one of many arguments in favor of the view that breast feeding of infants should be encouraged!).

more properly (Szentágothai et al., 1968), *the tuberoinfundibular* tract. The origin of these thin, unmyelinated fibers has been difficult to identify precisely. In Golgi sections they can be seen to come from small cells surrounding the basal part of the 3rd ventricle (particularly from the arcuate nucleus, see footnote 20). Using a special technique of surgically separating minute areas from the rest of the hypothalamus and studying the ensuing degeneration of nerve fibers and terminals, Réthely and Halász (1970) concluded that all fibers come from the medial basal hypothalamus, particularly from the arcuate nucleus but some also from the small suprachiasmatic nucleus (situated close to the midline) and part of the periventricular nucleus, in agreement with the Golgi studies of Szentágothai et al. (1968). Confirming evidence comes from recording of antidromic potentials in the hypothalamus following electrical stimulation of the median eminence (Makara, Harris, and Spyer, 1972). According to Bodoky and Réthely (1977), the arcuate nucleus contains two types of cells, both of which send their axons to the median eminence. In their combined Golgi and experimental study in the rat, these authors further demonstrated that the nucleus also gives off fibers to the ventromedial hypothalamic area, in partial agreement with physiological studies (see for example, Renaud, 1976).

While the major input to *the median eminence* comes from the tuberal region of the hypothalamus, recent studies indicate that it *also receives afferents from other regions*. In experimental electron microscopic studies, degenerating terminals were found in the median eminence following lesions of several brainstem structures, including the nucleus locus coeruleus, the dorsal raphe nucleus, and the nucleus of the solitary tract (Palkovits, Léránth, Záborszky, and Brownstein, 1977). (The latter findings tally with the presence of noradrenalin—as well as serotonin—containing terminals in the median eminence; see below).

Many studies have been devoted to the *structural organization of the median eminence*. In many respects it resembles that of the neural lobe. The median eminence can be subdivided into several zones. Against the 3rd ventricle (which extends as the infundibular recess), it is covered with a layer of ependymal cells of a particular type ("tanycytes"). Underneath the ependyma follows a layer of axons. Many of these are derived from the supraoptic and paraventricular nuclei and continue caudally into the neural lobe. Next to this "internal zone" is an "external zone" which contains densely packed unmyelinated axons, the distal (preterminal) parts of fibers of the tuberoinfundibular tract. The terminals of the fibers are collected in a so-called palisade zone, which in addition contains regularly arranged nonnervous (glial?) processes. The terminals are apposed to a basement membrane which abuts on a perivascular space. This contains collagen and is traversed by abundant capillaries belonging to the hypophysial portal vessel system (see below). (On the ventral side the perivascular zone is in contact with the pars tuberalis of the anterior lobe.)

Electron microscopic studies have clarified important details. Only a few will be mentioned. The majority of the cell processes found in the palisade layer are nervous terminals (Szentágothai et al., 1968; Raisman, 1973a; for an extensive review, see Kobayashi, Matsui, and Ishii, 1970). According to most authors, they do not contain neurosecretory material that can be made visible with the Gomori method. When studied with the electron microscope, the nerve terminals are found to contain abundant small vesicles of the "synaptic vesicle" type (about 300 Å in diameter, Bodian, 1966a; about 500 Å, Raisman, 1973a). In addition, other types of vesicles occur (see Rinne, 1966; Kobayashi, Matsui, and Ishii, 1970, for references), for example, dense-core vesicles ranging in size from 800 to 1600 Å (Raisman, 1973a). As in the neural lobe, the capillaries are fenestrated. Even if the small vesicles in the terminals of the fibers of the tuberoinfundibular tract resemble usual synaptic vesicles, they do not belong to synapses. Like the nervous terminals of fibers from the supraoptic and paraventricular nucleus in the neural lobe, those in the median eminence are part of *neurohemal contacts*. The electron microscopic analysis of the median eminence has contributed substantially to the understanding of how the hypothalamus may influence the hormonal production of the anterior lobe. Some functional aspects of these complex problems will be considered later.

It has been known for some time that the anterior lobe of the hypophysis is dependent for its function on connections with the hypothalamus. Since hypothalamic fibers do not reach the anterior lobe, this influence was difficult to understand until the discovery of the hypophysial portal vessel system, which provides a vascular link between the median eminence (where hypothalamic fibers terminate) and the anterior lobe.

The discovery of the *hypophysial portal vessel system* is attributed to Popa and Fielding (1930). In the hypophysial stalk and the median eminence those authors observed a net of sinusoid capillaries which were connected with vessels entering the anterior lobe as well as with larger vessels of the stalk. Those entering the anterior lobe were connected with the sinusoids there. The vessels connecting the two capillary networks (in the stalk and the anterior lobe) were called *portal vessels*.[33] In the following years divergent opinions were held, as to the course of the blood flow in these portal vessels, but the question could not be decided in purely anatomical studies. By direct observations of the portal vessels in living animals, Green and Harris (1949) and others have proved that the blood flow is from above downward, from the stalk and median eminence to the anterior lobe. This fact obviously is essential for the understanding of the functional role of the hypophysial portal vessel system. Before considering this subject, some other observations, extending those mentioned, will be reviewed with reference to Fig. 11-12, drawn on the basis of extensive studies in man by Xuereb, Prichard, and Daniel (1954a, b). These authors, in addition to the usual histologic techniques, employed the method of injecting the hypophysial arteries or veins with neoprene latex. When this hardens the tissue is macerated, and casts of the vascular system are obtained.

As seen in Fig. 11-12, the hypophysis is supplied by two sets of arteries, both arising from the internal carotid. The superior hypophysial artery (SHA in Fig. 11-12) runs in a posteromedial direction and forms an arterial ring around the uppermost part of the hypophysial stalk. Some branches from this artery supply the optic chiasma and some pass to the hypothalamus, but the majority enter the upper part of the infundibular stem. From each of the superior arteries a particular

---

[33] A hypophysial portal system has been found in all vertebrate species studied (see Green, 1951; Holmes and Ball, 1974), even in animals in which the anterior and posterior lobe are separated by a well-developed connective tissue septum, as for example in whales (Valsø, 1936; and others).

Hypothalamus

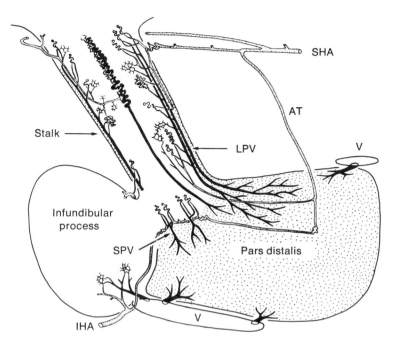

FIG. 11-12   Diagram illustrating the vessels of the hypophysial portal system in man. The portal vessels leading from the sinusoids in the hypophysial stalk to the anterior lobe are shown in black (see text). Abbreviations: *IHA* and *SHA:* Inferior and superior hypophysial arteries; *AT:* trabecular artery; *LPV* and *SPV:* long and short portal vessels, respectively; *V:*veins. From Adams, Daniel, and Prichard (1965-66).

branch, the so-called *trabecular artery* of Xuereb et al. (AT in Fig. 11-12), takes off in a descending direction and enters the anterior lobe. According to Xuereb, Prichard, and Daniel (1954a) it does not give off branches to the anterior lobe, while other authors maintain that it has a few small end arteries to this (see Stanfield, 1960). There is agreement that the trabecular artery (which has been given different names by other authors) finally enters the hypophysial stalk. The inferior hypophysial arteries (IHA) form a ring around the posterior lobe, from which branches are given off to this as well as to the lower part of the infundibular stem.

The most important finding in these studies is that *the anterior lobe receives all (or practically all) its blood supply via the hypophysial portal vessels.* When the hypophysial arteries enter the hypophysial stalk they break up into a number of coiled and looped sinusoid capillaries. Many of these are arranged in a very complex manner, as clearly shown in the photographs in the papers of Xuereb, Prichard, and Daniel (1954a, b; see also Daniel, 1966). The superior hypophysial artery distributes mainly to a superior net of sinusoids, the inferior to a net in the inferior part of the infundibular stem. Blood from the two nets of sinusoids collects into vessels which pass to the anterior lobe. These are the portal vessels, which after entering the anterior lobe give rise to a second capillary bed in this. A distinction may be made between the long portal vessels (LPV in Fig. 11-12) coming from the superior part of the stem and the short ones (SPV) from the inferior part. The latter supply chiefly the part of the anterior lobe adjacent to the lower infundibular stem. This different distribution is of some interest for the results of operative transection of the hypophysial stalk in man.[34]

This operation has been suggested and in some cases performed as an alternative to hypophysectomy for the treatment of malignant disease, especially mammary cancer, on the assumption that the anterior lobe will be inactivated by the transection. However, a transection of the stalk will interrupt only the blood supply of the anterior lobe coming via the superior portal vessels, while the inferior supply will be spared. In studies of hypophyses from 21 patients subjected to stalk section, Adams, Daniel, and Prichard (1966) found that the territory supplied by the inferior vessels had survived necrosis. Corresponding findings have been made in animals (see Daniel, 1966, for references). The functional activity of the remaining tissue may, however, not be normal. It is well known that the anterior lobe in order to function normally requires intact connections with the hypothalamus. (The hormone production of an autotransplanted hypophysis is much reduced.) In agreement with this, the epithelial cells in the part of the anterior lobe preserved after stalk section are much reduced in size and show other changes as well (see Adams, Daniel, and Prichard, 1966). The posterior lobe also shows changes after some time. Presumably these consequences of stalk section are due to the fact that neurohumors from the hypothalamus no longer act on the still viable part of the anterior lobe and the neural lobe since stalk section interrupts the neurosecretory hypothalamohypophysial fibers (see also below).

According to Daniel and Prichard (1975), the cells in a particular part of the pars distalis appear to be fed from the same few portal vessels. The authors are inclined to believe that this regional distribution of the vascular components of the neurovascular link between the hypothalamus and the pars distalis is paralleled by a corresponding arrangement in the neural elements of the link: the axons which innervate the capillary loops that drain into an individual portal vessel may be derived from nerve cells in a particular group, concerned in the production of a particular factor (see below).

There is abundant *physiological evidence* that the hypothalamus influences the hormonal activity of the anterior lobe. Thus it was shown early by Harris (1948) that stimulation of the posterior hypothalamus with the tuberal region yields secretion of the adrenocorticotrophic hormone, ACTH, and may induce ovulation. Stimulation of the anterior lobe directly does not have any of these effects. On the other hand, transection of the hypophysial stalk results in cessation of estrus in rabbits, and lesions of the posterior hypothalamic regions will abolish the secretion of ACTH which otherwise follows stress stimuli. A vast number of subsequent physiological studies have on the whole given support to the conclusions drawn by the early workers in the field. It appears to be established that hypothalamic hormones (see below) are released into the portal circulation from the median eminence and thus serve as final messengers for neural control of the anterior lobe hormone secretion (see diagram in Fig. 11-13). The mechanism has, however, turned out to be far more complex and differentiated than originally assumed. A vast literature has grown up, not least on the biochemistry of the substances involved. Information on these aspects will be found in books listed elsewhere in this chapter. The most recent, *The Hypothalamus* (1978), illustrates well the present state of complexity. Only a simplified presentation of some data can be given here.

---

[34] The account of the hypophysial portal system given here is highly simplified. According to studies of Török (see Szentágothai et al., 1968) there are other possible routes for a blood flow from the vascular loops of the hypophysial stalk in addition to the major one to the anterior lobe. See also Daniel and Prichard (1975).

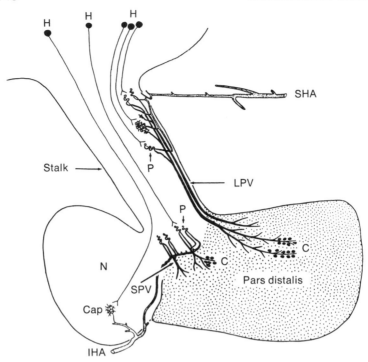

FIG. 11-13 Diagram showing the suggested pathways by which nerve cells in the hypothalamus (*H*) transmit neurohumors via their axons into loops of the primary capillary bed (*P*), and thence through the long (*LPV*) and short (*SPV*) portal vessels to control the output of hormone from cells (*C*) in a given area of the pars distalis. The innervation of the capillary bed (*Cap*) in the infundibular process by nerve cells in the hypothalamus is also shown. Compare with Fig. 11-12. From Adams, Daniel, and Prichard (1965–66).

An astonishingly large number of biologically active molecules have been identified in the median eminence: transmitters, enzymes, and hormones. Some of these substances have by some authors been found to be localized to particular structural elements (see, for example, Joseph and Knigge, 1978). The neurosecretory products of the cells in the arcuate (and possibly some other) nuclei, being released into the blood in the portal vessels, are generally referred to as *releasing factors* with reference to their action on the hormonal activity of anterior-lobe cells. It appears that the neurosecretory activity may be influenced by nonhypothalamic afferents to the median eminence. (Such afferent terminals, mentioned above, may establish axoaxonic contacts with neurosecretory terminals in the palisade zone.) This effect may be different for different transmitters, for example, dopaminergic fibers may influence the production of a particular releasing factor. Both cholinergic and monoaminergic mechanisms appear to be involved in the control of synthesis and release of neurosecretory substances.

In addition to neurosecretory substances acting as *releasing factors*, in recent years *inhibiting factors* have been identified. They are named according to the hormone they influence (for example, corticotrophin-, luteinizing hormone-, follicle-

stimulating hormone-, thyreotrophin-, and somatotrophin-*releasing factors;* soma-
totrophin- and prolactin-*inhibiting factors*). The chemistry of these factors is incompletely known, but they appear to be peptides of relatively small molecular
weight. There have been many attempts to ascertain whether the different factors
are synthesized in particular regions (nuclei) of the hypophysiotrophic area by undertaking chemical analyses of minute parts of this area or by studying changes in
hormone production or hormone-influenced behavior following differently placed,
isolated lesions of the hypothalamus or following electrical stimulation. Other
methods have been used as well, such as studies of morphological changes in the
target organs following lesions of the hypothalamus. Immunohistologic methods
and other recent developments have given much new information. Studies of these
problems meet with a number of difficulties, and there are many sources of error.
The anatomical situation is rather unfavorable. Many divergent findings have been
made, and there are still many open questions concerning the localization of the

FIG. 11-14 Diagrammatic representation of a sagittal section through the hypothalamus, showing the areas that have been implicated in the elaboration of some
releasing- and inhibiting factors (see key below figure for symbols). Some major
fiber bundles are indicated in black, and some of the main hypothalamic nuclei are
outlined. Based on illustrations of Bernardis (1974). Abbreviations: *AC:* anterior
commisure; *AH:* anterior hypothalamic area; *ARC:* arcuate nucleus; *DM:* dorsomedial nucleus; *FX:* fornix; *M:* mammillary nuclear complex; *MTT:* mammillothalamic tract; *OC:* optic chiasma; *PM:* premammillary nucleus; *PO:* preoptic
area; *PV:* paraventricular nucleus; *SC:* suprachiasmatic nucleus; *SEP:* septal nuclear
complex; *THAL:* thalamus; *VM:* ventromedial nucleus.

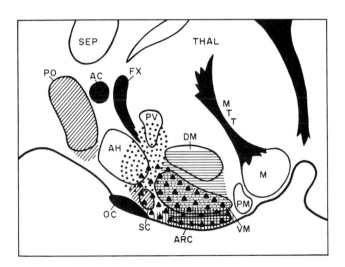

|||||||| Corticotrophin-releasing factor

∴∵ Follicle-stimulating hormone-releasing factor

/////, Luteinizing hormone-releasing factor

Growth hormone-releasing factor

▲▲ Thyreotrophin-releasing factor

≡ Prolactin-inhibiting factor

different releasing and inhibiting factors to particular areas of the hypothalamus. To some extent, however, topical differences have been described (see Fig. 11-14). Bernardis (1974) has given a concise review of the subject.[35]

> It appears from his review that the corticotrophin (ACTH)-releasing factor is manufactured in a fairly restricted diffusely outlined region in the basal hypothalamus which includes the arcuate nucleus. This area covers a small part of a rather large region, which extends through the anterior and medial hypothalamus and includes the supraoptic, suprachiasmatic, arcuate, and ventromedial hypothalamic nuclei, and which is related to the thyreotrophin-releasing factor. An important overlapping area, which covers the ventromedial, dorsomedial, and arcuate nuclei, has been attributed to the control of the prolactin-inhibiting factor. The hypothalamic area controlling growth hormone secretion is found in the vicinity of, and probably within, the ventromedial nucleus. The hypothalamic area that subserves the control of the luteinizing hormone has been concluded to be located in the preoptic-suprachiasmatic area, that for the follicle-stimulating hormone in a narrow region comprising the paraventricular nucleus and the area between this and the optic chiasma. (Figure 11-14 shows a simplified diagram.)

The overlapping between areas attributed to the control of different hormones produced by the anterior lobe is considerable, and many areas appear to be related to releasing factors for more than one hormone (for instance, growth hormone and thyreotrophin). In view of the integrative role played by the hypothalamus in endocrine control, this is indeed what might be expected. Likewise, available knowledge of the anatomy of the hypothalamus lends little support to the assumption that there is any marked degree of localization within this part of the nervous system, having chiefly diffuse dendritic and terminal axonal ramifications and abundant recurrent collaterals.[36]

It has been shown in some instances that a region is capable of producing a releasing factor, for example for corticotrophin, without afferent connections. However, the various regions responsible for the synthesis of releasing and inhibiting factors must obviously be under the influence of other parts of the nervous system. It has indeed been shown that stimulation of various nonhypothalamic regions (for example, the amygdala) may influence the production of releasing factors. Most likely this occurs by way of fibers from these regions to the hypothalamus.

> Recent research has brought forth a number of observations on the hormones and their action, illustrating the complexities present. The actions of many anterior lobe hormones have turned out to be less specific than was formerly believed. Several have been shown to affect functions other than those related to their main target organs. For example, lactation is known to be influenced by several hormones in addition to prolactin. Many functions in which hypothalamic hormones are involved concern several endocrine as well as other organs. Under normal circumstances such combined effects are produced by a variety of stimuli. The influence of the hormonal level of the target organs on hypothalamic–anterior lobe activity has attracted much interest. For example, the level of estrogens in the blood influences the secretion of gonadotrophic hormones (a high level of

[35] For further more extensive reviews see, for example, Mess, Zansi, and Tima, 1970; Blackwell and Guillemin, 1973; Porter, Mical, Ben-Jonathan, and Ondo, 1973; *The Hypothalamus,* 1978.

[36] When taken together, the hypothalamic areas that are considered to be related to releasing and inhibiting factors cover a larger territory than the arcuate nucleus, formerly often assumed to be the main or the only origin of the tuberoinfundibular tract. However, as was mentioned above, recent studies have shown that the fibers of this tract have a much wider origin. On the other hand, the arcuate nucleus gives off axons not only to the median eminence but also to the ventromedial nucleus and lateral hypothalamic areas. Findings like these concerning internuclear hypothalamic connections may contribute to explain the wide and overlapping zones for different releasing and inhibiting factors.

progesterone reduces the synthesis and release of the luteinizing hormone). Many feedback mechanisms have been demonstrated. For a fuller treatment of these subjects and references, special texts should be consulted, for example the monograph of Szentágothai, Flerkó, Mess, and Halász (1968), *The Pituitary Gland* (1966), McCann, Dhariwal, and Porter (1968), *The Hypothalamus* (1969), *The Hypothalamus* (1970), *The Anterior Pituitary* (1975), *The Pituitary. A Current Review* (1977), *The Hypothalamus* (1978).

The *pars intermedia* is a part of the hypophysis which in man appears to be of relatively little importance. In a series of vertebrates it has been known for some time to influence the color of the skin by producing expansion of the melanocytes or dispersion of the melanin granules within their cytoplasm. Its hormone, previously called *intermedin,* now generally referred to as *melanocyte-stimulating hormone* (MSH), has been shown to consist of two components, both peptides. As mentioned previously, the pars intermedia receives some nerve fibers from the hypophysial stalk. Their endings, according to most authors, contain neurosecretory granules. Whether the neurosecretory substance acts directly on the cells or by way of the vessels in the poorly vascularized pars intermedia appears uncertain. It has been suggested that in man the hormone may be of some importance for darkening of the skin as protection against sunlight. In patients suffering from adrenocortical insufficiency or some types of pituitary tumors, large amounts of MSH appear to be released and are presumably responsible for the marked darkening of the skin seen in these patients. (For an exhaustive treatment of the intermediate lobe see Volume 3 of *The Pituitary Gland,* 1966.)

In the simplified account given above, a wealth of particular data have not been considered, such as features concerning catecholaminergic as well as cholinergic terminals in the median eminence, and the suggested functional importance of the ependymal cells. It is beyond the scope of the present text to discuss these and other physiological and biochemical problems. (For an exhaustive account of these see Kobayashi, Matsui, and Ishii, 1970; concerning central monaminergic "systems" and their possible functional importance see Fuxe and Hökfelt, 1970).

Some functional aspects of central levels of autonomic integration. This subject merits some consideration in addition to the data concerning hypothalamoendocrine relations touched on above. It covers a vast field. Many observations have been made, but their interpretation is not easy, and many uncritical suggestions have been set forth. Some of these concern the question of "centers" for the regulation of particular functions influenced by the autonomic system. It is beyond the scope of the present text to treat this subject in any detail.

Mention has been made previously of the postulated "centers" in the brainstem (respiratory, cardiovascular, "vomiting," "micturition," and others). In the following discussion we shall mainly be concerned with the "centers" assumed to exist in the hypothalamus. As already mentioned, they appear to a large extent to act via the lower "centers," and to serve integration of autonomic functions of a more complex kind than the latter. Changes in a number of functions influenced by the autonomic system have been recorded in experimental animals following stimulation or destruction of various parts of the hypothalamus. Some information has been obtained from studies in man. It should be noted that many of the effects observed in the autonomic sphere represent fragments of more complex behavior changes.

Cardiovascular functions are influenced on stimulation of the hypothalamus, as noted by early workers in the field. Increase of blood pressure and acceleration of heart rate have been obtained on stimulation of the posterior lateral part of the hypothalamus. Vasodilatation and depressor responses have been seen chiefly from stimulation of anterior hypothalamic regions.

There is evidence that the latter effects depend on inhibition of the discharge of preganglionic sympathetic fibers to the cardiovascular system. Arterial baroreceptors and chemoreceptors have been shown to influence hypothalamic neurons. This and other data indicate that the hypothalamic control of cardiovascular functions does not occur exclusively via the brainstem "center," but that the descending influence from the hypothalamus on the cardiovascular system may be partly direct. Anatomical possibilities exist for this as well as for an action via the brainstem, but knowledge about the course of the pathways involved is fragmentary (see Calaresu, Faiers, and Mogenson, 1975, for a review. As mentioned above, fibers passing from the hypothalamus to the sympathetic intermediolateral cell column appear to exist).

The hypothalamus appears to play a decisive role in the *regulation of body temperature*. Early workers found that in cats increased loss of heat (which occurs mainly by panting since they only sweat on the pads of the feet) can be elicited by local heating of the anterior hypothalamus, suggesting that in this region there are cells which are susceptible to an increased temperature of the blood. By this mechanism an increase of the temperature of the body above the normal level is prevented. An animal with a bilateral lesion of this region will not start panting and sweating like a normal animal if it is brought into a heated room (Clark, Magoun, and Ranson, 1939), and its body temperature will rise. This reaction, or lack of reaction, is persistent, although some degree of adaptation takes place in the course of some weeks. Conversely, bilateral lesions in the posterior part of the hypothalamus, dorsolateral to the mammillary bodies, are followed by a deficient capacity to adjust to lowered environmental temperatures. Shivering and peripheral vasoconstriction do not occur on exposure to cold.

It appears from these and other data that the anterior, lateral part of the hypothalamus is concerned in regulation of the mechanism of heat loss, and in the lateral part of the posterior hypothalamus there is an area that regulates heat production and is able to increase this if necessary. However, none of the areas can be sharply circumscribed, and they apparently do not coincide with definite nuclei. Further research with more refined methods has extended the early observations.

Some cells in the hypothalamus have been shown to be sensitive to heating or cooling of the blood and to act as *thermoreceptors*. Such cells have been found only in the anterior hypothalamus. Pyrogens likewise produce fever only when injected into the anterior hypothalamus.[37] It appears that this hypothalamic region is the principal part of the body's "thermostat." It activates the posterior area when increased heat production is needed. (The posterior area appears to be "thermo-blind.") In recent years the chemical aspects of temperature regulation have received much attention. While the catecholamines adrenaline, noradrenaline, and dopamine lower body temperature when injected into the anterior hypothalamic area, serotonin produces a rise in temperature. It has been hypothesized (see Myers, 1969) that when a rise of body temperature is required, specific cells in the anterior hypothalamus release serotonin, which triggers the pathways for heat production; when body cooling is needed, another group of cells in the anterior hypothalamus releases a catecholamine, to activate the pathways that influence the mechanism for heat loss. There are still numerous open questions concerning the central regulation of body tempera-

[37] Recent investigations do not support the old conception that the thalamus and the striate body are concerned in the regulation of body temperature. Presumably, concomitant injury to the hypothalamus has been responsible for the changes in temperature observed. The antipyretic drugs must be assumed to act mainly on the hypothalamus.

ture. It appears that other parts of the nervous system are involved in addition to the hypothalamus, and that some hypothalamic units in the thermoregulatory territories react to various chemical stimuli (for example, endotoxins, morphine) from the blood as well as to cutaneous cold stimuli. The latter effect is presumably mediated along nervous pathways. For a review of the neural processes involved in thermoregulation, see Hensel (1973).

Clinical observations indicate that conditions are similar in man. *Postoperative hyperthermia* is not infrequently observed following operations in which the hypothalamus is apt to be injured. This applies particularly to operations for hypophysial tumors and suprasellar meningeomas. In some cases, which have taken a fatal course resulting in death from hyperthermia soon after the operation, lesions in the anterior part of the hypothalamus have been discovered. In the more common cases of moderate and transient postoperative hyperthermia, the condition is presumably due to a less severe damage to this part of the hypothalamus (edema or minute hemorrhages). Experience from cases of tumors (primary or metastatic) or hemorrhages into the hypothalamus lend further support to the view that the hypothalamus is essential for an effective regulation of body temperature (see, for example, Davison, 1940; Zimmerman, 1940). Disturbances in the regulation of body temperature are common in many affections of the hypothalamus, but may be seen in other diseases as well, for example myxedema. (For a recent survey, see Johnson and Spalding, 1974.)

Some other functions shown to be influenced from the hypothalamus may be mentioned. *Piloarrection* can be elicited on electrical stimulation of several parts of the hypothalamus. The hypothalamus is probably concerned in integrating the activity of the pilomotors with other autonomic functions in which piloarrection is brought into play, as, for example, in temperature regulation and in expressions of rage and fear.

Mention was made previously of the role of the hypothalamus in the regulation of *water metabolism* of the body and of the presence of a "drinking center." Correspondingly, certain regions of the hypothalamus appear to be closely related to the *control of food intake*. Certain medially placed lesions may result in hyperphagia and obesity, while lateral lesions are apt to produce aphagia and consequent emaciation. A lateral "feeding center" and a medial "satiety center" have been postulated. A vast literature has grown up on the "centers" in the hypothalamus related to feeding and drinking mechanisms. For recent surveys see, for example, Fitzsimons (1972), Grossman (1975), and *Hunger: Basic Mechanisms and Clinical Implications* (1976).

Quite recently Grossman et al. (1978) studied aphagia and adipsia following small injections of kainic acid into circumscribed regions of the lateral hypothalamus in rats. Kainic acid has been found to selectively destroy nerve cell bodies and dendrites in some regions, while nerve fibers are not affected. The findings are taken to indicate that aphagia and adipsia are the results of selective destruction of hypothalamic nerve cells, while other concomitant behavioral changes, usually seen following mechanic or electrolytic lesions of the same hypothalamic areas, are due to interruption of passing fibers.

Growth, bearing some relation to food intake, may likewise be influenced from the hypothalamus (presumably, in part at least, by way of the somatotrophic hormone from the anterior lobe). These findings are of interest in relation to the frequent occurrence of obesity in hypothalamic lesions in man, and fit in with the view that *dystrophia adiposogenitalis* is due to lesions of the hypothalamus.

Of clinical interest also is the demonstration that the hypothalamus may act on the *digestive organs*. It has been repeatedly demonstrated in animals that stimulation of the anterior hypothalamus produces increase of secretion and peristalsis in the stomach and gut. If the posterior part of the hypothalamus is stimulated, however, these functions are depressed, and symptoms of sympathetic activity occur. Lesions of the hypothalamus in animals have been shown by a number of students to be often followed by ulcerations and hemorrhages in the stomach and gut. Long, Leonard, Chou, and French (1962), making stereotatic hypothalamic lesions in cats, found that anterior lesions often resulted in acute gastric hemorrhage, while posterior lesions were followed either by hemorrhage, or more often, gastric ulceration. Chronic stimulation of the hypothalamus by implanted electrodes is likewise often followed by gastric ulceration or hemorrhage, apparently without reference to any particular localization of the electrode (see Long, Leonard, Story, and French, 1962). Lesions in other parts of the brain have also been shown to result in gastrointestinal hemorrhage and ulceration.

It is well known that melena and hematemesis occasionally may follow intracranial operations. In neurosurgical departments (as in animal laboratories), this is much more frequent after operations in which the hypothalamus is apt to be damaged than after other intracranial procedures. These changes occur very rapidly, in animals as well as in man, and in the course of a few days perforation may induce a fatal course.

It has been suggested that the gastrointestinal changes produced by hypothalamic lesions or stimulations result from an imbalance between the two components of the autonomic system. It has been maintained that stress situations are capable of producing gastric ulcers, presumably by influencing the hypothalamic control of the digestive tract. On the basis of experimental and clinical observations it has been suggested that changes in the hypothalamus may play a role in the development of gastric and duodenal ulcers, so frequent in human beings. Although the localization of the ulcers is different (the postoperative ulcerations are usually found on the great curvature or at the cardia or pylorus), it is generally admitted that hypothalamic factors may be of importance in the etiology of ordinary gastric ulcers, even if several factors, not only nervous, but also humoral and mechanical, are involved.

Other functions which appear to be influenced from the hypothalamus are the *sexual functions*. However, little is known regarding details. *Disturbances of sleep* have been observed following affections of the hypothalamus in animals and man as discussed in Chapter 6, and "waking" and "sleeping" centers have been postulated. The role of the hypothalamus in *emotional behavioral reactions* has been much discussed. In this sphere it obviously collaborates with other regions of the brain with which it has fiber connections, not least the amygdaloid nucleus (see Chap. 10).

It should be stressed that a distinction must be made between *emotional expressions and behavior* and *emotions*. The latter must be conceived as being primarily at least conscious feelings of a purely subjective nature. Emotional reactions, which involve autonomic as well as somatic phenomena, do not necessarily occur in emotions but may, at least to a large extent, be suppressed. On the other hand, findings in decorticate animals have shown that emotional *reactions* may be elicited when there can scarcely be any emotions. In animals in which the entire cerebral cortex, the basal ganglia, and the bulk of the thalamus have been removed, leaving only the hypothalamus in connection with the lower parts of the brain, the phenomenon of "sham rage" is observed (Bard, 1928). These animals react with severe expressions of rage, pupillary dilatation, erection of hairs, increased heart rate, scratching, biting, snarling, etc., to various stimuli, even stimuli which normally do not elicit rage. The phenomena have been explained as being due to an outburst of discharges from the posterior hypothalamus, since they cease when this also is removed. In contrast to normal emotionally induced reactions, "sham rage" behavior ceases almost immediately when the stimulation is discontinued.

Lesions of the hypothalamus in man have been observed in several instances to be followed by manic excitement or by depressive states. In cases in which the hypothalamus has been pressed upon or has been subjected to traction during operations, various symptoms have been noted. Anxiety feelings and expressions, flights of ideas and motor hyperactivity and talkativeness, severe depressive feelings, and other phenomena have been repeatedly observed. The emotional changes occurring in chronic epidemic encephalitis, especially in children, have also been claimed to be due to damage to the hypothalamus.

In addition to those mentioned, other functions, such as sweating, urinary secretion, and others, may be influenced from the hypothalamus. In fact, there is scarcely an activity of the body which cannot be influenced from this part of the brain. In spite of a number of studies, including recordings of potentials in the hypothalamus following stimulation of various peripheral and central sites and determination of regional variations in the distribution of substances such as noradrenaline, acetylcholine, and serotonin, we are far from understanding how the hypothalamus is able to perform its remarkably well integrated actions on a diversity of processes. The idea that the two divisions of the peripheral autonomic system are represented in different regions of the hypothalamus is difficult to uphold, even though, as we have seen, there is some evidence concerning certain phenomena that the anterior hypothalamus is related particularly to parasympathetic activity, the posterior to sympathetic. However, so far no clear borders have been defined between these two subdivisions of the hypothalamus, either morphologically or functionally. It should be recalled that it is virtually impossible to destroy or stimulate one of the divisions without involving fiber connections of the other.

"Centers" in the hypothalamus for the control of a number of functions have been postulated; some of them have already been mentioned. The concept of "centers" within the nervous system, i.e., morphologically fairly well circumscribed regions concerned with a particular "function," is becoming less and less attractive and less acceptable as our insight into the organization of the nervous system increases. As concerns the hypothalamus, it is obvious that several of the postulated "centers" should govern extremely complex processes (for example, regulation of body temperature or cardiovascular regulation), many of which involve not only autonomic influences on the target organs, but somatic and often endocrine influences as well (feeding behavior, sexual functions). It seems, *a priori,* to be extremely unlikely that a small spatially restricted collection of cells should be capable of the high degree of integration required for such tasks. The concept of "centers" is made even more dubious since several of them are postulated to have approximately the same location within the hypothalamus. (This indeed would seem to be a virtue of necessity, since even if each "center" is supposed to be rather small, it would be difficult to accommodate the large number of "centers" described within the small volume of the hypothalamus!)

It appears at present that *the only conclusion which can safely be drawn concerning the hypothalamic "centers" is that they represent areas of the central nervous system which are essential for the proper performance of certain functions.* Damage to these areas results in disturbances of these functions. However, the same functions which are controlled by the hypothalamus can be influenced from other regions of the brain as well, often likewise referred to as "centers." A number of such "centers" have been described, even for fairly elementary functions. For example, as concerns the neural control of sweating, Wang (1964) described "centers" in the reticular formation, the hypothalamus, the thalamus, and the striopallidum, and mentioned the cerebral cortex and the cerebellum as other regions that may influence sweating. In view of the fiber connections of the hypothalamus it is only to be expected that functions which are controlled by it may also be influenced from other parts of the brain. Accordingly, it is misleading to consider any of them as "centers." It appears that, in some way which is as yet poorly understood, an intimate collaboration between a number of particular regions of the brain takes place when autonomic functions are influenced and when these are integrated with each other as well as with somatic and endocrine changes, appearing as components in more complex reactions. It is interesting to note that in recent research on the hypothalamus, there is an increasing tendency to evaluate the different functional changes elicitable from the hypothalamus as parts of integrated behavioral patterns, engaging other parts of the brain as well (see, to mention one example, *Integrative Hypothalamic Activity,* 1974). This "integrative view" does not contradict the assumption that the hypothalamus is a particularly important region for the integration of autonomic functions. From a clinical standpoint the influence of the cerebral cortex on autonomic functions is of particular interest, in view of the obvious relationships between mental activity and autonomic changes, not least those occurring in emotional states. These relationships are far from being understood, in spite of a large body of observations, including behavioral studies of animals and man when lesions of certain parts of the brain are present. Only a few comments will be made here.

The cerebral cortex and autonomic functions. The assumption set forth by some older authors that the cerebral cortex influences autonomic functions has been confirmed. However, no cortical region has been found which can be regarded as influencing autonomic functions only. On the contrary, all evidence available seems to indicate that the "autonomic functions" of the cerebral cortex are taken care of by areas which are also concerned with somatic activities. Impulses from visceral receptors have been recorded from the cerebral cortex.

> For example, on stimulation of the splanchnic nerve, action potentials can be recorded in circumscribed parts of Sm I contralaterally and of Sm II bilaterally (see, for example, Newman, 1974). Vagal impulses traveling in the superior laryngeal nerve and gustatory impulses (relayed via the nucleus of the solitary tract and the VPM of the thalamus) have likewise been traced to the *sensorimotor cortex* (see Chap. 7, section c). In the thalamus as well as the cortex there appears to be considerable confluence of these pathways for somatic and visceral sensory input.
>
> Responses to visceral stimulation have been observed in other parts of the cerebral cortex, particularly the *posterior orbital cortex.* Both vagal and splanchnic electrical stimulations have been found to give rise to bilateral responses (see Newman, 1974, for references). The pathways utilized are incompletely known, but presumably the last link is made up of neurons in the MD of the thalamus, projecting to the frontal lobe (see Chap. 12).

Effects on autonomic functions can be obtained from various cortical areas. For example, *vasomotor changes* can be elicited on stimulation of areas 4 and 6, most clearly when the animals have been curarized and the somatic motor responses therefore suppressed. (A cortically elicited vasodilatation appears to be accompanied by a constriction of the renal vessels.) Extirpation of the precentral cortex is followed by vasomotor changes, and it is a well-known clinical experience that patients suffering from capsular hemiplegias commonly show vasomotor changes in the paretic extremities. The assumption that these changes are secondary and due to the paresis only is scarcely tenable. Following stimulation of various cortical areas, cardiac acceleration accompanied by increased blood pressure as well as slowing of the heart rate with decrease of arterial tension can be provoked. This has been found not only in experimental animals but also in man on electrical stimulation of different cortical regions, such as the frontal lobe, the temporal lobe tip, and the cingulate gyrus (see Appenzeller, 1976, for references).

Among other symptoms of an autonomic nature which can be elicited from the cortex may be mentioned *pupillary constriction* from area 19 and *pupillary dilatation* from area 8 (cf. Chap. 7). Furthermore, *sweat secretion* and *piloarrection* have been obtained from irritation of area 6 (the former, according to some authors, also from an area of the temporal lobe). The *secretory and motor functions of the stomach* are also influenced from the cortex. In experimental studies the posterior orbital cortex has been found to have a particularly potent effect on cardiovascular functions, respiration, digestive organs, temperature regulation, and other functions (see, for example, Kaada, 1960; Newman, 1974). The efferent pathways utilized are not known in detail, but projections from the orbitofrontal cortex to the brainstem may be involved (see Chap. 12).

The anatomical relationships of different parts of the cerebral cortex, and between some of these (especially the phylogenetically older paleo- and archicortex) and subcortical structures such as the amygdala and the hypothalamus, indicate that there is a functional collaboration, even if our knowledge of the connections involved are fragmentary. However, we have every reason to believe that what we call mental processes are particularly related to the cerebral cortex. It is well known that psychic activity is accompanied by changes in the autonomic sphere (for example, tachycardia and increased blood pressure in anxiety and excitation, salivation evoked by the imagination of palatable food, blushing in shyness and embarrassment). It appears likely that such changes are an expression of the close anatomical and functional relations between the cerebral cortex and the hypothalamus (and other subcortical regions related to emotional and autonomic responses). Corresponding mechanisms may be involved when changes in bodily functions accompany psychic disturbances, as is known to occur, not only in endogenous psychoses but also in response to psychic influences in mentally normal persons. The occurrence of so-called psychosomatic diseases may be understood from this point of view, even if the brain regions and the connections involved cannot be precisely indicated.[38]

The unduly simplified presentation above refers mainly to the influence of the cerebral cortex (serving primarily intellectual functions) on the hypothalamus

[38] Several bodily effects of mental processes are evidenced by changes in endocrine functions influenced by the hypothalamus via the hypophysis. Mention has been made of the psychic influence on milk letdown. Another well-known example is the cessation of the menstrual cycle following sorrow or other severe psychic stress. In young women even a change of the external milieu, for example when they start education in a new surrounding, may have this effect. The so-called anorexia nervosa is another example.

(engaged primarily in emotional and autonomic phenomena). On the other hand, the presence of (largely indirect) pathways from the hypothalamus to the cerebral cortex may in a correspondingly simplified scheme be taken to explain some other phenomena. It is a familiar observation that the emotional status of the organism is of importance for the intellectual functions of the brain, such as abstract reasoning and judgments of events and persons in the past, present, and future. In every conscious mental act our emotions are apt to interfere, thus making a completely objective judgment virtually impossible, at least when the object is of personal interest. The hypothalamus is presumably involved in some way in this mechanism. However, the hypothalamic influence is scarcely solely responsible. The hormonal status of the organism is probably of importance as well. Furthermore, as has been mentioned in Chapter 10 in reference to "the limbic system," several other parts of the brain appear to be particularly closely related to emotions and changes in functions of the autonomic nervous system. This is the case not only for subcortical structures, for example the septal nuclei and the amygdaloid nucleus, but also for certain parts of the cortex, as for example the cingulate gyrus and the orbitofrontal cortex. However, many authors have suggested that at least the main influence from these and other "limbic" structures occurs via the hypothalamus. This view appears to be in general agreement with our knowledge of their fiber connections (see Chap. 10). Nevertheless, it should be recalled that the hypothalamus is only one of many brain regions that influence and control the autonomic system and emotional reactions, and that all these regions must be assumed—and have in part been shown—to collaborate intimately.

The peripheral course of autonomic fibers.   As mentioned previously, in the peripheral parts of the autonomic nervous system, parasympathetic and sympathetic fibers (coming from the craniosacral and thoracolumbar divisions, respectively) usually run intermingled with each other. The large autonomic plexuses found in the body cavities contain fibers of both kinds, and in between there are usually ganglion cells, belonging in the peripheral plexuses largely to the parasympathetic system; in the upper abdominal plexuses, to the sympathetic. This situation presents considerable difficulties in anatomical and physiological studies and is of importance in surgical interventions on the autonomic system. No systematic description of the peripheral course of autonomic fibers will be given (see Mitchell, 1953; Pick, 1970, for complete accounts). The subject will be treated in broad outlines only, with emphasis on functional rather than topographical features. Only few references will be given. Since the parasympathetic fibers in the cranial nerves have been dealt with in Chapter 7, we shall here be concerned mainly with the peripheral sympathetic fibers, and the fibers from the sacral division of the parasympathetic.

The sympathetic fibers accompanying the peripheral nerves.   These are postganglionic fibers which reach the nerves from the ganglia of the sympathetic trunk via the gray rami communicantes (cf. Fig. 11-4). These fibers convey impulses to the *arrector pili muscles, the sweat glands, and the peripheral vessels.* Stimulation of these fibers (or the corresponding preganglionic fibers) produces *piloarrection, sweat secretion, and* (with some exceptions, see later) *vasoconstriction.*

It was shown relatively early in animals (Langley; and others) as well as in man (Foerster; André-Thomas; and others) that the *sympathetic fibers supplying the different organs and bodily regions are derived from distinct segments of the spinal cord within the levels between* Th_1 *and* L_2, *inclusive.*[39] Electrical stimulation of the ventral roots elicits piloarrection, sweat secretion, and vasoconstriction, and this is localized to the homolateral half of the face and neck when the 2nd to 4th upper thoracic roots are stimulated, and to the upper extremity when the 3rd to 6th thoracic segments are involved (Foerster; André-Thomas maintains the 4th to 8th; White and Smithwick include Th_{2-8}.) The lower extremity is supplied by the 10th thoracic to the 2nd lumbar segment (cf. Figs. 11-6 and 11-7 and the table on p. 765). Since fibers from more than one segment of the cord terminate in each ganglion, a single gray ramus communicans mediates impulses from several spinal cord segments. From this and from what has been said previously it is clear, therefore, that the effect of stimulation of a single ventral root will not be sharply limited to one dermatome, but the resulting changes will be more widespread. However, *a certain segmental distribution is retained,* although not entirely coincident with the somatic. This correspondence is closer in the distribution of the postganglionic fibers.

For example, Guttmann (1940) substantiated and amplified some earlier observations on this point by a study of the areas presenting loss of sweating when the body is exposed to external heat (thermoregulatory sweating) in patients in whom one or more of the cervical ganglia had been extirpated. When the superior cervical ganglion is removed, sweating is abolished in the face and the upper part of the neck corresponding to the dermatomes C_2 and C_3 approximately. When the middle cervical ganglion is extirpated the area also includes the lower part of the neck and the shoulder region (C_4 and part of C_5), and extirpation of the inferior cervical ganglion leads to anhidrosis also of the entire arm, with the exception of its medial aspect which is supplied by the uppermost thoracic nerves. (It is worthy of notice that when an area is thus made anhidrotic a compensatory hypersecretion of sweat is noticed in the adjacent areas.)

Richter and Woodruff (1945) made use of the increased electrical skin resistance which occurs in sympathetic denervated areas. They mapped areas showing increased resistance in a number of patients in whom lumbar and sacral ganglia of the sympathetic trunk had been surgically removed. The sympathetic dermatomes outlined in this way agreed closely with the sensory dermatomes. In previous studies the authors report corresponding findings for the face and trunk. The method has been used by several investigators, for example by Monro (1959) in an extensive study of the results of sympathectomies.

Peripherally, the sympathetic fibers accompany the somatic. *The piloarrection, sweating, and vasoconstriction which appear following irritation of the peripheral nerves after they have taken up the ramus communicans griseus will therefore present a distribution corresponding to the somatic sensory innervation.* In lesions of the plexus they will be of a segmental type, while in lesions of the peripheral nerves they will be found in the territory of the sensory distribution of the nerve in question. Some features of this sympathetic innervation deserve special attention.

The pilomotor impulses and pilomotor reflexes. Stimulation of the sympathetic fibers is followed by erection of the hairs (due to the oblique course of the arrector pili muscles) and the skin presents the well-known appearance of "goose-

---

[39] Different authors give somewhat varying limits, and in addition there are individual variations. For a complete account see Pick (1970).

flesh.'' An abolition of this aspect of the sympathetic innervation is of little or no importance in man, but the conditions of the pilomotors may give diagnostic information. (In animals piloarrection is a factor in temperature regulation.) Piloarrection may be induced with varying ease in different individuals. The appropriate stimulus is cold, but deep pressure, as well as disagreeable, sharp noises and psychic impressions may also be effective. In the latter cases pathways from the cortex-hypothalamus must be assumed to be involved, but with local, cutaneous stimuli the effect is mediated by *reflexes, having their reflex center in the spinal cord*. Consequently, *lesions of the spinal cord will be followed by changes in the pilomotor reflexes, if the lesion is situated between the 1st thoracic and the 2nd lumbar segment*. The pilomotor reflexes are most easily elicited from the shoulder region (the supra-spinous fossa) and the lumbar region. It is typical that this reflexly provoked piloarrection spreads from its starting point in an oral and caudal direction, and that it is confined to the homolateral half of the body. If a transverse lesion of the cord is present, this regular mode of spread will be interrupted, as the damage of the cord prevents the reflex impulses from passing downward or upward. However, it should be remembered that on account of the anatomical conditions (fibers from several segments to each vertebral ganglion) the border of the zone will not correspond exactly to the limit of the sensory changes. Thus, in a transverse lesion of the cord at the 9th thoracic segment, piloarrection elicited from the shoulder region will descend to approximately the 11th or 12th thoracic dermatome, whereas when starting from the lumbar region it will ascend to the 8th or 7th thoracic dermatome. Like other spinal reflexes the pilomotor reflexes cannot be provoked in the initial stages of a transverse lesion of the cord.

Lesions of the rami communicantes and the vertebral ganglia will also be accompanied by changes in the pilomotor reflexes. After extirpation of the stellate ganglion piloarrection will be absent in the face, neck, and arm on the operated side. When a peripheral nerve has been interrupted, a piloarrection progressing regularly will fail to involve the anesthetic area, for example, in a radial nerve lesion the piloarrection will be absent on the dorsal surface of the arm and forearm.[40]

When examining the pilomotor reflexes, care must be taken to avoid sources of error (the influence of room temperature, etc.), and the results obtained must be evaluated critically. There are also *local pilomotor reflexes* which may cause confusion. Such reflexes may occur even in anesthetic cutaneous areas. In many individuals the pilomotor reflexes are weak.

The innervation of the sweat glands.   The fibers to the sweat glands, as has been mentioned, accompany the pilomotor fibers, but in contradistinction to these they are *cholinergic*. In lesions of a peripheral nerve an irritation of the nerve will frequently be accompanied by an increased secretion of sweat in its area of distribution, whereas a more severe damage will interrupt the conduction and lead to anhidrosis. This is seen most clearly in lesions of the median nerve which

---

[40] On the dorsum of the hand the arrectores pili are rudimentary, and as will be recalled hairs are absent on the dorsal aspect of the two distal phalanges of fingers and toes, and on the palm of the hand and sole of the foot.

supplies the three radial fingers and the radial half of the fourth finger on the palmar side. The skin will become dry and desquamate in the anesthetic area.[41]

In transverse lesions of the cord reflex sweat secretion behaves in a similar manner to spinal reflex piloarrection. However, spinal sudomotor reflexes cannot be provoked at will. Sweat secretion is an important factor in the regulation of body temperature and appears to be regulated first and foremost by the hypothalamus. Thermoregulatory sweating appears on heating of the body and since it is initiated from the hypothalamus it will be absent below the transverse lesion, because the descending sudomotor fibers mentioned previously are interrupted. In lesions of the spinal roots or sympathetic ganglia, the anhidrosis will affect the area supplied by them.

> The sweat glands considered above are the coiled tubular eccrine glands occurring over most of the body surface. The *apocrine sweat glands,* present in some regions, notably in the axilla, differ from the eccrine sweat glands not only anatomically but also functionally. In the axilla cholinesterase has been identified histologically around the eccrine glands but not around the apocrine glands. The latter are adrenergic and respond to sympathiocomimetic drugs. They do not show thermoregulatory sweating but, as is well known, secrete in response to mental stress. It appears that they are influenced humorally by adrenaline. It has been claimed that sympathectomy in man does not abolish sweating from the apocrine glands. For some data see Evans (1957).

Vasomotor fibers and reflexes. More important than the sudomotor sympathetic fibers are the vasomotor fibers. The vasomotor fibers belonging to the sympathetic system in most instances act as vasoconstrictors. In their peripheral course they follow the pilomotor and sudomotor fibers, and thus like these show an approximate segmental distribution (cf. the table on p. 765). It appears to be settled that the cerebral vessels are also provided with sympathetic nerve fibers.

The sympathetic fibers are distributed to arteries, arterioles, and small veins.[42] It is of some importance that direct fibers from the sympathetic trunk accompany only the larger arteries in their proximal part (in the limbs approximately to the beginning of the brachial and femoral artery respectively). The arteries situated more peripherally in the limbs are supplied with sympathetic fibers by fine branches from the main nerve trunks which usually follow the arteries. Apart from plexuses of nerve fibers encircling the arteries, nerve fibers are present also in the arterial wall. In the adventitia these fibers are partly myelinated; in the muscular layer they are very fine and unmyelinated. Some of the fibers in the walls of the arteries are sensory, afferent. Like other visceral afferent fibers they have their perikarya in the dorsal root ganglia. Some of these fibers appear to mediate pain. In the adventitial coat of the arteries different types of terminal structures are found, for example some Vater-Pacinian corpuscles. These degenerate, as do also some of the fine fibers in the adventitia, after interruption of the sympathetic postganglionic supply. Finally, some cells described in the walls of the arteries have been interpreted as peripheral autonomic cells, interpolated between the incoming sympathetic fibers and the smooth-muscle cells. They have been assumed

---

[41] Guttmann (1940) has drawn attention to the occurrence of lesions of the cervical sympathetic trunk following its violent stretching, without concomitant injury to the spinal nerves or the skeleton. These lesions reveal themselves by local anhidrosis, and the location of this may give information on the site of the lesion.

[42] For an account of the innervation of particular vessels see Pick (1970).

to be engaged in the local vasomotor phenomena which may be observed after sympathectomy, but on the whole their significance is doubtful. Normally a local vascular tonus and local vasomotor reflexes are present, but are of little significance.

> The *capillaries* are accompanied by very fine nerve fibers, but the nature of these has not been established. The sympathetic nervous system and adrenaline appear to have no clear-cut effect on the capillaries, nor has acetylcholine. *The capillaries may change in caliber, independently of the arteries and arterioles.* This may be observed, for example, when a fever begins. The skin may be pale (contracted capillaries) but at the same time hot (dilated arterioles). Changes in the caliber of the capillaries have been studied directly under various circumstances, for example, in the frog in the mesentery or the web of the foot. The opinion held previously by some workers that capillary constriction is produced by contraction of the Rouget-Mayer cells, found close to the wall of the capillaries, appears to be untenable.

Arteriovenous anastomoses, establishing direct communications between arteries and veins, are extremely common. As more careful search for them has been made, they have been found in a large number of organs. They vary in dimensions and complexity, being perhaps most developed in the distal parts of the body, such as the fingers, toes, and ears. Usually they are rather coiled, and the arterial end is abundantly supplied with contractile cells and has an innervation of sympathetic fibers. The arteriovenous anastomoses obviously represent a potent mechanism for changing the pattern of blood flow through the capillary net, according to their variations in caliber. Opening up of the anastomoses in the skin provides an effective means for increasing the blood flow, for example in the external ear or fingers during exposure to cold.

The physiology of the peripheral circulation and the many factors, nervous and humoral, which influence it, is a vast and complex field of study. A number of reflexes are involved. Mention was made previously of some data on the central regulation, from which it is seen that vasodilator and vasoconstrictor effects of peripheral vessels can be obtained on stimulation of several regions of the central nervous system, especially the reticular formation of the lower brainstem, the hypothalamus, and certain regions of the cerebral cortex. Ultimately these central controlling and regulating mechanisms act on the peripheral autonomic nervous system.

As mentioned above the *main action of the sympathetic innervation of blood vessels is constrictor.* It appears to be established that the neurotransmitter responsible for this action is *noradrenaline.* The presence of *vasodilator fibers* has been the subject of much debate.[43] It appears to be generally agreed that such fibers are present in the parasympathetic chorda tympani and in the parasympathetic fibers to the external genital organs. The transmitter is assumed to be *acetylcholine.* The observation that vasodilatation may occur following stimulation of nerves led some early workers to postulate the presence of parasympathetic fibers in the peripheral nerves, in opposition to anatomical observations. This controversy is now resolved, since it has been shown, as mentioned previously, that some fibers

---

[43] It should be kept in mind that alterations of the caliber of blood vessels produced by nerve fibers can occur only by variations in a constrictor action. There is no structural mechanism for an *active* dilatation. Accordingly, dilatation of a vessel following changes in its innervation can occur only by a diminishing "tonus" of the constrictor influence which resists the intravascular hydrostatic pressure.

belonging anatomically to the sympathetic nervous system, are cholinergic. This is the case not only for the sympathetic fibers to the sweat glands, but also for some sympathetic fibers passing to vessels. It has been particularly well demonstrated for skeletal muscle (see Uvnäs, 1960). Whether such fibers are also present in the sympathetic supply of the heart, skin, and intestines has not been finally settled.

On stimulation of a bundle of sympathetic nerve fibers the constrictor effects on the vessels usually predominate and mask the vasodilator effects. The fact that vasodilatation can be clearly demonstrated following stimulation of certain regions of the central nervous system indicates that the two effects presumably may occur separately under normal circumstances. In clinical cases of affections of the peripheral parts of the sympathetic system the effects on the vasoconstrictors usually dominate the picture.

It was mentioned previously that irritation of the upper thoracic ventral roots is followed by vasoconstriction in the neck and head on the same side. The skin becomes cool and as a rule also pale. (Sweat secretion and piloarrection may appear concomitantly, but it should be pointed out that under normal as well as pathological conditions changes of the vasomotors, sudomotors, and pilomotors may occur independently of each other.) Following extirpation of the stellate ganglion a vasodilatation of face, neck, and lateral part of the arm is observed, betraying itself in an increased temperature, and reddening of the skin. This is due to interruption of the vasoconstrictor impulses to the areas mentioned, because they traverse the stellate ganglion. Some of the postganglionic fibers to the head accompany the arteries, but some leave the latter and enter the cranial nerves.

> On account of the sensitizing effect of sympathetic denervation, the vasodilatation following an extirpation, for example of the stellate ganglion, is apt to diminish somewhat in the course of a week or two. When instead the sympathetic fibers from the 2nd and 3rd thoracic segments in the white rami are severed, and the sympathetic trunk is divided between the 2nd and 3rd thoracic ganglia, the postganglionic neurons are left intact, sensitization is less pronounced, and the effect is more enduring. A Horner's syndrome is also avoided. Many neurosurgeons, therefore, prefer such *preganglionic* denervation of the upper extremity to extirpation of the stellate ganglion, which represents a postganglionic denervation. The role of sensitization in sympathectomies is still debated (see Monro, 1959, for some data).

In lesions of the peripheral nerves or the brachial or lumbosacral plexuses a paralysis of the vasoconstrictors also appears, but as a rule the increased temperature and reddening of the affected areas are not enduring. In irritative lesions of the peripheral nerves the ensuing vasoconstriction may, however, be rather conspicuous. Lesions affecting a limited part of the intermediolateral cell column will as a rule give no clear-cut symptoms, since each vertebral ganglion receives fibers from more than one segment of the cord.

The *vasomotor reflexes* are of supreme importance for the maintenance of the blood pressure and for the proper distribution of blood to different organs. They will only be briefly touched upon here. There are spinal vasomotor reflexes with a transverse reflex arc in the spinal cord. Other vasomotor reflexes utilize a longitudinal reflex arc, like the pilomotor (cf. above). They may consequently show alterations in transverse lesions of the cord. In the stage of "reorganization" after such lesions (see Chap. 4) the vasomotor reflexes are usually increased below the site of the lesion. The same may be observed in lesions of the brainstem. Lesions of the cord restricted to the region below the 2nd lumbar segment will, however, not be accompanied by vasomotor changes, since the intermediolateral cell column will escape injury.

The caliber of the *capillaries* appears to be controlled mainly by *chemical* means by changes due to local processes in the tissue. Thus an increase in the tension of *carbon dioxide* and a *lowered pH,* due to increased metabolic activity of the tissue, is followed by capillary dilatation, which secures a more abundant circulation of blood to the active tissues. Another substance which provokes a marked capillary dilatation is *histamine.* This phenomenon is closely related to the concept of *axon reflexes* and is of some clinical interest.

Following an intradermal injection of histamine, a conspicuous rubor develops around the place where the injection is made, due to capillary dilatation. This is soon followed by a local edema, changing the reddish spot into an urticarial papule. Surrounding this central spot where the capillaries are maximally dilated and where transudation of fluid from the blood occurs, a zone of active capillary hyperemia with reddening soon appears and has a tendency to widen (Lewis's triple response). The same reaction (but less pronounced) can be observed also after intradermal injection of irritating substances, such as distilled water, weak acids, and other substances. It has been assumed that in these cases the injected substance acts by producing a damage to the local tissue, whereby a substance which was originally thought to be histamine or a histamine-like compound (Lewis's H-substance) is liberated. It appears that the same mechanism underlies the dermographism, usually most pronounced in neurotic individuals. The role played by histamine in the triple reaction is not yet finally settled. Other substances, such as plasmakinins or serotonin, may also be involved.

It is remarkable that the reaction described above as following intradermal injection of histamine does not appear in cutaneous areas in which sensibility is abolished, for example in peripheral nerve lesions. In these cases only the urticarial papule develops. This indicates that nervous factors must be involved. The nerves which are of importance are the cutaneous ones. When a cutaneous sensory nerve is stimulated for a certain time, a strong vasodilatation develops in the area of distribution of the nerve. Irritation of a dorsal root yields vasodilatation in the corresponding dermatome.

It appears that the vasodilatation is mediated by fibers which leave the spinal cord in the dorsal roots and travel peripherally in the sensory nerves. There appears to be decisive evidence that the vasodilator impulses are mediated through the afferent dorsal root fibers, mainly the thinnest unmyelinated ones. Since the vasodilator impulses are transmitted in the opposite direction to the others, they are *antidromic impulses.*

Even if such impulses may arise on artificial irritation of dorsal roots and also may be a factor in the production of the eruption in herpes zoster, these facts do not prove that similar conditions prevail under normal circumstances. It appears that the vasodilatation which is observed following the application of local irritants to the skin, especially painful ones, is mediated by so-called *axon reflexes.* The fine sensory fibers are assumed to branch peripherally. On irritation of the skin, impulses are not only transmitted in a central direction and perceived as pain, but also pass in a peripheral direction through collateral branches. At their terminals histamine or another vasodilator substance (see above) is assumed to be liberated and to produce the capillary dilatation. If this is sufficiently strong, the permeability of the capillaries is increased and a local edema (the urticarial papule) develops. This theory explains why these vasodilator reflexes do not disappear immediately following transection of the dorsal roots distal to the ganglion, but only in the course of several days when the nerve fibers degenerate. It is assumed that the outer zone of vasodilatation which occurs on local application of histamine is due to axon reflexes of this type. When the nerve has degenerated, the usual rubor surrounding the point of injection is, therefore, absent. It appears that the triple response is influenced from the central nervous system. Thus the extension of the peripheral flare is different on the two sides following lesions, for example in the territory of the middle cerebral artery (Appenzeller, 1976).

The autonomic innervation of some organs. The general aspects of the functions of the autonomic nervous system have been considered above. It is appropriate, however, to deal with the innervation of some visceral organs in particular. The problems related to the autonomic innervation of visceral organs have turned out to be far more complex than was assumed a few decades ago. No attempt will be made to discuss this subject exhaustively. The main points in the autonomic innervation can be seen from the table on p. 765. For more detailed information see the tables of Bonica (1968), based in part on information following anesthetic block of autonomic fibers in man. It can be said that as a general rule the sympathetic and parasympathetic innervation of organs is antagonistic, although exceptions exist.

The autonomic innervation of the *eye* has been treated to some extent in Chapter 7 (see also Fig. 11-2). The parasympathetic fibers, as will be remembered, produce on stimulation contraction of the pupil and of the ciliary muscle (accommodation). The same effect is obtained by certain drugs, such as pilocarpine and eserine. A complete oculomotor palsy results in mydriasis and paralysis of accommodation (in addition to paralysis of most of the extrinsic muscles of the eye). Atropine produces mydriasis. Stimulation of the sympathetic fibers to the eye results in dilatation of the pupil, probably mainly by producing a contraction of the vessels of the iris. An interruption of the sympathetic fibers to the eye, on the other hand, produces a miosis. This is the case with lesions of the postganglionic fibers which originate in the superior cervical ganglion, or with lesions of the preganglionic fibers in their course from the 1st and 2nd thoracic segments of the cord through the inferior and middle cervical ganglion and the cervical sympathetic trunk. On account of this a lesion of the upper thoracic cord is usually followed by a miosis in the homolateral eye. The exact region that has to be damaged is the intermediolateral cell column in the 1st and 2nd thoracic segments, and this region is therefore frequently spoken of as the *"centrum ciliospinale."* However, the same result will also follow a lesion of the cervical cord or of the medulla and pons, if it encroaches on the tracts that descend in the lateral part of the medulla and the cord to the ciliospinal center. In all cases mentioned a *ptosis* occurs in addition to the *miosis* since the plain tarsal muscle is paralyzed, and frequently an *enophthalmos* is observed: the eyeball appears to be deeper in the orbit than on the other side. The latter phenomenon is usually attributed to a paralysis of the smooth muscles present in the inferior orbital fissure. However, these muscles are very sparse, and can scarcely be assumed to be responsible; furthermore, it appears that, when measurements are made by an exophthalmometer, the difference between the two sides is negligible. The enophthalmos, therefore, appears to mainly be only apparent, due to the ptosis. The combination of miosis, ptosis, and enophthalmos is commonly called *Horner's syndrome*. Its presence indicates a lesion of the structures enumerated above.[44]

If a Horner's syndrome is present, there is usually loss of sweating on the corresponding side of the face. This may be overlooked if not sought for. Occasionally *Raeder's (1924) paratrigeminal syndrome* is found. There is ptosis and miosis but no loss of sweating in the face, in-

[44] Some clinicians have utilized the reaction of the pupil to adrenaline instillation into the eye to differentiate between lesions of the pre- and postganglionic sympathetic neurons. If the postganglionic neuron is damaged, i.e. the lesion is peripheral to the superior cervical ganglion, the pupil dilates markedly when adrenaline is applied to the conjunctival sac. If the lesion is central to this, damaging the preganglionic fibers, the pupil reacts approximately normally, namely, very weakly or not at all. The difference is another example of sensitization following lesions of the postganglionic fibers.

dicating that the sympathetic innervation of the eye, but not of the face, is affected. This is usually accompanied by headaches above the eye, often beginning in the morning. Raeder found the syndrome in some cases of patients with head trauma or tumor, and explained it as being due to damage to the sympathetic fibers surrounding the internal carotid artery. Later studies indicate that the syndrome may also be related to migraine (see Ford and Walsh, 1958).

The *lacrimal gland* is supplied by parasympathetic fibers (cf. above and Fig. 11-2) which are secretory. If these fibers are damaged, for instance in fractures of the zygomatic bone, the reflex tear secretion of the corresponding eye is abolished. The reflex is most easily elicited by irritation of the conjunctiva or the nasal mucosa. The afferent path of the reflex arc passes through the trigeminal nerve.

The *digestive organs* on the whole get their secretory and motor impulses through the parasympathetic, their inhibitory through the sympathetic, although the latter statement requires some qualification. The course of the parasympathetic fibers to the *salivary glands* has been described previously (cf. Fig. 11-2). On stimulation of the chorda tympani, secretion takes place from the submandibular and sublingual glands. The sympathetic fibers appear to be of lesser importance. When the parasympathetic fibers are destroyed, reflex salivation cannot be elicited from the gland in question. Thus there occurs no salivation on olfactory or gustatory stimulation, and the secretion of saliva occurring in response to imagination of palatable food is also abolished. The latter reaction is probably initiated from the cerebral cortex.

The parasympathetic fibers to the *stomach and the gut* are secretory and motor. Thus stimulation of the vagus produces increased peristalsis and secretion of gastric and intestinal juices, and the sphincters relax. As remembered from Chapter 7, the rectum and the lower colon (from the splenic flexure) receive their parasympathetic supply from the sacral cord. The sympathetic influence on the digestive tube is, broadly speaking, inhibitory. Stimulation of the sympathetic is followed by reduced peristalsis and secretion, and increased tonus of the sphincters. The gastric secretion and peristalsis which occur after physic, gustatory, and olfactory stimuli are mediated through the vagus. However, considerable secretory and motor activity of the stomach is retained even if both vagi and both splanchnic nerves are cut, and in time such activity tends to become more normal. It is assumed that this remaining activity is mediated through the nerve cells and plexuses in the wall of the organs (Meissner's and Auerbach's plexuses).

The exocrine glands of the *pancreas* are supplied with secretory fibers from the vagus, although the sympathetic fibers may have some influence in this respect. Nerve fibers have also been traced in the electron microscope to the islands of Langerhans (see Stahl, 1963). It appears that β-cells receive both adrenergic and cholinergic nerve fibers, but their mutual role in influencing the secretion of insulin does not seem to be clear. The liver appears to receive only sympathetic fibers, but the vagus carries motor fibers to the gall bladder and bile duct, and stimulation of the vagus causes increased flow of bile and relaxation of the sphincter of Oddi. Because of the highly developed local "autonomy" of the digestive organs, symptoms from them will usually not be clearly evidenced in diseases of the autonomic nervous system.

Stimulation of the sympathetic fibers to the *heart* is followed by tachycardia, stimulation of the vagus by bradycardia. In addition, such stimulation has other ef-

THE PRINCIPAL FEATURES OF THE AUTONOMIC INNERVATION OF SOME ORGANS

Organ	SYMPATHETIC			PARASYMPATHETIC		
	Pregangl. neuron	Postgangl. neuron	Action	Pregangl. neuron	Postgangl. neuron	Action
Eye	Th_{1-2}	Sup. cerv. gangl.	Dilatation of pupil	Visc. oc. mot. nucl. n. III.	Ciliary gangl.	Pupillary constriction, accommodation
Lacrimal gland	Th_{1-2}	Sup. cerv. gangl.	?	Sup. saliv. nucl. n. VII	Sphenopalat. gangl.	Secretion of tears
Submandibular and sublingual glands	Th_{1-2}	Sup. cerv. gangl.	Vasoconstriction, secretion (?)	Sup. saliv. nucl. n. VII	Submandib. gangl.	Secretion of saliva, vasodilatation
Parotid gland	Th_{1-2}	Sup. cerv. gangl.	Vasoconstriction, secretion (?)	Inf. saliv. nucl. n. IX	Otic gangl.	Secretion of saliva, vasodilatation
Heart	$Th_{1-4}(5?)$	Sup., middle and inf. cerv. gangl., upper thor. gangl.	Acceleration. Dilatation of coronary arteries	Dors. mot. nucl. n. X	Cardiac plexus	Bradycardia
Bronchi, lungs	$Th_{2-7}(2-4?)$	Inf. cerv. gangl., upper thor. gangl.	Dilatation of bronchi, vasodilatation (?)	Dors. mot. nucl. n. X	Pulmonary and bronchial plexuses	Constriction of bronchi (secretion?)
Stomach	$Th_{6-10}(5-11?)$ Gr. splanchn. n.	Celiac gangl.	Inhibition of peristalsis and secretion. Contraction of pyloric sph.	Dors. mot. nucl. n. X	Gastric plexus	Secretion, peristalsis
Pancreas	$Th_{6-10}(5-11?)$ Gr. splanchn. n.	Celiac gangl.	Weak secretion (?)	Dors. mot. nucl. n. X	Periarterial plexuses	Secretion (pancreatic juice, insulin?)
Small intestine, colon asc., transv.	$Th_{6-10}(5-11)$ Gr. splanchn. n.	Celiac gangl., sup mesent. gangl. and other ganglia and plexuses	Inhibition of secretion and peristalsis	Dors. mot. nucl. n. X	Myenteric and submucous plexuses	Secretion, peristalsis
Colon desc. and sigm., rectum	L_{1-2}	Inf. mesent. gangl., hypogastr. plex. and other ganglia	Inhibition of secretion and peristalsis	S_{2-4}	Myenteric and submucous plexuses	Secretion, peristalsis
Kidney	$Th_{12}-L_1 (Th_{11}-L_2?)$	Celiac gangl., renal plexus	Vasomotor changes	?		
Ureter, bladder	$L_{1-2}(Th_{11-12}?)$	Hypogastr. and other plexuses	Vasoconstriction, contr. of int. sph. in ejaculation	?	Vesical plexus, uret. plexus	Contraction of bladder, (detrusor)
Adrenal medulla	$Th_{11}-L_1 (Th_{10}-L_2?)$ Lesser spl. n.	Cells of adrenal medulla	Secretion	S_{2-4}		
Head, neck (skin, muscles)	$Th_{2-4}(5?)$	Sup. and middle cerv. gangl.	Vasoconstriction. Sweat secretion. Piloarrection	No parasympathetic innervation		
Upper extremity (skin, muscles)	$Th_{3-6}(2-8?)$	Stellate gangl., upper thor. ganglia				
Lower extremity (skin, muscles)	$Th_{10}-L_2 (Th_{11}-L_2?)$	Lower lumbar and upper sacral ganglia				

fects, on the cardiac muscle and the coronary vessels (see *Nervous Control of the Heart*, 1965, for some data). The distribution of the postganglionic sympathetic and parasympathetic fibers to the heart is still not settled. It appears that the auricle and atrioventricular bundle are supplied with both vagal and sympathetic fibers, whereas the ventricles are supplied by the sympathetic only. The sinoauricular node is supplied mainly by the right vagus, a finding in harmony with the physiological effect of bradycardia following vagal stimulation. There has been some uncertainty concerning the action of the sympathetic nerves on the coronary vessels. The consensus of opinion has been that their effect is dilator and not constrictor as with other blood vessels. This is in agreement with the needs of the heart muscle under sympathetic stimulation. However, there has been discussion whether the dilator effect of sympathetic stimulation is secondary to changes in the activity of the myocardium. It is important to remember that the sympathetic supply to the heart is derived not only from cervical sympathetic ganglia (Fig. 11-5 and the table on p. 765), but also from the upper 4–5 thoracic ganglia. Most of the afferent fibers from the heart take the same course as the sympathetic. The afferent fibers from the heart which travel in the vagus nerves appear to be concerned in the mediation of cardiac reflexes such as the aortic reflex and the carotid sinus reflex. (Strictly speaking, however, it is probably mainly the glossopharyngeal fibers which are responsible for the carotid sinus reflex, cf. p. 464.

The *bronchi* and *lungs* are endowed with a rich supply of autonomic fibers, and fine nerve fibers can be traced along the arteries and arterioles up to the respiratory bronchi, where they innervate the smooth muscles. Stimulation of the sympathetic fibers, ultimately derived from the 2nd to 4th thoracic nerves, produces bronchodilatation; stimulation of the vagus results in constriction of the bronchioli and possibly an increased secretion from the bronchial glands. The beneficial effect of adrenaline and atropine in cases of bronchial asthma gives evidence of the action of the two types of autonomic nerves. The afferent fibers from the respiratory organs appear to pass mainly in the vagus, and they influence the reflex activity of the respiratory "centers" in the medulla. There is some uncertainty concerning the action of the nerves on the vessels of the lung, but the sympathetic influence appears to be predominantly vasoconstrictor, although this effect is not as pronounced as in many other organs. The thoracic organs (lungs, heart, and large arteries) are supplied with receptors of many kinds. They have mainly been studied electrophysiologically. Most of the information such organs provide is apparently not consciously perceived. (For a recent review see Paintal, 1977.)

The *kidneys* are supplied by sympathetic fibers which enter the organ with the renal vessels. Section of the splanchnic nerves is said to be followed by an increase of urinary secretion for some weeks, but this is probably due only to the increased blood flow. It is doubtful whether vagal fibers to the kidney exist. The *ureters* receive both sympathetic and parasympathetic fibers, but the peristalsis of the ureters is little influenced if all their nerves are sectioned.

The *autonomic innervation of the bladder* is of some practical importance to the neurologist. The parasympathetic fibers are derived from the 2nd to 4th sacral segments and are found in the visceral branches of the pudendal plexus, frequently called *nervi pelvici* or *nervi erigentes*. They fuse with the sympathetic fibers from the hypogastric plexus and form a vesical plexus, best developed at the base of the bladder. The preganglionic sympathetic fibers appear to be derived (in man) mainly from the intermediolateral cell column at L_1–L_2, perhaps from Th_{12} and Th_{11} as well (see Figs. 11-5 and 11-15, and the table on p. 765). Some of them descend in the sympathetic trunk to the level of the first four lumbar ganglia where, presumably, many of the postganglionic sympathetic neurons to the bladder are located. Others occur scattered in ganglia within the hypogastric plexus. Precise information on this point is not available. Postganglionic parasympathetic neurons are present beneath the serosa and between the muscle bundles of the bladder. *Afferent fibers* from the bladder proceed centrally along the routes taken by the efferents. Structures interpreted as receptors have been described in the bladder wall. The fibers from the stretch receptors, responding to distension of the bladder, are assumed to pass with the pelvic nerves, according to clinical and experimental findings (see Nyberg-Hansen, 1966c, for some references). The afferent fibers following the sympathetic outflow are believed to be concerned mainly with pain sensations.

The external, striated, urethral sphincter is supplied by somatic efferent fibers, being axons of motoneurons in S_{3-4}. They pass via the pudendal nerve. In a recent HRP study, Sato, Mizuno, and Konishi (1978) located the motoneurons supplying the external sphincter as well as other perineal striated muscles (external and sphincter, levator ani etc.) to S_1–S_2 in the cat. Figure 11-15 illustrates the main points in the innervation of the urinary bladder.

The physiology of micturition is a complex subject and the nervous factors in-

FIG. 11-15 A diagram of the main features of the innervation of the urinary bladder. Efferent nerve fibers are shown on the left, afferents on the right. See text. From Bors and Porter (1970).

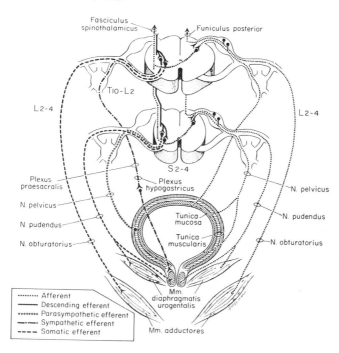

volved are not yet completely known. The primary stimulus for micturition is bladder distension. The basic reflex arc for micturition is a spinal one (afferent fibers from the bladder with perikarya in sensory ganglia S_{2-4}—a reflex center in these segments of the cord—preganglionic fibers from the intermediolateral cell column in S_{2-4} to the plexus in the bladder wall—postganglionic fibers from the latter to the muscles, see Fig. 11-15). However, this elementary reflex arc is influenced from higher levels, and under normal circumstances there is evidence that the spinal micturition "center" is under the control of a higher "center," presumably located in the pons (see below). Afferent impulses from the bladder reach even the cerebral cortex. In the following account, some data relevant to the act of micturition will be considered. It was formerly generally held that the internal sphincter was separate from the main mass of the bladder musculature. The latter is often collectively referred to as the *detrusor*. During micturition a parasympathetically induced contraction of the detrusor was thought to be combined with an inhibitory sympathetic action on the internal sphincter. However, the situation seems to be less schematic. According to later anatomical studies, there appears to be no true separate internal sphincter, but only a continuation of the longitudinal detrusor fibers around the bladder outlet, chiefly on its dorsal aspect. It appears that the opening of the internal urethral orifice on micturition is achieved simply by the contraction of the detrusor muscle and a separate inhibitory innervation of a sphincter is not necessary (Lapides, 1958). Studies of micturition in humans by "micturition urethro-cystography" (Petersén, Stener, Selldén, and Kollberg, 1962) lend support to this view.

Using the fluorescence method for demonstration of adrenergic nerves and terminals, Hamberger and Norberg (1965) found sympathetic fibers to be restricted to the muscle fibers of the vesical trigone, including the orifices of the ureters and the urethra. (It is of interest in this connection that, in contrast to the rest of the bladder which is of entodermal origin, the trigonum vesicae is derived from the mesoderm.)

While there is no dispute that stimulation of the parasympathetic fibers to the bladder produces micturition, the action of the sympathetic has been much debated, and opposing views have been held. In discussing the bladder it should be recalled that it has a dual function, that of passive collection of the urine delivered by the ureters, and that of an intermittent active expulsion of its contents. The two functions are related, insofar as expulsion is ordinarily elicited when the bladder is filled to a certain point. However, it is not the absolute volume which matters and which is the stimulus for evacuation, but the intravesical tension. When this reaches a certain level, afferent impulses from the stretch receptors travel centrally, mainly in the pelvic nerves, reach the sacral division of the cord, and by acting on the parasympathetic cells set up contraction of the detrusor muscle. The external sphincter opens later. The bladder wall possesses a certain "tonus" and shows a modest rhythmic contractile activity. While some authors have maintained that "bladder tone" is an intrinsic property of the bladder (see Gjone, 1965, for some references); others have held that it is influenced by the autonomic nervous system, the parasympathetic as well as the sympathetic. It appears from recent research (Gjone, 1965; Edvardsen, 1967; see also De Groat, 1975) that during the collecting phase there is little activity in the parasympathetic fibers to the bladder (the detrusor) because in this phase the sympathetic innervation has an inhibitory effect on the vesical ganglia and smooth muscles of the bladder. This allows increased filling of the bladder without increase in intravesical tension. With the onset of micturition, the sympathetic effect is reduced, and the micturition reflex with emptying of the bladder proceeds uninhibited.[45] All researchers agree that the parasympathetic innervation of the detrusor is essential for micturition. (The trigonal muscles appear to operate in the act of ejaculation in the male, mediated by sympathetic fibers to the ductus deferens, by preventing backflow of seminal fluid into the bladder.)

The role of the external sphincter, which can be voluntarily controlled, is mainly to contract in order to stop micturition when this is desirable. The sphincter can be closed very effectively, even if the detrusor contracts. Its opening when micturition begins appears to be reflexly induced. Electromyographic studies of the external sphincter in man by Petersen and his collaborators, and others, have given interesting information, especially when combined with "micturition urethrocystography" (see for example Petersén, Kollberg, and Dhunér, 1961; Petersén et al. 1962).

Even if the act of micturition is mediated by a basic spinal reflex, as described above, it is subject to *supraspinal influences,* a fact which is well known in man from experiences particularly with lesions of the spinal cord (to be considered later). Like most other autonomic functions, those of the bladder have been shown to be influenced from many regions of the brain, often somewhat light-heartedly referred to as "centers for micturition." Both excitation and inhibition of bladder activity have been observed. Among regions exhibiting such effects on electrical stimulation (see Ruch, 1960; Gjone, 1965) may be mentioned: the sensorimotor areas I and II, the orbital gyrus, the cingulate gyrus, the piriform cortex, the amygdaloid nucleus, the hypothalamus, the superior colliculi, and the reticular formation. Inhibition and facilitation have in part been obtained from different points (see Gjone, 1965).[46] Inhibition has further been reported to follow stimulation of the red nucleus, the substantia nigra, subthalamus, certain thalamic nuclei, and the pallidum (Lewin, Dillard, and Porter, 1967). To what extent the results of stimulation at many of these sites are due to stimulation of descending fibers appears to be unsettled, and the significance of the findings is far from clear. *In man* micturition (and defecation) have been observed to be affected following lesions of the superomedial part of the frontal lobe (Andrew and Nathan, 1964). While micturition as such occurs in a normal manner, the "higher control" of it (as well as of defecation) is disturbed with impaired sensation of bladder-filling and ex-

[45] The demonstration of an adrenergic inhibitory effect on the postganglionic parasympathetic perikarya in the wall of the bladder (see De Groat, 1975) appears to be in general agreement with data obtained in Hamberger and Norberg's (1965) fluorescence studies, referred to above. However, the observations cannot easily be correlated with the restricted anatomical distribution of sympathetic fibers to the bladder. The demonstration of the adrenergic effect is of interest from a clinical point of view, for example with regard to the beneficial effect of ephedrine in some cases of enuresis (see Edvardsen, 1967, for further correlations). The data mentioned above, like others, make it clear that the peripheral autonomic ganglia have a far more complex function than merely to serve as relays.

[46] In the amygdala, excitatory responses were obtained chiefly from the basolateral group, inhibitory responses from the corticomedial group (Gjone, 1965). The well-known urge to micturate in situations of fear but not of anger, may have a relation to the observation that fear reactions appear to be elicited most easily from the basolateral amygdaloid region (Ursin and Kaada, 1960).

treme urgency of micturition when the patient is awake, with incontinence when he is asleep. Micturition disturbances are not uncommon in frontal lobe tumors (Maurice-Williams, 1974).

On account of the practical importance of disturbances in the functions of the urinary bladder following injuries to the spinal cord, the descending pathways concerned in micturition have attracted much interest. In a number of patients subjected to chordotomy, Nathan and Smith (1958) found that "a bilateral lesion in the spinal cord which divides the region on an equatorial plane passing through the central canal has an effect upon micturition similar to that produced by a complete transection" (loc. cit., p. 188). It appears that this pathway is involved chiefly in the conscious, voluntary control of micturition, and that the fibers may belong to the corticospinal tract. If so, they apparently do not correspond to the descending vesicopressor fibers studied in animals.

The pathways for afferent impulses from the bladder within the central nervous system are incompletely known. According to Nathan and Smith (1951), the ascending fibers, mediating the sensation of fullness of the bladder and giving rise to the desire to micturate, pass in the superficial ventral part of the lateral funiculus. Fibers mediating sensations of pain from the bladder and the urethra ascend in the same region. These conclusions were made on the basis of bladder function in a series of patients subjected to chordotomies. The problems related to the central control of the bladder are still far from solved, however. There is reason to believe that the cortical influence may be more important in man than in animals. Most likely it is exerted by way of direct corticospinal fibers. However, close cooperation must exist between the voluntary control and the influence exerted by other regions of the brain, perhaps first and foremost the hypothalamus.

It is of theoretical as well as practical interest that the elementary micturition reflex appears not to be purely spinal, as described above, but utilizes a supraspinal arc. The reflex firing of sacral preganglionic fibers to the bladder following electrical stimulation of afferents from the bladder occurs after a long latency (80–120 msec in the intact cat). It is absent in the acute spinal animal. It is assumed that a reflex "center" in the pons, in the regions of the locus coeruleus, is involved (see De Groat, 1975). In chronic spinal animals, in which "automatic" bladder emptying occurs, the long-latency reflex is abolished while a short-latency reflex occurs, mediated via that sacral cord.

Lesions of various parts of the nervous connections involved in the normal activity of the bladder may cause disturbances. Bladder disturbances in man are seen most often in lesions of the spinal cord. The symptoms will differ according to the site of the lesion. In lesions above the sacral division voluntary control of the bladder will be abolished if the lesion affects both lateral funiculi. If the conduction of the fibers is completely abolished, control is entirely lost, and the bladder will empty automatically (reflexly) as soon as a certain amount of urine has accumulated. As a rule it will react with smaller quanitities of urine than normal. This is usually interpreted as being due to a decreased supraspinal inhibition. (In the case of a suddenly developing transverse lesion the bladder will at first be entirely paralyzed in the shock stage, as mentioned previously.) In partial lesions the phenomenon of precipitate micturition is common, betraying an imperfect, but not entirely lacking, central control, since the patient is able to suspend micturition for a certain time, although usually very short. This is not infrequently seen in disseminated sclerosis. The symptoms following lesions of the efferent and afferent fibers to the bladder will be considered later, in connection with lesions of the sacral cord.

The autonomic innervation of the *genital organs* is not known in detail. The gonads are supplied by sympathetic fibers following the vessels (the internal spermatic and ovarian arteries, respectively). The fibers are derived from the lower thoracic segments of the cord, and are mainly vasomotor. The sympathetic fibers to the seminal vesicles, prostate, and vas deferens are motor and vasoconstrictor. On stimulation of the hypogastric plexus (presacral nerves) ejaculation has been observed. The prostate gland appears, however, to be supplied also by parasympathetic fibers, but the main effect of the parasympathetic innervation is to be found in the vasodilatation of the erectile tissues of the genital organs, which follows stimulation of these nerves. (On account of the erection elicited in this case, the pelvic nerves used to be called the nervi erigentes.) It appears, therefore, that erection is brought about mainly by impulses through the parasympathetic, and an ''erection center'' has therefore been postulated in the sacral cord. Ejaculation, on the other hand, is initiated through the sympathetic fibers from the lower thoracic and upper lumbar segments, and consequently an ''ejaculation center'' has been located in this part of the cord.

The motor fibers to the *uterus* appear to be sympathetic; likewise those to the Fallopian tubes. The fibers are distributed mainly to the muscular coat. The presence of a parasympathetic innervation to the uterus is questionable. The action of the sympathetic on the uterus varies according to whether the uterus is pregnant or not, but on the whole the sympathetic innervation appears to be of no great importance. The uterus appears to possess considerable local autonomy. Thus the experimentally completely denervated uterus may still be able to function normally in parturition in animals, and human cases are known in which a normal childbirth has occurred with complete transverse lesions of the cord.

The importance of autonomic fibers to some of the *endocrine* glands is not yet completely understood. The endocrine gland that is most clearly under the influence of the autonomic system is the *adrenal medulla,* which receives an abundant supply of sympathetic fibers through the lesser splanchnic nerve and probably also fibers from the 12th thoracic and 1st lumbar segments. These fibers are, at least to a large extent, preganglionic, and their stimulation is followed by a marked liberation of adrenaline. This is of interest from a functional and clinical point of view, since conditions accompanied by an activity of the sympathetic will be apt to produce a sudden and massive liberation of adrenaline with ensuing symptoms of sympathetic activity. It appears that by this innervation a mechanism is provided for enabling a rapid and widespread sympathetic activity, useful under certain circumstances, particularly in ''emergency states.''

The influence of the autonomic system on the circulation in various bodily organs has been referred to above. It remains to consider briefly the *innervation of the cerebral blood vessels* and the role of the autonomic system in the control of blood circulation in the brain. Knowledge about both subjects is still rather insufficient. Fine nerve fibers were observed on small arteries of the brain by early anatomists, but were considered by some to be of little functional importance.

(For reviews, see Nelson and Rennels, 1970; Purves, 1972). These fibers were found to be continuations of the *sympathetic* periarterial plexuses surrounding the internal carotid and vertebral arteries and having their postganglionic perikarya in the superior cervical ganglion. Following the introduction of the Falck and Hillarp fluorescent technique (see Chap. 1), this was corroborated, since the periarterial intracranial plexuses, judged on the basis of their fluorescence, were found to be noradrenergic (Nielsen and Owman, 1967; and others), and the fluorescence of these nerve fibers disappeared after extirpation of the superior cervical ganglion. A *parasympathetic* innervation of intracranial vessels was described as coming via the intermediate nerve (Chorobski and Penfield, 1932). In histochemical studies, cholinergic fibers have been described along the cerebral vessels, intermingling with the noradrenergic ones (Edvinsson, Nielsen, Owman, and Sporrong, 1972).

> Electron microscopically, nerve terminals have been found in the walls of cerebral arteries. However, nerves have never been found to enter the muscular layer, but are restricted to the adventitia, in man (Dahl and Nelson, 1964) as well as in the cat and other animals (Sato, 1966; Nelson and Rennels, 1970; and others). The nerve fibers make contact with the most superficial muscle cells. The neuromuscular contacts appear largely to be of the "en passage" type. (Similar arrangements have been described in arteries in many other organs.) Synaptic "terminals" have been found to contain agranular (clear) as well as small and large dense-core vesicles. The presence of agranular and small granular (dense-core) vesicles has been taken to confirm the presence of cholinergic and noradrenergic synaptic formations, respectively (see Chap. 1).

There is generally good agreement between these anatomical studies and physiological investigations in animals and man. Sympathetic activity causes vasoconstriction; parasympathetic activity, elicited by way of the intermediate nerve, gives rise to vasodilatation. However, the role played in changes of cerebral vessels by the sympathetic and parasympathetic system is little known.

> Following local (on the pial surface) or intra-arterial administration of noradrenaline, vasoconstriction of the cerebral arteries occurs. In the dog D'Alecy and Feigl (1972) found evidence for cerebral vasoconstriction following electrical stimulation of the stellate ganglion. From studies in man (Skinhøj, 1972) it appears that vasoconstrictor fibers are not active under "normal conditions." On stimulation of the facial (intermediate) nerve, some authors have observed cerebral vasodilatation (for reviews, see Purves, 1972; and Lundberg and Jodal, 1975).

In addition to the efferent innervation, the cerebral vessels are provided with afferent fibers. Some of these mediate pain, as found on electrical stimulation of vessels in patients given only local anesthesia. The afferent fibers appear, at least most of them, to be branches of the trigeminal nerve, but little is actually known of the sensory innervation of cerebral vessels (see Purves, 1972). It is of clinical interest that trigeminal nerve fibers that supply the dural sinuses, the tentorium, and the falx, appear to be pain sensitive (for a brief review, see McNaughton and Feindel 1977).

Humoral factors, not least carbon dioxide tension, have long been known to play an important role in the regulation of cerebral circulation. A rise in pCO_2 elicits cerebral vasodilatation, a fall, vasoconstriction (cf. the dizziness that accompanies hyperventilation). Furthermore, the cerebral circulation has a considerable degree of autoregulation, enabling it to maintain its blood flow constant within wide ranges of blood pressure changes in the general circulation. Nervous factors appear to play a minor role. It has been suggested (Nelson and Rennels,

1970) that the neural influence on the cerebral circulation is superimposed on other, more basic, mechanisms and that it is concerned mainly in fine or rapid adjustments in cerebral circulation in response to local tissue requirements. Other explanations have been suggested (see Purves, 1972). As will be mentioned in Chapter 12, it has been shown that blood flow through a particular region of the cerebral cortex is related to the degree of its activity.

Visceral sensibility. Referred pain. From a practical point of view the afferent visceral fibers are of no less importance than the efferent. However, our knowledge concerning them is still far from complete.[47] Visceral afferent fibers are first and foremost of importance for the many *visceral reflexes,* some of which have been mentioned previously. However, it appears clear that some of the visceral afferent impulses may be transmitted to higher levels of the nervous system and ultimately reach consciousness. Under normal circumstances these impulses play a minor role and are for the most part not clearly recognized. They mediate impressions of a diffuse nature, such as the feelings of hunger, thirst, fullness of the bladder and bowels, etc. In certain pathological conditions, however, sensory impressions from the viscera are consciously perceived and are clinically important.

Our knowledge of *visceral sensibility* is incomplete. Some results of experimental research have been referred to in previous sections of this chapter. Most present-day information on this subject in man has been obtained from clinical observations. It has been known for some time that the viscera are themselves insensitive to touch, cutting, cold, and heat, and operations on visceral organs may on this account be performed under local anesthesia. In this type of operation the patients complain of pain or disagreeable sensations only if the parietal peritoneum, pleura, or pericardium are manipulated, or when the mesenteries are stretched. Many diseases of visceral organs, as is well known, are not accompanied by pain (e.g., endocarditis, pulmonary diseases such as tuberculosis, and renal tumors). On the other hand, the violent pains which accompany some diseases of visceral organs clearly demonstrate that these organs are not completely insensitive, but must be endowed with sensory fibers. Visceral pains have some characteristics in common which distinguish them to some extent from pain in superficial structures. They are as a rule *diffusely localized* by the patient and they have a tendency to *irradiate to cutaneous areas*. On this account they are often assumed by the patient to arise mainly or exclusively in surface regions of the body. This latter pain is termed "referred pain."

The conditions giving rise to visceral pain may be grouped under different headings. Most commonly the pain is due to distension of hollow viscera, probably mainly because the distension sets up forcible contractions of the smooth muscles of the walls, usually called "spasms." Of this type are the pains occurring in blockage of the ureter or the bile ducts by concrements ("renal" and "biliary colics" respectively) in which the pains commonly occur in bouts. Another cause of visceral pain is rapid stretching of the capsules of solid organs, such as the liver and the spleen. In the case of the heart a sudden anoxemia may give rise to severe pain and it is probable that the same mechanism may also be responsible in other organs. The pains occurring in some inflammatory diseases of visceral organs may be due to violent contractions of the muscular walls in some

---

[47] For some data, see Newman (1974).

cases (thus probably in the case of the pains seen in cystopyelitis), but in other cases they are probably due to irritation of the peritoneum.

As mentioned previously the visceral sensory fibers, like the somatic, have their perikarya in the spinal ganglia and enter the cord in the dorsal roots. Some of these fibers are fine: others are coarse. Several observations allow conclusions to be made concerning the course of the visceral afferent fibers, a point of supreme importance for the surgical treatment of visceral pain. *The bulk of the visceral pain fibers accompany the sympathetic fibers.* Thus transection of the splanchnic nerves or the corresponding communicant rami abolishes pain arising from the upper abdominal viscera, such as the stomach or the gut. Likewise, transection of the dorsal roots of the part of the cord which supplies the organ in question with sympathetic fibers will be effective. For example, the pains of biliary colic may be abolished by section of the dorsal roots of the 4th to 10th thoracic segments as well as by section of the splanchnic nerves. The visceral afferent fibers, at least the majority of them, according to observations of this kind, thus enter the cord through the dorsal roots. However, in some cases section of the dorsal roots has not been entirely successful, and the existence of other afferent pathways has been postulated.

The *only exception* to the rule that visceral pain fibers follow sympathetic fibers appears to be in the case of the fibers from the sigmoid colon and rectum, the neck of the bladder, the prostate, and the cervix of the uterus. These, according to the findings of neurosurgeons (for references see White and Sweet, 1955), enter the cord in the dorsal roots of the 2nd to 4th sacral segments (where the parasympathetic efferents originate). Similarly, in their peripheral course these visceral afferent fibers appear to accompany the parasympathetic nerves, being found in the pelvic nerves and the associated plexuses. (The pain fibers from the fundus of the bladder and the fundus of the uterus, however, pass through the hypogastric plexus to the dorsal roots of the 11th thoracic to the 1st lumbar segments.) There has been some dispute as to whether the vagus carries visceral pain fibers. According to surgical experience it apparently does not. Balchum and Weaver (1943) found that pain fibers from the stomach do not pass in the vagus but in the greater splanchnic nerves.

The peculiar phenomenon that visceral disease is frequently accompanied by pain which is referred to definite parts of the body surface, either alone or with concomitant sensation of pain conceived as of visceral origin, has been difficult to explain, and many features of this fact have not yet been clarified. The occurrence of *referred pain* is familiar to clinicians. As examples may be cited the pain in the right scapular region with diseases of the liver or gall bladder, the epigastric pain in gastric ulcer, the inguinal and testicular pain in renal colic. This *parietal pain* is localized quite regularly to definite cutaneous areas in affections of the various organs. In this cutaneous area the skin is frequently hyperesthetic, there may be vasomotor changes, and reflex rigidity of skeletal muscle (*défense,* "guarding") may be observed in the same area. These symptoms may occur also without real pain. The cutaneous areas presenting these changes are commonly termed Head zones, on account of the basic study of these zones by Henry Head. *The cutaneous areas or zones for the various viscera coincide roughly with the segmental distribution of the somatic sensory fibers which take origin from the same segments of the cord as the sympathetic fibers to the viscus in question.* Thus the zone for the stomach corresponds approximately to the dermatomes Th_{5-9}, i.e., the epigastrium, usually mainly on the left side, and within this area the cardiac region appears to be represented most cranially, the pyloric region most caudally. The zone for the liver and gall bladder comprises the dermatomes Th_{7-9} on the right side, that for the appendix and cecum, Th_{10-12}. In angina pectoris the upper thoracic dermatomes are involved, usually only on the left side, namely the upper parts of the thorax and the ulnar side of the arm. In the case of the viscera which send their pain fibers to the sacral cord (see above) the cutaneous areas coincide with the corresponding sacral dermatomes; they are found extending from the perineum (cf. Chap. 2 on the dermatomes).

The phenomenon of referred pain occurs not only in diseases of the viscera. It may be observed in affections of the surfaces of the body cavities, such as the pleura and peritoneum (''parietal referred pain''), as well as in injuries or diseases of deep somatic structures, such as muscles, joints, ligaments, and periosteum; reference of pain from one area of the skin to another is not uncommon. The many varieties of referred pain have many features in common, suggesting that the basic mechanism involved is essentially the same.

> Usually the area of reference becomes painful only when the initial pain, localized by the patient more or less precisely to the original focus, has lasted for some time. The referred pain may even persist after the local pain has vanished. The pain as well as the cutaneous changes, such as hyperesthesia, are not always distributed to the expected dermatomes, but may spread over wider areas. In this respect there are individual variations (assumed by some to depend upon anatomical differences in the distribution of sympathetic fibers). These individual differences have been particularly clearly demonstrated by workers (for example, Hockaday and Whitty, 1967) who have studied referred pain experimentally in man by injecting irritating substances, such as hypertonic saline solutions in the interspinous ligaments in healthy subjects, as done first by Lewis and Kellgren (1939). Hockaday and Whitty (1967), confirming the findings of other authors, found that anesthetizing the site of the stimulus consistently abolished referred pain and the accompanying changes, such as hyperesthesia and muscle spasm, but anesthetizing the site of reference did not consistently reduce or abolish the referred effects. The latter observations are of interest with reference to the role played by peripheral factors in the appearance of referred pain and especially in the changes accompanying the pain.

Many attempts have been made to explain the mechanism underlying referred pain. The occurrence of the cutaneous zones of Head seems to imply that, by one means or another, the visceral impulses must be propagated to the cutaneous sensory fibers or tracts. Mackenzie assumed that abnormally strong visceral impulses occurring in visceral disease produce some sort of increased irritability of the gray matter of the region of the cord where the impulses enter, and, on account of this, reinforce normally subliminal impulses from the somatic structures, leading to their conscious perception. An ''irritable focus'' of this type would explain the hyperesthesias that are so frequent in the cutaneous zones of referred pain. An abnormal irritation of the visceral efferent neurons would explain vasomotor, pilomotor, and sudomotor changes in the zone; and the reflex rigidity of somatic musculature might be due to a similar irritation of the ventral horn cells. Hinsey and Phillips (1940) tried to formulate the conceptions of Mackenzie in a more precise manner. They assumed that both visceral and somatic afferent fibers are capable of acting on a ''common pool'' of secondary neurons, which are subjected to summation and inhibition. The mechanisms involved are far from simple. Even though spinal mechanisms are undoubtedly involved, recent findings on the finer organization of the sensory pathways and their synaptic stations make it likely that important processes occur at supraspinal levels. Descending fibers, influencing the central transmission of afferent messages, may play a role. Reference has been made previously to the recent theory of pain mechanisms, the "gate theory" of Melzack and Wall (1965). This is of relevance also for the problems of referred pain, as discussed in their article, to which the reader is referred. For further comments on referred pain and discussions of possible central mechanisms, see White and Sweet (1955), Nathan (1956a), Nordenboos (1959), Appenzeller (1976), Nathan (1976), and Wall (1978). See also Chapter 2.

Although the mechanism of "referred pain" has not yet been adequately explained, it is of considerable importance for the clinician, and not only the neurologist, in the diagnosis of visceral disease. The information concerning visceral sensibility is also of relevance when therapeutic surgical measures are required to cope with pain originating from visceral organs.

In connection with the discussion of visceral pain, a special feature of it, namely *vascular sensibility*, should be briefly considered. It has been mentioned previously that the blood vessels are supplied with nerve fibers, some of them sensory, and clinical evidence substantiates this. Thus ligation of an unanesthetized artery is usually painful. Experimental injection of irritating substances (barium chloride, for example) into an artery likewise gives rise to pain and is followed by a general rise of blood pressure, if the nerve plexus of the artery is intact. These nerve fibers accompany the artery for a short distance, but soon leave it to join the main nerve trunks in their central course. The fibers have their perikarya in the spinal ganglia, like other visceral afferent fibers. Clinical experience seems to show that some of the vasosensible and probably also some of the other visceral afferent fibers accompany the arteries during virtually their entire course, to finally reach the spinal ganglia via the sympathetic trunk.

Symptoms in lesions of the autonomic system. Lesions of the hypothalamus and brainstem.   In view of the incomplete knowledge of the structure and particularly the function of the different parts of the autonomic system, it is no wonder that the clinical aspects of diseases of these structures are insufficiently understood. As a matter of fact, there are as yet not many clear-cut diseases or syndromes which can be unhesitatingly ascribed to lesions of the autonomic system, although there is good reason to believe that the autonomic system is involved in producing some of the symptoms in a multitude of diseases that do not primarily affect the system itself. The intimate interrelation between the hypothalamus and the cortex is worthy of emphasis, and should be especially borne in mind when so-called functional disorders and neuroses are treated.

In the preceding sections reference has repeatedly been made to the clinical aspects of disturbed autonomic functions and to symptoms following lesions of various parts of the autonomic system. The facts already considered will only be briefly mentioned here, and attention will be paid primarily to features which have not been dealt with. No complete account of all the symptoms which belong to this field will be attempted.

The symptoms following lesions of the *hypothalamus* have for the most part already been considered. One or more of the following symptoms are those most commonly observed: *Diabetes insipidus* (when the supraoptic nucleus or the hypophysial stalk is damaged), *disturbances in heat regulation* and *vasomotor disturbances, sleep disturbances* (most frequent as hypersomnia in lesions of the posterior hypothalamus), *dystrophia adiposogenitalis* (region of tuberal nuclei or ventromedial nucleus), *hemorrhages and ulcerations of the alimentary canal.* Furthermore, *Horner's syndrome* and *emotional changes* may occur. On account of the small size and the site of the hypothalamus, lesions of it will very frequently be associated with damage to other neighboring structures, particularly the optic chiasma or tracts and the hypophysis.

The hypothalamus may be affected by *fractures of the base of the skull.* In some of these cases diabetes insipidus develops. Hypothalamic lesions are frequently found after fatal *acute closed head injuries* (see Crompton, 1971). *Primary* and *metastatic tumors* may also affect the hypothalamus. Of primary tumors the craniopharyngeomas and the suprasellar meningeomas should be especially recalled (cf. Chap. 8). *Vascular incidents,* on the other hand, very rarely affect the hypothalamus, which has a very rich and efficient blood supply. Occasionally *aneurysms* of the internal carotid or the circulus arteriosus press on the hypothalamus. Far more frequently this region is affected by infections of the central nervous system. The regular involvement of the hypothalamus in *epidemic encephalitis* has been mentioned (cf. also Chap. 4). Another condition that commonly involves the hypothalamus is *syphilitic meningoencephalitis;* also in *general paresis* symptoms from this part of the brain may occur. In cases of *alcoholic encephalopathy* changes in the hypothalamus have been described, and in the amnestic confabulatory *Korsakoff's psychosis* the pathological alterations are often especially marked in the mammillary body (however see Chap. 10). *Progressive leukodystrophy* has been assumed to bear a relation to the hypothalamus. Multiple small hemorrhages, old and recent, have been found in the hypothalamus in cases of *chronic gastric and duodenal ulcers* (Vonderahe, 1940), and in *cancer of the stomach* changes resembling those found in chronic alcoholism have been described in the mammillary bodies (Neubürger, 1937). Baker, Cornwell, and Brown (1952) have drawn attention to the frequent affection of the hypothalamus in *poliomyelitis.* In a clinicopathological study of 115 cases they found clinical evidence of hypothalamic dysfunction (hyperthermia, hypothermia, gastric hemorrhage, and other symptoms) during the acute illness, and slighter changes that could persist for months or even years after recovery.

Lesions of the *pons and medulla oblongata* are apt to interrupt descending fibers to the autonomic preganglionic neurons of the brainstem and spinal cord. This is the case particularly with lesions in the lateral part of the pons and medulla (concerning other symptoms related to such lesions see Chaps. 2 and 4). Immediately after an acute interruption of the pathways the autonomic reflexes will be abolished below the lesion on the same side, and there will therefore be anhidrosis and lack of piloarrection and, as a rule, some vasodilatation in this half of the body. When the initial shock has passed, Horner's syndrome becomes evident on account of the interruption of the fibers to the centrum ciliospinale. Eventually this, as well as the other symptoms, becomes less conspicuous. Thus the difference in skin temperature and sweating in the two halves of the body will gradually diminish. The pilomotor reflexes are usually increased on the side of the lesion. This, as well as other features, has been explained as being caused by the lack of central regulation of the activity of the spinal autonomic cell groups (Foerster, Gagel, and Mahoney, 1939; List and Peet, 1939).

Lesions of the spinal cord. The conus syndrome. *In lesions of the spinal cord at levels above the 3rd lumbar segment* (stabbing wounds, bullet wounds, compressions due to fractures of the spine or intraspinal tumors, transverse myelitis), the autonomic descending pathways may be interrupted if the lesion involves the lateral funiculus. In these cases autonomic spinal reflexes, mediated via segments below the level of the lesion, will, after some time, increase, just as do the tendon reflexes. The pilomotor reflexes, as has been mentioned, are purely homolateral, whereas the vasomotor and sudomotor reflexes are also mediated by reflex arcs which cross the median plane of the cord. Changes in the latter reflexes may therefore be observed bilaterally even in unilateral lesions of the cord, and their diagnostic value is thus somewhat reduced. It has been mentioned previously that the pilomotor reflexes, especially, may give information on the site of the lesion in such cases, but it should be emphasized that the segmental level of the pilomotor disturbances does not correspond exactly with the level of the somatic symptoms (cf. above). Transverse lesions of the cord will abolish the voluntary control of the bladder, as referred to previously. In the following discussion we will return briefly to symptoms from the bladder in lesions of the cord above the sacral segments.

In addition to producing symptoms due to interruption of the descending autonomic pathways, lesions of the spinal cord between the levels of Th_1 and L_2 will as a rule also destroy some of the preganglionic neurons in the intermediolateral cell column. However, symptoms due to such a lesion tend to diminish after some time and, furthermore, an affection of only a limited part of the column will scarcely reveal itself by distinct symptoms.

Lesions in the upper part of the thoracic cord are most likely to produce symptoms due to involvement of preganglionic neurons. If, for example, the *two upper thoracic segments* are affected, a Horner's syndrome will develop on the side of the lesion. If also the following two or three thoracic segments are involved, the vasomotor, sudomotor, and pilomotor reflexes will be abolished not only in the face but also in the homolateral arm. Sweat secretion is abolished in this area, and during the first few days after the lesion has developed, a vasodilatation may be observed. Symptoms of this kind may be met with in *syringomyelia,* which frequently affects the lower cervical and upper thoracic cord. In the beginning, however, signs of sympathetic irritation may be observed, in the form of hyperhidrosis and vasoconstriction. In order for loss of spinal autonomic reflexes to become apparent, several segments have to be affected, since each autonomic ganglion receives fibers from several segments of the cord. If the *lower thoracic and upper two lumbar segments* are affected, symptoms such as those just described will occur in the lower extremity. It has been mentioned previously that loss of the spinal vasomotor and sudomotor reflexes tends to become less distinct after some time, since local autonomic reflexes become exaggerated. As mentioned above, ejaculation disturbances in the male may occur following affections of the lower thoracic and upper two lumbar segments.

Lesions of the sacral region of the spinal cord deserve particular consideration. The symptom complex which appears in these cases constitutes the so-called *"conus syndrome."* The *somatic* symptoms concern the middle and commonly also the lower sacral dermatomes and myotomes. There will be a *flaccid paralysis* of the muscles of the outlet of the pelvis, including the external anal and vesical sphincters and the ischiocavernosus and bulbocavernosus muscles. *Loss of sensation* will occur in the region of the perineum, around the anal orifice (S_{3-5}) and commonly also some of the posterior aspect of the thigh (S_2), giving the appearance of a so-called *saddle anesthesia*. If the 1st sacral segment is spared, the ankle jerk is usually retained. When pain is present in pure conus lesions, it is referred to the sacral dermatomes. The *destruction of the preganglionic parasympathetic perikarya and their fibers* will result in a *paralysis of the bladder*. The paralyzed bladder will be distended by the accumulating urine, and no automatic emptying occurs. However, when the distension is severe, some overflow will usually occur, with emptying of small quantities of urine at short intervals. This "dribbling incontinence" is not sufficient for emptying the bladder effectively. The voluntary influence on the bladder is, of course, also lost, and contrary to the condition with lesions of the cord at higher levels, which interfere mainly with the voluntary impulses to the bladder, peripheral stimuli will not elicit reflex contraction of the bladder in lesions of the conus. In addition to the paralysis of the bladder, there is also a *paralysis of the rectum and sigmoid colon*, which receive their parasympathetic motor innervation from the sacral cord. Consequently there is *no spontaneous defecation*, but on account of the paralysis of the voluntary, striped, external sphincter of the anus there will be *incontinentia alvi*. However, after some time a certain degree of automatism of the rectum develops, presumably mediated by the peripheral reflex mechanism (cf. above on the local autonomy of the gut). Finally the conus syndrome in the male is characterized by *loss of* the capacity of *erection and ejaculation*, on account of the destruction of the preganglionic parasympathetic neurons, and the somatic motor ventral horn cells. However, emission of semen occurs, since the motor fibers to the ductus deferens and seminal vesicle are sympathetic. (This emission of semen, however, does not occur as in ordinary ejaculation, since in this process the striped, somatic ischio- and bulbo-cavernosus muscles are also engaged.) During the development of a conus lesion, when some cells are still preserved, priapism may occur, due to irritation of the visceral and somatic motor cells of the sacral cord.[48] If the conus lesion is incomplete, it may also happen that not all symptoms usually included in the syndrome are present. Thus there may be loss of ejaculation and erection, but preserved or only moderately impaired function of the bladder, etc. (dissociated conus syndromes). It should be noted that the conus syndrome may be caused by spinal cord lesions that affect the segments S_{2-4} secondarily. Some authors maintain that the caudal end of a lesion must be above L_4 if the reflex activity of the bladder is to be fully preserved.

In connection with the paralysis of the bladder in lesions of the conus it should be stressed that identical disturbances will occur if the conus is intact, but the *efferent* fibers to the bladder in the pelvic nerves are damaged. On the other hand, a pure lesion of the *afferent* pathways from the bladder, damaging them in

---

[48] Priapism may be seen also in acute transverse lesions at higher levels of the cord.

the dorsal roots or in the cord, will also prevent reflex evacuation of the bladder and lead to a distended, *atonic* bladder. This is most commonly observed in tabes. From these facts it will be clear that *lesions of the cauda equina* usually will give rise to similar symptoms as the conus lesions. (The deliberations mentioned above concerning the bladder apply also to the rectum and sigmoid colon and the innervation of the genital organs.) However, since a lesion of the cauda will be apt to affect not only fibers of the sacral segments of the cord but also a varying number of the lumbar nerves, the sensory and motor losses will as a rule be more extensive and reach higher levels, and, furthermore, the distribution will usually be more irregular, since some of the roots of the cauda will avoid damage and be merely pushed aside, as in the case of a tumor developing intraspinally.

Brief mention has been made previously of the various disturbances in bladder function following transverse lesions of the cord above the sacral levels. On account of their frequency (traumatic cord injuries and multiple sclerosis being the most common causes) these disturbances have attracted much interest. Some additional remarks are therefore appropriate. (For more complete accounts, see for example *The Neurogenic Bladder,* 1966; Bors and Comarr, 1971). As mentioned previously, following an acute lesion of the cord above the sacral levels, "spinal shock" affects the bladder and there is retention of urine. When the "shock" declines, the bladder functions "automatically"; the spinal reflex apparatus in the paraplegic patient works without its normal supraspinal control.[49] The micturition reflex is usually more easily elicited than normally and, when started, micturition cannot be brought to a stop by voluntary innervation of the external sphincter. Various stimuli below the level of the lesion are apt to elicit the micturition reflex, for example, cooling, touching, or pricking of the skin of the perineum or the inside of the thigh. This is annoying but may be made use of by the paraplegic patient as a trick to elicit emptying of the bladder.[50] The increased reflex activity often involves the external sphincter. Electromyography has shown that in patients with a spastic paraplegia the external sphincter is often never completely relaxed, and may thus increase the outflow resistance during micturition with a resulting incomplete emptying of the bladder. (Operative interruption of the external sphincter by denervation or division of its fibers may be beneficial.)

In other cases of paraplegia, the bladder may be "overactive" or "uninhibited." This may be counteracted by drugs of the atropine group, which reduce the parasympathetic effect on the detrusor. On the other hand, cholinergic pharmaca are often used with good effect in transitory postoperative retention, and in chronic hypotonic function of the bladder when this is not due to physical obstruction. ("Ganglionic blocking agents" and "polysynaptic inhibitors" may be used with advantage in certain cases of bladder disturbances, as reviewed by Pedersen and Grynderup, 1966.) As mentioned in a previous section, there is some evidence for a sympathetic influence on micturition. Some slight differences in the

---

[49] As mentioned above, in the normal state, the micturition reflex appears to have its reflex center in the pons, while the automatic emptying of the bladder following a transverse lesion of the cord is taken care of by a spinal reflex arc with its center in S_{2-4}.

[50] In a number of patients Nathan (1956b) found that even if the spinal cord is completely transected the patient may have a certain awareness of bladder-filling, giving him some control of micturition. It appears that these sensory impulses are transmitted centrally via afferent fibers accompanying sympathetic nerves.

symptomatology of the bladder following cord lesions above or below the origin of the sympathetic fibers (but above the level of the sacral outflow) may perhaps be related to the presence or absence of sympathetic innervation (see Edvardsen, 1967, for some data).

Lesions of the peripheral parts of the autonomic system. Pathological processes affecting the *ganglia of the sympathetic trunk* will produce symptoms identical with those in lesions of the intermediolateral cell column, except that, in pure lesions of the trunk, somatic motor and sensory symptoms will be absent. However, in order to produce symptoms, such lesions have to involve several ganglia at most places, and are therefore not very frequently diagnosed. In the cervical and lumbar part of the trunk, however, even a small lesion will be able to interrupt fibers to ganglia situated cranially or caudally, respectively (cf. Fig. 11-5). Most common among lesions of peripheral parts of the sympathetic are perhaps lesions of the upper thoracic ganglia and inferior cervical ganglion or the stellate ganglion. As has been mentioned, these lesions are followed by abolished sweat secretion and piloarrection of the head and arm on the side of the lesion, accompanied by a vasodilatation. If the inferior cervical ganglion is included, Horner's syndrome is usually the most enduring symptom. Causes of such lesions include stabbing or gunshot wounds, most commonly seen in wartime, mediastinal tumors, aneurysms of the aorta, glandular abesses of the neck, and operative lesions. If the lesion develops gradually, irritative symptoms commonly mark the beginning, betraying themselves by vasoconstriction, sweat secretion, and piloarrection.

Interruption of the sympathetic innervation of the *abdominal and thoracic viscera* as a rule does not yield clear-cut symptoms, regardless of whether the vertebral ganglia or more peripheral structures are involved.

It has been remarked previously that *lesions of the plexuses of the spinal nerves or the peripheral nerves* themselves will be followed by disturbances of sympathetic nature which present a distribution corresponding to the arrangement of the somatic sensory fibers. In lesions of the plexuses, for example the brachial, a segmental vasodilatation, anhidrosis, and abolished piloarrection may be observed. In irritative lesions the symptoms may be more conspicuous, betraying the affection of the sympathetic fibers by the presence of vasoconstriction, increased sweating, and not infrequently also trophic disturbances even of the bones. In cases of polyneuritis irritative symptoms may occur together with deficiency symptoms. However, it is a peculiar circumstance that in some cases of lesions of the peripheral nerves with a complete anesthesia in the area of distribution, a certain diffuse, uncharacteristic sensibility may be retained. As a rule this sensibility is most clearly demonstrated by pinching or deep pressure. In some cases even the slightest touch or a slight pressure may give rise to severe pain, as in cases of causalgia. It has been assumed that this sensibility must be mediated through the sensory fibers which accompany the vessels.

No disease is known which can be attributed to a selective lesion of the *peripheral autonomic plexuses.* However, on the basis of animal experiments and findings made in operations on these structures in man, there is reason to believe that disturbances of the visceral innervation may be responsible for some symptoms seen in various diseases.

Operations on the autonomic nervous system. Most of these operations were originally made on the basis of theoretical speculations in cases where the etiology of the disease was unknown. Operations on the sympathetic system have been performed in nearly all types of diseases. The results have been of varying value. A considerable proportion of the operative procedures have as their principal aim the relief of otherwise intractable pain.

An interruption of the sympathetic pathways can be achieved by operations of various kinds. A *periarterial sympathectomy* (periarterial neurectomy) consists essentially in removing the periarterial sympathetic plexus of a part of an artery after it has been exposed. The adventitia is stripped off the artery for a certain length. Most neurosurgeons have found that the ensuing vasodilatation peripheral to the sympathectomized part of the artery is only transient and the effect on pain has not been convincing. This may perhaps be due to the fact, mentioned previously, that apart from the main arterial trunks of the proximal extremities, the vessels obtain their sympathetic fibers from the nerves in their immediate vicinity (see, for example, Stürup and Carmichael, 1935). The possibilities of regeneration of sympathetic fibers and of sensitization of denervated structures to humoral substances have been considered.

A complete interruption of sympathetic impulses to a certain region of the body can be accomplished by removing all the ganglia of the sympathetic trunk through which the impulses pass. This procedure is commonly called *ganglionectomy* (or sometimes sympathectomy). The same effect will be obtained also by cutting the white rami communicantes to the region in question. This *ramisection,* however, is a more delicate operation, and as a rule it will not be possible to cut only the white rami, the gray being also included. It will be understood that a ganglionectomy will in some cases represent an interruption mainly of the *preganglionic* fibers. This is the case, for instance, when the upper lumbar ganglia are removed in sympathectomy for the lower extremity (as a rule the 2nd to 4th ganglia are removed; the 1st should be spared in order to avoid sterility in the male by paralyzing the ductus deferens). In this case the postganglionic neurons in the lowest lumbar and sacral ganglia of the trunk will not be affected. When the stellate ganglion is removed in order to obtain a sympathetic denervation of the arm, this, in contrast, involves removing the *postganglionic* neurons. (In order to be effective the upper two thoracic ganglia have been included in the extirpation, since frequently the 2nd thoracic ganglion contains some postganglionic neurons to the arm.)

Following sympathectomies it has been noted by many neurosurgeons that some areas of the body are not sympathetically denervated. This subject has been thoroughly studied by Monro (1959).

In a number of patients he tested the effect of sympathectomies by means of the electrical skin resistance method. Following cervicodorsal sympathectomy (which includes the stellate and 2nd thoracic ganglion) there is an "escape area" in the central region of the face. Following a thoracolumbar sympathectomy (which includes the 4th thoracic to the 3rd lumbar ganglion) there is an area with retained sudomotor activity which comprises the 1st and 2nd lumbar dermatomes and often even extends above and below these dermatomes. In addition the perineum is an "escape area." On the basis of careful clinicoanatomical investigations and anatomical studies of normal material Monro (1959) concluded that the lumbar "escape area" following thoracolumbar sympathectomies is due to the presence of intermediate ganglia associated with the nerve roots of L_1 and L_2 whose connections have remained intact. The intermediate ganglia associated with the nerve roots of L_4 and L_5 will not cause sweating, because their preganglionic fibers have been sectioned below the lower level of the sympathetic outflow. (Monro found this to be at L_3 in 85% of his patients.) No satisfactory explanation can be given for the occurrence of the "escape areas" in the face and perineum. Monro's monograph (1959) contains a wealth of observations of clinical interest. He discussed the fact that results are usually more satisfactory following a thoracolumbar than cervicodorsal sympathectomies.

Before an operation is undertaken a *diagnostic injection of a local anaesthetic,* for example *procaine* into the ganglia in question is usually made. This procaine block is generally effective in interrupting the conduction of the fibers; but if its effects are unsatisfactory not much benefit can be expected from an operation.[51] On the other hand, sometimes the effect (for example on pain) of a procaine block is more enduring than would be expected, and in some cases repeated injections of procaine have been followed by complete relief. It is generally assumed that procaine in these cases acts by breaking a "vicious circle."

An effect similar to that obtained on extirpation of sympathetic ganglia may be achieved by *alcohol or phenol injections* into the ganglia. Although a minor procedure, this has in some cases a less enduring effect, and damage to neighboring structures cannot always be avoided. Thus damage of the intercostal nerves in injections into the stellate ganglion may give rise to neuralgias after some time.

Of other operative procedures may be mentioned *section of the splanchnic nerves, resection of the hypogastric plexus* (presacral neurectomy), and resections of the superior and inferior mesenteric plexuses.

In the first period of enthusiasm, surgery on the autonomic system was attempted in a variety of diseases. As more experience was gained operations of this kind have been limited to conditions where the results are generally satisfactory. In some instances advances in pharmacotherapy have abolished or minimized the need for surgical interventions, for example in the treatment of essential hypertension. In the following, reference will be made to some conditions where surgery on the autonomic nervous system has been found to be of benefit, even if in some of these disorders such operations are rarely performed nowadays. (For complete accounts see White and Smithwick, 1942; White and Sweet, 1955.)

Sympathetic denervation of one or more extremities has been frequently employed in diseases in which there is reason to believe that at least one factor responsible for the symptoms is an abnormal state of contraction of peripheral vessels. The most clear-cut disease of this type is the so-called *Raynaud's disease.*

This disorder occurs more frequently in women than in men, and is characterized mainly by vasomotor changes in peripheral parts of the body, most commonly fingers or toes, accompanied by pain. Cooling as well as emotional factors are apt to provoke the attacks, during which the fingers and hands or toes and feet become cold and frequently at the same time cyanotic. The cooling is due to constriction of the arterioles, the cyanosis to the slow circulation of the blood in

---

[51] As mentioned previously, the area of temporary sympathetic denervation produced by the block may be judged by determining the distribution of anhidrosis in tests of thermoregulatory sweating or by mapping the distribution of increased electrical skin resistance.

the capillaries which are dilated. Sometimes the capillaries are contracted and the skin accordingly pale. The pain is dull, deep, and may be accompanied by paresthesias. When the attack recedes, the affected parts grow warm and red, and this vasodilatation is usually followed by a severe burning pain. In long-lasting cases permanent changes are apt to develop. Necroses of the skin, particularly on the finger tips, and ulcerations and eventually atrophic changes in the subcutaneous tissues and the bones are not uncommon in long-lasting cases.

The etiology of the disease is not known. Probably several factors are responsible. That the sympathetic innervation of the blood vessels is of importance seems to be clear from the beneficial results obtained in sympathectomies. However, in this respect there is a marked difference between postganglionic procedures and preganglionic methods in the upper extremity. Those having experience with the preganglionic operative method have on the whole obtained satisfactory results in cases of Raynaud's disease (see, for example, White and Smithwick, 1942). For an evaluation of sympathectomy in vascular diseases see Boguslawski, Banach, and Dabrowski (1960).

Whether sympathectomies are of any use in peripheral vascular disorders appears to depend mainly on whether there is a component of vasospasm involved in the production of the symptoms. When organic changes in the vessels have advanced to a certain degree, no relief is, as a rule, to be expected. Among conditions which in some cases have benefited from sympathetic denervation may be mentioned *endarteritis obliterans* (Bürger's disease). In senile and diabetic gangrene some surgeons have tried sympathectomy, and some have found that demarcation is favored and that healing takes place more rapidly. The same holds true in cases of chronic ulcers of the extremities where there is reason to believe that the circulation is impaired by vasospasm.

Before the advent of modern hypotensive drugs, attempts were made to treat *essential hypertension* by extensive sympathetic denervation, and several authors reported favorable results, for example following resection of the splanchnic nerves.

Resection of the splanchnic nerves is not followed by definite symptoms from the abdominal viscera. This is in accord with other findings which have shown that these viscera possess a considerable degree of local autonomy. When disturbances in the functions of the abdominal viscera are present, removal of their sympathetic innervation has been attempted in some cases as a therapeutic measure, with most success in disorders in which there is reason to assume that overactivity of the sympathetic is responsible for abnormal states of contraction of smooth muscles. Such operations have been performed, for example in cases of *cardiospasm* and as a treatment for *pylorospasm*. When dealing with the digestive organs, it may be appropriate to mention the surgical interruption of their *parasympathetic* innervation. Section of the vagus nerve, *vagotomy,* has been attempted as treatment in patients with gastric or duodenal ulcers. The results have in general not been satisfactory.

The main aim of the operation is to transect the secretory fibers to the stomach. However, with a truncal vagotomy the motor innervation of the stomach and the vagal innervation of other abdominal organs are also interrupted. Complications of various kinds occur. In recent years, therefore, modifications of the surgical procedure have been developed, aiming at transecting only the branches of the nerve which supply the part of the stomach harboring the acid-producing parietal cells. The results of this *selective gastric vagotomy* are reported to be more promising than those of total vagotomy. However, recurrence of the ulcer is fairly common, perhaps due to regeneration of transected vagal fibers.

Although pain of visceral origin may be abolished by section of the appropriate dorsal roots or by chordotomy, it is obviously preferable to cope with visceral pain without producing loss of somatic sensibility. As a rule, therefore, visceral pain is treated by operative procedures involving the autonomic nervous system, even if in some cases the other methods have to be resorted to. Some conditions in which operations on the sympathetic nervous system have been employed to alleviate pain will be mentioned.

The pain of *angina pectoris,* which, as has been seen, must be assumed to travel via the afferent fibers from the heart, and which follow the sympathetic fibers, has been relieved by operations on the sympathetic, by alcohol injections in vertebral ganglia, or by ganglionectomy. Particularly after the recent advances in treatment of autonomic disorders with pharmaca, operations on the sympathetic system have, however, been little used.

It is essential that not only the stellate ganglion be removed, but that the upper two or (better) four thoracic ganglia are also included (see White and Sweet, 1955) since afferent fibers from the heart may reach the cord as low as the 4th (and in some cases probably even the 5th) thoracic ganglion. There appears to be no alteration in the function of the heart after the operation. Some authors have warned against the operation, claiming that when the pain impulses from the heart are abolished, the patient loses his warning signal. However, some diffuse feeling, sufficient to warn the patient of an attack, remains, such as a feeling of oppression, or palpitations (see, for example, Lindgren and Olivecrona, 1947).

Dysmenorrhea has been treated by interruption of the sympathetic supply of the uterus. This is done by removing the sympathetic plexus descending from the aorta to the pelvis as the hypogastric plexuses or ''presacral nerves.'' The operation is, therefore, commonly called *presacral neurectomy.* Since hormonal therapy has been found to be very effective in controlling dysmenorrhea, presacral neurectomy is now seldom used in this condition. The operation has been tried in cases of painful disorders of the bladder and prostate. However, the results have not been very satisfactory, as can be explained on the basis of the innervation of these organs (cf. above). In cases of malignant tumors, relief from pain can only be expected as long as the tumor has not infiltrated the surroundings, since somatic fibers will then be involved in addition.

Attempts to relieve the pain of *trigeminal neuralgia* by operations on the cervical sympathetic, particulary by removal of the superior cervical ganglion, have been made by some. There appears to be no sound basis for the view that the sympathetic is involved in producing the pain of trigeminal neuralgia, and the operation is not used any more.

Painful disorders of the extremities have been treated with operations on the sympathetic supply in many cases. The symptom complex called *causalgia* (meaning burning pain) by Weir Mitchell occurs in some cases of penetrating wounds with lesions of peripheral nerves, most often bullet wounds. As a rule the nerve is not entirely interrupted. Most frequently the symptoms are seen following injury to the median or sciatic nerves.

Causalgia or a closely similar clinical picture may, however, also be observed in irritative nerve lesions of other types. The symptoms often appear immediately after the injury or during the process of healing of the wound. The distressing pain, which is of a burning type, is localized in the area of distribution of the nerve peripheral to the lesion, and is often exaggerated by heat and diminished by cold. The affected area is extremely hyperesthetic, and the slightest stimulus of the skin is liable to provoke severe pain, as is also movement of the extremity, and in sensitive persons psychic stimuli may also start an attack of pain. In the hyperesthetic area, which frequently exceeds the cutaneous area of the nerve, thermal stimuli and deep pressure may also provoke pain. In typical cases objective alterations occur in the affected area. The skin is often hot and red, and may be glossy. A marked sweating may occur.

The pathogenesis of the syndrome is unknown. Many hypotheses have been set forth, but none of them is entirely satisfactory. According to Richards (1967), who reviews the literature, the most likely explanation for the appearance of the symptoms, especially the burning pain, is an abnormal interaction at the site of injury between efferent sympathetic and afferent sensory fibers.

In those cases of causalgia in which spontaneous recovery does not occur, several types of operations and treatment have been attempted. Intraneural injection of alcohol above the place of the lesion, resection or crushing of the nerve, periarterial sympathectomy, and other procedures have been tried without much success. The only satisfactory results are seen following interruption of the sympathetic nerve supply of the affected limb, by removal of the appropriate sympathetic ganglia, in the case of the upper limb cervicothoracic ganglionectomy. According to reports in the literature the results of such operations have been excellent in 77% of the cases, good in 19%, and poor in 4% (Richards, 1967).

Other disorders in which pain occurs in the extremities, and in which there is reason to believe that disturbance in the autonomic innervation is an important factor have also been more or less successfully treated by surgical procedures (see White and Sweet, 1955). However, little is known of the nature of these disorders, which commonly develop following closed traumatic injuries of an extremity, and are designated post-traumatic dystrophies, post-traumatic painful osteoporosis, etc. The distinction between these conditions and typical causalgia is not always clear.

Following amputations most patients say that they feel their amputated arm or foot is still present; they have what is called a *phantom-limb,* giving rise to different kinds of sensations. In some patients severe pain is felt in parts of the phantom limb for weeks or years. The pathogenesis of phantom-limb pain is very complex, but still little understood. For a recent discussion of possibly underlying neural mechanisms, see Melzack (1974). Psychological factors are apparently involved. Different types of operative treatment of this disorder have been attempted, but often with little success.

Recent advances in pharmacotherapy have gradually lessened the need for surgical interventions on the autonomic nervous system, and further advances are to be expected in this field in the future.

The account above deals with operations on the peripheral parts of the autonomic nervous system. However, neurosurgical interventions for therapeutic purposes have been directed also to its *central levels,* particularly the *hypothalamus.* Since the hypothalamus, as described above, has been known for some time to be related to emotional expressions and behavior, and even to harbor mutually opposing "centers," it is no wonder that, with the possibilities for stereotactic production of small lesions in the human brain, attempts have been made to use localized destructions of the hypothalamus as a therapeutic measure. These "*psy-*

chosurgical" operations have been undertaken chiefly in patients with a long-standing history of uncontrolled, violent, aggressive behavior, who have not responded to other types of therapy. Some authors hold that the operation is particularly well suited for mentally retarded patients who show aggressive behavior. The target for the destruction has most often been the *posterior* hypothalamus (bilaterally). This zone has been found in man to respond to stimulation with signs of sympathetic discharge.

Stereotactic *anterior* hypothalamotomy has been recommended in the treatment of "hedonia," a condition characterized by an uncontrollable urge to satisfy personal needs (with regard to food, drink, smoking, sexual activity). An abnormally strong emotional drive is supposed to be a basic feature in the condition. The operation has been claimed to be valuable in alcoholism and in pedophilia, and other sexual deviations. Interestingly, the main complication is said to be an increased appetite. Some authors have, rather naïvely, held that it is possible to influence particular aspects of sexual (and other) functions by destroying particular small areas of the hypothalamus, basing their belief on the presence of experimentally demonstrated localized "centers" for different aspects of sexual function. For some articles see *Neurosurgical Treatment in Psychiatry, Pain and Epilepsy* (1977).

Considering the main afferent and efferent connections of the hypothalamus, its composition of many minor, differently structured territories, and considering further that it is virtually not possible to correlate a circumscribed territory with a particular function, it appears that the rationale for "psychosurgical" hypothalamotomy is very weak, to say the least. We shall return to some general problems of "psychosurgery" in Chapter 12.

12

The Cerebral Cortex

THE CEREBRAL cortex has been touched upon repeatedly in the preceding chapters. It is appropriate at this point to consider the structure and function of the cortex more thoroughly, and to give a brief review of symptoms following lesions of the cortex apart from those already mentioned.

The cerebral cortex is developed from the telencephalon, forming the rostral part of the prosencephalon. However, the cortex cerebri is neither structurally nor functionally an entity. Some regions primarily have connections with fibers from the olfactory bulb, directly or through relay stations. These phylogenetically oldest parts of the cortex are often loosely called *the "rhinencephalon."* In lower vertebrates such as amphibians and reptiles, these parts constitute a considerable proportion of the telencephalon as has been described in Chapter 10. The other divisions of the cortex first undergo a marked progressive development in mammals, in the phylogenetic scale. The parts of the hemisphere appearing at this stage are usually called *neocortex* or *neopallium*. In the higher mammals and man the neocortex increases enormously in size. Its surface area, relatively, becomes increasingly greater than its volume, a condition made possible by the appearance of fissures and convolutions, necessary to accommodate the growing brain within the skull.

Structure of the cerebral cortex. Cytoarchitecture and areal parcellation. The general principles of structure of the neocortex are similar in all mammals, despite the progressive differentiation which reaches its peak in man. Some of the phylogenetic features will be touched upon below, after the description of the neocortex in man.

The neocortex is frequently also designated *isocortex* (O. Vogt) or *homogenetic cortex* (Brodmann) since its different parts develop in the same manner during ontogenesis. This isocortex, which makes up the bulk of the cerebral cortex in

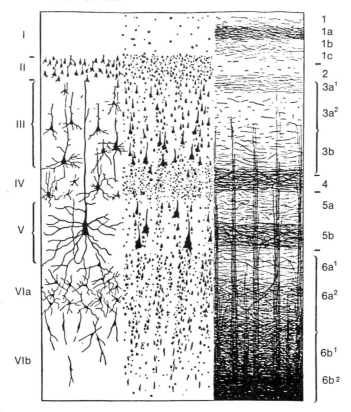

I		1, 1a, 1b, 1c
II		2, 3a¹
III		3a², 3b
IV		4
V		5a, 5b
VIa		6a¹, 6a²
VIb		6b¹, 6b²

FIG. 12-1 Diagram of the structure of the cerebral cortex. To the left, from a Golgi preparation; center, from a Nissl preparation; to the right, from a myelin sheath preparation. I: lamina zonalis; II: lamina granularis externa; III: lamina pyramidalis; IV: lamina granularis interna; V: lamina ganglionaris; VI: lamina multiformis. After Brodmann and O. Vogt.

man, presents several minor differences in various parts of the brain, to be discussed below, but with some modifications all regions can be made to fit into a general scheme of a six-layered cortex when examined in preparations stained for nerve cells. This is depicted in Fig. 12-1, the left column showing the cells as they appear in Golgi preparations, where axons and dendrites are made visible, the center column representing the picture seen in Nissl-stained sections. The separate layers are as follows:

I. *The molecular or zonal layer,* immediately beneath the pia, is rich in fibers but poor in cells. Scattered small nerve cells are present, the horizontal cells of Cajal, orientated with their longest axes parallel to the surface of the cortex. The fibers form a dense tangentially running plexus.

II. *The external granular layer* is composed of densely packed small cells, some of a pyramidal type, others round or star-shaped. The pyramidal cells in this layer, as in the other layers containing pyramids, have their apices directed towards the surface.

III. *The pyramidal layer* consists predominantly of pyramidal cells of medium size. The largest ones are found in the deeper parts of the layer.

IV. *The internal granular layer* contains chiefly small cells lying close together. Many of them are star-shaped, but at least as many are of the pyramidal type. This layer contains abundant horizontally directed fibers (the outer band of Baillarger, cf. below).

V. *The ganglionic layer* is composed mainly of pyramidal cells, mostly medium-sized and large. Like the other pyramidal cells, these have long apical dendrites, directed toward the molecular layer, and a generous amount of basal dendrites, leaving the cells at their bases and running more or less horizontally.

VI. *The multiform layer* contains predominantly spindle-shaped cells. It is frequently subdivided into an outer part VI*a* and an inner VI*b,* the latter at most places fusing gradually with the white matter of the hemisphere.

On closer examination it becomes apparent that, in spite of features common to all cortical regions, clear-cut differences can be ascertained between various smaller parts. This recognition is comparatively recent. The fundamental investigations in this field we owe to Campbell (1905), Brodmann (1909), and von Economo (1927). The cortex presents not only a varying thickness in different parts (ranging from 4.5 mm to 1.3 mm), but also microscopical differences. In cell-stained sections it is clearly seen that the different layers exhibit varying thicknesses, varying densities of cells, and also differences with regard to the sizes and types of cells which they contain. On the basis of these differences in structure, *the cytoarchitecture,* several more or less distinct small *areas* may be discerned within the cerebral cortex. Brodmann distinguished some 50 different areas, von Economo nearly twice this number.

Fig. 12-2 reproduces the cytoarchitectural map of the human cortex as worked out by Brodmann. Each area has been designated by a different figure. Von Economo employed other symbols in his map, but in the principal features the subdivisions of Brodmann and von Economo coincide. Since Brodmann's labeling is most extensively employed and simpler than that of von Economo, the latter will not be considered in more detail. A glance at the map of Brodmann will reveal that the borders between the different cortical areas to a large extent do not coincide with the sulci on the cerebral surface. Microscopically the limits between adjacent areas are also not quite sharp everywhere.

C. and O. Vogt have elaborated Brodmann's map in a somewhat more detailed manner, and proposed a further subdivision of some of Brodmann's areas. The parcellation derived by the Vogts in the monkey was transferred to the human brain by Foerster and employed by him in his presentation of clinical experiences in electrical stimulation of the cortex. However, objections may be raised to this, and although the Vogt-Foerster map was extensively employed by experimenters upon the cerebral cortex, recent research tends to show that a minute parcellation like that of the Vogts is not always justified.

Common features of several cytoarchitectural fields make it possible to collect some of them in larger groups. Von Economo arranges the different areas in five fundamental groups, as shown in Fig. 12-3. The cortical types labeled 2, 3, and 4 in the lower diagram of the figure contain the six typical layers, described above, although not all of them are developed to the same degree in the different types. These types are therefore *homotypical,* contrary to the *heterotypical* cortical types 1 and 5, in which six layers cannot be clearly discerned when the cortex is fully developed. (During fetal life, however, this heterotypical cortex, according to Brodmann, also passes a six-layered stage. Because of this it belongs to the homogenetic cortex or isocortex.)

It can be seen in the figure that the cortex of fundamental type no. 1 is distinguished by its lack of distinct granular layers (II and IV), whereas layers III and V are well developed. This *agranular cortex,* as can be seen from the upper drawing, is found first and foremost in the posterior parts of the frontal lobe, from the

FIG. 12-2 Brodmann's cytoarchitectural map of the human brain. The various areas are labeled with different symbols and their numbers indicated by figures.

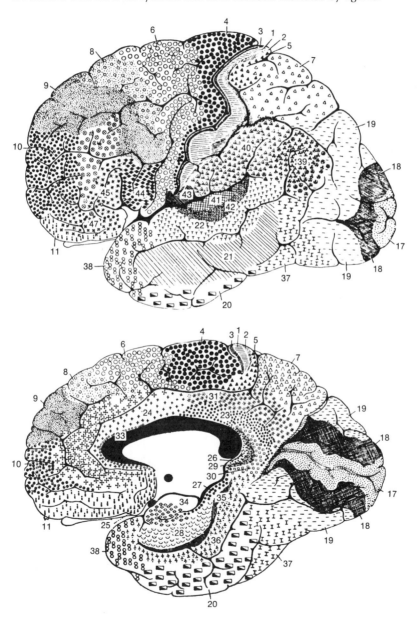

central sulcus forward. Areas 4 and 6 belong to this type, likewise part of areas 8 and 44. In addition, the same type of cortex is found on the medial surface of the hemisphere in the paracentral lobule, the anterior half of the gyrus cinguli, and anterior to the lamina terminalis, and in part of the hippocampus and uncus. (The latter regions belong to the *allocortex*, cf. Chap. 10.) A particular type of agranular cortex is found in area 4, where between the larger pyramids in the Vth layer the giant pyramids of Betz are found (cf. Chap. 4). Figure 8-7a is a photomicrograph from area 4 showing some of the principal features. The more massive efferent projection systems take origin preponderantly from agranular cortex. *The agranular cortex, therefore, with certain qualifications, can be considered the prototype of "motor" cortex.*

FIG. 12-3   The different principal types of cortex (below) and their distribution in the hemisphere (above). *1:* agranular cortex; *2:* frontal type of cortex; *3:* parietal type; *4:* polar type; *5:* granular cortex, koniocortex (see text). After von Economo from Kornmüller and Janzen (1939).

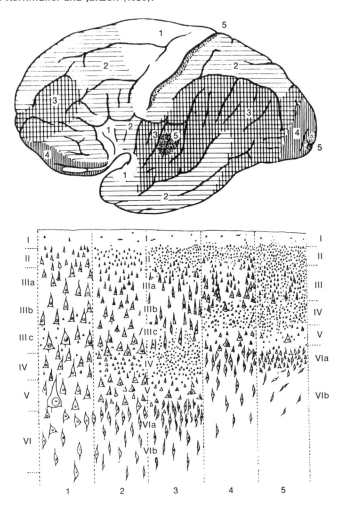

The other type of heterotypical cortex is labeled 5 in Fig. 12-3. This type is characterized by its richness in granular (or stellate) cells and the strongly developed granular layers II and IV. The layers III and V, on the other hand, are only poorly developed. Areas belonging to this type of cortex are found (cf. Fig. 12-3) in the anterior part of the postcentral convolution, along the calcarine fissure (only the dorsalmost end of this area is visible in Fig. 12-3; see also Fig. 12-4), in a limited part of the upper surface of the first temporal convolution, and finally in restricted parts of the hippocampal gyrus (not visible in Fig. 12-3). Cortex of this type, *granular cortex* or *koniocortex, is characteristic of those cortical areas which receive the primary afferent sensory systems.*[1] The most elaborate type of such koniocortex is found in the area striata, where, as can be recalled, the fibers of the optic radiation end. The number of granular cells is exceedingly high, and the granular layers are thick, the inner granular layer (IV) being also subdivided into two sublayers, IV*a* and IV*c* (see Fig. 8-7c). Layer IV*b* represents the intermediate layer, being poor in cells but abundant in myelinated fibers, which, as has been mentioned, form the line of Gennari or Vicq d'Azyr (cf. Fig. 8-6).

From Fig. 12-3 it can be seen that cortex of type 2 is found in the frontal lobe anterior to the distribution of cortex of the agranular type, but in addition it occupies parts of the parietal and temporal lobes. The pyramidal cells of layers III and V are well developed, giving this cortex a certain resemblance to the agranular cortex. On the other hand, the parietal type (3 in Fig. 12-3) has less massive pyramidal layers and more elaborate granular layers. The lamination is distinct in this type, which, however, is not limited to the parietal lobe. The last type, the polar, no. 4, is characterized *inter alia* by its thinness and high content of granular cells.

Just as a parcellation of the cerebral cortex can be made on the basis of differences in the amount, type, and pattern of distribution of the nerve cells, so detailed scrutiny of the myelinated fibers allows for a parcellation. The *myeloarchitectural areas* delimited in this way coincide fairly well with the cytoarchitectural areas.

In a myelin-sheath-stained section through the cerebral cortex of man (Fig. 12-1, right column) a wealth of myelinated fibers are seen. Some of these fibers ascend perpendicular to the surface of the cortex, more or less distinctly aggregated in radial bundles. Other fibers run parallel to the surface. In most parts of the cortex they are concentrated in three zones. The most superficial is found in the lamina zonalis or molecularis; the other two are situated more deeply, forming the so-called outer and inner lines or bands of Baillarger (4 and 5b in Fig. 12-1, to the right). Studies of myeloarchitecture have been particularly extensively performed by C. and O. Vogt and their pupils. The various areas present differences in the amount of fibers in the various layers as well as in the thickness of the fibers, their course, and terminal distribution. Following this line of investigation, a parcellation of the cortex even more detailed than the cytoarchitectural one has been found.

It has further been shown, particularly by Pfeifer (1940), that there are also regional variations in the vascular supply of the cortex. The density of the capil-

[1] Despite differences between areas with regard to architecture and thickness of the cortex, it appears that the *total* number of cells in a cylinder taken through the thickness of the neocortex is the same in all functional areas (except area 17) and in all species examined (Rockel, Hiorns, and Powell, 1974).

laries, their arrangement, caliber, and other features differ in various layers and various parts of the cortex. In many respects this *angioarchitectural parcellation* is in accord with the results obtained from the other methods of areal subdivision of the cortex.[2] In a corresponding manner the distribution and arrangement of the *glial cells* differ in different areas, but the details have not yet been completely elucidated. In recent years some studies have been made on the *chemoarchitecture* of the cortex. It turns out that there are differences between areas and between layers with regard to both chemical composition and amounts of different enzymes present. Such studies have been made in man as well as in animals, and attempts have been made in this way to obtain information on pathological processes in the cortex, for example in certain types of presenile dementia (see Pope, Hess, and Lewin, 1964). This field of research seems promising but is still at its beginning.

Studies of the chemical differences between minor parts of the cortex can be made with histochemical methods. More precise data can, however, be obtained by excising small cylinder-shaped vertical blocks from the cortex by biopsy or from autopsy material. The block is immediately frozen and subsequently serially sectioned. Alternate sections are used for cytoarchitectonic identifications and for microchemical determinations. The intracortical distribution of chemical constituents can then be related to cytoarchitecture.

Since the development of the Falck and Hillarp method (see Chap. 1), some authors have studied the distribution of monoamine-containing terminals within the cerebral cortex. Most studies have been made in the rat. Some topical differences have been found. While noradrenergic terminals appear to occur all over the cortex (see Chap. 6), dopaminergic terminals appear to have a more restricted distribution. To mention one example: according to Berger, Thierry, Tassin, and Moyne (1976), the areas in the frontal lobe which receive a dopaminergic innervation appear to correspond to the ''prefrontal'' cortex in primates. Much is still unclear in the interpretation of such findings. They will not be discussed further here; nor will reference be made to studies of regional variations of enzymes (acetylcholinesterase, succinate dehydrogenase, and numerous others).

The discovery that the cerebral cortex could be subdivided into structurally different areas strongly influenced the views on the problem of cerebral localization. The adherents to the view that the cerebral cortex contains ''centers'' for different specific functions found in this fact a solid foundation for their hypotheses concerning cerebral function. However, many facts are known which give clear evidence that a mutual interdependence exists between different areas. We will return later to this subject.

The total number of nerve cells in the human cerebral cortex is enormous. Several workers have made calculations of this figure, and values up to 14×10^9 (von Economo and Koskinas, 1925) have been reported. Most workers have not

---

[2] Presumably the regional differences in angioarchitecture are of functional importance. It should be mentioned in this connection that recent methods permit studies of regional blood flow in man. An inert radioactive gas, for example xenon-133, is injected intra-arterially, and the activity in various cortical regions is recorded by means of a series of detectors applied to the skull. For studies and references the reader should consult *Brain Work. The Coupling of Function, Metabolism and Blood Flow in the Brain* (1975). It has been shown, as might indeed have been expected, that neuronal activity in a cortical region is accompanied by a local increase of the blood flow in this region. For example, the activity as measured with this method increases markedly in the precentral hand region when movements are performed with the contralateral hand (see Oleson, 1971). Mental activities, such as reading, memorizing, and reasoning are likewise followed by regional changes in blood flow in particular cerebral regions (see Risberg and Ingvar, 1973). The regional blood flow pattern appears further to show characteristic changes in certain cerebral diseases, and the new methods may prove to be of diagnostic value in clinical neurology (for some data, see Ingvar, 1978).

taken into consideration the shrinkage of the brain that occurs during the histological procedures. Pakkenberg (1966), considering this, found a total number of 2.6
$\times$ 10^9 (2.6 billion) nerve cells in the brain of a young adult. The considerable
variations in the numbers calculated by different authors may be partly due to
technical factors. However, it does not seem unlikely that there may well be considerable individual variation among human brains in this respect.[3]

In recent years much valuable information has been obtained on the organization, structural and functional, of the neocortex. Before considering this subject it
is appropriate to review briefly some of the main features of the fiber connections
of the cerebral cortex.

The fiber connections of the cerebral cortex. Many of these connections
have been described in previous chapters. Here only some principal features will
be summarized. The fiber connections of the cerebral cortex may be subdivided
schematically into four groups: (1) corticofugal or efferent projection fibers; (2)
corticopetal or afferent fibers; (3) association fibers, connecting different regions
of the same hemisphere; and (4) commissural fibers, interconnecting the two
hemispheres.

Among *corticofugal projection fiber systems,* one of those most intensively
studied is the *pyramidal tract.* Its corticospinal and corticobular fibers arise in
various cortical areas, most of them in the pre- and postcentral gyrus (see Chap 4).
Other efferent fiber systems are the *corticopontine tracts,* likewise well developed
in man (see Chap. 5). The topically arranged projection from rather extensive cortical areas to *the caudate nucleus* and *the putamen* as well as corticofugal fibers to
the red nucleus and to *the reticular formation* have been considered in Chapters 4,
5, and 6. Minor contingents end in *the dorsal column nuclei* (see Chap. 2), *the inferior olive* (see Chap. 5), and *the subthalamic nucleus* (see Chap. 4). Other corticofugal fibers, especially from the occipital lobe, pass to *the colliculi* and *the
optic tectum* (see Chap. 7). Finally, mention should be made of the abundant
topically organized *corticothalamic fibers.* These pass largely to the thalamic
nucleus from which the cortical area emitting them receives its thalamic afferents.
When the various efferent fiber systems are analyzed, it becomes apparent that a
considerable number of the long, corticofugal fibers take origin from the central
region (areas 4, 3, 1, 2, and, in part, 6). It should be noted that fibers from one of
these cortical regions have regularly been traced to different subcortical stations,
for example from area 4 to the spinal cord, the red nucleus, the pontine nuclei, the
inferior olive, and some reticular nuclei. In physiological studies it is often assumed that cortical influences exerted on the subcortical stations are mediated by
collaterals of fibers proceeding to the cord. While corticospinal fibers have been
seen to give off collaterals to many subcortical structures, as mentioned in previous chapters, it is evident from recent anatomical studies tracing the retrograde
axonal transport, that such an assumption can be true only in part. Somata of cells

---

[3] The claim that the number of cortical neurons decreases with age (Brody, 1955) received no
support in a study by Cragg (1975). Nor did Cragg find any reduction in synaptic density. On the other
hand, Geinisman and Bondareff (1976) reported a significant decrease in the number of axodendritic
synapses in old rats (dentate gyrus), and Scheibel, Lindsay, Tomiyasu, and Scheibel (1975) found loss
of dendritic spines and dendritic arborizations in the aged cerebral cortex of man.

of origin of a particular "system" (corticospinal, corticostriate, corticothalamic, etc.) have been found to have a specific laminar or sublaminar distribution (see Humphrey and Rietz, 1976; Jones and Wise, 1977; Wise and Jones, 1977b). These data show that a cortical area is able to coactivate different parts of the brain in a far subtler way than would be the case if they were all supplied by collaterals of corticospinal fibers. There are relatively few subcortical gray masses (for example the pallidum and the vestibular nuclei) that do not receive projections from the cortex. This reflects the importance of the cortical influence on a number of functions (including the central transmission of sensory impulses; see Chap. 2).

In a corresponding manner *afferent fibers* appear to reach extensive regions of the neocortex. Most of these fibers are derived from the *thalamus*. In general, the thalamocortical projections are topically organized; a particular cortical area receives its fibers from a particular subdivision of the thalamus, as described in Chapter 2. The cortical projections from *the lateral and medial geniculate bodies* also belong to the thalamocortical projection (Chaps. 8 and 9, respectively). Some authors have suggested that direct cortical afferents arise in some subcortical structures other than the thalamus. Some of these connections have recently been definitely demonstrated. Of particular interest has been the demonstration of a widely dispersed cortical innervation by noradrenergic fibers from *the nucleus locus coeruleus* and from *the raphe nuclei* (presumably serotonergic) as mentioned in Chapter 6. (For two recent studies, see Gatter and Powell, 1977, and Bentivoglio, Macchi, Rossini, and Tempesta, 1978.) In general, precise information on the distribution of cortical afferents from nonthalamic sources is still scanty.

According to classical teaching the final links of the great sensory pathways (somatosensory, optic, and acoustic) end in cortical areas of the granular type (areas 3, 1, and 2; 17; and 41, respectively). However, as has been described in Chapters 2, 8, and 9, these projections are not restricted to granular cortex (type 5 in Fig. 12-3) only. For example, thalamic nuclei receiving the somatosensory pathways project not only onto areas 3, 1, and 2, but also to agranular cortex in front of the central sulcus and, in addition, to other cortical regions (Sm II and III). Likewise, there are several cortical acoustic areas, and visual impulses are relayed not only to visual area 17 (striate area), but also to the surrounding areas 18 and 19. Furthermore, whereas some cortical areas are related preponderantly to one type of sensory information, it appears that in many regions of the cortex there is a convergence of information of sensory impulses of different kinds, as has been referred to in previous chapters. The proportion of the total cortex receiving sensory information fairly directly from various kinds of receptors is, however, rather restricted. Other cortical regions can be outlined as the projection areas of thalamic nuclei which receive more or less well defined afferent connections (for example the VL from the cerebellum, the VA from the pallidum, the anterior thalamic nucleus from the mammillary body, the MD from the amygdaloid nucleus). As to other thalamic nuclei projecting onto the cortex, especially the pulvinar (see below) and the lateral nucleus (LP and LD, see Fig. 2-14), knowledge of their afferent connections has until recently been rather scanty. The cortical projection areas of such nuclei, covering chiefly parts of the temporal and parietal cortex, were formerly often included among the so-called *cortical association areas,* on the assumption that they are less influenced from other regions of the nervous system than the other parts of the cortex. These areas, as well as the thalamic nuclei projecting onto parts of them, increase along the phylogenetic scale and are particularly well developed in man. Their main task was considered to be the association and integration of impulses, and they were assumed to be involved chiefly in more complex functions, not least in processes related to mental activity. While this view may still be generally valid, more recent findings have shown that it is almost impossible to outline regions that can be considered as pure "association regions" in the classical sense. Furthermore, it has become apparent that in the highest integrative processes occurring in the brain the cortex collaborates in a complex way with a number of subcortical structures as well as with regions belonging to the archi- and paleocortex (for example the thalamus, the hippocampus, the amygdaloid nucleus), as must indeed be concluded from the extensive fiber connections interrelating the cortex and these regions. Integration of impulses from various sources occurs also in subcortical regions. The superior colliculus, for example, receives optic fibers, fibers from the visual and other areas of the cortex, fibers belonging to the somatosensory system, and some fibers of the acoustic pathways and from the cerebellum (see Chap. 7, section g).

A correlation of the activity of different cortical areas may occur by way of the *association fibers*. Such fibers are present in large numbers in the cerebral cortex. Some of the association fibers are very short, pass entirely within the gray matter of the cortex, and interconnect neighboring cortical areas. The *short association fibers* of the visual areas, the somatosensory areas, the motor cortex, and the acoustic cortex have been described in Chapters 8, 2, 4, and 9, respectively. Other association fibers are somewhat longer and course chiefly in the white matter. Some of the *long association fibers* connect cortical areas in the same cerebral hemisphere; others cross the midline and connect areas of the two hemispheres. Because they are functionally important and because recent anatomical studies have brought forth much new information on their organization, the long association fibers will be considered more closely below.

Long cortical intrahemispheric and interhemispheric (commissural) association connections and some notes on the cortical "association" areas. The bundles formed by these fibers were partly identified by the old anatomists by dissection, but it is doubtful that all fibers in some of these bundles are really association fibers, and long ones. The fibers in a bundle may differ with regard to their sites of origin and termination. The interhemispheric association fibers are often referred to as *commissural* fibers. Most of them pass in the corpus callosum. We shall first consider the intrahemispheric (ipsilateral) long association fibers, particularly emphasizing the interconnection of the so-called "association" areas of the cerebral cortex.

It is customary to distinguish several bundles of long ipsilateral association fibers. One is the *cingulum*, passing in the cingulate gyrus and mentioned in Chapter 10. The *superior longitudinal fasciculus* passes from the frontal to the occipital region; the *inferior longitudinal fasciculus* passes from the occipital lobe to the temporal pole; the *uncinate fasciculus* connects the temporal and frontal lobes. Other bundles have been described as well.

Numerous experimental anatomical studies have been devoted to these connections. It appears from early studies with the Marchi method (Polyak, 1927b;

Mettler, 1935; Jansen, 1937; and others) that, in general, as might indeed be expected, areas that have the closest functional interrelations are most closely linked by association fibers. This was confirmed in later studies with silver impregnation methods. In recent years much precise information on the *origin* of association fibers has been obtained with the HRP method. Most recent studies of cortical association fibers have been carried out in the monkey. Some experiments have been performed in cats. There appear to be some differences between the two species. For example, association fibers interconnecting the frontal and temporal lobes (see below) appear to be absent in the cat (see Kawamura, 1973c). Long association connections appear to be more developed in the monkey than in the cat. There is reason to believe that in man they are even more important, but little information is available.

It is obvious from experimental studies that, despite their ubiquitous occurrence, *the long intrahemispheric association fibers are precisely organized in an extremely complex pattern*. Thus, within a particular association bundle a specific pattern is usually present, for example in the temporofrontal projection (Jacobson and Trojanowski, 1977a). Different parts of the cortex of the superior temporal sulcus differ with regard to their input (Seltzer, and Pandya, 1978). Likewise, within the ''prefrontal'' granular cortex (see below) different areas are not identical with regard to their association connections. Future studies will certainly reveal further details. Our still imperfect knowledge of the long association connections cannot yet be correlated in a satisfactory manner with functional observations. Some general features, emerging chiefly from observations in monkeys, will be mentioned. Thus, to a large extent (see Fig. 12-4), the connections between cortical association areas are *reciprocally organized* (frontotemporal and temporofrontal connections, etc.). Association connections with allocortical regions exist. While long fibers seem to interconnect particularly the different ''association areas'' of the cortex, efferent fibers from these areas may pass to some more ''specific'' areas, particularly the ''premotor'' area 6.[4] This has been found to receive long association fibers from the ''peristriate belt'' (Pandya and Kuypers, 1969), from the parietal lobe, its area 5 (Jones and Powell, 1970a), and from the auditory cortex (Diamond, Jones, and Powell, 1968b). Even if it entails some repetition, it is deemed appropriate in the following description to consider the long association connections separately for each of the four major cerebral lobes (Fig. 12-4).

We shall start with the *frontal lobe,* and consider the part of its cortex that is generally considered as *''association cortex.''* This appears roughly to correspond to the so-called *''prefrontal'' granular cortex.*

In the following simplified presentation, distinction will not always be made between different parts of the frontal lobe. The cortex of the anterior part of the frontal lobe is often referred to as the *prefrontal cortex,* a rather illogical and not clearly defined designation. Because it is commonly used by authors studying the frontal lobe, the term will, however, be employed in the following description. The ''prefrontal'' cortex has no obvious posterior border. All authors

[4] It should be recalled that the primary sensory areas in the parietal, the temporal, and the occipital lobe project to areas in their immediate vicinity (as described in Chaps. 2, 10, and 8, respectively), and that, with some exception, the specific sensory or motor areas appear not to give off long association fibers.

include in it the anteriormost part of the convexity of the frontal lobe, where in primates the cortex is of the granular type (corresponding probably to Brodmann's areas 9, 10, 46, 45; see Fig. 12-2). Areas 11, 12, 13, 14, and 47 on the orbital surface likewise are of the granular type. Collectively they are often referred to more specifically as the *orbitofrontal region* or the posterior orbital cortex and considered part of the "limbic system." The transition from the granular "prefrontal" cortex to the agranular motor and premotor cortices (areas 4 and 6) is formed by area 8 (the frontal eye field) and area 44, having so-called "dysgranular cortex." Some authors appear to include these areas in the "prefrontal" cortex. On the medial surface of the frontal lobe, area 32 (see Fig. 12-2) is granular while area 24 is not. Some differences in the connections of the various parts of the "prefrontal" cortex will be mentioned in the consideration of the frontal lobe below.

In the monkey the *"prefrontal" granular cortex* receives a rather marked *afferent* projection from the ipsilateral temporal lobe (corresponding, at least in part, to the classical uncinate fasciculus), as determined in silver impregnation studies following lesions of the temporal lobe (Pandya and Kuypers, 1969; Chavis and Pandya, 1976; and others). Other afferents come from the peristriate cortex and from the parietal association areas (see below). In the HRP study of Jacobson and Trojanowski (1977a) the sites of origin were more precisely determined (chiefly the superior temporal gyrus in the temporal lobe, areas 19 and 7 in the occipital and parietal lobes, and areas 23 and 24 in the cingulate gyrus).

FIG. 12-4 A simplified sketch of some of the long association fiber connections of the "association areas" of the cerebral cortex, and their connections with parts of the allocortex, plotted on diagrams of the monkey's brain. Only some main features are shown. Fibers from the association areas to area 6 are not included (see text). Arrows point to sites of termination of fibers. Note that most association connections are reciprocally organized. (The fibers passing from the orbital surface of the frontal lobe to the entorhinal area are shown as passing in the cingulum bundle for the sake of clarity of the illustration, while they have been found to pass directly and to take off from the uncinate bundle (*U.b.*) between the orbitofrontal cortex and the temporal pole.) Numbers refer to approximate positions of some of Brodmann's areas.

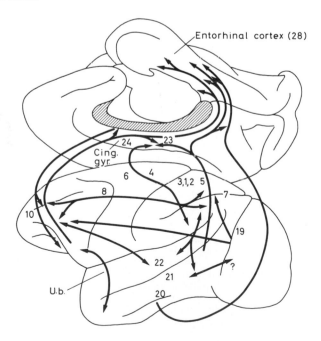

On the whole, the *efferent* long association projections of the prefrontal cortex reciprocate the afferent ones. Fibers have been traced to the temporal cortex (Nauta, 1964; Pandya and Kuypers, 1969; Kawamura and Otani, 1970; Pandya, Dye, and Butters, 1971; and others), the parietal lobe (see below), and the cingulate gyrus, chiefly its anterior part (see, for example, Pandya and Kuypers, 1969; Pandya, Dye, and Butters, 1971; Leichnetz and Astruc, 1976). In a study in the rhesus monkey Van Hosen, Pandya, and Butters (1975) mapped a projection from the orbitofrontal cortex to the entorhinal area (28) and adjoining cortex, confirmed by Leichnetz and Astruc (1976).

As concerns the *temporal lobe*, the association fibers arising from the acoustic cortex (see Chap. 9) and passing beyond this appear to originate from some of its subdivisions only and to pass chiefly to the cortex in the vicinity of the acoustic areas, particularly area 22 (Diamond, Jones, and Powell, 1968b; Kawamura, 1973a; and others; see also Chap. 9). Long association fibers (see Fig. 12-4) appear to arise, however, from the latter regions (which may be referred to as local acoustic association areas). These fibers have been traced to the prefrontal cortex (see above) and to the parietal lobe and cingulate gyrus (Mettler, 1935; Bignall, 1969; Pandya, Hallett, and Mukherjee, 1969; Pandya and Kuypers, 1969; and others). Other fibers, particularly from the inferior temporal gyrus, pass to the entorhinal and perirhinal cortex (see Van Hosen and Pandya, 1975a). The temporal lobe *receives* association fibers from the frontal lobe (see above); as well as from the parietal and the peristriate cortex (see below).

In the monkey, fibers from various regions of the temporal lobe have been found to be distributed in a topographical pattern in the amygdala, chiefly the basal and lateral nucleus (see Herzog and Van Hoesen, 1976).[5]

Most of the cortex of the *parietal lobe* is generally considered a typical "association cortex." This covers Brodmann's areas 5 and 7 above and 39 and 49 below in the parietal lobe (see Fig. 12-2). Posteriorly it borders on visual area 19, whereas the anteriorly situated part of the parietal lobe (areas 3, 1, and 2) represents the primary somatosensory cortex. The parietal association area has been found to *receive* long association fibers (see Fig. 12-4) from the frontal and temporal lobes (see above) and from the peristriate cortex (see below). As mapped with the HRP method (Mesulam, Van Hoesen, Pandya, and Geschwind, 1977), the afferent association connections of the inferior parietal lobe in the monkey appear to come chiefly from the "prefrontal" cortex (areas 8, 45, and 46), the superior temporal gyrus, the cingulate cortex (areas 24 and 23), and some other regions.[6] *Efferent* association fibers from different regions of the parietal association cortex have been traced to the superior temporal gyrus (Jones and Powell, 1970a), to the "prefrontal" cortex (see above), and to the cingulate gyrus (Jones

---

[5] There are differences with regard to association and other connections between parts of the temporal lobe. Not only the inferior temporal gyrus but particularly the temporal pole, is closely related to the "rhinencephalon," since it gives off fibers to the amygdala, prepiriform cortex, and entorhinal area, and to the posterior part of the cingulate gyrus (Akert, Gruesen, Woolsey, and Meyer, 1961). These neocortical projections to rhinencephalic structures have attracted interest in attempts to explain the symptoms seen following ablations of the temporal lobe (see Chap. 10).

[6] Mesulam et al. (1977) traced afferents from some subcortical regions, such as the pulvinar, intralaminar thalamic nuclei, the pretectal area, the nucleus locus coeruleus, and the raphe nuclei.

and Powell, 1970a; Pandya and Kuypers, 1969, in the monkey; Kawamura, 1973b, in the cat).

The *occipital lobe* is dominated by the visual areas 17, 18, and 19, which are amply interconnected (see Chap. 8). Long association fibers appear to arise preponderantly from area 19.[7] A fair number pass to the "prefrontal" cortex (Kuypers, Szwarcbart, Mishkin, and Rosvold, 1965; Chavis and Pandya, 1976). As shown in Fig. 12-4, other efferents have been traced to the temporal lobe (Kuypers et al., 1965; Pandya and Kuypers, 1969) and the inferior part of the parietal lobe in the monkey (Pandya and Kuypers, 1969), and to the parietal association area in the cat (Kawamura, 1973c). Few reciprocal *afferent* connections have been described. Pandya and Kuypers (1969) found fibers from the temporal lobe to the "peristriate belt" in the monkey, and there may be some from the "prefrontal" granular cortex (Garey, Jones, and Powell, 1968).

The preceding simplified summary of some main features in the organization of the long intrahemispheric association fibers and the diagram in Fig. 12-4 present only a few of the many detailed findings described in the rather abundant literature on this subject. It should be noted that almost all information stems from studies in animals, largely monkeys. Conditions in man are presumably even more complex. Below, some experimental data on the function of the association areas will be mentioned.

Single-unit recordings of responses to different kinds of stimuli have been undertaken. In behavioral studies a large variety of changes have been reported after lesions of these areas, or on stimulation in freely moving animals. As might be expected from what is known about the connections of the association areas, a wide array of disturbances has been observed. Their interpretation is extremely difficult, and there is considerable disagreement concerning details.[8]

It is clear from their fiber connections that the association cortices of the parietal, temporal, and occipital lobes (often collectively referred to as the parieto–temporo–occipital association cortex) are closely interrelated. Nevertheless, the different cortices are not identical in this respect and accordingly must be assumed to have somewhat different tasks. As described in other chapters, the primary sensory areas in the parietal, the temporal, and the occipital lobes project to areas in their immediate vicinity. Long association fibers emanate from these areas to association areas in the three other cerebral lobes (see Fig. 12-4). Information reaching the cerebral cortex via the optic, acoustic, and somatosensory systems may therefore obviously be integrated in all four association areas, even if the frontal lobe, in contrast to the three other lobes, does not harbor any primary sensory area. This is probably of functional relevance. The frontal lobe will be considered separately at a later junction. In the following discussion we will be concerned with the association areas in the other lobes.

It appears certain that in the monkey the so-called *inferotemporal* cortex (which covers the posterior regions of the inferior temporal gyrus and probably

[7] Some authors found a projection from area 17 to the cortex of the superior temporal sulcus (for a recent study see Martinez-Millán and Holländer, 1975).

[8] Many factors contribute to this situation. It may be difficult to decide exactly which part of an association area is stimulated or damaged. Even closely placed parts may not be identical. The association regions are related to the most complex brain functions. Much depends on the method chosen for study, the concepts used, and criteria chosen, for example, in distinguishing primary from secondary phenomena.

corresponds largely to Brodmann's area 20) is involved in visual pattern recognition.

Bilateral lesions produce severe deficits in visual discrimination. Single units in this region are reported to respond only to visual stimuli and have wide receptive fields. Most of them appear to be of a very complex type. The inferotemporal cortex in primates has therefore been spoken of as a "higher visual area." The units respond with a long latency (mean 120 msec) and appear to be activated, at least mainly, by way of association fibers from the striate cortex (for some relevant papers, see *The Neurosciences, Third Study Program*, 1974; see also Chap. 8). Another possibility would be that the units are activated by visual stimuli via fibers from the pulvinar since it (see below) receives afferents from the striate cortex and from the superior colliculus and possibly, at least in the cat, from the retina (Berman and Jones, 1977). Responses to visual stimuli have been recorded in the pulvinar which projects onto the parietotemporal association cortex.

Although acoustic input does not appear to reach the inferotemporal cortex, the *superior temporal cortex* (approximately Brodmann's area 22) appears to be rather purely related to acoustic functions. The more anterior parts of the temporal lobe (corresponding approximately to Brodmann's area 21) appear to be similar in many respects to the inferotemporal cortex with regard to higher visual functions.

The association areas of the *parietal cortex* comprise cytoarchitectonic areas posterior to the somatic sensory regions Sm I and Sm II. They appear to be concerned in higher sensory analysis and to be particularly related to somatosensory input from the contralateral half of the body. (Association fibers pass from Sm I to area 5, and from this to area 7.) There appears, however, to be extensive integration with visual information. Lesions of the posterior parietal cortex in the monkey have been found to impair somesthetic discrimination learning (compare astereognosis in man), and to produce disturbances of visually guided reaching and inattention to stimuli in the opposite visual field ("visual neglect"). There is growing evidence that there are notable differences between area 5 and area 7.[9]

In single-unit studies of area 7 in the monkey, Hyvärinen and Poranen (1974) found that some units respond to visual stimuli, whereas another type of cell is activated when the monkey tries to get hold of something. A fairly large group of cells reacts to both kinds of stimuli. Other authors found that in conscious monkeys, units in area 5 respond preponderantly to movements, whereas units in area 7 relate to visual targets (Mountcastle et al., 1975; Lynch et al., 1977). Most units recorded from in area 5 are related to active rather than passive movements of limb joints. Support for the assumption that there are functional differences between areas 5 and 7 comes from the recent studies of Stein (1978) in which local temporary cooling of the two cortical regions in the monkey was used. Stein concluded that area 5 is concerned in controlling contralateral movements in relation to visual targets in the contralateral visual field. The neuronal circuits involved in these actions of the parietal association areas are imperfectly known. For a discussion of possible pathways engaged, see Stein (1978) and the study of Leinonen, Hyvärinen, Nyman, and Linnankoski (1979) on functional properties of cells in area 7 in awake monkeys. According to Leinonen and Nyman (1979), the anterior part of area 7 is related particularly to receptive and motor functions of the face. The consequences of lesions of the parietal cortex in man are far more complex and more specific than in the monkey, particularly on account of the hemispheral dominance and the relation of the parietal lobe to language functions (see below).

The data mentioned above indicate, on the one hand, that different parts of the temporo–parieto–occipital cortical association area are not identical func-

[9] There is some disagreement among authors concerning the definition of cytoarchitectural areas of the posterior parietal lobe (posterior to area 5 and the somatosensory areas Sm I and II). The posterior parietal cortex, area 7, in the monkey may correspond to areas 7, 39, and 40 in man. A comparison of the consequences of lesions of this cortical region in monkey and in man was found to lend support to this concept (Petrides and Iversen, 1979).

tionally; on the other hand, in many respects they are similar. All of them appear, for example, to be concerned in the processes of high-level integration of different types of sensory information. Their afferent projections from the thalamus, different for different regions, play an important role in their functions—for example, the input from the pulvinar to parts of the temporoparietal association cortex (see below).

Attempts have been made to formulate integrative views on the function of the cortical association areas. Among these, the views set forth by Jones and Powell (1970a) on the basis of experimental studies offer attractive perspectives.

Jones and Powell (1970a) pointed out that within the somatosensory, the visual, and probably the acoustic system, each primary sensory area projects via association fibers to a local area in the neighborhood in the same cerebral lobe (parietal, occipital, and temporal, respectively) and also to a portion of the premotor cortex in the frontal lobe. The local association area sends fibers to a new local area (and also to the premotor area). Reciprocal connections exist between many areas (see above). Since most local association areas give off fibers to or receive fibers from other association areas, there must be almost unlimited possibilities for convergence of impulses originally set up by arrival of primary sensory messages. Particularly the cortex of the temporal lobe and the "prefrontal" granular cortex appear to be sites where such integration may occur.[10]

Particular attention has been devoted to *long association connections of neocortical areas with allocortical domains.* As mentioned above (see Fig. 12-4), the *cingulate gyrus* receives association fibers from the "prefrontal" areas, and the temporal and parietal lobes. In the cat the *entorhinal cortex* (see Chap. 10) receives fibers from the temporal lobe (Kawamura, 1973a), in the monkey from the inferior temporal gyrus (Van Hoesen and Pandya, 1975a) and the temporal pole (Akert, Gruesen, Woolsey, and Meyer, 1961) and also from the frontal lobe (the orbitofrontal cortex; Van Hoesen, Pandya, and Butters, 1975; see also Cragg, 1965, in the cat). These connections presumably mediate neocortical influences on the functions of the so-called "limbic system" (see Chap. 10). Other connections appear to serve the same purpose, such as the projections from the frontal and temporal neocortices to the amygdaloid nucleus (see Chap. 10) and the indirect pathway from the prefrontal cortex via the dorsomedial thalamic nucleus (MD) to the amygdaloid nucleus. On the other hand, many of the "limbic" structures receiving input from neocortical regions project back to these, in part in a rather precise reciprocal pattern. The presence of such connections serves as a reminder of the close functional interrelationship between the neocortex and the "limbic system," to which reference was made in Chapter 10.

The commissural connections between the two cerebral hemispheres are extensive. The vast majority of these fibers pass in the corpus callosum (Fig. 10-4); others pass in the hippocampal commissure and the anterior commissure. Those in the hippocampal commissure appear to be derived only from the paleo- and allocortex, as described in Chapter 10. Such fibers also utilize the anterior commissure, but this in addition carries (in its posterior limb) a fair number of commissural fibers from the temporal neocortex in the rat (Brodal, 1948) and in the monkey (Akert et al., 1961; Ebner and Myers, 1965; Pandya, Hallett, and Mukherjee, 1969). (The massa intermedia, which may be absent, contains interthalamic fibers.)

---

[10] It is, however, an undue oversimplification to consider the temporal cortex and the "prefrontal" cortex as entities. Minor parts of each are different.

The corpus callosum develops in proportion to the neocortex and reaches its greatest development in man.[11] Almost all information on the course of its fibers comes from animal studies. It appears that even if the majority of its fibers interconnect homotopic regions in the two hemispheres, there are also heterotopic connections, particularly with regard to different subdivisions of a sensory cortical field. The total distribution of commissural fibers was studied by Ebner and Myers (1965) in the cat and the racoon and by Myers (1965) and Karol and Pandya (1971) in the monkey after section of the corpus callosum. Largely corresponding findings were made. As seen in Fig. 12-5, there are clear regional variations with regard to the distribution of commissural afferents. As mentioned in other chapters, in the motor area Ms I (probably Ms II as well, Karol and Pandya, 1971) and the somatosensory areas Sm I and Sm II, commissural fibers are absent from the areas of representation of the distal parts of the limbs, and likewise from most of the visual area 17 (except near its border toward area 18; see Chap. 8) and parts of the auditory cortex (see, for example, Diamond, Jones, and Powell, 1968a). Some physiologists (see, for example, Robinson, 1973; Innocenti, Manzoni, and Spidalieri, 1974) disagree on certain points. In general, however, anatomical and physiological studies have given concordant results concerning the distribution pattern of callosal fibers (see Heimer, Ebner, and Nauta, 1967, for references). The projections between homotopic regions appear to be very precise (see, for example, Jones and Powell, 1969b, for the sensory cortical areas). The fibers from different lobes of the cerebrum have their particular sites in the corpus callosum (Pandya, Karol, and Heilbrunn, 1971, in the monkey). Those from the frontal

[11] The development of the cerebral commissures has been studied by a number of authors, most recently by Rakic and Yakovlev (1968). The number of callosal fibers is enormous. In the cat each square millimeter contains some 700,000 fibers (Myers, 1959). Not all fibers are commissural, however. Some pass to the striatum and other destinations. In man, the corpus callosum has been estimated to contain some 180 million fibers, 40% being unmyelinated (Tomasch, 1954).

FIG. 12-5   The dotted areas show the regions of the left hemisphere of the monkey which receive commissural fibers from the opposite hemisphere. Note that the striate area in the occipital cortex and the hand region of the first somatosensory area in the postcentral gyrus are free from commissural connections. From Myers (1965).

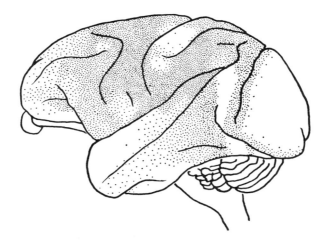

lobe pass in the rostral half of the corpus callosum; the others pass in the caudal half, with parietal lobe fibers anterior to those from the temporal lobe, and occipital lobe fibers most caudally. According to physiological studies (Innocenti, Manzoni, and Spidalieri, 1974), the callosal fibers from somatosensory regions even show some degree of somatotopic pattern. (The laminar distribution of terminal endings of commissural fibers and their cells of origin will be considered below.)

The problem of the topography of callosal connections has aroused renewed interest in recent years because of increasing insight into the function of the corpus callosum (see below). The patterns in these connections are extremely specific. This is well illustrated by such findings as those of Jones and Powell in their studies of the commissural connections of the somatosensory areas Sm I and Sm II in the cat (1968b) and in the monkey (1969b).

Intrinsic organization and synaptology of the cerebral cortex. When attempting to arrive at an understanding of the function of the cerebral cortex, we must avoid the pitfall of regarding the cortex cerebri simply as a two-dimensional sheet of mutually interconnected, structurally different areas. It would indeed be strange if the lamination—the regular arrangement of the cells in layers (varying in details from area to area)—should have no functional meaning. What is, for example, the functional reason for the extraordinary development of granule cells and granular layers in the areas receiving the "direct" sensory systems? Why are the "effector regions," to employ a somewhat unprecise designation, particularly rich in large pyramidal cells? The conclusion was drawn quite early by investigators in this field that there must be some correlation between "sensory" function and the development of the granular layers; and between "motor" function and pyramidal layers, particularly the Vth layer. Evidence supporting this view soon came from different areas of research, and it was generally accepted that the outer layers of the cortex, layers I to IV, have predominantly receptive and associative functions, whereas the effector functions of the cortex are taken care of by the deeper layers V and VI. This statement could be no more than a rough approximation as long as the details concerning the course and ramifications of the axons and dendrites within the cerebral cortex were not completely known. A clarification of these features and of the *synaptology* of the cortex requires minute studies by means of different methods. Lorente de Nó (1949) attacked these problems in systematic investigations using the Golgi method, supplementing and extending the now classical work of Cajal in this field. A summary of his results serves as a useful basis for the presentation of more recent data on the subject.[12]

According to the behavior of their axons, Lorente de Nó distinguished four fundamental types of nerve cells in the cerebral cortex (cf. Fig. 12-6). One type sends its axon toward the white matter, where it eventually continues as a projection fiber to deeper structures or to other parts of the cortex (1-5 in Fig. 12-6). Another type has only a very short axon, which arborizes in the immediate vicinity of the perikaryon (8, 9, 10 in Fig. 12-6). A third type of cell sends its axon toward the surface, and the axon and its collaterals are distributed to one or more cortical layers (11 in Fig. 12-6). The axons of the last type of cell take a horizon-

[12] The studies of Lorente de Nó do not include the frontal lobe. Presumably, however, conditions are similar to those found in the other three lobes.

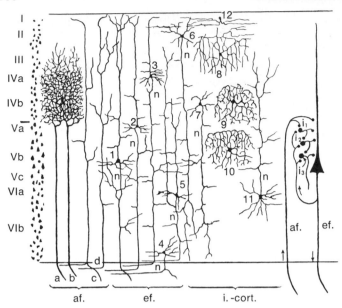

FIG. 12-6   A very simplified diagram of different types of neurons in the cerebral cortex, with axons, dendrites, and collaterals. *af:* axons of afferent fibers; *a, b,* and *c:* fibers from the thalamus; *d:* association fibers; *ef:* neurons sending their axons out of the cortex, from the cell labeled 1 as a projection fiber, from cells 2–5 as association fibers; *i.-cort:* cells whose axons arborize intracortically, those of cells 6 and 7 in the deeper layers, of cells 8 to 10 mainly in their own layer, cell 11 with its axon to more superficial layers, and cell 12 with a horizontal course of the axon (see text). To the right a simplified diagram of the passage of impulses within the cortex. *af.* and *ef.:* afferent and efferent fibers; i_1, i_2, and i_3: intercalated neurons. Redrawn and simplified from diagrams of Lorente de Nó (1949).

tal course (12 in Fig. 12-6). The nervous impulses to the cortical cells are derived from axons and collaterals of cells in the thalamus, partly from cells situated in other parts of the cortex on the same side (association fibers), or from the contralateral side (commissural fibers), and finally also from cortical cells in the immediate vicinity.[13]

Figure 12-6 shows some of the principal features in the intracortical arrangement of axons and dendrites in a very simplified fashion. Only a few of the numerous dendrites and collaterals of each cell are included in order to simplify the picture. On the extreme left in the figure are seen afferent fibers entering the cortex. According to Lorente de Nó, the bulk of the fibers from the thalamus, forming part of the direct sensory pathways, end with huge plexuses of terminal ramifications in the internal granular lamina, lamina IV (fibers a and b in Fig. 12-6). These "specific afferent fibers," as Lorente de Nó called them, do not give off collaterals in their passage through the deeper layers. Other thalamocortical fibers behave as diagrammed by the fiber c in the figure. They can be followed to the molecular lamina I and send some collaterals to several layers. Afferents of this type are often called "nonspecific" (cf. Chap. 2). In a similar manner association

---

[13] In recent years some afferent fibers have been found to originate from certain subcortical regions in addition to those from the thalamus (see below).

fibers from other cortical areas pass through all layers, but most of their collaterals appeared to Lorente de Nó to end on cells in the upper four layers, particularly layers II and III (d in Fig. 12-6).

Among cells sending their axons out of the gray matter to the cortex, some give origin to projection fibers traveling to remote nuclear masses, for example in the pyramidal tract. This holds true particularly for the cells in the Vth layer (1 in Fig. 12-6). The axons of other cells course as association fibers to other cortical areas in the same hemisphere, or as commissural fibers to the opposite hemisphere (2-5 in Fig. 12-6). Cells of this type are found particularly in layers III, V, and VI. It can be seen from the figure that the axons of all the latter cells have numerous collaterals which are given off during their course in the cortex and thus must be able to influence the nerve cells in the deeper layers through which the axons pass. The fiber plexuses of the Vth layer are, to a considerable extent, composed of collaterals from these efferent axons. Some of the cells in the deeper layers in addition send recurrent collaterals to the more superficial layers of the cortex (cell 5 in Fig. 12-6). The "effector" cells of layers V and VI will therefore be able to influence cells in more superficial layers.

Within all layers of the cortex there are cells whose axons and collaterals are only distributed within the cortex itself. Particularly abundant are cells of the second type of Golgi, often called stellate cells (cells 8, 9, 10 in Fig. 12-6). The axon arborizes extensively in the close vicinity of the perikaryon. All layers in addition contain cells (labeled 11 in Fig. 12-6) whose axons are directed toward the surface, where they are distributed to one or, more frequently, several layers. This implies that when cells in the deeper layers of the cortex are brought into action, they will be able to send impulses toward the surface, to the more superficial layers. Intracortical neurons with horizontal axons dominate in layer I (12 in Fig. 12-6).

It appears from these investigations that the afferent impulses to the cerebral cortex are, roughly speaking, distributed first and foremost to the superficial cortical layers. Particularly the IVth layer seems to be the terminal station for the direct sensory projection systems (the systems of the medial lemniscus, the spinothalamic tracts, and the brachium conjunctivum, optic, and acoustic systems). The other thalamocortical fibers and the incoming association fibers, on the other hand, exert their effect not only on the superficial layers but also on the deep ones. The findings support the idea that the deeper layers, particularly the Vth, are the most important sources of the efferent projection.

It is, however, equally clear from these investigations that functionally the deeper layers cannot be regarded as effector, the superficial as receptor, without qualifications. This is evidenced by the large number and variety of cells whose fibers arborize entirely wihin the cortex itself, some of which are provided with ascending axons, others with descending axons (6 and 7 in Fig. 12-6), whereas others are endowed with amply arborizing axons within their own layer. Every impulse reaching the cerebral cortex, by influencing these intracortical neurons, will be able to act upon cells in all cortical layers. Another feature contributing to this is the collaterals given off by the efferent neurons themselves.

An extremely simple path of a nervous impulse reaching the cortex may be imagined to be as shown in the diagram to the extreme right of Fig. 12-6. The afferent impulse entering through the fiber labeled af. is transmitted not only to an effector cell, sending its axon for example in the pyramidal tract, but will also reach the intracortical neuron i_1. The latter cell, when discharging, will not only be able to add another stimulus to the effector cell, but will also act on another association cell, i_2, which in turn excites a third cell, i_3, both the latter in turn influencing the effector cell. Finally the activity of these intercalated neurons will be influenced by the discharge of the effector cell through its recurrent collaterals.

A closer scrutiny of the mutual arrangement of the nerve cells in the cortex and the distribution of their dendrites, axons, and collaterals, makes it clear that there must be almost unlimited possibilities for variations in the impulse transmissions within the cortical gray matter. In all essentials Lorente de Nó's data on the general arrangement of dendrites, axons, collaterals, etc. in the cerebral cortex have been confirmed in further research. The use of new methods has revealed further details. The amount of new data and detailed information from recent years on the structure of the cerebral cortex is overwhelming.[14] Only some data will be considered here.

On the basis of studies of silver-impregnated sections investigators have doubted whether there are true terminal boutons in the neocortex. However, improved methods, particularly electron microscopy, have demonstrated that *in the cortex as elsewhere synaptic contacts are established by means of terminal boutons*. Both axosomatic and axodendritic synapses occur (Gray, 1959; Colonnier, 1967; and others). The number of synapses is enormous. In an electron microscope study Cragg (1967) estimated the number of synaptic contacts in the motor and striate areas in the mouse and the monkey. In the motor area he found on an average 60,000 synapses per neuron in the monkey, 13,000 in the mouse; in the striate area the figures were 5,600 and 7,000, respectively, for the two species.[15] These figures make it clear that a single cell must be subject to the influence of numerous others. Using rather conservative estimates, Cragg (1967) figured that some 600 *intracortical* neurons may contact a cell in the motor cortex of the monkey, 65 in the striate area. Another indication of the multitude of possible interconnections is found in the Golgi studies of the motor and striate area in the cat (Sholl, 1953), since the dendrites of one stellate cell extend over an area which contains 2,000 to 4,000 perikarya. Conversely, the branches of one afferent fiber encompass a territory which may contain 5,000 neurons (see Sholl, 1956).[16]

These figures give an impression of the wealth of possible interconnections in the cortex, but give no specific information about the lines of transmission or circuitry within the cortex. More specific information is needed. Before proceeding, it is appropriate to summarize some data on the origin and termination of the main fiber connections of the cortex. With regard to the *termination of afferents to the cortex* it is generally agreed, as a result of numerous experimental investigations and further Golgi studies, that the type called "specific" (fibers a and b in Fig.

[14] For some papers and references concerning general features of the architectonics and organization of the cerebral cortex, see for example *The Structural and Functional Organization of the Neocortex*, 1970; *Conceptual Models of Neural Organization*, 1974; *Architectonics of the Cerebral Cortex*, 1978.

[15] The marked increase in the number of synapses in the motor cortex in the monkey as compared to the mouse is correlated with a reduction in neuronal density. The neuronal density is even less in man than in the monkey (see Sholl, 1956; Cragg, 1967), indicating that still more of the space is taken up by axonal and dendritic branches.

[16] In extensive material Sholl (1956) also studied other quantitative aspects of cortical cells in the cat.

12-6; see also Chap. 2) and arising from the *thalamus* supply mainly layer IV, with some branches to layer III. This has been observed in numerous Golgi studies and likewise in experimental light- and electron microscopic studies.

However, there appear to be minor differences between areas. In the cat (Strick and Sterling, 1974) and monkey (Sloper, 1973) the afferents from the VL of the thalamus supply chiefly layer III in the motor cortex, but also in part layers I and VI, whereas in the primary sensory cortex layer IV is the principal target for "specific" afferents. Even between areas 3, 1, and 2 of the primary sensory cortex there are qualitative and quantitative differences in the pattern of afferents from the thalamus (Jones, 1975a).

The distribution within the cortex of the *"unspecific" afferents* to all layers has likewise been confirmed, although not all layers are equally well supplied.

In general, long (as well as short) *ipsilateral cortical association fibers* appear to be axons of small and medium-sized pyramidal cells, chiefly located in cortical layer III. Some have been found in adjoining parts of layers II and IV and in layer V, as learned from HRP studies (Jacobson and Trojanowski, 1977a; Mesulam, Van Hoesen, Pandya, and Geschwind, 1977). (There may be differences between areas in this respect.) The *terminal branches of the association fibers* are distributed preferentially to layers III and IV, even if there may be some differences between areas also here; less heavy terminations have been described in all layers from VI to III (Jones and Powell, 1968a; Diamond, Jones, and Powell, 1968b; Kawamura and Otani, 1970; Kawamura, 1973a, b, c; and others).

Commissural fibers appear to end in all cortical layers. However, there are differences between areas with regard to density of termination in various cortical laminae. There is some disagreement among authors concerning details. The fibers appear to end chiefly in supragranular and granular layers, particularly in layers III and IV, but terminations have been found by some in layer V, and even in layer VI. For studies of the cortical levels of terminations of callosal fibers, see for example, Ebner and Myers (1965), Jones and Powell (1970b), Kawamura and Otani (1970), Lund and Lund (1970), Karol and Pandya (1971), and Sloper (1973). There appear to be species differences (see, for example, Jacobson and Marcus, 1970). HRP studies have confirmed the assumption, based on studies with other methods (Shanks, Rockel, and Powell, 1975), that most, if not all, *callosal fibers* are axons of (chiefly pyramidal) cells in layer III, in part in layer V, in the visual cortex (Wong-Riley, 1974; Winfield, Gatter, and Powell, 1975) and in the prefrontal (Jacobson and Trojanowski, 1977b), the somatosensory (Jones, Burton, and Porter, 1975; Wise, 1975), and other parts of the cortex (Jacobson and Trojanowski, 1974; and others).

The cells of origin of *efferent, corticofugal fibers* have been referred to at various places in this book. The conception that most efferent axons from the cortex arise from neurons in layers rich in pyramidal cells (layers III and V) and in part also layer VI, has received support from HRP studies, which often permit exact identification of the perikarya of neurons of origin. There are regional and areal differences with regard to details. The origins of the association and commissural fibers were referred to above. Corticofugal *projections to subcortical regions* (considered in previous chapters) arise preferentially from certain cortical areas. As mentioned above (p. 796), fiber contingents arising from one cortical region but destined to supply different subcortical stations are to a large extent axons from different colonies of neurons, chiefly situated in layers III and V. This refutes the frequently assumed notion that subcortical stations are innervated, only or chiefly, by collaterals from axons descending further.

In recent years anatomical and physiological findings have made possible some novel ideas about the *fundamental organizational pattern of the cortex*. Particular interest has been devoted to the pattern of *columnar organization*, referred to in previous chapters. In Nissl- and Golgi-stained sections cut perpendicularly to the surface of the cortex the nerve cells can often be seen to be arranged in columns as shown in Fig. 8-7a and c, and the afferent and efferent myelinated fibers are to a large extent arranged in radially running bundles (Fig. 12-1, right column). However, it was only after the studies of Mountcastle and collaborators in the late 1950s that this columnar arrangement was considered to be of functional importance. As has been described (see Chap. 2), Mountcastle discovered that in the somatosensory cortex neurons located within vertical columns, extending from the pial surface to the white matter, respond to stimulation of the same type of receptor located at a particular site. The work of Hubel and Wiesel from the 1960s brought forward evidence of the presence of a columnar functional arrangement in the visual cortex (see Chap. 8). It appears to be present in the acoustic cortex (Chap. 9) and the motor cortex (Chap. 4) as well. A particularly striking example of a columnar organization was found when the "barrels," related to the mystacial vibrissae, were discovered in the somatosensory face area of some rodents (see Chap. 7, section f).

Many anatomical studies are compatible with the view that there is a *structural basis for the functional columnar organization*, although how structure and function are related is not clear (see below). An organization in the vertical plane (perpendicular to the surface) is revealed in many structural features. Thus, as mentioned above, many of the cortical nerve cells extend their dendrites and give off axons preferentially perpendicular to the surface (Fig. 12-6). This is the case not only for pyramidal cells but also for many cortical interneurons. Two types (not shown in Fig. 12-6) should be specially mentioned, the *"cellule fusiforme à double bouquet dendritique"* of Cajal, and the so-called *"chandelier" cell* of Szentágothai.

> The frequent occurrence of Cajal's (1909–11) *"cellule fusiforme à double bouquet dendritique"* was drawn attention to by Colonnier (1966, 1967). This cell (not shown in Fig. 12-6) gives off two dense bunches of dendrites, one extending toward the surface, the other toward the depth. The thin axons of these cells, present in layers II, III, and IV, pass vertically in either direction, give off several short, fine collaterals, and may extend through all layers of the cortex. Their number is considerable, especially in man. The *"chandelier"* cell of Szentágothai (1975) has a medium-sized cell body and is provided with vertically arranged dendrites. The axon, after an initial short vertical course, bends horizontally and gives off several vertical, ascending and descending, branches. The terminal varicosities of these branches form "a short strictly vertical cylinder of boutons of about 2 μm inner diameter" (Szentágothai, 1975). These branches occur in all layers except layer I (and appear to contact selectively apical dendrites of pyramidal cells).

Dendrite bundles, found in the spinal cord and in many other regions, occur also in the cerebral cortex where they were described by Fleischhauer, Petsche, and Witkowski (1972), Peters and Walsh (1972), and Scheibel, Davies, Lindsay, and Scheibel (1974). The bundles are vertically oriented and are particularly evident in layer IV, where each bundle consists of various numbers of thick, apical dendrites of pyramidal cells located in layer V, in addition to thinner dendrites, belonging apparently to other smaller pyramidal cells. However, even if such dendrite bundles may occur all over the cortex, their pattern is not identical in all parts of it (see Feldman and Peters, 1974; Fleischhauer, 1978). In man, bundles of Betz-cell dendrites (see also Chap. 4) have recently been described by Scheibel and Scheibel (1978).[17] Those authors, furthermore, emphasized that descending dendritic branches of many Betz cells occupy an approximately cylindrical volume of tissue. It has recently been noticed that ascending, recurrent collaterals of a pyramidal cell axon tend to be aggregated in the immediate surroundings of its apical dendrite. A final feature worth mentioning concerns the distribution within the cortex of its afferents, both thalamic and others (see Fig. 12-6). They are distributed preferentially in the vertical direction of the cortex.

Several morphological features thus support the presence of a vertical (perpendicular to the surface) organization of the cerebral cortex and are compatible with the notion of functional vertical columns. However, there are abundant anatomical possibilities for information occurring *across the columns,* in the horizontal plane, parallel to the surface.[18] Dendrites of many cortical neurons may extend for a considerable distance horizontally along their layer. The pyramids, for example, give off numerous basilar dendrites, largely within layers III and V. In man such dendrites of the Betz cells may have a length of 1–3 mm (Scheibel and Scheibel, 1978). In layer I the most conspicuous cell type is the so-called Cajal-Retzius cell, which gives off an axon that may run for a considerable distance in the horizontal direction (cell 12 in Fig. 12-6). However, in all layers of the cortex there are what is generally referred to as *horizontal neurons,* neurons whose dendrites are oriented preponderantly in the horizontal plane.

Colonnier (1964, 1966, 1967), studying Golgi sections cut parallel to the cortical surface, found the lateral distribution of the dendrites of most cortical nerve cells to be rather restricted. Many of the stellate cells may, however, have *axonal* branches, which extend for some distance in the horizontal plane. Some of these cells have been grouped together as so-called *basket cells.* They will be considered below.

The idea of a columnar organization of the cerebral cortex was received with enthusiasm, since a column might represent an "elementary functional unit." The concept of functionally discrete columns is based largely on recordings made with microelectrodes inserted perpendicularly or obliquely through the cerebral cortex. The types of sensory stimuli giving rise to responses, receptive fields, and latencies have been recorded. As has been mentioned at several places, microelectrodes permit recording of the activity of particular cells.

Until recently the cells studied could be identified only under conditions of antidromic activation. This requires that the axons can be stimulated separately. (Most studies of this kind have been made on the Betz cells contributing to the corticospinal tract.) What kind of cell in a column

---

[17] The functional role of dendrite bundles is so far entirely hypothetical (see Scheibel and Scheibel, 1978, for some data and references). Dendrodendritic synapses appear to occur at least in some cortical regions (for example, in the monkey's motor cortex, Sloper, 1971) and may perhaps establish contact also between dendrites within a bundle.

[18] Physiological evidence of a horizontal correlation between columns has been presented, for example for the striate area by Hubel and Wiesel (1968).

produces the afferent responses recorded is still uncertain. The method of identifying the cells by antidromic stimulation can scarcely be used. However, the method of identifying cells recorded from by subsequent injection of a marking dye through the microelectrode (see Chap. 1) might give some information when applied to the cerebral cortex (see, for example, Kelly and Van Essen, 1974).

The electrophysiological events in the cerebral cortex are obviously extremely complex. These problems will not be discussed here. Suffice it to mention that the contribution of the potentials of single cells to the composite changes recorded in the electroencephalogram (EEG) is still a disputed question. The recording and interpretation of the EEG have developed into a particular branch of neurology. Apart from the use of the EEG as a tool in the diagnosis and prognosis of diseases of the nervous system, the study of the EEG under various conditions has produced valuable information on questions of basic neurophysiological interest.

As alluded to above, a central problem for understanding the organization of the cortical columns is whether columns, functionally determined (largely by recordings with microelectrodes), can be correlated with particular anatomical structures. Obviously a column must consist of a large number of neurons, pyramidal and others. Another question concerns the impulse passage, the circuitry, within a functionally determined column.[19] The assumption that a column could be a unit in the anatomical sense is not immediately obvious. Physiologically the columns have been estimated to have a width (diameter) of some 350 to 450 μm. A single specific thalamic afferent fiber may spread over 400 μm (see, for example, Jones, 1975a); the basal dendrites of pyramidal cells, however, extend widely, in man some 1–3 mm. Other axons and dendrites, in part horizontally oriented, likewise cover a far more extensive field than what would correspond to a functionally outlined column. Such data indicate that the anatomical substrate of a functional cortical column can scarcely be clearly defined (Von Bonin and Mehler, 1971; and others). However, a functional column does not necessarily have to correspond to a particular anatomical unit. As expressed by Chow and Leiman (1970, p. 176), "the same cortical elements could serve as a component of one functional column at one time and of another column at another time. Functional columns could form and dissolve at different times." For a critical analysis of the hypothesis of columnar organization the reader should consult Towe (1975).[20]

A particular instance of columnar functional organization is the presence of so-called *barrels* in layer IV of the face region of the primary sensory cortex of some animals (Fig. 12-7). As was described in Chapter 7 (p. 516 ff.), each mystacial vibrissa has been found to project to one barrel. Whether an anatomically outlined barrel is equivalent to a functional longitudinal column appears, however, not yet to be decided.

The finer intrinsic structure of the barrels is so far incompletely known. Killackey and Leshin (1975) concluded from a degeneration study that each barrel is supplied by one thalamocortical af-

---

[19] The responses of the simple type obtainable from a column are largely found in layer IV, whereas more complex responses occur in layers above and below layer IV. As described above, the "specific" thalamic afferents terminate chiefly in layer IV (to some extent also in III and I). It is likely that the sensory impulses relayed to this layer, by way of internuncial cells, influence the activity of neurons in other layers belonging to this column, but little is known about this. For a study on the visual cortex see Kelly and Van Essen (1974).

[20] The subject of cortical columns has been most extensively studied in the visual cortex (see Chap. 8). Here the columnar pattern has been shown to be an important organizational feature and to be valid for cells subserving functions of different complexity (for example, ocular dominance).

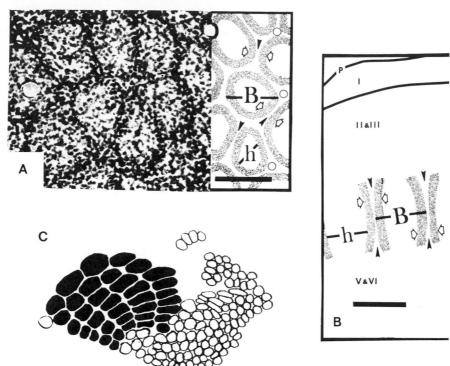

FIG. 12-7   Illustrations of some main features of the anatomy of the cortical barrels in the face region of the primary sensory cortex in the mouse.
A: A photomicrograph of a tangential section of cortical layer IV in the anterior barrel field. The explanatory diagram shows that each barrel (*B*) is made up of a ring of densely lying cells (stippled), which surrounds a less cellular area, the hollow (*h*). Septa (black arrowheads) separate the barrels. Bar = 100 μm.
B: A diagram of the somewhat larger barrels in the posteromedial barrel field, representing the mystacial vibrissae, as seen from the lateral aspect (constructed on the basis of coronal sections).
C: A camera lucida drawing of the barrel field of a mouse cerebral hemisphere. The barrels (black) in the posteromedial barrel field, related to the mystacial vibrissae are arranged in five distinct rows and are larger than the others. For descriptions see pp. 516–517. From Woolsey and Van der Loos (1970).

ferent (see also Donaldson, Hand, and Morrison, 1975). Feldman and Peters (1974) studied the distribution of pyramidal dendrite bundles in layer IV in the barrel field of the sensory cortex in the rat and other species. They found that numerous bundles occur within the territory of one barrel, and that the bundles are disposed equally in the walls and hollows of the barrel. White (1976), however, found that in the mouse the dendrite bundles occur preponderantly in barrel sides and septa. For a study of some connections, see White and DeAmicis (1977).

In spite of much research in recent years, our knowledge of the anatomy of the cerebral cortex is far from sufficient to understand properly the "nature" of the functional columns, not to speak of cortical organization in general. Here it will be possible to draw attention only to some main features out of a wealth of details. After a consideration of cell types, we shall briefly mention some data concerning synaptic contacts.

It appears that the original distinction between *pyramidal and stellate cells* (the two main types described in the cortex) is not sharp. Thus some pyramidal cells may have axons that do not leave the cortex (see, for example, Valverde, 1976; Tömböl, 1978). The stellate or granular cells (belonging to the heterogeneous group of Golgi II type cells) have been divided into several different varieties. The subdivisions made by various authors are not always concordant, probably an expression of the fact that there are fleeting transitions between types. (For recent studies, see, for example, Jones, 1975b; Valverde, 1976; Tömböl, 1978).

A general feature of stellate cells appears to be that they (at least the vast majority), are *interneurons* that do not send their axons out of the cortex. (Two of the particular types distinguished, the "cellule fusiforme à double bouquet dendritique" and the "chandelier" cell, were described above.) A third type, the *basket cell*, first described by Cajal, is characterized by having chiefly horizontal axonal branches that give off several fine, vertical collaterals with boutons. The collaterals surround the somata of more deeply lying pyramidal cells, like a nest or basket. The dendrites pass in the horizontal and vertical direction. The largest basket cells are found in layers III, IV, and V.

Such cells have been described also in the human motor cortex, where they are particularly frequent in layer IV (Marin-Padilla, 1969).[21] The basket cell must obviously be able to act on several pyramidal cells.

As to the *synaptic relations* within the cerebral cortex, there are many unsolved problems. Most direct information on synaptic relations can be obtained from electron microscopic studies of experimental material, in which a particular contingent of afferent fibers, arising outside or within the cortex, is destroyed.[22] As referred to above, the presence of both axosomatic and axodendritic synapses, suggested on the basis of light microscopic studies, has been confirmed with the electron microscope. Different types of synapses occur. It appears that the terminals of thalamic, commissural, and association fibers all end in synaptic contacts of the asymmetrical type and that their boutons contain spherical synaptic vesicles (see Jones and Powell, 1973). This type of contact (cf. Chap. 1) is generally taken to be excitatory. There are, however, also numerous synapses of the symmetric type. Here the terminals are often found to contain pleomorphic vesicles. These are generally interpreted as being inhibitory.

Dendrites of many cortical cells, pyramidal and stellate, are amply provided with spines, for example the apical dendrites of pyramidal cells. The number of spines increases exponentially with the distance from the cell body (Valverde, 1967). Many terminals establish synaptic contact with spines, others with smooth

[21] At least in man these basket cells appear to be "flat" neurons (150–200 μm thick), extending their branches perpendicular to the long axis of the precentral gyrus (Marin-Padilla and Stibitz, 1974). Baskets are formed around pyramidal cells in all layers. It appears that not all fibers in the baskets belong to basket cells, but that some are of extrinsic origin, in the motor cortex coming from the VL of the thalamus (Marin-Padilla, 1972).

[22] Attempts to elucidate the terminations of intracortical cells have been made by studying the structure in pieces of cortex which have been isolated from the underlying white matter for a relatively long period, so that afferent endings will have degenerated. The undercutting of the cortex may also be made between layers. In light and electron microscopic studies of this kind Szentágothai (1965b) confirmed that the tangential fibers in layer I are mainly derived from cells in the deeper layers. It further appears from such studies that intracortical cells take part in the pericellular baskets around the perikarya of the pyramidal cells, as has been concluded from Golgi studies (see also Colonnier, 1967).

parts of dendrites. These differences are presumably related to functional differences, but little definite is known about this problem.

There appear to be different types of spines (for example, some have a thick, others a thin stem). Spines of different types may occur on different parts of a cell's dendrites, for example those of pyramidal cells. This may correlate with the finding that afferents of different origin may establish synaptic contact with various parts of the dendritic tree. It appears that in the case of the pyramidal cells the thalamic "specific" afferents contact the middle part of the apical dendrite, and the "nonspecific" afferents all of it except its most proximal part; whereas commissural fibers contact chiefly oblique side branches of the apical dendrite, and recurrent collaterals its distal, apical branches and basal dendrites. For a discussion of these problems see, for example, Chow and Leiman (1970).

It appears from electron microscopic studies that contacts of the asymmetric type (excitatory?), for example from the thalamus and ipsi- and contralateral cortex, are established, at least preferentially, with spines. ("Specific" thalamic fibers appear to establish contact with spiny parts of dendrites of pyramidal as well as of stellate cells.) Running along the apical dendrites of the pyramidal cells there are fine axons that in Golgi sections appear to make series of asymmetric synaptic contacts with the spines of the dendrite along its course. This has been described also in the human brain (Marin-Padilla, 1968). These "climbing fibers" have been interpreted by some as collaterals of pyramidal cells, whereas other authors assume that they belong to commissural afferents or thalamic fibers or intracortical cells in the vicinity.

It appears that many of the numerous stellate (Golgi II type) cells in the cortex do establish synaptic contacts of the symmetric type and have terminals containing pleomorphic vesicles. They are often assumed to have an inhibitory effect on the postsynaptic cells. Particular interest has been devoted to the basket cells as mediators of an inhibition of the pyramidal cells.[23] Owing to the arrangement of its axon, one basket cell may be imagined to influence a series of pyramidal cells. The basket–pyramidal cell arrangement may be a feature in the fundamental pattern of cortical organization (the "basket–pyramidal system," studied especially in the motor cortex; see for example, Marin-Padilla, 1970, 1972).

Indirect information on synaptic relations in the cortex may be obtained by studying differences in its finer structure after experimentally induced alterations in its function. A few examples may be mentioned.

It has been observed that the spines on the apical dendrites of the pyramidal cells are reduced in number in mice which have been reared in complete darkness since birth (Valverde, 1967) and in rabbits (Globus and Scheibel, 1967a) and mice (Valverde, 1968) after eye enucleation or lesions of the lateral geniculate body. The lack of visual stimuli, normally transmitted to the striate area, thus produces morphological changes in its cells (as is known from previous studies). The disappearance of the spines on the central parts of the apical dendrites might indicate that the thin axons contacting the apical dendrites are fibers from the geniculate body which terminate in layer IV and, in part, in III. However, following enucleation, there are also specific variations in the orientation of the dendrites of stellate cells in layer IV (Valverde, 1968). The changes in the spines of the pyramidal cells may therefore be secondary.

Following transection of the corpus callosum in rabbits at birth, Globus and Scheibel (1967b) found loss of about one-third of the spines on the oblique branches only of the apical dendrites of the pyramidal cells in the parietal cortex. This suggests that callosal afferents are distributed to

---

[23] The largely corresponding arrangement of the basket cells in the cerebellar cortex and the hippocampus is taken to support the assumption of an inhibitory action of cortical basket cells.

these regions (i.e., laminae IV and III). This finding is compatible with the results of authors studying the termination of callosal fibers with silver impregnation methods.

Only a few of numerous observations on synaptic relations in the cerebral cortex have been referred to above. It should again be emphasized that there are differences among areas concerning the arrangement of cells, axons, dendrites, synapses, and other features. In addition there are differences among species. In spite of our still fragmentary factual information on these points, many researchers have been tempted to speculate upon possible common patterns of cortical organization which may serve as guidelines for further research. All attempts to formulate general views on cortical organization and function will have to be almost purely theoretical and hypothetical. It may be mentioned that central points in all theories are the synaptic arrangements involved in the relations between input and output of a functional unit in a module, for example the circuitry within a functional column, where thalamic afferents establish (excitatory?) contacts with stellate and pyramidal cells. The (inhibitory?) "basket–pyramidal" system, referred to above, appears to be an important part of almost all models of cortical function.

Even if far more data were known concerning the anatomical basis of cortical circuitry and microcircuitry, any formulation of ideas about the functional organization of the cerebral cortex would be seriously hampered by the difficulties inherent in studies of the functional aspects of synaptic transmission in general.[24] Furthermore, other factors related to transmitters and to general humoral conditions may be involved and show temporal changes. Speaking of the neuronal operation of a functional unit or module, Eccles writes, ". . . we have to envisage levels of complexity in the operation of a module far beyond anything yet conceived, and of a totally different order from any integrated microcircuits of electronics, . . ." (Popper and Eccles, 1977, p. 241).

It is scarcely appropriate to discuss the subject of principles of cortical organization further in this text and to consider the many hypothetical schemes set forth. The interested reader will find much relevant information, for example in the *Neurosciences Research Program Bulletins* of Chow and Leiman (1970) and Szentágothai and Arbib (1974). Szentágothai has given more extensive presentations of his "module concept" in separate articles (see Szentágothai, 1975). The Scheibels have likewise considered the problem extensively (see the preceding bulletins for some references). Other relevant articles may be found in *Architectonics of the Cerebral Cortex* (1978). For a brief account see Popper and Eccles (1977).

The cerebral cortex in phylogenesis. The higher an animal is in the phylogenetic scale, the more elaborate the transmission of nervous impulses in the cortex appears to be. The percentage of association cells, and particularly of intracortical neurons, increases markedly along the phylogenetic scale. In man such

---

[24] Even if in some inhibitory synapses the presynaptic part contains many flattened or ovoid vesicles, the presence of such vesicles in a bouton does not permit the conclusion that it has an inhibitory action (see Chap. 1). Nor does the presence of round vesicles in a bouton necessarily indicate its excitatory nature. The functional importance of morphological features such as membrane thickenings and spine apparatuses is conjectural.

cells make up a considerable part of the entire cell content of the cerebral cortex (Lorente de Nó). The cerebral cortex in mammals presents an increasing degree of cytoarchitectural and myeloarchitectural differentiation in phylogenesis. Although some of the more elementary areas can be recognized to be fairly uniform throughout the mammalian scale, for example the area striata, there can be little doubt that in the higher mammals a progressive differentiation of many areas has taken place. In the anthropoid apes and man, areas have been recognized which are missing in lower mammals. Particularly those areas of the cortex which are presumed to function as "association areas" show a progressive development in man. The parietal and temporal lobes appear to be particularly rich in areas of this type.[25] On the other hand, the areas giving rise to or receiving projection fibers do not increase correspondingly, and therefore occupy a *relatively* smaller part of the entire cerebral surface. Various aspects of cortical evolution have been critically reviewed by Holloway (1968).

Functions of the cerebral cortex. On account of the elaborate differentiation of the cerebral cortex in phylogenesis physiological experiments in animals can only to a certain extent give information concerning the function of the cortex in man. In general the results of experimental studies are in fairly good accord with observations in man, if allowance is made for the different development of the structures in question. Examples of this have been mentioned repeatedly in preceding chapters when discussing the results of stimulation or destruction of several cortical areas, especially of the sensorimotor, the visual and the acoustic areas. As far as these and some other cortical areas are concerned it seems reasonable to conclude that they are related predominantly to particular functions.[26] On the whole the cortical regions delimited by physiological methods as functionally different fields coincide fairly well with certain cytoarchitectural areas, even if the old notions that the area striata alone represents the visual cortex, area 41 (and 42?) the acoustic cortex, areas 3, 1, and 2 the somesthetic cortex, and area 4 the cortical motor area are no longer tenable. These questions have been considered in Chapters 8, 9, 2, and 4, respectively, to which the reader is referred.

Even if the regions of the cortex related more specifically to fairly well defined and relatively elementary functions occupy a much greater part of the entire neocortex than was previously assumed, extensive cortical regions are left. The functions of these regions appear to be more complex than those of the other areas and are difficult to investigate, especially since they are related to processes which in a loose manner may be characterized as "mental" (of various complexities) and which can, therefore, be properly studied only in man. (Many of these faculties are probably almost specific for man.) The functional observations concerning these parts of the neocortex will be considered briefly here, since they can be correlated with our present-day anatomical knowledge only to a very limited

[25] In this connection some findings on the ontogenetical development of the human cortex are of interest. Conel (1939) has shown that in newborn children area 4 is the area which has progressed farthest in differentiation. Least developed are the areas in the anterior part of the frontal lobe. The terminal areas of the large sensory projection systems are also rather well differentiated at birth. For some further data on cortical ontogenesis see Holloway (1968).

[26] The term *function* is here used in a purely colloquial manner. See also the last section of this chapter.

extent. To a certain degree it is possible to indicate certain regions of the cortex as related to particular, although rather complex functions. Following a consideration of the frontal lobes we will turn to the regions related to language and other so-called "psychosomatic" functions, and finally discuss some recent observations on the corpus callosum.

The frontal lobe.   The frontal lobe "has remained the most mystifying of the major subdivisions of the cerebral cortex" (Nauta, 1971). It is generally said to undergo an extensive differentiation in monkeys and particularly in man. This assumption has, however, been questioned (see Holloway, 1968). It has been generally assumed that the frontal lobe must be concerned in higher mental functions. Both animal experiments and clinical findings have strongly suggested that this is true, particularly for the anterior part of the frontal lobe. As described above (p. 789), these cortical regions, situated in front of areas 4, 6, and 8, usually are collectively referred to as the "prefrontal" granular cortex.[27] Before considering this cortex and its functions, mention should be made of a speech disturbance that may occur in lesions of the frontal lobe, namely *motor aphasia*. Like other disturbances in linguistic functions it can be observed and studied only in man.

Motor aphasia is seen in lesions of the dominant hemisphere involving the regions immediately in front of the cortical motor face area. This region is often called Broca's area, since it was first drawn attention to by Broca in 1861. Briefly, motor aphasia is characterized by the patient being incapable of expressing his thoughts in words. The peripheral apparatus involved in speech, the laryngeal, facial, and lingual muscles, are not paralyzed and can be used properly in other acts. There is, for example, no dysarthria. It is the *initiation and integration* of the movements necessary for performing articulate speech which are defective. In some cases of motor aphasia the patient has in addition lost the capacity to write, a condition called *agraphia,* in spite of the fact that his fingers are not paralyzed or paretic and are used normally in other performances. Motor aphasia is only one type of disturbance of the language functions which may be seen in lesions of the brain. We will return later to a brief consideration of these disorders.

We will deal with the "prefrontal" granular cortex and its functions below; the subject encompasses problems of theoretical as well as practical interest, to which a vast literature has been devoted.

It was found rather early by almost all experimenters that unilateral removal of the frontal lobe produces little change while bilateral ablations are followed by characteristic symptoms.

Among experiences from the early studies may be mentioned the fact that in monkeys the most conspicuous feature is a "hyperactivity" that is betrayed by the restless walking of the animal in its cage (Kennard, Spencer, and Fountain, Jr., 1941; and others). Ruch and Shenkin (1943) concluded that ablation of the posterior orbital cortex is responsible for the motor hyperac-

---

[27] Only in primates does the larger part of the prefrontal cortex have the cytoarchitectural features of a granular cortex; in subprimates it is of the agranular type. However, the consistent projection of the dorsomedial thalamic nucleus to the prefrontal areas in all species examined indicates that a "prefrontal cortex" is present in subprimates as well as in primates (see Nauta, 1971). Primate studies are of most interest from a clinical point of view.

tivity. In chimpanzees, Jacobsen and collaborators (1935, and later) also observed an increased distractibility and a reduced capacity for solving problems (as, for example, when the animals had to connect together several sticks in order to reach a piece of food). It has been suggested that the hyperactivity represents a release phenomenon, due to removal of a normally present inhibitory influence on motor activity by this part of the cortex (see Kaada, 1951, 1960).

On stimulation of the posterior orbital cortex a series of responses have been observed in cats and monkeys, such as inhibition of spontaneous and cortically and reflexly induced movements and inhibition or acceleration of respiratory movements; autonomic responses from the cardiovascular and digestive systems; and emotional reactions in the form of "arrest reaction," "attention," or "arousal" (see Kaada, 1960, for a review). As mentioned in Chapter 10, the posterior orbital cortex is usually included as part of the morphological basis of the "limbic system."

Since the appearance of the studies referred to above, several others have been performed. Extensive behavioral studies with complex tests have been undertaken of monkeys subjected to ablations of the anterior frontal areas (for some reports see *The Frontal Granular Cortex and Behavior,* 1964 and 1972). Gross behavioral changes have rarely been noted. (The hyperactivity referred to above appears to be a hyperreactivity rather than a hyperactivity.) While there appear to be few or no recognizable defects in tests of simple intellectual capacities and small changes in emotional behavior, more complex intellectual and emotional spheres seem to suffer. There are clear psychological changes, for example in the capacity of enduring attention and initiative. Thus deficits have been noted in tasks that require a rapid shift from one way of solving a problem to another. There are deficits in delayed-response tests. In studies of free-living monkeys it has been found that following bilateral prefrontal ablations the animals show an increased stereotyped locomotor activity (pacing), reduced aggressiveness, and reduced vocal activity. Furthermore, there are marked changes in their social behavior (Myers, Swett, and Miller, 1973) particularly loss of the animals' group affinity and neglect of offspring.[28]

Some of the disturbances found in animal studies resemble the changes seen in human patients. It might be expected that studies in man are better suited to give some insight into the functional role of the frontal lobe than experimental studies in monkeys.

In man several symptoms have long been recognized as being to a certain extent characteristic of lesions of the frontal lobe (the motor symptoms, including forced grasping, have been described in Chap. 4 and are due to lesions of the precentral and supplementary motor cortical areas). Most authors hold that in man the lesion must also be bilateral if the symptoms are to become distinct. However, Rylander (1939), in a study of the symptoms following unilateral partial ablations of the frontal lobe in 32 patients, found distinct postoperative changes in all these cases. Some data (see Milner, 1974) indicate that there may be functional differences between the right and left frontal lobes (apart from the relation of the latter to speech).

The symptoms most commonly described as following bilateral damage to or removal of the prefrontal areas in man are frequently grouped under two headings:

[28] Interestingly, the changes in social behavior seen in adult monkeys following bilateral prefrontal ablations do not develop following corresponding ablations in very young monkeys (up to 2 years) (see Franzen and Myers, 1973). This has been taken to indicate that a significant social maturation must be present for the prefrontal lobectomy to produce adverse effects. (Similar age-related differences have been seen following amygdalectomies.)

reduction of intellectual ability and of ethical standards. The patients present signs of complacency and self-satisfaction, and frequently also of boastfulness. Their powers of concentration, their initiative, and their endurance are reduced, and they are more distractible than normal. Often they are in high spirits, in a somewhat puerile manner. Memory for recent events suffers, and their capacity to solve problems requiring intellectual effort is reduced, particularly as regards abstract problems. Their power of judging their own situation is impaired: the patient's horizon is narrowed to the present and his own person. In syphilitic general paresis, where conspicuous anatomical changes are usually found in the cortex of the frontal lobe, and also in cases of *Pick's disease* with the same localization, similar symptoms are frequent.

Additional information concerning the functions of the prefrontal areas in man has been obtained from cases where a *frontal lobotomy* or a *prefrontal leucotomy* has been performed on patients suffering from mental disorders.

> Introduced by Egaz Moniz in 1936, this operation was for some years much used. The principle of the operation is to cut the fiber connections of the frontal "association areas." This was done by inserting an instrument through a burr hole and moving it in a coronal plane. Some surgeons preferred to make an open leucotomy, i.e., with the frontal lobe exposed by a craniotomy. As a rule these operations were performed bilaterally. Although they were made on patients suffering from one sort or another of mental disorder (schizophrenia, obsessions, etc.), the postoperative changes occurring in behavior and mind following a prefrontal leucotomy closely resembled those observed in cases of diseases of the frontal lobe. Changes in the emotional sphere appear to be most prominent, whereas there was some dispute as to whether purely intellectual functions suffered. However, the patient's capacity to judge the future consequences of his actions was as a rule reduced.

The techniques used in the classical prefrontal leucotomies, aiming at interrupting the connections of the prefrontal cortex, were rather crude, and the anatomical consequences of surgery were not predictable.[29] Attempts were soon made to ascertain whether less drastic surgical procedures could be used. Selective undercuttings of part of the projections of the prefrontal cortex, for example the orbital areas (*selective orbital leucotomy*), and excisions of particular and restricted cortical areas in open surgery (*topectomy*) were performed. Studies of numerous patients subjected to such operations gave rise to a number of publications. Not least, the influence of leucotomy in the emotional sphere attracted much interest. Freeman and Watts (1942) and others noted that delusions or obsessive ideas frequently do not disappear after operation, but they no longer disturb the patient. Still more striking was the observation (see Freeman and Watts, 1946) that patients subjected to leucotomy on account of unbearable pain of somatic origin are no longer troubled by the pain, even if it is still present.

[29] The fortuitous nature of the anatomical effects of leucotomy was convincingly demonstrated by Meyer and Beck (1945) in a study of nine brains from patients subjected to leucotomy. The place of the section in the brain varied greatly. This is to be expected in view of the considerable individual variations which exist within human brains; in addition there are variations between the brain and the surface markings of the skull which had to be relied upon when leucotomies were performed. Furthermore, it was evident that the depth and localization of the section also varied, even if the same technique was employed. Still more important, perhaps, is the fact that extensive secondary changes were found in certain of the cases, such as large cysts, probably due to intracerebral bleedings. Secondary changes (for example, bleedings) were sometimes found far from the place of the cut. (See also Meyer and Beck, 1954; Eie, 1954.)

The development of stereotactic methods in neurosurgery (shortly before 1950) opened up new possibilities for producing restricted lesions of the brain at chosen sites and for reducing the frequency of postoperative epilepsy. (For a historical account see Vaughan, 1975.) Chiefly on the basis of the assumption (see below) that the results would differ according to the region of the brain destroyed at operation, a rich variety of operations has been performed. To mention a few: *bilateral anterior stereotactic cingulotomy* (transection of the cingulum bundle in the white matter of area 24), *stereotactic anterior capsulotomy* (aiming at interruption of those fibers in the internal capsule which connect the anterior frontal cortex and the thalamus), *stereotactic limbic leucotomy* (a combination of lesions of the cingulum with transection of the fibers of the lower medial quadrant of the frontal lobe), so-called *stereotactic subcaudate tractotomy* (the section is placed rostral to the anterior limit of the 3rd ventricle), and an operation called *anterior mesoliviotomy* (transection of the most rostral part of the knee of the corpus callosum). In all these procedures the connections of only parts of the granular frontal cortex are transected.

The introduction of potent so-called *psychopharmaca* has meant a revolution in the treatment of mental disorders and has contributed to changes in prevalent views. Nevertheless, *psychosurgery* (operations on the intact brain for the alleviation of mental and behavioral disorders)[30] still has its adherents, and many variants of selective prefrontal leucotomy are carried out on patients who have not responded to any other kind of treatment. Other operations, such as *amygdalotomies* (see Chap. 10) and *destruction of the ventromedial hypothalamic nucleus* (see Chap. 11) likewise have been recommended and performed in the treatment of mental disorders. It is far beyond the scope of this text to consider these subjects fully.[31] Below a few data will be mentioned, and some general comments on psychosurgery will be made.

The original expectation that different kinds of mental disturbances could be cured or alleviated by destruction of particular areas of the brain had to be given up rather soon. Differences in the effect on mental functions can be correlated only to some extent with differences in the region destroyed at operation. Briefly stated, it appears that most patients subjected to one or another type of "prefrontal" leucotomy suffer from affective disorders, many of them from obsessive compulsive neurosis. The neurosurgical procedure found by most investigators to be best suited is one that involves *section of fibers in the cingulum bundle* (simple cingulotomy in area 24 or a variant; for some recent papers on cingulotomy see *Neurosurgical Treatment in Psychiatry, Pain, and Epilepsy,* 1977). Favorable results have been reported. Some authors have performed the same operation in patients with severe chronic depressive illness or anxiety states and describe good results in some cases. Relief from intractable pain has been obtained in many patients. It has been claimed that "personality changes" are not observed after cingulotomy, and that there is no reduction of intellectual functions. Severe obses-

---

[30] For some attempts to define the term *psychosurgery* more precisely, see Mark and Ordia (1976).

[31] A vast literature has grown up on the various aspects of prefrontal leucotomy and other spheres of psychosurgery (see, for example, *Selective Partial Ablation of the Frontal Cortex,* 1949; Valenstein, 1973; *Surgical Approaches in Psychiatry,* 1973; Robin and MacDonald, 1975; *Operating on the Mind,* 1975; *Neurosurgical Treatment in Psychiatry, Pain, and Epilepsy,* 1977).

sions, depressions, and anxiety have been the principal indications for another commonly used variant of ''prefrontal'' leucotomy, *orbital undercutting* (transection of the connections of the orbital surface of the prefrontal cortex). Good results have been reported, but side effects may follow more commonly than after cingulotomies.

Studies of the large number of patients subjected to selective leucotomies during the last 25 years have in general confirmed the early observations concerning the consequences of ''prefrontal'' leucotomies (referred to above), and additional information has been obtained, not least because many patients have been subjected to elaborate psychological tests.

> Attempts to differentiate between the consequences of lesions of various minor parts of the ''prefrontal'' granular cortex have given rather meager results. However, it appears to be generally accepted that destruction of the medial inferior (orbital) part of the frontal lobe is likely to result in emotional changes and to modify social behavior, while lesions of the dorsal and lateral convexity produce intellectual deficits and defects in the performance of delayed-response tasks. These differences can probably be related to differences in fiber connections of the two regions of the prefrontal cortex (see below).
>
> Valuable studies have recently been made in patients who have undergone unilateral prefrontal ablations for the treatment of epilepsy and have been subjected to various psychological tests (see for example Teuber, 1972; Milner, 1974). It appears from such studies that these patients do not have any linguistic disability. However, they tend to perseverate and show deficits in carrying out tasks requiring insight and flexibility (revealed, for example, in the so-called recency tests, where the task is to indicate which word or picture in a series the patient has seen most recently). The observations ''support the notion that the temporal ordering of events is disturbed after frontal lobectomy'' (Milner, 1974, p. 83). A remarkable feature following ablation appears to be a loss of foresight. This may be correlated with the results of the recency tests.

Many attempts have been made to understand the function of the ''prefrontal'' cortex and to ''explain'' the symptoms following ablations of the whole or parts of it. Theories and speculations abound. It has been suggested that mental disturbances that may be favorably influenced by leucotomy are due to abnormal activity (hyperexcitability or increased inhibition) of some other brain region, and that the result of surgery is to interrupt the lines of propagation of such abnormal activity. Our knowledge of these lines for impulse propagation to and from the prefrontal cortex is far from complete. Some salient points may be mentioned. (See Nauta, 1971, for an extensive review).

As noted above, the ''prefrontal'' cortex, in contrast to the three other cerebral lobes, does not receive fiber systems carrying primary sensory information (via thalamic relays). Nor do associational connections appear to reach it from the primary visual, acoustic, and sensory cortices. Nevertheless, the ''prefrontal'' cortex may be influenced by such sensory stimuli, since cortical regions adjoining the primary sensory fields and receiving association fibers from these (see Figs. 12-4 and 12-8A) project to the ''prefrontal'' cortex.[32] Even if important transformation

---

[32] As referred to previously, Jacobson and Trojanowski's HRP studies (1977a,) show that these projections are rather specific (see the original articles). The strongest input comes from the temporal lobe, particularly the superior temporal gyrus, while the contributions from the parietal lobe (area 7) and from area 19 are rather modest. Furthermore, Jacobson and Trojanowski (1977b) observed heterotopic callosal connections to the ''prefrontal'' granular cortex from the contralateral temporal and cingulate cortex.

of sensory messages presumably occurs during their passage along these multisynaptic routes, they are likely to influence the activity of the "prefrontal" cortex. The efferent connections from this cortex to the association areas of the three other cerebral lobes appear on the whole to be arranged reciprocally to the afferents from the latter to the "prefrontal" cortex (see Fig. 12-4). The existence of these connections strongly indicates that a close correlation must occur between the "prefrontal" cortex and those cortical areas which in man are concerned in speech and other complex mental functions. This is in agreement with the signs of defects in intellectual functions seen following (unilateral) "prefrontal" ablations, even if language disturbances appear not to occur (see above).

In addition to *receiving afferents from other (association) areas of the neocortex,* the "prefrontal" cortex receives fibers from many *subcortical brain regions* and from parts of the allocortex (see Fig. 12-8).

Thus the *dorsomedial thalamic nucleus* (MD, see Fig. 2-14), particularly its large-celled medial division, projects in an orderly topographical manner onto the "prefrontal" granular cortex. While some authors found this projection to end chiefly in the orbitofrontal cortex, others found a more widespread distribution (for some articles see Rose and Woolsey, 1948; Nauta, 1962; Roberts and Akert, 1963; Akert, 1964; Narkiewicz and Brutkowski, 1967; Leonard, 1969; Tobias, 1975). Kievit and Kuypers (1977) appear to have solved this dispute. The distribution of labeled cells in the thalamus was plotted following injection of HRP in different parts of the frontal cortex in 25 monkeys. Their study confirms the presence of a topographical order in the thalamic projection (a mediolateral sequence in the MD corresponds roughly to an anteroposterior sequence in the frontal cortex) and also clearly shows that the entire "prefrontal" cortex receives afferents from the MD. However, the site of origin of thalamic fibers to the "prefrontal" cortex is not limited to the MD, but forms a band that encroaches on neighboring thalamic nuclei as well. Among *afferent* connections to the MD (see Fig. 12-8B) are fibers from the amygdala and from the piriform cortex. As mentioned in Chapter 10, both these structures receive olfactory fibers. Even if fibers from the olfactory bulb have not been traced directly to the MD, these connections bear witness that the "prefrontal" cortex may be influenced by olfactory stimuli via the MD. In addition, such stimuli appear to have more immediate access to the prefrontal cortex by way of the recently established direct amygdaloprefrontal projection (see Chap. 10).

The MD is, however, much more than a relay station in routes for olfactory impulses to the "prefrontal" cortex. There is evidence that it may be involved in memory functions, as discussed in Chapter 10, and it is interesting to note that there appears to be an extensive convergence of afferents within this nucleus (see Fig. 12-8B; details of topographical distributions and synaptic relations will not be considered). Thus, fibers have been traced from the septal nuclei (see for example, Meibach and Siegel, 1977a; see also Chap. 10), and from the mesencephalic tegmentum (Guillery, 1959; and others). Some authors have advocated the view that there are fibers to the MD from the hypothalamus. According to Szentágothai et al. (1968) and Raisman (1966b) the evidence for the latter connection is not conclusive. Some other afferents to the MD have been described as well. The afferents listed above appear to end, at least mainly, in the medial division of the MD. The afferents to the lateral division are less well known. They have been claimed to arise in the cingulate gyrus and some other regions (see Siegel et al., 1977, for some references). It should be noted that the activity in the MD may be influenced from the "prefrontal" granular cortex, to which it projects by way of reciprocal corticothalamic projections (DeVito and Smith, 1964; Rinvik, 1968c; and others), even if these appear to supply mainly lateral parts of the MD, and the reciprocity is not complete.

Some of the subcortical regions, which are connected with the "prefrontal" cortex by an indirect route via the MD, also have a direct line of communication at their disposal. This is the case for the *amygdala* (see above) and the *septal nuclei* (see Chap. 10), possibly also for the *hypothalamus* and the *mesencephalic reticular formation.* Afferent fibers have also been traced

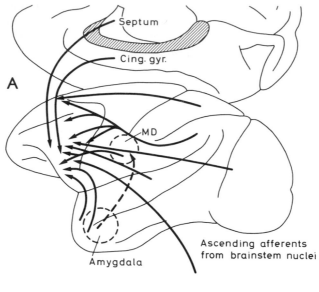

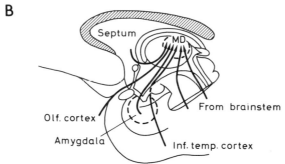

FIG. 12-8 *A:* Extremely simplified diagram of some main afferent connections of
the "prefrontal" cortex. Afferents come from other parts of the cortex and from
some subcortical regions (septum, amygdala, dorsomedial thalamic nucleus (*MD*),
brainstem nuclei, see text). Regional differences in origin and termination of particular connections are not shown. Fibers from the amygdala to the MD are included
in the diagram.
B: Schematic drawing showing main subcortical afferents of the dorsomedial thalamic nucleus (*MD*). This, particularly its medial part, receives afferents from the septum, from the olfactory cortex, from the amygdala, from the inferior temporal
region, and fibers ascending from the brainstem (particularly midbrain tegmentum).
Diagram *B* slightly altered after Nauta (1971).

from other structures such as the *raphe nuclei* and the *nucleus locus coeruleus* (see Chap. 6) to the
"prefrontal" cortex. It should finally be recalled, as described above (see Fig. 12-4), that parts of
the *allocortex*, the gyrus cinguli, the entorhinal cortex, and neighboring regions project onto the
"prefrontal" cortex.

It can be seen from the preceding incomplete summary that *the "prefrontal"
agranular cortex may be influenced from numerous other parts of the brain—
neocortical, allocortical, and subcortical.* Among the latter the dorsomedial thalamic nucleus (MD) appears to be particularly important. Just as the "prefrontal"

cortex receives information by way of afferents from many sources, the MD is provided with *reciprocal connections, enabling it to influence almost all regions that may act on it.*

As described above, the "prefrontal" granular cortex gives off intrahemispheric association fibers (see Figs. 12-4 and 12-9) to the neocortical temporal, parietal, and occipital association areas; to the allocortical cingulate gyrus; to the entorhinal area; and to at least some of the perirhinal cortex. In addition, there is a topographically arranged projection to the MD, to the amygdaloid nucleus, and to the septum (see Chap. 10); to the hypothalamus (see Chap. 11); to the mesencephalon, the raphe nuclei, and the nucleus locus coeruleus (see Chap. 6); and to some other regions (for example, caudate nucleus, intralaminar nuclei, probably the hippocampus). The pulvinar has been described as receiving some fibers (see Leichnetz and Astruc, 1976.

We may ask: Can the anatomical data outlined above be of any use in our attempts to understand the functional role of the frontal lobe, particularly of the "prefrontal" granular cortex? The answer seems to be: To a limited extent only. The ample and largely reciprocal interconnections between the "prefrontal" granular cortex and other cortical and subcortical regions and the interconnections between many of the latter regions leave us with a picture of an extremely com-

FIG. 12-9 Extremely simplified diagram of main efferent connections of the "prefrontal" cortex in the monkey. Efferent fibers pass to cortical and subcortical regions, to a large extent to regions from which the prefrontal cortex receives afferents (compare Fig. 12-8A). Not all contingents appear to be finally established.

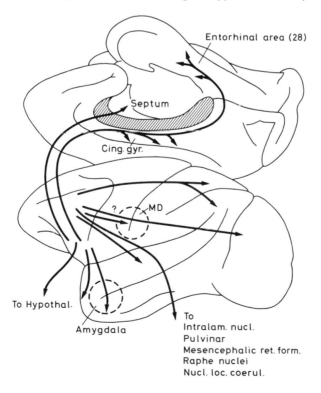

plex tangle of fiber connections. In the future more instances of precise topographical (and functional) relations will presumably be brought to light than are recognized today (only a few of them have been mentioned above).

The presence of architectonic, hodological, and other differences between minor parts of the "prefrontal" cortex must be taken to be correlated with functional differences. It would be an undue oversimplification to consider the "prefrontal" cortex as a unit. On the other hand, on the basis of present anatomical knowledge the functional differences between many minor regions must be assumed to be very subtle. Studies of such functional differences are extremely difficult and are beset with numerous pitfalls.

A few general features appear so far to be established. It is noteworthy that the "prefrontal" cortex, a part of the neocortex, has two-way connections with structures known to be involved in emotional and behavioral changes and in the regulation of the internal milieu of the organism. Among these are the amygdala, the septum, the hypothalamus, and the cingulate gyrus—structures that are all generally listed as belonging to the so-called "limbic system." On the basis of these connections it is no wonder that there are many similarities in the disturbances in emotional behavior following lesions of the "prefrontal" cortex and of the "limbic" structures, even if there are differences. On the other hand, the changes observed in intellectual functions following lesions of the "prefrontal" cortex are presumably related primarily to interruption of connections between this cortex and the temporo–parieto–occipital association cortex.

> As we have seen, to a large extent efferent and afferent connections of the "prefrontal" cortex are reciprocally arranged. A lesion or a surgical section will therefore in general interrupt both afferent and efferent links of two-way connections. Some authors have speculated about whether the functional disturbances occurring after surgical interventions are due chiefly to interruption of afferent or of efferent connections of the "prefrontal" lobe. At our present stage of knowledge (and presumably even if we knew the relations in all details) it appears to be of little benefit and rather artificial to hypothesize about this question.

It follows from the arrangement of the fiber connections that damage to any part of the frontal lobe will almost inevitably interrupt connections of the "prefrontal" cortex with both neocortical areas and with "limbic" structures. No wonder, therefore, that monkeys or humans subjected to lesions of the frontal lobe or "prefrontal" leucotomies show changes in both spheres. There is no conclusive evidence that different parts of the "prefrontal" cortex (and their connections) are related *exclusively* to either "intellectual" or "emotional" functional[33] spheres (an assumption that is a priori entirely unacceptable from a biological point of view). However, there is evidence that certain regions of the "prefrontal" lobe are *more closely* engaged in one than in the other of the two spheres of activity, and that there is an essential functional difference between the orbitofrontal cortex and the cortex on the convexity of the frontal lobe.[34] Anatomical data appear to support this distinction. This information is of interest, particularly with regard to types of procedures to be chosen in "psychosurgical" treatment. Reviewing the

---

[33] It is stressed here (as in other sections of this volume) that to speak of emotional and intellectual functions as separate spheres of activity is an unacceptable oversimplification. The distinction above is made only for the sake of convenience.

[34] This question has been discussed in several publications, some listed in footnote 31.

question, Teuber (1972, p. 617) concludes that "orbitofrontal lesions tend to create profound disturbances in emotional reactivity and to interfere with appropriate behavior in social groups." On the other hand, lesions of the dorsolateral cortex and of the inferior convexity result in deficits in delayed response and related tasks and in trouble with object-reversal tasks, respectively.

Experimental studies in monkeys (see Butter and Snyder, 1972) show that ablations restricted to the *orbitofrontal cortex* give rise to distinct changes in emotionality, especially aggressive and adversive behavior (in some ways resembling the effect of lesions of the amygdala). Of particular interest is the finding that lesions of the dorsomedial thalamic nucleus (MD) produce essentially the same results as ablations of the orbitofrontal cortex.

Anatomical data are in agreement with the functionally determined relation of the orbitofrontal cortex first and foremost to the "emotional sphere." The "prefrontal" cortex, as described above, has important connections with allocortical regions. The fibers passing in the cingulum bundle to the entorhinal area (and supplying the cingulate gyrus as well) originate largely in the orbitofrontal cortex (see Van Hoesen, Pandya, and Butters, 1975). Further, as described above, the medial part of the MD, projecting at least chiefly to the orbitofrontal cortex, receives a considerable proportion of its afferents from so-called "limbic" structures (see above and Fig. 12-8B). Correspondingly, the "prefrontal" afferents to the MD (its medial part) have been found to arise mainly in the orbitofrontal cortex.

As described above, lesions of the *convexity and dorsolateral parts of the "prefrontal" cortex* result in certain functional disturbances; changes in the "emotional sphere" are moderate. In agreement with this, it appears that these parts of the cortex are relatively little connected with "limbic" structures. Thus the afferents from the MD to these cortical regions appear to arise chiefly in its lateral, small-celled part, to which few fibers from "limbic" structures have been traced. Even if the majority of intrahemispheric connections of the dorsolateral part of the "prefrontal" cortex are established with the parietal and temporal association cortices, certain regions do give off fibers to the cingulate gyrus as well (see Pandya, Dye, and Butters, 1971).[35]

It appears from the discussion above that with regard to the "prefrontal" cortex there is general agreement between the results of behavioral studies in monkeys, anatomical findings, and clinical experience. However, in all three fields our knowledge is still rather crude. This has obvious practical consequences for the use of leucotomies as a therapeutic measure. The same is the case for other kinds of so-called psychosurgery such as amygdalotomy (see Chap. 10) and operations on the hypothalamus (see Chap. 11). It is appropriate at this junction to devote some comments to *psychosurgery in general,* particularly since in recent years it has met with opposition, in part quite furious.

Psychosurgical treatment meets with problems in many spheres—ethical, surgical, technical, and anatomical—as well as difficulties related to the psychiatric evaluation of patients before and shortly after psychosurgery, and studies of late effects.

It is of course essential that the *lesion* (produced electrolytically or by other means) be placed at the desired target as precisely as possible. Coordinates worked out for stereotaxic brain surgery and control of the position of the electrode (by means of X-ray examinations, results of stimulation, and other mea-

---

[35] It is of interest that projections to the cingulate gyrus appear to arise from only two of three areas of the dorsolateral cortex investigated. This illustrates the differences in patterns of connections between minor regions of the "prefrontal" cortex. On the other hand, the presence of efferents from this part of the "prefrontal" cortex to the cingulate cortex shows that it is not possible to distinguish between larger areas of the "prefrontal" cortex as being or not being related to "limbic" structures and emotional changes.

sures) usually permit this, even if there are individual variations among brains. It is important to realize that *any lesion of the brain will destroy not only neurons and their axons in a small patch of nervous tissue:* in addition numerous fibers passing through the area and often representing reciprocal connections will be transected. Even a minute destruction of brain tissue will, therefore, have consequences (functional and structural) for many other regions of the brain, sometimes far removed from the lesion. It is therefore not permissible to assume from behavioral and other functional disturbances following neurosurgery that these reflect interference with the function of the small destroyed part only. (Many hypotheses concerning "centers" for particular functions should be rejected for this reason.) These considerations are especially relevant in psychosurgery on the hypothalamus, where areas of nervous tissue claimed to serve particular functions are very small (see Chap. 11).

While even a very small lesion will inevitably have immediate consequences for the functions of many other brain regions, other factors may also play a role in the final result, e.g., bleeding from inadvertently damaged vessels and shrinkage of destroyed tissue with time. Further, there may occur some reorganization of structure and function with the establishment of new synaptic patterns (cf. Chap. 1).

The conditions enumerated above tend to limit the predictability of the consequences of *any* surgical intervention on the brain, not least "psychosurgery." Furthermore, the fact that psychosurgical operations aim at destroying parts of the brain concerned in its most elaborate functions entails particular difficulties in the evaluation of the patient's condition before and after the operation and the assessment of the results. As expressed by Valenstein (1973, p. 297), "in the case of psychosurgery, it is very difficult to know what the real relationship is between the excised brain tissue and any behavior change that might occur." In addition to what is considered above, there are other reasons for such statements. Thus Walsh (1977, p. 163) is probably right when he states, "the traditional diagnostic categories of psychiatry are inadequate as a basis for election of patients for psychosurgery." Special tests for particular functions must be elaborated, and the results must be very critically evaluated.

Robin and Mcdonald (1975) listed several factors that may influence the evaluation of the results of surgery and may be sources of fallacies in their assessment. Among these, some arise from the patient: spontaneous remissions of the disease process, recurrent disease, the psychological effect of the fact of treatment. Other factors are related to society's attitude (for example, its criteria for recovery and for what type of behavior is socially acceptable may change with time). Finally, there are fallacies arising from the therapist: errors in diagnostic reasoning, in psychiatric assessment (see above), and other problems.

Robin and Macdonald (1975), like many others, emphasized that controlled studies are essential for an adequate evaluation of the therapeutic efficiency of psychosurgical procedures. So far, relatively few studies of this kind have been undertaken. Justification for the different kinds of psychosurgery still rests almost exclusively on an empirical basis.

It follows from the anatomical and functional features referred to above that a psychosurgical operation of any kind, even if the damage inflicted is as desired, can never be expected to have effects on one mental "function" only. Side effects will always occur. The more carefully the patients who have undergone psychosurgery have been tested psychologically before the operation, the more deviations from the patient's preoperative (and premorbid) personality have been discovered. It has been said that in essence any psychosurgical operation results in changes in the patient's "personality" (a term difficult to define).[36] Another important point is that, in contrast to the changes that may follow psychotherapy and pharmacotherapy (which also may have undesirable side effects), the alterations caused by a psychosurgical procedure are irreversible.

The practice of psychosurgery raises problems of a social, ethical, and political nature. These aspects have been much discussed in recent years and have given rise to heated public debates. However, they cannot be evaluated properly unless the purely medical aspects of the problem are taken into consideration as well. This has not always been the case. Many of the most ardent personal antagonists to any kind of psychosurgery, and the organized groups of so-called antipsychiatrists do not belong to the medical profession or to any kind of health personnel. It is possible here to mention only a few arguments over a very complex problem.[37]

The opposition raised by the so-called antipsychiatrists concerns all kinds of psychosurgery. The use of such operations for mentally ill people is condemned as unethical. The main reasons for this opposition are the unpredictable side effects, particularly the postoperative changes in personality and the fact that the destruction of a part of the brain is irreparable; that the tissue destroyed is most often not diseased. Other arguments have been put forward, for example the problems involved in defining what is "normal" and what is not in the mental and behavioral spheres. Such objections may rightly be levied against psychosurgical treatment of mental disease. However, even some critical psychiatrists and neurosurgeons feel that in certain instances psychosurgery may still deserve consideration as a possible therapeutic procedure (see, for example, *Surgical Approaches in Psychiatry*, 1973; Mark and Ordia, 1976). To justify surgical procedures, it is required that certain conditions be fulfilled with regard to critical evaluation of the results of previous attempts at therapy, duration of the disease, etc. From an ethical point of view it is essential that the patient and his/her relatives be informed about possible therapeutic results and risks. A guiding principle must be to achieve the best for the patient. This evaluation may, of course, be extremely difficult.

If the patient's behavior is socially unacceptable, particular problems arise in evaluating the indications for psychosurgical treatment. This is the case with patients suffering from marked aggressive tendencies with uncontrollable bursts of violence directed against their immediate surroundings, and with sexual and other deviants. The problems are accentuated when such persons commit acts which are considered criminal and which cause them to be imprisoned. The problem of whether society has a moral right to "cure" these criminals by means of surgery of the brain has been discussed and has prompted very hard criticisms of medical professionals who have suggested this solution of a serious social problem (see, for example, the article by the ardent antipsychiatrist Chorover, 1976, and the balanced discussion of this topic by Valenstein, 1973, pp. 336–353).

[36] These personality changes and undesirable side effects appear to be more marked and easier to discover when the patient is living in his natural surroundings than when observed in an "artificial" hospital community. A corresponding difference has been found in groups of monkeys, observed in their natural surroundings or in the laboratory. For example, amygdalectomized monkeys who are "cured" of their aggression show marked changes in social behavior. They suffer from severe deficits in their adaptation to the ranking orders in the colony and are unnaturally submissive (see Valenstein, 1973, for a discussion).

[37] For some extensive treatments of the subject see, for example, Valenstein, 1973; *Operating on the Mind*, 1975; *Current Controversies in Neurosurgery*, 1976; *Neurosurgical Treatment in Psychiatry, Pain, and Epilepsy*, 1977.

It can be seen from the discussion above that any kind of operation on the human brain that may influence emotional, intellectual, and behavioral faculties must be judged with the greatest caution. Many of the arguments concerning social and ethical aspects brought forward by the "antipsychiatrists" are clearly relevant. However, most antipsychiatrists apparently possess limited insight into and lack practical experience with mentally ill people. Much of their reasoning is purely theoretical and, it appears, flavored by philosophical and political concepts.[38] The medical aspects of a psychosurgical operation will, however, always have to be taken into consideration in the particular concrete case. It appears that it would be premature to ban all so-called psychosurgical operations. There is reason to believe that in certain instances a psychosurgical operation is the most satisfactory therapeutic measure, in spite of all its shortcomings.

It is to be deplored that public interest in psychosurgery has fostered much emotionally colored and unrealistic animosity against neurosurgery and psychiatry and also against scientific medicine in general. The *theoretical* possibilities of manipulating the human brain by surgical (as well as chemical) means have inspired science fiction writers and may have contributed to a belief among the public that what they read is more than fiction. Not least have science fiction writers and antipsychiatrists found data of great interest among the possibilities for influencing mental functions and behavior by stimulation of the brain of the awake human being. A few data will be mentioned. For a balanced, well-documented, and critical analysis of these aspects the reader should consult Valenstein's monograph (1973). See also articles in *Neurosurgical Treatment in Psychiatry, Pain, and Epilepsy* (1977).

> Taken in a wider sense the field of psychosurgery includes the technique of implating electrodes into particular, selected areas of the brain when the intention is to produce mental and behavioral changes. These electrodes may be made to discharge in many ways, by the subject himself or by remote stimulation if connected with a radioreceiver in or on the body. In the monkey and in other animals, certain areas, for example, in the hypothalamus, on stimulation give rise to what appears to be pleasant sensations (so-called pleasure zones, belonging to so-called reward systems), and the animal, if allowed, may stimulate itself incessantly. On stimulation with electrodes inserted in corresponding brain sites in man, different kinds of sensations and reactions have been observed. The presence of particular "pleasure zones" in the human brain is questionable since, as emphasized by Valenstein (1973), the same stimulation may often evoke different responses, and identical responses may be evoked from different regions. This is indeed not surprising in view of the numerous direct and indirect interconnections between various parts of the brain and the incessantly changing patterns in the impulse traffic. While such factors are of importance in stimulation of all sites of the central nervous system (spinal cord, cerebellum, mesencephalon, basal ganglia; see other chapters of this book), it is likely that they play a particularly important role in regions related to the most complex nervous functions.

Cerebral dominance. The term "cerebral dominance"[39] is used to describe the fact that one of the cerebral hemispheres is the "leading" one in certain

---

[38] Mark and Ordia (1976, p. 727) comment, "The antipsychiatry campaigners have been in the vanguard of those who would precipitously dismantle our mental hospitals and discharge mentally ill patients into an unprepared community." These authors cite (loc. cit., p. 727) a political scientist "who suggests that this campaign is designed to keep mentally ill patients sick as a visible sign that our society has failed in its obligations and is indeed a sick society."

[39] The literature on this subject is overwhelming. For some recent relevant papers see, for example, *Hemisphere Function in the Human Brain* (1974), *Evolution and Lateralization of the Brain* (1977), *Lateralization in the Nervous System* (1977), *Asymmetrical Function of the Brain* (1978).

functions, the difference being most marked as concerns the complex language functions. The notion of a "leading hemisphere" appears to have been first presented by Hughlings Jackson in 1869 (see Zangwill, 1960) on the basis of the frequent occurrence of motor (and other kinds of) aphasia in right-handed persons when they suffer a hemiplegia of the right limbs, while a left-sided hemiplegia in such persons usually is not accompanied by aphasia. In left-handed individuals the situation is often said to be the reverse (a lesion of the right hemisphere, but not of the left, being likely to result in aphasia). However, this is no absolute rule. In left-handed persons most often the left hemisphere has turned out to be the dominant one with regard to speech functions. With increased insight it appears that the concept of a unilateral dominance of the left over the right hemisphere in man has to be abandoned and should be replaced by one of complementary specialization of the two hemispheres. The left hemisphere appears to be dominant or "leading" for certain functions, the right for others, particularly for what may loosely be referred to as "spatial functions" (see below).

It appears that with regard to language functions cerebral dominance is genetically determined (see Geschwind, 1974b; and below on anatomical asymmetries of the brain). Handedness, however, may at least in part, be environmentally determined. More or less ambidextrous persons are not rare. As phrased by Zangwill (1960, p. 27), "Handedness must be regarded as a graded characteristic; left-handedness, in particular, being less clear-cut than right-handedness and less regularly associated with dominance of either hemisphere." This conclusion is based on studies of the incidence of language disorders in left-handed patients who had sustained a strictly unilateral injury to the brain.[40] Corresponding conclusions have been made by Penfield and Roberts (1959) and others.

These results are of practical interest, for example in surgical therapy involving cortical excisions, particularly in the operative treatment of epilepsy. A method introduced by Wada in 1949 (see Wada and Rasmussen, 1960) has made it possible to identify the dominant hemisphere for speech in patients before they undergo operation. A solution of amylobarbital sodium is injected into one common carotid artery. The injection produces an immediate and temporary loss of function of the cerebral structures supplied by the artery.[41] Thus there ensues a transient contralateral hemiplegia. If the affected hemisphere is the dominant one, aphasia develops, as can be ascertained by various verbal and other tests. This method is now widely used. The results confirm that cerebral dominance cannot be judged on handedness alone. For example, Serafetinides, Hoare, and Driver (1965) found that of 12 patients who by this test had a left dominant hemisphere for speech, 8 were right-handed, 2 left-handed, and 2 were ambidextrous. Three left-handed patients had a right cerebral dominance, while three had no exclusively unilateral representation of speech.

There appears to be little doubt that so far as the perception of language and the performance of speech are concerned, in most persons one hemisphere is dom-

[40] The observations above are of interest in attempts to explain the occurrence of "congenital word blindness" in children (see Zangwill, 1960).

[41] Occasionally, however, even a unilateral injection may give bilateral clinical signs as well as bilateral changes in the EEG (see Perria, Rosadini, and Rossi, 1961). In such cases the test permits no conclusions as to cerebral dominance.

inant.[42] There are some reports that the two hemispheres differ in their influence on other functions as well. Injection of amylobarbital on the side of the dominant hemisphere has been accompanied by depressive emotional reaction, injection on the nondominant side by an euphoric reaction (Perria, Rosadini, and Rossi, 1961; and others). Serafetinides, Hoare, and Driver (1965) even suggested that there is a "cerebral dominance for consciousness," but Rosadini and Rossi (1967) failed to find support for this view. The question of whether consciousness is "represented" bilaterally in the human brain has been much discussed, but no final answer has been brought forward. For recent surveys see, for example, Zangwill (1974) and Pucetti (1977). See also Popper and Eccles (1977).

In recent years many studies have been devoted to the problems related to "cerebral dominance." Information obtained in studies of patients in whom the corpus callosum has been transected and corresponding experimental studies have greatly furthered our understanding of how the two hemispheres collaborate, and have shed light on the question of whether certain functions are related to one hemisphere only (see also below). In general, it may be said that the left hemisphere is particularly related to verbal skills, while noverbal skills depend more on the right hemisphere. The parts of the hemisphere showing these differences are first and foremost the lower part of the parietal lobe and the adjoining temporal cortex. These regions, as has been mentioned above, belong to the "association" areas of the cortex.

So far, little evidence for anatomical differences between the two hemispheres has been produced. Geschwind and Levitsky (1968) reported left–right asymmetries in the temporal speech region. The posterior region of the superior surface of the temporal lobe (the so-called "planum temporale") is larger on the left side than on the right in about two-thirds of the brains examined. This has been confirmed by Witelson and Pallie (1973), who also found that the left-sided area was best developed even in neonates. They suggested that this "indicates that the infant is born with a preprogrammed biological capacity to process speech sounds." A similar, but less marked, asymmetry may be present in the brain of the chimpanzee, but not in the macaque (Yeni-Komishian and Benson (1976).[43] Other asymmetries as well have been observed in human brains. As is well known to radiologists, the occipital horn of the lateral cerebral ventricle is typically longer on the left than on the right side. Side differences in venous and arterial patterns and other asymmetries have likewise been observed. For accounts of anatomical asymmetries of the brain, see Geschwind (1974b) and Galaburda, LeMay, Kemper, and Geschwind (1978).

The cerebral cortex and language functions.   Mention was made above of the fact that disturbances in language functions occur in lesions of the dominant

---

[42] Several authors have maintained that the nondominant hemisphere is not without importance for speech functions. In support of this view Larsen, Skinhøj, and Lassen (1978) reported alterations of local blood flow in the cortex of the nondominant as well as the dominant hemisphere during tests for automatic speech in normal persons. They studied the local blood flow following intracarotid injections of xenon-133 (see footnote 2).

[43] Most authors maintain that there is no reliable evidence in support of the presence of hemispheric specialization in infrahuman mammals. Some articles that discuss this problem and the question of asymmetries in the neonate may be found in *Evolution and Lateralization of the Brain*, 1977.

hemisphere only (if there is a clear dominance). These disturbances are usually called *aphasias* and have attracted much interest by clinicians, psychologists, and linguists. The literature dealing with disturbances of speech and the perception of language (spoken or written) is enormous, and crowded with classifications and designations. It is just as crowded with hypotheses, presenting to one unfamiliar with the matter a veritable jungle, difficult to penetrate even to a limited extent.[44] Various aphasic disturbances which may be seen have been given particular names. A crude distinction is often made between disorders in which the production of speech suffers while the understanding of spoken and written language is intact (motor aphasia, see preceding section), and aphasias where primarily the receptive functions suffer (sensory aphasia and alexia). The defects in understanding in these cases may result in difficulties in expression. The two groups are also referred to as *expressive* and *receptive* aphasias, respectively. The era of classifications of aphasias into a number of different specific kinds or types appears now to have passed. It seems to be generally agreed that when patients suffering from one or another "type" of aphasia are closely examined, one never encounters the pure types. In all cases features are present which, according to classical schemas, are peculiar to other types of aphasia. Scarcely two patients show identical symptoms. The disturbances in the language functions appear to be determined not only by the site and size of the lesion, but also by the premorbid personality of the person, his intellectual standard, his sphere of knowledge, etc. In their now classical monograph on aphasia Weisenburg and McBride (1935, p. 442) state: "A study of the essential nature of the aphasic disorder shows individual differences which are so great that no single hypothesis will acount for the changes."

The study of aphasias has several aspects, psychological, physiological, and anatomical. The latter aspect concerns the question whether a lesion of a particular part of the brain will result in language disturbances and whether differently situated lesions give rise to different aphasic phenomena. At one time it was customary to assign particular locations to different types of aphasic disturbances, and to speak of "speech areas" in the cortex, on the assumption that the different "types" of aphasia were caused by affections of rather circumscribed parts of the cerebral cortex, and that these areas were "centers" for particular components of the complex language functions. In keeping with the realization that pure types of aphasia scarcely exist, these classical detailed "psychoanatomical" correlations have had to be discarded. This has become clear from careful investigations of the speech disorders occurring in cases of controlled surgical ablations of parts of the cortex and from postmortem studies of aphasic patients whose language performances have been carefully studied prior to death. Additional valuable information has been produced in studies of the results of electrical stimulations of the "speech areas" in conscious patients. Such studies, performed especially by Penfield and his collaborators, have given information of other aspects of the psycho-

---

[44] Of works dealing with the problems of aphasia a few may be mentioned: Weisenburg and McBride, 1935; Nielsen, 1946; Penfield and Roberts, 1959; Brain, 1961, 1965; *Disorders of Language*, 1964; *Brain Mechanisms Underlying Speech and Language*, 1967; Brown, 1972. In the journal *Cortex* the reader will find many articles related to language functions and the problem of cerebral dominance. Critchley's monograph on aphasiology (1970) is interesting reading. For a recent study of recovery and prognosis in aphasia (and references) see Kertesz and McCabe (1977).

physical functions of the brain as well. Some of these observations will be briefly mentioned. Most of the observations have been made in patients suffering from temporal lobe epilepsy.

On stimulation of the temporal lobe the patient may experience what Penfield has called an *experiential response* (see Penfield and Perot, 1963). This may be visual, auditory, combined visual and auditory, or more difficult to classify. The patient on stimulation recalls a scene from his past: hears a familiar voice, sees some of his relatives, etc.[45] Responses of these kinds were obtained only from the temporal lobe. As described above, association fibers conveying information about visual and acoustic input converge in the temporal lobe. The temporal lobe experiential responses tend to occur more often on stimulation of the nondominant temporal lobe than of the dominant lobe. Other phenomena were also noted, collectively designated as *"interpretive responses,"* or *"interpretive illusions"* (see Mullan and Penfield, 1959). The sounds heard may seem louder or fainter; things seen may seem clearer, blurred, larger, or smaller; present experience may seem familiar (*déjà vue* sensation), stronger, altered, or unreal. Other sensations also occurred. The visual responses were obtained from the temporal lobe, predominantly of the nondominant hemisphere, the acoustic responses from the superior temporal gyrus on either side. Stimulation of other parts of the cortex does not give rise to such phenomena.

From the findings, briefly summarized above, it is concluded that the temporal lobe, especially the first or superior temporal gyrus, plays a particular role in the recall of perceptions from the past (chiefly auditory and visual) and in the interpretation of acoustic and visual signals. For these reasons this cortical region is called by Penfield the *"interpretive cortex."* It is stressed that this cortex *is separate from the cortical areas which appear to be concerned in language mechanisms.*

Language mechanisms are influenced by stimulation of rather extensive regions of the cortex (Fig. 12-10). An anterior region approximately covering Broca's area gives rise to motor aphasia when destroyed. A posterior region occupies the inferior and posterior parts of the parietal lobe and the adjoining posterior part of the temporal lobe, except the first temporal gyrus (see above). This region covers approximately the area described long ago by Wernicke as the area whose destruction resulted in "sensory aphasia." In addition, changes in speech may be obtained from a third, superior region covering the supplementary motor area.

On electrical stimulation of the brain two kinds of effects on speech may be obtained (see Penfield and Roberts, 1959). Electrical stimulation may produce vocalization, or it may result in inability to vocalize or to use words properly. Vocalization, in the form of a sustained or interrupted vowel cry, can be elicited from the face subdivisions of the first and the supplementary motor areas in both hemispheres. These effects are presumably due to interference with the motor control of the organs of speech.

Arrest of speech or inability to vocalize spontaneously follows stimulation of the "motor areas," but may also be seen on stimulation of more extensive cortical areas, particularly those outlined in Fig. 12-10. From these areas in addition other effects have been observed: slurring of speech, distortion and repetition of words and syllables, confusion of numbers while counting, inability to name objects,

---

[45] Corresponding phenomena may occur during spontaneous seizures in these cases (then called "experiential hallucinations"). In the experiential response the patient is fully aware of his present situation.

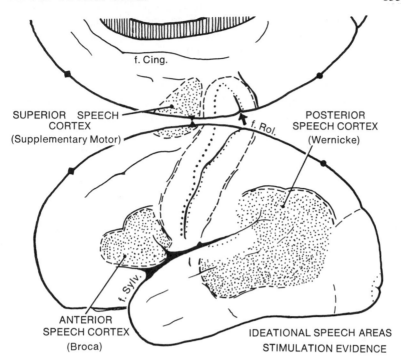

FIG. 12-10 Map of the cortex showing (dots) the areas in the dominant hemisphere of man from which interference with speech has been obtained on electrical stimulation. Note that the anterior part of the superior temporal gyrus is not included (see text). From Penfield and Roberts (1959).

with retained ability to speak. Occasionally other effects, such as difficulty in reading, have been observed. No significant differences have been found between the results of stimulation of the anterior, posterior, and superior area. Several of the effects listed above, which are obtained from the dominant hemisphere (usually the left) regardless of handedness, can scarcely be interpreted as due to disturbances in the control of muscles used in speech. The observations strongly suggest that the areas are involved in more integrated aspects of speech and language functions. This is in agreement with the studies of the consequences of cortical lesions involving these regions. (For some experimental studies of brain areas involved in vocalization, see Chap. 7, section c.)

This subject has been studied most extensively by Penfield and his collaborators, and has been discussed in some detail by Penfield and Roberts (1959). It appears from their material that transient aphasic disturbances may follow upon excision of various parts of the cortex. Persistent aphasia or dysphasia will result only when the removal involves the areas which on stimulation give disturbances in language functions (see Fig. 12-10). These areas are all assumed to be concerned in the "ideational mechanism of speech." The posterior area is considered indispensable to normal speech; the superior is the least important. No conclusive evidence that particular regions of the posterior area are related to different aspects of language functions could be produced, although there are some indications that lesions of the posterior parts are more likely to give rise to aphasic disorders most pronounced in the visual sphere. On the whole, corresponding results have been reported by Hécaen and Angelergues (1964) in a study of patients with local brain damage.

Even if there appears to be convincing evidence that certain regions of the dominant cerebral hemisphere (see Fig. 12-10) are particularly closely related to the complex functions involved in understanding and using language, some important qualifications should be borne in mind. In the first place, these areas do not function in isolation but in collaboration with other parts of the brain (the "interpretive cortex" of Penfield in the superior temporal gyrus may be especially important). Aphasic disturbances may indeed be seen with lesions outside these parts of the cortex.[46] Secondly, there is no reason to assume that particular aphasic disturbances can be more precisely linked with certain regions of these areas. Finally, it should be realized that when a certain type of speech disturbance is observed following a circumscribed cerebral lesion, this does not mean that the destroyed region is a "center" for a certain component of the mechanism of speech. What is observed is really quite another thing, namely, to what extent the remaining intact parts of the brain are able to perform such functions.

Lord Brain, in a review of "the neurology of language" (1961, p. 164) expressed his views as follows: "We have seen that the physiological organizations upon which speech depends are of great complexity, extend over considerable areas of the brain, and are organized in time as well as in space. These serve psychological functions, but the breakdown of a physiological schema does not necessarily, or indeed usually, disturb speech in a way that corresponds to a single type of psychological defect. Moreover, the anatomical organization of the schemas means that only rarely will a single type of schema be disturbed in isolation from others."

The *anatomical connections of the cortical areas related more closely to speech* are not yet sufficiently known. Associational connections with other regions are well developed as described above (p. 797 ff.). The regions receive (in part) abundant commissural connections (see Fig. 12-5). Of particular interest are the thalamic projections to these areas, since these projections and their nuclei of origin have been shown to be related to speech mechanisms in man. Thus Penfield and Roberts in 1959 (p. 215) described a patient with a small hemorrhagic lesion in the pulvinar of the dominant hemisphere, who had severe aphasia. Following stimulation or stereotactic lesions involving the pulvinar in man aphasic disturbances have been reported by several authors (see, for example, Ojemann, Fedio, and Van Buren, 1968). Language disturbances have been elicited not only from the pulvinar but from more anterior parts of the thalamus as well. A methodo-

---

[46] It should also be remembered that language functions are closely related to the emotional conditions of the person at the moment. This may be even more true for musical faculties. From an extensive clinical study of patients suffering from amusia, Ustvedt (1937) arrived at the conclusion that subcortical centers (hypothalamus, thalamus, and corpus striatum) are also important for the musical functions, particularly for the more emotional components of them. It has been suggested (Scheid and Eccles, 1975) that the right temporal lobe, particularly the planum temporale, is more closely related to the "musical sense" than the left, which is involved in linguistic functions. Readers interested in the relations between the brain and musical functions should consult *Music and the Brain,* 1977.

logically important point in such studies is the exact localization of the stimulating electrode (for some notes on methodology see Ojemann, Fedio, and Van Buren, 1968). It is characteristic that such disturbances have been seen only with left and not with right thalamic lesions, and that the language defects do not seem to be identical with any of those described following cortical lesions. Most often there is anomia (failure to name objects correctly) or dysnomia and perseveration. Short-term verbal memory may be defective. Observations of language functions following electrical stimulation of the thalamus during stereotactic operations for dyskinesias have confirmed that certain parts of the left thalamus are intimately related to language and short-term verbal memory. There appear, however, to be some differences with regard to influences on linguistic functions between the pulvinar and the ventrolateral parts of the thalamus. It appears that the latter regions are particularly related to the changes in motor and autonomic functions accompanying speech. For a recent report on the relation of the thalamus to speech functions, see Ojemann (1977).

Our knowledge of the thalamic projections to the "posterior speech cortex in man" is fragmentary. The most important source appears to be the *pulvinar* (see Fig. 2-14). Retrograde cell changes have been found in patients with speech disorders and in whom lesions were present in the posterior cortical speech regions (Van Buren and Borke, 1969). Great care must be exercised in drawing conclusions about the projections in man from experimental studies in animals, because these parts of the cortex are far more developed in primates than in other mammals and are particularly large in man. The corresponding thalamic projection areas show parallel differences. Thus, while the pulvinar is a large nucleus in man and is well developed in monkeys, what constitutes its homologue in lower mammals has been disputed (see, for example, Harting, Hall, and Diamond, 1972). In the following discussion, therefore, reference will be made chiefly to some general points emerging from experimental studies in monkeys, and references to particular subdivisions of the thalamic nuclei will largely be omitted. There are ample possibilities for confusion. Thus the borders between nuclei, for example, between the pulvinar and the nucleus lateralis posterior (LP in Fig. 2-14), are not drawn identically by different authors. Furthermore, there are differences between Old and New World monkeys. (For recent studies in the cat and references, see Graybiel, 1972b; Niimi, Kadota, and Matsushita, 1974.)

The *pulvinar* can be subdivided into subnuclei (commonly a medial, lateral, and inferior nucleus). It appears to be the main, but not the only, site of origin of thalamic fibers to the parieto–temporo–occipital association cortex in monkeys. In addition, there is a contribution from the LP (nucleus lateralis posterior). The distribution in the thalamus of retrograde cellular changes following lesions of the temporal, parietal, and occipital cortex has been studied by many authors (see, for example, Le Gros Clark and Boggen, 1935; Le Gros Clark, 1936b; Le Gros Clark and Northfield, 1937; Chow, 1950; Simpson, 1952a; Locke, 1960; Akert et al., 1961). Although there are several discrepancies with regard to details, the projection was found to be fairly precise, various subnuclei having their particular projection area in the cortex (see Chow, 1950; Simpson, 1952a). Best known is the projection of the pulvinar.

Mappings of the cortical projections of the *pulvinar* have recently been undertaken with the HRP method. In monkeys the fibers to the parietal lobe have been found to pass particularly to area 7 (Baleydier and Mauguière, 1977; Divac, LaVail, Rakic, and Winston, 1977; Pearson, Brodal, and Powell, 1978). Mesulam et al. (1977) reported fibers to the inferior parietal lobule. Autoradiographic (Benevento and Rezek, 1976), HRP (Trojanowski and Jacobson, 1975), and other studies have confirmed projections from the pulvinar to the temporal lobe, particularly to areas 20 and 21. It is of interest that in addition to fibers to the so-called association areas, a projection, topographically arranged, passes from the pulvinar (medial nucleus) to the frontal eye field, area 8 (Trojanowski and Jacobson, 1974). Other fibers appear to pass to the primary somatosensory cortex, Sm I and Sm II (DeVito, 1978), and to the peristriate region of the occipital lobe (see Winfield, Gatter, and Powell, 1975). In the monkey a projection to the striate area is disputed (see Ogren and Hendrickson, 1976).

Like the pulvinar, the *nucleus lateralis posterior* (*LP*) of the thalamus projects to the temporoparietal association area. From HRP studies some authors concluded that it projects to area 7 (Baleydier and Mauguière, 1977), but most authors found its main cortical projection to pass to area 5 (DeVito, 1978; and others). The projection appears to be topographically organized (Pearson, Brodal, and Powell, 1978).

In view of the functional observations on the *pulvinar* (see below) its *afferents* deserve attention. In addition to fibers from the cortex, it receives afferents from several subcortical structures, suggesting a role in integrating mechanisms.

An important, apparently rather massive contingent consists of fibers from the *temporoparietal cortex;* others come from the *peristriate region* of the occipital lobe (see, for example, Mathers, 1972, in the monkey; Kawamura et al., 1974, in the cat). These connections appear to be arranged reciprocally to the projection of the pulvinar onto the cortex (Trojanowski and Jacobson, 1975; DeVito and Simmons, 1976; Ogren and Hendrickson, 1976; DeVito, 1978). Fibers have been traced from the "prefrontal" cortex (Leichnetz and Astruc, 1976), a finding confirmed in HRP studies (Kievit and Kuypers, 1977). Autoradiographically a projection to the pulvinar was traced from the *somatosensory areas* Sm I and Sm II (DeVito, 1978). Fibers have further been traced from the *pretectal area* (Carpenter and Pierson, 1973; Benevento, Rezek, and Santos-Anderson 1977) and the *superior colliculus* in the monkey (see Mathers, 1971; Benevento and Fallon, 1975; Benevento and Rezak, 1976) and in other species, such as the cat (see Graham, 1977), tupaia (Harting, Hall, Diamond, and Martin, 1973), and the gray squirrel (Robson and Hall, 1977; see this paper for references). There are probably some afferents from the *retina* (see Campos-Ortega, Hayhow, and Clüver, 1970; Berman and Jones, 1977); and a projection from the *striate cortex* has been shown by several authors (see Campos-Ortega and Hayhow, 1972; Kawamura, Sprague, Niimi, 1974). It is of interest that there is anatomical as well as physiological evidence for a visuotopic representation in the pulvinar of the monkey (in the division referred to as the inferior pulvinar; see for example, Campos-Ortega and Hayhow, 1972; Kawamura, Sprague, and Niimi, 1974). Some other afferents have been reported, including some from the *spinal cord* (DeVito and Simmons, 1976). Rather surprising is the indication that there are some afferents from the *hypothalamus* (HRP studies in the cat; Yoshii, Fujii, and Mizokami, 1978). The afferents to the LP in part correspond to those to the pulvinar. Thus the LP has been found to receive afferents from the superior colliculus (see Benevento and Fallon, 1975). See Chapter 2 for further information.

It can be seen from the description above that the pulvinar, in addition to receiving fibers from the temporoparietal cortex, receives afferents from structures that receive visual input, such as the superior colliculus, the pretectum, and (in some species at least) the striate cortex. This applies particularly to the "inferior" pulvinar. The relation of the pulvinar to (complex) visual functions is likewise brought out in physiological studies, as is the case for the parieto–temporo–occipital association cortex to which it projects (see above).

Responses to visual stimuli have been recorded in monkeys in a great number of the cells of the pulvinar (and in a lesser proportion of cells in the LP). The electrophysiological properties of the responding units resemble those of cells in the superior colliculus and the association cortex which respond to visual stimuli (see Mathers and Rapisardi, 1973; Robson and Hall, 1977; for recent studies in the cat, see Chalupa and Fish, 1978; Mason, 1978). In monkeys destruction of the pulvinar has been found to produce loss of retention in visual discrimination problems and some other difficulties related to vision.

In man, disturbances of complex visual functions appear to have been little studied following affections of the pulvinar, whereas such disturbances have been observed following lesions of the parietal and temporal lobes (see below) in rather good agreement with findings in monkeys, to which reference was made above. No information can be obtained from animal studies concerning speech. However, the data on the close relation between the pulvinar and the parieto–temporo–occipital association cortex found in monkeys are in agreement with the observations in man, referred to above, that the pulvinar, and possibly in part the LP, is functionally closely linked with the "posterior speech cortex." [47]

The parietal lobe. Since particular symptoms have been correlated with lesions of the parietal lobe, this will be considered under a separate heading. In Chapter 2 the *sensory symptoms* resulting from irritation or lesions of the anterior parts of the parietal lobe (the postcentral gyrus and the superior parietal lobule) were discussed at some length. Among other symptoms which may be seen in lesions of the parietal lobe (considered fully in Critchley's monograph, 1953) mention is made of the occurrence of *muscular wasting in the contralateral half of the body*. Usually the atrophy is most marked in the upper limb. It may occur in slowly developing lesions (tumors, atrophies) and in acute cerebrovascular affections. There may be an accompanying paresis with some spasticity, which is often explained as being due to concomitant involvement of the precentral region. No satisfactory explanation of the muscular atrophy following lesions of the parietal lobe has so far been given, but its appearance may be of some diagnostic value. *Ataxia* of "sensory" type is often seen as a consequence of parietal lobe lesions.

Symptoms indicating derangement of more complex functions, involving somatic as well as psychic elements, have often been observed following lesions of the parietal lobe, especially the right. The literature on this subject is almost as confusing and the terminology as manifold as is the case for aphasia. Some of these disturbances are collectively designated as *agnosias* (lack of awareness), and a number of different types have been described. The common feature is a defective recognition of sensory impressions. There is a neglect of stimuli impinging upon the opposite half of the body and in the opposite half of the visual space. In general it appears that deficits are especially marked in tasks which require appreciation of spatial relations.

[47] As mentioned above, the pulvinar consists of different subnuclei, and the various contingents of its connections differ more or less with regard to their sites of termination or origin, a question not considered in the highly simplified presentation above. For example, it appears that the division most closely related to visual functions is the inferior nucleus of the pulvinar, while the lateral and the medial pulvinar are more closely related to the parietotemporal association cortex. Much remains before clear concepts can be formulated on the function of the pulvinar and its collaboration with the cortex in speech and other complex functions.

In *visual agnosia* the vision of the patient may be intact, but he fails to recognize known objects or persons. In *tactile agnosia* the patient lacks the power of synthesizing and associating the sensations experienced, for example, when handling an object with his eyes closed. In spite of normal sensation and absence of astereognosia (the patient is, for example, able to decide if the object is rough or smooth), he cannot realize which object he is holding. More rarely so-called *finger agnosia* is seen. Here the patient is unable to name his fingers or to indicate individual fingers on request and often has difficulties in distinguishing right from left. Even stranger is a condition called *anosognosia* (lack of awareness of disease) where the patient, for example, does not recognize that his limbs (usually the left) are paralyzed, or denies that he is blind (Anton's syndrome). Relatively common in affections of the parietal lobe is apraxia. Different types have been described. One is a condition called *constructional apraxia,* where in its simplest form the patient has difficulties ''in putting together one-dimensional units so as to form two-dimensional figures or patterns'' (Critchley, 1953, p. 172). This may become evident when the patient tries to copy a drawing.[48] Many variations of this disorder have been described, and it is considered by many to be closely related to agnostic disturbances. Constructional apraxia as well as agnostic disorders have been considered also as elements in a disturbance of the ''body scheme'' or ''body image.'' These terms refer to the individual's awareness of his own body and of his spatial relationships to the surroundings. The patient may have difficulties in route-finding and orientation. It scarcely needs stressing that the study of these complex functions requires special and often rather complex tests, if conclusive results are to be obtained. For a survey of parietal lobe syndromes, see Jewesbury (1969).

The preceding list of disturbances which may be seen in lesions of the parietal lobe is far from complete. There has been much discussion of whether the particular symptoms and symptom constellations are variants of disturbances of a common basic function or represent separate entities. For example, it has been held by many that finger agnosia, being a component of Gerstmann's syndrome,[49] is a specific disturbance caused by a lesion of the angular gyrus of the parietal lobe. This view appears to be based on very meager evidence (see Critchley, 1966). Likewise, for example, the disorder called ''visual agnosia'' is not as specific as was previously assumed. The inability to identify objects visually is usually combined with some degree of mental defect, and the brain damage does not appear to be restricted to a specific region (see Bender and Feldman, 1972). Further points of discussion have been the relations between agnostic and aphasic disorders, and the question whether the symptoms mentioned above occur only with lesions of the right parietal lobe (in the nondominant hemisphere) as advocated by some authors. In several agnostic disorders there is obviously a defect in symbolic thinking, closely related to language functions. In general it appears that disorders of ''spatial thought,'' such as anosognosia and constructional apraxia are most often observed following affections of the right parietal lobe. (Finger agnosia, however, appears to occur with almost equal frequency in lesions of the left and right parietal lobes.) From a detailed study of 17 patients subjected to localized ablations of the parietal lobe, Hécaen, Penfield, Bertrand, and Malmo (1956) concluded that agnosia of various kinds (including apractognosia) and disturbances in the body scheme appeared only when the ablation involved the lower and posterior parts of the parietal lobe and the upper part of the temporal lobe. The area coincides approximately with the ''posterior speech area'' (see Fig.

[48] The term *apractognosia* appears to be somewhat more extensive and to cover defects in practical tasks (dressing, etc.) as well as ideational difficulties.

[49] A fully developed Gerstmann's syndrome as generally defined includes, in addition to finger agnosia, dysgraphia, dyscalculia, and right–left disorientation (see Critchley, 1966).

12-10), but while the latter is regularly found in the dominant (usually the left) hemisphere, the other disturbances are most often observed following ablations of the right parietal lobe. They are then contralateral. However, according to Hécaen et al., corresponding ablations of the left parietal lobe are followed by bilateral disturbances in gnostic functions and topographical relationships (body scheme), in addition to aphasic disturbances. It has been argued by some that preponderance of such disturbances observed in patients with right parietal lesions is due to the fact that the presence of an aphasia in left-sided lesions obscures agnostic disturbances or at least make them difficult to notice, unless especially searched for. This view receives some support from a study by Weinstein, Cole, Mitchell, and Lyerly (1964), who further found that in lesions of the left parietal lobe there is usually an inverse relationship between the degree of aphasia and anosognosia.

The many and complex questions touched upon are far from solved. On the whole, the symptomatology of parietal lesions in man appears to be compatible with evidence of the function of the parietal lobe obtained in experimental studies in monkeys (see above). However, the information which can be obtained from such studies concerns only major functional features. The functional tasks of the parietal lobe are obviously far more complex in man than in monkeys. One feature that illustrates this is the lateralization of function in the human brain. With regard to this, however, and as referred to above, the distinction between a dominant (major) and a nondominant (minor) hemisphere appears to be less definite than was formerly believed. The observations available may suggest that a collaboration between the two hemispheres, particularly between the parietal lobes, is of major importance for the functions. Further support for this view comes from recent studies of the functional importance of the commissural connections of the cerebral cortex.

Functions of the corpus callosum. As described in a previous section, the corpus callosum is the main commissural connection between the two neocortices. The phylogenetic increase in the size of the neocortex in mammals is followed by a parallel increase in the size of the corpus callosum. It seems a likely assumption that the main function of this great fiber mass would be to secure cooperation between the two cerebral hemispheres, and that this function should be especially important in man, but until recently little was known of the function of the corpus callosum. It has been known for a long time that in some individuals the corpus callosum is not developed (agenesis of the corpus callosum). Loeser and Alvord Jr. (1968) studied the brains of 12 patients with this anomaly. A complete account of all published cases of agenesis of the corpus callosum and of 33 of their own cases has been given by Unterharnscheidt, Jachnik, and Gött (1968). The condition can now be diagnosed during life by air encephalography. Although in some such cases there is evidence of other developmental deficits, usually there are no overt neurological symptoms (see Ettlinger, Blakemore, Milner, and Wilson, 1974, for a recent study). In patients whose corpus callosum has been sectioned in attempts to treat spread of epileptic convulsions, no marked behavioral deficits were observed (see, for example, Smith and Akelaitis, 1942; Bridgman and Smith, 1945). Observations of this kind led some to conclude that the corpus callosum could scarcely be of great importance for the functioning of the human brain. It is now clear that the reason for the negative findings is that appropriate tests were not employed to unravel the function of the corpus callosum.[50]

The first conclusive evidence of the functional role of the corpus callosum came from the *experimental studies* in cats and monkeys by Myers, Sperry, and others and concerned visual functions. Advantage was taken of the presence of crossed as well as uncrossed connections in the visual pathways (see Chap. 8). Following transection of the optic chiasm, visual impulses from one eye are transmitted to the ipsilateral striate area only. With one eye occluded the cats were trained to discriminate between different figures, for example, a square and a circle (Myers, 1956). This discrimination must be due to processes occurring in the hemisphere ipsilateral to the eye used. If this eye is then covered, and the other eye used, the animals can discriminate equally well, even if the sensory input is then only to the other hemisphere which had so far not been used. Obviously, there must be a highly developed transmission of visual experiential information between the two hemispheres. If, in a cat with transected optic chiasm, the corpus callosum and anterior commissure are also transected the situation is completely different. The animal is not able to recall through one eye the tasks learned through the other eye. If it is trained subsequently to use the second eye, no evidence of benefit from the previous training of the first eye is found. Obviously, the commissural connections have been responsible for transmitting the information related to visual pattern discrimination from the hemisphere receiving the information to the other. Similar results have been obtained in studies of visually guided behavior in monkeys with transection of the commissures and the optic chiasm (Downer, 1959).

The importance of the corpus callosum for information transmission has been demonstrated also in other spheres. Myers and Henson (1960) studied tactuokinesthetic learning in chimpanzees. A chimpanzee with its corpus callosum and anterior commissure transected is taught to solve a latch box problem by using only one of its hands. If the same problem is later presented to it in a situation which requires that it use the other hand for sensory information the chimpanzee does not know how to solve the problem. In contrast, under the same conditions a normal animal will immediately solve the problem equally well with the other hand. Again it appears that the information of the test situation has not been transmitted to the opposite hemisphere. A number of studies on "splitbrain" monkeys have been performed in recent years in order to get a better understanding of the interhemispheric transfer. Other approaches have been tried as well. By making use of the inactivation of the cortex produced by "spreading cortical depression," blocking of one hemisphere during training can be obtained. From studies of this kind in rats, Russell and Ochs (1963) concluded that a memory trace may be restricted to the "trained" hemisphere for two weeks or more. Transfer to the other hemisphere occurred, however, if the animal was allowed one reinforced response with bilateral cortical function.

In agreement with what is known of the anatomy of the commissural connec-

---

[50] For some data see *Interhemispheric Relations and Cerebral Dominance*, 1962; *Functions of the Corpus Callosum*, 1965; Gazzaniga, 1970. Relevant papers may further be found in *Hemisphere Function in the Human Brain*, 1974; *Lateralization in the Nervous System*, 1977; *Evolution and Lateralization of the Brain*, 1977.

tions (see p. 803), different regions of the corpus callosum are involved in the transfer of visual and somatosensory information. The fibers responsible for transfer of information of visual experience are found posteriorly. Section of the posterior part of the corpus callosum in cats prevents the transfer, while following section of the anterior part the animals react as normal ones in these tests (Myers, 1959). From largely corresponding experiments in monkeys (see Gazzaniga, 1970), it appears that visual transfer may involve to some extent also more anterior parts of the corpus callosum. In the monkey, transfer of learned tactual discrimination appears to occur via the posterior part of the trunk of the corpus callosum (Myers and Ebner, 1976; see, however, Hunter, Ettlinger, and Maccabe, 1975). Concerning the cortical regions involved, it is of relevance to recall the anatomical distribution of the commissural fibers (see Fig. 12-5). The striate area, the hand region, and to a lesser extent the foot region, of the primary sensorimotor area and the primary acoustic area appear to be devoid of commissural connections. Thus, the transfer must occur between other areas, presumably first and foremost those adjoining the specific sensory areas assumed to be concerned in more complex perceptual processes. What is transferred via commissural fibers cannot be elementary sensory information. It appears a likely assumption that the interhemispheric "cross talk" involves other commissural fibers in addition to those from the areas immediately surrounding the specific sensory areas. This conclusion receives some support from other experimental findings as well as from observations in man (see Geschwind, 1965).

Possibilities for the study of *interhemispheral relationships in man* have been greatly increased following the introduction of transection of the corpus callosum as a therapeutic measure in certain cases of epilepsy. Several investigations on such patients have been performed in recent years. In addition to confirming the findings in animals, human studies have given some new information on problems concerning the function of the cortex in language mechanisms and on the relative role played by the right and left hemispheres in these functions. Studies of this kind require very elaborate tests, and conclusions must be drawn with caution. Only a few data will be mentioned. Gazzaniga, Bogen, and Sperry (1965), studying two patients with complete transection of the commissures, concluded that "activities that involved speech and writing were well preserved, but only insofar as they could be governed by the left hemisphere" (loc. cit. p. 236). In a further study Gazzaniga and Sperry (1967) found, in agreement with the above statement, that sensory information which entered the left hemisphere could be reported through speech and writing without difficulty. But the patients were totally unable to give accurate spoken or written reports for even the simplest kind of sensory information reaching the right hemisphere. They were, for example, unable to name objects palpated with the left hand. These studies thus are in agreement with conclusions made from stimulation and ablation experiments of the parietal lobe, that *verbal expression* is a matter almost exclusively of the left hemisphere. The same appears to be the case for calculation. As concerns *comprehension of language,* Gazzaniga and Sperry (1967) found evidence for its being related to both hemispheres, in contrast to observations by others. For other interesting observations the reader is referred to their paper and to the publications listed in footnote 50.

There seems to be agreement among those studying the subject that section-ing of the corpus callosum is not followed by changes in intellect, behavior, or emotions. The condition is briefly characterized as follows by Sperry (1966, p. 299): "Everything we have seen so far indicates that the surgery has left these people with two separate minds, that is, two separate spheres of consciousness. What is experienced in the right hemisphere seems to be entirely outside the realm of awareness of the left hemisphere. This mental division has been demonstrated in regard to perception, cognition, volition, learning, and memory. One of the hemispheres, the left, dominant or major hemisphere, has speech and is normally talkative and conversant. The other, the minor hemisphere, however, is mute or dumb, being able to express itself only through nonverbal reactions." While overt intellectual deficits are usually not noted in patients subjected to commissurotomy, some slighter disturbances have been found on close examination. Several authors have reported that there appears to be a loss of sustained attention. Sperry (1974) had the impression that "their overall mental potential is affected." Dimond (1976) in a special study of this problem found a "gross depletion of attentional capacity" and suggested that this is due to a defective hemispheric cooperation. Further, it appears that patients subjected to commissurotomy show marked and persistent difficulties with short-term memory (Zaidel and Sperry, 1974). Dis-orders of memory ("autopragmatic amnesia") and a general simplification of lan-guage were observed by Dimond, Scammell, Brouwers, and Weeks (1977) in a patient whose corpus callosum was transected in the central section (trunk) only.

The question whether each hemisphere in these split-brain patients "has its own consciousness" is likely to interest philosophers. As mentioned above, this question has been much debated, but is still unsolved. For a review see Zangwill, 1974. (See also Gazzaniga, 1970; Popper and Eccles, 1977; Pucetti, 1977.) However, it appears from the studies undertaken that in daily life there is only rarely evidence of conflict between the two hemispheres. Experimental situa-tions may, however, be arranged where this can be observed. Animal studies have likewise shown that the two hemispheres may learn different tasks, which may be performed simultaneously under appropriate experimental conditions. In the nor-mal brain it appears that any information reaching one hemisphere is regularly communicated to the other, largely to corresponding regions. Some of the findings made in patients with transected commissures indicate that the distinction between the two hemispheres with regard to language functions and body scheme may not be as marked as was previously assumed. Some of the aphasic disturbances may perhaps be easier to understand in the light of the new findings. Geschwind (1965), who discussed these problems, stressed the importance of intracortical connections between the regions involved in these functions, in addition to the commissural ones. He has coined the name "disconnection syndromes" as a com-mon denominator for "the effects of lesions of association pathways, either those which lie exclusively within a single cerebral hemisphere or those which join the two halves of the brain" (loc. cit., p. 242). More knowledge of the anatomy of the connections is greatly needed. Even with present knowledge, however, it ap-pears from the analysis of Geschwind (1965) that an explanation may be found for the occurrence of similar functional disturbances following differently situated lesions.

Recent findings indicate that at least most of the changes observed after sectioning of the corpus callosum in man are due to transection of the fibers in the posterior third. Thus Gordon, Bogen, and Sperry (1971) found that when commissurotomy is restricted to the anterior commissure and the anterior portion of the corpus callosum, highly detailed information of different sensory modalities can still be transferred between the two hemispheres. What functions are mediated by the large frontal section of the corpus callosum is so far not known.

The findings made in split-brain animals and man are of great interest for the understanding of several aspects of brain functions, such as the learning of language, as well as other kinds of learning and behavior. A vast literature has grown up on these subjects, which will not, however, be considered here. Several symposia have been devoted to discussion of these problems,[51] but much is still conjectural. It should not be overlooked in the treatment of these subjects that the cortex must collaborate with many other parts of the nervous system. Particularly, one should not minimize the emotional aspects of the functions, betrayed in everyday language by terms such as interest, motivation, etc.

Evaluation of clinical findings in lesions of the cerebral cortex. It is apparent from the preceding account that lesions of the cerebral cortex in man may result in a multitude of different symptoms: motor, sensory, autonomic, behavioral, and purely mental. Lesions of various parts of the cortex give rise to fairly characteristic symptoms, and from a knowledge of the anatomical and functional organization of the cortex the site of the lesion can in many cases be determined quite accurately.[52] If a proper and careful examination of the patient is made, this might be expected to be a relatively simple affair. Unfortunately, this is often not the case. Several circumstances contribute to this, some of which will be mentioned below.

A tumor may cause widespread secondary changes in other parts of the brain, which may progress even after the tumor has been removed. In the case of the surgical removal of a malignant tumor it is difficult to exclude the possibility that some of the mass is left and that this may be responsible for some of the symptoms. Secondary, even remote, bleedings may occasionally occur, and damage may be inflicted on subcortical structures. At operation it is not always easy to be sure which gyri are removed and which have been left, on account of the deformation of the convolutional pattern, edema, and dislocation of the brain which so frequently occur. With regard to mental changes, their evaluation is invalidated by the fact that, as a rule, comparable data from the patient's normal life before the onset of disease are lacking. Some information may be gained only concerning major behavioral features. These factors urge that the utmost care should be taken in interpreting certain symptoms as due to removal of a particular part of the cortex. For the same reason anatomical inferences—often too lightheartedly drawn

[51] See, for example, *Brain Mechanisms and Learning,* 1961; *Brain and Conscious Experience,* 1966; *Hemisphere Function in the Human Brain,* 1974; *Lateralization in the Nervous System,* 1977; *Evolution and Lateralization in the Human Brain,* 1977.

[52] A cortical lesion, particularly a tumor, will only rarely be restricted to the cortex. As a rule more or less of the underlying white matter and often even central gray masses are involved. The symptoms commonly seen in cortical affections of different locations can, therefore, not all be attributed to affections of the cortex itself.

from clinical observations—should be evaluated very critically. If reliable conclusions are to be made, a complete postmortem examination of the brain is essential. Hebb (1945) stresses these facts, and points particularly to the possibility of failure arising from the effect which the presence of a pathological process in the brain may have on the function of other parts. On the basis of the findings in a patient in whom the anterior third of both frontal lobes were removed on account of epilepsy following traumatic injury (Hebb and Penfield, 1940) and who showed moderate postoperative clinical changes, Hebb maintains that the presence of a pathological process in the brain may affect the mental state of a patient more than the absence of a part of the brain.

As has been mentioned at several occasions in other chapters, a pathological process as a rule will not be limited to certain structural and functional "units" (insofar as such are present at all), but it will affect several structures. However, it is fortunate that some diseases show a marked predilection for occurring in or affecting predominantly certain regions of the brain. This is the case for many tumors. This circumstance may allow the experienced clinician to determine not only the site of the lesion, but to calculate with a fair degree of accuracy also the nature of the disease, apart from information which the case history may yield on this point. On the other hand, it should be remembered that the lesion may be found in other places than expected. One of the classical examples is the occurrence of a frontal lobe tumor which clinically imitates a cerebellar tumor (see Chap. 5). Indeed, "false localizing signs" are not uncommon, as appears from the exhaustive account of Gassel (1961) based on observations in a series of 250 patients with intracranial meningcomas. Diffuse processes may cause diagnostic difficulties, since they may give rise to focal symptoms, in spite of being widespread. Thus some cortical atrophies may simulate the presence of a tumor (see, for example, Jackson, 1946; Fleminger, 1946). When a process starts in the so-called "mute regions," it may be without focal symptoms for a considerable time, making diagnosis difficult. It is, on the whole, important to bear in mind the frequent occurrence of secondary changes in order to avoid overestimation of "focal" signs.

The considerations above refer mainly to the problems encountered in clinical examinations of the patients. Many of the difficulties mentioned can be overcome or at least reduced when use is made of ancillary methods, such as X-ray investigations, pneumoencephalography, ventriculography, angiography, computerized axial tomography, electroencephalography, or other procedures (localization of certain tumors by intravascular injections of radioactive isotopes, and other methods). However, even with the use of all available methods, there are still cases where a precise diagnosis of the type and site of a cerebral lesion cannot be made.

In recent years a wealth of psychological test methods have been developed for studies of mental capacities, such as language functions, perception, etc. Tests of these types may give valuable diagnostic information on the lateralization of a disease process. So far, however, most valuable appears to be the information such tests are able to yield about normal brain function. For some comments on the use of psychological methods, whose results are often difficult to interpret, see Goldstein (1974).

Cortical localization. This subject has been repeatedly touched upon in the present chapter. In concluding the account of the cerebral cortex it is appropriate to consider whether any conclusions can be reached concerning the problem of cortical localization.[53] It is legitimate to assume, as those who adhere to the theory of extreme localization do, that the various cytoarchitectural areas each subserve a special function? Or does the cerebral cortex always function as a whole, as the most ardent opponents of the theory of cortical localization maintain?

Before proceeding, it is essential to devote some thought to *the concept of "function."* In the first place we should not overlook the fact that the classification of events and processes in the organism into various "functions" is arbitrary. When we speak of movements, secretion, sensory perception, etc., as functions, this is a convention adopted for practical purposes. In the living organism activity is scarcely ever restricted to one of these "man-made" categories of function. Even the slightest movement of a finger is accompanied by, for example, vascular alterations in it, and by stimulation of sensory receptors. At the other extreme, in behavioral reactions a multitude of different functions are always integrated, such as motor, secretory, cardiovascular, respiratory, alterations in the sensitivity of sensory mechanisms, etc. As one passes from more elementary functions to the more complex ones it becomes increasingly impossible to relate a particular part of the nervous system (a nucleus or a fiber tract) to a particular function. Examples of this have been mentioned in several chapters of this book, and the fallacies inherent in considering a particular region of the central nervous system as a "center" for one or another "function" have repeatedly been emphasized. The fact that the stimulation or ablation of a brain region gives rise to alterations in certain functions should not induce us to forget that what it is possible to observe in the experimental situation are only fragments of the totality of changes which actually take place. These must of necessity be intimately linked in order to secure an integrated reaction from the organism, adapted to the incessant changes in the organism itself and in its surroundings.

As concerns the cerebral cortex we have seen that *the classical conception of a clear-cut functional localization cannot be upheld.* For example, no clear distinction can be made between "motor" and "somatosensory" cortical areas. The striate area is not purely sensory, since it gives rise to pathways which mediate ocular movements. Damage to rather extensive cortical regions may result in disturbances of the extremely complex language functions. Particular cortical areas cannot be related to specific "functions," when this word is taken in the usual sense. On the other hand, certain cortical areas are more or less closely related to some of the functions. Indeed, recent research has made it clear that the functional role of, for example, the three somatosensory cortical regions is not identical; areas 17, 18, and 19 are not equivalent in the function of vision.

---

[53] A vast literature has grown up on this subject. The comments made in the following do not pretend to be more than a brief statement of some principal points of view. For this reason definition of the many terms employed, such as localization, activity, etc., which should rightly precede any discussion of a subject like this, is omitted. For a recent review on the problem of cortical localization, see Powell (1973).

The parcellation of the cerebral cortex into a number of cytoarchitecturally different areas has been criticized severely by many students (see for example, Lashley and Clark, 1946), and it is probably rightly said that the parcellations have been overdone by some. It is further obvious that the borders between areas (with a few exceptions, the striate area being the most notable) are not sharp. With these qualifications, however, the principle of a parcellation of the cortex into minor regions which differ structurally in certain respects is sound. The cytoarchitectonic differences between areas parallel other structural features, as described previously in this and other chapters. Where recent studies of fiber connections of the cortex have been performed accurately, they have demonstrated the correctness of this view, with the above qualification that the borders are not sharp. As examples of this, one may mention the afferent and efferent projections of the first sensorimotor region considered in Chapters 2 and 4, of the visual areas 17, 18, and 19 (Chap. 8), and of the thalamocortical and corticothalamic projections. Correspondingly, as has been seen, the terminal aborizations of cortical association fibers and of commissural fibers adhere to the cyto-architectonic patterns. With increasingly precise studies, more and more details are brought to light supporting the validity of the concept of a cortical morphological parcellation. It was mentioned in a preceding section of this chapter that the internal organization of the radially oriented cortical "units" differs between areas, and that the total number of boutons differs. The presence of chemical differences should also be mentioned. On almost every point where the problem has been investigated, even neighboring cortical areas have also been found to differ physiologically in certain respects (for example areas 3, 1, and 2 in the sensorimotor cortex).

When taken together, the available anatomical information, although far from complete, strongly supports the idea that *the various cortical areas differ with regard to their finer structural organization,* even if they all present variations of a common pattern. This leads to the assumption supported by an increasing number of physiological observations, that the cortical areas have different tasks, or different functions, provided one takes care not to use the word function in the usual sense (movement, sensation, etc.). In any of the latter functions in which the cortex is involved, a number of cortical areas must participate in an extensive mutual collaboration, and they must collaborate with subcortical structures. If considered in this way, and if account is taken of the different connotations of the word "function," there appears to be no real incompatibility between the view that the cortex is a mosaic of units, each with its particular function, and the holistic view of the cortex and the brain functioning as a whole. The problem, as so many others, is reduced to one of semantics.

Epilogue
Some General Considerations

THE READER WHO has had the patience to follow the account given in the preceding twelve chapters may have been impressed, overwhelmed, and perhaps frustrated by the amount of anatomical detail presented in the text, even though in many places the anatomical features have been treated rather summarily. Some questions have probably arisen in the minds of many readers: Do we need all this detailed anatomy? What is the use of it? Can it help us in the diagnosis and treatment of disorders of the nervous system? Some general comments bearing on these questions may be appropriate.

As we have seen, by combining various methods of study it is today possible to determine the anatomical organization of the brain in rather great detail, but in spite of considerable advances in recent years there are still vast areas of the brain which are insufficiently known. As progress has been made, one feature has become increasingly clear, and will have struck any student of the subject: *In the anatomical organization of the central nervous system there is an extremely high degree of order. This is true even with regard to the most minute structural features*. Many examples have been described in the text. A particular cell type may present a regular topographic distribution as well as a highly specific arrangement with regard to its dendritic and axonal processes. Strict topographical localization is found in the interconnections between many nuclei as well as in the associational and commissural connections of the cerebral cortex. Somatotopic patterns have been demonstrated in fiber projections where they were previously unknown. The various contingnents of afferent fibers to a nucleus may show a differential and precisely organized distribution within this. Afferents from different sources may establish synaptic contacts at different sites of the dendrites or perikarya of a category of nerve cells. Many varieties of synaptic structures are known, and different presynaptic formations have been shown to belong to afferent fibers of different origin. Even regions generally considered as being rather diffusely organized, such as the reticular formation and the "nonspecific" thalamic nuclei, have been shown to possess a fair degree of specificity in their anatomical organization and further research may well produce further evidence in support of this. There can be little doubt that if sufficiently meticulous studies are made, in the future more and more data will be brought forward and will demonstrate the extremely precise and orderly anatomical arrangements existing in the central nervous system.

The recent data alluded to above lead us to *consider the central nervous system as being composed of a multitude of minor units, each with its particular structural organization, specific as regards its finer intrinsic organization as well as its connections with other units.* Scarcely two regions are identically organized. If the structure of living matter is related to its function, an axiom which is generally accepted, it follows that *there are in the nervous system as many minor functional units as there are structural ones.* Provided that the term functional is not applied to "function" ("sensory," "motor," etc.) as discussed in the consideration of cortical localization (p. 847), there is nothing to contradict this view. In fact, wherever anatomical information is precise enough and sufficiently detailed physiological analyses have been performed, evidence in support of it has been found, as for example in the studies of the somatosensory fiber systems, of the vestibular nuclei, the visual system, and in other places. Studies of single unit potentials especially have in many instances shown a complete agreement with anatomical data, as concerns for example the behavior of units in response to afferent impulses from different sources. In some instances chemical and pharmacological studies have demonstrated differences between minor subdivisions of a structure, but on the whole evidence of this kind is so far relatively sparse, because a sufficiently close correlation of the findings with the structural organization has not been possible or has not been performed.

For the total behavior of the organism collaboration of the many structural and functional "units" of the nervous system is essential. This collaboration occurs by way of fiber connections between the "units." The more precisely these connections are studied, the more obvious it is that there are far more connections than previously assumed, and that they are much more multifarious and complex than was believed some decades ago. In fact, it is possible to envisage connections which may—via more or less circuitous routes—lead from almost any "unit" to almost any other "unit." While this does not permit the conclusion that these connections are all always utilized when a "unit" is active, it demonstrates the *possibilities* for a widespread and varied impulse conduction under different circumstances. In a general way it explains the well-known variability of responses to a particular stimulus, depending on the concomitant activity of other parts of the nervous system than those immediately activated by the stimulus, produced by other external or internal stimuli. Other features influencing the response are the chemical and hormonal conditions in the organism at the moment. The considerations above generally apply even to "simple" spinal reflexes, but are especially relevant with regard to more complex behavioral and mental responses to a variety of stimuli, as is well known to everyone from experiences in daily life.

The more we know of the structural and functional organization of the central nervous system, the more obvious its extreme complexity becomes. So tremendous is this complexity that there may be reason to doubt that the human brain is good enough to fully understand the human brain. Attempts have been made to make models of parts of the brain, of particular systems or fragments of it, and to apply cybernetic principles and computer analysis. While such attempts may throw light on certain aspects, it would be naïve to believe that they can really tell us much, for the very reason that they are yet too simple. They suffer from the disadvantage that a number of factors at work in the living brain are left out of consideration. The important aspect of the collaboration between "units" is unduly oversimplified, in part because too little is known of the structure and function of the parts included in the model, and in part because taking all relevant parts into consideration would amount almost to making a model of the entire brain, an obviously impossible undertaking. Attempts to analogize the human brain with electronic computers has induced some, particularly nonbiologists, to adopt a rigid mechanistic view of the brain.

Even if our prospects of really understanding our most complex organ are indeed small, there is no doubt that we may advance much further than we have so far. It is important to be aware, however, that if progress is to be made, studies of the nervous system have to be concerned with the most minute details of its function and structure. It is only by knowing these that we may hope to put bits of evidence together and to increase our understanding. As Whitfield (1967, p. 192) phrases the problem: "Each time we examine a yet smaller sub-unit of the nervous system we find it to consist of some more interconnected 'black-boxes,' and always we find that a behavioural model of the black-box makes little contribution to our understanding of the mechanism. There is no substitute for taking the lid off."

It may be added that *the investigation of the structure of the nervous system in all its detail is a prerequisite for progress in studies of its function*. Every function is carried out by a structure, and conclusions about functions without reference to the structures involved will remain incomplete and easily may lead us astray.

The view of the nervous system outlined above may seem pessimistic and may be particularly disheartening to the clinical neurologist, since it obviously runs counter to the traditional tendency to correlate a particular clinical symptom to damage to—or dysfunction of—a certain part or "system" of the brain. It is well known to critical clinical neurologists that, more often than not, the symptoms observed in a patient do not correspond to the traditional schematic presentations in the textbooks. But all too often features in the clinical picture which do not fit the schema are disregarded as being of little importance. In the preceding chapters of this book numerous examples have been mentioned of symptoms which according to traditional concepts are considered as specific to involvement of one or another part of the brain, but which have turned out to be far from "specific," the so-called pyramidal tract syndrome being perhaps the most impressive.

The complexity and multitude of interconnections in the central nervous system may help us to understand the remarkable degree of functional restitution or "compensation" which is often seen in lesions of the brain. An understanding of the complexities in the organization of the nervous system will relieve the neurologist of the "moral obligation" to make his findings fit preconceived traditional conceptions. It will encourage him to penetrate deeper in his analysis of the particular case. Even if he will often have to admit that a final conclusion as to the site and nature of the disturbance may not be possible, it will make clinical neurology more fascinating and engaging. An approach to the subject in this way will encourage observations which may increase our understanding of disturbances of the nervous system of man. The story of medicine contains many examples of how careful investigation of a single patient has given more information of value than a collective analysis of a number of cases showing (apparently) identical symptoms. If clinical neurological work in the future is to bring results of value, it is essential that the neurologist understand the major principles in the organization of the nervous system, and that he have a fair knowledge of its structure and function. In particular a knowledge of the lines of communication between the "units" of the nervous system is essential. It is also essential that improved and precise methods for the study of functional disturbances of the nervous system in man be developed. As has been seen, in recent years considerable advances have been made in this field, and additional work along this line should be encouraged.

The reader will have realized that in spite of all the research devoted to the nervous system, our knowledge is still rather limited. On almost all points there are unsolved problems. Our present-day concepts are still to a large extent based upon assumptions and hypotheses, built upon a modest body of factual knowledge. If further progress is to be made, it is important to be aware of this situation, and to make every effort to distinguish between observations and interpretations. Working hypotheses are important and necessary tools in research, but, as the history of science shows, they have a tendency to become accepted as truths and to hamper instead of promoting progress. The author has done his best to stress the relatively weak foundations upon which many of our present concepts of the central nervous system are built. Accordingly, few final conclusions can be found in this text, perhaps to the disappointment of some students who would like to have clear-cut and final statements. It is the author's hope that the present book, by emphasizing many unsolved problems, may contribute to making the reader aware of the challenge and fascination of the study of the nervous system, and to fostering a fruitful and rewarding approach to the problems which face the clinical neurologist in his daily work.

References

Abdel-Kader, G. A. (1968). The organization of the corticopontine system of the rabbit. *J. Anat. (Lond.) 102*, 165–181.

Abrahams, V. C., and P. K. Rose (1975). Projections of extraocular, neck muscle, and retinal afferents to superior colliculus in the cat: their connections to cells of origin of tectospinal tract. *J. Neurophysiol. 38*, 10–18.

Abzug, C., M. Maeda, B. W. Peterson, and V. J. Wilson (1974). Cervical branching of lumbar vestibulospinal axons. *J. Physiol. (Lond.) 243*, 499–522.

Active Touch. The Mechanism of Recognition of Objects by Manipulation. A Multi-disciplinary Approach (1978). (G. Gordon, ed.). Pergamon Press, Oxford.

Adal, M. N., and D. Barker (1965). Intramuscular branching of fusimotor fibres. *J. Physiol. (Lond.) 177*, 288–299.

Adam, J., C. D. Marsden, P. A. Merton, and H. B. Morton (1976). The effect of lesions in the internal capsule and the sensorimotor cortex on servo action in the human hand. *J. Physiol. (Lond.) 254*, 27–28P.

Adams, C. B. T., and V. Logue (1971). Studies of cervical spondylotic myelopathy. 1. Movement of the cervical roots, dura and cord, and their relation to the course of the extrathecal roots. *Brain 94*, 557–568.

Adams, J. C. (1977). Organization of the margins of the anteroventral cochlear nucleus. *Anat. Rec. 187*, 520.

Adams, J. C. (1979). Ascending projections to the inferior colliculus. *J. Comp. Neurol. 183*, 519–538.

Adams, J. C., and B. W. Warr (1976). Origins of axons in the cat's acoustic striae determined by injection of horseradish peroxidase into severed tracts. *J. Comp. Neurol. 170*, 107–122.

Adams, J. H., P. M. Daniel, and M. M. L. Prichard (1956–66). Observations on the portal circulation of the pituitary gland. *Neuroendocrinology 1*, 193–213.

Adams, J. H., P. M. Daniel, and M. M. L. Prichard (1966). Transection of the pituitary stalk in man: anatomical changes in the pituitary glands of 21 patients. *J. Neurol. Neurosurg. Psychiat. 29*, 545–555.

Adams, R. D. (1975). *Diseases of Muscle. A Study in Pathology*, 2nd ed. Harper & Row, Hagerstown, Md.

Addison, W. H. F. (1915). On the rhinencephalon of *Delphinus delphis*. *J. Comp. Neurol. 25*, 497–522.

Ades, H. W. (1944). Midbrain auditory mechanisms in cats. *J. Neurophysiol. 7*, 415–424.

Ades, H. W., and R. Felder (1942). The acoustic area of the monkey *(Macaca mulatta). J. Neurophysiol. 5*, 49–54.

Adey, W. R. (1951). An experimental study of the hippocampal connexions of the cingulate cortex in the rabbit. *Brain 74*, 233–247.

Adey, W. R. (1953). An experimental study of the central olfactory connexions in a marsupial *(Trichosurus vulpecula). Brain 76*, 311–330.

Adey, W. R., and M. Meyer (1952a). An experimental study of hippocampal afferent pathways from prefontal and cingulate areas in the monkey. *J. Anat. (Lond.) 86*, 58–74.

Adey, W. R., and M. Meyer (1952b). Hippocampal and hypothalamic connexions of the temporal lobe in the monkey. *Brain 75*, 358–384.

Adie, W. J. (1932). Tonic pupils and absent tendon reflexes: a benign disorder *sui generis;* its complete and incomplete forms. *Brain 55*, 98–113.

Adie, W. J., and M. Critchley (1927). Forced grasping and groping. *Brain 50*, 142–170.

Adinolfi, A. M., and G. D. Pappas (1968). The fine structure of the caudate nucleus of the cat. *J. Comp. Neurol. 133*, 167–184.

Adler, A. (1934). Zur Topik des Verlaufes der Geschmackssinnsfasern und anderer afferenter Bahnen im Thalamus. *Z. ges. Neurol. Psychiat. 149*, 208–220.

Adrian, E. D. (1941). Afferent discharges to the cerebral cortex from peripheral sense organs. *J. Physiol. (Lond.) 100*, 159–191.

Adrian, E. D. (1943). Afferent areas in the cerebellum connected with the limbs. *Brain 66*, 289–315.

Adrian, E. D. (1950a). Sensory discrimination. With some recent evidence from the olfactory organ. *Brit. Med. Bull. 6*, 330–332.

Adrian, E. D. (1950b). The electrical activity of the mammalian olfactory bulb. *Electroencephal. Clin. Neurophysiol. 2*, 377–388.

Advances in Pain Research and Therapy, Vol. 1, Proceedings of the First World Congress on Pain (1976). (J. J. Bonica and D. G. Albe-Fessard, eds.). Raven Press, New York.

Advances in Pain Research and Therpy, Vol. 2, International Symposium on Pain of Advanced Cancer (1979). (J. J. Bonica and V. Ventafridda, eds.). Raven Press, New York.

Advances in Sleep Research, Vol. 1 (1974) Vol. 2 (1976) (E. D. Weitzman, ed.). Spectrum, New York.

Advances in Stereoencephalotomy, III (1967). Third International Symposium on Stereoencephalotomy, Madrid (E. A. Spiegel and H. T. Wycis, eds.) *Confin. Neurol. (Basel) 29*, 65–282.

Afifi, A., and W. W. Kaelber (1964). Efferent connections of the substantia nigra in the cat. *Exp. Neurol. 11*, 474–482.

Agarashi, M., J. L. Cranford, Y. Nakai, and B. R. Alford (1979). Behavioral auditory function after transsection of crossed olivo-cochlear bundle in the cat. IV. Study on pure-tone frequency discrimination. *Acta Otolaryng. (Stockh.) 87*, 79–83.

Aggression and Defense. Neural Mechanisms and Social Patterns, Brain Function, Vol. V (1967). (C. D. Clemente and D. B. Lindsley, eds.). Univ. of California Press, Berkeley and Los Angeles, Calif.

Aghajanian, G. K., and D. W. Gallager (1975). Raphe origin of serotonergic nerves terminating in the cerebral ventricles. *Brain Research 88*, 221–231.

Aghajanian, G. K., and R. Y. Wang (1977). Habenular and other midbrain raphe afferents demonstrated by a modified retrograde tracing technique. *Brain Research 122*, 229–242.

Aitken, J. T., and J. E. Bridger (1961). Neuron size and neuron population density in the lumbosacral region of the cat's spinal cord. *J. Anat. (Lond.) 95*, 38–53.

Aitkin, L. M., D. W. Blake, S. Fryman, and G. R. Bock (1972). Responses of neurons in the rabbit inferior colliculus. II. Influence of binaural tonal stimulation. *Brain Research 47*, 91–101.

Aitkin, L. M., and J. Boyd (1978). Acoustic input to the lateral pontine nuclei. *Hearing Res. 1*, 67–77.

Aitkin, L. M., and W. R. Webster (1972). Medial geniculate body of the cat: organization and responses to tonal stimuli of neurons in ventral division. *J. Neurophysiol. 35,* 365–380.

Ajmone Marsan, C. (1965). The thalamus. Data on its functional anatomy and on some aspects of thalamocortical integration. *Arch. Ital. Biol. 103,* 847–882.

Akert, K. (1964). Comparative anatomy of frontal cortex and thalamofrontal connections. In *The Frontal Granular Cortex and Behavior* (J. M. Warren and K. Akert, eds.), pp. 372–394. McGraw-Hill, New York.

Akert, K. (1965). The anatomical substrate of sleep. In *Sleep Mechanisms, Progress In Brain Research,* Vol. 18 (K. Akert, C. Bally, and J. P. Schadé, eds.), pp. 9–19. Elsevier, Amsterdam.

Akert, K., and B. Andersson (1951). Experimenteller Beitrag zur Physiologie des Nucleus caudatus. *Acta Physiol. Scand. 22,* 281–298.

Akert, K., R. A. Gruesen, C. N. Woolsey, and D. R. Meyer (1961). Klüver-Bucy syndrome in monkeys with neocortical ablations of temporal lobe. *Brain 84,* 480–498.

Akert, K., C. N. Woolsey, I. T. Diamond, and W. D. Neff (1959). The cortical projection area of the posterior pole of the medial geniculate body in *Macaca mulatta. Anat. Rec. 133,* 242.

Akil, H., and D. J. Mayer (1972). Antagonism of stimulation-produced analgesia by p-CPA, a serotonin synthesis inhibitor. *Brain Research 44,* 692–697.

Albe-Fessard, D. (1968). Central nervous mechanisms involved in pain and analgesia. In *Pharmacology of Pain,* Vol 9. Proceedings 3rd Internal Pharmacological Meeting 1966, pp. 131–168. Pergamon Press, Oxford.

Albe-Fessard, D., G. Arfel, and G. Guiot (1963). Activités électriques caractéristiques de quelques structures cérébrales chez l'homme. *Ann. Chir. 17,* 1185–1214.

Albe-Fessard, D., G. Arfel, G. Guiot, P. Derome, and G. Guilbaud (1967). Thalamic unit activity in man. *Electroenceph. Clin. Neurophysiol. Suppl. 25,* 132–142.

Albe-Fessard, D., G. Arfel, G. Guiot, P. Derome, E. Hertzog, G. Vourc'h, H. Brown, P. Aleonard, J. de la Herran, and J. C. Trigo (1966). Electrophysiological studies of some deep cerebral structures in man. *J. Neurol. Sci. 3,* 37–51.

Albe-Fessard, D., J. Boivie, G. Grant, and A. Levante (1975). Labelling of cells in the medulla oblongata and the spinal cord of the monkey after injections of horseradish peroxidase in the thalamus. *Neurosci. Letters 1,* 75–80.

Albe-Fessard, D., and D. Bowsher (1965). Responses of monkey thalamus to somatic stimuli under chloralose anaesthesia. *Electroenceph. Clin. Neurophysiol. 19,* 1–15.

Albe-Fessard, D., and J. Delacour (1968). Notions anatomo-physiologiques sur les voies et les centres d'intégration des messages douloureux. *J. Psychol. Norm. Path. 65,* 1–44.

Albe-Fessard, D., G. Guiot, Y. Lamarre, and G. Arfel (1966). Activation of thalamocortical projections related to tremorogenic processes. In *The Thalamus* (D. P. Purpura and M. D. Yahr, eds.), pp. 237–249. Columbia Univ. Press, New York.

Albe-Fessard, D., A. Levante, and Y. Lamour (1974). Origin of spinothalamic and spinoreticular pathways in cats and monkeys. *Advan. Neurol. 4,* 157–166.

Albe-Fessard, D., and J. Liebeskind (1966). Origine des messages somato-sensitifs activant les cellules du cortex moteur chez le singe. *Exp. Brain Res. 1,* 127–246.

Albe-Fessard, D., J. Massion, R. Hall, and W. Rosenblith (1964). Modifications au cour de la veille et du sommeil des valeurs moyennes de réponses nerveuses centrales induites par des stimulations somatiques chez le Chat libre. *C. R. Acad. Sci. (Paris) 258,* 353–356.

Albus, K., and F. Donate-Oliver (1977). Cells of origin of the occipito-pontine projection in the cat: functional properties and intracortical location. *Exp. Brain Res. 28,* 167–174.

Aldskogius, H. (1974). Indirect and direct Wallerian degeneration in the intramedullary root fibres of the hypoglossal nerve. An electron microscopical study in the kitten. *Advan. Anat. Embryol. Cell Biol. 50,* 1–78.

Aleksic, S. N., and A. E. George (1973). Pure motor hemiplegia with occlusion of the extracranial carotid artery. *J. Neurol. Sci. 19,* 331–339.

Alema, G., L. Perria, G. Rosadini, G. F. Rossi, and J. Zattoni (1966). Functional inactivation of the human brain stem related to the level of consciousness. *J. Neurosurg. 24,* 629–639.

Alema, G., and G. Rosadini (1964). Données cliniques et EEG de l'introduction d'amytal sodium dans la circulation encephalique concernant l'état de conscience. *Acta Neurochir. 12,* 240–257.

Alexander, L. (1942a). The vascular supply of the strio-pallidum. *Res. Publ. Ass. Res. Nerv. Ment. Dis. 21,* 77–132.

Alexander, L. (1942b). The fundamental types of histopathological changes encountered in cases of athetosis and paralysis agitans. *Res. Publ. Ass. Res. Nerv. Ment. Dis. 21,* 334–493.

Alexander, R. S. (1946). Tonic and reflex functions of medullary sympathetic cardiovascular centers. *J. Neurophysiol. 9,* 205–217.

Alexander, W. F. (1940). The innervation of the biliary system. *J. Comp. Neurol. 72,* 357–370.

Alksne, J. F., T. W. Blackstad, F. Walberg, and L. E. White, Jr. (1966). Electron microscopy of axon degeneration: a valuable tool in experimental neuroanatomy. *Ergebn. Anat. Entwickl.-Gesch. 39,* 1–32.

Allen, G. I., C. B. Azzena, and T. Ohno (1972). Contribution of the cerebro-reticulo-cerebellar pathway to the early mossy fibre response in the cerebellar cortex. *Brain Research 44,* 670–675.

Allen, G. I., C. B. Azzena, and T. Ohno (1974). Somatotopically organized inputs from fore- and hindlimb areas of sensorimotor cortex to cerebellar Purkyně cells. *Exp. Brain Res. 20,* 255–272.

Allen, G. I., P. F. C. Gilbert, R. Marini, W. Schultz, and T. C. T. Yin (1977). Integration of cerebral and peripheral inputs by interpositus neurons in monkey. *Exp. Brain Res. 27,* 81–99.

Allen, G. I., H. Korn, and T. Oshima (1975). The mode of synaptic linkage in the cerebro-ponto-cerebellar pathway of the cat. I. Responses in the brachium pontis. *Exp. Brain Res. 24,* 1–14.

Allen, G. I., H. Korn, T. Oshima, and K. Toyama (1975). The mode of synaptic linkage in the cerebro-ponto-cerebellar pathway of the cat. II. Responses of single cells in the pontine nuclei. *Exp. Brain Res. 24,* 15–36.

Allen, G. I., T. Oshima, and K. Toyama (1977). The mode of synaptic linkage in the cerebro-ponto-cerebellar pathway investigated with intracellular recording from pontine nuclei cells of the cat. *Exp. Brain Res. 29,* 123–136.

Allen, G. I., and N. Tsukahara (1974). Cerebro-cerebellar communication systems. *Physiol. Rev. 54,* 957–1006.

Allen, M. W. Van (1967). Transient recurring paralysis of ocular abduction. A syndrome of intracranial hypertension. *Arch. Neurol. (Chic.) 17,* 81–88.

Allen, W. F. (1924). Localization in the ganglion semilunare in the cat. *J. Comp. Neurol. 38,* 1–25.

Allen, W. F. (1940). Effect of ablating the frontal lobes, hippocampi and occipito-parieto-temporal (excepting pyriform areas) lobes on positive and negative olfactory conditioned reflexes. *Am. J. Physiol. 128,* 754–771.

Allen, W. F. (1941). Effect of ablating the pyriform-amygdaloid areas and hippocampi on positive and negative olfactory conditioned reflexes and on conditioned olfactory differentiation. *Am. J. Physiol. 132,* 81–91.

Alley, K., R. Baker, and J. I. Simpson (1975). Afferents to the vestibulo-cerebellum and the origin of the visual climbing fibers in the rabbit. *Brain Research 98,* 582–589.

Allison, A. C. (1953a). The structure of the olfactory bulb and its relationship to the olfactory pathways in the rabbit and the rat. *J. Comp. Neurol. 98,* 309–353.

Allison, A. C. (1953b). The morphology of the olfactory system in the vertebrates. *Biol. Rev. 28,* 195–244.

Allison, A. C. (1954). The secondary olfactory areas in the human brain. *J. Anat. (Lond.) 88,* 481–488.

Allison, A. C., and R. T. T. Warwick (1949). Quantitative observations on the olfactory system of the rabbit. *Brain 72,* 186–197.

Altman, J. (1962). Some fiber projections to the superior colliculus in the cat. *J. Comp. Neurol. 119,* 77–95.

Altman, J., and S. A. Bayer (1977). Time of origin and distribution of a new cell type in the cerebellar cortex. *Exp. Brain Res. 29,* 265–274.

Altman, J., and M. B. Carpenter (1961). Fiber projections of the superior colliculus in the cat. *J. Comp. Neurol. 116,* 157–178.

Altman, J. A., N. N. Bechterev, E. A. Radionova, G. N. Shmigidina, and J. Syka (1976). Electrical responses of the auditory area of the cerebellar cortex to acoustic stimulation. *Exp. Brain Res. 26,* 285–298.

Alvarado, J. A., and C. Van Horn (1975). Muscle cell types of the cat inferior oblique. In *Basic Mechanisms of Ocular Motility and their Clinical Implications* (G. Lennerstrand and P. Bach-y-Rita, eds.), pp. 15–45. Pergamon Press, Oxford.

Alvarado-Mallart, R. M., C. Batini, C. Buisseret, J. P. Gueritaud, and G. Horcholle-Bossavit (1975). Mesencephalic projections of the rectus lateralis muscle afferents in the cat. *Arch. Ital. Biol. 113,* 1–20.

Amaral, D. G. (1978). A Golgi study of cell types in the hilar region of the hippocampus in the rat. *J. Comp. Neurol. 182,* 851–914.

Amassian, V. E. (1961). Microelectrode studies of the cerebral cortex. *Int. Rev. Neurobiol. 3,* 67–136.

Amato, G., V. La Grutta, and F. Enida (1969). The control exerted by the auditory cortex on the activity of the medial geniculate body and inferior colliculus. *Arch. Sci. Biol. (Bologna) 53,* 291–313.

Amici, R., G. Avanzini, and L. Pacini (1976). *Cerebellar Tumors. Clinical Analysis and Physiopathologic Correlations.* S. Karger, Basel.

Amoore, J. E. (1970). Computer correlation of molecular shape with odour: a model for structure-activity relationships. In *Taste and Smell in Vertebrates* (G. E. W. Wolstenholme and J. Knight, eds.), pp. 293–306. J. & A. Churchill, London.

Amoore, J. E. (1975). Four primary odor modalities of man: experimental evidence and possible significance. In *Olfaction and Taste V* (D. A. Denton and J. P. Coghlan, eds.), pp. 283–289. Academic Press, New York.

Amoore, J. E., J. W. Johnston Jr., and M. Rubin (1964). The stereochemical theory of odor. *Scientific Amer. 210,* 42–49.

Amoore, J. E., and D. Venstrom (1967). Correlations between stereochemical assessments and organoleptic analysis of odorus compounds. In *Olfaction and Taste II.* Proceedings Second International Symposium Tokyo 1965, pp. 3–17. Pergamon Press, Oxford.

Andén, N.-E., A. Carlsson, A. Dahlström, K. Fuxe, N. A. Hillarp, and K. Larsson (1964). Demonstration and mapping out of nigro-neostriatal dopamine neurons. *Life Sci. 3,* 523–530.

Andén, N. E., K. Fuxe, B. Hamberger, and T. Hökfelt (1966). A quantitative study on the nigro-neostriatal dopamine neuron system in the rat. *Acta Physiol. Scand. 67,* 306–312.

Andersen, P. (1960). Interhippocampal impulses. IV. A correlation of functional and structural properties of the interhippocampal fibres in cat, rabbit and rat. *Acta Physiol. Scand. 48,* 329–351.

Andersen, P. (1975). Organization of hippocampal neurons and their interconnections. In *The Hippocampus,* Vol. 1 (R. L. Isaacson and K. H. Pribram, eds.), pp. 155–175. Plenum Press, New York.

Andersen, P., and S. A. Andersson (1968). *The Physiological Basis of the Alpha Rhythm.* Appleton-Century-Crofts, New York.

Andersen, P., T. W. Blackstad, and T. Lømo (1966). Location and identification of excitatory synapses on hippocampal pyramidal cells. *Exp. Brain Res. 1,* 236–248.

Andersen, P., T. V. P. Bliss, and K. K. Skrede (1971). Lamellar organization of hippocampal excitatory pathways. *Exp. Brain Res. 13,* 222–238.

Andersen, P., C. McC. Brooks, J. C. Eccles, and T. A. Sears (1964). The ventro-basal nucleus of the thalamus: potential fields, synaptic transmission and excitability of both presynaptic and post-synaptic components. *J. Physiol. (Lond.) 174,* 348–369.

Andersen, P., H. Bruland, and B. R. Kaada (1961). Activation of the field CAI of the hip-
    pocampus by septal stimulation. *Acta Physiol. Scand. 51*, 29–40.
Andersen, P., J. C. Eccles, and Y. Løyning (1964a). Location of postsynaptic inhibitory
    synapses on hippocampal pyramids. *J. Neurophysiol. 27*, 592–607.
Andersen, P., J. C. Eccles, and Y. Løyning (1964b). Pathway of postsynaptic inhibition in
    the hippocampus. *J. Neurophysiol. 27*, 608–619.
Andersen, P., J. C. Eccles, R. F. Schmidt, and T. Yokota (1964a). Depolarization of
    presynaptic fibers in the cuneate nucleus. *J. Neurophysiol. 27*, 92–106.
Andersen, P., J. C. Eccles, R. F. Schmidt, and T. Yokota (1964b). Identification of relay
    cells and interneurons in the cuneate nucleus. *J. Neurophysiol. 27*, 1080–1095.
Andersen, P., J. C. Eccles, and T. A. Sears (1964a). Cortically evoked depolarization of
    primary afferent fibers in the spinal cord. *J. Neurophysiol. 27*, 63–77.
Andersen, P., J. C. Eccles, and T. A. Sears (1964b). The ventro-basal complex of the
    thalamus: types of cells, their responses and their functional organization. *J. Physiol.*
    (Lond.) 174, 370–399.
Andersen, P., P. J. Hagan, C. G. Phillips, and T. P. S. Powell (1975). Mapping by
    microstimulation of overlapping projections from area 4 to motor units of the baboon's
    hand. *Proc. Roy. Soc. B 188*, 31–60.
Andersen, P., B. Holmqvist, and P. E. Voorhoeve (1966). Excitatory synapses on hip-
    pocampal apical dendrites activated by entorhinal stimulation. *Acta Physiol. Scand.*
    66, 461–472.
Andersen, P., and T. Lømo (1965). Excitation of hippocampal pyramidal cells by dendritic
    synapses. *J. Physiol. (Lond.) 181*, 39–40P.
Anderson, F. D. (1963). The structure of a chronically isolated segment of the cat spinal
    chord. *J. Comp. Neurol. 120*, 279–316.
Anderson, F. D., and C. M. Berry (1959). Degeneration studies of long ascending fiber
    systems in the cat brain stem. *J. Comp. Neurol. 111*, 195–229.
Anderson, M. E. (1971). Cerebellar and cerebral inputs to physiologically identified ef-
    ferent cell groups in the red nucleus of the cat. *Brain Research 30*, 49–66.
Anderson, M., and M. Yoshida (1977). Electrophysiological evidence for branching nigral
    projections to the thalamus and the superior colliculus. *Brain Research 137*, 361–364.
Anderson, M. E., M. Yoshida, and V. J. Wilson (1971). Influence of the superior
    colliculus on cat neck motoneurons. *J. Neurophysiol. 34*, 898–907.
Anderson, S. D., A. I. Basbaum, and H. L. Fields (1977). Response of medullary raphe
    neurons to peripheral stimulation and to systemic opiates. *Brain Research 123*,
    363–368.
Andersson, B. (1957). Polydipsia, antidiuresis and milk ejection caused by hypothalamic
    stimulation. In *Neurohypophysis* (H. Heller, ed.), pp. 131–140. Butterworths, Lon-
    don.
Andersson, G., and O. Oscarsson (1978). Climbing fibre microzones in cerebellar vermis
    and their projection to different groups of cells in the lateral vestibular nucleus. *Exp.*
    Brain Res. 32, 565–579.
Andersson, S. A. (1962). Projection of different spinal pathways to the second somatic
    sensory area in cat. *Acta Physiol. Scand. 56, Suppl. 194*, 1–74.
Andersson, S., and B. E. Gernandt (1956). Ventral root discharge in response to vestibular
    and proprioceptive stimulation. *J. Neurophysiol. 19*, 524–543.
Andres, K. H. (1965). Der Feinbau des Bulbus olfactorius der Ratte unter besonderer
    Berücksichtigung der synaptischen Verbindungen. *Z. Zellforsch. 65*, 530–561.
Andres, K. H., and M. von Düring (1973). Morphology of cutaneous receptors. In *Hand-*
    book of Sensory Physiology, Vol. II (A. Iggo, ed.), pp. 3–28. Springer-Verlag, Berlin.
Andrew, J., and P. W. Nathan (1964). Lesions of the anterior frontal lobes and distur-
    bances of micturition and deflæcation. *Brain 87*, 233–262.
Andy, O. J., and M. Jurko (1977). The human amygdala: excitability state and aggression.
    In *Neurosurgical Treatment in Psychiatry, Pain, and Epilepsy* (W. H. Sweet, S.
    Obrador, and J. G. Martín-Rodríguez, eds.), pp. 417–427. Univ. Park Press, Bal-
    timore.

Andy, O. J., and H. Stephan (1968). The septum in the human brain. *J. Comp. Neurol.* *133*, 383–410.

Angaut, P. (1969a). Étude anatomique expérimentale des efférences cérébelleuses ascendantes. Analyse électro-anatomique des projections cérébelleuses sur le noyau ventral latéral du thalamus. (Thèse, Faculté des Sciences, Paris).

Angaut, P. (1969b). The fastigio-tectal projections. An anatomical experimental study. *Brain Research 13*, 186–189.

Angaut, P. (1970). The ascending projections of the nucleus interpositus posterior of the cat cerebellum: an experimental anatomical study using silver impregnation methods. *Brain Research 24*, 377–394.

Angaut, P. (1973). Bases anatomo-fonctionelles des interrelations cérébello-cérébrales. *J. Physiol. (Paris) 67*, 53A–116A.

Angaut, P., and D. Bowsher (1970). Ascending projections of the medial cerebellar (fastigial) nucleus: an experimental study in the cat. *Brain Research 24*, 49–68.

Angaut, P., and A. Brodal (1967). The projection of the "vestibulocerebellum" onto the vestibular nuclei in the cat. *Arch. Ital. Biol. 105*, 441–479.

Angaut, P., G. Guilbaud, and M.-C. Reymond (1968). An electrophysiological study of the cerebellar projections to the nucleus ventralis lateralis of thalamus in the cat. I. Nuclei fastigii et interpositus. *J. Comp. Neurol. 134*, 9–20.

Angaut, P., and C. Sotelo (1973). The fine structure of the cerebellar nuclei in the cat. II. Synaptic organization. *Exp. Brain Res. 16*, 431–454.

Angaut-Petit, D. (1975). The dorsal column system: II. Functional properties and bulbar relay of the postsynaptic fibres of the cat's fasciculus gracilis. *Exp. Brain Res. 22*, 471–493.

Angel, R. W., and W. W. Hofmann (1963). The H reflex in normal, spastic, and rigid subjects. *Arch. Neurol. (Chic.) 8*, 591–596.

Angeletti, P. U., R. Levi-Montalcini, and F. Caramia (1972). Structural and ultra-structural changes in developing sympathetic ganglia induced by guanethidine. *Brain Research 43*, 515–525.

Angevine, J. B., Jr. (1965). Time of neuron origin in the hippocampal region. An autoradiographic study in the mouse. *Exp. Neurol. Suppl. 2*, 1–70.

Angevine, J. B., Jr., S. Locke, and P. I. Yakovlev (1962). Limbic nuclei of thalamus and connections of limbic cortex. IV. Thalamocortical projection of the ventral anterior nucleus in man. *Arch. Neurol. (Chic.) 7*, 518–528.

Angevine, J. B., Jr., E. L. Mancall, and P. I. Yakovlev (1961). *The Human Cerebellum. An Atlas of Gross Topography in Serial Sections*. J. & A. Churchill, London.

Anterior Pituitary, The (1975). (A. Tixier-Vidal and M. G. Farquhar, eds.). Academic Press, New York.

Antonetty, C. M., and K. E. Webster (1975). The organization of the spinotectal projection. An experimental study in the rat. *J. Comp. Neurol. 163*, 449–466.

Appelberg, B., T. Jeneskog, and H. Johansson (1975). Rubrospinal control of static and dynamic fusimotor neurones. *Acta Physiol. Scand. 95*, 431–440.

Appelberg, B., and I. Z. Kosary (1963). Excitation of flexor fusimotor neurons by electrical stimulation in the red nucleus. *Acta Physiol. Scand. 59*, 445–453.

Appenzeller, O. (1976). *Autonomic Nervous System: An Introduction to Basic and Clinical Concepts*, 2nd ed. North-Holland, Amsterdam.

Applebaum, A. E., J. E. Beall, R. D. Foreman, and W. D. Willis (1975). Organization and receptive fields of primate spinothalamic tract neurons. *J. Neurophysiol. 38*, 572–586.

Applebaum, M. L., G. L. Clifton, R. E. Coggeshall, J. D. Coulter, W. H. Vance, and W. D. Willis (1976). Unmyelinated fibres in the sacral 3 and caudal 1 ventral roots of the cat. *J. Physiol. (Lond.) 256*, 557–572.

Apter, J. T. (1945). Projection of the retina on superior colliculus of cats. *J. Neurophysiol. 8*, 123–134.

Apter, J. T. (1946). Eye movements following strychninization of the superior colliculus of cats. *J. Neurophysiol. 9*, 73–86.

Apter, J. T. (1954). The significance of the unilateral Argyll Robertson pupil. Part I, a report of 13 cases. *Amer. J. Ophthalm. 38,* 34–43.

Architectonics of the Cerebral Cortex (1978). (M. A. B. Brazier and H. Petsche, eds.). Int. Brain Res. Org. Monogr. Ser. (IBRO), Vol. 3. Raven Press, New York.

Arey, L. B., and M. Gore (1942). The numerical relation between the ganglion cells of the retina and the fibers in the optic nerve of the dog. *J. Comp. Neurol. 77,* 609–617.

Arezzo, J., A. Pickoff, and H. G. Vaughan, Jr. (1975). The sources and intracerebral distribution of auditory evoked potentials in the alert rhesus monkey. *Brain Research 90,* 57–73.

Aring, C. D. (1944). Clinical symptomatology. In *The Precentral Motor Cortex* (P. C. Bucy, ed.), pp. 409–423. Univ. of Illinois Press, Urbana.

Armand, J., J. Padel, and A. M. Smith (1974). Somatotopic organization of the corticospinal tract in the cat motor cortex. *Brain Research 74,* 209–229.

Armstrong, D. M. (1974). Functional significance of the inferior olive. *Physiol. Rev. 54,* 358–417.

Armstrong, D. M., and R. J. Harvey (1966). Responses in the inferior olive to stimulation of the cerebellar and cerebral cortices in the cat. *J. Physiol. (Lond.) 187,* 553–574.

Armstrong, D. M., R. J. Harvey, and R. F. Schild (1971). Distribution in the anterior lobe of the cerebellum of branches from climbing fibres to the paramedian lobule. *Brain Research 25,* 203–206.

Armstrong, D. M., R. J. Harvey, and R. F. Schild (1973a). Branching of inferior olivary axons to terminate in different folia, lobules or lobes of the cerebellum. *Brain Research 54,* 365–371.

Armstrong, D. M., R. J. Harvey, and R. F. Schild (1973b). Spino-olivocerebellar pathways to the posterior lobe of the cat cerebellum. *Exp. Brain Res. 18,* 1–18.

Armstrong, D. M., R. J. Harvey, and R. F. Schild (1974). Topographical localization in the olivocerebellar projection: an electrophysiological study in the cat. *J. Comp. Neurol. 154,* 287–302.

Armstrong, D. M., and R. F. Schild (1978a). An investigation of the cerebellar corticonuclear projections in the rat using an autoradiographic tracing method. I. Projections from the vermis. *Brain Research 141,* 1–19.

Armstrong, D. M., and R. F. Schild (1978b). An investigation of the cerebellar corticonuclear projections in the rat using an autoradiographic tracing method. II. Projections from the hemisphere. *Brain Research 141,* 235–249.

Arnesen, A. R., and K. K. Osen (1978). The cochlear nerve in the cat: topography, cochleotopy, and fiber spectrum. *J. Comp. Neurol. 178,* 661–678.

Arnesen, A. R., and K. K. Osen (1980). Fibre spectrum of the vestibulo-cochlear anastomosis in the cat. *Acta Otolaryng. (Stockh.).* In press.

Arnesen, A. R., K. K. Osen, and E. Mugnaini (1978). Temporal and spatial sequence of anterograde degeneration in the cochlear nerve fibers of the cat. A light microscopic study. *J. Comp. Neurol. 178,* 679–696.

Arthur, R. P., and W. B. Shelley (1959). The peripheral mechanism of itch in man. In *Pain and Itch. Nervous Mechanisms.* Ciba Foundation Study Group, No. 1 (G. E. W. Wolstenholme and M. O'Connor, eds.), pp. 84–95. J. & A. Churchill, London.

Arvidsson, J. (1975). Location of cat trigeminal ganglion cells innervating dental pulp of upper and lower canines studied by retrograde transport of horseradish peroxidase. *Brain Research 99,* 135–139.

Arvidsson, J., and G. Grant (1979). Further observations on transganglionic degeneration in trigeminal primary sensory neurons. *Brain Research 162,* 1–12.

Asanuma, H. (1975). Recent developments in the study of the columnar arrangement of neurons within the motor cortex. *Physiol. Rev. 55,* 143–156.

Asanuma, H., J. Fernandez, M. Scheibel, and A. B. Scheibel (1974). Characteristics of projections from the nucleus ventralis lateralis to the motor cortex in the cats: an anatomical and physiological study. *Exp. Brain Res. 20,* 315–330.

Asanuma, H., and R. W. Hunsperger (1975). Functional significance of projection from the cerebellar nuclei to the motor cortex in the cat. *Brain Research 98,* 73–92.

Asanuma, H., and I. Rosén (1972). Topographical organization of cortical efferent zones projecting to distal forelimb muscles in the monkey. *Exp. Brain Res. 14,* 243–256.

Asanuma, H., and I. Rosén (1973). Spread of mono- and polysynaptic connections within cat's motor cortex. *Exp. Brain Res. 16,* 507–520.

Asanuma, H., and H. Sakata (1967). Functional organization of a cortical efferent system examined with focal depth stimulation in cats. *J. Neurophysiol. 30,* 35–54.

Asanuma, H., S. D. Stoney, Jr., and C. Abzug (1968). Relationship between afferent input and motor outflow in cat motorsensory cortex. *J. Neurophysiol. 31,* 670–681.

Asanuma, H., P. Zarzecki, E. Jankowska, T. Hongo, and S. Marcus (1979). Projection of individual pyramidal tract neurons to lumbar motor nuclei of the monkey. *Exp. Brain Res. 34,* 73–89.

Aschan, G. (1958). Different types of alcohol nystagmus. *Acta Otolaryng. (Stockh.) Suppl. 140,* 69–78.

Aschan, G. (1964). Nystagmography and caloric testing. In *Neurological Aspects of Auditory and Vestibular Disorders* (W. S. Fields and B. R. Alford, eds.), pp. 216–246. C. C. Thomas, Springfield, Ill.

Aschan, G., M. Bergstedt, and L. Goldberg (1964). Positional alcohol nystagmus in patients with unilateral and bilateral labyrinthine destructions. *Confin. Neurol. (Basel) 24,* 80–102.

Aschan, G., M. Bergstedt, and J. Stahle (1956). Nystagmography. Recording of nystagmus in clinical neuro-otological examinations. *Acta Otolaryng. (Stockh.) Suppl. 129,* 1–103.

Aschan, G., L. Ekvall, and G. Grant (1964). Nystagmus following stimulation in the central vestibular pathways using permanently implanted electrodes. International Vestibular Symposium, Uppsala, 1963. *Acta Otolaryng. (Stockh.) Suppl. 192,* 63–77.

Aspects of Cerebellar Anatomy (1954). (J. Jansen and A. Brodal, eds.). Johan Grundt Tanum Forlag, Oslo.

Astruc, J. (1971). Corticofugal connections of area 8 (frontal eye field) in *Macaca mulatta. Brain Research 33,* 241–256.

Asymmetrical Function of the Brain (1978). (M. Kinsbourne, ed.). Cambridge Univ. Press, Cambridge.

Atkinson, D. H., J. J. Seguin, and M. Wiesendanger (1974). Organization of corticofugal neurones in somatosensory area II of the cat. *J. Physiol. (Lond.) 236,* 663–679.

Atweh, S. F., and M. J. Kuhar (1977). Autoradiographic localization of opiate receptors in rat brain. II. The brain stem. *Brain Research 129,* 1–12.

Avanzino, G. L., P. B. Bradley, and J. H. Wolstencroft (1975). Pharmacological and electrophysiological characteristics of neurones in the paramedian reticular nucleus. *Arch. Ital. Biol. 113,* 193–204.

Avendaño, C. (1976). Proyecciones desde el núcleo de Darkschewitsch al gyrus sigmoideus en el gato, determinadas mediante el transporte axonal retrógrado de horseradish peroxidase. *Ann. Anat. 25,* 27–35.

Avendaño, C., F. Reinoso-Suarez, and A. Llamas (1976). Projections to gyrus sigmoideus from the substantia nigra in the cat, as revealed by the horseradish peroxidase retrograde transport technique. *Neurosci. Letters 2,* 61–65.

Bach, L. M. N. (1952). Relationships between bulbar respiratory, vasomotor and somatic facilitatory and inhibitory areas. *Amer. J. Physiol. 171,* 417–435.

Bach-y-Rita, P. (1975). Structural-functional correlations in eye muscle fibers. Eye muscle proprioception. In *Basic Mechanisms of Ocular Motility and their Clinical Implications* (G. Lennerstrand and P. Bach-y-Rita, eds.), pp. 91–111. Pergamon Press, Oxford.

Bach-y-Rita, P. (1980). Brain plasticity as a basis for therapeutic procedures. In *Recovery of Function. Theoretical Considerations for Brain Injury Rehabilitation,* pp. 225–263. (P. Bach-y-Rita, ed.). Huber, Bern.

Bach-y-Rita, P., and F. Ito (1966). Properties of stretch receptors in cat extra-ocular muscles. *J. Physiol. (Lond.) 186,* 663–688.

Bailey, P., G. von Bonin, H. W. Garol, and W. S. McCulloch (1943). Functional organization of temporal lobe of monkey (*Macaca mulatta*) and chimpanzee (*Pan Satyrus*). *J. Neurophysiol. 6*, 121–128.

Bailey, P., and H. Cushing (1925). Medulloblastoma cerebelli: a common type of mid-cerebellar glioma of childhood. *Arch. Neurol. Psychiat.* (*Chic.*) *14*, 192–223.

Bak, I. J., W. B. Choi, R. Hassler, K. G. Usunoff, and A. Wagner (1975). Fine structural synaptic organization of the corpus striatum and substantia nigra in rat and cat. In *Advances in Neurology*, Vol. 9, *Dopaminergic Mechanisms* (T. N. Chase and A. Barbeau, eds.), pp. 25–41. Raven Press, New York.

Bak, I. J., C. H. Markham, M. L. Cook, and J. G. Stevens (1977). Intraaxonal transport of *Herpes simplex* virus in the rat central nervous system. *Brain Research 136*, 415–429.

Baker, A. B., S. Cornwell, and I. A. Brown (1952). Poliomyelitis. VI. The hypothalamus. *Arch. Neurol. Psychiat.* (*Chic.*) *68*, 16–36.

Baker, A. B., H. A. Matzke, and J. R. Brown (1950). Poliomyelitis. III. Bulbar poliomyelitis, a study of medullary function. *Arch. Neurol. Psychiat.* (*Chic.*) *63*, 257–281.

Baker, G. S., and F. W. L. Kerr (1963). Structural changes in the trigeminal system following compression procedures. *J. Neurosurg. 20*, 181–184.

Baker, R. (1977). The nucleus prepositus hypoglossi. In *Eye Movements* (B. A. Brooks and F. J. Bajandas, eds.), pp. 145–178. Plenum Press, New York.

Baker, R., and A. Berthoz (1975). Is the prepositus hypoglossi nucleus the source of another vestibulo-ocular pathway? *Brain Research 86*, 121–127.

Baker, R., A. Berthoz, and J. Delgado-García (1977). Monosynaptic excitation of trochlear motoneurons following electrical stimulation of the prepositus hypoglossi nucleus. *Brain Research 121*, 157–161.

Baker, R., M. Gresty, and A. Berthoz (1976). Neuronal activity in the prepositus hypoglossi nucleus correlated with vertical and horizontal eye movement in the cat. *Brain Research 101*, 366–371.

Baker, R., and S. M. Highstein (1975). Physiological identification of interneurons and motoneurons in the abducens nucleus. *Brain Research 91*, 292–298.

Baker, R., W. Precht, and R. Llinás (1972). Mossy and climbing fiber projections of extraocular muscle afferents to the cerebellum. *Brain Research 38*, 440–445.

Balchum, O. J., and H. M. Weaver (1943). Pathways for pain from the stomach of the dog. *Arch. Neurol. Psychiat.* (*Chic.*) *49*, 739–753.

Baldissera, F., and G. ten Bruggencate (1976). Rubrospinal effects on ventral spinocerebellar tract neurones. *Acta Physiol. Scand. 96*, 233–249.

Baldissera, F., and W. J. Roberts (1976). Effects from the vestibulospinal tract on transmission from primary afferents to ventral spinocerebellar tract neurones. *Acta Physiol. Scand. 96*, 217–232.

Baleydier, C., and F. Mauguière (1977). Pulvinar-latero posterior afferents to cortical area 7 in monkeys demonstrated by horseradish peroxidase tracing technique. *Exp. Brain Res. 27*, 501–507.

Baleydier, C., and F. Mauguière (1978). Projections of the ascending somesthetic pathways in the cat superior colliculus visualized by the horseradish peroxidase technique. *Exp. Brain Res. 31*, 43–50.

Ballesteros, M. L. F., F. Buchthal, and P. Rosenfalck (1965). The pattern of muscular activity during the arm swing of natural walking. *Acta Physiol. Scand. 63*, 296–310.

Baloh, R. W., H. R. Konrad, D. Dirks, and V. Honrubia (1976). Cerebellar-pontine angle tumors. Results of quantitative vestibulo-ocular testing. *Arch. Neurol.* (*Chic.*). *33*, 507–512.

Balthasar, K. (1952). Morphologie der spinalen Tibialis- und Peronaeus-Kerne bei der Katze: Topograhie, Architekonik, Axon- und Dendritenverlauf der Motoneurone und Zwischenneurone in den Segmenten L_6-S_2. *Arch. Psychiat. Nervenkr. 188*, 345–378.

Ban, M., and T. Ohno (1977). Projection of cerebellar nuclear neurones to the inferior olive by descending collaterals of ascending fibres. *Brain Research 133*, 156–161.

Ban, T., and F. Omukai (1959). Experimental studies on the fiber connections of the amygdaloid nuclei in the rabbit. *J. Comp. Neurol. 113*, 245–279.

Ban, T., and K. Zyo (1962). Experimental studies on the fiber connections of the rhinencephalon. I. Albino rat. *Med. J. Osaka Univ. 12*, 385–424.

Bannister, L. H. (1976). Sensory terminals of peripheral nerves. In *The Peripheral Nerve* (D. N. Landon, ed.), pp. 396–463. Chapman and Hall, London.

Bantli, H., and J. R. Bloedel (1976). Characteristics of the output from the dentate nucleus to spinal neurons via pathways which do not involve the primary sensorimotor cortex. *Exp. Brain Res. 25*, 199–220.

Barbeau, A. (1973). Biology of the striatum. In *Biology of Brain Dysfunction*, Vol. 2 (G. E. Gaull, ed.), pp. 333–350. Plenum Press, New York.

Barbeau, A. (1976a). Parkinson's disease: etiological considerations. In *The Basal Ganglia* (M. D. Yahr, ed.), pp. 281–292. Raven Press, New York.

Barbeau, A. (1976b). The nonsurgical treatment of "Parkinson's disease": a personal view. In *Current Controversies in Neurosurgery* (T. P. Morley, ed.), pp. 419–428. W. B. Saunders, Philadelphia.

Barbeau, A. (1976c). Recent developments in Parkinson's disease and Huntington's chorea. *Int. J. Neurol. (Montevideo) 11*, 17–27.

Barber, R. P., J. E. Vaughn, K. Saito, B. J. McLaughlin, and E. Roberts (1978). GABAergic terminals are presynaptic to primary afferent terminals in the substantia gelatinosa of the rat spinal cord. *Brain Research 141*, 35–55.

Barbizet, J. (1963). Defect of memorizing of hippocampal-mammillary origin: a review. *J. Neurol. Neurosurg. Psychiat. 26*, 127–135.

Bard, Ph. (1928). A diencephalic mechanism for the expression of rage with special reference to the sympathetic nervous system. *Amer. J. Physiol. 84*, 490–515.

Bard, Ph., and D. McK. Rioch (1937). A study of four cats deprived of neocortex and additional portions of the forebrain. *Bull. Johns Hopk. Hosp. 60*, 73–147.

Bargmann, W. (1949). Über die neurosekretorische Verknüpfung von Hypothalamus und Neurohypophyse. *Z. Zellforsch. 34*, 610–634.

Bargmann, W. (1954). *Das Zwischenhirnhypophysensystem*. Springer-Verlag, Berlin.

Bargmann, W., and A. Knoop (1957). Elektronenmikroskopische Beobachtungen an der Neurohypophyse. *Z. Zellforsch. 46*, 242–251.

Barilari, M. G., and H. G. J. M. Kuypers (1969). Propriospinal fibers interconnecting the spinal enlargements in the cat. *Brain Research 14*, 321–330.

Barker, D. (1948). The innervation of the muscle-spindle. *Quart. J. Micr. Sci. 89*, 143–186.

Barker, D. (1962). The structure and distribution of muscle receptors. In *Symposium on Muscle Receptors* (D. Barker, ed.), pp. 227–240. Hong Kong Univ. Press; Oxford Univ. Press, London.

Barker, D. (1966). The motor innervation of the muscle spindle. In *Nobel Symposium I: Muscular Afferents and Motor Control* (R. Granit, ed.), pp. 51–58. Almqvist & Wiksell, Stockholm; Wiley, New York.

Barker, D. (1974). The morphology of muscle receptors. In *Handbook of Sensory Physiology*, Vol. III/2: *Muscle Receptors* (C. C. Hunt, ed.), pp. 1–190. Springer-Verlag, Berlin.

Barker, D., R. W. Banks, D. W. Harker, A. Milburn, and M. J. Stacey (1976). Studies of the histochemistry, ultrastructure, motor innervation and regeneration of mammalian muscle spindles. *Progr. Brain Res. 44*, 67–87.

Barker, D., P. Bessou, E. Jankowska, B. Paqès, and M. J. Stacey (1978). Identification of intrafusal muscle fibres activated by single fusimotor axons and injected with fluorescent dye in cat tenuissimus spindles. *J. Physiol. (Lond.) 275*, 149–165.

Barker, D., F. Emonet-Dénand, D. W. Harker, L. Jami, and Y. Laporte (1976). Distribution of fusimotor axons to intrafusal muscle fibres in cat tenuissimus spindles as determined by the glycogen depletion method. *J. Physiol. (Lond.) 266*, 49–69.

Barker, D., F. Emonet-Dénand, D. W. Harker, L. Jami, and Y. Laporte (1977). Types of intra- and extrafusal muscle fibre innervated by dynamic skeleto-fusimotor axons in cat peroneus brevis and tenuissimus muscles, as determined by the glycogen-depletion method. *J. Physiol. (Lond.) 266*, 713-726.

Barnard, J. W. (1936). A phylogenetic study of the visceral afferent areas associated with the facial, glossopharyngeal and vagus nerves, and their fiber connexions. The efferent facial nucleus. *J. Comp. Neurol. 65,* 503–602.

Barnard, J. W. (1940). The hypoglossal complex of vertebrates. *J. Comp. Neurol. 72,* 489–524.

Barnard, J. W., and C. N. Woolsey (1956). A study of localization in the corticospinal tracts of monkey and rat. *J. Comp. Neurol. 105,* 25–50.

Barnard, J. W., C. N. Woolsey, and R. A. Lende (1953). Localization in the cortico-spinal tract. (Abstract.) *Anat. Rec. 115,* 279.

Barnes, W. T., H. W. Magoun, and S. W. Ranson (1943). The ascending auditory pathway in the brain-stem of the monkey. *J. Comp. Neurol. 79,* 129–152.

Barnhart, M., R. Rhines, J. C. McCarter, and H. W. Magoun (1948). Distribution of lesions of the brain stem in poliomyelitis. *Arch. Neurol. Psychiat. (Chic.) 59,* 368–377.

Barricks, M. E., D. B. Traviesa, J. S. Glaser, and I. S. Levy (1977). Ophthalmoplegia in cranial arteritis. *Brain 100,* 209–221.

Barris, R. W. (1936). A pupillo-constrictor area in the cerebral cortex of the cat and its relationship to the pretectal area. *J. Comp. Neurol. 63,* 353–368.

Barris, R. W., W. R. Ingram, and S. W. Ranson (1935). Optic connexions of the midbrain and thalamus of the cat. *J. Comp. Neurol. 62,* 117–153.

Barron, D. H. (1936). A note on the course of the proprioceptor fibers from the tongue. *Anat. Rec. 66,* 11–15.

Basal Ganglia, The (1976). (M. D. Yahr, ed.). Raven Press, New York.

Basbaum, A. I., C. H. Clanton, and H. L. Fields (1976). Opiate and stimulus-produced analgesia: functional anatomy of a medullospinal pathway. *Proc. Nat. Acad. Sci. (Wash.) 73,* 4685–4688.

Basbaum, A. J., C. H. Clanton, and H. L. Fields (1978). Three bulbospinal pathways from the rostral medulla of the cat: an autoradiographic study of pain modulating systems. *J. Comp. Neurol. 178,* 209–224.

Basic Aspects of Central Vestibular Mechanisms, Progress in Brain Research, Vol. 37 (1972). (A. Brodal and O. Pompeiano, eds.). Elsevier, Amsterdam.

Basic Mechanisms of the Epilepsies (1969). (H. H. Jasper, A. A. Ward, Jr., and A. Pope, eds.). Little, Brown, Boston.

Basic Mechanisms in Hearing (1973). (Aa. R. Møller and P. Boston, eds.). Academic Press, New York.

Basic Mechanisms of Ocular Motility and their Clinical Implications (1975). (G. Lennerstrand and P. Bach-y-Rita, eds.). Pergamon Press, Oxford.

Basmajian, J. V. (1974). *Muscles Alive. Their Functions Revealed by Electromyography,* 3rd ed. Williams & Wilkins, Baltimore.

Bates, J. A. V. (1953). A comparison between movements produced by stimulation of the motor cortex and the internal capsule in the same individual. *J. Physiol. (Lond.) 123,* 49–50P.

Bates, J. A. V. (1957). Observations on the excitable cortex in man. *Lect. Sci. Basis Med. 5,* 333–347.

Bates, J. A. V. (1960). The individuality of the motor cortex. *Brain 83,* 654–667.

Bates, J. A. V. (1963). The significance of complex motor patterns in the response to cortical stimulation. *Int. J. Neurol. (Montevideo) 4,* 92–99.

Bates, J. A. V. (1969). The significance of tremor phasic units in the human thalamus. In *Third Symposium on Parkinson's Disease* (F. J. Gillingham and I. M. L. Donaldson, eds.), pp. 118–124. Livingstone, Edinburgh.

Bates, J. A. V. (1973). Electrical recording from the thalamus in human subjects. In *Handbook of Sensory Physiology,* Vol. II (A. Iggo, ed.), pp. 561–578. Springer-Verlag, Berlin.

Bates, J. A. V., and G. Ettlinger (1960). Posterior biparietal ablations in the monkey. Changes to neurological and behavioral testing. *Arch. Neurol. (Chic.) 3,* 177–192.

Batini, C., P. Buisseret, and C. Buisseret-Delmas (1975). Trigeminal pathway of the extrinsic eye muscle afferents in cat. *Brain Research 85,* 74–78.

Batini, C., P. Buisseret, and R. T. Kado (1974). Extraocular proprioceptive and trigeminal projections to the Purkinje cells of the cerebellar cortex. *Arch. Ital. Biol. 112*, 1–17.

Batini, C., J. Corvisier, J. Destombes, H. Gioanni, and J. Everett (1976). The climbing fibers of the cerebellar cortex, their origin and pathways in the cat. *Exp. Brain Res. 26*, 407–422.

Batini, C., F. Magni, M. Palestini, G. F. Rossi, and A. Zanchetti (1959). Neural mechanisms underlying the enduring EEG and behavioral activation in the midpontine pretrigeminal cat. *Arch. Ital. Biol. 97*, 13–25.

Batini, C., G. Moruzzi, and O. Pompeiano (1957). Cerebellar release phenomena. *Arch. Ital. Biol. 95*, 71–95.

Batini, C., and O. Pompeiano (1958). Effects of rostro-medial and rostro-lateral fastigial lesions on decerebrate rigidity. *Arch. Ital. Biol. 96*, 315–329.

Batton III, R. R., A Jayaraman, D. Ruggiero, and M. B. Carpenter (1977). Fastigial efferent projections in the monkey: an autoradiographic study. *J. Comp. Neurol. 174*, 281–306.

Beall, J. E., R. F. Martin, A. E. Applebaum, and W. D. Willis (1976). Inhibition of primate spinothalamic tract neurons by stimulation in the region of the nucleus raphe magnus. *Brain Research 114*, 328–333.

Beatty, R. A., O. Sugar, and T. A. Fox (1968). Protrusion of the posterior longitudinal ligament simulating herniated lumbar intervertebral disc. *J. Neurol. Neurosurg. Psychiat. 31*, 61–66.

Beck, E. (1950). The origin, course and termination of the prefronto-pontine tract in the human brain. *Brain 73*, 368–391.

Beck, E., and A. Bignami (1968). Some neuro-anatomical observations in cases with stereotactic lesions for the relief of Parkinsonism. *Brain 91*, 589–618.

Beckstead, R. M. (1978). Afferent connections of the entorhinal area in the rat as demonstrated by retrograde cell-labeling with horseradish peroxidase. *Brain Research 152*, 249–264.

Bédard, P., L. Larochelle, A. Parent, and L. J. Poirier (1969). The nigro-striatal pathway: a correlative study based on neuroanatomical and neurochemical criteria in the cat and the monkey. *Exp. Neurol. 25*, 365–377.

Beitz, A. J. (1976). The topographical organization of the olivo-dentate and dentato-olivary pathways in the cat. *Brain Research 115*, 311–317.

Békésy, G. von (1960). *Experiments in Hearing* (Research articles from 1928–1958). McGraw-Hill, New York.

Ben-Ari, Y., R. E. Zigmond, and K. E. Moore (1975). Regional distribution of tyrosine hydroxylase, norepinephrine, and dopamine within the amygdaloid complex of the rat. *Brain Research 87*, 96–101.

Ben-Ari, Y., R. E., Zigmond, C. C. D. Shute, and P. R. Lewis (1977). Regional distribution of choline acetyltransferase and acetylcholinesterase within the amygdaloid complex and stria terminalis system. *Brain Research 120*, 435–445.

Bender, M. B., and M. Feldman (1972). The so-called "visual agnosias". *Brain 95*, 173–186.

Bender, M. B., and E. A. Weinstein (1944). Effects of stimulation and lesion of the medial longitudinal fasciculus in the monkey. *Arch. Neurol. Psychiat. (Chic.) 52*, 106–113.

Beneson, F. D., C. D. Marsden, and J. C. Meadows (1974). The amnesic syndrome of posterior cerebral artery occlusion. *Acta Neurol. Scand. 50*, 113–145.

Benevento, L. A., and J. H. Fallon (1975). The ascending projections of the superior colliculus in the rhesus monkey (*Macaca mulatta*). *J. Comp. Neurol. 160*, 339–362.

Benevento, L. A., and M. Rezak (1976). The cortical projections of the inferior pulvinar and adjacent lateral pulvinar in the rhesus monkey (*Macaca mulatta*): an autoradiographic study. *Brain Research 108*, 1–24.

Benevento, L. A., M. Rezak, and R. Santos-Anderson (1977). An autoradiographic study of the projections of the pretectum in the rhesus monkey (*Macaca mulatta*): evidence for sensorimotor links to the thalamus and oculomotor nuclei. *Brain Research 127*, 197–218.

Benjamin, R. M. (1962). Some thalamic and cortical mechanisms of taste. In *Olfaction and Taste*. Proceedings First International Symposium Wenner-Gren Center 1962 (Y. Zotterman, ed.), pp. 309–329. Pergamon Press, Oxford.

Benjamin, R. M., and K. Akert (1959). Cortical and thalamic areas involved in taste discrimination in the albino rat. *J. Comp. Neurol. 111*, 231–259.

Benjamin, R. M., and J. C. Jackson (1974). Unit discharges in the mediodorsal nucleus of the squirrel monkey evoked by electrical stimulation of the olfactory bulb. *Brain Research 75*, 181–191.

Benjamin, R. M., and W. I. Welker (1957). Somatic receiving areas of cerebral cortex of squirrel monkey (*Saimiri sciureus*). *J. Neurophysiol. 20*, 286–299.

Bennet, M. R. (1972). *Autonomic Neuromuscular Transmission*. Cambridge Univ. Press, Cambridge.

Bennett, G. J., H. Hayashi, M. Abdelmoumene, and R. Dubner (1979). Physiological properties of stalked cells of the substantia gelatinosa intracellularly stained with horseradish peroxidase. *Brain Research 164*, 285–289.

Bentivoglio, M., D. vander Kooy, and H. G. J. M. Kuypers (1979). The organization of the efferent projections of the substantia nigra in the rat. A retrograde fluorescent double labeling study. *Brain Research, 174*, 1–17.

Bentivoglio, M., G. Macchi, P. Rossini, and E. Tempesta (1978). Brain stem neurons projecting to neocortex: a HRP study in the cat. *Exp. Brain Res. 31*, 489–498.

Beran, R. L., and G. F. Martin (1971). Reticulo-spinal fibers of the opossum, *Didelphis virginiana*. I. Origin. *J. Comp. Neurol. 141*, 453–466.

Bergmans, J., and S. Grillner (1968). Monosynaptic control of static α-motoneurones from the lower brain stem. *Experientia (Basel) 24*, 146–147.

Berger, B., A. M. Thierry, J. P. Tassin, and M. A. Moyne (1976). Dopaminergic innervation of the rat prefrontal cortex: a fluorescence histochemical study. *Brain Research 106*, 133–145.

Berke, J. J. (1960). The claustrum, the external capsule and the extreme capsule of *Macaca mulatta*. *J. Comp. Neurol. 115*, 297–331.

Berkley, K. J. (1975). Different targets of different neurons in nucleus gracilis of the cat. *J. Comp. Neurol. 163*, 285–304.

Berkley, K. J., and P. J. Hand (1978a). Projections to the inferior olive of the cat. II. Comparisons of input from the gracile, cuneate and the spinal trigeminal nuclei. *J. Comp. Neurol. 180*, 253–264.

Berkley, K. J., and P. J. Hand (1978b). Efferent projections of the gracile nucleus in the cat. *Brain Research 153*, 263–283.

Berkley, K. J., and R. Parmer (1974). Somatosensory cortical involvement in responses to noxious stimulation in the cat. *Exp. Brain Res. 20*, 363–374.

Berkley, K. J., and I. G. Worden (1978). Projections to the inferior olive of the cat. I. Comparisons of input from the dorsal column nuclei, the lateral cervical nucleus, the spino-olivary pathways, the cerebral cortex and the cerebellum. *J. Comp. Neurol. 180*, 237–252.

Berlucchi, G., L. Maffei, G. Moruzzi, and P. Strata (1964). EEG and behavioral effects elicited by cooling of medulla and pons. *Arch. Ital. Biol. 102*, 372–392.

Berman, N. (1977). Connections of the pretectum in the cat. *J. Comp. Neurol. 174*, 227–254.

Berman, N., and E. G. Jones (1977). Retino-pulvinar projection in the cat. *Brain Research 134*, 237–248.

Bernardis, L. L. (1974). Localization of neuroendocrine functions within the hypothalamus. *Canad. J. Neurol. Sci. 1*, 29–39.

Bernhard, C. G., E. Bohm, and I. Petersen (1953). Investigation on the organization of the cortico-spinal system in monkeys. *Acta Physiol. Scand. 29*, 79–103.

Berrevoets, C. E., and H. G. J. M. Kuypers (1975). Pericruciate cortical neurons projecting to brain stem reticular formation, dorsal column nuclei and spinal cord in the cat. *Neurosci. Letters 1*, 257–262.

Berson, D. M., and A. M. Graybiel (1978). Parallel thalamic zones in the LP-pulvinar complex of the cat identified by their afferent and efferent connections. *Brain Research 147,* 139–148.

Bertrand, C., and S. N. Martinez (1962). Localization of lesions, mostly with regard to tremor and rigidity. *Confin. Neurol. (Basel) 22,* 274–282.

Bertrand, F., A. Hugelin, and J. F. Vibert (1973). Quantitative study of anatomical distribution of respiration related neurons in the pons. *Exp. Brain Res. 16,* 383–399.

Bertrand, G. (1956). Spinal efferent pathways from the supplementary motor area. *Brain, 79,* 461–473.

Bertrand, G. H. Jasper, and A. Wong (1967). Microelectrode study of the human thalamus: functional organization in the ventro-basal complex. *Confin. Neurol. (Basel) 29,* 81–86.

Besson, J. M., C. Conseiller, K.-F. Hamann, and M.-C. Maillard (1972). Modifications of dorsal horn cell activities in the spinal cord, after intra-arterial injection of bradykinin. *J. Physiol. (Lond.) 221,* 189–205.

Bessou, P., and Y. Laporte (1962). Responses from primary and secondary endings of the same neuromuscular spindle of the tenuissimus muscle of the cat. In *Symposium on Muscle Receptors* (D. Barker, ed.), pp. 105–119. Hong Kong Univ. Press; Oxford Univ. Press, London.

Bessou, P., and E. R. Perl (1969). Responses of cutaneous sensory units with unmyelinated fibers to noxious stimuli. *J. Neurophysiol. 32,* 1025–1043.

Bettag, W., and T. Yoshida (1960). Über stereotaktische Schmerzoperationen. *Acta Neurochir. (Wien) 8,* 299–317.

Bienfang, D. C. (1968). Location of the cell bodies of the superior rectus and inferior oblique motoneurons in the cat. *Exp. Neurol. 21,* 455–466.

Bienfang, D. C. (1978). The course of direct projections from the abducens nucleus to the contralateral medial rectus subdivision of the oculomotor nucleus in the cat. *Brain Research 145,* 277–289.

Bignall, K. E. (1969). Bilateral temporofrontal projections in the squirrel monkey: origin, distribution and pathways. *Brain Research 13,* 319–327.

Billings-Gagliardi, S., V. Chan-Palay, and S. L. Palay (1974). A review of lamination in area 17 of the visual cortex of *Macaca mulatta. J. Neurocytol. 3,* 619–629.

Billingsley, P. R., and S. W. Ranson (1918). On the number of nerve cells in the ganglion cervicale superius and of nerve fibers in the cephalic end of the truncus sympathicus in the cat, and on the numerical relations of preganglionic and postganglionic neurons. *J. Comp. Neurol. 29,* 359–366.

Binder, M. D., J. S. Kroin, G. P. Moore, E. K. Stauffer, and D. G. Stuart (1976). Correlation analysis of muscle spindle responses to single motor unit contractions. *J. Physiol. (Lond.) 257,* 325–336.

Biochemistry and Pharmacology of the Basal Ganglia (1966). (E. Costa, L. J. Côte, and M. D. Yahr, eds.). Raven Press, New York.

Biology of Memory (1970). (K. H. Pribram and D. E. Broadbent, eds.). Academic Press, New York.

Bird, E. D. (1976). Biochemical studies on γ-aminobutyric acid metabolism in Huntington's chorea. In *Biochemistry and Neurology* (H. F. Bradford and C. D. Marsden, eds.), pp. 83–92. Academic Press, New York.

Bird, E. D., and L. L. Iversen (1974). Huntington's chorea—Postmortem measurement of glutamic acid decarboxylase, choline acetyl transferase and dopamine in basal ganglia. *Brain 97,* 457–472, 1974.

Birkemayer, W., and O. Hornykiewicz (1961). Der L-Dioxyphenylalanin (=L-Dopa)-Effekt bei der Parkinson-Akinese. *Wien. Klin. Wschr. 73,* 787–788.

Biscoe, T. J., and S. R. Sampson (1970). Field potentials evoked in the brain stem of the cat by stimulation of the carotid sinus, glossopharyngeal, aortic and superior laryngeal nerves. *J. Physiol. (Lond.) 209,* 341–358.

Bishop, G. A., R. A. McCrea, and S. T. Kitai (1976). A horseradish peroxidase study of the cortico-olivary projection in the cat. *Brain Research 116,* 306–311.

Bishop, P. O. (1953). Synaptic transmission. An analysis of the electrical activity of the lateral geniculate nucleus in the cat after optic nerve stimulation. *Proc. Roy. Soc. B 141*, 362–392.

Bishop, P. O. (1973). Neurophysiology of binocular single vision and stereopsis. In *Handbook of Sensory Physiology*, Vol. VII/3 (R. Jung, ed.), pp. 256–305. Springer-Verlag, Berlin–New York.

Bizzi, E. (1974). The coordination of eye-head movement. *Sci. Amer. 231*, 100–106.

Bizzi, E., and P. H. Schiller (1970). Single unit activity in the frontal eye fields of unanesthetized monkeys during eye and head movements. *Exp. Brain Res. 10*, 151–158.

Björk, A., and E. Kugelberg (1953a). Motor unit activity in the human extraocular muscles. *Electroenceph. Clin. Neurophysiol. 5*, 271–278.

Björk, A., and E. Kugelberg (1953b). The electrical activity of the muscles of the eye and eyelids in various positions and during movement. *Electroenceph. Clin. Neurophysiol. 5*, 595–602.

Björklund, A., B. Falck, and U. Stenevi (1971). Classification of monoamine neurones in the rat mesencephalon: distribution of a new monoamine neurone system. *Brain Research 32*, 269–285.

Björklund, A., and O. Lindvall (1978). The meso-telencephalic dopamine neuron system: a review of its anatomy. In *Limbic Mechanisms* (K. E. Livingston and O. Hornykiewicz, eds.), pp. 307–333. Plenum Press, New York.

Björkman, A., and G. Wohlfart (1936). Faseranalyse der Nn. oculomotorius, trochlearis und abducens des Menschen und des N. abducens verschiedener Tiere. *Z. mikr.-anat. Forsch. 39*, 631–647.

Black, A. H. (1975). Hippocampal electrical activity and behavior. In *The Hippocampus*, Vol. 2 (R. L. Isaacson and K. H. Pribram, eds.), pp. 129–167. Plenum Press, New York.

Blackstad, T. W. (1956). Commissural connections of the hippocampal region in the rat, with special reference to their mode of termination. *J. Comp. Neurol. 105*, 417–537.

Blackstad, T. W. (1958). On the termination of some afferents to the hippocampus and fascia dentata. *Acta Anat. (Basel) 35*, 202–214.

Blackstad, T. W. (1967). Cortical grey matter. A correlation of light and microscopic data. In *The Neuron* (H. Hydén, ed.), pp. 49–118. Elsevier, Amsterdam.

Blackstad, T. W. (1975a). Golgi preparations for electron microscopy: controlled reduction of the silver chromate by ultraviolet illumination. In *Golgi Centennial Symposium: Perspectives in Neurobiology* (M. Santini, ed.), pp. 123–132. Raven Press, New York.

Blackstad, T. W. (1975b). Electron microscopy of experimental axonal degeneration in photochemically modified Golgi preparations: a procedure for precise mapping of nervous connections. *Brain Research 95*, 191–210.

Blackstad, T. W. (1977). Notes sur l'hodologie comparative de structures rhinencéphaliques (limbiques). Quelques relations mutuelles et néocorticales. In *Rhinencéphale, Neurotransmetteurs et Psychoses* (J. de Ajuriaguerra and R. Tissot, eds.), pp. 17–60. (Symp. Bel-Air V, Genève, Sept. 1976). Georg & Cie S. A., Genève; Masson & Cie, Paris.

Blackstad, T. W., K. Brink, J. Hem, and B. Jeune (1970). Distribution of hippocampal mossy fibers in the rat. An experimental study with silver impregnation methods. *J. Comp. Neurol. 138*, 433–450.

Blackstad, T., A. Brodal, and F. Walberg (1951). Some observations on normal and degenerating terminal boutons in the inferior olive of the cat. *Acta Anat. (Basel) 11*, 461–477.

Blackstad, T. W., and P. R. Flood (1963). Ultrastructure of hippocampal axosomatic synapses. *Nature (Lond.) 198*, 542–543.

Blackstad, T. W., and Å. Kjaerheim (1961). Special axo-dendritic synapses in the hippocampal cortex: electron and light microscopic studies on the layer of mossy fibers. *J. Comp. Neurol. 117*, 113–159.

Blackwell, R. E., and R. Guillemin (1973). Hypothalamic control of adenohypophysial secretions. *Amer. Rev. Physiol. 35*, 357–390.

Blakemore, C. (1974). Development of functional connexions in the mammalian visual system. *Brit. Med. Bull. 30,* 152–156.

Blakeslee, G. A., I. S. Freiman, and S. E. Barrera (1938). The nucleus lateralis medullae. An experimental study of its anatomic connections in *Macacus rhesus. Arch. Neurol. Psychiat.* (*Chic.*) *39,* 687–704.

Blanks, R. H. I., R. Volkind, W. Precht, and R. Baker (1977). Responses of cat prepositus hypoglossi neurons to horizontal angular acceleration. *Neuroscience 2,* 391–403.

Bleier, R. (1969). Retrograde transsynaptic cellular degeneration in mammillary and ventral tegmental nuclei following limbic decortication in rabbits of various ages. *Brain Research 15,* 365–393.

Blevins, C. E. (1964). Studies on the innervation of the stapedius muscle of the cat. *Anat. Rec. 149,* 157–172.

Blom, S. (1963). Tic douloureux treated with new anticonvulsant. *Arch. Neurol.* (*Chic.*) *9,* 285–290.

Blomquist, A., R. Fink, D. Bowsher, S. Griph, and J. Westman (1978). Tectal and thalamic projections of dorsal column and lateral cervical nuclei: a quantitative study in the cat. *Brain Research 141,* 335–341.

Blomquist, A., and J. Westman (1976). Interneurons and initial axon collaterals in the feline gracile nucleus demonstrated with the rapid Golgi technique. *Brain Research 111,* 407–410.

Blomquist, A. J., R. M. Benjamin, and R. Emmers (1962). Thalamic localization of afferents from the tongue in squirrel monkeys (*Saimiri sciureus*). *J. Comp. Neurol. 118,* 77–87.

Blomquist, A. J., and C. A. Lorenzini (1965). Projection of dorsal roots and sensory nerves to cortical sensory motor regions of squirrel monkey. *J. Neurophysiol. 28,* 1195–1205.

Bloom, F. E. (1972). Localization of neurotransmitters by electron microscopy. In *Neurotransmitters, Res. Publ. Ass. Nerv. Ment. Dis.,* Vol 50 (I. J. Kopin, ed.), pp. 25–57. Williams & Wilkins, Baltimore.

Bloom, F. E., B. J. Hoffer, and G. R. Siggins (1971). Studies on norepinephrine-containing afferents to Purkinje cells of rat cerebellum. I. Localization of the fibers and their synapses. *Brain Research 25,* 501–521.

Blum, P. S., L. D. Abraham, and S. Gilman (1979). Vestibular, auditory, and somatic input to the posterior thalamus of the cat. *Exp. Brain Res. 34,* 1–9.

Bobillier, P., F. Petitjean, D. Salvert, M. Ligier, and S. Seguin (1975). Differential projections of the nucleus raphe dorsalis and nucleus raphe centralis as revealed by autoradiography. *Brain Research 85,* 205–210.

Bobillier, P., S. Seguin, F. Petitjean, D. Salbert, M. Touret, and M. Jouvet (1976). The raphe nuclei of the cat brain stem: a topographical atlas of their efferent projections as revealed by autoradiography. *Brain Research 113,* 449–486.

Bodian, D. (1937). An experimental study of the optic tracts and retinal projection of the Virginia opossum. *J. Comp. Neurol. 66,* 113–144.

Bodian, D. (1942). Studies on the diencephalon of the Virginia opossum. III. The thalamocortical projection. *J. Comp. Neurol. 77,* 525–575.

Bodian, D. (1946). Spinal projections of brainstem in rhesus monkey, deduced from retrograde chromatolysis. (Abstract.) *Anat. Rec. 94,* 512–513.

Bodian, D. (1947). Nucleic acid in nerve-cell regeneration. In *Nucleic Acid, Symp. Soc. Exp. Biol.* (*N.Y.*) *1,* 163–178.

Bodian, D. (1949). Poliomyelitis: pathologic anatomy. In *Poliomyelitis. Papers and Discussions Presented at the First International Poliomyelitis Conference,* pp. 62–84. Lippincott, Philadelphia.

Bodian, D. (1951). Nerve endings, neurosecretory substance and lobular organization of the neurohypophysis. *Bull. Johns Hopk. Hosp. 89,* 354–376.

Bodian, D. (1963). Cytological aspects of neurosecretion in opossum neurohypophysis. *Bull. Johns Hopk. Hosp. 113,* 57–93.

Bodian, D. (1964). An electron-microscopic study of the monkey spinal cord. I. Fine structure of normal motor column. II. Effects of retrograde chromatolysis. III. Cytologic effects of mild and virulent poliovirus infection. *Bull. Johns Hopk. Hosp. 114,* 13–119.

Bodian, D. (1966a). Herring bodies and neuro-apocrine secretion in the monkey. An electron microscopic study of the fate of the neurosecretory product. *Bull. Johns Hopk. Hosp. 118,* 282–326.

Bodian, D. (1966b). Synaptic types on spinal motoneurons. An electron microscopic study. *Bull. Johns Hopk. Hosp. 119,* 16–45.

Bodian, D. (1975). Origin of specific synaptic types in the motoneuron neuropil of the monkey. *J. Comp. Neurol. 159,* 225–244.

Bodoky, M., and M. Réthelyi (1977). Dendritic arborization and axon trajectory of neurons in the hypothalamic arcuate nucleus of the rat. *Exp. Brain Res. 28,* 543–555.

Boesten, A. J. P., and J. Voogd (1975). Projections of the dorsal column nuclei and the spinal cord on the inferior olive in the cat. *J. Comp. Neurol. 161,* 215–238.

Bogaert, L. van, and I. Bertrand (1932). Étude anatomo-clinique d'un syndrome alterne du noyau rouge avec mouvements involontaires rythmés de l'hémiface et de l'avant-bras. *Rev. Neurol. 1,* 38–45.

Boguslawski, S., S. Banach, and M. Dabrowski (1960). Clinical and statistical evaluation of the results obtained by sympathectomy in 285 cases of various peripheral vascular diseases. *J. Neurosurg. 17,* 824–829.

Bohm, E. (1953). An electro-physiological study of the ascending spinal antero-lateral fibre system connected to coarse cutaneous afferents. *Acta Physiol. Scand. 29, Suppl. 106:8,* 1–35.

Bohm, E., and R. R. Strang (1962). Glossopharyngeal neuralgia. *Brain 85,* 371–388.

Boisacq-Schepens, N., and M. Hanus (1972). Motor cortex vestibular responses in the chloralosed cat. *Exp. Brain Res. 14,* 539–549.

Boivie, J. (1970). The termination of the cervicothalamic tract in cat. An experimental study with silver impregnation methods. *Brain Research 19,* 333–360.

Boivie, J. (1971a). The termination in the thalamus and the zona incerta of fibres from the dorsal column nuclei (DCN) in the cat. An experimental study with silver impregnation methods. *Brain Research 28,* 459–490.

Boivie, J. (1971b). The termination of the spinothalamic tract in the cat. An experimental study with silver impregnation methods. *Exp. Brain Res. 12,* 331–353.

Boivie, J. (1978). Anatomical observations on the dorsal column nuclei, their thalamic projection and the cytoarchitecture of some somatosensory thalamic nuclei in the monkey. *J. Comp. Neurol. 178,* 17–48.

Boivie, J. (1979). An anatomical reinvestigation of the termination of the spinothalamic tract in the monkey. *J. Comp. Neurol. 186,* 343–370.

Boivie, J., G. Grant, D. Albe-Fessard, and A. Levante (1975). Evidence for a projection to the thalamus from the external cuneate nucleus in the monkey. *Neurosci. Letters. 1,* 3–8.

Boivie, J. J. G., and E. R. Perl (1975). Neural substrates of somatic sensation. In *MTP International Review of Science. Physiology Series One,* Vol. 3 (C. C. Hunt, ed.), pp. 303–411. Butterworths, London.

Bok, S. T. (1922). Die Entwicklung von Reflexen und Reflexbahnen. II. Die Ontogenese des Rückenmarkreflexapparates mit den zentralen Verhältnissen des Nervus sympathicus. *Psychiat. Neurol. Bl. (Amst.) 26,* 174–233.

Bolk, L. (1906). *Das Cerebellum der Säugetiere.* De Erven F. Bohn, Haarlem; G. Fischer, Jena.

Bonica, J. J. (1968). Autonomic innervation of the viscera in relation to nerve block. *Anesthesiology 29,* 793–813, 1968.

Bonin, G. von (1944). Architecture of the precentral motor cortex and some adjacent areas. In *The Precentral Motor Cortex* (P. C. Bucy, ed.), pp. 7–82. Univ. of Illinois Press, Urbana.

Bonin, G. von, and W. R. Mehler (1971). On columnar arrangement of nerve cells in cerebral cortex. *Brain Research 27,* 1–9.

Bonnevie, K. (1943). Hereditary hydrocephalus in the house mouse. I. Manifestation of the *hy*-mutation after birth in embryos 12 days old or more. *Skr. Norske Vidensk.-Akad., I. Mat.-Nat. Kl.* No. 4.

Bonnevie, K., and A. Brodal (1946). Hereditary hydrocephalus in the house mouse. IV. The development of the cerebellar anomalies during foetal life with notes on the normal development of the mouse cerebellum. *Skr. Norske Vidensk.-Akad., I. Mat.-Nat. Kl.* No. 4.

Borg, E. (1973a). A neuroanatomical study of the brainstem auditory system of the rabbit. Part II. Descending connections. *Acta Morph. Neerl.-Scand. 11,* 49–62.

Borg, E. (1973b). On the neuronal organization of the acoustic middle ear reflex. A physiological and anatomical study. *Brain Research 49,* 101–123.

Börnstein, W. S. (1940). Cortical representation of taste in man and monkey. II. The localization of the cortical taste area in man and a method of measuring impairment of taste in man. *Yale J. Biol. Med. 13,* 133–156.

Bors, E., and A. E. Comarr (1971). *Neurological Urology. Physiology of Micturition. Its Neurological Disorders and Sequelae.* Karger, Basel.

Bors, E., and R. W. Porter (1970). Neurosurgical considerations in bladder dysfunction. *Urol. Int. (Basel) 25,* 114–133.

Bortolami, R., A. Veggetti, E. Callegari, M. L. Lucchi, and G. Palmieri (1977). Afferent fibers and sensory ganglion cells within the oculomotor nerve in some mammals and man. I. Anatomical investigations. *Arch. Ital. Biol. 115,* 355–385.

Botterell, E. H., and J. F. Fulton (1938). Functional localization in the cerebellum of primates. *J. Comp. Neurol. 69,* 31–46, 47–62, and 63–87.

Boudreau, J. C., B. E. Bradley, P. R. Bierer, S. Kruger, and C. Tsuchitani (1971). Single unit recordings from the geniculate ganglion of the facial nerve of the cat. *Exp. Brain Res. 13,* 461–488.

Boudreau, J. C., and C. Tsuchitani (1973). *Sensory Neurophysiology with Special Reference to the Cat.* Van Nostrand, New York.

Bourk, T. R. (1976). Electrical responses of neural units in the anteroventral cochlear nucleus of the cat. Thesis, M.I.T., Boston.

Bowden, D. M., D. C. German, and W. D. Poynter (1978). An autoradiographic, semi-stereotaxic mapping of major projections from locus coeruleus and adjacent nuclei in *Macaca mulatta. Brain Research 145,* 257–276.

Bowsher, D. (1957). Termination of the central pain pathway in man: the conscious appreciation of pain. *Brain 80,* 606–622.

Bowsher, D. (1958). Projection of the gracile and cuneate nuclei in *Macaca mulatta:* an experimental degeneration study. *J. Comp. Neurol. 110,* 135–156.

Bowsher, D. (1961). The termination of secondary somatosensory neurons within the thalamus of *Macaca mulatta:* an experimental degeneration study. *J. Comp. Neurol. 117,* 213–227.

Bowsher, D. (1962). The topographical projection of fibres from the anterolateral quadrant of the spinal cord to the subdiencephalic brain stem in man. *Psychiat. et Neurol. (Basel) 143,* 75–99.

Bowsher, D. (1975). Diencephalic projections from the midbrain reticular formation. *Brain Research 95,* 211–220.

Bowsher, D. (1976). Role of the reticular formation in responses to noxious stimulation. *Pain 2,* 361–378.

Bowsher, D., A. Mallart, D. Petit, and D. Albe-Fessard (1968). A bulbar relay to the centre median. *J. Neurophysiol. 31,* 288–300.

Bowsher, D., and J. Westman (1970). The gigantocellular reticular region and its spinal afferents: a light and electron microscope study in the cat. *J. Anat. (Lond.) 106,* 23–36.

Boycott, B. B., and J. E. Dowling (1969). Organization of the primate retina: light microscopy. *Phil. Trans. B 255,* 109–176.

Boycott, B. B., and H. Kolb (1973). The connections between bipolar cells and photoreceptors in the retina of the domestic cat. *J. Comp. Neurol. 148,* 91–114.

Boycott, B. B., and H. Wässle (1974). The morphological types of ganglion cells of the domestic cat's retina. *J. Physiol. (Lond.) 240,* 397–419.

Boyd, I. A. (1962). The structure and innervation of the nuclear bag muscle fibre system and the nuclear chain muscle fibre system in mammalian muscle spindles. *Phil. Trans. B 245*, 81–136.

Boyd, I. A. (1976). The response of fast and slow nuclear bag fibres and nuclear chain fibres in isolated cat muscle spindles to fusimotor stimulation, and the effect of intrafusal contraction on the sensory endings. *Quart. J. Exp. Physiol. 61*, 203–284.

Boyd, I. A., M. H. Oladden, P. N. McWilliam, and J. Ward (1977). Control of dynamic and statis nuclear bag fibres and nuclear chain fibres by gamma and beta axons in isolated cat muscle spindles. *J. Physiol. (Lond.) 265*, 133–162.

Boyd, J. D. (1937). Observations on the human carotid sinus and its nerve supply. *Anat. Anz. 84*, 386–399.

Boyd, J. D. (1941). The sensory component of the hypoglossal nerve in the rabbit. *J. Anat. (Lond.) 75*, 330–344.

Bradley, P. B., B. N. Dhawan, and J. H. Wolstencroft (1964). Some pharmacological properties of cholinoceptive neurones in the medulla and pons of the cat *J. Physiol. (Lond.) 170*, 59–60P.

Bradley, P. B., and J. H. Wolstencroft (1962). Excitation and inhibition of brain-stem neurones by noradrenaline and acetylcholine. *Nature (Lond.) 196*, 840 and 873.

Brain, R. (1961). The neurology of language. *Brain 84*, 145–166.

Brain, R., and M. Wilkinson (1959). Observations on the extensor plantar reflex and its relationship to the functions of the pyramidal tract. *Brain 82*, 297–320.

Brain, W. R. (1965). *Speech Disorders. Aphasia, Apraxia, and Agnosia*, 2nd. ed. Butterworths, London.

Brain and Conscious Experience (1966). (J. C. Eccles, ed.). Springer-Verlag, Berlin.

Brain Mechanisms and Consciousness (1954). CIOMS Symposium (J. F. Delafresnaye, ed.). Blackwell, Oxford.

Brain Mechanisms and Learning (1961). CIOMS Symposium (J. F. Delafresnaye, ed.). Blackwell, Oxford.

Brain Mechanisms Underlying Speech and Language (1967). Proceedings of a conference held at Princeton, New Jersey, November 9–12, 1965 (C. H. Millikan and F. L. Darley, eds.). Grune and Stratton, New York.

Brain Work. The Coupling of Function, Metabolism and Blood Flow in the Brain (1975). (D. H. Ingvar and N. A. Lassen, eds.). Munksgaard, Copenhagen.

Braitenberg, V., and R. P. Atwood (1958). Morphological observations on the cerebellar cortex. *J. Comp. Neurol. 109*, 1–33.

Brask, T. (1978). Extratympanic manometry in man. Clinical and experimental investigations of the acoustic stapedius and tensor tympani contractions in normal subjects and in patients. *Scand. Audiol., Suppl. 7*, 1–199.

Brawer, J. R., and D. Kent Morest (1975). Relations between auditory nerve endings and cell types in the cat's anteroventral cochlear nucleus seen with the Golgi method and Nomarski optics. *J. Comp. Neurol. 160, 491*–506.

Brawer, J. R., D. K. Morest, and E. C. Kane (1974). The neuronal architecture of the cochlear nucleus of the cat. *J. Comp. Neurol. 155*, 251–300.

Bray, G. M., and A. J. Aguayo (1974). Regeneration of peripheral unmyelinated nerves. Fate of axonal sprouts which develop after injury. *J. Anat. (Lond.) 117*, 517–529.

Bredberg, G. (1968). Cellular pattern and nerve supply of the human organ of Corti. *Acta Oto-laryng. (Stockh.), Suppl. 236*, 1–135.

Bredberg, G., H. W. Ades, and H. Engstrøm (1972). Scanning electron microscopy of the normal and pathologically altered organ of Corti. *Acta Oto-laryng. (Stockh.), Suppl. 301*, 3–48.

Breig, A. (1960). *Biomechanics of the Central Nervous System. Some Basic Normal and Pathologic Phenomena*. Year Book Publishers, Chicago.

Breig, A. (1978). *Adverse Mechanical Tension in the Central Nervous System. An Analysis of Cause and Effect. Relief by Functional Neurosurgery*. Almqvist & Wiksell, Uppsala.

Bremer, F. (1935a). Cerveau isolé et physiologie du sommeil. *C. R. Soc. Biol. (Paris)* *118*, 1235–1242.

Bremer, F. (1935b). Le cervelet. In *Traité de physiologie normale et pathologique*, Vol. 10: *Physiologie nerveuse* (G. H. Roger and L. Binet, eds.), pt. 1–2. Masson & Cie, Paris.

Bremer, F., and R. S. Dow (1939). The acoustic area of the cerebral cortex of the cat. A combined oscillographic and cyto-architectonic study. *J. Neurophysiol. 2*, 308–318.

Bridgman, C. S., and K. U. Smith (1945). Bilateral neural integration in visual perception after section of the corpus callosum. *J. Comp. Neurol. 83*, 57–68.

Brightman, M. W., and S. L. Palay (1963). The fine structure of ependyma in the brain of the rat. *J. Cell Biol. 19*, 415–439.

Brindley, G. S. (1973). Sensory effects of electrical stimulation of the visual and paravisual cortex in man. In *Handbook of Sensory Physiology*, Vol VII/3, Central Processing of Visual Information, Part B, Morphology and Function of Visual Centers in the Brain (R. Jung, ed.), pp. 583–594. Springer-Verlag, Berlin.

Brindley, G. S., and P. A. Merton (1960). The absence of position sense in the human eye. *J. Physiol. (Lond.) 153*, 127–130.

Brinkman, C., and R. Porter (1979). Supplementary motor area in the monkey: activity of neurons during performance of a learned motor task. *J. Neurophysiol. 42*, 681–709.

Brion, S., and G. Guiot (1964). Topographie des faisceaux de projection du cortex dans la capsule interne et dans le pédoncule cérébral. Étude des dégénérescences secondaires dans la sclérose latérale amyotrophique et la maladie de Pick. *Rev. Neurol. 110*, 123–144.

Broadwell, R. D. (1975a). Olfactory relationships of the telencephalon and diencephalon in the rabbit. I. An autoradiographic study of the efferent connections of the main and accessory olfactory bulbs. *J. Comp. Neurol. 163*, 329–345.

Broadwell, R. D. (1975b). Olfactory relationships of the telencephalon and diencephalon in the rabbit. II. An autoradiographic and horseradish peroxidase study of the efferent connections of the anterior olfactory nucleus. *J. Comp. Neurol. 164*, 389–410.

Brocklehurst, R. J., and F. H. Edgeworth (1940). The fibre components of the laryngeal nerves of *Macaca mulatta J. Anat. (Lond.) 74*, 386–389.

Brodal, A. (1939). Experimentelle Untersuchungen über retrograde Zellveränderungen in der unteren Olive nach Läsionen des Kleinhirns. *Z. ges. Neurol. Psychiat. 166*, 624–704.

Brodal, A. (1940a). Modification of Gudden method for study of cerebral localization. *Arch. Neurol. Psychiat. (Chic.) 43*, 46–58.

Brodal, A. (1940b). Experimentelle Untersuchungen über die olivo-cerebellare Lokalisation. *Z. ges. Neurol. Psychiat. 169*, 1–153.

Brodal, A. (1941). Die Verbindungen des Nucleus cuneatus externus mit dem Kleinhirn beim Kaninchen und bei der Katze. Experimentelle Untersuchungen. *Z. ges. Neurol. Psychiat. 171*, 167–199.

Brodal, A. (1943). The cerebellar connections of the nucleus reticularis lateralis (nucleus funiculi lateralis) in the rabbit and the cat. Experimental investigations. *Acta Psychiat. (Kbh.) 18*, 171–233.

Brodal, A. (1945). Defective development of the cerebellar vermis (partial agenesis) in a child. *Skr. Norske Vidensk.-Akad., I. Mat.-Nat. Kl.* No. 3.

Brodal, A. (1947a). Central course of afferent fibers for pain in facialis, glossopharyngeal, and vagus nerves. Clinical observations. *Arch. Neurol. Psychiat. (Chic.) 57*, 292–306.

Brodal, A. (1947b). The hippocampus and the sense of smell. A review. *Brain 70*, 179–222.

Brodal, A. (1947c). The amygdaloid nucleus in the rat. *J. Comp. Neurol. 87*, 1–16.

Brodal, A. (1948). The origin of the fibers of the anterior commissure in the rat. Experimental studies. *J. Comp. Neurol. 88*, 157–205.

Brodal, A. (1949). Spinal afferents to the lateral reticular nucleus of the medulla oblongata in the cat. An experimental study. *J. Comp. Neurol. 91*, 259–295.

Brodal, A. (1952). Experimental demonstration of cerebellar connexions from the peri-hypoglossal nuclei (nucleus intercalatus, nucleus praepositus hypoglossi and nucleus of Roller) in the cat. *J. Anat. (Lond.) 86*, 110–129.

Brodal, A. (1953). Reticulo-cerebellar connections in the cat. An experimental study. *J. Comp. Neurol. 98*, 113–154.

Brodal, A. (1956). Anatomical aspects of the reticular formation of the pons and medulla oblongata. In *Progress in Neurobiology*. (J. Ariëns Kappers, ed.), pp. 240–255. Elsevier, Amsterdam.

Brodal, A. (1957). *The Reticular Formation of the Brain Stem. Anatomical Aspects and Functional Correlations*. The Henderson Trust Lecture. Oliver and Boyd, Edinburgh.

Brodal, A. (1962). Spasticity—Anatomical aspects. *Acta Neurol. Scand. 38, Suppl. 3*, 9–40.

Brodal, A. (1963). Some data and perspectives on the anatomy of the so-called "extrapyramidal system." *Acta Neurol. Scand. 39, Suppl. 4*, 17–38.

Brodal, A. (1964). Anatomical points of view on the alleged morphological basis of consciousness. *Acta Neurochir. (Wien) 12*, 166–186.

Brodal, A. (1965a). Experimental anatomical studies of the corticospinal and cortico-rubrospinal connections in the cat. *Symp. Biol. Hung. 5*, 207–217.

Brodal, A. (1965b). *The Cranial Nerves. Anatomy and Anatomico-Clinical Correlations*, 2nd ed. Blackwell, Oxford.

Brodal, A. (1967a). Anatomical studies of cerebellar fibre connections with special reference to problems of functional localization. In *The Cerebellum, Progress in Brain Research*, Vol. 25 (J. P. Schadé, ed.), pp. 135–173. Elsevier, Amsterdam.

Brodal, A. (1967b). Anatomical organization of cerebello-vestibulo-spinal pathways. In *Myotatic, Kinesthetic and Vestibular Mechanisms*. Section III. Vestibular Mechanisms: Nervous Pathways. Ciba Foundation Symposium (A. V. D. de Reuck and Julie Knight, eds.), pp. 148–166. J. & A. Churchill, London.

Brodal, A. (1972a). Cerebrocerebellar pathways. Anatomical data and some functional implications. *Acta Neurol. Scand., Suppl. 51*, 153–196.

Brodal, A. (1972b). Some features in the anatomical organization of the vestibular nuclear complex in the cat. In *Basic Aspects of Central Vestibular Mechanisms, Progress in Brain Research*, Vol. 37 (A. Brodal and O. Pompeiano, eds.), pp. 31–53. Elsevier, Amsterdam.

Brodal, A. (1972c). Organization of the commissural connections: anatomy. In *Basic Aspects of Central Vestibular Mechanisms, Progress in Brain Research*, Vol. 37 (A. Brodal and O. Pompeiano, eds.), pp. 167–176. Elsevier, Amsterdam.

Brodal, A. (1972d). Vestibulocerebellar input in the cat: anatomy. In *Basic Aspects of Central Vestibular Mechanisms, Progress in Brain Research*, Vol. 37 (A. Brodal and O. Pompeiano, eds.), pp. 315–327. Elsevier, Amsterdam.

Brodal, A. (1972e). Anatomy of the vestibuloreticular connections and possible 'ascending' vestibular pathways from the reticular formation. In *Basic Aspects of Central Vestibular Mechanisms, Progress in Brain Research*, Vol. 37 (A. Brodal and O. Pompeiano, eds.), pp. 553–565. Elsevier, Amsterdam.

Brodal, A. (1973). Self-observations and neuro-anatomical considerations after a stroke. *Brain 96*, 675–694.

Brodal, A. (1974). Anatomy of the vestibular nuclei and their connections. In *Handbook of Sensory Physiology*, Vol. VI/1 (H. H. Kornhuber, ed.), pp. 240–352. Springer-Verlag, Berlin.

Brodal, A. (1976). The olivocerebellar projection in the cat as studied with the method of retrograde axonal transport of horseradish peroxidase. II. The projection to the uvula. *J. Comp. Neurol. 166*, 417–426.

Brodal, A., and P. Angaut (1967). The termination of spinovestibular fibres in the cat. *Brain Research 5*, 494–500.

Brodal, A., and P. Brodal (1971). The organization of the nucleus reticularis tegmenti pontis in the cat in the light of experimental anatomical studies of its cerebral cortical afferents. *Exp. Brain Res. 13*, 90–110.

Brodal, A., S. Bøyesen, and A. G. Frøvig (1953). Progressive neuropathic (peroneal) muscular atrophy (Charcot-Marie-Tooth Disease). *Arch. Neurol. Psychiat. (Chic.) 70,* 1–29.

Brodal, A., and J. Courville (1973). Cerebellar corticonuclear projection in the cat. Crus II. An experimental study with silver methods. *Brain Research 50,* 1–23.

Brodal, A., J. Destombes, A. M. Lacerda, and P. Angaut (1972). A cerebellar projection onto the pontine nuclei. An experimental anatomical study in the cat. *Exp. Brain Res. 16,* 115–139.

Brodal, A., and P. A. Drabløs (1963). Two types of mossy fiber terminals in the cerebellum and their regional distribution. *J. Comp. Neurol. 121,* 173–187.

Brodal, A., and L. Fegersten Saugstad (1965). Retrograde cellular changes in the mesencephalic trigeminal nucleus in the cat following cerebellar lesions. *Acta Morph. Neerl.-Scand. 6,* 147–159.

Brodal, A., and A. C. Gogstad (1954). Rubro-cerebellar connections. An experimental study in the cat. *Anat. Rec. 118,* 455–486.

Brodal, A., and A. C. Gogstad (1957). Afferent connexions of the paramedian reticular nucleus of the medulla oblongata in the cat. An experimental study. *Acta Anat. (Basel) 30,* 133–151.

Brodal, A., and G. Grant (1962). Morphology and temporal course of degeneration in cerebellar mossy fibers following transection of spinocerebellar tracts in the cat. An experimental study with silver methods. *Exp. Neurol. 5,* 67–87.

Brodal, A., and E. Hauglie-Hanssen (1959). Congenital hydrocephalus with defective development of the cerebellar vermis (Dandy-Walker syndrome). *J. Neurol. Neurosurg. Psychiat. 22,* 99–108.

Brodal, A., and G. H. Hoddevik (1978). The pontocerebellar projection to the uvula in the cat. *Exp. Brain Res. 32,* 105–116.

Brodal, A., and B. Høivik (1964). Site and mode of termination of primary vestibulocerebellar fibres in the cat. An experimental study with silver impregnation methods. *Arch. Ital. Biol. 102,* 1–21.

Brodal, A., and J. Jansen (1941). Beitrag zur Kenntnis der spino-cerebellaren Bahnen beim Menschen. *Anat. Anz. 91,* 185–195.

Brodal, A., and J. Jansen (1946). The ponto-cerebellar projection in the rabbit and cat. Experimental investigations. *J. Comp. Neurol. 84,* 31–118.

Brodal, A., and K. Kawamura (1980). The olivocerebellar projection. A review. *Advan. Anat. Embryol. Cell Biol.* In press.

Brodal, A., A. M. Lacerda, J. Destombes, and P. Angaut (1972). The pattern in the projection of the intracerebellar nuclei onto the nucleus reticularis tegmenti pontis in the cat. An experimental anatomical study. *Exp. Brain Res. 16,* 140–160.

Brodal, A., and O. Pompeiano (1957). The vestibular nuclei in the cat. *J. Anat. (Lond.) 91,* 438–454.

Brodal, A., and O. Pompeiano (1958). The origin of ascending fibres of the medial longitudinal fasciculus from the vestibular nuclei. An experimental study in the cat. *Acta Morph. Neerl.-Scand. 1,* 306–328.

Brodal, A., O. Pompeiano, and F. Walberg (1962). *The Vestibular Nuclei and their Connections. Anatomy and Functional Correlations.* The Henderson Trust Lectures. Oliver and Boyd, Edinburgh.

Brodal, A., and B. Rexed (1953). Spinal afferents to the lateral cervical nucleus in the cat. An experimental study. *J. Comp. Neurol. 98,* 179–212.

Brodal, A., and G. F. Rossi (1955). Ascending fibers in brain stem reticular formation of cat. *Arch. Neurol. Psychiat. (Chic.) 74,* 68–87.

Brodal, A., and L. F. Saugstad (1965). Retrograde cellular changes in the mesencephalic trigeminal nucleus in the cat following cerebellar lesions. *Acta Morph. Neerl.-Scand. 6,* 147–159.

Brodal, A., T. Szabo, and A. Torvik (1956). Corticofugal fibers to sensory trigeminal nuclei and nucleus of solitary tract. An experimental study in the cat. *J. Comp. Neurol. 106,* 527–556.

Brodal, A., and G. Szikla (1972). The termination of the brachium conjunctivum descendens in the nucleus reticularis tegmenti pontis. An experimental anatomical study in the cat. *Brain Research 39,* 337–351.

Brodal, A., E. Taber, and F. Walberg (1960). The raphe nuclei of the brain stem in the cat. II. Efferent connections. *J. Comp. Neurol. 114,* 239–259.

Brodal, A., and A. Torvik (1954). Cerebellar projection of paramedian reticular nucleus of medulla oblongata in cat. *J. Neurophysiol. 17,* 484–495.

Brodal, A., and A. Torvik (1957). Über den Ursprung der sekundären vestibulo-cerebellaren Fasern bei der Katze. Eine experimentell-anatomische Studie. *Arch. Psychiat. Nervenkr. 195,* 550–567.

Brodal, A., and F. Walberg (1977a). The olivocerebellar projection in the cat studied with the method of retrograde axonal transport of horseradish peroxidase. IV. The projection to the anterior lobe. *J. Comp. Neurol. 172,* 85–108.

Brodal, A., and F. Walberg (1977b). The olivocerebellar projection in the cat studied with the method of retrograde axonal transport of horseradish peroxidase. VI. The projection onto longitudinal zones of the paramedian lobule. *J. Comp. Neurol. 176,* 281–294.

Brodal, A., F. Walberg, and T. Blackstad (1950). Termination of spinal afferents to inferior olive in cat. *J. Neurophysiol. 13,* 431–454.

Brodal, A., F. Walberg, and G. H. Hoddevik (1975). The olivocerebellar projection in the cat studied with the method of retrograde axonal transport of horseradish peroxidase. I. The projection to the paramedian lobule. *J. Comp. Neurol. 164,* 449–470.

Brodal, A., F. Walberg, and E. Taber (1960). The raphe nuclei of the brain stem in the cat. III. Afferent connections. *J. Comp. Neurol. 114,* 261–281.

Brodal, P. (1968a). The corticopontine projection in the cat. I. Demonstration of a somatotopically organized projection from the primary sensorimotor cortex. *Exp. Brain Res. 5,* 212–237.

Brodal, P. (1968b). The corticopontine projection in the cat. II. Demonstration of a somatotopically organized projection from the second somatosensory cortex. *Arch. Ital. Biol. 106,* 310–332.

Brodal, P. (1971a). The corticopontine projection in the cat. I. The projection from the proreate gyrus. *J. Comp. Neurol. 142,* 127–140.

Brodal, P. (1971b). The corticopontine projection in the cat. II. The projection from the orbital gyrus. *J. Comp. Neurol. 142,* 141–152.

Brodal, P. (1972a). The corticopontine projection from the visual cortex in the cat. I. The total projection and the projection from area 17. *Brain Research 39,* 297–317.

Brodal, P. (1972b). The corticopontine projection from the visual cortex in the cat. II. The projection from areas 18 and 19. *Brain Research 39,* 319–335.

Brodal, P. (1972c). The corticopontine projection in the cat. The projection from the auditory areas. *Arch. Ital. Biol. 110,* 119–144.

Brodal, P. (1975). Demonstration of a somatotopically organized projection onto the paramedian lobule and the anterior lobe from the lateral reticular nucleus: an experimental study with the horseradish peroxidase method. *Brain Research 95,* 221–239.

Brodal, P. (1978a). Principles of organization of the moneky corticopontine projection. *Brain Research 148,* 214–218.

Brodal, P. (1978b). The corticopontine projection in the rhesus monkey. Origin and principles of organization. *Brain 101,* 251–283.

Brodal, P. (1979). The ponto-cerebellar projection in the rhesus monkey: an experimental study with retrograde axonal transport of horseradish peroxidase. *Neuroscience 4,* 193–208.

Brodal, P. (1980a). The cortical projection to the nucleus reticularis tegmenti pontis in the rhesus monkey. *Exp. Brain Res. 38,* 19–27.

Brodal, P. (1980b). The projection from the nucleus reticularis tegmenti pontis to the cerebellum in the rhesus monkey. *Exp. Brain Res. 38,* 29–36.

Brodal, P., J. Maršala, and A. Brodal (1967). The cerebral cortical projection to the lateral reticular nucleus in the cat, with special reference to the sensorimotor cortical areas. *Brain Research 6,* 252–274.

Brodal, P., and F. Walberg (1977). The pontine projection to the cerebellar anterior lobe. An experimental study in the cat with retrograde transport of horseradish peroxidase. *Exp. Brain Res. 29*, 233–248.

Brodmann, K. (1909). *Vergleichende Lokalisationslehre der Grosshirnrinde,* J. A. Barth, Leipzig.

Brody, H. (1955). Organization of the cerebral cortex. III. A study of aging in the human cerebral cortex. *J. Comp. Neurol. 102*, 511–556.

Brooke, M. H., and K. K. Kaiser (1974). The use and abuse of muscle histochemistry. In *Trophic Functions of the Neuron* (D. B. Drachman, ed.). *Ann. N.Y. Acad. Sci. 228*, 121–144.

Brooke, R. N. L., J. de C. Downer, and T. P. S. Powell (1965). Centrifugal fibres to the retina in the monkey and cat. *Nature (Lond.) 207*, 1365–1367.

Brooks, V. B. (1975). Roles of cerebellum and basal ganglia in initiation and control of movements. *J. Can. Sci. Neurol. 2*, 265–277.

Brooks, V. B., I. B. Kozlovskaya, A. Atkin, F. E. Horvath, and M. Uno (1973). Effects of cooling dentate nucleus on tracking-task performance in monkeys. *J. Neurophysiol. 36*, 974–995.

Brooks, V. B., and S. D. Stoney, Jr. (1971). Motor mechanisms: the role of the pyramidal system in motor control. *Ann. Rev. Physiol. 33*, 337–392.

Brouwer, B., and W. P. C. Zeeman (1926). The projection of the retina in the primary optic neuron in the monkey. *Brain 49*, 1–35.

Brouwer, B., W. P. C. Zeeman, and A. W. Houwer (1923). Experimentell-anatomische Untersuchungen über die Projektion der Retina auf die primären Opticuszentren. *Schweiz. Arch. Neurol. Psychiat. 13*, 118–138.

Brown, A. G. (1973). Ascending and long spinal pathways: dorsal columns, spinocervical tract and spinothalamic tract. In *Handbook of Sensory Physiology,* Vol. II: *Somatosensory System* (A. Iggo, ed.), pp. 315–338. Springer-Verlag, Berlin.

Brown, A. G. (1977). Cutaneous axons and sensory neurones in the spinal cord. *Brit. Med. Bull. 33*, 109–112.

Brown, A. G., J. D. Coulter, P. K. Rose, A. D. Short, and P. J. Snow (1977). Inhibition of spinocervical tract discharges from localized areas of the sensorimotor cortex in the cat. *J. Physiol. (Lond.) 264*, 1–16.

Brown, A. G., and R. E. Fyffe (1978). The morphology of group Ia afferent fibre collaterals in the spinal cord of the cat. *J. Physiol. (Lond.) 274*, 111–127.

Brown, A. G., and G. Gordon (1977). Subcortical mechanisms concerned in somatic sensation. *Brit. Med. Bull. 33*, 121–128.

Brown, A. G., P. K. Rose, and P. J. Snow (1977a). The morphology of spinocervical tract neurones revealed by intracellular injection of horseradish peroxidase. *J. Physiol. (Lond.). 270*, 747–764.

Brown, A. G., P. K. Rose, and P. J. Snow (1977b). The morphology of hair follicle afferent fibre collaterals in the spinal cord of the cat. *J. Physiol. (Lond.) 272*, 779–797.

Brown, A. G., P. K. Rose, and P. J. Snow (1978). Morphology and organization of axon collaterals from afferent fibres of slowly adapting type I units in cat spinal cord. *J. Physiol. (Lond.) 277*, 15–27.

Brown, J. R. (1949). Localizing cerebellar syndromes. *J. Amer. Med. Ass. 141*, 518–521.

Brown, J. R., F. L. Darley, and A. E. Aronson (1970). Ataxic dysarthria. *Int. J. Neurol. (Montevideo) 7*, 302–318.

Brown, J. T., V. Chan-Palay, and S. L. Palay (1977). A study of afferent input to the inferior olivary complex in the rat by retrograde axonal transport of horseradish peroxidase. *J. Comp. Neurol. 176*, 1–22.

Brown, J. W. (1967). Physiology and phylogenesis of emotional expression. *Brain Research 5*, 1–14.

Brown, J. W. (1972). *Aphasia, Apraxia and Agnosia.* C. C. Thomas, Springfield, Ill.

Brown, L. T. (1974a). Corticorubral projections in the rat. *J. Comp. Neurol. 154*, 149–168.

Brown, L. T. (1974b). Rubrospinal projections in the rat. *J. Comp. Neurol. 154*, 169–188.

Brown, M. C., J. K. S. Jansen, and D. van Essen (1976). Polyneural innervation of skeletal muscle in new-born rats and its elimination during maturation. *J. Physiol. (Lond.)* *261*, 387–422.

Brown, P. B., and J. L. Fuchs (1975). Somatotopic representation of hindlimb skin in cat dorsal horn. *J. Neurophysiol. 38*, 1–9.

Brown, W. J., and H. C. H. Fang (1961). Spastic hemiplegia in man. Lack of flaccidity in lesion of the pyramidal tract. *Neurology (Minneap.) 11*, 829–835.

Brownson, R. H., R. Uusitalo, and A. Palkama (1977). Intraaxonal transport of horseradish peroxidase in the sympathetic nervous system. *Brain Research 120*, 407–422.

Bruckmoser, P., M.-C. Hepp-Reymond, and M. Wiesendanger (1970). Effects of peripheral, rubral, and fastigial stimulation on neurons of the lateral reticular nucleus of the rat. *Exp. Neurol. 27*, 388–398.

Bruesch, S. R. (1944). The distribution of myelinated afferent fibers in the branches of the cat's facial nerve. *J. Comp. Neurol. 81*, 169–191.

Bruesch, S. R., and L. B. Arey (1942). The number of myelinated and unmyelinated fibers in the optic nerve of vertebrates. *J. Comp. Neurol. 77*, 631–665.

Brugge, J., and M. Merzenich (1973). Responses of neurons in auditory cortex of the macaque monkey to monaural and binaural stimulation. *J. Neurophysiol. 36*, 1138–1158.

Bryan, R. N., J. D. Coulter, and N. D. Willis (1974). Cells of origin of the spinocervical tract in the monkey. *Exp. Neurol. 42*, 574–586.

Bryan, R. N., D. L. Trevino, J. D. Coulter, and W. D. Willis (1973). Location and somatotopic organization of the cells of origin of the spino-cervical tract. *Exp. Brain Res. 17*, 177–189.

Bryden, M. P. (1963). Ear preference in auditory perception. *J. Exp. Psychol. 65*, 103–105.

Buchtel, H. A., G. Iosif, G. F. Marchesi, L. Provini, and P. Strata (1972). Analysis of the activity evoked in the cerebellar cortex by stimulation of the visual pathways. *Exp. Brain Res. 15*, 278–288.

Buchthal, F. (1961). The general concept of the motor unit. In *Neuromuscular Disorders, Res. Publ. Ass. Nerv. Ment. Dis. 38*, 3–30.

Buchthal, F. (1962). The electromyogram. Its value in the diagnosis of neuromuscular disorders. *Wld. Neurol. 3*, 16–34.

Buchthal, F. (1965). Electromyography in paralysis of the facial nerve. *Arch. Otolaryng. 81*, 463–469.

Buchthal, F., and P. Rosenfalck (1973). On the structure of motor units. In *New Developments in Electromyography and Clinical Neurophysiology*, Vol. 1 (J. E. Desmedt, ed.) pp. 71–85. Karger, Basel.

Buchthal, F. and H. Schmalbruch (1970). Contraction times and fibre types in intact human muscle. *Acta Physiol. Scand. 79*, 435–452.

Buchwald, N. A., and C. D. Hull (1967). Some problems associated with interpretation of physiological and behavioral responses to stimulation of caudate and thalamic nuclei. *Brain Research 6*, 1–11.

Bucy, P. C. (1944a). Effects of extirpation in man. In *The Precentral Motor Cortex* (P. C. Bucy, ed.), pp. 353–394. Univ. of Illinois Press. Urbana, Ill.

Bucy, P. C. (1944b). Relation to abnormal involuntary movements. In *The Precentral Motor Cortex* (P. C. Bucy, ed.), pp. 395–408. Univ. of Illinois Press, Urbana, Ill.

Bucy, P. C. (1957). Is there a pyramidal tract? *Brain 80*, 376–392.

Bucy, P. C., and J. E. Keplinger (1961). Section of the cerebral peduncles. *Arch. Neurol. (Chic.) 5*, 132–139.

Bucy, P. C., J. E. Keplinger, and E. B. Siqueira (1964). Destruction of the "pyramidal tract" in man. *J. Neurosurg. 21*, 385–398.

Bucy, P. C., and H. Klüver (1955). An anatomical investigation of the temporal lobe in the monkey (*Macaca mulatta*). *J. Comp. Neurol. 103*, 151–252.

Bucy, P. C., R. Ladpli, and A. Ehrlich (1966). Destruction of the pyramidal tract in the monkey. The effects of bilateral section of the cerebral peduncles. *J. Neurosurg. 25,* 1–20.

Bunney, B. S., and G. K. Aghajanian (1976a). The precise localization of nigral afferents in the rat as determined by a retrograde tracing technique. *Brain Research 117,* 423–435.

Bunney, B. S., and G. K. Aghajanian (1976b). Dopaminergic influence in the basal ganglia: evidence for striatonigral feedback regulation. In *The Basal Ganglia* (M. D. Yahr, ed.), pp. 249–266. Raven Press, New York.

Bunt, A. H., A. E. Hendrickson, J. S. Lund, R. D. Lund, and A. F. Fuchs (1975). Monkey retinal ganglion cells: morphometric analysis and tracing of axonal projections with a consideration of the peroxidase technique. *J. Comp. Neurol. 164,* 265–286.

Burandt, D. C., G. M. French, and K. Akert (1961). Relationships between the caudate nucleus and the frontal cortex in *Macaca mulatta. Confin. Neurol. (Basel) 21,* 289–306.

Buren, J. M. van, and M. Baldwin (1958). The architecture of the optic radiation with the temporal lobe in man. *Brain 81,* 15–40.

Buren, J. M. van, and R. C. Borke (1969). Alterations in speech and the pulvinar. A serial section study of cerebrothalamic relationships in cases of acquired speech disorders. *Brain 92,* 255–284.

Buren, J. M. van, and R. C. Borke (1972). *Variations and Connections of the Human Thalamus. 1. The Nuclei and the Cerebral Connections of the Human Thalamus.* Springer-Verlag, Berlin.

Burger, L. J., N. H. Kalvin, and J. L. Smith (1970). Acquired lesions of the fourth cranial nerve. *Brain 93,* 567–574.

Burgess, P. R., and F. J. Clark (1969). Characteristics of knee joint receptors in the cat. *J. Physiol. (Lond.) 203,* 317–335.

Burgess, P. R., and E. R. Perl (1973). Cutaneous mechanoreceptors and nociceptors. In *Handbook of Sensory Physiology,* Vol. II: *Somatosensory System* (A. Iggo, ed.), pp. 30–78. Springer-Verlag, Berlin.

Bürgi, S. (1957). Das Tectum Opticum. Seine Verbindungen bei der Katze und seine Bedeutung beim Menschen. *Dtsch. Z. Nervenheilk. 176,* 701–729.

Burke, D., and G. Eklund (1977). Muscle spindle activity in man during standing. *Acta Physiol. Scand. 100,* 187–199.

Burke, D., K.-E. Hagbarth, L. Löfstedt, and B. G. Wallin (1976a). Responses of human spindle endings to vibration of non-contracting muscles. *J. Physiol. (Lond.) 261,* 673–693.

Burke, D., K.-E. Hagbarth, L. Löfstedt, and B. G. Wallin (1976b). The responses of human muscle spindle endings to vibration during isometric contraction. *J. Physiol. (Lond.) 261,* 695–711.

Burke, R., A. Lundberg, and F. Weight (1971). Spinal border cell origin of the ventral spinocerebellar tract. *Exp. Brain Res. 12,* 283–294.

Burke, R. E., P. L. Strick, K. Kanda, C. C. Kim, and B. Walmsley (1977). The anatomy of the medial gastrocnemius and soleus motor nuclei in the cat spinal cord. *J. Neurophysiol. 40,* 667–680.

Burke, R. E., and P. Tsairis (1973). Anatomy and innervation ratios in motor units of cat gastronemius. *J. Physiol. (Lond.) 234,* 749–765.

Burke, R. E., and P. Tsairis (1974). The correlation of physiological properties with histochemical characteristics in single muscle units. In *Trophic Functions of the Neuron* (D. B. Drachman, ed.). *Ann. N.Y. Acad. Sci. 228,* 145–158.

Burke, R. E., B. Walmsley, and J. A. Hodgson (1979). HRP anatomy of group Ia afferent contacts on alpha motoneurones. *Brain Research 160,* 347–352.

Burne, R. A., M. A. Eriksson, J. A. Saint-Cyr, and D. J. Woodward (1978). The organization of the pontine projection to lateral cerebellar areas in the rat: dual zones in the pons. *Brain Research 139,* 340–347.

Burnstock, G. (1976). Do some nerve cells release more than one transmitter? *Neuroscience 1,* 239–248.

Burnstock, G., B. Evans, B. J. Gannon, J. W. Heath, and V. James (1971). A new method of destroying adrenergic nerves in adult animals using guanethidine. *Brit. J. Pharmacol. 43*, 295–301.

Burnstock, G., and M. E. Holman (1961). The transmission of excitation from autonomic nerve to smooth muscle. *J. Physiol. (Lond.) 155*, 115–133.

Burrows, G. R., and W. R. Hayhow (1971). The organization of the thalamo-cortical visual pathways in the cat. *Brain Behav. Evol. 4*, 220–272.

Burton, H., and E. G. Jones (1976). The posterior thalamic region and its cortical projection in New world and Old world monkeys. *J. Comp. Neurol. 168*, 249–302.

Burton, H., and A. D. Loewy (1976). Descending projections from the marginal cell layer and other regions of the monkey spinal cord. *Brain Research 116*, 485–491.

Burton, H., and A. D. Loewy (1977). Projections to the spinal cord from medullary somatosensory relay nuclei. *J. Comp. Neurol. 173*, 773–792.

Burton, J. E., and N. Onoda (1977). Interpositus neuron discharge in relation to a voluntary movement. *Brain Research 121*, 167–172.

Busch, H. F. M. (1961). *An Anatomical Analysis of the White Matter in the Brain Stem of the Cat*. Van Gorcum & Co. N. V., Assen, The Netherlands.

Buser, P. (1966). Subcortical controls of pyramidal activity. In *The Thalamus* (D. P. Purpura and M. D. Yahr, eds.), pp. 323–347. Columbia Univ. Press, New York.

Buser, P., G. Pouderoux, and J. Mereaux (1974). Single unit recording in the caudate nucleus during sessions with elaborate movements in the awake monkey. *Brain Research 71*, 337–344.

Buskirk, C. van (1945). The seventh nerve complex. *J. Comp. Neurol. 82*, 303–333.

Butcher, L. L., and G. J. Giesler, Jr. (1977). Nigro–neostriato–nigral relationships demonstrated by horseradish peroxidase (HRP) histochemistry. *Neurosci. Abstracts III*, 35.

Butcher, S. G., and L. L. Butcher (1974). Origin and modulation of acetylcholine activity in the neostriatum. *Brain Research 71*, 167–171.

Butler, R. A., I. T. Diamond, and W. D. Neff (1957). Role of auditory cortex in discrimination of changes in frequency. *J. Neurophysiol. 20*, 108–120.

Butson, A. R. C. (1950). Regeneration of the cervical sympathetic. *Brit. J. Surg. 38*, 223–239.

Butter, C. M., and D. R. Snyder (1972). Alterations in adversive and aggressive behaviors following orbital frontal lesions in monkeys. *Acta Neurobiol. Exp. 32*, 525–565.

Büttner, U., J. A. Büttner-Ennever, and V. Henn (1977). Vertical eye movement related activity in the rostral mesencephalic reticular formation of the alert monkey. *Brain Research 130*, 239–252.

Büttner, U., and V. Henn (1976). Thalamic unit activity in the alert monkey during natural vestibular stimulation. *Brain Research 103*, 127–132.

Büttner-Ennever, J. A. (1977). Pathways from the pontine reticular formation to structures controlling horizontal and vertical eye movements in the moneky. In *Control of Gaze by Brain Stem Neurons* (R. Baker and A. Berthoz, eds.), pp. 89–98. Elsevier/North Holland Biomedical Press, Amsterdam—New York.

Büttner-Ennever, J. A., and U. Büttner (1978). A cell group associated with vertical eye movements in the rostral mesencephalic reticular formation. *Brain Research 151*, 31–47.

Büttner-Ennever, J. A., and V. Henn (1976). An autoradiographic study of the pathways from the pontine reticular formation involved in horizontal eye movements. *Brain Research 108*, 155–164.

Büttner-Ennever, J. A., and W. Lang (1978). Vestibular afferents to an eye movement area of the monkey mesencephalon. *Neurosci. Letters, Suppl. 1*, p. 110 (Abstr.).

Büttner-Ennever, J. A., T. A. Miles, and V. Henn (1975). The role of the pontine reticular formation in oculomotor function. *Exp. Brain Res.* Suppl. to Vol. 23, p. 31.

Buxton, D. F., and D. C. Goodman (1967). Motor function and the corticospinal tracts in the dog and raccoon. *J. Comp. Neurol. 129*, 341–360.

Bystrzycka, E., B. S. Nail, and M. Rowe (1977). Inhibition of cuneate neurones: its afferent source and influence on dynamically sensitive tactile neurones. *J. Physiol. (Lond.) 268*, 251–270.

Cabot, J. B., J. M. Wild, and D. H. Cohen (1979). Raphe inhibition of sympathetic pre-ganglionic neurons. *Science 203*, 184–186.

Cain, D. P., and D. Bindra (1972). Responses of amygdala single units to odors in the rat. *Exp. Neurol. 35*, 98–110.

Cairns, H. R. (1952). Disturbances of consciousness with lesions of the brainstem and diencephalon. *Brain 75*, 109–146.

Cajal, S. Ramon y (1909–11). *Histologie du Système Nerveux de l'Homme et des Vertébrés*. Maloine, Paris.

Cajal, S. Ramon y (1954). *Neuron Theory or Reticular Theory?* (W. U. Purkiss and C. A. Fox, trans.). Consejo Superior de Investigaciones Cientificas Instituto "Ramon y Cajal," Madrid.

Calaresu, F. R., A. A. Faiers, and G. J. Mogenson (1975). Central neural regulation of heart and blood vessels in mammals. *Neurobiology 5*, 1–35.

Calne, D. B., and C. A. Pallis (1966). Vibratory sense: a critical review. *Brain 89*, 723–746.

Calvino, B., J. Lagowska, and Y. Ben-Ari (1979). Morphine withdrawal syndrome: differential participation of structures located within the amygdaloid complex and striatum of the rat. *Brain Research 177*, 19–34.

Cammermeyer, J. (1960). The post-mortem origin and mechanism of neuronal hyperchromatosis and nuclear pyknosis. *Exp. Neurol. 2*, 379–405.

Cammermeyer, J. (1962). An evaluation of the significance of the "dark" neuron. *Ergebn. Anat. Entwickl.-Gesch. 36*, 1–61.

Cammermeyer, J. (1963). Peripheral chromatolysis after transection of mouse facial nerve. *Acta Neuropath. (Berl.) 2*, 213–230.

Cammermeyer, J. (1969). Species differences in acute retrograde neuronal reaction of the facial and hypoglossal nuclei. *J. Hirnforsch. 11*, 13–29.

Cammermeyer, J. (1972). Nonspecific changes of the central nervous system in normal and experimental material. In *The Structure and Function of Nervous Tissue*, Vol. VI (G. H. Bourne, ed.), pp. 131–251. Academic Press, New York.

Cammermeyer, J. (1975). The effect of postmortem trauma on neuronal cell types stained histochemically for phospholipids. *Exp. Neurol. 46*, 616–633.

Campbell, A. W. (1905). *Histological Studies on the Localisation of Cerebral Function*. Cambridge Univ. Press, Cambridge.

Campbell, S. K., T. D. Parker, and W. Welker (1974). Somatotopic organization of the external cuneate nucleus in albino rats. *Brain Research 77*, 1–23.

Campos-Ortega, J. A., and W. R. Hayhow (1972). On the organization of the visual cortical projection to the pulvinar in *Macaca mulatta*. *Brain Behav. Evol. 6*, 394–423.

Campos-Ortega, J. A., W. R. Hayhow, and P. F. de V. Clüver (1970). A note of the problem of retinal projections to the inferior pulvinar nucleus of primates. *Brain Research 22*, 126–130.

Caplan, L. R. (1974). Ptosis. *J. Neurol. Neurosurg. Psychiat. 37*, 1–7.

Capps, M. J., and H. W. Ades (1968). Auditory frequency discrimination after transection of the olivocochlear bundle in squirrel monkeys. *Exp. Neurol. 21*, 147–158.

Car, A. (1970). La commande corticale du centre déglutiteur bulbaire. *J. Physiol. (Paris) 62*, 361–386.

Car, A. (1971). Étude macrophysiologique et microphysiologique de la zone déglutitrice du cortex frontal. *J. Physiol. (Paris) 63*, 183A.

Car, A., A. Jean, and C. Roman (1975). A pontine primary relay for ascending projections in the superior laryngeal nerve. *Exp. Brain Res. 22*, 197–210.

Carli, G., K. Diete-Spiff, and O. Pompeiano (1967a). Responses of the muscle spindles and of the extrafusal fibres in an extensor muscle to stimulation of the lateral vestibular nucleus in the cat. *Arch. Ital. Biol. 105*, 209–242.

Carli, G., K. Diete-Spiff, and O. Pompeiano (1967b). Mechanisms of muscle spindle excitation. *Arch. Ital. Biol. 105*, 273–289.

Carman, J. B., W. M. Cowan, and T. P. S. Powell (1963). The organization of the cortico-striate connexions in the rabbit. *Brain 86*, 525–562.

Carman, J. B., W. M. Cowan, and T. P. S. Powell (1964a). The cortical projection upon the claustrum. *J. Neurol. Neurosurg. Psychiat. 27*, 46–51.

Carman, J. B., W. M. Cowan, and T. P. S. Powell (1964b). Cortical connexions of the thalamic reticular nucleus. *J. Anat. (Lond.) 98*, 587–598.

Carman, J. B., W. M. Cowan, T. P. S. Powell, and K. E. Webster (1965). A bilateral cortico-striate projection. *J. Neurol. Neurosurg. Psychiat. 28*, 71–77.

Carmel, P., and A. Starr (1963). Acoustic and nonacoustic factors modifying middle-ear muscle activity in waking cats. *J. Neurophysiol. 26*, 598–616.

Carmichael, E. A., M. R. Dix, and C. S. Hallpike (1954). Lesions of the cerebral hemispheres and their effects upon optokinetic and caloric nystagmus. *Brain 77*, 345–372.

Carmichael, E. A., M. R. Dix, and C. S. Hallpike (1965). Observations upon the neurological mechanism of directional preponderance of caloric nystagmus resulting from vascular lesions of the brain stem. *Brain 88*, 51–74.

Carmichael, E. A., and H. H. Woollard (1933). Some observations on the fifth and seventh cranial nerves. *Brain 56*, 109–125.

Carney, L. R. (1967). Considerations on the cause and treatment of trigeminal neuralgia. *Neurology (Minneap.) 17*, 1143–1151.

Carpenter, D., A. Lundberg, and U. Norrsell (1963). Primary afferent depolarization evoked from the sensorimotor cortex. *Acta Physiol. Scand. 59*, 126–142.

Carpenter, M. B. (1957). The dorsal trigeminal tract in the rhesus monkey. *J. Anat. (Lond.) 91*, 82–90.

Carpenter, M. B. (1960). Experimental anatomical-physiological studies of the vestibular nerve and cerebellar connections. In *Neural Mechanisms of the Auditory and Vestibular Systems* (G. L. Rasmussen and W. Windle, eds.), pp. 279–323. C. C. Thomas, Springfield, Ill.

Carpenter, M. B. (1961). Brain stem and infratentorial neuraxis in experimental dyskinesia. *Arch. Neurol. (Chic.) 5*, 504–524.

Carpenter, M. B. (1964). Ascending vestibular projections and conjugate horizontal eye movements. In *Neurological Aspects of Auditory and Vestibular Disorders* (W. S. Fields and B. R. Alford, eds.), pp. 150–189. C. C. Thomas, Springfield, Ill.

Carpenter, M. B. (1976). Anatomical organization of the corpus striatum and related nuclei. In *The Basal Ganglia* (M. D. Yahr, ed.), pp. 1–35. Raven Press, New York.

Carpenter, M. B., D. S. Bard, and F. A. Alling (1959). Anatomical connections between the fastigial nuclei, the labyrinth and the vestibular nuclei in the cat. *J. Comp. Neurol. 111*, 1–25.

Carpenter, M. B., G. M. Brittin, and J. Pines (1958). Isolated lesions of the fastigial nuclei in the cat. *J. Comp. Neurol. 109*, 65–89.

Carpenter, M. B., and C. S. Carpenter (1951). Analysis of somatotopic relations of the corpus Luysi in man and monkey. *J. Comp. Neurol. 95*, 349–370.

Carpenter, M. B., R. A. R. Fraser, and J. E. Shriver (1968). The organization of pallido-subthalamic fibers in the monkey. *Brain Research 11*, 522–559.

Carpenter, M. B., W. Glinsmann, and H. Fabrega (1958). Effects of secondary pallidal and striatal lesions upon cerebellar dyskinesia in the rhesus monkey. *Neurology (Minneap.) 8*, 352–358.

Carpenter, M. B., and G. R. Hanna (1961). Fiber projections from the spinal trigeminal nucleus in the cat. *J. Comp. Neurol. 117*, 117–131.

Carpenter, M. B., and G. R. Hanna (1962). Lesions of the medial longitudinal fasciculus in the cat. *Amer. J. Anat. 110*, 307–332.

Carpenter, M. B., J. W. Harbison, and P. Peter (1970). Accessory oculomotor nuclei in the monkey: projections and effects of discrete lesions. *J. Comp. Neurol. 140*, 131–154.

Carpenter, M. B., and R. E. McMasters (1963). Disturbances of conjugate horizontal eye movements in the monkey. II. Physiological effects and anatomical degeneration resulting from lesions in the medial longitudinal fasciculus. *Arch. Neurol. (Chic.) 8*, 347–368.

Carpenter, M. B., and R. E. McMasters (1964). Lesions of the substantia nigra in the Rhesus monkey. Efferent fiber degeneration and behavioral observations. *Amer. J. Anat. 114*, 293–320.

Carpenter, M. B., R. E. McMasters, and G. R. Hanna (1963). Disturbances of conjugate horizontal eye movements in the monkey. I. Physiological effects and anatomical degeneration resulting from lesions of the abducens nucleus and nerve. *Arch. Neurol.* (*Chic.*) *8*, 231–247.

Carpenter, M. B., and F. A. Mettler (1951). Analysis of subthalamic dyskinesia in the monkey, with special reference to ablations of agranular cortex. *J. Comp. Neurol. 95*, 125–158.

Carpenter, M. B., K. Nakano, and R. Kim (1976). Nigrothalamic projections in the monkey demonstrated by autoradiographic technics. *J. Comp. Neurol. 165*, 401–416.

Carpenter, M. B., and H. R. Nova (1960). Descending division of the brachium conjunctivum in the cat: a cerebelloreticular system. *J. Comp. Neurol. 114*, 295–305.

Carpenter, M. B., and P. Peter (1970/71). Accessory oculomotor nuclei in the monkey. *J. Hirnforsch. 12*, 405–418.

Carpenter, M. B., and P. Philip (1972). Nigrostriatal and nigrothalamic fibers in the rhesus monkey. *J. Comp. Neurol. 144*, 93–116.

Carpenter, M. B., and R. J. Pierson (1973). Pretectal region and the pupillary light reflex. An anatomical analysis in the monkey. *J. Comp. Neurol. 149*, 271–300.

Carpenter, M. B., B. M. Stein, and P. Peter (1972). Primary vestibulocerebellar fibers in the monkey: distribution of fibers arising from distinctive cell groups of the vestibular ganglia. *Amer. J. Anat. 135*, 221–250.

Carpenter, M. B., and N. L. Strominger (1964). Cerebello-oculomotor fibers in the rhesus monkey. *J. Comp. Neurol. 123*, 211–230.

Carpenter, M. B., and N. L. Strominger (1965). The medial longitudinal fasciculus and disturbances of conjugate horizontal eye movements in the monkey. *J. Comp. Neurol. 125*, 41–66.

Carpenter, M. B., and N. L. Strominger (1966). Corticostriate encephalitis and paraballism in the monkey. *Arch. Neurol.* (*Chic.*) *14*, 241–253.

Carpenter, M. B., and N. L. Strominger (1967). Efferent fibers of the subthalamic nucleus in the monkey. A comparison of the efferent projections of the subthalamic nucleus, substantia nigra and globus pallidus. *Amer. J. Anat. 121*, 41–72.

Carrea, R. M. E., and F. A. Mettler (1947). Physiologic consequences following extensive removals of the cerebellar cortex and deep cerebellar nuclei and effect of secondary cerebral ablations in the primate. *J. Comp. Neurol. 87*, 169–288.

Carrea, R. M. E., M. Reissig, and F. A. Mettler (1947). The climbing fibers of the simian and feline cerebellum. Experimental inquiry into their origin by lesions of the inferior olives and deep cerebellar nuclei. *J. Comp. Neurol. 87*, 321–365.

Carreras, M., and S. A. Andersson (1963). Functional properties of neurones of the anterior ectosylvian gyrus of the cat. *J. Neurophysiol. 26*, 100–126.

Carrier, O. (1972). *Pharmacology of the Peripheral Autonomic System*. Year Book Medical Publishers, Chicago.

Carstens, E., D. Klumpp, and M. Zimmermann (1980). Differential inhibitory effects of medial and lateral midbrain stimulation on spinal neuronal discharges to noxious skin heating in the cat. *J. Neurophysiol. 43:*332–342.

Carstens, E., and D. L. Trevino (1978a). Laminar origins of spinothalamic projections in the cat as determined by the retrograde transport of horseradish peroxidase. *J. Comp. Neurol. 182*, 151–166.

Carstens, E., and D. L. Trevino (1978b). Anatomical and physiological properties of ipsilaterally projecting spinothalamic neurons in the second cervical segment of the cat's spinal cord. *J. Comp. Neurol. 182*, 167–184.

Casseday, J. H., I. T. Diamond, and J. K. Harting (1976). Auditory pathways to the cortex in Tupaia glis. *J. Comp. Neurol. 166*, 303–340.

Castiglioni, A. J., M. C. Gallaway, and J. D. Coulter (1978). Spinal projections from the midbrain in monkey. *J. Comp. Neurol. 178*, 329–346.

Caughell, K. A., and B. A. Flumerfelt (1977). The organization of the cerebellorubral projection: an experimental study in the rat. *J. Comp. Neurol. 176*, 295–306.

Cauna, N. (1954). Nature and functions of the papillary ridges of the digital skin. *Anat. Rec. 119*, 449–468.

Cauna, N., and G. Mannan (1958). The structure of human digital Pacinian corpuscles (*Corpuscula lamellosa*) and its functional significance. *J. Anat. (Lond.)* 92, 1–20.

Cawthorne, E. A., G. Fitzgerald, and C. S. Hallpike (1942). Studies in human vestibular function. II. Observations on directional preponderance of caloric nystagmus ('Nystagmusbereitschaft') resulting from unilateral labyrinthectomy. *Brain* 65, 138–160.

Cedarbaum, J. M., and G. K. Aghajanian (1978). Afferent projections to the rat locus coeruleus as determined by a retrograde tracing technique. *J. Comp. Neurol.* 178, 1–16.

Celesia, G. G. (1963). Segmental organization of cortical afferent areas in the cat. *J. Neurophysiol.* 26, 193–206.

Celesia, G. G. (1976). Organization of auditory cortical areas in man. *Brain* 99, 403–414.

Celesia, G. G., R. J. Broughton, T. Rasmussen, and C. Branch (1968). Auditory evoked responses from the exposed human cortex. *Electroenceph. Clin. Neurophysiol.* 24, 458–466.

Central Organization of the Autonomic Nervous System (1975). (A. Sato, ed.). *Brain Research 87*, 137–437.

Central Rhythmic and Regulation; Circulation, Respiration, Extrapyramidal Motor System (1974). (W. Umbach and H. P. Koepchen, eds.). Hippokrates Verlag, Stuttgart.

Cerebellar Stimulation in Man (1978). (I. S. Cooper, ed.). Raven Press, New York, 222 pp.

Cerebellum, Epilepsy, and Behavior, The (1974). (I. S. Cooper, M. Riklan, and R. S. Snider, eds.). Plenum Press, New York.

Cerf, J. A., and L. W. Chacko (1958). Retrograde reaction in motoneuron dendrites following ventral root section in the frog. *J. Comp. Neurol.* 109, 205–219.

Chacko, L. W. (1948). The laminar pattern of the lateral geniculate body in the primates. *J. Neurol. Neurosurg. Psychiat.* 11, 211–224.

Chalupa, L. M., and S. E. Fish (1978). Response characteristics of visual and extravisual neurons in the pulvinar and lateral posterior nuclei of the cat. *Exp. Neurol.* 61, 96–120.

Chambers, M. R., K. H. Andres, M. von Düring, and A. Iggo (1972). The structure and function of the slowly adapting type II mechanoreceptor in hairy skin. *Quart. J. Exp. Physiol.* 57, 417–445.

Chambers, W. W. (1947). Electrical stimulation of the interior of the cerebellum in the cat. *Amer. J. Anat.* 80, 55–93.

Chambers, W. W., and C-N. Liu (1957). Cortico-spinal tract of the cat. An attempt to correlate the pattern of degeneration with deficits in reflex activity following neocortical lesions. *J. Comp. Neurol.* 108, 23–55.

Chambers, W. W., C-N. Liu, G. P. McCouch, and E. d'Aquili (1966). Descending tracts and spinal shock in the cat. *Brain* 89, 377–390.

Chambers, W. W., and J. M. Sprague (1955a). Functional localization in the cerebellum. I. Organization in longitudinal cortico-nuclear zones and their contribution to the control of posture, both extrapyramidal and pyramidal. *J. Comp. Neurol.* 103, 105–130.

Chambers, W. W., and J. M. Sprague (1955b). Functional localization in the cerebellum. II. Somatotopic organization in cortex and nuclei. *Arch. Neurol. Psychiat. (Chic.)* 74, 653–680.

Chandler, W. F., and E. C. Crosby (1975). Motor effects of stimulation and ablation of the caudate nucleus of the monkey. *Neurology (Minneap.)* 25, 1160–1163.

Chang, H.-T., and T. C. Ruch (1947). Topographical distribution of spinothalamic fibres in the thalamus of the spider monkey. *J. Anat. (Lond.)* 81, 150–164.

Chang, H.-T., T. C. Ruch, and A. A. Ward, Jr. (1947). Topographical representation of muscles in motor cortex of monkeys. *J. Neurophysiol.* 10, 39–56.

Chan-Palay, V. (1973a). On certain fluorescent axon terminals containing granular synaptic vesicles in the cerebellar nucleus lateralis. *Z. Anat. Entwickl.-Gesch.* 142, 239–258.

Chan-Palay, V. (1973b). Neuronal circuitry in the nucleus lateralis of the cerebellum. *Z. Anat. Entwickl.-Gesch.* 142, 259–265.

Chan-Palay, V. (1975). Fine structure of labelled axons in the cerebellar cortex and nuclei of rodents and primates after intraventricular infusions with tritiated serotonin. *Anat. Embryol. 148*, 235–265.

Chan-Palay, V. (1977). *Cerebellar Dentate Nucleus. Organization, Cytology and Transmitters.* Springer-Verlag, Berlin.

Chase, M. H., Y. Nakamura, C. D. Clemente, and M. B. Sterman (1967). Afferent vagal stimulation: neurographic correlates of induced EEG synchronization and desynchronization. *Brain Research 5*, 236–249.

Chatrian, G. E., L. E. White, Jr., and C. M. Shaw (1964). EEG pattern resembling wakefulness in unresponsive decerebrate state following traumatic brain-stem infarct. *Electroenceph. Clin. Neurophysiol. 16*, 285–289.

Chavis, D. A., and D. N. Pandya (1976). Further observations on corticofrontal connections in the rhesus monkey. *Brain Research 117*, 369–386.

Chemical Pathways in the Brain. Handbook of Psychopharmacology, Vol. 9 (1978), (L. L. Iversen, S. D. Iversen, and S. H. Snyder, eds.). Plenum Press, New York.

Chi, C. C. (1970). An experimental silver study of the ascending projections of the central gray substance and adjacent tegmentum in the rat with observations in the cat. *J. Comp. Neurol. 139*, 259–272.

Chokroverty, S., and F. A. Rubino (1975). "Pure" motor hemiplegia. *J. Neurol. Neurosurg. Psychiat. 38*, 896–899.

Chokroverty, S., F. A. Rubino, and C. Haller (1975). Pure motor hemiplegia due to pyramidal infarction. *Arch. Neurol. (Chic.) 32*, 647–648.

Chorobski, J., and W. Penfield (1932). Cerebral vasodilator nerves and their pathway from the medulla oblongata. With observations on the pial and intracerebral vascular plexus. *Arch. Neurol. Psychiat. (Chic.) 28*, 1257–1289.

Chorover, S. L. (1976). The pacification of the brain: from phrenology to psychosurgery. In *Current Controversies in Neurosurgery* (T. P. Morley, ed.), pp. 730–767. W. B. Saunders, Philadelphia.

Chouchov, Ch. (1978). Cutaneous receptors. *Advan. Anat. Embryol. Cell Biol. 54*, 5–62.

Chow, K. L. (1950). A retrograde cell degeneration study of the cortical projection field of the pulvinar in the monkey. *J. Comp. Neurol. 93*, 313–340.

Chow, K. L., and A. L. Leiman (1970). The structural and functional organization of the neocortex. *Neurosci. Res. Progr. Bull. 8*, 153–220.

Chow, K. L., and K. H. Pribram (1956). Cortical projection of the thalamic ventrolateral nuclear group in monkeys. *J. Comp. Neurol. 104*, 57–75.

Christ, J. F. (1969). Derivation and boundaries of the hypothalamus, with atlas of hypothalamic grisea. In *The Hypothalamus* (W. Haymaker, E. Anderson, and W. J. H. Nauta, eds.), pp. 13–60. C. C. Thomas, Springfield, Ill.

Christensen, B. N., and E. R. Perl (1970). Spinal neurons specifically excited by noxious or thermal stimuli: marginal zone of the dorsal horn. *J. Neurophysiol. 33*, 293–307.

Christoff, N. (1974). A clinicopathologic study of vertical eye movements. *Arch. Neurol. (Chic.) 31*, 1–8.

Chronister, R. B., R. W. Sikes, and L. E. White, Jr. (1976). The septo-hippocampal system: significance of the subiculum. In *The Septal Nuclei* (J. F. DeFrance, ed.), pp. 115–132. Plenum Press, New York.

Chu, N.-S., and F. E. Bloom (1974). The catecholamine-containing neurons in the cat dorsolateral pontine tegmentum: distribution of the cell bodies and some axonal projections. *Brain Research 66*, 1–21.

Chung, J. M., K. Chung, and R. D. Wurster (1975). Sympathetic preganglionic neurons of the cat spinal cord: horseradish peroxidase study. *Brain Research 91*, 126–131.

Clare, M. H., and G. H. Bishop (1954). Responses from an association area secondarily activated from optic cortex. *J. Neurophysiol. 17*, 271–277.

Clark, D. A. (1931). Muscle counts of motor units: a study in innervation ratios. *Amer. J. Physiol. 96*, 296–304.

Clark, F. J. (1972). Central projection of sensory fibers from the cat knee joint. *J. Neurobiol. 3*, 101–110.

Clark, F. J. (1975). Information signaled by sensory fibers in medial articular nerve. *J. Neurophysiol. 38*, 1464–1472.

Clark, F. J., and P. R. Burgess (1975). Slowly adapting receptors in cat knee joint: can they signal joint angle? *J. Neurophysiol. 38*, 1448–1463.

Clark, F. J., S. Landgren, and H. Silfvenius (1973). Projections to the cat's cerebral cortex from low threshold joint afferents. *Acta Physiol. Scand. 89*, 504–521.

Clark, G., H. W. Magoun, and S. W. Ranson (1939). Hypothalamic regulation of body temperature. *J. Neurophysiol. 2*, 61–80.

Clark, G., and J. W. Ward (1948). Responses elicited from the cortex of monkeys by electrical stimulation through fixed electrodes. *Brain 71*, 332–342.

Clark, W. E. Le Gros (1926). The mammalian oculomotor nucleus. *J. Anat. (Lond.) 60*, 426–448.

Clark, W. E. Le Gros (1936a). The termination of ascending tracts in the thalamus of the macaque monkey. *J. Anat. (Lond.) 71*, 7–40.

Clark, W. E. Le Gros (1936b). The thalamic connections of the temporal lobe of the brain in the monkey. *J. Anat. (Lond.) 70*, 447–464.

Clark, W. E. Le Gros (1937). The connections of the arcuate nucleus of the thalamus. *Proc. Roy. Soc. B 123*, 166–176.

Clark, W. E. Le Gros (1941a). Observations on the association fibre system of the visual cortex and the central representation of the retina. *J. Anat. (Lond.) 75*, 225–235.

Clark, W. E. Le Gros (1941b). The laminar organisation and cell content of the lateral geniculate body in the monkey. *J. Anat. (Lond.) 75*, 419–433.

Clark, W. E. Le Gros (1942). The cells of Meynert in the visual cortex of the monkey. *J. Anat. (Lond.) 76*, 369–375.

Clark, W. E. Le Gros (1951). The projection of the olfactory epithelium on the olfactory bulb in the rabbit. *J. Neurol. Neurosurg. Psychiat. 14*, 1–10.

Clark, W. E. Le Gros (1956). Observations on the structure and organization of olfactory receptors in the rabbit. *Yale J. Biol. Med. 29*, 83–95.

Clark, W. E. Le Gros (1957). Inquiries into the anatomical basis of olfactory discrimination. *Proc. Roy. Soc. B 146*, 299–319.

Clark, W. E. Le Gros, J. Beattie, G. Riddoch, and N. M. Dott (1938). *The Hypothalamus. Morphological, Functional, Clinical, and Surgical Aspects*. Oliver and Boyd, Edinburgh.

Clark, W. E. Le Gros, and R. H. Boggon (1933). On the connections of the anterior nucleus of the thalamus. *J. Anat. (Lond.) 67*, 215–226.

Clark, W. E. Le Gros, and R. H. Boggon (1935). The thalamic connections of the parietal and frontal lobes of the brain in the monkey. *Phil. Trans. B 224*, 313–359.

Clark, W. E. Le Gros, and M. Meyer (1947). The terminal connexions of the olfactory tract in the rabbit. *Brain 70*, 304–328.

Clark, W. E. Le Gros, and D. W. C. Northfield (1937). The cortical projection of the pulvinar in the macaque monkey. *Brain 60*, 126–142.

Clark, W. E. Le Gros, and G. G. Penman (1934). The projection of the retina in the lateral geniculate body. *Proc. Roy. Soc. B 114*, 291–313.

Clark, W. E. Le Gros, and T. P. S. Powell (1953). On the thalamo-cortical connexions of the general sensory cortex of Macaca. *Proc. Roy. Soc. B 141*, 467–487.

Clarke, A. M. (1967). Effect of the Jendrassik maneuvre on a phasic stretch reflex in normal human subjects during experimental control over supraspinal influences. *J. Neurol. Neurosurg. Psychiat. 30*, 34–42.

Clarke, W. B., and D. Bowsher (1962). Terminal distribution of primary afferent trigeminal fibers in the rat. *Exp. Neurol. 6*, 372–383.

Clemente, C. D., and M. H. Chase (1973). Neurological substrates of aggressive behaviour. *Ann. Rev. Physiol. 35*, 329–356.

Clemente, C. D., and M. B. Sterman (1963). Cortical synchronization and sleep patterns in acute restrained and chronic behaving cats induced by basal forebrain stimulations. *Electroenceph. Clin. Neurophysiol. Suppl. 24*, 172–187.

Clendenin, M., C.-F. Ekerot, and O. Oscarsson (1974). The lateral reticular nucleus in the cat. III. Organization of component activated from ipsilateral forelimb tract. *Exp. Brain Res. 21*, 501–513.

Clendenin, M., C.-F. Ekerot, and O. Oscarsson (1975). The lateral reticular nucleus in the cat. IV. Activation from dorsal funiculus and trigeminal afferents. *Exp. Brain Res. 24*, 131–144.

Clendenin, M., C.-F. Ekerot, O. Oscarsson, and I. Rosén (1974a). The lateral reticular nucleus in the cat. I. Mossy fibre distribution in cerebellar cortex. *Exp. Brain Res. 21*, 473–486.

Clendenin, M., C.-F. Ekerot, O. Oscarsson, and I. Rosén (1974b). The lateral reticular nucleus in the cat. II. Organization of component activated from bilateral ventral flexor reflex tract (bVFRT). *Exp. Brain Res. 21*, 487–500.

Clendenin, M., C.-F. Ekerot, O. Oscarsson, and I. Rosén (1974c). Functional organization of two spinocerebellar paths relayed through the lateral reticular nucleus in the cat. *Brain Research 69*, 140–143.

Clifton, G. L., R. E. Coggeshall, W. H. Vance, and W. D. Willis (1976). Receptive fields of unmyelinated ventral root afferent fibres in the cat. *J. Physiol. (Lond.) 256*, 573–600.

Clopton, B. M., J. A. Winfield, and F. J. Flammino (1974). Tonotopic organization: review and analysis. *Brain Research 76*, 1–20.

Clough, J. F. M., C. G. Phillips, and J. D. Sheridan (1971). The short-latency projection from the baboon's motor cortex to fusimotor neurones of the forearm and hand. *J. Physiol. (Lond.) 216*, 257–279.

Coërs, C., and A. L. Woolf (1959). *The Innervation of Muscle. A Biopsy Study*. Blackwell, Oxford.

Cogan, D. C. (1974). Paralysis of down-gaze. *Arch. Ophthal. 91*, 192–199.

Cogan, D. G. (1966). *Neurology of the Visual System*. C. C. Thomas, Springfield, Ill.

Coggeshall, R. E., M. L. Applebaum, M. Frazen, T. B. Stubbs III, and M. T. Sykes (1975). Unmyelinated axons in human ventral roots, a possible explanation for the failure of dorsal rhizotomy to relieve pain. *Brain 98*, 157–166.

Coggeshall, R. E., and S. L. Galbraith (1978). Categories of axons in mammalian rami communicantes. Part II. *J. Comp. Neurol. 181*, 349–360.

Cohen, B. (1971). Vestibulo-ocular relations. In *The Control of Eye Movements* (P. Bach-y-Rita, C. C. Collins, and J. E. Hyde, eds.), pp. 105–148. Academic Press, New York.

Cohen, B. (1974). The vestibulo-ocular reflex arc. In *Handbook of Sensory Physiology*, Vol. VI/1 (H. H. Kornhuber, ed.), pp. 477–540. Springer-Verlag, Berlin.

Cohen, B., K. Goto, S. Shanzer, and A. H. Weiss (1965). Eye movements induced by electrical stimulation of the cerebellum in the alert cat. *Exp. Neurol. 13*, 145–162.

Cohen, B., and S. M. Highstein (1972). Cerebellar control of the vestibular pathways to oculomotor neurons. In *Basic Aspects of Central Vestibular Mechanisms, Progress in Brain Research*, Vol. 37 (A. Brodal and O. Pompeiano, eds.), pp. 411–425. Elsevier, Amsterdam.

Cohen, B., and A. Komatsuzaki (1972). Eye movements induced by stimulation of the pontine reticular formation: evidence for integration in oculomotor pathways. *Exp. Neurol. 36*, 101–117.

Cohen, B., A. Komatsuzaki, and M. B. Bender (1968). Electrooculographic syndrome in monkeys after pontine reticular formation lesions. *Arch. Neurol. 18*, 78–92.

Cohen, B., J.-I. Suzuki, and M. B. Bender (1964). Eye movements from semicircular canal nerve stimulation in the cat. *Ann. Otol. (St. Louis) 73*, 153–169.

Cohen, D., W. W. Chambers, and J. M. Sprague (1958). Experimental study of the efferent projections from the cerebellar nuclei to the brain stem of the cat. *J. Comp. Neurol. 109*, 233–259.

Cohen, L. A. (1961). Role of eye and neck proprioceptive mechanisms in body orientation and motor coordination. *J. Neurophysiol. 24*, 1–11.

Cohen, M. J., S. Landgren, L. Strøm, and Y. Zotterman (1957). Cortical reception of touch and taste in the cat. A study of single cortical cells. *Acta Physiol. Scand. 40,* *Suppl. 135,* 1–50.

Coimbra, A., B. P. Sodré-Borges, and M. M. Magalhães (1974). The substantia gelatinosa Rolandi of the rat. Fine structure, cytochemistry (acid phosphatase) and changes after dorsal root section. *J. Neurocytol. 3,* 199–217.

Cole, M., W. J. H. Nauta, and W. R. Mehler (1964). The ascending efferent projections of the substantia nigra. *Trans. Amer. Neurol. Ass. 89,* 74–78.

Collier, J., and E. F. Buzzard (1901). Descending mesencephalic tracts in cat, monkey and man; Monakow's bundle; the dorsal longitudinal bundle; the ventral longitudinal bundle; the ponto-spinal tracts, lateral and ventral; the vestibulo-spinal tract; the central tegmental tract (centrale Haubenbahn); descending fibres of the fillet. *Brain 24,* 177–221.

Collins, C. C. (1975). The human oculomotor control system. In *Basic Mechanisms of Ocular Motility and their Clinical Implications* (G. Lennerstrand and P. Bach-y-Rita, eds.), pp. 145–180. Pergamon Press, Oxford.

Collins, W. F., Jr., F. E. Nulsen, and C. T. Randt (1960). Relation of peripheral nerve fiber size and sensation in man. *Arch. Neurol. (Chic.) 3,* 381–385.

Colonnier, M. (1964). The tangential organization of the visual cortex. *J. Anat. (Lond.) 98,* 327–344.

Colonnier, M. L. (1966). The structural design of the neocortex. In *Brain and Conscious Experience* (J. C. Eccles, ed.), pp. 1–18. Springer-Verlag, Berlin.

Colonnier, M. (1967). The fine structural arrangement of the cortex. Arch. Neurol. (*Chic.*) *16,* 651–657.

Colonnier, M., and E. G. Gray (1962). Degeneration in the cerebral cortex. In *Electron Microscopy. Fifth International Congress for Electron Microscopy.* Vol. 2, U-3 (S. S. Breese, Jr., ed.). Academic Press, New York.

Colonnier, M., and R. W. Guillery (1964). Synaptic organization in the lateral geniculate nucleus of the monkey. *Z. Zellforsch. 62,* 333–355.

Colonnier, M., and S. Rossignol (1969). Heterogeneity of the cerebral cortex. In *Basic Mechanisms of the Epilepsies* (H. H. Jasper et al., eds.), pp. 29–40. Little, Brown, Boston.

Combs, C. M. (1956). Bulbar regions related to localized cerebellar afferent impulses. *J. Neurophysiol. 19,* 285–300.

Conceptual Models of Neural Organization (1974). (J. Szentágothai and M. A. Arbib, eds.). *Neurosci. Res. Progr. Bull.,* Vol. 12, 305–510

Condé, H., and P. Angaut (1970). An electrophysiological study of the cerebellar projections to the nucleus ventralis lateralis thalami in the cat. II. Nucleus lateralis. *Brain Research 20,* 107–119.

Conel, J. Le Roy (1939). *The Postnatal Development of the Human Cerebral Cortex,* Vol. I. The Cortex of the Newborn. Harvard Univ. Press, Cambridge, Mass.

Conrad, B., and V. B. Brooks (1974). Effects of dentate cooling on rapid alternating arm movements. *J. Neurophysiol. 37,* 792–804.

Conrad, L. C. A., C. M. Leonard, and D. W. Pfaff (1974). Connections of the median and dorsal raphe nuclei in the rat: an autoradiographic and degeneration study. *J. Comp. Neurol. 156,* 179–206.

Conrad, L. C. A., and D. W. Pfaff (1976a). Efferents from medial basal forebrain and hypothalamus in the rat. I. An autoradiographic study of the medial preoptic area. *J. Comp. Neurol. 169,* 185–220.

Conrad, L. C. A., and D. W. Pfaff (1976b). Efferents from the medial basal forebrain and hypothalamus in the rat. II. An autoradiographic study of the anterior hypothalamus. *J. Comp. Neurol. 169,* 221–261.

Conradi, S. (1969). On motoneuron synaptology in adult cats. An electron microscopic study of the structure and location of neuronal and glial elements on cat lumbosacral motoneurons in the normal state and after dorsal root section. *Acta Physiol. Scand. Suppl. 332.*

Consciousness and the Brain. A Scientific and Philosophical Inquiry (1976). (G. G. Globus, G. Maxwell and I. Savodnik, eds.), Plenum Press, New York.

Contemporary Research Methods in Neuroanatomy (1970). (W. J. H. Nauta and S. O. E. Ebbesson, eds.). Springer-Verlag, Berlin.

Contributions to Sensory Physiology, Vol. 4 (1970). (W. D. Neff, ed.). Academic Press, New York.

Control of Eye Movements, The (1972). (P. Bach-y-Rita, C. C. Collins, and J. E. Hyde, eds.). Academic Press, New York.

Control of Gaze by Brain Stem Neurons (1977). (R. Baker and A. Berthoz, eds.). Elsevier/North-Holland, Amsterdam.

Cook, A. W., and E. J. Browder (1965). Function of posterior columns in man. *Arch. Neurol. (Chic.) 12*, 72–79.

Cook, J. R., and M. Wiesendanger (1976). Input from trigeminal cutaneous afferents to neurones of the inferior olive in rats. *Exp. Brain Res. 26*, 193–202.

Cook, W. H., J. H. Walker, and M. L. Barr (1951). A cytological study of transneuronal atrophy in the cat and rabbit. *J. Comp. Neurol. 94*, 267–292.

Cooke, J. D., B. Larson, O. Oscarsson, and B. Sjölund (1971a). Origin and termination of cuneocerebellar tract. *Exp. Brain Res. 13*, 339–358.

Cooke, J. D., Larson, O. Oscarsson, and B. Sjölund (1971b). Organization of afferent connections to cuneocerebellar tract. *Exp. Brain Res. 13*, 359–377.

Cooke, J. D., O. Oscarsson, and B. Sjölund (1972). Termination areas of climbing fibre paths in paramedian lobule. *Acta Physiol. Scand. 84*, 37A–38A.

Cooke, P. M., and R. S. Snider (1953). Some cerebellar effects on the electrocorticogram. *Electroenceph. Clin. Neurophysiol. 5*, 563–569.

Cooke, P. M., and R. S. Snider (1955). Some cerebellar influences on electrically induced cerebral seizures. *Epilepsia 4*, 19–28.

Cooper, I. S. (1959). Dystonia musculorum deformans alleviated by chemopallidectomy and chemopallidothalamectomy. *Arch. Neurol. (Chic.) 81*, 5–19.

Cooper, I. S. (1965a). Clinical and physiologic implications of thalamic surgery for disorder of sensory communication. Part 1. Thalamic surgery for intractable pain. *J. Neurol. Sci. 2*, 493–519.

Cooper, I. S. (1965b). Clinical and physiologic implications of thalamic surgery for disorders of sensory communication. Part 2. Intention tremor, dystonia, Wilson's disease and torticollis. J. Neurol. Sci. 2, 520–553.

Cooper, I. S. (1976). Dystonia: surgical approaches to treatment and physiologic implications. In *The Basal Ganglia* (M. D. Yahr, ed.), pp. 369–383. Raven Press, New York.

Cooper, I. S., L. L. Bergmann, and A. Caracalos (1963). Anatomic verification of the lesion which abolishes Parkinsonian tremor and rigidity. *Neurology (Minneap.) 13*, 779–787.

Cooper, K. E. (1966). Temperature regulation and the hypothalamus. *Brit. Med. Bull. 22*, 238–242.

Cooper, M. H., and J. A. Beal (1978). The neurons and the synaptic endings in the primate basilar pontine gray. *J. Comp. Neurol. 180*, 17–42.

Cooper, M. H., and C. A. Fox (1976). The basilar pontine gray in the adult monkey (*Macaca mulatta*): A Golgi study. *J. Comp. Neurol. 168*, 145–174.

Cooper, S. (1929). The relations of active and inactive fibres in fractional contraction of muscle. *J. Physiol. (Lond.) 67*, 1–13.

Cooper, S. (1953). Muscle spindles in the intrinsic muscles of the human tongue. *J. Physiol. (Lond.) 122*, 193–202.

Cooper, S. (1960). Muscle spindles and other muscle receptors. In *Structure and Function of Muscle*, Vol. I, Structure (G. H. Bourne, ed.), pp. 381–420. Academic Press, New York.

Cooper, S., and P. M. Daniel (1949). Muscle spindles in human extrinsic eye muscles. *Brain 72*, 1–24.

Cooper, S., and P. M. Daniel (1963). Muscle spindles in man; their morphology in the lumbricals and the deep muscles of the neck. *Brain 86,* 563–586.

Cooper, S., P. M. Daniel, and D. Whitteridge (1951). Afferent impulses in the oculomotor nerve from the extrinsic eye muscles. *J. Physiol. (Lond.) 113,* 463–474.

Cooper, S., P. M. Daniel, and D. Whitteridge (1953a). Nerve impulses in the brainstem of the goat. Short latency responses obtained by stretching the extrinsic eye muscles and the jaw muscles. *J. Physiol. (Lond.) 120,* 471–491.

Cooper, S., P. M. Daniel, and D. Whitteridge (1953b). Nerve impulses in the brainstem of the goat. Responses with long latencies obtained by stretching the extrinsic eye muscles. *J. Physiol. (Lond.) 120,* 491–513.

Cooper, S., P. M. Daniel, and D. Whitteridge (1955). Muscle spindles and other sensory endings in the extrinsic eye muscles; the physiology and anatomy of these receptors and of their connexions with the brain-stem. *Brain 78,* 564–583.

Cooper, S., and C. S. Sherrington (1940). Gower's tract and spinal border cells. *Brain 63,* 123–134.

Coote, J. H. (1975). Physiological significance of somatic afferent pathways from skeletal muscle and joints with reflex effects on the heart and circulation. *Brain Research 87* 139–144.

Coquery, J.-M. (1978). Role of active movement in control of afferent input from skin in cat and man. In *Active Touch. The Mechanism of Recognition of Objects by Manipulation* (G. Gordon, ed.), pp. 161–169. Pergamon Press, Oxford.

Corazza, R., E. Fadiga, and P. L. Parmeggiani (1963). Patterns of pyramidal activation of cat's motoneurons. *Arch. Ital Biol. 101,* 337–364.

Corbin, K. B. (1940). Observations on the peripheral distribution of fibers arising in the mesencephalic nucleus of the fifth cranial nerve. *J. Comp. Neurol. 73,* 153–177.

Corbin, K. B., and F. Harrison (1940). Function of mesencephalic root of the fifth cranial nerve. *J. Neurophysiol. 3,* 423–435.

Corsellis, J. A. N. (1970). The neuropathology of temporal lobe epilepsy. In *Modern Trends in Neurology,* Vol. 5 (D. Williams, ed.), pp. 254–270. Butterworths, London.

Corticothalamic Projections and Sensorimotor Activities (1972). (T. L. Frigyesi, E. Rinvik, and M. D. Yahr, eds.). Raven Press, New York.

Corvaja, N., I. Grofová, O. Pompeiano, and F. Walberg (1977a). The lateral reticular nucleus in the cat. I. An experimental anatomical study of its spinal and supraspinal afferent connections. *Neuroscience. 2,* 537–553.

Corvaja, N., I. Grofová, O. Pompeiano, and F. Walberg (1977b). The lateral reticular nucleus in the cat. II. Effects of lateral reticular lesions on posture and reflex movements. *Neuroscience 2,* 929–943.

Cottle, M. K. W., and F. R. Calaresu (1975). Projections from the nucleus and tractus solitarius in the cat. *J. Comp. Neurol. 161,* 143–158.

Cotzias, G. C., M. H. Van Woert, and L. M. Schiffer (1967). Aromatic amino acids and modification of parkinsonism. *New Engl. J. Med. 280,* 337–345.

Coulter, J. D. (1974). Sensory transmission through lemniscal pathway during voluntary movement in the cat. *J. Neurophysiol. 37,* 831–845.

Coulter, J. D., L. Ewing, and C. Carter (1976). Origin of primary sensorimotor cortical projections to lumbar spinal cord of cat and monkey. *Brain Research 103,* 366–372.

Coulter, J. D., and E. G. Jones (1977). Differential distribution of corticospinal projections from individual cytoarchitectonic fields in the monkey. *Brain Research 129,* 335–340.

Coulter, J. D., R. A. Maunz, and W. D. Willis (1974). Effects of stimulation of sensorimotor cortex on primate spinothalamic neurons. *Brain Research 65,* 351–356.

Courville, J. (1966a). Rubrobulbar fibres to the facial nucleus and the lateral reticular nucleus (nucleus of the lateral funiculus). An experimental study in the cat with silver impregnation methods. *Brain Research 1,* 317–337.

Courville, J. (1966b). The nucleus of the facial nerve; the relation between cellular groups and peripheral branches of the nerve. *Brain Research 1,* 338–354.

Courville, J. (1966c). Somatotopical organization of the projection from the nucleus interpositus anterior of the cerebellum to the red nucleus. An experimental study in the cat with silver impregnation methods. *Exp. Brain Res. 2,* 191–215.

Courville, J. (1975). Distribution of olivocerebellar fibers demonstrated by a radioautographic tracing method. *Brain Research 95,* 253–263.

Courville, J., J. R. Augustine, and P. Martel (1977). Projections from the inferior olive to the cerebellar nuclei in the cat demonstrated by retrograde transport of horseradish peroxidase. *Brain Research 130,* 405–419.

Courville, J., and A. Brodal (1966). Rubrocerebellar connections in the cat. An experimental study with silver impregnation methods. *J. Comp Neurol. 126,* 471–485.

Courville, J., and C. W. Cooper (1970). The cerebellar nuclei of *Macaca mulatta.* A morphological study. *J. Comp. Neurol. 140,* 241–254.

Courville, J., and N. Diakiw (1976). Cerebellar corticonuclear projection in the cat. The vermis of the anterior and posterior lobes. *Brain Research 110,* 1–20.

Courville, J., and N. Diakiw, and A. Brodal (1973). Cerebellar corticonuclear projection in the cat. The paramedian lobule. An experimental study with silver methods. *Brain Research 50,* 25–45.

Courville, J., and S. Otabe (1974). The rubroolivary projection in the macaque: an experimental study with silver impregnation methods. *J. Comp Neurol. 158,* 479–494.

Cowan, W. M. (1970). Anterograde and retrograde transneuronal degeneration in the central and peripheral nervous system. In *Contemporary Research Methods in Neuroanatomy* (W. J. H. Nauta and S. O. E. Ebbesson, eds.), pp. 217–249. Springer-Verlag, Berlin.

Cowan, W. M., L. Adamson, and T. P. S. Powell (1961). An experimental study of the avian visual system. *J. Anat. (Lond.) 95,* 545–563.

Cowan, W. M., and M. Cuénod (1975). The use of axonal transport for the study of neural connections: a retrospective survey. In *The Use of Axonal Transport for Studies of Neuronal Connectivity* (W. M. Cowan and M. Cuénod, eds.), pp. 1–24. Elsevier, Amsterdam.

Cowan, W. M., D. I. Gottlieb, A. E. Hendrickson, J. L. Price, and T. A. Woolsey (1972). The autoradiographic demonstration of axonal connections in the central nervous system. *Brain Research 37,* 21–51.

Cowan, W. M., R. W. Guillery, and T. P. S. Powell (1964). The origin of the mamillary peduncle and other hypothalamic connexions from the midbrain. *J. Anat. (Lond.) 98,* 345–363.

Cowan, W. M., and T. P. S. Powell (1954). An experimental study of the relation between the medial mammillary nucleus and the cingulate cortex. *Proc. Roy. Soc. B 143,* 114–125.

Cowan, W. M., and T. P. S. Powell (1955). The projection of the midline and intralaminar nuclei of the thalamus of the rabbit. *J. Neurol. Neurosurg. Psychiat. 18,* 266–279.

Cowan, W. M., and T. P. S. Powell (1963). Centrifugal fibers in the avian visual system. *Proc. Roy. Soc. B 158,* 232–252.

Cowan, W. M., and T. P. S. Powell (1966). Striopallidal projection in the monkey. *J. Neurol. Neurosurg. Psychiat. 29,* 426–439.

Cowan, W. M., G. Raisman, and T. P. S. Powell (1965). The connexions of the amygdala. *J. Neurol. Neurosurg. Psychiat. 28,* 137–151.

Cowey, A. (1964). Projection of the retina on to striate and prestriate cortex in the squirrel monkey, *Saimiri sciureus. J. Neurophysiol. 27,* 366–393.

Cowley, L. H., and J. M. Shackleford (1970). An ultrastructural study of the submandibular glands of the squirrel monkey, *Saimiri sciureus. J. Morph. 132,* 117–136.

Coxe, W. S., and W. M. Landau (1965). Observations upon the effect of supplementary motor cortex ablation in the monkey. *Brain 88,* 763–772.

Coyle, J. T., and R. Schwarcz (1976). Lesion of striatal neurons with kainic acid provides a model for Huntington's chorea. *Nature 263,* 244–246.

Cracco, R. Q., and R. C. Bickford (1968). Somatomotor and somatosensory evoked responses. *Arch. Neurol. (Chic.) 18,* 52–68.

Cragg, B. G. (1960). Responses of the hippocampus to stimulation of the olfactory bulb and of various afferent nerves in five mammals. *Exp. Neurol. 2*, 547–572.

Cragg, B. G. (1961). Olfactory and other afferent connections of the hippocampus in the rabbit, rat, and cat. *Exp. Neurol. 3*, 588–600.

Cragg, B. G. (1962). Centrifugal fibers to the retina and olfactory bulb, and composition of the supraoptic commissures in the rabbit. *Exp. Neurol. 5*, 406–427.

Cragg, B. G. (1965). Afferent connexions of the allocortex. *J. Anat. (Lond.) 99*, 339–357.

Cragg, B. G. (1967). The density of synapses and neurones in the motor and visual areas of the cerebral cortex. *J. Anat. (Lond.) 101*, 639–654.

Cragg, B. G. (1970). What is the signal for chromatolysis? *Brain Research 23*, 1–21.

Cragg, B. G. (1974). Plasticity of synapses. *Brit. Med. Bull. 30*, 141–144.

Cragg, B. G. (1975). The density of synapses and neurons in normal, mentally defective and ageing human brains. *Brain 98*, 81–90.

Cragg, B. G., and L. H. Hamlyn (1959). Histologic connections and electrical and autonomic responses evoked by stimulation of the dorsal fornix in the rabbit. *Exp. Neurol. 1*, 187–213.

Craggs, M. D., and D. N. Rushton (1976). The stability of the electrical stimulation map of the motor cortex of the anæsthetized baboon. *Brain 99*, 575–600.

Craig, A. D., Jr. (1978). Spinal and medullary input to the lateral cervical nucleus. *J. Comp. Neurol. 181*, 728–744.

Craig, A. D., Jr., and H. Burton (1979). The lateral cervical nucleus in the cat: anatomic organization of cervicothalamic neurons. *J. Comp. Neurol. 185*, 329–346.

Craig, A. D., Jr., and D. N. Tapper (1978). Lateral cervical nucleus in the cat: functional organization and characteristics. *J. Neurophysiol. 41*, 1511–1534.

Cravioto, H., J. Silberman, and I. Feigin (1960). A clinical and pathologic study of akinetic mutism. *Neurology (Minneap.) 10*, 10–21.

Creutzfeldt, O., and H. Akimoto (1958). Konvergenz und gegenseitige Beeinflussung von Impulsen aus der Retina und den unspezifischen Thalamuskernen an einzelnen Neuronen des optischen Cortex. *Arch. Psychiat. Nervenkr. 196*, 520–538.

Creutzfeldt, O. D., H. D. Lux, and S. Watanabe (1966). Electrophysiology of cortical nerve cells. In *The Thalamus* (D. P. Purpura and M. D. Yahr, eds.), pp. 209–230. Columbia Univ. Press, New York.

Crill, W. E. (1970). Unitary multiple-spiked responses in cat inferior olive nucleus. *J. Neurophysiol. 33*, 199–209.

Crill, W., and T. T. Kennedy (1967). Inferior olive of the cat: intracellular recording. *Science 157*, 717–718.

Critchley, M. (1953). *The Parietal Lobes*. Edward Arnold, London.

Critchley, M. (1966). The enigma of Gerstmann's syndrome. *Brain 89*, 183–198.

Critchley, M. (1970). *Aphasiology and other Aspects of Language*. Edward Arnold, London.

Crompton, M. R. (1971). Hypothalamic lesions following dorsal head injury. *Brain 94*, 165–172.

Crosby, E. C. (1953). Relations of brain centers to normal and abnormal eye movements in the horizontal plane. *J. Comp. Neurol. 99*, 437–480.

Crosby, E. C., and J. W. Henderson (1948). The mammalian midbrain and isthmus régions. Part II. Fiber connections of the superior colliculus. B. Pathways concerned in automatic eye movements, *J. Comp. Neurol. 88*, 53–92.

Crosby, E. C., and T. Humphrey (1941). Studies of the vertebrate telencephalon. II. The nuclear pattern of the anterior olfactory nucleus, tuberculum olfactorium and the amygdaloid complex in adult man. *J. Comp. Neurol. 74*, 309–352.

Crosby, E. C., R. C. Schneider, B. R. DeJonge, and P. Szonyi (1966). The alterations of tonus and movements through the interplay between the cerebral hemispheres and the cerebellum. *J. Comp. Neurol. 127, Suppl. 1*, 1–91.

Cross, B. A., and I. A. Silver (1966). Electrophysiological studies on the hypothalamus. *Brit. Med. Bull. 22*, 254–258.

Crowe, S. J., S. R. Guild, and L. M. Polvogt (1934). Observations on the pathology of high-tone deafness. *Bull. Johns Hopk. Hosp. 54*, 315–379.

Crowell, R. M., E. Perret, J. Siegfried, and J. P. Villoz (1968). Movement units and tremor phasic units in the human thalamus. *Brain Research 11*, 481–488.

Cruce, J. A. F. (1975). An autoradiographic study of the projections of the mammillothalamic tract in the rat. *Brain Research 85*, 211–219.

Cruce, J. A. F. (1977). An autoradiographic study of the descending connections of the mammillary nuclei of the rat. *J. Comp. Neurol. 176*, 631–644.

Cuénod, M., and W. M. Cowan (1975). Some future developments in the use of axonal transport mechanisms for tracing pathways in the central nervous system. In *The Use of Axonal Transport for Studies of Neuronal Connectivity* (W. M. Cowan and M. Cuénod, eds.), pp. 337–346. Elsevier, Amsterdam.

Culberson, J. L., and L. M. N. Bach (1973). Limbic projections to the ventromedial hypothalamus of the opossum. *Exp. Neurol. 41*, 683–689.

Culberson, J. L., and D. L. Kimmel (1972). Central distribution of primary afferent fibers of the glossopharyngeal and vagal nerves in the opossum, *Didelphis virginiana. Brain Research 44*, 325–335.

Cullheim, S., and J. O. Kellerth (1976). Combined light and electron microscopical tracing of neurones, including axons and synaptic terminals, after intracellular injection of horseradish peroxidase. *Neurosci. Letters 2*, 307–313.

Cullheim, S., J.-O. Kellerth, and S. Conradi (1977). Evidence for direct synaptic interconnections between cat spinal α-motoneurons via the recurrent axon collaterals: a morphological study using intracellular injection of horseradish peroxidase. *Brain Research 132*, 1–10.

Current Controversies in Neurosurgery (1976). (T. P. Morley, ed.). W. B. Saunders, Philadelphia.

Curry, M. J. (1972). The exteroceptive properties of neurones in the somatic part of the posterior group (PO). *Brain Research 44*, 439–462.

Curzon, G. (1967). The biochemistry of dyskinesias. *Int. Rev. Neurobiol. 10*, 323–370.

Cushing, H. (1904). The sensory distribution of the fifth cranial nerve. *Bull. Johns. Hopk. Hosp. 15*, 213–232.

Cushing, H. (1922). The field defects produced by temporal lobe lesions. *Brain 44*, 341–396.

Cutaneous Innervation. Advances in Biology of Skin, Vol. 1 (1960). (W. Montagna, ed.). Pergamon Press, Oxford.

Czarkowska, J., E. Jankowska, and E. Sybirska (1976). Axonal projection of spinal interneurones excited by group I afferents in the cat, revealed by intracellular staining with horseradish peroxidase. *Brain Research 118*, 115–118.

Dahl, E., and E. Nelson (1964). Electron microscopic observations on human intracranial arteries. *Arch. Neurol. (Chic.) 10*, 158–164, 1964.

Dahlström, A., and K. Fuxe (1964). Evidence for the existence of monoamine-containing neurons in the central nervous system. I. Demonstration of monoamines in the cell bodies of brain stem neurons. *Acta Physiol. Scand. 62, Suppl. 232*. 1–55.

Dahlström, A., and K. Fuxe (1965). Evidence for the existence of monoamine neurons in the central nervous system. *Acta Physiol. Scand. 64, Suppl. 247*, 1–36.

Daitz, H. M., and T. P. S. Powell (1954). Studies of the connexions of the fornix system. *J. Neurol. Neurosurg. Psychiat. 17*, 75–82.

D'Alecy, L. G., and E. O. Feigl (1972). Sympathetic control of cerebral blood flow in dogs. *Circulat. Res. 31*, 267–283.

Daniel, P. (1946). Spiral nerve endings in the extrinsic eye muscles of man. *J. Anat. (Lond.) 80*, 189–193.

Daniel, P. M. (1966). The blood supply of the hypothalamus and pituitary gland. *Brit. Med. Bull. 22*, 202–208.

Daniel, P. M., and M. L. Prichard (1975). Studies on the hypothalamus and the pituitary gland. *Acta Endocr. (Kbh.) 80, Suppl. 201,* 1–216.

Danta, G., R. C. Hilton, and P. G. Lynch (1975). Chronic progressive external ophthalmoplegia. *Brain 98,* 473–492.

Darian-Smith, I. (1965). Presynaptic component in the afferent inhibition observed within trigeminal brain-stem nuclei in the cat. *J. Neurophysiol. 28,* 695–709.

Darian-Smith, I. (1966). Neural mechanisms of facial sensation. *Int. Rev. Neurobiol. 9,* 301–395.

Darian-Smith, I. (1973). The trigeminal system. In *Handbook of Sensory Physiology,* Vol. II, *Somatosensory System* (A. Iggo, ed.), pp. 271–314. Springer-Verlag, Berlin.

Darian-Smith, I., J. Isbister, H. Mok, and T. Yokota (1966). Somatic sensory cortical projection areas excited by tactile stimulation of the cat: a triple representation. *J. Physiol. (Lond.) 182,* 671–689.

Darian-Smith, I., and K. O. Johnson (1977). Temperature sense in the primate. *Brit. Med. Bull. 33,* 143–148.

Darian-Smith, I., and G. Phillips (1964). Secondary neurones within a trigemino-cerebellar projection to the anterior lobe of the cerebellum in the cat. *J. Physiol. (Lond.) 170,* 53–68.

Darian-Smith, I., and T. Yokota (1966). Cortically evoked depolarization of trigeminal cutaneous afferent fibers in the cat. *J. Neurophysiol. 29,* 170–184.

Daval, G., and J. Leveteau (1974). Electrophysiological studies of centrifugal and centripetal connections of the anterior olfactory nucleus. *Brain Research 78,* 395–410.

Davis, H. (1961). Some principles of sensory receptor action. *Physiol. Rev. 41,* 391–416.

Davis, R. (1969). Brachio-rubral and rubro-brachial activity in the cat. *Brain Research 15,* 157–173.

Davison, C. (1940). Disturbances of temperature regulation in man. *Res. Publ. Ass. Nerv. Ment. Dis. 20,* 774–823.

Davison, C. (1942). The role of the globus pallidus and substantia nigra in the production of rigidity and tremor. A clinico-pathological study of paralysis agitans. *Res. Publ. Ass. Nerv. Ment. Dis. 21,* 267–333.

Dawson, B. H. (1958). The blood vessels of the human optic chiasma and their relation to those of the hypophysis and hypothalamus. *Brain 81,* 207–217.

deAndrade, J. R., C. Grant, and A. St. J. Dixon (1965). Joint distension and reflex muscle inhibition in the knee. *J. Bone Jt. Surg. 47-A,* 313–321.

DeArmond, S. J., M. M. Fusco, and M. M. Dewey (1976). *Structure of the Human Brain. A Photographic Atlas,* 2nd ed. Oxford Univ. Press, New York.

Deecke, L., P. Scheid, and H. H. Kornhuber (1969). Distribution of readiness potential, pre-motion positivity, and motor potential of the human cerebral cortex preceding voluntary finger movements. *Exp. Brain Res. 7,* 158–168.

Deecke, L., D. W. F. Schwarz, and J. M. Fredrickson (1974). Nucleus ventroposterior inferior (VPI) as the vestibular thalamic relay in the rhesus monkey. I. Field potential investigation. *Exp. Brain Res. 20,* 88–100.

Defaria, C. R., and K. Toyonaga (1978). Motor unit estimation in a muscle supplied by the radial nerve. *J. Neurol. Neurosurg. Psychiat. 41,* 794–797.

Defendini, R., and E. A. Zimmerman (1978). The magnocellular neurosecretory system of the mammalian hypothalamus. In *The Hypothalamus* (S. Reichlin, R. J. Baldessarini, and J. B. Martin, eds.), pp. 137–152. Raven Press, New York.

DeFrance, J. F., S. T. Kitai, and T. Shimono (1973). Electrophysiological analysis of the hippocampal-septal projections: I. Response and topographical characteristics. *Exp. Brain Res. 17,* 447–462.

De Gail, P., J. W. Lance, and P. P. Neilson (1966). Differential effects on tonic and phasic stretch reflex mechanisms produced by vibration of muscle in man. *J. Neurol. Neurosurg. Psychiat. 29,* 1–11.

De Groat, W. D. (1975). Nervous control of the urinary bladder of the cat. *Brain Research 87,* 201–211.

Déjérine, J. (1914). *Sémiologie des Affections du Système Nerveux.* Massion et Cie, Paris.

Dekaban, A. (1953). Human thalamus: an anatomical developmental and pathological study. *J. Comp. Neurol. 99*, 639–683.

Dekker, J. J., and H. G. J. M. Kuypers (1976). Quantitative EM study of projection terminals in the rat's AV thalamic nucleus. Autoradiographic and degeneration techniques compared. *Brain Research 117*, 399–422.

Delgado, J. M. R. (1963). Effect of brain stimulation on task-free situations. *Electroenceph. Clin. Neurophysiol. 24*, 260–280.

Delgado, J. M. R. (1964). Free behavior and brain stimulation. *Int. Rev. Neurobiol. 6*, 349–449.

Dell, P., M. Bonvallet, and H. Hugelin (1961). Mechanisms of reticular deactivation. In *The Nature of Sleep*. Ciba Foundation Symposium (G. E. W. Wolstenholme and M. O'Connor, eds.), pp. 86–102. J. & A. Churchill, London.

DeLong, M. R. (1971). Activity of pallidal neurons during movement. *J. Neurophysiol. 34*, 414–427.

DeLong, M. R. (1972). Activity of basal ganglia neurons during movement. *Brain Research 40*, 127–135.

DeLong, M. R. (1973). Putamen: activity of single units during slow and rapid arm movements. *Science 179*, 1240–1242.

DeLong, M. R., and A. P. Georgopoulos (1979). Functional organization of the substantia nigra, globus pallidus and subthalamic nucleus in the monkey. In *The Extrapyramidal System and its Disorders* (L. J. Poirier, T. L. Sourkes, and P. J. Bédard, eds.), pp. 131–140. Raven Press, New York.

DeLong, M. R., and P. L. Strick (1974). Relation of basal ganglia, cerebellum, and motor cortex units to ramp and ballistic limb movements. *Brain Research 71*, 327–335.

De Lorenzo, A. J. D. (1963). Studies on the ultrastructure and histophysiology of cell membranes, nerve fibers and synaptic junctions in chemoreceptors. In *Olfaction and Taste* (Y. Zotterman, ed.), pp. 5–17. Pergamon Press, Oxford.

De Lorenzo, A. J. D. (1970). The olfactory neuron and the blood–brain barrier. In *Taste and Smell in Vertebrates* (G. E. Wolstenholme and J. Knight, eds.), pp. 151–173. J. & A. Churchill, London.

Delwaide, P. J. (1973). Human monosynaptic reflexes and presynaptic inhibition. An interpretation of spastic hyperreflexia. In *New Developments in Electromyography and Clinical Neurophysiology*, Vol. 3 (J. E. Desmedt, ed.), pp. 508–522. Karger, Basel.

Dempsey, E. W., and R. S. Morison (1942). The production of rhythmically recurrent cortical potentials after localized thalamic stimulation. *Amer. J. Physiol. 135*, 293–300.

DeMyer, W. (1959). Number of axons and myelin sheaths in adult human medullary pyramids. Study with silver impregnation and iron hematoxylin staining methods. *Neurology (Minneap.) 9*, 42–47.

Deniau, J. M., J. Feger, and C. Le Guyader (1976). Striatal evoked inhibition of identified nigro-thalamic neurons. *Brain Research 104*, 152–156.

Deniau, J. M., C. Hammond, A. Riszk, and J. Feger (1978). Electrophysiological properties of identified output neurons of the rat substantia nigra (pars compacta and pars reticulata): evidence for the existence of branched neurons. *Exp. Brain Res. 32*, 409–422.

Dennis, B. J., and D. I. B. Kerr (1975). Olfactory bulb connections with basal rhinencephalon in the ferret: an evoked potential and neuroanatomical study. *J. Comp. Neurol. 159*, 129–148.

Dennis, B. J., and D. I. B. Kerr (1976). Origins of olfactory bulb centrifugal fibres in the cat. *Brain Research 110*, 593–600.

Denny-Brown, D. (1962). *The Basal Ganglia and Their Relation to Disorders of Movement*. Oxford Univ. Press, London.

Denny-Brown, D. (1968). Clinical symptomatology of diseases of the basal ganglia. In *Diseases of the Basal Ganglia*, Vol. 6, *Handbook of Clinical Neurology* (P. J. Vinken and G. W. Bruyn, eds.), pp. 133–172. North-Holland, Amsterdam.

Denny-Brown, D. (1977). Spasm of visual fixation. In *Physiological Aspects of Clinical Neurology* (F. Clifford Rose, ed.), pp. 43–75. Blackwell, Oxford.

Denny-Brown, D., E. J. Kirke, and N. Yanagisawa (1973). The tract of Lissauer in relation to sensory transmission in the dorsal horn of spinal cord in the Macaque monkey. *J. Comp. Neurol. 151*, 175–200.

Denny-Brown, D., and N. Yanagisawa (1973). The function of the descending root of the fifth nerve. *Brain 96*, 783–814.

Densert, O. (1974). Adrenergic innervation in the rabbit cochlea. *Acta Otolaryng. (Stockh.) 78*, 345–356.

de Olmos, J. S. (1972). The amygdaloid projection field in the rat as studied with the cupric-silver method. In *The Neurobiology of the Amygdala* (B. E. Elftheriou, ed.), pp. 145–204. Plenum Press, New York.

de Olmos, J., H. Hardy, and L. Heimer (1978). The afferent connections of the main and the accessory olfactory bulb formation in the rat. An experimental HRP-study. *J. Comp. Neurol. 181*, 213–244.

De Robertis, E. (1967). Ultrastructure and cytochemistry of the synaptic region. *Science 156*, 907–914.

Deschênes, M., P. Landry, and A. Labelle (1979). The comparative effectiveness of the 'brown and blue reactions' for tracing neuronal processes of cells injected intracellularly with horseradish peroxidase. *Neurosci. Letters 12*, 9–15.

Desmedt, J. E. (1962). Auditory-evoked potentials from cochlea to cortex as influenced by activation of the efferent olivo-cochlear bundle. *J. Acoust. Soc. Amer. 34*, 1478–1496.

Desmedt, J. E. (1973). The neuromuscular disorder in myasthenia gravis. In *New Developments in Electromyography and Clinical Neurophysiology*, Vol. 1 (J. E. Desmedt, ed.), pp. 241–304 and 305–342. Karger, Basel.

Desmedt, J. E., and E. Godaux (1977). Ballistic contraction in man: characteristic recruitment pattern of single motor units of the tibialis anterior muscle. *J. Physiol. (Lond.) 264*, 673–693.

Destombes, J. (1971). Étude anatomique expérimentale des projections cérébello-pontiques. (Thèse, Faculté des Sciences, Paris).

Development and Regeneration in the Nervous System (1974). *Brit. Med. Bull. 30*, 105–189 (R. M. Gaze and M. J. Keating, scientific eds.).

DeVito, J. L. (1967). Thalamic projection of the anterior ectosylvian gyrus (somatic area II) in the cat. *J. Comp. Neurol. 131*, 67–78.

DeVito, J. L. (1978). A horseradish peroxidase-autoradiographic study of parietopulvinar connections in *Saimiri sciureus*. *Exp. Brain Res. 32*, 581–590.

DeVito, J. L., K. W. Clausing, and O. A. Smith (1974). Uptake and transport of horseradish peroxidase by cut end of the vagus nerve. *Brain Research 82*, 269–271.

DeVito, J. L., and D. M. Simmons (1976). Some connections of the posterior thalamus in monkey. *Exp. Neurol. 51*, 347–362.

DeVito, J. L., and O. A. Smith (1959). Projections from the mesial frontal cortex (supplementary motor area) to the cerebral hemispheres and brain stem of the *Macaca mulatta*. *J. Comp. Neurol. 111*, 261–278.

DeVito, J. L., and O. A. Smith, Jr. (1964). Subcortical projections of the prefrontal lobe of the monkey. *J. Comp. Neurol. 123*, 413–424.

DeVito, J. L., and L. E. White, Jr. (1966). Projections from the fornix to the hippocampal formation in the squirrel monkey. *J. Comp. Neurol. 127*, 389–398.

Dhanarajan, P., D. G. Rüegg, and M. Wiesendanger (1977). An anatomical investigation of the corticopontine projection in the primate (*Saimiri sciureus*). The projection from motor and somatosensory areas. *Neuroscience 2*, 913–922.

Diamantopoulos, E., and P. Z. Olsen (1967). Excitability of motor neurones in spinal shock in man. *J. Neurol. Neurosurg. Psychiat. 30*, 427–431.

Diamond, I. T., K. L. Chow, and W. D. Neff (1958). Degeneration of caudal medial geniculate body following cortical lesion ventral to auditory area II in cat. *J. Comp. Neurol. 109*, 349–362.

Diamond, I. T., E. G. Jones, and T. P. S. Powell (1968a). Interhemispheric fiber connections of the auditory cortex of the cat. *Brain Research 11*, 177–193.

Diamond, I. T., E. G. Jones, and T. P. S. Powell (1968b). The association connections of auditory cortex of the cat. *Brain Research 11*, 560–579.

Diamond, I. T., E. G. Jones, and T. P. S. Powell (1969). The projection of the auditory cortex upon the diencephalon and brain stem in the cat. *Brain Research 15*, 305–340.

Diamond, I. T., and W. D., Neff (1957). Ablation of temporal cortex and discrimination of auditory patterns. *J. Neurophysiol. 20*, 300–315.

Diamond, I. T., and J. D. Utley (1963). Thalamic retrograde degeneration of sensory cortex in opossum. *J. Comp. Neurol. 120*, 129–160.

Dichgans, J., and R. Jung (1975). Oculomotor abnormalities due to cerebellar lesions. In *Basic Mechanisms of Ocular Motility and their Clinical Implications* (G. Lennerstrand and P. Bach-y-Rita, eds.), pp. 281–298. Pergamon Press, Oxford.

Dierssen, G., L. Bergmann, G. Gioni, and I. S. Cooper (1962). Surgical lesions affecting parkinsonian symptomatology: A clinico-anatomical discussion of two cases. *Acta Neurochir. (Wien) 10*, 125–133.

Diete-Spiff, K., G. Carli, and O. Pompeiano (1967). Comparison of the effects of stimulation of the VIIIth cranial nerve, the vestibular nuclei or the reticular formation on the gastrocnemius muscle and its spindles. *Arch. Ital. Biol. 105*, 243–272.

Dietrichs, E., and F. Walberg (1979). The cerebellar projection from the lateral reticular nucleus as studied with retrograde transport of horseradish peroxidase. *Anat. Embryol. 155*, 273–290.

Dietrichson, P. (1971). *The Role of the Fusimotor System in Spasticity and Parkinsonian Rigidity*. Universitetsforlaget, Oslo.

DiFiglia, M., P. Pasik, and T. Pasik (1976). A Golgi study of neuronal types in the neostriatum of monkeys. *Brain Research 114*, 245–256.

Dila, C. J. (1971). A midbrain projection to the centre median nucleus of the thalamus. A neurophysiological study. *Brain Research 25*, 63–74.

Dimitrijevic, M. R., and P. W. Nathan (1967a). Studies of spasticity in man. 1. Some features of spasticity. *Brain 90*, 1–30.

Dimitrijevic, M. R., and P. W. Nathan (1967b). Studies of spasticity in man. 2. Analysis of stretch reflexes in spasticity. *Brain 90*, 333–358.

Dimond, S. J. (1976). Depletion of attentional capacity after total commissurotomy in man. *Brain 99*, 347–356.

Dimond, S. J., R. E. Scammell, E. Y. M. Brouwers, and R. Weeks (1977). Functions of the centre section (trunk) of the corpus callosum in man. *Brain 100*, 543–562.

Discussion on Subcellular and Macromolecular Aspects of Synaptic Transmission, A (1971). *Phil. Trans. B 261*, 273–440.

Diseases of the Basal Ganglia, Vol. 6, *Handbook of Clinical Neurology* (1968). (J. P. Vinken and G. W. Bruyn, eds.). North-Holland, Amsterdam.

Disorders of Language (1964). A Ciba Foundation Symposium (A. V. S. de Reuck and M. O'Connor, eds.). J & A. Churchill, London.

Divac, I., J. H. LaVail, P. Rakic, and K. R. Winston (1977). Heterogenous afferents to the inferior parietal lobule of the rhesus monkey revealed by the retrograde transport method. *Brain Research 123*, 197–207.

Dixon, J. S., and J. A. Gosling (1972). The distribution of autonomic nerves in the musculature of the rat vas deferens. A light and electron microscope investigation. *J. Comp. Neurol. 146*, 175–188.

Djupesland, G. (1980). The acoustic reflex. In *Handbook of Clinical Impedence Audiometry*, 2nd ed. (J. F. Jerger and J. L. Northern, eds.). American Electromedics Corporation, New York.

Dobry, P. J. K., and K. L. Casey (1972). Roughness discrimination in cats with dorsal column lesions. *Brain Research 44*, 385–397.

Dodt, E. (1956). Centrifugal impulses in rabbit's retina. *J. Neurophysiol. 19*, 301–307.

Dom, R., J. S. King, and G. F. Martin (1973). Evidence for two direct cerebello-olivary connections. *Brain Research 57*, 498–501.

Domesick, V. B. (1969). Projections from the cingulate cortex in the rat. *Brain Research 12*, 296–320.

Domesick, V. B. (1977). The topographical organization of the striatonigral connection in the rat. *Anat. Rec. 187*, 567.

Domesick, V. B., R. M. Beckstead, and W. J. H. Nauta (1976). Some ascending and descending projections of the substantia nigra and ventral tegmental area in the rat. *Neurosci. Abstr. II*, 61.

Domino, E. F. (1958). A pharmacologic analysis of some reticular and spinal cord systems. In *Reticular Formation of the Brain* (H. H. Jaspers, et al., eds.), pp. 285–312. Little, Brown, Boston.

Donaldson, L., P. J. Hand, and A. R. Morrison (1975). Cortical-thalamic relationships in the rat. *Exp. Neurol. 47*, 448–458.

Dostrovsky, J., J. Millar, and P. D. Wall (1976). The immediate shift of afferent drive of dorsal column nucleus cells following deafferentation: a comparison of acute and chronic deafferentation in gracile nucleus and spinal cord. *Exp. Neurol. 52*, 480–495.

Douek, E. (1974). *The Sense of Smell and its Abnormalities*. Churchill, Livingstone, Edinburgh–London.

Douglas, P. R., D. G. Ferrington, and M. Rowe (1978). Coding of information about tactile stimuli by neurones of the cuneate nucleus. *J. Physiol. (Lond.) 285*, 493–513.

Douglas, R. J. (1967). The hippocampus and behavior. *Psychol. Bull. 67*, 416–442.

Døving, K. B. (1967). Problems in the physiology of olfaction. In *Chemistry and Physiology of Flavors*. Symposium on Foods (H. W. Schultz, E. A. Day, and L. M. Libbey, eds.), pp. 52–94. Avi, Westport, Conn.

Døving, K. B., and A. J. Pinching (1973). Selective degeneration of neurons in the olfactory bulb following prolonged odour exposure. *Brain Research 52*, 115–129.

Dow, R. S. (1936). The fiber connections of the posterior parts of the cerebellum in the cat and rat. *J. Comp. Neurol. 63*, 527–548.

Dow, R. S. (1938a). Efferent connections of the flocculo-nodular lobe in *Macaca mulatta*. *J. Comp. Neurol. 68*, 297–305.

Dow, R. S. (1938b). The effects of unilateral and bilateral labyrinthectomy in monkey, baboon and chimpanzee. *Amer. J. Physiol. 121*, 392–399.

Dow, R. S. (1938c). Effects of lesions in the vestibular part of the cerebellum in primates. *Arch. Neurol. Psychiat. (Chic.) 40*, 500–520.

Dow, R. S. (1939). Cerebellar action potentials in response to stimulation of various afferent connections. *J. Neurophysiol. 2*, 543–555.

Dow, R. S. (1942a). The evolution and anatomy of the cerebellum. *Biol. Rev. 17*, 179–220.

Dow, R. S. (1942b). Cerebellar action potentials in response to stimulation of the cerebral cortex in monkeys and cats. *J. Neurophysiol. 5*, 121–136.

Dow, R. S. (1969). Cerebellar syndromes. In *Handbook of Clinical Neurology*, Vol. 2 (P. J. Vinken and G. W. Bruyn, eds.), pp. 392–431. North-Holland, Amsterdam.

Dow, R. S., and G. Moruzzi (1958). *The Physiology and Pathology of the Cerebellum*. Univ. of Minnesota Press, Minneapolis.

Dowling, J. E. (1970). Organization of vertebrate retinas. *Invest. Ophthal. 9*, 655–680.

Dowling, J. E., and B. B. Boycott (1966). Organization of the primate retina: electron microscopy. *Proc. Roy. Soc. B, 166*, 80–111.

Downer, J. L. de C. (1959). Changes in visually guided behaviour following mid-sagittal division of optic chiasm and corpus callosum in monkey (*Macaca mulatta*). *Brain 82*, 251–259.

Downman, C. B. B., C. N. Woolsey, and R. A. Lende (1960). Auditory areas I, II and Ep: cochlear representation, afferent paths and interconnections. *Bull. Johns Hopk. Hosp. 106*, 127–142.

Drachman, D. A., and J. Arbit (1966). Memory and hippocampal complex. II. Is memory a multiple process? *Arch. Neurol (Chic.) 15*, 52–61.

Drachman, D. A., and A. K. Ommaya (1964). Memory and the hippocampal complex. *Arch. Neurol. (Chic.) 10*, 411–425.

Drachman, D. A., N. Wetzel, M. Wasserman, and H. Naito (1969). Experimental denervation of ocular muscles. A critique of the concept of "Ocular myopathy." *Arch. Neurol. (Chic.) 21*, 170–183.

Dreyer, D. A., P. R. Loe, C. B. Metz, and B. L. Whitsel (1975). Representation of head and face in postcentral gyrus of the Macaque. *J. Neurophysiol. 38*, 714–733.

Droogleever-Fortuyn, J. (1950). On the configuration and the connections of the medioventral area and the midline cells in the thalamus of the rabbit. *Folia Psychiat. Neerl. 53*, 213–254.

Droogleever-Fortuyn, J., and R. Stefens (1951). On the anatomical relations of the intralaminar and midline cells of the thalamus. *Electroenceph. Clin. Neurophysiol. 3*, 393–400.

Druckman, R. (1952). A critique of "suppression," with additional observations in the cat. *Brain 75*, 226–243.

Druga, R. (1968). Cortico-claustral connections. II. Connectioss from the parietal, temporal and occipital cortex to the claustrum. *Folia Morph. (Praha) 16*, 142–149.

Dublin, W. B. (1976). *Fundamentals of Sensorineural Auditory Pathology*. C. C. Thomas, Springfield, Ill.

Dubner, R. (1967). Interaction of peripheral and central input in the main sensory trigeminal nucleus of the cat. *Exp. Neurol. 17*, 186–202.

Dubner, R., and B. J. Sessle (1971). Presynaptic excitability changes of primary afferent and corticofugal fibers projecting to trigeminal brain stem nuclei. *Exp. Neurol. 30*, 223–238.

DuBois, F. S., and J. O. Foley (1937). Quantitative studies of the vagus nerve in the cat. II. The ratio of jugular to nodose fibers. *J. Comp. Neurol. 67*, 69–87.

Duensing, F., and K.-P. Schaefer (1957a). Die Neuronaktivität in der Formatio reticularis des Rhombencephalons beim vestibulären Nystagmus. *Arch. Psychiat. Nervenkr. 196*, 265–290.

Duensing, F. and K.-P. Schaefer (1957b). Die "locker gekoppelten" Neurone der Formatio reticularis des Rhombencephalons beim vestibulären Nystagmus. *Arch. Psychiat. Nervenkr. 196*, 402–420.

Duensing, F., and K.-P. Schaefer (1958). Die Aktivität einzelner Neurone im Bereich der Vestibulariskerne bei Horizontalbeschleunigungen unter besonderer Berücksichtigung des vestibulären Nystagmus. *Arch. Psychiat. Nervenkr. 198*, 225–252.

Duensing, F., and K.-P. Schaefer (1959). Über die Konvergenz verschiedener labyrinthärer Afferenzen auf einzelne Neurone des Vestibulariskerngebietes. *Arch. Psychiat. Nervenkr. 199*, 345–371.

Duggan, A. W., and C. J. A. Game (1975). Spontaneous and synaptic excitation of paramedian reticular neurons in the decerebrate cat. *J. Physiol. (Lond.) 247*, 1–24.

Duncan, D., and R. Morales (1978). Relative numbers of several types of synaptic connections in the substantia gelatinosa of the cat spinal cord. *J. Comp. Neurol. 182*, 601–610.

Dusser de Barenne, J. G. (1933). 'Corticalization' of function and functional localization in the cerebral cortex. *Arch. Neurol. Psychiat. (Chic.) 30*, 884–901.

Dusser de Barenne, J. G., H. W. Garol, and W. S. McCulloch (1941). The 'motor' cortex of the chimapanzee. *J. Neurophysiol. 4*, 287–303.

Dusser de Barenne, J. G., H. W. Garol, and W. S. McCulloch (1942). Physiological neuronography of the cortico-striatal connections. *Res. Publ. Ass. Nerv. Ment. Dis. 21*, 246–266.

Dusser de Barenne, J. G., and W. S. McCulloch (1938). Functional organization in the sensory cortex of the monkey (*Macaca mulatta*). *J. Neurophysiol. 1*, 69–85.

Duvoisin, R. C., M. D. Yahr, M. D. Schweitzer, and H. H. Merritt (1963). Parkinsonism before and since the epidemic of encephalitis lethargica. *Arch. Neurol. (Chic.) 9*, 232–236.

Dyachkova, L. N., P. G. Kostyuk, and N. C. Pogorelaya (1971). An electron microscopic analysis of pyramidal tract terminations in the spinal cord of the cat. *Exp. Brain Res. 12*, 105–119.

Dyhre-Poulsen, P. (1978). Perception of tactile stimuli before ballistic and during tracking movements. In *Active Touch. The Mechanism of Recognition of Objects by Manipulation* (G. Gordon, ed.). pp. 171–176. Pergamon Press, Oxford.

Dyson, C., and G. S. Brindley (1966). Strength-duration curves for the production of cutaneous pain by electrical stimuli. *Clin. Sci. 30*, 237–241.

Eager, R. P. (1963a). Efferent cortico-nuclear pathways in the cerebellum of the cat. *J. Comp. Neurol. 120*, 81–103.

Eager, R. P. (1963b). Cortical association pathways in the cerebellum of the cat. *J. Comp. Neurol. 121*, 381–394.

Eager, R. P. (1966). Patterns and mode of termination of cerebellar corticonuclear pathways in the monkey (*Macaca mulatta*). *J. Comp. Neurol. 126*, 551–565.

Eager, R. P., and R. J. Barrnett (1966). Morphological and chemical studies of Nauta-stained degenerating cerebellar and hypothalamic fibers. *J. Comp. Neurol. 126*, 487–510.

Earle, A. M., and H. A. Matzke (1974). Efferent fibers of the deep cerebellar nuclei in hedgehogs. *J. Comp. Neurol. 154*, 117–132.

Earle, K. M. (1952). The tract of Lissauer and its possible relation to the pain pathway. *J. Comp. Neurol. 96*, 93–111.

Ebbesson, S. O. E. (1968a). A connection between the dorsal column nuclei and the dorsal accessory olive. *Brain Research 8*, 393–397.

Ebbesson, S. O. E. (1968b). Quantitative studies of the superior cervical sympathetic ganglia in a variety of primates including man. I. The ratio of preganglionic fibers to ganglionic neurons. *J. Morph. 124*, 117–132.

Ebner, F. F., and R. E. Myers (1965). Distribution of corpus callosum and anterior commissure in cat and raccoon. *J. Comp. Neurol. 124*, 353–365.

Eccles, J. C. (1957). *The Physiology of Nerve Cells.* Johns Hopkins Press, Baltimore.

Eccles, J. C. (1964). *The Physiology of Synapses.* Springer-Verlag, Berlin.

Eccles, J. C. (1966a). Functional organization of the cerebellum in relation to its role in motor control. In *Muscular Afferents and Motor Control.* First Nobel Symposium, Stockholm, Sweden, 1965 (R. Granit, ed.), pp. 19–36. Almqvist and Wiksell, Stockholm; Wiley, New York.

Eccles, J. C. (1966b). Cerebral synaptic mechanisms. In *Brain and Conscious Experience* (J. C. Eccles, ed.), pp. 24–50. Springer-Verlag, Berlin.

Eccles, J. C. (1973). *The Understanding of the Brain.* McGraw-Hill, New York.

Eccles, J. C., R. M. Eccles, I. Iggo, and A. Lundberg (1960). Electrophysiological studies on gamma motoneurons. *Acta Physiol. Scand. 50*, 32–40.

Eccles, J. C., R. M. Eccles, and A. Lundberg (1957). The convergence of monosynaptic excitatory afferents on to many different species of alpha motoneurons. *J. Physiol. (Lond.) 137*, 22–50.

Eccles, J. C., R. M. Eccles, and A. Lundberg (1960). Types of neurone in and around the intermediate nucleus of the lumbosacral cord. *J. Physiol. (Lond.) 154*, 89–114.

Eccles, J. C., M. Ito, and J. Szentágothai (1967). *The Cerebellum as a Neuronal Machine,* Springer-Verlag, Berlin.

Eccles, J. C., O. Oscarsson, and W. D. Willis (1961). Synpatic action of group I and II afferent fibres of muscle on the cells of the dorsal spinocerebellar tract. *J. Physiol. (Lond.) 158*, 517–543.

Eccles, J. C., T. Rantucci, N. H. Sabah, and H. Táboríková (1974). Somatotopic studies on cerebellar fastigial cells. *Exp. Brain Res. 19*, 100–118.

Eccles, J. C., I. Rosén, P. Scheid, and H. Táboríková (1972). Cutaneous afferent responses in interpositus neurones of the cat. *Brain Research 42*, 207–211.

Eccles, J. C., and C. S. Sherrington (1930). Numbers and contraction-values of individual motor-units examined in some muscles of the limb. *Proc. Roy. Soc. B 106*, 326–357.

Eccles, R., and A. Lundberg (1959). Synaptic actions in motoneurons by efferents which may evoke the flexion reflex. *Arch. Ital. Biol. 97*, 199–221.

Economo, C. von (1927). *Zellaufbau der Grosshirnrinde des Menschen,* Julius Springer, Berlin.

Economo, C. von (1920). *Die Encephalitis lethargica, ihre Nachkrankheiten und ihre Behandlung.* Urban & Schwarzenberg, Berlin.

Economo, C. von, and L. Horn (1930). Über Windungsrelief, Masse und Rindenarchitektonik der Supratemporalfläche, ihre individuellen und ihre Seitenunterschiede. *Z. ges. Neurol. Psychiat. 130*, 678–757.

Economo, C. von, and G. N. Koskinas (1925). *Die Cytoarchitektonik der Hirnrinde des erwachsenen Menschen.* Springer-Verlag, Berlin.

Edgar, M. A., and S. Nundy (1966). Innervation of the spinal dura mater. *J. Neurol. Neurosurg. Psychiat. 29,* 530–534.

Edvardsen, P. (1967). Nervous control of urinary bladder in cat. A survey of recent experimental results and their relation to clinical problems. *Acta Neurol. Scand. 43,* 543–563.

Edvinsson, L., K. C. Nielsen, C. Owman, and B. Sporrong (1972). Cholinergic mechanisms in pial vessels. Histochemistry, electron microscopy and pharmacology. *Z. Zellforsch. 134,* 311–325.

Edström, L., and E. Kugelberg (1968). Histochemical composition, distribution of fibres and fatiguability of single motor units. Anterior tibial muscle of the rat. *J. Neurol. Neurosurg. Psychiat. 31,* 424–433.

Edwards, S. B. (1972). The ascending and descending projections of the red nucleus in the cat: an experimental study using an autoradiographic tracing method. *Brain Research 48,* 45–63.

Edwards, S. B. (1975). Autoradiographic studies of the midbrain reticular formation: descending projections of nucleus cuneiformis. *J. Comp. Neurol. 161,* 341–358.

Edwards, S. B. (1977). The commissural projection of the superior colliculus in the cat. *J. Comp. Neurol. 173,* 23–40.

Edwards, S. B., C. L. Ginsburgh, C. K. Henkel, and B. E. Stein (1979). Sources of subcortical projections to the superior colliculus in the cat. *J. Comp. Neurol. 184,* 309–330.

Edwards, S. B., and C. K. Henkel (1978). Superior colliculus connections with the extraocular motor nuclei in the cat. *J. Comp. Neurol. 179,* 451–467.

Edwards, S. B., A. C. Rosenquist, and L. A. Palmer (1974). An autoradiographic study of ventral lateral geniculate projections in the cat. *Brain Research 72,* 282–287.

Ehringer, H., and O. Hornykiewicz (1960). Verteilung von Noradrenalin und Dopamin (3-hydroxydyramine) in Gehirn des Menschen und ihr Verhalten bei Erkrankungen des extrapyramidalen Systems. *Klin. Wschr. 38,* 1236–1239.

Eie, N. (1954). Macroscopical investigations of twenty-nine brains subjected to frontal leucotomy. With some observations on clinicopathological correlations. *Acta Psychiat. Scand., Suppl. 90,* 40 pp.

Eisenman, J., S. Landgren, and D. Novin (1963). Functional organizatin in the main sensory trigeminal nucleus and in the rostral subdivision of the nucleus of the spinal trigeminal tract in the cat. *Acta Physiol. Scand. 59, Suppl. 214,* 1–44.

Eklund, G. (1972). General features of vibration-induced effects on balance. *Uppsala J. Med. Sci. 77,* 112–124.

Elfvin, L.-G., and C. J. Dalsgaard (1977). Retrograde axonal transport of horseradish peroxidase in afferent fibers of the inferior mesenteric ganglion of the guinea pig. Identification of the cells of origin in dorsal root ganglia. *Brain Research 126,* 149–153.

Ellaway, P. H., and J. R. Trott (1976). Reflex connections from muscle stretch receptors to their own fusimotor neurons. *Progr. Brain Res. 44,* 113–120.

Ellaway, P. H., and J. R. Trott (1978). Autogenetic reflex action on to gamma motoneurones by stretch of triceps surae in the decerebrated cat. *J. Physiol. (Lond.) 276,* 49–66.

Eller, T., and V. Chan-Palay (1976). Afferents to the cerebellar lateral nucleus. Evidence from retrograde transport of horseradish peroxidase after pressure injections through micropipettes. *J. Comp. Neurol. 166,* 285–301.

Elliot, K. A. S. (1965). γ-aminobutyric acid and other inhibitory substances. *Brit. Med. Bull. 21,* 70–75.

Ellison, J. P., and G. M. Clark (1975). Retrograde axonal transport of horseradish peroxidase in peripheral autonomic nerves. *J. Comp. Neurol. 161,* 103–114.

Elsberg, C. A. (1935). The sense of smell. VIII. Olfactory fatigue. *Bull. Neurol. Inst. N. Y. 4,* 479–495.

Elsberg, C. A. (1936). The sense of smell. XII. The localization of tumors of the frontal lobe of the brain by quantitative olfactory tests. *Bull. Neurol. Inst. N.Y. 4*, 535–543.

Elsberg, C. A., E. D. Brewer, and I. Levy (1935). The sense of smell. IV. Concerning conditions which may temporarily alter normal olfactory acuity. *Bull. Neurol. Inst. N.Y. 4*, 31–34.

Elsberg, C. A., and I. Levy (1935). The sense of smell. I. A new and simple method of quantitative olfactometry. *Bull. Neurol. Inst. N.Y. 4*, 5–19.

Elsberg, C. A., and H. Spotnitz (1942). Value of quantitative olfactory tests for localization of supratentorial disease. *Arch. Neurol. Psychiat. (Chic.) 48*, 1–12.

Elverland, H. H. (1977). Descending connections between the superior olivary and cochlear nuclear complexes in the cat studied by autoradiographic and horseradish peroxidase methods. *Exp. Brain Res. 27*, 397–412.

Elverland, H. H. (1978). Ascending and intrinsic projections of the superior olivary complex in the cat. *Exp. Brain Res. 32*, 117–134.

Emmers, R., R. M. Benjamin, and A. J. Blomquist (1962). Thalamic localization of afferents from the tongue in albino rat. *J. Comp. Neurol. 118*, 43–48.

Emson, P. C. (1979). Peptides as neurotransmitter candidates in the mammalian CNS. *Progr. Neurobiol. 13*, 61–116.

Engel, A. G., and T. Santa (1973). Motor endplate fine structure. In *New Developments in Electromyography and Clinical Neurophysiology*, Vol. 1 (J. E. Desmedt, ed.), pp. 196–228. Karger, Basel.

Engel, J., M. V. Driver, and M. A. Falconer (1975). Electrophysiological correlates of pathology and surgical results in temporal lobe epilepsy. *Brain 98*, 129–156.

Engel, W. K. (1965). Histochemistry of neuromuscular disease. Significance of muscle fiber types. In *Proceedings of the VIII International Congress of Neurology, Vienna*, pp. 67–101, *Elsevier Excerpta Medica*, Amsterdam.

English, D. T., and C. E. Blevins (1969). Motor units of laryngeal muscles. *Arch. Otolaryng. (Chic.) 89*, 778–784.

Engström, H. (1958). On the double innervation of the sensory epithelia of the inner ear. *Acta Otolaryng. (Stockh.) 49*, 109–118.

Engström, H. (1967). The ultrastructure of the sensory cells of the cochlea. *J. Laryng. 81*, 687–715.

Engström, H., H. W. Ades, and A. Andersson (1966). *Structural Pattern of the Organ of Corti. A Systematic Mapping of Sensory Cells and Neural Elements.* Almqvist & Wiksell, Stockholm.

Engström, H., and H. Wersäll (1958). The ultrastructural organization of the organ of Corti and of the vestibular sensory epithelia. *Exp. Cell Res., Suppl. 5*, 460–492.

Enoch, D. M., and F. W. L. Kerr (1967a). Hypothalamic vasopressor and vesicopressor pathways. I. Functional studies. *Arch. Neurol. (Chic.) 16*, 290–306.

Enoch, D. M., and F. W. L. Kerr (1967b). Hypothalamic vasopressor and vesicopressor pathways. II. Anatomic study of their course and connections. *Arch. Neurol. (Chic.) 16*, 307–320.

Erickson, T. C. (1945). Erotomania (nymphomania) as an expression of cortical epileptiform discharge. *Arch. Neurol. Psychiat. (Chic.) 53*, 226–230.

Erickson, T. C., and C. N. Woolsey (1951). Observations on the supplementary motor area of man. *Trans. Amer. Neurol. Ass. 76*, 50–52.

Erulkar, S. D., and M. Fillenz (1960). Single-unit activity in the lateral geniculate body of the cat. *J. Physiol. (Lond.) 154*, 206–218.

Ervin, F. R., and V. H. Marks (1960). Stereotactic thalamotomy in the human. Part II. Physiologic observations on the human thalamus. *Arch. Neurol. (Chic.) 3*, 368–380.

Escobar, A., E. D. Sampedro, and R. S. Dow (1968). Quantitative data on the inferior olivary nucleus in man, cat and vampire bat. *J. Comp. Neurol. 132*, 397–404.

Essick, C. R. (1912). The development of the nuclei pontis and the nucleus arcuatus in man. *Amer. J. Anat. 13*, 25–54.

Ettlinger, G., C. B. Blakemore, A. D. Milner, and J. Wilson (1974). Agenesis of the corpus callosum: a further behavioural investigation. *Brain 97*, 225–234.

Euler, U. S. von (1959). Autonomic neuroeffector transmission. In *Handbook of Physiology, Section I, Neurophysiology, Vol. I* (J. Field, H. W. Magoun and V. E. Hall, eds.), pp. 215–237. American Physiological Society, Washington, D.C.

Evans, C. L. (1957). Sweating in relation to sympathetic innervation. *Brit. Med. Bull. 13*, 197–201.

Evans, E. F., H. F. Ross, and I. C. Whitfield (1965). The spatial distribution of unit characteristic frequency in the primary auditory cortex of the cat. *J. Physiol. (Lond.) 179*, 238–247.

Evarts, E. V. (1965). Relation of discharge frequency to conduction velocity in pyramidal tract neurons. *J. Neurophysiol. 28*, 216–228.

Evarts, E. V. (1966). Pyramidal tract activity associated with conditioned hand movement in the monkey. *J. Neurophysiol. 29*, 1011–1027.

Evarts, E. V. (1968). Relation of pyramidal tract activity to force exerted during voluntary movement. *J. Neurophysiol. 31*, 14–27.

Evarts, E. V. (1972). Contrasts between activity of precentral and postcentral neurons of cerebral cortex during movements in the monkey. *Brain Research 40*, 25–31.

Evarts, E. V., and R. Granit (1976). Relations of reflexes and intended movements. *Progr. Brain Res. 44*, 1–14.

Evoked Electrical Activity in the Auditory Nervous System (1978). (R. F. Naunton and C. Fernández, eds.). Academic Press, New York.

Evolution and Lateralization of the Brain (1977). *Ann. N.Y. Acad. Sci.*, Vol. 299 (S. J. Dimond and D. A. Blizard, eds.). New York Academy of Sciences, New York.

Extrapyramidal System and its Disorders, The (1979). *Advanc. Neurol.*, Vol. 24 (L. J. Poirier, T. L. Sourkes, and P. J. Bédard, eds.). Raven Press, New York.

Faaborg-Andersen, K. (1957). Electromyographic investigation of intrinsic laryngeal muscles in humans. *Acta Physiol. Scand. 41, Suppl. 140*, 1–149.

Faber, D. S., and J. T. Murphy (1969). Axonal branching in the climbing fiber pathway to the cerebellum. *Brain Research 15*, 262–267.

Faden, A. I., and J. M. Petras (1978). An intraspinal sympathetic preganglionic pathway: anatomic evidence in the dog. *Brain Research 144*, 358–362.

Fadiga, E., T. Manzoni, S. Sapienza, and A. Urbano (1968). Synchronizing and desynchronizing fastigial influences on the electrocortical activity of the cat, in acute experiments. *Electroenceph. Clin. Neurophysiol. 24*, 330–342.

Fadiga, E., and G. C. Pupilli (1964). Teleceptive components of the cerebellar function, *Physiol. Rev. 44*, 432–486.

Falck, B. (1962). Observations on the possibilities of the cellular localization of monoamines by a fluorescence method. *Acta Physiol. Scand. 56, Suppl. 197*, 1–25.

Falck, B., N. Å. Hillarp, G. Thieme, and A. Torp (1962). Fluorescence of catecholamines and related compounds condensed with formaldehyde. *J. Histochem. Cytochem. 10*, 348–354.

Falconer, M. A. (1949). Intramedullary trigeminal tractotomy and its place in the treatment of facial pain. *J. Neurol. Neurosurg. Psychiat. 12*, 297–311.

Falconer, M. A., E. A. Serafetinides, and J. A. N. Corsellis (1964). Etiology and pathogenesis of temporal lobe epilepsy. *Arch. Neurol. (Chic.) 10*, 233–248.

Falconer, M. A., and J. L. Wilson (1958). Visual field changes following anterior, temporal lobectomy, their significance in relation to "Meyer's loop" of the optic radiation. *Brain 81*, 1–14.

Fallon, J. H., J. N. Riley, and R. Y. Moore (1978). Substantia nigra dopamine neurons: separate populations project to neostriatum and allocortex. *Neurosci. Letters 7*, 157–162.

Famiglietti, E. V., Jr. (1970). Dendro-dendritic synapses in the lateral geniculate nucleus of the cat. *Brain Research 20*, 181–191.

Famiglietti, E. V., and H. Kolb (1975). A bistratified amacrine cell and synaptic circuitry in the inner plexiform layer of the retina. *Brain Research 84*, 293–300.

Famiglietti, E. V., and A. Peters (1972). The synaptic glomerulus and the intrinsic neuron in the dorsal lateral geniculate nucleus of the cat. *J. Comp. Neurol. 144*, 285–334.

Farley, I. J., and O. Hornykiewicz (1977). Noradrenaline distribution in subcortical areas of the human brain. *Brain Research 126*, 53–62.

Farrell, G., L. F. Fabre, and E. W. Rauschkolb (1966). The neurohypophysis. *Ann. Rev. Physiol. 30*, 557–588.

Faull, R. L. M., and J. B. Carman (1968). Ascending projections of the substantia nigra in the rat. *J. Comp. Neurol. 132*, 73–92.

Faull, R. L. M., and W. R. Mehler (1979). The cells of origin of nigrotectal, nigrothalamic and nigrostriatal projections in the rat. *Neuroscience 3*, 989–1002.

Favale, E., C. Loeb, G. F. Rossi, and G. Sacco (1961). EEG synchronization and behavioral signs of sleep following low frequency stimulation of the brain stem reticular formation. *Arch. Ital. Biol. 99*, 1–22.

Fay, T. (1927). Observations and results from intracranial section of the glossopharyngeus and vagus nerves in man. *J. Neurol. Psychopath. 8*, 110–123.

Fedio, P., and A. K. Ommaya (1970). Bilateral cingulum lesions and stimulation in man with lateralized impairment in short-term verbal memory. *Exp. Neurol. 29*, 84–91.

Feindel, W. (1954). Anatomical overlap of motor-units. *J. Comp. Neurol. 101*, 1–14.

Feindel, W., J. R. Hinshaw, and G. Weddell (1952). The pattern of motor innervation in mammalian striated muscle. *J. Anat. (Lond.) 86*, 35–48.

Feinstein, B., B. Lindegård, E. Nyman, and G. Wohlfart (1954). Morphologic studies of motor units in normal human muscles. *Acta Anat. (Basel) 23*, 127–142.

Feldman, M. L., and J. M. Harrison (1971). The superior olivary complex in primates. In *Medical Primatology 1970. 2nd Conference on Experimental Medicine and Surgery in Primates, New York 1969* (E. I. Goldsmith and J. Moor-Jankowski, eds.), pp. 329–340. S. Karger, Basel.

Feldman, M. L., and A. Peters (1974). A study of barrels and pyramidal dendritic clusters in the cerebral cortex. *Brain Research 77*, 55–76.

Feldon, S., P. Feldon, and L. Kruger (1970). Topography of the retinal projection upon the superior colliculus in the cat. *Vision Res. 10*, 135–143.

Feltz, P., and J. de Champlain (1972). Persistence of caudate unitary responses to nigral stimulation after destruction and functional impairment of the striatal dopaminergic terminals. *Brain Research 43*, 595–600.

Feremutsch, K. (1955). Strukturanalyse des menschlichen Hypothalamus. *Mschr. Psychiat. Neurol. 130*, 1–85.

Feremutsch, K., and K. Simma (1959). Beitrag zur Kenntnis der "Formatio reticularis medullae oblongatae et pontis" des Menschen. *Z. Anat. Entwickl.-Gesch. 121*, 271–291.

Ferin, M., R. A. Grigorian, and P. Strata (1971). Mossy and climbing fibre activation in the cat cerebellum by stimulation of the labyrinth. *Exp. Brain Res. 12*, 1–17.

Fernández, C. (1951). The innervation of the cochlea (guinea pig). *Laryngoscope (St. Louis) 61*, 1152–1172.

Fernández, C., and J. M. Fredrickson (1964). Experimental cerebellar lesions and their effect on vestibular function. International Vestibular Symposium, Uppsala 1963. *Acta Otolaryng. (Stockh.) Suppl. 192*, 52–62.

Fernandez de Molina, A., and R. W. Hunsperger (1962). Organization of the subcortical system governing defence and flight reactions in the cat. *J. Physiol. (Lond.) 160*, 200–213.

Ferraro, A., and S. E. Barrera (1935a). The nuclei of the posterior funiculi in *Macacus rhesus*. An anatomic and experimental investigation. *Arch. Neurol. Psychiat. (Chic.) 33*, 262–275.

Ferraro, A., and S. E. Barrera (1935b). Posterior column fibers and their termination in *Macacus rhesus*. *J. Comp. Neurol. 62*, 507–530.

Ferraro, A., and S. E. Barrera (1936). Lamination of the medial lemniscus in *Macacus rhesus*. *J. Comp. Neurol. 64*, 313–324.

Ferrer, N. G. (1969). Efferent projections of the anterior olfactory nucleus. *J. Comp. Neurol. 137*, 309–320.

Fetz, E. E. (1968). Pyramidal tract effects on interneurons in the cat lumbar dorsal horn. *J. Neurophysiol. 31*, 69–80.

Fex, J. (1967). The olivocochlear feedback system. In *Sensorineural Hearing Processes and Disorders*. Henry Ford Hospital International Symposium (A. B. Graham, ed.), pp. 77–86. J. & A. Churchill, London.

Fex, J. (1974). Neural excitatory processes of the inner ear. In *Handbook of Sensory Physiology*, Vol. V/1 (W. D. Keidel and W. D. Neff, eds.), pp. 585–646. Springer-Verlag, Berlin.

Fibiger, H. C., R. E. Pudritz, P. L. McGeer, and E. G. McGeer (1972). Axonal transport in nigro-striatal and nigro-thalamic neurons: effects of medial 6-hydroxydopamine. *J. Neurochem. 19,* 1697–1708.

Field, P. M. (1972). A quantitative ultrastructural analysis of the distribution of amygdaloid fibres in the preoptic area and the ventromedial hypothalamic nucleus. *Exp. Brain Res. 14*, 527–538.

Fields, H. L., and A. I. Basbaum (1978). Brainstem control of spinal pain-transmission neurons. *Ann. Rev. Physiol. 40*, 217–248.

Fields, H. L., A. I. Basbaum, C. H. Clanton, and S. D. Anderson (1977). Nucleus raphe magnus inhibition of spinal cord dorsal horn neurons. *Brain Research 126*, 441–453.

Fields, H. L., C. H. Clanton, and S. D. Anderson (1977). Somatosensory properties of spinoreticular neurons in the cat. *Brain Research 120*, 49–66.

Fillenz, M. (1955). Responses in the brainstem of the cat to stretch of extrinsic ocular muscles. *J. Physiol. (Lond.) 128*, 182–199.

Fillenz, M., and R. M. Pollard (1976). Quantitative differences between sympathetic nerve terminals. *Brain Research 109*, 443–454.

Fink, R. P., and L. Heimer (1967). Two methods for selective silver impregnation of degenerating axons and their synaptic endings in the central nervous system. *Brain Research 4*, 369–374.

Finley, K. H. (1940). Angio-architecture of the hypothalamus and its peculiarities. *Res. Publ. Ass. Nerv. Ment. Dis. 20*, 286–309.

Fisher, C., W. R. Ingram, W. K. Hare, and S. W. Ranson (1935). The degeneration of the supra-optico-hypophyseal system in diabetes insipidus. *Anat. Rec. 63*, 29–52.

Fisher, C. M. (1965). Pure sensory stroke involving face, arm and leg. *Neurology (Minneap.) 15*, 76–80.

Fisher, C. M. (1967). Some neuro-ophthalmological observations. *J. Neurol. Neurosurg. Psychiat. 30*, 383–392.

Fisher, C. M., and R. D. Adams (1956). Diphtheritic polyneuritis—A pathological study. *J. Neuropath. Exp. Neurol. 15*, 243–268.

Fisher, C. M., and H. B. Curry (1965). Pure motor hemiplegia of vascular origin. *Arch. Neurol. (Chic.) 13*, 30–44.

Fisher, S. K., and B. B. Boycott (1974). Synaptic connexions made by horizontal cells within the outer plexiform layer of the retina of the cat and the rabbit. *Proc. Roy. Soc. B 186*, 317–331.

Fisken, R. A., L. J. Garey, and T. P. S. Powell (1975). The intrinsic association and commissural connections of area 17 of the visual cortex. *Phil. Trans. B 272*, 487–536.

Fitzgerald, G., and C. S. Hallpike (1942). Studies in human vestibular function. I. Observations on the directional preponderance ("Nystagmusbereitschaft") of caloric nystagmus resulting from cerebral lesions. *Brain 65*, 115–137.

Fitzpatrick, D., I. T. Diamond, and D. Raczkowski (1978). Descending projections from auditory cortex to the thalamus and tectum of *Galago senegalensis*. *Abstr. Soc. Neurosci. 4*, 6.

Fitzpatrick, K. A. (1975). Cellular architecture and topographic organization of the inferior colliculus of the squirrel monkey. *J. Comp. Neurol. 164*, 185–208.

Fitzpatrick, K. A., and T. J. Imig (1976). Projections of the auditory cortex in the owl monkey. *Anat. Rec. 184*, 403.

Fitzsimons, J. T. (1966). The hypothalamus and drinking. *Brit. Med. Bull. 22*, 232–237.

Fitzsimons, J. T. (1972). Thirst. *Physiol. Rev. 52*, 468–561.

Fleischhauer, K. (1978). Cortical architectonics: the last 50 years and some problems of today. In *Architectonics of the Cerebral Cortex* (M. A. B. Brazier and H. Petsche, eds.), IBRO 3, pp. 99–117. Raven Press, New York.

Fleischhauer, K., H. Petsche, and W. Wittkowski (1972). Vertical bundles of dendrites in the neocortex. *Z. Anat. Entwickl.-Gesch. 136*, 213–223.

Fleminger, J. F. (1946). Discussion on cortical atrophy. *Proc. Roy. Soc. Med. (Lond.) 39*, 427–430.

Flindt-Egebak, P. (1977). Autoradiographical demonstration of the projections from the limb areas of the feline sensorimotor cortex to the spinal cord. *Brain Research 135*, 153–156.

Flock, Å. (1964). Structure of the macula utriculi with special reference to directional interplay of sensory responses as revealed by morphological polarization. *J. Cell Biol. 22*, 413–431.

Flock, Å., and A. J. Duvall (1965). The ultrastructure of the kinocilium of the sensory cells in the inner ear and lateral line organs. *J. Cell Biol. 25*, 1–8.

Flood, S., and J. Jansen (1961). On the cerebellar nuclei in the cat. *Acta Anat. (Basel) 46*, 52–72.

Flood, S., and J. Jansen (1966). The efferent fibres of the cerebellar nuclei and their distribution on the cerebellar peduncles in the cat. *Acta Anat. (Basel) 63*, 137–166.

Flumerfelt, B. A., S. Otabe, and J. Courville (1973). Distinct projections to the red nucleus from the dentate and interposed nuclei in the monkey. *Brain Research 50*, 408–414.

Fluur, Erik (1959). Influences of semicircular ducts on extraocular muscles. *Acta Otolaryng. (Stockh.) Suppl. 149*, 1–46.

Foerster, O. (1911). Resection of the posterior nerve roots of the spinal cord. *Lancet 2*, 76–79.

Foerster, O. (1929). Beiträge zur Pathophysiologie der Sehbahn und der Sehsphäre. *J. Psychol. Neurol. (Lpz.) 39*, 463–485.

Foerster, O. (1933). The dermatomes in man. *Brain 56*, 1–39.

Foerster, O. (1936a). Symptomatologie der Erkrankungen des Rückenmarks und seiner Wurzeln. In *Hdb. d. Neurol.*, Vol. 5 (Bumke and Foerster, eds.), pp. 1–403. Springer-Verlag, Berlin.

Foerster, O. (1936b). Motorische Felder und Bahnen. In *Hdb. d. Neurol.*, Vol. 6 (Bumke and Foerster, eds.), pp. 1–357. Springer-Verlag, Berlin.

Foerster, O., and O. Gagel (1932). Die Vorderseitenstrangdurchschneidung beim Menschen. Eine klinisch-patho-physiologisch-anatomische Studie. *Z. ges. Neurol. Psychiat, 138*, 1–92.

Foerster, O., O. Gagel, and W. Mahoney (1939). Die encephalen Tumoren des verlängerten Markes, der Brücke und des Mittelhirns. *Arch. Psychiat. Nervenkr. 110*, 1–74.

Foix, Ch., and J. Nicolesco (1925). *Anatomie cérébrale. Les noyaux gris centraux et la région mésencéphalo-sous-optique, suivi d'un appendice sur l'anatomie pathologique de la maladie de Parkinson.* Masson et Cie, Paris.

Foley, J. O., and F. DuBois (1937). Quantitative studies of the vagus nerve in the cat. I. The ratio of sensory to motor fibers. *J. Comp. Neurol. 67*, 49–67.

Foley, J. O., and F. DuBois (1943). An experimental study of the facial nerve. *J. Comp. Neurol. 79*, 79–105.

Foltz, E. L., and L. E. White, Jr. (1962). Pain "relief" by frontal cingulumotomy. *J. Neurosurg. 19*, 89–100.

Fonnum, F., Z. Gottesfeld, and I. Grofová (1978). Distribution of glutamate decarboxylase, choline acetyltransferase and aromatic amino acid decarboxylase in the basal ganglia of normal and operated rats. Evidence for striatopallidal, striatoentopeduncular and striatonigral GABAergic fibers. *Brain Research 143*, 125–138.

Fonnum, F., I. Grofová, E. Rinvik, J. Storm-Mathisen, and F. Walberg (1974). Origin and distribution of glutamate decarboxylase in substantia nigra of the cat. *Brain Research 71*, 77–92.

Fonnum, F., I. Grofová, and E. Rinvik (1978). Origin and distribution of glutamate decarboxylase in the nucleus subthalamicus of the cat. *Brain Research 153*, 370–374.

Fonnum, F., and I. Walaas (1979). Localization of neurotransmitter candidates in neostriatum. In *The Neostriatum* (I. Divac and R. G. E. Öberg, eds.), pp. 53–69. Pergamon Press, Oxford.

Fonnum, F., and F. Walberg (1973). An estimation of the concentration of γ-aminobutyric acid and glutamate decarboxylase in the inhibitory Purkinje axon terminals in the cat. *Brain Research 54,* 115–127.

Forbes, B. F., and N. Moskowitz (1974). Projection of auditory responsive cortex in the squirrel monkey. *Brain Research 67,* 239–254.

Ford, F. R., and F. B. Walsh (1958). Raeder's paratrigeminal syndrome. A benign disorder, possibly a complication of migraine. *Bull. Johns Hopk. Hosp. 103,* 296–298.

Foreman, R. D., A. E. Applebaum, J. E. Beall, D. L. Trevino, and W. D. Willis (1975). Responses of primate spinothalamic tract neurons to electrical stimulation of hindlimb peripheral nerves. *J. Neurophysiol. 38,* 132–145.

Foreman, R. D., J. E. Beall, A. E. Applebaum, J. D. Coulter, and W. D. Willis (1976). Effects of dorsal column stimulation on primate spinothalamic tract neurons. *J. Neurophysiol. 39,* 534–546.

Foreman, R. D., R. F. Schmidt, and W. D. Willis (1979). Effects of mechanical and chemical stimulation of fine muscle afferents upon primate spinothalamic tract cells. *J. Physiol. (Lond.) 286,* 215–231.

Forman, D., and J. W. Ward (1957). Responses to electrical stimulation of caudate nucleus in cats in chronic experiments. *J. Neurophysiol. 20,* 230–244.

Fox, C. A. (1943). The stria terminals, longitudinal association bundle and precommissural fornix fibers in the cat. *J. Comp. Neurol. 79,* 277–295.

Fox, C. A. (1949). Amygdalo-thalamic connections in *Macaca mulatta* (Abstract). *Anat. Rec. 103,* 537.

Fox, C. A., A. N. Andrade, D. E. Hillman, and R. C. Schwyn (1971). The spiny neurons in the primate striatum: a Golgi and electron microscopic study. *J. Hirnforsch. 13,* 181–201.

Fox, C. A., A. N. Andrade, I. J. Lu Qui, and J. A. Rafols (1974). The primate globus pallidus: a Golgi and electron microscopic study. *J. Hirnforsch. 15,* 75–93.

Fox, C. A., A. Andrade, and R. C. Schwyn (1969). Climbing fiber branching in the granular layer. In *Neurobiology of Cerebellar Evolution and Development* (R. Llinás, ed.), pp. 603–611, Amer. Med. Ass., Chicago.

Fox, C. A., A. N. Andrade, R. C. Schwyn, and J. A. Rafols (1971/72). The aspiny neurons and the glia in the primate striatum: a Golgi and electron microscopic study, *J. Hirnforsch. 13,* 341–361.

Fox, C. A., and J. W. Barnard (1957). A quantitative study of the Purkinje cell dendritic branchlets and their relationship to afferent fibres. *J. Anat. (Lond.) 91,* 299–313.

Fox, C. A., D. E. Hillman, K. A. Siegesmund, and C. R. Dutta (1967). The primate cerebellar cortex: a Golgi and electron microscopic study. In *The Cerebellum, Progress in Brain Research,* Vol. 25 (C. A. Fox and R. S. Snider, eds.), pp. 174–225. Elsevier Publishing Co., Amsterdam.

Fox, C. A., D. E. Hillman, K. A. Siegesmund, and L. A. Sether (1966). The primate globus pallidus and its feline and avian homologues: a Golgi and electron microscopic study. In *Evolution of the Forebrain. Phylogenesis and Ontogenesis of the Forebrain* (R. Hassler and H. Stephan, eds.), pp. 237–248. Georg Thieme Verlag, Stuttgart.

Fox, C. A., I. J. Lu Qui, and J. A. Rafols (1974). Further observations on Ramón y Cajal's "dwarf" or "neurogliaform" neurons and the oligodendroglia in the primate striatum. *J. Hirnforsch. 15,* 517–527.

Fox, C. A., and J. A. Rafols (1975). The radial fibers in the globus pallidus. *J. Comp. Neurol. 159,* 177–200.

Fox, C. A., and J. A. Rafols (1976). The striatal efferents in the globus pallidus and the substantia nigra. In *The Basal Ganglia* (M. D. Yahr, ed.), pp. 37–55). Raven Press, New York.

Fox, C. A., J. A. Rafols, and W. M. Cowan (1975). Computer measurements of axis cylinder diameters of radial fibers and comb bundle fibers. *J. Comp. Neurol. 159,* 201–224.

Fox, C. A., and J. T. Schmitz (1943). A Marchi study of the distribution of the anterior commissure in cat. *J. Comp. Neurol. 79*, 297–314.

Fox, C. A., K. A. Siegesmund, and C. R. Dutta (1964). The Purkinje cell dendritic branchlets and their relation with the parallel fibers: light and electron microscopic observations. In *Morphological and Biochemical Correlates of Neural Activity* (M. M. Cohen and R. S. Snider, eds.), pp. 112–141. Harper & Row, New York.

Fox, J. L., and J. F. Kurtzke (1966). Trauma-induced intention tremor relieved by stereotaxic thalamotomy. *Arch. Neurol. (Chic.) 15*, 247–251.

Fraioli, B., and B. Guidetti (1977). Posterior partial rootlet section in the treatment of spasticity. *J. Neurosurg. 46*, 618–626.

Frank, E., J. K. S. Jansen, T. Lømo, and R. H. Westgaard (1975). The interaction between foreign and original motor nerves innervating the soleus muscle of rats. *J. Physiol. (Lond.) 247*, 725–743.

Frankfurter, A., J. T. Weber, and J. K. Harting (1977). Brainstem projections to lobule VII of the posterior vermis in the squirrel monkey: as demonstrated by the retrograde axonal transport of tritiated horseradish peroxidase. *Brain Research 124*, 135–139.

Frankfurter, A., J. T. Weber, G. J. Royce, N. L. Strominger, and J. K. Harting (1976). An autoradiographic analysis of the tecto-olivary projection in primates. *Brain Research 118*, 245–257.

Franzen, E. A., and R. E. Meyers (1973). Age effects on social behavior deficits following prefrontal lesions in monkeys. *Brain Research 54*, 277–286.

Frazier, C. H., and S. N. Rowe (1934). Certain observations upon localization in fifty-one verified tumors of the temporal lobe. *Res. Publ. Ass. Nerv. Ment. Dis. 13*, 251–258.

Fredrickson, J. M., U. Figge, P. Scheid, and H. H. Kornhuber (1966). Vestibular nerve projection to the cerebral cortex of the rhesus monkey. *Exp. Brain Res. 2*, 318–327.

Fredrickson, J. M., D. Schwarz, and H. H. Kornhuber (1965). Convergence and interaction of vestibular and deep somatic afferents upon neurons in the vestibular nuclei of the cat. *Acta Otolaryng. (Stockh.) 61*, 168–186.

Freedman, W., S. Minassian, and R. Herman (1976). Functional stretch reflex (FSR)—a cortical reflex? *Progr. Brain Res. 44*, 487–490.

Freeman, M. A. R., and B. Wyke (1967). The innervation of the knee joint. An anatomical and histological study in the cat. *J. Anat. (Lond.) 101*, 505–532.

Freeman, W., and J. W. Watts (1942). *Psychosurgery. Intelligence, Emotion and Social Behavior following Prefrontal Lobotomy for Mental Disorders.* C. C. Thomas, Springfield, Ill.

Freeman, W., and J. W. Watts (1946). Pain of organic disease relieved by prefrontal lobotomy. *Proc. Roy. Soc. Med. 39*, 445–447.

Freeman, W., and J. W. Watts (1947). Retrograde degeneration of the thalamus following prefrontal lobotomy. *J. Comp. Neurol. 86*, 65–93.

French, J. D. (1952). Brain lesions associated with prolonged unconsciousness. *Arch. Neurol. (Chic.) 68*, 722–740.

French, J. D., F. K. von Amerongen, and H. W. Magoun (1952). An activating system in brain stem of monkey. *Arch. Neurol. Psychiat. (Chic.) 68*, 557–590.

Frequency Analysis and Periodicity Detection in Hearing (1970). (R. Plomp and G. F. Smoorenburg, eds.) A. W. Sijthoff, Leiden.

Freund, H.-J. (1973). Neuronal mechanisms of the lateral geniculate body. In *Handbook of Sensory Physiology*, Vol. VII/3, Central Processing of Visual Information, part B, Morphology and Function of Visual Centers in the Brain (R. Jung, ed.), pp. 177–246. Springer-Verlag, Berlin.

Frezik, J. (1963). Associative connections established by Purkinje axon collaterals between different parts of the cerebellar cortex. *Acta Morph. Acad. Sci. Hung. 12*, 9–14.

Fricke, R., and N. M. Cowan (1978). An autoradiographic study of the commissural and ipsilateral hippocampodentate projections in the adult rat. *J. Comp. Neurol. 181*, 253–270.

Friedman, A. P., D. H. Harter, and H. H. Merritt (1962). Ophthalmoplegic migraine. *Arch. Neurol. (Chic.) 7*, 320–327.

Fritsch, G., and E. Hitzig (1870). Ueber die elektrische Erregbarkeit des Grosshirns. *Arch. Anat. Physiol. Wiss. Med. 37*, 300–332.

Frontal Granular Cortex and Behavior, The (1964). (J. M. Warren and K. Akert, eds.). McGraw-Hill, New York.

Frontal Granular Cortex and Behavior, The (1972). (J. Konorski, H-L. Teuber, and B. Zernicki, eds.). *Acta Neurobiol. Exp. 32*, 115–656.

Fry, F. J., and W. M. Cowan (1972). A study of retrograde cell degeneration in the lateral mammillary nucleus of the cat, with special reference to the role of axonal branching in the preservation of the cell. *J. Comp. Neurol. 144*, 1–24.

Fry, W. J., R. Krumins, F. J. Fry, G. Thomas, S. Borbely, and H. Ades (1963). Origins and distributions of some efferent pathways from the mammillary nuclei of the cat. *J. Comp. Neurol. 120*, 195–257.

Fuchs, A. F., and H. H. Kornhuber (1969). Extraocular muscle afferents to the cerebellum of the cat. *J. Physiol. (Lond.) 200*, 713–722.

Fukushima, T., and F. W. L. Kerr (1979). Organization of trigeminothalamic tracts and other thalamic afferent systems of the brainstem in the rat: presence of gelatinosa neurons with thalamic connections. *J. Comp. Neurol. 183*, 169–184.

Fukushima, K., B. W. Peterson, Y. Uchino, J. D. Coulter, and V. J. Wilson (1977). Direct fastigiospinal fibers in the cat. *Brain Research 126*, 538–542.

Fuller, J. H. (1975). Brain stem reticular units: some properties of the course and origin of the ascending trajectory. *Brain Research 83*, 349–367.

Fullerton, B. C. (1975). The organization of the inferior colliculus of the rhesus monkey. *Anat. Rec. 181*, 359.

Fulton, J. F. (1934). Forced grasping and groping in relation to the syndrome of the premotor area. *Arch. Neurol. Psychiat. (Chic.) 31*, 221–235.

Fulton, J. F., C. F. Jacobsen, and M. A. Kennard (1932). A note concerning the relation of the frontal lobes to posture and forced grasping in monkeys. *Brain 55*, 524–536.

Fulton, J. F., and A. D. Keller (1932). *The Sign of Babinski. A Study of the Evolution of Cortical Dominance*. C. Thomas, Springfield, Ill.

Fulton, J. F., and M. A. Kennard (1934). A study of flaccid and spastic paralyses produced by lesions of the cerebral cortex in primates. *Res. Publ. Ass. Nerv. Ment. Dis. 13*, 158–210.

Functional Recovery after Lesions of the Nervous System (1974). (E. Eidelberg and D. G. Stein, eds.). *Neurosci. Res. Progr. Bull. Vol. 12*, 189–303.

Functions of the Corpus Callosum (1965). Ciba Foundation Study Group, No. 20 (E. G. Ettlinger, A. V. S. de Reuck, and R. Porter, eds.). J. & A. Churchill,,London.

Functions of the Septo-Hippocampal System. (1978). Ciba Foundation Symposium 58, new series. Elsevier/ Excerpta Medica, North-Holland, Amsterdam.

Furness, J. B., and M. Costa (1973). The ramifications of adrenergic nerve terminals in the rectum, anal sphincter and anal accessory muscles of the guinea-pig. *Z. Anat. Entwickl.-Gesch. 140*, 109–128.

Fuxe, K. (1965). Evidence for the existence of monoamine neurons in the central nervous system. IV. Distribution of monoamine terminals in the central nervous system. *Acta Physiol. Scand. 64, Suppl. 247*, 37–85.

Fuxe, K., and T. Hökfelt (1970). Central monoaminergic systems and hypothalamic function. In *The Hypothalamus* (L. Martini, M. Motta, and F. Fraschini, eds.), pp. 123–152. Academic Press, New York.

Fuxe, K., T. Hökfelt, O. Johansson, G. Jonsson, P. Lidbrink, and Å. Ljungdahl (1974). The origin of the dopamine nerve terminals in limbic and frontal cortex. Evidence for meso-cortico dopamine neurons. *Brain Research 82*, 349–355.

Fuxe, K., T. Hökfelt, and U. Ungerstedt (1970). Morphological and functional aspects of central monoamine neurons. *Int. Rev. Neurobiol. 13*, 93–126. (C. C. Pfeiffer and J. R. Smythies, eds.). Academic Press, New York.

Gabella, G. (1976). *Structure of the Autonomic Nervous System*. Chapman and Hall, London.

Gacek, R. R. (1960). Efferent component of the vestibular nerve. In *Neural Mechanisms of the Auditory and Vestibular Systems* (G. L. Rasmussen and W. F. Windle, eds.), pp. 276–284. C. C. Thomas, Springfield, Ill.

Gacek, R. R. (1961). The efferent cochlear bundle in man. *Arch. Otolaryng. 74,* 690–694.

Gacek, R. R. (1969). The course and central termination of first order neurons supplying vestibular endorgans in the cat. *Acta Otolaryng. (Stockh.), Suppl. 254,* 1–66.

Gacek, R. R. (1971). Anatomical demonstration of the vestibulo-ocular projections in the cat. *Acta Otolaryng. (Stockh.), Suppl. 293,* 1–63.

Gacek, R. R. (1974). Localization of neurons supplying the extrocular muscles in the kitten using horseradish peroxidase. *Exp. Neurol. 44,* 381–403.

Gacek, R. R. (1977). Location of brain stem neurons projecting to the oculomotor nucleus in the cat. *Exp. Neurol. 57,* 725–749.

Gacek, R. R., and M. Lyon (1974). The localization of vestibular efferent neurons in the kitten with horseradish peroxidase. *Acta Otolaryng. (Stockh.) 77,* 92–101.

Gacek, R. R., and G. L. Rasmussen (1961). Fiber analysis of the statoacoustic nerve of guinea pig, cat, and monkey. *Anat. Rec. 139,* 455–463.

Gala, K., J.-S. Hong and A. Guidotti (1977). Presence of substance P and GABA in separate striatonigral neurons. *Brain Research 136,* 371–375.

Galaburda, A. M., M. LeMay, T. L. Kemper, and N. Geschwind (1978). Right-left asymmetries in the brain. *Science 199,* 852–856.

Galambos, R. (1956). Suppression of auditory nerve activity by stimulation of efferent fibers to cochlea. *J. Neurophysiol. 19,* 424–437.

Galambos, R., and J. E. Rose (1952). Microelectrode studies on medial geniculate body of cat. III. Response to pure tones. *J. Neurophysiol. 15,* 381–400.

Ganchrow, D., and R. P. Erickson (1972). Thalamocortical relations in gustation. *Brain Research 36,* 289–305.

Gandevia, S. C., and D. I. McCloskey (1976). Joint sense, muscle sense, and their combination as position sense, measured at the distal interphalangeal joint of the middle finger. *J. Physiol. (Lond.) 260,* 387–407.

Ganes, T., B. R. Kaada, and R. Nyberg-Hansen (1966). Failure to produce postural tremor by mesencephalic lesions in cats. *J. Comp. Neurol. 128,* 127–132.

Garcin, R., and J. Lapresle (1954). Syndrome sensitif de type thalamique et a topographie chéiro-orale par lésion localisée du thalamus. *Rev. Neurol. 90,* 124–129.

Gardner, E. (1944). The distribution and termination of nerves in the knee joint of the cat. *J. Comp. Neurol. 80,* 11–32.

Gardner, E. (1967). Spinal cord and brain stem pathways for afferents from joints. In *Myotatic and Vestibular Mechanisms.* Ciba Foundation Symposium (A. V. D. de Reuck and J. Knight, eds.), pp. 56–76. J. & A. Churchill, London.

Gardner, E., and H. M. Cuneo (1945). Lateral spinothalamic tract and associated tracts in man. *Arch. Neurol. Psychiat. (Chic.) 53,* 423–430.

Garey, L. J., E. G. Jones, and T. P. S. Powell (1968). Interrelationships of striate and extrastiate cortex with the primary relay sites of the visual pathway. *J. Neurol. Neurosurg. Psychiat. 31,* 135–157.

Garey, L. J., and T. P. S. Powell (1967). The projection of the lateral geniculate nucleus upon the cortex in the cat. *Proc. Roy. Soc. B 169,* 107–126.

Garey, L. J., and T. P. S. Powell (1968). The projection of the retina in the cat. *J. Anat. (Lond.) 102,* 189–222.

Garey, L. J., and T. P. S. Powell (1971). An experimental study of the termination of the lateral geniculo-cortical pathway in the cat and monkey. *Proc. Roy. Soc. B 179,* 41–63.

Garver, D. L., and J. R. Sladek, Jr. (1975). Monamine distribution in primate brain. I. Catecholamine-containing perikarya in the brain stem of *Macaca speciosa. J. Comp. Neurol. 159,* 289–304.

Gassel, M. M. (1961). False localizing signs. A reveiw of the concept and analysis of the occurrence in 250 cases of intracranial meningioma. *Arch. Neurol. (Chic.) 4,* 526–554.

Gassel, M. M. (1969). Occipital lobe syndromes (excluding hemianopia). In *Handbook of Clinical Neurology*, Vol. 2 (P. J. Vinken and G. W. Bruyn, eds.), pp. 640–699. North-Holland, Amsterdam; Wiley, New York.

Gassel, M. M., and E. Diamantopoulos (1964). The Jendrassik maneuver. II. An analysis of the mechanism. *Neurology (Minneap.) 14*, 640–642.

Gassel, M. M., and D. Williams (1963). Visual function in patients with homonymous hemianopia. Part III. The completion phenomenon; insight and attitude to the defect; and visual functional efficiency. *Brain 86*, 229–260.

Gasser, H. S. (1935). Conduction in nerves in relation to fiber types. *Res. Publ. Ass. Nerv. Ment. Dis. 15*, 35–56.

Gastaut, H. (1954). The brain stem and cerebral electrogenesis in relation to consciousness. In *Brain Mechanisms and Consciousness*. A Ciba Foundation Symposium (J. F. Delafresnaye, ed.), pp. 249–279. Blackwell, Oxford.

Gatter, K. C., and T. P. S. Powell (1977). The projection of the locus coeruleus upon the neocortex in the macaque monkey. *Neuroscience. 2*, 441–445.

Gatter, K. C., and T. P. S. Powell (1978). The intrinsic connections of the cortex of area 4 of the monkey. *Brain 101*, 513–541.

Gatter, K. C., J. J. Sloper, and T. P. S. Powell (1978). An electron microscopic study of the termination of intracortical axons upon Betz cells in area 4 of the monkey. *Brain 101*, 543–553.

Gautier, C., and J. Lereboullet (1927). Syndrome inférieur du noyau rouge. *Rev. Neurol. 1*, 57–61.

Gaze, R. M., F. J. Gillingham, S. Kalyanaraman, R. W. Porter, A. A. Donaldson, and I. M. L. Donaldson (1964). Microelectrode recordings from the human thalamus. *Brain 87*, 691–706.

Gaze, R. M., and G. Gordon (1954). The representation of cutaneous sense in the thalamus of the cat and monkey. *Quart. J. Exp. Physiol. 39*, 279–304.

Gazzaniga, M. S. (1970). *The Bisected Brain*. Appleton-Century-Crofts, New York.

Gazzaniga, M. S., J. E. Bogen, and R. W. Sperry (1965). Observations on visual perception after disconnexion of the cerebral hemispheres in man. *Brain 88*, 221–236.

Gazzaniga, M. S., and R. W. Sperry (1967). Language after section of the cerebral commissures. *Brain 90*, 131–148.

Geffen, L. B., and B. G. Livett (1971). Synaptic vesicles in sympathetic neurons. *Physiol. Rev. 51*, 98–157.

Geinisman, Y., and W. Bondareff (1976). Decrease in the number of synapses in the senescent brain: a quantitative electron microscopic analysis of the dentate gyrus molecular layer in the rat. *Mech. Age Dev. 5*, 11–23.

Geis, G. S., and R. D. Wurster (1980). Horseradish peroxidase localization of cardiac vagal preganglionic somata. *Brain Research 182*, 19–30.

Geisert, E. E., Jr. (1976). The use of tritiated horseradish peroxidase for defining neuronal pathways: a new application. *Brain Research 117*, 130–135.

Gelfan, S., and A. F. Rapisarda (1964). Synaptic density on spinal neurons of normal dogs and dogs with experimental hind-limb rigidity. *J. Comp. Neurol. 123*, 73–96.

Geniec, P., and D. K. Morest (1971). The neuronal architecture of the human posterior colliculus. A study with the Golgi method. *Acta Otolaryng. (Stockh.), Suppl. 295*, 1–33.

George, J. M., and D. M. Jacobowitz (1975). Localization of vasopressin in discrete areas of the rat hypothalamus. *Brain Research 93*, 363–366.

Gerebtzoff, M. A. (1939). Les voies centrales de la sensibilité et du goût et leurs terminaisons thalamiques. *Cellule 48*, 91–146.

Gerebtzoff, M. A. (1939). Contribution à l'étude des voies afférentes de l'olive inférieure. *J. Belge Neurol. Psychiat.*, 719–728.

German, D. C., and D. M. Bowden (1975). Locus ceruleus in rhesus monkey (*Macaca mulatta*): a combined histochemical fluorescence, Nissl and silver study. *J. Comp. Neurol. 161*, 19–30.

German, W. J., and B. S. Brody (1940). The external geniculate bodies. Degeneration studies following occipital lobectomy. *Arch. Neurol. Psychiat. (Chic.) 43*, 997–1003.

Gernandt, B. (1949). Response of mammalian vestibular neurons to horizontal rotation and caloric stimulation. *J. Neurophysiol. 12*, 173–184.

Gernandt, B. E., and S. Gilman (1960). Interactions between vestibular, pyramidal, and cortically evoked extrapyramidal activities. *J. Neurophysiol. 23*, 516–533.

Gernandt, B. E., M. Iranyi, and R. B. Livingston (1959). Vestibular influences on spinal mechanisms. *Exp. Neurol. 1*, 248–273.

Geschwind, N. (1965). Disconnexion syndromes in animals and man. *Brain 88*, 237–294 and 585–644.

Geschwind, N. (1974a). Late changes in the nervous system: an overview. In *Plasticity and Recovery of Function in the Central Nervous System* (D. G. Stein, J. J. Rosen, and N. Butters, eds.), pp. 467–508. Academic Press, New York.

Geschwind, N. (1974b). The anatomical basis of hemispheric differentiation. In *Hemisphere Function in the Human Brain* (S. J. Dimond and J. G. Beaumont, eds.), pp. 7–24. Wiley, New York.

Geschwind, N., and W. Levitsky (1968). Human brain: left-right asymmetries in temporal speech region. *Science 161*, 186–187.

Getz, B. (1952). The termination of spinothalamic fibres in the cat as studied by the method of terminal degeneration. *Acta. Anat. (Basel) 16*, 271–290.

Getz, B., and T. Sirnes (1949). The localization within the dorsal motor vagal nucleus. An experimental investigation. *J. Comp. Neurol. 90*, 95–110.

Ghelarducci, B., O. Pompeiano, and K. M. Spyer (1974). Activity of precerebellar reticular neurones as a function of head position. *Arch. Ital. Biol. 112*, 98–125.

Ghez, C., and M. Pisa (1972). Inhibition of afferent transmission in cat lemniscal system during voluntary movement. *Brain Research 40*, 145–151.

Giannazzo, E., T. Manzoni, R. Raffaele, S. Sapienza, and A. Urbano (1969). Effect of chronic fastigial lesions on the sleep–wakefulness rhythm in the cat. *Arch. Ital. Biol. 107*, 1–18.

Giesler, G. J., Jr., J. T. Cannon, G. Urca, and J. C. Liebeskind (1978). Long ascending projections from substantia gelatinosa Rolandi and the subjacent dorsal horn in the rat. *Science 202*, 984–986.

Giesler, G. J., Jr., D. Menétrey, and A. I. Basbaum (1979). Differential origins of spinothalamic tract projections to medial and lateral thalamus in the rat. *J. Comp. Neurol. 184*, 107–126.

Gijn, J. van (1975). Babinski response: stimulus and effector. *J. Neurol. Neurosurg. Psychiat. 38*, 180–186.

Gijn, J. van (1976). Equivocal plantar responses: a clinical and electromyographic study. *J. Neurol. Neurosurg. Psychiat. 39*, 275–282.

Gijn, J. van (1977). The plantar reflex. A historical, clinical and electromyographic study. Thesis. Krips Repro-Meppel, Rotterdam.

Gijn, J. van (1978). The Babinski sign and the pyramidal syndrome. *J. Neurol. Neurosurg. Psychiat. 41*, 865–873.

Gilbert, C. D., and J. P. Kelly (1975). The projections of cells in different layers of the cat's visual cortex. *J. Comp. Neurol. 163*, 81–106.

Gildenberg, P. L. (1972). Physiologic observations concerned with percutaneous cordotomy. In *Neurophysiology Studied in Man. Int. Congr. Ser.*, Vol. 253 (G. G. Somjen, ed.), pp. 231–236. Elsevier Excerpta Medica, Amsterdam.

Gildenberg, P. L., and R. Hassler (1971). Influence of stimulation of the cerebral cortex on vestibular nuclei units in the cat. *Exp. Brain Res. 14*, 77–94.

Gillingham, F. J. (1962). Small localized surgical lesion of the internal capsule in the treatment of the dyskinesias. *Confin. Neurol. (Basel) 22*, 385–392.

Gilman, S. (1969). The mechanism of cerebellar hypotonia. An experimental study in the monkey. *Brain 92*, 621–638.

Gilman, S. (1973). A cerebello-thalamo-cortical pathway controlling fusimotor activity. *Advan. Behav. Biol. 7*, 309–329.

Gilman, S. (1976). Patterns of motoneuron responses to natural stimuli. In *Mechanisms in Transmission of Signals for Conscious Behaviour* (T. Desiraju, ed.), pp. 23–40. Elsevier, Amsterdam.

Gilman, S., and D. Denny-Brown (1966). Disorders of movement and behaviour following dorsal column lesions. *Brain 89*, 397–418.

Gilman, S., J. S. Lieberman, and L. A. Marco (1974). Spinal mechanisms underlying the effects of unilateral ablation of areas 4 and 6 in monkeys. *Brain 97*, 49–64.

Gilman, S., and L. A. Marco (1971). Effects of medullary pyramidotomy in the monkey. I. Clinical and electromyographic abnormalities. *Brain 94*, 495–514.

Gilman, S., L. A. Marco, and H. C. Ebel (1971). Effects of medullary pyramidotomy in the monkey. II. Abnormalities of spindle afferent responses. *Brain 94*, 515–530.

Gilman, S., and J. P. van der Meulen (1966). Muscle spindle activity in dystonic and spastic monkeys. *Arch. Neurol. (Chic.) 14*, 553–563.

Giolli, R. A. (1961). An experimental study of the accessory optic tracts (transpeduncular tracts and anterior accessory optic tracts) in the rabbit. *J. Comp. Neurol. 117*, 77–95.

Giolli, R. A. (1963). An experimental study of the accessory optic system in the Cynomolgus monkey. *J. Comp. Neurol. 121*, 89–108.

Gisbergen, van, J. A. M., J. L. Grashuis, P. I. M. Johannesma, and A. J. H. Vendrik (1975). Statistical analysis and interpretation of the initial response of cochlear nucleus neurons to tone bursts. *Exp. Brain Res. 23*, 407–423.

Gjone, R. (1965). A dual peripheral and supraspinal autonomic influence on the urinary bladder. *J. Oslo Cy. Hosp. 15*, 173–182.

Glasgow, E. F., and D. C. Sinclair (1962). Dissociation of thermal sensibility in procaine nerve block. *Brain 85*, 791–798.

Glees, P. (1943). The Marchi reaction: its use on frozen sections and its time limit. *Brain 66*, 229–232.

Glees, P. (1944). The anatomical basis of cortico-striate connections. *J. Anat. (Lond.) 78*, 47–51.

Glees, P. (1945). The interrelation of the strio-pallidum and the thalamus in the macaque monkey. *Brain 68*, 331–346.

Glees, P. (1946). Terminal degeneration within the central nervous system as studied by a new silver method. *J. Neuropath. Exp. Neurol. 5*, 54–59.

Glees, P. (1952). Der Verlauf und die Endigung des Tractus spinothalamicus und der medialen Schleife, nach Beobachtungen beim Menschen und Affen. *Verh. anat. Ges. (Jena) 50*, 48–58.

Glees, P., and W. E. Le Gros Clark (1941). The termination of optic fibres in the lateral geniculate body of the monkey. *J. Anat. (Lond.) 75*, 295–308.

Glees, P., J. Cole, C. W. M. Whitty, and H. Cairns (1950). The effects of lesions in the cingular gyrus and adjacent areas in monkeys. *J. Neurol. Neurosurg. Psychiat. 13*, 178–190.

Glees, P., and H. B. Griffith (1952). Bilateral destruction of the hippocampus (Cornu Ammonis) in a case of dementia. *Mth. Rev. Psychiat. Neurol. 123*, 193–204.

Glees, P., R. B. Livingston, and J. Soler (1951). Der intraspinale Verlauf und die Endigungen der sensorischen Wurzeln in den Nucleus Gracilis und Cuneatus. *Arch. Psychiat. Nervenkr. 187*, 190–204.

Glees, P., and P. D. Wall (1946). Fibre connections of the subthalamic region and the centro-median nucleus of the thalamus. *Brain 69*, 195–208.

Glickstein, M., R. A. King, J. Miller, and M. Berkley (1967). Cortical projections from the dorsal lateral geniculate nucleus of cats. *J. Comp. Neurol. 130*, 55–76.

Globus, A., and A. B. Scheibel (1967a). Synaptic loci on visual cortical neurons of the rabbit: the specific afferent radiation. *Exp. Neurol. 18*, 116–131.

Globus, A., and A. B. Scheibel (1967b). Synaptic loci on parietal cortical neurons: termination of corpus callosum fibers. *Science 156*, 1127–1129.

Gloor, P. (1955). Electrophysiological studies on the connections of the amygdaloid nucleus in the cat. *Electroenceph. Clin. Neurophysiol. 7*, 243–264.

Gloor, P. (1960). Amygdala. In *Handbook of Physiology*, Vol. II, Section I, Neurophysiology (J. Field, H. W. Magoun, and V. E. Hall, eds.), pp. 1395–1420. American Physiological Society, Washington, D.C.

Gobel, S. (1971). Structural organization in the main sensory trigeminal nucleus. In *Oral-Facial Sensory and Motor Mechanisms* (R. Dubner and Y. Kawamura, eds.), pp. 183–202. Appleton-Century-Crofts, New York.

Gobel, S. (1974). Synaptic organization of the substantia gelatinosa glomeruli in the spinal trigeminal nucleus of the adult cat. *J. Neurocytol. 3*, 219–243.

Gobel, S. (1975a). Golgi studies of the substantia gelatinosa neurons in the spinal trigeminal nucleus. *J. Comp. Neurol. 162*, 397–416.

Gobel, S. (1975b). Neurons with two axons in the substantia gelatinosa layer of the spinal trigeminal nucleus of the adult cat. *Brain Research 88*, 333–338.

Gobel, S., and J. M. Brinck (1977). Degenerative changes in primary trigeminal axons and in neurons in nucleus caudalis following tooth pulp extirpations in the cat. *Brain Research 132*, 347–354.

Gobel, S., and R. Dubner (1969). Fine structural studies of the main sensory trigeminal nucleus in the cat and rat. *J. Comp. Neurol. 137*, 459–494.

Gobel, S., and M. B. Purvis (1972). Anatomical studies of the organization of the spinal V nucleus: the deep bundles and the spinal tract. *Brain Research 48*, 27–44.

Godaux, E., and J. E. Desmedt (1975a). Human masseter muscle: H- and tendon reflexes. *Arch. Neurol. (Chic.) 32*, 229–234.

Godaux, E., and J. E. Desmedt (1975b). Exteroceptive suppression and motor control of the masseter and temporalis muscles in normal man. *Brain Research 85*, 447–458.

Godfrey, D. A., N. Y. S. Kiang, and B. E. Norris (1975a). Single unit activity in the posteroventral cochlear nucleus of the cat. *J. Comp. Neurol. 162*, 247–268.

Godfrey, D. A., N. Y. S. Kiang, and B. E. Norris (1975b). Single unit activity in the dorsal cochlear nucleus of the cat. *J. Comp. Neurol. 162*, 269–284.

Goebel, H. H., A. Komatsuzaki, M. B. Bender, and B. Cohen (1971). Lesions of the pontine tegmentum and conjugate gaze paralysis. *Arch. Neurol. (Chic.) 24*, 431–440.

Goldberg, J. M., and P. B. Brown (1968). Functional organization of the dog superior olivary complex: an anatomical and electrophysiological study. *J. Neurophysiol. 31*, 639–656.

Goldberg, J. M., and R. Y. Moore (1967). Ascending projections of the lateral lemniscus in the cat and monkey. *J. Comp. Neurol. 129*, 143–156.

Goldberg, J. M., and W. D. Neff (1961). Frequency discrimination after bilateral section of the brachium of the inferior colliculus. *J. Comp. Neurol. 116*, 265–290.

Goldberg, M. E., and R. H. Wurtz (1972). Activity of superior colliculus in behaving monkey. I. Visual receptive fields of single neurons. *J. Neurophysiol. 35*, 542–559.

Goldberger, M. E. (1974a). Recovery of movement after CNS lesions in monkeys. In *Plasticity and Recovery of Function in the Central Nervous System* (D. G. Stein, J. J. Rosen, and N. Butters, eds.), pp. 265–337. Academic Press, New York.

Goldberger, M. E. (1974b). Recovery of function and collateral sprouting in cat spinal cord. In *Functional Recovery after Lesions of the Nervous System* (E. Eidelberg and D. Stein, eds.), pp. 235–239. *Neurosci. Res. Progr. Bull., Vol. 12.*

Goldberger, M. E., and J. H. Growden (1971). Tremor at rest following cerebellar lesions in monkeys: effect of L-DOPA administration. *Brain Research 27*, 183–187.

Goldie, I., and M. Wellisch (1969). The presence of nerves in original and regenerated synovial tissue in patients synovectomised for rheumatoid arthritis. *Acta Orthop. Scand. 40*, 143–152.

Goldman, P. S. (1978). Neuronal plasticity in primate telencephalon: anomalous projections induced by prenatal removal of frontal cortex. *Science 202*, 768–770.

Goldman, P. S., and W. J. H. Nauta (1976). Autoradiographic demonstration of a projection from the prefrontal association cortex to the superior colliculus in the rhesus monkey. *Brain Research 116*, 145–149.

Goldman, P. S., and W. J. H. Nauta (1977). An intricately patterned prefronto-caudate projection in the rhesus monkey. *J. Comp. Neurol. 171*, 369–386.

Goldring, S., E. Aras, and P. C. Weber (1970). Comparative study of sensory input to motor cortex in animals and man. *Electroenceph. Clin. Neurophysiol. 29,* 537–550.

Goldstein, G. (1974). The use of clinical neurophysiological methods in the lateralisation of brain lesions. In *Hemisphere Function in the Human Brain* (S. J. Dimond and J. G. Beaumont, eds.), pp. 279–310. Wiley, New York.

Goldstein, M. H., Jr., and M. Abeles (1975). Note on tonotopic organization of primary auditory cortex in the cat. *Brain Research 100,* 188–191.

Goldstein, M. H., Jr., M. Abeles, R. L. Daly, and J. McIntosh (1970). Functional architecture in cat primary auditory cortex: tonotopic organization. *J. Neurophysiol. 33,* 188–197.

Golgi Centennial Symposium: Perspectives in Neurobiology (1975). (M. Santini, ed.). Raven Press, New York.

Goode, G. E., and M. Sreesai (1978). An electron microscopic study of rubrospinal projections to the lumbar spinal cord of the opossum. *Brain Research 143,* 61–70.

Goodman, D. C., R. E. Hallett, and R. B. Welch (1963). Patterns of localization in the cerebellar corticonuclear projections of the albino rat. *J. Comp. Neurol. 121,* 51–67.

Goodwin, G. M., D. I. McCloskey, and P. B. C. Matthews (1972). The contribution of muscle afferents to kinesthesia shown by vibration induced illusions of movement and by the effects of paralysing joint afferents. *Brain 95,* 705–748.

Gordon, B. (1973). Receptive fields in deep layers of cat superior colliculus. *J. Neurophysiol. 36,* 157–178.

Gordon, G., and D. Horrobin (1967). Antidromic and synaptic responses in the cat's gracile nucleus to cerebellar stimulation. *Brain Research 5,* 419–421.

Gordon, G., and M. G. M. Jukes (1964a). Dual organization of the exteroceptive components of the cat's gracile nucleus. *J. Physiol. (Lond.) 173,* 263–290.

Gordon, G., and M. G. M. Jukes (1964b). Descending influence on the exteroceptive organizations of the cat's gracile nucleus. *J. Physiol. (Lond.) 173,* 291–319.

Gordon, G., and C. H. Paine (1960). Functional organization in nucleus gracilis of the cat. *J. Physiol. (Lond.) 153,* 331–349.

Gordon, H. W. (1970). Hemispheric asymmetries in the perception of musical chords. *Cortex 6,* 387–398.

Gordon, H. W., J. E. Bogen, and R. W. Sperry (1971). Absence of deconnexion syndrome in two patients with partial section of the neocommissures. *Brain 94,* 327–336.

Gordon, T., R. Jones, and G. Vrbová (1976). Changes in chemosensitivity of skeletal muscles as related to endplate formation. *Progr. Neurobiol. 6,* 103–136.

Gottlieb, D. J., and W. M. Cowan (1973). Autoradiographic studies of the commissural and ipsilateral association connections of the hippocampus and dentate gyrus of the rat. I. The commissural connections. *J. Comp. Neurol. 149,* 393–422.

Gould, B. B., and A. M. Graybiel (1976). Afferents to the cerebellar cortex in the cat: evidence for an intrinsic pathway leading from the deep nuclei to the cortex. *Brain Research 110,* 601–611.

Grabow, J. D., M. J. Ebersold, J. W. Albers, and E. M. Schima (1974). Cerebellar stimulation for the control of seizures. *Mayo Clin. Proc. 49,* 759–774.

Grafstein, B. (1971). Transneuronal transfer of radioactivity in the central nervous system. *Science 172,* 177–179.

Graham, J. (1977). An autoradiographic study of the efferent connections of the superior colliculus in the cat. *J. Comp. Neurol. 173,* 629–654.

Grand, W. (1971). Positional nystagmus: an early sign in medulloblastoma. *Neurology (Minneap.) 21,* 1157–1159.

Granit, R. (1950). The organization of the vertebrate retinal elements. *Ergebn. Physiol. 46,* 31–70.

Granit, R. (1955a). *Receptors and Sensory Perception.* Yale Univ. Press, New Haven, Conn.

Granit, R. (1955b). Centrifugal and antidromic effects on ganglion cells of retina. *J. Neurophysiol. 18,* 388–411.

Granit, R. (1970). *The Basis of Motor Control.* Academic Press, New York.

Granit, R. (1975). The functional role of the muscle spindles—facts and hypotheses. *Brain* *98*, 531–556.

Granit, R. (1977). Reconsidering the "alpha-gamma switch" in cerebellar action. In *Physiological Aspects of Clinical Neurology* (F. Clifford Rose, ed.), pp. 201–213. Blackwell, Oxford.

Granit, R., and H.-D. Henatsch (1956). Gamma control of dynamic properties of muscle spindles. *J. Neurophysiol. 19,* 356–366.

Granit, R., H.-D. Henatsch, and G. Steg (1956). Tonic and phasic ventral horn cells differentiated by post-tetanic potentiation in cat extensors. *Acta Physiol. Scand. 37,* 114–126.

Granit, R., and B. Holmgren (1955). Two pathways from brain stem to gamma ventral horn cells. *Acta Physiol. Scand. 35,* 93–108.

Granit, R., B. Holmgren, and P. A. Merton (1955). The two routes for excitation of muscle and their subservience to the cerebellum. *J. Physiol. (Lond.) 130,* 213–224.

Granit, R., C. Job, and B. R. Kaada (1952). Activation of muscle spindles in pinna reflex. *Acta Physiol. Scand. 27,* 161–168.

Granit, R., and B. R. Kaada (1952). Influence of stimulation of central nervous structures on muscle spindles in cat. *Acta Physiol. Scand. 27,* 130–160.

Grant, F. C., and L. M. Weinberger (1941). Experiences with intra-medullary tractotomy. I. Relief of facial pain and summary of operative results. *Arch. Surg. 42,* 681–692.

Grant, G. (1962a). Spinal course and somatotopically localized termination of the spinocerebellar tracts. An experimental study in the cat. *Acta Physiol. Scand. 56, Suppl. 193,* 1–45.

Grant, G. (1962b). Projection of the external cuneate nucleus onto the cerebellum in the cat: an experimental study using silver methods. *Exp. Neurol. 5,* 179–195.

Grant, G. (1970). Neuronal changes central to the site of axon transection. A method for the identification of retrograde changes in perikarya, dendrites and axons by silver impregnation. In *Contemporary Research Methods in Neuroanatomy* (W. J. H. Nauta and S. O. E. Ebbesson, eds.), pp. 173–185. Springer-Verlag, Berlin.

Grant, G. (1975). Retrograde neuronal degeneration. In *Golgi Centennial Symposium: Perspectives in Neurobiology* (M. Santini, ed.), pp. 195–200. Raven Press, New York.

Grant, G., and H. Aldskogius (1967). Silver impregnation of degenerating dendrites, cells and axons central to axonal transection. I. A Nauta study on the hypoglossal nerve in kittens. *Exp. Brain Res. 3,* 150–162.

Grant, G., and J. Arvidsson (1975). Transganglionic degeneration in trigeminal primary sensory neurons. *Brain Research 95,* 265–279.

Grant, G., J. Arvidsson, B. Robertson, and J. Ygge (1979). Transganglionic transport of horseradish peroxidase in primary sensory neurons. *Neurosci. Letters 12,* 23–28.

Grant, G., G. Aschan, and L. Ekvall (1964). Nystagmus produced by localized cerebellar lesions. International Vestibular Symposium, Uppsala 1963. *Acta Otolaryng. (Stockh.) Suppl. 192,* 78–84.

Grant, G., J. Boivie, and A. Brodal (1968). The question of a cerebellar projection from the lateral cervical nucleus re-examined. *Brain Research 9,* 95–102.

Grant, G., J. Boivie, and H. Silfvenius (1973). Course and termination of fibres from the nucleus z of the medulla oblongata. An experimental light microscopical study in the cat. *Brain Research 55,* 55–70.

Grant, G., S. Landgren, and H. Silfvenius (1975). Columnar distribution of U-fibres from the postcruciate cerebral projection area of the cat's group I muscle afferents. *Exp. Brain Res. 24,* 57–74.

Grant, G., and O. Oscarsson (1966). Mass discharges evoked in the olivocerebellar tract on stimulation of muscle and skin nerves. *Exp. Brain Res. 1,* 329–337.

Grant, G., O. Oscarsson, and I. Rosén (1966). Functional organization of the spinoreticulocerebellar path with identification of its spinal component. *Exp. Brain Res. 1,* 306–319.

Grant, G., and B. Rexed (1958). Dorsal spinal root afferents to Clarke's column. *Brain 81*, 567–576.

Grant, G., and F. Walberg (1974). The light and electron microscopical appearance of anterograde and retrograde neuronal degeneration. In *Dynamics of Degeneration and Growth in Neurons* (K. Fuxe, L. Olson, and Y. Zotterman, eds.), pp. 5–18. Pergamon Press, Oxford.

Gray, E. G. (1959). Axo-somatic and axo-dendritic synapses of the cerebral cortex: an electron microscope study. *J. Anat. (Lond.) 93*, 420–433.

Gray, E. G. (1961). The granule cells, mossy synapses and Purkinje spine synapses of the cerebellum: light and electron microscope observations. *J. Anat. (Lond.) 95*, 345–356.

Gray, E. G. (1963). Electron microscopy of presynaptic organelles of the spinal cord. *J. Anat. (Lond.) 97*, 101–106.

Gray, E. G. (1966). Problems of interpreting the fine structure of vertebrate and invertebrate synapses. *Int. Rev. Gen. Exp. Zool. 2*, 139–170.

Gray, E. G., and R. W. Guillery (1966). Synaptic morphology in the normal and degenerating nervous system. *Int. Rev. Cytol. 19*, 111–182.

Gray, E. G., and L. H. Hamlyn (1962). Electron microscopy of experimental degeneration in the avian optic tectum. *J. Anat. (Lond.) 96*, 309–316.

Gray, J. A. B., and P. B. C. Matthews (1951). Response of Pacinian corpuscles in the cat's toe. *J. Physiol. (Lond.) 113*, 475–482.

Graybiel, A. M. (1972a). Some extrageniculate visual pathways in the cat. *Invest. Ophthal. 11*, 322–332.

Graybiel, A. M. (1972b). Some ascending connections of the pulvinar and nucleus lateralis posterior of the thalamus in the cat. *Brain Research 44*, 99–125.

Graybiel, A. M. (1973). The thalamo-cortical projection of the so-called posterior nuclear group: a study with anterograde degeneration methods in the cat. *Brain Research 49*, 229–244.

Graybiel, A. M. (1974). Visuo-cerebellar and cerebello-visual connections involving the ventral lateral geniculate nucleus. *Exp. Brain Res. 20*, 303–306.

Graybiel, A. M. (1975a). Anatomical organization of retinotectal afferents in the cat: an autoradiographic study. *Brain Research 96*, 1–23.

Graybiel, A. M. (1975b). Wallerian degeneration and anterograde tracing methods. In *The Use of Axonal Transport for Studies of Neuronal Connectivity* (W. M. Cowan and M. Cuénod, eds.), pp. 173–216. Elsevier, Amsterdam.

Graybiel, A. M. (1977). Direct and indirect preoculomotor pathways of the brainstem: an autoradiographic study of the pontine reticular formation in the cat. *J. Comp. Neurol. 175*, 37–78.

Graybiel, A. M. (1978). Organization of the nigrotectal connection: an experimental tracer study in the cat. *Brain Research 143*, 339–348.

Graybiel, A. M., and E. A. Hartwieg (1974). Some afferent connections of the oculomotor complex in the cat. *Brain Research 81*, 543–551.

Graybiel, A. M., H. J. W. Nauta, R. J. Lasek, and W. J. H. Nauta (1973). A cerebello-olivary pathway in the cat: an experimental study using autoradiographic tracing techniques. *Brain Research 58*, 205–211.

Green, J. D. (1951). The comparative anatomy of the hypophysis, with special reference to its blood supply and innervation. *Amer. J. Anat. 88*, 225–311.

Green, J. D., and A. A. Arduini (1954). Hippocampal electrical activity in arousal. *J. Neurophysiol. 17*, 533–557.

Green, J. D., and G. W. Harris (1949). Observation of the hypophysio-portal vessels of the living rat. *J. Physiol. (Lond.) 108*, 359–361.

Greenwood, L. F., and B. J. Sessle (1976). Inputs to trigeminal brain stem neurones from facial, oral, tooth pulp and pharyngolaryngeal tissues. II. Role of trigeminal nucleus caudalis in modulating responses to innocuous and noxious stimuli. *Brain Research 117*, 227–238.

Gresty, M., and R. Baker (1976). Neurons with visual receptive field, eye movement and neck displacement sensitivity within and around the nucleus prepositus hypoglossi in the alert cat. *Exp. Brain Res. 24,* 429–433.

Greving, R. (1928). Die zentralen Anteile des vegetativen Nervensystems. *In v. Möllendorff's Handbuch der mikroskopischen Anatomie des Menschen,* IV/1, 917–1060. Springer-Verlag, Berlin.

Grigg, P. (1975). Mechanical factors influencing responses of joint afferent neurons from cat knee joint. *J. Neurophysiol. 38,* 1473–1484.

Grigg, P., G. A. Finerman, and L. H. Riley (1973). Joint position sense after total hip replacement. *J. Bone Jt. Surg. 55A,* 1016–1025.

Grigg, P., and B. J. Greenspan (1977). Response of primate joint afferent neurons to mechanical stimulation of knee joint. *J. Neurophysiol. 40,* 1–8.

Grigg, P., E. P. Harrigan, and K. E. Fogarty (1978). Segmental reflexes mediated by joint afferent neurons in cat knee. *J. Neurophysiol. 41,* 9–14.

Grigg, P., and J. B. Preston (1971). Baboon flexor and extensor fusimotor neurons and their modulation by motor cortex. *J. Neurophysiol. 34,* 428–436.

Grillner, S. (1974). A consideration of stretch and vibration data in relation to the tonic stretch reflex. In *Control of Posture and Locomotion* (R. B. Stein, K. B. Pearson, R. S. Smith, and J. B. Redford, eds.), pp. 397–405. Plenum Press, New York.

Grillner, S. (1975). Locomotion in vertebrates: central mechanisms and reflex interaction. *Physiol. Rev. 55,* 247–304.

Grillner, S., T. Hongo, and S. Lund (1966a). Interaction between the inhibitory pathways from the Deiters' nucleus and IA afferents to flexor motorneurones. (Abstract). *Acta Physiol. Scand. 68, Suppl. 277,* 61.

Grillner, S., T. Hongo, and S. Lund (1966b). Monosynaptic excitation of spinal α-motoneurones from the brain stem. *Experientia (Basel) 22,* 691.

Grillner, S., T. Hongo, and S. Lund (1970). The vestibulospinal tract. Effects on alpha motoneurones in the lumbosacral spinal cord in the cat. *Exp. Brain Res. 10,* 94–120.

Grillner, S., T. Hongo, and S. Lund (1971). Convergent effects on alpha motoneurones from the vestibulospinal tract and a pathway descending in the medial longitudinal fasciculus. *Exp. Brain Res. 12,* 457–479.

Grillner, S., and P. Zangger (1979). On the central generation of locomotion in the low spinal cat. *Exp. Brain Res. 34,* 241–261.

Grillo, M. A. (1966). Electron microscopy of sympathetic tissues. *Pharmacol. Rev. 18,* 387–399.

Grim, M. (1967). Muscle spindles in the posterior cricoarytenoid muscle of the human larynx. *Folia Morph. (Praha) 15,* 124–131.

Grimby, L. (1963a). Normal plantar response: integration of flexor and extensor reflex components. *J. Neurol. Neurosurg. Psychiat. 26,* 39–50.

Grimby, L. (1963b). Pathological plantar response: disturbances of the normal integration of flexor and extensor reflex components. *J. Neurol. Neurosurg. Psychiat. 26,* 314–321.

Grimby, L. (1965). Pathological plantar response. Part I. Flexor and extensor components in early and late reflex parts. Part II. Loss of significance of stimulus site. *J. Neurol. Neurosurg. Psychiat. 28,* 469–481.

Grimby, L., and J. Hannerz (1975). Disturbances in the voluntary recruitment order of anterior tibial motor units in ataxia. *J. Neurol. Neurosurg. Psychiat. 38,* 46–51.

Grimby, L., and J. Hannerz (1977). Firing rate and recruitment order of toe extensor motor units in different modes of voluntary contraction. *J. Physiol. (Lond.) 264,* 865–879.

Grimby, L., E. Kugelberg, and B. Löfström (1966). The plantar response in narcosis. *Neurology (Minneap.) 16,* 139–144.

Grimm, R. J., and D. S. Rushmer (1974). The activity of dentate neurons during an arm movement sequence. *Brain Research 71,* 309–326.

Groenewegen, H. J., A. J. P. Boesten, and J. Voogd (1975). The dorsal column nuclear projections to the nucleus ventralis posterior lateralis thalami and the inferior olive in the cat: an autoradiographic study. *J. Comp. Neurol. 162,* 505–518.

Groenewegen, H. J., and J. Voogd (1977). The parasagittal zonation within the olivocerebellar projection. I. Climbing fiber distribution in the vermis of cat cerebellum. *J. Comp. Neurol. 174,* 417–488.

Groenewegen, H. J., J. Voogd, and S. L. Freedman (1979). The parasagittal zonal organization within the olivocerebellar projection. II. Climbing fiber distribution in the intermediate and hemispheric parts of cat cerebellum. *J. Comp. Neurol. 183,* 551–602.

Grofová, I. (1969). Experimental demonstration of a topical arrangement of the pallido-subthalamic fibers in the cat. *Psychiat. Neurol. Neurochir. (Amst.) 72,* 53–59.

Grofová, I. (1975). The identification of striatal and pallidal neurons projecting to substantia nigra. An experimental study by means of retrograde axonal transport of horseradish peroxidase. *Brain Research 91,* 286–291.

Grofová, I. (1979). Extrinsic connections of the neostriatum. In *The Neostriatum* (I. Divac and R. E. Öberg, eds.), pp. 37–51. Pergamon Press, Oxford.

Grofová, I., and J. Maršala (1959). Tvar a struktura nucleus ruber u člověka. (Form and structure of the nucleus ruber in man.) *Cs. Morfol. 8,* 215–237.

Grofová, I., O. P. Ottersen, and E. Rinvik (1978). Mesencephalic and diencephalic afferents to the superior colliculus and periaqueductal gray substance demonstrated by retrograde axonal transport of horseradish peroxidase in the cat. *Brain Research 146,* 205–220.

Grofová, I., and E. Rinvik (1974). Light and electron microscopical studies of the normal structure and main afferent connections of the nuclei ventralis lateralis and ventralis anterior thalami of the cat. *Confin. Neurol. (Basel) 36,* 256–272.

Groos, W. P., L. K. Ewing, C. M. Carter, and J. D. Coulter (1978). Organization of corticospinal neurons in the cat. *Brain Research 143,* 393–419.

Gros, C., G. Ouaknine, B. Vlahovitch, and Ph. Frerebeau (1966). Selective posterior radicotomy in the treatment of pyramidal hypertony. *Neuro-Chirurgie 3,* 505–518.

Gross, N. B., W. S. Lifschitz, and D. J. Anderson (1974). The tonotopic organization of the auditory thalamus of the squirrel monkey (*Saimiri sciureus*). *Brain Research 65,* 323–332.

Grossman, S. P. (1975). Role of the hypothalamus in the regulation of food and water intake. *Psychol. Rev. 82,* 200–224.

Grossman, S. P. (1976). Behavioral functions of the septum: a re-analysis. In *The Septal Nuclei* (J. F. DeFrance, ed.), pp. 361–422. Plenum Press, New York.

Grossman, S. P., D. Dacey, A. E. Halaris, T. Collier, and A. Routtenberg (1978). Aphagia and adipsia after preferential destruction of nerve cell bodies in the hypothalamus. *Science 202,* 537–539.

Growdon, J. H., W. W. Chambers, and C. N. Liu (1967). An experimental study of cerebellar dyskinesia in the rhesus monkey. *Brain 90,* 603–632.

Grünbaum, A. S. F., and C. S. Sherrington (1902). Observations on the physiology of the cerebral cortex of some of the higher apes. *Proc. Roy. Soc. B 69,* 206–209.

Grünbaum, A. S. F., and C. S. Sherrington (1903). Observations on the physiology of the anthropoid apes. *Proc. Roy. Soc. B 72,* 152–155.

Grundfest, H., and B. Campbell (1942). Origin, conduction, and termination of impulses in the dorsal spino-cerebellar tracts. *J. Neurophysiol. 5,* 275–294.

Gudden, B. (1870). Über einen bisher nicht beschriebenen Nervenfasernstrang im Gehirne der Säugetiere und des Menschen. *Arch. Psychiat. Nervenkr. 2,* 364–366.

Gudden, B. (1881). Über den Tractus peduncularis transversus. *Arch. Psychiat. Nervenkr. 11,* 415–423.

Guilbaud, G., D. Caille, J. M. Besson, and G. Benelli (1977). Single units activities in ventral posterior and posterior group thalamic nuclei during nociceptive stimulations in the cat. *Arch. Ital. Biol. 115,* 38–56.

Guilbaud, G., J. L. Oliveras, G. Giesler, Jr., and J. M. Besson (1977). Effects induced by stimulation of the centralis inferior nucleus of the raphe on dorsal horn interneurons in cat's spinal cord. *Brain Research 126,* 355–360.

Guild, S. R. (1932). Correlations of histologic observations and the acuity of hearing. *Acta Otolaryng. (Stockh.) 17,* 207–249.

Guild, S. R., S. J. Crowe, C. C. Bunch, and L. M. Polvogt (1931). Correlation of differences in the density of innervation of the organ of Corti with differences in the acuity of hearing. *Acta Otolaryng. (Stockh.) 15*, 269–308.

Guillery, R. W. (1956). Degeneration in the post-commissural fornix and the mammillary peduncle of the rat. *J. Anat. (Lond.) 90*, 350–370.

Guillery, R. W. (1957). Degeneration in the hypothalamic connexions of the albino rat. *J. Anat. (Lond.) 91*, 91–115.

Guillery, R. W. (1959). Afferent fibres to the dorso-medial thalamic nucleus in the cat. *J. Anat. (Lond.) 93*, 403–419.

Guillery, R. W. (1966). A study of Golgi preparations from the dorsal lateral geniculate nucleus of the adult cat. *J. Comp. Neurol. 128*, 21–50.

Guillery, R. W. (1969a). The organization of synaptic interconnections in the laminae of the dorsal lateral geniculate nucleus of the cat. *Z. Zellforsch. 96*, 1–38.

Guillery, R. W. (1969b). A quantitative study of synaptic interconnections in the dorsal lateral geniculate nucleus of the cat. *Z. Zellforsch. 96*, 39–49.

Guillery, R. W. (1970). The laminar distribution of retinal fibres in the dorsal lateral geniculate nucleus of the cat: a new interpretation. *J. Comp. Neurol. 138*, 339–369.

Guillery, R. W., H. O. Adrian, C. N. Woolsey, and J. E. Rose (1966). Activation of somatosensory areas I and II of cat's cerebral cortex by focal stimulation of the ventrobasal complex. In *The Thalamus* (D. P. Purpura and M. D. Yahr, eds.), pp. 197–206. Columbia Univ. Press, New York.

Guillery, R. W., and M. Colonnier (1970). Synaptic patterns in the dorsal lateral geniculate nucleus of the monkey. *Z. Zellforsch. 103*, 90–108.

Guillery, R. W., and H. J. Ralston (1964). Nerve fibers and terminals: electron microscopy after Nauta staining. *Science 143*, 1331–1332.

Guillery, R. W., and G. L. Scott (1971). Observations on synaptic patterns in the dorsal lateral geniculate nucleus of the cat: the C laminae and the perikaryal synapses. *Exp. Brain Res. 12*, 184–203.

Guiot, G., M. Sachs, E. Hertzog, S. Brion, J. Rougerie, J. C. Dalloz, and F. Napoléone (1959). Stimulation électrique et lésions chirurgicales de la capsule interne. Déductions anatomiques et pysiologiques. *Neuro-chirurgie 5*, 17–36.

Gulley, R. L., and R. L. Wood (1971). The fine structure of the neurons in the rat substantia nigra. *Tissue & Cell 3*, 675–690.

Gurfinkel, V. S., M. I. Lipshits, S. Mori, and K. E. Popov (1976). The state of stretch reflex during quiet standing in man. *Progr. Brain Res. 44*, 473–486.

Guth, P. S., C. H. Norris, and R. P. Bobbin (1976). The pharmacology of transmission in the peripheral auditory system. *Pharmacol. Rev. 28*, 95–125.

Gutman, E. (1976). The multiple regulation of contractile and histochemical properties of cross-striated muscle. In *The Motor System: Neurophysiology and Muscle Mechanisms* (M. Shahani, ed.), pp. 1–13. Elsevier, Amsterdam.

Guttmann, L. (1940). The distribution of disturbances of sweat secretion after extirpation of certain sympathetic cervical ganglia in man. *J. Anat. (Lond.) 74*, 537–549.

Gwyn, D. G., and B. A. Flumerfelt (1974). A comparison of the distribution of cortical and cerebellar afferents in the red nucleus of the rat. *Brain Research 69*, 130–135.

Ha, H. (1971). The cervicothalamic tract in the rhesus monkey. *Exp. Neurol. 33*, 205–212.

Ha, H., and C.-N. Liu (1966). Organization of the spino-cervicothalamic system. *J. Comp. Neurol. 127*, 445–470.

Ha, H., and C.-N. Liu (1968). Cell origin of the ventral spinocerebellar tract. *J. Comp. Neurol. 133*, 185–206.

Ha, H., and F. Morin (1964). Comparative anatomical observations of the cervical nucleus, n. cervicalis lateralis, of some primates. (Abstract.) *Anat. Rec. 148*, 374.

Haberly, L. B., and J. L. Price (1977). The axonal projections of the mitral and tufted cells of the olfactory bulb in the rat. *Brain Research 129*, 152–157.

Hagbarth, K.-E. (1973). The effect of muscle vibration in normal man and in patients with motor disorders. In *New Developments in Electromyography and Clinical Neurophysiology,* Vol. 3 (J. E. Desmedt, ed.), pp. 428–443. Karger, Basel.

Hagbarth, K.-E., and G. Eklund (1966). Motor effects of vibratory muscle stimuli in man. In *Muscular Afferents and Motor Control* (R. Granit, ed.), pp. 177–186. Almqvist & Wiksell, Stockholm.

Hagbarth, K.-E., and J. Fex (1959). Centrifugal influences on single unit activity in spinal sensory paths. *J. Neurophysiol. 22,* 321–338.

Hagbarth, K.-E., A. Hongell, and B. G. Wallin (1970). The effect of gamma fibre block on afferent muscle nerve activity during voluntary contractions. *Acta Physiol. Scand. 79,* 27A–28A.

Hagbarth, K.-E., and D. I. B. Kerr (1954). Central influences on spinal afferent conduction. *J. Neurophysiol. 17,* 295–307.

Hagbarth, K.-E., and E. Kugelberg (1958). Plasticity of the human abdominal skin reflex. *Brain 81,* 305–318.

Hagbarth, K.-E., G. Wallin, D. Burke, and L. Löfstedt (1975). Effects of the Jendrassik manoeuvre on muscle spindle activity in man. *J. Neurol. Neurosurg. Psychiat. 38,* 1143–1153.

Hagbarth, K.-E., G. Wallin, and L. Löfstedt (1973). Muscle spindle responses to stretch in normal and spastic subjects. *Scand. J. Rehab. Med. 5,* 156–159.

Hagbarth, K.-E., G. Wallin, and L. Löfstedt (1975). Muscle spindle activity in man during voluntary fast alternating movements. *J. Neurol. Neurosurg. Psychiat. 38,* 625–635.

Hagbarth, K.-E., G. Wallin, L. Löfstedt, and S. M. Aquilonius (1975). Muscle spindle activity in alternating tremor of Parkinsonism and in clonus. *J. Neurol. Neurosurg. Psychiat. 38,* 636–641.

Hagen, E. (1949/50). Neurohistologische Untersuchungen an der menschlichen Hypophyse. *Z. Anat. Entwickl.-Gesch. 114,* 640–679.

Hagen, E., H. Knoche, D. C. Sinclair, and G. Weddell (1953). The role of specialized nerve terminals in cutaneous sensibility. *Proc. Roy. Soc. B 141,* 279–287.

Häggqvist, G. (1936). Analyse der Faserverteilung in einem Rückenmarkquerschnitt (Th3). *Z. Mikr.-Anat. Forsch. 39,* 1–34.

Häggqvist, G. (1937). Faseranalytische Studien über die Pyramidenbahn. *Acta Psychiat. Neurol. Scand. 12,* 457–466.

Haines, D. E. (1975a). Cerebellar corticovestibular fibers of the posterior lobe in a prosimian primate, the lesser bushbaby (*Galago senegalensis*). *J. Comp. Neurol. 160,* 363–398.

Haines, D. E. (1975b). Cerebellar cortical efferents of the posterior lobe vermis in a prosimian primate (*Galago*) and the tree shrew (*Tupaia*). *J. Comp. Neurol. 163,* 21–40.

Haines, D. E. (1976). Cerebellar corticonuclear and corticovestibular fibers of the anterior lobe vermis in a prosimian primate (*Galago senegalensis*). *J. Comp. Neurol. 170,* 67–96.

Haines, D. E. (1977a). Cerebellar corticonuclear and corticovestibular fibers of the flocculonodular lobe in a prosimian primate (*Galago senegalensis*). J. Comp. Neurol. 174, 607–630.

Haines, D. E. (1977b). A proposed functional significance of parvicellular regions of the lateral and medial cerebellar nuclei. *Brain Behav. Evol. 14,* 328–340.

Haines, D. E., J. L. Culberson, and G. F. Martin (1976). Laterality and topography of cerebellar cortical efferents in the opossum (*Didelphis marsupialis virginiana*). *Brain Research 106,* 152–158.

Haines, D. E., and J. A. Rubertone (1977). Cerebellar corticonuclear fibers: evidence of zones in the primate anterior lobe. *Neurosci. Letters 6,* 231–236.

Haines, D. E., and R. H. Whitworth (1978). Cerebellar cortical efferent fibers of the paraflocculus of tree shrew (Tupaia). *J. Comp. Neurol. 182,* 137–150.

Halász, N., Å. Ljungdahl, and T. Hökfelt (1978). Transmitter histochemistry of the rat olfactory bulb. II. Fluorescence histochemical, autoradiographic and electron microscopic localization of monoamines. *Brain Research 154,* 253–271.

Halban, H. V., and M. Infeld (1902). Zur Pathologie der Hirnschenkelhaube mit besonderer Berücksichtigung der posthemiplegischen Bevegungserscheinungen. *Arb. Neurol. Inst. Univ. Wien 9,* 329–404.

Halgren, E., R. Rausch, and P. H. Crandall (1977). Neurons in the human basolateral amygdala and hippocampal formation do not respond to odors. *Neurosci. Letters 4,* 331–335.

Halgren, E., R. D. Walter, D. G. Cherlow, and P. H. Crandall (1978). Mental phenomena evoked by electrical stimulation of the human hippocampal formation and amygdala. *Brain 101,* 83–117.

Hall, E. A. (1963). Efferent connections of the basal and lateral nuclei of the amygdala in the cat. *Amer. J. Anat. 113,* 139–145.

Hall, E. (1972). Some aspects of the structural organization of the amygdala. In *The Neurobiology of the Amygdala* (B. E. Eleftheriou, ed.), pp. 95–121. Plenum Press, New York.

Hall, E. (1975). The anatomy of the limbic system. In *Neural Integration of Physiological Mechanisms and Behavior* (G. J. Mogenson and F. R. Calaresu, eds.), pp. 68–94. Univ. of Toronto Press, Toronto.

Hall, E., and F. A. Geneser-Jensen (1971). Distribution of acetylcholinesterase and monoamine oxidase in the amygdala of the guinea pig. *Z. Zellforsch. 120,* 204–221.

Hall, J. G. (1964). The cochlea and the cochlear nuclei in neonatal asphyxia. *Acta Otolaryng. (Stockh.), Suppl. 194,* 1–93.

Hallett, M., B. T. Shahani, and R. R. Young (1975a). EMG analysis of stereotyped voluntary movements in man. *J. Neurol. Neurosurg. Psychiat. 38,* 1154–1162.

Hallett, M., B. T. Shahani, and R. R. Young (1975b). EMG analysis of patients with cerebellar deficits. *J. Neurol. Neurosurg. Psychiat. 38,* 1163–1169.

Halperin, J. J., and J. H. LaVail (1975). A study of the dynamics of retrograde transport and accumulation of horseradish peroxidase in injured neurons. Brain Research 100, 253–269.

Halstead, W. C., A. E. Walker, and P. C. Bucy (1940). Sparing and nonsparing of 'macular' vision associated with occipital lobectomy in man. *Arch. Ophthal. 24,* 948–966.

Hamberger, B., and K. A. Norberg (1965). Adrenergic synaptic terminals and nerve cells in bladder ganglia of the cat. *Int. J. Neuropharmacol. 4,* 41–45.

Hamilton, B. L. (1973a). Cytoarchitectural subdivisions of the periaqueductal gray matter in the cat. *J. Comp. Neurol. 149,* 1–28.

Hamilton, B. L. (1973b). Projections of the nuclei of the periaqueductal gray matter in the cat. *J. Comp. Neurol. 152,* 45–58.

Hamilton, B. L., and F. M. Skultety (1970). Efferent connections of the periaqueductal gray matter in the cat. *J. Comp. Neurol. 139,* 105–114.

Hamilton, L. W. (1976). *Basic Limbic System Anatomy of the Rat.* Plenum Press, New York.

Hamlyn, L. H. (1962). The fine structure of the mossy fibre endings in the hippocampus of the rabbit. *J. Anat. (Lond.) 96,* 112–120.

Hamlyn, L. H. (1963). An electron microscope study of pyramidal neurons in the Ammon's horn of the rabbit. *J. Anat. (Lond.) 97,* 189–201.

Hammond, P. H. (1956). The influence of prior instruction to the subject on an apparently involuntary neuromuscular response. *J. Physiol. (Lond.) 132,* 17–18P.

Hámori, J., and E. Mezey (1977). Serial and triadic synapses in the cerebellar nuclei of the cat. *Exp. Brain Res. 30,* 259–273.

Hámori, J., T. Pasik, P. Pasik, and J. Szentágothai (1974). Triadic synaptic arrangements and their possible significance in the lateral geniculate nucleus of the monkey. *Brain Research 80,* 379–393.

Hámori, J., and J. Szentágothai (1966a). Identification under the electron microscope of climbing fibers and their synaptic contacts. *Exp. Brain Res. 1,* 65–81.

Hámori, J., and J. Szentágothai (1966b). Participation of Golgi neuron processes in the cerebellar glomeruli: an electron microscope study. *Exp. Brain Res. 2,* 35–48.

Hampson, J. L. (1949). Relationship between cat cerebral and cerebellar cortices. *J. Neurophysiol. 12*, 37–50.

Hampson, J. L., C. R. Harrison, and C. N. Woolsey (1952). Cerebro-cerebellar projections and the somatotopic localization of motor function in the cerebellum. *Res. Publ. Ass. Nerv. Ment. Dis. 30*, 299–316.

Hanaway, J., and R. R. Young (1977). Localization of the pyramidal tract in the internal capsule of man. *J. Neurol. Sci. 34*, 63–70.

Hanbery, J., C. Ajmone-Marsan, and M. Dilworth (1954). Pathways of nonspecific thalamocortical projection systems. *Electroenceph. Clin. Neurophysiol. 6*, 103–118.

Hancock, M. B. (1976). Cells of origin of hypothalamo-spinal projections in the cat. *Neurosci. Letters 3*, 179–184.

Hancock, M. B., and C. L. Fougerousse (1976). Spinal projections from the nucleus locus coeruleus and nucleus subcoeruleus in the cat and monkey as demonstrated by the retrograde transport of horseradish peroxidase. *Brain Res. Bull. 1*, 229–234.

Hand, P. J. (1966). Lumbosacral dorsal root terminations in the nucleus gracilis of the cat. Some observations on terminal degeneration in other medullary sensory nuclei. *J. Comp. Neurol. 126*, 137–156.

Hand, P., and C.-N. Liu (1966). Efferent projections of the nucleus gracilis. (Abstract). *Anat. Rec. 154*, 353–354.

Hand, P. J., and A. R. Morrison (1970). Thalamocortical projections from the ventrobasal complex to somatic sensory areas I and II. *Exp. Neurol. 26*, 291–308.

Hand, P. J., and A. R. Morrison (1972). Thalamocortical relationships in the somatic sensory system as revealed by silver impregnation techniques. *Brain Behav. Evol. 5*, 273–302.

Hand, P. J., and T. van Winkle (1977). The efferent connections of the feline nucleus cuneatus. *J. Comp. Neurol. 171*, 83–110.

Handbook of Sensory Physiology Vol. V/1, Auditory System, (1974) (W. D. Keidel and W. D. Neff, eds.). Springer-Verlag, Berlin.

Handbook of Sensory Physiology Vol. VI/1, Vestibular System Part 1: Basic Mechanisms (1974) (H. H. Kornhuber, ed.). Springer-Verlag, Berlin.

Hansebout, R. R., and J. B. R. Cosgrove (1966). Effects of intrathecal phenol in man. *Neurology (Minneap.) 16*, 277–282.

Hansen, K., and H. Schliack (1962). *Segmentale Innervation. Ihre Bedeutung für Klinik und Praxis*. Georg Thieme Verlag, Stuttgart.

Hansen, S., and J. P. Ballantyne (1978). A quantitative electrophysiological study of motor neurone disease. *J. Neurol. Neurosurg. Psychiat. 41*, 773–783.

Hanström, B. (1952). Transportation of colloid from the neurosecretory hypothalamic centres in the brain into the blood vessels of the neural lobe of the hypophysis. *Kungl. Fysiogr. Sällsk. Lund Förh. 22*, 31–35.

Hare, W. K., H. W. Magoun, and S. W. Ranson (1935). Pathways for pupillary constriction. *Arch. Neurol. Psychiat. (Chic.) 34*, 1188–1194.

Harik, S. I., and P. L. Morris (1973). The effects of lesions in the head of the caudate nucleus on spontaneous and L-DOPA induced activity in the cat. *Brain Research 62*, 279–285.

Harkmark, W. (1954). The rhombic lip and its derivatives in relation to the theory of neurobiotaxis. In *Aspects of Cerebellar Anatomy* (J. Jansen and A. Brodal, eds.), pp. 264–284. Johan Grundt Tanum, Oslo.

Harkmark, W., C. H. Köhler, and B. Srebro (1978). Afferent connections from the hippocampus to the septum telencephali in the rat as studied by a retrograde transport of horseradish peroxidase. *Israel J. Med. Sci. 14*, 894–895.

Harlem, O. K., and A. Lönnum (1957). A clinical study of the abdominal skin reflexes in newborn infants. *Arch. Dis. Childh. 32*, 127–130.

Harper, J. W., and R. G. Heath (1973). Anatomic connections of the fastigial nucleus to the rostral forebrain in the cat. *Exp. Neurol. 39*, 285–292.

Harper, J. W., and R. G. Heath (1974). Ascending projections of the cerebellar fastigial nuclei: connections to the ectosylvian gyrus. *Exp. Neurol. 42*, 241–247.

Harriman, D. G. F., and H. Garland (1968). The pahtology of Adie's syndrome. *Brain 91,* 401–418.

Harrington, R. B., R. W. Hollenhorst, and G. P. Sayre (1966). Unilateral internuclear ophthalmoplegia. *Arch. Neurol. (Chic.) 15,* 29–34.

Harris, A. J., R. Hodes, and H. W. Magoun (1944). The afferent path of the pupillodilator reflex in the cat. *J. Neurophysiol. 7,* 231–243.

Harris, D. A., and E. Henneman (1977). Identification of two species of alpha motoneurons in cat's plantaris pool. *J. Neurophysiol. 40,* 16–25.

Harris, G. W. (1947). The innervation and actions of the neuro-hypophysis; an investigation using the method of remote-control stimulation. *Phil. Trans. B 232,* 385–441.

Harris, G. W. (1948). Electrical stimulation of the hypothalamus and the mechanism of neural control of the adenohypophysis. *J. Physiol. (Lond.) 107,* 418–429.

Harris, M. C., and M. Sanghera (1974). Projection of medial basal hypothalamic neurones to the preoptic anterior hypothalamic areas and the paraventricular nucleus in the rat. *Brain Research 81,* 401–411.

Harrison, F., and K. B. Corbin (1942). Oscillographic studies on the spinal tract of the fifth cranial nerve. *J. Neurophysiol. 5,* 465–482.

Harrison, J. M. (1974). The auditory system of the medulla and localization. *Fed. Proc. 33,* 1901–1903.

Harrison, J. M., and M. E. Howe (1974a). Anatomy of the afferent auditory nervous system in mammals. In *Handbook of Sensory Physiology, Vol. V/1* (W. D. Keidel and W. D. Neff, eds.), pp. 283–336. Springer-Verlag, Berlin.

Harrison, J. M., and M. E. Howe (1974b). Anatomy of the descending auditory system (Mammalian). In *Handbook of Sensory Physiology, Vol. V/1* (W. D. Keidel and W. D. Neff, eds.), pp. 363–388. Springer-Verlag, Berlin.

Harrison, J. M., and R. Irving (1964). Nucleus of the trapezoid body. Dual afferent innervation. *Science 143,* 473–474.

Harrison, J. M., and R. Irving (1965). Anterior ventral cochlear nucleus. *J. Comp. Neurol. 124,* 15–21.

Harrison, J. M., and R. Irving (1966a). Ascending connections of the anterior ventral cochlear nucleus in the rat. *J. Comp. Neurol. 126,* 51–63.

Harrison, J. M., and R. Irving (1966b). The organization of the posterior ventral cochlear nucleus in the rat. *J. Comp. Neurol. 126,* 391–401.

Harrison, M. S. (1962). "Epidemic vertigo"—"vestibular neuronitis". A clinical study. *Brain 85,* 613–620.

Harting, J. K. (1977). Descending pathways from the superior colliculus: an autoradiographic analysis in the rhesus monkey (*Macaca mulatta*). *J. Comp. Neurol. 173,* 583–612.

Harting, J. K., and R. W. Guillery (1976). Organization of retinocollicular pathways in the cat. *J. Comp. Neurol. 166,* 133–144.

Harting, J. K., W. C. Hall, and I. T. Diamond (1972). Evolution of the pulvinar. *Brain Behav. Evol. 6,* 424–452.

Harting, J. K., W. C. Hall, I. T. Diamond, and G. F. Martin (1973). Anterograde degeneration study of the superior colliculus in *Tupaia glis:* evidence for a subdivision between superficial and deep layers. *J. Comp. Neurol. 148,* 361–386.

Harting, J. K., and G. F. Martin (1970). Neocortical projections to the pons and medulla of the nine banded armadillo. *J. Comp. Neurol. 138,* 483–500.

Harting, J. K., and C. R. Noback (1971). Subcortical projections from the visual cortex in the tree shrew. *Brain Research 25,* 21–33.

Hartmann-von Monakow, K., K. Akert, and H. Künzle (1979). Projections of precentral and premotor cortex to the red nucleus and other midbrain areas in *Macaca fascicularis. Exp. Brain Res. 34,* 91–105.

Hashikawa, T., and K. Kawamura (1977). Identification of cells of origin of tectopontine fibers in the cat superior colliculus: an experimental study with the horseradish perioxidase method. *Brain Research 130,* 65–79.

Hassel, H. J. van, M. A. Biedenbach, and A. C. Brown (1972). Cortical potentials evoked by tooth pulp stimulation in rhesus monkeys. *Arch. Oral Biol. 17,* 1059–1066.

Hassler, R. (1949). Über die afferenten Bahnen und Thalamuskerne des motorischen Systems des Grosshirns. I. Mitteilung. Bindearm und Fasciculus thalamicus. II. Mitteilung. Weitere Bahnen aus Pallidum, Ruber, vestibulärem System zum Thalamus; Übersicht und Besprechung der Ergebnisse. *Arch. Psychiat. Nervenkr. 182,* 759–785, 786–818.

Hassler, R. (1950). Projections of the cerebellum to the midbrain and the thalamus. *Dtsch. Z. Nervenheilk. 163,* 629–671.

Hassler, R. (1956). Die zentralen Apparate der Wendebewegungen. II. Die neuronalen Apparate der vestibulären Korrekturwendungen und der Adversivbewegungen. *Arch. Psychiat. Nervenkr. 194,* 481–516.

Hassler, R. (1959). Anatomy of the thalamus. In *Introduction to Stereotaxis with an Atlas of the Human Brain,* Vol. I (G. Schaltenbrand and P. Bailey, eds.), pp. 230–290. Grune & Stratton, New York.

Hassler, R. (1960). Die zentralen Systeme des Schmerzes. *Acta. Neurochir. (Wien) 8,* 353–423.

Hassler, R. (1961). Motorische und sensible Effekte umschriebener Reizungen und Ausschaltungen im menschlichen Zwischenhirn. *Dtsch. Z. Nervenheilk. 183,* 148–171.

Hassler, R. (1964). Spezifische und unspezifische Systeme des menschlichen Zwischenhirns. In *Lectures on the Diencephalon, Progress in Brain Research,* Vol. 5 (W. Bargmann and J. P. Schadé, eds.), pp. 1–32. Elsevier, Amsterdam.

Hassler, R. (1966a). Thalamic regulation of muscle tone and the speed of movements. In *The Thalamus* (D. P. Purpura and M. D. Yahr, eds.), pp. 419–436. Columbia Univ. Press. New York.

Hassler, R. (1966b). Extrapyramidal motor areas of cat's frontal lobe: their functional and architectonic differentiation. *Int. J. Neurol. (Montevideo) 5,* 301–316.

Hassler, R., and G. Dieckmann (1967). Arrest reaction, delayed inhibition and unusual gaze behaviour resulting from stimulation of the putamen in awake, unrestrained cats. *Brain Research 5,* 504–507.

Hassler, R., and K. Muhs-Clement (1964). Architektonischer Aufbau des sensorimotorischen und parietalen Cortex der Katze. *J. Hirnforsch. 6,* 377–420.

Hassler, R., and T. Riechert (1959). Klinische und anatomische Befunde bei stereotaktischen Schmerzoperationen im Thalamus. *Arch. Psychiat. Nervenkr. 200,* 93–122.

Hast, M. H., J. M. Fischer, A. B. Wetzel, and V. E. Thompson (1974). Cortical motor representation of the laryngeal muscles in *Macaca mulatta. 73,* 229–240.

Hattori, T., H. C. Fibiger, and P. L. McGeer (1975). Demonstration of a pallidonigral projection innervating dopaminergic neurons. *J. Comp. Neurol. 162,* 487–504.

Hattori, T., P. L. McGeer, H. C. Fibiger, and E. G. McGeer (1973). On the source of gaba-containing terminals in the substantia nigra. Electron microscopic autoradiographic and biochemical studies. *Brain Research 54,* 103–114.

Haug, F.-M. Š. (1974). Light microscopical mapping of the hippocampal region, the pyriform cortex and the corticomedial amygdaloid nuclei of the rat with Timm's sulphide silver method. I. Area dentata, hippocampus and subiculum. *Z. Anat. Entwickl.-Gesch. 145,* 1–27.

Haug, F.-M. Š. (1976). Sulphide silver pattern and cytoarchitectonics of parahippocampal areas in the rat. Special reference to the subdivision of area entorhinalis (area 28) and its demarcation from the pyriform cortex. *Advan. Anat. Embryol. Cell Biol. 52,* Fasc. 4, 73 pp.

Hauge, T. (1954). Catheter vertebral angiography. *Acta Radiol. (Stockh.) Suppl. 109,* 1–219.

Hauglie-Hanssen, E. (1968). Intrinsic neuronal organization of the vestibular nuclear complex in the cat. A Golgi study. *Ergebn. Anat. Entwickl.-Gesch. 40, Heft 5,* 1–105.

Hayashi, M. (1924). Einige wichtige Tatsachen aus der ontogenetischen Entwicklung des menschlichen Kleinhirns. *Dtsch. Z. Nervenheilk. 81,* 74–82.

Hayes, N. L., and A. Rustioni (1979). Dual projections of single neurons are visualized simultaneously: use of enzymatically inactive [³H] HRP. *Brain Research 165,* 321–326.

Hayhow, W. R. (1958). The cytoachitecture of the lateral geniculate body in the cat in relation to the distribution of crossed and uncrossed optic fibers. *J. Comp. Neurol. 110,* 1–64.

Hayhow, W. R. (1959). An experimental study of the accessory optic fiber system in the cat. *J. Comp. Neurol. 113,* 281–313.

Haymaker, W., and B. Woodhall (1945). *Peripheral Nerve Injuries. Principles of Diagnosis.* W. B. Saunders, Philadelphia.

Hayward, J. N. (1975). Neural control of the posterior pituitary. *Ann. Rev. Physiol. 37,* 191–210.

Head, H. (1920). *Studies in Neurology,* 2 Vols. Oxford Univ. Press, London.

Head, H., and G. Holmes (1911–1912). Sensory disturbances from cerebral lesions. *Brain 34,* 102–254.

Headley, P. M., D. Lodge, and A. W. Duggan (1976). Drug-induced rhythmical activity in the inferior olivary complex of the rat. *Brain Research 101,* 461–478.

Hearing Mechanisms in Vertebrates. A Ciba Foundation Symposium (1968) (A. V. S. de Reuck and J. Knight, eds.). J. & A. Churchill, London.

Heath, C. J., J. Hore, and C. G. Philips (1976). Inputs from low threshold muscle and cutaneous afferents of hand and forearm to Areas 3a and 3b of baboon's cerebral cortex. *J. Physiol. (Lond.) 257,* 199–227.

Heath, C. J., and E. G. Jones (1971a). An experimental study of ascending connections from the posterior group of the thalamic nuclei in the cat. *J. Comp. Neurol. 141,* 397–426.

Heath, C. J., and E. G. Jones (1971b). The anatomical organization of the suprasylvian gyrus of the cat. *Ergebn. Anat. Entwickl.-Gesch. 45,* 1–64.

Hebb, D. A. (1945). Man's frontal lobe. A critical review. *Arch. Neurol. Psychiat. (Chic.) 54,* 10–24.

Hebb, D. A. (1954). The problem of consciousness and introspection. In *Brain Mechanisms and Consciousness.* CIOMS Symposium (J. F. Delafresnaye, ed.), pp. 402–417. Blackwell, Oxford.

Hebb, D. A., and W. Penfield (1940). Human behaviour after extensive bilateral removal from the frontal lobes. *Arch. Neurol. Psychiat. (Chic.) 44,* 421–438.

Hécaen, H., and R. Angelergues (1964). Localization of symptoms in aphasia. In *Disorders of Language.* Ciba Foundation Symposium (A. V. S. de Reuck and M. O'Connor, eds.), pp. 223–246. J. & A. Churchill, London.

Hécaen, H., W. Penfield, C. Bertrand, and R. Malmo (1956). The syndrome of apractognosia due to lesions of the minor hemisphere. *Arch. Neurol. Psychiat. (Chic.) 75,* 400–434.

Hees, J. van, and J. M. Gybels (1972). Pain related to single afferent fibers from human skin. *Brain Research 48,* 397–400.

Heidary, H., and J. Tomasch (1969). Neuron numbers and perikaryon areas in the human cerebellar nuclei. *Acta Anat. (Basel) 74,* 290–296.

Heilman, K. M., and G. W. Sypert (1977). Korsakoff's syndrome resulting from bilateral fornix lesions. *Neurology (Minneap.) 27,* 490–493.

Heimer, L. (1967). Silver impregnation of terminal degeneration in some forebrain fiber systems: a comparative evaluation of current methods. *Brain Research 5,* 86–108.

Heimer, L. (1968). Synaptic distribution of centripetal and centrifugal nerve fibres in the olfactory system of the rat. An experimental anatomical study. *J. Anat. (Lond.) 103,* 413–432.

Heimer, L. (1972). The olfactory connections of the diencephalon in the rat. An experimental light- and electron-microscope study with special emphasis on the problem of terminal degeneration. *Brain Behav. Evol. 6,* 484–523.

Heimer, L., F. F. Ebner, and W. J. H. Nauta (1967). A note on the termination of commissural fibers in the neocortex. *Brain Research 5,* 171–177.

Heimer, L., and W. J. H. Nauta (1969). The hypothalamic distribution of the stria terminalis in the rat. *Brain Research 13,* 284–297.

Heimer, L., and P. D. Wall (1968). The dorsal root distribution to the substantia gelatinosa of the rat with a note on the distribution in the cat. *Exp. Brain Res. 6,* 89–99.

Heinbecker, P., G. H. Bishop, and J. O'Leary (1933). Pain and touch fibers in peripheral nerves. *Arch. Neurol. (Chic.) 29,* 771–789.

Held, M., and F. Kleinknecht (1927). Die lokale Entspannung der Basilarmembran und ihre Hörlücken. *Pflügers Arch. Ges. Physiol. 216,* 1–31.

Heller, H. (1966). The hormone content of the vertebrate hypothalamo-neurohypophysial system. *Brit. Med. Bull. 22,* 227–231.

Helmholtz, H. L. F. (1863). Die Lehre von den Tonempfindungen als physiologische Grundlage für die Theorie der Musik. *On the Sensations of Tone* (1875). English translation of 3rd edition (A. J. Ellis, trans.). Longmans, Green, London.

Hemisphere Function in the Human Brain (1974). (S. J. Dimond and J. G. Beaumont, eds.). Wiley, New York.

Henderson, J. L. (1939). The congenital facial diplegia syndrome, clinical features, pathology and aetiology. *Brain 62,* 381–403.

Hendrickson, A. (1969). Electron microscopic radioautography: identification of origin of synaptic terminals in normal nervous tissue. *Science 165,* 194–196.

Hendrickson, A. E., N. Wagoner, and W. M. Cowan (1972). An autoradiographic and electron microscopic study of retino-hypothalamic connections. *Z. Zellforsch. 135,* 1–26.

Hendrickson, A., M. E. Wilson, and M. J. Toyne (1970). The distribution of optic nerve fibers in *Macaca mulatta. Brain Research 23,* 425–427.

Henn, V., and B. Cohen (1976). Coding of information about rapid eye movements in the pontine reticular formation of alert monkeys. *Brain Research 108,* 307–325.

Henneman, E. (1974). Principles governing distribution of sensory input to motor neurons. In *The Neurosciences, Third Study Program* (F. O. Schmidt and F. G. Worden, eds.), pp. 281–291. The MIT Press, Cambridge, Mass.

Henneman, E., and C. B. Olson (1965). Relations between structure and function in the design of skeletal muscles. *J. Neurophysiol. 28,* 581–598.

Henneman, E., G. Somjen, and D. O. Carpenter (1965a). Functional significance of cell size in spinal motoneurons. *J. Neurophysiol. 28,* 560–580.

Henneman, E., G. Somjen, and D. O. Carpenter (1965b). Excitability and inhibitibility of motoneurons of different sizes. *J. Neurophysiol. 28,* 599–620.

Henriksson, N. G., G. Johansson, and H. Østlund (1967). New techniques of otoneurological diagnosis. II. Vestibulo-spinal and postural patterns. In *Myotatic and Vestibular Mechanisms.* Ciba Foundation Symposium (A. V. D. de Reuck and J. Knight, eds.), pp. 231–237. J. & A. Churchill, London.

Henry, J. L., and F. R. Calaresu (1972). Topography and numerical distribution of neurons of the thoraco-lumbar intermediolateral nucleus in the cat. *J. Comp. Neurol. 144,* 205–214.

Henry, J. L., and F. R. Calaresu (1974a). Excitatory and inhibitory inputs from medullary nuclei projecting to spinal cardioacceleratory neurons in the cat. *Exp. Brain Res. 20,* 485–504.

Henry, J. L., and F. R. Calaresu (1974b). Pathways from medullary nuclei to spinal cardioacceleratory neurons in the cat. *Exp. Brain Res. 20,* 505–514.

Henschen, F. (1907). Seröse Zyste und partieller Defekt des Kleinhirns. *Z. Klin. Med. 63,* 115–152.

Hensel, A. (1973). Cutaneous thermoreceptors. In *Handbook of Sensory Physiology,* Vol II, *Somatosensory System* (A. Iggo, ed.), pp. 79–110. Springer-Verlag, Berlin.

Hensel, H. (1973). Neural processes in thermo-regulation. *Physiol. Rev. 53,* 948–1017.

Hensel, H., and K. K. A. Boman (1960). Afferent impulses in cutaneous sensory nerves in human subjects. *J. Neurophysiol. 23,* 564–578.

Hensel, H., A. Iggo, and I. Witt (1960). Quantitative study of sensitive cutaneous thermoreceptors with C afferent fibres. *J. Physiol. (Lond.) 153,* 113–126.

Henson, R. A. (1977). Henry Head: his influence on the development of ideas of sensation. *Brit. Med. Bull. 33,* 91–96.

Hentall, I. (1977). A novel class of unit in the substantia gelatinosa of the spinal cat. *Exp. Neurol. 57,* 792–806.

Hepp-Reymond, M.-C., E. Trouche, and M. Wiesendanger (1974). Effects of unilateral and bilateral pyramidotomy on a conditioned rapid precision grip in monkeys (*Macaca fascicularis*). *Exp. Brain Res. 21*, 519–527.

Herkenham, M., and W. J. H. Nauta (1977). Afferent connections of the habenular nucleus in the rat. A horseradish peroxidase study, with a note on the fiber-of-passage problem. *J. Comp. Neurol. 173*, 123–146.

Herman, R. (1970). The myotatic reflex. *Brain 93*, 273–312.

Herman, R., W. Freedman, and S. M. Meeks (1973). Physiological aspects of hemiplegic and paraplegic spasticity. In *New Developments in Electromyography and Clinical Neurophysiology*, Vol. 3 (J. E. Desmedt, ed.), pp. 579–588. Karger, Basel.

Hern, J. E. C., C. G. Phillips, and R. Porter (1962). Electrical thresholds of unimpaled corticospinal cells in the cat. *Quart. J. Exp. Physiol. 47*, 134–140.

Hernández-Peón, R., and G. Chávez-Ibarra (1963). Sleep induced by electrical or chemical stimulation of the forebrain. *Electroenceph. Clin. Neurophysiol. Suppl. 24*, 188–198.

Hernández-Peón, R., and K.-E. Hagbarth (1955). Interaction between afferent and cortically induced reticular responses. *J. Neurophysiol. 18*, 44–55.

Herrick, C. J. (1933). The functions of the olfactory parts of the cerebral cortex. *Proc. Nat. Acad. Sci. (Wash.) 19*, 7–14.

Herzog, A. G., and G. W. Van Hoesen (1976). Temporal neocortical afferent connections to the amygdala in the rhesus monkey. *Brain Research 115*, 57–69.

Hess, W. R. (1944). Das Schlafsyndrom als Folge diencephaler Reizung. *Helv. Physiol. Pharmacol. Acta 2*, 305–344.

Hickey, T. L., and R. W. Guillery (1974). An autoradiographic study of retinogeniculate pathways in the cat and the fox. *J. Comp. Neurol. 156*, 239–254.

Highstein, S. M. (1971). Organization of the inhibitory and excitatory vestibulo-ocular reflex pathways to the third and fourth nuclei in the rabbit. *Brain Research 32*, 218–224.

Highstein, S. M., B. Cohen, and K. Matsunami (1974). Monosynaptic projections from the pontine reticular formation to the IIIrd nucleus in the cat. *Brain Research 75*, 340–344.

Highstein, S. M., K. Maekawa, A. Steinacker, and B. Cohen (1976). Synaptic input from the pontine reticular nuclei to abducens motoneurons and internuclear neurons in the cat. *Brain Research 112*, 162–167.

Hild, W. (1954). Das morphologische, kinetische und endokrinologische Verhalten von hypothalamischem und neurohypophysärem Gewebe in vitro. *Z. Zellforsch. 40*, 257–312.

Hild, W., and G. Zetler (1951). Über das Vorkommen der drei sogenannten "Hypophysenhinterlappenhormone" Adiuretin, Vasopressin und Oxytocin im Zwischenhirn als wahrscheinlicher Ausdruck einer neurosekretorischen Leistung der Ganglienzellen der *Nuclei supraopticus* und *paraventricularis*. *Experientia (Basel) 7*, 189–191.

Hild, W., and G. Zetler (1953). Experimenteller Beweis für die Entstehung der sog. Hypophysenhinterlappenwirkstoffe im Hypothalamus. *Pflügers Arch. ges. Physiol. 257*, 169–201.

Hillarp, N.-Å. (1946). Structure of the synapse and the innervation apparatus of the automatic nervous system. *Acta Anat. (Basel) Suppl. IV*, 1–153.

Hillarp, N.-Å. (1960). Peripheral autonomic mechanisms. In *Handbook of Physiology, Section I Neurophysiology, Vol. II* (J. Field, H. W. Magoun, and V. E. Hall, eds.), pp. 979–1006. American Physiological Society, Washington, D.C.

Hilton, S. M. (1975). Ways of viewing the central nervous control of the circulation—old and new. *Brain Research 87*, 213–219.

Hines, M. (1937). The 'motor' cortex. *Bull. Johns Hopk. Hosp. 60*, 313–336.

Hines, M. (1944). Significance of the precentral motor cortex. In *The Precentral Motor Cortex* (P. C. Bucy, ed.), pp. 459–494. Univ. of Illinois Press, Urbana.

Hinman, A., and M. B. Carpenter (1959). Efferent fiber projections of the red nucleus in the cat. *J. Comp. Neurol. 113*, 61–82.

Hinojosa, R., and J. D. Robertson (1967). Ultrastructure of the spoon type synaptic endings in the nucleus vestibularis tangentialis of the chick. *J. Cell Biol. 34*, 421–430.

Hinsey, J. C., and R. A. Phillips (1940). Observations upon diaphragmatic sensation. *J. Neurophysiol. 3*, 175–181.

Hippocampus, The (1975), Vol 1: *Structure and Development*, Vol. 2: *Neurophysiology and Behavior* (R. L. Isaacson and K. H. Pribram, eds.). Plenum Press, New York.

Hirai, N., Y. Uchino, and S. Watanabe (1977). Neuronal organization of the fastigio-trochlear pathway in the cat. *Brain Research 131*, 362–366.

Hiraoka, M., and M. Shimamura (1977). Neural mechanisms of the corneal blinking reflex in cats. *Brain Research 125*, 265–275.

Hirayama, K., T. Tsubaki, Y. Toyokura and S. Okinaya (1962). The representation of the pyramidal tract in the internal capsule and basis pedunculi. A study based on three cases of amyotrophic sclerosis. *Neurology (Minneap.) 12*, 337–342.

Hiura, T. (1977). Salivatory neurons innervate the submandibular and sublingual glands in the rat: horseradish peroxidase study. *Brain Research 137*, 145–149.

Hjorth-Simonsen, A. (1971). Hippocampal efferents to the ipsilateral entorhinal area: an experimental study in the rat. *J. Comp. Neurol. 142*, 417–438.

Hjorth-Simonsen, A. (1972). Projection of the lateral part of the entorhinal area to the hippocampus and fascia dentata. *J. Comp. Neurol. 146*, 219–232.

Hjorth-Simonsen, A. (1973). Some intrinsic connections of the hippocampus in the rat: an experimental analysis. *J. Comp. Neurol. 147*, 145–162.

Hjorth-Simonsen, A., and B. Jeune (1972). Origin and termination of the hippocampal perforant path in the rat studied by silver impregnation. *J. Comp. Neurol. 144*. 215–232.

Hockaday, J. M., and C. W. M. Whitty (1967). Patterns of referred pain in the normal subject. *Brain 90*, 481–496.

Hockfield, S., and S. Gobel (1978). Neurons in and near nucleus caudalis with long ascending projection axons demonstrated by retrograde labeling with horseradish peroxidase. *Brain Research 139*, 333–339.

Hoddevik, G. H. (1975). The pontocerebellar projection onto the paramedian lobule in the cat: an experimental study with the use of horseradish peroxidase as a tracer. *Brain Research 95*, 291–307.

Hoddevik, G. H. (1977). The pontine projection to the flocculonodular lobe and the paraflocculus studied by means of retrograde axonal transport of horseradish peroxidase in the rabbit. *Exp. Brain Res. 30*, 511–526.

Hoddevik, G. H. (1978). The projection from nucleus reticularis tegmenti pontis onto the cerebellum in the cat. A study using the methods of anterograde degeneration and retrograde axonal transport of horseradish peroxidase. *Anat. Embryol. 153*, 227–242.

Hoddevik, G. H., and A. Brodal (1977). The olivocerebellar projection studied with the method of retrograde axonal transport of horseradish peroxidase. V. The projections to the flocculonodular lobe and the paraflocculus in the rabbit. *J. Comp. Neurol. 176*, 269–280.

Hoddevik, G. H., A. Brodal, K. Kawamura, and T. Hashikawa (1977). The pontine projection to the cerebellar vermal visual area studied by means of the retrograde axonal transport of horseradish peroxidase. *Brain Research 123*, 209–227.

Hoddevik, G. H., A. Brodal, and F. Walberg (1975). The reticulovestibular projection in the cat. An experimental study with silver impregnation methods. *Brain Research 94*, 383–399.

Hoddevik, G. H., A. Brodal, and F. Walberg (1976). The olivocerebellar projection in the cat studied with the method of retrograde axonal tranport of horseradish peroxidase. III. The projection to the vermal visual area. *J. Comp. Neurol. 169*, 155–170.

Hoddevik, G. H., and F. Walberg (1979). The pontine projection onto longitudinal zones of the paramedian lobule in the cat. *Exp. Brain Res. 34*, 233–240.

Hoff, E. C., and H. E. Hoff (1934). Spinal termination of the projection fibres from the motor cortex of primates. *Brain 57*, 454–474.

Hoffer, B. J., G. R. Siggins, and F. E. Bloom (1971). Studies on norepinephrine-containing afferents to Purkinje cells of rat cerebellum. II. Sensitivity of Purkinje cells to norepinephrine and related substances administered by microiontophoresis. *Brain Research 25*, 523–534.

Hoffman, D. L., and J. R. Sladek, Jr. (1973). The distribution of catecholamines within the inferior olivary complex of the gerbil and rabbit. *J. Comp. Neurol. 151,* 101–112.

Hoffmann, K.-P., J. Stone, and S. M. Sherman (1972). Relay of receptive field properties in dorsal lateral geniculate nucleus of the cat. *J. Neurophysiol. 35,* 518–531.

Hoffmann, P. (1918). Über die Beziehungen der Sehnenreflexe zur wilkürlichen Bewegung und zum Tonus. *Z. Biol. 68,* 351–370.

Hoffmann, W. (1933). Thalamussyndrom auf Grund einer kleinen Läsion. *J. Psychol. Neurol. (Lpz.) 45,* 362–374.

Hogg, I. D. (1944). The development of the nucleus dorsalis (Clarke's column). *J. Comp. Neurol. 81,* 69–95.

Hökfelt, T., R. Elde, O. Johansson, R. Luft, G. Nilsson, and A. Arimura (1976). Immunohistochemical evidence for separate populations of somatostatin-containing and substance P-containing primary afferent neurons in the rat. *Neuroscience 1,* 131–136.

Hökfelt, T., J.-O. Kellerth, G. Nilsson, and B. Pernow (1975). Experimental immunohistochemical studies on the localization and distribution of substance P in cat primary sensory neurons. *Brain Research 100,* 235–252.

Hökfelt, T., and Å. Ljungdahl (1975). Uptake mechanisms as a basis for the histochemical identification and tracing of transmitter-specific neuron populations. In *The Use of Axonal Transport for Studies of Neuronal Connectivity* (W. M. Cowan and M. Cuénod, eds.), pp. 249–305. Elsevier, Amsterdam.

Hökfelt, T., and U. Ungerstedt (1969). Electron and fluroescence microscopical studies on the nucleus caudatus putamen of the rat after unilateral lesions of ascending nigroneostriatal dopamine neurons. *Acta Physiol. Scand. 76,* 415–426.

Holländer, H. (1974a). On the origin of the corticotectal projections in the cat. *Exp. Brain Res. 21,* 433–439.

Holländer, H. (1974b). Projections from the striate cortex to the diencephalon in the squirrel monkey (*Saimiri sciureus*). A light microscopic radioautographic study following intracortical injection of H^3 leucine. *J. Comp. Neurol. 155,* 425–440.

Holländer, H., P. Brodal, and F. Walberg (1969). Electronmicroscopic observations on the structure of the pontine nuclei and the mode of termination of the corticopontine fibres. An experimental study in the cat. *Exp. Brain Res. 7,* 95–110.

Holländer, H., and L. Martinez-Millán (1975). Autoradiographic evidence for a topographically organized projection from the striate cortex to the lateral geniculate nucleus in the rhesus monkey (*Macaca mulatta*). *Brain Research 100,* 407–411.

Holländer, H., and H. Vanegas (1977). The projection from the lateral geniculate nucleus onto the visual cortex in the cat. A quantitative study with horseradish-peroxidase. *J. Comp. Neurol. 173,* 519–536.

Holloway, R. L., Jr. (1968). The evolution of the primate brain: some aspects of quantitative relations. *Brain Research 7,* 121–172.

Holm, P., and P. Flindt-Egebak (1976). The absence of horseradish peroxidase uptake by cerebellar afferents to the red nucleus of the cat. *Neurosci. Letters 2,* 315–318.

Holmes, G. (1917). The symptoms of acute cerebellar injuries due to gunshot injuries. *Brain 40,* 461–535.

Holmes, G. (1938). The cerebral integration of ocular movements. *Brit. Med. J. II,* 107–112.

Holmes, G. (1939). The cerebellum of man. *Brain 62,* 1–30.

Holmes, G., and W. T. Lister (1916). Disturbances of vision from cerebral lesions, with special reference to the cortical representation of the macula. *Brain 39,* 34–73.

Holmes, G. L., and W. P. May (1909). On the exact origin of the pyramidal tracts in man and other mammals. *Brain 32,* 1–42.

Holmes, G., and T. G. Stewart (1908). On the connection of the inferior olives with the cerebellum in man. *Brain 31,* 125–137.

Holmes, R. L., and J. N. Ball (1974). *The Pituitary Gland. A Comparative Account.* Cambridge Univ. Press, London.

Holmqvist, B., A. Lundberg, and O. Oscarsson (1960). Supraspinal inhibitory control of transmission to three ascending spinal pathways influenced by the flexion reflex afferents. *Arch. Ital. Biol. 98,* 60–80.

Holmqvist, B., O. Oscarsson, and I. Rosén (1963). Functional organization of the cuneo-cerebellar tract in the cat. *Acta Physiol. Scand. 58,* 216–235.

Hong, J. S., H.-Y. T. Yang, G. Racagni, and E. Casta (1977). Projections of substance P containing neurons from neostriatum to substantia nigra. *Brain Research 122,* 541–544.

Hongo, T., and E. Jankowska (1967). Effects from the sensorimotor cortex on the spinal cord in cats with transected pyramids. *Exp. Brain Res. 3,* 117–134.

Hongo, T., E. Jankowska, and A. Lundberg (1965). Effects evoked from the rubrospinal tract in cats. *Experientia (Basel) 21,* 525–526.

Hongo, T., E. Jankowska, and A. Lundberg (1969). The rubrospinal tract. I. Effects on alpha-motoneurones innervating hindlimb muscles in the cat. *Exp. Brain Res. 7,* 344–364.

Hongo, T., E. Jankowska, and A. Lundberg (1972). The rubrospinal tract. IV. Effects on interneurones. *Exp. Brain Res. 15,* 54–78.

Hongo, T., and Y. Okada (1967). Cortically evoked pre- and postsynaptic inhibition of impulse transmission to the dorsal spinocerebellar tract. *Exp. Brain Res., 3,* 163–177.

Hongo, T., Y. Okada, and M. Sato (1967). Corticofugal influences on transmission to the dorsal spinocerebellar tract from hindlimb primary afferents. *Exp. Brain Res. 3,* 135–149.

Hopkins, D. A. (1975). Amygdalotegmental projections in the rat, cat and rhesus monkey. *Neurosci. Letters 1,* 263–270.

Hopkins, D. A., and G. Holstege (1978). Amygdaloid projections to the mesencephalon, pons and medulla oblongata in the cat. *Exp. Brain Res. 32,* 529–547.

Hopkins, D. A., and D. G. Lawrence (1975). On the absence of a rubrothalamic projection in the monkey with observations on some ascending mesencephalic projections. *J. Comp. Neurol. 161,* 269–294.

Hopkins, D. A., and L. W. Niessen (1976). Substantia nigra projections to the reticular formation, superior colliculus and central gray in the rat, cat and monkey. *Neurosci. Letters 2,* 253–259.

Horch, K. W., P. R. Burgess, and D. Whitehorn (1976). Ascending collaterals of cutaneous neurons in the fasciculus gracilis of the cat. *Brain Research 117,* 1–17.

Horch, K. W., F. J. Clark, and P. R. Burgess (1975). Awareness of knee joint angle under static conditions. *J. Neurophysiol. 38,* 1436–1447.

Hore, J., J. B. Preston, R. G. Durkovic, and P. D. Cheney (1976). Responses of cortical neurons (areas 3a and 4) to ramp stretch of hindlimb muscles in the baboon. *J. Neurophysiol. 39,* 484–500.

Horel, J. A. (1978). The neuroanatomy of amnesia. A critique of the hippocampal memory hypothesis. *Brain 101,* 403–445.

Horel, J. A., E. G. Keating, and L. J. Misantone (1975). Partial Klüver Bucy syndrome produced by destroying temporal neocortex or amygdala. *Brain Research 94,* 347–359.

Horizons in Neuro-psychopharmacology, Progress in Brain Research, Vol. 16 (1965) (W. A. Himwich and J. P. Schadé, eds.). Elsevier, Amsterdam.

Hornykiewicz, O. (1966). Metabolism of brain dopamine in human parkinsonism: neurochemical and clinical aspects. In *Biochemistry and Pharmacology of the Basal Ganglia* (E. Costa, L. J. Côte, and M. D. Yahr, eds.), pp. 171–185. Raven Press, New York.

Hornykiewicz, O. (1971). Neurochemical pathology and pharmacology of brain dopamine and acetylcholine: rational basis for the current drug treatment of parkinsonism. In *Recent Advances in Parkinson's Disease* (F. H. McDowell and C. H. Markham, eds.), pp. 34–65. F. A. Davis, Philadelphia.

Hornykiewicz, O. (1976). Neurohumoral interactions and basal ganglia function and dysfunction. In *The Basal Ganglia* (M. D. Yahr, ed.), pp. 269–278. Raven Press, New York.

Horrobin, D. F. (1966). The lateral cervical nucleus of the cat: an electrophysiological study. *Quart. J. Exp. Physiol. 51,* 351–371.

Hosoba, M., T. Bando, and N. Tsukahara (1978). The cerebellar control of accommodation of the eye in the cat. *Brain Research 153*, 495–505.

Hosobuchi, Y., J. E. Adams, and R. Linchitz (1977). Pain relief by electrical stimulation of the central gray matter in humans and its reversal by naloxone. *Science 197*, 183–186.

Hosoya, Y., and M. Matsushita (1979). Identification and distribution of the spinal and hypophyseal projection neurons in the paraventricular nucleus of the rat. A light and electron microscopic study with the horseradish peroxidase method. *Exp. Brain Res. 35*, 315–331.

Houk, J., and E. Henneman (1976). Responses of Golgi tendon organs to active contractions of the soleus muscle of the cat. *J. Neurophysiol. 30*, 466–481.

Howe, J. F., J. D. Loeser, and W. H. Calvin (1977). Mechanosensitivity of dorsal root ganglia and chronically injured axons: a physiological basis for the radicular pain of nerve root compression. *Pain 3*, 25–41.

Hoyt, W. F., and R. B. Daroff (1971). Supranuclear disorders of ocular control systems in man. Clinical, anatomical and physiological correlations. In *The Control of Eye Movements* (P. Bach-y-Rita, C. C. Collins, and J. E. Hyde, eds.), pp. 175–236. Academic Press, New York.

Hubbard, J. E., and V. Di Carlo (1973). Fluorescence histochemistry of monoamine-containing cell bodies in the brain stem of the squirrel monkey (*Saimiri sciureus*). I. The locus coeruleus. *J. Comp. Neurol. 147*, 553–566.

Hubbard, J. E., and V. Di Carlo (1974a). Fluorescence histochemistry of monoamine-containing cell bodies in the brain stem of the squirrel monkey (*Saimiri sciureus*). II. Catecholamine-containing groups. *J. Comp. Neurol. 153*, 369–384.

Hubbard, J. E., and V. Di Carlo (1947b). Fluorescence histochemistry of monoamine-containing cell bodies in the brain stem of the squirrel monkey (*Saimiri sciureus*). III. Serotonin-containing groups. *J. Comp. Neurol. 153*, 385–398.

Hubbard, J. I., and O. Oscarsson (1962). Localization of the cell bodies of the ventral spinocerebellar tract in lumbar segments of the cat. *J. Comp. Neurol. 118*, 199–204.

Hubel, D. H., S. LeVay, and T. N. Wiesel (1975). Mode of termination of retinotectal fibers in macaque monkey: an autoradiographic study. *Brain Research 96*, 25–40.

Hubel, D. H., and T. N. Wiesel (1961). Integrative action in the cat's lateral geniculate body. *J. Physiol. (Lond.) 155*, 385–398.

Hubel, D. H., and T. N. Wiesel (1962). Receptive fields, binocular interaction and functional architecture in the cat's visual cortex. *J. Physiol. (Lond.) 160*, 106–154.

Hubel, D. H., and T. N. Wiesel (1965). Receptive fields and functional architecture in two non-striate visual areas (18 and 19) of the cat. *J. Neurophysiol. 28*, 229–289.

Hubel, D. H., and T. N. Wiesel (1968). Receptive fields and functional architecture of monkey striate cortex. *J. Physiol. (Lond.) 195*, 215–243.

Hubel, D. H., and T. N. Wiesel (1969a). Anatomical demonstration of columns in the monkey striate cortex. *Nature 221*, 747–750.

Hubel, D. H., and T. N. Wiesel (1969b). Visual area of the lateral suprasylvian gyrus (Clare-Bishop area) of the cat. *J. Physiol. (Lond.) 202*, 251–260.

Hubel, D. H., and T. N. Wiesel (1970). Cells sensitive to binocular depth in area 18 of the macaque monkey cortex. *Nature 225*, 41–42.

Hubel, D. H., and T. N. Wiesel (1972). Laminar and columnar distribution of geniculo-cortical fibers in the macaque monkey. *J. Comp. Neurol. 146*, 421–450.

Hubel, D. H., and T. N. Wiesel (1974a). Sequence regularity and geometry of orientation columns in the monkey striate cortex. *J. Comp. Neurol. 158*, 267–294.

Hubel, D. H., and T. N. Wiesel (1974b). Uniformity of monkey striate cortex: a parallel relationship between field size, scatter, and magnification factor. *J. Comp. Neurol. 158*, 295–306.

Huber, E. (1930). Evolution of facial musculature and cutaneous fields of trigeminus. *Quart. Rev. Biol. 5*, 389–347.

Hugelin, A., M. Bonvallet, and P. Dell (1953). Topographie des projections cortico-motrices au niveau du télencéphale, du diencéphale, du tronc cérébral et du cervelet chez le chat. *Rev. Neurol. 89*, 419–425.

Hughes, J., and H. W. Kosterlitz (1977). Opioid peptides. *Brit. Med. Bull. 33,* 157–161.

Hukuhara, T., Jr. (1974). Functional organization of brain stem respiratory neurons and rhythmogenesis. In *Central Rhythmic and Regulation* (W. Umbach and H. P. Koepchen, eds.), pp. 35–49. Hippokrates Verlag, Stuttgart.

Hultborn, H., K. Mori, and N. Tsukahara (1978). Cerebellar influence on parasympathetic neurones innervating intra-ocular muscles. *Brain Research 159,* 269–278.

Humphrey, D. R., and W. S. Corrie (1978). Properties of pyramidal tract neuron system within a functionally defined subregion of primate motor cortex. *J. Neurophysiol. 41,* 216–243.

Humphrey, D. R., and R. R. Rietz (1976). Cells of origin of corticorubral projections from the arm area of primate motor cortex and their synaptic actions in the red nucleus. *Brain Research 110,* 162–169.

Humphrey, T. (1972). The development of the human amygdaloid complex. In *The Neurobiology of the Amygdala* (B. E. Elftheriou, ed.), pp. 21–80. Plenum Press, New York.

Hunger: Basic Mechanisms and Clinical Applications (1976). (D. Novin, W. Wyrwicka, and G. A. Bray, eds.), Raven Press, New York.

Hunt, C. C. (1961). On the nature of vibration receptors in the hind limb of the cat. *J. Physiol. (Lond.) 155,* 175–186.

Hunt, J. R. (1907). On herpetic inflammations of the geniculate ganglion: a new syndrome and its complications. *J. Nerv. Ment. Dis. 34,* 73–96.

Hunt, J. R. (1937). Geniculate neuralgia (neuralgia of the nervus facialis). *Arch. Neurol. Psychiat. (Chic.) 37,* 253–285.

Hunter, M., G. Ettlinger, and J. J. Maccabe (1975). Intermanual transfer in the monkey as a function of amount of callosal sparing. *Brain Research 93,* 223–240.

Hursh, J. B. (1939). Conduction velocity and diameter of nerve fibers. *Amer. J. Physiol. 127,* 131–139.

Hyde, J. E., and R. G. Eason (1959). Characteristics of ocular movements evoked by stimulation of brainstem of cat. *J. Neurophysiol. 22,* 666–678.

Hyde, J. E., and S. G. Eliasson (1957). Brainstem induced eye movements in cats. *J. Comp. Neurol. 108,* 139–172.

Hyndman, O. R., and C. von Epps (1939). Possibility of differential section of the spinothalamic tract: a clinical and histological study. *Arch. Surg. 38,* 1036–1053.

Hypothalamus, The (1969). (W. Haymaker, E. Anderson, and W. J. H. Nauta, eds.). C. C. Thomas, Springfield, Ill.

Hypothalamus, The (1970). (L. Martini, M. Motta, and F. Fraschini, eds.). Academic Press, New York.

Hypothalamus, The (1978). (S. Reichlin, R. J. Baldessarini, and J. B. Martin, eds.). *Ass. Res. Nerv. Dis. Res. Publ.,* Vol. 56. Raven Press, New York.

Hyvärinen, J., and A. Poranen (1974). Function of the parietal association area 7 as revealed from cellular discharges in alert monkeys. *Brain 97,* 673–692.

Ibata, Y., Y. Nojyo, T. Matsuura, and Y. Sano (1973). Nigro–neostriatal projection. A correlative study with Fink-Heimer impregnation, fluorescence histochemistry and electron microscopy. *Z. Zellforsch. 138,* 333–344.

Iggo, A. (1962a). New specific sensory structures in hairy skin. *Acta Neuroveg. (Wien) 24,* 175–180.

Iggo, A. (1926b). An electrophysiological analysis of afferent fibres in primate skin. *Acta Neuroveg. (Wien) 24,* 225–240.

Iggo, A. (1966). Cutaneous receptors with a high sensitivity to mechanical displacement. In *Touch, Heat and Pain.* Ciba Foundation Symposium (A. V. S. de Reuck and J. Knight, eds.), pp. 237–256. J. & A. Churchill, London.

Iggo, A. (1974). Cutaneous receptors (1974). In *The Peripheral Nervous System* (J. I. Hubbard, ed.), pp. 347–420. Plenum Press, New York.

Iggo, A. (1977). Cutaneous and subcutaneous sense organs (1977). *Brit. Med. Bull. 33,* 97–102.

Iggo, A., and A. R. Muir (1969). The structure and function of a slowly adapting touch corpuscle in hairy skin. *J. Physiol. (Lond.) 200*, 763–796.

Iles, J. F. (1976). Central terminations of muscle afferents on motoneurones in the cat spinal cord. *J. Physiol. (Lond.) 262*, 91–117.

Illert, M., and M. Gabriel (1972). Descending pathways in the cervical cord of cats affecting blood pressure and sympathetic activity. *Pflügers Arch. 335*, 109–124.

Illert, M., A. Lundberg, Y. Padel, and R. Tanaka (1978). Integration in descending motor pathways controlling the forelimb in the cat. 5. Properties of and monosynaptic excitatory convergence on C3–C4 propriospinal neurones. *Exp. Brain Res. 33*, 101–130.

Illert, M., A. Lundberg, and R. Tanaka (1974). Disynaptic corticospinal effects in forelimb motoneurones in the cat. *Brain Research 75*, 312–315.

Illert, M., A. Lundberg, and R. Tanaka (1975). Integration in a disynaptic corticomotoneuronal pathway to the forelimb in the cat. *Brain Research 93*, 525–529.

Illert, M., A. Lundberg, and R. Tanaka (1976). Integration in descending motor pathways controlling the forelimb in the cat. 1. Pyramidal effects on motoneurones. *Exp. Brain Res. 26*, 509–519.

Illert, M., A. Lundberg, and R. Tanaka (1977). Integration in descending motor pathways controlling the forelimb in the cat. 3. Convergence on propriospinal neurones transmitting disynaptic excitation from the corticospinal tract and other descending tracts. *Exp. Brain Res. 29*, 323–346.

Illert, M., and R. Tanaka (1978). Integration in descending motor pathways controlling the forelimb in the cat. 4. Corticospinal inhibition of forelimb motoneurones mediated by short propriospinal neurones. *Exp. Brain Res. 31*, 131–141.

Illis, L. (1964a). Spinal cord synapses in the cat: the normal appearances by the light microscope. *Brain 87*, 543–554.

Illis, L. (1964b). Spinal cord synapses in the cat: the reaction of the boutons termineaux at the motoneurone surface to experimental denervation. *Brain 87*, 555–572.

Illis, L. S. (1967). The motor neuron surface and spinal shock. In *Modern Trends in Neurology, 4* (D. Williams, ed.), pp. 53–68. Butterworths, London.

Illis, L. S. (1973a). Experimental model of regeneration in the central nervous system. I. Synaptic changes. *Brain 96*, 47–60.

Illis, L. S. (1973b). Experimental model of regeneration in the central nervous system. II. The reaction of glia in the synaptic zone. *Brain 96*, 61–68.

Imai, Y., and T. Kusama (1968). Distribution of the dorsal root fibres in the cat. *Brain Research 13*, 338–359.

Imig, T. J., and H. O. Adrián (1977). Binaural columns in the primary field (AI) of cat auditory cortex. *Brain Research 138*, 241–257.

Imig. T. J., M. A. Ruggero, L. M. Kitzes, E. Javel, and J. F. Brugge (1977). Organization of auditory cortex in the owl monkey (*Aotus trivirgatus*). *J. Comp. Neurol. 171*, 111–128.

Ingle, D., and J. M. Sprague (1975). Sensorimotor function of the midbrain tectum. *Neurosci. Res. Progr. Bull. 13*, 169–288.

Ingram, W. R. (1940). Nuclear organization and chief connections of the primate hypothalamus. *Res. Publ. Ass. Nerv. Ment. Dis. 20*, 195–244.

Ingvar, D. H. (1978). Localisation of cortical functions by multiregional measurements of the cerebral blood flow. In *Architectonics of the Cerebral Cortex* (M. A. B. Brazier and H. Petsche, eds.), IBRO 3, pp. 235–243, Raven Press, New York.

Inner Ear Studies (1972). (H. W. Ades and H. Engström, eds.) *Acta Otolaryng. (Stockh.), Suppl. 301*, 1–126.

Innnocenti, G. M., T. Manzoni, and G. Spidalieri (1974). Patterns of the somesthetic messages transferred through the corpus callosum. *Exp. Brain Res. 19*, 447–466.

Integrative Hypothalamic Activity. Progress in Brain Research, Vol. 41 (1974) (D. F. Swaab and J. P. Schadé, eds.). Elsevier, Amsterdam.

Interhemispheric Relations and Cerebral Dominance (1962). (V. B. Mountcastle, ed.). The Johns Hopkins Press, Baltimore.

International Vestibular Symposium (1964). Uppsala 1963, in memory of Robert Bárány.

(A. Sjöberg, G. Aschan & J. Stahle, eds.). Almqvist & Wiksell, Uppsala. (Acta Otolaryng. Suppl. 192).

Irving, R., and J. M. Harrison (1967). The superior olivary complex and audition: a comparative study. *J. Comp. Neurol. 130*, 77–86.

Isaacson, R. L. (1974). *The Limbic System*. Plenum Press, New York.

Isaacson, R. L. (1975a). The myth of recovery from early brain damage. In *Aberrant Development in Infancy: Human and Animal Studies* (N. R. Ellis, ed.), pp. 1–26. Laurence Erlbaum Assoc., Potomac, Md.

Isaacson, R. L. (1975b). Memory processes and the hippocampus. In *Short-Term Memory* (D. Deutsch and J. A. Deutsch, eds.), pp. 313–337. Academic Press, New York.

Isaacson, R. L. (1976). Experimental brain lesions and memory. In *Neural Mechanisms of Learning and Memory* (M. R. Rosenzweig and E. L. Bennett, eds.), pp. 521–543. MIT Press, Cambridge, Mass.

Ishii, T., Y. Murakami, and K. Balogh (1967). Acetylcholinesterase activity in the efferent nerve fibers of the human inner ear. *Ann. Otol. (St. Louis) 76*, 69–82.

Ishizuka, N., H. Mannen, T. Hongo, and S. Sasaki (1979). Trajectory of group Ia afferent fibers stained with horseradish peroxidase in the lumbosacral spinal cord of the cat: three dimensional reconstructions from serial sections. *J. Comp. Neurol. 186*, 189–212.

Ito, M., T. Hongo, M. Yoshida, Y. Okada, and K. Obata (1964). Antidromic and trans-synaptic activation of Deiters' neurones induced from the spinal cord. *Jap. J. Physiol. 14*, 638–658.

Ito, M., M. Udo, N. Mano, and N. Kawai (1970). Synaptic action of the fastigiobulbar impulses upon neurones in the medullary reticular formation and vestibular nuclei. *Exp. Brain Res. 11*, 29–47.

Ito, M., and M. Yoshida (1966). The origin of cerebellar-induced inhibition of Deiters neurones. I. Monosynaptic initiation of the inhibitory postsynaptic potentials. *Exp. Brain Res. 2*, 330–349.

Itoh, K. (1977). Efferent projections of the pretectum in the cat. *Exp. Brain Res. 30*, 89–105.

Iurato, S. (1962). Efferent fibers to the sensory cells of the Corti's organ. *Exp. Cell Res. 27*, 162–164.

Iurato, S. (1974). Efferent innervation of the cochlea. In *Handbook of Sensory Physiol.*, Vol. V/1 (W. D. Keidel and W. D. Neff, eds.), pp. 259–282. Springer-Verlag, Berlin.

Iurato, S., C. A. Smith, D. H. Eldredge, D. Henderson, C. Carr, Y. Ueno, S. Cameron, and R. Richter (1978). Distribution of the crossed olivocochlear bundle in the chinchilla's cochlea. *J. Comp. Neurol. 182*, 57–76.

Iversen, S. D. (1974). 6-hydroxydopamine: a chemical lesion technique for studying the role of amine neurotransmitters in behavior. In *The Neurosciences, 3rd Study Program* (F. O. Schmitt and F. G. Worden, eds.), pp. 705–711. MIT Press, Cambridge, Mass.

Iversen, L. L. (1975). How do antipsychotic drugs work? *Neurosci. Res. Progr. Bull. 13 (Suppl.)*, 29–51.

Iwata, N., S. T. Kitai, and S. Olson (1972). Afferent component of the facial nerve: its relation to the spinal trigeminal and facial nucleus. *Brain Research 43*, 662–667.

Izquierdo, I. (1975). The hippocampus and learning. *Progr. Neurobiol. 5*, 37–75.

Jabbur, S. J., M. A. Baker, and A. L. Towe (1972). Wide-field neurons in thalamic nucleus ventralis posterolateralis of the cat. *Exp. Neurol. 36*, 213–238.

Jabbur, S. J., S. I. Harik, and J. A. Hush (1976). Caudate influence on transmission in the cuneate nucleus. *Brain Research 120*, 559–563.

Jabbur, S. J., and A. L. Towe (1961). Cortical excitation of neurons in dorsal column nuclei of cat, including an analysis of pathways. *J. Neurophysiol. 24*, 499–509.

Jackson, H. (1946). Discussion on cortical atrophy. *Proc. Roy. Soc. Med. 39*, 423–427.

Jackson, H. C. II, R. K. Winkelmann, and W. Bickel (1966). Nerve endings in human lumbar spinal column and related structures. *J. Bone Jt. Surg. 48A,* 1272–1281.

Jackson, J. H., and C. E. Beevor (1890). Case of tumour of the right temporosphenoidal lobe bearing on the localisation of the sense of smell and on the interpretation of a particular variety of epilepsy. *Brain 12,* 346–357.

Jacobi, H. M., H. M. Krott, and M. B. Poremba (1970). Recording of proprioceptive muscle afferents in man by various disorders of tendon reflex. *Electromyography 10,* 307–316.

Jacobs, B. L., W. D. Wise, and K. M. Taylor (1974). Differential behavioral and neurochemical effects following lesions of the dorsal or median raphe nuclei in rats. *Brain Research 79,* 353–361.

Jacobs, L., P. J. Anderson, and M. B. Bender (1973). The lesions producing paralysis of downward but not upward gaze. *Arch. Neurol. (Chic.) 28,* 319–323.

Jacobsen, C. F. (1934). Influence of motor and premotor area lesions upon the retention of acquired skilled movements in monkeys and chimpanzees. *Res. Publ. Ass. Nerv. Ment. Dis. 13,* 225–247.

Jacobsen, C. F. (1935). Functions of frontal association areas in primates. *Arch, Neurol. Psychiat. (Chic.) 33,* 558–569.

Jacobson, S., and E. M. Marcus (1970). The laminar distribution of fibers of the corpus callosum: a comparative study in the rat, cat, rhesus monkey and chimpanzee. *Brain Research 24,* 517–520.

Jacobson, S., and J. Q. Trojanowski (1974). The cells of origin of the corpus callosum in rat, cat and rhesus monkey. *Brain Research 74,* 149–155.

Jacobson, S., and J. Q. Trojanowski (1975a). Amygdaloid projections to prefrontal granular cortex in rhesus monkey demonstrated with horseradish peroxidase. *Brain Research 100,* 132–139.

Jacobson, S., and J. Q. Trojanowski (1975b). Corticothalamic neurons and thalamocortical terminal fields: an investigation in rat using horseradish peroxidase and autoradiography. *Brain Research 85,* 385–401.

Jacobson, S., and J. Q. Trojanowski (1977a). Prefrontal granular cortex of the rhesus monkey. I. Intrahemispheric cortical afferents. *Brain Research 132,* 209–233.

Jacobson, S., and J. Q. Trojanowski (1977b). Prefrontal granular cortex of the rhesus monkey. II. Interhemispheric cortical afferents. *Brain Research 132,* 235–246.

Jakob, H. (1955). Zur Analyse konsekutiver Olivenschäden bei vasculär bedingten Kleinhirndefekten. *Arch. Psychiat. Nervenkr. 193,* 583–600.

Jami, L., and J. Petit (1976). Frequency of tendon organ discharges elicited by the contraction of motor units in cat leg muscles. *J. Physiol. (Lond.) 261,* 633–645.

Jampel, R. S. (1960). Convergence, divergence, pupillary reactions and accommodation of the eye from faradic stimulation of the macaque brain. *J. Comp. Neurol. 115,* 371–399.

Jampel, R. S., and J. Mindel (1967). The nucleus for accommodation in the midbrain of the macaque. *Invest. Ophthal. 6,* 40–50.

Jampolsky, A. (1970). What can electromyography do for the ophthalmologist? *Invest. Ophthal. 9,* 570–599.

Jankowska, E. (1975). Identification of interneurons interposed in different spinal reflex pathways. In *Proceedings of Golgi Centennial Symposium* (M. Santini, ed.), pp. 235–246. Raven Press, New York.

Jankowska, E. (1978). Some problems of projections and actions of cortico- and rubrospinal fibers. *J. Physiol. (Paris) 74,* 209–214.

Jankowska, E., and S. Lindström (1970). Morphological identification of physiologically defined neurones in the cat spinal cord. *Brain Research 20,* 323–326.

Jankowska, E., and S. Lindström (1972). Morphology of interneurones mediating Ia reciprocal inhibition of motoneurones in the spinal cord of the cat. *J. Physiol. (Lond.) 226,* 805–823.

Jankowska, E., S. Lund, A. Lundberg, and O. Pompeiano (1968). Inhibitory effects evoked through ventral reticulospinal pathways. *Arch. Ital. Biol. 106,* 124–140.

Jankowska, E., A. Lundberg, W. J. Roberts, and D. Stuart (1974). A long propriospinal system with direct effect on motoneurones and on interneurones in the cat lumbosacral cord. *Exp. Brain Res. 21,* 169–194.

Jankowska, E., Y. Padel, and R. Tanaka (1975). Projections of pyramidal tract cells to α-motoneurones innervating hind-limb muscles in the monkey. *J. Physiol. (Lond.) 249,* 637–667.

Jankowska, E., Y. Padel, and R. Tanaka (1976). Disynaptic inhibition of spinal motoneurones from the motor cortex in the monkey. *J. Physiol. (Lond.) 258,* 467–487.

Jankowska, E., J. Rastad, and J. Westman (1976). Intracellular application of horseradish peroxidase and its light and electron microscopical appearance in spinocervical tract cells. *Brain Research 105,* 557–562.

Jansen, J. (1933). Experimental studies on the intrinsic fibers of the cerebellum. I. The arcuate fibers. *J. Comp. Neurol. 57,* 369–400.

Jansen, J. (1937). Experimental investigations of the associational connections of the cerebral cortex, with special reference to the conditions in the frontal lobe. *Skr. Norske Vidensk.-Akad., I. Mat.-nat. Kl. No. 1,* 1–21.

Jansen, J. (1950). The morphogenesis of the cetacean cerebellum. *J. Comp. Neurol. 93,* 341–400.

Jansen, J. (1954). On the morphogenesis and morphology of the mammalian cerebellum. In *Aspects of Cerebellar Anatomy* (J. Jansen and A. Brodal, eds.), pp. 13–81. Johan Grundt Tanum, Oslo.

Jansen, J., and A. Brodal (1940). Experimental studies on the intrinsic fibers of the cerebellum. II. The corticonuclear projection. *J. Comp. Neurol. 73,* 267–321.

Jansen, J., and A. Brodal (1942). Experimental studies on the intrinsic fibers of the cerebellum. III. The corticonuclear projection in the rabbit and the monkey. *Skr. Norske Vidensk.-Akad., I. Mat.-nat. Kl. No. 3,* 1–50.

Jansen, J., and A. Brodal (1958). Das Kleinhirn. In *Möllendorff's Handbuch der mikroskopischen Anatomie des Menschen IV/8.* Springer-Verlag, Berlin.

Jansen, J., and J. Jansen Jr. (1955). On the efferent fibers of the cerebellar nuclei in the cat. *J. Comp. Neurol. 102,* 607–632.

Jansen, J., and F. Walberg (1964). Cerebellar corticonuclear projection studied experimentally with silver impregnation methods. *J. Hirnforsch. 6,* 338–354.

Jansen, J., Jr. (1957). Afferent impulses to the cerebellar hemispheres from the cerebral cortex and certain subcortical nuclei. An electroanatomical study in the cat. *Acta Physiol. Scand. 41, Suppl. 143,* 1–99.

Jansen, J. K. S. (1962). Spasticity—functional aspects. *Acta Neurol. Scand. 38, Suppl. 3,* 41–51.

Jansen, J. K. S., and D. C. van Essen (1975). Re-innervation of rat skeletal muscle in the presence of α-bungarotoxin. *J. Physiol. (Lond.) 250,* 651–667.

Jansen, J. K. S., T. Lømo, K. Nicolaysen, and R. H. Westgaard (1973). Hyperinnervation of skeletal muscle fibres: dependence on muscle activity. *Science 181,* 559–561.

Jansen, J. K. S., and P. B. C. Matthews (1962a). The central control of the dynamic response of muscle spindle receptors. *J. Physiol. (Lond.) 161,* 357–378.

Jansen, J. K. S., and P. B. C. Matthews (1962b). The effects of fusimotor activity on the static responsiveness of primary and secondary endings of muscle spindles in the decerebrate cat. *Acta Physiol. Scand. 55,* 376–386.

Jansen, J. K. S., K. Nicolaysen, and T. Rudjord (1966). Discharge pattern of neurons of the dorsal spinocerebellar tract activated by static extension of primary endings of muscle spindles. *J. Neurophysiol. 29,* 1061–1086.

Jansen, J. K. S., and T. Rudjord (1964). On the silent period and Golgi tendon organs of the soleus muscle of the cat. *Acta Physiol. Scand. 62,* 364–379.

Jasper, H. H. (1949). Diffuse projection systems: the integrative action of the thalamic reticular system. *Electroenceph. Clin. Neurophysiol. 1,* 405–420.

Jasper, H. H., and G. Bertrand (1966). Thalamic units involved in somatic sensation and voluntary and involuntary movements in man. In *The Thalamus* (D. P. Purpura and M. D. Yahr, eds.), pp. 365–384. Columbia Univ. Press, New York.

Jasper, H. H., and J. Drooglever-Fortuyn (1947). Experimental studies on the functional anatomy of petit mal epilepsy. *Res. Publ. Ass. Nerv. Ment. Dis. 26*, 272–298.

Jasper, H., R. Lende, and T. Rasmussen (1960). Evoked potentials from the exposed somato-sensory cortex in man. *J. Nerv. Ment. Dis. 130*, 526–537.

Jayaraman, A., R. R. Batton, and M. B. Carpenter (1977). Nigrotectal projections in the monkey: an autoradiographic study. *Brain Research 135*, 147–152.

Jean-Baptiste, M., and D. K. Morest (1975). Transneuronal changes of synaptic endings and nuclear chromatin in the trapezoid body following cochlear ablations in cats. *J. Comp. Neurol. 162*, 111–134.

Jefferson, A. (1963). Trigeminal root and ganglion injections using phenol in glycerine for the relief of trigeminal neuralgia. *J. Neurol. Neurosurg. Psychiat. 26*, 345–352.

Jefferson, G. (1938). On the saccular aneurysms of the internal carotid artery in the cavernous sinus. *Brit. J. Surg. 26*, 267–302.

Jefferson, G. (1958). The reticular formation and clinical neurology. In *Reticular Formation of the Brain*. Henry Ford Hospital International Symposium (H. H. Jasper, L. D. Proctor, R. S. Knighton, and R. T. Costello, eds.), pp. 729–738. Little, Brown, Boston.

Jefferson, G. (1960). *Selected Papers*. Pitman, London.

Jefferson, M. (1952). Altered consciousness associated with brain-stem lesions. *Brain 75*, 55–67.

Jenkner, F. L. (1961). Selective anterolateral chordotomy for upper extremity pain. *Arch. Neurol. (Chic.) 4*, 660–662.

Jerge, C. R. (1963a). Organization and function of the trigeminal mesencephalic nucleus. *J. Neurophysiol. 26*, 379–392.

Jerge, C. R. (1963b). The function of the nucleus supratrigeminalis. *J. Neurophysiol. 26*, 393–402.

Jerger, J., L. Lovering, and M. Wertz (1972). Auditory disorder following bilateral temporal lobe insult: report of a case. *J. Speech Dis. 37*, 523–535.

Jervis, G. A. (1963). Huntington's chorea in childhood. *Arch. Neurol. (Chic.) 9*, 244–257.

Jessel, T. M., and L. L. Iversen (1977). Opiate analgesics inhibit substance P release from rat trigeminal nucleus. *Nature 268*, 549–551.

Jewesbury, E. C. O. (1969). Parietal lobe syndromes. In *Handbook of Clinical Neurology*, Vol. 2, *Localization in Clinical Neurology* (P. J. Vinken and G. W. Bruyn, eds.), pp. 680–699. North-Holland, Amsterdam; Wiley, New York.

Johansson, H., and H. Silfvenius (1977a). Axon-collateral activation by dorsal spinocerebellar tract fibres of group I relay cells of nucleus Z in the cat medulla oblongata. *J. Physiol. (Lond.) 265*, 341–369.

Johansson, H., and H. Silfvenius (1977b). Input from ipsilateral proprio- and exteroceptive hindlimb afferents to nucleus Z of the cat medulla oblongata. *J. Physiol. (Lond.) 265*, 371–393.

Johansson, R. S., and K. Å. Olsson (1976). Microelectrode recordings from human oral mechanoreceptors. *Brain Research 118*, 307–311.

Johansson, R. S., and Å. B. Vallbo (1979). Tactile sensibility in the human hand: relative and absolute densities of four types of mechanoreceptive units in glabrous skin. *J. Physiol. (Lond.) 286*, 283–300.

Johnson, J. L. (1977). Glutamic acid as a synaptic transmitter candidate in the dorsal sensory neuron: reconsiderations. *Life Sci. 20*, 1637–1644.

Johnson, R. H., and D. M. K. Spalding (1974). *Disorders of the Autonomic Nervous System*. Blackwell, Oxford.

Johnson, T. N., and C. D. Clemente (1959). An experimental study of the fiber connections between the putamen, globus pallidus, ventral thalamus, and midbrain tegmentum in cat. *J. Comp. Neurol. 113*, 83–101.

Jones, B. E., S. T. Harper, and A. E. Halaris (1977). Effects of locus coeruleus lesions upon cerebral monoamine content sleep-wakefulness states and the response to amphetamine in the cat. *Brain Research 124*, 473–496.

Jones, B. E., and R. Y. Moore (1974). Catecholamine-containing neurons of the nucleus locus coeruleus in the cat. *J. Comp. Neurol., 157*, 43–52.

Jones, B. E., and R. Y. Moore (1977). Ascending projections of the locus coeruleus in the rat. II. Autoradiographic study. *Brain Research 127*, 23–53.

Jones, D. G. (1975). *Synapses and Synaptosomes. Morphological Aspects*. Chapman and Hall, London.

Jones, D. G. (1978). Some Current Concepts of Synaptic Organization. *Advan. Anat. Embryol. Cell Biol., Vol. 55*, Fasc. 4, 69 pp.

Jones, D. R., J. H. Casseday, and I. T. Diamond (1976). Further study of parallel auditory pathways in the tree shrew, *Tupaia glis. Anat. Rec. 184*, 438.

Jones, E. G. (1967). Pattern of cortical and thalamic connexions of the somatic sensory cortex. *Nature (Lond.) 216*, 704–705.

Jones, E. G. (1974). The anatomy of extrageniculostriate visual mechanisms. In *The Neurosciences*, 3rd Study Program (F. O. Schmitt and F. G. Worden, eds-in-chief), pp. 215–227. MIT Press, Cambridge, Mass.

Jones, E. G. (1975a). Lamination and differential distribution of thalamic afferents within the sensory motor cortex of the squirrel monkey. *J. Comp. Neurol. 160*, 167–204.

Jones, E. G. (1975b). Varieties and distribution of nonpyramidal cells in the somatic sensory cortex of the squirrel monkey. *J. Comp. Neurol. 160*, 205–268.

Jones, E. G. (1975c). Some aspects of the organization of the thalamic reticular complex. *J. Comp. Neurol. 162*, 285–308.

Jones, E. G., and H. Burton (1974). Cytoarchitecture and somatic sensory connectivity of thalamic nuclei other than the ventrobasal complex in the cat. *J. Comp. Neurol. 154*, 395–432.

Jones, E. G., and H. Burton (1976a). A projection from the medial pulvinar to the amygdala in primates. *Brain Research 104*, 142–147.

Jones, E. G., and H. Burton (1976b). Areal differences in the laminar distribution of thalamic afferents in cortical fields of the insular, parietal and temporal regions of primates. *J. Comp. Neurol. 168*, 197–248.

Jones, E. G., H. Burton, and R. Porter (1975). Commissural and corticocortical "columns" in the somatic sensory cortex of primates. *Science 190*, 572–574.

Jones, E. G., J. D. Coulter, H. Burton, and R. Porter (1977). Cells of origin and terminal distribution of corticostriatal fibers arising in the sensory–motor cortex of monkeys. *J. Comp. Neurol. 173*, 53–80.

Jones, E. G., J. D. Coulter, and S. H. C. Hendry (1978). Intracortical connectivity of architectonic fields in the somatic sensory, motor and parietal cortex of monkeys. *J. Comp. Neurol. 181*, 291–348.

Jones, E. G., and R. Y. Leavitt (1973). Demonstration of thalamo-cortical connectivity in the cat somato-sensory system by retrograde axonal transport of horseradish peroxidase. *Brain Research 63*, 414–418.

Jones, E. G., and R. Y. Leavitt (1975). Retrograde axonal transport and the demonstration of non-specific projections to the cerebral cortex and striatum from thalamic intralaminar nuclei in the rat, cat and monkey. *J. Comp. Neurol. 154*, 349–378.

Jones, E. G., and T. P. S. Powell (1968a). The ipsilateral cortical connexions of the somatic sensory areas in the cat. *Brain Research 9*, 71–94.

Jones, E. G., and T. P. S. Powell (1968b). The commissural connexions of the somatic sensory cortex in the cat. *J. Anat. (Lond.) 103*, 433–455.

Jones, E. G., and T. P. S. Powell (1968c). The projections of the somatic sensory cortex upon the thalamus in the cat. *Brain Research 10*, 369–391.

Jones, E. G., and T. P. S. Powell (1969a). Connexions of the somatic sensory cortex in the rhesus monkey. I. Ipsilateral cortical connexions. *Brain 92*, 477–502.

Jones, E. G., and T. P. S. Powell (1969b). Connexions of the somatic sensory cortex of the rhesus monkey. II. Contralateral cortical connexions. *Brain 92*, 717–730.

Jones, E. G., and T. P. S. Powell (1969c). The cortical projections of the ventroposterior nucleus of the thalamus in the cat. *Brain Research 13*, 298–318.

Jones, E. G., and T. P. S. Powell (1969d). Electron microscopy of synaptic glomeruli in the thalamic relay nuclei of the cat. *Proc. Roy. Soc. B 172*, 153–171.

Jones, E. G., and T. P. S. Powell (1970a). An anatomical study of converging sensory pathways within the cerebral cortex of the monkey. *Brain 93*, 793–820.

Jones, E. G., and T. P. S. Powell (1970b). An electron microscopic study of the laminar pattern and mode of termination of afferent fibre pathways in the somatic sensory cortex of the cat. *Phil. Trans. B 257*, 45–62.

Jones, E. G., and T. P. S. Powell (1970c). Connexions of the somatic sensory cortex of the rhesus monkey. III. Thalamic connexions. *Brain 93*, 37–56.

Jones, E. G., and T. P. S. Powell (1971). An analysis of the posterior group of thalamic nuclei on the basis of its afferent connections. *J. Comp. Neurol. 143*, 185–216.

Jones, E. G., and Powell (1973). Anatomical organization of the somatosensory cortex. In *Handbook of Sensory Physiology*, Vol. II (A. Iggo, ed.), pp. 579–620. Springer-Verlag, Berlin.

Jones, E. G., and S. P. Wise (1977). Size, laminar and columnar distribution of efferent cells in the sensory-motor cortex of monkeys. *J. Comp. Neurol. 175*, 391–438.

Jones, G. M., and D. G. D. Watt (1971). Observations on the control of stepping and hopping movements in man. *J. Physiol. (Lond.) 219*, 709–727.

Jongkees, L. B. W., and A. J. Philipszoon (1960). Some nystagmographical methods for the investigation of the effect of drugs upon the labyrinth. The influence of cinnarazine, hyoscine, largactil and nembutal on the vestibular system. *Acta Physiol. Pharmacol. Neerl. 9*, 240–275.

Jordan, H., and H. Holländer (1972). The structure of the ventral part of the lateral geniculate nucleus. A cyto- and myeloarchitectonic study in the cat. *J. Comp. Neurol. 145*, 259–272.

Jordan, L. M., D. R. Kenshalo, Jr., R. F. Martin, L. H. Haber, and W. D. Willis (1979). Two populations of spinothalamic tract neurons with opposite responses to 5-hydroxytryptamine. *Brain Research 164*, 342–346.

Joseph, J. W., G. M. Shambes, J. M. Gibson, and W. Welker (1978). Tactile projections to granule cells in caudal vermis of the rat's cerebellum. *Brain Behav. Evol. 15*, 141–149.

Joseph, S. A., and K. M. Knigge (1978). The endocrine hypothalamus: recent anatomical studies. In *The Hypothalamus* (S. Reichlin, R. J. Baldessarini, and J. B. Martin, eds.), pp. 15–47. Raven Press, New York.

Jouvet, M. (1964). Étude neurophysiologique clinique des troubles de la conscience. *Acta Neurochir. (Wien) 12*, 258–269.

Jouvet, M. (1965). Paradoxical sleep—A study of its nature and mechanisms. *Progr. Brain Res. 18*, 20–62.

Jouvet, M. (1972). The role of monoamines and acetylcholine containing neurons in the regulation of the sleep-waking cycle. *Rev. Physiol. Biochem. Exp. Pharmacol. 64*, 166–307.

Jouvet, M., and D. Jouvet (1963). A study of the neurophysiological mechanisms of dreaming. *Electroenceph. Clin. Neurophysiol. Suppl. 24*, 133–157.

Joynt, R. J. (1966). Verney's concept of the osmoreceptor. *Arch. Neurol. (Chic.) 14*, 331–344.

Jung, R. (1958). Coordination of specific and nonspecific afferent impulses at single neurons of the visual cortex. In *Reticular Formation of the Brain*. Henry Ford Hospital International Symposium. (H. H. Jasper, L. D. Proctor, R. S. Knighton, W. C. Noshay, and R. J. Costello, eds.), pp. 423–434. Little, Brown, Boston.

Jung, R., and R. Hassler (1960). The extrapyramidal motor systems. In *Handbook of Physiology*, Section 1, Vol. II (J. Field, H. W. Magoun, and V. E. Hall, eds.), pp. 863–927. American Physiological Society, Washington, D.C.

Juraska, J. M., C. J. Wilson, and P. M. Groves (1977). The substantia nigra of the rat: a Golgi study. *J. Comp. Neurol. 172*, 585–600.

Jürgens, U. (1976). Projections from the cortical larynx area in the squirrel monkey. *Exp. Brain Res. 25*, 401–411.

Jürgens, U., and D. Ploog (1970). Cerebral representation of vocalization in the squirrel monkey. *Exp. Brain Res. 10*, 532–554.

Jürgens, U., and R. Pratt (1979). Role of the periaqueductal grey in vocal expression of emotion. *Brain Research 167*, 367–378.

Kaada, B. R. (1951). Somato-motor, autonomic and electrocorticographic responses to electrical stimulation of 'rhinencephalic' and other structures in primates, cat and dog.—A study of responses from the limbic, subcallosal, orbito-insular, piriform and temporal cortex, hippocampus-fornix and amygdala. *Acta Physiol. Scand. 23, Suppl. 83,* 1–285.

Kaada, B. (1960). Cingulate, posterior orbital, anterior insular and temporal pole cortex. In *Handbook of Physiology,* Section 1. Neurophysiology, Vol. II (J. Field, H. W. Magoun, and V. E. Hall, eds.), pp. 1345–1372. American Physiological Society, Washington, D.C.

Kaada, B. (1963). Pathophysiology of Parkinson tremor, rigidity and hypokinesia. *Acta Neurol. Scand. 39, Suppl. 4,* 39–51.

Kaada, B. (1967). Brain mechanisms related to aggressive behavior. In *Aggression and Defense. Neural Mechanisms and Social Patterns* (C. D. Clemente and D. B. Lindsley, eds.), pp. 95–216. Univ. of California Press, Berkeley.

Kaada, B. R. (1972). Stimulation and regional ablation of the amygdaloid complex with reference to functional representations. In *The Neurobiology of the Amygdala* (B. E. Elftheriou, ed.), pp. 205–281. Plenum Press, New York.

Kaada, B. R., P. Andersen, and J. Jansen, Jr. (1954). Stimulation of the amygdaloid nuclear complex in unanesthetized cats. *Neurology (Minneap.) 4,* 48–64.

Kaada, B. R., W. Harkmark and O. Stokke (1961). Deep coma associated with desynchronization in EEG. *Electroenceph. Clin. Neurophysiol. 13,* 785–789.

Kaada, B. R., J. Jansen, Jr., and P. Andersen (1953). Stimulation of the hippocampus and medial cortical areas in unanesthetized cats. *Neurology (Minneap.) 3,* 844–857.

Kaelber, W. W., and C. L. Mitchell (1967). The centrum medianum—central tegmental fasciculus complex. A stimulation, lesion and degeneration study in the cat. *Brain 90,* 83–100.

Kalil, M. (1976). Motor and sensory regions of the rhesus monkey ventral thalamic nuclei defined by their afferent and efferent connections. *Proc. Soc. Neurosci. 6,* 544.

Kalil, K. (1978). Patch-like termination of thalamic fibers in the putamem of the rhesus monkey: an autoradiograhic study. *Brain Research 140,* 333–339.

Kanaseki, T., and J. M. Sprague (1974). Anatomical organization of pretectal nuclei and tectal laminae in the cat. *J. Comp. Neurol. 158,* 319–338.

Kanazawa, I., E. Bird, R. O'Connell, and D. Powell (1977). Evidence for a decrease in substance P content of substantia nigra in Huntington's chorea. *Brain Research 120,* 387–392.

Kanazawa, I., P. C. Emson, and A. C. Cuello (1977). Evidence for the existence of substance P-containing fibres in striato-nigral and pallido-nigral pathways in rat brain. *Brain Research 119,* 447–453.

Kanazawa, I., G. R. Marshall, and J. S. Kelly (1976). Afferents to the rat substantia nigra studied with horseradish peroxidase with special reference to fibers from the subthalamic nucleus. *Brain Research 115,* 485–491.

Kane, E. S. (1974). Patterns of degeneration in the caudal cochlear nucleus of the cat after cochlear ablation. *J. Comp. Neurol. 179,* 67–92.

Kane, E. S. (1976). Descending projections to specific regions of cat cochlear nucleus: a light microscopic study. *Exp. Neurol. 52,* 372–388.

Kane, E. S. (1977a). Descending inputs to the cat dorsal cochlear nucleus: an electron microscopic study. *J. Neurocytol. 6,* 583–605.

Kane, E. S. (1977b). Descending inputs to the octopus cell area of the cat cochlear nucleus: an electron microscopic study. *J. Comp. Neurol. 173,* 337–354.

Kane, E. S., and R. C. Finn (1977). Descending and intrinsic inputs to dorsal cochlear nucleus of cats: a horseradish peroxidase study. *Neuroscience 2,* 897–912.

Kaneko, C. R. S., N. Steinacker, B. Cohen, R. Maciewicz, and S. M. Highstein (1975). Synaptic linkage of the reticulo-ocular pathway in the cat. *Neurosci. Abstracts I,* 225. Society for Neurosci., Bethesda, Md.

Kanki, S., and T. Ban (1952). Cortico-fugal connections of frontal lobe in man. *Med. J. Osaka Univ. 3,* 201–222.

Kappers, C. U. A., G. C. Huber and E. Crosby (1936). *The Comparative Anatomy of the Nervous System of Vertebrates, including Man*, Macmillan, New York.

Karamanlidis, A. (1968). Trigemino-cerebellar fiber connections in the goat studied by means of the retrograde cell degeneration method. *J. Comp. Neurol. 133*, 71–88.

Karamanlidis, A. N., H. Michaloudi, O. Mángana, and R. P. Saigal (1978). Trigeminal ascending projections in the rabbit, studied with horseradish peroxidase. *Brain Research 156*, 110–116.

Karamanlidis, A. N., and J. Voogd (1970). Trigemino-thalamic fibre connections in the goat. An experimental anatomical study. *Acta Anat. (Basel) 75*, 596–622.

Karol, E. A., and D. N. Pandya (1971). The distribution of the corpus callosum in the rhesus monkey. *Brain 94*, 471–486.

Kato, M., and J. Tanji (1971). The effects of electrical stimulation of Deiters' nucleus upon hindlimb γ-motoneurons in the cat. *Brain Research 30*, 385–395.

Kato, M., and J. Tanji (1972). Volitionally controlled single motor units in human finger muscles. *Brain Research 40*, 345–357.

Kawamura, K. (1973a). Corticocortical fiber connections of the cat cerebrum. I. The temporal region. *Brain Research 51*, 1–21.

Kawamura, K. (1973b). Corticocortical fiber connections of the cat cerebrum. II. The parietal region. *Brain Research 51*, 23–40.

Kawamura, K. (1973c). Corticocortical fiber connections of the cat cerebrum. III. The occipital region. *Brain Research 51*, 41–60.

Kawamura, K. (1975). The pontine projection from the inferior colliculus in the cat. An experimental anatomical study. *Brain Research 95*, 309–322.

Kawamura, K., and A. Brodal (1973). The tectopontine projection in the cat: an experimental anatomical study with comments on pathways for teleceptive impulses to the cerebellum. *J. Comp. Neurol. 149*, 371–390.

Kawamura, K., A. Brodal, and G. Hoddevik (1974). The projection of the superior colliculus onto the reticular formation of the brainstem. An experimental anatomical study in the cat. *Exp. Brain Res. 19*, 1–19.

Kawamura, K., and T. Hashikawa (1975). Studies of the tecto-ponto-cerebellar pathway in the cat by anterograde degeneration and axonal transport techniques. *J. Physiol. Soc. Jap. 37*, 370–372.

Kawamura, K., and T. Hashikawa (1978). Cell bodies of origin of reticular projections from the superior colliculus in the cat. An experimental study with the use of horseradish peroxidase as a tracer. *J. Comp. Neurol. 182*, 1–16.

Kawamura, K., and T. Hashikawa (1979). Olivocerebellar projections in the cat studied by means of anterograde axonal transport of labeled amino acids as tracers. *Neuroscience 4*, 1615–1633.

Kawamura, K., and K. Otani (1970). Corticocortical fiber connections in the cat cerebrum: the frontal region. *J. Comp. Neurol. 139*, 423–448.

Kawamura, S. (1974). Topical organization of the extrageniculate visual system in the cat. *Exp. Neurol. 45*, 451–461.

Kawamura, S., J. M. Sprague, and K. Niimi (1974). Corticofugal projections from the visual cortices to the thalamus, pretectum and superior colliculus in the cat. *J. Comp. Neurol. 158*, 339–362.

Kawana, E. (1969). Projections of the anterior ectosylvian gyrus to the thalamus, the dorsal column nuclei, the trigeminal nuclei and the spinal cord in cats. *Brain Research 14*, 117–136.

Keegan, J. J., and F. D. Garrett (1948). The segmental distribution of the cutaneous nerves in the limbs of man. *Anat. Rec. 102*, 409–437.

Keele, C. A. (1966). Measurement of responses to chemically induced pain. In *Touch, Heat and Pain*. A Ciba Foundation Symposium (A. V. S. de Reuck and J. Knight, eds.), pp. 57–72. J. & A. Churchill, London.

Keele, K. D. (1957). *Anatomies of Pain*. Blackwell, Oxford.

Keller, E. L. (1974). Participation of medial pontine reticular formation in eye movement generation in monkey. *J. Neurophysiol. 37*, 316–332.

Keller, J. H., and P. J. Hand (1970). Dorsal root projections to nucleus cuneatus of the cat. *Brain Research 20,* 1–17.

Keller, J. H., and B. C. Moffett, Jr. (1968). Nerve endings in the temporomandibular joint of the Rhesus macaque. *Anat. Rec. 160,* 587–594.

Kelly, D. L., S. Goldring, and J. L. O'Leary (1965). Averaged evoked somatosensory responses from exposed cortex of man. *Arch. Neurol. (Chic.) 13,* 1–9.

Kelly, J. B. (1973). The effects of insular and temporal lesions in cats on two types of auditory pattern discrimination. *Brain Research 62,* 71–87.

Kelly, J. B., and I. C. Whitfield (1971). Effects of auditory cortical lesions on discriminations of rising and falling frequency-modulated tones. *J. Neurophysiol. 34,* 802–816.

Kelly, J. P., and D. C. van Essen (1974). Cell structure and function in the visual cortex of the cat. *J. Physiol. (Lond.) 238,* 515–547.

Kelly, P. H. (1975). Unilateral 6-hydroxydopamine lesions of nigrostriatal or mesolimbic dopamine-containing terminals and the drug-induced rotation of rats. *Brain Research 100,* 163–169.

Kemp, J. M. (1970). The termination of strio-pallidal and strio-nigral fibres. *Brain Research 17,* 125–128.

Kemp, J. M., and T. P. S. Powell (1970). The corticostriate projection in the monkey. *Brain 93,* 525–546.

Kemp, J. M., and T. P. S. Powell (1971a). The structure of the caudate nucleus of the cat: light and electron microscopy. *Phil. Trans. B 262,* 383–402.

Kemp, J. M., and T. P. S. Powell (1971b). The site of termination of afferent fibers in the caudate nucleus. *Phil. Trans. B 262,* 403–412.

Kemper, T. L., S. J. Wright, Jr., and S. Locke (1972). Relationship between the septum and the cingulate gyrus in *Macaca mulatta. J. Comp. Neurol. 146,* 465–478.

Kennard, M. A. (1944). Experimental analysis of the functions of the basal ganglia in monkeys and chimpanzees. *J. Neurophysiol. 7,* 127–148.

Kennard, M. A., S. Spencer, and G. Fountain (1941). Hyperactivity in monkeys following lesions of the frontal lobes. *J. Neurophysiol. 4,* 512–524.

Kernell, D. (1966). Input resistance, electrical excitability, and size of ventral horn cells in cat spinal cord. *Science 152,* 1637–1640.

Kerr, D. I. B., and K.-E. Hagbarth (1955). An investigation of olfactory centrifugal fiber system. *J. Neurophysiol. 18,* 362–374.

Kerr, D. I., F. P. Haugen, and R. Melzack (1955). Responses evoked in the brain stem by tooth stimulation. *Amer. J. Physiol. 183,* 253–258.

Kerr, F. W. L. (1961). Structural relation of the trigeminal spinal tract to upper cervical roots and the solitary nucleus in the cat. *Exp. Neurol. 4,* 134–148.

Kerr, F. W. L. (1962). Facial, vagal and glossopharyngeal nerves in the cat. Afferent connections. *Arch. Neurol. (Chic.) 6,* 264–281.

Kerr, F. W. L. (1963a). The divisional organization of afferent fibres of the trigeminal nerve. *Brain 86,* 721–732.

Kerr, F. W. L. (1963b). The etiology of trigeminal neuralgia. *Arch. Neurol. (Chic.) 8,* 15–25.

Kerr, F. W. L. (1966). On the question of ascending fibers in the pyramidal tract: with observations on spinotrigeminal and spinopontine fibers. *Exp. Neurol. 14,* 77–85.

Kerr, F. W. L. (1969). Preserved vagal visceromotor function following destruction of the dorsal motor nucleus. *J. Physiol. (Lond.) 202,* 755–769.

Kerr, F. W. L. (1970a). The fine structure of the subnucleus caudalis of the trigeminal nerve. *Brain Research 23,* 129–145.

Kerr, F. W. L. (1970b). The organization of primary afferents in the subnucleus caudalis of the trigeminal. A light and electron microscope study of degeneration. *Brain Research 23,* 147–165.

Kerr, F. W. L. (1971). Electronmicroscopic observations on primary deafferentation of the subnucleus caudalis of the trigeminal nerve. In *Oral-Facial Sensory and Motor Mech-*

anisms (R. Dubner and Y. Kawamura, eds.), pp. 159–181. Appleton-Century-Crofts, New York.

Kerr, F. W. L. (1975). The ventral spinothalamic tract and other ascending systems of the ventral funiculus of the spinal cord. *J. Comp. Neurol. 159*, 335–356.

Kerr, F. W. L., and S. Alexander (1964). Descending autonomic pathways in the spinal cord. *Arch. Neurol. (Chic.) 10*, 249–261.

Kerr, F. W. L., and J. A. Brown (1964). Pupillomotor pathways in the spinal cord. *Arch. Neurol. (Chic.) 10*, 262–270.

Kerr, F. W. L., N. Hendler, and P. Bowron (1970). Viscerotopic organization of the vagus. *J. Comp. Neurol. 138*, 279–290.

Kerr, F. W. L., and O. W. Hollowell (1964). Location of pupillomotor and accommodation fibres in the oculomotor nerve: experimental observations on paralytic mydriasis. *J. Neurol. Neurosurg. Psychiat. 27*, 473–481.

Kerr, F. W. L., and R. A. Olafson (1961). Trigeminal and cervical volleys: convergence on single units in the spinal gray at C-1 and C-2. *Arch. Neurol. (Chic.) 5*, 171–178.

Kertesz, A., and P. McCabe (1977). Recovery patterns and prognosis in aphasia. *Brain 100*, 1–18.

Khayyat, G. F., Y. J. Yu, and R. B. King (1975). Response patterns to noxious and nonnoxious stimuli in rostral trigeminal relay nuclei. *Brain Research 97*, 47–60.

Kiang, N. Y.-S. (1965). *Discharge Patterns of Single Fibers in the Cat's Auditory Nerve.* MIT Press, Cambridge, Mass.

Kiang, N. Y.-S., D. A. Godfrey, B. E. Norris, and S. E. Moxon (1975). A block model of the cat cochlear nucleus. *J. Comp. Neurol. 162*, 221–246.

Kievit, J., and H. G. J. M. Kuypers (1972). Fastigial cerebellar projections to the ventrolateral nucleus of the thalamus and the organization of the descending pathways. In *Corticothalamic Projections and Sensorimotor Activities*. (T. L. Frigyesi, E. Rinvik, and M. D. Yahr, eds.), pp. 91–111. Raven Press, New York.

Kievit, J., and H. G. J. M. Kuypers (1977). Organization of the thalamocortical connexions to the frontal lobe in the rhesus monkey. *Exp. Brain Res. 29*, 299–322.

Killackey, H. P., and S. Leshin (1975). The organization of specific thalamocortical projections to the posteromedial barrel subfield in the rat somatic sensory cortex. *Brain Research 86*, 469–472.

Kiloh, L. G., and S. Nevin (1951). Progressive dystrophy of the external ocular muscles (ocular myopathy). *Brain 74*, 115–143.

Kim, R., K. Nakano, A. Jayaraman, and M. B. Carpenter (1976). Projections of the globus pallidus and adjacent structures: an autoradiographic study in the monkey. *J. Comp. Neurol. 169*, 263–290.

Kimmel, D. L. (1941). Development of the afferent components of the facial, glossopharyngeal and vagus nerves in the rabbit embryo. *J. Comp. Neurol. 74*, 447–471.

Kimura, D. (1961). Cerebral dominance and the perception of verbal stimuli. *Canad. J. Psychol. 15*, 166–171.

Kimura, D. (1964). Left-right differences in the perception of melodies. *Quart. J. Exp. Psychol. 16*, 355–358.

Kimura, R., and J. Wersäll, Jr. (1962). Termination of the olivo-cochlear bundle in relation to the outer hair cells of the organ of Corti in the guinea-pig. *Acta Otolaryng. (Stockh.) 55*, 11–32.

King, J. S. (1976). The synaptic cluster (glomerulus) in the inferior olivary nucleus. *J. Comp. Neurol. 165*, 387–400.

King, J. S., J. A. Andrezik, W. M. Falls, and G. F. Martin (1976). The synaptic organization of the cerebello-olivary circuit. *Exp. Brain Res. 26*, 159–170.

King, J. S., M. H. Bowman, and G. F. Martin (1971). The red nucleus of the opossum (*Didelphis marsupialis virginiana*): a light and electron microscopic study. *J. Comp. Neurol. 143*, 157–184.

King, J. S., R. M. Dom, J. B. Conner, and G. F. Martin (1973). An experimental light and electron microscopic study of cerebellorubral projections in the opossum, *Didelphis marsupialis virginiana. Brain Research 52*, 61–78.

King, J. S., G. F. Martin, and T. Biggert (1968). The basilar pontine gray of the opossum. I. Morphology. *J. Comp. Neurol. 133,* 439–446.

King, J. S., G. F. Martin, and M. H. Bowman (1975). The direct spinal area of the inferior olivary nucleus: an electron microscopic study. *Exp. Brain Res. 22,* 13–24.

King, J. S., G. F. Martin, and J. B. Conner (1972). A light and electron microscopic study of cortico-rubral projections in the opossum, *Didelphis marsupialis virginiana. Brain Research 38,* 251–265.

King, J. S., R. C. Schwyn, and C. A. Fox (1971). The red nucleus in the monkey (*Macaca mulatta*): a Golgi and an electron microscopic study. *J. Comp. Neurol. 142,* 75–108.

Kirkwood, P. A., and T. A. Sears (1974). Monosynaptic excitation of motoneurones from secondary endings of muscle spindles. *Nature 252,* 243–244.

Kirsche, W. (1958). Synaptische Formationen in den Ganglia lumbalia des Truncus sympathicus vom Menschen einschliesslich Bemerkungen über den heutigen Stand der Neuronenlehre. *Z. Mikr. Anat. Forsch. 64,* 707–772.

Kitai, S. T., J. F. DeFrance, K. Hatada, and D. T. Kennedy (1974). Electrophysiological properties of lateral reticular nucleus cells. II. Synaptic activation. *Exp. Brain Res. 21,* 419–432.

Kitai, S. T., J. D. Kocsis, and T. Kiyohara (1976). Electrophysiological properties of nucleus reticularis tegmenti pontis cells: antidromic and synaptic activation. *Exp. Brain Res. 24,* 295–309.

Kitai, S. T., R. A. McCrea, R. J. Preston, and G. A. Bishop (1977). Electrophysiological and horseradish peroxidase studies of precerebellar afferents to the nucleus interpositus anterior. I. Climbing fiber system. *Brain Research 122,* 197–214.

Kitai, S. T., T. Tanaka, N. Tsukahara, and H. Yu (1972). The facial nucleus of cat: antidromic and synaptic activation and peripheral nerve representation. *Exp. Brain Res. 16,* 161–183.

Kitai, S. T., A. Wagner, W. Precht, and T. Ohno (1975). Nigro-caudate and caudato-nigral relationship: an electrophysiological study. *Brain Research 85,* 44–48.

Kite, W. C., Jr., R. D. Whitfield, and E. Campbell (1957). The thoracic herniated intervertebral disc syndrome. *J. Neurosurg. 14,* 61–67.

Kleitman, N. (1963). *Sleep and Wakefulness,* revised and enlarged edition. The Univ. of Chicago Press, Chicago.

Kling, A. (1972). Effects of amygdalectomy on social affective behavior in non-human primates. In *The Neurobiology of the Amygdala* (B. E. Eleftheriou, ed.), pp. 511–536. Plenum Press, New York.

Klinke, R., and N. Galley (1974). Efferent innervation of vestibular and auditory receptors. *Physiol. Rev. 54,* 316–357.

Klüver, H. (1958). "The temporal lobe syndrome" produced by bilateral ablations. In *Neurological Basis of Behaviour.* Ciba Foundation Symposium (E. E. Wolstenholme and C. M. O'Connor, eds.), pp. 175–182. J. & A. Churchill, London.

Klüver, H., and P. C. Bucy (1937). Psychic blindness and other symptoms following bilateral temporal lobectomy in rhesus monkeys. *Amer. J. Physiol. 119,* 352–353.

Kneisley, L. W., M. P. Biber, and J. H. LaVail (1978). A study of the origin of brain stem projections to the monkey spinal cord using the retrograde transport method. *Exp. Neurol. 60,* 116–139.

Knighton, R. S. (1950). Thalamic relay nucleus for the second somatic sensory receiving area in the cerebral cortex of the cat. *J. Comp. Neurol. 92,* 183–191.

Knook, H. L. (1965). *The Fibre Connections of the Forebrain,* Thesis. Van Gorcum, Assen.

Knutsson, E., V. Lindblom, and A. Mårtensson (1973). Differences in effects in gamma and alpha spasticity induced by the GABA derivative baclofen (Lioresal). *Brain 96,* 29–46.

Kobayashi, H., T. Matsui, and S. Ishii (1970). Functional electron microscopy of the hypothalamic median eminence. *Int. Rev. Cytol. 29,* 282–382.

Kocsis, J. D., and C. P. Vandermaelen (1977). The caudate projection of the retrorubral nucleus and its relationship to the substantia nigra in the cat. *Anat. Rec. 187,* 628–629.

Köhler, C., M. T. Shipley, B. Srebro, and W. Harkmark (1978). Some retrohippocampal afferents to the entorhinal cortex. Cells of origin as studied by the HRP method in the rat and mouse. *Neurosci. Letters 10*, 115–120.

Koikegami, H. (1957). On the correlation between cellular and fibrous patterns of the human brain stem reticular formation with some cytoarchitectonic remarks on the other mammals. *Acta. Med. Biol. (Niigata) 5*, 21–72.

Kolb, H. (1974). The connections between horizontal cells and photoreceptors in the retina of the cat: electron microscopy of Golgi preparations. *J. Comp. Neurol. 155*, 1–14.

Koppang, K. (1962). Intrathecal phenol in the treatment of spastic conditions. *Acta Neurol. Scand. 38, Suppl. 3*, 63–68.

Korn, H., C. Sotelo, and F. Crepel (1972/73). Electronic coupling between neurons in the rat lateral vestibular nucleus. *Exp. Brain Res. 16*, 255–275.

Korneliussen, H. K. (1967). Cerebellar corticogenesis in Cetacea, with special reference to regional variations. *J. Hirnforsch. 9*, 151–185.

Korneliussen, H. K. (1968). On the ontogenetic development of the cerebellum (nuclei, fissures, and cortex) of the rat, with special reference to regional variations in corticogenesis. *J. Hirnforsch. 10*, 379–412.

Korneliussen, H. K. (1969). Cerebellar organization in the light of cerebellar nuclear morphology and cerebellar corticogenesis. In *Neurobiology of Cerebellar Evolution and Development* (R. Llinás, ed.), pp. 515–523. Education & Research Foundation, Chicago.

Korneliussen, H. K., and J. Jansen (1965). On the early development and homology of the central cerebellar nuclei in Cetacea. *J. Hirnforsch. 8*, 47–56.

Korner, P. I. (1971). Integrative neural cardiovascular control. *Physiol. Rev. 51*, 312–367.

Kornmüller, A. E., and R. Janzen (1939). Über die normalen bioelektrischen Erscheinungen des menschlichen Gehirns. *Arch. Psychiat. Nervenkr. 110*, 224–252.

Korte, G. E. (1979). The brainstem projection of the vestibular nerve in the cat. *J. Comp. Neurol. 184*, 279–292.

Korte, G. E., and E. Mugnaini (1979). The cerebellar projection of the vestibular nerve in the cat. *J. Comp. Neurol. 184*, 265–278.

Kostowski, W., E. Giacalone, S. Garattini, and L. Valzelli (1968). Studies on behavioural and biochemical changes in rats after lesions of midbrain raphe. *Europ. J. Pharmacol. 4*, 371–376.

Kostyuk, P. G., and G. G. Skibo (1975). An electron microscopic analysis of rubrospinal tract termination in the spinal cord of the cat. *Brain Research 85*, 511–516.

Kotby, M. N., and L. K. Haugen (1970a). The mechanics of laryngeal function. *Acta Otolaryng. (Stockh.) 70*, 203–211.

Kotby, M. N., and L. K. Haugen (1970b). Attempts at evaluation of the function of various laryngeal muscles in the light of muscle and nerve stimulation experiments in man. *Acta Otolaryng. (Stockh.) 70*, 419–427.

Kotby, M. N., and L. K. Haugen (1970c). Clinical application of electromyography in vocal fold mobility disorders. *Acta Otolaryng. (Stockh.) 70*, 428–437.

Kotchabhakdi, N., G. H. Hoddevik, and F. Walberg (1978). Cerebellar afferent projections from the perihypoglossal nuclei: an experimental study with the method of retrograde axonal transport of horseradish peroxidase. *Exp. Brain Res. 31*, 13–29.

Kotchabhakdi, N., and F. Walberg (1977). Cerebellar afferents from neurons in motor nuclei of cranial nerves demonstrated by retrograde axonal transport of horseradish peroxidase. *Brain Research 137*, 158–163.

Kotchabhakdi, N., and F. Walberg (1978a). Primary vestibular afferent projections to the cerebellum as demonstrated by retrograde axonal transport of horseradish peroxidase. *Brain Research 142*, 142–146.

Kotchabhakdi, N., and F. Walberg (1978b). Cerebellar afferent projections from the vestibular nuclei in the cat: an experimental study with the method of retrograde axonal transport of horseradish peroxidase. *Exp. Brain Res. 31*, 591–604.

Kotchabhakdi, N., F. Walberg, and A. Brodal (1978). The olivocerebellar projection in the cat studied with the method of retrograde transport of horseradish peroxidase. VII. The projection to lobulus simplex, crus I and II. *J. Comp. Neurol. 182,* 293–314.

Krauthamer, G. M., and D. Albe-Fessard (1964). Electrophysiologic studies of the basal ganglia and striopallidal inhibition of non-specific afferent activity. *Neuropsychologia* 2, 73–83.

Krettek, J. E., and J. L. Price (1974). A direct input from the amygdala to the thalamus and the cerebral cortex. *Brain Research 67,* 169–174.

Krettek, J. E., and J. L. Price (1977a). The cortical projection of the medio-dorsal nucleus and adjacent thalamic nuclei in the rat. *J. Comp. Neurol. 171,* 157–192.

Krettek, J. E., and J. L. Price (1977b). Projections from the amygdaloid complex to the cerebral cortex and thalamus in the rat and cat. *J. Comp. Neurol. 172,* 687–722.

Krettek, J. E., and J. L. Price (1977c). Projections from the amygdaloid complex and adjacent olfactory structures to the entorhinal cortex and to the subiculum in the rat and cat. *J. Comp. Neurol. 172,* 723–752.

Krettek, J. E., and J. L. Price (1978). Amygdaloid projections to subcortical structures within the basal forebrain and brainstem in the rat and cat. *J. Comp. Neurol. 178,* 225–254.

Kreutzberg, G. W., and P. Schubert (1975). The cellular dynamics of intraneuronal transport. In *The Use of Axonal Transport for Studies of Neuronal Connectivity* (W. M. Cowan and M. Cuénod, eds.), pp. 83–112. Elsevier, Amsterdam.

Krishnamurti, A., F. Sanides, and W. I. Welker (1976). Microelectrode mapping of modality-specific somatic sensory cerebral neocortex in slow loris. *Brain Behav. Evol. 13,* 367–383.

Kristensson, K. (1975). Retrograde axonal transport of protein tracers. In *The Use of Axonal Transport for Studies of Neuronal Connectivity* (W. M. Cowan and M. Cuénod, eds.), pp. 69–82. Elsevier, Amsterdam.

Kristensson, K., and Y. Olsson (1976). Retrograde transport of horseradish peroxidase in transected axons. 3. Entry into injured axons and subsequent localization in perikaryon. *Brain Research 115,* 201–213.

Kristiansen, K. (1949). Neurological investigations of patients with acute injuries of the head. *Skr. Norske Vidensk.-Akad., I. Mat.-Nat. Kl.* No. 1, 1–169.

Kristiansen, K. (1964). Neurosurgical considerations on the brain mechanisms of consciousness. *Acta Neurochir. (Wien) 12,* 289–314.

Krnjević, K. (1974). Chemical nature of synaptic transmission in vertebrates. *Physiol. Rev. 54,* 418–540.

Kruger, L., and D. Albe-Fessard (1960). Distribution of responses to somatic afferent stimuli in the diencephalon of the cat under chloralose anesthesia. *Exp. Neurol. 2,* 442–467.

Kruger, L., and F. Michel (1962a). A morphological and somatotopic analysis of single unit activity in the trigeminal sensory complex of the cat. *Exp. Neurol. 5,* 139–156.

Kruger, L., and F. Michel (1962b). Reinterpretation of the representation of pain based on physiological excitation of single neurons in the trigeminal sensory complex. *Exp. Neurol. 5,* 157–178.

Kruger, L., R. Siminoff, and P. Witkovsky (1961). Single neuron analysis of dorsal column nuclei and spinal nucleus of trigeminal in cat. *J. Neurophysiol. 24,* 333–349.

Kubie, L. S. (1954). Psychiatric and psychoanalytic considerations of the problem of consciousness. In *Brain Mechanisms and Consciousness* (J. F. Delafresnaye, ed.), pp. 444–467. Blackwell, Oxford.

Kubik, C. S., and R. D. Adams (1946). Occlusion of the basilar artery.—A clinical and pathological study. *Brain 69,* 73–121.

Kudo, M., and K. Niimi (1978). Ascending projections of the inferior colliculus onto the medial geniculate body in the cat studied by anterograde and retrograde tracing techniques. *Brain Research 155,* 113–117.

Kugelberg, E. (1976). Adaptive transformation of rat soleus motor units during growth. *J. Neurol. Sci. 27,* 269–289.

Kugelberg, E., K. Eklund, and L. Grimby (1960). An electromyographic study of the nociceptive reflexes of the lower limb. Mechanism of the plantar responses. *Brain 83*, 394–410.

Kugelberg, E., and K. E. Hagbarth (1958). Spinal mechanism of the abdominal and erector spinae skin reflexes. *Brain 81*, 290–304.

Kugelberg, E., and U. Lindblom (1959). The mechanism of pain in trigeminal neuralgia. *J. Neurol. Neurosurg. Psychiat. 22*, 36–43.

Kuhar, M. J., G. K. Aghajanian, and R. H. Roth (1972). Tryptophan hydroxylase activity and synaptosomal uptake of serotonin in discrete brain regions after midbrain raphe lesions: correlations with serotonin levels and histochemical fluorescence. *Brain Research 44*, 165–176.

Kuhar, M. J., R. H. Roth, and G. K. Aghajanian (1971). Selective reduction of tryptophan hydroxylase activity in rat forebrain after midbrain raphe lesions. *Brain Research 35*, 167–176.

Kuhlenbeck, H. (1954). The human diencephalon. A summary of development, structure, function and pathology. *Confin. Neurol. (Basel) Suppl. 14*, 1–230.

Kuhlenbeck, H. (1957). *Brain and Consciousness*. S. Karger, Basel.

Kuhlenbeck, H. (1959). Further remarks on brain and consciousness: the brain-paradox and the meaning of consciousness. *Confin. Neurol. (Basel) 19*, 462–485.

Kuhn, R. A. (1950). Functional capacity of the isolated human cord. *Brain 73*, 1–51.

Kumar, S., and P. R. Davis (1973). Lumbar vertebral innervation and intraabdominal pressure. *J. Anat. (Lond.) 114*, 47–53.

Kunc, Z. (1965). Treatment of essential neuralgia of the 9th nerve by selective tractotomy. *J. Neurosurg. 23*, 494–500.

Kunc, Z. (1970). Significant factors pertaining to the results of trigeminal tractotomy. In *Trigeminal Neuralgia* (R. Hassler and A. E. Walker, eds.). Georg Thieme Verlag, Stuttgart.

Kunc, Z., and J. Marsala (1962). La localisation et la termination des voies afférentes des nerfs IX et X dans le bulbe. *Acta Neurochir. (Wien) 19*, 512–522.

Kuno, M., E. J. Muñoz-Martinez, and M. Randić (1973). Sensory inputs to neurones in Clarke's column from muscle, cutaneous and joint receptors. *J. Physiol. (Lond.) 228*, 327–342.

Kuo, J.-S., and M. B. Carpenter (1973). Organization of pallidothalamic projections in the rhesus monkey. *J. Comp. Neurol. 151*, 201–236.

Kuntz, A. (1920). The development of the sympathetic nervous system in man. *J. Comp. Neurol. 32*, 173–229.

Kuntz, A. (1938). The structural organization of the coeliac ganglia. *J. Comp. Neurol. 69*, 1–12.

Kuntz, A. (1940). The structural organization of the inferior mesenteric ganglia. *J. Comp. Neurol. 72*, 371–382.

Kuntz, A., and D. I. Farnsworth (1931). Distribution of afferent fibers via the sympathetic trunks and gray communicating rami to the brachial and lumbosacral plexuses. *J. Comp. Neurol. 53*, 389–399.

Kuntz, A., and C. A. Richins (1946). Reflex pupillodilator mechanisms. An experimental analysis. *J. Neurophysiol. 9*, 1–7.

Künzle, H. (1973). The topographic organization of spinal afferents to the lateral reticular nucleus of the cat. *J. Comp. Neurol. 149*, 103–116.

Künzle, H. (1975a). Autoradiographic tracing of the cerebellar projections from the lateral reticular nucleus in the cat. *Exp. Brain Res. 22*, 255–266.

Künzle, H. (1975b). Bilateral projections from precentral motor cortex to the putamen and other parts of the basal ganglia. An autoradiographic study in Macaca fascicularis. *Brain Research 88*, 195–210.

Künzle, H. (1976). Thalamic projections from the precentral motor cortex in Macaca fascicularis. *Brain Research 105*, 253–267.

Künzle, H. (1977). Projections from the primary somatosensory cortex to basal ganglia and thalamus in the monkey. *Exp. Brain Res. 30*, 481–492.

Künzle, H. (1978). An autoradiographic analysis of the efferent connections from premotor and adjacent prefrontal regions (areas 6 and 9) in Macaca fascicularis. *Brain Behav. Evol. 15*, 185–234.

Künzle, H., and K. Akert (1977). Efferent connections of cortical area 8 (frontal eye field) in Macaca fascicularis. A reinvestigation using the autoradiographic technique. *J. Comp. Neurol. 173*, 147–164.

Künzle, H., and M. Cuénod (1973). Differential uptake of (^{3}H) proline and (^{3}H) leucine by neurons: its importance for the autoradiographic tracing of pathways. *Brain Research 62*, 213–217.

Künzle, H., and M. Wiesendanger (1974). Pyramidal connections to the lateral reticular nucleus in the cat: a degeneration study. *Acta Anat. (Basel) 88*, 105–114.

Kupfer, C. (1962). The projection of the macula in the lateral geniculate nucleus of man. *Amer. J. Ophthal. 54*, 597–609.

Kupfer, C., L. Chumbley, and J. de C. Downer (1967). Quantitative histology of optic nerve, optic tract and lateral geniculate nucleus of man. *J. Anat. (Lond.) 101*, 393–401.

Kusama, T., K. Otani, and E. Kawana (1966). Projections of the motor, somatic sensory, auditory and visual cortices in cats. In *Progress in Brain Research*, Vol. 21, part A (T. Tokizane and J. P. Schadé, eds.), pp. 292–322. Elsevier, Amsterdam.

Kuypers, H. G. J. M. (1958a). An anatomical analysis of cortico-bulbar connexions to the pons and lower brain stem in the cat. *J. Anat. (Lond.) 92*, 198–218.

Kuypers, H. G. J. M. (1958b). Some projections from the peri-central cortex to the pons and lower brain stem in monkey and chimpanzee. *J. Comp. Neurol. 110*, 221–255.

Kuypers, H. G. J. M. (1958c). Cortico-bulbar connexions to the pons and lower brain stem in man. An anatomical study. *Brain 81*, 364–388.

Kuypers, H. G. J. M. (1960). Central cortical projections to motor and somatosensory cell groups. *Brain 83*, 161–184.

Kuypers, H. G. J. M. (1966). Discussion. In *The Thalamus* (D. P. Purpura and M. D. Yahr, eds.), pp. 122–126. Columbia Univ. Press, New York.

Kuypers, H. G. J. M., M. Bentivoglio, D. van der Kooy, and C. E. Catsman-Berrevoets (1979). Retrograde transport of bisbenzimide and propidium iodide through axons to their parent cell bodies. *Neurosci. Letters 12*, 1–8.

Kuypers, H. G. J. M., and J. Brinkman (1970). Precentral projections to different parts of the spinal intermediate zone in the rhesus monkey. *Brain Research 24*, 29–48.

Kuypers, H. G. J. M., W. R. Fleming, and J. W. Farinholt (1962). Subcorticospinal projections in the rhesus monkey. *J. Comp. Neurol. 118*, 107–137.

Kuypers, H. G. J. M., J. Kievit, and A. C. Groenklevant (1974). Retrograde axonal transport of horseradish peroxidase in rat's forebrain. *Brain Research 67*, 211–218.

Kuypers, H. G. J. M., and D. G. Lawrence (1967). Cortical projections to the red nucleus and the brain stem in the rhesus monkey. *Brain Research 4*, 151–188.

Kuypers, H. G. J. M., and V. A. Maisky (1975). Retrograde axonal transport of horseradish peroxidase from spinal cord to brain stem cell groups in the cat. *Neurosci. Letters 1*, 9–14.

Kuypers, H. G. J. M., M. K. Szwarcbart, M. Mishkin, and H. E. Rosvold (1965). Occipitotemporal corticocortical connections in the rhesus monkey. *Exp. Neurol. 11*, 245–262.

Kuypers, H. G. J. M., and J. D. Tuerk (1964). The distribution of the cortical fibres within the nuclei cuneatus and gracilis in the cat. *J. Anat. (Lond.) 98*, 143–162.

Ladpli, R., and A. Brodal (1968). Experimental studies of commissural and reticular formation projections from the vestibular nuclei in the cat. *Brain Research 8*, 65–96.

Laemle, L. K. (1975). Cell populations of the lateral geniculate nucleus of the cat as determined with horseradish peroxidase. *Brain Reserch 100*, 650–656.

Lafleur, J., J. De Lean, and L. J. Poirier (1974). Physiopathology of the cerebellum in the monkey. Part 1. Origin of cerebellar afferent nervous fibers from the spinal cord and brain stem. *J. Neurol. Sci. 22,* 471–490.

Lamarre, Y., A. J. Joffroy, M. Dumont, C. De Montigny, F. Grou, and J. P. Lund (1975). Central mechanisms of tremor in some feline and primate models. *Canad. J. Neurol. Sci. 2,* 227–233.

Lammers, H. J. (1972). The neural connections of the amygdaloid complex in mammals. In *The Neurobiology of the Amydala* (B. E. Eleftheriou, ed.), pp. 123–144. Plenum Press, New York.

La Motte, C. (1977). Distribution of the tract of Lissauer and the dorsal root fibers in the primate spinal cord. *J. Comp. Neurol. 172,* 529–562.

La Motte, C., C. B. Pert, and S. H. Snyder (1976). Opiate receptor binding in primate spinal cord: distribution and changes after dorsal root section. *Brain Research 112,* 407–412.

Lance, J. W., R. S. Schwab, and E. A. Peterson (1963). Action tremor and the cogwheel phenomenon in Parkinson's disease. *Brain 86,* 95–110.

Land, L. J. (1973). Localized projection of olfactory nerves to rabbit olfactory bulb. *Brain Research 63,* 153–166.

Landau, W. M., and M. H. Clare (1959). The plantar reflex in man, with special reference to some conditions where the extensor response is unexpectedly absent. *Brain 82,* 321–355.

Landau, W. M., and M. H. Clare (1964). Fusimotor function. Part VI. H reflex, tendon jerk, and reinforcement in hemiplegia. *Arch. Neurol. (Chic.) 10,* 128–134.

Landgren, S., A. Nordwall, and C. Wengström (1965). The location of the thalamic relay in the spino-cervico-lemniscal path. *Acta Physiol. Scand. 65,* 164–175.

Landgren, S., and K. Å. Olsson (1976). Localization of evoked potentials in the digastric, masseteric, supra- and intertrigeminal subnuclei of the cat. *Exp. Brain Res. 26,* 299–318.

Landgren, S., and K. Å. Olsson (1977). The effect of electrical stimulation in the defense attack area of the hypothalamus on the monosynaptic jaw closing and the disynaptic jaw opening reflexes in the cat. In *Pain in the Trigeminal Region* (D. J. Anderson and B. Matthews, eds.), pp. 385–394. Elsevier, Amsterdam.

Landgren, S., C. G. Phillips, and R. Porter (1962). Cortical fields of origin of the monosynaptic pyramidal pathways to some alpha motoneurones of the baboon's hand and forearm. *J. Physiol. (Lond.) 161,* 112–125.

Landgren, S., and H. Silfvenius (1969). Projection to cerebral cortex of group I muscle afferents from the cat's hind limb. *J. Physiol. (Lond.) 200,* 353–372.

Landgren, S., and H. Silfvenius (1971). Nucleus Z, the medullary relay in the projection path to the cerebral cortex of group I muscle afferents from the cat's hind limb. *J. Physiol. (Lond.) 218,* 551–571.

Landgren, S., and D. Wolsk (1966). A new cortical area receiving input from group I muscle afferents. *Life Sciences 5,* 75–79.

Landis, S. C., and F. E. Bloom (1975). Ultrastructural identification of noradrenergic boutons in mutant and normal mouse cerebellar cortex. *Brain Research 96,* 299–305.

Lang, W., J. A. Büttner-Ennever, and U. Büttner (1978). Vestibulothalamic pathways in the monkey. *Neurosci. Letters, Suppl. 1,* p. 354 (Abstr.)

Lange, W. (1972). Regionale Unterschiede in der Cytoarchitektonik der Kleinhirnrinde bei Mensch, Rhesusaffe und Katze. *Z. Anat. Entwickl.-Gesch. 138,* 329–346.

Langer, T. P., and R. D. Lund (1974). The upper layers of the superior colliculus in the rat: a Golgi study. *J. Comp. Neurol. 158,* 405–436.

Langfitt, T. W., K. Kamei, G. Y. Koff, and S. M. Peacock, Jr. (1963). Gamma neuron control by thalamus and globus pallidus. *Arch. Neurol. (Chic.) 9,* 593–606.

Langhorst, P., and M. Werz (1974). Concept of functional organization of the brain stem cardiovascular center. In *Central Rhythmic and Regulation* (W. Umbach and H. P. Koepchen, eds.), pp. 238–255. Hippokrates Verlag, Stuttgart.

Langworth, E. P., and D. Taverner (1963). The prognosis in facial palsy. *Brain 86,* 465–480.

Langworthy, O. R., and L. Ortega (1943). The iris. Innervation of the iris of the albino rabbit as related to its function. Theoretical discussion of abnormalities of the pupils observed in man. *Medicine (Baltimore) 22,* 287–362.

Lanoir, J., and J. Schlag (1976). Le thalamus du chat et du singe. Données anatomiques, hodologiques et fonctionnelles. 1re partie: Les noyaux spécifiques sensoriels et le groupe des noyaux ventraux. *J. Physiol. (Paris) 72,* 1–170.

Lapides, J. (1958). Structure and function of the internal vesical sphincter. *J. Urol. (Baltimore) 80,* 341–353.

Laporte, Y., and F. Emonet-Dénand (1976). The skeleto-fusimotor innervation of cat muscle spindle. *Progr. Brain Res. 44,* 99–105.

Larsell, O. (1934). Morphogenesis and evolution of the cerebellum. *Arch. Neurol. Psychiat. (Chic.) 31,* 373–395.

Larsell, O. (1937). The cerebellum. A review and interpretation. *Arch. Neurol. Psychiat. (Chic.) 38,* 580–607.

Larsell, O. (1945). Comparative neurology and present knowledge of the cerebellum. *Bull. Minn. Med. Found. 5,* 73–112.

Larsell, O. (1952). The morphogenesis and adult pattern of the lobules and fissures of the cerebellum of the white rat. *J. Comp. Neurol. 97,* 281–356.

Larsell, O. (1953). The anterior lobe of the mammalian and the human cerebellum. *Anat. Rec. 115,* 341.

Larsell, O. (1967). *The Comparative Anatomy and Histology of the Cerebellum from Myxinoids through Birds* (J. Jansen, ed.). Univ. of Minnesota Press, Minneapolis.

Larsell, O. (1970). *The Comparative Anatomy and Histology of the Cerebellum from Monotremes through Apes* (J. Jansen, ed.). Univ. of Minnesota Press, Minneapolis.

Larsell, O., and R. A. Fenton (1928). The embryology and neurohistology of the sphenopalatine ganglion connections; a contribution to the study of otalgia. *Laryngoscope (St. Louis) 38,* 371–389.

Larsell, O., and J. Jansen (1972). *The Comparative Anatomy and Histology of the Cerebellum. The Human Cerebellum, Cerebellar Connections, and Cerebellar Cortex.* Univ. of Minnesota Press, Minneapolis.

Larsen, B., E. Skinhöj, and N. A. Lassen (1978). Variations in regional cortical blood flow in the right and left hemispheres during automatic speech. *Brain 101,* 193–209.

Larson, C. R., D. Sutton, and R. C. Lindeman (1978). Cerebellar regulation of phonation in rhesus monkey (*Macaca mulatta*). *Exp. Brain Res. 33,* 1–18.

Laruelle, L. (1937). La structure de la moelle épinière en coupes longitudinales. *Rev. Neurol. 67,* 695–725.

Lashley, K. S. (1934). The mechanism of vision. VII. The projection of the retina upon the primary optic centers in the rat. *J. Comp. Neurol. 59,* 341–373.

Lashley, K. S. (1941). Thalamo-cortical connections of the rat's brain. *J. Comp. Neurol. 75,* 67–121.

Lashley, K. S., and G. Clark (1946). The cytoarchitecture of the cerebral cortex of Ateles: a critical examination of architectonic studies. *J. Comp. Neurol. 85,* 223–306.

Lassek, A. M. (1940). The human pyramidal tract. II. A numerical investigation of the Betz cells of the motor area. *Arch. Neurol. Psychiat. (Chic.) 44,* 718–724.

Lassek, A. M. (1941). The human pyramidal tract. III. Magnitude of the large cells of the motor area (area 4). *Arch. Neurol. Psychiat. (Chic.) 45,* 964–972.

Lassek, A. M. (1942a). The pyramidal tract. The effect of pre- and postcentral cortical lesions on the fiber components of the pyramids in monkey. *J. Nerv. Ment. Dis. 95,* 721–729.

Lassek, A. M. (1942b). The human pyramidal tract. IV. A study of the mature, myelinated fibers of the pyramid. *J. Comp. Neurol. 76,* 217–225.

Lassek, A. M. (1944). The human pyramidal tract. X. The Babinski sign and destruction of the pyramidal tract. *Arch. Neurol. Psychiat. (Chic.) 52,* 484–494.

Lassek, A. M. (1945). The human pyramidal tract. XI. Correlations of the Babinski sign and the pyramidal syndrome. *Arch. Neurol. Psychiat. (Chic.) 53,* 375–377.

Lassek, A. M., and G. L. Rasmussen (1939). The human pyramidal tract. A fiber and numerical analysis. *Arch. Neurol. Psychiat. (Chic.) 42,* 872–876.

Lassek, A. M., and M. D. Wheatley (1945). The pyramidal tract. An enumeration of the large motor cells of area 4 and the axons in the pyramids of chimpanzee. *J. Comp. Neurol. 82*, 299–302.

Lateralization in the Nervous System (1977). (S. Harnad, R. W. Doty, L. Goldstein, J. Jaynes, and G. Krauthamer, eds.). Academic Press, New York.

Laties, A. M., and J. M. Sprague (1966). The projection of optic fibers to the visual centers in the cat. *J. Comp. Neurol. 127*, 35–70.

Laursen, A. M. (1963). Corpus striatum. *Acta Physiol. Scand. 59, Suppl. 211*, 1–106.

Laursen, A. M., and M. Wiesendanger (1967). The effect of pyramidal lesions on response latency in cats. *Brain Research 5*, 207–220.

LaVail, J. H. (1975). Retrograde cell degeneration and retrograde transport technique. In *The Use of Axonal Transport for Studies of Neuronal Connectivity* (W. M. Cowan and M. Cuénod, eds.), pp. 217–248. Elsevier, Amsterdam.

LaVail, J. H., and M. M. LaVail (1972). Retrograde axonal transport in the central nervous system. *Science 176*, 1416–1417.

LaVelle, A., and F. W. LaVelle (1958). Neuronal swelling and chromatolysis as influenced by the state of cell development. *Amer. J. Anat. 102*, 219–241.

LaVelle, A., and F. W. LaVelle (1959). Neuronal reaction to injury during development. Severance of the facial nerve *in utero. Exp. Neurol. 1*, 82–95.

Lawn, A. M. (1966). The localization, in the nucleus ambiguus of the rabbit, of the cells of origin of motor nerve fibers in the glossopharyngeal nerve and various branches of the vagus nerve by means of retrograde degeneration. *J. Comp. Neurol. 127*, 293–305.

Lawrence, D. G., and D. A. Hopkins (1976). The development of motor control in the rhesus monkey: evidence concerning the role of corticomotoneuronal connections. *Brain 99*, 235–254.

Lawrence, D. G., and H. G. J. M. Kuypers (1968a). The functional organization of the motor system in the monkey. I. The effects of bilateral pyramidal lesions. *Brain 91*, 1–14.

Lawrence, D. G., and H. G. J. M. Kuypers (1968b). The functional organization of the motor system in the monkey. II. The effects of lesions of the descending brain-stem pathways. *Brain 91*, 15–36.

Le Beau, J., M. Dondey, and D. Albe-Fessard (1962). Détermination de la fonction de certaines structures cérébrales profondes par refroidissement localisé et réversible. (Principles de la méthode, premières applications animales et humaines). *Rev. Neurol. 107*, 485–499.

Le Beau, J., and A. Petrie (1953). A comparison of the personality changes after (1) prefrontal selective surgery for the relief of intractable pain and for the treatment of mental cases; (2) cingulectomy and topectomy. *J. Ment. Sci. 99*, 53–61.

Leek, B. F. (1977). Abdominal and pelvic visceral receptors. *Brit. Med. Bull. 33*, 163–168.

Lees, F., and L. A. Liversedge (1962). Descending ocular myopathy. *Brain 85*, 701–710.

Leestma, J. E., and A. Noronha (1976). Pure motor hemiplegia, medullary pyramid lesion, and olivary hypertrophy. *J. Neurol. Neurosurg. Psychiat. 39*, 877–884.

Leibowitz, U. (1969). Epidemic incidence of Bell's palsy. *Brain 92*, 109–114.

Leichnetz, G. R., and J. Astruc (1976). The efferent projections of the medial prefrontal cortex in the squirrel monkey (*Saimiri sciureus*). *Brain Research 109*, 455–472.

Leionen, L., J. Hyvärinen, G. Nyman, and I. Linnankoski (1979). I. Functional properties of neurons in lateral part of associative area 7 in awake monkeys. *Exp. Brain Res. 34*, 299–320.

Leinonen, L., and G. Nyman (1979). II. Functional properties of cells in anterolateral part of area 7 associative face area of awake monkeys. *Exp. Brain Res. 34*, 321–333.

Leksell, L. (1945). The action potential and excitatory effects of the small ventral root fibres to skeletal muscle. *Acta Physiol. Scand. 10, Suppl. 31*, 1–84.

Lele, P. P., and G. Weddell (1956). The relationship between neurohistology and corneal sensibility. *Brain 79*, 119–154.

Lele, P. P., G. Weddell, and C. Williams (1954). The relationship between heat transfer, skin temperature and cutaneous sensibility. *J. Physiol. (Lond.) 126*, 206–234.

Lenn, N. J., and T. S. Reese (1966). The fine structure of nerve endings in the nucleus of the trapezoid body and the ventral cochlear nucleus. *J. Anat. (Lond.) 118,* 375–390.

Leonard, C. M. (1969). The prefrontal cortex of the rat. I. Cortical projection of the medio-dorsal nucleus. II. Efferent connections. *Brain Research 12,* 321–343.

Leonard, C. M., and J. W. Scott (1971). Origin and distribution of the amygdalofugal pathways in the rat: an experimental neuroanatomical study. *J. Comp. Neurol. 141,* 313–330.

Leontovich, T. A., and G. P. Zhukova (1963). The specificity of the neuronal structure and topography of the reticular formation in the brain and spinal cord of carnivora. *J. Comp. Neurol. 121,* 347–379.

Léránth, C., L. Záborszky, J. Marton, and M. Palkovits (1975). Quantitative studies on the supraoptic nucleus in the rat. I. Synaptic organization. *Exp. Brain Res. 22,* 509–523.

LeVay, S. (1971). On the neurons and synapses of the lateral geniculate nucleus of the monkey, and the effects of eye enucleation. Z. Zellforsch. 113, 369–419.

LeVay, S., D. H. Hubel, and T. N. Wiesel (1975). The pattern of ocular dominance columns in macaque visual cortex revealed by a reduced silver stain. *J. Comp. Neurol. 159,* 559–576.

Levin, P. M. (1936). The efferent fibers of the frontal lobe of the monkey, macaca mulatta. *J. Comp. Neurol. 63,* 369–419.

Levin, P. M., and F. K. Bradford (1938). The exact origin of the cortico-spinal tract in the monkey. *J. Comp. Neurol. 68,* 411–422.

Levitt, L. P., D. J. Selkoe, B. Frankenfield, and W. Schoene (1975). Pure motor hemiplegia secondary to brain-stem tumour. *J. Neurol. Neurosurg. Psychiat. 38,* 1240–1243.

Levitt, M., M. Carreras, C. N. Liu, and W. W. Chambers (1964). Pyramidal and extrapyramidal modulation of somatosensory activity in gracile and cuneate nuclei. *Arch. Ital. Biol. 102,* 197–229.

Lewin, R. J., G. V. Dillard, and R. W. Porter (1967). Extrapyramidal inhibition of the urinary bladder. *Brain Research 4,* 301–307.

Lewin, W. (1961). Observations on selective leucotomy. *J. Neurol. Neurosurg. Psychiat. 24,* 37–44.

Lewin, W., and C. G. Phillips (1952). Observations on partial removal of the post-central gyrus for pain. *J. Neurol. Neurosurg. Psychiat. 15,* 143–147.

Lewis, D., and W. E. Dandy (1930). The course of the nerve fibers transmitting sensation of taste. *Arch. Surg. 21,* 249–288.

Lewis, P. R., and C. C. D. Shute (1967). The cholinergic limbic system: projections to hippocampal formation, medial cortex, nuclei of the ascending cholinergic reticular system and the subfornical organ and supra-optic crest. *Brain 90,* 521–540.

Lewis, R., and G. S. Brindley (1965). The extrapyramidal cortical motor map. *Brain 88,* 397–406.

Lewis, T. (1942). *Pain.* Macmillan, New York.

Lewis, T., and J. H. Kellgren (1939). Observations relating to referred pain, visceromotor reflexes and other associated phenomena. *Clin. Sci. 4,* 47–71.

Lewis, V. A., and G. F. Gebhart (1977). Evaluation of the periaqueductal central gray (PAG) as a morphine-specific locus of action and examination of morphine-induced and stimulation-produced analgesia at coincident PAG loci. *Brain Research 124,* 283–303.

Lewy, F. H., and H. Kobrak (1936). The neural projection of the cochlear spirals on the primary acoustic centers. *Arch. Neurol. Psychiat. (Chic.) 35,* 839–852.

Leyton, A. S. F., and C. S. Sherrington (1917). Observations on the excitable cortex of the chimpanzee, orang-utan, and gorilla. Quart. J. Exp. Physiol. 11, 135–222.

Libet, B. (1973). Electrical stimulation of cortex in human subjects, and conscious sensory aspects. In *Handbook of Sensory Physiology,* Vol. II (A. Iggo, ed.), pp. 743–790. Springer-Verlag, Berlin.

Libet, B., W. W. Alberts, E. W. Wright, Jr., and B. Feinstein (1967). Responses of human somatosensory cortex to stimuli below threshold for conscious sensation. *Science 158,* 1597–1600.

Libet, B., W. W. Alberts, E. W. Wright, Jr., M. Lewis, and B. Feinstein (1975). Cortical representation of evoked potentials relative to conscious sensory responses, and of somatosensory qualities in man. In *The Somatosensory System* (H. H. Kornhuber, ed.), pp. 291–308. Georg Thieme Verlag, Stuttgart.

Liddell, E. G. T., and C. G. Phillips (1944). Pyramidal section in the cat. *Brain 67,* 1–9.

Liddell, E. G. T., and C. G. Phillips (1950). Thresholds of cortical representation. *Brain 73,* 125–140.

Lieberman, A. R. (1971). The axon reaction: a review of the principal features of peri-karyal responses to axon injury. *Int. Rev. Neurobiol. 14,* 49–124.

Lieberman, A. R., and K. E. Webster (1974). Aspects of the synaptic organization of intrinsic neurons in the dorsal lateral geniculate nucleus. An ultrastructural study of the normal and of the experimentally deafferented nucleus in the rat. *J. Neurocytol. 3,* 677–710.

Liebeskind, J. C., G. Guilbaud, G. Besson, and J.-L. Oliveras (1973). Analgesia from electrical stimulation of the periaqueductal gray matter in the cat: behavioral observations and inhibitory effects on spinal cord interneurons. *Brain Research 50,* 441–446.

Liedgren, S. R. C., K. Kristensson, B. Larsby, and L. M. Ødkvist (1976). Projection of thalamic neurons to cat primary vestibular cortical fields studied by means of retrograde axonal transport of horseradish peroxidase. *Exp. Brain Res. 24,* 237–243.

Light, A. R., and E. R. Perl (1979a). Reexamination of the dorsal root projection to the spinal dorsal horn including observations on the differential termination of coarse and fine fibers. *J. Comp. Neurol. 186,* 117–132.

Light, A. R., and E. R. Perl (1979b). Spinal termination of functionally identified primary afferent neurons with slowly conducting myelinated fibers. *J. Comp. Neurol. 186,* 133–150.

Light, A. R., D. L. Trevino, and E. R. Perl (1979). Morphological features of functionally defined neurons in the marginal zone and substantia gelatinosa of the spinal dorsal horn. *J. Comp. Neurol. 186,* 151–172.

Limbic and Autonomic Nervous Systems Research (1974). (L. V. DiCara, ed.). Plenum Press, New York.

Limbic System Mechanics and Autonomic Function (1972). (Ch. C. Hockman, ed.). C. C. Thomas, Springfield, Ill.

Lin, C.-S., M. M. Merzenich, M. Sur, and J. H. Kaas (1979). Connections of areas 3b and 1 of the parietal somatosensory strip with the ventroposterior nucleus in the owl monkey (*Aotus trivirgatus*). *J. Comp. Neurol. 185,* 355–372.

Lin, C. S., E. Wagor, and J. H. Kaas (1974). Projections from the pulvinar to the middle temporal visual area (MT) in the owl monkey, *Actus trivirgatus. Brain Research 76,* 145–149.

Linauts, M., and G. F. Martin (1978a). The organization of olivo-cerebellar projections in the opposum, *Didelphis virginiana,* as revealed by the retrograde transport of horseradish peroxidase. *J. Comp. Neurol. 179,* 355–382.

Linauts, M., and G. F. Martin (1978b). An autoradiographic study of midbrain-dien-cephalic projections to the inferior olivary nucleus in the opossum (*Didelphis virginiana*). *J. Comp. Neurol. 179,* 325–354.

Lindblom, K., and B. Rexed (1948). Spinal nerve injury in dorsolateral protrusions of lumbar discs. *J. Neurosurg. 5,* 413–432.

Lindblom, U. F., and J. O. Ottosson (1956). Bulbar influence on spinal cord dorsum potentials and ventral root reflexes. *Acta Physiol. Scand. 35,* 203–214.

Lindblom, U. F., and J. O. Ottosson (1957). Influence of pyramidal stimulation upon the relay of coarse cutaneous afferents in the dorsal horn. *Acta Physiol. Scand. 38,* 309–318.

Lindeman, H. H. (1969). Studies on the morphology of the sensory regions of the vestibular apparatus. *Ergebn. Anat. Entwickl.-Gesch. 42,* 1–113.

Lindeman, H. H. (1973). Anatomy of the otolith organs. *Advan. Oto-rhino-laryng. 20,* 405–433.

Lindeman, H. H., H. W. Ades, G. Bredberg, and H. Engström (1971). The sensory hairs and the tectorial membrane in the development of the cat's organ of Corti. *Acta Otolaryng. (Stockh.) 72,* 229–242.

Lindgren, I., and H. Olivecrona (1947). Surgical treatment of angina pectoris. *J. Neurosurg. 4,* 19–39.

Lindsley, D. B., L. H. Schreiner, W. B. Knowles, and H. W. Magoun (1950). Behavioral and EEG changes following chronic brain stem lesions in the cat. *Electroenceph. Clin. Neurophysiol. 2,* 483–498.

Lindström, S., and M. Takata (1972). Monosynaptic excitation of dorsal spinocerebellar tract neurones from low threshold joint afferents. *Acta Physiol. Scand. 84,* 430–432.

Lindvall, O., A. Björklund, R. Y. Moore, and U. Stenevi (1974). Mesencephalic dopamine neurons projecting to neocortex. *Brain Research 81,* 325–331.

List, C. F., and M. M. Peet (1939). Sweat secretion in man. V. Disturbances of sweat secretion with lesions of the pons, medulla and cervical portion of the cord. *Arch. Neurol. Psychiat. (Chic.) 42,* 1098–1127.

List, C. F., and J. R. Williams (1957). Pathogenesis of trigeminal neuralgia. A review. *Arch. Neurol. Psychiat. (Chic.) 77,* 36–43.

Liu, C-N. (1956). Afferent nerves to Clarke's and the lateral cuneate nuclei in the cat. *Arch. Neurol. Psychiat. (Chic.) 75,* 67–77.

Liu, C-N., and W. W. Chambers (1958). Intraspinal sprouting of dorsal root axons. *Arch. Neurol. Psychiat. (Chic.) 79,* 46–61.

Liu, C-N., and W. W. Chambers (1964). An experimental study of the corticospinal system in the monkey (*Macaca mulatta*). The spinal pathways and preterminal distribution of degenerating fibers following discrete lesions of the pre- and postcentral gyri and bulbar pyramid. *J. Comp. Neurol. 123,* 257–284.

Ljungberg, T., and U. Ungerstedt (1976). Sensory inattention produced by 6-hydroxy-dopamine-induced degeneration of ascending dopamine neurons in the brain. *Exp. Neurol. 53,* 585–600.

Ljungdahl, Å., T. Hökfelt, M. Goldstein, and D. Park (1975). Retrograde peroxidase tracing of neurons, combined with transmitter histochemistry. *Brain Research 84,* 313–319.

Llamas, A., C. Avendaño, and F. Reinoso-Suárez (1977). Amygdaloid projections to prefrontal and motor cortex. *Science 195,* 794–796.

Llamas, A., F. Reinoso-Suárez, and E. Martinez-Moreno (1975). Projections to the gyrus proreus from the brain stem tegmentum (locus coeruleus, raphe nuclei) in the cat, demonstrated by retrograde transport of horseradish peroxidase. *Brain Research 89,* 331–336.

Llinás, R., R. Baker, and C. Sotelo (1974). Electrotonic coupling between neurons in cat inferior olive. *J. Neurophysiol. 37,* 560–571.

Llinás, R., and R. A. Volkind (1973). The olivocerebellar system: functional properties as revealed by harmaline-induced tremor. *Exp. Brain Res. 18,* 69–87.

Lloyd, D. P. C. (1941). The spinal mechanism of the pyramidal system in cats. *J. Neurophysiol. 4,* 525–546.

Locke, S. (1960). The projection of the medial pulvinar of the macaque. *J. Comp. Neurol. 115,* 155–167.

Locke, S. (1961). The projection of the magnocellular medial geniculate body. *J. Comp. Neurol. 116,* 179–193.

Locke, S. (1967). Thalamic connections to insular and opercular cortex of monkey. *J. Comp. Neurol. 129,* 219–240.

Locke, S., J. B. Angevine, Jr., and P. I. Yakovlev (1964). Limbic nuclei of thalamus and connections of limbic cortex. VI. Thalamocortical projection of lateral dorsal nucleus in cats and monkeys. *Arch. Neurol. (Chic.) 11,* 1–12.

Loe, P. R., B. L. Whitsel, D. A. Dreyer, and C. B. Metz (1977). Body representation in ventrobasal thalamus of macaque: a single-unit analysis. *J. Neurophysiol. 40,* 1339–1355.

Loeb, G. E. (1976). Ventral root projections of myelinated dorsal root ganglion cells in the cat. *Brain Research 106,* 159–165.

Loeb, C., and G. Poggio (1953). Electroencephalograms in a case with pontomesen-cephalic haemorrhage. *Electroenceph. clin. Neurophysiol. 5*, 295–296.

Loeser, J. D., and E. C. Alvord, Jr. (1968). Agenesis of the corpus callosum. *Brain 91*, 553–570.

Loewy, A. D., J. C. Araujo, and F. W. L. Kerr (1973). Pupillodilator pathways in the brain stem of the cat. Anatomical and electrophysiological identification of a central autonomic pathway. *Brain Research 60*, 65–91.

Loewy, A. D., and H. Burton (1978). Nuclei of the solitary tract: efferent projections to the lower brain stem and spinal cord of the cat. *J. Comp. Neurol. 181*, 421–450.

Loewy, A. D., and C. B. Saper (1978). Edinger-Westphal nucleus: projections to the brain stem and spinal cord in the cat. *Brain Research 150*, 1–27.

Loewy, A. D., C. B. Saper, and N. D. Yamodis (1978). Re-evaluation of the efferent projections of the Edinger-Westphal nucleus in the cat. *Brain Research 141*, 153–159.

Lohman, A. H. M. (1963). The anterior olfactory lobe of the guinea pig. A descriptive and experimental anatomical study. *Acta. Anat. (Basel) 53, Suppl. 49*, 1–109.

Lohman, A. H. M., and G. M. Mentink (1969). The lateral olfactory tract, the anterior commissure and the cells of the olfactory bulb. *Brain Research 12*, 396–413.

Løken, Aa. C., and A. Brodal (1970). A somatotopical pattern in the human lateral ves-tibular nucleus. *Arch. Neurol. (Chic.) 23*, 350–357.

Long, D. M., A. S. Leonard, S. N. Chou, and L. A. French (1962). Hypothalamus and gastric ulceration. I. Gastric effects of hypothalamic lesions. *Arch. Neurol. (Chic.) 7*, 167–175.

Long, D. M., A. S. Leonard, J. Story, and L. A. French (1962). Hypothalamus and gastric ulceration. II. Production of gastrointestinal ulceration by chronic hypothalamic stimu-lation. *Arch. Neurol. (Chic.) 7*, 176–183.

Lopes da Silva, F. H., and D. E. A. T. Arnolds (1978). Physiology of the hippocampus and related structures. *Ann. Rev. Physiol. 40*, 185–216.

Lorens, S. A., and H. C. Guldberg (1974). Regional 5-hydroxytryptamine following selec-tive midbrain raphe lesions in the rat. *Brain Research 78*, 45–56.

Lorente de Nó, R. (1933a). Anatomy of the eighth nerve. I. The central projection of the nerve endings of the internal ear. *Laryngoscope (St. Louis) 43*, 1–38.

Lorente de Nó, R. (1933b). Anatomy of the eighth nerve. III. General plan of structure of the primary cochlear nuclei. *Laryngoscope (St. Louis) 43*, 327–350.

Lorente de Nó, R. (1933c). Studies on the structure of the cerebral cortex. I. The area en-torhinalis. *J. Psychol. Neurol. (Lpz.) 45*, 381–438.

Lorente de Nó, R. (1934). Studies on the structure of the cerebral cortex. II. Continuation of the study of the Amnonic system. *J. Psychol. Neurol. (Lpz.) 46*, 117–177.

Lorente de Nó, R. (1949). Cerebral cortex: architecture, intracortical connections, motor projections. In Fulton's *Physiology of the Nervous System*, 3rd ed., pp. 288–312. Ox-ford Univ. Press, New York.

Lorente de Nó, R. (1976). Some unresolved problems concerning the cochlear nerve. *Ann. Otol. (St. Louis) 85*, 1–28.

Lowenstein, O. (1966). The functional significance of the ultrastructure of the vestibular end organs. In *NASA SP-115*. Second symposium on the role of the vestibular organs in space exploration, pp. 73–87. Ames Research Center, Moffett Field, Calif.

Lowenstein, O. (1967). Functional aspects of vestibular structure. In *Myotatic and Ves-tibular Mechanisms*, Ciba Foundation Symposium (A. V. D. de Reuck and J. Knight, eds.), pp. 121–128. J & A. Churchill, London.

Löwenstein, O., and A. Sand (1940). The mechanism of the semicircular canal. A study of the responses of single-fibre preparations to angular accelerations and to rotation at constant speed. *Proc. Roy. Soc. B 129*, 256–275.

Lucas Keene, M. F. (1961). Muscle spindles in the human laryngeal muscles. *J. Anat. (Lond.) 95*, 25–29.

Lund, J. S. (1973). Organization of neurons in the visual cortex, area 17, of the monkey. *J. Comp. Neurol. 147*, 455–496.

Lund, J. S., and R. D. Lund (1970). The termination of callosal fibers in the paravisual cortex of the rat. *Brain Research 17*, 25–45.

Lund, J. S., R. D. Lund, A. E. Hendrickson, A. H. Bunt, and A. F. Fuchs (1975). The origin of efferent pathways from the primary visual cortex, area 17, of the macaque monkey as shown by retrograde transport of horseradish peroxidase. *J. Comp. Neurol. 164*, 287–304.

Lund, R. D. (1972). Synaptic patterns in the superficial layers of the superior colliculus of the monkey, *Macaca mulatta*. *Exp. Brain Res. 15*, 194–211.

Lund, R. D., and J. S. Lund (1971). Synaptic adjustment after deafferentation of the superior colliculus. *Science 171*, 804–807.

Lund, R. D., and K. E. Webster (1967a). Thalamic afferents from the dorsal column nuclei. An experimental anatomical study in the rat. *J. Comp. Neurol. 130*, 301–312.

Lund, R. D., and K. E. Webster (1967b). Thalamic afferents from the spinal cord and trigeminal nuclei. An experimental anatomical study in the rat. *J. Comp. Neurol. 130*, 313–327.

Lund, R. D., and L. E. Westrum (1966). Neurofibrils and the Nauta method. *Science 151*, 1397–1399.

Lund, S., and O. Pompeiano (1965). Descending pathways with monosynaptic action on motoneurones. *Experientia (Basel) 21*, 602–603.

Lund, S., and O. Pompeiano (1968). Monosynaptic excitation of alpha motoneurones from supraspinal structures in the cat. *Acta Physiol. Scand. 73*, 1–21.

Lundberg, A. (1966). Integration in the reflex pathway. In *Nobel Symposium I: Muscular Afferents and Motor Control* (R. Granit, ed.), pp. 275–305. Almqvist & Wiksell, Stockholm; Wiley, New York.

Lundberg, A. (1967). The supraspinal control of transmission in spinal reflex pathways. In *Recent Advances in Clinical Neurophysiology* (L. Widén, ed.), *Electroenceph. Clin. Neurophysiol. Suppl. 25*, 35–46.

Lundberg, A. (1975). Control of spinal mechanisms from the brain. In *The Nervous System*, Vol 1: *The Basic Neurosciences* (D. B. Tower, ed.), pp. 253–265. Raven Press, New York.

Lundberg, A., U. Norsell, and P. Voorhoeve (1962). Pyramidal effects on lumbo-sacral interneurones activated by somatic afferents. *Acta Physiol. Scand. 56*, 220–229.

Lundberg, A., and O. Oscarsson (1960). Functional organization of the dorsal spino-cerebellar tract in the cat. VII. Identification of units by antidromic activation from the cerebellar cortex with recognition of five functional subdivisions. *Acta Physiol. Scand. 50*, 356–374.

Lundberg, A., and P. Voorhoeve (1962). Effects from the pyramidal tract on spinal reflex arcs. *Acta Physiol. Scand. 56*, 201–219.

Lundberg, P. O. (1957). A study of neurosecretory and related phenomena in the hypothalamus and pituitary of man. *Acta Morph. Neerl.-Scand. 1*, 256–285.

Lundberg, P. O. (1960). Cortico-hypothalamic connexions in the rabbit. An experimental neuro-anatomical study. *Acta Physiol. Scand. 49, Suppl. 171*, 1–80.

Lundervold, A., T. Hauge, and Aa. C. Løken (1956). Unusual EEG in unconscious patient with brain stem atrophy. *Electroenceph. Clin. Neurophysiol. 8*, 665–670.

Lundgren, O., and M. Jodal (1975). Regional blood flow. *Ann. Rev. Physiol. 37*, 395–414.

Lynch, G., S. Deadwyler, and C. Cotman (1973). Post-lesion axonal growth produces permanent functional connections. *Science 180*, 1364–1366.

Lynch, G., C. Gall, P. Mensah, and C. W. Cotman (1974). Horseradish peroxidase histochemistry: a new method for tracing efferent projections in the central nervous system. *Brain Research 65*, 373–380.

Lynch, G., R. L. Smith, and R. Robertson (1973). Direct projections from brainstem to telencephalon. *Exp. Brain Res. 17*, 221–228.

Lynch, G., B. Stanfield, and C. W. Cotman (1973). Developmental differences in post-lesion axonal growth in the hippocampus. *Brain Research 59*, 155–168.

Lynch, G., B. Stanfield, Th. Parks, and C. W. Cotman (1974). Evidence for selective postlesion axonal growth in the dentate gyrus of the rat. *Brain Research 69*, 1–11.

Lynch, J. C., V. B. Mountcastle, N. H. Talbot, and T. C. T. Yin (1977). Parietal lobe mechanisms for directed visual attention. *J. Neurophysiol. 40*, 362–389.

Lynn, B. (1975). Somatosensory receptors and their CNS connections. *Ann. Rev. Physiol. 37*, 105–127.

Lynn, B. (1977). Cutaneous hyperalgesia. *Brit. Med. Bull. 33*, 103–108.

Lyon, M. J. (1975). Localization of the efferent neurons to the tensor tympani muscle of the newborn kitten using horseradish peroxidase. *Exp. Neurol. 49*, 439–455.

Lyon, M. J. (1978). The central location of the motor neurons to the stapedius muscle in the cat. *Brain Research 143*, 437–444.

Mabuchi, M., and T. Kusama (1966). The corticorubral projection in the cat. *Brain Research 2*, 254–273.

Mabuchi, M., and T. Kusama (1970). Mesodiencephalic projections to the inferior olive and the vestibular and perihypoglossal nuclei. *Brain Research 17*, 133–136.

Macchi, G. (1958). Organizzazione morfologica delle connessioni thalamo-corticali. *Monit. Zool. Ital. Suppl. 66*, 25–121.

Macchi, G., F. Angeleri, and G. Guazzi (1959). Thalamo-cortical connections of the first and second somatic sensory areas in the cat. *J. Comp. Neurol. 111*, 387–405.

Macchi, G., M. Bentivoglio, C. D'Atena, P. Rossini, and E. Tempesta (1977). The cortical projections of the thalamic intralaminar nuclei restudied by means of the HRP retrograde axonal transport. *Neurosci. Letters 4*, 121–126.

Macchi, G., M. Bentivoglio, D. Minciacchi, P. Rossini, and E. Tempesta (1978). The claustro-cortical projections: a HRP study in the cat. *Neurosci. Letters. Suppl. 1*, 166.

Macchi, G., and E. Rinvik (1976). Thalamotelencephalic circuits: a neuroanatomical survey. In *Handbook of Electroencephalography and Clinical Neurophysiology. Vol. 2, Electrical Activity from the Neuron to the EEG and EMG. Part A. Morphological Basis of EEG Mechanisms* (O. Creutzfeldt, ed., A. Rémond, ed.-in-chief), pp. 86–133. Elsevier, Amsterdam.

Machne, X., and J. P. Segundo (1956). Unitary responses to afferent volleys in amygdaloid complex. *J. Neurophysiol. 19*, 232–240.

Maciewicz, R. J. (1975). Thalamic afferents to areas 17, 18 and 19 of cat cortex traced with horseradish peroxidase. *Brain Research 84*, 308–312.

Maciewicz, R. J., K. Eagen, C. R. S. Kaneko, and S. M. Highstein (1977). Vestibular and medullary brain stem afferents to the abducens nucleus in the cat. *Brain Research 123*, 229–240.

Maciewicz, R. J., C. R. S. Kaneko, S. M. Highstein, and R. Baker (1975). Morphophysiological identification of interneurons in the oculomotor nucleus that project to the abducens nucleus in the cat. *Brain Research 96*, 60–65.

McBride, R. L., and J. Sutin (1977). Amygdaloid and pontine projections to the ventromedial nucleus of the hypothalamus. *J. Comp. Neurol. 174*, 377–396.

McComas, A. J., P. R. W. Fawcett, M. J. Campbell, and R. E. P. Sica (1971). Electrophysiological estimation of number of motor units within a human muscle. *J. Neurol. Neurosurg. Psychiat. 34*, 121–131.

McComas, A. J., R. E. P. Sica, A. R. M. Upton, D. Longmire, and M. R. Caccia (1972). Physiological estimates of the numbers and sites of motor units in man. *Advanc. Behav. Biol. 7*, 55–72.

McCouch, G. P., J. D. Deering, and T. H. Ling (1951). Location of receptors for tonic neck reflexes. *J. Neurophysiol. 14*, 191–195.

McCouch, G. P., C-N. Liu, and W. W. Chambers (1966). Descending tracts and spinal shock in the monkey (*Macaca mulatta*). *Brain 89*, 359–376.

McCouch, G. P., C.-N. Liu, W. W. Chambers, and J. Yu (1970). Recurrent inhibition and facilitation in the monkey, *Macaca mulatta*, in relation to spinal shock. *Exp. Neurol. 29*, 92–100.

McCrea, R. A., G. A. Bishop, and S. T. Kitai (1976). Intracellular staining of Purkinje cells and their axons with horseradish peroxidase. *Brain Research 118*, 132–136.

McCrea, R. A., G. A. Bishop, and S. T. Kitai (1977). Electrophysiological and horseradish peroxidase studies of precerebellar afferents to the nucleus interpositus anterior. II. Mossy fiber system. *Brain Research 122*, 215–228.

McCrea, R. A., G. A. Bishop, and S. T. Kitai (1978). Morphological and electrophysiological characteristics of projection neurons in the nucleus interpositus of the cat cerebellum. *J. Comp. Neurol. 181*, 397–420.

McCulloch, W. S. (1944). Cortico-cortical connections. In *The Precentral Motor Cortex* (P. C. Bucy, ed.), pp. 211–242. Univ. of Illinois Press, Urbana.

McDonald, D. M., and G. L. Rasmussen (1971). Ultrastructural characteristics of synaptic endings in the cochlear nucleus having acetylcholinesterase activity. *Brain Research 28*, 1–18.

McGeer, E. G., P. L. McGeer, D. S. Grewaal, and V. K. Singh (1975). Striatal cholinergic interneurons and their relation to dopaminergic nerve endings. *J. Pharmacol. (Paris) 6*, 143–152.

McGeer, P. L., and E. G. McGeer (1976). Enzymes associated with the metabolism of catecholamines, acetylcholine and GABA in human controls and patients with Parkinson's disease and Huntington's chorea. *J. Neurochem. 26*, 65–76.

McGeer, P. L., E. G. McGeer, and H. C. Fibiger (1973). Cholineacetylase and glutamic acid decarboxylase in Huntington's chorea. A preliminary study. *Neurology (Minneap.) 23*, 912–917.

McGeer, P. L., E. G. McGeer, H. C. Fibiger, and V. Wickson (1971). Neostriatal choline acetylase and acetylcholinesterase following selective brain lesions. *Brain Research 35*, 308–314.

McGeer, P. L., T. Hattori, V. K. Singh, and E. G. McGeer (1976). Cholinergic systems in extrapyramidal function. In *The Basal Ganglia* (M. D. Yahr, ed.), pp. 213–222. Raven Press, New York.

McIlwain, J. T. (1973). Topographic relationships in projection from striate cortex to superior colliculus of the cat. *J. Neurophysiol. 36*, 690–701.

McIlwain, J. T., and P. Buser (1968). Receptive fields of single cells in the cat's superior colliculus. *Exp. Brain Res. 5*, 314–325.

McIntyre, A. K. (1962). Cortical projection of impulses in the interosseus nerve of the cat's hind limb. *J. Physiol. (Lond.) 163*, 46–60.

McIntyre, A. K., M. E. Holman, and J. L. Veale (1967). Cortical responses to impulses from single Pacinian corpuscles in the cat's hind limb. *Exp. Brain Res. 4*, 243–255.

MacKay, D. M. (1967). The human brain. *Science Journal 3*, 43–47.

McKenzie, J. S., D. M. Gilbert, and D. K. Rogers (1971). Hippocampal and neostriatal inhibition of extra-lemniscal thalamic unitary responses in the cat. *Brain Research 27*, 382–395.

McKinley, W. A., and H. W. Magoun (1942). The bulbar projection of the trigeminal nerve. *Amer. J. Physiol. 137*, 217–224.

McKissock, W., and K. W. E. Paine (1958). Primary tumours of the thalamus. *Brain 81*, 41–63.

McLardy, T. (1948). Projection of the centromedian nucleus of the human thalamus. *Brain 71*, 290–303.

McLardy, T. (1950). Thalamic projections to frontal cortex in man. *J. Neurol. Neurosurg. Psychiat. 13*, 198–202.

McLardy, T. (1955a). Observations on the fornix of the monkey. I. Cell studies. *J. Comp. Neurol. 103*, 305–326.

McLardy, T. (1955b). Observations on the fornix of the monkey. II. Fiber studies. *J. Comp. Neurol. 103*, 327–344.

McLaughlin, B. J. (1972a). The fine structure of neurons and synapses in the motor nuclei of the cat spinal cord. *J. Comp. Neurol. 144*, 429–460.

McLaughlin, B. J. (1972b). Dorsal root projections to the motor nuclei in the cat spinal cord. *J. Comp. Neurol. 144*, 461–474.

McLaughlin, B. J. (1972c). Propriospinal and supraspinal projections to the motor nuclei in the cat spinal cord. *J. Comp. Neurol. 144,* 475–500.

MacLean, P. D. (1975). An ongoing analysis of hippocampal inputs and outputs: microelectrode and neuroanatomical findings in squirrel monkeys. In *The Hippocampus,* Vol. 1 (R. L. Isaacson and K. H. Pribram, eds.), pp. 177–211. Plenum Press, New York.

McLeod, J. G., and S. H. Wray (1967). Conduction velocity and fibre diameter of the median and ulnar nerves of the baboon. *J. Neurol. Neurosurg. Psychiat. 30,* 240–247.

McMahan, U. J. (1967). Fine structure of synapses in the dorsal nucleus of the lateral geniculate body of normal and blinded rats. *Z. Zellforsch. 76,* 116–146.

McMasters, R. E., A. H. Weiss, and M. B. Carpenter (1966). Vestibular projections to the nuclei of the extraocular muscles. Degeneration resulting from discrete partial lesions of the vestibular nuclei in the monkey. *Amer. J. Anat. 118,* 163–194.

McNaughton, F. L., and W. H. Feindel (1977). Innervation of intracranial structures. A reappraisal. In *Physiological Aspects of Clinical Neurology* (F. Clifford Rose, ed.), pp. 279–293. Blackwell, London.

Maeda, T., and N. Shimizu (1972). Projections ascendantes du locus coeruleus et d'autres neurones aminergiques pontiques au niveau du prosencéphale du rat. *Brain Research 36,* 19–35.

Maekawa, K., and J. I. Simpson (1973). Climbing fiber responses evoked in vestibulocerebellum of rabbit from visual system. *J. Neurophysiol. 36,* 649–666.

Maekawa, K., and T. Takeda (1975). Mossy fiber responses evoked in the cerebellar flocculus of rabbits by stimulation of the optic pathway. *Brain Research 98,* 590–595.

Maffei, L., and O. Pompeiano (1962a). Cerebellar control of flexor motoneurons. An analysis of the postural responses to stimulation of the paramedian lobule in the decerebrate cat. *Arch. Ital. Biol. 100,* 476–509.

Maffei, L., and O. Pompeiano (1962b). Effects of stimulation of the mesencephalic tegmentum following interruption of the rubrospinal tract. *Arch. Ital. Biol. 100,* 510–525.

Magalhães-Castro, H. H., A. D. de Lima, P. E. S. Saraiva, and B. Magalhães-Castro (1978). Horseradish peroxidase labeling of cat tectotectal cells. *Brain Research 148,* 1–13.

Magalhães-Castro, H. H., P. E. S. Saraiva, and B. Magalhães-Castro (1975). Identification of corticotectal cells of the visual cortex of cats by means of horseradish peroxidase. *Brain Research 83,* 474–479.

Magladery, J. W., and R. D. Teasdall (1961). Corneal reflexes. An electromyographic study in man. *Arch. Neurol. (Chic.) 5,* 269–274.

Magnes, J., G. Moruzzi, and O. Pompeiano (1961). Synchronization of the EEG produced by low-frequency electrical stimulation of the region of the solitary tract. *Arch. Ital. Biol. 99,* 33–67.

Magni, F., R. Melzack, G. Moruzzi, and C. J. Smith (1959). Direct pyramidal influences on the dorsal-column nuclei. *Arch. Ital. Biol. 97,* 357–377.

Magni, F., F. Moruzzi, G. F. Rossi, and Z. Zanchetti (1959). EEG arousal following inactivation of the lower brain stem by selective injection of barbiturate into the vertebral circulation. *Arch. Ital. Biol. 97,* 33–46.

Magni, F., and W. D. Willis (1963). Identification of reticular formation neurons by intracellular recording. *Arch. Ital. Biol. 101* 681–702.

Magni, F., and W. D. Willis (1964a). Cortical control of brain stem reticular neurons. *Arch. Ital. Biol. 102,* 418–433.

Magni, F., and W. D. Willis (1964b). Subcortical and peripheral control of brain stem reticular neurones. *Arch. Ital. Biol. 102,* 434–448.

Magoun, H. W., and W. A. McKinley (1942). The termination of ascending trigeminal and spinal tracts in the thalamus of the cat. *Amer. J. Physiol. 137,* 409–416.

Magoun, H. W., and S. W. Ranson (1935). The afferent path of the light reflex. A review of the literature. *Arch. Ophthal. 13,* 862–874.

Magoun, H. W., and S. W. Ranson (1939). Retrograde degeneration of the supraoptic nuclei after section of the infundibular stalk in the monkey. *Anat. Rec. 75,* 107–124.

Magoun, H. W., and R. Rhines (1946). An inhibitory mechanism in the bulbar reticular formation. *J. Neurophysiol. 9*, 165–171.

Majorossy, K., and A. Kiss (1976a). Specific patterns of neuron arrangement and of synaptic articulation in the medial geniculate body. *Exp. Brain Res. 26*, 1–17.

Majorossy, K., and A. Kiss (1976b). Types of interneurons and their participation in the neuronal network of the medial geniculate body. *Exp. Brain Res. 26*, 19–37.

Makara, G. B., M. C. Harris, and M. Spyer (1972). Identification and distribution of tuberoinfundibular neurones. *Brain Research 40*, 283–290.

Makara, G. B., and L. Hodács (1975). Rostral projections from the hypothalamic arcuate nucleus. *Brain Research 84*, 23–29.

Makous, W., S. Nord, B. Oakley, and C. Pfaffmann (1963). The gustatory relay in the medulla. In *Olfaction and Taste*. Proceedings First International Symposium, Wenner-Gren Center 1962 (Y. Zotterman, ed.), pp. 381–393. Pergamon Press, Oxford.

Malamud, N. (1966). The epileptogenic focus in temporal lobe epilepsy from a pathological standpoint. *Arch. Neurol. (Chic.) 14*, 190–195.

Maler, L., H. C. Fibiger, and P. L. McGeer (1973). Demonstration of the nigrostriatal projection by silver staining after nigral injections of 6-hydroxydopamine. *Exp. Neurol. 40*, 505–515.

Malis, L. I., K. H. Pribram, and L. Kruger (1953). Action potentials in motor cortex evoked by peripheral nerve stimulation. *J. Neurophysiol. 16*, 161–167.

Mannen, H. (1975). Reconstruction of axonal trajectory of individual neurons in the spinal cord using Golgi-stained serial sections. *J. Comp. Neurol. 159*, 357–374.

Mannen, H., and Y. Sugiura (1976). Reconstruction of neurons of dorsal horn proper using Golgi-stained serial sections. *J. Comp. Neurol. 168*, 303–312.

Manni, E. (1950). Localizzazioni cerebellari corticali nella cavia. Nota 2a: Effetti di lesioni della "parti vestibolari" del cervelletto. *Arch. Fisiol. 50*, 1–14.

Manni, E., G. Palmieri, and R. Marini (1972). Pontine trigeminal termination of proprioceptive afferents from the eye muscles. *Exp. Neurol. 36*, 310–318.

Manni, E., G. Palmieri, and R. Marini (1974). Central pathway of the extraocular muscle proprioception. *Exp. Neurol. 42*, 181–190.

Manni, E., and V. E. Pettorossi (1976). Somatotopic localization of the eye muscle afferents in the semilunar ganglion. *Arch. Ital. Biol. 114*, 178–187.

Marani, E., J. Voogd, and A. Boekee (1977). Acetylcholinesterase staining in subdivisions of the cat's inferior olive. *J. Comp. Neurol. 174*, 209–226.

Marchesi, G. F., and P. Strata (1970). Climbing fibers of cat cerebellum: modulation of activity during sleep. *Brain Research 17*, 145–148.

Marg, E. (1973). Neurophysiology of the accessory optic system. In *Handbook of Sensory Physiology, Vol. VII/3 Central Processing of Visual Information, part B: Morphology and Function of Visual Centers in the Brain* (R. Jung, ed.), pp. 103–111. Springer Verlag, Berlin.

Marie, P., C. Foix, and T. Alajouanine (1922). De l'atrophie cérébelleuse tardive à prédominance corticale. *Rev. Neurol. 38*, 849–885, 1082–1111.

Marie, P., and G. Guillain (1903). Lésion ancienne du noyau rouge. Dégénérations secondaires. *Nouv. Iconogr. Salpêt. 16*, 80–83.

Marin, O. S. M., J. B. Angevine, Jr., and S. Locke (1962). Topographical organization of the lateral segment of the basis pedunculi in man. *J. Comp. Neurol. 118*, 165–183.

Marini, G., L. Provini, and A. Rosina (1975). Macular input to the cerebellar nodulus. *Brain Research 99*, 367–371.

Marin-Padilla, M. (1967). Number and distribution of the apical dendritic spines of the layer V pyramidal cells in man. *J. Comp. Neurol. 131*, 475–489.

Marin-Padilla, M. (1968). Cortical axo-spinodentric synapses in man: a Golgi study. *Brain Research 8*, 196–200.

Marin-Padilla, M. (1969). Origin of the pericellular basket of the pyramidal cells of the human motor cortex. A Golgi study. *Brain Research 14*, 633–646.

Marin-Padilla, M. (1970). Prenatal and early postnatal ontogenesis of the human motor cortex: a Golgi study. II. The basket-pyramidal system. *Brain Research 23*, 185–191.

Marin-Padilla, M. (1972). Double origin of the pericellular baskets of the pyramidal cells of the human motor cortex: a Golgi study. *Brain Research 38,* 1–12.

Marin-Padilla, M., and G. R. Stibitz (1974). Three-dimensional reconstruction of the basket cell of the human motor cortex. *Brain Research 70,* 511–514.

Mark, R. (1974). *Memory and Nerve Cell Connections. Criticisms and Contributions from Developmental Neurophysiology.* Clarendon Press, Oxford.

Mark, V. H., F. R. Ervin, and T. P. Hackett (1960). Clinical aspects of stereotactic thalamotomy. Part I. The treatment of severe pain. *Arch. Neurol. (Chic.) 3,* 351–367.

Mark, V. H., F. R. Ervin, and P. I. Yakovlev (1963). Stereotactic thalamotomy. III. The verification of anatomical lesion sites in the human thalamus. *Arch. Neurol. (Chic.) 8,* 528–538.

Mark, V. H., and I. J. Ordia (1976). The controversies over the use of neurosurgery in aggressive states and an assessment of the critics of this kind of surgery. In *Current Controversies in Neurosurgery* (T. P. Morley, ed.), pp. 722–729. W. B. Saunders, Philadelphia.

Markham, C. H., W. J. Brown, and R. W. Rand (1966). Stereotaxic lesions in Parkinson's disease. Clinicopathological correlations. *Arch. Neurol. (Chic.) 15,* 480–497.

Markham, C. H., and R. W. Rand (1963). Stereotactic surgery in Parkinson's disease. *Arch. Neurol. (Chic.) 8,* 621–631.

Marquardsen, J., and B. Harvald (1964). The electroencephalogram in acute vascular lesions of the brain stem and the cerebellum. *Acta Neurol. Scand. 40,* 58–68.

Marsden, C. D. (1976). Dystonia: the spectrum of the disease. In *The Basal Ganglia* (M. D. Yahr, ed.), pp. 351–367. Raven Press, New York.

Marsden, C. D., J. C. Meadows, and H. J. F. Hodgson (1969). Observations on the reflex response to muscle vibration and its voluntary control. *Brain 92,* 829–946.

Marsden, C. D., P. A. Merton, and H. B. Morton (1972). Servo action in human voluntary movement. *Nature 238,* 140–143.

Marsden, C. D., P. A. Merton, and H. B. Morton (1973). Is the human stretch reflex cortical rather than spinal? *Lancet 1,* 759–761.

Marsden, C. D., P. A. Merton, and H. B. Morton (1976a). Servo action of the human thumb. *J. Physiol. (Lond.) 257,* 1–44.

Marsden, C. D., P. A. Merton, and H. B. Morton (1976b). Stretch reflex and servo action in a variety of human muscles. *J. Physiol. (Lond.) 259,* 531–560.

Marsden, C. D., P. A. Merton, and H. B. Morton (1977). The sensory mechanism of servo action in human muscle. *J. Physiol. (Lond.) 265,* 521–535.

Marsden, C. D., P. A. Merton, H. B. Morton, and J. Adam (1977). The effect of posterior column lesions on servo responses from the human long thumb flexor. *Brain 100,* 185–200.

Marsden, C. D., P. A. Merton, H. B. Morton, M. Hallett, J. Adam, and D. N. Rushton (1977). Disorders of movement in cerebellar disease in man. In *Physiological Aspects of Clinical Neurology* (F. Clifford Rose, ed.), pp. 179–199. Blackwell, Oxford.

Marshall, J. (1951). Sensory disturbances in cortical wounds with special reference to pain. *J. Neurol. Neurosurg. Psychiat. 14,* 187–204.

Martin, G. F., M. S. Beattie, H. C. Hughes, M. Linauts, and M. Panneton (1977). The organization of reticulo-olivo-cerebellar circuits in the North American opossum. *Brain Research 137.* 253–266.

Martin, G. F., and R. Dom (1971). Reticulospinal fibers of the opossum, *Didelphis virginiana.* II. Course, caudal extent and distribution. *J. Comp. Neurol. 141,* 467–484.

Martin, G. F., R. Dom, S. Katz, and J. S. King (1974). The organization of projection neurons in the opossum red nucleus. *Brain Research 78,* 17–34.

Martin, G. F., and J. S. King (1968). The basilar pontine gray of the opossum (*Didelphis virginiana*). II. Experimental determination of neocortical input. *J. Comp. Neurol. 133,* 447–462.

Martin, G. F., J. S. King, and R. Dom (1974). The projections of the deep cerebellar nuclei of the opossum, *Didelphis marsupialis virginiana. J. Hirnforsch. 15,* 545–573.

Martin, J. P. (1967). Role of the vestibular system in the control of posture and movement in man. In *Myotatic, Kinesthetic and Vestibular Mechanisms*. Ciba Foundation Symposium (A. V. S. de Reuck and J. Knight, eds.), pp. 92–96. J. & A. Churchill, London.

Martin, J. P., and J. R. McCaul (1959). Acute hemiballismus treated by ventrolateral thalamolysis. *Brain 82,* 104–108.

Martin, R. F., L. H. Haber, and W. D. Willis (1979). Primary afferent depolarization of identified cutaneous fibers following stimulation of medial brain stem. *J. Neurophysiol. 42,* 779–790.

Martin, R. F., L. M. Jordan, and W. D. Willis (1978). Differential projections of cat medullary raphe neurons demonstrated by retrograde labelling following spinal cord lesions. *J. Comp. Neurol. 182,* 77–88.

Martinez, A. (1955). Some efferent connexions of the human frontal lobe. *J. Neurosurg. 12,* 18–25.

Martinez, A. (1961). Fiber connections of the globus pallidus in man. *J. Comp. Neurol. 117,* 37–41.

Martinez, S. N., C. Bertrand, and C. Botana-Lopez (1967). Motor fiber distribution within the cerebral peduncle. *Confin. Neurol. (Basel) 29,* 117–122.

Martinez-Millán, L., and H. Holländer (1975). Cortico-cortical projections from striate cortex of the squirrel monkey (*Saimiri sciureus*). A radioautographic study. *Brain Research 83,* 405–417.

Martner, J. (1975). Cerebellar influences on autonomic mechanisms. An experimental study in the cat with special reference to the fastigial nucleus. *Acta Physiol. Scand., Suppl. 425,* 1–42.

Mascitti, T. A., and S. N. Ortega (1966). Efferent connections of the olfactory bulb in the cat. An experimental study with silver impregnation methods. *J. Comp. Neurol. 127,* 121–136.

Mason, R. (1978). Functional organization in the cat's pulvinar complex. *Exp. Brain Res. 31,* 51–66.

Massion, J. (1967). The mammalian red nucleus. *Physiol. Rev. 47,* 383–436.

Massion, J. (1973). The intervention des voies cérébello-corticale et cortico-cérébelleuses dans l'organisation et la regulation du movements. *J. Physiol. (Paris) 67,* 117A–170A.

Massopust, L. C. Jr., and H. J. Daigle (1960). Cortical projection of the medial and spinal vestibular nuclei in the cat. *Exp. Neurol. 2,* 179–185.

Massopust, L. C., and J. M. Ordy (1962). Auditory organization of the inferior colliculi in the cat. *Exp. Neurol. 6,* 465–477.

Masterton, B., J. A. Jane, and I. T. Diamond (1967). Role of brainstem auditory structures in sound localization. I. Trapezoid body, superior olive, and lateral lemniscus. *J. Neurophysiol. 30,* 341–359.

Masterton, R. B., J. A. Jane, and I. T. Diamond (1968). Role of brain-stem auditory structures in sound localization. II: Inferior colliculus and its brachium. *J. Neurophysiol. 31,* 96–108.

Matano, S., and T. Ban (1967). Some observations on the efferent fibers from the superior and inferior colliculi to the midbrain central grey matter. *Med. J. Osaka Univ. 18,* 25–31.

Mathers, L. H. (1971). Tectal projections to the posterior thalamus of the squirrel monkey. *Brain Research 35,* 255–298.

Mathers, L. H. (1972). The synaptic organization of the cortical projection to the pulvinar of the squirrel monkey. *J. Comp. Neurol. 146,* 43–59.

Mathers, L. H., and S. C. Rapisardi (1973). Visual and somatosensory receptive fields of neurons in the squirrel monkey pulvinar. *Brain Research 64,* 65–83.

Matsushita, M., and Y. Hosoya (1978). The location of spinal projection neurons in the cerebellar nuclei (cerebellospinal tract neurons) of the cat. A study with the horseradish peroxidase technique. *Brain Research 142,* 237–248.

Matsushita, M., and Y. Hosoya (1979). Cells of origin of the spinocerebellar tract in the rat, studied with the method of retrograde transport of horseradish peroxidase. *Brain Research 173,* 185–200.

Matsushita, M., and M. Ikeda (1970a). Olivary projections to the cerebellar nuclei in the cat. *Exp. Brain Res. 10,* 488–500.

Matsushita, M., and M. Ikeda (1970b). Spinal projections to the cerebellar nuclei in the cat. *Exp. Brain Res. 10,* 501–511.

Matsushita, M., and M. Ikeda (1973). Propriospinal fiber connections of the cervical motor nuclei in the cat: a light and electron microscope study. *J. Comp. Neurol. 150,* 1–32.

Matsushita, M., and M. Ikeda (1975). The central cervical nucleus as cell origin of a spinocerebellar tract arising from the cervical cord: a study in the cat using horseradish peroxidase. *Brain Research 100,* 412–417.

Matsushita, M., and M. Ikeda (1976). Projections from the lateral reticular nucleus to the cerebellar cortex and nuclei in the cat. *Exp. Brain Res. 24,* 403–421.

Matsushita, M., M. Ikeda, and Y. Hosoya (1979). The location of spinal neurons with long descending axons (long descending propriospinal tract neurons) in the cat: a study with the horseradish peroxidase technique. *J. Comp. Neurol. 184,* 63–80.

Matsushita, M., and N. Iwahori (1971a). Structural organization of the fastigial nucleus. I. Dendrites and axonal pathways. *Brain Research 25,* 597–610.

Matsushita, M., and N. Iwahori (1971b). Structural organization of the fatigial nucleus. II. Afferent fiber systems. *Brain Research 25,* 611–624.

Matsushita, M., and N. Iwahori (1971c). Structural organization of the interpositus and the dentate nuclei. *Brain Research 35,* 17–36.

Matsushita, M., and T. Ueyama (1973a). Projections from the spinal cord to the cerebellar nuclei in the rabbit and rat. *Exp. Neurol. 38,* 438–448.

Matsushita, M., and T. Ueyama (1973b). Ventral motor nucleus of the cervical enlargement in some mammals: its specific afferents from the lower cord levels and cytoarchitecture. *J. Comp. Neurol. 150,* 33–52.

Matthews, B. H. C. (1933). Nerve endings in mammalian muscle. *J. Physiol. (Lond.) 78,* 1–53.

Matthews, D. A., C. Cotman, and G. Lynch (1976). An electron microscopic study of lesion-induced synaptogenesis in the dentate gyrus of the adult rat. I. Magnitude and time course of degeneration. *Brain Research 115,* 1–21.

Matthews, M. A., W. D. Willis, and V. Williams (1971). Dendrite bundles in lamina IX of cat spinal cord: a possible source for electrical interaction between motoneurons? *Anat. Rec. 171,* 313–328.

Matthews, M. R., W. M. Cowan, and T. P. S. Powell (1960). Transneuronal cell degeneration in the lateral geniculate nucleus of the macaque monkey. *J. Anat. (Lond.) 94,* 145–169.

Matthews, M. R., and T. P. S. Powell (1962). Some observations on transneuronal cell degeneration in the olfactory bulb of the rabbit. *J. Anat. (Lond.)* 96, 89–102.

Matthews, P. B. C. (1962). The differentiation of two types of fusimotor fibre by their effects on the dynamic response of muscle spindle primary endings. *Quart. J. Exp. Physiol. 47,* 324–33.

Matthews, P. B. C. (1964). Muscle spindles and their motor control. *Physiol. Rev. 44,* 219–288.

Matthews, P. B. C. (1966). The reflex excitation of the soleus muscle of the decerebrate cat caused by vibration applied to its tendon. *J. Physiol. (Lond.) 184,* 450–472.

Matthews, P. B. C. (1969). Evidence that the secondary as well as the primary endings of the spindles may be responsible for the tonic stretch reflex of the decerebrate cat. *J. Physiol. (Lond.) 204,* 365–393.

Matthews, P. B. C. (1972). *Mammalian Muscle Receptors and their Central Actions.* Monogr. Physiol. Soc. No. 23. Edward Arnold, London.

Matthews, P. B. C. (1977). Muscle afferents and kinaesthesia. *Brit. Med. Bull. 33,* 137–142.

Matthysse, S. (1973). Antipsychotic drug actions: a clue to the neuropathology of schizophrenia? *Fed. Proc. 32,* 200–205.

Matzke, H. A. (1951). The course of the fibers arising from the nucleus gracilis and cuneatus of the cat. *J. Comp. Neurol. 94,* 439–452.

Maunz, R. A., N. G. Pitts, and B. W. Peterson (1978). Cat spinoreticular neurons: location, responses and changes in responses during repetitive stimulation. *Brain Research 148*, 365–379.

Maurice-Williams, R. S. (1974). Micturition symptoms in frontal tumours. *J. Neurol. Neurosurg. Psychiat. 37*, 431–436.

Mayer, D. J., and J. C. Liebeskind (1974). Pain reduction by focal electrical stimulation of the brain: an anatomical and behavioral analysis. *Brain Research 68*, 73–93.

Mayer, D. J., and D. D. Price (1976). Central nervous system mechanisms of analgesia. *Pain 2*, 379–404.

Mayer, D. J., D. D. Price, and D. P. Becker (1975). Neurophysiological characterization of the anterolateral spinal cord neurons contributing to pain perception in man. *Pain 1*, 51–58.

Mayer, D. J., T. L. Wolfe, H. Akil, B. Carder, and J. C. Liebeskind (1971). Analgesia from electrical stimulation in the brainstem of the cat. *Science 174*, 1351–1354.

Maynard, C. W., R. B. Leonard, J. D. Coulter, and R. E. Coggeshall (1977). Central connections of ventral root afferents as demonstrated by the HRP method. *J. Comp. Neurol. 172*, 601–608.

Meadows, J. C. (1973). Dysphagia in unilateral cerebral lesions. *J. Neurol. Neurosurg. Psychiat. 36*, 853–860.

Meessen, H., and J. Olszewski (1949). *A Cytoarchitectonic Atlas of the Rhombencephalon of the Rabbit*. S. Karger, Basel.

Mehler, W. R. (1962). The anatomy of the so-called "pain tract" in man: an analysis of the course and distribution of the ascending fibers of the fasciculus anterolateralis. In *Basic Research in Paraplegia* (J. D. French and R. W. Porter, eds.), pp. 26–55. C. C. Thomas, Springfield, Ill.

Mehler, W. R. (1966a). Some observations on secondary ascending afferent systems in the central nervous system. In *Pain* (R. S. Knighton and P. R. Dumke, eds.), pp. 11–32. Little, Brown, Boston.

Mehler, W. R. (1966b). Further notes on the centre médian, nucleus of Luys. In *The Thalamus* (D. P. Purpura and M. D. Yahr, eds.), pp. 109–127. Columbia Univ. Press, New York.

Mehler, W. R. (1969). Some neurological species differences. A posteriori. *Ann. N.Y. Acad. Sci. 169*, 424–468.

Mehler, W. R. (1971). Idea of a new anatomy of the thalamus. *J. Psychiat. Res. 8*, 203–217.

Mehler, W. R. (1974). Central pain and the spinothalamic tract. *Advan. Neurol. 4*, 127–146.

Mehler, W. R., M. E. Feferman, and W. J. H. Nauta (1960). Ascending axon degeneration following anterolateral cordotomy. An experimental study in the monkey. *Brain 83*, 718–750.

Mei, N. (1970). Disposition anatomique et propriétés électrophysiologiques des neurones sensitifs vagaux chez le chat. *Exp. Brain Res. 11*, 465–479.

Meibach, R. C., and A. Siegel (1975). The origin of fornix fibers which project to the mammillary bodies in the rat: a horseradish peroxidase study. *Brain Research 88*, 508–512.

Meibach, R. C., and A. Siegel (1977a). Efferent connections of the septal area in the rat: an analysis utilizing retrograde and anterograde transport methods. *Brain Research 119*, 1–20.

Meibach, R. C., and A. Siegel (1977b). Efferent connections of the hippocampal formation in the rat. *Brain Research 124*, 197–224.

Meibach, R. C., and A. Siegel (1977c). Thalamic projections of the hippocampal formation: evidence for an alternate pathway involving the internal capsule. *Brain Research 134*, 1–12.

Meikle, T. H., Jr., and J. M. Sprague (1964). The neural organization of the visual pathways in the cat. *Int. Rev. Neurobiol. 6*, 149–189.

Mellgren, S. I., W. Harkmark, and B. Srebro (1977). Some enzyme histochemical characteristics of the human hippocampus. *Cell Tiss. Res. 181*, 459–471.

Mellgren, S. J., and B. Srebro (1973). Changes in acetylcholinesterase and distribution of degenerating fibers in the hippocampal region after septal lesions in the rat. *Brain Research 52*, 19–36.

Meltzer, G. E., R. S. Hunt, and W. M. Landau (1963). Fusimotor function. Part III. The spastic monkey. *Arch. Neurol. (Chic.) 9*, 133–136.

Melvill-Jones, G., and D. G. D. Watt (1971). Muscular control of landing from unexpected falls in man. *J. Physiol. (Lond.) 219*, 729–737.

Melzack, R. (1974). Central neural mechanisms in phantom limb pain. In *Advan. Neurol.*, Vol. 4, *Pain* (J. J. Bonica, ed.), pp. 319–328. Raven Press, New York.

Melzack, R., and F. P. Haugen (1957). Responses evoked at the cortex by tooth stimulation. *Amer. J. Physiol. 190*, 570–574.

Melzack, R., and P. D. Wall (1962). On the nature of cutaneous sensory mechanisms. *Brain 83*, 331–356.

Melzack, R., and P. D. Wall (1965). Pain mechanisms: a new theory. *Science 150*, 971–979.

Mergner, T., O. Pompeiano, and N. Corvaja (1977). Vestibular projections to the nucleus intercalatus of Staderini mapped by retrograde transport of horseradish peroxidase. *Neurosci. Letters 5*, 309–313.

Merrill, E. G. (1970). The lateral respiratory neurones of the medulla: their associations with nucleus ambiguus, nucleus retroambigualis, the spinal accessory nucleus and the spinal cord. *Brain Research 24*, 11–28.

Merrillees, N. C. R., S. Sunderland, and W. Hayhow (1950). Neuromuscular spindles in the extraocular muscles in man. *Anat. Rec. 108*, 23–30.

Merzenich, M. M., and J. F. Brugge (1973). Representation of the cochlear partition on the superior temporal plane of the macaque monkey. *Brain Research 50*, 275–296.

Merzenich, M. M., J. H. Kaas, and G. L. Roth (1976). Auditory cortex in the grey squirrel: tonotopic organization and architectonic fields. *J. Comp. Neurol. 166*, 387–402.

Merzenich, M. M., J. H. Kaas, M. Sur, and C.-S. Lin (1978). Double representation of the body surface within cytoarchitectonic areas 3b and 1 in "SI" in the owl monkey (*Actus trivirgatus*). *J. Comp. Neurol. 181*, 41–74.

Merzenich, M. M., P. L. Knight, and G. L. Roth (1975). Representation of cochlea within primary auditory cortex in the cat. *J. Neurophysiol. 38*, 231–249.

Merzenich, M. M., and M. D. Reid (1974). Representation of the cochlea within the inferior colliculus of the cat. *Brain Research 77*, 397–415.

Merzenich, M. M., G. L. Roth, R. A. Andersen, P. L. Knight, and S. A. Colwell (1977). Some basic features of organization of the central auditory nervous system. In *Psychophysics and Physiology of Hearing* (E. F. Evans and J. P. Wilson, eds.), pp. 485–497. Academic Press, New York.

Mess, B., M. Zansi, and L. Tima (1970). Site of production of releasing and inhibiting factors. In *The Hypothalamus* (L. Martini, M. Motta, and F. Fraschini, eds.), pp. 259–276. Academic Press, New York.

Mesulam, M.-M. (1976). The blue reaction product in horseradish peroxidase neurohistochemistry: incubation parameters and visibility. *J. Histochem. Cytochem. 24*, 1273–1280.

Mesulam, M.-M. (1978). Tetramethylbenzidine for horseradish peroxidase neurohistochemistry. A non-carcinogenic blue reaction-product with superior sensitivity for visualizing neural afferents and efferents. *J. Histochem. Cytochem. 26*, 106–117.

Mesulam, M.-M., and D. N. Pandya (1973). The projections of the medial geniculate complex within the sylvian fissure of the rhesus monkey. *Brain Research 60*, 315–333.

Mesulam, M.-M., G. W. Van Hoesen, D. N. Pandya, and N. Geschwind (1977). Limbic and sensory connections of the inferior parietal lobule (area PG) in the rhesus monkey: a study with a new method for horseradish histochemistry. *Brain Research 136*, 393–414.

Mettler, F. A. (1935). Corticofugal fiber connections of the cortex of *Macaca mulatta*. The occipital region. *J. Comp. Neurol. 61*, 221–256. The frontal region. Ibid. 509–542. The parietal region. Ibid. *62*, 263–292. The temporal region. Ibid. *63*, 25–48.

Mettler, F. A. (1942). Relation between pyramidal and extrapyramidal function. *Res. Publ. Ass. Nerv. Ment. Dis. 21*, 150–227.

Mettler, F. A. (1944). Physiologic consequences and anatomic degenerations following lesions of the primate brain-stem: plantar and patellar reflexes. *J. Comp. Neurol. 80*, 69–148.

Mettler, F. A. (1945). Fiber connections of the corpus striatum of the monkey and baboon. *J. Comp. Neurol. 82*, 169–204.

Mettler, F. A. (1964). Substantia nigra and parkinsonism. *Arch. Neurol. (Chic.) 11*, 529–542.

Mettler, F. A. (1966). Experimentally produced tremor. Temporal factors in its development and disappearance in the monkey. *Arch. Neurol. (Chic.) 15*, 241–246.

Mettler, F. A., H. W. Ades, E. Lipman, and E. A. Culler (1939). The extrapyramidal system. An experimental demonstration of function. *Arch. Neurol. Psychiat. (Chic.) 41*, 984–995.

Mettler, F. A., and A. J. Lubin (1942). Termination of the brachium pontis. *J. Comp. Neurol. 77*, 391–397.

Mettler, F. A., and G. M. Stern (1962). Somatotopic localization in rhesus subthalamic nucleus. *Arch. Neurol. (Chic.) 7*, 328–329.

Meyer, A. (1957). Hippocampal lesions in epilepsy. In *Modern Trends in Neurology* (D. Williams, ed.), pp. 301–306. Butterworths, London.

Meyer, A., and E. Beck (1945). Neuropathological problems arising from prefrontal leucotomy. *J. Ment. Sci. 91*, 411–422.

Meyer, A., and E. Beck (1954). *Prefrontal Leucotomy and Related Operations: Anatomical Aspects of Success and Failure*. William Ramsay Henderson Trust Lecture. Oliver and Boyd, Edinburgh—London.

Meyer, J. S., and R. M. Herndon (1962). Bilateral infarction of the pyramidal tracts in man. *Neurology (Minneap.) 12*, 637–642.

Meyer, M. (1949). A study of efferent connexions of the frontal lobe in the human brain after leucotomy. *Brain 72*, 265–296.

Meyer, M., and A. C. Allison (1949). An experimental investigation of the connexions of the olfactory tracts in the monkey. *J. Neurol. Neurosurg. Psychiat. 12*, 274–286.

Meyer-Lohmann, J., B. Conrad, K. Matsunami, and V. Brooks (1975). Effects of dentate cooling on precentral unit activity following torque pulse injections into elbow movements. *Brain Research 94*, 237–251.

Meyer-Lohmann, J., W. Riebord, and D. Robrecht (1974). Mechanical influence of the extrafusal muscle on the behaviour of deefferented primary muscle spindle endings in cat. *Pflügers Arch. Ges. Physiol. 352*, 267–278.

Meyers, R. (1953). The extrapyramidal system. An inquiry into the validity of the concept. *Neurology (Minneap.) 3*, 627–655.

Meyers, R. (1958). Historical background and personal experiences in the surgical relief of hyperkinesia and hypertonus. In *Pathogenesis and Treatment of Parkinsonism* (W. S. Fields, ed.), pp. 229–270. C. C. Thomas, Springfield, Ill.

Michail, S., and A. N. Karamanlidis (1970). Trigemino-thalamic fibre connexions in the dog and pig. *J. Anat. (Lond.) 107*, 557–566.

Middlebrooks, J. C., R. W. Dykes, and M. M. Merzenich (1978). Binaural response-specific bands within AI in the cat: specialization within isofrequency contours. *Abstr. Soc. Neurosci. 4*, p. 8.

Mihailoff, G. A., and J. S. King (1975). The basilar pontine gray of the opossum: a correlated light and electron microscopic analysis. *J. Comp. Neurol. 159*, 521–552.

Millar, J. (1975). Flexion-extension sensitivity of elbow joint afferents in cat. *Exp. Brain Res. 24*, 209–214.

Millar, J. (1979). Loci of joint cells in the cuneate and external cuneate nuclei of the cat. *Brain Research 167*, 385–390.

Miller, J. J., T. L. Richardson, H. C. Fibiger, and H. McLennan (1975). Anatomical and electrophysiological identification of a projection from the mesencephalic raphe to the caudate-putamen in the rat. *Brain Research 97*, 133–138.

Miller, R. A., and N. L. Strominger (1973). Efferent connections of the red nucleus in the brainstem and spinal cord of the rhesus monkey. *J. Comp. Neurol. 152,* 327–346.

Miller, R. A., and N. L. Strominger (1977). An experimental study of the efferent connections of the superior peduncle in the rhesus monkey. *Brain Research 133,* 237–250.

Miller, S., N. Nezlina, and O. Oscarsson (1969). Projection and convergence patterns in climbing fibre paths to cerebellar anterior lobe activated from cerebral cortex and spinal cord. *Brain Research 14,* 230–233.

Millhouse, O. E. (1973). The organization of the ventromedial hypothalamic nucleus. *Brain Research 55,* 71–87.

Milner, B. (1974). Hemispheric specialization: scope and limits. In *The Neurosciences. Third Study Program* (F. O. Schmitt and F. G. Worden, eds.), pp. 75–89. MIT Press, Cambridge, Mass.

Minckler, J. (1940). The morphology of the nerve terminals of the human spinal cord as seen in block silver preparations, with estimates of the total number per cell. *Anat. Rec. 77,* 9–25.

Minckler, J., R. M. Klemme, and D. Minckler (1944). The course of efferent fibers from the human premotor cortex. *J. Comp. Neurol. 81,* 259–277.

Minderhoud, J. M. (1971). An anatomical study of the efferent connections of the thalamic reticular nucleus. *Exp. Brain Res. 12,* 435–446.

Mirsky, A. F., H. E. Rosvold, and K. H. Pribram (1957). Effects of cingulectomy on social behavior in monkeys. *J. Neurophysiol. 20,* 588–601.

Miskolczy, D. (1931). Ueber die Endigungsweise der spinocerebellaren Bahnen. *Z. Anat. Entwickl.-Gesch. 96,* 537–542.

Miskolczy, D. (1934). Die Endigungsweise der olivo-cerebellaren Faserung. *Arch. Psychiat. Nervenkr. 102,* 197–201.

Mitchell, D., and R. F. Hellon (1977). Neuronal and behavioural responses in rats during noxious stimulation of the tail. *Proc. Roy. Soc. B 197,* 169–194.

Mitchell, G. A. G. (1953). *Anatomy of the Autonomic Nervous System.* E. and S. Livingstone, Edinburgh.

Mitchell, G. A. G., and R. Warwick (1955). The dorsal vagal nucleus. *Acta Anat. (Basel) 25,* 371–395.

Miura, M. (1975). Postsynaptic potentials recorded from nucleus of the solitary tract and its subjacent reticular formation elicited by stimulation of the carotid sinus nerve. *Brain Research 100,* 437–440.

Mizuno, N. (1966). An experimental study of the spino-olivary fibers in the rabbit and the cat. *J. Comp. Neurol. 127,* 267–291.

Mizuno, N. (1970). Projection fibers from the main sensory trigeminal nucleus and the supra-trigeminal region. *J. Comp. Neurol. 139,* 457–472.

Mizuno, N., A. Konishi, and M. Sato (1975). Localization of masticatory motoneurons in the cat and rat by means of retrograde axonal transport of horseradish peroxidase. *J. Comp. Neurol. 164,* 105–115.

Mizuno, N. K. Mochizuki, C. Akimoto, and R. Matsushima (1973). Pretectal projections to the inferior olive in the rabbit. *Exp. Neurol. 39,* 498–506.

Mizuno, N., K. Mochizuki, C. Akimoto, R. Matsushima, and Y. Nakamura (1973). Rubrobulbar projections in the rabbit. A light and electron microscopic study. *J. Comp. Neurol. 147,* 267–280.

Mizuno, N., K. Mochizuki, C. Akimoto, R. Matsushima, and K. Sasaki (1973). Projections from the parietal cortex to the brain stem nuclei in the cat, with special reference to the parietal cerebrocerebellar system. *J. Comp. Neurol. 147,* 511–522.

Mizuno, N., and Y. Nakamura (1970). Direct hypothalamic projections to the locus coeruleus. *Brain Research 19,* 160–162.

Mizuno, N., and Y. Nakamura (1973). An electron microscope study of spinal afferents to the lateral reticular nucleus of the medulla oblongata in the cat. *Brain Research 53,* 187–191.

Mizuno, N., Y. Nakamura, and N. Iwahori (1974). An electron microscope study of the dorsal cap of the inferior olive in the rabbit, with special reference to the pretecto-olivary fibers. *Brain Research 77,* 385–395.

Mizuno, N., K. Nakano, M. Imaizumi, and M. Okamoto (1967). The lateral cervical nucleus of the Japanese monkey (*Macaca fuscata*). *J. Comp. Neurol. 129*, 375–384.

Mizuno, N., E. K. Sauerland, and C. D. Clemente (1968). Projections from the orbital gyrus in the cat. I. To brain stem structures. *J. Comp. Neurol. 133*, 463–476.

Mlonyeni, M. (1973). The number of Purkinje cells and inferior olivary neurones in the cat. *J. Comp. Neurol. 147*, 1–10.

Moatamed, F. (1966). Cell frequencies in the human inferior olivary nuclear complex. *J. Comp. Neurol. 128*, 109–116.

Molenaar, I., and H. G. J. M. Kuypers (1978). Cells of origin of propriospinal fibers and of fibers ascending to supraspinal levels. A HRP study in cat and rhesus monkey. *Brain Research 152*, 429–450.

Mollaret, B., and M. Goulon (1959). Le coma dépassé. *Rev. Neurol. 101*, 3–15.

Monrad-Krohn, G. H. (1927). Clinical neurology of the facial nerve. *Trans. Amer. Neurol. Ass.*, pp. 190–193.

Monrad-Krohn, G. H. (1939). On facial dissociation. *Acta Psychiat. (Kbh). 14*, 557–566. (See also *Brain 47*, 1924.)

Monrad-Krohn, G. H. (in cooperation with S. Refsum) (1964). *The Clinical Examination of the Nervous System*, 12th ed. H. K. Lewis, London.

Monro, P. A. G. (1959). *Sympathectomy. An Anatomical and Physiological Study with Clinical Applications*. Oxford Univ. Press, London.

Moore, C. N., J. H. Casseday, and W. D. Neff (1974). Sound localization: the role of the commissural pathways of the auditory system of the cat. *Brain Research 82*, 13–26.

Moore, J. K. (1980). The primate cochlear nuclei: loss of lamination as a phylogenetic process. *J. Comp. Neurol.* In press.

Moore, J. K., and R. Y. Moore (1971). A comparative study of the superior olivary complex in the primate brain. *Folia Primat. 16*, 35–51.

Moore, J. K., and K. K. Osen (1979). The cochlear nuclei in man. *Amer. J. Anat. 154*, 393–418.

Moore, R. Y. (1973). Retinohypothalamic projection in mammals: a comparative study. *Brain Research 49*, 403–409.

Moore, R. Y., R. K. Bhatnagar, and A. Heller (1971). Anatomical and chemical studies of a nigroneostriatal projection in the cat. *Brain Research 30*, 119–136.

Moore, R. Y., A. Björklund, and U. Stenevi (1971). Plastic changes in the adrenergic innervation of the rat septal area in response to denervation. *Brain Research 33*, 13–35.

Moore, R. Y., and J. M. Goldberg (1963). Ascending projections of the inferior colliculus in the cat. *J. Comp. Neurol. 121*, 109–136.

Moore, R. Y., and J. M. Goldberg (1966). Projections of the inferior colliculus in the monkey. *Exp. Neurol. 14*, 429–438.

Moore, R. Y., and A. E. Halaris (1975). Hippocampal innervation by serotonin neurons of the midbrain raphe in the rat. *J. Comp. Neurol. 164*, 171–183.

Moore, R. Y., and N. J. Lenn (1972). A retinohypothalamic projection in the rat. *J. Comp. Neurol. 146*, 1–14.

Morest, D. K. (1964a). The laminar structure of the inferior colliculus of the cat. *Anat. Rec. 148*, 314.

Morest, D. K. (1964b). The neuronal architecture of the medial geniculate body of the cat. *J. Anat. (Lond.) 98*, 611–630.

Morest, D. K. (1965a). The laminar structure of the medial geniculate body of the cat. *J. Anat. (Lond.) 99*, 143–160.

Morest, D. K. (1965b). The lateral tegmental system of the midbrain and the medial geniculate body: study with Golgi and Nauta methods in cat. *J. Anat. (Lond.) 99*, 611–634.

Morest, D. K. (1967). Experimental study of the projections of the nucleus of the tractus solitarius and the area postrema in the cat. *J. Comp. Neurol. 130*, 277–300.

Morest, D. K. (1971). Dendrodendritic synapses of cells that have axons: the fine structure of the Golgi type II cell in the medial geniculate body of the cat. *Z. Anat. Entwickl.-Gesch. 133*, 216–246.

Morest, D. K. (1975). Synaptic relationships of Golgi type II cells in the medial geniculate body of the cat. *J. Comp. Neurol. 162,* 157–194.

Morgane, P. J., and W. C. Stern (1974). Chemical anatomy of brain circuits in relation to sleep and wakefulness. In *Advances in Sleep Research,* Vol. 1 (E. D. Weitzman, ed.), pp. 1–131. Spectrum, New York.

Morillo, A., and D. Baylor (1963). Electrophysiological investigation of lemniscal and paraleminsical input to the midbrain reticular formation. *Electroenceph. Clin. Neurophysiol. 15,* 455–464.

Morin, F., and J. V. Catalano (1955). Central connections of a cervical nucleus (nucleus cervicalis lateralis of the cat). *J. Comp. Neurol. 103,* 17–32.

Morin, F., D. T. Kennedy, and E. Gardner (1966). Spinal afferents to the lateral reticular nucleus. I. An histological study. *J. Comp. Neurol. 126,* 511–522.

Morin, F., S. T. Kitai, H. Portnoy, and C. Demirjian (1963). Afferent projections to the lateral cervical nucleus: a microelectrode study. *Amer. J. Phsyiol. 204,* 667–672.

Morin, F., H. G. Schwartz, and J. L. O'Leary (1951). Experimental study of the spinothalamic and related tracts. *Acta Psychiat. Neurol. Scand. 26,* 371–396.

Morison, R. S., and E. W. Dempsey (1942). A study of thalamo-cortical relations. *Amer. J. Physiol. 135,* 281–292.

Morris, C. J. (1969). Human skeletal muscle fibre type grouping and collateral reinnervation. *J. Neurol. Neurosurg. Psychiat. 32,* 440–444.

Mortimer, E. M., and K. Akert (1961). Cortical control and representation of fusimotor neurons. *Amer. J. Phys. Med. 40,* 228–248.

Mortimer, J. A. (1975). Cerebellar responses to teleceptive stimuli in alert monkeys. *Brain Research 83,* 369–390.

Morton, A. (1969). A quantitative analysis of the normal neuron population of the hypothalamic magnocellular nuclei in man and of their projections to the neurohypophysis. *J. Comp. Neurol. 136,* 143–158.

Morton, A. (1970). The time course of retrograde neuron loss in the hypothalamic magnocellular nuclei in man. *Brain 93,* 329–336.

Moruzzi, G. (1940). Palaeocerebellar inhibition of vasomotor and respiratory carotid sinus reflexes. *J. Neurophysiol. 3,* 20–32.

Moruzzi, G. (1950). *Problems in Cerebellar Physiology.* C. C. Thomas, Springfield, Ill.

Moruzzi, G. (1963). Active processes in the brain stem during sleep. *Harvey Lect. 58,* 233–297. Academic Press, New York.

Moruzzi, G. (1972). The sleep-waking cycle. *Rev. Physiol. Biochem. Exp. Parmacol. 64,* 1–165.

Moruzzi, G., and H. W. Magoun (1949). Brain stem reticular formation and activation of the EEG. *Electroenceph. Clin. Neurophysiol. 1,* 455–473.

Moruzzi, G., and O. Pompeiano (1956). Crossed fastigial influence on decerebrate rigidity. *J. Comp. Neurol. 106,* 371–392.

Mosko, S. S., D. Haubrich, and B. L. Jacobs (1977). Serotonergic afferents to the dorsal raphe nucleus: evidence from HRP and synaptosomal uptake studies. *Brain Research 119,* 269–290.

Mosko, S., G. Lynch, and C. W. Cotman (1973). The distribution of septal projections to the hippocampus of the rat. *J. Comp. Neurol. 152,* 163–174.

Mosso, J. A., and L. Kruger (1972). Spinal trigeminal neurons excited by noxious and thermal stimuli. *Brain Research 38,* 206–210.

Mosso, J. A., and L. Kruger (1973). Receptor categories represented in spinal trigeminal nucleus caudalis. *J. Neurophysiol. 36,* 472–488.

Motokizawa, F. (1947). Electrophysiological studies of olfactory projection to the mesencephalic reticular formation. *Exp. Neurol. 44,* 135–144.

Mountcastle, V. B. (1957). Modality and topographic properties of single neurons of cat's somatic sensory cortex. *J. Neurophysiol. 20,* 408–434.

Mountcastle, V. B., M. R. Covian, and C. R. Harrison (1952). The central representation of some forms of deep sensibility. *Ass. Res. Nerv. Dis. Proc. 30,* 339–370.

Mountcastle, V., and E. Henneman (1949). Pattern of tactile representation in thalamus of cat. *J. Neurophysiol. 12,* 85–100.

Mountcastle, V. B., and E. Henneman (1952). The representation of tactile sensibility in the thalamus of the monkey. *J. Comp. Neurol. 97*, 409–440.

Mountcastle, V. B., J. C. Lynch, A. Georgopoulos, H. Sakata, and C. Acuna (1975). Posterior parietal association cortex of the monkey: command functions for operations within extrapersonal space. *J. Neurophysiol. 38*, 871–908.

Mountcastle, V. B., G. F. Poggio, and G. Werner (1963). The relation of thalamic cell response to peripheral stimuli varied over an intensive continuum. *J. Neurophysiol. 26*, 807–834.

Mountcastle, V. B., and T. P. S. Powell (1959a). Central nervous mechanisms subserving position sense and kinesthesis. *Bull. Johns Hopk. Hosp. 105*, 173–200.

Mountcastle, V. B., and T. P. S. Powell (1959b). Neural mechanisms subserving cutaneous sensibility, with special reference to the role of afferent inhibition in sensory perception and discrimination. *Bull. Johns Hopk. Hosp. 105*, 201–232.

Moushegian, G., A. L. Rupert, and T. L. Langford (1967). Stimulus coding by medial superior olivary neurons. *J. Neurophysiol. 30*, 1239–1261.

Mugnaini, E. (1972). The histology and cytology of the cerebellar cortex. In *The Comparative Anatomy and Histology of the Cerebellum. The Human Cerebellum, Cerebellar Connections, and Cerebellar Cortex* (O. Larsell and J. Jansen, eds.), pp. 201–262. Univ. of Minnesota Press, Minneapolis.

Mugnaini, E., and A.-L. Dahl (1975). Mode of distribution of aminergic fibers in the cerebellar cortex of the chicken. *J. Comp. Neurol. 162*, 417–432.

Mugnaini, E., K. K. Osen, A.-L. Dahl, V. L. Friedrich, Jr., and G. Korte (1980). Fine structure of granule cells and related interneurones (termed Golgi cells) in the cochlear nuclear complex of cat, rat and mouse. *J. Neurocytol.* In press.

Mugnaini, E., and F. Walberg (1967). An experimental electron microscopical study on the mode of termination of cerebellar corticovestibular fibres in the cat lateral vestibular nucleus (Deiters' nucleus). *Exp. Brain Res. 4*, 212–236.

Mugnaini, E., F. Walberg, and A. Brodal (1967). Mode of termination of primary vestibular fibres in the lateral vestibular nucleus. An experimental electron microscopical study in the cat. *Exp. Brain Res. 4*, 187–211.

Mugnaini, E., F. Walberg, and E. Hauglie-Hanssen (1967). Observations on the fine structure of the lateral vestibular nucleus (Deiters' nucleus) in the cat. *Exp. Brain Res. 4*, 146–186.

Mugnaini, E., W. B. Warr, and K. K. Osen (1980). Distribution and light microscopic features of granule cells in the cochlear nuclei of cat, rat and mouse. *J. Comp. Neurol.* In press.

Mullan, S., and W. Penfield (1959). Illusions of comparative interpretation and emotion. *Arch. Neurol. Psychiat. (Chic.) 81*, 269–284.

Müller, R., and G. Wohlfart (1947). Om tumörer i corpus pineale (Pineal tumours). *Nord. Med. 33*, 15–21.

Murakami, F., Y. Fujito, and N. Tsukahara (1976). Physiological properties of the newly formed cortico-rubral synapses of red nucleus neurons due to collateral sprouting. *Brain Research 103*, 147–151.

Murphy, J. T., Y. C. Wong, and H. C. Kwan (1974). Distributed feedback systems for muscular control. *Brain Research 71*, 495–505.

Murray, E. A., and J. D. Coulter (1976). Corticospinal projections from the medial cerebral hemisphere in monkey. *Abstr. Soc. Neurosci. 3*, 275.

Murray, H. M., and D. E. Haines (1975). The rubrospinal tract in a prosimian primate (*Galago senegalensis*). *Brain Behav. Evol. 12*, 311–333.

Murray, M. (1966). Degeneration of some intralaminar thalamic nuclei after cortical removals in the cat. *J. Comp. Neurol. 127*, 341–367.

Murray, M., and M. E. Goldberger (1974). Restitution of function and collateral sprouting in the cat spinal cord: the partially hemisected animal. *J. Comp. Neurol. 158*, 19–36.

Muscle Receptors. Handbook of Sensory Physiology, Vol. III/2 (1974). (C. C. Hunt, ed.). Springer-Verlag, Berlin.

Music and the Brain. Studies in the Neurology of Music (1977). (C. MacDonald and R. A. Henson, eds.). W. Heinemann, London.

Myers, R. D. (1969). Temperature regulation: neurochemical systems in the hypothalamus. In *The Hypothalamus* (W. Haymaker, E. Anderson, and W. J. H. Nauta, eds.), pp. 506–523. C. C. Thomas, Springfield, Ill.

Myers, R. E. (1956). Function of corpus callosum in interocular transfer. *Brain 79,* 358–363.

Myers, R. E. (1959). Localization of function in the corpus callosum. Visual gnostic transfer. *Arch. Neurol. (Chic.) 1,* 74–77.

Myers, R. E. (1962). Commissural connections between occipital lobes of the monkey. *J. Comp. Neurol. 118;* 1–16.

Myers, R. E. (1965). Phylogenetic studies of commissural connexions. In *Functions of the Corpus Callosum.* Ciba Foundation Study Group, No. 20 (E. G. Ettlinger, A. V. S. de Reuck, and R. Porter, eds.), pp. 138–142. J. & A. Churchill, London.

Myers, R. E., and F. F. Ebner (1976). Localization of function in corpus callosum: tactual information transmission in *Macaca mulatta. Brain Research 103,* 455–462.

Myers, R. E., and C. O. Henson (1960). Role of corpus callosum in transfer of tactuokinesthetic learning in chimpanzee. *Arch. Neurol. (Chic.) 3,* 404–409.

Myers, R. E., and C. Swett (1970). Social behavior deficits of free-ranging monkeys after anterior temporal cortex removal: a preliminary report. *Brain Research 18,* 551–556.

Myers, R. E., C. Swett, and M. Miller (1973). Loss of social group affinity following prefrontal lesions in free-ranging macaques. *Brain Research 64,* 257–269.

Myotatic, Kinesthetic and Vestibular Mechanisms. Ciba Foundation Symposium (1967). (A. V. D. de Reuck and J. Knight, eds.). J. & A. Churchill, London.

Nádvorník, P., M. Sramka, L. Lisý, and I. Svička (1972). Experiences with dentatotomy. *Confin. Neurol. (Basel) 34,* 320–324.

Nafstad, P. M. J. (1967). An electron microscope study on the termination of the perforant path fibres in the hippocampus and the fascia dentata. *Z. Zellforsch. 76,* 532–542.

Naito, H., K.-J. Tanimura, N. Taga, and Y. Hosoya (1974). Microelectrode study on the subnuclei of the oculomotor nucleus in the cat. *Brain Research 81,* 215–231.

Nakai, Y. (1970). Electron microscopic observations on synapse-like contacts between pituicytes and different types of nerve fibers in the *anuran* pars nervosa. *Z. Zellforsch. 110,* 27–39.

Nakamura, J., and N. Mizuno (1971). An electron microscopic study of the interpositorubral connections in the cat and rabbit. *Brain Research 35,* 283–286.

Nakashima, M., K. Mori, and S. F. Takagi (1978). Centrifugal influence on olfactory bulb activity in the rabbit. *Brain Research 154,* 301–316.

Nansen, F. (1886). The structure and combination of the histological elements of the central nervous system. *Bergen Museums Årsberetn.*

Narabayashi, H., T. Nagao, Y. Saito, M. Yoshida, and M. Nagahata (1963). Stereotaxic amygdalotomy for behavior disorders. *Arch. Neurol. (Chic.) 9,* 1–16.

Narkiewicz, O., and S. Brutkowski (1967). The organization of projections from the thalamic mediodorsal nucleus to the prefrontal cortex of the dog. *J. Comp. Neurol. 129,* 361–374.

Nashner, L. M. (1973). Vestibular and reflex control of normal standing. In *Control of Posture and Locomotion* (Stein, Pearson, Smith and Redford, eds.). *Advan. Behav. Biol. 7,* 291–308.

Nashold, B., G. Somjen, and H. Friedman (1972). Paresthesias and EEG potentials evoked by stimulation of the dorsal funiculi in man. *Exp. Neurol. 36,* 273–287.

Nashold, Jr., B. S., and D. G. Slaughter (1969). Effects of stimulating or destroying the deep cerebellar regions in man. *J. Neurosurg. 31,* 172–186.

Nathan, P. W. (1956a). Reference of sensation at the spinal level. *J. Neurol. Neurosurg. Psychiat. 19,* 88–100.

Nathan, P. W. (1956b). Awareness of bladder filling with divided sensory tract. *J. Neurol. Neurosurg. Psychiat. 19,* 101–105.

Nathan, P. W. (1963). Results of antero-lateral cordotomy for pain in cancer. *J. Neurol. Neurosurg. Psychiat. 26,* 353–362.

Nathan, P. W. (1976). The gate-control theory of pain. A critical review. *Brain 99,* 123–158.

Nathan, P. W., and T. A. Sears (1960). Effects of phenol on nervous conduction. *J. Physiol. (Lond.) 150,* 565–580.

Nathan, P. W., T. A. Sears, and M. C. Smith (1965). Effects of phenol solutions on the nerve roots of the cat: an electrophysiological and histological study. *J. Neurol. Sci. 2,* 7–29.

Nathan, P. W., and M. C. Smith (1950). Normal mentality associated with a maldeveloped "rhinencephalon." *J. Neurol. Neurosurg. Psychiat. 13,* 191–197.

Nathan, P. W., and M. C. Smith (1951). The centripetal pathway from the bladder and urethra within the spinal cord. *J. Neurol. Neurosurg. Psychiat. 14,* 262–280.

Nathan, P. W., and M. C. Smith (1955a). Long descending tracts in man. I. Review of present knowledge. *Brain 78,* 248–303.

Nathan, P. W., and M. C. Smith (1955b). The Babinski response: a review and new observations. *J. Neurol. Neurosurg. Psychiat. 18,* 250–259.

Nathan, P. W., and M. C. Smith (1958). The centrifugal pathway for micturition within the spinal cord. *J. Neurol. Neurosurg. Psychiat. 21,* 177–189.

Nathan, P. W., and M. C. Smith (1959). Fasciculi proprii of the spinal cord in man: review of present knowledge. *Brain 82,* 610–668.

Nathan, P. W., and J. W. A. Turner (1942). The efferent pathway for pupillary contraction. *Brain 65,* 343–351.

Nature of Sleep, The. Ciba Foundation Symposium (1961). (G. E. W. Wolstenholme and M. O'Connor, eds.). J. & A. Churchill, London.

Nauta, H. J. W. (1974). Efferent projections of the caudate nucleus, pallidal complex, and subthalamic nucleus in the cat. (Thesis), Case Western Reserve University, Cleveland, Ohio.

Nauta, H. J. W., and M. Cole (1978). Efferent projections of the subthalamic nucleus: an autoradiographic study in monkey and cat. *J. Comp. Neurol. 180,* 1–16.

Nauta, H. J. W., M. B. Pritz, and R. J. Lasek (1974). Afferents to the rat caudoputamen studied with horseradish peroxidase. An evaluation of a retrograde neuroanatomical research method. *Brain Research 67,* 219–238.

Nauta, W. J. H. (1946). Hypothalamic regulation of sleep in rats. An experimental study. *J. Neurophysiol. 9,* 285–314.

Nauta, W. J. H. (1956). An experimental study of the fornix in the rat. *J. Comp. Neurol. 104,* 247–272.

Nauta, W. J. H. (1957). Silver impregnation of degenerating axons. In *New Research Techniques of Neuroanatomy* (W. F. Windle, ed.), pp. 17–26. C. C. Thomas, Springfield, Ill.

Nauta, W. J. H. (1958). Hippocampal projections and related neural pathways to the midbrain in the cat. *Brain 81,* 319–340.

Nauta, W. J. H. (1961). Fibre degeneration following lesions of the amygdaloid complex in the monkey. *J. Anat. (Lond.) 95,* 515–531.

Nauta, W. J. H. (1962). Neural associations of the amygdaloid complex in the monkey. *Brain 85,* 505–520.

Nauta, W. J. H. (1964). Some efferent connections of the prefrontal cortex in the monkey. In *The Frontal Granular Cortex and Behavior* (J. M. Warren and K. Akert, eds), pp. 397–407. McGraw-Hill, New York.

Nauta, W. J. H. (1971). The problem of the frontal lobe: a reinterpretation. *J. Psychiat. Res. 8,* 167–187.

Nauta, W. J. H., and P. A. Gygax (1954). Silver impregnation of degenerating axons in the central nervous system. A modified technic. *Stain Technol. 29,* 91–93.

Nauta, W. J. H., and W. Haymaker (1969). Hypothalamic nuclei and fiber connections. In *The Hypothalamus* (W. Haymaker, E. Anderson, and W. J. H. Nauta, eds.), pp. 136–209. C. C. Thomas, Springfield, Ill.

Nauta, W. J. H., and H. G. J. M. Kuypers (1958). Some ascending pathways in the brain stem reticular formation. In *Reticular Formation of the Brain*. Henry Ford Hospital Symposium (H. H. Jasper and L. D. Proctor, eds.), pp. 3–30. Little, Brown, Boston.

Nauta, W. J. H., and W. R. Mehler (1966). Projections of the lentiform nucleus in the monkey. *Brain Research 1,* 3–42.

Nauta, W. J. H., and D. G. Whitlock (1954). An anatomical analysis of the nonspecific thalamic projection system. In *Brain Mechanisms and Consciousness* (J. F. Delafres-naye, ed.), pp. 81–104. Blackwell, Oxford.

Navaratnam, V., and P. R. Lewis (1970). Cholinesterase-containing neurones in the spinal cord of the rat. *Brain Research 18,* 411–425.

Neff, W. D. (1960). Role of the auditory cortex in sound discrimination. In *Neural Mechanisms of the Auditory and Vestibular Systems* (G. L. Rasmussen and W. F. Windle, eds.), pp. 211–216. C. C. Thomas, Springfield, Ill.

Neff, W. D., J. F. Fisher, I. T. Diamond, and M. Yela (1956). Role of auditory cortex in discrimination requiring localization of sound in space. *J. Neurophysiol. 19,* 500–512.

Nelson, E., and M. Rennels (1970). Innervation of intracranial arteries. *Brain 93,* 475–490.

Nelson, P. G. (1966). Interaction between spinal motoneurons of the cat. *J. Neurophysiol. 29,* 275–287.

Nervous Control of the Heart (1965). (W. C. Randall, ed.). Williams & Wilkins, Baltimore.

Nervous System, The, Vol. 3, *Human Communication and its Disorders* (1975). (D. B. Tower, ed.-in-chief; E. L. Eagles, vol. ed.). Raven Press, New York.

Neubürger, K. (1937). Über die nichtalkoholische Wernickesche Krankheit, insbesondere über ihr Vorkommen beim Krebsleiden. *Virchows Arch. Path. Anat. 298,* 68–86.

Neural Integration of Physiological Mechanisms and Behaviour (1975). (G. J. Mogenson and F. R. Calaresu, eds.). Univ. of Toronto Press, Toronto.

Neural Mechanisms of the Auditory and Vestibular Systems (1960). (G. L. Rasmussen and W. Windle, eds.). C. C. Thomas, Springfield, Ill.

Neuroanatomy of the Auditory System. Report on Workshop (1973). (B. W. Konigsmark, ed.). *Arch. Otolaryng. 98,* 397–413.

Neurobiology of the Amygdala, The (1972). (B. E. Eleftheriou, ed.). Plenum Press, New York.

Neurobiology of Peptides (1978). (L. L. Iversen, R. A. Nicoll, and N. N. Vale, eds.). *Neurosci. Res. Progr. Bull.,* Vol. 16, MIT Press, Cambridge, Mass.

Neurogenic Bladder, The (1966). (E. Pedersen, ed.). *Acta Neurol. Scand. 42, Suppl. 20,* 1–186. (Proc. Third Scand. Symp. Multiple Sclerosis, Aarhus 1965).

Neurological Aspects of Auditory and Vestibular Disorders (1964). (W. S. Fields and B. R. Alford, eds.). C. C. Thomas, Springfield, Ill.

Neurological Basis of Behavior. Ciba Foundation Symposium (1958). (E. E. W. Wolstenholme and C. M. O'Connor, eds.). J. & A. Churchill, London.

Neuron-Target Cell Interactions (1976). *Neurosci. Res. Progr. Bull.,* Vol. 14 (B. H. Smith and G. W. Kreutzberg, eds.).

Neurophysiologie und Psychophysik des visuellen Systems (1961). Symposium Freiburg 1960 (R. Jung and H. Kornhuber, eds.) Springer-Verlag, Berlin.

Neurosciences, The. Third Study Program (1974). (F. O. Schmitt and F. G. Worden, eds.). MIT Press, Cambridge, Mass.

Neurosurgical Treatment in Psychiatry, Pain, and Epilepsy (1977). (W. H. Sweet, S. Obrador, and J. G. Martín-Rodríguez, eds.). Univ. Park Press, Baltimore.

Neurotransmitters (1972). *Res. Publ. Ass. Nerv. Ment. Dis., Vol. 50* (I. J. Kopin, ed.). Williams & Wilkins, Baltimore.

Neurotransmitters and Metabolic Regulation (1972). Biochem. Soc. Symp. No. 36 (R. M. Smellie, ed.). Biochemical Society, London.

New Developments in Electromyography and Clinical Neurophysiology, Vol. 1–3 (1973). (J. E. Desmedt, ed.). Karger, Basel.

Newman, P. P. (1974). *Visceral Afferent Functions of the Nervous System.* Edward Arnold, London.

Nicolaissen, B., and A. Brodal (1959). Chronic progressive external ophthalmoplegia. *Arch. Ophthal. 61,* 202–210.

Nielsen, J. M. (1946). *Agnosia, Apraxia, Aphasia. Their Value in Cerebral Localization,* 2nd ed. Hoeber, New York.

Nielsen, K. C., and C. Owman (1967). Adrenergic innervation of pial arteries related to the circle of Willis in the cat. *Brain Research 6,* 773–776.

Nieoullon, A., and Y. Gahéry (1978). Influence of pyramidotomy on limb flexion movements induced by cortical stimulation and associated postural adjustment in the cat. *Brain Research 149,* 39–52.

Nieoullon, A., and L. Rispal-Padel (1976). Somatotopic localization in cat motor cortex. *Brain Research 105,* 405–422.

Nieuwenhuys, R., J. Voogd, and C. van Huijzen (1978). *The Human Central Nervous System. A synopsis and Atlas.* Springer-Verlag, Berlin.

Niimi, K., M. Kadota, and Y. Matsushita (1974). Cortical projections of the pulvinar nuclear group of the thalamus in the cat. *Brain Behav. Evol. 9,* 422–457.

Niimi, K., S. Kawamura, and S. Ishimaru (1971). Projections of the visual cortex to the lateral geniculate and posterior thalamic nuclei in the cat. *J. Comp. Neurol. 143,* 279–312.

Niimi, K. and H. Matsuoka (1979). Thalamo-cortical organization of the auditory system in the cat studied by retrograde axonal transport of horseradish peroxidase. *Advan. Anat. Embryol. Cell Biol. 57,* 1–56.

Niimi, K., M. Niimi, and Y. Okada (1978). Thalamic afferents to the limbic cortex in the cat studied with the method of retrograde axonal transport of horseradish peroxidase. *Brain Research 145,* 225–238.

Niimi, K., and J. M. Sprague (1970). Thalamocortical organization of the visual system in the cat. *J. Comp. Neurol. 138,* 219–250.

Niimi, M. (1978). Cortical projections of the anterior thalamic nuclei in the cat. *Exp. Brain Res. 31,* 403–416.

Nijensohn, D. E., and F. W. L. Kerr (1975). The ascending projections of the dorsolateral funiculus of the spinal cord in the primate. *J. Comp. Neurol. 161,* 459–470.

Nishida, S., and M. Sears (1969). Dual innervation of the iris sphincter muscle of the albino guinea pig. *Exp. Eye Res. 8,* 467–469.

Nisimaru, N., and M. Yamamoto (1977). Depressant action of the posterior lobe of the cerebellum upon renal sympathetic nerve activity. *Brain Research 133,* 371–375.

Nissl, F. (1892). Über die Veränderungen der Ganglienzellen am Facialiskern des Kaninchens nach Ausreissung der Nerven. *Allg. Z. Psychiat. 48,* 197–198.

Nissl, F. (1908). Experimentalergebnisse zur Frage der Hirnrindenschichtung. *Mschr. Psychiat. Neurol. 23,* 186–188.

Noack, W., L. Dumitrescu, and J. U. Schweichel (1972). Scanning and electron microscopical investigations of the surface structures of the lateral ventricles in the cat. *Brain Research 46,* 121–129.

Nomura, Y., and H. F. Schuknecht (1965). The efferent fibers in the cochlea. *Ann. Otol. (St. Louis) 74,* 289–302.

Noordenbos, W. (1959). *Pain. Problems Pertaining to the Transmission of Nerve Impulses which give Rise to Pain.* Elsevier, Amsterdam.

Norberg, K. A. (1967). Transmitter histochemistry of the sympathetic adrenergic nervous system. *Brain Research 5,* 125–170.

Norberg, K. A., and L. Olson (1965). Adrenergic innervation of the salivary glands in the rat. *Z. Zellforsch. 68,* 183–189.

Nord, S. G., and G. S. Ross (1973). Responses of trigeminal units in the monkey bulbar lateral reticular formation to noxious and non-noxious stimulation of the face: experimental and theoretical considerations. *Brain Research 58,* 385–399.

Nord, S. G., and R. F. Young (1975). Projection of tooth pulp afferents to the cat trigeminal nucleus caudalis. *Brain Research 90,* 195–204.

Norgren, R. (1970). Gustatory responses in the hypothalamus. *Brain Research 21,* 63–77.

Norgren, R. (1974). Gustatory afferents to ventral forebrain. *Brain Research 81,* 285–295.

Norgren, R. (1978). Projections from the nucleus of the solitary tract in the rat. *Neurosci. 3,* 207–218.

Norgren, R., and C. M. Leonard (1973). Ascending central gustatory pathways. *J. Comp. Neurol. 150,* 217–238.

Norgren, R., and C. Pfaffmann (1975). The pontine taste area in the rat. *Brain Research 91,* 99–117.

Norgren, R., and G. Wolf (1975). Projections of thalamic gustatory and lingual areas in the rat. *Brain Research 92,* 123–129.

Nørholm, T., and I. Tygstrup (1960). Correlation between the clinical effect of stereotactic operations and brain-autopsy findings. *Acta Neurol. Scand. 39, Suppl. 4,* 196–203.

Norrsell, U. (1967a). A conditioned reflex study of sensory defects caused by cortical somatosensory ablations. *Physiol. Behav. 2,* 73–81.

Norrsell, U. (1967b). Afferent pathways of a tactile conditioned reflex after cortical somatosensory ablations. *Physiol. Behav. 2,* 83–86.

Norsell, U. (1979). Thermosensory defects after cervical spinal cord lesions in cat. *Exp. Brain Res. 35,* 479–494.

Norsell, U. (1980). Behavioural studies of the somatosensory system. *Physiol. Rev.* In press.

Norrsell, U., and P. Voorhoeve (1962). Tactile pathways from the hindlimb to the cerebral cortex in cat. *Acta Physiol. Scand. 54,* 9–17.

Noort, J. van (1969). The structure and connections of the inferior colliculus. An investigation of the lower auditory system. Thesis, van Gorcum & Co. N.V., Leiden.

Nyberg-Hansen, R. (1964a). Origin and termination of fibers from the vestibular nuclei descending in the medial longitudinal fasciculus. An experimental study with silver impregnation methods in the cat. *J. Comp. Neurol. 122,* 355–367.

Nyberg-Hansen, R. (1964b). The location and termination of tectospinal fibers in the cat. *Exp. Neurol. 9,* 212–227.

Nyberg-Hansen, R. (1965a). Sites and mode of termination of reticulospinal fibers in the cat. An experimental study with silver impregnation methods. *J. Comp. Neurol. 124,* 71–100.

Nyberg-Hansen, R. (1965b). Anatomical demonstration of gamma motoneurons in the cat's spinal cord. *Exp. Neurol. 13,* 71–81.

Nyberg-Hansen, R. (1966a). Functional organization of descending supraspinal fibre systems to the spinal cord. Anatomical observations and physiological correlations. *Ergebn. Anat. Entwickl.-Gesch. 39, Heft 2,* 1–48.

Nyberg-Hansen, R. (1966b). Sites of termination of interstitiospinal fibers in the cat. An experimental study with silver impregnation methods. *Arch. Ital. Biol. 104,* 98–111.

Nyberg-Hansen, R. (1966c). Innervation and nervous control of the urinary bladder. *Acta Neurol. Scand. 42, Suppl. 20,* 7 24.

Nyberg-Hansen, R. (1969a). Corticospinal fibres from the medial aspect of the cerebral hemisphere in the cat. An experimental study with the Nauta method. *Exp. Brain Res. 7,* 120–132.

Nyberg-Hansen, R. (1969b). Further studies on the origin of corticospinal fibres in the cat. An experimental study with the Nauta method. *Brain Research 16,* 39–54.

Nyberg-Hansen, R., and A. Brodal (1963). Sites of termination of corticospinal fibers in the cat. An experimental study with silver impregnation methods. *J. Comp. Neurol. 120,* 369–391.

Nyberg-Hansen, R., and A. Brodal (1964). Sites and mode of termination of rubrospinal fibres in the cat. An experimental study with silver impregnation methods. *J. Anat. (Lond.) 98,* 235–253.

Nyberg-Hansen, R., and J. Horn (1972). Functional aspects of cerebellar signs in clinical neurology. *Acta Neurol. Scand., Suppl. 51,* 219–245.

Nyberg-Hansen, R., and T. Mascitti (1964). Sites and mode of termination of fibers of the vestibulospinal tract in the cat. An experimental study with silver impregnation methods. *J. Comp. Neurol. 122,* 369–387.

Nyberg-Hansen, R., and E. Rinvik (1963). Some comments on the pyramidal tract, with special reference to its individual variations in man. *Acta Neurol. Scand. 39,* 1–30.

Nyby, O., and J. Jansen (1951). An experimental investigation of the corticopontine projection in *Macaca mulatta. Skr. Norske Vidensk.-Akad., I. Mat.-nat. Kl. No. 3*, 1–47.

Nygren, L.-G., and L. Olson (1977). A new major projection from locus coeruleus: the main source of noradrenergic nerve terminals in the ventral and dorsal columns of the spinal cord. *Brain Research 132*, 85–93.

Nylén, C. O. (1950). Positional nystagmus. A review and further prospects, *J. Laryng. 64*, 295–318.

Nyquist, B., S. Refsum, and A. Torkildsen (1939). Det infraclinoide carotis interna-aneurysme—hemicrania ophthalmoplegica. *Nord. Med. 3*, 2325–2335.

O'Connell, J. E. A. (1973). The anatomy of the optic chiasma and heteronymous hemianopia. *J. Neurol. Neurosurg. Psychiat. 36*, 710–723.

O'Connell, J. E. A. (1978). Trigeminal false localizing signs and their causation. *Brain 101*, 119–142.

O'Connell, J. E. A., and E. P. G. H. Du Boulay (1973). Binasal hemianopia. *J. Neurol. Neurosurg. Psychiat. 36*, 697–709.

Oculomotor System, The (1964). (M. B. Bender, ed.). Harper & Row, New York.

Oculomotor System and Brain Functions, The (1973). (V. Zikmund, ed.). Butterworths, London.

Ödkvist, L. M., D. W. F. Schwarz, J. M. Fredrickson, and R. Hassler (1974). Projection of the vestibular nerve to the area 3a arm field in the squirrel monkey (*Saimiri sciureus*). *Exp. Brain Res. 21*, 97–105.

Ödkvist, L. M., S. R. C. Liedgren, B. Larsby, and J. Jerlvall (1975). Vestibular and somatosensory inflow to the vestibular projection area in the post cruciate dimple region of the cat cerebral cortex. *Exp. Brain Res. 22*, 185–196.

Odutola, A. B. (1977). On the location of reticular neurons projecting to the cuneo-gracile nuclei in the rat. *Exp. Neurol. 54*, 54–59.

Ogawa, T., and S. Mitomo (1938). Eine experimentell-anatomische Studie über zwei merkwürdige Faserbahnen im Hirnstamm des Hundes: Tractus mesencephalo-olivaris medialis (Economo et Karplus) und Tractus tecto-cerebellaris. *Jap. J. Med. Sci., Trans. I. Anat. 7*, 77–94.

Ogren, M., and A Hendrickson (1976). Pathways between striate cortex and subcortical regions in *Macaca mulatta* and *Saimiri sciureus:* evidence for a reciprocal pulvinar connection. *Exp. Neurol. 53*, 780–800.

Ogura, J. H., and R. L. Lam (1953). Anatomical and physiological correlations on stimulating the human superior laryngeal nerve. *Laryngoscope (St. Louis) 63*, 947–959.

Ohye, C., R. Bouchard, L. Larochelle, P. Bédard, R. Boucher, B. Raphy, and L. J. Poirier (1970). Effect of dorsal rhizotomy on postural tremor in the monkey. *Exp. Brain Res. 10*, 140–150.

Ohye, C., K. Kubota, T. Hongo, T. Nagao, and H. Narabayashi (1964). Ventrolateral and subventrolateral thalamic stimulation. *Arch. Neurol. (Chic.) 11*, 427–434.

Ojemann, G. A. (1977). Asymmetric function of the thalamus in man. In *Evolution and Lateralization in the Brain* (S. J. Dimond and D. A. Blizard, eds.), pp. 380–396. New York Academy of Sciences, New York.

Ojemann, G. A., P. Fedio, and J. M. van Buren (1968). Anomia from pulvinar and subcortical parietal stimulation. *Brain 91*, 99–116.

Oka, H., and K. Jinnai (1978). Common projection of the motor cortex to the caudate nucleus and the cerebellum. *Exp. Brain Res. 31*, 31–42.

Oka, H., K. Sasaki, Y. Matsuda, T. Yasuda, and N. Mizuno (1975). Responses of pontocerebellar neurones to stimulation of the parietal association and the frontal motor cortices. *Brain Research 93*, 399–407.

O'Keefe, J., and L. Nadel (1978). *The Hippocampus as a Cognitive Map*. Clarendon Press, Oxford.

O'Leary, J. L. (1940). A structural analysis of the lateral geniculate nucleus of the cat. *J. Comp. Neurol. 73*, 405–430.

O'Leary, J. (1941). Structure of the area striata of the cat. *J. Comp. Neurol. 75*, 131–164.

O'Leary, J. L., D. S. B. Smith, J. M. Inukai, and M. O'Leary (1970). Termination of olivocerebellar system in the cat. *Arch. Neurol. (Chic.) 22*, 193–206.

Oleson, J. (1971). Contralateral focal increase of cerebral blood flow in man during arm work. *Brain 94*, 635–646.

Olfaction and Taste. V. Proceedings of the Fifth International Symposium (1975). (D. A. Denton and J. P. Coghlan, eds.). Academic Press, New York.

Olivecrona, Hans (1957). Paraventricular nucleus and pituitary gland. *Acta Physiol. Scand. 40, Suppl. 136*, 1–178.

Olivecrona, Herbert (1942). Tractotomy for relief of trigeminal neuralgia. *Arch. Neurol. Psychiat. (Chic.) 47*, 544–564.

Oliver, D. L., and W. C. Hall (1975). Subdivisions of the medial geniculate body in the tree shrew *(Tupaia glis)*. *Brain Research 86*, 217–227.

Oliver, D. L., and W. C. Hall (1978a). The medial geniculate body of the tree shrew, *Tupaia glis*. I. Cytoarchitecture and midbrain connections. *J. Comp. Neurol. 182*, 423–458.

Oliver, D. L., and W. C. Hall (1978b). The medial geniculate body of the tree shrew, *Tupaia glis*. II. Connections with the neocortex. *J. Comp. Neurol. 182*, 459–494.

Oliver, D. L., M. M. Merzenich, G. L. Roth, W. C. Hall, and J. H. Kaas (1976). Tonotopic organization and connections of primary auditory cortex in the tree shrew, *Tupaia glis*. *Anat. Rec. 184*, p. 491.

Oliver, J. E., Jr., W. E. Bradley, and T. F. Fletcher (1969a). Identification of preganglionic parasympathetic neurons in the sacral spinal cord of the cat. *J. Comp. Neurol. 137*, 321–328.

Oliver, J. E., Jr., W. E. Bradley, and T. F. Fletcher (1969b). Spinal cord representation of the micturition reflex. *J. Comp. Neurol. 137*, 329–346.

Oliver, L. C. (1952). The supranuclear arc of the corneal reflex. *Acta Psychiat. Scand. 27*, 329–333.

Oliveras, J. L., J. M. Besson, G. Guilbaud, and J. C. Liebeskind (1974). Behavioral and electrophysiological evidence of pain inhibition from midbrain stimulation in the cat. *Exp. Brain Res. 20*, 32–44.

Oliveras, J. L., F. Redjemi, G. Guilbaud, and J. M. Besson (1975). Analgesia induced by electrical stimulation of the inferior centralis nucleus of the raphe in the cat. *Pain 1*, 139–145.

Olmstead, C. E., J. R. Villablanca, R. J. Marcus, and D. L. Avery (1976). Effects of caudate nuclei or frontal cortex ablations in cats. IV. Bar pressing, maze learning, and performance. *Exp. Neurol. 53*, 670–693.

Olpe, H. R., and W. P. Koella (1977). The response of striatal cells upon stimulation of the dorsal and median raphe nuclei. *Brain Research 122*, 357–360.

Olson, L., and K. Fuxe (1971). On the projections from the locus coeruleus noradrenaline neurons: the cerebellar innervation. *Brain Research 28*, 165–171.

Olson, L., and K. Fuxe (1972). Further mapping out of central noradrenaline neuron systems: projections of the 'subcoeruleus' area. *Brain Research 43*, 289–295.

Olson, L., and T. Malmfors (1970). Growth characteristics of adrenergic nerves in the adult rat. *Acta Physiol. Scand., Suppl. 348*, 1–112.

Olszewski, J. (1950). On the anatomical and functional organization of the spinal trigeminal nucleus. *J. Comp. Neurol. 92*, 401–413.

Olszewski, J. (1952). *The Thalamus of the Macaca Mulatta. An Atlas for Use with the Stereotaxic Instrument*. S. Karger, Basel.

Olszewski, J. (1954). The cytoarchitecture of the human reticular formation. In *Brain Mechanisms and Consciousness* (J. F. Delafresnaye, ed.), pp. 54–80. Blackwell, Oxford.

Olszewski, J., and D. Baxter (1954). *Cytoarchitecture of the Human Brain Stem*. S. Karger, New York.

Olton, D. S., J. A. Walker, and F. H. Gage (1978). Hippocampal connections and spatial discrimination. *Brain Research 139*, 295–308.

O'Neal, J. T., and L. E. Westrum (1973). The fine structural synaptic organization of the cat lateral cuneate nucleus. A study of sequential alterations in degeneration. *Brain Research 51*, 97–124.

Operating on the Mind. The Psychosurgery Conflict. (1975). (W. M. Gaylin, J. S. Meister, and R. C. Neville, eds.). Basic Books, New York.

Oral-Facial Sensory and Motor Mechanisms (1971). (R. Dubner and Y. Kawamura, eds.). Appleton-Century-Crofts, New York.

Orbach, J., and K. L. Chow (1959). Differential effects of resections of somatic areas I and II in monkeys. *J. Neurophysiol. 22*, 195–203.

Orioli, F. L., and F. A. Mettler (1956). The rubro-spinal tract in *Macaca mulatta*. *J. Comp. Neurol. 106*, 299–318.

Ortmann, R. (1951). Über experimentelle Veränderungen der Morphologie des Hypophysenzwischenhirnsystems und die Beziehung der sog. "Gomorisubstanz" zum Adiuretin. *Z. Zellforsch. 36*, 92–140.

Ortmann, R. (1960). Neurosecretion. In *Handbook of Physiology, Section 1: Neurophysiology*, Vol. II. (J. Field, H. W. Magoun, and V. E. Hall, eds.), pp. 1039–1065. Amer. Physiol. Soc., Washington, D.C.

Oscarsson, O. (1965). Functional organization of the spino- and cuneocerebellar tracts. *Physiol. Rev. 45*, 495–522.

Oscarsson, O. (1967). Termination and functional organization of a dorsal spino-olivocerebellar path. *Brain Research 5*, 531–534.

Oscarsson, O. (1973). Functional organization of spinocerebellar paths. In *Handbook of Sensory Physiology*, Vol. II, *Somatosensory System* (A. Iggo, ed.), pp. 339–380. Springer-Verlag, Berlin.

Oscarsson, O., and I. Rosén (1963). Cerebral projection of group I afferents in fore-limb muscle nerves of cat. *Experientia (Basel) 19*, 206–207.

Oscarsson, O., and I. Rosén (1966a). Short-latency projections to the cat's cerebral cortex from skin and muscle afferents in the contralateral forelimb. *J. Physiol. (Lond.) 182*, 164–184.

Oscarsson, O., and I. Rosén (1966b). Response characteristics of reticulocerebellar neurones activated from spinal afferents. *Exp. Brain Res. 1*, 320–328.

Oscarsson, O., and B. Sjölund (1977). The ventral spino-olivocerebellar system in the cat. I. Identification of five paths and their termination in the cerebellar anterior lobe. *Exp. Brain Res. 28*, 469–486.

Oscarsson, O., and N. Uddenberg (1964). Identification of a spinocerebellar tract activated from forelimb afferents in the cat. *Acta Physiol. Scand. 62*, 125–136.

Oscarsson, O., and N. Uddenberg (1966). Somatotopic termination of spino-olivocerebellar path. *Brain Research 3*, 204–207.

Osen, K. K. (1969). Cytoarchitecture of the cochlear nuclei in the cat. *J. Comp. Neurol. 136*, 453–484.

Osen, K. K. (1970). Course and termination of the primary afferents in the cochlear nuclei of the cat. An experimental anatomical study. *Arch. Ital. Biol. 108*, 21–51.

Osen, K. K. (1972). The projection of the cochlear nuclei to the inferior colliculus in the cat. *J. Comp. Neurol. 144*, 355–372.

Osen, K. K., and J. Jansen (1965). The cochlear nuclei in the common porpoise, *Phocaena phocaena*. *J. Comp. Neurol. 125*, 223–257.

Osen, K. K., and K. Roth (1969). Histochemical localization of cholinesterases in the cochlear nuclei of the cat, with notes on the origin of acetylcholinesterase-positive afferents and the superior olive. *Brain Research 16*, 165–185.

O'Steen, W. K., and G. M. Vaughan (1968). Radioactivity in the optic pathway and hypothalamus of the rat after intraocular injection of tritiated 5-hydroxy-tryptophan. *Brain Research 8*, 209–212.

Ostertag, B. (1936). *Einteilung und Charakteristik der Hirngewächse*. G. Fischer, Jena.

Otophysiology Advan. Oto-rhino-laryng., Vol. 20 (1973). (C. R. Pfaltz, ser. ed.; J. E. Hawkins, M. Lawrence, and W. P. Work, vol. eds.). Karger, Basel.

Otsuka, R., and R. Hassler (1962). Über Aufbau und Gliederung der corticalen Sehsphäre bei der Katze. *Arch. Psychiat. Z. ges. Neurol. 203*, 212–234.

Ottersen, O. P., and Y. Ben-Ari (1978a). Demonstration of a heavy projection of midline thalamic neurons upon the lateral nucleus of the amygdala of the rat. *Neurosci. Letters 9,* 147–152.

Ottersen, O. P., and Y. Ben-Ari (1978b). Pontine and mesencephalic afferents to the central nucleus of the amygdala of the rat. *Neurosci. Letters 8,* 329–334.

Ottersen, O. P., and Y. Ben-Ari (1979). Afferent connections to the amygdaloid complex of the rat and cat. I. Projections from the thalamus. *J. Comp. Neurol. 187,* 401–424.

Outcome of Severe Damage to the Central Nervous System (1975). Ciba Foundation Symposium 34 (new series). (R. Porter and D. W. Fitzsimons, eds.). Elsevier, Excerpta Medica, Amsterdam.

Padel, Y., J. Armand, and A. M. Smith (1972). Topography of rubrospinal units in the cat. *Exp. Brain Res. 14,* 363–371.

Padel, Y., A. M. Smith, and J. Armand (1973). Topography of projections from the motor cortex to rubrospinal units in the cat. *Exp. Brain Res. 17,* 315–332.

Pain. Res. Publ. Ass. Res. Nerv. Dis., Vol. 23 (1943). (H. G. Wolff, H. S. Gasser, and J. C. Hinsey, eds.). Williams & Wilkins, Baltimore.

Pain (1978). (F. W. L. Kerr and K. L. Casey, eds.). *Neurosci. Res. Progr. Bull.,* Vol. 16. MIT Press, Cambridge, Mass.

Pain, International Symposium on (1974). (J. J. Bonica, ed.). *Advan. Neurol.,* Vol. 4. Raven Press, New York.

Pain in the Trigeminal Region (1977). (D. J. Anderson and B. Matthews, eds.). Elsevier/North-Holland Biomedical Press, Amsterdam.

Paintal, A. S. (1977). Thoracic receptors connected with sensation. *Brit. Med. Bull. 33,* 169–174.

Pakkenberg, H. (1966). The number of nerve cells in the cerebral cortex of man. *J. Comp. Neurol. 128,* 17–20.

Palay, L. (1955). An electron microscope study of the neurohypophysis in normal, hydrated and dehydrated rats. (Abstract.) *Anat. Rec. 121,* 348.

Palay, S. L. (1957). The fine structure of the neurohypophysis. In *Ultrastructure and Cellular Chemistry of Neural Tissue* (H. Waelsch, ed.), pp. 31–44. Hoeber, New York.

Palay, S. L., and V. Chan-Palay (1974). *Cerebellar Cortex. Cytology and Organization.* Springer-Verlag, Berlin.

Palkovits, M., M. Brownstein, J. M. Saavedra, and J. Axelrod (1974). Norepinephrine and dopamine content of hypothalamic nuclei of the rat. *Brain Research 77,* 137–149.

Palkovits, M., M. Fekete, G. B. Makara, and J. P. Herman (1977). Total and parietal hypothalamic deafferentiations for topographical identification of catecholaminergic innervations of certain preoptic and hypothalamic nuclei. *Brain Research 127,* 127–136.

Palkovits, M., C. Léránth, L. Záborszky, and M. J. Brownstein (1977). Elecron microscopic evidence of direct neuronal connections from the lower brain stem to the median eminence. *Brain Research 136,* 339–344.

Palkovits, M., P. Magyar, and J. Szentágothai (1971a). Quantitative histological analysis of the cerebellar cortex in the cat. I. Number and arrangement in space of the Purkinje cells. *Brain Research 32,* 1–13.

Palkovits, M., P. Magyar, and J. Szentágothai (1971b). Quantitative histological analysis of the cerebellar cortex in the cat. III. Structural organization of the molecular layer. *Brain Research 34,* 1–18.

Palkovits, M., E. Mezey, J. Hámori, and J. Szentágothai (1977). Quantitative histological analysis of the cerebellar nuclei in the cat. I. Numerical data on cells and on synapses. *Exp. Brain Res. 28,* 189–209.

Palmer, L. A., and A. C. Rosenquist (1974). Visual receptive fields of single striate cortical units projecting to the superior colliculus in the cat. *Brain Research 67,* 27–42.

Pandya, D. N., P. Dye, and N. Butters (1971). Efferent corticocortical projections of the prefrontal cortex in the rhesus monkey. *Brain Research 31,* 35–46.

Pandya, D. N., M. Hallett, and S. K. Mukherjee (1969). Intra- and interhemispheric connections of the neocortical auditory system in the rhesus monkey. *Brain Research 14,* 49–65.

Pandya, D. N., E. A. Karol, and D. Heilbronn (1971). The topographical distribution of interhemispheric projections in the corpus callosum of the rhesus monkey. *Brain Research 32,* 31–43.

Pandya, D. N., and H. G. J. M. Kuypers (1969). Corticocortical connections in the rhesus monkey. *Brain Research 13,* 13–36.

Pandya, D. N., and F. Sanides (1973). Architectonic parcellation of the temporal operculum in rhesus monkey and its projection pattern. *Z. Anat. Entwickl.-Gesch. 139,* 127–161.

Pandya, D. N., G. W. Van Hoesen, and V. B. Domesick (1973). A cingulo-amygdaloid projection in the rhesus monkey. *Brain Research 61,* 369–373.

Pandya, D. N., and L. A. Vignolo (1968). Interhemispheric neocortical projections of somatosensory areas I and II in the rhesus monkey. *Brain Research 7,* 300–303.

Pandya, D. N., and L. A. Vignolo (1971). Intra- and interhemispheric projections of the precentral, premotor and arcuate areas in the rhesus monkey. *Brain Research 26,* 217–233.

Papez, J. W. (1937). A proposed mechanism of emotion. *Arch. Neurol. Psychiat. (Chic.) 38,* 725–743.

Pappas, G. D. (1975). The fine structure of electrotonic synapses. In *Golgi Centennial Symposium: Perspectives in Neurobiology* (M. Santini, ed.), pp. 339–345. Raven Press, New York.

Pappas, G. D., E. B. Cohen, and D. P. Purpura (1966). Fine structure and nonsynaptic neuronal relations in the thalamus of the cat. In *The Thalamus* (D. P. Purpura and M. D. Yahr, eds.), pp. 47–71. Columbia Univ. Press, New York.

Pappas, G. D., and S. G. Waxman (1972). Synaptic fine structure—morphological correlates of chemical and electrotonic transmission. In *Structure and Function of Synapses* (G. D. Pappas and D. P. Purpura, eds.), pp. 1–43. Raven Press, New York.

Partlow, G. D., M. Colonnier, and J. Szabo (1977). Thalamic projections of the superior colliculus in the rhesus monkey, *Macaca mulatta.* A light and electron microscopic study. *J. Comp. Neurol. 171,* 285–378.

Pasik, P., T. Pasik, and M. DiFiglia (1976). Quantitative aspects of neuronal organization in the neostriatum of the macaque monkey. In *The Basal Ganglia* (M. D. Yahr, ed.), pp. 57–89. Raven Press, New York.

Pasik, T., P. Pasik, and J. Hámori (1973). Nucleus of the accessory optic tract. Light and electron microscopic study in normal monkeys and after eye enucleation. *Exp. Neurol. 41,* 612–627.

Pasik, P., T. Pasik, J. Hámori, and J. Szentágothai (1973). Golgi type II interneurons in the neuronal circuit of the monkey lateral geniculate nucleus. *Exp. Brain Res. 17,* 18–34.

Pasquier, D. A., and F. Reinoso-Suarez (1976). Direct projections from hypothalamus to hippocampus in the rat demonstrated by retrograde transport of horseradish peroxidase. *Brain Research 108,* 165–169.

Pasquier, D. A., and F. Reinoso-Suarez (1977). Differential efferent connections of the brain stem to the hippocampus in the cat. *Brain Research 120,* 540–548.

Pass, I. J. (1933). Anatomic and functional relationship of the nucleus dorsalis (Clarke's column) and of the dorsal spinocerebellar tract (Flechsig's). *Arch. Neurol. Psychiat. (Chic.) 30,* 1025–1045.

Patton, H. D., T. C. Ruch, and A. E. Walker (1944). Experimental hypogeusia from Horsley-Clarke lesions of the thalamus in *Macaca mulatta. J. Neurophysiol. 7,* 171–184.

Paul, R. L., H. Goodman, and M. Merzenich (1972). Alterations in mechanoreceptor input to Brodmann's areas 1 and 3 of the postcentral hand area of *Macaca mulatta* after nerve section and regeneration. *Brain Research 39,* 1–19.

Paul, R. L., M. Merzenich, and H. Goodman (1972). Representation of slowly and rapidly adapting cutaneous mechanoreceptors of the hand in Brodmann's areas 3 and 1 of *Macaca mulatta. Brain Research 36,* 229–249.

Paul, S. M., R. G. Heath, and J. P. Ellison (1973). Histochemical demonstration of a direct pathway from the fastigial nucleus to the septal region. *Exp. Neurol. 40,* 798–805.

Paula-Barbosa, M. M., and A. Sousa-Pinto (1973). Auditory cortical projections to the superior colliculus in the cat. *Brain Research 50,* 47–61.

Paulsen, K. (1958). Über Vorkommen und Zahl von Muskelspindeln in inneren Kehlkopfmuskeln des Menschen (M. cricoarytaenoideus dorsalis, M. cricothyreoideus). *Z. Zellforsch. 48,* 349–355.

Peachey, L. (1971). The structure of the extra-ocular muscle fibers of mammals. In *The Control of Eye Movements* (P. Bach-y-Rita, C. C. Collins, and J. H. Hyde, eds.), pp. 45–66. Academic Press, New York.

Pearce, G. W. (1960). Some cortical projections to the midbrain reticular formation. In *Structure and Function of the Cerebral Cortex* (D. B. Tower and J. P. Schadé, eds.), pp. 131–137. Elsevier, Amsterdam.

Pearson, A. A. (1938). The spinal accessory nerve in human embryos. *J. Comp. Neurol. 68,* 243–266.

Pearson, A. A. (1939). The hypoglossal nerve in human embryos. *J. Comp. Neurol. 71,* 21–39.

Pearson, A. A. (1944). The oculomotor nucleus in the human fetus. *J. Comp. Neurol. 80,* 47–68.

Pearson, A. A. (1945). Observations on the root of the facial nerve in human fetuses. (Abstract.) *Anat. Rec. 91,* 294–295.

Pearson, A. A. (1949a). The development and connections of the mesencephalic root of the trigeminal nerve in man. *J. Comp. Neurol. 90,* 1–46.

Pearson, A. A. (1949b). Further observations on the mesencephalic root of the trigeminal nerve. *J. Comp. Neurol. 91,* 147–194.

Pearson, A. A. (1952). Role of gelatinous substance of spinal cord in conduction of pain. *Arch. Neurol. Psychiat. (Chic.) 68,* 515–529.

Pearson, R. C. A., P. Brodal, and T. P. S. Powell (1978). The projection of the thalamus upon the parietal lobe in the monkey. *Brain Research 144,* 143–148.

Pecci Saavedra, J., and O. L. Vaccarezza (1968). Synaptic organization of the glomerular complexes in the lateral geniculate nucleus of cebus monkey. *Brain Research 8,* 389–393.

Pedersen, E. (1959). Epidemic vertigo. Clinical picture, epidemiology and relation to encephalitis. *Brain 82,* 566–580.

Pedersen, E., and V. Grynderup (1966). Clinical pharmacology of the neurogenic bladder. *Acta Neurol. Scand. 42, Suppl. 20,* 111–120.

Pedersen, E., and P. Juul-Jensen (1965). Treatment of spasticity by subarachnoid phenolglycerin. *Neurology 15,* 256–262.

Peele, T. L. (1942). Cytoarchitecture of individual parietal areas in the monkey (*Macaca mulatta*) and the distribution of the efferent fibers. *J. Comp. Neurol. 77,* 693–737.

Peitersen, E. (1965). *Vestibulo-spinale reflexer. Kliniske og eksperimentelle undersögelser ved hjælp af steppingtesten.* Munksgaard, København.

Pellegrini, M., O. Pompeiano, and N. Corvaja (1977). Identification of different size motoneurons labeled by the retrograde axonal transport of horseradish peroxidase. *Pflügers Arch. ges. Physiol. 368,* 161–163.

Penfield, W. (1958). Centrencephalic integrating system. *Brain 81,* 231–234.

Penfield, W. G., and E. Boldrey (1937). Somatic motor and sensory representation in the cerebral cortex of man as studied by electrical stimulation. *Brain 60,* 389–443.

Penfield, W., and T. C. Erickson (1941). *Epilepsy and Cerebral Localization. A Study of the Mechanism, Treatment and Prevention of Epileptic Seizures.* C. C. Thomas, Springfield, Ill.

Penfield, W., and M. E. Faulk, Jr. (1955). The insula. Further observations on its function. *Brain 78,* 445–470.

Penfield, W., and H. Jasper (1954). *Epilepsy and the Functional Anatomy of the Human Brain*. Little, Brown, Boston.

Penfield, W., and K. Kirstiansen (1951). *Epileptic Seizure Patterns*. C. C. Thomas, Springfield, Ill.

Penfield, W., and B. Milner (1958). Memory deficit produced by bilateral lesions in the hippocampal zone. *Arch. Neurol. Psychiat. (Chic.) 79*, 475–497.

Penfield, W., and P. Perot (1963). The brain's record of auditory and visual experience. A final summary and discussion. *Brain 86*, 595–696.

Penfield, W., and T. Rasmussen (1950). *The Cerebral Cortex of Man*. Macmillan, New York.

Penfield, W., and L. Roberts (1959). *Speech and Brain Mechanisms*. Princeton Univ. Press, Princeton.

Penfield, W., and K. Welch (1951). The supplementary motor area of the cerebral cortex. A clinical and experimental study. *Arch. Neurol. Psychiat. (Chic.) 66*, 289–317.

Perl, E. R. (1976). Sensitization of nociceptors and its relation to sensation. In *Advances in Pain Research and Therapy*, Vol. 1 (J. J. Bonica and D. Albe-Fessard, eds.), pp. 17–28. Raven Press, New York.

Perl, E. R., and D. G. Whitlock (1961). Somatic stimuli exciting spinothalamic projections to thalamic neurons in cat and monkey. *Exp. Neurol. 3*, 256–296.

Perl, E. R., D. G. Whitlock, and J. R. Gentry (1962). Cutaneous projection to second-order neurons of the dorsal column system. *J. Neurophysiol. 25*, 337–358.

Perria, L., G. Rosadini, and G. F. Rossi (1961). Determination of side of cerebral dominance with amobarbital. *Arch. Neurol. (Chic.) 4*, 172–181.

Perry, T. L., S. Hansen, and M. Kloster (1973). Huntington's chorea—deficiency of γ-aminobutyric acid in brain. *New Engl. J. Med. 288*, 337–342.

Peters, A., and S. L. Palay (1966). The morphology of laminae A and A_1 of the dorsal nucleus of the lateral geniculate body of the cat. *J. Anat. (Lond.) 100*, 451–486.

Peters, A., and J. Saldanha (1976). The projection of the lateral geniculate nucleus to area 17 of the rat cerebral cortex. III. Layer VI. *Brain Research 105*, 533–537.

Peters, A., and T. M. Walsh (1972). A study of the organization of apical dendrites in the somatic sensory cortex of the rat. *J. Comp. Neurol. 144*, 253–268.

Petersén, I., S. Kollberg, and K.-G. Dhunér (1961). The effect of the intravenous injection of succinylcholine on micturition. An electromyographic study. *Brit. J. Urol. 33*, 392–396.

Petersén, I., and B. Stener (1959). Experimental evaluation of the hypothesis of ligamento-muscular protective reflexes. III. A study in man using the medial collateral ligament of the knee joint. *Acta Physiol. Scand. 48, Suppl. 166*, 51–61.

Petersén, I., I. Stener, U. Selldén, and S. Kollberg (1962). Investigation of urethral sphincter in women with simultaneous electromyography and micturition urethro-cystography. *Acta Neurol. Scand. 38, Suppl., 3*, 145–151.

Peterson, B. W. (1970). Distribution of neural responses to tilting within vestibular nuclei of the cat. *J. Neurophysiol. 33*, 750–767.

Peterson, B. W. (1977). Identification of reticulo-spinal projections that may participate in gaze control. In *Control of Gaze by Brain Stem Neurons* (R. Baker and A. Berthoz, eds.), pp. 143–152. Elsevier/North-Holland, Amsterdam.

Peterson, B. W., M. E. Anderson, and M. Filion (1974). Responses of ponto-medullary reticular neurons to cortical, tectal and cutaneous stimuli. *Exp. Brain Res. 21*, 19–44.

Peterson, B. W., and J. D. Coulter (1977). A new long spinal projection from the vestibular nuclei in the cat. *Brain Research 122*, 351–356.

Peterson, B. W., and L. P. Felpel (1971). Excitation and inhibition of reticulospinal neurons by vestibular, cortical and cutaneous stimulation. *Brain Research 27*, 373–376.

Peterson, B. W., M. Filion, L. P. Felpel, and C. Abzug (1975). Responses of medial reticular neurons to stimulation of the vestibular nerve. *Exp. Brain Res. 22*, 335–350.

Peterson, B. W., R. A. Maunz, N. G. Pitts, and R. G. Mackel (1975). Patterns of projection and branching of reticulospinal neurons. *Exp. Brain Res. 23*, 333–351.

Peterson, B. W., N. G. Pitts, and K. Fukushima (1979). Reticulospinal connections with limb and axial motoneurons. *Exp. Brain Res. 36,* 1–20.

Petras, J. M. (1967). Cortical, tectal and tegmental fiber connections in the spinal cord of the cat. *Brain Research 6,* 275–324.

Petras, J. M. (1972). Corticostriate and corticothalamic connections in the chimpanzee. In *Corticothalamic Projections and Sensorimotor Activities* (T. L. Frigyesi, E. Rinvik, and M. D. Yahr, eds.), pp. 201–216. Raven Press, New York.

Petras, J. M., and J. F. Cummings (1972). Autonomic neurons in the spinal cord of the rhesus monkey: a correlation of the findings of cytoarchitectonics and sympathectomy with fiber degeneration following dorsal rhizotomy. *J. Comp. Neurol. 146,* 189–218.

Petras, J. M., and J. F. Cummings (1977). The origin of spinocerebellar pathways. II. The nucleus centrobasalis of the cervical enlargement and the nucleus dorsalis of the thoracolumbar spinal cord. *J. Comp. Neurol. 173,* 693–716.

Petras, J. M., and A. I. Faden (1978). The origin of sympathetic preganglionic neurons in the dog. *Brain Research 144,* 353–357.

Petrén, K. (1910). Ueber die Bahnen der Sensibilität im Rückenmarke, besonders nach den Fällen von Stichverletzung studiert. *Arch. Psychiat. 47,* 495–557.

Petrides, M., and S. D. Iversen (1979). Restricted posterior parietal lesions in the rhesus monkey and performance on visuospatial tasks. *Brain Research 161,* 63–77.

Petrovicky, P. (1966). A comparative study of the reticular formation of the guinea pig. *J. Comp. Neurol. 128,* 85–108.

Pfaffmann, C. (1939). Afferent impulses from the teeth due to pressure and noxious stimulation. *J. Physiol. (Lond.) 97,* 207–219.

Pfeifer, R. A. (1920). Myelogenetisch-anatomische Untersuchungen über das kortikale Ende der Hörleitung. *Abh. Math.-Physik. Kl. Sächs. Akad. Wiss.,* Vol. *37,* No. II, 1–54.

Pfeifer, R. A. (1925). Myelogenetisch-anatomische Untersuchungen über den zentralen Abschnitt der Sehleitung. *Monogr. Gesamtgeb. Neurol. Psychiat.,* Heft *43,* 1–149.

Pfeifer, R. A. (1936). Pathologie der Hörstrahlung und der corticalen Hörsphäre. In *Handbuch der Neurologie* (Bumke-Foerster, eds.), Vol. *6,* pp. 533–626. Berlin.

Pfeifer, R. A. (1940). *Die angioarchitektonische areale Gliederung der Grosshirnrinde.* Georg Thieme, Leipzig.

Pfenninger, K. H. (1973). Synaptic Morphology and Cytochemistry. *Progr. Histochem. Cytochem.,* Vol. 5, No. 1. Fisher Verlag, Stuttgart-Portland.

Phillips, C. G. (1967). Corticomotoneuronal organization. Projection from the arm area of the Baboon's motor cortex. *Arch. Neurol. (Chic.) 17,* 188–195.

Phillips, C. G. (1969). Motor apparatus of the baboon's hand. The Ferrier Lecture 1968. *Proc. Roy. Soc. B 173,* 141–174.

Phillips, C. G., and R. Porter (1964). The pyramidal projection to motoneurones of some muscle groups of the baboon's forelimb. In *Physiology of Spinal Neurons, Progress in Brain Research,* Vol. 12 (J. C. Eccles and J. P. Schadé, eds.), pp. 222–242. Elsevier, Amsterdam.

Phillips, C. G., and R. Porter (1977). *Corticospinal Neurones. Their Role in Movement.* Monograph of the Physiol. Society, No. 34. Academic Press, London.

Phillips, C. G., T. P. S. Powell, and M. Wiesendanger (1971). Projection from low-threshold muscle afferents of hand and forearm to area 3a of baboon's cortex. *J. Physiol. (Lond.) 217,* 419–446.

Physiological Basis of Mental Activity, The (1963). (R. Hernández-Peón, ed.). *Electroenceph. Clin. Neurophysiol. Suppl. 24.* Elsevier, Amsterdam.

Physiology of the Auditory System. A Workshop (1971). (M. B. Sachs, ed.). National Educational Consultants, Inc., Baltimore.

Pick, J. (1970). *The Autonomic Nervous System. Morphological, Comparative, Clinical and Surgical Aspects.* Lippincott, Philadelphia.

Pick, J., and D. Sheehan (1946). Sympathetic rami in man. *J. Anat. (Lond.) 80,* 12–20.

Pickel, V. M., M. Segal, and F. E. Bloom (1974). A radioautographic study of the efferent pathways of the nucleus locus coeruleus. *J. Comp. Neurol. 155,* 15–42.

Pickering, B. T. (1976). The molecules of neurosecretion: their formation, transport and release. In *Perspectives in Brain Research, Progress in Brain Research,* Vol. 45 (M. A. Corner and D. F. Swaab, eds.), pp. 161–179. Elsevier, Amsterdam.

Piercy, M. (1967). Studies of the neurological basis of intellectual functions. Clinical studies. In *Modern Trends in Neurology,* Vol. 4 (D. Williams, ed.), pp. 106–124. Butterworths, London.

Pierson, R. J., and M. B. Carpenter (1974). Anatomical analysis of pupillary reflex pathway. *J. Comp. Neurol. 158,* 121–144.

Pilgrim, C. (1974). Histochemical differentiation of hypothalamic areas. In *Integrative Hypothalamic Activity, Progress in Brain Research,* Vol. 41 (D. F. Swaab and J. P. Schadé, eds.), pp. 97–110. Elsevier, Amsterdam.

Pilyavsky, A. (1975). Characteristics of fast and slow corticobulbar fibre projections to reticulospinal neurones. *Brain Research 85,* 49–52.

Pinching, A. J. (1977). Clinical testing of olfaction reassessed. *Brain 100,* 377–388.

Pinching, A. J., and K. B. Døving (1974). Selective degeneration in the rat olfactory bulb following exposure to different odours. *Brain Research 82,* 195–204.

Pinching, A. J., and T. P. S. Powell (1971). The neuropil of the glomeruli of the olfactory bulb. *J. Cell Sci. 9,* 347–377.

Pitts, R. F. (1940). The respiratory center and its descending pathways. *J. Comp. Neurol. 72,* 605–625.

Pitts, R. F., H. W. Magoun, and S. W. Ranson (1939). Localization of the medullary respiratory centers in the cat. *Amer. J. Physiol. 126,* 673–688.

Pituitary Gland, The, Vols. 1–3 (1966). (G. W. Harris and B. T. Donovan, eds.). Butterworths, London.

Pituitary, The. A Current Review (1977). (M. B. Allen, Jr. and V. B. Mahesh, eds.). Academic Press, New York.

Plasticity and Recovery of Function in the Central Nervous System (1974). (D. G. Stein, J. J. Rosen, and N. Butters, eds.). Academic Press, New York.

Poggio, G. F., and V. B. Mountcastle (1960). A study of the functional contributions of the lemniscal and spinothalamic systems to somatic sensibility. *Bull. Johns Hopk. Hosp. 106,* 266–316.

Poggio, G. F., and V. B. Mountcastle (1963). The functional properties of ventrobasal thalamic neurons studied in unanesthetized monkeys. *J. Neurophysiol. 26,* 775–806.

Poirier, L. J. (1960). Experimental and histological study of midbrain dyskinesias. *J. Neurophysiol. 23,* 534–551.

Poirier, L. J., and G. Bouvier (1966). The red nucleus and its efferent nervous pathways in the monkey. *J. Comp. Neurol. 128,* 223–244.

Poirier, L. J., G. Bouvier, P. Bédard, R. Boucher, L. Larochelle, A. Olivier, and P. Singh (1969). Essai sur les circuits neuronaux impliqués dans le tremblement postural et l'hypokinésie. *Rev. Neurol. 120,* 15–40.

Poirier, L. J., M. Filion, L. Larochelle, and J. C. Péchadre (1975). Physiopathology of experimental parkinsonism in the monkey. *Canad. J. Neurol. Sci. 2,* 255–263.

Poirier, L. J., E. G. McGeer, L. Larochelle, P. L. McGeer, P. Bédard, and R. Boucher (1969). The effect of brain stem lesions on tyrosine and tryptophan hydroxylase in various structures of the telencephalon of the cat. *Brain Research 14,* 147–155.

Poirier, L. J., P. Singh, R. Boucher, G. Bouvier, A. Olivier, and P. Larochelle (1967). Effect of brain lesions on the concentration of striatal dopamine and serotonin in the cat. *Arch. Neurol. (Chic.) 17,* 601–608.

Poirier, L. J., and T. L. Sourkes (1965). Influence of the substantia nigra on the catecholamine content of the striatum. *Brain 88,* 181–192.

Poirier, L. J., T. L. Sourkes, G. Bouvier, R. Boucher, and S. Carabin (1966). Striatal amines, experimental tremor and the effect of harmaline in the monkey. *Brain 89,* 37–52.

Pola, J., and D. A. Robinson (1976). An explanation of eye movements seen in internuclear ophthalmoplegia. *Arch. Neurol. (Chic.) 33,* 447–452.

Polácek, P. (1966). Receptors of the joints. Their structure, variability and classification. *Acta Fac. Med. Univ. Brun, 23,* 1–107 + 60 Plates.

Poletti, C. E., and G. Creswell (1977). Fornix system efferent projection in the squirrel monkey: an experimental degeneration study. *J. Comp. Neurol. 175,* 101–128.

Poljak, S. (1924). Die Struktureigentümlichkeiten des Rückenmarkes bei den Chiropteren. Zugleich ein Beitrag zu der Frage über die spinalen Zentren des Sympathicus. *Z. Anat. Entwickl.-Gesch. 74,* 509–576.

Pollack, S. L. (1960). The grasp response in the neonate. Its characteristics and interaction with the tonic neck reflex. *Arch. Neurol. (Chic.) 3,* 574–588.

Pollock, L. J., and L. Davis (1930). Muscle tone in Parkinsonian states. *Arch. Neurol. Psychiat. (Chic.) 23,* 303–319.

Pollock, M., and R. W. Hornabrook (1966). The prevalence, natural history and dementia of Parkinson's disease. *Brain 89,* 429–448.

Polyak, S. (1927a). Über den allgemeinen Bauplan des Gehörsystems und über seine Bedeutung für die Physiologie, für die Klinik, und für die Psychologie. *Z. Neurol. 110,* 1–49.

Polyak, S. (1927b). An experimental study of the association, callosal and projection fibers of the cerebral cortex of the cat. *J. Comp. Neurol. 44,* 197–258.

Polyak, S. (1932). *The Main Afferent Fiber Systems of the Cerebral Cortex in Primates. Univ. Calif. Publ. Anat.,* Vol. 2. Berkeley.

Polyak, S. (1933). A contribution to the cerebral representation of the retina. *J. Comp. Neurol. 57,* 541–617.

Polyak, S. (1936). Minute structure of the retina in monkeys and in apes. *Arch. Ophthal. 15,* 477–519.

Polyak, S. (1941). *The Retina. The Anatomy and Histology of the Retina in Man, Ape and Monkey, including the Consideration of Visual Functions, the History of Physiological Optics, and the Laboratory Technique.* Univ. of Chicago Press, Chicago.

Polyak, S. (1957). *The Vertebrate Visual System* (H. Klüver, ed.). Univ. of Chicago Press, Chicago.

Pomeranz, B., P. D. Wall, and W. V. Weber (1968). Cord cells responding to fine myelinated afferents from viscera, muscle and skin. *J. Physiol. (Lond.) 199,* 511–532.

Pompeiano, O. (1957). Analisi degli effetti della stimolazione elettrica del nucleo rosso nel Gatto decerebrato. *Rend. Acc. Naz. Lincei, Cl. Sci. Fis., Mat. Nat. 22,* 100–103.

Pompeiano, O. (1959). Organizzazione somatotopica delle risposte flessorie alla stimolazione elettrica del nucleo interposito nel Gatto decerebrato. *Arch. Sci. Biol. (Bologna) 43,* 163–176.

Pompeiano, O. (1960). Organizzazione somatotopica delle risposte posturali alla stimolazione elettrica del nucleo di Deiters nel Gatto decerebrato. *Arch. Sci. Biol. (Bologna) 44,* 497–511.

Pompeiano, O. (1962). Somatotopic organization of the postural responses to stimulation and destruction of the caudal part of the fastigial nucleus. *Arch. Ital. Biol. 100,* 259–271.

Pompeiano, O. (1967). Functional organization of the cerebellar projections to the spinal cord. In *The Cerebellum. Progress in Brain Research,* Vol. 25. (J. P. Schadé, ed.), pp. 282–321. Elsevier, Amsterdam.

Pompeiano, O. (1972a). Vestibulospinal relations: vestibular influences on gamma motoneurons and primary afferents. In *Basic Aspects of Central Vestibular Mechanisms, Progress in Brain Research,* Vol. 37 (A. Brodal and O. Pompeiano, eds.), pp. 197–232. Elsevier, Amsterdam.

Pompeiano, O. (1972b). Spinovestibular relations: anatomical and physiological aspects. In *Basic Aspects of Central Vestibular Mechanisms. Progress in Brain Research,* Vol. 37 (A. Brodal and O. Pompeiano, eds.), pp. 263–296. Elsevier, Amsterdam.

Pompeiano, O. (1972c). Reticular control of the vestibular nuclei: physiology and pharmacology. In *Basic Aspects of Central Vestibular Mechanisms* (A. Brodal and O. Pompeiano, eds.). *Progress in Brain Research,* Vol. 37, 601–618. Elsevier, Amsterdam.

Pompeiano, O. (1973). Reticular formation. In *Handbook of Sensory Physiology,* Vol. II: *Somatosensory System* (A. Iggo, ed.), pp. 381–488. Springer-Verlag, Berlin.

Pompeiano, O. (1974). Cerebello-vestibular interrelations. In *Handbook of Sensory Physiology,* Vol. VI/1 (H. H. Kornhuber, ed.), pp. 417–476. Springer-Verlag, Berlin.

Pompeiano, O., and A. Brodal (1957a). The origin of vestibulospinal fibres in the cat. An experimental-anatomical study, with comments on the descending medial longitudinal fasciculus. *Arch Ital. Biol. 95,* 166–195.

Pompeiano, O., and A. Brodal (1957b). Spino-vestibular fibers in the cat. *J. Comp. Neurol. 108,* 353–382.

Pompeiano, O., and A. Brodal (1957c). Experimental demonstration of a somatotopical origin of rubrospinal fibers in the cat. *J. Comp. Neurol. 108,* 225–252.

Pompeiano, O., and E. Cotti (1959). Analisi microelettrodica delle proiezioni cerebello-deitersiane. *Arch. Sci. Biol. (Bologna) 43,* 57–101.

Pompeiano, O., T. Mergner, and N. Corvaja (1978). Commissural, perihypoglossal and reticular afferent projections to the vestibular nuclei in the cat. *Arch. Ital. Biol. 116,* 130–172.

Pompeiano, O., and A. R. Morrison (1966). Vestibular influences during sleep. III. Dissociation of the tonic and phasic inhibition of spinal reflexes during desynchronized sleep following vestibular lesions. *Arch. Ital. Biol. 104,* 231–246.

Pompeiano, O., and J. E. Swett (1962a). EEG and behavioral manifestations of sleep induced by cutaneous nerve stimulation in normal cats. *Arch Ital. Biol. 100,* 311–342.

Pompeiano, O., and J. E. Swett (1962b). Identification of cutaneous and muscular afferent fibers producing EEG synchronization or arousal in normal cats. *Arch. Ital. Biol. 100,* 343–380.

Pompeiano, O., and J. E. Swett (1963). Actions of graded cutaneous and muscular afferent volleys on brain stem units in the decerebrate, cerebellectomized cat. *Arch Ital. Biol. 101,* 552–582.

Pompeiano, O., and F. Walberg (1957). Descending connections to the vestibular nuclei. An experimental study in the cat. *J. Comp. Neurol. 108,* 465–502.

Pontes, C., F. F. Reis, and A. Sousa-Pinto (1975). The auditory cortical projections onto the medial geniculate body in the cat. An experimental anatomical study with silver and autoradiographic methods. *Brain Research 91,* 43–63.

Popa, G. T., and U. Fielding (1930). A portal circulation from the pituitary to the hypothalamic region. *J. Anat. (Lond.) 65,* 88–91.

Pope, A. (1967). Microchemical architecture of human isocortex. *Arch. Neurol. (Chic.) 16,* 351–356.

Pope, A., H. H. Hess, and E. Lewin (1964). Studies on the microchemical pathology of human cerebral cortex. In *Morphological and Biochemical Correlates of Neural Activity* (M. M. Cohen and R. S. Snider, eds.), pp. 98–111. Hoeber Medical Division, Harper & Row, New York.

Popper, K. R., and J. C. Eccles (1977). *The Self and Its Brain.* Springer-Verlag, Berlin.

Porter, J. C., R. S. Mical, N. Ben-Jonathan, and J. G. Ondo (1973). Neurovascular regulation of the anterior hypophysis. *Recent Progr. Hormone Res. 29,* 161–198.

Porter, L., and J. Semmes (1974). Preservation of cutaneous temperature sensitivity after ablation of sensory cortex in monkeys. *Exp. Neurol. 42,* 206–219.

Porter, R. (1972). Relationship of the discharges of cortical neurons to movement in free-to-move monkeys. *Brain Research 40,* 39–43.

Porter, R. (1976). Influences of movement detectors on pyramidal tract neurons in primates. *Physiol. Rev. 76,* 121–137.

Porter, R., and McD. Lewis (1975). Relationship of neuronal discharges in the precentral gyrus of monkeys to the performance of arm movements. *Brain Research 98,* 21–36.

Portugal, J. R., and M. Brock (1962). On the pathogenesis of the Dandy-Walker-Brodal syndrome. *Zbl. Neurochir. 23,* 80–97.

Poulos, D. A., and J. T. Molt (1977). Thermosensory mechanisms in the spinal trigeminal nucleus of cats. In *Pain in the Trigeminal Region* (D. J. Anderson and B. Matthews, eds.), pp. 213–224. Elsevier/North-Holland, Amsterdam.

Powell, E. W. (1973). Limbic projections to the thalamus. *Exp. Brain Res. 17,* 394–401.

Powell, E. W., K. Akagi, and J. B. Hatton (1974). Subcortical projections of the cingulate gyrus in cat. *J. Hirnforsch. 15,* 269–278.

Powell, E. W., and J. B. Hatton (1969). Projections of the inferior colliculus in the cat. *J. Comp. Neurol. 136,* 183–192.

Powell, E. W., and P. F. Robinson (1975). Cinguloseptal projections in the squirrel monkey. *Brain Research 96,* 310–316.

Powell, T. P. S. (1952). Residual neurons in the human thalamus following hemidecortication. *Brain 75,* 571–584.

Powell, T. P. S. (1973a). The organization of the major functional areas of the cerebral cortex. *Symp. Zool. Soc. Lond. 33,* 235–252.

Powell, T. P. S. (1973b). Sensory convergence in the cerebral cortex. In *Surgical Approaches in Psychiatry* (L. V. Laitinen and K. E. Livingston, eds.), pp. 266–281. Medical Technical, Lancaster, England.

Powell, T. P. S. (1976). Bilateral cortico-tectal projection from the visual cortex in the cat. *Nature 260,* 526–527.

Powell, T. P. S., and W. M. Cowan (1954a). The connexions of the midline and intralaminar nuclei of the thalamus of the rat. *J. Anat. (Lond.) 88,* 307–319.

Powell, T. P. S., and W. M. Cowan (1954b). The origin of the mamillothalamic tract in the rat. *J. Anat. (Lond.) 88,* 489–497.

Powell, T. P. S., and W. M. Cowan (1956). A study of thalamo-striate relations in the monkey. *Brain 79,* 364–390.

Powell, T. P. S., and W. M. Cowan (1963). Centrifugal fibres in the lateral olfactory tract. *Nature 199,* 1296–1297.

Powell, T. P. S., and W. M. Cowan (1964). A note on retrograde fibre degeneration. *J. Anat. (Lond.) 98,* 579–585.

Powell, T. P. S., W. M. Cowan, and G. Raisman (1963). Olfactory relationships of the diencephalon. *Nature 199,* 710–712.

Powell, T. P. S., W. M. Cowan, and G. Raisman (1965). The central olfactory connexions. *J. Anat. (Lond.) 99,* 791–813.

Powell, T. P. S., and S. D. Erulkar (1962). Transneuronal cell degeneration in the auditory relay nuclei of the cat. *J. Anat. (Lond.) 96,* 249–268.

Powell, T. P. S., R. W. Guillery, and W. M. Cowan (1957). A quantitative study of the fornix-mamillo-thalamic system. *J. Anat. (Lond.) 91,* 419–437.

Powell, T. P. S., and V. B. Mountcastle (1959a). The cytoarchitecture of the postcentral gyrus of the monkey macaca mulatta. *Bull. Johns Hopk. Hosp. 106,* 108–131.

Powell, T. P. S., and V. B. Mountcastle (1959h). Some aspects of the functional organization of the postcentral gyrus of the monkey: a correlation of findings obtained in a single unit analysis with cytoarchitecture. *Bull. Johns Hopk. Hosp. 105,* 133–162.

Precentral Motor Cortex, The (1944). (P. C. Bucy, ed.). Univ. of Illinois Press, Urbana.

Precht, W. (1974a). Physiological aspects of the efferent vestibular system. In *Handbook of Sensory Physiology,* Vol. VI/1 (H. H. Kornhuber, ed.), pp. 221–236. Springer-Verlag, Berlin.

Precht, W. (1974b). The physiology of the vestibular nuclei. In *Handbook of Sensory Physiology,* Vol. VI/1 (H. H. Kornhuber, ed.), pp. 353–416. Springer-Verlag, Berlin.

Precht, W. (1975). Cerebellar influences on eye movements. In *Basic Mechanisms of Ocular Motility and their Clinical implications* (G. Lennerstrand and P. Bach-y-Rita, eds.), pp. 261–280. Pergamon Press, Oxford.

Precht, W. (1978). Neuronal operations in the vestibular system. *Studies of Brain Function,* Vol. 2 (H. B. Barlow, E. Florey, O.-J. Grüsser, and H. van der Loos, eds.), 225 pp. Springer-Verlag, Berlin.

Precht, W., and R. Llinás (1969). Functional organization of the vestibular afferents to the cerebellar cortex of frog and cat. *Exp. Brain Res. 9,* 30–52.

Precht, W., P. C. Schwindt, and P. C. Magherini (1974). Tectal influences on cat ocular motoneurons. *Brain Research 82,* 27–40.

Precht, W., and H. Shimazu (1965). Functional connections of tonic and kinetic vestibular neurons with primary vestibular afferents. *J. Neurophysiol. 28*, 1014–1028.

Precht, W., J. I. Simpson, and R. Llinás (1976). Response of Purkinje cells in rabbit nodulus and uvula to natural vestibular and visual stimuli. *Pflügers Arch. ges. Physiol. 367*, 1–6.

Precht, W., R. Volkind, and R. H. I. Blanks (1977). Functional organization of the vestibular input to the anterior and posterior cerebellar vermis of cat. *Exp. Brain Res. 27*, 143–160.

Precht, W., and M. Yoshida (1971). Blockage of caudate-evoked inhibition of neurons in the substantia nigra by picrotoxin. *Brain Research 32*, 229–233.

Prelević, S., W. McIntyre Burnham, and P. Gloor (1976). A microelectrode study of amygdaloid afferents: temporal neocortical inputs. *Brain Research 105*, 437–457.

Prestige, M. C. (1966). Initial collaterals of motor axons within the spinal cord of the cat. *J. Comp. Neurol. 126*, 123–136.

Preston, J. B., and D. G. Whitlock (1961). Intracellular potentials recorded from motoneurons following precentral gyrus stimulation in primate. *J. Neurophysiol. 24*, 91–100.

Pribram, K. H., K. L. Chow, and J. Semmes (1953). Limit and organization of the cortical projection from the medial thalamic nucleus in monkey. *J. Comp. Neurol. 98*, 433–448.

Price, D. D. (1972). Characteristics of second pain and flexion reflexes indicative of prolonged central summation. *Exp. Neurol. 37*, 371–387.

Price, D. D., R. Dubner, and J. W. Hu (1976). Trigeminothalamic neurons in nucleus caudalis responsive to tactile, thermal, and nociceptive stimulation of monkey's face. *J. Neurophysiol. 39*, 936–953.

Price, D. D., and D. J. Mayer (1975). Neurophysiological characterization of the anterolateral quadrant neurons subserving pain in *M. mulatta*. *Pain 1*, 59–72.

Price, J. L. (1968). The termination of centrifugal fibres in the olfactory bulb. *Brain Research 7*, 483–486.

Price, J. L. (1973). An autoradiographic study of complementary laminar patterns of termination of afferent fibers to the olfactory cortex. *J. Comp. Neurol. 150*, 87–108.

Price, J. L., and T. P. S. Powell (1970a). An experimental study of the origin and the course of the centrifugal fibres to the olfactory bulb in the rat. *J. Anat. (Lond.) 107*, 215–237.

Price, J. L., and T. P. S. Powell (1970b). An electron-microscopic study of the termination of the afferent fibres to the olfactory bulb from the cerebral hemisphere. *J. Cell Sci. 7*, 157–187.

Price, J. L., and T. P. S. Powell (1970c). The mitral and short axon cells of the olfactory bulb. *J. Cell Sci. 7*, 631–651.

Price, J. L., and T. P. S. Powell (1971). Certain observations on the olfactory pathway. *J. Anat. (Lond.) 110*, 105–126.

Price, T. R., and K. E. Webster (1972). The corticothalamic projection from the primary somatosensory cortex of the rat. *Brain Research 44*, 636–640.

Provini, L., S. Redman, and P. Strata (1968). Mossy and climbing fibre organization in the anterior lobe of the cerebellum activated by forelimb and hindlimb areas of the sensorimotor cortex. *Exp. Brain Res. 6*, 216–233.

Provins, K. A. (1958). The effect of peripheral nerve block on the appreciation and execution of finger movements. *J. Physiol. (Lond.) 143*, 55–67.

Pubols, B. H., Jr. (1968). Retrograde degeneration study of somatic sensory thalamocortical connections in brain of *Virginia opossum*. *Brain Research 7*, 232–251.

Pubols, B. H., and L. M. Pubols (1971). Somatotopic organization of spider monkey somatic sensory cerebral cortex. *J. Comp. Neurol. 141*, 63–76.

Pubols, B. H., and L. M. Pubols (1972). Neural organization of somatic sensory representation in the spider monkey. *Brain Behav. Evol. 5*, 342–366.

Pubols, B. H., Jr., W. I. Welker, and C. I. Johnson, Jr. (1965). Somatic sensory representation of forelimb in dorsal root fibers of raccoon, coatimundi, and cat. *J. Neurophysiol. 28*, 312–341.

Pucetti, R. (1977). Bilateral organization of consciousness in man. In *Evolution and Lateralization of the Brain* (S. J. Dimond and D. A. Blizard, eds.), pp. 448–457. New York Academy of Sciences, New York.

Pugh, W. W., and I. H. Wagman (1977). Responses to somatic stimulation in N. parafascicularis (Pf) and in the posterior group of nuclei (PO) in the unanesthetized cat. *Proc. Soc. Neurosci. 7*, 490.

Purves, H. D. (1966). Cytology of the adenohypophysis. In *The Pituitary Gland,* Vol. 1 (G. W. Harris and B. T. Donovan, eds.), pp. 147–232. Butterworths, London.

Purves, M. J. (1972). *The Physiology of the Cerebral Circulation.* Cambridge Univ. Press, Cambridge.

Purpura, D. P. (1972). Synaptic mechanisms in coordination of activity in thalamic internuncial common paths. In *Corticothalamic Projections and Sensorimotor Activities* (T. Frigyesi, E. Rinvik, and M. D. Yahr, eds.), pp. 21–51. Raven Press, New York.

Putnam, T. J. (1926). Studies on the central visual system. IV. The details of the organization of the geniculo-striate system in man. *Arch. Neurol. Psychiat. (Chic.) 16*, 683–707.

Putnam, T. J., and S. Liebman (1942). Cortical representation of the macula lutea with special reference to the theory of bilateral representation. *Arch. Ophthal. 28*, 415–443.

Putnam, T. J., and I. K. Putnam (1926). Studies on the central visual system. I. The anatomic projection of the retinal quadrants on the striate cortex of the rabbit. *Arch. Neurol. Psychiat. (Chic.) 16*, 1–20.

Rafols, J. A., and C. A. Fox (1976). The neurons in the primate subthalamic nucleus: a Golgi and electron microscopic study. *J. Comp. Neurol. 168*, 75–111.

Rafols, J. A., and H. A. Matzke (1970). Efferent projections of the superior colliculus in the opossum. *J. Comp. Neurol. 138*, 147–160.

Rafols, J. A., and F. Valverde (1973). The structure of the dorsal lateral geniculate nucleus in the mouse. A Golgi and electron microscopic study. *J. Comp. Neurol. 150*, 303–332.

Raaf, J., and J. W. Kernohan (1944). Relation of abnormal collections of cells in posterior medullary velum of cerebellum to origin of medulloblastoma. *Arch. Neurol. Psychiat. (Chic.) 52*, 163–169.

Raeder, J. G. (1924). Paratrigeminal paralysis of the oculopupillary sympathetic. *Brain 47*, 149–158.

Raisman, G. (1966a). The connexions of the septum. *Brain 89*, 317–348.

Raisman, G. (1966b). Neural connections of the hypothalamus. *Brit. Med. Bull. 22*, 197–201.

Raisman, G. (1969a). Neuronal plasticity in the septal nuclei of the adult rat. *Brain Research 14*, 25–48.

Raisman, G. (1969b). A comparison of the mode of termination of the hippocampal and hypothalamic afferents to the septal nuclei as revealed by electron microscopy of degeneration. *Exp. Brain Res. 7*, 317–343.

Raisman, G. (1972). An experimental study of the projection of the amygdala to the accessory olfactory bulb and its relationship to the concept of a dual olfactory system. *Exp. Brain Res. 14*, 395–408.

Raisman, G. (1973a). Electron microscopic studies of the development of new neurohaemal contacts in the median eminence of the rat after hypophysectomy. *Brain Research 55*, 245–261.

Raisman, G. (1973b). An ultrastructural study of the effects of hypophysectomy on the supraoptic nucleus of the rat. *J. Comp. Neurol. 147*, 181–208.

Raisman, G. (1975). An experimental study of the projection of the amygdala to the accessory olfactory bulb and its relationship to the concept of a dual olfactory system. *Exp. Brain Res. 14*, 395–408.

Raisman, G., W. M. Cowan, and T. P. S. Powell (1965). The extrinsic afferent, commissural and association fibres of the hippocampus. *Brain 88*, 963–996.

Raisman, G., W. M. Cowan, and T. P. S. Powell (1966). An experimental analysis of the efferent projection of the hippocampus. *Brain 89*, 83–108.

Raisman, G., and P. M. Field (1973). A quantitative investigation of the development of collateral reinnervation after partial deafferentation of the septal nuclei. *Brain Research 50*, 241–264.

Raisman, G., P. M. Field, A. J. C. Ostberg, L. L. Iversen, and R. E. Ziegmond (1974). A quantitative ultrastructural and biochemical analysis of the process of reinnervation of the superior cervical ganglion in the adult cat. *Brain Research 71, 1–16*.

Rakic, P. (1975). Local circuit neurons. *Neurosci. Res. Progr. Bull. 13*, 291–446.

Rakic, P., and P. I. Yakovlev (1968). Development of the corpus callosum and cavum septi in man. *J. Comp. Neurol. 132*, 45–72.

Rall, W., G. M. Shepherd, T. S. Reese, and M. W. Brightman (1966). Dendrodendritic synaptic pathway for inhibition in the olfactory bulb. *Exp. Neurol. 14*, 44–56.

Ralston, H. J. (1965). The organization of the substantia gelatinosa Rolandi in the cat lumbosacral cord. *Z. Zellforsch. 67*, 1–23.

Ralston, H. J. III (1968a). The fine structure of neurons in the dorsal horn of the cat spinal cord. *J. Comp. Neurol. 132*, 275–302.

Ralston, H. J. III (1968b). Dorsal root projections to dorsal horn neurons in the cat spinal cord. *J. Comp. Neurol. 132*, 303–330.

Ralston, H. J. III (1969). The synaptic organization of lemniscal projections to the ventrobasal thalamus of the cat. *Brain Research 14*, 99–115.

Ralston, H. J. III (1971a). Evidence for presynaptic dendrites and a proposal for their mechanism of action. *Nature 230*, 585–587.

Ralston, H. J. III (1971b). The synaptic organization in the dorsal horn of the spinal cord and in the ventrobasal thalamus in the cat. In *Oral-Facial Sensory and Motor Mechanisms* (R. Dubner and Y. Kawamura, eds.), pp. 229–250. Appleton-Century-Crofts, New York.

Ralston, H. J. III (1979). The fine structure of laminae I, II and III of the Macaque spinal cord. *J. Comp. Neurol. 184*, 619–642.

Ralston, H. J. III, and M. M. Herman (1969). The fine structure of neurons and synapses in the ventrobasal thalamus of the cat. *Brain Research 14*, 77–97.

Ralston, H. J. III, and D. D. Ralston (1979). The distribution of dorsal root axons in laminae I, II and III of the Macaque spinal cord: a quantititive electron microscope study. *J. Comp. Neurol. 184*, 643–684.

Ralston, H. J. III, and P. V. Sharp (1973). The identification of thalamocortical relay cells in the adult cat by means of retrograde axonal transport of horseradish peroxidase. *Brain Research 62*, 273–278.

Ramón-Moliner, E., and W. J. H. Nauta (1966). The isodendritic core of the brain stem. *J. Comp. Neurol. 126*, 311–335.

Randolph, M., and J. Semmes (1974). Behavioral consequences of selective subtotal ablations in the postcentral gyrus of *Macaca mulatta. Brain Research 70*, 55–70.

Ranieri, F., J. Crousillat, and N. Mei (1975). Étude électrophysiologique et histologique des fibres afférentes splanchniques. *Arch. Ital. Biol. 113*, 354–373.

Ranieri, F., N. Mei, and J. Crousillat (1973). Les afférences splanchniques provenant des mécanorécepteurs gastro-intestinaux et péritonéaux. *Exp. Brain Res. 16*, 276–290.

Ranson, S. W. (1943). *The Anatomy of the Nervous System*, 7th ed. W. B. Saunders, Philadelphia.

Ranson, S. W., and P. R. Billingsley (1918). The superior cervical ganglion and the cervical portion of the sympathetic trunk. *J. Comp. Neurol. 29*, 313–358.

Ranson, S. W., H. K. Davenport, and E. A. Doles (1932). Intramedullary course of the dorsal root fibers of the first three cervical nerves. *J. Comp. Neurol. 54*, 1–12.

Ranson, S. W., W. H. Droegemueller, H. K. Davenport, and C. Fisher (1935). Number, size and myelination of the sensory fibers in the cerebrospinal nerves. *Res. Publ. Ass. Nerv. Ment. Dis. 15*, 3–28.

Ranson, S. W., and H. W. Magoun (1933). The central path of the pupilloconstriction reflex in response to light. *Arch. Neurol. Psychiat. (Chic.) 30,* 1193–1202.

Ranson, S. W., S. W. Ranson, Jr., and M. Ranson (1941). Fiber connections of corpus striatum as seen in Marchi preparations. *Arch. Neurol. Psychiat. (Chic.) 46,* 230–249.

Rao, G. S., J. E. Breazile, and R. L. Kitchell (1969). Distribution and termination of spinoreticular afferents in the brain stem of sheep. *J. Comp. Neurol. 137,* 185–196.

Raphan, T., and B. Cohen (1978). Brainstem mechanisms for rapid and slow eye movements. *Ann. Rev. Physiol. 40,* 527–552.

Rasmussen, A. T. (1932). *The Principal Nervous Pathways.* Macmillan, New York.

Rasmussen, A. T. (1938). The proportions of the various subdivisions of the normal adult human hypophysis cerebri and the relative number of the different types of cells in pars distalis, with biometric evaluation of age and sex differences and special consideration of basophilic invasion into the infundibular process. *Res. Publ. Ass. Nerv. Ment. Dis. 17,* 118–150.

Rasmussen, A. T. (1940a). Studies of the VIIIth cranial nerve of man. *Laryngoscope (St. Louis) 50,* 67–83.

Rasmussen, A. T. (1940b). Effects of hypophysectomy and hypophysial stalk resection on the hypothalamic nuclei of animals and man. *Res Publ. Ass. Nerv. Ment. Dis. 20,* 245–269.

Rasmussen, A. T., and W. T. Peyton (1948). The course and termination of the medial lemniscus in man. *J. Comp. Neurol. 88,* 411–424.

Rasmussen, G. L. (1946). The olivary peduncle and other fiber projections of the superior olivary complex. *J. Comp. Neurol. 84,* 141–220.

Rasmussen, G. L. (1953). Further observations of the efferent cochlear bundle. *J. Comp. Neurol. 99,* 61–74.

Rasmussen, G. L. (1955). Descending or "feed-back" connections of auditory system of the cat. *Amer. J. Physiol. 183,* 653.

Rasmussen, G. L. (1960). Efferent fibers of the cochlear nerve and cochlear nucleus. In *Neural Mechanisms of the Auditory and Vestibular Systems* (G. L. Rasmussen and W. Windle, eds.), pp. 105–115. C. C. Thomas, Springfield, Ill.

Rasmussen, G. L. (1961). Distribution of fibers originating from the inferior colliculus (Abstract.) *Anat. Rec. 139,* 266.

Rasmussen, G. L. (1964). Anatomic relationships of the ascending and descending auditory systems. In *Neurological Aspects of Auditory and Vestibular Disorders.* (W. S. Fields and B. R. Alford, eds.), pp. 5–23. C. C. Thomas, Springfield, Ill.

Rasmussen, G. L. (1967). Efferent connections of the cochlear nerve. In *Sensorineural Hearing Processes and Disorders.* Henry Ford Hospital International Symposium (A. B. Graham, ed.), pp. 61–75. Little, Brown, Boston.

Rasmussen, G. L, R. R. Gacek, E. P. McCrane, and C. C. Baker (1960). Model of cochlear nucleus (cat) displaying its afferent and efferent connections. (Abstract.) *Anat. Rec. 136,* 344.

Rasmussen, T. (1976). The place of surgery in the treatment of epilepsy. In *Current Controversies in Neurosurgery* (T. P. Morley, ed.), pp. 465–477. W. B. Saunders, Philadelphia.

Ravizza, R., and I. T. Diamond (1974). Role of auditory cortex in sound localization: a comparative ablation study of hedgehog and bushbaby. *Fed. Proc. 33,* 1917–1919.

Ravizza, R. J., R. B. Straw, and P. D. Long (1976). Laminar origin of efferent projections from auditory cortex in the golden Syrian hamster. *Brain Research 114,* 497–500.

Raymond, F., and R. Cestan (1902). Sur un cas de papillome épithélioïde du noyau rouge. Contribution à l'étude des fonctions du noyau rouge. *Arch. Neurol. (Paris) Sér. 2, 14,* 81–100.

Raymond, J., A. Sans, and R. Marty (1974). Projections thalamiques des voyaux vestibulaires: étude histologique chez le chat. *Exp. Brain Res. 20,* 273–283.

Recent Advances in Parkinson's Disease (1971). (F. H. McDowell and C. H. Markham, eds.). F. A. Davis, Philadelphia.

Recovery from Brain Damage. Research and Theory (1978). (S. Finger, ed.). Plenum Press, New York.

Recovery of Function. Theoretical Considerations for Brain Injury Rehabilitation (1980). (P. Bach-y-Rita, ed.). Huber, Bern. In press.

Reese, T. S., and M. W. Brightman (1970). Olfactory surface and central olfactory connexions in some vertebrates. In *Taste and Smell in Vertebrates* (G. E. Wolstenholme and J. Knight, eds.), pp. 115–143. J. & A. Churchill, London.

Reese, T. S., and G. M. Shepherd (1972). Dendrodendritic synapses in the central nervous system. In *Structure and Function of Synapses* (G. D. Pappas and D. P. Purpura, eds.), pp. 121–136. Raven Press, New York.

Reichert, F. L. (1934). Neuralgias of the glossopharyngeal nerve. With particular reference to the sensory, gustatory and secretory functions of the nerve. *Arch. Neurol. Psychiat. (Chic.) 32*, 1030–1037.

Reinking, R. M., J. A. Stephens, and D. G. Stuart (1975). The tendon organs of cat medial gastrocnemius: significance of motor unit type and size for the activation of Ib afferents. *J. Physiol. (Lond.) 250*, 491–512.

Remmel, R. S., R. D. Skinner, and J. Pola (1977). Cat pontomedullary neurons projecting to the regions of the ascending MLF and the vestibular nuclei. In *Control of Gaze by Brain Stem Neurons* (R. Baker and A. Berthoz, eds.), pp. 163–166. Elsevier/North-Holland, Amsterdam.

Renaud, L. P. (1976). An electrophysiological study of amygdalo-hypothalamic projections to the ventromedial nucleus of the rat. *Brain Research 105*, 45–58.

Renaud, L. P., and D. A. Hopkins (1977). Amygdala afferents from the mediobasal hypothalamus: an electrophysiological and neuroanatomical study in the rat. *Brain Research 121*, 201–213.

Renaud, L. P., and J. B. Martin (1975). Electrophysiological studies of connections of hypothalamic ventromedial nucleus neurons in the rat: evidence for a role in neuroendocrine regulation. *Brain Research 93*, 145–151.

Réthelyi, M. (1970). Ultrastructural synaptology of Clarke's column. *Exp. Brain Res. 11*, 159–174.

Réthelyi, M. (1972). Cell and neuropil architecture of the intermediolateral (sympathetic) nucleus of cat spinal cord. *Brain Research 46*, 203–213.

Réthelyi, M. (1977). Preterminal and terminal axon arborizations in the substantia gelatinosa of cat's spinal cord. *J. Comp. Neurol. 172*, 511–528.

Réthelyi, M., and B. Halász (1970). Origin of the nerve endings in the surface zone of the median eminence of the rat hypothalamus. *Exp. Brain Res. 11*, 145–158.

Réthelyi, M., and J. Szentágothai (1969). The large synaptic complex of the substantia gelatinosa. *Exp. Brain Res. 7*, 258–274.

Réthelyi, M., and J. Szentágothai (1973). Distribution and connections of afferent fibers in the spinal cord. In *Handbook of Sensory Physiology*, Vol. II, *Somatosensory System* (A. Iggo, ed.), pp. 207–252. Springer-Verlag, Berlin.

Reticular Formation of the Brain (1958). (H. H. Jasper, L. D. Proctor, R. S. Knighton, W. C. Noshay, and R. T. Costello, eds.). Henry Ford Hospital International Symposium. Little, Brown, Boston.

Reuben, R. N., and C. Gonzalez (1964). Ocular electromyography in brain stem dysfunction. *Arch. Neurol. (Chic.) 11*, 265–272.

Rexed, B. (1944). Contributions to the knowledge of the post-natal development of the peripheral nervous system in man. *Acta Psychiat. (Kbh.) Suppl. 33*, 1–206.

Rexed, B. (1952). The cytoarchitectonic organization of the spinal cord in the cat. *J. Comp. Neurol. 96*, 415–495.

Rexed, B. (1954). A cytoarchitectonic atlas of the spinal cord in the cat. *J. Comp. Neurol. 100*, 297–379.

Rexed, B., and A. Brodal (1951). The nucleus cervicalis lateralis. A spinocerebellar relay nucleus. *J. Neurophysiol. 14*, 399–407.

Rexed, B., and G. Strøm (1952). Afferent nervous connexions of the lateral cervical nucleus. *Acta Physiol. Scand. 25*, 219–229.

Rexed, B., and K. G. Wennerström (1959). Arachnoidal proliferation and cystic formation in the spinal nerveroot pouches of man. *J. Neurosurg. 16*, 73–84.

Reynolds, D. V. (1969). Surgery in the rat during electrical analgesia induced by focal brain stimulation. *Science 164*, 444–445.

Rhinencephalon and Related Structures, The. Progress in Brain Research, Vol. 3 (1963). (W. Bargmann and J. P. Schadé, eds.). Elsevier, Amsterdam.

Rhines, R., and H. W. Magoun (1946). Brain stem facilitation of cortical motor response. *J. Neurophysiol. 9*, 219–229.

Rhoton, A. L. (1968). Afferent connections of the facial nerve. *J. Comp. Neurol. 133*, 89–100.

Rhoton, A. L., J. L. O'Leary, and J. P. Ferguson (1966). The trigeminal, facial, vagal and glossopharyngeal nerves in the monkey. *Arch. Neurol. (Chic.) 14*, 530–540.

Ribak, C. E., and A. Peters (1975). An autoradiographic study of the projections from the lateral geniculate body of the rat. *Brain Research 92*, 341–368.

Ribak, C. E., J. E. Vaughn, K. Saito, R. Barber, and E. Roberts (1976). Immunocytochemical localization of glutamate decarboxylase in rat substantia nigra. *Brain Research 116*, 287–298.

Ricardo, J. A., and E. T. Koh (1978). Anatomical evidence of direct projections from the nucleus of the solitary tract to the hypothalamus, amygdala, and other forebrain structures in the rat. *Brain Research 153*, 1–26.

Richards, J. G., H. P. Lorez, and J. P. Tranzer (1973). Indolealkylamine nerve terminals in cerebral ventricles: identification by electron microscopy and fluorescence histochemistry. *Brain Research 57*, 277–288.

Richards, R. L. (1967). Causalgia. A centennial review. *Arch. Neurol. (Chic.) 16*, 339–350.

Richardson, D. E., and H. Akil (1977a). Pain reduction by electrical brain stimulation in man. Part 1: Acute administration in periaqueductal and periventricular sites. *J. Neurosurg. 47*, 178–183.

Richardson, D. E., and H. Akil (1977b). Pain reduction by electrical brain stimulation in periaqueductal and periventricular sites. Part 2: Chronic self-administration in the periventricular gray matter. *J. Neurosurg. 47*, 184–194.

Richardson, K. C. (1958). Electronmicroscopic observations on Auerbach's plexus in the rabbit, with special reference to the problem of smooth muscle innervation. *Amer. J. Anat. 103*, 99–135.

Richardson, K. C. (1962). The fine structure of autonomic nerve endings in smooth muscle of the rat vas deferens. *J. Anat. (Lond.) 96*, 427–442.

Richardson, K. C. (1964). The fine structure of the albino rabbit iris with special reference to the identification of adrenergic and cholinergic nerves and nerve endings in its intrinsic muscles. *Amer. J. Anat. 114*, 173–205.

Richmond, F. J. R., and V. C. Abrahams (1975). Morphology and distribution of muscle spindles in dorsal muscles of the cat neck. *J. Neurophysiol. 38*, 1322–1339.

Richter, C. P., and M. Hines (1934). The production of the 'grasp reflex' in adult macaques by experimental frontal lobe lesions. *Res. Publ. Ass. Nerv. Ment. Dis. 13*, 211–224.

Richter, C. P., and B. G. Woodruff (1945). Lumbar sympathetic dermatomes in man determined by the electrical skin resistance method. *J. Neurophysiol. 8*, 323–338.

Richter, D. W., and H. Seller (1975). Baroreceptor effects on medullary respiratory neurones of the cat. *Brain Research 86*, 168–171.

Ries, E. A., and O. R. Langworthy (1937). A study of the surface structure of the brain of the whale. (*Balaenoptera physalus* and *Physeter catadon*). *J. Comp. Neurol. 68*, 1–47.

Rinne, U. K. (1966). Ultrastructure of the median eminence of the rat. *Z. Zellforsch. 74*, 98–122.

Rinvik, E. (1965). The corticorubral projection in the cat. Further observations. *Exp. Neurol. 12*, 278–291.

Rinvik, E. (1966). The cortico-nigral projection in the cat. An experimental study with silver impregnation methods. *J. Comp. Neurol. 126*, 241–254.

Rinvik, E. (1968a). The corticothalamic projection from the pericruciate and coronal gyri in the cat. An experimental study with silver-impregnation methods. *Brain Research 10*, 79–119.

Rinvik, E. (1968b). The corticothalamic projection from the second somatosensory cortical area in the cat. An experimental study with silver impregnation methods. *Exp. Brain Res. 5*, 153–172.

Rinvik, E. (1968c). The corticothalamic projection from the gyrus proreus and the medial wall of the rostral hemisphere in the cat. An experimental study with silver impregnation methods. *Exp. Brain Res. 5*, 129–152.

Rinvik, E. (1968d). A re-evaluation of the cytoarchitecture of the ventral nuclear complex of the cat's thalamus on the basis of the corticothalamic connections. *Brain Research 8*, 237–254.

Rinvik, E. (1972). Organization of thalamic connections from motor and somatosensory cortical areas in the cat. In *Corticothalamic Projections and Sensorimotor Activities* (T. L. Frigyesi, E. Rinvik, and M. D. Yahr, eds.), pp. 57–88. Raven Press, New York.

Rinvik, E. (1975). Demonstration of nigrothalamic connections in the cat by retrograde axonal transport of horseradish peroxidase. *Brain Research 90*, 313–318.

Rinvik, E., and I. Grofová (1970). Observations on the fine structure of the substantia nigra in the cat. *Exp. Brain Res. 11*, 229–248.

Rinvik, E., and I. Grofová (1974a). Light and electron microscopical studies of the normal nuclei ventralis lateralis and ventralis anterior thalami in the cat. *Anat. Embryol. 146*, 57–93.

Rinvik, E., and I. Grofová (1974b). Cerebellar projections to the nuclei ventralis lateralis and ventralis anterior thalami. Experimental electron microscopical and light microscopical studies in the cat. *Anat. Embryol. 146*, 95–111.

Rinvik, E., I. Grofová, and O. P. Ottersen (1976). Demonstration of nigrotectal and nigroreticular projections in the cat by axonal transport of protein. *Brain Research 112*, 388–394.

Rinvik, E., I. Grofová, C. Hammond, J. M. Deniau, and J. Féger (1979). Afferent connections to the subthalamic nucleus in the monkey and the cat studied with the HRP technique. In *The Extrapyramidal System and its Disorders* (L J. Poirier, T. L. Sourkes, and P. J. Bédard, eds.), pp. 53–70. Raven Press, New York.

Rinvik, E., and F. Walberg (1963). Demonstration of a somatotopically arranged corticorubral projection in the cat. An experimental study with silver methods. *J. Comp. Neurol. 120*, 393–407.

Rinvik, E., and F. Walberg (1969). Is there a cortico-nigral tract? A comment based on experimental electron microscopic observations in the cat. *Brain Research 14*, 742–744.

Rinvik, E., and F. Walberg (1975). Studies on the cerebellar projections from the main and external cuneate nuclei in the cat by means of retrograde axonal transport of horseradish peroxidase. *Brain Research 95*, 371–381.

Risberg, J., and D. H. Ingvar (1973). Patterns of activation in the grey matter of the dominant hemisphere during memorizing and reasoning. *Brain 96*, 737–756.

Rispal-Padel, L., and A. Grangetto (1977). The cerebello-thalamo-cortical pathway. Topographical investigation at the unitary level in the cat. *Exp. Brain Res. 28*, 101–123.

Rispal-Padel, L., and J. Latreille (1974). The organization of projections from the cerebellar nuclei to the contralateral motor cortex in the cat. *Exp. Brain Res. 19*, 36–60.

Rispal-Padel, L., and J. Massion (1970). Relations between the ventrolateral nucleus and the motor cortex in the cat. *Exp. Brain Res. 10*, 331–339.

Rioch, D. M. (1929). Studies on the diencephalon of carnivora. I. Nuclear configuration of thalamus, epithalamus and hypothalamus of dog and cat. *J. Comp. Neurol. 49*, 1–94.

Riva-Sanseverino, E., and A. Urbano (1965). Electrical activity of paraflocculus and other cerebellar lobuli following vestibular rotatory stimulation in the cat. *Arch. Sci. Biol. (Bologna) 49*, 83–96.

Rivers, W. H. R., and H. Head (1908). A human experiment in nerve division. *Brain 31*, 323–450.

Roberts, P. A., and H. A. Matzke (1971). Projections of the subnucleus caudalis of the trigeminal nucleus in the sheep. *J. Comp. Neurol. 141*, 273–282.

Roberts, T. D. M. (1967). *Neurophysiology of Postural Mechanisms*. Butterworths, London.

Roberts, T. S., and K. Akert (1963). Insular and opercular cortex and its thalamic projection in *Macaca mulatta*. *Schweiz. Arch. Neurol. Neurochir. Psychiat. 92,* 1–43.

Robertson, R. T. (1977). Thalamic projections to parietal cortex. *Brain Behav. Evol. 14,* 161–184.

Robertson, R. T., G. S. Lynch, and R. F. Thompson (1973). Diencephalic distributions of ascending reticular systems. *Brain Research 55,* 309–322.

Robertson, R. T., and E. Rinvik (1973). The corticothalamic projections from parietal regions of the cerebral cortex. Experimental degeneration studies in the cat. *Brain Research 51,* 61–79.

Robin, A., and D. MacDonald (1975). *Lessons of Leucotomy.* Henry Kimpton, London.

Robinson, D. A. (1972). Eye movements evoked by collicular stimulation in the alert monkey. *Vision Res. 12,* 1795–1808.

Robinson, D. L. (1973). Electrophysiological analysis of interhemispheric relations in the second somatosensory cortex of the cat. *Exp. Brain Res. 18,* 131–144.

Robson, J. A., and W. C. Hall (1977). The organization of the pulvinar in the grey squirrel (*Sciureus carolinenus*). I. Cytoarchitecture and connections. *J. Comp. Neurol. 173,* 355–388.

Rockel, A. J., R. W. Hiorns, and T. P. S. Powell (1974). Numbers of neurons through full depth of neocortex. *J. Anat. (Lond.) 118,* 371.

Rockel, A. J., and E. G. Jones (1973). The neuronal organization of the inferior colliculus of the adult cat. I. The central nucleus. *J. Comp. Neurol. 147,* 11–60.

Rogers, D. (1972). Ultrastructural identification of degenerating boutons of monosynaptic pathways to the lumbosacral segments in the cat after spinal hemisection. *Exp. Brain Res. 14,* 293–311.

Rogers, D. C., and G. Burnstock (1966). The interstitial cell and its place in the concept of the autonomic ground plexus. *J. Comp. Neurol. 126,* 255–284.

Rogers, K. D., and J. S. McKenzie (1973). Regional differences within the caudate nucleus for suppression of extralemniscal thalamic units. *Brain Research 56,* 345–349.

Romanes, G. J. (1953). The motor cell groupings of the spinal cord. In *The Spinal Cord.* Ciba Symposium. (J. L. Malcolm, J. A. B. Gray, and G. E. Wolstenholme, eds.), pp. 24–42. J. & A. Churchill, London.

Ron, S., and D. A. Robinson (1973). Eye movements evoked by cerebellar stimulation in alert monkey. *J. Neurophysiol. 36,* 1004–1022.

Roofe, P. G. (1940). Innervation of annulus fibrosus and posterior longitudinal ligament. *Arch. Neurol. Psychiat. (Chic.) 44,* 100–103.

Rosadini, G., and G. F. Rossi (1967). On the suggested cerebral dominance for consciousness. *Brain 90,* 101–112.

Rose, A. M., T. Hattori, and H. C. Fibiger (1976). Analysis of the septo-hippocampal pathway by light and electron microscopic autoradiography. *Brain Research 108,* 170–174.

Rose, J. E. (1949). The cellular structure of the auditory region of the cat. *J. Comp. Neurol. 91,* 409–439.

Rose, J. E. (1952). The cortical connections of the reticular complex of the thalamus. *Res. Publ. Ass. Nerv. Ment. Dis. 30,* 454–479.

Rose, J. E., R. Galambos, and J. R. Hughes (1959). Microelectrode studies of the cochlear nuclei of the cat. *Bull. Johns Hopk. Hosp. 104,* 211–251.

Rose, J. E., D. D. Greenwood, J. M. Goldberg, and J. E. Hind (1963). Some discharge characteristics of single neurons in the inferior colliculus of the cat. I. Tonotopical organization, relation of spike counts to tone intensity, and firing patterns of single elements. *J. Neurophysiol. 26,* 294–320.

Rose, J. E., N. B. Gross, C. D. Geisler, and J. E. Hind (1966). Some neural mechanisms in the inferior colliculus of the cat which may be relevant to the localization of a sound source. *J. Neurophysiol. 29,* 288–314.

Rose, J. E., and L. I. Malis (1965a). Geniculo-striate connections in the rabbit. I. Retrograde changes in the dorsal lateral geniculate body after destruction of cells in various layers of the striate region. *J. Comp. Neurol. 125,* 95–120.

Rose, J. E., and L. I. Malis (1965b). Geniculo-striate connections in the rabbit. II. Cytoarchitectonic structure of the striate region and of the dorsal lateral geniculate body; organization of the geniculo-striate projections. *J. Comp. Neurol. 125,* 121–140.

Rose, J. E., and V. B. Mountcastle (1952). The thalamic tactile region in rabbit and cat. *J. Comp. Neurol. 97,* 441–489.

Rose, J. E., and C. N. Woolsey (1948). The orbitofrontal cortex and its connections with the mediodorsal nucleus in rabbit, sheep and cat. *Res. Publ. Ass. Nerv. Ment. Dis. 27,* 210–232.

Rose, J. E., and C. N. Woolsey (1949a). Organization of the mammalian thalamus and its relationships to the cerebral cortex. *Electroenceph. Clin. Neurophysiol. 1,* 391–403.

Rose, J. E., and C. N. Woolsey (1949b). The relations of thalamic connections, cellular structure and evocable electrical activity in the auditory region of the cat. *J. Comp. Neurol. 97,* 441–466.

Rose, J. E., and C. N. Woolsey (1958). Cortical connections and functional organization of the thalamic auditory system of the cat. In *Biological and Biochemical Bases of Behavior* (H. F. Harlow and C. N. Woolsey, eds.), pp. 127–150. Univ. of Wisconsin Press, Madison.

Rose, M. (1935). Cytoarchitektonik und Myeloarchitektonik der Grosshirnrinde. In *Handbuch der Neurologie* (Bumke-Foerster, eds.), Vol. 1, pp. 588–778. Julius Springer, Berlin.

Rosén, I. (1969). Localization in caudal brain stem and cervical spinal cord of neurones activated from forelimb group I afferents in the cat. *Brain Research 16,* 55–71.

Rosén, I., and P. Scheid (1973a). Patterns of afferent input to the lateral reticular nucleus of the cat. *Exp. Brain Res. 18,* 242–255.

Rosén, I., and P. Scheid (1973b). Responses to nerve stimulation in the bilateral ventral flexor reflex tract (bVFRT) of the cat. *Exp. Brain Res. 18,* 256–267.

Rosén, I., and P. Scheid (1973c). Responses in the spino-reticular-cerebellar pathway to stimulation of cutaneous mechanoreceptors. *Exp. Brain Res. 18,* 268–278.

Rosén, I., and B. Sjölund (1973). Organization of group I activated cells in the main and external cuneate nuclei in the cat: identification of muscle receptors. *Exp. Brain Res. 16,* 221–237.

Rosene, D. L., and G. W. Van Hoesen (1977). Hippocampal efferents reach widespread areas of cerebral cortex and amygdala in the rhesus monkey. *Science 198,* 315–317.

Rosenquist, A. C., S. B. Edwards, and L. A. Palmer (1974). An autoradiographic study of the projections of the dorsal lateral geniculate nucleus and the posterior nucleus in the cat. *Brain Research 80,* 71–93.

Rosenstock, J., T. D. Field, and E. Greene (1977). The role of mammillary bodies in spatial memory. *Exp. Neurol. 55,* 340 352.

Ross, A. T., and W. E. DeMyer (1966). Isolated syndrome of the medial longitudinal fasciculus in man. *Arch. Neurol. (Chic.) 15,* 203–205.

Ross, M. D. (1969). The general visceral efferent component of the eighth cranial nerve. *J. Comp. Neurol. 135,* 453–478.

Ross, M. D. (1973). Autonomic components of the VIIIth nerve. *Advan. Oto-rhino-laryng. 20,* 316–336.

Ross, R. T. (1972). Corneal reflex in hemisphere disease. *J. Neurol. Neurosurg. Psychiat. 35,* 877–880.

Rossi, G., and G. Cortesina (1965a). The "efferent cochlear and vestibular system" in *Lepus cuniculus* L. *Acta Anat. 60,* 362–381.

Rossi, G., and C. Cortesina (1965b). Morphological study of the laryngeal muscles in man. *Acta Otolaryng. (Stockh.) 59,* 575–592.

Rossi, G. F. (1964). A hypothesis on the neural basis of consciousness. *Acta Neurochir. 12,* 187–197.

Rossi, G. F. (1965). Brain stem facilitating influences on EEG synchronization. Experimental findings and observations in man. *Acta Neurochir. 13,* 257–288.

Rossi, G. F., and A. Brodal (1956a). Corticofugal fibres to the brain stem reticular formation. An experimental study in the cat. *J. Anat. (Lond.) 90,* 42–62.

Rossi, G. F., and A. Brodal (1956b). Spinal afferents to the trigeminal sensory nuclei and the nucleus of the solitary tract. *Confin. Neurol. (Basel) 16*, 321–332.

Rossi, G. F., and A. Brodal (1957). Terminal distribution of spinoreticular fibers in the cat. *Arch. Neurol. Psychiat. (Chic.) 78*, 439–453.

Rossi, G. F., and Z. Zanchetti (1957). The brain stem reticular formation. *Arch. Ital. Biol. 95*, 199–435.

Rossignol, S., and M. Colonnier (1971). A light microscopic study of degeneration patterns in cat cortex after lesions of the lateral geniculate nucleus. *Vision Res. Suppl. 3*, 329–338.

Rothman, S. (1960). Pathophysiology of itch sensation. In *Cutaneous Innervation* (W. Montagna, ed.), pp. 189–200. Pergamon Press, Oxford.

Roucoux, A., and M. Crommelinck (1976). Eye movements evoked by superior colliculus stimulation in the alert cat. *Brain Research 106*, 349–363.

Rowbotham, G. F. (1939). Observations on the effect of trigeminal denervation. *Brain 62*, 364–380.

Rowe, M. J. (1977). Cerebral cortical areas associated with the activation of climbing fibre input to cerebellar Purkinje cells. *Arch. Ital. Biol. 115*, 79–93.

Rowe, M. J., and B. J. Sessle (1972). Responses of trigeminal ganglion and brain stem neurones in the cat to mechanical and thermal stimulation of the face. *Brain Research 42*, 367–384.

Royce, G. J. (1978a). Autoradiographic evidence for a discontinuous projection to the caudate nucleus from the centromedian nucleus in the cat. *Brain Research 146*, 145–150.

Royce, G. J. (1978b). Cells of origin of subcortical afferents to the caudate nucleus: a horseradish peroxidase study in the cat. *Brain Research 153*, 465–475.

Rubia, F. J. (1970). The projection of visceral afferents to the cerebellar cortex of the cat. *Pflügers Arch. Ges. Physiol. 320*, 97–110.

Ruch, T. C. (1960). Central control of the bladder. In *Handbook of Physiology. Section 1: Neurophysiology, Vol. II.* (J. Field, H. W. Magoun, and V. E. Hall, eds.), pp. 1207–1223. Amer. Physiol. Soc., Washington, D.C.

Ruch, T. C., J. F. Fulton, and W. J. German (1938). Sensory discrimination in monkey, chimpanzee and man after lesions of the parietal lobe. *Arch. Neurol. Psychiat. (Chic.) 39*, 919–938.

Ruch, T. C., and H. A. Shenkin (1943). The relation of area 13 on orbital surface of frontal lobes to hyperactivity and hyperphagia in monkeys. *J. Neurophysiol. 6*, 349–360.

Rüegg, D. G., E. Eldred, and M. Wiesendanger (1978). Spinal projection to the dorsolateral nucleus of the caudal basilar pons in the cat. *J. Comp. Neurol. 179*, 383–393.

Rüegg, D. G., J. J. Séguin, and M. Wiesendanger (1977). Effects of electrical stimulation of somatosensory and motor areas of the cerebral cortex on neurones of the pontine nuclei in squirrel monkeys. *Neurosci. 2*, 923–927.

Rüegg, D., and M. Wiesendanger (1975). Corticofugal effects from sensorimotor area I and somatosensory area II on neurones of the pontine nuclei in the cat. *J. Physiol. (Lond.) 247*, 745–757.

Ruggiero, D., R. R. Batton III, A. Jayaraman, and M. B. Carpenter (1977). Brainstem afferents to the fastigial nucleus in the cat demonstrated by transport of horseradish peroxidase. *J. Comp. Neurol. 172*, 189–210.

Rushworth, G. (1960). Spasticity and rigidity: an experimental study and review. *J. Neurol. Neurosurg. Psychiat. 23*, 99–118.

Ruska, H., and C. Ruska (1961). Licht- und Elektronenmikroskopie des peripheren neurovegetativen Systems im Hinblick auf die Funktion. *Dtsch. Med. Wschr. 86*, 1697–1701; 1770–1772.

Russell, G. V. (1955). The nucleus locus coeruleus (dorsolateralis tegmenti). *Tex. Rep. Biol. Med. 13*, 939–988.

Russell, I. S., and S. Ochs (1963). Localization of a memory trace in one cortical hemisphere and transfer to the other hemisphere. *Brain 86*, 37–54.

Russell, J. R., and W. DeMeyer (1961). The quantitative cortical origin of pyramidal axons of Macaca rhesus. *Neurology 11*, 96–108.

Rustioni, A. (1973). Non-primary afferents to the nucleus gracilis from the lumbar cord of the cat. *Brain Research 51*, 81–95.

Rustioni, A. (1977). Spinal neurons project to the dorsal column nuclei of rhesus monkeys. *Science 196*, 656–658.

Rustioni, A., N. L. Hayes, and S. O'Neill (1979). Dorsal column nuclei and ascending spinal afferents in macaques. *Brain 102*, 95–125.

Rustioni, A., and Macchi, G. (1968). Distribution of dorsal root fibers in the medulla oblongata in the cat. *J. Comp. Neurol. 134*, 113–126.

Rustioni, A., and I. Molenaar (1975). Dorsal column nuclei afferents in the lateral funiculus of the cat: distribution pattern and absence of sprouting after chronic deafferentation. *Exp. Brain Res. 23*, 1–12.

Rutherford, J. G., and D. G. Gwyn (1977). Gap junctions in the inferior olivary nucleus of the squirrel monkey, *Saimiri sciureus*. *Brain Research 128*, 374–378.

Rutherford, W. (1886). A new theory of hearing. *J. Anat. Physiol. (Lond.) 21*, 166–168.

Rylander, G. (1939). Personality changes after operations on the frontal lobes. A clinical study of 32 cases. *Acta Psychiat. Scand. Suppl. 20*, 1–327.

Saavedra, J. M., M. Brownstein, and M. Palkovits (1974). Serotonin distribution in the limbic system of the rat. *Brain Research 79*, 437–441.

Saavedra, J. M., M. Palkovits, M. J. Brownstein, and J. Axelrod (1974). Serotonin distribution in the nuclei of the rat hypothalamus and preoptic region. *Brain Research 77*, 157–165.

Sadjadpour, K. (1975). Postfacial palsy phenomena: faulty nerve regeneration or ephaptic transmission? *Brain Research 95*, 403–406.

Sadjadpour, K., and A. Brodal (1968). The vestibular nuclei in man. A morphological study in the light of experimental findings in the cat. *J. Hirnforsch. 10*, 299–323.

Sadun, A. (1975). Differential distribution of cortical terminations in the cat red nucleus. *Brain Research 99*, 145–151.

Saint-Cyr, J. A., and J. Courville (1979). Projection from the vestibular nuclei to the inferior olive in the cat: an autoradiographic and horseradish peroxidase study. *Brain Research 165*, 189–200.

Sakai, K., M. Touret, D. Salvert, L. Leger, and M. Jouvet (1977). Afferent projections to the cat locus coeruleus as visualized by the horseradish peroxidase technique. *Brain Research 119*, 21–41.

Sakumoto, T., M. Tohyama, K. Satoh, Y. Kimoto, T. Kinugasa, O. Tanizawa, K. Kurachi, and N. Shimizu (1978). Afferent fiber connections from lower brainstem to hypothalamus studied by the horseradish peroxidase method with special reference to noradrenaline innervation. *Exp. Brain Res. 31*, 81–94.

Salmoiraghi, C. C., and F. A. Steiner (1963). Acetylcholine sensitivity of cat's medullary neurons. *J. Neurophysiol. 26*, 581–597.

Samuel, E. P. (1952). The autonomic and somatic innervation of the articular capsule. *Anat. Rec. 113*, 53–70.

Sanchez-Longo, L. P., and F. M. Forster (1958). Clinical significance of impairment of sound localization. *Neurology 8*, 119–125.

Sand, P. G. (1970). *The Human Lumbo-sacral Vertebral Column. An Osteometric Study*. Universitetsforlaget, Oslo.

Sando, I. (1965). The anatomical interrelationships of the cochlear nerve fibers. *Acta Otolaryng. (Stockh.) 59*, 417–436.

Sanford, H. S., and H. L. Bair (1939). Visual disturbances associated with tumours of the temporal lobe. *Arch. Neurol. Psychiat. (Chic.) 42*, 21–43.

Sanides, D. (1975). The retinal projection to the ventral lateral geniculate nucleus of the cat. *Brain Research 85*, 313–316.

Sanides, D., W. Fries, and K. Albus (1978). The corticopontine projection from the visual cortex of the cat: an autoradiographic investigation. *J. Comp. Neurol. 179*, 77–88.

Sanides, F., and J. Hoffmann (1969). Cyto- and myeloarchitecture of the visual cortex of the cat and the surrounding integration cortices. *J. Hirnforsch. 11*, 79–104.

Sans, A., J. Raymond, and R. Marty (1970). Résponses thalamiques et corticales à la stimulation électrique du nerf vestibulaire chez le chat. *Exp. Brain Res. 10*, 265–275.

Sans, A., J. Raymond, and R. Marty (1972). Projections des crêtes ampullaires et de l'utricule dans les noyaux vestibulaires primaires. Étude microphysiologique et corrélations anatomofonctionelles. *Brain Research 44*, 337–355.

Saper, C. B., A. D. Loewy, L. W. Swanson, and W. M. Cowan (1976). Direct hypothalamo-autonomic connections. *Brain Research 117*, 305–312.

Sar, M., W. E. Stumpf, R. J. Miller, K.-J. Chang, and P. Cuatrecasas (1978). Immunohistochemical localization of enkephalin in rat brain and spinal cord. *J. Comp. Neurol. 182*, 17–38.

Sasaki, K., A. Namikawa, and S. Hashiramoto (1960). The effect of midbrain stimulation upon alpha motoneurons in lumbar spinal cord of the cat. *Jpn. J. Physiol. 10*, 303–316.

Sasaki, K., H. Oka, S. Kawaguchi, K. Jinnai, and T. Yasuda (1977). Mossy fibre and climbing fibre responses produced in the cerebellar cortex by stimulation of the cerebral cortex in monkeys. *Exp. Brain Res. 29*, 419–428.

Sasaki, K., H. Oka, Y. Matsuda, T. Shimono, and N. Mizuno (1975). Electrophysiological studies of the projections from the parietal association area to the cerebellar cortex. *Exp. Brain Res. 23*, 91–102.

Sato, A., and R. F. Schmidt (1973). Somatosympathetic reflexes: afferent fibers, central pathways, discharge characteristics. *Physiol. Rev. 53*, 916–947.

Sato, M., N. Mizuno, and A. Konishi (1978). Localization of motoneurons innervating perineal muscles: a HRP study in cat. *Brain Research 140*, 149–154.

Sato, S. (1966). An electron microscopic study on the innervation of the intracranial artery of the rat. *Amer. J. Anat. 118*, 873–889.

Satomi, H., T. Yamamoto, H. Ise, and H. Takatama (1978). Origins of the parasympathetic preganglionic fibers to the cat intestine as demonstrated by the horseradish peroxidase method. *Brain Research 151*, 571–578.

Sauerland, E. K., Y. Nakamura, and C. D. Clemente (1967). The role of the lower brainstem in cortically induced inhibition of somatic reflexes in the cat. *Brain Research 6*, 164–180.

Sawaki, Y. (1977). Retinohypothalamic projection: electrophysiological evidence for the existence in female rats. *Brain Research 120*, 336–341.

Scalia, F. (1966). Some olfactory pathways in the rabbit brain. *J. Comp. Neurol. 126*, 285–310.

Scalia, F. (1972). The termination of retinal axons in the pretectal region of mammals. *J. Comp. Neurol. 145*, 223 258.

Scalia, F., and S. S. Winans (1975). The differential projections of the olfactory bulb and accessory olfactory bulb in mammals. *J. Comp. Neurol. 161*, 31–56.

Scharlock, D. P., W. D. Neff, and N. L. Strominger (1965). Discrimination of tone duration after bilateral ablation of cortical auditory areas. *J. Neurophysiol. 28*, 673–681.

Scharrer, B. (1976). Neurosecretion—comparative and evolutionary aspects. In *Perspectives in Brain Research, Progress in Brain Research*, Vol. 45 (M. A. Corner and D. F. Swaab, eds.), pp. 125–137. Elsevier, Amsterdam.

Scharrer, E., and R. Gaupp (1933). Neuere Befunde am Nucleus supraopticus und Nucleus paraventricularis des Menschen. *Z. ges Neurol. Psychiat. 148*, 766–772.

Scharrer, E., and B. Scharrer (1954). Neurosekretion. In v. Möllendorff's *Handbuch der mikroskopischen Anatomie des Menschen*, Bd. VI/5, pp. 953–1066. Springer, Berlin.

Schechter, P. J., and R. I. Henkin (1974). Abnormalities of taste and smell after head trauma. *J. Neurol. Neurosurg. Psychiat. 37*, 802–810.

Scheibel, A. B. (1980). The brainstem reticular core and sensory function. In *American Handbook of Physiology*, Vol. II, 2nd ed. (I. Darian-Smith, ed.), American Physiological Society, Washington, D.C. In press.

Scheibel, M. E., T. L. Davies, R. D. Lindsay, and A. B. Scheibel (1974). Basilar dendrite bundles of giant pyramidal cells. *Exp. Neurol. 42*, 307–319.

Scheibel, M. E., T. L. Davies, and A. B. Sheibel (1972). On dendrodendritic relations in the dorsal thalamus of the adult cat. *Exp. Neurol. 36*, 519–529.

Scheibel, M. E., R. D. Lindsay, U. Tomiyasu, and A. B. Scheibel (1975). Progressive dendritic changes in aging human cortex. *Exp. Neurol. 47*, 392–403.

Scheibel, M. E., and A. B. Scheibel (1954). Observations on the intracortical relations of the climbing fibers of the cerebellum. A Golgi study. *J. Comp. Neurol. 101*, 733–764.

Scheibel, M. E., and A. B. Scheibel (1955). The inferior olive. A Golgi study. *J. Comp. Neurol. 102*, 77–132.

Scheibel, M. E., and A. B. Scheibel (1958). Structural substrates for integrative patterns in the brainstem reticular core. In *Reticular Formation of the Brain* (Henry Ford Hosp. Symposium), (H. H. Jasper, L. D. Proctor et al., eds.), pp. 31–55. Little, Brown, Boston.

Scheibel, M. E., and A. B. Scheibel (1965). Periodic sensory nonresponsiveness in reticular neurons. *Arch. Ital. Biol. 103*, 300–316.

Scheibel, M. E., and A. B. Scheibel (1966a). Spinal motorneurons, interneurons and Renshaw cells. A Golgi study. *Arch. Ital. Biol. 104*, 328–353.

Scheibel, M. E., and A. B. Scheibel (1966b). Terminal axonal patterns in cat spinal cord. I. The lateral corticospinal tract. *Brain Research 2*, 333–350.

Scheibel, M. E., and A. B. Scheibel (1966c). The organization of the ventral anterior nucleus of the thalamus. A Golgi study. *Brain Research 1*, 250–268.

Scheibel, M. E., and A. B. Scheibel (1966d). The organization of the nucleus reticularis thalami: a Golgi study. *Brain Research 1*, 43–62.

Scheibel, M. E., and A. B. Scheibel (1966e). Patterns of organization in specific and nonspecific thalamic fields. In *The Thalamus* (D. P. Purpura and M. D. Yahr, eds.), pp. 13–46. Columbia Univ. Press, New York.

Scheibel, M. E., and A. B. Scheibel (1967). Structural organization of nonspecific thalamic nuclei and their projection toward cortex. *Brain Research 6*, 60–94.

Scheibel, M. E., and A. B. Scheibel (1968). Terminal axonal patterns in cat spinal cord. II. The dorsal horn. *Brain Research 9*, 32–58.

Scheibel, M. E., and A. B. Scheibel (1969). Terminal patterns in cat spinal cord. III. Primary afferent collaterals. *Brain Research 13*, 417–443.

Scheibel, M. E., and A. Scheibel (1970). Organization of spinal motoneuron dendrites in bundles. *Exp. Neurol. 28*, 106–112.

Scheibel, M. E., and A. B. Scheibel (1972a). Specialized organizational pattern within the nucleus reticularis thalami of the cat. *Exp. Neurol. 34*, 316–322.

Scheibel, M. E., and A. B. Scheibel (1972b). Elementary processes in selected thalamic and cortical subsystem—the structural substrates. In *The Neurosciences. Second Study Program* (F. O. Schmitt, ed.), pp. 443–457. Rockefeller Univ. Press, New York.

Scheibel, M. E., and A. B. Scheibel (1975a). Dendrite bundles, central programs and the olfactory bulb. *Brain Research 95*, 407–421.

Scheibel, M. E., and A. B. Scheibel (1975b). Dendrites as neuronal couplers: the dendrite bundle. In *Golgi Centennial Symposium: Perspectives in Neurobiology* (M. Santini, ed.), pp. 347–354. Raven Press, New York.

Scheibel, M. E., and A. B. Scheibel (1977). The anatomy of constancy. *Ann. N.Y. Acad. Sci. 290*, 421–435.

Scheibel, M. E., and A. B. Scheibel (1978). The dendritic structures of the human Betz cell. In *Architectonics of the Cerebral Cortex* (M. A. B. Brazier and H. Petsche, eds.), pp. 43–57. Raven Press, New York.

Scheibel, M. E., A. B. Scheibel, and T. L. Davies (1972). Some substrates for centrifugal control over thalamic cell ensembles. In *Corticothalamic Projections and Sensorimotor Activities* (T. Frigyesi, E. Rinvik, and M. D. Yahr, eds.), pp. 131–156. Raven Press, New York.

Scheibel, M., A. Scheibel, A. Mollica, and G. Moruzzi (1955). Convergence and interaction of afferent impulses on single units of reticular dormation. *J. Neurophysiol. 18*, 309–331.

Scheibel, M., A. Scheibel, F. Walberg, and A. Brodal (1956). Areal distribution of axonal and dendritic patterns in inferior olive. *J. Comp. Neurol. 106*, 21–49.

Scheibel, M. E., U. Tomiyasu, and A. B. Scheibel (1975). Do raphe nuclei of the reticular formation have a neurosecretory or vascular sensor function? *Exp. Neurol. 47,* 316–329.

Scheid, P., and J. C. Eccles (1975). Music and speech: artistic functions of the human brain. *Psychol. Music 3,* 21–35.

Schimert, J. S. (1938). Die Endigungsweise des Tractus vestibulospinalis. *Z. Anat. Entwickl. Gesch. 108,* 761–767.

Schimert, J. S. (1939). Das Verhalten der Hinterwurzelkollateralen im Rückenmark. *Z. anat. Entwickl. Gesch. 109,* 665–687.

Schmitt, F. O., P. Dev, and B. H. Smith (1976). Electrotonic processing of information by brain cells. *Science 193,* 114–120.

Schneider, R. J., A. T. Kulics, and T. B. Ducker (1977). Proprioceptive pathways in the spinal cord. *J. Neurol. Neurosurg. Psychiat. 40,* 417–433.

Schoen, J. H. R. (1964). Comparative aspects of the descending fibre systems in the spinal cord. In *The Organization of the Spinal Cord, Progress in Brain Research,* Vol. 11 (J. C. Eccles and J. P. Schadé, eds.), pp. 203–222. Elsevier, Amsterdam.

Scholten, J. M. (1946). *De Plaats van den Paraflocculus in het Geheel der Cerebellaire Correlaties.* Acad. Proefschr., N. V. Noord-Hollandsche Uitgevers Maatschappij, Amsterdam.

Schoultz, T. W., and J. E. Swett (1972). The fine structure of the Golgi tendon organ. *J. Neurocytol. 1,* 1–26.

Schoultz, T. W., and J. E. Swett (1974). Ultrastructural organization of the sensory fibers innervating the Golgi tendon organ. *Anat. Rec. 179,* 147–162.

Schroeder, D. M., and J. A. Jane (1971). Projection of dorsal column nuclei and spinal cord to brainstem and thalamus in the tree shrew, *Tupaia glis. J. Comp. Neurol. 142,* 309–350.

Schroeder, K. F., A. Hopf, H. Lange, and G. Thörner (1975). Morphometrisch-statistische Strukturanalysen des Striatum, Pallidum und Nucleus subthalamicus beim Menschen. I. Striatum. *J. Hirnforsch. 16,* 333–350.

Schubert, P. (1974). Transport in Dendriten einzelner Motoneurone. *Bull. Schweiz. Akad. Med. Wiss. 30,* 56–65.

Schubert, P., and H. Holländer (1975). Methods for the delivery of tracers to the central nervous system. In *The Use of Axonal Transport for Studies of Neuronal Connectivity* (W. M. Cowan and M. Cuénod, eds.), pp. 113–125. Elsevier, Amsterdam.

Schuknecht, H. F. (1960). Neuroanatomical correlates of auditory sensitivity and pitch discrimination in the cat. In *Neural Mechanisms of the Auditory and Vestibular Systems* (G. L. Rasmussen and W. F. Windle, eds.), pp. 76–90. C. C. Thomas, Springfield, Ill.

Schultz, W., E. B. Montgomery, Jr., and R. Marini (1976). Stereotyped flexion of forelimb and hindlimb to microstimulation of dentate nucleus in cebus monkeys. *Brain Research 107,* 151–155.

Schulze, M. L. (1955–56). Die absolute und relative Zahl der Muskelspindeln in den kurzen Daumenmuskeln des Menschen. *Anat. Anz. 102,* 290–291.

Schumacher, P. (1940). Zur Prognose der Fazialislähmung bei Poliomyelitis. *Münch. Med. Wschr. 87,* 591–593.

Schürmann, K., M. Butz, and M. Brock (1972). Temporal retrogasserian resection of trigeminal root versus controlled elective percutaneous electrocoagulation of the ganglion of Gasser in the treatment of trigeminal neuralgia. *Acta Neurochir. (Wien) 26,* 33–53.

Schwab, M., Y. Agid, J. Glowinski, and H. Thoenen (1977). Retrograde axonal transport of ^{125}I-tetanus toxin as a tool for tracing fiber connections in the central nervous system; connections of the rostral part of the rat neostriatum. *Brain Research 126,* 211–224.

Schwarcz, R., J. P. Bennett, and J. T. Coyle (1977). Inhibitors of GABA metabolism: implications for Huntington's disease. *Ann. Neurol. 2,* 299–303.

Schwartz, H. G., and J. L. O'Leary (1941). Section of the spinothalamic tract in the medulla with observations on the pathway for pain. *Surgery 9,* 183–193.

Schwartz, H. G., G. E. Roulhac, R. L. Lam, and J. O'Leary (1951). Organization of the fasciculus solitarius in man. *J. Comp. Neurol. 94*, 221–237.

Schwartz, H. G., and G. Weddell (1938). Observations on the pathways transmitting the sensation of taste. *Brain 61*, 99–115.

Schwartz, M. S., E. Stålberg, H. Schiller, and B. Thiele (1976). The reinnervated motor unit in man. A single fibre EMG multielectrode investigation. *J. Neurol. Sci. 27*, 303–312.

Schwartzkroin, P. A. (1975). Characteristics of CA1 neurons recorded intracellularly in the hippocampal *in vitro* slice preparation. *Brain Research 85*, 423–436.

Schwartzman, R. J., and M. D. Bogdonoff (1969). Proprioception and vibration sensibility discrimination in the absence of the posterior columns. *Arch. Neurol. (Chic.) 20*, 349–353.

Schwarz, D. W. F., and R. D. Tomlinson (1977). Neuronal responses to eye muscle stretch in cerebellar lobule VI of the cat. *Exp. Brain Res. 27*, 101–111.

Schwarz, G. A., and L. J. Barrows (1960). Hemiballism without involvement of Luys' body. *Arch. Neurol. (Chic.) 2*, 420–434.

Schwyn, R. C., and C. A. Fox (1974). The primate substantia nigra: a Golgi and electron microscopic study. *J. Hirnforsch. 15*, 95–126.

Scollo-Lavizzari, G., and K. Akert (1963). Cortical area 8 and its thalamic projection in *Macaca mulatta. J. Comp. Neurol. 121*, 259–270.

Scott, J. W., and B. R. Chafin (1975). Origin of olfactory projections to lateral hypothalamus and nuclei gemini of the rat. *Brain Research 88*, 64–68.

Scott, J. W., and C. M. Leonard (1971). The olfactory connections of the lateral hypothalamus in the rat, mouse and hamster. *J. Comp. Neurol. 141*, 331–344.

Scoville, W. B., and B. Milner (1957). Loss of recent memory after bilateral hippocampal lesions. *J. Neurol. Neurosurg. Psychiat. 20*, 11–21.

Second Symposium on the Role of the Vestibular Organs in Space Exploration (1966). NASA SP-115, Ames Research Center, Moffett Field, Calif.

Sedgwick, E. M., and T. D. Williams (1967). Responses of single units in the inferior olive to stimulation of the limb nerves, peripheral skin receptors, cerebellum, caudate nucleus and motor cortex. *J. Physiol. (Lond.) 189*, 261–279.

Segal, M. (1977). Afferents to the entorhinal cortex of the rat studied by the method of retrograde transport of horseradish peroxidase. *Exp. Neurol. 57*, 750–765.

Segal, M. (1979). A potent inhibitory monosynaptic hypothalamo-hippocampal connection. *Brain Research 162*, 137–141.

Segal, M., and S. Landis (1974a). Afferents to the hippocampus of the rat studied with the method of retrograde transport of horseradish peroxidase. *Brain Research 78*, 1–15.

Segal, M., and S. C. Landis (1974b). Afferents to the septal area of the rat studied with the method of retrograde axonal transport of horseradish peroxidase. *Brain Research 82*, 263–268.

Selective Partial Ablation of the Frontal Cortex (1949). (F. A. Mettler, ed.). Hoeber, New York.

Selhorst, J. B., L. Stark, A. L. Ochs, and W. F. Hoyt (1976). Disorders in cerebellar ocular motor control. I. Saccadic overshoot dysmetria. An oculographic, control system and clinico-anatomical analysis. *Brain 99*, 497–508.

Seltzer, B., and D. N. Pandya (1978). Afferent cortical connections and architectonics of the superior temporal sulcus and surrounding cortex in the rhesus monkey. *Brain Research 149*, 1–24.

Selzer, M., and W. A. Spencer (1969a). Convergence of visceral and cutaneous afferent pathways in the lumbar spinal cord. *Brain Research 14*, 331–348.

Selzer, M., and W. A. Spencer (1969b). Interactions between visceral and cutaneous afferents in the spinal cord: reciprocal primary afferent fiber depolarization. *Brain Research 14*, 349–366.

Sem-Jacobsen, C. W., M. C. Petersen, H. W. Dodge, Q. D. Jacks, J. A. Lazarte, and C. B. Holman (1956). Electric activity of the olfactory bulb in man. *Amer. J. Med. Sci. 232*, 243–251.

Semmes, J. (1969). Protopathic and epicritic sensation: a reappraisal. In *Contributions to Clinical Neurophysiology* (A. L. Benton, ed.), pp. 142–171. Aldine, Chicago.

Semmes, J. (1973). Somesthetic effects of damage to the central nervous system. In *Handbook of Sensory Physiology*, Vol. II (A. Iggo, ed.), pp. 719–742. Springer-Verlag, Berlin.

Semmes, J., and M. Mishkin (1965). Somatosensory loss in monkeys after ipsilateral cortical ablation. *J. Neurophysiol. 28*, 473–486.

Sensorineural Hearing Loss (1970). (E. G. Wolstenholme and J. Knight, eds.). J. & A. Churchill, London.

Sensorineural Hearing Processes and Disorders. (Henry Ford Hosp. Symp.) (1967). (A. B. Graham, ed.), J. & A. Churchill, London.

Sensory Functions of the Skin in Primates. With Special Reference to Man (1976). Wenner-Gren Center International Symposium Series, Vol. 27 (Y. Zotterman, ed.). Pergamon Press, Oxford.

Septal Nuclei, The, (1976). In *Advan. Behav. Biol.,* Vol. 20 (J. F. DeFrance, ed.). Plenum Press, New York.

Serafetinides, E. A., R. D. Hoare, and M. V. Driver (1965). Intracarotid sodium amylobarbitone and cerebral dominance for speech and consciousness. *Brain 88,* 107–130.

Sessle, B. J., and L. F. Greenwood (1976). Inputs to trigeminal brain stem neurones from facial, oral, tooth pulp and pharyngolaryngeal tissues: I. Responses to innocuous and noxious stimuli. *Brain Research 117,* 211–226.

Seyffarth, H. (1940). The behaviour of motor units in voluntary contraction. *Norske Vid.-Akad. Avh. I, Math.-Nat. Kl.* No. 4, and *Acta Psychiat. Scand. 16,* 79–109 and 261–278 (1941).

Seyffarth, H., and D. Denny-Brown (1948). The grasp reflex and the instinctive grasp reaction. *Brain 71,* 109–183.

Shambes, G. M., J. M. Gibson, and W. Welker (1978). Fractured somatotopy in granule cell tactile areas of rat cerebellar hemispheres revealed by micromapping. *Brain Behav. Evol. 15,* 94–140.

Shanks, M. F., A. J. Rockel, and T. P. S. Powell (1975). The commissural fibre connections of the primary somatic sensory cortex. *Brain Research 98,* 166–171.

Shanzer, S., and M. B. Bender (1959). Oculomotor responses on vestibular stimulation of monkeys with lesions of the brain stem. *Brain 82,* 669–682.

Sharrard, W. J. W. (1955). The distribution of the permanent paralysis in the lower limb in poliomyelitis. *J. Bone Jt. Surg. 37,* 540–558.

Sheehan, D. (1941). Spinal autonomic outflows in man and monkey. *J. Comp. Neurol. 75,* 341–370.

Sheehan, D., and J. Pick (1943). The rami communicantes in the rhesus monkey. *J. Anat. (Lond.) 77,* 125–139.

Shelden, C. H., R. H. Pundenz, D. B. Freshwater, and B. L. Crue (1955). Compression rather than decompression for trigeminal neuralgia. *J. Neurosurg. 12,* 123–126.

Shenkin, H. A., and F. H. Lewey (1944). Taste aura preceding convulsions in a lesion of the partetal operculum. *J. Nerv. Ment. Dis. 100,* 352–354.

Shepherd, G. M. (1972). Synaptic organization of the mammalian olfactory bulb. *Physiol. Rev. 52,* 864–917.

Shepherd, G. M. (1974). *The Synaptic Organization of the Brain. An Introduction.* Oxford Univ. Press, New York.

Shepherd, G. M., T. V. Getchell, and J. S. Kauer (1975). Analysis of structure and function in the olfactory pathway. In *The Nervous System,* Vol. 1: *The Basic Neurosciences* (D. B. Tower, ed.), pp. 207–220. Raven Press, New York.

Sheppard, R. N., and N. H. Wadia (1956). Atypical features in acoustic neuroma. *Brain 79,* 282–318.

Sheps, J. G. (1945). The nuclear configuration and cortical connections of the human thalamus. *J. Comp. Neurol. 83,* 1–56.

Sherlock, D. A., P. M. Field, and G. Raisman (1975). Retrograde transport of horseradish peroxidase in the magnocellular neurosecretory system of the rat. *Brain Research 88,* 403–414.

Shigenaga, Y., A. Sakai, and K. Okada (1976). Effect of tooth pulp stimulation in trigeminal nucleus caudalis and adjacent reticular formation in rat. *Brain Research 103,* 400–406.

Shimazu, H. (1972). Organization of the commissural connections: physiology. In *Basic Aspects of Central Vestibular Mechanisms, Progress in Brain Research,* Vol. 37 (A. Brodal and O. Pompeiano, eds.), pp. 177–190. Elsevier, Amsterdam.

Shimazu, H., T. Hongo, and K. Kubota (1962). Two types of central influences on gamma motor system. *J. Neurophysiol. 25,* 309–323.

Shimazu, H., T. Hongo, K. Kubota, and H. Narabayashi (1962). Rigidity and spasticity in man. Electromyographic analysis with reference to the role of the globus pallidus. *Arch. Neurol. (Chic.) 6,* 10–17.

Shimazu, H., and W. Precht (1965). Tonic and kinetic responses of cat's vestibular neurons to horizontal angular acceleration. *J. Neurophysiol. 28,* 991–1013.

Shimazu, H., and W. Precht (1966). Inhibition of central vestibular neurons from the contralateral labyrinth and its mediating pathway. *J. Neurophysiol. 29,* 467–492.

Shimizu, N., S. Ohnishi, M. Tohyama, and T. Maeda (1974). Demonstration by degeneration silver methods of the ascending projection from the locus coeruleus. *Exp. Brain Res. 21,* 181–192.

Shinnar, S., R. J. Maciewicz, and R. J. Shofer (1975). A raphe projection to cat cerebellar cortex. *Brain Research 97,* 139–143.

Shinoda, Y., A. P. Arnold, and H. Asanuma (1976). Spinal branching of corticospinal axons in the cat. *Exp. Brain Res. 26,* 215–234.

Shinoda, Y., G. Ghez, and A. Arnold (1977). Spinal branching of rubrospinal axons in the cat. *Exp. Brain Res. 30,* 203–218.

Shinoda, Y., P. Zarzecki, and H. Asanuma (1979). Spinal branching of pyramidal tract neurons in the monkey. *Exp. Brain Res. 34,* 59–72.

Shipley, M. T. (1975). The topographical and laminar organization of the presubiculum's projection to the ipsi- and contralateral entorhinal cortex in the guinea pig. *J. Comp. Neurol. 160,* 127–146.

Sholl, D. A. (1953). Dendritic organization in the neurons of the visual and motor cortices of the cat. *J. Anat. (Lond.) 87,* 387–406.

Sholl, D. A. (1955). The organization of the visual cortex in the cat. *J. Anat. (Lond.) 89,* 33–46.

Sholl, D. A. (1956). *The Organization of the Cerebral Cortex.* Methuen, London; Wiley, New York.

Short-Term Memory (1975). (D. Deutsch and J. A. Deutsch, eds.). Academic Press, New York.

Shriver, J. E., and C. R. Noback (1967). Cortical projections to the lower brain stem and spinal cord in the tree shrew (*Tupaia glis*). *J. Comp. Neurol. 130,* 25–54.

Shriver, J. E., B. M. Stein, and M. B. Carpenter (1968). Central projections of spinal dorsal roots in the monkey. I. Cervical and upper thoracic dorsal roots. *Amer. J. Anat. 123,* 27–74.

Shute, C. C. D., and P. R. Lewis (1966). Cholinergic and monoaminergic pathways in the hypothalamus. *Brit. Med. Bull. 22,* 221–226.

Shute, C. C. D., and P. R. Lewis (1967). The ascending cholinergic reticular system: neocortical, olfactory and subcortical projections. *Brain 90,* 497–520.

Sica, R. E. P., A. J. McComas, A. R. M. Upton, and D. Longmira (1974). Motor unit estimations in small muscles of the hand. *J. Neurol. Neurosurg. Psychiat. 37,* 55–67.

Siegel, A., H. Edinger, and S. Ohgami (1974). The topographical organization of the hippocampal projection to the septal area: a comparative neuroanatomical analysis in the gerbil, rat, rabbit and cat. *J. Comp. Neurol. 157,* 359–378.

Siegel, A., H. Edinger, and R. Troiano (1973). The pathway from the mediodorsal nucleus of thalamus to the hypothalamus in the cat. *Exp. Neurol. 38*, 202–217.

Siegel, A., T. Fukushima, R. Meibach, L. Burke, H. Edinger, and S. Weiner (1977). The origin of the afferent supply to the mediodorsal thalamic nucleus: enhancement of HRP transport by selective lesions. *Brain Research 135*, 11–23.

Siegel, A., S. Ohgami, and H. Edinger (1975). Projections of the hippocampus to the septal area in the squirrel monkey. *Brain Research 99*, 247–260.

Siegel, A., R. Troiano, and A. Royce (1973). Differential projections of the anterior and posterior cingulate gyrus to the thalamus in the cat. *Exp. Neurol. 38*, 192–201.

Siegfried, J., F. R. Ervin, Y. Miyazaki, and V. H. Mark (1962). Localized cooling of the central nervous system. I. Neurophysiological studies in experimental animals. *J. Neurosurg. 19*, 840–852.

Siegfried, J., E. Esslen, U. Gretener, E. Ketz, and E. Perret (1970). Functional anatomy of the dentate nucleus in the light of stereotaxic operations. *Confin. Neurol. (Basel) 32*, 1–10.

Sikes, R. W., R. B. Chronister, and L. E. White, Jr. (1977). Origin of the direct hippocampus—anterior thalamic bundle in the rat: a combined horseradish peroxidase–Golgi analysis. *Exp. Neurol. 57*, 379–395.

Sillito, A. M., and A. W. Zbrozyna (1970). The localization of pupilloconstrictor function within the mid-brain of the cat. *J. Physiol. (Lond.) 211*, 461–477.

Silverman, S. M., P. S. Bergman, and M. B. Bender (1961). The dynamics of transient cerebral blindness. Report of nine episodes following vertebral angiography. *Arch. Neurol. (Chic.) 4*, 333–348.

Simantov, R., M. J. Kuhar, G. W. Pasternak, and S. H. Snyder (1976). The regional distribution of a morphine-like factor enkephalin in monkey brain. *Brain Research 106*, 189–197.

Simma, K. (1951). Zur Projektion des Centrum medianum und Nucleus parafascicularis thalami beim Menschen. *Mschr. Psychiat. Neurol. 122*, 32–46, 1951.

Simmons, H. J., and E. W. Powell (1972). Septomammillary projections in the squirrel monkey. *Acta Anat. (Basel) 82*, 159–178.

Simpson, D. A. (1952a). The projection of the pulvinar to the temporal lobe. *J. Anat. (Lond.) 86*, 20–28.

Simpson, D. A. (1952b). The efferent fibres of the hippocampus in the monkey. *J. Neurol. Neurosurg. Psychiat. 15*, 79–92.

Simpson, J. I., and K. E. Alley (1974). Visual climbing fiber input to rabbit vestibulocerebellum: a source of direction-specific information. *Brain Research 82*, 302–308.

Simpson, J. I., W. Precht, and R. Llinás (1974). Sensory separation in climbing and mossy fiber inputs to cat vestibulocerebellum. *Pflügers Arch. ges. Physiol. 351*, 183–193.

Sinclair, D. C. (1967). *Cutaneous Sensation.* Oxford Univ. Press, London.

Sinclair, D. C., and B. A. R. Stokes (1964). The production and characteristics of "second pain." *Brain 87*, 609–618.

Sinclair, D. C., G. Weddell, and W. H. Feindel (1948). Referred pain and associated phenomena. *Brain 71*, 184–211.

Sinclair, D. C., G. Weddell, and E. Zander (1952). The relationship of cutaneous sensibility to neurohistology in the human pinna. *J. Anat. (Lond.) 86*, 402–411.

Singleton, M. C., and T. L. Peele (1965). Distribution of optic fibers in the cat. *J. Comp. Neurol. 125*, 303–328.

Siqueira, E. B. (1965). The temporo-pulvinar connections in the rhesus monkey. *Arch. Neurol. (Chic.) 13*, 321–330.

Sjöberg, A. (1943). Något om centrala och perifera läsioner av nionde och tionde hjärnnerverna. *Nord. Med. 19*, 1398–1402.

Sjöqvist, O. (1938). Studies on the pain conduction in the trigeminal nerve. *Acta Psychiat. Scand. Suppl. 17*, 1–139.

Sjöstrand, F. (1961). Electron microscopy of the retina. In *The Structure of the Eye* (G. K. Smelser, ed.), pp. 1–28. Academic Press, New York.

Skeen, L. C., and W. C. Hall (1977). Efferent projections of the main and the accessory olfactory bulb in the tree shrew (*Tupaia glis*). *J. Comp. Neurol. 172*, 1–36.

Skinhöj, E. (1972). The sympathetic nervous system and the regulation of cerebral blood flow in man. *Stroke 3*, 711–716.

Skinner, J. E., and D. B. Lindsley (1967). Electrophysiological and behavioral effects of blockade of the nonspecific thalamocortical system. *Brain Research 6*, 95–118.

Skoglund, S. (1956). Anatomical and physiological studies of knee joint innervation in the cat. *Acta Physiol. Scand. 36, Suppl. 124*, 1–101.

Skoglund, S. (1973). Joint receptors and kinaesthesis. In *Handbook of Sensory Physiology*, Vol. II, *Somatosensory System* (A. Iggo, ed.), pp. 111–136. Springer-Verlag, Berlin.

Skrede, K. K., and R. H. Westgaard (1971). The transverse hippocampal slice: a well defined cortical structure maintained *in vitro. Brain Research 35*, 589–593.

Sladek, J., J. R., and J. P. Bowman (1975). The distribution of catecholamines within the inferior olivary complex of the cat and rhesus monkey. *J. Comp. Neurol. 163*, 203–214.

Slauck, A. (1921). Beiträge zur Kenntnis der Muskelpathologie. *Z. Neurol. 71*, 352–356.

Sleep Mechanisms. Progress in Brain Research, Vol. 18 (1965). (K. Akert, C. Bally, and J. P. Schadé, eds.). Elsevier, Amsterdam.

Sloan, N., and H. Jasper (1950). The identity of spreading depression and "suppression." *Electroenceph. Clin. Neurophysiol. 2*, 59–78.

Sloper, J. C. (1966). Hypothalamic neurosecretion. *Brit. Med. Bull. 22*, 209–215.

Sloper, J. J. (1971). Dendro-dendritic synapses in the primate motor cortex. *Brain Research 34*, 186–192.

Sloper, J. J. (1973). An electron microscope study of the termination of afferent connections to the primate motor cortex. *J. Neurocytol. 2*, 361–368.

Sloper, J. J., and T. P. S. Powell (1978). Dendro-dendritic and reciprocal synapses in the primate motor cortex. *Proc. Roy. Soc. B 203*, 23–38.

Sloper, J. J., and T. P. S. Powell (1979a). Ultrastructural features of the sensori-motor cortex of the primate. *Phil. Trans. B 285*, 123–139.

Sloper, J. J., and T. P. S. Powell (1979b). An experimental electron microscopic study of afferent connections to the primate motor and somatic sensory cortices. *Phil. Trans. B 285*, 199–226.

Smith, A. D. (1971). Summing up: some implications of the neuron as a secreting cell. *Phil. Trans. B 261*, 423–437.

Smith, C. A. (1961). Innervation pattern of the cochlea. The internal hair cell. *Ann. Oto-rhino-laryng. 70*, 504–527.

Smith, C. A. (1975). The inner ear: its embryological development and microstructure. In *The Nervous System*, Vol. 3: *Human Communication and Its Disorders* (D. B. Tower, ed.), pp. 1–18. Raven Press, New York.

Smith, C. A. (1978). Structure of the cochlear duct. In *Evoked Electrical Activity in the Auditory Nervous System* (R. F. Naunton and C. Fernández, eds.), pp. 3–19. Academic Press, New York.

Smith, C. A., and G. L. Rasmussen (1963). Recent observations of the olivocochlear bundle. *Ann. Otol. (St. Louis) 72*, 489–506.

Smith, C. A., and G. L. Rasmussen (1965). Degeneration in the efferent nerve endings in the cochlea after axonal section. *J. Cell Biol. 26*, 63–77.

Smith, C. A., and F. S. Sjöstrand (1961). Structure of the nerve endings on the external hair cells of the guinea pig cochlea as studied by serial sections. *J. Ultrastruct. Res. 5*, 523–556.

Smith, K. U., and A. J. Akelaitis (1942). Studies on the corpus callosum. I. Laterality in behavior and bilateral motor organization in man before and after section of the corpus callosum. *Arch. Neurol. Psychiat. (Chic.) 47*, 519–543.

Smith, M. C. (1951). The use of Marchi staining in the later stages of human tract degeneration. *J. Neurol. Neurosurg. Psychiat. 14*, 222–225.

Smith, M. C. (1956a). Observations on the extended use of the Marchi method. *J. Neurol. Neurosurg. Psychiat. 19*, 67–73.

Smith, M. C. (1956b). The recognition and prevention of artefacts of the Marchi method. *J. Neurol. Neurosurg. Psychiat. 19*, 74–83.

Smith, M. C. (1957). The anatomy of the spino-cerebellar fibers in man. 1. The course of the fibers in the spinal cord and brain stem. *J. Comp. Neurol. 108*, 285–352.

Smith, M. C. (1960). Nerve fibre degeneration in the brain in amyotrophic lateral sclerosis. *J. Neurol. Neurosurg. Psychiat. 23*, 269–282.

Smith, M. C. (1961). The anatomy of the spino-cerebellar fibers in man. II. The distribution of the fibers in the cerebellum. *J. Comp. Neurol. 117*, 329–354.

Smith, M. C. (1962). Location of stereotactic lesions confirmed at necropsy. *Brit. Med. J. 1*, 900–906.

Smith, M. C. (1967). Stereotactic operations for Parkinson's disease—anatomical observations. In *Modern Trends in Neurology* (D. Williams, ed.), pp. 21–52. Butterworths, London.

Smith, M. C., S. J. Strich, and P. Sharp (1956). The value of the Marchi method for staining tissue stored in formalin for prolonged periods. *J. Neurol. Neurosurg. Psychiat. 19*, 62–64.

Smith, O. A. (1974). Reflex and central mechanisms involved in the control of the heart and the circulation. *Ann. Rev. Physiol. 36*, 93–123.

Smith, R. L. (1973). The ascending fiber projections from the principal sensory trigeminal nucleus in the rat. *J. Comp. Neurol. 148*, 423–446.

Smith, R. L. (1975). Axonal projections and connections of the principal sensory trigeminal nucleus in the monkey. *J. Comp. Neurol. 163*, 347–376.

Smith, W. K. (1944a). The frontal eye fields. In *The Precentral Motor Cortex* (Paul C. Bucy, ed.), pp. 307–342. Univ. of Illinois Press, Urbana.

Smith, W. K. (1944b). The results of ablation of the cingular region of the cerebral cortex. *Fed. Proc. 3*, 42.

Smyth, G. E. (1939). The systematization and central connections of the spinal tract and nucleus of the trigeminal nerve. A clinical and pathological study. *Brain 62*, 41–87.

Snider, R. S. (1936). Alterations which occur in mossy terminals of the cerebellum, following transection of the brachium pontis. *J. Comp. Neurol. 64*, 417–431.

Snider, R. S. (1940). Morphology of the cerebellar nuclei in the rabbit and the cat. *J. Comp. Neurol. 72*, 399–415.

Snider, R. S. (1945). Electro-anatomical studies on a tecto-cerebellar pathway. (Abstract.) *Anat. Rec. 91*, 299.

Snider, R. S. (1950). Recent contributions to the anatomy and physiology of the cerebellum. *Arch. Neurol. Psychiat. (Chic.) 64*, 196–219.

Snider, R. S. (1975). A cerebellar-ceruleus pathway. *Brain Research 88*, 59–63.

Snider, R. S., and E. Eldred (1948). Cerebral projections to the tactile, auditory and visual areas of the cerebellum. (Abstract.) *Anat. Rec. 100*, 714.

Snider, R., and E. Eldred (1951). Electro-anatomical studies on cerebro-cerebellar connections in the cat. *J. Comp. Neurol. 95*, 1–16.

Snider, R. S., and E. Eldred (1952). Cerebro-cerebellar relationships in the monkey. *J. Neurophysiol. 15*, 27–40.

Snider, R. S., and A. Stowell (1942). Evidence of a representation of tactile sensibility in the cerebellum of the cat. *Fed. Proc. 1*, 82–83.

Snider, R. S., and A. Stowell (1944). Receiving areas of the tactile, auditory and visual systems in the cerebellum. *J. Neurophysiol. 7*, 331–357.

Snider, R. S., and N. Wetzel (1965). Electroencephalographic changes by stimulation of the cerebellum of man. *Electroenceph. Clin. Neurophysiol. 18*, 176–183.

Snow, P. J., P. K. Rose, and A. Brown (1976). Tracing axons and axon collaterals of spinal neurones using intracellular injection of horseradish peroxidase. *Science 191*, 312–313.

Snyder, S. H., S. P. Banerjee, H. I. Yamamura, and D. Greenberg (1974). Drugs, neurotransmitters and schizophrenia. *Science 184*, 1243–1253.

Soffin, G., M. Feldman, and M. B. Bender (1968). Alterations of sensory levels in vascular lesions of lateral medulla. *Arch. Neurol. (Chic.) 18*, 178–190.

Somana, R., and F. Walberg (1978). Cerebellar afferents from the paramedian reticular nucleus studied with retrograde transport of horseradish peroxidase. *Anat. Embryol. 154*, 353–368.

Somana, R., and F. Walberg (1979). Cerebellar afferents from the nucleus of the solitary tract. *Neurosci. Letters 11*, 41–47.

Somatic and Visceral Sensory Mechanisms (1977). (G. Gordon, ed.). *Brit. Med. Bull. 33*, 89–182.

Somatosensory System. Handbook of Sensory Physiology, Vol. II (1973). (A. Iggo, ed.). Springer-Verlag, Berlin.

Somatosensory System, The (1975). (H. H. Kornhuber, ed.). Georg Thieme Verlag, Stuttgart.

Somatosensory and Visceral Receptor Mechanisms. Progr. Brain Res., Vol. 43 (1976). (A. Iggo and O. B. Ilyinsky, eds.). Elsevier, Amsterdam.

Somjen, G., D. O. Carpenter, and E. Henneman (1965). Responses of motoneurons of different sizes to graded stimulation of supraspinal centres of the brain. *J. Neurophysiol. 28*, 958–965.

Sotelo, C. (1975). Morphological correlates of electronic coupling between neurons in mammalian nervous system. In *Golgi Centennial Symposium: Perspectives in Neurobiology* (M. Santini, ed.), pp. 355–365. Raven Press, New York.

Sotelo, C., and P. Angaut (1973). The fine structure of the cerebellar central nuclei in the cat. I. Neurons and neuroglial cells. *Exp. Brain Res. 16*, 410–430.

Sotelo, C., R. Llinás, and R. Baker (1974). Structural study of inferior olivary nucleus of the cat: morphological correlates of electronic coupling. *J. Neurophysiol. 37*, 541–559.

Sotelo, C., and S. L. Palay (1970). The fine structure of the lateral vestibular nucleus in the rat. II. Synaptic organization. *Brain Research 18*, 93–115.

Sotelo, C., and D. Riche (1974). The smooth endoplasmic reticulum and the retrograde and fast orthograde transport of horseradish peroxidase in the nigro-striato-nigral loop. *Anat. Embryol. 146*, 209–218.

Sotgui, M. L., and M. Margnelli (1976). Electrophysiological identification of pontomedullary reticular neurons directly projecting into dorsal column nuclei. *Brain Research 103*, 443–453.

Sotgiu, M. L., and G. Marini (1977). Reticulo-cuneate projections as revealed by horseradish peroxidase axonal transport. *Brain Research 128*, 341–345.

Sousa-Pinto, A. (1969). Experimental anatomical demonstration of a cortico-olivary projection from area 6 (supplementary motor area?) in the cat. *Brain Research 16*, 73–83.

Sousa-Pinto, A. (1970a). The cortical projection onto the paramedian reticular and perihypoglossal nuclei (nucleus praepositus hypoglossi, nucleus intercalatus and nucleus of Roller) of the medulla oblongata of the cat. An experimental study. *Brain Research 18*, 77–91.

Sousa-Pinto, A. (1970b). Electron microscopic observations on the possible retinohypothalamic projection in the rat. *Exp. Brain Res. 11*, 528–538.

Sousa-Pinto, A. (1973). Cortical projections of the medial geniculate body in the cat. *Advan. Anat. Embryol. Cell Biol. 48*, 1–42.

Sousa-Pinto, A., and A. Brodal (1969). Demonstration of a somatotopical pattern in the cortico-olivary projection in the cat. An experimental-anatomical study. *Exp. Brain Res. 8*, 364–386.

Špaček, J., A. R. Lieberman (1974). Ultrastructure and three-dimensional organization of synaptic glomeruli in rat somatosensory thalamus. *J. Anat. (Lond.) 3*, 487–516.

Spalding, J. M. K. (1952a). Wounds of the visual pathway. Part I: The visual radiation. *J. Neurol. Neurosurg. Psychiat. 15*, 99–109.

Spalding, J. M. K. (1952b). Wounds of the visual pathway. Part II: The striate cortex. *J. Neurol. Neurosurg. Psychiat. 15*, 169–183.

Specht, S., and B. Grafstein (1973). Accumulation of radioactive protein in mouse cerebral cortex after injection of ³H-fucose into the eye. *Exp. Neurol. 41,* 705–722.

Sperry, R. W. (1966). Brain bisection and mechanisms of consciousness. In *Brain and Conscious Experience* (J. C. Eccles, ed.), pp. 298–308. Springer-Verlag, Berlin.

Sperry, R. W. (1974). Lateral specialization in the surgically separated hemispheres. In *The Neurosciences. Third Study Program* (F. O. Schmitt and F. G. Worden, eds.), pp. 5–19. MIT Press, Cambridge, Mass.

Spiegel, E. A., and E. G. Szekely (1961). Prolonged stimulation of the head of the caudate nucleus. *Arch. Neurol. (Chic.) 4,* 55–65.

Spiegel, E. A., H. T. Wycis, M. Marks, and A. J. Lee (1947). Stereotaxic apparatus for operations on the human brain. *Science 106,* 349–350.

Spiller, W. G., and C. H. Frazier (1901). The division of the sensory root of the trigeminus for the relief of tic douloureux; an experimental, pathological and clinical study, with a preliminary report of one surgically successful case. *Philad. Med. J. 8,* 1039–1049.

Spoendlin, H. (1956). Elektronenmikroskopische Untersuchungen am Cortischen Organ. *Pract. Oto-rhino-laryng. (Basel) 18,* 246.

Spoendlin, H. (1960). Submikroskopische Strukturen im Cortischen Organ der Katze. *Acta Otolaryng. (Stockh.) 52,* 111–130.

Spoendlin, H. H. (1964). Organization of the sensory hairs in the gravity receptors in utricle and saccule of the squirrel monkey. *Z. Zellforsch. 62,* 701–716.

Spoendlin, H. H. (1970). Structural basis for peripheral frequency analysis. In *Frequency Analysis and Periodicity Detection in Hearing* (R. Plomp and G. F. Smoorenburg, eds.), pp. 2–40. A. W. Sijthoff, Leiden.

Spoendlin, H. (1972). Innervation densities of the cochlea. *Acta Otolaryng. (Stockh.) 73,* 235–248.

Spoendlin, H. (1973). The innervation of the cochlear receptor. In *Basic Mechanisms in Hearing* (A. R. Möller, ed.), pp. 185–234. Academic Press, New York.

Spoendlin, H. H., and R. R. Gacek (1963). Electron microscopic study of the efferent and afferent innervation of the organ of Corti in the cat. *Ann. Otol. (St. Louis) 72,* 660–686.

Spoendlin, H., and W. Lichtensteiger (1966). The adrenergic innervation of the labyrinth. *Acta Otolaryng. (Stockh.) 61,* 423–434.

Spoendlin, H., and W. Lichtensteiger (1967). The sympathetic nerve supply to the inner ear. *Arch. klin. exp Ohr.-Nas.-Kehlkopfheilk. 189,* 346–359.

Sprague, J. M. (1953). Spinal "border cells" and their role in postural mechanism (Schiff-Sherrrington phenomenon). *J. Neurophysiol. 16,* 464–474.

Sprague, J. M. (1958). The distribution of dorsal root fibres on motor cells in the lumbosacral spinal cord of the cat, and the site of excitatory and inhibitory terminals in monosynaptic pathways. *Proc. Roy. Soc. B 149,* 534–556.

Sprague, J. M. (1975). Mammalian tectum: intrinsic organization, afferent inputs, and integrative mechanisms. Anatomical substrate. *Neurosci. Res. Progr. Bull. 13,* 204–213.

Sprague, J. M., and W. W. Chambers (1954). Control of posture by reticular formation and cerebellum in the intact, anesthetized and unanesthetized and in the decerebrated cat. *Amer. J. Physiol. 176,* 52–64.

Sprague, J. M., and H. Ha (1964). The terminal fields of dorsal root fibers in the lumbosacral cord of the cat, and the dendritic organization of the motor nuclei. In *Organization of the Spinal Cord* (J. C. Eccles and J. P. Schadé, eds.), pp. 120–152. Elsevier, Amsterdam.

Sprague, J. M., M. Levitt, K. Robson, C. N. Liu, E. Stellar, and W. W. Chambers (1963). A neuroanatomical and behavioral analysis of the syndromes resulting from midbrain lemniscal and reticular lesions in the cat. *Arch. Ital. Biol. 101,* 225–295.

Sprague, J. M., and T. H. Meikle (1965). The role of the superior colliculus in visually guided behavior. *Exp. Neurol. 11,* 115–146.

Spurling, R. G., and E. G. Grantham (1940). Neurologic picture of herniations of the nucleus pulposus in the lower part of the lumbar region. *Arch. Surg. 40,* 375–388.

Spyer, K. M. (1975). Organisation of baroreceptor pathways in the brain. *Brain Research* *87*, 221–226.

Spyer, K. M., and J. H. Wolstencroft (1971). Problems of the afferent input to the paramedian reticular nucleus, and the central connections of the sinus nerve. *Brain Research* *26*, 411–414.

Srebro, B., and S. A. Lorens (1975). Behavioral effects of selective midbrain raphe lesions in the rat. *Brain Research 89*, 303–325.

Sreesai, M. (1974). Cerebellar cortical projections of the opossum (*Didelphis marsupialis virginiana*). *J. Hirnforsch. 15*, 529–544.

Stahl, M. (1963). Elektronenmikroskopische Untersuchungen über die vegetative Innervation der Bauchspeicheldrüse. *Z. Mikr.-Anat. Forsch. 70*, 62–102.

Stanfield, J. P. (1960). The blood supply of the human pituitary gland. *J. Anat. (Lond.) 94*, 257–273.

Starzl, T. E., and D. G. Whitlock (1952). Diffuse thalamic projection system in monkey. *J. Neurophysiol. 15*, 449–468.

Stauffer, E. K., D. G. D. Watt, A. Taylor, R. M. Reinking, and D. G. Stuart (1976). Analysis of muscle receptor connections by spike-triggered averaging. 2. Spindle group II afferents. *J. Neurophysiol. 39*, 1393–1402.

Steg, G. (1966). Efferent muscle control in rigidity. In *Muscular Afferents and Motor Control*. Proceedings 1st Nobel Symposium, Stockholm 1965 (R. Granit, ed.), pp. 437–443. Almqvist & Wiksell, Stockholm; Wiley, New York.

Steiger, H.-J., and J. Büttner-Ennever (1978). Relationship between motoneurons and internuclear neurons in the abducens nucleus: a double retrograde tracer study in the cat. *Brain Research 148*, 181–188.

Steiger, H.-J., and J. A. Büttner-Ennever (1979). Oculomotor nucleus afferents in the monkey demonstrated with horseradish peroxidase. *Brain Research 160*, 1–15.

Stein, B. E., S. J. Goldberg, and H. P. Clamann (1976). The control of eye movements by the superior colliculus in the alert cat. *Brain Research 118*, 469–474.

Stein, B. E., B. Magalhães-Castro, and L. Kruger (1976). Relationship between visual and tactile representation in cat superior colliculus. *J. Neurophysiol. 39*, 401–419.

Stein, B. M., and M. B. Carpenter (1967). Central projections of portions of the vestibular ganglia innervating specific parts of the labyrinth in the rhesus monkey. *Amer. J. Anat. 120*, 281–318.

Stein, J. (1978). Effects of parietal lobe cooling on manipulative behaviour in the conscious monkey. In *Active Touch* (G. Gordon, ed.), pp. 79–90. Pergamon Press, Oxford.

Stein, R. B. (1974). Peripheral control of movement. *Physiol. Rev. 54*, 215–243.

Steinhausen, W. (1931). Über den Nachweis der Bewegungen der Cupula in der intakten Bogengangampulle des Labyrinthes bei der natürlichen rotatorischen und calorischen Reizung. *Pflügers Arch. ges. Physiol. 228*, 322–328.

Stell, W. K. (1972). The morphological organization of the vertebrate retina. In *Handbook of Sensory Physiology, Vol. VII/2, Physiology of Photoreceptor Organs* (M. G. F. Fuortes, ed.), pp. 111–213. Springer Verlag, Berlin.

Stephan, H. (1975). In *Allocortex. Handbuch der mikroskopischen Anatomie des Menschen* (W. Bargman, ed.), Vol. 4: *Nervensystem*, Part 9. Springer-Verlag, Berlin.

Stephan, H., and O. J. Andy (1962). The septum (A comparative study on its size in insectivores and primates). *J. Hirnforsch. 5*, 229–244.

Stepien, L. S., J. P. Cordeau, and T. Rasmussen (1960). The effect of temporal lobe and hippocampal lesions on auditory and visual recent memory in monkeys. *Brain 83*, 470–489.

Sterling, P. (1973). Quantitative mapping with the electron microscope: retinal terminals in the superior colliculus. *Brain Research 54*, 347–354.

Sterling, P., and H. G. J. M. Kuypers (1967a). Anatomical organization of the brachial spinal cord of the cat. I. The distribution of dorsal root fibers. *Brain Research 4*, 1–15.

Sterling, P., and H. G. J. M. Kuypers (1967b). Anatomical organization of the brachial spinal cord of the cat. II. The motoneuron plexus. *Brain Research 4*, 16–32.

Stern, G. (1966). The effects of lesions in the substantia nigra. *Brain 89*, 449–478.

Stern, J., and A. Ward, Jr. (1960). Inhibition of the muscle spindle discharge by ventrolateral thalamic stimulation. Its relation to Parkinsonism. *Arch. Neurol. (Chic.) 3*, 193–204.

Stern, K. (1938). Note on the nucleus ruber magnocellularis and its efferent pathway in man. *Brain 61*, 284–289.

Stevens, J. R., C. Kim, and P. D. MacLean (1961). Stimulation of caudate nucleus. Behavioral effects of chemical and electrical excitation. *Arch. Neurol. (Chic.) 4*, 47–65.

Steward, O. (1976). Topographic organization of the projections from the entorhinal area to the hippocampal formation of the rat. *J. Comp. Neurol. 167*, 285–314.

Steward, O., C. W. Cotman, and G. S. Lynch (1974). Growth of a new fiber projection in the brain of adult rats: re-innervation of the dentate gyrus by the contralateral entorhinal cortex following ipsilateral entorhinal lesions. *Exp. Brain Res. 20*, 45–66.

Steward, O., S. A. Scoville, and S. L. Vinsant (1977). Analysis of collateral projections with a double retrograde labeling technique. *Neurosci. Letters 5*, 1–5.

Stewart, D. H., Jr., C. J. Scibetta, and R. B. King (1967). Presynaptic inhibition in the trigeminal relay nuclei. *J. Neurophysiol. 30*, 135–153.

Stewart, R. M. (1939). Arhinencephaly. *J. Neurol. Psychiat. 2*, 303–312.

Stewart, W. A., and R. B. King (1963). Fiber projections from the nucleus caudalis of the spinal trigeminal nucleus. *J. Comp. Neurol. 121*, 271–286.

Stibbe, E. P. (1939). Surgical anatomy of the subtentorial angle with special reference to the acoustic and trigeminal nerves. *Lancet ii*, 859–862.

Stilwell, D. L., Jr. (1956). The nerve supply of the vertebral column and its associated structures in the monkey. *Anat. Rec. 125*, 139–170.

Stilwell, D. L., Jr. (1957a). The innervation of deep structures of the foot. *Amer. J. Anat. 101*, 59–74.

Stilwell, D. L., Jr. (1957b). The innervation of deep structures of the hand. *Amer. J. Anat. 101*, 75–99.

Stone, J. (1972). Morphology and physiology of the geniculo-cortical synapse in the cat: the question of parallel input to the striate cortex. *Invest. Ophthal. 11*, 338–346.

Stone, J., and B. Dreher (1973). Projection of X- and Y-cells of the cat's lateral geniculate nucleus to area 17 and 18 of visual cortex. *J. Neurophysiol. 36*, 551–567.

Stone, J., and R. B. Freeman (1973). Neurophysiological mechanisms in the visual discrimination of form. In *Handbook of Sensory Physiology, Vol. VII/3, Part A, Integrative Functions and Comparative Data* (R. Jung, ed.), pp. 153–207. Springer-Verlag, Berlin.

Stone, J., and Y. Fukuda (1974). Properties of cat retinal ganglion cells: a comparison of W-cells with X- and Y-cells. *J. Neurophysiol. 37*, 722–748.

Stone, J., and S. M. Hansen (1966). The projection of the cat's retina on the lateral geniculate nucleus. *J. Comp. Neurol. 126*, 601–624.

Stone, J., and K.-P. Hoffmann (1971). Conduction velocity as a parameter in the organization of the afferent relay in the cat's lateral geniculate nucleus. *Brain Research 32*, 454–459.

Stookey, B., and J. Ransohoff (1959). *Trigeminal Neuralgia. Its History and Treatment.* C. C. Thomas, Springfield, Ill.

Stopford, J. S. B. (1922). The nerve supply of the interphalangeal and metacarpophalangeal joints. *J. Anat. (Lond.) 56*, 1–11.

Storm-Mathiesen, J. (1978). Localization of putative transmitters in the hippocampal formation. With a note on the connections to septum and hypothalamus. In *Functions of the Septo-Hippocampal System.* Ciba Foundation Symposium 58 (new series), pp. 49–79. Elsevier Excerpta Medica, Amsterdam.

Stotler, W. A. (1953). An experimental study of the cells and connections of the superior olivary complex of the cat. *J. Comp. Neurol. 98*, 401–432.

Straschill, M., and P. Rieger (1973). Eye movements evoked by focal stimulation of the cat's superior colliculus. *Brain Research 59*, 211–227.

Strata, P. (1975). The dual input to the cerebellar cortex. In *Golgi Centennial Symposium, Proc.* (M. Santini, ed.), pp. 273–280. Raven Press, New York.

Strick, P. L. (1973). Light microscopic analysis of the cortical projection of the thalamic ventrolateral nucleus in the cat. *Brain Research 55*, 1–24.

Strick, P. L. (1975). Multiple sources of thalamic input to the primate motor cortex. *Brain Research 88, 372–377*.

Strick, P. L., and C. C. Kim (1978). Input to primate motor cortex from posterior parietal cortex (area 5). I. Demonstration by retrograde transport. *Brain Research 157*, 325–331.

Strick, P. L., and J. B. Preston (1978). Multiple representation in the primate motor cortex. *Brain Research 154*, 366–370.

Strick, P. L., and P. Sterling (1974). Synaptic termination of afferents from the ventrolateral nucleus of the thalamus in the cat motor cortex. A light and electron microscope study. *J. Comp. Neurol. 153*, 77–106.

Strominger, N. L., and M. B. Carpenter (1965). Effects of lesions in the substantia nigra upon subthalamic dyskinesia in the monkey. *Neurology 15*, 587–594.

Strominger, N. L., L. R. Nelson, and W. J. Dougherty (1977). Second order auditory pathways in the chimpanzee. *J. Comp. Neurol. 172*, 349–366.

Strong, O. S., and A. Elwyn (1943). *Human Neuroanatomy*. Williams & Wilkins, Baltimore.

Structural and Functional Organization of the Neocortex, The. (1970) (K. L. Chow and A. L. Leiman, eds.). *Neurosci. Res. Progr. Bull. 8*, 155–220.

Structure and Function of the Limbic System (1967). Progress in Brain Research, Vol. 27 (W. R. Adey and T. Tokizane, eds.). Elsevier, Amsterdam.

Structure and Function of Synapses (1972). (G. D. Pappas and D. P. Purpura, eds.). Raven Press, New York.

Structure of Human Memory, The (1976). (C. N. Cofer, ed.). Freeman, San Francisco.

Stryker, M. P., and P. H. Schiller (1975). Eye and head movements evoked by electrical stimulation of monkey superior colliculus. *Exp. Brain Res. 23*, 103–112.

Stürup, G. K., and E. A. Carmichael (1935). Pain: The peripheral pathway. *Brain 58*, 216–219.

Stålberg, E., M. S. Schwartz, B. Thiele, and H. H. Schiller (1976). The normal motor unit in man. A single fibre EMG multielectrode investigation. *J. Neurol. Sci. 27*, 291–301.

Suga, N., and W. E. O'Neill (1978). Mechanisms of echolocation in bats—comments on the neuroethology of the biosonar system of "CF-FM" bats. *Trends Neurosci. 1*, 35–38.

Sugiura, Y. (1975). Three dimensional analysis of neurons in the substantia gelatinosa Rolandi. *Proc. Jpn. Acad. 51*, 336–341.

Sullivan, P. R., J. Kuten, M. S. Atkinson, J. B. Angevine, and P. I. Yakovlev (1958). Cell count in the lateral geniculate nucleus of man. *Neurology 8*, 566–567.

Sumner, D. (1964). Post-traumatic anosmia. *Brain 87*, 107–120.

Sumner, D. (1967). Post-traumatic ageusia. *Brain 90*, 187–202.

Sunderland, S. (1940). The projection of the cerebral cortex on the pons and cerebellum in the macaque monkey. *J. Anat. (Lond.) 74*, 201–226.

Sunderland, S. (1948). Neurovascular relations and anomalies at the base of the brain. *J. Neurol. Neurosurg. Psychiat. 11*, 243–257.

Sunderland, S., and K. C. Bradley (1953). Disturbances of oculomotor function accompanying extradural haemorrhage. *J. Neurol. Neurosurg. Psychiat. 16*, 35–46.

Sunderland, S., and D. F. Cossar (1953). The structure of the facial nerve. *Anat. Rec. 116*, 147–165.

Sunderland, S., and E. S. R. Hughes (1946). The pupillo-constrictor pathway and the nerves to the ocular muscles in man. *Brain 69*, 301–309.

Sunderland, S., and J. O. Lavarack (1953). The branching of nerve fibres. *Acta Anat. 17*, 46–61.

Sunderland, S., and W. E. Swaney (1952). The intraneural topography of the recurrent laryngeal nerve in man. *Anat. Rec. 114*, 411–426.

Surgical Approaches in Psychiatry (1973). (L. V. Laitinen and K. E. Livingston, eds.). Univ. Park Press, Baltimore.

Suzuki, J.-I., and B. Cohen (1964). Head, eye, body and limb movements from semicircular canal nerves. *Exp. Neurol. 10*, 393–405.

Svendgaard, N. A., A. Björklund, and U. Stenevi (1975). Regenerative properties of central monoamine neurons. *Advan. Anat. Embryol. Cell Biol. 51/4*, 1–77.

Swanson, A. G., G. C. Buchan, and E. C. Alvord, Jr. (1965). Anatomic changes in congenital insensitivity to pain. *Arch. Neurol. (Chic.) 12*, 12–18.

Swanson, L. W. (1976). The locus coeruleus: a cytoarchitectonic, Golgi and immunohistochemical study in the albino rat. *Brain Research 110*, 39–56.

Swanson, L. W. (1977). Immunohistochemical evidence for a neurophysin-containing autonomic pathway arising in the paraventricular nucleus of the hypothalamus. *Brain Research 128*, 346–353.

Swanson, L. W. (1978). The anatomical organization of septo-hippocampal projections. In *Functions of the Septo-Hippocampal System*. Ciba Foundation Symposium 58 (new series) pp. 25–43. Elsevier Excerpta Medica, Amsterdam.

Swanson, L. W., and W. M. Cowan (1975). A note on the connections and development of the nucleus accumbens. *Brain Research 92*, 324–330.

Swanson, L. W., and W. M. Cowan (1976). Autoradiographic studies of the development and connections of the septal area in the rat. In *The Septal Nuclei* (J. F. De France, ed.), pp. 37–64. Plenum Press, New York.

Swanson, L. W., and W. M. Cowan (1977). An autoradiographic study of the organization of the efferent connections of the hippocampal formation in the rat. *J. Comp. Neurol. 172*, 49–84.

Swanson, L. W., W. M. Cowan, and E. G. Jones (1974). An autoradiographic study of the efferent connections of the ventral lateral geniculate nucleus in the albino rat and the cat. *J. Comp. Neurol. 156*, 143–164.

Swash, M., and K. P. Fox (1972). Muscle spindle innervation in man. *J. Anat. (Lond.) 112*, 61–80.

Swett, J. E., and E. Eldred (1960). Distribution and numbers of stretch receptors in medial gastrocnemius and soleus muscles of the cat. *Anat. Rec. 137*, 453–460.

Swett, W. H., V. H. Mark, and H. Hamlin (1960). Radiofrequency lesions in the central nervous system of man and cat. *J. Neurosurg. 17*, 213–225.

Synapse, The (1976). *Cold Spring Harbor Symposia on Quantitative Biology*, Vol. 40. Cold Spring Harbor Laboratory, New York.

Szabo, J. (1962). Topical distribution of the striatal efferents in the monkey. *Exp. Neurol. 5*, 21–36.

Szabo, J. (1967). The efferent projections of the putamen in the monkey. *Exp. Neurol. 19*, 463–476.

Szabo, J. (1970). Projections from the body of the caudate nucleus in the rhesus monkey. *Exp. Neurol. 27*, 1–15.

Szabo, J. (1971). A silver impregnation study of nigrostriate projections in the cat. *Anat. Rec. 169*, 441.

Szabo, J. (1972). The course and distribution of the efferents from the tail of the caudate nucleus in the monkey. *Exp. Neurol. 37*, 562–572.

Szabo, J. (1977). Horseradish peroxidase study of the striatal and nigral efferents. *Anat. Rec. 187*, 726.

Szabo, T., and M. Dussardier (1964). Les noyaux d'origine du nerf vague chez le mouton. *Z. Zellforsch. 63*, 247–276.

Szentágothai, J. (1942). Die zentrale Leitungsbahn des Lichtreflexes der Pupillen. *Arch. Psychiat. Nervenkr. 115*, 136–156.

Szentágothai, J. (1943a). Die Lokalisation der Kehlkopfmuskulatur in den Vaguskernen. *Z. Anat. Entwickl. Gesch. 112*, 704–710.

Szentágothai, J. (1943b). Die zentrale Innervation der Augenbewegungen. *Arch. Psychiat. Nervenkr. 116*, 721–760.

Szentágothai, J. (1948). Anatomical considerations of monosynaptic reflex arcs. *J. Neurophysiol. 11*, 445–454.

Szentágothai, J. (1949). Functional representation in the motor trigeminal nucleus. *J. Comp. Neurol. 90*, 111–120.

Szentágothai, J. (1951). Short propriospinal neurons and intrinsic connections of the spinal gray matter. *Acta Morph. Acad. Sci. Hung. 1,* 81–94.

Szentágothai, J. (1952a). *Die Rolle der einzelnen Labyrinthrezeptoren bei der Orientation von Augen und Kopf im Raume.* Akadémiai Kiado, Budapest.

Szentágothai, J. (1952b). The general visceral efferent column of the brain stem. *Acta Morph. Acad. Sci. Hung. 2,* 313–328.

Szentágothai, J. (1958). The anatomical basis of synaptic transmission of excitation and inhibition in motoneurons. *Acta Morph. Acad. Sci. Hung. 8,* 287–309.

Szentágothai, J. (1961). Somatotopic arrangement of synapses of primary sensory neurones in Clarke's column. *Acta Morph. Acad. Sci. Hung, 10,* 307–311.

Szentágothai, J. (1963). The structure of the synapse in the lateral geniculate body. *Acta Anat. (Basel) 55,* 166–186.

Szentágothai, J. (1964a). Neuronal and synaptic arrangement in the substantia gelatinosa Rolandi. *J. Comp. Neurol. 122,* 219–239.

Szentágothai, J. (1964b). Pathways and synaptic articulation patterns connecting vestibular receptors and oculomotor nuclei. In *The Oculomotor System* (Morris B. Bender, ed.), pp. 205–223. Hoeber Medical Division, Harper & Row, New York.

Szentágothai, J. (1964c). The structure of the autonomic interneuronal synapse. *Acta Neuroveg. (Wien) 26,* 338–359.

Szentágothai, J. (1965a). Complex synapses. In *Aus der Werkstatt der Anatomen* (W. Bargmann, ed.), pp. 147–167. Georg Thieme Verlag, Stuttgart.

Szentágothai, J. (1965b). The use of degeneration methods in the investigation of short neuronal connexions. In *Degeneration Patterns in the Nervous System.* Progress in Brain Research, Vol. 14 (M. Singer and J. P. Schadé, eds.), pp. 1–32. Elsevier, Amsterdam.

Szentágothai, J. (1967). Synaptic architecture of the spinal motoneuron pool. In *Recent Advances in Clinical Neurophysiology. Electroenceph. Clin. Neurophysiol.,* Suppl. 25 (L. Widén, ed.), pp. 4–19. Elsevier, Amsterdam.

Szentágothai, J. (1973a). Neuronal and synaptic architecture of the lateral geniculate nucleus. In *Handbook of Sensory Physiology, Vol. VII/3, Central Processing of Visual Information, Part B, Morphology and Function of Visual Centers in the Brain* (R. Jung, ed.), pp. 141–176. Springer-Verlag, Berlin.

Szentágothai, J. (1973b). Synaptology of the visual cortex. In *Handbook of Sensory Physiology, Vol. VII/3, Central Processing of Visual Information, Part B, Morphology and Function of Visual Centers in the Brain* (R. Jung, ed.), pp. 269–324. Springer-Verlag, Berlin.

Szentágothai, J. (1975). The 'module-concept' in cerebral cortex architecture. *Brain Research 95,* 475–496.

Szentágothai, J., and A. Albert (1955). The synaptology of Clarke's column. *Acta Morph. Acad. Sci. Hung. 5,* 43–51.

Szentágothai, J., and M. A. Arbib (1974). Conceptual models of neural organization. *Neurosci. Res. Progr. Bull. 12,* 305–510.

Szentágothai, J., A. Donhoffer, and K. Rajkovits (1954). Die Lokalisation der Cholinesterase in der interneuronalen Synapse. *Acta Histochem. (Jena) 1,* 272–281.

Szentágothai, J., B. Flerkó, B. Mess, and B. Halász (1968). *Hypothalamic Control of the Anterior Pituitary,* 2nd ed. Akadémiai Kiadó, Budapest.

Szentágothai, J., J. Hámori, and T. Tömböl (1966). Degeneration and electron microscope analysis of the synaptic glomeruli in the lateral geniculate body. *Exp. Brain. Res. 2,* 283–301.

Szentágothai, J., and K. Rajkovits (1958). Der Hirnnervenanteil der Pyramidenbahn und der prämotorische Apparat motorischer Hirnnervenkerne. *Arch. Psychiat. Nervenkr. 197,* 335–354.

Szentágothai, J., and K. Rajkovits (1959). Über den Ursprung der Kletterfasern des Kleinhirns. *Z. Anat. Entwickl. Gesch. 121,* 130–141.

Szentágothai-Schimert, J. (1941a). Die Endigungsweise der absteigenden Rückenmarksbahnen. *Z. Anat. Entwickl. Gesch. 111,* 322–330.

Szentágothai-Schimert, J. (1941b). Die Bedeutung des Faserkalibers und der Markscheidendicke im Zentralnervensystem. *Z. Anat. Entwickl. Gesch. 111,* 201–223.

Szumski, A. J., D. Burg, A. Struppler, and F. Velko (1974). Activity of muscle spindles during muscle twitch and clonus in normal and spastic humans subjects. *Electroenceph. Clin. Neurophysiol. 37,* 589–597.

Taarnhøj, P. (1952). Decompression of the trigeminal root and the posterior part of the ganglion as treatment in trigeminal neuralgia. Preliminary communication. *J. Neurosurg. 9,* 288–290.

Taber, E. (1961). The cytoarchitecture of the brain stem of the cat. I. Brain stem nuclei of cat. *J. Comp. Neurol. 116,* 27–70.

Taber, E., A. Brodal, and F. Walberg (1960). The raphe nuclei of the brain stem in the cat. I. Normal topography and cytoarchitecture and general discussion. *J. Comp. Neurol. 114,* 161–187.

Taber Pierce, E., W. E. Foote, and J. Allen Hobson (1976). The efferent connection of the nucleus raphe dorsalis. *Brain Research 107,* 137–144.

Taber Pierce, E., G. H. Hoddevik, and F. Walberg (1977). The cerebellar projection from the raphe nuclei in the cat as studied with the method of retrograde transport of horseradish peroxidase. *Anat. Embryol. 152,* 73–87.

Takeda, T., and K. Maekawa (1976). The origin of the pretecto-olivary tract. A study using the horseradish peroxidase method. *Brain Research 117,* 319–325.

Talbot, W. H., I. Darian-Smith, H. H. Kornhuber, and V. B. Mountcastle (1968). The sense of flutter vibration: comparison of the human capacity with response patterns of mechanoreceptive afferents from the monkey hand. *J. Neurophysiol. 31,* 301–334.

Tanaka, T. (1975). Afferent projections in the hypoglossal nerve to the facial neurons of the cat. *Brain Research 99,* 140–144.

Tanaka, T. (1976). Pyramidal activation of the facial nucleus in the cat. *Brain Research 103,* 389–393.

Tanaka, T., Y. Takeuchi, and K. Nakano (1978). Cells of origin of the spino-facial pathway in the cat: a horseradish peroxidase study. *Brain Research 142,* 580–585.

Tandon, P. N., and K. Kristiansen (1966). Clinico-pathological observations on brain stem dysfunction in cranio-cerebral injuries. Proc. IIIrd Int. Congr. Neurol. Surg., Copenhagen, 1965. *Excerpta Med. Int. Congr. Ser. 110,* 126–129.

Tanji, J. (1975). Activity of neurons in cortical area 3a during maintenance of steady postures by the monkey. *Brain Research 88,* 549–553.

Tanji, J., and E. V. Evarts (1976). Anticipatory activity of motor cortex neurons in relation to direction of an intended movement. *J. Neurophysiol. 39,* 1062–1068.

Taren, J. A., and E. A. Kahn (1962). Anatomic pathways related to pain in face and neck. *J. Neurosurg. 19,* 116–119.

Tarlov, E. (1969). The rostral projections of the primate vestibular nuclei. An experimental study in macaque, baboon and chimpanzee. *J. Comp. Neurol. 135,* 27–56.

Tarlov, E. (1970). Organization of vestibulo-oculomotor projections in the cat. *Brain Research 20,* 159–179.

Tarlov, E., and S. R. Tarlov (1971). The representation of extraocular muscles in the oculomotor nuclei: experimental studies in the cat. *Brain Research 34,* 37–52.

Tarlov, E. C., and R. Y. Moore (1966). The tecto-thalamic connections in the brain of the rabbit. *J. Comp. Neurol. 126,* 403–422.

Tasker, R. R. (1976). Surgery for Parkinson's disease. In *Current Controversies in Neurosurgery* (T. P. Morley, ed.), pp. 411–418. W. B. Saunders, Philadelphia.

Taste and Smell in Vertebrates. A Ciba Foundation Symposium (1970). (G. E. W. Wolstenholme and J. Knight, eds.). W. & A. Churchill, London.

Taverner, D. (1955). Bell's palsy. A clinical and electromyographic study. *Brain 78,* 209–228.

Taverner, D. (1969). The localization of isolated cranial nerve lesions. In *Handbook of Clinical Neurology,* Vol. 2 (P. J. Vinken and G. W. Bruyn, eds.), pp. 52–85. North-Holland, Amsterdam; Wiley, New York.

Taxi, J. (1965). Contribution a l'étude des connexions des neurones moteurs du système nerveux autonome. *Ann. Sci. Nat. Zool. 7*, 413–674.

Teasdall, R. D., and J. W. Magladery (1959). Superficial abdominal reflexes in man. *Arch. Neurol. Psychiat. (Chic.) 81*, 28–36.

Teasdall, R. D., and M. L. Sears (1960). Ocular myopathy. Clinical and electromyographic considerations. *Arch. Neurol. (Chic.) 2*, 281–292.

Terayama, Y., K. Yamamoto, and T. Sakamoto (1969). The efferent olivo-cochlear bundle in the guinea pig cochlea. *Ann. Otol. (St. Louis) 78*, 1254–1268.

Ternaux, J. P., F. Héry, S. Bourgoin, J. Adrien, J. Glowinski, and M. Hamon (1977). The topographical distribution of serotoninergic terminals in the neostriatum of the rat and the caudate nucleus of the cat. *Brain Research 121*, 311–326.

Terzian, H., and G. D. Ore (1955). Syndrome of Klüver-Bucy reproduced in man by bilateral removal of the temporal lobes. *Neurology 5*, 373–380.

Terzuolo, C. A., and R. Llinás (1966). Distribution of synaptic inputs in the spinal motoneurone and its functional significance. In *Muscular Afferents and Motor Control*. Proceedings 1st Nobel Symposium, Stockholm (R. Granit, ed.), pp. 373–384. Almqvist & Wiksell, Stockholm; Wiley, New York.

Terzuolo, C. A., J. F. Soechting, and P. Viviani (1973). Studies on motor control of some simple motor tasks. I. Relations between parameters of movements and EMG activities. *Brain Research 58*, 212–216.

Testa, C. (1964). Functional implications of the morphology of spinal ventral horn neurons of the cat. *J. Comp. Neurol. 123*, 425–444.

Teuber, H.-L. (1972). Unity and diversity of frontal lobe functions. *Acta Neurobiol. Exp. 32*, 615-656.

Teuber, H. L. (1976). Complex functions of basal ganglia. In *The Basal Ganglia* (M.D. Yahr, ed.), pp. 151–168. Raven Press, New York.

Thach, W. T. (1967). Somatosensory receptive fields of single units in cat cerebellar cortex. *J. Neurophysiol. 30*, 675–696.

Thach, W. T. (1970a). Discharge of cerebellar neurons related to two maintained postures and two prompt movements. I. Nuclear cell output. *J. Neurophysiol. 33*, 527–536.

Thach, W. T. (1970b). Discharge of cerebellar neurons related to two maintained postures and two prompt movements. II. Purkinje cell output and input. *J. Neurophysiol. 33*, 537–547.

Thach, W. T. (1975). Timing of activity in cerebellar dentate nucleus and cerebral motor cortex during prompt volitional movement. *Brain Research 88*, 233–241.

Thage, O. (1965). The myotomes L2—S2 in man. *Acta Neurol. Scand. 41, Suppl. 13*, 241–243.

Thage, O. (1974). *Quadriceps Weakness and Wasting. A Neurological, Electrophysiological and Histological Study*. FADLs forlag, København.

Thalamus, The (1966). (D. P. Purpura and M. D. Yahr, eds.). Columbia Univ. Press, New York.

Thierry, A. M., G. Blanc, A. Sobel, L. Stinus, and J. Glowinski (1973). Dopamine terminals in the rat cortex. *Science 182*, 499–501.

Thoenen, H. (1972). Surgical, immunological and chemical sympathectomy. Their application in the investigation of the physiology and pharmacology of the sympathetic nervous system. In *Handbook of Experimental Pharmacology, Catecholamines*, Vol. 33 (H. Blaschko and E. Muscholl, eds.), pp. 813–844. Springer-Verlag, Berlin.

Thomas, D. M., R. P. Kaufman, J. M. Sprague, and W. W. Chambers (1956). Experimental studies of the vermal cerebellar projections in the brain stem of the cat (fastigiobulbar tract). *J. Anat. (Lond.) 90*, 371–385.

Thomas, M. R., and F. R. Calaresu (1974). Localization and function of medullary sites mediating vagal bradycardia in the cat. *Amer. J. Physiol. 226*, 1344–1349.

Thompson, G. C. (1978). Afferent auditory projections to the inferior colliculus of bush baby (*Galago senegalensis*). Abstr. Soc. Neurosci. 4, p. 11.

Thörner, G., H. Lange, and A. Hopf (1975). Morphometrisch-statistische Strukturanalyse des Striatum, Pallidum und Nucleus subthalamicus beim Menschen. II. Pallidum. *J. Hirnforsch. 16*, 401–413.

Tigges, J., and W. K. O'Steen (1974). Termination of retinofugal fibers in squirrel monkey: a re-investigation using autoradiographic methods. *Brain Research 79*, 489–495.

Tiwari, R. K., and R. B. King (1974). Fiber projections from trigeminal nucleus caudalis in primate (squirrel monkey and baboon). *J. Comp. Neurol. 158*, 191–206.

Tobias, T. J. (1975). Afferents to prefrontal cortex from the thalamic mediodorsal nucleus in the rhesus monkey. *Brain Research 83*, 191–212.

Todo, K., T. Yamamoto, H. Satomi, H. Ise, H. Takatama, and K. Takahashi (1977). Origins of vagal preganglionic fibers to the sino-atrial and atrio-ventricular node regions in the cat heart as studied by the horseradish peroxidase method. *Brain Research 130*, 545–550.

Tolbert, D. L., H. Bantli, and J. R. Bloedel (1976). Anatomical and physiological evidence for a cerebellar nucleocortical projection in the cat. *Neurosci. 1*, 205–217.

Tolbert, D. L., H. Bantli, and J. R. Bloedel (1977). The intracerebellar nucleocortical projection in a primate. *Exp. Brain Res. 30*, 425–434.

Tolbert, D. L., H. Bantli, and J. R. Bloedel (1978). Multiple branching of cerebellar efferent projections in cats. *Exp. Brain Res. 31*, 305–316.

Tolbert, D. L., L. C. Massopust, M. G. Murphy, and P. A. Young (1977). The anatomical organization of the cerebello-olivary projection in the cat. *J. Comp. Neurol. 170*, 525–544.

Tolbert, L. P., and D. K. Morest (1978). Patterns of synaptic organization in the cochlear nucleus of the cat. *Abstr. Soc. Neurosci. 4*, p. 11.

Tomasch, J. (1954). Size, distribution, and number of fibres in the human corpus callosum. *Anat. Rec. 119*, 119–135.

Tomasch, J. (1963). Über die Zahl and Grösse der Zellen in den motorischen Hirnnervenkernen des Menschen. *Z. Mikr.-Anat. Forsch. 70*, 298–314.

Tomasch, J. (1968). The overall information carrying capacity of the major afferent and efferent cerebellar cell and fiber systems. *Confin. Neurol. (Basel) 30*, 359–367.

Tomasch, J. (1969). The numerical capacity of the human cortico-ponto-cerebellar system. *Brain Research 13*, 476–484.

Tömböl, T. (1966–67). Short neurons and their synaptic relations in the specific thalamic nuclei. *Brain Research 3*, 307–326.

Tömböl, T. (1969). Two types of short axon (Golgi 2nd) interneurons in the specific thalamic nuclei. *Acta Morph. Acad. Sci. Hung. 17*, 285–297.

Tömböl, T. (1978). Comparative data on the Golgi architecture of interneurons of different cortical areas in cat and rabbit. In *Architectonics of the Cerebral Cortex* (M.A.B. Brazier and H. Petsche, eds.), pp. 59–76. Raven Press, New York.

Torebjörk, H. E., K.-E. Hagbarth, and G. Eklund (1978). Tonic finger flexion reflex induced by vibratory activation of digital mechanoreceptors. In *Active Touch, The Mechanism of Recognition of Objects by Manipulation. A Multi-disciplinary Approach* (G. Gordon, ed.), pp. 197–203. Pergamon Press, Oxford.

Torebjörk, H. E., and R. G. Hallin (1973). Perceptual changes accompanying controlled preferential blocking of A and C fibre responses in intact human skin. *Exp. Brain Res. 16*, 321–332.

Torebjörk, H. E., and R. G. Hallin (1974). Identification of afferent units in intact human skin nerves. *Brain Research 67*, 387–403.

Torvik, A. (1956a). Transneuronal changes in the inferior olive and pontine nuclei in kittens. *J. Neuropath. Exp. Neurol. 15*, 119–145.

Torvik, A. (1956b). Afferent connections to the sensory trigeminal nuclei, the nucleus of the solitary tract and adjacent structures. An experimental study in the rat. *J. Comp. Neurol. 106*, 51–142.

Torvik, A. (1957a). The ascending fibers from the main trigeminal sensory nucleus. An experimental study in the cat. *Amer. J. Anat. 100*, 1–16.

Torvik, A. (1957b). The spinal projection from the nucleus of the solitary tract. An experimental study in the cat. *J. Anat. (Lond.) 91*, 314–322.

Torvik, A. (1957c). Die Lokalisation des "Speichelzentrums" bei der Katze. *Z. Mikrosk.-Anat. Forsch. 63*, 317–326.

Torvik, A. (1959). Sensory, motor, and reflex changes in two cases of intractable pain after stereotactic mesencephalic tractotomy. *J. Neurol. Neurosurg. Psychiat. 22*, 299–305.

Torvik, A. (1976). Central chromatolysis and the axon reaction: a reappraisal. *Neuropath. Appl. Neurobiol. 2*, 423–432.

Torvik, A., and A. Brodal (1954). The cerebellar projection of the peri-hypoglossal nuclei (nuclei intercalatus, nucleus praepositus hypoglossi and nucleus of Roller) in the cat. *J. Neuropath. Exp. Neurol. 13*, 515–527.

Torvik, A., and A. Brodal (1957). The origin of reticulospinal fibers in the cat. An experimental study. *Anat. Rec. 128*, 113–137.

Touch, Heat and Pain. Ciba Foundation Symposium (1966). (A. V. de Reuck and J. Knight, eds.). J. & A. Churchill, London.

Towe, A. L. (1973). Somatosensory cortex: descending influences on ascending systems. In *Handbook of Sensory Physiology*, Vol. 2, *Somatosensory System* (A. Iggo, ed.), pp. 701–718. Springer-Verlag, Berlin.

Towe, A. L. (1975). Notes on the hypothesis of columnar organization in somatosensory cerebral cortex. *Brain. Behav. Evol. 11*, 16–47.

Towe, A. L., and S. J. Jabbur (1961). Cortical inhibition of neurons in dorsal column nuclei of cat. *J. Neurophysiol. 24*, 488–498.

Tower, S. S. (1940). Pyramidal lesions in the monkey. *Brain 63*, 36–90.

Tower, S. S. (1944). The pyramidal tract. Definition and structure. In *The Precentral Motor Cortex* (P. C. Bucy, ed.), pp. 151–172. Univ. of Illinois Press, Urbana.

Tower, S., D. Bodian, and H. Howe (1941). Isolation of intrinsic and motor mechanism of the monkey's spinal cord. *J. Neurophysiol. 4*, 388–397.

Toyama, K., N. Tsukahara, K. Kosaka, and K. Matsunami (1970). Synaptic excitation of red nucleus neurones by fibres from interpositus nucleus. *Exp. Brain Res. 11*, 187–198.

Travis, A. M. (1955). Neurological deficiencies following supplementary motor area lesions in *Macaca mulatta. Brain 78*, 174–198.

Trevino, D. L. (1976). The origin and projections of a spinal nociceptive and thermoreceptive pathway. In *Sensory Functions of the Skin* (Y. Zotterman, ed.), pp. 367–377. Pergamon Press, Oxford.

Trevino, D. L., R. A. Maunz, R. N. Bryan, and W. D. Willis (1972). Location of cells of origin of the spinothalamic tract in the lumbar enlargement of cat. *Exp. Neurol. 34.*, 64–77.

Trigeminal Neuralgia. Pathogenesis and Pathophysiology (1970). (R. Hassler and A. E. Walker, eds.). Georg Thieme Verlag, Stuttgart.

Trojaborg, W., and F. Buchthal (1965). Malignant and benignant fasciculations. *Acta Neurol. Scand. 41*, Suppl. 13, 251–254.

Trojanowski, J. Q., and S. Jacobson (1974). Medial pulvinar afferents to frontal eye fields in rhesus monkey demonstrated by horseradish peroxidase. *Brain Research 80*, 395–411.

Trojanowski, J. Q., and S. Jacobson (1975). A combined horseradish peroxidase-autoradiographic investigation of reciprocal connections between superior temporal gyrus and pulvinar in squirrel monkey. *Brain Research 85*, 347–353.

Trotter, W., and H. M. Davies (1909). Experimental studies in the innervation of the skin. *J. Physiol. (Lond.) 38*, 134–246.

Truex, R. C., M. J. Taylor, M. Q. Smythe, and P. L. Gildenberg (1970). The lateral cervical nucleus of cat, dog and man. *J. Comp. Neurol. 139*, 93–104.

Tsuchitani, C. (1977). Functional organization of lateral cell groups of cat superior olivary complex. *J. Neurophysiol. 40*, 296–318.

Tsuchitani, C., and J. C. Boudreau (1966). Single unit analysis of cat superior olive S segment with tonal stimuli. *J. Neurophysiol. 29*, 684–697.

Tsukahara, N., and T. Bando (1970). Red nuclear and interposate nuclear excitation of pontine nuclear cells. *Brain Research 19*, 295–298.

Tsukahara, N., and K. Kosaka (1968). The mode of cerebral excitation of red nucleus neurons. *Exp. Brain Res. 5*, 102–117.

Tsukahara, N., K. Toyama, and K. Kosaka (1964). Intracellulary recorded responses of red nucleus neurones during antidromic and orthodromic activation. *Experientia (Basel) 20*, 632.

Tsukahara, N., K. Toyama, and K. Kosaka (1967). Electrical activity of red nucleus neurones investigated with intracellular microelectrodes. *Exp. Brain Res. 4*, 18–33.

Tulloch, I. F., G. W. Arbuthnott, and A. K. Wright (1978). Topographical organization of the striatonigral pathway revealed by anterograde and retrograde neuroanatomical tracing techniques. *J. Anat. (Lond.) 127*, 425–441.

Tunturi, A. R. (1950). Physiological determination of the arrangement of the afferent connections to the middle ectosylvian auditory area in the dog. *Amer. J. Physiol. 162*, 489–502.

Tunturi, A. R. (1952). A difference in the representation of auditory signals for the left and right ears in the iso-frequency contours of the right middle ectosylvian auditory cortex of the dog. *Amer. J. Physiol. 168*, 712–727.

Türck, L. (1851). Über secundäre Erkrankung einzelner Rückenmarksstränge und ihrer Fortsetzung zum Gehirne. *S. B. Akad. Wiss. Wien 6*, 288. (Cited from Nathan and Smith, 1955a.)

Turner, B. H., K. C. Gupta, and M. Mishkin (1978). The locus and cytoarchitecture of the olfactory bulb projection areas in *Macaca mulatta*. *J. Comp. Neurol. 177*, 381–396.

Uddenberg, N. (1968). Differential localization in dorsal funiculus of fibres originating from different receptors. *Exp. Brain Res. 4*, 367–376.

Udo, M., and N. Mano (1970). Discrimination of different spinal monosynaptic pathways converging onto reticular neurons. *J. Neurophysiol. 33*, 227–238.

Uemura, T., and B. Cohen (1973). Effects of vestibular nuclei lesions on vestibulo-ocular reflexes and posture in monkeys. *Acta Otolaryng. (Stockh.), Suppl. 315*, 1–71.

Understanding the Stretch Reflex (1976) (S. Homma, ed.). *Progr. Brain Res.*, Vol. 44. Elsevier, Amsterdam.

Ungerstedt, U. (1971). Stereotaxic mapping of the monoamine pathways in the rat brain. *Acta Physiol Scand., Suppl. 367*, 1–48.

Ungerstedt, U. (1974). Brain dopamine neurons and behavior. In *The Neurosciences, The Third Study Program* (F. O. Schmitt and F. G. Worden, eds.), pp. 695–703. MIT Press, Cambridge, Mass.

Uno, M., I. B. Kozlovskaya, and V. B. Brooks (1973). Effects of cooling interposed nuclei on tracking-task performance in monkeys. *J. Neurophysiol. 36*, 996–1003.

Uno, M., K. Kubota, C. Ohye, T. Nagao, and H. Narabayashi (1967). Topographical arrangement between thalamic ventro-lateral nucleus and precentral motor cortex in man. *Electroenceph. Clin. Neurophysiol. 22*, 437–443.

Uno, M., M. Yoshida, and I. Hirota (1970). The mode of cerebello-thalamic relay transmission investigated with intracellular recording from cells of the ventrolateral nucleus of cat's thalamus. *Exp. Brain Res. 10*, 121–139.

Unterharnscheidt, F., D. Jachnik, and H. Gött (1968). *Der Balkenmangel. Monographien aus dem Gesamtgebiete der Neurologie und Psychiatrie*. Heft *128*, pp. 1–232. Springer-Verlag, Berlin.

Urquhart, N., T. L. Perry, S. Hansen, and J. Kennedy (1975). GABA content and glutamic acid decarboxylase activity in brain of Huntington's chorea patients and control subjects. *J. Neurochem. 24*, 1071–1075.

Ursin, H. (1965). Effect of amygdaloid lesions on avoidance behavior and visual discrimination in cats. *Exp. Neurol. 11*, 298–317.

Ursin, H., and B. R. Kaada (1960). Functional localization within the amygdaloid complex in the cat. *Electroenceph. Clin. Neurophysiol. 12*, 1–20.

Use of Axonal Transport for Studies of Neuronal Connectivity, The (1975). (W. M. Cowan and M. Cuénod, eds.). Elsevier, Amsterdam.

Ustvedt, H. J. (1937). Über die Untersuchung der musikalischen Funktionen bei Patienten mit Gehirnleiden, besonders bei Patienten mit Aphasie. *Acta Med. Scand., Suppl. 86*, 1–737.

Usunoff, K. G., R. Hassler, K. Romansky, R. P. Usunova, and A. Wagner (1976). The nigrostriatal projection in the cat. I. Silver impregnation study. *J. Neurol. Sci. 28,* 265–288.

Usunoff, K. G., R. Hassler, A. Wagner, and I. J. Bak (1974). The efferent connections of the head of the caudate nucleus in the cat: an experimental morphological study with special reference to a projection to the raphe nuclei. *Brain Research 74,* 143–148.

Uvnäs, B. (1960). Central cardiovascular control. In *Handbook of Physiology. Section 1: Neurophysiology, Vol. II* (J. Field, H. W. Magoun, and V. E. Hall, eds.), pp. 1131–1162. Amer. Physiol. Soc., Washington, D.C.

Vachananda, B. (1959). The major spinal afferent systems to the cerebellum and the cerebellar corticonuclear connections in *Macaca mulatta J. Comp. Neurol. 112,* 303–351.

Valenstein, E. S. (1973). *Brain Control: A Critical Examination of Brain Stimulation and Psychosurgery.* Wiley, New York.

Valenstein, E. S., and W. J. H. Nauta (1959). A comparison of the distribution of the fornix system in the rat, guinea pig, cat, and monkey. *J. Comp. Neurol. 113,* 337–363.

Vallbo, Å. B. (1970a). Slowly adapting muscle receptors in man. *Acta Physiol. Scand. 78,* 315–333.

Vallbo, Å. B. (1970b). Discharge patterns in human muscle spindle afferents during isometric voluntary contractions. *Acta Physiol. Scand. 80,* 552–566.

Vallbo, Å. B. (1974a). Afferent discharge from human muscle spindles in non-contracting muscles. Steady state frequency as a function of joint angle. *Acta Physiol. Scand. 90,* 303–318.

Vallbo, Å. B. (1974b). Human muscle spindle discharge during isometric voluntary contractions. Amplitude relations between spindle frequency and torque. *Acta Physiol. Scand. 90,* 319–336.

Vallbo, Å. B., and R. S. Johansson (1976). Skin mechanoreceptors in the human hand: neural and psychophysical thresholds. In *Sensory Functions of the Skin in Primates. With Special Reference to Man.* Wenner-Gren Center International Symposium Series, Vol. 27 (Y. Zotterman, ed.), pp. 185–199. Pergamon Press, Oxford.

Vallbo, Å. B., and R. S. Johansson (1978). The tactile sensory innervation of the glabrous skin of the human hand. In *Active Touch. The Mechanism of Recognition of Objects by Manipulation. A Multi-disciplinary Approach* (G. Gordon, ed.), pp. 29–54. Pergamon Press, Oxford.

Valsø, J. (1936). Die Hypophyse des Blauwals (*Balaenoptera sibbaldii*). Makroskopische und mikroskopische Anatomie. *Z. Anat. Entwickl.-Gesch. 105,* 713–719.

Valverde, F. (1961a). Reticular formation of the pons and medulla oblongata. A Golgi study. *J. Comp. Neurol. 116,* 71–100.

Valverde, F. (1961b). A new type of cell in the lateral reticular formation of the brain stem. *J. Comp. Neurol. 117,* 189–195.

Valverde, F. (1962). Reticular formation of the albino rat's brain stem; cytoarchitecture and corticofugal connections. *J. Comp. Neurol. 119,* 25–53.

Valverde, F. (1965). *Studies on the Piriform Lobe.* Harvard Univ. Press, Cambridge, Mass.

Valverde, F. (1966). The pyramidal tract in rodents. A study of its relations with the posterior column nuclei, dorsolateral reticular formation of the medulla oblongata, and cervical spinal cord (Golgi and electron microscopic observations). *Z. Zellforsch. 71,* 297–363.

Valverde, F. (1967). Apical dendritic spines of the visual cortex and light deprivation in the mouse. *Exp. Brain Res. 3,* 337–353.

Valverde, F. (1968). Structural changes in the area striata of the mouse after enucleation. *Exp. Brain Res. 5,* 274–292.

Valverde, F. (1973). The neuropil in superficial layers of the superior colliculus of the mouse. A correlated Golgi and electron microscopic study. *Z. Anat. Entwickl.-Gesch. 142,* 117–147.

Valverde, F. (1976). Aspects of cortical organization related to the geometry of neurons with intracortical axons. *J. Neurocytol. 5*, 509–529.

Vanegas, H., H. Holländer, and H. Distel (1978). Early stages of uptake and transport of horseradish peroxidase by cortical structures and its use for the study of local neurons and their processes. *J. Comp. Neurol. 177*, 193–211.

Van Buskirk, R. L., and R. P. Erickson (1977). Odorant responses in taste neurons of the rat NTS. *Brain Research 135*, 287–303.

VanGilder, J. C., and J. L. O'Leary (1970). Topical projection of the olivocerebellar system in the cat: an electrophysiological study. *J. Comp. Neurol. 140*, 69–80.

Van Hoesen, G. W., M.-M. Mesulam, and R. Haaxma (1976). Temporal cortical projections to the olfactory tubercle in the rhesus monkey. *Brain Research 109*, 375–381.

Van Hoesen, G. W., and D. N. Pandya (1975a). Some connections of the entorhinal (area 28) and perirhinal (area 35) cortices of the rhesus monkey. I. Temporal lobe afferents. *Brain Research 95*, 1–24.

Van Hoesen, G. W., and D. N. Pandya (1975b). Some connections of the entorhinal (area 28) and perirhinal (area 35) cortices of the rhesus monkey. III. Efferent connections. *Brain Research 95*, 39–59.

Van Hoesen, G. W., D. N. Pandya, and N. Butters (1975). Some connections of the entorhinal (area 28) and perirhinal (area 35) cortices of the rhesus monkey. II. Frontal lobe afferents. *Brain Research 95*, 25–38.

Van Rossum, J. (1969). Corticonuclear and corticovestibular projections of the cerebellum: an experimental investigation of the anterior lobe, simple lobule and the caudal vermis in the rabbit. Thesis. Van Gorcum, Assen.

Vaughan, H. G., Jr. (1975). Psychosurgery and brain stimulation in historical perspective. In *Operating on the Mind* (W. M. Gaylin, J. S. Meister, and R. C. Neville, eds.), pp. 24–72. Basic Books, New York.

Veale, J. L., S. Rees, and R. F. Mark (1973). Renshaw cell activity in normal and spastic man. In *New Developments in Electromyography and Clinical Neurophysiology*, Vol. 3 (J. E. Desmedt, ed.), pp. 523–537. Karger, Basel.

Vedel, J. P., and J. Mouillac-Baudevin (1969). Etude fonctionelle du contrôle de l'activité des fibres fusimotrices dynamiques et statiques par les formations réticulées mésencéphalique, pontique et bulbaire chez le chat. *Exp. Brain Res. 9*, 325–345.

Veening, I. G. (1978a). Cortical afferents of the amygdaloid complex in the rat. An HRP study. *Neurosci. Letters 8*, 191–195.

Veening, I. G. (1978b). Subcortical afferents of the amygdaloid complex in the rat. *Neurosci. Letters 8*, 197–202.

Verbiest, H. (1954). A radicular syndrome from developmental narrowing of the lumbar vertebral canal. *J. Bone Jt. Surg. 36*, 230–237.

Verbiest, H. (1977). *Neurogenic Intermittent Claudication.* North-Holland, Amsterdam.

Verhaart, W. J. C. (1950). Fiber analysis of the basal ganglia. *J. Comp. Neurol. 93*, 425–440.

Verney, E. B. (1947). The antidiuretic hormone and the factors which determine its release. *Proc. Roy. Soc. B. 135*, 25–106.

Vestibular System, The (1975). (R. F. Naunton, ed.). Academic Press, New York.

Vibert, J. F., F. Bertrand, M. Denavit-Saubié, and A. Hugelin (1976a). Discharge patterns of bulbo-pontine respiratory unit populations in cat. *Brain Research 114*, 211–225.

Vibert, J. F., F. Bertrand, M. Denavit-Saubié, and A. Hugelin (1976b). Three dimensional representation of bulbo-pontine respiratory networks architecture from unit density maps. *Brain Research 114*, 227–244.

Victor, D. J. (1977). The diagnosis of congenital unilateral third-nerve palsy. *Brain 99*, 711–718.

Victor, M., R. D. Adams, and G. H. Collins (1971). *The Wernicke-Korsakoff Syndrome.* Blackwell, Oxford.

Victor, M., R. D. Adams, and E. L. Mancall (1959). A restricted form of cerebellar cortical degeneration occurring in alcoholic patients. *Arch. Neurol. (Chic.) 1*, 579–688.

Victor, M., J. B. Angevine, E. L. Mancall, and C. M. Fisher (1961). Memory loss with lesions of hippocampal formation. *Arch. Neurol. (Chic.) 5*, 244–263.

Vierck, C. J. (1966). Spinal pathways mediating limb position sense. *Anat. Rec. 154,* 437.

Vierck, C. J., Jr. (1978a). Interpretations of the sensory and motor consequences of dorsal column lesions. In *Active Touch. The Mechanism of Recognition of Objects by Manipulation. A Multi-disciplinary Approach* (G. Gordon, ed.), pp. 139–159. Pergamon Press, Oxford.

Vierck, C. J., Jr. (1978b). Comparison of forelimb and hindlimb motor deficits following dorsal column section in monkeys. *Brain Research 146,* 279–294.

Vierck, C. J., Jr., and M. M. Luck (1979). Loss and recovery of reactivity to noxious stimuli in monkeys with primary spinothalamic cordotomies, followed by secondary and tertiary lesions of other cord sectors. *Brain 102,* 233–248.

Villablanca, J. R., R. J. Marcus, and C. E. Olmstead (1976b). Effects of caudate nuclei on frontal cortical ablations in cats. I. Neurology and gross behavior. *Exp. Neurol. 52,* 389–420.

Villablanca, J. R., R. J. Marcus, and C. E. Olmstead (1976b). Effects of caudate nuclei on frontal cortex ablations in cats. II. Sleep—wakefulness, EEG and motor activity. *Exp. Neurol. 53,* 31–50.

Villablanca, J. R., R. J. Marcus, C. E. Olmstead, and D. L. Avery (1976). Effects of caudate nuclei or frontal cortex ablations in cats. III. Recovery of limb placing reactions, including observations in hemispherectomized animals. *Exp. Neurol. 53,* 289–303.

Vincent, J. D., E. Arnaud, and A. Nicolescu-Catargi (1972). Osmoceptors and neurosecretory cells in the supraoptic complex of the unanesthetized monkey. *Brain Research 45,* 278–281.

Vincent, S. R., T. Hattori, and E. G. McGeer (1978). The nigrotectal projection: a biochemical and ultrastructural characterization. *Brain Research 151,* 159–164.

Vogt, B. A., and D. N. Pandya (1978). Cortico-cortical connections of somatic sensory cortex (areas 3, 1 and 2) in the rhesus monkey. *J. Comp. Neurol. 177,* 179–191.

Vogt, C., and O. Vogt (1919). Allgemeinere Ergebnisse unserer Hirnforschung. *J. Psychol. Neurol. (Lpz.) 25,* 279–462.

Vogt, C., and O. Vogt (1920). Zur Lehre der Erkrankungen des striären Systems. *J. Psychol. Neurol. (Lpz.) 25, Ergänzungsheft 3,* 627–846.

Vogt, C., and O. Vogt (1941). Thalamusstudien I–III. *J. Psychol. Neurol. (Lpz.) 50,* 31–154.

Vonderahe, A. R. (1940). Changes in the hypothalamus in organic disease. *Res. Publ. Ass. Nerv. Ment. Disc. 20,* 689–712.

Voneida, T. J. (1960). An experimental study of the course and destination of fibers arising in the head of the caudate nucleus in the cat and monkey. *J. Comp. Neurol. 115,* 75–87.

Voogd, J. (1964). *The Cerebellum of the Cat. Structure and Fibre Connexions.* Proefschr. Van Gorcum & Co. N.V., Assen.

Voogd, J. (1969). The importance of fiber connections in the comparative anatomy of the mammalian cerebellum. In *Neurobiology of Cerebellar Evolution and Development,* (R. Llinás, ed.), pp. 493–514. Amer. Med. Ass., Chicago.

Voss, H. (1956). Zahl und Anordnung der Muskelspindeln in den oberen Zungenbeinmuskeln, im M. trapezius and M. latissimus dorsi. *Anat. Anz. 103,* 443–446.

Vraa-Jensen, G. (1942). *The Motor Nucleus of the Facial Nerve.* Munksgaard, Copenhagen.

Vyklický, L., O. L. Keller, G. Brożek, and S. M. Butkhuzi (1972). Cortical potentials evoked by stimulation of tooth pulp afferents in the cat. *Brain Research 41,* 211–213.

Wachs, H., and B. Boshes (1961). Tremor studies in normals and in Parkinsonism. *Arch. Neurol. (Chic.) 4,* 66–82.

Wada, J., and T. Rasmussen (1960). Intracarotid injection of sodium amytal for the lateralization of cerebral speech dominance: experimental and clinical observations. *J. Neurosurg. 17,* 266–282.

Wagman, I. H. (1964). Eye movements induced by electrical stimulation of cerebrum in monkeys and their relationship to bodily movements. In *The Oculomotor System* (M. B. Bender, ed.), pp. 18–39. Hoeber, New York.

Waite, P. M. E. (1973). Somatotopic organization of vibrissal responses in the ventro-basal complex of the rat thalamus. *J. Physiol. (Lond.) 228*, 527–540.

Walberg, F. (1952). The lateral reticular nucleus of the medulla oblongata in mammals. A comparative-anatomical study. *J. Comp. Neurol. 96*, 283–344.

Walberg, F. (1956). Descending connections to the inferior olive. An experimental study in the cat. *J. Comp. Neurol. 104*, 77–174.

Walberg, F. (1957a). Corticofugal fibres to the nuclei of the dorsal columns. An experimental study in the cat. *Brain 80*, 273–287.

Walberg, F. (1957b). Do the motor nuclei of the cranial nerves receive corticofugal fibres? An experimental study in the cat. *Brain 80*, 597–605.

Walberg, F. (1958). Descending connections to the lateral reticular nucleus. An experimental study in the cat. *J. Comp. Neurol. 109*, 363–390.

Walberg, F. (1960). Further studies on the descending connections to the inferior olive. Reticulo-olivary fibers: an experimental study in the cat. *J. Comp. Neurol. 114*, 79–87.

Walberg, F. (1964). The early changes in degenerating boutons and the problem of argyrophilia. Light and electron microscopic observations. *J. Comp. Neurol. 122*, 113–137.

Walberg, F. (1965a). An electron microscopic study of terminal degeneration in the inferior olive of the cat. *J. Comp. Neurol. 125*, 205–222.

Walberg, F. (1965b). Axoaxonic contacts in the cuneate nucleus, probable basis for presynaptic depolarization. *Exp. Neurol. 13*, 218–231.

Walberg, F. (1966). The fine structure of the cuneate nucleus in normal cats and following interruption of afferent fibres. An electron microscopical study with particular reference to findings made in Glees and Nauta sections. *Exp. Brain Res. 2*, 107–128.

Walberg, F. (1974a). Descending connections from the mesencephalon to the inferior olive: an experimental study in the cat. *Exp. Brain Res. 20*, 145–156.

Walberg, F. (1974b). Crossed reticulo-reticular projections in the medulla, pons and mesencephalon. An autoradiographic study in the cat. *Z. Anat. Entwickl.-Gesch. 143*, 127–134.

Walberg, F., D. Bowsher, and A. Brodal (1958). The termination of primary vestibular fibers in the vestibular nuclei in the cat. An experimental study with silver methods. *J. Comp. Neurol. 110*, 391–419.

Walberg, F., and A. Brodal (1953a). Pyramidal tract fibers from temporal and occipital lobes. An experimental study in the cat. *Brain 76*, 491–508.

Walberg, F., and A Brodal (1953b). Spino-pontine fibers in the cat. An experimental study. *J. Comp. Neuro. 99*, 251–288.

Walberg, F., and A. Brodal (1979). The longitudinal zonal pattern in the paramedian lobule of the cat's cerebellum: an analysis based on a correlation of recent HRP data with results of studies with other methods. *J. Comp. Neurol. 187*, 581–588.

Walberg, F., A. Brodal, and G. H. Hoddevik (1976). A note on the method of retrograde transport of horseradish peroxidase as a tool in studies of afferent cerebellar connections, particularly those from the inferior olive; with comments on the orthograde transport in Purkinje cell axons. *Exp. Brain Res. 24*, 383–401.

Walberg, F., H. Holländer, and I. Grofová (1976). An autoradiographic identification of Purkinje axon terminals in the cat. *J. Neurocytol. 5*, 157–169.

Walberg, F., and J. Jansen (1961). Cerebellar corticovestibular fibers in the cat. *Exp. Neurol. 3*, 32–52.

Walberg, F., and J. Jansen (1964). Cerebellar corticonuclear projection studied experimentally with silver impregnation methods. *J. Hirnforsch. 6*, 338–354.

Walberg, F., N. Kotchabhakdi, and G. H. Hoddevik (1979). The olivocerebellar projections to the flocculus and paraflocculus in the cat, compared to those in the rabbit. A study using horseradish peroxidase as a tracer. *Brain Research 161*, 389–398.

Walberg, F., and O. Pompeiano (1960). Fastigiofugal fibers to the lateral reticular nucleus: An experimental study in the cat. *Exp. Neurol. 2*, 40–53.

Walberg, F., O. Pompeiano, A. Brodal, and J. Jansen (1962). The fastigiovestibular projection in the cat. An experimental study with silver impregnation methods. *J. Comp. Neurol. 118*, 49–76.

Walberg, F., O. Pompeiano, L. E. Westrum, and E. Hauglie-Hanssen (1962). Fastigioreticular fibers in cat. An experimental study with silver methods. *J. Comp. Neurol. 119*, 187–199.

Walker, A. E. (1934). The thalamic projection to the central gyri in *Macacus rhesus*. *J. Comp. Neurol. 60*, 161–184.

Walker, A. E. (1936). An experimental study of the thalamocortical projection of the macaque monkey. *J. Comp. Neurol. 64*, 1–39.

Walker, A. E. (1937). Experimental anatomical studies of the topical localization within the thalamus of the chimpanzee. *Proc. Kon. Ned. Akad. Wet. 40*, 198–206.

Walker, A. E. (1938a). The thalamus of the chimpanzee. I. Terminations of the somatic afferent systems. *Confin. Neurol. (Basel) 1*, 99–127.

Walker, A. E. (1938b). The thalamus of the chimpanzee. IV. Thalamic projections to the cerebral cortex. *J. Anat. (Lond.) 73*, 37–93.

Walker, A. E. (1938c). *The Primate Thalamus*. Univ. of Chicago Press, Chicago.

Walker, A. E. (1939). The origin, course and terminations of the secondary pathways of the trigeminal nerve in primates. *J. Comp. Neurol. 71*, 59–89.

Walker, A. E. (1940a). The spinothalamic tract in man. *Arch. Neurol. Psychiat. (Chic.) 43*, 284–298.

Walker, A. E. (1940b). The medial thalamic nucleus. A comparative anatomical, physiological and clinical study of the nucleus medialis dorsalis thalami. *J. Comp. Neurol. 73*, 87–115.

Walker, A. E. (1942a). Somatotopic localization of spinothalamic and secondary trigeminal tracts in mesencephalon. *Arch. Neurol. Psychiat. (Chic.) 48*, 884–889.

Walker, A. E. (1942b). Relief of pain by mesencephalic tractotomy. *Arch. Neurol. Psychiat. (Chic.) 48*, 865–880.

Walker, A. E. (1943). Central representation of pain. In *Pain., Res. Publ. Ass. Nerv. Ment. Dis.*, Vol. 23. (H. G. Wolff, H. S. Gasser, and J. C. Hinsey, eds.), pp. 63–85. Williams & Wilkins, Baltimore.

Walker, A. E., and E. H. Botterell (1937). The syndrome of the superior cerebellar peduncle in the monkey. *Brain 60*, 329–353.

Walker, A. E., and J. F. Fulton (1938). The thalamus of the chimpanzee. III. Methathalamus. Normal structure and cortical connections. *Brain 61*, 250–268.

Walker, A. E., and H. Richter (1966). Section of the cerebral peduncle in the monkey. *Arch. Neurol. (Chic.) 14*, 231–240.

Walker, A. E., and T. A. Weaver, Jr. (1940). Ocular movements from the occipital lobe in the monkey. *J. Neurophysiol. 3*, 353–357.

Walker, A. E., and T. A. Weaver, Jr. (1942). The topical organization and termination of the fibers of the posterior columns in *Macaca mulatta*. *J. Comp. Neurol. 76*, 145–158.

Wall, P. D. (1960). Cord cells responding to touch, damage, and temperature of skin. *J. Neurophysiol. 23*, 197–210.

Wall, P. D. (1967). The laminar organization of dorsal horn and effects of descending impulses. *J. Physiol. (Lond.) 188*, 403–423.

Wall, P. D. (1970). The sensory and motor role of impulses travelling in the dorsal columns towards cerebral cortex. *Brain 93*, 505–524.

Wall, P. D. (1973). Dorsal horn electrophysiology. In *Handbook of Sensory Physiology*, Vol. II, *Somatosensory System* (A. Iggo, ed.), pp. 253–270. Springer-Verlag, Berlin.

Wall, P. D. (1978). The gate control theory of pain mechanisms. A re-examination and restatement. *Brain 101*, 1–18.

Wall, P. D. (1980). Mechanisms of plasticity of connection following damage in adult mamalian nervous system. In *Recovery of Function. Theoretical Considerations for Brain Injury Rehabilitation*, pp. 91–105. (P. Bach-y-Rita, ed.). Huber, Bern.

Wall, P. D., and J. R. Cronly-Dillon (1960). Pain, itch and vibration. *Arch. Neurol. (Chic.) 2*, 365–375.

Wall, P. D., and M. D. Egger (1971). Formation of new connexions in adult rat brains after partial deafferentation. *Nature 232*, 542–545.

Wall, P. D., E. G. Merrill, and T. L. Yaksh (1979). Responses of single units in laminae 2 and 3 of cat spinal cord. *Brain Research 160*, 245–260.

Wall, P. D., and W. Noordenbos (1977). Sensory functions which remain in man after complete transections of dorsal columns. *Brain 100*, 641–653.

Wall, P. D., and A. Taub (1962). Four aspects of the trigeminal nucleus and a paradox. *J. Neurophysiol. 25*, 110–126.

Wall, P. D., S. Waxman, and A. I. Basbaum (1974). Ongoing activity in peripheral nerve: injury discharge. *Exp. Neurol. 45*, 576–589.

Wall, P. D., and R. Werman (1976). The physiology and anatomy of long ranging afferent fibres within the spinal cord. *J. Physiol. (Lond.) 255*, 321–334.

Wallgren, A. (1929). Zur Aetiologie der 'rheumatischen' Facialisparese im Kindesalter. *Acta Med. Scand. 71*, 21–28.

Wallin, B. G., A. Hongell, and K.-E. Hagbarth (1973). Recording from muscle afferents in Parkinsonian rigidity. In *New Developments in Electromyography and Clinical Neurophysiology*, Vol. 3 (J. E. Desmedt, ed.), pp. 263–272. Karger, Basel.

Walsh, E. G. (1957). An investigation of sound localization in patients with neurological abnormalities. *Brain 80*, 222–250.

Walsh, F. B., and W. F. Hoyt (1969). *Clinical Ophthalmology*, 3 volumes. 3rd ed. Williams and Wilkins, Baltimore.

Walsh, K. W. (1977). Neuropsychological aspects of modified leucotomy. In *Neurosurgical Treatment in Psychiatry, Pain, and Epilepsy* (W. H. Sweet, S. Obrador, and J. G. Martin-Rodriguez, eds.), pp. 163–174. Univ. Park Press, Baltimore.

Walshe, F. (1956). The Babinski plantar response, its forms and its physiological and pathological significance. *Brain 79*, 529–556.

Walshe, F. M. R. (1924). Observations on the nature of the muscular rigidity of paralysis agitans and its relationship to tremor. *Brain 47*, 159–177.

Walshe, F. M. R. (1942a). The anatomy and physiology of cutaneous sensibility: a critical review. *Brain 65*, 48–114.

Walshe, F. M. R. (1942b). The giant cells of Betz, the motor cortex and the pyramidal tract: a critical review. *Brain 65*, 409–461.

Walshe, F. M. R. (1943). On the mode of representation of movements in the motor cortex, with special reference to "convulsions beginning unilaterally" (Jackson). *Brain 66*, 104–139.

Walter, R. D., R. W. Rand, P. H. Crandall, C. H. Markham, and W. R. Adey (1963). Depth electrode studies of thalamus and basal ganglia. Results in movement disorders in man. *Arch. Neurol. (Chic.) 8*, 388–397.

Wang, G. H. (1964). *The Neural Control of Sweating*. Univ. of Wisconsin Press, Madison.

Wang, R. Y., and G. K. Aghajanian (1977). Inhibition of neurons in the amygdala by dorsal raphe stimulation: mediation through a direct serotonergic pathway. *Brain Research 120*, 85–102.

Wang, S. C., and S. W. Ranson (1939). Autonomic responses to electrical stimulation of the lower brain stem. *J. Comp. Neurol. 71*, 437–455.

Ward, A. A., Jr. (1948). The cingular gyrus: area 24. *J. Neurophysiol. 11*, 13–23.

Ward, D. G., and C. G. Gunn (1976). Locus coeruleus complex: elicitation of a pressor response and a brain stem region necessary for its occurrence. *Brain Research 107*, 401–406.

Ware, C. B., and E. J. Mufson (1979). Spinal cord projections from the medial cerebellar nucleus in the tree shrew (*Tupaia glis*). *Brain Research 171*, 383–400.

Warmolts, J. R., and W. K. Engel (1972). Open biopsy electromyography. I. Correlation of motor unit behavior with histochemical muscle fiber type in human limb muscle. *Arch. Neurol. (Chic.) 27*, 512–517.

Warr, W. B. (1966). Fiber degeneration following lesions in the anterior ventral cochlear nucleus off the cat. *Exp. Neurol. 14*, 453–474.

Warr, W. B. (1969). Fiber degeneration following lesions in the posteroventral cochlear nucleus of the cat. *Exp. Neurol. 23*, 140–155.

Warr, W. B. (1975). Olivocochlear and vestibular efferent neurons of the feline brain stem: their location, morphology and number determined by retrograde axonal transport and acetylcholinesterase histochemistry. *J. Comp. Neurol. 161,* 159–182.

Warr, W. B. (1978). The olivocochlear bundle: its origins and terminations in the cat. In *Evoked Electrical Activity in the Auditory Nervous System* (R. F. Naunton and C. Fernández, eds.), pp. 43–65. Academic Press, New York.

Wartenberg, R. (1954). Cerebellar signs. *J. Amer. Med. Ass. 156,* 102–105.

Warwick, R. (1953). Representation of the extra-ocular muscles in the oculomotor nuclei of the monkey. *J. Comp. Neurol. 98,* 449–503.

Warwick, R. (1954). The ocular parasympathetic nerve supply and its mesencephalic sources. *J. Anat. (Lond.) 88,* 71–93.

Warwick, R. (1955). The so-called nucleus of convergence. *Brain 78,* 92–114.

Watt, D. G. D., E. K. Stauffer, A. Taylor, R. M. Reinking, and D. G. Stuart (1976). Analysis of muscle receptor connections by spike-triggered averaging. 1. Spindle primary and tendon organ afferents. *J. Neurophysiol. 39,* 1375–1392.

Weaver, R., W. M. Landau, and J. F. Higgins (1963). Fusimotor function. Part II. Evidence of fusimotor depression in human spinal shock. *Arch. Neurol. (Chic.) 9,* 127–132.

Weaver, T. A., Jr., and A. E. Walker (1941). Topical arrangement within the spinothalamic tract of the monkey. *Arch. Neurol. Psychiat. (Chic.) 46,* 877–883.

Webber, R. H. (1958). A contribution on the sympathetic nerves in the lumbar region. *Anat. Rec. 130,* 581–604.

Weber, J. T., G. D. Partlow, and J. K. Harting (1978). The projection of the superior colliculus upon the inferior olivary complex of the cat: an autoradiographic and horseradish peroxidase study. *Brain Research 144,* 369–377.

Webster, K. E. (1961). Cortico-striate interrelations in the albino rat. *J. Anat. (Lond.) 95,* 532–544.

Webster, K. E. (1965). The cortico-striatal projection in the cat. *J. Anat. (Lond.) 99,* 329–337.

Webster, K. E. (1977). Somaesthetic pathways. *Brit. Med. Bull. 33,* 113–120.

Weddell, G. (1941a). The multiple innervation of sensory spots in the skin. *J. Anat. (Lond.) 75,* 441–446.

Weddell, G. (1941b). The pattern of cutaneous innervation in relation to cutaneous sensibility. *J. Anat. (Lond.) 75,* 346–368.

Weddell, G., B. Feinstein, and R. E. Pattle (1944). The electrical activity of voluntary muscle in man under normal and pathological conditions. *Brain 67,* 178–257.

Weddell, G., J. A. Harpman, D. G. Lambley, and L. Young (1940). The innervation of the musculature of the tongue. *J. Anat. (Lond.) 74,* 255–267.

Weinberger, L. N., and F. C. Grant (1942). Experiences with intramedullary tractotomy. III. Studies in sensation. *Arch. Neurol. Psychiat. (Chic.) 48,* 355–381.

Weinberger, N. M., M. Velasco, and D. B. Lindsley (1965). Effects of lesions upon thalamically induced electrocortical desynchronization and recruiting. *Electroenceph. Clin. Neurophysiol. 18,* 369–377.

Weinstein, E. A., M. Cole, M. S. Mitchell, and O. G. Lyerly (1964). Anosognosia and aphasia. *Arch. Neurol. (Chic.) 10,* 376–386.

Weinstein, S., J. Semmes, L. Ghent, and H.-L. Teuber (1958). Roughness discrimination after penetrating brain injury in man: analysis according to locus of lesion. *J. Comp. Physiol. Psychol. 51,* 269–275.

Weisberg, J. A., and A. Rustioni (1976). Cortical cells projecting to the dorsal column nuclei of cats. An anatomical study with the horseradish peroxidase technique. *J. Comp. Neurol. 168,* 425–437.

Weisberg, J. A., and A. Rustioni (1977). Cortical cells projecting to the dorsal column nuclei of Rhesus monkey. *Exp. Brain Res. 28,* 521–528.

Weisenburg, T., and K. E. McBride (1935). *Aphasia. A Clinical and Psychological Study.* Commonwealth Fund, New York.

Welch, W. K., and M. A. Kennard (1944). Relation of cerebral cortex to spasticity and flaccidity. *J. Neurophysiol. 7,* 255–268.

Welker, C. (1971). Microelectrode delineation of fine grain somatotopic organization of SmI cerebral neocortex in albino rat. *Brain Research 26*, 259–275.

Welker, C. (1976). Receptive fields of barrels in the somatosensory neocortex in the rat. *J. Comp. Neurol. 166*, 173–189.

Welker, C., and T. A. Woolsey (1974). Structure of layer IV in the somatosensory neocortex of the rat: description and comparison with the mouse. *J. Comp. Neurol. 158*, 437–454.

Welker, W. I., R. M. Benjamin, R. C. Milles, and C. N. Woolsey (1957). Motor effects of stimulation of cerebral cortex of squirrel monkey (*Saimiri sciureus*). *J. Neurophysiol. 20*, 347–364.

Welker, W. I., and J. I. Johnson, Jr. (1965). Correlation between nuclear morphology and somatotopic organization in ventro-basal complex of the raccoon's thalamus. *J. Anat. (Lond.) 99*, 761–790.

Welker, W. I., and S. Seidenstein (1959). Somatic sensory representation in the cerebral cortex of the raccoon (*Procyon lotor*). *J. Comp. Neurol. 111*, 469–502.

Wendt, R., and D. Albe-Fessard (1962). Sensory responses of the amygdala with special reference to somatic afferent pathways. In *Physiologie de l'hippocampe*, pp. 171–200. Ed. Centre National de la Recherche Scientifique, Paris.

Werner, G., and B. L. Whitsel (1968). Topology of body representation in somatosensory area I of primates. *J. Neurophysiol. 31*, 856–869.

Werner, G., and B. L. Whitsel (1973). Functional organization of the somatosensory cortex. In *Handbook of Sensory Physiology*, Vol. II, *Somatosensory System* (A. Iggo, ed.), pp. 621–700. Springer-Verlag, Berlin.

Wersäll, J. (1956). Studies on the structure and innervation of the sensory epithelium of the cristae ampullares in the guinea pig. A light and electron microscopic investigation. *Acta Otolaryng. (Stockh.), Suppl. 126*, 1–85.

Wersäll, J., L. Gleisner, and P.-G. Lundquist (1967). Ultrastructure of the vestibular end organs. In *Myotatic and Vestibular Mechanisms*. Ciba Foundation Symposium. (A. V. S. de Reuck and J. Knight, eds.), pp. 105–116. J. & A. Churchill, London.

Wersäll, J., and P.-G Lundquist (1966). Morphological polarization of the mechanoreceptors of the vestibular and acoustic system. In *NASA SP-115: Second Symposium on the Role of the Vestibular Organs in Space Exploration*. Ames Research Center, Moffett Field, California, pp. 57–71.

Westergaard, E. (1972). The fine structure of nerve fibers and endings in the lateral cerebral ventricles of the rat. *J. Comp. Neurol. 144*, 345–354.

Weston, J. K. (1939). Notes on the comparative anatomy of the sensory areas of the vertebrate inner ear. *J. Comp. Neurol. 70*, 355–394.

Westrum, L. E., and T. W. Blackstad (1962). An electron microscopic study of the stratum radiatum of the rat hippocampus (regio superior, CA 1) with particular emphasis on synaptology. *J. Comp. Neurol. 119*, 281–309.

Westrum, L. E., R. C. Canfield, and R. G. Black (1976). Transganglionic degeneration in the spinal trigeminal nucleus following removal of tooth pulps in adult cats. *Brain Research 101*, 137–140.

Wever, E. G. (1964). The physiology of the peripheral hearing mechanism. In *Neurological Aspects of Auditory and Vestibular Disorders* (W. S. Fields, ed.), pp. 24–50. C. C. Thomas, Springfield, Ill.

White, E. L. (1973). Synaptic organization of the mammalian olfactory glomerulus: new findings including an intraspecific variation. *Brain Research 60*, 299–313.

White, E. L. (1976). Ultrastructure and synaptic contacts in barrels of mouse SI cortex. *Brain Research 105*, 229–251.

White, E. L., and R. A. DeAmicis (1977). Afferent and efferent projections of the region in mouse SmI cortex which contains the posteromedial barrel subfield. *J. Comp. Neurol. 175*, 455–482.

White, J. C., and R. H. Smithwick (1942). *The Autonomic Nervous System. Anatomy, Physiology, and Surgical Application*, 2nd ed., Henry Kimpton, London.

White, J. C., and W. H. Sweet (1955). *Pain. Its Mechanisms and Neurosurgical Control*. C. C. Thomas, Springfield. Ill.

White, J. C., and W. H. Sweet (1969). *Pain and the Neurosurgeon. A Forty-Year Experience*. C. C. Thomas, Springfield, Ill.

White, L. E., Jr. (1959). Ipsilateral afferents to the hippocampal formation in the albino rat. I. Cingulum projections. *J. Comp. Neurol. 113*, 1–42.

White, L. E., Jr. (1965a). Olfactory bulb projections of the rat. *Anat. Rec. 152*, 465–479.

White, L. E., Jr. (1965b). A morphological concept of the limbic lobe. *Int. Rev. Neurobiol. 8*, 1–34.

White, L. E., Jr. W. M. Nelson, and E. L. Foltz 1960). Cingulum fasciculus study by evoked potentials. *Exp. Neurology 2*, 406–421.

Whitfield, I. C. (1967). *The Auditory Pathway*. Edward Arnold, London.

Whitfield, I. C., and E. F. Evans (1965). Responses of auditory cortical neurons to changing frequency. *J. Neurophysiol. 28*, 655–672.

Whitlock, D. G. (1952). A neurohistological and neurophysiological study of afferent fiber tracts and receptive areas of the avian cerebellum. *J. Comp. Neurol. 97*, 567–636.

Whitlock, D. G., and W. J. H. Nauta (1956). Subcortical projections from the temporal neocortex in *Macaca mulatta*. *J. Comp. Neurol. 106*, 183–212.

Whitlock, D. G., and E. R. Perl (1961). Thalamic projections of spinothalamic pathways in monkey. *Exp. Neurol. 3*, 240–255.

Whitsel, B. L., D. A. Dreyer, and J. R. Roppolo (1971). Determinants of body representation in postcentral gyrus of Macaques. *J. Neurophysiol. 34*, 1018–1034.

Whitsel, B. L., L. M. Petrucelli, and G. Sapiro (1969). Modality representation in the lumbar and cervical fasciculus gracilis of squirrel monkeys. *Brain Research 15*, 67–78.

Whitsel, B. L., L. M. Petrucelli, G. Sapiro, and H. Ha (1970). Fiber sorting in the fasciculus gracilis of squirrel monkeys. *Exp. Neurol. 29*, 227–242.

Whitsel, B. L., L. M. Petrucelli, and G. Werner (1969). Symmetry and connectivity in the map of the body surface in somatosensory area II of primates. *J. Neurophysiol. 32*, 170–183.

Whitsel, B. L., A. Rustioni, D. A. Dreyer, P. R. Loe, E. E. Allen, and C. B. Metz (1978). Thalamic projections to S-I in macaque monkey. *J. Comp. Neurol. 178*, 385–410.

Whitteridge, D. (1960). Central control of eye movements. In *Handbook of Physiology. Section 1: Neurophysiology*, Vol. II (J. Field, H. W. Magoun, and V. E. Hall, eds.), pp. 1089–1109. Amer. Physiol. Soc. Washington, D.C.

Whittier, J. R. (1947). Ballism and the subthalamic nucleus (nucleus hypothalamicus; corpus Luysi). Review of the literature and study of thirty cases. *Arch. Neurol. Psychiat. (Chic.) 58*, 672–692.

Whittier, J. R., and F. A. Mettler (1949). Studies on the subthalamus of the rhesus monkey. I. Anatomy and fiber connections of the subthalamic nucleus of Luys. *J. Comp. Neurol. 90*, 281–317.

Whitty, C. W. M., and W. Lewin (1960). A Korsakoff syndrome in the postcingulectomy confusional state. *Brain 83*, 648–653.

Widén, L., and C. Ajmone Marsan (1961). Action of afferent and corticofugal impulses on single elements in the dorsal lateral geniculate nucleus. In *Neurophysiologie und Psychophysik des visuellen Systems* (R. Jung and H. Kornhuber, eds.), pp. 125–132. Springer-Verlag, Berlin.

Wiesel, T. N., and D. H. Hubel (1965). Extent of recovery from the effects of visual deprivation in kittens. *J. Neurophysiol. 28*, 1060–1072.

Wiesel, T. N., D. H. Hubel, and D. M. K. Lam (1974). Autoradiographic demonstration of ocular-dominance in the monkey striate cortex by means of transneuronal transport. *Brain Research 79*, 273–279.

Wiesendanger, M. (1973). Input from muscle and cutaneous nerves of the hand and forearm to neurones of the precentral gyrus of baboons and monkeys. *J. Physiol. (Lond.) 228*, 203–219.

Wiesendanger, M., and D. Felix (1969). Pyramidal excitation of lemniscal neurons and facilitation of sensory transmission in the spinal trigeminal nucleus of the cat. *Exp. Neurol. 25*, 1–17.

Wiesendanger, M., J. J. Séguin, and H. Kunzle (1974). The supplementary motor area—a control system for posture. *Advan. Behav. Biol. 7*, 331–346.

Wiesman, G. G., D. S. Jones, and W. C. Randall (1966). Sympathetic outflows from cervical spinal cord in the dog. *Science 152*, 381–382.

Wiitanen, J. T. (1969). Selective impregnation of degenerating axons and axon terminals in the central nervous system of the monkey. *Brain Research 14*, 546–548.

Wiklund, L., A. Björklund, and B. Sjölund (1977). The indolaminergic innervation of the inferior olive. 1. Convergence with the direct spinal afferents in the areas projecting to the cerebellar anterior lobe. *Brain Research 131*, 1–21.

Wiksten, B. (1975). The central cervical nucleus—a source of spinocerebellar fibres, demonstrated by retrograde transport of horseradish peroxidase. *Neurosci. Letters 1*, 81–84.

Wiksten, B. (1979a). The central cervical nucleus in the cat. II. The cerebellar connections studied with retrograde transport of horseradish peroxidase. *Exp. Brain Res. 36*, 155–173.

Wiksten, B. (1979b). The central cervical nucleus in the cat. III. The cerebellar connections studied with anterograde transport of ^{3}H-leucine. *Exp. Brain Res. 36*, 175–189.

Willer, J. C., and Y. Lamour (1977). Electrophysiological evidence for a facio-facial reflex in the facial muscles in man. *Brain Research 119*, 459–464.

Williams, M., and J. Pennybacker (1954). Memory disturbances in third ventricle tumours. *J. Neurol. Neurosurg. Psychiat. 17*, 115–123.

Willis, W. D., L. H. Haber, and R. F. Martin (1977). Inhibition of spinothalamic tract cells and interneurons by brain stem stimulation in the monkey. *J. Neurophysiol. 40*, 968–981.

Willis, W. D., R. B. Leonard, and D. R. Kenshalo, Jr. (1978). Spinothalamic tract neurons in the substantia gelatinosa. *Science 202*, 986–988.

Willis, W. D., R. A. Maunz, R. D. Foreman, and J. D. Coulter (1975). Static and dynamic responses of spinothalamic tract neurons to mechanical stimuli. *J. Neurophysiol. 38*, 587–600.

Willis, W. D., D. L. Trevino, J. D. Coulter, and R. A. Maunz (1974). Responses of primate spinothalamic tract neurons to natural stimulation of the hindlimb. *J. Neurophysiol. 37*, 358–372.

Willis, W. D., and J. C. Willis (1966). Properties of interneurons in the ventral spinal cord. *Arch. Ital. Biol. 104*, 354–386.

Wilson, M. E., and B. G. Cragg (1967). Projections from the lateral geniculate nucleus in the cat and monkey. *J. Anat. (Lond.) 101*, 677–692.

Wilson, M. E., and M. Toyne (1970). Retino-tectal and cortico-tectal projections in *Macaca mulatta*. *Brain Research 24*, 395–406.

Wilson, S. A. K. (1914). An experimental research into the anatomy and physiology of the corpus striatum. *Brain 36*, 427–492.

Wilson, V. J., and G. M. Jones (1979). *Mammalian Vestibular Physiology*. Plenum Press, New York.

Wilson, V. J., M. Kato, B. W. Peterson, and R. M. Wylie (1967). A single-unit analysis of the organization of Deiters' nucleus. *J. Neurophysiol. 30*, 603–619.

Wilson, V. J., M. Kato, R. C. Thomas, and B. W. Peterson (1966). Excitation of lateral vestibular neurons by peripheral afferent fibers. *J. Neurophysiol. 29*, 508–529.

Wilson, V. J., Y. Uchino, A. Susswein, and K. Fukushima (1977). Properties of direct fastigiospinal fibers in the cat. *Brain Research 126*, 543–546.

Wilson, V. J., and M. Yoshida (1969a). Comparison of effects of stimulation of Deiters' nucleus and medial longitudinal fasciculus on neck, forelimb, and hindlimb motoneurons. *J. Neurophysiol. 32*, 743–758.

Wilson, V. J., and M. Yoshida (1969b). Monosynaptic inhibition of neck motoneurons by the medial vestibular nucleus. *Exp. Brain. Res. 9*, 365–380.

Windle, W. F. (1926). Non-bifurcating nerve fibers of the trigeminal nerve. *J. Comp. Neurol. 40*, 229–240.

Windle, W. F. (1931). The sensory component of the spinal accessory nerve. *J. Comp. Neurol. 53*, 115–127.

Winer, J. A., I. T. Diamond, and D. Raczkowski (1977). Subdivisions of the auditory cortex of the cat: the retrograde transport of horseradish peroxidase to the medial geniculate body and posterior thalamic nuclei. *J. Comp. Neurol. 176*, 387–418.

Winfield, D. A., K. C. Gatter, and T. P. S. Powell (1975). Certain connections of the visual cortex of the monkey shown by the use of horseradish peroxidase. *Brain Research 92*, 456–461.

Winfield, D. A., and T. P. S. Powell (1976). The termination of thalamo-cortical fibres in the visual cortex of the cat. *J. Neurocytol. 5*, 269–281.

Winter, D. L. (1965). N. gracilis of cat. Functional organization and corticofugal effects. *J. Neurophysiol. 28*, 48–70.

Wise, S. P. (1975). The laminar organization of certain afferent and efferent fiber systems in the rat somatosensory cortex. *Brain Research 90*, 139–142.

Wise, S. P., and E. G. Jones (1977a). Somatotopic and columnar organization in the corticotectal projection of the rat somatic sensory cortex. *Brain Research 133*, 223–235.

Wise, S. P., and E. G. Jones (1977b). Cells of origin and terminal distribution of descending projections of the rat somatic sensory cortex. *J. Comp. Neurol. 175*, 129–158.

Witelson, S. F., and W. Pallie (1973). Left hemispheric specialization for language in the newborn. Neuroanatomical evidence of asymmetry. *Brain 96*, 641–646.

Wohlfahrt, S. (1932). Die vordere Zentralwindung bei Pyramidenbahnläsionen verschiedener Art. *Acta Med. Scand., Suppl. 46*, 1–234.

Wohlfahrt, S., and G. Wohlfart (1935). Mikroskopische Untersuchungen an progressiven Muskelatrophien. *Acta Med. Scand., Suppl. 63*, 1–137.

Wohlfart, G., and K. G. Henriksson (1960). Observations on the distribution, number and innervation of Golgi musculo-tendinous organs. *Acta Anat. 41*, 192–204.

Wold, J. E., and A. Brodal (1973). The projection of cortical sensorimotor regions onto the trigeminal nucleus in the cat. An experimental anatomical study. *Neurobiol. 3*, 353–375.

Wold, J. E., and A. Brodal (1974). The cortical projection of the orbital and proreate gyri to the sensory trigeminal nuclei in the cat. An experimental anatomical study. *Brain Research 65*, 381–395.

Wolf, G., and J. Sutin (1966). Fiber degeneration after lateral hypothalamic lesions in the rat. *J. Comp. Neurol. 127*, 137–155.

Wolf, G. A., Jr. (1941). The ratio of preganglionic neurons to postganglionic neurons in the visceral nervous system. *J. Comp. Neurol. 75*, 235–243.

Wolfe, J. W. (1971). Relationship of cerebellar potentials to saccadic eye movements. *Brain Research 30*, 204–206.

Wolstencroft, J. H. (1964). Reticulospinal neurones. *J. Physiol. (Lond.) 174*, 91–108.

Wong-Riley, M. T. T. (1972a). Neuronal and synaptic organization of the normal dorsal lateral geniculate nucleus of the squirrel monkey, *Saimiri sciureus*. *J. Comp. Neurol. 144*, 25–60.

Wong-Riley, M. T. T. (1972b). Terminal degeneration and glial reactions in the lateral geniculate nucleus of the squirrel monkey after eye removal J. Comp. Neurol. 144, 61–92.

Wong-Riley, M. T. T. (1972c). Changes in the dorsal lateral geniculate nucleus of the squirrel monkey after unilateral ablation of the visual cortex. *J. Comp. Neurol. 146*, 519–548.

Wong-Riley, M. T. T. (1974). Demonstration of geniculocortical and callosal projection neurons in the squirrel monkey by means of retrograde axonal transport of horseradish peroxidase. *Brain Research 79*, 267–272.

Wong-Riley, M. T. T. (1976a). Endogenous peroxidatic activity in brainstem neurons as demonstrated by their staining with diaminobenzidine in normal squirrel monkeys. *Brain Research 108*, 257–277.

Wong-Riley, M. T. T. (1976b). Projections from the dorsal lateral geniculate nucleus to prestriate cortex in the squirrel monkey as demonstrated by retrograde transport of horseradish peroxidase. *Brain Research 109*, 595–600.

Wood, J. G., B. J. McLaughlin, and J. E. Vaughn (1976). Immunocytochemical localization of GAD in electron miscroscopic preparations of rodent CNS. In *GABA in Ner-*

vous System Function (E. Roberts, T. N. Chase, and D. B. Tower, eds.), pp. 133–148.

Wood, J. H., C. R. Lake, M. G. Ziegler, and J. M. van Buren (1977). Neurophysiological and neurochemical alterations during electrical stimulation of human caudate nucleus. *J. Neurosurg. 46,* 361–368.

Woodburne, R. T. (1936). A phylogenetic consideration of the primary and secondary centers and connections of the trigeminal complex in a series of vertebrates. *J. Comp. Neurol. 65,* 403–502.

Woodburne, R. T. (1939). Certain phylogenetic anatomical relations of localizing significance for the mammalian nervous system. *J. Comp. Neurol. 71,* 215–257.

Woodburne, R. T., E. C. Crosby, and R. E. McCotter (1946). The mammalian midbrain and isthmus regions. Part II. The fiber connections. A. The relations of the tegmentum of the midbrain with the basal ganglia in *Macaca mulatta. J. Comp. Neurol. 85,* 67–92.

Woollard, H. H. (1935). Observations on the termination of cutaneous nerves. *Brain 58,* 352–367.

Woollard, H. H., and J. H. Harpman (1939). The cortical projection of the medial geniculate body. *J. Neurol. Psychiat. 2,* 35–44.

Woollard, H. H., and J. A. Harpman (1940). The connexions of the inferior colliculus and of the dorsal nucleus of the lateral lemniscus. *J. Anat. (Lond.) 74,* 441–458.

Woollard, H. H., G. Weddell, and J. A. Harpman (1940). Observations on the neurohistological basis of cutaneous pain. *J. Anat. (Lond.) 74,* 413–440.

Woolsey, C. N. (1947). Patterns of sensory representation in the cerebral cortex. *Fed. Proc. 6,* 437–441.

Woolsey, C. N. (1958). Organization of somatic sensory and motor areas of the cerebral cortex. In *Biological and Biochemical Bases of Behavior* (H. F. Harlow and C. N. Woolsey, eds.), pp. 63–81. Univ. Wisconsin Press, Madison.

Woolsey, C. N. (1960). Organization of cortical auditory system: a review and a synthesis. In *Neural Mechanisms of the Auditory and Vestibular Systems* (G. L. Rasmussen and W. F. Windle, eds.), pp. 165–180. C. C. Thomas, Springfield, Ill.

Woolsey, C. N. (1964). Cortical localization as defined by evoked potential and electrical stimulation studies. In *Cerebral Localization and Organization* (G. Schaltenbrand and C. N. Woolsey, eds.), pp. 17–32. Univ. Wisconsin Press, Madison.

Woolsey, C. N. (1971). Tonotopic organization of the auditory cortex. In *Physiology of the Auditory System.* A Workshop (M. B. Sachs, ed.), pp. 271–282. National Educational Consultants, Inc., Baltimore.

Woolsey, C. N., and D. Fairman (1946). Contralateral, ipsilateral, and bilateral representation of cutaneous receptors in somatic areas I and II of the cerebral cortex of pig, sheep, and other mammals. *Surgery 19,* 684–702.

Woolsey, C. N., W. H. Marshall, and P. Bard (1942). Representation of cutaneous tactile sensibility in the cerebral cortex of the monkey as indicated by evoked potentials. *Bull. Johns Hopk. Hosp. 70,* 399–441.

Woolsey, C. N., W. H. Marshall, and P. Bard (1943). Note on the organization of the tactile sensory area of the cerebral cortex of the chimpanzee. *J. Neurophysiol. 6,* 287–291.

Woolsey, C. N., P. H. Settlage, D. R. Meyer, W. Spencer, T. P. Hamuy, and A. M. Travis (1952). Patterns of localization in precentral and "supplementary" motor areas and their relation to the concept of a premotor area. *Res. Publ. Ass. Nerv. Ment. Dis. 30,* 238–264.

Woolsey, C. N., and E. M. Walzl (1942). Topical projection of nerve fibers from local regions of the cochlea to the cerebral cortex of the cat. *Bull. John Hopk. Hosp. 71,* 315–344.

Woolsey, R. M., and J. S. Nelson (1975). Asymptomatic destruction of the fornix in man. *Arch. Neurol. (Chic.) 32,* 566–568.

Woolsey, T. A., and H. Van der Loos (1970). The structural organization of layer IV in the somatosensory region (SI) of mouse cerebral cortex. The description of a cortical field composed of discrete cytoarchitectonic units. *Brain Research 17,* 205–242.

Wrete, M. (1930). Morphogenetische und anatomische Untersuchungen über die Rami communicates der Spinalnerven beim Menschen. *Uppsala Lak.-Fören. Förh. 35,* 219–380.

Wrete, M. (1935). Die Entwicklung der intermediären Ganglien beim Menschen. *Morph. Jb. 75,* 229–268.

Wyke, B. D. (1947). Clinical physiology of the cerebellum. *Med. J. Aust. 2/18,* 533–540.

Wyke, B. D. (1967). The neurology of joints. Arris and Gale Lecture. *Ann. Roy. Coll. Surg. Engl. 41,* 25–50.

Xuereb, G. P., M. M. L. Prichard, and P. M. Daniel (1954a). The arterial supply and venous drainage of the human hypophysis cerebri. *Quart. J. Exp. Physiol. 39,* 199–217.

Xuereb, G. P., M. M. L. Prichard, and P. M. Daniel (1954b). The hypophysial portal system of vessels in man. *Quart. J. Exp. Physiol. 39,* 219–230.

Yakovlev, P. I., S. Locke, D. Y. Koskoff, and R. A. Patton (1960). Limbic nuclei of thalamus and connections of limbic cortex. I. Organization of the projections of the anterior group of nuclei and of the midline nuclei of the thalamus to the anterior cingulate gyrus and hippocampal rudiment in the monkey. *Arch. Neurol. (Chic.) 3,* 621–641.

Yamamoto, M., I. Shimoyama, and S. M. Highstein (1978). Vestibular nucleus neurons relaying excitation from the anterior canal to the oculomotor nucleus. *Brain Research 148,* 31–42.

Yamamoto, T., H. Satomi, H. Ise, and K. Takahashi (1977). Evidence of the dual innervation of the cat stomach by the vagal dorsal motor and medial solitary nuclei as demonstrated by the horseradish peroxidase method. *Brain Research 122,* 125–131.

Yamamoto, T., K. Takahashi, H. Satomi, and H. Ise (1977). Origins of primary afferent fibers in the spinal ventral roots in the cat as demonstrated by the horseradish peroxidase method. *Brain Research 126,* 350–354.

Yanagisawa, N., R. Tanaka, and Z. Ito (1976). Reciprocal inhibition in spastic hemiplegia in man. *Brain 99,* 555–574.

Yee, J., F. Harrison, and K. B. Corbin (1939). The sensory innervation of the spinal accessory and tongue musculature in the rabbit. *J. Comp. Neurol. 70,* 305–314.

Yeni-Komishian, G. H., and D. A. Benson (1976). Anatomical study of cerebral asymmetry in the temporal lobe of humans, chimpanzees, and rhesus monkeys. *Science 192,* 387–389.

Yeterian, E. H., and G. W. van Hoesen (1978). Cortico-striate projections in the rhesus monkey: the organization of certain corticocaudate connections. *Brain Research 139,* 43–63.

Yokota, T., and N. Nishikawa (1977). Somatotopic organization of trigeminal neurons within caudal medulla oblongata. In *Pain in the Trigeminal Region* (D. J. Anderson and B. Matthews, eds.), pp. 243–257. Elsevier/North-Holland. Amsterdam.

York, D. H., and J. E. Faber (1977). An electrophysiological study of nigrotectal relationships: a possible role in turning behavior. *Brain Research 130,* 383–386.

Yoshida, M., and W. Precht (1971). Monosynaptic inhibition of neurons of the substantia nigra by caudato-nigral fibers. *Brain Research 32,* 225–228.

Yoshii, N., M. Fujii, and T. Mizokami (1978). Hypothalamic projection to the pulvinar-LP complex in the cat; a study by the HRP-method. *Brain Research 155,* 343–346.

Yoss, R. E. (1952). Studies of the spinal cord. Part I. Topographic localization within the dorsal spino-cerebellar tract in *Macaca mulatta. J. Comp. Neurol. 97,* 5–20.

Yoss, R. E. (1953). Studies of the spinal cord. Part II. Topographic localization within the ventral spino-cerebellar tract in the macaque. *J. Comp. Neurol. 99,* 613–638.

Yuen, H., R. M. Dom, and G. F. Martin (1974). Cerebellopontine projections in the

American opossum. A study of their origin, distribution and overlap with fibers from the cerebral cortex. *J. Comp. Neurol. 154*, 257–286.

Yumiya, H., K. Kubota, and H. Asanuma (1974). Activities of neurons in area 3a of the cerebral cortex during voluntary movements in the monkey. *Brain Research 78*, 169–177.

Záborszky, L., C. Léranth, G. B. Makara, and M. Palkovits (1975). Quantitative studies on the supraoptic nucleus in the rat. II. Afferent fiber connections. *Exp. Brain Res. 22*, 525–549.

Zaidel, D., and W. R. Sperry (1974). Memory impairment after commissurotomy in man. *Brain 97*, 263–272.

Zander Olsen, P., and E. Diamantopoulos (1967). Excitability of spinal motor neurons in normal subjects and patients with spasticity, Parkinsonian rigidity, and cerebellar hypotonia. *J. Neurol. Neurosurg. Psychiat. 30*, 325–331.

Zangwill, O. L. (1960). *Cerebral Dominance and its Relation to Psychological Function.* William Ramsay Henderson Trust Lecture. Oliver and Boyd, Edinburgh.

Zangwill, O. L. (1974). Consciousness and the cerebral hemispheres. In *Hemisphere Function in the Human Brain* (S. J. Dimond and J. G. Beaumont, eds.), pp. 264–278. Wiley, New York.

Zarzecki, P., Y. Shinoda, and H. Asanuma (1978). Projection from area 3a to the motor cortex by neurons activated from group I muscle afferents. *Exp. Brain Res. 33*, 269–282.

Zarzecki, P., P. L. Strick, and H. Asanuma (1978). Input to primate motor cortex from posterior parietal cortex (area 5). II. Identification by antidromic activation. *Brain Research 157*, 331–336.

Zeki, S. M. (1971). Interhemispheric connections of prestriate cortex in monkey. *Brain Research 19*, 63–75.

Zeki, S. M. (1974). Functional organization of a visual area in the posterior bank of the superior temporal sulcus of the rhesus monkey. *J. Physiol. (Lond.) 236*, 549–573.

Zeman, W. (1970). Pathology of the torsion dystonias (dystonia musculorum deformans). *Neurology (Minneap.) 20*, 79–88.

Zentay, Paul J. (1937). Motor disorders of the central nervous system and their significance for speech. Part I. Cerebral and cerebellar dysarthrias. *Laryngoscope (St. Louis) 47*, 147–156.

Zervas, N. T., F. A. Horner, and K. S. Pickren (1967). The treatment of dyskinesia by stereotaxic dendatectomy. *Conf. Neurol. (Basel) 29*, 93–100.

Zilstorff-Pedersen, K. (1964). Determinations and variations of olfactory thresholds. *Arch. Otolaryng. 79*, 412–417.

Zimmer, J. (1973). Extended commissural and ipsilateral projections in postnatally de-entorhinated hippocampus and fascia dentata demonstrated in rats by silver impregnation. *Brain Research 64*, 293–311.

Zimmer, J. (1974a). Proximity as a factor in the regulation of aberrant axonal growth in postnatally deafferented fascia dentata. *Brain Research 72*, 137–142.

Zimmer, J. (1974b). Long term synaptic reorganization in rat fascia dentata deafferented at adolescent and adult stages: observations with the Timm method. *Brain Research 76*, 336–342.

Zimmerman, H. M. (1940). Temperature disturbances and the hypothalamus. *Res. Publ. Ass. Nerv. Ment. Dis. 20*, 824–840.

Zimmerman, I. D. (1968). A triple representation of the body surface in the sensorimotor cortex of the squirrel monkey. *Exp. Neurol. 20*, 415–431.

Zimny, R., T. Sobusiak, and Z. Matlosz (1970/71). The afferent components of the hypoglossal nerve. An experimental study with toluidine blue and silver impregnation methods. *J. Hirnforsch. 12*, 83–100.

Zotterman, Y. (1959). The peripheral nervous mechanism of pain: a brief review. In *Pain and Itch. Nervous Mechanisms.* Ciba Foundation Study Group no. 1 (G. E. Wolstenholme and M. O'Connor, eds.), pp. 13–24. J. & A. Churchill, London.

Zotterman, Y. (1975). The physiology of taste. Electrophysiological experiments on human taste nerves. *Bull. Acad. Roy. Méd. Belg. 130,* 132–147.

Zucker, E., and W. I. Welker (1969). Coding of somatic sensory input by vibrissae neurons in the rat's trigeminal ganglion. *Brain Research 12,* 138–156.

Åström, K. E. (1953). On the central course of afferent fibers in the trigeminal, facial, glossopharyngeal, and vagal nerves and their nuclei in the mouse. *Acta Physiol. Scand. 29, Suppl. 106,* 209–320.

Index

Where a subject is treated on more than one page, ff. after the page number indicates the place where it is most completely described. n. indicates reference to a footnote.